ADAPTIVE &
DIGITAL
SIGNAL
PROCESSING

with Digital Filtering Applications

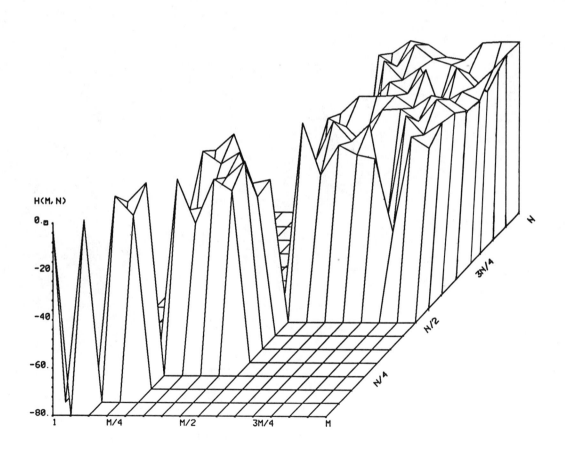

ADAPTIVE & DIGITAL SIGNAL PROCESSING

with Digital Filtering Applications

CLAUDE S. LINDQUIST

Professor of Electrical & Computer Engineering
University of Miami
Coral Gables, Florida 33124

Volume 2

INTERNATIONAL SERIES IN SIGNAL PROCESSING AND FILTERING

Cover: The unsmoothed gain matrix of a sixteen-point Hadamard vector filter used to detect a cosine-triangle pulse in white noise at a signal-to-noise ratio of 10 dB.

ADAPTIVE & DIGITAL SIGNAL PROCESSING
Copyright ©1989 by Claude S. Lindquist

Printed in the United States of America
10 9 8 7 6 5 4 3 2 1

Published by
Steward & Sons
P.O. Box 24-8583
Miami, FL 33124

Library of Congress Catalog Card Number 87-061772
ISBN 0-917144-03-1

To my father and mother,
Claude R. and Eva E. Lindquist

"And now abides faith, hope, love, these three;
but the greatest of these is love." (1 Cor. 13:13)

PREFACE

Samuel Johnson said *"The two most engaging powers of an author are to make new things familiar, and familiar things new."* Making new things familiar and familiar things new is our goal in this book on adaptive and digital signal processing.

This is a challenging goal because current research is rapidly expanding an already massive amount of existing theory. Digital signal processing is being used by most disciplines throughout universities. Digital signal processing is also being extensively used by all types of engineers in all areas of engineering in industry. The precursor of this wide popularity was the introduction of the microcomputer chip in the late 1970's. The microcomputer forms the basic building block of most digital signal processing systems. The small size and low cost of the microcomputer overcame the general utilization barriers imposed by the large and costly mainframe computers of previous decades.

Adaptive and Digital Signal Processing with Digital Filtering Applications is a textbook for undergraduate and graduate level courses in both adaptive and digital signal processing, and digital filtering. It is also an excellent reference and self-study text for engineers. This book is unique compared to other digital signal processing texts for a variety of reasons. It uses an unusual chapter sequence to maximize the cohesive development of material. This development carefully balances theoretical complexity and practical usefulness. Lengthy and unnecessary algebraic derivations are avoided to provide room for practical simulations and discussion. An extremely comprehensive collection of topics, design methods, and algorithms are discussed and integrated together. Many design examples are presented to illustrate concepts and to demonstrate system implementations. Another unusual feature of this book is that it includes adaptive filters which are routinely developed using our frequency domain filter design approach. We believe that the reader will find this digital signal processing book to be one of his most useful and comprehensive texts.

This book is composed of twelve chapters. Chapters 1–6 are concerned with analysis and Chapters 7–12 are concerned with synthesis or design. Chapter 1 presents the basic concepts in digital signal and system analysis. Chapter 2 introduces frequency domain analysis using generalized transforms and windowing functions. Chapter 3 discusses time domain analysis using generalized inverse transforms and interpolators. Chapters 2 and 3 complement one another and should be studied together. Chapter 4 classifies and formulates classical, optimum, and adaptive filters. Homomorphic and amplitude domain filters are also discussed. Chapter 5 presents digital transformations which convert the analog lowpass filters of Chapter 4 into digital lowpass filters. Chapter 6 derives frequency transformations which convert the analog lowpass filters of Chapter 4 into digital highpass, bandpass, bandstop, and allpass filters.

Chapter 7 classifies digital filters into fast transform, nonrecursive, and recursive types and discusses their advantages and disadvantages. Chapter 8 investigates fast transform filter design using scalar and vector filters and generalized transforms. Chapter 9 demonstrates the usefulness of these concepts by designing a variety of FFT filters which meet specialized applications. Chapter 10 presents nonrecursive filter design methods and Chapter 11 presents recursive filter design methods. Chapter 12 discusses hardware implementation of digital filters using a variety of realization forms.

vii

Complete problem sets are presented at the end of each chapter. The student should solve these problems to test, enhance, and develop his conceptual understanding. Answers to selected problems are found at the end of the book. These answers will help him verify that he has properly grasped concepts. A complete and detailed index is included to enable him to easily and quickly locate information.

Different chapter sequences can be followed by the instructor in one- and two-semester courses. When analysis and design are to be taught in separate courses, use:

- Analysis: Chapters 1–6;
- Design: Chapters 7–12.

When analysis and design are to be integrated and taught in the same course, use:

- FIR and IIR Filter Design: Chapters 1–3 (selected sections), 4.2, 5, 7, 10–12;
- Fast Transform Filter Design: Chapters 1–4 (remaining sections), 6, 7, 8, 9.

Other chapter combinations are possible but we have found these sequences to be very suitable for teaching and also for self-study programs. Graduate courses cover the material in detail while undergraduate courses cover selected topics in abbreviated form.

Those familiar with our previous book *Active Network Design with Signal Filtering Applications* will already feel familiar with this book. Both books use almost the same chapter sequence (same first half, analogous second half) and have been organized to be complementary. This complementary feature enhances the cohesiveness between digital filter courses and active filter courses that use both books.

It is always an author's pleasure to thank those people who have helped him. Special thanks go to my students who have patiently studied from partially completed manuscripts. They provided many useful comments and suggestions. Some especially motivated students pursued research, thesis, and special projects which have been incorporated into this book and help make it unique. These students include M. Almeer, M. Bavay, E. Chiang, L. Conboy, S. Cox, S. Graham, W. Hilligoss, H. Lin, J. Reis, N. Rizq, H. Rozran, K. Severence, T. Wongterarit, S. Yhann, C. Yu, and D. Yuen. Some helped with the emergency typing of manuscripts including Howard Rozran, Nader Rizq, and even my father C.R. Lindquist. Celestino Corral drew many of the figures. Clinton Powell, Alan Ravitz, and Ming Zhao helped write software and run numerous simulations. I am also pleased to recognize many years of research with William Haas.

The enthusiasm of my colleagues and secretaries in the Department of Electrical and Computer Engineering of the University of Miami has been greatly welcomed. Some of these faculty members include B. Avanic, A. Condom, J. Duncan, N. Einspruch, B. Furht, G. Gonzalez, T. Giuma, E. Jury, M. Kabuka, A. Katbab, R. Lask, E. Lee, P. Liu, A. Recio, M. Tapia, F. Urban, J.P. Violette, N. Weinberg, K. Yacoub, and T. Young. My ever helpful secretaries are B. Braverman, N. Perez, M. Prohias, and C. Williamson. I especially want to thank Kamal Yacoub, past department chairman, for fostering this congenial department atmosphere. Fellow authors Guillermo Gonzalez and Eli Jury provided me with much encouragement. My acknowledgements would not be complete without also thanking Gene Hostetter and Samuel Stearns who have been faithful friends and supportive colleagues for many years.

Claude S. Lindquist
Miami, Florida
July 4, 1988

CONTENTS

1 GENERAL CONCEPTS

In the beginning God created the heaven and the earth.
And the earth was without form, and void;
and darkness was upon the face of the deep ...
And God said, Let there be light: and there was light.
And God saw the light, that it was good:
and God divided the light from the darkness.

Genesis 1:1–4

This book presents a thorough study of the digital filter. Although digital networks have been utilized for decades, digital filters are relatively new. They have led to unique design approaches and concepts that are industrially useful. Digital filters were originally built using shift registers, multipliers, and adders. With the advent of microprocessors, assembly language implementations became possible. Entire microcomputer systems using high-level software are now practical. These hardware and software developments are responsible for today's explosive interest in digital filters and digital signal processing. This will be known as the era of the microprocessor and the digital filter.

A number of different filter types are available to the design engineer:[1]

1. *Digital filters:* Composed of microprocessor-based systems or shift registers, multipliers, and adders.

2. *Active filters:* Composed of resistors, capacitors, and operational amplifiers.

3. *LC filters:* Composed of inductors and capacitors.

4. *Mechanical filters:* Composed of mechanical resonators (bars or discs) coupled by rods or wires and driven by electromechanical transducers (magnetostrictive ferrite or piezoelectric).

5. *Crystal or ceramic filters:* Composed of crystals, inductors, and capacitors; or of crystals (electroacoustic) or ceramic (piezoelectric) resonators and driven directly.

6. *Monolithic crystal or ceramic filters:* Composed of metallic electrodes deposited on a single crystal or ceramic substrate and driven directly.

Each filter type has a particular frequency band and fractional bandwidth (or Q) range where it excels. These are shown in Fig. 1.0.1. It is clear that digital filters are useful at frequencies up to 100 KHz and Q's as high as 100,000. This is the frequency and Q range where the bulk of industrial applications exist. Thus, digital filters are invaluable to the design engineer.

1

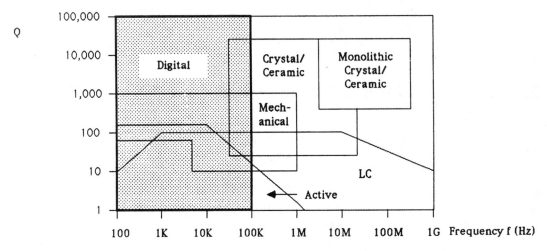

Fig. 1.0.1 Frequency and Q range for various filter types.

However, this book is much more than simply a study of digital filters as a perusal of the index will show. Broadly speaking, it is a complete theoretical and applied guide to digital signal processing and filtering. The theory presented has been carefully selected on the basis of industrial usefulness. All theory is rigorously derived and thoroughly interpreted using a myriad of practical design examples. A number of standard design curves have been collected for easy use. Many new concepts and design techniques are introduced. A variety of useful references are included. The design engineer will find this book to be perhaps his most useful book on digital signal processing and filtering.

This first chapter is a collection of ideas and concepts upon which much of our later work will be based. Some of the material may appear to be of only theoretical interest. However this is not the case as later development will show.

1.1 DIGITAL VERSUS ANALOG FILTERS

This book is concerned with digital filters. Digital filters offer several distinct advantages over analog filters. These are summarized in Table 1.1.1 which will now be discussed. Analog filters process signals that are continuous in time and have a continuum of values. However digital filters process signals that are sampled only at certain times and are quantized to have only certain values as shown in Fig. 1.1.1. Analog filters use passive (R, L, and C) components and active components (e.g., op amps) which have imprecise values (initial tolerances) that drift with temperature and time (component drift). Digital filters are implemented with digital elements (shift registers, multipliers, and adders) that do not drift. However there is coefficient rounding since digital signals are represented by finite length digital words. When analog filter signals become too large (positive or negative), they usually limit and become distorted. In digital filters, the digital word simply overflows the register.

Noise in analog filters is produced by a variety of sources which gives rise to different noise types like thermal, shot, $1/f$, etc. Although this noise is present in the

Table **1.1.1** Comparisons of digital and analog filters.

Property	Digital Filter	Analog Filter
Input/output signals	Discrete magnitudes (quantized)	Continuous magnitudes
Time	Discrete time	Continuous time
Component tolerance	Coefficient rounding	Initial tolerance
Component drift	No coefficient drift	Component drift
Excessive signal levels	Overflow	Overload, nonlinearity
Sampling	Aliasing	No aliasing
Noise	Quantizing, aliasing, limit cycles	Thermal, shot, $1/f$, etc.
Cost	High – decreasing rapidly	Low
Speed	To MHz range	To GHz range
Applications	Very wide	Limited
Flexibility	Excellent	Limited

digital electronic circuits used to form digital signals, noise is not present in the digital words themselves. Unfortunately digital filter noise arises from other sources like signal quantization and signal sampling (aliasing).

The cost of digital filters is usually considerably higher than their analog filter counterparts but cost continues to decrease. Digital filters cannot operate at higher frequencies like some analog filters. However newer parallel processing techniques with higher-speed logic and multipliers make much faster operation possible.

Perhaps the major advantage of digital filters is their flexibility, especially for digital filters implemented in software. Digital filter transfer functions can be easily optimized and updated in real time using adaptive algorithms. Nonlinear operations often used in signal processing can also be incorporated directly into the program. Higher-level languages further reduce the designer's burden for easily implementing digital systems. Also since many new devices generate signals directly in digital form using digital hardware, digital filters are system compatible while analog filters are not.[2]

1.2 DIGITAL FILTER COMPONENTS

Now let us consider the relation between analog and digital filters in more detail using the block diagram of Fig. 1.2.1. In general terms, the analog filter accepts input

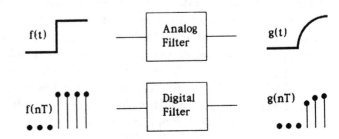

Fig. **1.1.1** Relation between analog and digital filters.

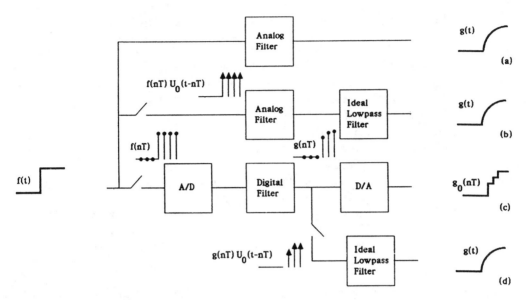

Fig. **1.2.1** Relation between analog and digital filters showing (a) analog filter, (b) sampled-data analog filter with ideal lowpass interpolator, (c) digital filter with zero-order interpolator, and (d) digital filter with ideal lowpass interpolator.

$f(t)$ and forms output $g(t)$. The digital filter accepts a sequence of input samples $\ldots f(-T), f(0), f(T) \ldots$ and forms a sequence of output samples $\ldots g(-T), g(0), g(T) \ldots$ The digital input $f(nT)$ equals the analog input $f(t)$ for times $t = nT$ where $n = \ldots -1, 0, 1 \ldots$. If the sample rate f_s is sufficiently high (above the Nyquist rate) then the digital output $g(nT)$ equals, on the average, the analog output $g(t)$ for times $t = nT$. Using interpolation formulas, the analog output $g(t)$ can be exactly reconstructed from the sampled output $g(nT)$. Ideal lowpass filters can also be used to provide the same analog output $g(t)$ from $g(nT)$.

Notice that two different types of switches are used in Fig. 1.2.1. The switch before the A/D block samples the analog signal $f(t)$ and outputs pulses (numbers) $f(nT)$ (or $f(nT)D_0(t - nT)$ whose amplitudes equal the analog samples $f(nT)$). The switch before the analog filter and ideal lowpass filter instead outputs impulses $f(nT)U_0(t - nT)$ whose areas equal the analog samples $f(nT)$. We will refer to these as pulse and impulse switches, respectively. Pulse switches (impulse switches) output numbers (output impulses) and must be used ahead of digital filters (analog filters). Digital filters and impulse switches are incompatable and vice versa.

Sampling in digital filters is usually performed by *A/D and D/A converters*. The analog-to-digital converter (A/D or ADC) processes voltages (or currents converted to voltages) in some range (e.g., –10 volts $\leq V \leq$ 10 volts).[3] It samples $f(t)$, holds the sample $f(nT)$ in a *sample-and-hold (S/H)*, and outputs a digital word composed of k bits (usually $k = 8, 12,$ or 16). The digital word ranges between $(00\ldots 0)$ for the minimum analog value and $(11\ldots 1)$ for the maximum analog value. Other coding schemes like sign-magnitude, two's-complement, and Gray codes can also be used. The code is selected to simplify the hardware implementation or improve system performance. In any event, the input is coded into one of 2^k output levels. The characteristics of linear (nonlinear) A/D and D/A converters are shown in Fig. 1.2.2a (Fig. 1.2.2b). In nonlinear converters, the A/D is sometimes called the *compressor* and the D/A is called the *expandor*. Together they are referred to as a *compandor*. Assuming a linear A/D converter is used (i.e., no companding or nonlinear characteristic as might be used in applications like voice processing), the output step size is $1/2^k$ and the input step size is $(f_{max} - f_{min})/2^k$. The error never exceeds $\pm 1/2^{k+1}$ between the linear A/D's input and output for $f_{max} - f_{min} = 1$.

These *quantized* input samples are applied to the digital filter which operates on them according to some algorithm and outputs the results to a digital-to-analog converter (D/A or DAC). The D/A takes the k-bit digital word and converts it to one of the 2^k quantized levels as shown in Fig. 1.2.2 where we invert the plot by interchanging the f-axis and g-axis. If companding is used, then the A/D and D/A characteristics are nonlinear but are still inverses of one another. The D/A includes a sample-and-hold which holds this value until the next sample is converted. A lowpass filter is often used to smooth the steps in the D/A output $g_0(t)$. Alternatively the digital filter output can be converted to impulses whose areas equal $g_0(t)$ at the sample times $t = nT$. Applying these impulses to an ideal lowpass filter in Fig. 1.2.1 results in an output $g(t)$ identical to that from the analog filter (assuming we sample above the Nyquist rate).

Digital filters are composed of *delay elements, multipliers,* and *adders* as shown in Fig. 1.2.3. These components process k-bit words and are interconnected in a variety of ways to obtain the desired filter characteristics. When the digital filters are implemented as high-level software programs rather than discrete hardware, the words are stored in arrays instead of delay elements. The dimensions of the arrays equal the required number of unit delays plus one. Simple single-line statements like $y(n) = ax(n) + bx(n-1)$ perform both multiplication and addition. Usually implementation in software is simpler, faster, and more flexible than hardware. However execution speed is much slower because there is considerable software overhead (e.g., operating system, high-level language support, device handlers, etc.).

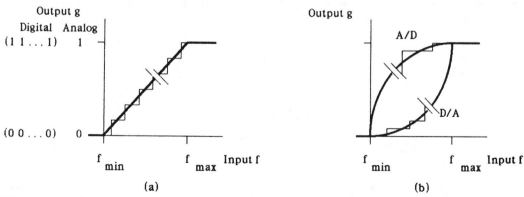

Fig. 1.2.2 Transfer characteristics of (a) linear and (b) nonlinear A/D converter and D/A converter.

Component	Equation	Symbol
Unit Delay	$y(nT) = x(nT - T)$	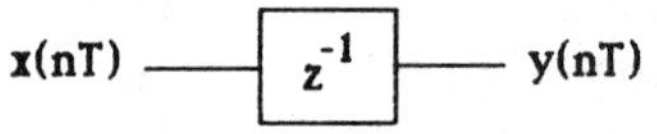 $x(nT)$ — z^{-1} — $y(nT)$
Adder	$y(nT) = \sum_{i=1}^{k} x_i(nT)$	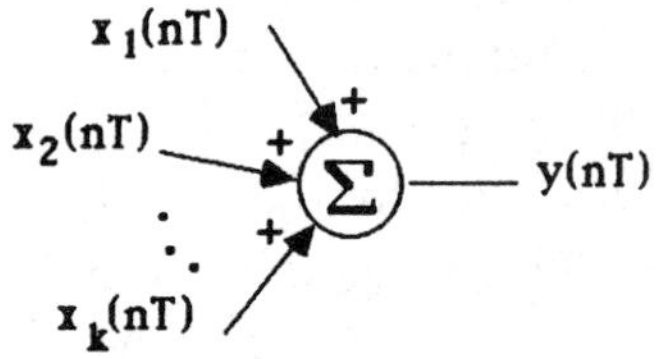 $x_1(nT)$, $x_2(nT)$ ... $x_k(nT)$ → Σ → $y(nT)$
Multiplier	$y(nT) = Kx(nT)$	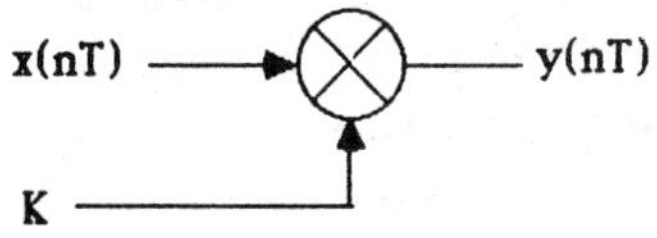 $x(nT)$ → $\otimes$ → $y(nT)$, K

Fig. 1.2.3 Digital filter components.

1.3 SAMPLING THEOREM

It is important to mathematically describe the sampling process represented by the impulse switches in Fig. 1.2.1. The switch can be replaced by a multiplier having the two inputs $f(t)$ and $s(t)$ and the output $f_s(t)$ shown in Fig. 1.3.1.

1.3.1 UNIFORM SAMPLING

For *uniform sampling*, the sampling signal $s(t)$ is an impulse train of infinite length. When $s(t)$ is synchronized to $t = 0$ so that an impulse occurs at $t = 0$, then

$$s(t) = \sum_{n=-\infty}^{\infty} U_0(t - nT) \tag{1.3.1}$$

Pulse switches use $D_0(t - nT)$ terms instead (see Eq. 3.1.11). The sampled signal $f_s(t)$

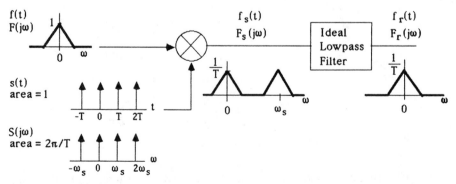

Fig. 1.3.1 Relation between unsampled and sampled signal spectra.

equals

$$f_s(t) = f(t)s(t) = f(t) \sum_{n=-\infty}^{\infty} U_0(t - nT) = \sum_{n=-\infty}^{\infty} f(nT)U_0(t - nT) \qquad (1.3.2)$$

The Laplace transform of $f_s(t)$ can be expressed from Table 1.5.2 as

$$F_s(s) = \frac{1}{2\pi}F(s) * S(s) = \frac{1}{2\pi}F(s) * \omega_s \sum_{n=-\infty}^{\infty} U_0(s - jn\omega_s) = \frac{\omega_s}{2\pi}\sum_{n=-\infty}^{\infty} F(s - jn\omega_s)$$
$$(1.3.3)$$

where the sampling frequency ω_s equals $2\pi/T$ in radians/second or $1/T$ Hertz.

When $s(t)$ is delayed by one-half clock period $T/2$, the impulse stream $s(t - T/2)$ is centered on either side of $t = 0$. $s(t - T/2)$ is still described by Eq. 1.3.1 but the impulse term is rewritten as $U_0(t - (n - 1/2)T)$ which accounts for the $T/2$ second delay. $F_s(s)$ must be multiplied by $e^{-sT/2}$ in Eq. 1.3.3.

EXAMPLE 1.3.1 Determine the Laplace transform for a unit step signal sampled every T seconds.
SOLUTION The unit step has a Laplace transform which equals

$$F(s) = \frac{1}{s}, \qquad \text{Re } s > 0 \qquad (1.3.4)$$

(see Eq. 1.5.6). Therefore the sampled unit step has a transform

$$F_s(s) = \frac{1}{T}\sum_{n=-\infty}^{\infty} F(s - jn\omega_s) = \frac{1}{T}\sum_{n=-\infty}^{\infty} \frac{1}{s - jn\omega_s}, \qquad \text{Re } s > 0 \qquad (1.3.5)$$

using Eq. 1.3.3. The pole-zero pattern of the unit step is shown in Fig. 1.3.2a. It has a pole at the origin and a zero at infinity. The sampled unit step has the pole-zero pattern shown in Fig. 1.3.2b. It is a replicated version of Fig. 1.3.2a obtained by vertically shifting the entire pole-zero pattern by $n\omega_s$ for $n = \ldots, -1, 0, 1, \ldots$ and summing the results. ∎

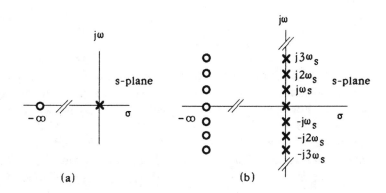

Fig. 1.3.2 Pole-zero
pattern of (a) unit step
and (b) sampled unit
step in Example 1.3.1.

To show the meaning of $F_s(s)$, assume that $F(s)$ has a region of convergence which includes the $j\omega$-axis so its Fourier transform is defined. Setting $s = j\omega$ in Eq. 1.3.3,

$$F_s(j\omega) = \frac{\omega_s}{2\pi} \sum_{n=-\infty}^{\infty} F(j(\omega - n\omega_s)), \qquad \frac{\omega_s}{2\pi} = \frac{1}{T} \tag{1.3.6}$$

Thus the sampled spectrum $F_s(j\omega)$ is the superposition of an infinite number of frequency signal spectrums $F(j\omega)$ as shown in Fig. 1.3.1. When $F(j\omega)$ is band-limited so $F(j\omega) = 0$ for $\omega > B$, then the spectrums centered about $\ldots, -\omega_s, 0, \omega_s, 2\omega_s, \ldots$ do not overlap if the sampling frequency $\omega_s > 2B$. Otherwise the spectral terms overlap causing what is called *frequency aliasing*.[4] Without frequency aliasing, the original spectrum $F(j\omega)$ can be recovered by filtering the sampled spectrum $F_s(j\omega)$ with an ideal lowpass filter having a gain $H(j\omega)$ of

$$\begin{aligned} H(j\omega) &= 1, \qquad |\omega| \leq \omega_s/2 = \omega_N \\ &= 0, \qquad |\omega| > \omega_s/2 \end{aligned} \tag{1.3.7}$$

It is rather remarkable but not surprising that band-limited signals can be reconstructed from their samples if the samples are taken sufficiently close together. The closeness is determined by the highest frequency present in the signal. The sampled signal $f_s(t)$ equals the inverse Fourier transform of the $F_s(j\omega)$ sampled spectrum as (see Eq. 1.6.1)

$$f_s(t) = \frac{1}{2\pi} \int_{-\infty}^{\infty} F_s(j\omega)e^{j\omega t}\, d\omega = \frac{f(t)}{T} \sum_{m=-\infty}^{\infty} e^{jm\omega_s t} = \frac{f(t)}{T}\left[1 + 2\sum_{m=1}^{\infty} \cos\left(\frac{2\pi m t}{T}\right)\right] \tag{1.3.8}$$

This shows the sampled signal and the signal itself are always identical within a constant $\omega_s/2\pi = 1/T$ at times t which are integer multiples of T. At all other times, they are generally not equal. To obtain the interpolated signal $f_r(t)$, the lowpass filter output in Fig. 1.3.1 can be expressed in convolutional form as

$$\begin{aligned} f_r(t) = h(t) * f_s(t) &= \frac{\omega_s}{\pi}\frac{\sin(\omega_s t/2)}{\omega_s t/2} * f(t) \sum_{n=-\infty}^{\infty} U_0(t - nT) \\ &= \frac{\omega_s}{\pi} \sum_{n=-\infty}^{\infty} f(nT)\frac{\sin\left[(\omega_s/2)(t - nT)\right]}{(\omega_s/2)(t - nT)} \end{aligned} \tag{1.3.9}$$

Half the sampling frequency $\omega_s/2$ is defined to equal the *Nyquist frequency* ω_N. No frequency aliasing occurs when the signal bandwidth $B < \omega_N$.

To illustrate these concepts, consider a signal which is band-limited for frequencies $|\omega| < 1$. For convenience, the spectrum $F(j\omega)$ is assumed to be real-valued as shown in Fig. 1.3.3a. Using sampling frequencies ω_s of 3, 2, and 1.5 radians/second (or sampling intervals $T = 2\pi/\omega_s$ of $2\pi/3, 2\pi/2$, and $2\pi/1.5$ seconds), $F(j\omega)$ is shifted integer multiples of ω_s, added to itself, and scaled by $1/T$. The resulting spectra are shown in Figs. 1.3.3b, 1.3.3c, and 1.3.3d, respectively. In Fig. 1.3.3b the sampling

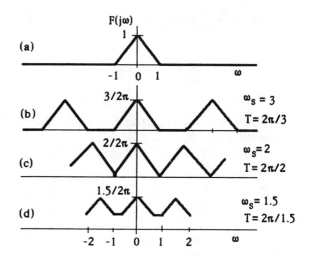

Fig. 1.3.3 (a) Band-limited spectrum $F(j\omega)$ and its sampled versions at sampling frequencies ω_s of (b) 3, (c) 2, and (d) 1.5 rad/sec.

frequency $\omega_s = 3$ is above the double bandwidth frequency of 2 so no spectral overlap occurs and there is no aliasing. However in Fig. 1.3.3d, ω_s is only 1.5 so there is spectral overlap and aliasing. Passing these spectra through the ideal lowpass filter given by Eq. 1.3.7 recovers $1/T$ times the original signal in Fig. 1.3.3b and not in Fig. 1.3.3d. The signal in Fig. 1.3.3d is said to have *aliasing distortion* which cannot be eliminated.

Now consider the case where sampling occurs at exactly the double bandwidth rate and spectral overlap occurs only at the band-edge as shown in Fig. 1.3.3c. Unless there is a discrete spectral component at B (i.e., a term in $F(j\omega)$ of $kU_0(\omega \pm B)$), the sampled spectrum contains zero energy at $\omega = \pm B$ (see Eq. 1.6.18). Such a zero spectrum is called a *null spectrum*. It is analogous to a *null signal* in the time domain which equals zero except at one time where it is nonzero but nonimpulsive. Null spectra and null signals have transform inverses (Eq. 1.6.1) that are zero everywhere and are of little practical interest. Thus when signals are sampled at the double the band-limit frequency $2B$, signals having no impulsive spectra at $\pm B$ can be perfectly recovered with no aliasing distortion by ideal lowpass filtering. This is the case in Fig. 1.3.3c. Those having impulsive spectra cannot be recovered without causing aliasing distortion.

Notice that if the signal spectrum $F(j\omega)$ being summed in Eq. 1.3.6 is complex-valued, then the shifted versions cannot be directly added to form the sampled spectrum as was done in Fig. 1.3.3. However a bound on the sampled spectrum magnitude $|F_s(j\omega)|$ can be found by manipulating Eq. 1.3.6. Since the magnitude of a complex sum can never exceed the sum of magnitudes (Schwartz's inequality), then

$$|F_s(j\omega)| = \left| \frac{1}{T} \sum_{n=-\infty}^{\infty} F(j(\omega - n\omega_s)) \right| \le \frac{1}{T} \sum_{n=-\infty}^{\infty} \left| F(j(\omega - n\omega_s)) \right| \qquad (1.3.10)$$

Thus, by summing shifted magnitude spectra, an upper bound is formed on the magnitude of the sampled spectrum. All spectra having the same magnitude but different phases have the same upper bound in Eq. 1.3.10. This equation will often be used.

EXAMPLE 1.3.2 Consider a signal $f(t)$ with a spectrum $F(j\omega)$ where

$$f(t) = 3e^{-t}\sin(3t)\, U_{-1}(t)$$

$$F(j\omega) = \frac{9}{10 - \omega^2 + j2\omega} \tag{1.3.11}$$

Determine the sampling rate for a spectral overlap of −20 dB. Compute the approximate aliasing error.

SOLUTION The magnitude of $F(j\omega)$ equals

$$|F(j\omega)| = \frac{9}{\sqrt{(10 - \omega^2)^2 + (2\omega)^2}} = \frac{9}{\sqrt{\omega^4 - 16\omega^2 + 100}} \tag{1.3.12}$$

which is plotted in Fig. 1.3.4a. The sampled spectrum $F_s(j\omega)$ is plotted in Fig. 1.3.4b. The corner frequency $\omega_n = \sqrt{10} \cong 3$ and the asymptotic roll-off is −40 dB/dec. The response is down −20 dB from its dc value at a frequency about 0.5 decade $= \sqrt{10}$ above ω_n. Therefore the −20 dB Nyquist frequency equals $\sqrt{10}\,\omega_n = 10$ rad/sec. Then the minimum sampling frequency ω_s equals $2\omega_N = 20$ rad/sec $\cong 3$ Hz. One measure of the approximate frequency aliasing error is the normalized energy in the signal spectrum $F(j\omega)$ lying beyond the Nyquist frequency ω_N. Mathematically the normalized energy error equals

Energy Error =

$$\frac{\int_{\omega_N}^{\infty}|F(j\omega)|^2\,d\omega}{\int_0^{\infty}|F(j\omega)|^2\,d\omega} = \frac{\int_{\omega_N}^{\infty}|F(j\omega)|^2\,d\omega}{\pi\int_{-\infty}^{\infty}|f(t)|^2\,dt} \cong \frac{\int_{\omega_N}^{\infty}(9/\omega^2)^2\,d\omega}{\pi\int_0^{\infty}9e^{-2t}\sin^2(3t)\,dt} = \frac{0.0315}{\pi(2.025)} = 0.5\%$$

$$\tag{1.3.13}$$

The approximate aliasing error is about 0.5% which is acceptably small. Eq. 1.3.13 has been simplified by using Parseval's theorem to convert the denominator integral into its time integral equivalent, and approximating $|F(j\omega)|$ by its stopband asymptote of $9/\omega^2$. ∎

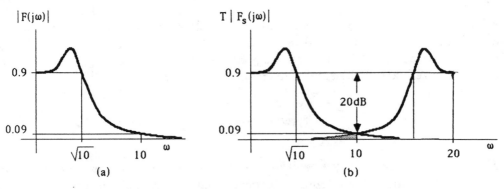

Fig. 1.3.4 Spectrum of (a) signal and (b) sampled signal in Example **1.3.2**.

1.3.2 MULTIRATE SAMPLING

Signals can be sampled at various uniform rates to achieve better results. For example, small bandwidth signals need be sampled only very slowly before processing. After digital processing, we would like to sample very fast to improve signal reconstruction. This is called *multirate sampling*. A multirate system is shown in Fig. 1.3.5. The input is sampled at time $t = nT$ while the output is sampled at $t = nT/k$ where $k > 0$. This corresponds to sampling frequencies of $f_s = 1/T$ and $kf_s = k/T$ Hz, respectively. The two impulse sampling streams, when synchronized to $t = 0$, therefore equal

$$s_1(t) = \sum_{n=-\infty}^{\infty} U_0(t - nT), \qquad s_2(t) = \sum_{n=-\infty}^{\infty} U_0\left(t - n\frac{T}{k}\right) \qquad (1.3.14)$$

When the new sampling rate is higher than the original sample rate ($k > 1$), the conversion process is called *interpolation*. When the new rate is lower ($0 < k < 1$), the conversion process is called *decimation*. The number of samples is increased in interpolation since additional samples are being created. The size is decreased in decimation since samples are being discarded. Interpolation and decimation are dual processes. Interpolators can be converted into decimators and vice versa.

By the time scaling theorem of Table 1.6.2, if a signal $f(t)$ sampled every T seconds has a spectrum $F_{s1}(j\omega)$, then

$$f_{s2}(t) = f_{s1}(t/k), \qquad F_{s2}(j\omega) = kF_{s1}(jk\omega) \qquad (1.3.15)$$

where $F_{s2}(j\omega)$ is the spectrum of $f(t)$ sampled every T/k seconds. The F_{s1} spectrum

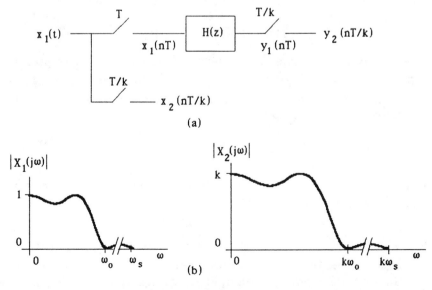

Fig. 1.3.5 (a) Multirate digital filter and its (b) ac steady-state response.

is frequency scaled by k and magnitude scaled by k to obtain F_{s2}. Therefore F_{s1} and F_{s2} are simply scaled versions of one another.

1.3.3 NONUNIFORM SAMPLING

Almost all digital filters use uniform sampling where $f(t)$ is sampled every T seconds at a $1/T$ Hz sampling rate. This was the case in Fig. 1.2.1. Multirate sampling systems use at least two uniform sampling rates such as $1/T$ and k/T Hz. *Nonuniform sampling* is a more general case where there is nonuniform time between samples. Samples are collected only when needed. When there is little information in $f(t)$ and its short-term spectrum is narrow-band, only a few samples are collected. When $f(t)$ contains much information and its short-term spectrum becomes wide-band, many more samples are collected. In other words, the sampling rate is not constant but varies to accommodate the short-term spectrum being processed. Multirate sampling systems are the same as those used in Fig. 1.2.1 but the switches have variable sampling rate.

The impulse sampling stream for nonuniform sampling equals

$$s(t) = \sum_{n=-\infty}^{\infty} U_0(t - t_n) \qquad (1.3.16)$$

In uniform sampling, the sampling time $t_n = nT$. In multirate sampling, at least two sampling times are used as $t_{n1} = nT$ and $t_{nk} = nT/k$. Nonuniform sampling uses different t_n functions and can be adaptive or nonadaptive. In *nonadaptive sampling*, t_n is only a function of time and not a function of the signal being sampled. For example, logarithmic nonuniform sampling sets $t_n = k \ln nT$ while power law nonuniform sampling sets $t_n = k^{nT}$. If logarithmic nonuniform sampling is used for an exponential function $f(t) = e^{-at}$, then $f(t_n)$ equals

$$t_n = k \ln nT, \qquad f(t_n) = e^{-ak \ln nT} = (nT)^{-ak} = 1/nT \text{ for } k = 1/a \qquad (1.3.17)$$

The nonuniformly sampled exponential function $f(t_n)$ decreases as $1/nT$ rather than exponentially as $\left(e^{-aT}\right)^n$ if uniform sampling was used. In general, nonuniform sampling is used only if the nature of the signal is known apriori so t_n can be properly selected. Otherwise adaptive sampling is used.

1.3.4 ADAPTIVE SAMPLING

In *adaptive sampling*, the sampling time t_n is a function g of the signal f being sampled and not of time. Therefore

$$t_n = g[f(t)] \qquad (1.3.18)$$

For bandwidth compression, $f(t)$ is sampled only when it is fluctuating. The more rapid the fluctuations, the faster the sampling.

In zero-order adaptive sampling, the signal is sampled only when the absolute change of $f(t)$ exceeds ϵ. Therefore,

$$f(t_n) = f(t_{n-1}) \pm \epsilon \tag{1.3.19}$$

in which case t_n equals

$$t_n = f^{-1}\left[f(t_{n-1}) \pm \epsilon\right] \tag{1.3.20}$$

f^{-1} is the inverse of f. Zero-order sampling is usually used in delta modulation. For the exponential signal, $f(t_{n-1}) = e^{-at_{n-1}}$ so that the sample times equal

$$t_n = (-1/a)\ln\left[e^{-at_{n-1}} \pm \epsilon\right] = t_{n-1} + (-1/a)\ln\left[1 \pm \epsilon e^{at_{n-1}}\right] \tag{1.3.21}$$

$+\epsilon$ is used for increasing exponentials and $-\epsilon$ is used for decreasing exponentials.

First-order adaptive sampling sets the sampling frequency proportional to the absolute value of the time rate of change of $f(t)$ (i.e., derivative) so that

$$t_n = t_{n-1} + \frac{1}{k|f'(t_{n-1})|} \quad \text{since} \quad \frac{1}{t_n - t_{n-1}} = k|f'(t_{n-1})| \tag{1.3.22}$$

For the exponential signal $f(t) = e^{-at}$, then $f'(t) = -ae^{-at} = -af(t)$ and

$$t_n = t_{n-1} + \frac{1}{k|af(t_{n-1})|} = t_{n-1} + (1/k|a|)e^{at_{n-1}} \tag{1.3.23}$$

The derivative in Eq. 1.3.22 can be approximated in a variety of ways. Using a backward difference leads to a recursion equation for t_n as

$$t_n \cong t_{n-1} + \frac{1}{k}\left|\frac{t_{n-1} - t_{n-2}}{f(t_{n-1}) - f(t_{n-2})}\right|, \quad f(t_{n-1}) \neq f(n_{n-2}) \tag{1.3.24}$$

and $t_n = t_{n-1} + \Delta t_{max}$ when $f(t_{n-1}) = f(t_{n-2})$. Defining the initial starting points as $t_{-1} = -T$ and $t_0 = 0$, then the next sample is taken at $t_1 = T/k|f(0) - f(-T)|$ or Δt_{max}. The sampling process continues indefinitely using Eq. 1.3.24.

Although the zero-order and first-order adaptive sampling algorithms given by Eqs. 1.3.20 and 1.3.22 are somewhat involved mathematically, they are easily implemented in hardware.[4]

1.4 CONTINUOUS TRANSFORMS, SERIES TRANSFORMS, AND DISCRETE TRANSFORMS

We shall now discuss generalized transforms of which there are an infinite number. Some of the best known and most often used are listed in Table 1.4.1. Generalized transforms fall into three categories – continuous, series, and discrete. From the most general viewpoint, continuous transforms are used to describe continuous time, continuous frequency signals. Series transforms are used to describe continuous time, discrete frequency signals. Discrete transforms are used to describe discrete time, discrete frequency signals.

Continuous transforms or *integral transforms* have the integral form[5]

$$F(p) = \int_{t_1}^{t_2} K(p,t)f(t)dt, \qquad p \in \mathcal{R}_p$$

$$f(t) = \int_{p_1}^{p_2} K^{-1}(p,t)F(p)dp, \qquad t \in \mathcal{R}_t$$

$$(1.4.1)$$

$K(p,t)$ is called the *kernel* and $K^{-1}(p,t)$ is the *inverse kernel*. The kernel and inverse kernel are orthogonal where $\int_{p_1}^{p_2} K(p,t)K^{-1}(p,\tau)dp = U_0(t-\tau)$. The transform converts a function $f(t)$ of time t into a function (or generalized spectrum) $F(p)$ of complex variable p (or generalized frequency). $F(p)$ is only defined in its region of convergence $\mathcal{R}_p$. The spectral frequencies are all the p values in $\mathcal{R}_p$. The spectral coefficient $F(p)$ equals the amount, level, or contribution of the kernel $K(p,t)$ in $f(t)$. The inverse transform converts $F(p)$ back into $f(t)$. The times are all the t values in $\mathcal{R}_t$. Laplace and Fourier transforms will be discussed in Chaps. 1.5 and 1.6, respectively.

The properties of a transform are determined by the properties of its kernel. The inverse kernels are called *basis functions* and are used to linearly approximate $f(t)$. They describe or span the function space of $f(t)$. If the kernel set is *complete*, $f(t)$ can be expressed as a linear combination (i.e., integral) of the basis functions with an integral-squared error of zero.

When $F(p)$ is nonzero for all p in $\mathcal{R}_p$, $f(t)$ is said to have a *continuous spectrum*. When $F(p)$ is nonzero at only certain p (such as p_m) in $\mathcal{R}_p$, then $f(t)$ is said to have a *discrete* or *line spectrum*. In this case, $F(p)$ becomes impulsive and the continuous transform reduces to a series transform. *Series transforms* have the infinite series form

$$F(p) = \sum_{m=1}^{\infty} F(p_m)U_0(p - p_m) \quad \text{where} \quad F(p_m) = \int_{t_1}^{t_2} K(p_m,t)f(t)dt, \ p_m \in \mathcal{R}_p$$

$$f(t) = c \sum_{m=1}^{\infty} F(p_m)K^{-1}(p_m,t), \quad t \in \mathcal{R}_t$$

$$(1.4.2)$$

Integral transforms describe continuous signals having continuous spectra. Series transforms describe continuous signals having discrete spectra. The discrete spectrum given by Eq. 1.4.2a can also be viewed as a zero-order approximation of the continuous spectrum of Eq. 1.4.1a where the integral is approximated by a sum. The approximation is exact when $F(p)$ is impulsive. The relation between these transforms will be discussed further in Chap. 3.8. Fourier series will be discussed in Chap. 1.7.

Discrete transforms have the finite series form

$$F(m) = \sum_{n=1}^{N} K(m,n)f(n), \qquad m = 1,\ldots,M \quad (\text{or } 0,\ldots,M-1)$$

$$f(n) = \sum_{m=1}^{M} K^{-1}(n,m)F(m), \qquad n = 1,\ldots,N \quad (\text{or } 0,\ldots,N-1)$$

$$(1.4.3)$$

Table 1.4.1 Table of continuous and discrete transform kernels.

Continuous Type	$K(p,t)$	$K^{-1}(p,t)$
Laplace–LT	e^{-st}	$(1/j2\pi)e^{st}$
Fourier–FT	$e^{-j\omega t}$	$(1/2\pi)e^{j\omega t}$
Cosine–CosT	$\cos(\omega t)$	$(1/2\pi)\cos(\omega t)$
Sine–SinT	$\sin(\omega t)$	$(1/2\pi)\sin(\omega t)$
Paley–PT	$Pal(p,t)$	$Pal^{-1}(p,t)$, Fig. 2.8.1c
Hadamard–HT	$Had(p,t)$	$Had^{-1}(p,t)$, Fig. 2.8.1d
Haar–HaT	$Ha(p,t)$	$Ha^{-1}(p,t)$, Fig. 2.8.1e
Slant–ST	$S(p,t)$	$S^{-1}(p,t)$, Fig 2.8.1f

Discrete Type	$K(m,n)$	$K^{-1}(n,m)$
z–ZT	z^{-n}	$(1/j2\pi T)z^{n-1}$
Fourier–DFT	$e^{-j2nm\pi/N}$	$(1/N)e^{j2nm\pi/N}$
Cosine–DCosT	$\cos(2nm\pi/N)$	$(1/N)\cos(2nm\pi/N)$
Sine–DSinT	$\sin(2nm\pi/N)$	$(1/N)\sin(2nm\pi/N)$
Paley–DPT	$(1/N)(-1)^{\sum_{i=0}^{L-1} b_i(n)b_{L-1-i}(m)}$	$(-1)^{\sum_{i=0}^{L-1} b_i(n)b_{L-1-i}(m)}$
Hadamard–DHT	$(1/N)(-1)^{\sum_{i=0}^{L-1} b_i(n)b_i(m)}$	$(-1)^{\sum_{i=0}^{L-1} b_i(n)b_i(m)}$
Haar–DHaT	Eq. 2.8.28	Eq. 3.3.13
Slant–DST	Eq. 2.8.33	Eq. 3.3.17

Discrete transforms are used when the signal spectrum is discrete and band-limited (i.e., the infinite series of Eq. 1.4.2a is truncated to finite length) and the signal is impulsively sampled in time. Both discrete transforms of Eq. 1.4.3 can be viewed as zero-order approximations of the continuous transforms of Eq. 1.4.1. Time sampling $f(t)$ at $t = t_n$ gives $f(n)$ while frequency sampling $F(p)$ at $p = p_m$ gives $F(p_m)$. $K(m,n)$ and $K^{-1}(m,n)$ are determined by time and frequency sampling the continuous $K(p,t)$ and $K^{-1}(p,t)$, respectively. The $K^{-1}(n,m)$ matrix is the transpose of the $K^{-1}(m,n)$ matrix. Uniform sampling is generally used. The discrete Fourier transform and z-transform will be discussed in Chaps. 1.8 and 1.9–1.10, respectively. The discrete Paley, Hadamard, Haar, and slant transforms will be discussed in Chaps. 2.8 and 3.3.

Regardless of which transform type is used, only a few spectral terms are nonzero when $f(t)$ is closely *matched* to (i.e., resembles) its basis functions. When $f(t)$ is *mis-*

matched, many spectral terms are present. *Generalized bandwidth* is measured by the number of terms containing significant energy. Bandwidth is minimized by selecting a transform that has only one (or a few) basis function which resembles the signal.

1.5 LAPLACE TRANSFORM (LT) AND INVERSE TRANSFORM (ILT)

The *Laplace transform (LT)* is a mathematical operation that transforms a time domain function $f(t)$ into a frequency domain function $F(s)$. It equals[6]

$$F(s) = \mathcal{L}\big[f(t)\big] = \int_{-\infty}^{\infty} f(t)e^{-st}dt, \qquad s \in \mathcal{R} \tag{1.5.1}$$

where $\mathcal{L}[\ ]$ denotes the Laplace transform operation. Thus, a function f of time t is converted into a function F of complex frequency $s = \sigma + j\omega$. s has a real part σ and an imaginary part ω. A particular complex frequency s_o represents a point in the *s-plane*. Although $f(t)$ is depicted in a 2-dimensional plot (f versus t), $F(s)$ must be depicted in one 4-dimensional plot ($|F|$ and arg F or Re F and Im F versus σ and ω) or two 3-dimensional plots. In general, $F(s)$ will only exist (be defined) over some restricted region $\mathcal{R}$ (i.e., values of s) of the s-plane. This is called the *region of convergence $\mathcal{R}$*. F is analytic inside $\mathcal{R}$. Eq. 1.5.1 describes the *two-sided* Laplace transform. The *one-sided* LT uses integration limits of $(0, \infty)$. It is used only for signals that are zero for negative time (i.e., causal).

$f(t)$ is obtained from $F(s)$ using the *inverse Laplace transform (ILT)* where

$$f(t) = \mathcal{L}^{-1}\big[F(s)\big] = \frac{1}{j2\pi} \int_{c-j\infty}^{c+j\infty} F(s)e^{st}ds, \qquad c \in \mathcal{R} \tag{1.5.2}$$

where $\mathcal{L}^{-1}[\ ]$ denotes the inverse Laplace transform operation. Thus, a function F of complex frequency s is converted into a function of time t. The integration takes place along a vertical line Re $s = c$ in the s-plane. This line must lie in the region of convergence $\mathcal{R}$ of F. To minimize notational confusion, lower-case (small) letters are reserved for time domain functions and upper case (capital letters) for s-domain functions. Thus, f represents a time domain function and F represents a frequency domain function.

EXAMPLE 1.5.1 Determine the Laplace transform and its region of convergence for a unit step function $U_{-1}(t)$ shown in Fig. 1.5.1a. Sketch $f(t)$ and $F(s)$.
SOLUTION Using Eq. 1.5.1, the Laplace transform $F(s)$ of $f(t)$ must equal

$$F(s) = \int_{-\infty}^{\infty} U_{-1}(t)e^{-st}dt = \int_{0}^{\infty} e^{-st}dt = \frac{e^{-st}}{-s}\bigg|_{t=0}^{\infty} = \frac{1}{s}\big(1 - \lim_{t\to\infty} e^{-st}\big) \tag{1.5.3}$$

Since $s = \sigma + j\omega$, the second term can be bounded by

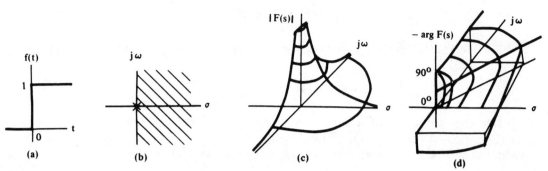

Fig. 1.5.1   (a) Signal $f(t)$, (b) region of convergence, (c) magnitude $|F(s)|$, and (d) phase arg $F(s)$ of Example 1.5.1.

$$|e^{-st}| = |e^{-(\sigma+j\omega)t}| = |e^{-\sigma t}||e^{-j\omega t}| = e^{-\sigma t} \qquad (1.5.4)$$

The limit is therefore

$$\lim_{t\to\infty} |e^{-st}| = \lim_{t\to\infty} e^{-\sigma t} = 0, \qquad \sigma > 0 \qquad (1.5.5)$$

The limit is zero only if Re $s = \sigma > 0$. Otherwise the limit is undefined. This establishes the region of convergence in the s-plane where the Laplace transform exists. Thus, the Laplace transform of the unit step equals

$$F(s) = \mathcal{L}\big[U_{-1}(t)\big] = \frac{1}{s}, \qquad \text{Re } s > 0 \qquad (1.5.6)$$

In any other region, $F(s)$ is undefined. Since

$$|F(s)| = \frac{1}{|s|}, \qquad \text{arg } F(s) = -\text{arg } s, \qquad \text{Re } s > 0 \qquad (1.5.7)$$

the magnitude and angle of F are shown in Fig. 1.5.1. Since we cannot draw a 4-dimensional sketch to characterize F, we must use two 3-dimensional sketches instead. Notice the Laplace transform is only defined (or exists) in the right-half s-plane (or simply right-half-plane) which is its region of convergence. This region is often denoted by shading the appropriate area in the s-plane. ■

EXAMPLE 1.5.2 Determine the Laplace transform for $-U_{-1}(-t)$. Compare its region of convergence with that in Example 1.5.1.
SOLUTION   Its Laplace transform equals

$$F(s) = -\int_{-\infty}^{\infty} U_{-1}(-t)e^{-st}dt = \int_{\infty}^{0} e^{st}dt = \frac{e^{st}}{s}\bigg|_{t=\infty}^{0} = \frac{1}{s}(1 - \lim_{t\to\infty} e^{st}) \qquad (1.5.8)$$

Evaluating the limit as before in Eq. 1.5.5,

$$\lim_{t\to\infty} e^{st} \leq \lim_{t\to\infty} |e^{st}| = \lim_{t\to\infty} e^{\sigma t} = 0, \qquad \sigma < 0 \tag{1.5.9}$$

which is equal to zero only if Re $s = \sigma < 0$. Thus, the Laplace transform equals

$$F(s) = \mathcal{L}\big[-U_{-1}(-t)\big] = \frac{1}{s}, \qquad \text{Re } s < 0 \tag{1.5.10}$$

Notice that the only difference between this transform and that in Example 1.5.1 is its region of convergence. This transform converges only in the left-half-plane, while that of the previous example converged in the right-half-plane. ■

A number of Laplace transforms are tabulated in Table 1.5.1 for future reference. These are derived using the definition of the Laplace transform. Several noncausal functions are included as a matter of interest.

The Laplace transform exists for those values of s for which $F(s)$ is finite. Since

$$|F(s)| = \left| \int_{-\infty}^{\infty} f(t)e^{-st}\,dt \right| \leq \int_{-\infty}^{\infty} |f(t)e^{-st}|\,dt = \int_{-\infty}^{\infty} |f(t)|e^{-\sigma t}\,dt, \qquad \text{Re } s = \sigma \tag{1.5.11}$$

it is sufficient for $F(s)$ to exist if

$$\int_{-\infty}^{\infty} |f(t)|e^{-\sigma t}\,dt < \infty \tag{1.5.12}$$

(is finite). All exponential-order functions satisfy this requirement. A function $f(t)$ is of *exponential order k* if

$$\begin{aligned} |f(t)| &\leq M_1 e^{k_1 t}, \quad t > 0 \\ &\leq M_2 e^{k_2 t}, \quad t < 0 \end{aligned} \tag{1.5.13}$$

In other words, function $f(t)$ does not grow more rapidly than its associated exponential bound. Substituting these exponential bounds into Eq. 1.5.12 shows that

$$\int_{-\infty}^{\infty} |f(t)|e^{-\sigma t}\,dt \leq \int_{-\infty}^{0} M_2 e^{(k_2-\sigma)t}\,dt + \int_{0}^{\infty} M_1 e^{(k_1-\sigma)t}\,dt \tag{1.5.14}$$

$$= \frac{M_2}{k_2 - \sigma} - \frac{M_1}{k_1 - \sigma} < \infty \qquad \text{for} \qquad k_1 < \sigma \text{ and } k_2 > \sigma$$

Therefore, the Laplace transform $F(s)$ for $f(t)$ has a region of convergence no less than

$$k_1 < \text{Re } s < k_2 \tag{1.5.15}$$

and which may be even greater. A number of examples are shown in Fig. 1.5.2 which the engineer should verify.

Note that when $f(t)$ is zero for $t < 0$, it is bounded by e^{kt} where $k = \infty$. Thus, Re $s = \infty$ forms the right-hand boundary on the region of convergence $\mathcal{R}$. The right-hand boundary is determined by the form of $f(t)$ for $t < 0$. Conversely, when $f(t)$ is zero for $t > 0$, it is bounded by e^{kt} where $k = -\infty$. Thus, Re $s = -\infty$ forms the left-

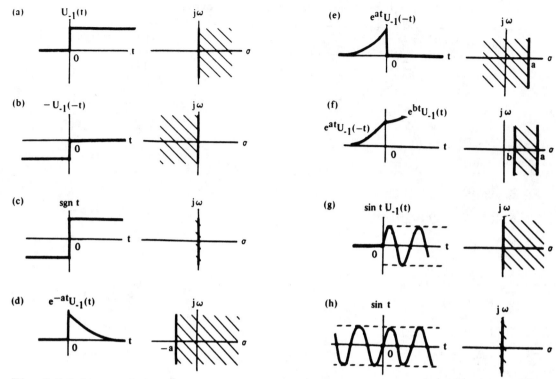

Fig. 1.5.2 Assorted time domain signals and their (minimum) regions of convergence.

hand boundary of $\mathcal{R}$. The left-hand boundary then depends upon the form of $f(t)$ for $t > 0$. Of course, there is an infinite number of functions which are not of exponential order such as $t^t, \exp(t^2)$, and $1/t$. They simply vary too rapidly to be bounded by an exponential. However, they are of little practical interest.

The Laplace transform possesses a number of interesting properties. These properties are listed in Table 1.5.2 for the two-sided Laplace transform. They are easily proved using the transformation definition. The engineer will find it interesting to see how many of the entries in the Laplace transform table (Table 1.5.1) can be derived using these properties and the unit step or unit impulse transform. It will be found that they can all be written very simply which shows the great utility of Table 1.5.2.

In situations where the Laplace transform appears to be undefined, generalized function theory is used. Often integrals that do not exist in the classical sense do exist in the generalized sense. Several examples are tabulated at the end of Table 1.5.1. They are extremely useful as we shall see in the next section when their physical meaning will be interpreted. It is useful to note that these transforms can be combined with others from Table 1.5.1 using superposition to extend their region of convergence.

When describing the properties of Laplace transforms $F(s)$, and simply functions in general, several mathematical definitions are used. Recall that a function $F(s)$ is *analytic* at a point s_o if it is single-valued and has a unique finite derivative at s_o and in

Table 1.5.1 Table of Laplace transforms.

f(t)	F(s)	Convergence Region
1. $U_0(t)$	1	all s
2. $U_1(t)$	s	all s
3. $U_n(t)$	s^n	all s
4. $U_{-1}(t)$	$\dfrac{1}{s}$	Re s $>$ 0
5. $tU_{-1}(t) = U_{-2}(t)$	$\dfrac{1}{s^2}$	Re s $>$ 0
6. $(t^{n-1}/(n-1)!)U_{-1}(t) = U_{-n}(t)$	$\dfrac{1}{s^n}$	Re s $>$ 0
7. sgn t	$\dfrac{2}{s}$	Re s $=$ 0
8. $-U_{-1}(-t)$	$\dfrac{1}{s}$	Re s $<$ 0
9. $e^{-\sigma t}U_{-1}(t)$	$\dfrac{1}{s+\sigma}$	Re s $>-$ Re σ
10. $te^{-\sigma t}U_{-1}(t)$	$\dfrac{1}{(s+\sigma)^2}$	Re s $>-$ Re σ
11. $(t^n/n!)e^{-\sigma t}U_{-1}(t)$	$\dfrac{1}{(s+\sigma)^{n+1}}$	Re s $>-$ Re σ
12. $(1-e^{-\sigma t})U_{-1}(t)$	$\dfrac{\sigma}{s(s+\sigma)}$	Re s $>$ max $(0, -$ Re $\sigma)$
13. $e^{-\sigma|t|}$	$\dfrac{2\sigma}{\sigma^2-s^2}$	$-$ Re $\sigma <$ Re s $<$ Re σ
14. $2 \sinh \sigma t\ U_{-1}(t)$	$-\dfrac{2\sigma}{\sigma^2-s^2}$	Re s $>$ max (Re $\sigma, -$ Re σ)
15. $\cos \omega t\ U_{-1}(t)$	$\dfrac{s}{s^2+\omega^2}$	Re s $>$ 0
16. $\sin \omega t\ U_{-1}(t)$	$\dfrac{\omega}{s^2+\omega^2}$	Re s $>$ 0
17. $e^{-\sigma t}\cos \omega t\ U_{-1}(t)$	$\dfrac{s+\sigma}{(s+\sigma)^2+\omega^2}$	Re s $>-$ Re σ
18. $e^{-\sigma t}\sin \omega t\ U_{-1}(t)$	$\dfrac{\omega}{(s+\sigma)^2+\omega^2}$	Re s $>-$ Re σ
19. 1	$2\pi U_0(-js)$	Re s $=$ 0
20. t^n	$2\pi j^n U_n(-js)$	Re s $=$ 0
21. $e^{j\omega_0 t}$	$2\pi U_0(-js-\omega_0)$	Re s $=$ 0
22. $\cos \omega_0 t$	$\pi[U_0(-js+\omega_0) + U_0(-js-\omega_0)]$	Re s $=$ 0
23. $\sin \omega_0 t$	$j\pi[U_0(-js+\omega_0) - U_0(-js-\omega_0)]$	Re s $=$ 0
24. sgn t	$2/s$	Re s $=$ 0

Table 1.5.2 Properties of Laplace transforms.

	$f(t)$	$\mathcal{L}[f(t)] = F(s) = \int_{-\infty}^{\infty} f(t)e^{-st}\,dt$	Region of Convergence $a < \mathrm{Re}\,s < \beta$
Definition			
Linear Operations			
Homogeneity	$Kf(t)$	$KF(s)$	$a < \mathrm{Re}\,s < \beta$
Superposition	$f_1(t) \pm f_2(t)$	$F_1(s) \pm F_2(s)$	$\left.\begin{array}{l} \max(a_1, a_2) < \mathrm{Re}\,s \\ < \min(\beta_1, \beta_2) \end{array}\right\}$
Both operations	$a_1 f_1(t) \pm a_2 f_2(t)$	$a_1 F_1(s) \pm a_2 F_2(s)$	
Symmetry	$F(t)$	$2\pi f(-s)$	$a < \mathrm{Re}\,s = 0 < \beta$
Scaling			
Time scaling	$f(at)$	$\dfrac{1}{|a|} F(\dfrac{s}{a})$	$a\,a < \mathrm{Re}\,s < a\,\beta$
Magnitude scaling	$af(t)$	$aF(s)$	$a < \mathrm{Re}\,s < \beta$
Shifting			
Time shifting	$f(t - a)$	$e^{-as}F(s)$	$a < \mathrm{Re}\,s < \beta$
Frequency shifting	$e^{-at}f(t)$	$F(s + a)$	$a - \mathrm{Re}\,a < \mathrm{Re}\,s < \beta - \mathrm{Re}\,a$
Periodic Functions (for $t > 0$ only)	$f(t) = \sum\limits_{n=0}^{\infty} f_1(t + nT)$	$\dfrac{F_1(s)}{1 - e^{-Ts}} = F_1(s)\sum\limits_{n=0}^{\infty} e^{-nTs}$	$a < \mathrm{Re}\,s$
		where $F_1(s) = \int_0^T f(t)e^{-st}\,dt$	$a < \mathrm{Re}\,s$
Convolution			
Time convolution (Frequency product)	$f_1(t)*f_2(t) = \int_{-\infty}^{\infty} f_1(\tau)f_2(t - \tau)\,d\tau$	$F_1(s)F_2(s)$	$\left\{\begin{array}{l} \max(a_1, a_2) < \mathrm{Re}\,s \\ < \min(\beta_1, \beta_2) \end{array}\right.$
Frequency convolution (Time product)	$f_1(t)f_2(t)$	$\dfrac{1}{2\pi} F_1(s)*F_2(s) = \dfrac{1}{2\pi j}\int_{c - j\infty}^{c + j\infty} F_1(p)F_2(s - p)\,dp$	$\begin{array}{l} a_1 + a_2 < \mathrm{Re}\,s < \beta_1 + \beta_2 \\ a_1 < c < a_2 \end{array}$
Time Differentiation *&			
First derivative	$df(t)/dt$	$sF(s) - f(0^+)$	$a < \mathrm{Re}\,s < \beta$
Second derivative	$d^2 f(t)/dt^2$	$s^2 F(s) - sf(0^+) - df(0^+)/dt$	$a < \mathrm{Re}\,s < \beta$
nth derivative	$d^n f(t)/dt^n$	$s^n F(s) - s^{n-1}f(0^+) - s^{n-2}df(0^+)/dt$ $- \ldots - d^{n-1}f(0^+)/dt^{n-1}$	$a < \mathrm{Re}\,s < \beta$
Time Integration *&			
First integral	$\int_{-\infty}^{t} f(t)\,dt = f^{(-1)}(t)$	$F(s)/s + f^{(-1)}(0^+)/s$	
Second integral	$\int_{-\infty}^{t}\int_{-\infty}^{t} f(t)\,dt\,dt = f^{(-2)}(t)$	$F(s)/s^2 + f^{(-1)}(0^+)/s^2 + f^{(-2)}(0^+)/s$	$\left.\begin{array}{l} \max(a, 0) \\ < \mathrm{Re}\,s < \beta \end{array}\right\}$
nth integral	$f^{(-n)}(t)$	$F(s)/s^n + f^{(-1)}(0^+)/s^n + f^{(-2)}(0^+)/s^{n-1}$ $+ \ldots + f^{(-n)}(0^+)/s$	
Frequency Differentiation			
First derivative	$tf(t)$	$-dF(s)/ds$	$a < \mathrm{Re}\,s < \beta$
nth derivative	$t^n f(t)$	$(-1)^n d^n F(s)/ds^n$	$a < \mathrm{Re}\,s < \beta$
Frequency Integration			
nth integral	$t^{-n}f(t)$	$(-1)^n F^{(-n)}(s)$	$a < \mathrm{Re}\,s < \beta$
Initial Value *	$f(0^+) = \lim\limits_{t \to 0} f(t)$	$= \lim\limits_{s \to \infty} sF(s)$	
Final Value *	$f(\infty) = \lim\limits_{t \to \infty} f(t)$	$= \lim\limits_{s \to 0} sF(s)$	providing poles of $sF(s)$ in left-half-plane
Initial Slopes *	$d^{m-1}f(0+)/dt^{m-1}$	$= \lim\limits_{s \to \infty} s^m F(s)$	

*For one-sided transforms. &In two-sided case, drop all terms involving $f^{(i)}(0+)$ in F(s) and retain only first term.

the neighborhood surrounding s_0. Any analytic function can be described by a *Taylor series* about s_0. If $s_0 = 0$ which is the origin of the s-plane, then such a series is called the *Maclaurin series*. The Taylor and Maclaurin series have the respective forms

$$F(s) = F(s_0) + F'(s_0)(s - s_0) + F''(s_0)\frac{(s - s_0)^2}{2!} + \cdots,$$

$$F(s) = F(0) + sF'(0) + \frac{s^2}{2!}F''(0) + \cdots, \qquad |s| < \infty$$

(1.5.16)

Singularities or, more precisely, *singular points* are points in the s-plane (i.e., values of s) where $F(s)$ fails to be analytic. *Isolated singularities* are singular points around which nonoverlapping circles can be drawn. Every $F(s)$ having isolated singularities can be described by a *Laurent series* which has the form

$$F(s) = \sum_{n=1}^{N} \frac{F^{-n}(s_0)}{(s - s_0)^n} + \sum_{n=0}^{\infty} \frac{(s - s_0)^n F^n(s_0)}{n!}, \qquad |s - s_0| \leq r \qquad (1.5.17)$$

in the vicinity of every isolated singularity s_0. The vicinity or region where $F(s)$ is so defined is bounded by a circle with center s_0 and radius r. The radius is equal to the distance between s_0 and its nearest isolated singularity. To further classify such $F(s)$, we consider the descending part of the series. If N is finite, then the isolated singular point s_0 of $F(s)$ is said to be a *pole* of order N. When N is infinite, the singularity s_0 is said to be an *essential singularity* or an *accumulation point* of $F(s)$. If $F(s)$ has only poles as singularities, it is said to be a *rational function*. Alternatively, it is sometimes called a *meromorphic function*. However, meromorphic functions may have an infinite number of poles while rational functions may have only a finite number.

When the reciprocal function $1/F(s)$ is considered, the same terminology is used. When the poles of $1/F(s)$ are established, these are said to be *zeros* of $F(s)$. Usually the functions which we shall consider will have only poles and zeros with no other types of singularities.

Two-dimensional (2-D) Laplace transforms. These are used to describe signals that are functions of two independent variables rather than one. The 2-D Laplace transform $F(s_1, s_2)$ and its inverse equal

$$F(s_1, s_2) = \mathcal{L}[f(t_1, t_2)] = \int_{-\infty}^{\infty} \int_{-\infty}^{\infty} f(t_1, t_2)e^{-(s_1 t_1 + s_2 t_2)} dt_1 dt_2, \quad s_1, s_2 \in \mathcal{R}$$

$$f(t_1, t_2) = \mathcal{L}^{-1}[F(s_1, s_2)]$$

$$= \frac{1}{(j2\pi)^2} \int_{c_1 - j\infty}^{c_1 + j\infty} \int_{c_2 - j\infty}^{c_2 + j\infty} F(s_1, s_2)e^{(s_1 t_1 + s_2 t_2)} ds_1 ds_2, \quad c_1, c_2 \in \mathcal{R}$$

(1.5.18)

This is a simple generalization of the 1-D Laplace transform $F(s)$ of Eq. 1.5.1. It can be expanded to N dimensions by extending the power of the exponential to include the $s_N t_N$ term and integrating over all N dimensions. This produces $F(s_1, s_2, \ldots, s_N)$ which has a region of convergence $s_1, s_2, \ldots, s_N \in \mathcal{R}$ in the N-dimensional space.

It should be remembered that this transform maps a function f of N independent variables like time or space into N independent complex frequency variables s_i. Some 2-D examples are time delay analysis of $f(t, T)$, transmission line analysis of $f(x, t)$, and image analysis of $f(x, y)$. Although the two physical dimensions involve different combinations of time and spacial variables (with different units), they are all transformed using a 2-D Laplace transform.

EXAMPLE 1.5.3 Determine the Laplace transform for an image having an intensity function $f(x, y)$ of

$$f(x, y) = e^{-y} \left[U_{-1}(x) - U_{-1}(x - 1) \right] U_{-1}(y) \tag{1.5.19}$$

SOLUTION $f(x, y)$ is defined in the first quadrant with a constant intensity in the x-direction and exponential decrease in the y-direction. A value of unity represents blackness and zero represents whiteness. From Eq. 1.5.18, $f(x, y)$ has a 2-D Laplace transform which equals

$$F(s_1, s_2) = \int_0^\infty \int_0^1 e^{-y} e^{-(s_1 x + s_2 y)} dx\, dy = \int_0^1 e^{-s_1 x} dx \int_0^\infty e^{-(s_2+1)y} dy$$

$$= \frac{1 - e^{-s_1}}{s_1(s_2 + 1)}, \quad \text{all } s_1,\ \text{Re } s_2 > -1$$

$$\tag{1.5.20}$$

The transform $F(s_1, s_2)$ has no pole at $s_1 = 0$, a single pole at $s_2 = -1$, and a zero at infinity. Its region of convergence is the intersection of all s_1 and Re $s_2 > -1$. ∎

The Laplace transform of a sampled signal $f_s(t)$ was given by Eq. 1.3.3 as

$$F_s(s) = \frac{\omega_s}{2\pi} \sum_{n=-\infty}^{\infty} F(s - jn\omega_s), \qquad f_s = \frac{\omega_s}{2\pi} = \frac{1}{T} \tag{1.5.21}$$

f_s is the sampling frequency in Hz. The Laplace transform $F_s(s)$ is obtained from the

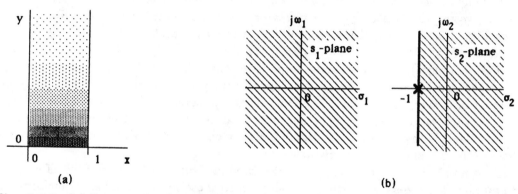

(a)   (b)

Fig. 1.5.3 (a) Image $f(x, y)$ and (b) its region of convergence of Example 1.5.3.

Fourier transform $F_s(j\omega)$ using analytic continuation. This involves setting $\omega = s/j = -js$ (since $s = j\omega$) as will be discussed in the next section. $F_s(s)$ is the sum of an infinite number of frequency-shifted $F(s)$, the Laplace transform of the unsampled signal. This is illustrated in Fig. 1.16.2 and is discussed in Example 1.16.2.

1.6 FOURIER TRANSFORM (FT) AND INVERSE TRANSFORM (IFT)

The Fourier transform is a special case of the Laplace transform. Mathematically, the *Fourier transform (FT)* and the *inverse Fourier transform (IFT)* equal[7]

$$F(j\omega) = \mathcal{F}[f(t)] = \int_{-\infty}^{\infty} f(t)e^{-j\omega t}dt, \qquad \omega \in (-\infty, \infty)$$

$$f(t) = \mathcal{F}^{-1}[F(j\omega)] = \frac{1}{2\pi} \int_{-\infty}^{\infty} F(j\omega)e^{j\omega t}d\omega, \qquad t \in (-\infty, \infty)$$

$$\text{(1.6.1)}$$

$\mathcal{F}[\,]$ and $\mathcal{F}^{-1}[\,]$ denote the Fourier transform and inverse Fourier transform operations, respectively. Thus, a function f of time t is converted into a function F of frequency ω. Interpreting this transformation physically, if $f(t)$ describes a signal, then $F(j\omega)$ describes the ac steady-state frequency content or *spectrum* of the signal. $|F(j\omega)|$ is called the *magnitude spectrum* and arg $F(j\omega)$ is called the *phase spectrum*. However, if $F(j\omega)$ describes the ac steady-state gain of a system, then $F(j\omega)$ describes its output spectrum in response to a *flat* input spectrum (which deterministically is an impulse or stochastically is white noise in time).

Comparing the Laplace and Fourier transforms (Eqs. 1.5.1 and 1.6.1), we see that the Fourier transform is indeed a special case of the Laplace transform where

$$F(j\omega) = F(s)\Big|_{s=j\omega} \qquad \text{and} \qquad f(t) = \mathcal{F}^{-1}\left(\mathcal{L}[f(t)]\Big|_{s=j\omega}\right) \qquad \text{(1.6.2)}$$

Recall, in general, that the complex frequency $s = \sigma + j\omega$ where σ is the real part of s (nepers/sec) and ω is the imaginary part of s (radians/sec) or the ac steady-state frequency. Since we set $s = j\omega$ in relating the two transforms, this is equivalent to investigating the ac steady-state behavior along the $j\omega$-axis where $\sigma = \text{Re } s = 0$ (e.g., see Fig. 1.5.1). Thus, if we vertically slice the $F(s)$ plot (or alternatively, the $|F(s)|$ and arg $F(s)$ plots) along the $j\omega$-axis, we obtain the $F(j\omega)$ plot (or the $|F(j\omega)|$ and arg $F(j\omega)$ plots). This viewpoint emphasizes that $F(j\omega)$ is less general than $F(s)$.

We can obtain $F(j\omega)$ directly from $F(s)$ by substituting $s = j\omega$ whenever the region of convergence $\mathcal{R}$ for $F(s)$ contains the $j\omega$-axis, i.e., $\sigma = 0 \in \mathcal{R}$. If $\mathcal{R}$ does not contain the $j\omega$-axis, then $F(j\omega)$ does not exist. In Table 1.5.1 of the previous section, some Laplace transforms were tabulated that only existed along the $j\omega$-axis. In such cases, the FT and LT are equivalent.

A number of Fourier transforms are listed in Table 1.6.1. Fourier transforms satisfy a number of properties which are analogous to those for the Laplace transform. These are listed in Table 1.6.2 for the two-sided case. They allow the engineer to easily expand a Fourier transform table. The FT and IFT are summarized in Table 1.6.1 for later reference. Approximating the integrals of Eq. 1.6.1 by finite sums result in the discrete Fourier transform and inverse discrete Fourier transform. They will be introduced in Chap. 1.8 and discussed at length in Chaps. 2 and 3.

EXAMPLE 1.6.1 Determine the Fourier transform for a dc signal of unity. Interpret the result.

SOLUTION We find from Table 1.5.1 that the Laplace transform of unity equals

$$F(s) = \mathcal{L}[1] = 2\pi U_0(-js), \qquad \text{Re } s = 0 \tag{1.6.3}$$

Since the Fourier transform is evaluated letting $s = j\omega$, the $j\omega$-axis must be included in the region of convergence $\mathcal{R}$ for $F(s)$. Since this is the case in Eq. 1.6.3, the Fourier transform of the dc signal equals

$$F(j\omega) = F(s)\Big|_{s=j\omega} = 2\pi U_0(\omega) \tag{1.6.4}$$

The spectrum is zero everywhere except at the origin as shown in Fig. 1.6.1. This must be the case since the signal contains only dc frequency (i.e., $\omega = 0$). The 2π factor will be explained later. ∎

EXAMPLE 1.6.2 Determine the Fourier transform for a signum function sgn t where, from Table 1.5.1,

$$F(s) = \frac{2}{s}, \qquad \text{Re } s = 0 \tag{1.6.5}$$

SOLUTION The region of convergence for $F(s)$ contains the $j\omega$-axis (actually it consists only of the $j\omega$-axis). Thus, letting $s = j\omega$ then

$$F(j\omega) = F(s)\Big|_{s=j\omega} = \frac{2}{j\omega}, \qquad \text{all } \omega \tag{1.6.6}$$

The spectrum of the signum function is plotted in Fig. 1.6.2. It is composed of all frequencies from 0 to $+\infty$ (and image frequencies from $-\infty$ to 0). Notice that although $|F| \to \infty$ as $\omega \to 0$, $f(t)$ contains no dc component (has zero average value) since F contains no $U_0(\omega)$. ∎

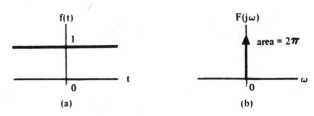

Fig. 1.6.1 (a) DC signal and its **(b)** spectrum of Example 1.6.1.

Table 1.6.1 Table of Fourier transforms.

f(t)	F(jω)	f(t)	F(jω)
1. $U_0(t)$	1	8. t^n	$2\pi j^n U_n(\omega)$
2. $U_1(t)$	$j\omega$	9. $\|t\|$	$-2/\omega^2$
3. $U_n(t)$	$(j\omega)^n$	10. sgn t	$2/j\omega$
4. $U_{-1}(t)$	$\pi U_0(\omega) + 1/j\omega$	11. $e^{-\sigma t}U_{-1}(t)$	$\dfrac{1}{\sigma + j\omega}$
5. $tU_{-1}(t) = U_{-2}(t)$	$j\pi U_1(\omega) - 1/\omega^2$		
6. $t^n U_{-1}(t)$	$\pi j^n U_n(\omega) + n!/(j\omega)^{n+1}$	12. $te^{-\sigma t}U_{-1}(t)$	$\dfrac{1}{(\sigma + j\omega)^2}$
7. 1	$2\pi U_0(\omega)$	13. $(t^n/n!)e^{-\sigma t}U_{-1}(t)$	$\dfrac{1}{(\sigma + j\omega)^{n+1}}$

f(t)	F(jω)
14. $e^{-\sigma\|t\|}$	$\dfrac{2\sigma}{\sigma^2 + \omega^2}$
15. $e^{-\sigma t^2}$	$\sqrt{\dfrac{\pi}{\sigma}}\, e^{-\omega^2/4\sigma}$
16. $\dfrac{1}{\sigma^2 + t^2}$	$\dfrac{\pi}{\sigma}\, e^{-\sigma\|\omega\|}$
17. $U_{-1}(t + T/2) - U_{-1}(t - T/2)$	$T\,\dfrac{\sin(\omega T/2)}{\omega T/2}$
18. $\cos \omega_0 t$	$\pi[U_0(\omega + \omega_0) + U_0(\omega - \omega_0)]$
19. $\sin \omega_0 t$	$j\pi[U_0(\omega + \omega_0) - U_0(\omega - \omega_0)]$
20. $\cos \omega_0 t\, U_{-1}(t)$	$\dfrac{\pi}{2}[U_0(\omega + \omega_0) + U_0(\omega - \omega_0)] + \dfrac{j\omega}{\omega_0^2 - \omega^2}$
21. $\sin \omega_0 t\, U_{-1}(t)$	$\dfrac{j\pi}{2}[U_0(\omega + \omega_0) - U_0(\omega - \omega_0)] + \dfrac{\omega_0}{\omega_0^2 - \omega^2}$
22. $e^{-\sigma t}\cos \omega_0 t\, U_{-1}(t)$	$\dfrac{\sigma + j\omega}{(\sigma + j\omega)^2 + \omega_0^2}$
23. $e^{-\sigma t}\sin \omega_0 t\, U_{-1}(t)$	$\dfrac{\omega_0}{(\sigma + j\omega)^2 + \omega_0^2}$

Table 1.6.2 Properties of Fourier transforms.

Definition	$f(t)$	$F(j\omega) = \int_{-\infty}^{\infty} f(t)e^{-j\omega t}\,dt$
Linear Operations		
Homogeneity	$Kf(t)$	$KF(j\omega)$
Superposition	$f_1(t) \pm f_2(t)$	$F_1(j\omega) \pm F_2(j\omega)$
Both operations	$a_1 f_1(t) \pm a_2 f_2(t)$	$a_1 F_1(j\omega) \pm a_2 F_2(j\omega)$
Symmetry	$F(t)$	$2\pi f(-j\omega)$
Scaling		
Time scaling	$f(at)$	$\dfrac{1}{|a|} F(\dfrac{j\omega}{a})$
Magnitude scaling	$af(t)$	$aF(j\omega)$
Shifting		
Time shifting	$f(t-a)$	$e^{-ja\omega}F(j\omega)$
Frequency shifting	$e^{-at}f(t)$	$F(j\,\omega + a)$
Periodic Functions (for $t > 0$ only)	$f(t) = \sum\limits_{n=0}^{\infty} f_1(t+nT)$	$\dfrac{F_1(j\omega)}{1 - e^{-j\omega T}} = F_1(j\omega)\sum\limits_{n=0}^{\infty} e^{-jn\omega T}$ where $F_1(j\omega) = \int_0^T f(t)e^{-j\omega t}dt$
Convolution		
Time convolution (Frequency product)	$f_1(t)*f_2(t) = \int_{-\infty}^{\infty} f_1(\tau)f_2(t-\tau)\,d\tau$	$F_1(j\omega)F_2(j\omega)$
Frequency convolution (Time product)	$f_1(t)f_2(t)$	$\dfrac{1}{2\pi} F_1(j\omega)*F_2(j\omega) = \dfrac{1}{2\pi j}\int_{-j\infty}^{j\infty} F_1(p)F_2(j\omega - p)\,dp$
Time Differentiation		
First derivative	$df(t)/dt$	$j\omega F(j\omega)$
Second derivative	$d^2 f(t)/dt^2$	$(j\omega)^2 F(j\omega)$
nth derivative	$d^n f(t)/dt^n$	$(j\omega)^n F(j\omega)$
Time Integration		
First integral	$\int_{-\infty}^{t} f(t)\,dt = f^{(-1)}(t)$	$F(j\omega)/j\omega$
Second integral	$\int_{-\infty}^{t}\int_{-\infty}^{t} f(t)\,dt\,dt = f^{(-2)}(t)$	$F(j\omega)/(j\omega)^2$
nth integral	$f^{(-n)}(t)$	$F(j\omega)/(j\omega)^n$
Frequency Differentiation		
First derivative	$tf(t)$	$-dF(j\omega)/d(j\omega)$
nth derivative	$t^n f(t)$	$(-1)^n\, d^n F(j\omega)/d(j\omega)^n$
Frequency Integration		
nth integral	$t^{-n}f(t)$	$(-1)^n F^{(-n)}(j\omega)$
Initial Value *	$f(0^+) = \lim\limits_{t \to 0} f(t)$	$= \lim\limits_{\omega \to \infty} j\omega F(j\omega)$
Final Value *	$f(\infty) = \lim\limits_{t \to \infty} f(t)$	$= \lim\limits_{\omega \to 0} j\omega F(j\omega)$ providing poles of $sF(s)$ in in left-half-plane
Initial Slopes *	$d^{m-1}f(0+)/dt^{m-1}$	$= \lim\limits_{\omega \to \infty} (j\omega)^m F(j\omega)$

*For one-sided transforms.

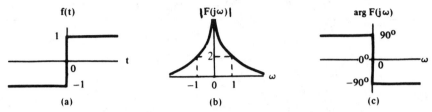

Fig. 1.6.2 (a) Signum function and its (b) magnitude and
(c) phase spectra of Example 1.6.2.

EXAMPLE 1.6.3 Find the Fourier transform for a step function $U_{-1}(t)$ where

$$F(s) = \mathcal{L}[f(t)] = \frac{1}{s}, \qquad \text{Re } s > 0 \qquad (1.6.7)$$

Interpret the result.

SOLUTION Since the region of convergence $\mathcal{R}$ for $F(s)$ does not include the $j\omega$-axis, we cannot evaluate $F(j\omega)$ from this result. However, rewrite $f(t)$ as

$$f(t) = U_{-1}(t) = 0.5 + 0.5 \text{ sgn } t \qquad (1.6.8)$$

Then the Laplace transform for each term can be found in Table 1.5.1. Summing the two transforms gives

$$F(s) = \mathcal{L}[0.5] + \mathcal{L}[0.5 \text{ sgn } t] = \pi U_0(-js) + \frac{1}{s}, \qquad \text{Re } s \geq 0 \qquad (1.6.9)$$

This extends the definition of $F(s)$ to include the $j\omega$-axis. The process of enlarging the region of convergence is called *analytic continuation*. Then $F(j\omega)$ equals

$$F(j\omega) = \pi U_0(\omega) + \frac{1}{j\omega} \qquad (1.6.10)$$

The first term corresponds to a dc term of $1/2$ which agrees with the dc (average) value of $f(t)$. Thus, $|F|$ contains an impulse of area π at $\omega = 0$. The remaining spectrum allows $f(t)$ to make a unit transition at $t = 0$. ∎

EXAMPLE 1.6.4 Describe the difference in the spectrum between the cosine signal and the gated-cosine signal whose Laplace transforms equal

$$\mathcal{L}[\cos \omega_o t] = \pi [U_0(-js + \omega_o) + U_0(-js - \omega_o)], \qquad \text{Re } s = 0$$

$$\mathcal{L}[\cos \omega_o t \, U_{-1}(t)] = \frac{\pi}{2}[U_0(-js + \omega_o) + U_0(-js - \omega_o)] + \frac{s}{s^2 + \omega_o^2}, \qquad \text{Re } s \geq 0$$

$$(1.6.11)$$

SOLUTION The Fourier transforms equal

$$\mathcal{F}[\cos \omega_o t] = \pi [U_0(\omega + \omega_o) + U_0(\omega - \omega_o)]$$

$$\mathcal{F}[\cos \omega_o t \, U_{-1}(t)] = \frac{\pi}{2}[U_0(\omega + \omega_o) + U_0(\omega - \omega_o)] + \frac{j\omega}{\omega_o^2 - \omega^2} \qquad (1.6.12)$$

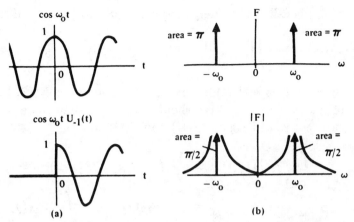

Fig. 1.6.3 (a) Cosine signals and their (b) magnitude spectra of Example 1.6.4.

The continuous cosine signal exists for all time. Its spectrum consists of only two frequency components at $\pm\omega_o$. The gated-cosine has the same two frequency components but all other frequencies besides. This is shown in Fig. 1.6.3. ∎

These examples illustrate the general result that signals which are repetitive or periodic consist of a countably infinite number of *discrete* frequencies. A countably infinite number means that every number has a one-to-one correspondence with an integer n, such as $\omega_n = n\omega_o$. Thus, their spectra consist entirely of lines or impulses at distinct frequencies. The representation of these signals will be discussed more in the next section. Nonrepetitive signals consist of a continuum or band of frequencies (in addition perhaps to discrete frequencies). If the signal lasts only for a finite time, then it contains all frequencies from $\omega = -\infty$ to $+\infty$. However, if it exists for all time, then it *may* contain only a finite band of frequencies. This is called a *band-limited* signal.

The *energy* E contained in a signal $f(t)$ is defined to equal

$$E = \int_{-\infty}^{\infty} |f(t)|^2 dt \qquad (1.6.13)$$

If $f(t)$ is real-valued, then $|f(t)|^2 = f(t)f^*(t) = f^2(t)$. The power P contained in a signal $f(t)$ is defined to equal

$$P = \lim_{T \to \infty} \frac{1}{T} \int_{-T/2}^{T/2} |f(t)|^2 dt \qquad (1.6.14)$$

Parseval's theorem relates the energy contained within a signal to that contained within its spectrum as

$$E = \int_{-\infty}^{\infty} |f(t)|^2 dt = \frac{1}{2\pi} \int_{-\infty}^{\infty} F(j\omega)F^*(j\omega)d\omega = \frac{1}{2\pi} \int_{-\infty}^{\infty} |F(j\omega)|^2 d\omega$$

$$= \int_{-\infty}^{\infty} |F(jf)|^2 df = \int_{0}^{\infty} S(j\omega)d\omega = \int_{0}^{\infty} S(jf)df \qquad (1.6.15)$$

$S(j\omega)$ is called the *energy spectrum* or *energy spectral density* of $f(t)$ and equals

$$S(j\omega) = \frac{1}{\pi}|F(j\omega)|^2 = \frac{1}{\pi}F(j\omega)F^*(j\omega), \qquad S(jf) = 2|F(jf)|^2 = 2F(jf)F^*(jf)$$
$$(1.6.16)$$

When $f(t)$ is real-valued, its spectrum satisfies $F^*(j\omega) = F(-j\omega)$. $S(j\omega)$ has the units of joules/(rad/sec). We should also note that the energy contained in the product of two real-valued signals $f_1(t)$ and $f_2(t)$ equals

$$\int_{-\infty}^{\infty} f_1(t)f_2^*(t)dt = \frac{1}{2\pi}\int_{-\infty}^{\infty} F_1(j\omega)F_2^*(j\omega)d\omega = \int_{-\infty}^{\infty} F_1(jf)F_2^*(jf)df$$
$$= \int_0^{\infty} S_{12}(j\omega)d\omega = \int_0^{\infty} S_{12}(jf)df$$
$$(1.6.17)$$

$S_{12}(j\omega)$ is called the *cross-energy spectrum* of $f_1(t)$ and $f_2(t)$.

The energy in $f(t)$ contained within a band of frequencies between ω_1 and ω_2 equals

$$E(\omega_1, \omega_2) = E(f_1, f_2) = \int_{\omega_1}^{\omega_2} S(j\omega)d\omega = \int_{f_1}^{f_2} S(jf)df \qquad (1.6.18)$$

When E is continuous, the energy spectrum of the signal contains no spectral impulses. That is, the signal contains no discrete frequencies. In this case, the energy $E(\omega, \omega + \Delta\omega) \rightarrow 0$ as $\Delta\omega \rightarrow 0$ so there is zero energy in a bandwidth of zero.

EXAMPLE 1.6.5 Determine the energy spectrum of a unit pulse of duration T seconds centered about $t = 0$. The unit pulse can be expressed as

$$f(t) = U_{-1}(t + T/2) - U_{-1}(t - T/2) \qquad (1.6.19)$$

SOLUTION Using Eq. 1.6.9, its Laplace transform must therefore equal

$$F(s) = \left[\pi U_0(-js) + \frac{1}{s}\right]\left[e^{sT/2} - e^{-sT/2}\right] = \frac{e^{sT/2} - e^{-sT/2}}{s}, \qquad \text{all } s \qquad (1.6.20)$$

from Example 1.6.3 and the time-shifting theorem of Table 1.5.2. Therefore, its Fourier transform equals

$$F(j\omega) = F(s)\Big|_{s=j\omega} = \frac{e^{j\omega T/2} - e^{-j\omega T/2}}{j\omega} \qquad (1.6.21)$$

$F(j\omega)$ can be rewritten as

$$F(j\omega) = \frac{\sin(\omega T/2)}{\omega/2} = T\frac{\sin(\omega T/2)}{\omega T/2} = T\operatorname{sinc}(\omega T/2) \qquad (1.6.22)$$

which has a $\sin x/x$ (denoted as sinc x) variation with frequency as shown in Fig. 1.6.4b. The energy density spectrum of the pulse equals

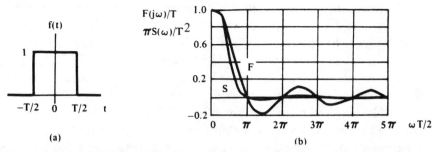

Fig. 1.6.4 (a) Pulse signal and its (b) spectrum of Example 1.6.5.

$$S(j\omega) = \frac{1}{\pi}|F(j\omega)|^2 = \frac{1}{\pi}\big[T\,\text{sinc}(\omega T/2)\big]^2 \tag{1.6.23}$$

which is also plotted in Fig. 1.6.4b. We see the bulk of the energy in the signal is contained in the frequency range $0 \le f \le 1/T$. Using Eq. 1.6.18, numerical integration shows that 92.2 percent of the total output energy lies within the main lobe, and 4.8, 1.7, 0.8, and 0.5 percent within the first, second, third, and fourth side-lobes, respectively. The total energy contained within the pulse equals

$$E(\infty) = \int_{-\infty}^{\infty} |f(t)|^2 dt = \int_{-T/2}^{T/2} dt = T \tag{1.6.24}$$

■

EXAMPLE 1.6.6 Now determine the energy spectrum of a gated-cosine or unit cosine pulse of length T given by

$$f(t) = \cos(\omega_o t)\,\big[U_{-1}(t + T/2) - U_{-1}(t - T/2)\big] \tag{1.6.25}$$

SOLUTION The spectrum of the gated-cosine is easily determined using the frequency convolution theorem of Table 1.6.2 as

$$F(j\omega) = \mathcal{F}\big[\cos(\omega_o t)\big] \;*\; \frac{T}{2\pi}\,\text{sinc}(\omega T/2) \tag{1.6.26}$$

From Example 1.6.4, the Fourier transform of the cosine signal was given by Eq. 1.6.12a. Therefore evaluating Eq. 1.6.26 gives

$$F(j\omega) = \frac{T}{2}\left(\text{sinc}\left[\frac{(\omega + \omega_o)T}{2}\right] + \text{sinc}\left[\frac{(\omega - \omega_o)T}{2}\right]\right) \tag{1.6.27}$$

Thus, the signal spectrum is the sum of two $\sin x/x$ characteristics centered about $\pm\omega_o$. Now the energy is concentrated in the frequency band $|\Delta f| \le 1/T$ centered about $\pm\omega_o$. The total energy in the gated-cosine pulse equals

$$E(\infty) = \frac{T}{2}\big[1 + \text{sinc}(\omega_o T)\big] \tag{1.6.28}$$

after some calculation. $E(\infty)$ approaches $T/2$ as $\omega_o T \to \infty$.

■

Signals and impulse response functions $f(t)$ are often described in terms of their 3 dB bandwidth, asymptotic roll-off, highest side-lobe level, and noise bandwidth of their magnitude spectrums $|F(j\omega)|$. Several signals are listed in Table 2.6.2. The *3 dB bandwidth* of the spectrum is the frequency where $|F(j\omega)|$ is reduced 3 dB from its maximum value. The *asymptotic roll-off* of the spectrum is the high frequency slope of the envelope of $|F(j\omega)|$ and is measured in dB/dec or dB/oct. The *highest side-lobe level* is the maximum level attained by $|F(j\omega)|$ in its side-lobes (i.e., excluding the main lobe). The *noise bandwidth* is equal to the total energy E contained within the signal, or equivalently, the bandwidth B of an ideal rectangular filter (i.e., having $|F(j\omega)| = 1$ for $|\omega| < B$ and zero elsewhere) whose impulse response contains the same energy. The engineer can determine these parameters for the pulse of Example 1.6.5 using the rectangular entry in Table 2.6.2 as we will show in Chap. 2.6.

Two-dimensional (2-D) Fourier transforms describe the spectrum of signals that are functions of two independent variables. The 2-D FT and its inverse equal

$$
\begin{aligned}
F(j\omega_1, j\omega_2) &= \mathcal{F}\big[f(t_1, t_2)\big] \\
&= \int_{-\infty}^{\infty} \int_{-\infty}^{\infty} f(t_1, t_2) e^{-j(\omega_1 t_1 + \omega_2 t_2)} dt_1 dt_2, \quad \omega_1, \omega_2 \in (-\infty, \infty) \\
f(t_1, t_2) &= \mathcal{F}^{-1}\big[F(j\omega_1, j\omega_2)\big] = \frac{1}{(2\pi)^2} \int_{-\infty}^{\infty} \int_{-\infty}^{\infty} F(j\omega_1, j\omega_2) e^{j(\omega_1 t_1 + \omega_2 t_2)} d\omega_1 d\omega_2
\end{aligned}
$$
$$(1.6.29)$$

2-D Fourier transforms are obtained from their Laplace transforms (Eq. 1.5.18) as

$$
\begin{aligned}
F(j\omega_1, j\omega_2) &= F(s_1, s_2)\Big|_{s_1 = j\omega_1, s_2 = j\omega_2} \\
f(t_1, t_2) &= \mathcal{F}^{-1}\left(\mathcal{L}\big[f(t_1, t_2)\big]\Big|_{s_1 = j\omega_1, s_2 = j\omega_2} \right)
\end{aligned}
$$
$$(1.6.30)$$

A variety of 2-D signals and spectrums are shown in Table 1.6.3.

Fourier transforms are easily generalized to any length N by extending the power of the exponential to include the $\omega_N t_N$ term and integrating over all the dimensions in Eq. 1.6.29. Alternatively, the N-dimensional Laplace transform may be converted as shown by Eq. 1.6.30 which is evaluated from $s_1 = j\omega_1$ to $s_N = j\omega_N$. 2-D Fourier transforms are now routinely used in such areas as seismic, sonar, and image processing while 3-D Fourier transforms are used in radar and tomography. They are implemented using fast Fourier transforms (FFT) which are extensively discussed in Chap. 2. Examples of 2-D transforms are presented in Chap. 2.10 so they will not be discussed further here. It should be obvious that these higher-dimensioned transforms will be very valuable to the engineer so he will want to master them.

EXAMPLE 1.6.7 Find the Fourier transform for an image having

$$
f(x, y) = e^{-y}\big[U_{-1}(x) - U_{-1}(x - 1)\big]U_{-1}(y) \tag{1.6.31}
$$

Table 1.6.3 Table of 2-D Fourier transforms. (From R.N. Bracewell, *The Fourier Transform and Its Applications*, 2nd ed., rev., pp. 246–247, McGraw-Hill, NY, 1986.)

	Spacial Domain	Frequency Domain

1. Single impulse
$U_0(x, y)$

2. Impulse stream
$\sum_{n=-\infty}^{\infty} U_0(x, y - n)$

3. Constant–impulse product
$U_0(y)$

4. Cosine–impulse product
$\cos(\pi y) U_0(x)$

5. Cosine
$\cos(\pi x)$

6. Cosine rotated θ
$\cos\left[2\pi(x\cos\theta + y\sin\theta)\right]$

7. Cosine product
$\cos(2\pi x)\cos(2\pi y)$

8. Sinc product
$\operatorname{sinc}(x)\operatorname{sinc}(y)$

9. Sinc squared product
$\operatorname{sinc}^2(x)\operatorname{sinc}^2(y)$

10. Sinc squared–sinc product
$\operatorname{sinc}^2(x)\operatorname{sinc}(y)$

11. Gaussian – circular
$\exp\left[-\pi(x^2 + y^2)\right]$

12. Gaussian – elliptical
$\exp\left[-\pi(x^2/A^2 + y^2/a^2)\right]$

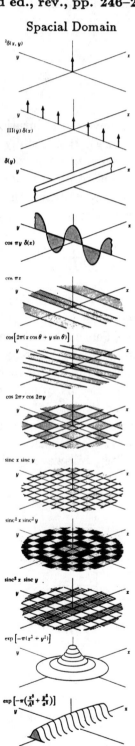
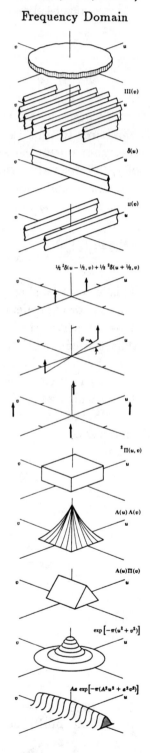

SOLUTION The Laplace transform $F(s_1, s_2)$ for $f(x, y)$ is given by Eq. 1.5.20. Since its region of convergence (all s_1 and Re $s_2 > -1$) includes the $j\omega_1$-axis and $j\omega_2$-axis as shown in Fig. 1.5.3b, its Fourier transform exists. Setting $s_1 = j\omega_1$ and $s_2 = j\omega_2$,

$$F(j\omega_1, j\omega_2) = F(s_1, s_2)\Big|_{s_1=j\omega_1,\ s_2=j\omega_2} = \frac{1 - e^{-j\omega_1}}{j\omega_1(1 + j\omega_2)} \qquad (1.6.32)$$

∎

In practice, all actual signals have finite duration and change mathematical forms from time to time (are nonstationary). It is usually of interest to determine the signal spectrum during these short time periods where its statistics are stationary instead of its average spectrum over the entire observation period. Thus short-term Fourier transforms and analysis are often more desirable than long-term transforms. The *short-term* or *running Fourier transform* equals

$$F(j\omega; T_o) = \int_{T_o - NT/2}^{T_o + NT/2} f(t) e^{-j\omega t} dt = \int_{-\infty}^{\infty} h(T_o - t) f(t) e^{-j\omega t} dt = h(t) * \left[f(t) e^{-j\omega t} \right]$$

$$(1.6.33)$$

This is accomplished by using a data window of size or duration NT seconds centered at time T_o, or equivalently, a lowpass filter with impulse response $h(t) = 0$ for $|t| > NT/2$. This truncates the data and the Fourier transform of Eq. 1.6.1. Eq. 1.6.33 shows that the convolution of $h(t)$ and $f(t) e^{-j\omega t}$ produces the running Fourier transform. The spectrum of a running transform is shown in Fig. 2.11.1. These matters will be discussed extensively in Chap. 2. Let us briefly consider the implementation of Eq. 1.6.33 which will help motivate and explain later work.

Short-term Fourier transforms are measured using spectrum analyzers. A spectrum analyzer consists of a bank of analyzer channels. As shown in Figs. 1.6.5a, each analyzer channel consists of a bandpass filter, full-wave rectifier, and lowpass filter. Each filter

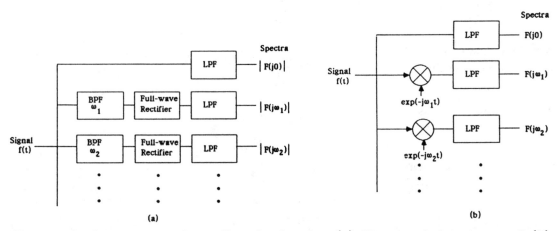

Fig. 1.6.5 Spectrum analyzer filter bank using (a) filters and detectors and (b) filters and multipliers.

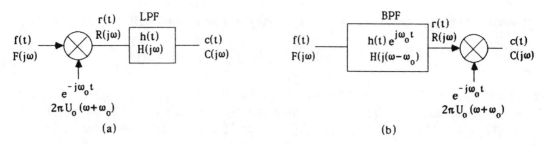

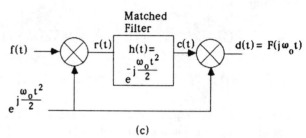

Fig. 1.6.6 Spectrum analyzer using multipliers and (a) lowpass filter, (b) bandpass filter, and (c) matched filter.

output measures the signal magnitude in a narrow frequency band centered at frequency ω_i. Alternatively, each channel consists of a multiplier and lowpass filter. In this case, each filter output measures both the signal magnitude and phase in the narrow frequency band around ω_i. Fig. 1.6.5a is a noncoherent (no phase recovery) approach while Fig. 1.6.5b is a coherent (phase recovery) approach. Now let us consider the coherent channel in more detail.

In the lowpass filter system of Fig. 1.6.6a, the defining equations are

$$r(t) = f(t)e^{-j\omega_o t}, \qquad R(j\omega) = F(j\omega) * 2\pi U_0(\omega + \omega_o) = F(j(\omega + \omega_o)),$$
$$C(j\omega) = H(j\omega)R(j\omega) = H(j\omega)F(j(\omega + \omega_o)), \qquad c(t) = h(t) * \left[f(t)e^{-j\omega_o t}\right]$$
$$(1.6.34)$$

The bandpass filter system of Fig. 1.6.6b (one channel in Fig. 1.6.5b) has

$$r(t) = f(t) * 2\pi\left[h(t)e^{j\omega_o t}\right], \qquad R(j\omega) = F(j\omega)H(j(\omega - \omega_o)),$$
$$C(j\omega) = R(j\omega) * 2\pi U_0(\omega + \omega_o) = R(j(\omega + \omega_o)) = H(j\omega)F(j(\omega + \omega_o)), \qquad (1.6.35)$$
$$c(t) = r(t)e^{-j\omega_o t} = \left[h(t)e^{j\omega_o t} * f(t)\right]e^{-j\omega_o t}$$

These two systems are equivalent since their $C(j\omega)$ are identical. It is interesting to verify this by showing that their $c(t)$ are equal even though they appear not to be. The lowpass filter has impulse response $h(t)$ or gain $H(j\omega)$ and the bandpass filter has impulse response $h(t)e^{j\omega_o t}$ or gain $H(j(\omega - \omega_o))$. The two gain characteristics are simply frequency-shifted versions of one another. The lowpass filter response is centered at $\omega = 0$; the bandpass filter response is centered at ω_o.

Now if $H(j\omega)$ is an ideal lowpass filter having a zero-bandwidth gain response centered at $\omega = 0$ or dc, then

$$H(j\omega) = U_0(\omega) \text{ (unity gain bandwidth)}, \qquad h(t) = \frac{1}{2\pi} \qquad (1.6.36)$$

Substituting this into Eq. 1.6.34 shows that $c(t)$ is a dc output with complex value

$$c(t) = \mathcal{F}^{-1}\left[F(j(\omega + \omega_o))U_0(\omega)\right] = \mathcal{F}^{-1}\left[F(j\omega_o)U_0(\omega)\right] = F(j\omega_o) \qquad (1.6.37)$$

$c(t)$ equals the signal spectrum $F(j\omega)$ evaluated at the modulation frequency $\omega = \omega_o$.

Suppose instead that $H(j\omega)$ is a matched filter (i.e., matched to the modulating signal $s(t) = e^{-j\omega_o t}$ so that $H(j\omega) = S(-j\omega)$ or $h(t) = s(-t)$). Then modifying Eq. 1.6.34 gives

$$h(t) = e^{j\omega_o t}, \qquad H(j\omega) = 2\pi U_0(\omega - \omega_o),$$
$$C(j\omega) = 2\pi F(j(\omega + \omega_o))U_0(\omega - \omega_o) = 2\pi F(j2\omega_o)U_0(\omega - \omega_o), \qquad (1.6.38)$$
$$c(t) = [f(t)e^{-j\omega_o t}] * e^{j\omega_o t} = F(j2\omega_o)e^{j\omega_o t}$$

In this case, $c(t)$ equals a complex exponential with complex amplitude $F(j2\omega_o)$ and frequency ω_o. Following the system in Fig. 1.6.6a with another multiplier with input $e^{-j\omega_o t}$ yields an output $d(t) = F(j2\omega_o)$.

These results suggest that if the frequency modulating signal $s(t)$ is made to vary or sweep linearly, then the output of Fig. 1.6.6a will sweep also. Since $s(t) = \exp(j\omega_o t^2/2)$ has phase $\theta = \omega_o t^2/2$ and instantaneous frequency $\omega_i = d\theta/dt = \omega_o t$, then $\exp(-j\omega_o t^2/2)$ can be used as the modulating signal rather than $\exp(-j\omega_o t)$. In radar, this type of signal is called a *frequency chirp* or simply a *chirp signal*.

Such a matched filter system is shown in Fig. 1.6.6c. It generates the Fourier transform $F(j\omega)$ of $f(t)$ in continuous time. The defining equations are

$$r(t) = f(t)e^{-j\omega_o t^2/2}, \qquad R(j\omega) = F(j\omega) * \frac{\sqrt{\pi}}{\omega_o}e^{j(\omega^2 + \pi/4)},$$

$$C(j\omega) = H(j\omega)R(j\omega) = \frac{\sqrt{\pi}}{\omega_o}e^{j(\omega^2 + \pi/4)}R(j\omega), \qquad (1.6.39)$$

$$c(t) = h(t) * r(t) = e^{j\omega_o t^2/2} * \left[f(t)e^{-j\omega_o t^2/2}\right] = e^{j\omega_o t^2/2}F(j\omega_o t),$$

$$d(t) = c(t) * e^{-j\omega_o t^2/2} = F(j\omega_o t) \equiv F(j\omega)$$

We see from Eq. 1.6.39 that the $d(t)$ output exactly equals the $f(t)$ signal spectrum $F(j\omega_o t)$ where time and frequency are related as $\omega = \omega_o t$. The matched filter output $c(t)$ equals the signal spectrum augmented by phase $\omega_o t^2/2$. The second modulator cancels this phase distortion by introducing a negative phase of $-\omega_o t^2/2$. If only the magnitude spectrum is required, the second multiplier is unnecessary since

$$|c(t)| = |F(j\omega_o t)|, \qquad \arg c(t) = \arg F(j\omega_o t) + \frac{\omega_o t^2}{2} \qquad (1.6.40)$$

1.7 FOURIER SERIES (FS)

Fourier series (FS) describe periodic signals. A Fourier series is a special case of the Fourier transform and thus it is also a special case of the Laplace transform. Fourier series result when the Fourier or Laplace transforms are composed entirely of impulse functions. Since rigorous derivation of such functions requires generalized function theory which is not widely known, Fourier series are almost always treated as a different topic. This should not be the case. Fourier series of periodic signals can be obtained directly from the Fourier transform of a single period or cycle of that signal as will now be shown.

Consider Fig. 1.7.1 where $f(t)$ is a nonperiodic truncated signal which is repeated every T seconds to form the periodic signal $f_p(t)$ as

$$f_p(t) = \sum_{n=-\infty}^{\infty} f(t - nT) \tag{1.7.1}$$

For mathematical simplicity, $f(t)$ is arbitrarily defined in the interval $(-T/2, T/2)$ which is centered at $t = 0$. To make $f(t)$ periodic with period T, $f(t)$ is convolved with an impulse stream or train $p(t)$ of period T as

$$f_p(t) = f(t) * p(t) = f(t) * \sum_{n=-\infty}^{\infty} U_0(t - nT) \tag{1.7.2}$$

Since convolution in time corresponds to multiplication in frequency from the convolution property in Table 1.6.2, the Fourier transform of $f_p(t)$ equals

$$F_p(j\omega) = F(j\omega)P(j\omega) = \omega_o F(j\omega) \sum_{m=-\infty}^{\infty} U_0(\omega - m\omega_o) = \omega_o \sum_{m=-\infty}^{\infty} F(j\omega_o)U_0(\omega - m\omega_o) \tag{1.7.3}$$

where $\omega_o = 2\pi/T$. $P(j\omega)$ is obtained from $p(t)$ using Eqs. 1.3.2, 1.3.3, and 1.6.4.

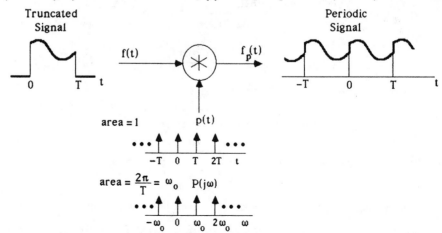

Fig. 1.7.1 Signal processing system which relates Fourier transform and Fourier series.

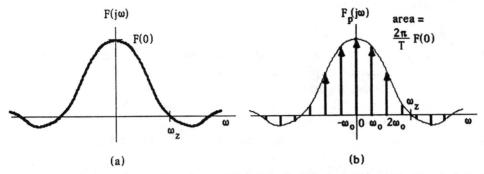

$F(j\omega)$ $F_p(j\omega)$

(a) (b)

Fig. 1.7.2 Spectra of (a) nonperiodic signal $f(t)$ and (b) periodic signal $f_p(t)$.

This spectral multiplication operation is shown in Fig. 1.7.2. Multiplying ω_o times the Fourier transform $F(j\omega)$ of the nonperiodic signal forms the envelope of the discrete spectra $F_p(j\omega)$. This spectrum is an infinite series of impulses separated by ω_o rad/sec or $1/T$ Hz.

EXAMPLE 1.7.1 Consider a unit pulse. Use its spectrum to find the Fourier series of a unit-amplitude square wave having frequency $f_o = 1/T$ Hz.

SOLUTION A unit pulse with duration $T/2$ seconds has a spectrum $F(j\omega)$ of

$$f(t) = U_{-1}(t + T/4) - U_{-1}(t - T/4)$$

$$F(j\omega) = \frac{T}{2} \frac{\sin(\omega T/4)}{\omega T/4} \tag{1.7.4}$$

from Eqs. 1.6.19 and 1.6.22. This spectrum is shown in Fig. 1.6.4. (Use $T/4$ rather than $T/2$.) Convolving $f(t)$ with an impulse stream of period T produces a square wave where

$$f_p(t) = U_{-1}(\cos(2\pi f_o))$$

$$F_p(j\omega) = \frac{2\pi}{T} \frac{T}{2} \sum_{m=-\infty}^{\infty} \frac{\sin(\omega T/4)}{\omega T/4} U_0\left(\omega - m\frac{2\pi}{T}\right) \tag{1.7.5}$$

$$= \pi U_0(\omega) + 2 \sum_{\substack{m=1 \\ m \text{ odd}}}^{\infty} \frac{(-1)^{\frac{m-1}{2}}}{m} \left[U_0(\omega - 2\pi m f_o) + U_0(\omega + 2\pi m f_o) \right]$$

Since $e^{j\omega_o t}$ has a spectrum of $2\pi U_0(\omega - \omega_o)$ from Table 1.6.1, then $f_p(t)$ equals the IFT of $F_p(j\omega)$ which taken term by term yields

$$f_p(t) = \mathcal{F}^{-1}[F_p(j\omega)] = \frac{1}{2} + \frac{1}{\pi} \sum_{\substack{m=1 \\ m \text{ odd}}}^{\infty} \frac{(-1)^{\frac{m-1}{2}}}{m} \left[e^{j2\pi m f_o t} + e^{-j2\pi m f_o t} \right] \tag{1.7.6}$$

Expressing $\frac{1}{2}(e^x + e^{-x}) = \cos(x)$ gives the Fourier series for the square wave as

$$f_p(t) = \frac{1}{2} + \frac{2}{\pi} \sum_{\substack{m=1 \\ m \text{ odd}}}^{\infty} \frac{(-1)^{\frac{m-1}{2}}}{m} \cos(2\pi m f_o t) \tag{1.7.7}$$

■

Now let us derive Fourier series using the conventional approach. The Fourier series of a periodic function $f_p(t)$ of period T is[9]

$$f_p(t) = \sum_{m=-\infty}^{\infty} c_m e^{jmw_o t} = c_0 + \sum_{m=1}^{\infty} 2|c_m| \cos(mw_o t + \arg c_m)$$

$$= a_0 + \sum_{m=1}^{\infty} [a_m \cos(mw_o t) + b_m \sin(mw_o t)]$$

(1.7.8)

The first equation is the complex exponential series form, the second is the cosine series (magnitude and angle) form, while the third is the trigonometric series form. The c_m are the complex coefficients of the Fourier series which equal

$$c_m = \frac{1}{T} \int_{-T/2}^{T/2} f_p(t) e^{-jmw_o t} dt = \frac{1}{2}(a_m - jb_m), \qquad c_0 = a_0 \qquad (1.7.9)$$

They satisfy $c_{-m} = c_m^*$ (complex conjugates for negative indices). It is clear from Eq. 1.7.9 that $a_m = \text{Re}\,(2c_m)$ and $b_m = -\text{Im}\,(2c_m)$. The fundamental frequency w_o is related to the signal period T as $w_o = 2\pi/T$ rad/sec or $f_o = 1/T$ Hz. Taking the Fourier transform of Eq. 1.7.8 gives the Fourier series spectrum as

$$F_p(j\omega) = 2\pi \left(c_0 U_0(\omega) + \sum_{m=1}^{\infty} [c_m U_0(\omega - mw_o) + c_m^* U_0(\omega + mw_o)] \right)$$

$$= 2\pi \left[a_0 + \sum_{m=1}^{\infty} \left(\frac{a_m}{2} [U_0(\omega - mw_o) + U_0(\omega + mw_o)] \right. \right.$$

(1.7.10)

$$\left. \left. - j\frac{b_m}{2} [U_0(\omega - mw_o) - U_0(\omega + mw_o)] \right) \right]$$

Entries 7, 18, and 19 of Table 1.6.1 are used to obtain this result. $F_p(j\omega)$ consists of a sequence of impulses located at frequencies $\omega = mw_o$ with complex areas having amplitude $|c_m|$ and phase $-\arg(c_m)$.

Now compare this $F_p(j\omega)$ with that found in Eq. 1.7.3 using the continuous spectrum approach. It is clear that the Fourier series coefficients c_m of the periodic signal are related to the original nonperiodic signal spectrum $F(j\omega)$ as

$$2\pi c_m = w_o F(j\omega) \Big|_{\omega = mw_o} = w_o F(m) \qquad \text{or} \qquad c_m = \frac{1}{T} F(m) \qquad (1.7.11)$$

Therefore the FS coefficients can be obtained from $F(j\omega)$ directly as

$$a_0 = \frac{1}{T} F(0), \qquad a_m = \frac{2}{T} \text{Re } F(m), \qquad b_m = \frac{-2}{T} \text{Im } F(m) \qquad (1.7.12)$$

with no need to use the usual FS approach and calculations. Eq. 1.7.12 shows that $f_p(t)$ is composed of a dc component $F(0)/T$ and an infinite number of harmonically related ac components with peak value $2|F(m)|/T$, phase $-\arg F(m)$, and fundamental frequency $w_o = 2\pi/T$. The Fourier series equations are summarized in Table 1.8.1.

Approximating the c_m integral (Eq. 1.7.9) by a finite sum and the $f_p(t)$ infinite series by a finite series produces a discrete Fourier series (DFS). The DFS will be discussed in Chap. 1.8.

Let us apply this *solution by inspection* approach to reanalyze Example 1.7.1. The pulse spectrum is evaluated at frequencies $\omega = m\omega_o$ so $\omega T/4 = m\pi/2$ and

$$F(jm\omega_o) = \frac{T}{2} \quad (m = 0), \qquad \frac{T(-1)^{(m-1)/2}}{\pi m} \quad (m \text{ odd}), \qquad 0 \quad (\text{all other } m) \tag{1.7.13}$$

from Eq. 1.7.4b. Then Eqs. 1.7.11 and 1.7.12 give the square wave coefficients as

$$c_m = 0.5 \quad (m = 0), \qquad \frac{(-1)^{(m-1)/2}}{\pi m} \quad (m \text{ odd}), \qquad 0 \quad (\text{all other } m)$$

$$a_m = 0.5 \quad (m = 0), \qquad \frac{2(-1)^{(m-1)/2}}{\pi m} \quad (m \text{ odd}), \qquad 0 \quad (\text{all other } m) \tag{1.7.14}$$

where $b_m = 0$. Since c_m is real, the Fourier series has the cosine form. These coefficients match those in Eq. 1.7.7.

1.8 DISCRETE FOURIER TRANSFORM (DFT) AND SERIES (DFS)

The Fourier transform (FT) and Fourier series (FS) can also be used to analyze sampled signals. This produces the discrete Fourier transform (DFT) and discrete Fourier series (DFS), respectively. They have an interesting relation to the z-transform and its inverse which are described in the next two sections.

The *discrete Fourier transform* or *DFT* equals the FT of a sampled signal $f(nT)$. Mathematically, the DFT equals

$$F_s(j\omega) = \int_{-\infty}^{\infty} \Big[\sum_{n=-\infty}^{\infty} f(t)U_0(t - nT) \Big] e^{-j\omega t} dt$$

$$= \sum_{n=-\infty}^{\infty} \Big[\int_{-\infty}^{\infty} f(t)e^{-j\omega t}U_0(t - nT)dt \Big] = \sum_{n=-\infty}^{\infty} f(nT)e^{-jn\omega T} \tag{1.8.1}$$

which is obtained by combining Eqs. 1.3.2, 1.5.1, and 1.6.2. $F_s(j\omega)$ describes the spectrum of the sampled signal $f_s(t)$. The DFT may also be viewed as the zero-order approximation to the FT of the continuous (unsampled) signal $f(t)$ as

$$F(j\omega) = \int_{-\infty}^{\infty} f(t)e^{-j\omega t}dt \cong T \sum_{n=-\infty}^{\infty} f(nT)e^{-jn\omega T} = TF_s(j\omega) \tag{1.8.2}$$

where

$$f(t) \cong T \sum_{n=-\infty}^{\infty} f(nT)U_0(t - nT) = Tf_s(t) \tag{1.8.3}$$

Here the signal $f(t)$ is approximated by a series of impulses as shown in Fig. 1.8.1. T times the DFT of a signal approximates its FT. The approximation error approaches

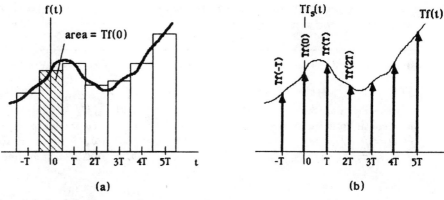

Fig. 1.8.1   (a) Zero-order approximation of signal $f(t)$ and (b) sampled signal $Tf_s(t)$ where $T = 2\pi/\omega_s = 1/f_s$.

zero as the sampling interval $T \to 0$. If the FT is band-limited to frequencies less than $\omega_s/2$, then $T \times$ DFT equals the FT for frequencies $|\omega| < \omega_s/2$ (no frequency aliasing error). Properties of the FT and DFT are listed in Table 1.8.1.

In DFT's, $F_s(j\omega)$ is usually computed at N equally-spaced frequencies in the interval between 0 and ω_s. Since the frequency separation between spectral components is ω_s/N, Eq. 1.8.1 is evaluated at frequencies $\omega = m(\omega_s/N)$ for $m = 0, 1, \ldots, N-1$. Defining $F(m) \equiv F_s(jm\omega_s/N)$ and $f(n) \equiv f(nT)$, then Eq. 1.8.1 can be rewritten in shorthand notation as

$$F(m) = \sum_{n=-\infty}^{\infty} f(n)e^{-j2\pi nm/N} = \sum_{n=-\infty}^{\infty} f(n)W_N^{nm} \tag{1.8.4}$$

where the constant $W_N = \exp(-j2\pi/N)$. This is the usual form of the DFT. Notice that $F(m) = F(m+N)$ so the DFT spectrum is periodic in m with period N.

For real-time analysis, the signal $f(t)$ can only be observed over a finite interval or time window such as $0 \le t < T_w$. Also only a finite number of samples can be used for numerical evaluation. Assuming that N equally-spaced samples are collected, then the DFT given by Eq. 1.8.4 reduces to the short-term DFT where

$$F(m) = \sum_{n=0}^{N-1} f(n)W_N^{nm} \tag{1.8.5}$$

The interval between samples is $T = T_w/N$. Since only N data points are used, then only N of the coefficients $F(m)$ are independent. The sampled signal $f_s(t)$ is converted back (or reconstructed) into a continuous time signal $f_r(t)$ by passing the sampled signal spectrum $F_s(j\omega)$ through the ideal lowpass filter described by Eq. 1.3.7. The filter bandwidth $B = \omega_s/2 = \omega_N$ equals the Nyquist frequency. From Eq. 1.3.6, the band-limited spectrum $F_r(j\omega)$ of $f_r(t)$ equals

$$F_r(j\omega) = \frac{\omega_s}{2\pi} \sum_{n=-\infty}^{\infty} F(j(\omega - n\omega_s)) = F_s(j\omega), \quad |\omega| \le \omega_s/2$$
$$= 0, \quad\quad\quad |\omega| > \omega_s/2 \tag{1.8.6}$$

Table 1.8.1 Comparisons of continuous and discrete Fourier transforms and series.

FT	DFT
Transform (Eq. 1.6.1)	**(Eq. 1.8.5)** $W_N = e^{-j\frac{2\pi}{N}}$
$F(j\omega) = \int_{-\infty}^{\infty} f(t)e^{-j\omega t}dt$	$F(m) = \sum_{n=0}^{N-1} f(n)W_N^{nm}$
t = continuous time	$T = T_w/N$ = sampling interval
ω = continuous frequency	$T_w = NT$ = sampling window
$f(n) = f(nT)$	$\Delta\omega = \omega_s/N$ = interval between bins
$TF(m) = F(j\omega)\|_{\omega=m\Delta\omega}$	$\omega_s = N\Delta\omega$ = sampling frequency
Inverse Transform (Eq. 1.6.1)	**(Eq. 1.8.10)**
$f(t) = \frac{1}{2\pi}\int_{-\infty}^{\infty} F(j\omega)e^{j\omega t}d\omega$	$f(n) = \frac{1}{N}\sum_{n=0}^{N-1} F(m)W_N^{-nm}$
Orthogonality (Table 1.5.1, no. 22)	
$\int_{-\infty}^{\infty} e^{-j(\omega-\omega_o)t}dt = 2\pi U_0(\omega - \omega_o)$	$\frac{1}{N}\sum_{m=0}^{N-1} W_N^{m(n-k)} = 1,$
	$\qquad n = k \pmod{N}$ and 0 otherwise
Parseval's Theorem (Eqs. 1.6.15, 1.6.17)	**(Eq. 1.8.12)**
$\int_{-\infty}^{\infty} \|f(t)\|^2 dt = \frac{1}{2\pi}\int_{-\infty}^{\infty} \|F(j\omega)\|^2 d\omega$	$\sum_{n=0}^{N-1} \|f(n)\|^2 = \frac{1}{N}\sum_{m=0}^{N-1} \|F(m)\|^2$
$\int_{-\infty}^{\infty} f_1(t)f_2^*(t)dt$	$\sum_{n=0}^{N-1} f_1(n)f_2^*(n)$
$\qquad = \frac{1}{2\pi}\int_{-\infty}^{\infty} F_1(j\omega)F_2^*(j\omega)d\omega$	$\qquad = \frac{1}{N}\sum_{m=0}^{N-1} F_1(m)F_2^*(m)$

FS	DFS
Transform (Eq. 1.7.9)	**(Eq. 1.8.15)**
$c_m = \frac{1}{NT}\int_{-NT/2}^{NT/2} f(t)e^{-jm\omega_o t}dt$	Same as DFT
$\omega_o = 1/NT$ = repetition frequency	
Inverse Transform (Eq. 1.7.8)	**(Eq. 1.8.15)**
$f(t) = \sum_{m=-\infty}^{\infty} c_m e^{jm\omega_o t}$	Same as IDFT
Orthogonality	
$\int_{-T/2}^{T/2} e^{-j(m-n)\omega_o t}dt = 2\pi,$	Same as IDFT
$\qquad m = n \pmod{N}$ and 0 otherwise	
Parseval's Theorem	
$\int_{-\infty}^{\infty} \|f(t)\|^2 dt = \sum_{m=-\infty}^{\infty} \|c_m\|^2$	Same as DFT
$\int_{-\infty}^{\infty} f_1(t)f_2^*(t)dt = \sum_{m=-\infty}^{\infty} c_{1m}c_{2m}^*$	Same as DFT

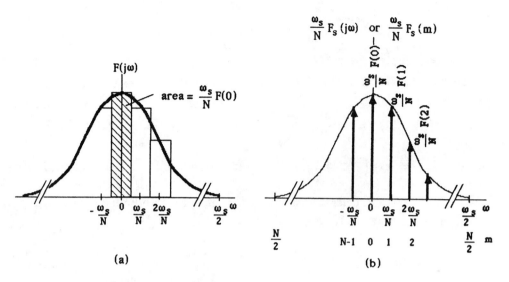

Fig. 1.8.2 (a) Zero-order approximation of band-limited spectrum $F(j\omega)$ and (b) sampled spectrum $F_s(j\omega)$ where $\omega_s = 2\pi/T$.

Then the inverse Fourier transform (IFT) equals

$$f_r(nT) = \frac{1}{2\pi} \int_{-\infty}^{\infty} F_r(j\omega)e^{j\omega nT}\,d\omega = \frac{1}{2\pi} \int_{-\omega_s/2}^{\omega_s/2} F_s(j\omega)e^{j\omega nT}\,d\omega \qquad (1.8.7)$$

As shown in Fig. 1.8.2, the spectrum $F_s(j\omega)$ of the sampled signal $f_s(t)$ can be zero-order approximated as

$$F_s(j\omega) \cong \frac{\omega_s}{N} \sum_{m=-\infty}^{\infty} F_s\left(\frac{jm\omega_s}{N}\right) U_0\left(\omega - \frac{m\omega_s}{N}\right) \qquad (1.8.8)$$

When there is no frequency aliasing, $F(j\omega)/T = F_s(j\omega)$. $F_r(j\omega)$ is also given by Eq. 1.8.8 where the summation limits extend between $(-N/2, (N/2) - 1)$. $F_r(j\omega)$ can be numerically evaluated since it involves only a finite number of terms unlike $F_s(j\omega)$. Notice the upper limit is $(N/2) - 1$ and not $N/2$. The extra half-bin below $-N/2$ compensates for (equals) the missing half-bin below $N/2$ as shown in Fig. 1.8.2.

Substituting Eq. 1.8.8 into Eq. 1.8.7 yields the *inverse discrete Fourier transform* or *IDFT* which equals

$$f_r(nT) = \frac{1}{2\pi} \int_{-\omega_s/2}^{\omega_s/2} \frac{2\pi}{NT} \left[\sum_{m=-\infty}^{\infty} F_s\left(\frac{jm\omega_s}{N}\right) U_0\left(\omega - \frac{m\omega_s}{N}\right) \right] e^{j\omega nT}\,d\omega$$

$$= \sum_{m=-\infty}^{\infty} \frac{1}{NT} \left[\int_{-\omega_s/2}^{\omega_s/2} F_s\left(\frac{jm\omega_s}{N}\right) e^{j\omega nT} U_0\left(\omega - \frac{m\omega_s}{N}\right) \right] d\omega \qquad (1.8.9)$$

$$= \frac{1}{NT} \sum_{m=-N/2}^{(N/2)-1} U_0\left(\omega - \frac{m\omega_s}{N}\right) e^{jmn\omega_s T/N}$$

Rewriting Eq. 1.8.9 in the simpler shorthand notation used for Eq. 1.8.4,

$$f(n) = Tf_r(n) = \frac{1}{N} \sum_{m=-N/2}^{(N/2)-1} F_s(m) W_N^{-nm} = \frac{1}{N} \sum_{m=0}^{N-1} F_s(m) W_N^{-nm} \qquad (1.8.10)$$

where $W_N = \exp(-j2\pi/N)$. The summation limits can be changed because of the periodicity of $F_s(m)$ and W_N.

The discrete Fourier transform and inverse discrete Fourier transform pair are given by Eqs. 1.8.5 and 1.8.10 as

$$F(m) = F_s(m) = \sum_{n=0}^{N-1} f(n) W_N^{nm}, \qquad m = 0, 1, \ldots, N-1$$

$$\qquad (1.8.11)$$

$$f(n) = Tf_r(n) = \frac{1}{N} \sum_{m=0}^{N-1} F(m) W_N^{-nm}, \qquad n = 0, 1, \ldots, N-1$$

Parseval's theorem was given by Eqs. 1.6.15 and 1.6.17 for the continuous Fourier transform. In the DFT case, Parseval's theorem is

$$\sum_{n=0}^{N-1} |f(n)|^2 = \frac{1}{N} \sum_{m=0}^{N-1} |F(m)|^2$$

$$\qquad (1.8.12)$$

$$\sum_{n=0}^{N-1} f_1(n) f_2^*(n) = \frac{1}{N} \sum_{m=0}^{N-1} F_1(m) F_2^*(m)$$

If $f_2^*(n)$ is real-valued, then $F_2^*(m) = F_2(-m) = F_2(N - m)$.

EXAMPLE 1.8.1 Find the 4-point DFT and IDFT of a sawtooth signal having samples $[f] = [0, 1, 2, 3]$.
SOLUTION Using Eq. 1.8.11a, the DFT spectrum equals

$$\begin{aligned} F(0) &= 0(+1) + 1(+1) + 2(+1) + 3(+1) = +6 \\ F(1) &= 0(+1) + 1(-j) + 2(-1) + 3(+j) = -2 + j2 \\ F(2) &= 0(+1) + 1(-1) + 2(+1) + 3(-1) = -2 \\ F(3) &= 0(+1) + 1(+j) + 2(-1) + 3(-j) = -2 - j2 \end{aligned}$$

$$\qquad (1.8.13)$$

From Eq. 1.8.11b, the IDFT reconstructed signal equals

$$\begin{aligned} f(0) &= [6(+1) + (-2+j2)(+1) - 2(+1) + (-2-j2)(+1)]/4 = 0 \\ f(1) &= [6(+1) + (-2+j2)(+j) - 2(-1) + (-2-j2)(-j)]/4 = 1 \\ f(2) &= [6(+1) + (-2+j2)(-1) - 2(+1) + (-2-j2)(-1)]/4 = 2 \\ f(3) &= [6(+1) + (-2+j2)(-j) - 2(-1) + (-2-j2)(+j)]/4 = 3 \end{aligned}$$

$$\qquad (1.8.14)$$

■

The DFT may also be viewed as a Fourier series (FS) approximation as follows. Using Fourier series from Chap. 1.7, a periodic function $f(t)$ of period T_w is written as an infinite sum where[10]

$$c_m = \frac{1}{T_w} \int_0^{T_w} f(t)e^{-jmw_ot}dt \cong \frac{T}{NT} \sum_{n=0}^{N-1} f(nT)e^{-jnmw_oT} = F_s(jmw_o)$$

$$f(t) = \sum_{m=-\infty}^{\infty} c_m e^{jmw_ot} \cong \sum_{m=-M}^{M} c_m e^{jmw_ot}$$

(1.8.15)

using Eqs. 1.7.8 and 1.7.9. The fundamental frequency w_o is related to the width of the data window T_w as $w_o = 2\pi/T_w = 2\pi/NT$ radians/second or $1/NT$ Hz. The discrete Fourier series results when we approximate the c_m time integral by a discrete sum and the infinite sum of $f(t)$ by a finite sum. The DFS given by the right-hand side of Eq. 1.8.15 is identical to the DFT given by Eq. 1.8.11 when $F(m) = Nc_m$, $w_oT = 2\pi/N$, $W_N = \exp(-j2\pi/N)$, and $N = 2M$. Properties of the DFS are listed in Table 1.8.1. These FS approximations are good if $f(t)$ is almost constant over any T interval, or equivalently, almost band-limited beyond the frequency $Mw_o = 2\pi M/NT$. Sampling $f(t)$ forces the c_m to also be periodic as $c(m+N) = c(m)$.

The DFT can be related to the z-transform to be discussed in Chap. 1.9. Rewriting Eq. 1.8.4 in the z-transform notation of Eq. 1.9.2 gives

$$F(m) = \sum_{n=-\infty}^{\infty} f(n)W_N^{nm} = F(z)\Big|_{z=W_N^{-m}=e^{j2\pi m/N}}$$

(1.8.16)

Thus sampling the z-transform $F(z)$ at N equally-spaced points counterclockwise around the unit circle produces the DFT spectral coefficients. The IDFT can be obtained from the inverse z-transform given by Eq. 1.10.1 as

$$f_r(nT) = \frac{1}{j2\pi T} \oint_C F(z)z^{n-1}dz$$

(1.8.17)

$f(nT)$ is evaluated around the unit circle where $z = e^{j\theta} = W_N^{-m}$ and $W_N = \exp(-j2\pi/N)$. Approximating the integral by a sum where $dz = je^{j\theta}d\theta \cong jW_N^{-m}(2\pi/N)$ then

$$Tf_r(n) = \frac{1}{j2\pi} \sum_{m=0}^{N-1} F(m)W_N^{-m(n-1)} jW_N^{-m}\frac{2\pi}{N}$$

$$= \frac{1}{N} \sum_{m=0}^{N-1} F(m)W_N^{-mn} = \frac{1}{N} \sum_{m=0}^{N-1} F(z)z^{-n}\Big|_{z=W_N^{-m}}$$

(1.8.18)

which is identical to Eq. 1.8.11b.

Because of the symmetry of these equations, the DFT and IDFT can be easily implemented on the computer and numerically evaluated. The ZT and IZT cannot be

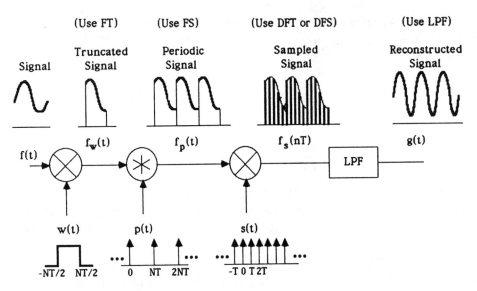

Fig. 1.8.3 Signal processing system which relates the Fourier transform, Fourier series, and their discrete versions.

numerically evaluated because they involve an infinite number of points. By limiting their evaluation to a finite number of points around the unit circle, the ZT and IZT reduce to the DFT (or DFS) and IDFT (or IDFS), respectively. Because of their importance, the DFT and IDFT will be extensively investigated in Chaps. 2.1 and 3.2, respectively, along with a variety of other discrete transforms.

To summarize the results of this section and to gain greater perspective, the DFT, DFS, and FS are related in Fig. 1.8.3. This is a generalization of Fig. 1.7.1. Let $f(t)$ represent the signal and $f_w(t)$ represent the truncated or windowed signal. Then

$$f_w(t) = f(t)w(t)$$
$$F_w(j\omega) = \frac{1}{2\pi}F(j\omega) * W(j\omega) \tag{1.8.19}$$

where $w(t)$ is a rectangular window with width T_w centered at any time we wish to use. $W(j\omega)$ is given by Eq. 1.6.22. $F_w(j\omega)$ is the Fourier transform of $f_w(t)$. To make $f_w(t)$ periodic with period T_w, $f_w(t)$ is convolved with a sampling pulse train $p(t)$ of period T_w as was done in Fig. 1.7.1 which gives

$$f_p(t) = f_w(t) * p(t) = f_w(t) * \sum_{n=-\infty}^{\infty} U_0(t - nT_w)$$
$$F_p(j\omega) = F_w(j\omega)P(j\omega) = \frac{2\pi}{T_w}F_w(j\omega)\sum_{n=-\infty}^{\infty} U_0\left(\omega - m\frac{2\pi}{T_w}\right) \tag{1.8.20}$$

$P(j\omega)$ is given by Eq. 1.7.3. This operation converts a Fourier transform to a Fourier series $F_p(j\omega)$. Finally sampling $f_p(t)$ at a frequency $1/T$ yields

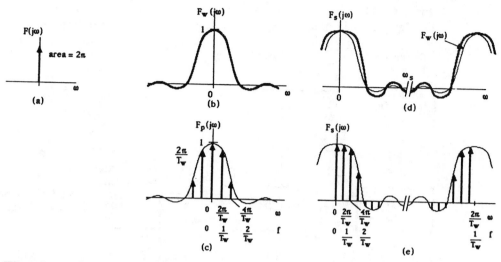

Fig. 1.8.4 Spectra of (a) signal $f(t)$, (b) windowed signal $f_w(t)$, (c) periodic signal $f_p(t)$, and (d)–(e) sampled signal $f_s(t)$.

$$f_s(t) = f_p(t)s(t) = f_p(t) \sum_{n=-\infty}^{\infty} U_0(t - nT)$$

$$F_s(j\omega) = \frac{1}{2\pi} F_p(j\omega) * P(j\omega) = \frac{1}{T} \sum_{n=-\infty}^{\infty} F_p\left[j\left(\omega - m\frac{2\pi}{T}\right)\right] \tag{1.8.21}$$

It is these windowing, convolving, and sampling operations of Eqs. 1.8.19–1.8.21 that convert continuous transforms into discrete transforms.

The results of these operations are shown in Fig. 1.8.4. The signal spectrum $F(j\omega)$ is shown in Fig. 1.8.4a. Windowing the signal for T_w seconds, or equivalently, truncating data introduces spectral distortion in $F_w(j\omega)$ which is shown in Fig. 1.8.4b and discussed in Chap. 2.5. Convolving this windowed signal $f_w(t)$ with an impulse stream $p(t)$ of period T_w produces the sampled spectrum $F_p(j\omega)$ shown in Fig. 1.8.4c. $F_w(j\omega)$ is the Fourier transform of the nonperiodic signal $f_w(t)$ and $F_p(j\omega)$ is the Fourier series coefficient of the periodic signal $f_p(t)$ formed from $f_w(t)$. The spectral coefficients of $F_p(j\omega)$ equal those of $F_w(j\omega)$ scaled by $2\pi/T_w$ when evaluated at frequencies $\omega = m2\pi/T_w$. Sampling the periodic signal $f_p(t)$ produces $f_s(t)$ whose spectrum is shown in Figs. 1.8.4d and 1.8.4e.

A number of transform and inverse transform relations have been discussed. For convenience and later reference, these relations are summarized in Table 1.8.2. The center column lists time domain functions. The left-hand column lists complex frequency domain functions and the right-hand column lists ac steady-state functions. The first row are continuous time relations, the second row are sampled-time relations, while the last rows are discrete time relations. The equation number used to convert one form into another are listed between the blocks.

Table **1.8.2** Summary of equations relating continuous and discrete time signals to Laplace, Fourier, z, and generalized transforms.

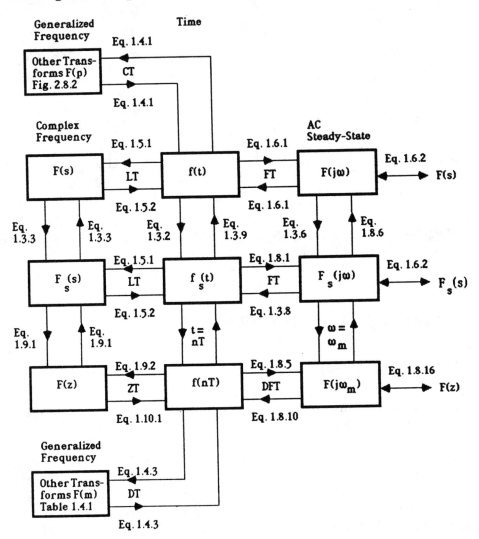

1.9 Z-TRANSFORM (ZT)

Digital filters process pulse streams and sampled-data analog filters process impulse streams as described by Eq. 1.3.2. Unfortunately the Laplace transform is not well-suited to describing such signals because it generates infinite sums involving 0 or e^{sT}, respectively. However another transform called the *z-transform (ZT)* is well-suited.[11] The ZT is related to the LT as

$$F(z) = Z\left[f_s(t)\right] = \mathcal{L}\left[f_s(t)\right]\Big|_{z=e^{sT}} , \qquad z \in \mathcal{R} \qquad (1.9.1)$$

where $Z[\ ]$ denotes the z-transform operation. By setting $z = e^{sT}$, the infinite sums involve the rational function z rather than the irrational function e^{sT}. From Eq. 1.3.2, the z-transform of $f_s(t)$ equals

$$F(z) = F(s) * \sum_{n=-\infty}^{\infty} e^{-nsT}\bigg|_{z=e^{sT}} = F\left(\frac{\ln z}{T}\right) * \sum_{n=-\infty}^{\infty} z^{-n} = \sum_{n=-\infty}^{\infty} f(nT)z^{-n} \quad (1.9.2)$$

The z-transform $F(z)$ is an infinite series in z^{-1}. It converts a function f_s of time t into a function F of complex variable $z = x+jy$. A particular complex variable z_o represents a point in the z-plane. In general, $F(z)$ will only exist (be summable) in some region $\mathcal{R}$ of the z-plane which is its region of convergence. A number of z-transforms are listed in Table 1.9.1 for future reference.

EXAMPLE 1.9.1 Find the z-transform for $e^{-\sigma t}U_{-1}(t)$ which is an exponential or geometric step. This function reduces to a unit step when $\sigma = 0$.
SOLUTION Using Eq. 1.9.2, the z-transform $F(z)$ of $f(nT)$ must equal

$$F(z) = \sum_{n=-\infty}^{\infty} e^{-\sigma nT}U_{-1}(n)z^{-n} = \sum_{n=0}^{\infty}(ze^{\sigma T})^{-n} = \sum_{n=0}^{\infty}\left((ze^{\sigma T})^{-1}\right)^n$$

$$= \frac{1}{1 - (ze^{\sigma T})^{-1}} = \frac{z}{z - e^{-\sigma T}}, \qquad \frac{1}{|ze^{\sigma T}|} < 1 \quad \text{or} \quad |z| > |e^{-\sigma T}| \qquad (1.9.3)$$

The infinite summation is expressed in closed-form using the binomial expansion which converges for $|z| > |e^{-\sigma T}|$. Outside this region $F(z)$ is undefined. $F(z)$ has a pole at $z = e^{-\sigma T}$ and a zero at the origin. All this information is summarized in Fig. 1.9.1b. The magnitude and phase of $F(z)$ are also plotted in Figs. 1.9.1c and 1.9.1d, respectively. Notice that the sampled exponential signal $e^{-\sigma nT}U_{-1}(n)$ can be more easily expressed as the sampled geometric signal $a^n U_{-1}(n)$ where $a = e^{-\sigma T}$. This further simplifies Eq. 1.9.3 since $F(z) = z/(z - a), |z| > |a|$. ∎

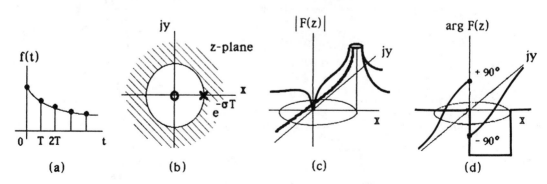

Fig. 1.9.1 (a) Signal $f(nT)$, (b) region of convergence, (c) magnitude $|F(z)|$, and (d) phase arg $F(z)$ of Example 1.9.1.

Table 1.9.1   Table of z-transforms.

$f(t)$	$f(n)$	$F(z)$	Convergence Region
1. $U_0(t)$	$U_0(n)$ or $D_0(n)$	1	all z
2. $U_{-1}(t)$	$U_{-1}(n)$	$\dfrac{z}{z-1}$	$|z| > 1$
3. $U_{-2}(t)$	$U_{-2}(n) = nU_{-1}(n)$	$\dfrac{z}{(z-1)^2}$	$|z| > 1$
4. $U_{-3}(t)$	$U_{-3}(n) = n^2 U_{-1}(n)$	$\dfrac{z(z+1)}{(z-1)^3}$	$|z| > 1$
5. $U_{-4}(t)$	$U_{-4}(n) = n^3 U_{-1}(n)$	$\dfrac{z(z^2 + 4z + 1)}{(z-1)^4}$	$|z| > 1$
6. $a^{t/T} U_{-1}(t)$	$a^n U_{-1}(n)$	$\dfrac{z}{z-a}$	$|z| > |a|$
7. $(t/T)a^{t/T} U_{-1}(t)$	$na^n U_{-1}(n)$	$\dfrac{az}{(z-a)^2}$	$|z| > |a|$
8. $(t/T)^2 a^{t/T} U_{-1}(t)$	$n^2 a^n U_{-1}(n)$	$\dfrac{az(z+a)}{(z-a)^3}$	$|z| > |a|$
9. $\dfrac{(t/T + k - 1)!}{(t/T)!(k-1)!} a^{t/T} U_{-1}(t)$	$\dfrac{(n+k-1)!}{n!(k-1)!} a^n U_{-1}(n)$	$\dfrac{z^k}{(z-a)^k}$	$|z| > |a|$
10. $\cos(bt/T)\, U_{-1}(t)$	$\cos nb\, U_{-1}(n)$	$\dfrac{z(z - \cos b)}{z^2 - 2z\cos b + 1}$	$|z| > |1|$
11. $\sin(bt/T)\, U_{-1}(t)$	$\sin nb\, U_{-1}(n)$	$\dfrac{z \sin b}{z^2 - 2z\cos b + 1}$	$|z| > |1|$
12. $a^{t/T} \cos(bt/T)\, U_{-1}(t)$	$a^n \cos nb\, U_{-1}(n)$	$\dfrac{z(z - a\cos b)}{z^2 - 2az\cos b + a^2}$	$|z| > |a|$
13. $a^{t/T} \sin(bt/T)\, U_{-1}(t)$	$a^n \sin nb\, U_{-1}(n)$	$\dfrac{az \sin b}{z^2 - 2az\cos b + a^2}$	$|z| > |a|$
14. $-U_{-1}(-t - T)$	$-U_{-1}(-n - 1)$	$\dfrac{z}{z-1}$	$|z| < 1$
15. $-(t/T)U_{-1}(-t - T)$	$-nU_{-1}(-n - 1)$	$\dfrac{z}{(z-1)^2}$	$|z| < 1$
16. $-(t/T)^2 U_{-1}(-t - T)$	$-n^2 U_{-1}(-n - 1)$	$\dfrac{z(z+1)}{(z-1)^3}$	$|z| < 1$
17. $-(t/T)^3 U_{-1}(-t - T)$	$-n^3 U_{-1}(-n - 1)$	$\dfrac{z(z^2 + 4z + 1)}{(z-1)^4}$	$|z| < 1$
18. $-a^{t/T} U_{-1}(-t - T)$	$-a^n U_{-1}(-n - 1)$	$\dfrac{z}{z-a}$	$|z| < |a|$
19. $-(t/T)a^{t/T} U_{-1}(-t - T)$	$-na^n U_{-1}(-n - 1)$	$\dfrac{az}{(z-a)^2}$	$|z| < |a|$
20. $-(t/T)^2 a^{t/T} U_{-1}(-t - T)$	$-n^2 a^n U_{-1}(-n - 1)$	$\dfrac{az(z+a)}{(z-a)^3}$	$|z| < |a|$
21. $-\cos(bt/T)\, U_{-1}(-t - T)$	$-\cos nb\, U_{-1}(-n - 1)$	$\dfrac{z(z - \cos b)}{z^2 - 2z\cos b + 1}$	$|z| < |1|$
22. $-\sin(bt/T)\, U_{-1}(-t - T)$	$-\sin nb\, U_{-1}(-n - 1)$	$\dfrac{z \sin b}{z^2 - 2z\cos b + 1}$	$|z| < |1|$
23. $-a^{t/T} \cos(bt/T)\, U_{-1}(-t - T)$	$-a^n \cos nb\, U_{-1}(-n - 1)$	$\dfrac{z(z - a\cos b)}{z^2 - 2az\cos b + a^2}$	$|z| < |a|$
24. $-a^{t/T} \sin(bt/T)\, U_{-1}(-t - T)$	$-a^n \sin nb\, U_{-1}(-n - 1)$	$\dfrac{az \sin b}{z^2 - 2az\cos b + a^2}$	$|z| < |a|$

Table 1.9.2 Properties of z-transforms.

Property	Discrete Time	z-transform	Region of Convergence
Definition	$f(nT)$	$F(z) = Z\left[f(nT)\right]$ $= \sum_{n=-\infty}^{\infty} f(nT)z^{-n}$	$\gamma < \|z\| < \delta$
Linear Operations			
Homogeneity	$Kf(nT)$	$KF(z)$	$\gamma < \|z\| < \delta$
Superposition	$f_1(nT) \pm f_2(nT)$	$F_1(z) \pm F_2(z)$	$\max(\gamma_1, \gamma_2) < \|z\| <$ $\min(\delta_1, \delta_2)$
Both Operations	$a_1 f_1(nT) \pm a_2 f_2(nT)$	$a_1 F_1(z) \pm a_2 F_2(z)$	$\max(\gamma_1, \gamma_2) < \|z\| <$ $\min(\delta_1, \delta_2)$
Symmetry	$F(nT)$	$f(1/z)$	$\delta < \|1/z\| < \gamma$
Scaling			
Time Scaling	$f(anT)$ $f(-anT)$	$F(z)$ with $T \to aT$ $F(1/z)$ with $T \to aT$	$\gamma < \|z\| < \delta$ $\delta < \|1/z\| < \gamma$
Magnitude Scaling	$af(nT)$	$aF(z)$	$\gamma < \|z\| < \delta$
Shifting			
Time Shifting	$f(nT - n_oT)$	$z^{-n_o} F(z)$	$\gamma < \|z\| < \delta$
Time Shifting*	$f(nT - n_oT)$	$z^{-n_o}[F(z) + zf(-1) +$ $z^2 f(-2) + \ldots + z^{n_o} f(-n_o)]$	$\gamma < \|z\| < \delta$
Frequency Shifting	$a^n f(nT)$	$F(z/a)$	$\gamma\|a\| < \|z\| <$ $\delta\|a\|$
Periodic Functions* (for $t > 0$ only)	$f(nT) =$ $\sum_{n=0}^{\infty} f_1((n + n_o)T)$	$\frac{F_1(z)}{1 - z^{-1}} = F_1(z)\sum_{n=0}^{\infty} z^{-n},$ $F_1(z) = \sum_{n=0}^{n_o-1} f(nT)z^{-n}$	$\gamma < \|z\|$
Convolution			
Time convolution (Frequency product)	$f_1(nT) * f_2(nT)$	$F_1(z)F_2(z)$	$\max(\gamma_1, \gamma_2) < \|z\| <$ $\min(\delta_1, \delta_2)$
Frequency convolution (Time product)	$f_1(nT)f_2(nT)$	$\frac{1}{j2\pi} F_1(z) * F_2(z) =$ $\frac{1}{j2\pi} \oint_c F_1(v) F_2(z/v) \frac{dv}{v}$	$\gamma_1\gamma_2 < \|z\| < \delta_1\delta_2$
Zero Padding	$f(nT/k),\ k = \text{integer}$ 0, otherwise	$F(z^k)$	$\gamma^{1/k} < \|z\| < \delta^{1/k}$
Time Integration*	$\sum_{k=-\infty}^{n} f(kT)$	$\frac{z}{z-1} F(z)$	$\max(1, \gamma) < \|z\| < \delta$
Frequency Differentiation			
First derivative	$nTf(nT)$	$(-Tz)dF(z)/dz$	$\gamma < \|z\| < \delta$
kth derivative	$(nT)^k f(nT)$	$(-Tz)^k d^k F(z)/dz^k$	$\gamma < \|z\| < \delta$
Frequency Integration			
First integral	$f(nT)/nT$	$(1/Tz)\int_z^{\infty} F(z)dz$ or $(-1/T)\int_0^z F(z)/z\, dz$	$\gamma < \|z\| < \delta$ $\gamma < \|z\| < \delta$
Initial Value*	$f(0^+) = \lim_{n \to 0} f(nT)$	$\lim_{z \to \infty} F(z)$	
Final Value*	$f(\infty) = \lim_{n \to \infty} f(nT)$	$\lim_{z \to 1} \frac{z-1}{z} F(z)$	if poles of $\frac{z-1}{z}F(z)$ inside unit circle
Initial Slope*	$\Delta f(0^+)/T =$ $\lim_{n \to 0} \frac{f((n+1)T) - f(nT)}{T}$	$\lim_{z \to \infty} \frac{(z-1)F(z) - zf(0)}{T}$	

*For one-sided transforms

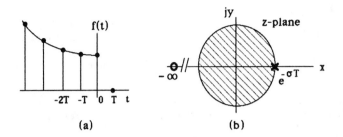

Fig. 1.9.2 (a) Exponential step signal and (b) region of convergence of Example 1.9.2.

EXAMPLE 1.9.2 Determine the z-transform for the sequence $e^{-\sigma t}U_{-1}(-t)$.
SOLUTION Using Eq. 1.9.2 where $t = nT$, $F(z)$ must equal

$$F(z) = \sum_{n=-\infty}^{0} e^{-\sigma nT}z^{-n} = \sum_{n=-\infty}^{0}(ze^{\sigma T})^{-n} = \sum_{n=0}^{\infty}(ze^{\sigma T})^n$$

$$= \frac{1}{1 - ze^{\sigma T}}, \qquad |ze^{\sigma T}| < 1 \tag{1.9.4}$$

which converges for $|z| < |e^{-\sigma T}|$ as shown in Fig. 1.9.2. ∎

EXAMPLE 1.9.3 Determine the conditions for the two-sided geometric step function $e^{-\sigma t}U_{-1}(t) - e^{-\beta t}U_{-1}(-t - T)$ to have a z-transform.
SOLUTION The transform for the $e^{-\beta T}U_{-1}(-t-T)$ term is found from Eq. 1.9.4 by simply subtracting the $n = 0$ term (which is one) from $F(z)$. Then combining this transform with that of Eq. 1.9.3, we obtain

$$F(z) = \frac{z}{z - e^{-\sigma T}} - \frac{1}{1 - ze^{\beta T}} + 1 = \frac{z}{z - e^{-\sigma T}} + \frac{z}{z - e^{-\beta T}}$$

$$= \frac{z(2z - e^{-\sigma T} - e^{-\beta T})}{(z - e^{-\sigma T})(z - e^{-\beta T})}, \qquad |e^{-\sigma T}| < |z| < |e^{-\beta T}| \tag{1.9.5}$$

$F(z)$ converges only if Re σ > Re β and then only for values of z lying inside the annular region shown in Fig. 1.9.3b. ∎

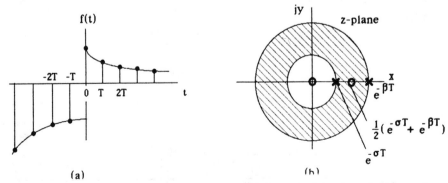

Fig. 1.9.3 (a) Doubled-sided exponential step signal and (b) region of convergence of Example 1.9.3.

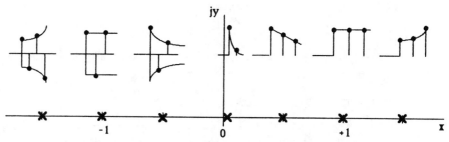

Fig. 1.9.4 Relation of exponential step $e^{-n\sigma T}\,U_{-1}(n)$ to real-axis pole locations.

EXAMPLE 1.9.4 Determine the z-transform of an exponential step with alternating sign $(-1)^n e^{-n\sigma T} U_{-1}(n)$.
SOLUTION Using Eq. 1.9.2, the z-transform equals

$$F(z) = \sum_{n=0}^{\infty}(-1)^n e^{-\sigma nT} z^{-n} = \sum_{n=0}^{\infty}(-1)^n (ze^{\sigma T})^{-n}$$

$$= \frac{1}{1+(ze^{\sigma T})^{-1}} = \frac{z}{z+e^{-\sigma T}}, \qquad |z| > |e^{-\sigma T}| \tag{1.9.6}$$

The pole of $F(z)$ lies at $z = -e^{-\sigma T}$ unlike the pole of $F(z)$ in Example 1.9.1 which lies at $z = +e^{-\sigma T}$. The difference between the two examples is sign alternation. This results when the pole is rotated by 180 degrees in the z-plane (by $e^{j\pi} = -1$). This is summarized in Fig. 1.9.4 which shows the real-axis location as a function of the time constant $1/\sigma$ of the exponential step. ∎

EXAMPLE 1.9.5 Determine the z-transform of a sinusoidal step $\sin(\omega t)U_{-1}(t)$.
SOLUTION The z-transform can be indirectly derived using Euler's identity $\sin(\omega t) = (e^{j\omega t} - e^{-j\omega t})/j2$. Then using the $e^{-\sigma nT}$ transform found in Example 1.9.1, we can write directly that

$$F(z) = \frac{1/j2}{1-(ze^{j\omega T})^{-1}} - \frac{1/j2}{1-(ze^{-j\omega T})^{-1}}$$

$$= \frac{z\sin(\omega T)}{z^2 - 2z\cos(\omega T) + 1}, \qquad \frac{1}{|ze^{\pm j\omega T}|} < 1 \quad \text{or} \quad |z| > 1 \tag{1.9.7}$$

Here the two poles are complex. They lie on the unit circle at an angle of ωT radians with respect to the positive real axis. One zero lies at the origin and the other lies at infinity. Therefore given the sinusoidal frequency ω, then the sampling time T determines the pole position as shown in Fig. 1.9.5. A sinusoidal step with alternating sign $(-1)^n$ has the same pole-zero pattern but it is rotated by 180°. ∎

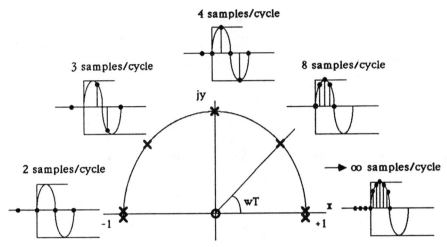

Fig. 1.9.5 Relation of sinusoidal step $\sin(nwT)\, U_{-1}(n)$ of frequency ω to unit circle pole locations. Conjugate poles not shown.

The z-transform satisfies a number of interesting properties. These properties are listed in Table 1.9.2 for the two-sided z-transform. They are easily proved using the z-transform definition. The engineer should practice using these relations by deriving all the entries in Table 1.9.1 utilizing only the first entry and Table 1.9.2. This exercise will show the great usefulness of these properties.

The two-dimensional (2-D) z-transform $F(z_1, z_2)$ and its inverse equal

$$F(z_1, z_2) = Z\left[f(n_1T_1, n_2T_2)\right] = \sum_{n_1=-\infty}^{\infty} \sum_{n_2=-\infty}^{\infty} f(n_1T_1, n_2T_2)z_1^{-n_1} z_2^{-n_2}, \quad z_1, z_2 \in R$$

$$f(n_1T_1, n_2T_2) = Z^{-1}\left[F(z_1, z_2)\right]$$

$$= \frac{1}{(j2\pi)^2 T_1 T_2} \oint_{C_1} \oint_{C_2} F(z_1, z_2)z_1^{n_1-1} z_2^{n_2-1}, \quad C_1, C_2 \in R$$

$$(1.9.8)$$

The 2-D z-transform $F(z_1, z_2)$ is related to the 2-D Laplace transform $F(s_1, s_2)$ and 2-D Fourier transform $F(j\omega_1, j\omega_2)$ as

$$F(z_1, z_2) = Z\left[f_s(t_1, t_2)\right] = \mathcal{L}\left[f_s(t_1, t_2)\right]\Big|_{\substack{z_1 = e^{s_1 T_1} \\ z_2 = e^{s_2 T_2}}} = \mathcal{F}\left[f_s(t_1, t_2)\right]\Big|_{\substack{z_1 = e^{j\omega_1 T_1} \\ z_2 = e^{j\omega_2 T_2}}} \quad (1.9.9)$$

by analogy with Eqs. 1.6.2 and 1.9.1.

EXAMPLE 1.9.6 Determine the z-transform of impulse, step, exponential, and complex exponential 2-D signals $f(n_1, n_2)$.

SOLUTION Applying the 2-D z-transform from Eq. 1.9.8 gives

$$Z\left[U_0(n_1, n_2)\right] = \sum_{n_1=-\infty}^{\infty} \sum_{n_2=-\infty}^{\infty} U_0(n_1, n_2)z_1^{-n_1} z_2^{-n_2} = 1, \quad \text{all } (z_1, z_2)$$

$$Z\left[U_{-1}(n_1, n_2)\right] = \sum_{n_1=0}^{\infty} z_1^{-n_1} \sum_{n_2=0}^{\infty} z_2^{-n_2} = \frac{1}{(1 - z_1^{-1})(1 - z_2^{-1})}, \qquad |z_1|, |z_2| > 1$$

$$Z\left[a_1^{n_1} a_2^{n_2} U_{-1}(n_1, n_2)\right] = \frac{1}{(1 - a_1 z_1^{-1})(1 - a_2 z_2^{-1})}, \qquad |z_1| > |a_1|, |z_2| > |a_2|$$

$$Z\left[e^{j(n_1\omega_1 T_1 + n_2\omega_2 T_2)} U_{-1}(n_1, n_2)\right]$$

$$= \frac{1}{(1 - e^{j\omega_1 T_2} z_1^{-1})(1 - e^{j\omega_2 T_2} z_2^{-1})}, \qquad |z_1|, |z_2| > 1$$

$$(1.9.10)$$

∎

Multirate sampling was discussed in Chap. 1.3 and illustrated in Fig. 1.3.5. The spectral relation between a signal $f(t)$ sampled at two different rates ($1/T$ and k/T Hz) was given in Eq. 1.3.15. Now let us relate their z-transforms. Denote the z-transform of $f(t)$ sampled at a $1/T$ Hz rate as $F_1(z)$ where

$$F_1(z) = \sum_{n=-\infty}^{\infty} f(nT) z^{-n}, \qquad |z| \in \mathcal{R} \qquad (1.9.11)$$

Then the z-transform of $f(t)$ sampled at a k/T Hz rate equals

$$F_2(z) = \sum_{n=-\infty}^{\infty} f(nT/k) z^{-n/k} = \sum_{n=-\infty}^{\infty} f(nT/k)\left(z^{1/k}\right)^{-n}$$

$$= \sum_{n=-\infty}^{\infty} f(nT/k)(z^{1/k})^{-n} = F_1(z^{1/k}), \qquad |z^{1/k}| \in \mathcal{R} \qquad (1.9.12)$$

Therefore the z-transform of the signal sampled every nT/k seconds can be obtained from the z-transform of the signal sampled every nT seconds by replacing z by $z^{1/k}$. In ac steady-state, then

$$F_2(e^{j\omega T}) = F_1(e^{j\omega T/k}) \qquad (1.9.13)$$

which is the digital equivalent of Eq. 1.3.15. The F_2 spectrum is translated by a factor k in frequency. Thus the Nyquist frequency $\omega T = \pi$ for F_1 maps to $k\pi$ for F_2. All else remains unchanged.

The transformation $z = e^{sT}$ is of paramount importance in digital filters. Let us briefly investigate its properties. The z-transform pair equal

$$z = e^{sT} \qquad \text{and} \qquad sT = \ln z \qquad (\text{or} \qquad s = \frac{1}{T}\ln z) \qquad (1.9.14)$$

Expressing s and z in rectangular form as $s = \sigma + j\omega$ and $z = x + jy$, then Eq. 1.9.14 can be manipulated as

$$sT = \ln(|z|e^{j \arg z}) = \ln|z| + j \arg z = (\sigma + j\omega)T \qquad (1.9.15)$$

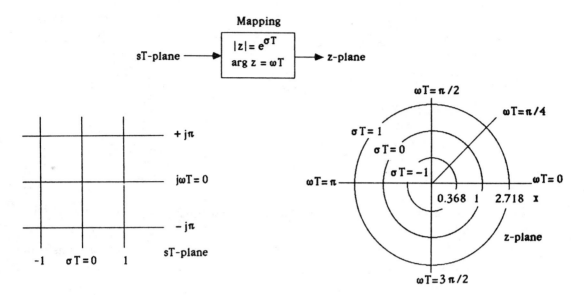

Fig. 1.9.6 Mapping from the s-plane to the z-plane in rectangular form.

Therefore the real and imaginary parts of sT equal

$$\sigma T = \ln |z|$$
$$\omega T = \arg z \pm 2k\pi, \qquad k = 0, 1, 2, \ldots \tag{1.9.16}$$

The mapping from the s-plane to the z-plane is single-valued but the mapping from the z-plane to the s-plane is multi-valued. That is, a single point in the z-plane maps into an infinite number of points in the s-plane as will be illustrated in Example 1.9.7. From Eq. 1.9.15, the magnitude and phase of z equal

$$|z| = e^{\sigma T}, \qquad \arg z = \omega T \tag{1.9.17}$$

Thus, vertical root loci having real part σ in the s-plane map into circles of radius $e^{\sigma T}$ centered about the origin in the z-plane. Root loci of imaginary part ω in the s-plane map into radial lines of angle ωT radians in the z-plane. This is shown in Fig. 1.9.6. If s is expressed in polar form rather than rectangular form as $s = \omega_n e^{j(\pi-\theta)}$ where $\theta = \cos^{-1} \varsigma$, then the root loci for constant ω_n and ς map into the z-plane as shown in Fig. 1.9.7.

EXAMPLE 1.9.7 Consider an integrator having a transfer function $H(s) = 1/s$. Find the digital transfer function using z-transforms and plot its pole-zero pattern. **SOLUTION** The z-transform equals the transform of the sampled impulse response $h(t) = U_{-1}(t)$ of the integrator. Using Eq. 1.9.3 where $\sigma = 0$,

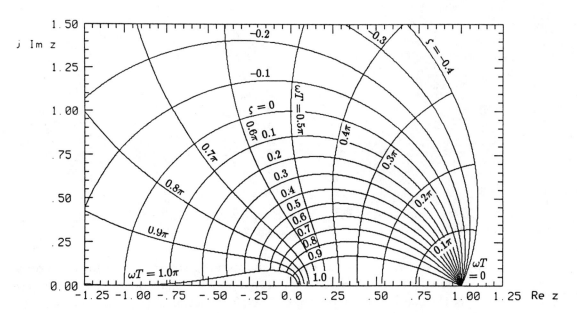

Fig. 1.9.7 Mapping from the s-plane to the z-plane in polar form.

$$H(z) = \frac{z}{z-1}, \qquad |z| > 1 \tag{1.9.18}$$

so there is one pole at $z = 1$ and one zero at the origin. The sampled-domain transfer function equals

$$H_s(s) = \frac{1}{T} \sum_{n=-\infty}^{\infty} \frac{1}{s - jn\omega_s}, \qquad \text{Re } s > 0, \quad \omega_s = \frac{2\pi}{T} \tag{1.9.19}$$

as found in Example 1.3.1 (see Eq. 1.3.10). Then

$$H(z) = H_s(s)\Big|_{\substack{z=e^{sT},\\ sT=\ln z}} = \frac{1}{T} \sum_{n=-\infty}^{\infty} \frac{T}{\ln z - jn\omega_s T} = \frac{z}{z-1} \tag{1.9.20}$$

The poles of $H(z)$ equal

$$\ln z = sT = jn\omega_s T = \ln|z| + j \, \arg z \tag{1.9.21}$$

Solving Eq. 1.9.21 for $|z|$ and arg z yields

$$\begin{aligned}
|z| &= 1 \qquad \text{since} \qquad \ln|z| = 0 \\
\arg z &= n\omega_s T = 2n\pi, \qquad n = \ldots, -2, -1, 0, 1, 2, \ldots
\end{aligned} \tag{1.9.22}$$

This shows that all poles lie on the unit circle at an angle of $\omega_s T = 2\pi$ radians. Thus all the s-plane poles and zeros map into a single pole and zero in the z-plane. This is shown in Fig. 1.9.8. ∎

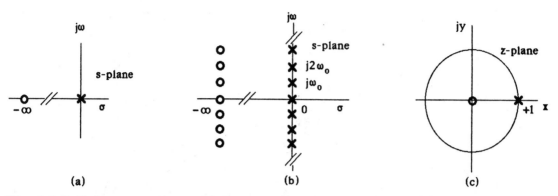

Fig. 1.9.8 Pole-zero pattern of integrator transfer function for (a) unsampled $H(s)$, (b) sampled $H(s)$, and (c) $H(z)$ in Example 1.9.7.

It is important to remember that the transformation $z = e^{sT}$ is used for transforming the *sampled* transfer function $H_s(s)$ from the s-domain to the z-domain. It is *not used* for transforming the continuous (unsampled) transfer function $H(s)$. This is a common conceptual misunderstanding. For example, if $H(s) = 1/s$ then $H(z) \neq T/\ln z$. Example 1.9.7 illustrates the proper conversion method for manipulating sampled transfer functions. Unsampled transfer functions are often converted using s-to-z transformations to be discussed in Chap. 5. The next example illustrates the technique.

EXAMPLE 1.9.8 Convert the analog integrator of Example 1.9.7 into a digital integrator using the s-to-z transform $s = 1 - z^{-1}$ (see Chap. 5.4). Determine its ac steady-state gain.

SOLUTION The digital integrator has the transfer function

$$H(z) = H(s)\Big|_{s=1-z^{-1}} = \frac{1}{s}\Big|_{s=\frac{z-1}{z}} = \frac{z}{z-1} \tag{1.9.23}$$

which is identical to $H(z)$ found in Eq. 1.9.20. Setting $s = j\omega$ or $z = e^{j\omega T}$ in Eq. 1.9.23 yields the ac steady-state gain as

$$H(e^{j\omega T}) = \frac{z}{z-1}\Big|_{z=e^{j\omega T}} = \frac{e^{j\omega T/2}}{j\,\sin(\omega T/2)} \tag{1.9.24}$$

The magnitude and phase of H are plotted in Fig. 1.9.9b. At low frequencies where $\omega T/2 \cong 0$, Eq. 1.9.24 reduces to (since $\exp(j\theta) \cong 1 + j\theta$ and $\sin\theta \cong \theta$)

$$H(e^{j\omega T}) \cong \frac{1 + j\omega T/2}{j\omega T} \cong \frac{1}{j\omega T} \tag{1.9.25}$$

The digital integrator well approximates the ac steady-state gain of the ideal analog integrator having $H(s) = 1/s$ as shown in Fig. 1.9.9a. Shifting $H(j\omega)$ to multiples of the sampling frequency $\omega_s = 2\pi/T$ and adding as required by superposition in Eq. 1.3.3 converts Fig. 1.9.9a into Fig. 1.9.9b. The magnitude response is bounded by Eq. 1.3.10. ∎

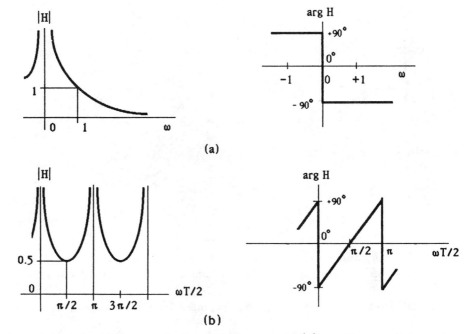

Fig. 1.9.9 Magnitude and phase responses of (a) continuous integrator and (b) discrete integrator in Example 1.9.7.

1.10 INVERSE Z-TRANSFORM (IZT)

The *z-transform (ZT)* and *inverse z-transform (IZT)* are related as

$$F(z) = Z\big[f(nT)\big] = \sum_{n=-\infty}^{\infty} f(nT)z^{-n}, \quad z \in \mathcal{R}$$

$$f(nT) = Z^{-1}\big[F(z)\big] = \frac{1}{j2\pi T} \oint_C F(z)z^{n-1}dz, \quad C \in \mathcal{R} \tag{1.10.1}$$

The integration path along the closed contour C must lie in the region of convergence of $F(z)$ as shown in Fig. 1.10.1. The z-transform pair result by substituting the sampled time function $f_s(t)$ given by Eq. 1.3.2 in the Laplace transform pair of Eqs. 1.5.1 and 1.5.2, and making the change of variable $z = \exp(sT)$ so $dz = T\exp(sT)ds$ or $ds = (1/Tz)dz$.

The inverse z-transforms of $F(z)$ can be determined using three methods: (1) partial fraction expansion (PFE), (2) power (Taylor) series expansion, (3) and contour integration using Eq. 1.10.1b. We choose to use the partial fraction expansion method since we consider only rational $F(z)$. The power series method, although simple, does not yield a closed form solution for $f(nT)$. The contour method is computationally complex and needed only for irrational $F(z)$ which are not considered.

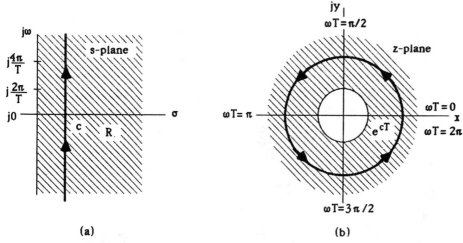

Fig. 1.10.1 (a) Integration path Re $s = c$ in s-plane and the equivalent path $z = e^{cT}$ in z-plane. Both paths lie inside the regions of convergence.

$f(nT)$ results by computing the inverse z-transform of $F(z)/z$ where

$$F(z) = \frac{cz \prod_{j=1}^{M-1}(z - z_j)}{\prod_{i=1}^{N}(z - p_i)}, \qquad |z| > k \tag{1.10.2}$$

As with Laplace transforms, this is facilitated by expressing $F(z)/z$ in a partial fraction expansion (PFE) as[12]

$$F(z) = \sum_{i=1}^{N} \frac{K_i z}{z - p_i} \qquad \text{or} \qquad \frac{F(z)}{z} = \sum_{i=1}^{N} \frac{K_i}{z - p_i} \tag{1.10.3}$$

K_i is the residue of $F(z)/z$ evaluated at the pole $z = p_i$. Such an expansion or representation is valid for any rational $F(z)$ having: (1) $N \geq M$ (denominator degree $\geq$ numerator degree) and (2) only simple or first-order poles. The expansion is valid for all values of z. These two restrictions can be removed by modifying the PFE form as will be discussed in a moment.

The residue K_i is obtained by multiplying $F(z)/z$ by $(z - p_i)$ and evaluating the product at $z = p_i$ as

$$K_i = (z - p_i)\frac{F(z)}{z}\bigg|_{z=p_i} = K_i + \sum_{\substack{q=1 \\ q \neq i}}^{N} \frac{(z - p_i)K_q}{z - p_q}\bigg|_{z=p_i} = K_i \tag{1.10.4}$$

All the terms within the summation equal zero since the poles are simple where $p_q \neq p_i$. Expressing $F(z)/z = N(z)/D(z)$, then K_i can also be evaluated as

$$K_i = \lim_{z \to p_i} (z - p_i)\frac{N(z)}{D(z)} = \lim_{z \to p_i} \frac{N(z)}{D(z)/(z - p_i)} = \frac{N(z)}{dD(z)/dz}\bigg|_{z=p_i} \tag{1.10.5}$$

which is a useful form in many situations. The K_i residue measures the contribution to the total response due to the pole at $z = p_i$. A pole close to the other poles but far from the zeros has a large residue and vice versa.

The K_i residue has the well-known graphical interpretation shown in Fig. 1.10.2. Combining Eqs. 1.10.2 and 1.10.4, K_i is a complex number equal to

$$K_i = (z - p_i)\frac{F(z)}{z}\bigg|_{z=p_i} = \frac{c\prod_{j=1}^{M-1}(p_i - z_j)}{\prod_{\substack{q=1\\q\neq i}}^{N}(p_i - p_q)} = \frac{c\prod_{j=1}^{M-1}A_j e^{j\phi_j}}{\prod_{\substack{q=1\\q\neq i}}^{N}B_q e^{j\theta_q}} = |K_i|e^{j\,\arg K_i} \quad (1.10.6)$$

The magnitude and angle of K_i equal

$$|K_i| = \frac{c\prod_{j=1}^{M-1}A_j}{\prod_{\substack{q=1\\q\neq i}}^{N}N_q}, \qquad \arg K_i = \sum_{j=1}^{M-1}\phi_j - \sum_{\substack{q=1\\q\neq i}}^{N}\theta_q \quad (1.10.7)$$

The magnitude of K_i is equal to c times the product of the *zero* vector magnitudes A_j divided by the product of the *pole* vector magnitudes B_q. The zero vectors originate from each zero and extend to the test pole $z = p_i$. The pole vectors originate from each pole and extend to $z = p_i$ except for the test pole itself. The phase of K_i is equal to the sum of zero vector angles ϕ_j minus the pole vector angles θ_q.

Since the polynomials in the numerator and denominator of $F(z)$ are assumed to have real coefficients, all poles and zeros exist in conjugate pairs when complex. Thus, the net angle contribution from a complex pole or zero pair must be zero when p_i is real. Therefore all real poles have positive or negative real residues. Since the residues of a complex pole pair must be complex conjugates, the two terms in the partial fraction expansion arising from these complex poles can be rewritten as

$$\frac{K}{z - x - jy} + \frac{K^*}{z - x + jy} = 2\frac{(z - x)\,\mathrm{Re}\,K - y\,\mathrm{Im}\,K}{(z - x)^2 + y^2} \quad (1.10.8)$$

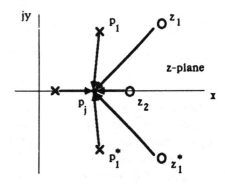

Fig. 1.10.2 Graphical interpretation for residue K_i.

Thus, the partial fraction expansion form for $F(z)/z$ having only N simple poles (of which P pairs are complex) and $N \geq M$ is

$$\frac{F(z)}{z} = \sum_{i=P+1}^{N} \frac{K_i}{z - p_i} + 2\sum_{i=1}^{P} \frac{(z - x_{2i-1})\,\text{Re}\,K_{2i-1} - y_{2i-1}\text{Im}\,K_{2i-1}}{(z - x_{2i-1})^2 + y_{2i-1}^2}, \qquad |z| > z_o$$

$$(1.10.9)$$

where the K_i are given by Eqs. 1.10.4 and 1.10.5. If $F(z)/z$ fails to have a pole at some z value where a residue is evaluated, then $K_i = 0$ at that point. This situation arises when there is a pole-zero cancellation in $F(z)/z$ which is not recognized. It is encouraging to see that residue evaluation is so forgiving. Now consider the more general situation.

Suppose that $F(z)$ has a pole p_1 of order k and $N \geq M$. Then the partial fraction expansion of $F(z)/z$ equals

$$\frac{F(z)}{z} = \frac{c\prod_{j=1}^{M}(z - z_j)}{(z - p_1)^k \prod_{i=k+1}^{N}(z - p_i)} = \sum_{i=1}^{k} \frac{K_{1,i}}{(z - p_1)^i} + \sum_{i=k+1}^{N} \frac{K_i}{z - p_i} \qquad (1.10.10)$$

The residue $K_{q,i}$ is subscripted so that q equals the pole number and i equals the order of pole being considered. Note that if the pole is simple, then $k = 1$ and all the $K_{1,i} = 0$ except $K_{1,1}$. The $K_{1,i}$ residue equals

$$K_{1,i} = \frac{1}{(k - i)!} \frac{d^{k-i}}{dz^{k-i}} \left[\frac{(z - p_1)^k F(z)}{z} \right]\Bigg|_{z=p_1}, \qquad i = 1, 2, \ldots, k \qquad (1.10.11)$$

using the same reasoning as that resulting in Eq. 1.10.5. As before, this equation is true for both real and complex multiple-order poles. Thus for every multiple-order pole in $F(z)/z$, the additional summation terms given in Eq. 1.10.10 must be added to the partial fraction expansion for $F(z)/z$.

Now suppose that $F(z)$ has more (finite) zeros than poles so $M > N$. Then additional terms in ascending powers of z must be included in the partial fraction expansion of $F(z)/z$ given by Eq. 1.10.3 as

$$\frac{F(z)}{z} = \sum_{i=0}^{M-N-1} K_{-i} z^i + X(z) \qquad (1.10.12)$$

The remainder $X(z)$ is given by Eq. 1.10.10. The K_{-i} residues of these terms equal

$$K_{-i} = \frac{1}{(M - N - i)!} \frac{d^{M-N-i}}{d(1/z)^{M-N-i}} \left[\frac{F(z)}{z^{M-N}} \right]\Bigg|_{|z|=\infty}, \qquad i = 1, 2, \ldots, M - N - 1$$

$$(1.10.13)$$

using the same reasoning as that used to obtain Eq. 1.10.5 and noting that $R(z) \to 0$ as $|z| \to \infty$. The partial fraction expansion of Eq. 1.10.12 can be more easily obtained by dividing the denominator of $F(z)/z$ into its numerator.

The response $f(nT)$ of a filter can now be easily determined. $f(nT)$ is equal to the sum of responses due to each individual term in the partial fraction expansion of

$F(z)$. The necessary transform pairs are found in Table 1.9.1 as (assume causality so $f(nT) = 0$ for $n < 0$)

$$Z^{-1}\left[\frac{z}{z-p}\right] = p^n U_{-1}(n), \qquad |z| > p$$

$$Z^{-1}\left[\frac{z(z-x)}{(z-x)^2+y^2}\right] = p^n \cos(n\omega T) U_{-1}(n), \qquad |z| > |p| \qquad (1.10.14)$$

$$Z^{-1}\left[\frac{zy}{(z-x)^2+y^2}\right] = p^n \sin(n\omega T) U_{-1}(n), \qquad |z| > |p|$$

where $p = \sqrt{x^2 + y^2}$ and $\omega T = \tan^{-1}(y/x)$. Assuming that $F(z)$ has $N \geq M$ and only first-order poles) then from Eq. 1.10.9, $f(nT)$ equals

$$f(nT) = Z^{-1}[F(z)] = \left[\sum_{i=k+1}^{N} K_i p_i^n\right.$$

$$\left. + 2\sum_{i=1}^{k} p_{2i-1}^n \left[\text{Re } K_{2i-1}\cos(n\omega_{2i-1}T) - \text{Im } K_{2i-1}\sin(n\omega_{2i-1}T)\right]\right] U_{-1}(nT)$$

$$= \left[\sum_{i=k+1}^{N} K_i p_i^n + 2\sum_{i=1}^{k} |K_{2i-1}| p_{2i-1}^n \cos(n\omega_{2i-1}T + \arg K_{2i-1})\right] U_{-1}(nT)$$

$$(1.10.15)$$

The response form of $f(nT)$ depends upon the pole locations. As shown in Fig. 1.9.4, a simple real-axis pole gives rise to an exponential response whose time constant depends upon its location on the real axis. Locations inside the unit circle give rise to decreasing responses (stable); locations outside the unit circle to increasing responses (unstable); and unit circle locations to constant responses (conditionally stable). The closer the pole is to $z = 1$, the smaller is the time constant and the slower is the response. Alternating sign sequences have left-half-plane locations while non-alternating sign sequences have right-half-plane locations.

Complex pole pairs (of first-order) give rise to oscillating responses as shown in Fig. 1.9.5. The frequency of oscillation ω (radians/second) is equal to the pole angle ωT divided by the time T between samples. The number of samples over one period of oscillation (samples/cycle) equals 2π divided by the pole angle. The amplitude of the response is bounded by positive and negative exponential envelopes. The envelope corresponds to the exponential response of a real-axis pole equal to the radius of the complex pole being considered. All pole pairs on a constant radius circle have the same exponential response bounds. Poles inside the unit circle have oscillating responses with exponentially-decreasing bounds (stable); poles outside the unit circle have oscillating responses with exponentially-increasing bounds (unstable); and poles on the unit circle have oscillating responses with constant bounds (conditionally stable).

Now let us consider the effects that multiple-order poles have upon the impulse response. If $F(z)$ contains a pole p of order k, then there are additional terms in the partial fraction expansion. From Eq. 1.10.10 and Table 1.9.1, these produce additional

$n^{k-1}p^n U_{-1}(n)$ type terms in the impulse response. The response decays to zero for $|p| < 1$ but rises to infinity for $|p| \geq 1$. Thus multiple-poles on or outside the unit circle cause the system to be unstable. Therefore, stable causal systems can have only first-order (simple) poles inside the unit circle.

When the numerator of $F(z)$ is of greater degree than the denominator, the partial fraction expansion of $F(z)/z$ must be augmented by the terms given by Eq. 1.10.12. These additional terms in the partial fraction expansion produce a response

$$f_2(nT) = Z^{-1}[F(z)] = \sum_{i=0}^{M-N-1} K_{-i}U_0((n+i)T) \qquad (1.10.16)$$

using Table 1.9.1. These terms involve the impulse U_0 shifted in time. Since these terms equal zero as $n \to \infty$, they do not affect system stability.

We see that partial fraction expansions allow the response of any filter having a rational transfer function to be determined. Although this is a conceptually simple process, it can become tedious and time consuming for systems having more than several poles. In practice, the engineer will want to utilize complete transform tables where the partial fraction expansions have already been tabulated.

EXAMPLE 1.10.1 Find the impulse response of a filter having gain

$$H(z) = \frac{1}{(z - 0.5)(z - 1)}, \qquad |z| > 1 \qquad (1.10.17)$$

SOLUTION The impulse response $h(nT)$ is the inverse z-transform of $H(z)$ (see Eq. 1.12.6). Since the denominator of $H(z)$ is of higher degree than the numerator, the partial fraction expansion is given by Eq. 1.10.3 as

$$\frac{H(z)}{z} = \frac{1}{z(z - 0.5)(z - 1)} = \frac{K_0}{z} + \frac{K_1}{z - 0.5} + \frac{K_2}{z - 1}, \qquad |z| > 1 \qquad (1.10.18)$$

Evaluating the residue of each pole using Eq. 1.10.4 gives

$$K_0 = z\left[\frac{H(z)}{z}\right]\Big|_{z=0} = \frac{1}{(z - 0.5)(z - 1)}\Big|_{z=0} = 2$$

$$K_1 = (z - 0.5)\left[\frac{H(z)}{z}\right]\Big|_{z=0.5} = \frac{1}{z(z - 1)}\Big|_{z=0.5} = -4 \qquad (1.10.19)$$

$$K_2 = (z - 1)\left[\frac{H(z)}{z}\right]\Big|_{z=1} = \frac{1}{z(z - 0.5)}\Big|_{z=1} = 2$$

Substituting these residues into Eq. 1.10.18 gives the PFE of $H(z)$ as

$$H(z) = 2 - \frac{4z}{z - 0.5} + \frac{2z}{z - 1}, \qquad |z| > 1 \qquad (1.10.20)$$

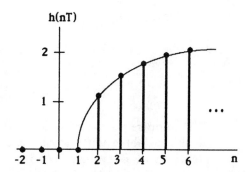

Fig. 1.10.3 Impulse response of digital filter in Example 1.10.1.

Taking the inverse of $H(z)$ term by term using Table 1.9.1 yields

$$h(nT) = 2U_0(n) + \left[-4(0.5)^n + 2\right] U_{-1}(n) = 2U_0(n) + \left[2 - 0.5^{n-2}\right] U_{-1}(n) \quad (1.10.21)$$

where $U_0(n)$ is the unit impulse and $U_{-1}(n)$ is the unit step. The response is plotted in Fig. 1.10.3. It rises monotonically from 0 to 2 in step sizes of 0.5^n after a two clock period delay.

This response can be verified by expressing $H(z)$ in a Taylor series in $1/z$ about $z = \infty$ as

$$H(z) = \frac{1}{z^2 - 1.5z + 0.5} = z^{-2} + 1.5z^{-3} + 1.75z^{-4} + 1.875z^{-5} + \ldots, \qquad |z| > 1$$
$$(1.10.22)$$

This form can also be obtained by simply performing long division on $H(z)$. Taking the inverse z-transform of Eq. 1.10.22 gives

$$h(nT) = U_0(n-2) + 1.5U_0(n-3) + 1.75U_0(n-4) + 1.875U_0(n-5) + \ldots \quad (1.10.23)$$

This is equivalent to the impulse response given by Eq. 1.10.21. ∎

EXAMPLE 1.10.2 Find the impulse response of

$$H(z) = \frac{z^3}{(z - 0.5)(z - 1)}, \qquad |z| > 1 \qquad (1.10.24)$$

SOLUTION In this case, the numerator of $H(z)$ is of higher degree than the denominator. Therefore, long division must first be performed so $H(z)$ has the form of Eq. 1.10.12. This results in

$$H(z) = z + \frac{1.5z^2 - 0.5z}{z^2 - 1.5z + 0.5}, \qquad |z| > 1 \qquad (1.10.25)$$

Now the remainder $X(z)$ can be expressed in the standard PFE as

$$\frac{X(z)}{z} = \frac{1.5z - 0.5}{(z - 0.5)(z - 1)} = \frac{-0.5}{z - 0.5} + \frac{2}{z - 1}, \qquad |z| > 1 \qquad (1.10.26)$$

Using Table 1.9.1, the transform pair equal

$$H(z) = z - \frac{0.5z}{z - 0.5} + \frac{2z}{z - 1}, \qquad |z| > 1$$

$$h(nT) = U_0(n+1) + \left[-0.5(0.5)^n + 2\right] U_{-1}(n) \tag{1.10.27}$$

This impulse response is identical to that plotted in Fig. 1.10.3 except $h(nT)$ must be shifted three units to the left (to $n = -2$). Alternatively viewed, the transfer functions of Examples 1.10.1 and 1.10.2 are related as $H_2(z) = z^3 H_1(z)$. Using the time shifting theorem of Table 1.9.2, then the impulse responses are related as $h_2(nT) = h_1((n+3)T)$. Replacing n by $(n+3)$ in $h_2(nT)$ given by Eq. 1.10.21 and manipulating the result produces Eq. 1.10.27.   ■

EXAMPLE 1.10.3 Find the step response of

$$H(z) = \frac{1}{z^2 - \sqrt{2}z + 1}, \qquad |z| > 1 \tag{1.10.28}$$

SOLUTION The transformed step response equals

$$R(z) = \frac{z}{(z - 1)(z^2 - \sqrt{2}z + 1)}$$

$$= \frac{K_1 z}{z - 1} + \frac{K_2 z}{z - 0.707 - j0.707} + \frac{K_3 z}{z - 0.707 + j0.707}, \qquad |z| > 1 \tag{1.10.29}$$

by taking the product of $H(z)$ and $Z[U_{-1}(nT)] = z/(z - 1)$. Evaluating residues,

$$K_1 = (z - 1)\left[\frac{R(z)}{z}\right]\Big|_{z=1} = \frac{1}{z^2 - \sqrt{2}z + 1}\Big|_{z=1} = 1.707$$

$$K_2 = (z - 0.707 - j0.707)\left[\frac{R(z)}{z}\right]\Big|_{\substack{z=0.707 \\ +j0.707}} = \frac{1}{(z - 1)(z - 0.707 + j0.707)}\Big|_{\substack{z=0.707 \\ +j0.707}}$$

$$= -0.924e^{-j22.5°}$$

$$K_3 = (z - 0.707 + j0.707)\left[\frac{R(z)}{z}\right]\Big|_{\substack{z=0.707 \\ -j0.707}} = \frac{1}{(z - 1)(z - 0.707 - j0.707)}\Big|_{\substack{z=0.707 \\ -j0.707}}$$

$$= -0.924e^{j22.5°} \tag{1.10.30}$$

Therefore, finding the inverse z-transform of Eq. 1.10.29 in Table 1.9.1 gives

$$r(nT) = \left[K_1(1)^n + K_2(0.707 + j0.707)^n + K_2^*(0.707 - j0.707)^n\right] U_{-1}(n)$$

$$= \left[1.707 + 2\,\text{Re}\,(-0.924e^{-22.5°})(e^{j45n°})\right] U_{-1}(n)$$

$$= \left[1.707 - 1.848\cos(45n° - 22.5°)\right] U_{-1}(n) \tag{1.10.31}$$

It is useful to observe several important points. First, complex poles (and zeros) always occur in conjugate pairs so $p_3 = p_2^*$ (see Eq. 1.10.29). Second, their residues are always complex conjugates so $K_3 = K_2^*$ (see Eq. 1.10.30). Third, when inverse z-transforms are taken of each term, the resulting terms are still complex conjugates so $K_3 p_3^n = K_2^* p_2^{*n}$. Fourth, such terms combine into a sum where the imaginary parts cancel leaving twice the real part as $2\,\text{Re}(K_2 p_2^n)$. Eq. 1.10.31 clearly shows these facts. Eq. 1.10.31 can be written directly using the general inverse transform expression of Eq. 1.10.15.   ■

EXAMPLE 1.10.4 Find the step response of

$$H(z) = \frac{1}{(z-1)(z-2)^2}, \qquad |z| > 2 \qquad (1.10.32)$$

SOLUTION Multiplying $H(z)$ by $z/(z-1)$ forms the transformed step response $R(z)$. Since $R(z)$ has double poles at $z = 1$ and 2, the partial fraction expansion of Eq. 1.10.10 must be used as

$$\frac{R(z)}{z} = \frac{1}{(z-1)^2(z-2)^2} = \frac{K_{1,2}}{(z-1)^2} + \frac{K_{1,1}}{z-1} + \frac{K_{2,2}}{(z-2)^2} + \frac{K_{2,1}}{z-2}, \qquad |z| > 2$$

$$(1.10.33)$$

Evaluating the residues using Eq. 1.10.11 gives

$$K_{1,2} = (z-1)^2 \left[\frac{R(z)}{z}\right]\Big|_{z=1} = G_1(z)\Big|_{z=1} = \frac{1}{(z-2)^2}\Big|_{z=1} = 1$$

$$K_{1,1} = \frac{dG_1(z)}{d(z-1)}\Big|_{z=1} = \frac{-2(z-2)}{(z-2)^4}\Big|_{z=1} = 2$$

$$K_{2,2} = (z-2)^2 \left[\frac{R(z)}{z}\right]\Big|_{z=2} = G_2(z)\Big|_{z=1} = \frac{1}{(z-1)^2}\Big|_{z=2} = 1 \qquad (1.10.34)$$

$$K_{2,1} = \frac{dG_2(z)}{d(z-2)}\Big|_{z=2} = \frac{-2(z-1)}{(z-1)^4}\Big|_{z=2} = -2$$

Therefore the transform pair become

$$R(z) = \frac{z}{(z-1)^2} + \frac{2z}{z-1} + \frac{z}{(z-2)^2} - \frac{2z}{z-2}, \qquad |z| > 2$$

$$r(nT) = [1^n n + 2 + 0.5(2^n)n - 2(2^n)]\,U_{-1}(n) = [(2^{n-1}+1)n - 2(2^n - 1)]\,U_{-1}(n)$$

$$(1.10.35)$$

   ■

EXAMPLE 1.10.5 Find the inverse z-transform of

$$H(z) = \frac{1}{(z-0.5)(z-1)}, \qquad 0.5 < |z| < 1 \qquad (1.10.36)$$

SOLUTION The partial fraction of $H(z)$ was found in Example 1.10.1 (see Eq. 1.10.20). However the region of convergence is different in this transform. To proceed, express $H(z)$ as the sum of two transforms (H_1, H_2) each having its own region of convergence $(\mathcal{R}_1, \mathcal{R}_2)$. Their intersection $(\mathcal{R}_1 \cap \mathcal{R}_2)$ forms the region of

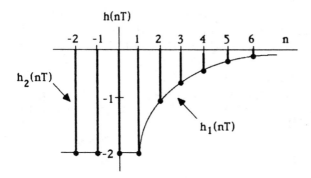

Fig. 1.10.4 Impulse response of digital filter in Example 1.10.5.

convergence of $H(z)$. From Eq. 1.10.20,

$$H(z) = 2 - \frac{4z}{z - 0.5} + \frac{2z}{z - 1} = H_1(z) + H_2(z), \qquad |z| > 0.5 \quad \text{and} \quad |z| < 1$$

$$(1.10.37)$$

We choose to assign

$$H_1(z) = 2 - \frac{4z}{z - 0.5}, \quad |z| > 0.5 \quad \text{and} \quad H_2(z) = \frac{2z}{z - 1}, \quad |z| < 1 \quad (1.10.38)$$

Taking the inverse z-transform from Table 1.9.1 gives

$$h_1(nT) = 2U_0(n) - 4(0.5)^n U_{-1}(n), \qquad h_2(nT) = -2(1)^n U_{-1}(-n-1) \quad (1.10.39)$$

These responses are plotted in Fig. 1.10.4. $h_1(nT)$ is a decaying exponential for positive time. $h_2(nT)$ is a step for negative time. It should be mentioned that assigning the term 2 to $H_1(z)$ in Eq. 1.10.38 was arbitrary. It could just as well have been assigned to $H_2(z)$ or split between $H_1(z)$ and $H_2(z)$.   ■

1.11 LINEARITY

A general digital *MIMO* system has multiple inputs and multiple outputs as shown in Fig. 1.11.1a. The inputs to the system (independent variables) are $x_1, x_2, \ldots, x_n$. The system outputs (dependent variables) are $y_1, y_2, \ldots, y_m$. The simplest possible system is the single input and single output *SISO* case shown in Fig. 1.11.1b. Suppose that the system transforms inputs into outputs using transformation $\mathbf{h}$ where

$$\begin{bmatrix} y_1 \\ y_2 \\ \vdots \\ y_m \end{bmatrix} = \mathbf{h} \begin{bmatrix} x_1 \\ x_2 \\ \vdots \\ x_n \end{bmatrix} \qquad \text{or} \qquad y = \mathbf{h}[x] \qquad (1.11.1)$$

$\mathbf{h}$ is an $m \times n$ matrix containing operators. Note that in the single input/single output case, $y = \mathbf{h}[x]$ so that the vector equation reduces to a scalar equation. The system is *linear* if **h** is a *linear transformation*. **h** is a linear transform if it satisfies homogeneity and superposition. Transformation **h** satisfies the *homogeneity condition* if

$$y(nT) = \mathbf{h}\big[ax(nT)\big] = a\mathbf{h}\big[x(nT)\big] \qquad (1.11.2)$$

for an input ax. Transformation **h** satisfies the *superposition condition* if

$$y(nT) = \mathbf{h}\big[x(nT) + \Delta x(nT)\big] = \mathbf{h}\big[x(nT)\big] + \mathbf{h}\big[\Delta x(nT)\big] \qquad (1.11.3)$$

for an input $x + \Delta x$. This transformation or operator notation is especially convenient for mathematically defining some other system categories.

A system is *time-invariant* if for all inputs $x(nT - D)$, then

$$y(nT - D) = \mathbf{h}\big[x(nT - D)\big] \qquad (1.11.4)$$

for any discrete time nT and time delay D. Thus, delaying an input by time D simply delays the output by time D in time-invariant systems. In other words, the output is independent of the time origin and depends only upon the input shape.

A system is *memoryless*, if for all inputs $x(nT)$, the outputs can be expressed as

$$y(nT) = \mathbf{h}\big[x(nT), nT\big] \qquad (1.11.5)$$

This equation stipulates that the output at discrete time nT of a memoryless system depends only upon the input at that particular time, and not upon the input at any other time. We will use these terms to describe several examples in a moment.[13]

Before proceeding, consider a linear time-varying SISO system having the Eq. 1.11.1 form of $y = \mathbf{h}[x]$ where

$$y(n) = \sum_{k=-\infty}^{\infty} h(n, k)x(k), \qquad \text{all } n \qquad (1.11.6)$$

Writing $y(n)$ in matrix form gives

$$
\begin{bmatrix} \vdots \\ y(-1) \\ y(0) \\ y(1) \\ \vdots \end{bmatrix}
=
\begin{bmatrix}
\vdots & \vdots & \vdots & \vdots & \vdots \\
\cdots & h(-1,-1) & h(-1,0) & h(-1,1) & \cdots \\
\cdots & h(0,-1) & h(0,0) & h(0,1) & \cdots \\
\cdots & h(1,-1) & h(1,0) & h(1,1) & \cdots \\
\vdots & \vdots & \vdots & \vdots & \vdots
\end{bmatrix}
\begin{bmatrix} \vdots \\ x(-1) \\ x(0) \\ x(1) \\ \vdots \end{bmatrix}
\qquad (1.11.7)
$$

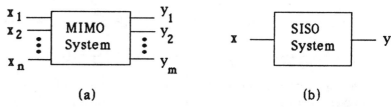

(a) (b)

Fig. 1.11.1 (a) Multiple input/multiple output system and (b) single input/single output system.

Here $[x]$ is the input vector, $[y]$ is the output vector, and $\mathbf{h}$ is the transmission matrix. The input sequence needed to produce a specified output sequence equals $x = \mathbf{h}^{-1}[y]$ where $\mathbf{h}^{-1}$ is the inverse of $\mathbf{h}$ and must be nonsingular (i.e., determinant $|\mathbf{h}| \neq 0$). When the system $\mathbf{h}$ is time-invariant, then $h(k, n) = h(k - n)$. For this very common situation, Eqs. 1.11.6 and 1.11.7 reduce to

$$y(n) = \sum_{k=-\infty}^{\infty} h(n - k)x(k) = \sum_{k=-\infty}^{\infty} h(k)x(n - k), \qquad \text{all } n \tag{1.11.8}$$

or

$$\begin{bmatrix} \vdots \\ y(-1) \\ y(0) \\ y(1) \\ \vdots \end{bmatrix} = \begin{bmatrix} \ddots & \ddots & \ddots & \ddots & \ddots \\ \ddots & h(0) & h(-1) & h(-2) & \ddots \\ \ddots & h(1) & h(0) & h(-1) & \ddots \\ \ddots & h(2) & h(1) & h(0) & \ddots \\ \ddots & \ddots & \ddots & \ddots & \ddots \end{bmatrix} \begin{bmatrix} \vdots \\ x(-1) \\ x(0) \\ x(1) \\ \vdots \end{bmatrix} \tag{1.11.9}$$

These matrix equations will be discussed more fully in the next section.

1.12 CAUSALITY

A system is *causal* if its response to an input at discrete time nT does not depend on future input values.[14] Mathematically, a causal system satisfies

$$y(nT) = \mathbf{h}\big[x(mT)\big], \quad n \geq m \tag{1.12.1}$$

Noncausal or anticipatory systems are said to be *physically nonrealizable*. Such systems generate outputs before inputs are applied. The name stems from the fact that noncausal systems cannot perform controlled functions in real time. They can be implemented in nonreal time.

EXAMPLE 1.12.1 A variety of systems are described by their operators in Fig. 1.12.1. Classify the systems in terms of linearity, time-invariance, memory, and causality.
SOLUTION In the first system, operator $H[\ ] = k(nT)[\ ]$ and the input/output relation equals
$$y(nT) = k(nT)x(nT) \tag{1.12.2}$$

Setting $x = ax_1$ gives $y = kax_1$. Setting $x = bx_1 + cx_2$, we find

$$y = kx = k(bx_1 + cx_2) = k(bx_1) + k(cx_2) = kb(x_1) + kc(x_2) = by_1 + cy_2 \tag{1.12.3}$$

so the system is linear since Eqs. 1.11.2 and 1.11.3 are satisfied. We test for time-invariance by replacing nT by $(nT - D)$ in Eq. 1.12.2 and compare results. Since

$$y(nT - D) = k(nT - D)x(nT - D) \neq k(nT)x(nT - D) \tag{1.12.4}$$

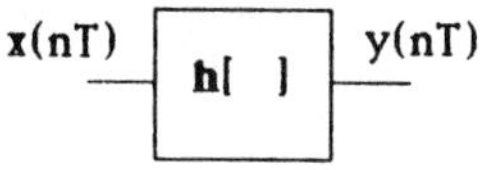

$x(nT)$ — $\boxed{h[\ ]}$ — $y(nT)$

(a) $h[\ ] = k(nT)[\ ]$
Linear
Time-varying
Memoryless
Causal

(b) $h[\ ] = [\ ]^2$
Nonlinear
Time-invariant
Memoryless
Causal

(c) $h[\ ] = \displaystyle\sum_{m=-\infty}^{k} [\ ]$
Linear
Time-invariant
Memory
Causal

(d) $h[\ ] = \displaystyle\sum_{m=-\infty}^{\infty} [\ ]\, h(nT, mT)$
Linear
Time-varying
Memory
Noncausal

(e) $h[\ ] = \displaystyle\sum_{m=-\infty}^{\infty} [\ ]\, h((n-m)T)$
Linear
Time-invariant
Memory
Noncausal

Fig. 1.12.1 A variety of systems and their classifications in Example 1.12.1.

the system is time-varying because Eq. 1.11.4 is not satisfied. Testing for memory, we see that an instantaneous input/output exists. Thus, the system is memoryless since Eq. 1.11.5 is satisfied. Testing for causality, we input an $x(nT)$ beginning at time $nT = 0$. Since no output occurs for $n < 0$, the system is causal and Eq. 1.12.1 is satisfied. These system characteristics are listed in Fig. 1.12.1a. The other examples follow in the same way so we shall not discuss them in detail. ∎

Now consider a single input/single output time-invariant linear system having a transfer function $H(z)$. The z-transformed output $Y(z)$ of the system equals

$$Y(z) = H(z)X(z) \tag{1.12.5}$$

where $X(z)$ is the z-transformed input. A convenient analytical input for testing the system is an impulse whose z-transform is $X(z) = 1$ (see Chap. 3.1). Then the system output $Y(z)$ equals $H(z)$ so

$$y(nT) = Z^{-1}[Y(z)] = Z^{-1}[H(z)] = h(nT) \tag{1.12.6}$$

$h(nT)$ is called the *impulse response* of the system. It depends only upon the gain $H(z)$ of the system. Since the input occurs at $n = 0$, causal systems must have impulse responses which satisfy

$$h(nT) = 0, \qquad n < 0 \tag{1.12.7}$$

This alternative causality definition allows $h(nT)$ to be tested directly.

EXAMPLE 1.12.2 Consider the following digital filter transfer functions and their inverses. Do they represent physically realizable systems? Remember that such transfer functions are periodic about $n\omega_o$ for all n (see Eq. 1.3.3).

(a) $|H_1(e^{j\omega T})| = 1, \quad |\omega T| \leq \omega_o T$

$\qquad\qquad = 0, \quad \omega_o T < |\omega T| < \pi$

(b) $|H_2(e^{j\omega T})| = \dfrac{1}{T} \displaystyle\sum_{n=-\infty}^{\infty} e^{-(\omega - n\omega_o)^2}, \qquad |\omega T| < \pi \tag{1.12.8}$

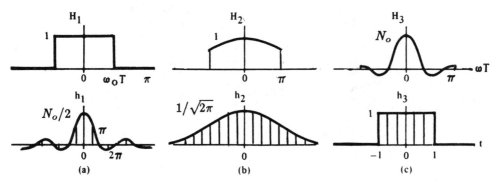

Fig. 1.12.2 Several examples of noncausal systems in Example 1.12.2.

$$\text{(c)} \quad |H_3(e^{j\omega T})| = \frac{\sin(\omega N_o T/2)}{\sin(\omega T/2)}, \qquad |\omega T| < \pi$$

SOLUTION H_1 represents the gain of an ideal low-pass filter while H_2 is that of an ideal Gaussian filter. It can be shown that the impulse responses equal

$$\text{(a)} \quad h_1(nT) = \frac{\sin(n\pi N_o/N)}{\sin(2n\pi/N)}, \qquad N_o = \frac{\omega_o T}{\pi}$$

$$\text{(b)} \quad h_2(nT) = \frac{1}{\sqrt{2\pi}} e^{-(nT)^2} \qquad\qquad\qquad (1.12.9)$$

$$\text{(c)} \quad h_3(nT) = 1, \quad |nT| < N_o T/2$$
$$\qquad\qquad = 0, \quad \text{elsewhere}$$

These impulse responses are drawn in Fig. 1.12.2. Since they are all nonzero for some negative time, all the transfer functions are physically nonrealizable.   ∎

Transfer functions can be tested directly for causality (rather than their impulse responses) using the generalized *Paley-Wiener criterion*. For a magnitude function $|H(e^{j\omega T})|$ to be realizable, a necessary and sufficient condition is that

$$PWC = \int_{-\pi/T}^{\pi/T} \big| \log|H(e^{j\omega T})| \big|\, d\omega = \int_{-\pi}^{\pi} \big| \log|H(j\theta)| \big|\, d\theta < \infty \qquad (1.12.10)$$

Thus, not only must $|H(e^{j\omega T})|$ exist but the area under the plot of $\big| \log|H(e^{j\omega T})| \big|$ between the Nyquist frequencies must be finite. This further implies that the magnitude characteristics $|H(e^{j\omega T})|$ cannot be zero over any band of frequencies. However, $|H(e^{j\omega T})|$ may possess a countably infinite number of zeros over the frequency band without violating the test condition.

The converse of the Paley-Wiener criterion says that $h(nT) = 0$ for $n < 0$ if the condition of Eq. 1.12.10 is satisfied. This corresponds to Eq. 1.12.7. Thus, realizability can be tested in either the time domain or the frequency domain depending upon which is more convenient.

EXAMPLE 1.12.3 Test the three filters of Example 1.12.2 for realizability using the Paley-Wiener criterion.

SOLUTION To test H_1, we must evaluate

$$PWC_1 = 2 \int_0^{\pi/T} \big| \log |H_1| \big| \; d\omega = 2 \Big[|\log 1| \int_0^{\omega_o T} d(\omega T) + |\log 0| \int_{\omega_o T}^{\pi} d(\omega T) \Big]$$

$$= 0 + \infty = \infty \qquad\qquad (1.12.11)$$

Thus, $|H_1|$ does not describe a realizable filter. Likewise testing $|H_2|$ gives

$$PWC_2 = \frac{2}{T} \int_0^{\pi/T} \Big| \log \Big| \sum_{n=-\infty}^{\infty} e^{-(\omega - n\omega_e)^2} \Big| \Big| \; d\omega = \infty \qquad\qquad (1.12.12)$$

$|H_2|$ describes a nonrealizable filter. Finally, testing $|H_3|$ in the same manner gives

$$PWC_3 = 2 \int_0^{\pi} \Big| \log \Big| \frac{\sin(\omega N_o T/2)}{\sin(\omega T/2)} \Big| \Big| \; d(\omega T) < \infty \qquad\qquad (1.12.13)$$

Thus, $|H_3|$ describes a realizable filter. In fact, it is clear from the Paley-Wiener test that any $[\sin(\omega T)/\omega T]^n$ magnitude function described a realizable filter ($n \geq 1$).

We found in Example 1.12.2 that $|H_3|$ described a nonrealizable filter, and therefore, are faced by an apparent conflict and dilemma. There is no conflict, however, when it is noted that an appropriate phase $\arg H$ must be chosen for H if it is to be realizable. Recall that the gain H equals

$$H(j\omega) = |H(j\omega)| e^{j \; \arg H(j\omega)} \qquad\qquad (1.12.14)$$

and that $\arg H$ does not affect magnitude. We shall see that choosing a linear phase $\arg H = \omega$ is sufficient to make H_3 realizable. This has the effect of delaying $h_3(t)$ in Fig. 1.12.2 by one second which forces $h_3(t) = 0$ for $t < 0$. ∎

Reconsider the most common linear SISO system discussed in Eqs. 1.11.6–1.11.9 where $y = \mathbf{h}x$. Causal time-varying systems have $h(k,n) = 0$ for $k < n$ so

$$\begin{bmatrix} \vdots \\ y(-1) \\ y(0) \\ y(1) \\ \vdots \end{bmatrix} = \begin{bmatrix} \vdots & \vdots & \vdots & \vdots & \vdots \\ \cdots & h(-1,-1) & 0 & 0 & \cdots \\ \cdots & h(0,-1) & h(0,0) & 0 & \cdots \\ \cdots & h(1,-1) & h(1,0) & h(1,1) & \cdots \\ \vdots & \vdots & \vdots & \vdots & \vdots \end{bmatrix} \begin{bmatrix} \vdots \\ x(-1) \\ x(0) \\ x(1) \\ \vdots \end{bmatrix} \qquad (1.12.15)$$

Causal time-invariant systems satisfy Eq. 1.12.7 so Eq. 1.12.15 reduces to

$$\begin{bmatrix} \vdots \\ y(-1) \\ y(0) \\ y(1) \\ \vdots \end{bmatrix} = \begin{bmatrix} \ddots & \ddots & \ddots & \ddots & \ddots \\ \ddots & h(0) & 0 & 0 & \ddots \\ \ddots & h(1) & h(0) & 0 & \ddots \\ \ddots & h(2) & h(1) & h(0) & \ddots \\ \ddots & \ddots & \ddots & \ddots & \ddots \end{bmatrix} \begin{bmatrix} \vdots \\ x(-1) \\ x(0) \\ x(1) \\ \vdots \end{bmatrix} \qquad (1.12.16)$$

Notice the matrices are triangular in either case because their upper-triangular sections are filled with zeros. We see that the elements of H, which are $h(k,n)$ or $h(n)$, represent the unit impulse response samples of the system. The simplest causal, time-invariant case has $x(n) = 0$ for $n < 0$. Eq. 1.12.16 then reduces to

$$\begin{bmatrix} y(0) \\ y(1) \\ y(2) \\ \vdots \end{bmatrix} = \begin{bmatrix} h(0) & 0 & 0 & \cdot_\cdot \\ h(1) & h(0) & 0 & \cdot_\cdot \\ h(2) & h(1) & h(0) & \cdot_\cdot \\ \cdot_\cdot & \cdot_\cdot & \cdot_\cdot & \cdot_\cdot \end{bmatrix} \begin{bmatrix} x(0) \\ x(1) \\ x(2) \\ \vdots \end{bmatrix} \qquad (1.12.17)$$

For example, $h(k,n) = a^{k-n}U_{-1}(k-n)$ is causal and time-invariant; $h(k,n) = a^{k}U_{-1}(k-n)$ is causal but time-varying; $h(k,n) = a^{k-n}$ is noncausal and time-invariant; while $h(k,n) = a^{k}$ is noncausal and time-varying. Time-invariant (time-varying) discrete filter matrices must also have circular symmetry where $h(k,n) = h(\mathrm{mod}_N(k+r), \mathrm{mod}_N(n+r))$, $r = 1, 2, \ldots, N-1$ (almost arbitrary symmetry) regardless of causality.

In design, the input $[x]$ and output $[y]$ are specified and $[h]$ must be found. Assuming N samples of both x and y are available (N total equations), then N^2 values of $h(\ )$ must be found. Since $N^2 > N$, the problem is underconstrained and there is no unique solution for $[h]$ in the noncausal case. In the causal time-varying case given by Eq. 1.12.15, there are $\frac{1}{2}N(N+1)$ nonzero $[h]$ matrix elements. Since $\frac{1}{2}N(N+1) > N$ there is also no unique solution. However in the causal, time-invariant case given by Eqs. 1.12.16 and 1.12.17, there are only N nonzero matrix elements so the solution is unique. Solving for $h(k)$ and progressing row by row in Eq. 1.12.17 yields the recursion equation

$$h(0) = \frac{y(0)}{x(0)}, \qquad h(n) = \frac{1}{x(0)}\left[y(n) - \sum_{k=0}^{n-1} x(n-k)h(k) \right], \qquad n = 1, 2, \ldots \qquad (1.12.18)$$

This process of generating or synthesizing $[h]$ from $[x]$ and $[y]$ is called *deconvolution*.

1.13 STABILITY

In assessing system behavior, we are often interested in its stability. System stability depends upon its time domain response. Several definitions of stability exist. Roughly speaking, a system is stable if its output is bounded for every bounded input. The following is a precise definition that can be tested mathematically.[15]

Absolutely stable system: A system whose impulse response $|h(nT)| \to 0$ as $|n| \to \infty$.
Conditionally stable system: A system whose impulse response $|h(nT)|$ is bounded (by a constant) as $|n| \to \infty$.
Unstable system: A system whose impulse response $|h(nT)| \to \infty$ as $|n| \to \infty$.

It should be noted that stability and causality are different concepts. Let us investigate the stability of the systems whose impulse response is shown in Fig. 1.13.1. The systems can be classified in terms of stability and causality as:

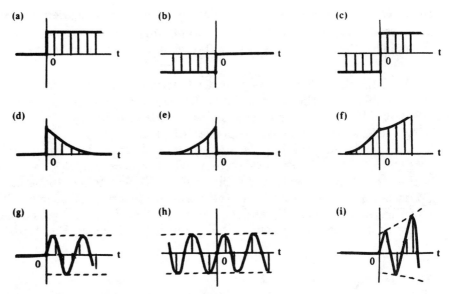

Fig. **1.13.1** Assorted impulse response functions.

a. Conditionally stable, causal   f. Unstable, noncausal
b. Conditionally stable, noncausal   g. Conditionally stable, causal
c. Conditionally stable, noncausal   h. Conditionally stable, noncausal
d. Absolutely stable, causal   i. Unstable, causal
e. Absolutely stable, noncausal

We see that systems (f) and (i) are unstable and systems (d) and (e) are absolutely stable. The remaining systems are conditionally stable since their impulse responses approach bounded levels rather than zero.

From the causality standpoint, the systems of (a), (d), (g), and (i) are causal and therefore realizable. Therefore, these responses can be generated using real systems. The other systems are nonrealizable and cannot be implemented in real-time.

If we are concerned with *causal* systems, then such systems are *absolutely stable* if all their poles lie inside the unit circle. They will be *conditionally stable* if they have any additional simple poles on the unit circle. They will be *unstable* if they have multiple-order poles on the unit circle or any poles outside the unit circle. The system zeros have no effect on system stability. The statement that stable systems are systems whose transfer function poles lie inside the unit circle is incorrect unless realizability is also required.

Another test for stability uses the gain function $H(j\omega)$ and the final value theorem from Table 1.9.2. Since the final value of the impulse response $h(t)$ satisfies

$$h(\infty) - h(-\infty) = \lim_{s \to 0} sH(s) = \lim_{\omega \to 0} \omega H(j\omega) = \lim_{z \to 1} \frac{z-1}{z} H(z) \qquad (1.13.1)$$

then we can sometimes test $H(j\omega)$ directly to determine system stability. It must be remembered however, that this limit can only be performed if the region of convergence of H includes the s-plane origin ($s = 0$ or $\omega = 0$) or $z = 1$ in the z-plane.

EXAMPLE 1.13.1 Three transfer functions were tested for causality in Example 1.12.2. Now test them for stability directly in the frequency domain.

SOLUTION Since all three gains approach a constant as $\omega \to 0$ as shown in Fig. 1.12.2, they all have responses which satisfy $\omega H \to 0$. Therefore from Eq. 1.13.1, they all have impulse responses with equal limiting (endpoint) values as $h(+\infty) = h(-\infty)$. For the systems which are causal, $h(-\infty) = 0$ which insures $h(+\infty) = 0$ and absolute stability. Since H_3 was found to be causal in Example 1.12.2, this shows H_3 is also a stable transfer function. This is confirmed by $h_3(nT)$ in Fig. 1.12.2. Since H_1 and H_2 are noncausal, $h(-\infty)$ does not necessarily equal zero so no conclusion about stability can be made using Eq. 1.13.1. ∎

1.14 SIGNAL FLOW GRAPHS

A signal flow graph is a pictorial representation of a set of equations. Once the equations describing a system have been written, they can be represented in signal flow graph form. The flow graph can be easily manipulated and simplified using standard rules. Much greater insight is gained by manipulating the graph of the equations, rather than the equations themselves, as we shall see.[16]

Mason introduced signal flow graph theory. A *signal flow graph* is a graphical representation of the relationships between the variables of a set of linear algebraic equations. The graph consists of oriented branches connected to nodes. The nodes represent the signals or variables of the system. A branch leaving node x_i and terminating on node x_j represents the linear dependance of variable x_j on variable x_i, but not vice versa. Each branch has a unidirectional gain or transmittance T_{ij}. The direction is denoted by an arrowhead. A signal is transmitted through a branch in the arrowhead direction and is multiplied by the branch gain. Each node algebraically sums the incoming signals, and then transmits the resulting signal to every outgoing branch. A number of standard definitions and terminologies are used in signal flow graph theory. The most important for our uses are the following:

1. *Signal Flow Graph:* A network of nodes connected by (directed) branches.
2. *Node:* Represents a signal or variable (dependent/independent) in the system.
3. *Branch:* Represents the dependence of a dependent variable upon an independent variable.
4. *Input (Source) Node:* A node having only outgoing branches. This represents a signal input or excitation.
5. *Output (Sink) Node:* A node having only incoming branches. This represents a signal output or response.
6. *Path:* Any continuous succession or connected set of branches transversed in their indicated directions.

7. *Forward Path (Open Path):* Any path connecting an input node to an output node, along which no node is encountered more than once.
8. *Feedback Path (Loop, Closed Path):* Any path which originates and terminates on the same node, in which no node is transversed more than once.
9. *Path Gain:* The product of the branch gains of the branches forming the path.
10. *Loop Gain:* The product of the branch gains forming the loop.
11. *Residual Graph:* A graph containing only sources and sinks. Residual graphs are obtained from the original signal flow graph using reduction techniques.
12. *Signal Flow Graph Gain:* The signal appearing at the sink per unit signal applied at the source. For multiple input-output graphs, every sink-source combination has its own gain.
13. *Essential Nodes:* Those nodes which must be removed to eliminate all feedback loops.
14. *Flow Graph Order (Index):* Equals the minimum number of essential nodes.
15. *Node Splitting:* A node that has been separated into a source node and a sink node. Consists of grouping input branches on one side of a node, output branches on the opposite side, and splitting the node in half.
16. *Loop Gain of Node:* Signal returned to a node per unit signal transmitted by that node; or gain between the source and sink created by splitting the node.
17. *Loop Difference (Return Difference):* Difference between a unit signal injected at a node and the signal returned to the node; or one minus the loop gain of the node.
18. *Nontouching Loops:* Loops having no common nodes.

One of the great benefits of signal flow graphs is the ease with which gain expressions can be determined. In fact, Mason described signal flow graph by saying, *"A way to enhance, writing gain at a glance, ..."* . The general expression for the gain between any pair of sink and source nodes in a graph is

$$H = \frac{\sum T_k D_k}{D} \qquad (1.14.1)$$

where

T_k = gain of the kth forward path
$D = 1 - \sum P_{m_1} + \sum P_{m_2} - \sum P_{m_3} + \cdots \qquad (1.14.2)$
P_{m_r} = gain product of the mth possible combination of r nontouching loops
D_k = the value of D for that part of the graph not touching the kth forward path

Sometimes D is called the *determinant of the graph* and D_k is called the *cofactor of the T_k path.* The engineer may prefer to express the determinant D as

$D = 1-$ (sum of all loop gains)
   $+$ (sum of loop gain products of all combinations of 2 nontouching loops)
   $-$ (sum of loop gain products of all combinations of 3 nontouching loops) $+\ldots$
$$(1.14.3)$$

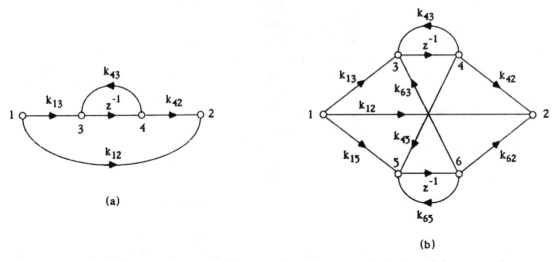

Fig. 1.14.1 (a) First-order and (b) second-order canonical signal flow graphs.

Two loops are nontouching if they have no common nodes. After some practice, Mason's gain formula can be written by inspection so it is an especially convenient analysis tool.

Recall from our flow graph definitions that the graph order is equal to the minimum number of nodes, which if removed, eliminate all feedback loops. Any flow graph of order n can be reduced to a so-called *canonical flow graph* having $(n+1)^2$ branches. The first- and second-order canonical graphs are shown in Fig. 1.14.1 (z^{-1} branches must be removed using the feedback loop equivalence to yield the minimum number of nodes). Digital filter structures which are described by these flow graphs will be analyzed in Chap. 12.

Let us consider signal flow graphs from the viewpoint of linear algebra. A flow graph pictorially represents a system of n linear equations as $y = hx$. However, the flow graph is not unique. There are $n!$ signal flow graphs which describe the same set of n equations. The particular graph obtained depends upon the order in which the dependent variables are chosen. Nevertheless, all the graphs contain the same information although they have different signal flows.

EXAMPLE 1.14.1 The two-point discrete Fourier transform (DFT) takes two data points in the time domain (f_0, f_1) and converts them to two data points in the frequency domain (F_0, F_1) as

$$F_0 = f_0 + f_1, \qquad F_1 = f_0 - f_1 \qquad (1.14.4)$$

Draw the flow graph representation of the equations.
SOLUTION Since f_0 and f_1 are independent variables, they are source nodes. Since F_0 and F_1 are dependent variables, they are sink nodes. The signal flow graph of the equations is shown in Fig. 1.14.2. This is an example of the MIMO linear system drawn in Fig. 1.11.1a. There are 2^2 transfer functions (four combinations

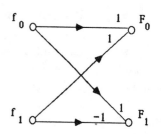

Fig. 1.14.2 Signal flow graph for 2-point DFT.

of f_i and F_j) which equal

$$H_{00} = \frac{F_0}{f_0} = 1, \qquad H_{01} = \frac{F_1}{f_0} = 1, \qquad H_{10} = \frac{F_0}{f_1} = 1, \qquad H_{11} = \frac{F_1}{f_1} = -1$$

$$(1.14.5)$$

where $H_{ij} = F_j/f_i$. Each transfer function is represented by a branch in Fig. 1.14.2. i represents the source node number and j represents the sink node number. Although most DFT's use at least 1024 data points, flow graphs facilitate their understanding. 8-point DFT's are shown in Figs. 2.3.4–2.3.8 and 2.4.4–2.4.7. ■

EXAMPLE 1.14.2 The canonical first- and second-order flow graphs are shown in Fig. 1.14.1. Determine their overall gain using Mason's gain formula.
SOLUTION The first-order canonical graph has one delay element (z^{-1}). Applying Mason's gain equation given by Eq. 1.14.1, we see that there are two forward paths with gains $k_{13}k_{42}z^{-1}$ and k_{12}. There is one feedback path with gain $k_{43}z^{-1}$. The k_{12} forward path does not touch this path. The overall gain equals

$$H(z) = k_{12} + \frac{k_{13}k_{42}z^{-1}}{1 - k_{43}z^{-1}} \qquad (1.14.6)$$

Repeating the same procedure for the second-order canonical graph which has two delay elements, there are five forward paths with gains $k_{13}k_{42}z^{-1}$, $k_{12}, k_{15}k_{62}z^{-1}$, $k_{13}k_{45}k_{62}/z^{-2}$, and $k_{15}k_{63}k_{42}/z^{-2}$, and three feedback loops with gains $k_{43}z^{-1}$, $k_{65}z^{-1}$, and $k_{45}k_{63}z^{-2}$. The overall gain equals

$$H(z) = k_{12} +$$
$$\frac{k_{13}k_{42}(1 - k_{65}z^{-1})z^{-1} + k_{15}k_{62}(1 - k_{43}z^{-1})z^{-1} + (k_{13}k_{45}k_{62} + k_{15}k_{42}k_{63})z^{-2}}{D(z)}$$

$$(1.14.7)$$

where the system determinant $D(z)$ equals

$$D(z) = 1 - k_{43}z^{-1} - k_{65}z^{-1} + (k_{43}k_{65} - k_{45}k_{63})z^{-2} \qquad (1.14.8)$$

It is rather remarkable that transfer functions can be written with such ease. ■

1.15 SENSITIVITY

Sensitivity is used to estimate the percentage drift of a dependent variable y which results from some percentage drift of an independent variable x. Mathematically, the *sensitivity* of y with respect to x equals[17]

$$S_x^y = \frac{dy/y}{dx/x} = \frac{dy/dx}{y/x} = \frac{d\ln y}{d\ln x} \tag{1.15.1}$$

For a change of Δy in y due to a drift of Δx in x, the percentage drifts or tolerances in y and x equal

$$T_y = \frac{\Delta y}{y}, \qquad T_x = \frac{\Delta x}{x} \tag{1.15.2}$$

Therefore, the sensitivity S_x^y can be related to the tolerances T_x and T_y as

$$S_x^y = \lim_{\Delta x \to 0} \frac{T_y}{T_x} \tag{1.15.3}$$

Sensitivity is equal to the ratio of the two percentage drifts for differential drift in x. For small percentage drifts in x

$$T_y \cong S_x^y T_x \tag{1.15.4}$$

Thus, when some percentage drift in x occurs, and the sensitivity of y with respect to x is known, the resulting percentage drift in y can be found.

Greater insight into the relation between tolerance and sensitivity can be gained by considering the dependent variable $y = y_o + \Delta y$ expressed as a Taylor series where

$$y(x) = y_o + \Delta y = y(x_o) + \frac{dy}{dx}\bigg|_{x_o} \Delta x + \frac{d^2 y}{dx^2}\bigg|_{x_o} \frac{\Delta x^2}{2!} + \dots \tag{1.15.5}$$

Then the change Δy in y due to a change Δx in x is given by

$$\Delta y = (y_o + \Delta y) - y_o = \frac{dy}{dx}\bigg|_{x_o} \Delta x + \frac{d^2 y}{dx^2}\bigg|_{x_o} \frac{\Delta x^2}{2!} + \dots \tag{1.15.6}$$

Using Eq. 1.15.6, the ratio of changes can be rewritten as

$$\frac{\Delta y}{\Delta x} = y'(x_o) + y''(x_o)\frac{\Delta x}{2!} + y'''(x_o)\frac{\Delta x^2}{3!} + \dots \tag{1.15.7}$$

Forming the ratio T_y/T_x from Eq. 1.15.2, substituting Eq. 1.15.7, and manipulating the result gives

$$\frac{T_y}{T_x} = \frac{\Delta y/\Delta x}{y/x} = S_x^y + \frac{y''(x_o)}{2!\,(y_o/x_o)}\Delta x + \frac{y'''(x_o)}{3!\,(y_o/x_o)}\Delta x^2 + \dots \tag{1.15.8}$$

Table 1.15.1 Table of sensitivity relations.

$S_x^{kx} = 1$	$S_x^{y+k} = \dfrac{y}{y+k} S_x^y$
$S_x^{kx^n} = n$	$S_x^{k-y} = -\dfrac{y}{k-y} S_x^y$
$S_x^{y(x)} = S_x^y$	$S_x^{yz} = S_x^y + S_x^z$
$S_x^{1/y} = -S_x^y$	$S_x^{y/z} = S_x^y - S_x^z$
$S_x^{y^n} = n S_x^y$	$S_x^{y+z+\cdots} = \dfrac{1}{y+z+\ldots} (y S_x^y + z S_x^z + \ldots)$
$S_x^{ky} = S_x^y$	$S_x^z = S_x^y \, S_y^z$

Eq. 1.15.8 reduces to Eq. 1.15.3 when $\Delta x \to 0$. This shows that the sensitivity is equal to the tolerance ratio only for very small Δx and for small second-order (and higher-order) derivatives of y relative to y_o/x_o.

When dependent variable y is a function of two or more independent variables x_i (for $i = 1, \ldots, n$), then the sensitivity of y with respect to x_i equals

$$S_{x_i}^y = \frac{\partial y/y}{\partial x_i/x_i} = \frac{\partial y/\partial x_i}{y/x_i} = \frac{\partial \ln y}{\partial \ln x_i}, \qquad i = 1, \ldots, n \qquad (1.15.9)$$

This is a generalization of Eq. 1.15.1. Thus, when more than a single independent variable is present, partial differentiation replaces normal differentiation. All independent variable values are held constant except the one being considered for differentiation. For sufficiently small drifts in the independent variables, the change Δy in y due to changes Δx_i in the x_i is

$$T_y \cong S_{x_1}^y T_{x_1} + S_{x_2}^y T_{x_2} + \ldots + S_{x_n}^y T_{x_n} \qquad (1.15.10)$$

If the x_i tolerances are uncorrelated, then the maximum and minimum tolerances equal

$$\begin{aligned}
T_{y\,max} &\cong \left|S_{x_1}^y\right| T_{x_1 max} + \left|S_{x_2}^y\right| T_{x_2 max} + \ldots + \left|S_{x_n}^y\right| T_{x_n max} \\
T_{y\,min} &\cong \left|S_{x_1}^y\right| T_{x_1 min} + \left|S_{x_2}^y\right| T_{x_2 min} + \ldots + \left|S_{x_n}^y\right| T_{x_n min}
\end{aligned} \qquad (1.15.11)$$

When the x_i tolerances have equal positive and negative values, Eq. 1.15.11 reduces to

$$\pm T_y = \pm \left(\left|S_{x_1}^y\right| \left|T_{x_1}\right| + \left|S_{x_2}^y\right| \left|T_{x_2}\right| + \ldots + \left|S_{x_n}^y\right| \left|T_{x_n}\right| \right) \qquad (1.15.12)$$

Eq. 1.15.12 is used to estimate worst-case drifts in y as will be discussed momentarily.

A number of useful relationships involving the sensitivities of various functions of x can be easily derived using the sensitivity definition. These are listed in Table 1.15.1 for ready reference. Here k and n are constants and y and z are differentiable functions of x. The reader should verify these relations for they allow complicated sensitivities to be written almost by inspection.

EXAMPLE 1.15.1   Consider the first-order canonical flow graph of Fig. 1.14.2. Assuming that $k_{12} = 0$, then the transfer function $H(z)$ equals

$$H(z) = \frac{k_{13}k_{42}}{z - k_{43}} \tag{1.15.13}$$

Compute the sensitivities of H with respect to all the k_{ij} coefficients and z. Also determine the drift in H that results from drifts in the k_{ij} and z.

SOLUTION   Since $H = H(z, k_{13}, k_{42}, k_{43})$, there are four sensitivities to compute. These are found using Eq. 1.15.1 or they can be written by inspection using the identities in Table 1.15.1. Using the $S_z^{kx}, S_x^{1/y}$, and S_x^{y+k} identities in conjunction with Eq. 1.15.13, it can be written by inspection that

$$S_{k_{13}}^H = S_{k_{42}}^H = 1, \qquad S_{k_{43}}^H = \frac{k_{43}}{z - k_{43}}, \qquad S_z^H = \frac{-z}{z - k_{43}} = -(1 + S_{k_{43}}^H) \tag{1.15.14}$$

The normalized drift T_H in H is given by Eq. 1.15.10 as

$$T_H = T_{k_{13}} + T_{k_{42}} + \frac{k_{43}}{z - k_{43}} T_{k_{43}} - \frac{z}{z - k_{43}} T_z \tag{1.15.15}$$

where $T_{k_{ij}}$ represents the tolerances of the k_{ij} constants and T_z is the tolerance of z. The worst-case normalized drift in H is given by Eq. 1.15.12 as

$$|T_H| = |T_{k_{13}}| + |T_{k_{42}}| + \left| \frac{k_{43}}{z - k_{43}} \right| |T_{k_{43}}| + \left| \frac{z}{z - k_{43}} \right| |T_z| \tag{1.15.16}$$

The k_{ij} constants are generally numbers having any value in the continuous range between $\pm\infty$. However as discussed in Chap. 1.2, the constants are stored in binary words of length L in the computer. The maximum normalized word error $T_{k_{ij}}$ is therefore $\pm 1/2^{L+1}$. In ac steady-state, $z = e^{j\omega T}$ where $\omega T = \theta$ is the frequency-time product. Since $T_z = S_\theta^z T_\theta = j\theta T_\theta$, the z tolerance equals $j\theta = S_\theta^z$ times the angle tolerance T_θ. T_θ is equal to the clock period tolerance T_T or minus the clock frequency tolerance T_{f_o}. Substituting these results into Eqs. 1.15.15 and 1.15.16 along with k_{43} and z yields the tolerance on the gain H due to coefficient truncation and clock frequency variations.   ∎

EXAMPLE 1.15.2   Consider the transfer function of Eq. 1.15.13 where $k_{13} = 1$ and $k_{42} = k_{43} = k$ so

$$H(z) = \frac{k}{z - k} \tag{1.15.17}$$

This is one digital approximation to an analog first-order lowpass filter (see Chap. 5). Compute the tolerance of $H(z)$ where $T_{k_{13}} = 0$ and $T_{k_{42}} = T_{k_{43}} = T_k$.

SOLUTION   Evaluating Eq. 1.15.15 gives the tolerance of H as

$$T_H = 0 + \left(1 + \frac{k}{z - k}\right) T_k - \frac{z}{z - k} T_z = \frac{z}{z - k}(T_k - T_z) = (1 + H)(T_k - T_z) \tag{1.15.18}$$

Eq. 1.15.16 gives the worst-case tolerance as

$$|T_H| = 0 + \left(1 + \left|\frac{k}{z-k}\right|\right)|T_k| + \left|\frac{z}{z-k}\right||T_z| = (1 + |H|)|T_k| + |1 + H||T_z| \quad (1.15.19)$$

If k and z have equal tolerances so $T_k = T_z$, then $T_H = 0$ but this is seldom the case. Generally $T_k = \pm 1/2^{L+1}$ and $T_z = j\theta T_\theta$. Therefore the tolerance T_H equals

$$T_H = \frac{e^{j\theta}}{e^{j\theta} - k}(T_k - T_z) = \frac{1}{1 - ke^{-j\theta}}\left(\frac{\pm 1}{2^{L+1}} - j\theta T_\theta\right) \quad (1.15.20)$$

$\theta = \omega T$ is the normalized frequency of interest.   ∎

1.16 ROOT LOCUS

The z-transform allowed us to transform sum and difference equations in time into algebraic equations in z. Signal flow graphs gave us a convenient tool for simply representing these equations and the means by which to calculate the filter gains. We saw that (rational) gain expressions were characterized by their pole and zero locations in the z-plane. Generally, when system parameters are changed, their pole and zero locations also change. The *root locus* is simply a plot of the pole or zero locations of filter gain as a parameter K varies continuously from $-\infty$ to $+\infty$. Thus, root locus allows the engineer to predict how system poles and zeros change when a given parameter changes. Then using the frequency domain and time domain results to be developed in Chaps. 2 and 3, respectively, it can then be assessed how these root changes affect system performance. Therefore, the root locus concept is very important to the designer.[18]

Let us express the general gain (or whatever function of z) as

$$H(z) = \frac{C(z)}{R(z)} \quad (1.16.1)$$

To determine the finite zeros and poles of $H(z)$, we find the z values which satisfy

$$C(z) = 0 \quad \text{and} \quad R(z) = 0 \quad (1.16.2)$$

respectively. Very often, these equations can be expressed in the simple linear form

$$N(z) + KM(z) = 0 \quad (1.16.3)$$

K is a system parameter of interest which may take on values in the range $(-\infty, \infty)$. Note that this equation can be rearranged as

$$1 + K\frac{M(z)}{N(z)} = 1 + KG(z) = 0 \quad (1.16.4)$$

Whenever an equation can be rearranged into this form, we can readily construct the locus of its roots as K varies from $-\infty$ to $+\infty$ using well-known rules.

The root locus construction rules are listed in Table 1.16.1. Although these rules are usually developed only for $K \in (0, \infty)$, these have been generalized for $K \in (-\infty, \infty)$. When determining the roots of Eq. 1.16.3, or equivalently Eq. 1.16.4, we define $G(z) = M(z)/N(z)$. Then the m finite zeros of $G(z)$ satisfy

$$M(z) = 0 \qquad (1.16.5)$$

while the n finite poles of $G(z)$ satisfy

$$N(z) = 0 \qquad (1.16.6)$$

The poles and zeros of $G(z)$ are easily located and facilitate the drawing of the root locus. Generally, they bear no direct relation (in themselves) to the poles and zeros of $H(z)$ in Eq. 1.16.1, except when $K = 0$ or $1/K = 0$.

EXAMPLE 1.16.1 An exponentially-decaying sinusoidal step $e^{-n\sigma T} \cos n\omega T \times U_{-1}(n)$ has a z-transform

$$F(z) = \frac{ze^{\sigma T}\left(ze^{\sigma T} - \cos \omega T\right)}{z^2 e^{2\sigma T} - 2ze^{\sigma T} \cos \omega T + 1} \qquad (1.16.7)$$

from Table 1.9.1. Draw the root locus for the zeros and poles of $F(z)$ as a function of the frequency-time product ωT.
SOLUTION Defining $K = \cos \omega T$ and $p = ze^{\sigma T}$, the standard root locus equation for the (one finite) zero of $F(z)$ equals

$$1 - K\,\frac{1}{p} = 0 \qquad (1.16.8)$$

while that for the poles of $F(z)$ equals

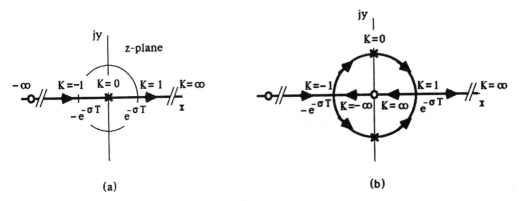

(a)   (b)

Fig. 1.16.1 Root locus of (a) zero and (b) poles of signal in Example 1.16.1.

Table 1.16.1 Construction rules for root locus.

1. *Number of loci:* The number of separate loci equals the number of poles or zeros of $G(z)$ when critical frequencies at infinity are included.

2. *Loci end points:* Each branch of the loci starts at a zero (for $k = -\infty$), passes through a pole (for $k = 0$), and terminates at a zero ($k = +\infty$)

3. *Symmetry of loci:* The loci are symmetrical about the real axis (i.e., complex roots exist in conjugate pairs).

4. *Loci on the real axis:* The parts of the real axis which comprise sections of the loci are to the left of an *even* number of poles and zeros for $k < 0$, and to the left of an *odd* number of poles and zeros for $k > 0$.

5. *Loci near infinity (asymptotes of loci):*
   a. The loci near infinity (where $|z| \to \infty$) approach asymptotic lines having angle

   $$\theta_A = \pm \frac{p\pi}{n-m}, \quad p \text{ an odd integer for } k > 0 \text{ and even integer for } k < 0$$

   b. The asymptotes intersect the real axis at σ_A where

   $$\sigma_A = \left(\sum_n \text{poles} - \sum_m \text{zeros} \right) \Big/ (n-m)$$

   where $n - m$ is the difference between the number of finite poles and zeros of $G(z)$.

6. *Loci intersecting real axis:* The points of intersection of the loci with the real axis (breakaway points) are determined by solving

   $$\frac{dK}{dz} = \frac{d}{dz}\left(-\frac{1}{G(z)} \right) = 0$$

   The tangents to the loci at the breakaway points are equally spaced over 360°.

7. *Loci intersecting imaginary axis:* The points of intersection of the loci with the imaginary axis is determined using the Routh-Hurwitz test.

8. *Loci intersecting unit circle:* The points of intersection of the loci with the unit circle is determined using the Schur-Cohn test or Jury test.

9. *Loci near poles and zeros:*
   a. The angle of departure ψ_d from the zero z_x equals

   $$\psi_d = \sum_{i=1}^{n} \arg(z_x - p_i) - \sum_{\substack{j=1 \\ j \neq x}}^{n} \arg(z_x - z_j)$$

   b. The angle of arrival ψ_a to a zero z_x equals $\psi_a = \psi_d \pm 180^\circ$.
   c. The angle of arrival ϕ_a to a pole p_x equals

   $$\phi_a = \sum_{j=1}^{m} \arg(p_x - z_j) - \sum_{\substack{i=1 \\ i \neq x}}^{n} \arg(p_x - p_i)$$

   d. The angle of departure ϕ_d from a pole p_x equals $\phi_d = \phi_a \pm 180^\circ$

$$1 - K \frac{2p}{1+p^2} = 0 \qquad (1.16.9)$$

using Eq. 1.16.4. Using the root locus construction rules in Table 1.16.1, we can easily draw and label the root locus as shown in Fig. 1.16.1 for $-\infty < K < \infty$. Now since $K = \cos \omega T$, then for $-\infty < \omega T < \infty$ we see that $-1 \le K \le 1$ is the K range that should be used. Notice the poles always lie on a circle centered about the origin with radius $e^{-\sigma T}$. Varying K is identical to varying ωT. Although Fig. 1.16.1 is labelled in terms of K, it can also be labelled in terms of $\omega T = \cos^{-1} K$. K's of $(-1, 0, 1)$ correspond to ωT's of $(180°, \pm 90°, 0°)$ as shown in Fig. 1.9.5.   ■

EXAMPLE 1.16.2 The s-plane locus for the roots of a sampled $H(s)$ may be converted to the z-plane locus for the roots of $H(z)$ using the z-transformation $z = e^{sT}$. Convert the root locus shown in Fig. 1.16.2a.

SOLUTION The unsampled analog transfer function $H(s)$ having the root locus shown in Fig. 1.16.2a is

$$1 - K \frac{s^2}{(s+1)^2} = 0 \qquad (1.16.10)$$

The root locus for the sampled analog transfer function is shown in Fig. 1.16.2b which is obtained by superposition of frequency shifted versions of Fig. 1.16.2a according to Eq. 1.5.21. The root locus of the digital transfer function is obtained by mapping points in Fig. 1.16.2b into equivalent points in Fig. 1.16.2c using Figs. 1.9.6 or 1.9.7.

This can also be done analytically by noting that the circular root locus in the s-plane is described parametrically by

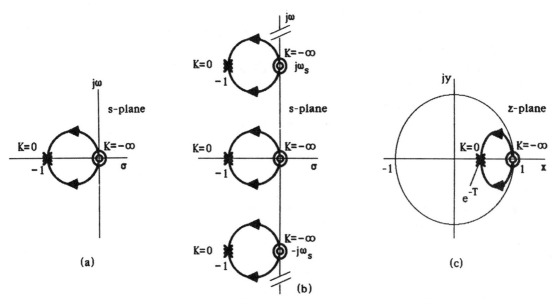

Fig. 1.16.2 Root locus of poles of (a) unsampled analog gain, (b) sampled analog gain, and (c) digital gain in Example 1.16.2.

$$(\alpha - 0.5)^2 + \omega^2 = 0.5^2 \qquad \text{or} \qquad \omega = \pm\sqrt{\alpha - \alpha^2} \qquad (1.16.11)$$

where $s = \alpha + j\omega$. Then transforming as $z = e^{sT}$,

$$z = e^{sT} = e^{\sigma T} e^{j\omega T} = e^{\sigma T} e^{\pm jT\sqrt{\alpha - \alpha^2}} \qquad (1.16.12)$$

The z-plane root locus has

$$|z| = e^{\sigma T}, \qquad \arg z = \pm T\sqrt{\alpha - \alpha^2} \qquad (1.16.13)$$

for $-1 \leq \alpha \leq 0$ corresponding to $-\infty < K \leq 0$.   ∎

EXAMPLE 1.16.3 Convert the s-plane root locus of $F(s)$ shown in Fig. 1.16.3a into the equivalent z-plane root locus. Use different sampling rates.
SOLUTION   The characteristics of the $z = e^{sT}$ transformation were discussed in Chap. 1.9 and certain mapping contours were drawn in Fig. 1.9.6 (rectangular form) and 1.9.7 (polar form). Notice that the s-plane locus of Fig. 1.16.3a has a constant radius ω_n and a variable angle $\theta = \cos^{-1}\varsigma$. Therefore the polar plot of Fig. 1.9.7 contains the information required to solve this problem. For $K = 0$, then $s = -\omega_n$ which under the transformation maps into $z = e^{-\omega_n T}$. Since the region $0 \leq \omega_n < \infty$ is equivalent to $0 < z \leq 1$, the origination point for the z-plane locus is the positive real axis contained within the unit circle. Choosing normalized sampling times $\omega_n T = \pi, \pi/2$, and $\pi/5$, then $z = e^{-\pi}, e^{-\pi/2}$, and $e^{-\pi/5}$ are the z-plane starting points as shown in Figs. 1.16.3b, 1.16.3c, and 1.16.3d, respectively.

The loci termination points at $s = \pm j\omega_n$ map to $z = e^{\pm j\omega_n T}$. These points lie on the unit circle of the z-plane at an angle $\omega_n T$ radians from the positive real axis. This is also shown in Figs. 1.16.3a–d. Between these endpoints, the loci must be computed. Fortunately, because the s-plane loci has polar coordinates (ω_n, θ) where ω_n is constant, the loci have already been determined in Fig. 1.9.7. All that needs be done is to use the proper $\omega_n T$ curve where $\omega_n T = \pi, \pi/2$, and $\pi/5$. This completes the loci in Fig. 1.16.3.   ∎

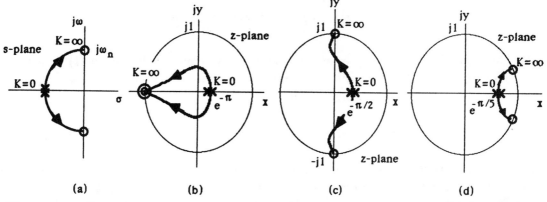

Fig. 1.16.3   Root locus of poles of (a) unsampled analog gain and (b) digital gain for $\omega_n T = \pi$, (c) $\pi/2$, and (d) $\pi/5$ in Example 1.16.3.

1.17 RANDOM SIGNALS

There are many situations when signals are not known exactly and must be described statistically.[19] In these cases, a signal $v(t)$ is viewed as a random variable which is characterized by a probability density function $f(v(t))$ or $f(v;t)$. When the probability density functions are time-invariant as $f(v)$, the statistical process is said to be *stationary*. The probability that a signal $v(t)$ does not exceed V_0 is determined using the *probability density function (pdf)* $f(v;t)$ as

$$Pr(v(t) \leq V_0) = \int_{-\infty}^{V_0} f(v;t)dv = F(V_0;t) \qquad (1.17.1)$$

$F(v;t)$ is called the *cumulative probability function* or the *probability distribution function (df)* of v. $F(v;t)$ and $f(v;t)$ are related as

$$F(v;t) = \int_{-\infty}^{v} f(v;t)dv, \qquad f(v;t) = \frac{dF(v;t)}{dv} \qquad (1.17.2)$$

The probability that a signal $v(t)$ lies between V_1 and V_2 therefore equals

$$Pr(V_1 \leq v(t) \leq V_2) = \int_{V_1}^{V_2} f(v;t)dt = F(V_2;t) - F(V_1;t) \qquad (1.17.3)$$

using Eq. 1.17.1. A variety of signals and their probability density and distribution functions are shown in Table 1.17.1.

Certain statistics are usually used to describe random signals. These often involve an *expected value* of some function $g(v)$ of a random variable v. Expected values are denoted as $E\{\ \}$ and are defined to equal

$$E\Big\{g\big[v(t)\big]\Big\} = \int_{-\infty}^{\infty} g(v)f(v;t)dv \qquad (1.17.4)$$

We shall now discuss several important statistics. Consider $v_1(t)$ and $v_2(t)$ which are random variables and are described by probability density functions $f(v_1;t)$ and $f(v_2;t)$.

The *mean* or *average* of a signal equals

$$\eta_1(t) = E\{v_1(t)\}, \qquad \eta_2(t) = E\{v_2(t)\} \qquad (1.17.5)$$

It is called a *dc value* when η_1 or η_2 are independent of t. Stationary processes have constant (i.e. dc) means. Their *mean square* values equal $E\{v_1^2\}$ and $E\{v_2^2\}$, respectively. Their *variances* equal $E\{(v_1 - \eta_1)^2\}$ and $E\{(v_2 - \eta_2)^2\}$, respectively.

The *autocorrelation* of a signal equals

Table 1.17.1 Statistical characterization of signals.
(From Monograph No. 1, "An introduction to correlation," pp. 6–7,
Rockland Scientific Corp. (formerly Federal Scientific Corp.), NJ, 5-1-72.)

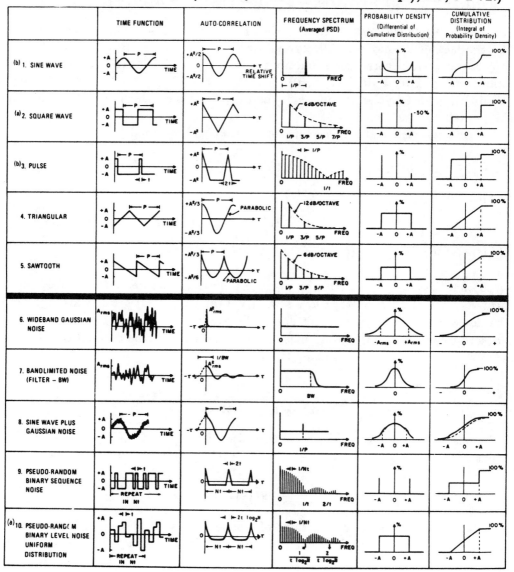

$$R_{11}(t,\tau) = E\{v_1(t+\tau)v_1^*(t)\} = E\{v_1(t)v_1^*(t-\tau)\}$$
$$R_{22}(t,\tau) = E\{v_2(t+\tau)v_2^*(t)\} = E\{v_2(t)v_2^*(t-\tau)\} \qquad (1.17.6)$$

When $\tau = 0$, $R_{11}(t,0)$ and $R_{22}(t,0)$ equal the average powers in v_1 and v_2, respectively. The *cross-correlation* between two signals equals

$$R_{12}(t,\tau) = E\{v_1(t+\tau)v_2^*(t)\} = E\{v_1(t)v_2^*(t-\tau)\}$$

$$R_{21}(t,\tau) = E\{v_2(t+\tau)v_1^*(t)\} = E\{v_1^*(t-\tau)v_2(t)\} = R_{12}^*(t,-\tau) \tag{1.17.7}$$

When $\tau = 0$, $R_{12}(t,0) = R_{21}^*(t,0)$ and the cross-correlations equal the average power in the product of v_1 and v_2. All correlation functions satisfy $R_{xy}(t,\tau) = R_{yx}^*(t,-\tau)$ for any x and y. Autocorrelation functions satisfy $R_{vv}(0) \geq |R_{vv}(t)|$. Cross-correlation functions satisfy $\frac{1}{2}[R_{v_1v_1}(0) + R_{v_2v_2}(0)] \geq \sqrt{R_{v_1v_1}(0)R_{v_2v_2}(0)} \geq |R_{v_1v_2}(t)|$. The equality condition holds only for periodic $v_1(t)$ and $v_2(t)$ at times which are integer multiples of the period.

The *autocovariance* and *cross-covariance* of v_1 and v_2 equal

$$C_{11}(t,\tau) = E\left\{[v_1(t+\tau) - \eta_1(t+\tau)][v_1^*(t) - \eta_1^*(t)]\right\} = R_{11}(t,\tau) - \eta_1(t+\tau)\eta_1^*(t)$$

$$C_{22}(t,\tau) = R_{22}(t,\tau) - \eta_2(t+\tau)\eta_2^*(t)$$

$$C_{12}(t,\tau) = E\left\{[v_1(t+\tau) - \eta_1(t+\tau)][v_2^*(t) - \eta_2^*(t)]\right\} = R_{12}(t,\tau) - \eta_1(t+\tau)\eta_2^*(t)$$

$$C_{21}(t,\tau) = R_{21}(t,\tau) - \eta_2(t+\tau)\eta_1^*(t) = C_{12}^*(t,-\tau) \tag{1.17.8}$$

$C_{11}(t,0)$ and $C_{22}(t,0)$ represent the average powers in the derivations of v_1 and v_2, respectively, about their average levels. $C_{12}(t,0)$ and $C_{21}(t,0)$ are the average powers in the deviation of the v_1v_2 product about the average v_1v_2 level. In other words, the R's describe correlations including means while the C's describe correlations excluding means. The correlation coefficient $\rho_{12}(t,\tau)$ equals

$$\rho_{12}(t,\tau) = \frac{C_{12}(t,\tau)}{\sqrt{C_{11}(t,\tau)C_{22}(t,\tau)}} = \rho_{21}^*(t,-\tau) \tag{1.17.9}$$

Signals v_1 and v_2 are said to be *independent* if the pdf of their product equals the product of their pdf's as $f(v_1,v_2;t+\tau,t) = f(v_1;t+\tau)f(v_2;t)$. In this situation,

$$E\left\{g_1[v_1(t+\tau)]g_2[v_2^*(t)]\right\} = E\left\{g_1[v_1(t+\tau)]\right\}E\left\{g_2[v_2(t)]\right\} \tag{1.17.10}$$

The expected cross-product of functions of independent variables equals the product of the expected functions of each variable. Two processes v_1 and v_2 are *orthogonal* when

$$R_{12}(t,\tau) = 0, \quad \text{all } (t,\tau) \tag{1.17.11}$$

They are *uncorrelated* when

$$C_{12}(t,\tau) = 0, \quad \text{all } (t,\tau) \tag{1.17.12}$$

or equivalently from Eq. 1.17.9

$$R_{12}(t,\tau) = E\{v_1(t+\tau)v_2^*(t)\} = \eta_1(t)\eta_2^*(t) = E\{v_1(t+\tau)\}E\{v_2^*(t)\}, \quad \text{all } (t,\tau) \tag{1.17.13}$$

The cross-correlation between v_1 and v_2 equals the product of their means. Independent signals are uncorrelated but uncorrelated signals are not necessarily independent (except for two Gaussian signals).

The average correlation time T_{11} of a process (or between two processes T_{12}) is defined as

$$T_{11}(t) = \frac{\int_0^\infty C_{11}(t,\tau)d\tau}{C_{11}(t,0)}, \qquad T_{12}(t) = \frac{\int_0^\infty C_{12}(t,\tau)d\tau}{C_{12}(t,0)} \qquad (1.17.14)$$

A process $v_1(t)$ is heavily correlated ($\rho \cong 1$) with itself (or another process $v_2(t)$) for shift periods less than T_{11} (or T_{12}). Otherwise it is almost uncorrelated ($\rho \cong 0$).

The *power spectrum* or *power spectral density PSD* S_{11} of v_1 and S_{22} of v_2 is found by taking the Fourier transform of their autocorrelations. Thus, when the R's are functions of τ only (wide-sense stationary)

$$S_{11}(j\omega) = \int_{-\infty}^{\infty} R_{11}(\tau)e^{-j\omega\tau}d\tau, \qquad S_{22}(j\omega) = \int_{-\infty}^{\infty} R_{22}(\tau)e^{-j\omega\tau}d\tau \qquad (1.17.15)$$

Power spectra are always real and even (for real v). The spectral density of the product of v_1 and v_2, or *cross-spectral density*, equals

$$S_{12}(j\omega) = \int_{-\infty}^{\infty} R_{12}(\tau)e^{-j\omega\tau}d\tau = S_{21}^*(j\omega) \qquad (1.17.16)$$

If R_{12} and R_{21} are real-valued which is usually the case, then $S_{21}^*(j\omega) = S_{21}(-j\omega)$. The cross-power spectra S_{12} and S_{21} are generally complex even when v_1 and v_2 are real. When the R's are functions of both t and τ, then two-dimensional Fourier transforms are used which give $S_{11}(j\omega_1, j\omega_2)$, $S_{22}(j\omega_1, j\omega_2)$, and $S_{12}(j\omega_1, j\omega_2)$, respectively. S_{xy} is the mean square voltage density (volts^2/rad/sec) at frequency ω. Examples of autocorrelations and power spectra are shown in Table 1.17.1. The coherence function $\Gamma_{12}(j\omega)$ between the two signals is

$$\Gamma_{12}(j\omega) = \frac{|S_{12}(j\omega)|}{\sqrt{S_{11}(j\omega)S_{22}(j\omega)}} \qquad (1.17.17)$$

It is real-valued, bounded as $-1 \leq \Gamma_{12}(j\omega) \leq 1$, and generally varies with frequency.

Very often, we are concerned only with the rms signal contained between frequencies ω_1 and ω_2 with bandwidth $B = \omega_2 - \omega_1$. In this situation, the rms signal values equal

$$V_{1rms} = \sqrt{\frac{1}{\pi}\int_{\omega_1}^{\omega_2} S_{11}(j\omega)d\omega}, \qquad V_{2rms} = \sqrt{\frac{1}{\pi}\int_{\omega_1}^{\omega_2} S_{22}(j\omega)d\omega} \qquad (1.17.18)$$

In most situations, the pdf $f(v;t)$ is not known for the random signal v. Usually only a single $v(t)$ is available from all those possible (called the *ensemble*). In such cases, time averages (denoted by $< >$) must be used to form estimates (denoted by $\hat{\ }$) of ensemble averages (denoted by $E\{ \}$). When time averages equal ensemble averages, the random process which generates $v(t)$ is said to be *ergodic*. Only stationary processes can be ergodic. For further details, see Ref. 19.

The estimate of the mean of a signal v over an interval of 2δ seconds equals

$$\hat{\eta}(t) = \hat{E}\{v(t)\} = \frac{1}{2\delta}\int_{-\delta}^{\delta} v(t)dt \qquad (1.17.19)$$

The estimates of the autocorrelation and cross-correlation between two signals (having infinite energy but finite power) equal

$$\hat{R}_{11}(t,\tau) = \hat{E}\{v_1(t+\tau)v_1^*(t)\} = \frac{1}{2\delta}\int_{-\delta}^{\delta} v_1(t+\tau)v_1^*(t)dt$$

$$\hat{R}_{12}(t,\tau) = \hat{E}\{v_1(t+\tau)v_2^*(t)\} = \frac{1}{2\delta}\int_{-\delta}^{\delta} v_1(t+\tau)v_2^*(t)dt \qquad (1.17.20)$$

The estimates of the power spectrum, cross-power spectrum, and coherence function when the R's depend only on τ are

$$\hat{S}_{11}(j\omega) = \int_{-\infty}^{\infty} \hat{R}_{11}(\tau)e^{-j\omega\tau}d\tau, \qquad \hat{S}_{12}(j\omega) = \int_{-\infty}^{\infty} \hat{R}_{12}(\tau)e^{-j\omega\tau}d\tau$$

$$\hat{\Gamma}_{12}(j\omega) = |\hat{S}_{12}(j\omega)|\Big/\sqrt{\hat{S}_{11}(j\omega)\ \hat{S}_{22}(j\omega)} \qquad (1.17.21)$$

2-D Fourier transforms are used when the R's depend on both t and τ. In practice, the integration period $(-\delta, \delta)$ is made short relative to the stationarity period of the process. Over this time, a nonstationary process appears to be almost stationary so its statistics estimated by Eqs. 1.17.19–1.17.21 are constant. In effect, they become *running estimates* which are discussed in Chap. 2.11. These concepts will be used in Chaps. 4 and 9.

For finite energy signals, the autocorrelation and cross-correlation estimates are defined to equal

$$\hat{R}_{11}(\tau) = \hat{E}\{v_1(t+\tau)v_1^*(t)\} = \int_{-\infty}^{\infty} v_1(t+\tau)v_1^*(t)dt = v_1(-\tau) * v_1^*(\tau)$$

$$\hat{R}_{12}(\tau) = \hat{E}\{v_1(t+\tau)v_2^*(t)\} = \int_{-\infty}^{\infty} v_1(t+\tau)v_2^*(t)dt = v_1(-\tau) * v_2^*(\tau) = \hat{R}_{21}^*(-\tau)$$

$$(1.17.22)$$

which are convolution products. Their power spectrum, cross-power spectrum, and coherence functions are still given by Eq. 1.17.21 and reduce to

$$\hat{S}_{11}(j\omega) = V_1(j\omega)V_1^*(j\omega) = |V_1(j\omega)|^2, \qquad \hat{S}_{22}(j\omega) = V_2(j\omega)V_2^*(j\omega) = |V_2(j\omega)|^2$$

$$\hat{S}_{12}(j\omega) = V_1(j\omega)V_2^*(j\omega), \qquad \hat{\Gamma}_{12}(j\omega) = 1$$

$$(1.17.23)$$

When the signals have infinite energy but finite power, then the correlation functions of Eq. 1.17.20 are used instead. Very often, nonperiodic signals have finite energy while periodic signals have infinite energy but finite power. As a practical matter, most engineers define correlation using Eq. 1.17.22 rather than Eqs. 1.17.6–1.17.7 because of the convenient time-convolution and frequency-multiplication relation of Table 1.6.2.

PROBLEMS

1.1   Some filter applications with their frequencies and Q ranges are listed below. What implementation forms could be used? (a) submarine and long-range navigation having $f = 3-30$ KHz and $Q = 50$; (b) frequency-division multiplexed communications system having $f = 60-108$ KHz, $Q = 1$; (c) marine communication and radio beacons with $f = 30 - 300$ KHz and $Q = 50$; (d) maritime radio and AM broadcasting having $f = 300-3000$ KHz and $Q = 20$; (e) telephone, telegraph, and facsimile systems with $f = 3 - 30$ MHz and $Q = 100$; (f) VHF television and FM radio with $f = 30 - 300$ MHz and $Q = 200$; (g) UHF television and microwave links having $f = 0.3 - 3$ GHz with $Q = 500$; (h) satellite communication and radar microwave links with $f = 3 - 30$ GHz and $Q = 1000$.

1.2   Draw and compare the block diagrams of an analog and digital filter. (a) What are their similarities and what are their differences? (b) How must the transfer characteristics of the A/D and D/A be related? (c) Where are S/H blocks needed in the system and what is their function?

1.3   Nonlinear A/D and D/A blocks can be modeled as linear A/D and D/A blocks cascaded between nonlinear analog blocks called *compandors*. If this input *compressor* block has the following characteristics, what gain must the output *expandor* block have? (a) μ-law compression

$$|w_2(t)| = \frac{\ln(1 + \mu|w_1(t)|)}{\ln(1 + \mu)}, \qquad |\pm w_1(t)| \le 1$$

($\mu = 0$ results in linear mode, $\mu = 255$ is used in Bell System PCM); (b) A-law compression

$$|w_2(t)| = \frac{A|w_1(t)|}{1 + \ln A}, \ 0 \le |w_1(t)| \le \frac{1}{A} \quad \text{and} \quad \frac{1 + \ln(A|w_1(t)|)}{1 + \ln A}, \ \frac{1}{A} \le |w_1(t)| \le 1$$

($A = 87.6$ is used in United Kingdom).

1.4   Consider a signal $f(t) = e^{-t}\sin(t)U_{-1}(t)$. (a) Determine its Fourier transform. (b) Draw its magnitude spectrum. (c) How fast must $f(t)$ be sampled so that the spectrum is -20 dB at the Nyquist frequency ω_N? Compute the aliasing error. (d) Repeat part c for -40 dB gain at ω_N.

1.5   Repeat Problem 1.4 when the signal $f(t) = e^{-2t}\sin(3t)U_{-1}(t)$.

1.6   Repeat Problem 1.4 when the signal $f(t) = e^{-t}\cos(t)U_{-1}(t)$.

1.7   Repeat Problem 1.4 when the signal $f(t)$ equals (a) $e^{-|t|}$, (b) $e^{-t}U_{-1}(t)$, (c) e^{-t^2}.

1.8   Find the Laplace transforms of the following signals and their regions of convergence using Tables 1.5.1 and 1.5.2: (a) $(2t^3 - 2\cos(3t) + 3e^{-2t})U_{-1}(t)$, (b) $(t - 3)^2U_{-1}(t)$, (c) $\int_0^t \sin(3x)dx\, U_{-1}(t)$, (d) $t^3 e^{-t}U_{-1}(t)$, (e) $[\sin(t)/t]U_{-1}(t)$.

1.9   Find the Laplace transform of the signal (including its region of convergence) shown in: (a) Fig. P1.9a, (b) Fig. P1.9b, (c) Fig. P1.9c, (d) Fig. P1.9d.

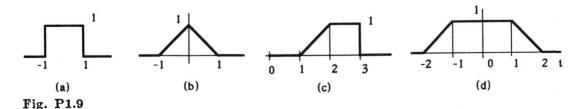

(a)   (b)   (c)   (d)

Fig. P1.9

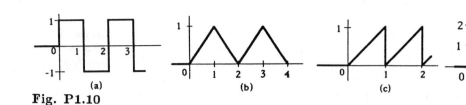

Fig. P1.10

1.10   Find the Laplace transform of the semiperiodic signal (including its region of convergence) shown in: (a) Fig. P1.10a, (b) Fig. P1.10b, (c) Fig. P1.10c, (d) Fig. P1.10d.

1.11   Make the semiperiodic signal of: (a) Fig. P1.10a, (b) Fig. P1.109b, (c) Fig. P1.10c, and (d) Fig. P1.10d periodic by setting $f(t) = f(t+T)$, $t < 0$. Find the Laplace transforms of these periodic signals (including their regions of convergence).

1.12   Find four different signals all having $1/[(s+1)(s-1)(s-2)]$ as their Laplace transform. Show the region of convergence for each of the signals.

1.13   Find the 2-D Laplace transforms of the following spatial functions $f(x,y)$ (including their regions of convergence): (a) $U_0(y)$, (b) $U_0(x,y)$, (c) $\sum_{n=-\infty}^{\infty} U_0(x)$, (d) $U_{-1}(y)$, (e) $U_{-1}(x,y)$, (f) $\sum_{n=0}^{\infty} U_{-1}(x-n)$, (g) $\exp(-x)U_{-1}(x)$, (h) $\exp(-x)U_{-1}(y)$, (i) $\exp(-x)$, (j) $\exp[-\pi(x^2+y^2)]$, (k) $\cos(\pi y)U_0(x)$, (l) $\cos(\pi y)U_{-1}(x)$, (m) $\cos(\pi x)$, (n) $\cos[2\pi(x\cos\theta + y\sin\theta)]$, (o) $\cos(2\pi x)\cos(2\pi y)$

1.14   Find the Fourier transform of the signal shown in: (a) Fig. P1.9a, (b) Fig. P1.9b, (c) Fig. P1.9c, (d) Fig. P1.9d.

1.15   If the Fourier transform of $f(t)$ equals $F(j\omega)$, show that (a) $\mathcal{F}[f(t)\cos(\omega_o t)] = [F(\omega + \omega_o) + F(\omega - \omega_o)]/2$, (b) $\mathcal{F}[f(t)\sin(\omega_o t)] = j[F(\omega + \omega_o) - F(\omega - \omega_o)]/2$.

1.16   Use the symmetry property in Table 1.6.2 to find the Fourier transform of: (a) $\mathrm{sinc}(t) = \sin(t)/t$, (b) $\mathrm{sinc}^2(t)$.

1.17   Show that the Fourier transform of a Gaussian pulse $f(t) = \exp(-at^2)$ is also Gaussian as $\sqrt{\pi/a}\exp(-\omega^2/4a)$.

1.18   Use Parseval's theorem to prove: (a) $\int_{-\infty}^{\infty} \exp(-\pi t^2)\cos(\omega_o t)dt = \exp(-\omega_o^2/4\pi)$, (b) $\int_{-\infty}^{\infty} \mathrm{sinc}^2(t)\cos(\pi t)dt = 0.5$.

1.19   Find the Fourier series of the signal shown in: (a) Fig. P1.19a, (b) Fig. P1.19b, (c) Fig. P1.19c.

1.20   Make the semiperiodic signal of: (a) Fig. P1.10a, (b) Fig. P1.10b, (c) Fig. P1.10c, and (d) Fig. P1.10d periodic by setting $f(t) = f(t+T)$, $t < 0$. Find the Fourier series these periodic functions.

1.21   Consider the infinite impulse stream $p(t) = \sum_{n=-\infty}^{\infty} U_0(t - nT)$. $p(t)$ is a periodic function with period T. (a) Show that the Fourier series of $p(t) = (1/T)\sum_{n=-\infty}^{\infty} e^{jn\omega_o t}$ where $\omega_o = 2\pi/T$. Use this result to show that the Fourier transform of $p(t)$ is $\mathcal{F}[p(t)] = \omega_o \sum_{m=-\infty}^{\infty} U_0(\omega - m\omega_o)$.

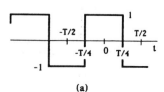

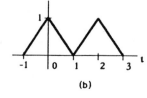

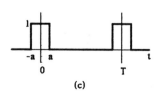

Fig. P1.19

1.22 Find the 4-point DFT of the signal $f(t) = e^{-t}U_{-1}(t)$ using a sampling rate of: (a) 1, (b) 2, (c) 4, (d) 8, (e) 16 Hz.

1.23 Given a signal sequence $[f(n)] = [0, 1, 2, 3]$. (a) Find the DFT of the signal sequence. Use this result to find the DFT of: (b) $f(n)\cos(n\pi)$, (c) $f(n)\sin(n\pi)$.

1.24 Perform an 8-point DFT over the nonzero interval on the signal shown in: (a) Fig. P1.9a, (b) Fig. P1.9b, (c) Fig. P1.9c, (d) Fig. 1.9d.

1.25 Find the z-transforms for the following $f(nT)$ signals (including their regions of convergence): (a) $(1/2)^n U_{-1}(n)$, (b) $-(1/2)^n U_{-n-1}(n)$, (c) $(1/2)^n [U_{-1}(n) - U_{-1}(-1 - n)]$, (d) $nTe^{-nT}U_{-1}(n)$, (e) $e^{-2n}\cos(n)U_{-1}(n)$, (f) $e^{-2|n|}\cos(n)$.

1.26 Find the z-transforms for the following $f(nT)$ signals (including their regions of convergence): (a) $a^{|n|}$ for $0 < |a| < 1$; (b) $a^{-n}\cos(n\omega_o T + \theta)U_{-1}(n)$ for $0 < a < 1$; (c) 1 for $0 \le n \le N - 1$ and 0 otherwise; (d) n for $0 \le n \le N$, $(2N - n)$ for $N + 1 \le n \le 2n$, and 0 otherwise.

1.27 Determine the inverse z-transforms of the following $X(z)$: (a) $1/(1 + z^{-1}/2)$, $|z| > 1/2$; (b) $1/(1 + z^{-1}/2)$, $|z| < 1/2$; (c) $(1 - z^{-1}/2)/(1 + z^{-1}/2)$, $|z| > 1/2$; (d) $(1 - z^{-1}/2)/(1 + 3z^{-1}/4 + z^{-2}/8)$, $|z| > 1/2$; (e) $(1 - az^{-1}/2)/(z^{-1} - a)$, $|z| > 1/|a|$.

1.28 Find three different signals all having $X(z)$ as their z-transforms. Specify the region of convergence for each one: (a) $1/[(z-1)(z-2)]$, (b) $z/[(z-0.25)(z-0.5)]$, (c) $z/[(z-2)(z^2+1)]$.

1.29 Let $x(n) = a^n U_{-1}(n)$ be the input signal to a linear time-invariant system having impulse response $h(n) = b^n U_{-1}(n)$. Find the: (a) z-transforms of $x(n)$ and $h(n)$ including their regions of convergence, (b) z-transform of the system output $Y(z)$, (c) system output $y(n)$.

1.30 Consider a linear time-invariant discrete system having $y(n) + 0.5y(n - 1) = x(n)$. Use the z-transform approach to find two possible impulse responses for the system.

1.31 Consider the digital filter shown in Fig. P1.31. Assume $K = 1$. (a) Write its transfer function. (b) Compute and plot its impulse response.

1.32 Consider the digital filter shown in Fig. P1.32. Assume $K = 1$. (a) Write its transfer function. (b) Compute and plot its impulse response.

1.33 Consider the digital filter shown in Fig. P1.33. Suppose that $H_1(z) = z/(z - 0.5)$, $H_2(z) = (z - 2)/(z - 0.5)$, and $K = 1$. (a) Write its transfer function. (b) Compute and plot its impulse response.

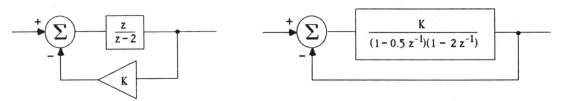

Fig. P1.31   Fig. P1.32

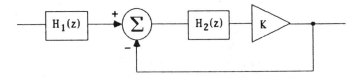

Fig. P1.33

1.34   Let $U_0(m, n)$ and $U_{-1}(m, n)$ be the 2-D impulse and step sequences, respectively. (a) Express $U_{-1}(m, n)$ in terms of $U_0(m, n)$ and vice versa. (b) Find their z-transforms. (c) Sketch their pole-zero patterns and show their regions of convergence.

1.35   Let $x(m, n), h(m, n)$, and $y(m, n)$ be the input, impulse response, and output of a 2-D linear, time-invariant system, respectively. (a) Relate y to x and h in both the time and frequency domains. (b) Show that if x and h are separable, then y is also separable. (c) Why is this useful?

1.36   Let $x(n, m) = a^n U_0(n - m), 0 \le n, m < \infty$ and $h(n, m) = b^n U_0(n - m), 0 \le n, m < \infty$. Find the z-transform of: (a) $x(n, m)$, (b) $h(n, m)$, (c) $y(n, m)$.

1.37   Suppose the impulse response of a 2-D system equals $h(n, m) = 1$ for $0 \le n < N$ and $0 \le m < M$. $h(n, m)$ is zero elsewhere. (a) Find $H(z_1, z_2)$ and the (b) frequency response of the system.

1.38   Determine if the following operators $\mathbf{h}[x(n)]$ are linear, causal, stable, time-invariant: (a) $f(n)x(n)$, (b) $\sum_{k=n_o}^{n} x(k)$, (c) $\sum_{k=n-n_o}^{n+n_o} x(k)$, (d) $x(n + n_o)$, (e) $x(n - n_o)$, (f) $a^{x(n)}$, (g) $[ax(n) + b]$, (h) $ax(n)$.

1.39   Consider the system $y(n) - ay(n - 1) = x(n)$ where $y(0) = 0$. (a) Determine if the system is linear or time-invariant. (b) Repeat part a if $y(0) = 1$.

1.40   Determine if the systems having the following transfer functions $H(z)$ are causal and stable: (a) $z/(z - 1), |z| > 1$; (b) $z/(z - 1), |z| < 1$; (c) $(z - a)/(z - 1), |z| > 1$.

1.41   Use the Paley-Wiener criterion to determine if the following gain magnitude functions $|H(e^{j\omega T})|$ are realizable (causal). $|H|$ is defined for $|\omega| \le \omega_N = \pi/T$ and is periodic outside this interval. (a) $\exp(-0.347\omega^2)$, (b) $\exp(-|\omega|)$, (c) $\cos^2(0.572\omega)$ for $|\omega| < \Omega$ and 0 elsewhere; (d) 1 for $|\omega| \le \Omega$ and 0 elsewhere.

1.42   Second-order digital filters have the transfer function denominators $\Delta(z)$ listed below. Draw the root locus for the filter poles as gain K varies: (a) $z^2 + Kz + 1$, (b) $(z + a_1)(z + a_2) - K(z + a_3)$, (c) $(z + a_1)(z + a_2) - Kz(z + a_3)$, (d) $(z + a_1)(z + a_2) - K(z^2 + a_3^2)$, (e) $(z + a_1)(z + a_2) - K(z + a_3)(z + a_4)$.

1.43   Consider the digital system shown in Fig. P1.31. (a) Determine the overall transfer function of the system. (b) Draw the root locus for the system poles. (c) For what gain values K are the system stable?

1.44   Consider the digital system shown in Fig. P1.32. (a) Determine the overall transfer function of the system. (b) Draw the root locus for the system poles. (c) For what gain values K are the system stable?

1.45   Consider the digital system shown in Fig. P1.33. (a) Determine the overall transfer function of the system. (b) What equation do the poles satisfy? (c) Suppose that $H_1(z) = z/(z - 0.5)$, $H_2(z) = (z - 2)/(z - 0.5)$, and $K = 1$. For what gain values K are the system stable?

1.46   Suppose a bandpass analog filter has $H(s) = s_n/(s_n^2 + s_n/Q + 1)$ where $s_n = s/\omega_c$. It is converted into a digital filter using the s-to-z transform $s = (2/T)(z - 1)/(z + 1)$. (a) Determine the overall transfer function of the system. (b) Draw the root locus for the system poles. (c) For what Q values are the system stable?

1.47   Repeat Problem 1.46 using the s-to-z transform $s = (z - 1)/T$. (a) Determine the overall transfer function of the system. (b) Draw the root locus for the system poles. (c) For what Q values are the system stable?

1.48   Consider a deterministic signal $s(t)$ of: (a) $U_{-1}(t)$, (b) $e^{-t}U_{-1}(t)$, (c) $\cos(\omega t)U_{-1}(t)$, (d) $U_{-1}(t) - U_{-1}(t - T)$. Random noise $n(t)$ has a Gaussian pdf with zero mean and variance σ^2. Find the mean and variance of $s(t) + n(t)$.

1.49   Find the autocorrelation of the signal and noise and their cross-correlation in Prob. 1.48.

1.50 Find the autocovariance of the signal and noise and their cross-covariance in Prob. 1.48. Find their correlation coefficient.

1.51 Show whether or not the signal and noise of Prob. 1.48 are: (a) independent, (b) orthogonal, (c) uncorrelated.

1.52 Find the power spectral densities and cross-spectral densities of signal and noise in Prob. 1.48. Find the coherence function.

REFERENCES

1. Lindquist, C.S., *Active Network Design with Signal Filtering Applications*, Chap. 1.0, Steward and Sons, CA, 1977.
2. Rabiner, L.R. and B. Gold, *Theory and Application of Digital Signal Processing*, Chaps. 1.1–1.2, Prentice-Hall, NJ, 1975.
3. Graeme, J.G., G.E. Tobey, and L.P. Huelsman, *Operational Amplifiers*, Chap. 9, McGraw-Hill, NY, 1971.
4. Oppenheim, A.V. and R.W. Schafer, *Digital Signal Processing*, Chap. 1.7, Prentice-Hall, NJ, 1975.
   Steele, R., *Delta Modulation Systems*, Pentech Press, London, 1975.
5. Wylie, C.R., *Advanced Engineering Mathematics*, 4th ed., McGraw-Hill, NY, 1975.
6. Ref. 1. Chap. 1.8.
7. Ref. 1. Chap. 1.9.
8. Oppenheim, A.V., A.S. Willsby, and I.T. Young, *Signals and Systems*, pp. 511–512, Prentice-Hall, NJ, 1983.
9. Van Valkenburg, M.E., *Network Analysis*, 3rd ed., Chap. 15, Prentice-Hall, NJ, 1974.
10. Papoulis, A., *Circuits and Systems: A Modern Approach*, Chap., 7, Holt, Rinehart, and Winston, NY, 1980.
11. Ref. 8, Chaps. 10.1–10.2.
12. Oppenheim, A.V. and R.W. Schafer, *Digital Signal Processing*, Chap. 2.2, Prentice-Hall, NJ, 1975.
13. Ref. 1, Chap. 1.4.
    Ref. 2, Chaps. 2.4–2.5.
14. Ref. 1, Chap. 1.5.
    Jong, M.J., *Methods of Discrete Signal and System Analysis*, Chap. 3, McGraw-Hill, NY, 1982.
15. Ref. 1, Chap. 1.6.
16. Ref. 1, Chap. 1.14.
    Mason, S.J., "Feedback theory–some properties of signal flow graphs," Proc. IRE, vol. 41, pp. 1142–1156, Sept. 1953.
    —, "Feedback theory–further properties of signal flow graphs," Proc. IRE, vol. 44, pp. 920–926, July 1956.
17. Ref. 1, Chap. 1.15.
18. Ref. 1, Chap. 1.16.
19. Papoulis, A., *Probability, Random Variables, and Stochastic Processes*, 2nd ed., Chaps. 9–10, McGraw-Hill, NY, 1980.
    —, *The Fourier Integral and its Applications*, Chap. 12, McGraw-Hill, NY, 1962.

2 FREQUENCY DOMAIN ANALYSIS

And God said, Let us make man in our image ...
and let them have ... dominion ... over all the earth ...
So God created man in his own image...
And the Lord God commanded the man, saying,
Of every tree of the garden thou mayest freely eat:
But of the tree of knowledge of good and evil, thou shalt not eat of it:
for in the day that thou eatest thereof thou shall surely die.

Genesis 1:26–27, 2:16–17

Frequency and time domain characteristics are very important to the engineer when designing digital filters and assessing their performance. This chapter concentrates on frequency domain analysis while the next chapter discusses time domain analysis.

A general digital signal processing system is shown in Fig. 2.0.1. A signal $r(t)$ is input and band-limited by a lowpass filter to obtain $f(t)$. $f(t)$ is sampled to form $f(nT)$ and then transformed into $F(m)$. The transformed signal $F(m)$ is processed by a filter with generalized transfer function $H(m)$. This yields the transformed output signal $G(m) = H(m)F(m)$. Then $G(m)$ is inverse transformed to obtain the sampled output $g(nT)$. Finally $g(nT)$ is interpolated to obtain the continuous output signal $g(t)$.

This chapter is concerned with implementing the *transform block* in Fig. 2.0.1. First the concept of the discrete Fourier transform (DFT) is reviewed. Next the fast Fourier transform (FFT) is introduced. The FFT reduces the computational time of the DFT. Window functions are presented which improve the spectral information generated by the FFT. Other useful discrete transforms including Walsh, Paley, Hadamard, Haar, slant, and their fast versions are discussed. They further reduce computational load and are more suitable for characterizing certain types of signals. The various discrete transforms will be interrelated.

Two-dimensional and higher-order transforms are discussed. These are important because the 1-D digital signal processing system in Fig. 2.0.1 can be used for higher-dimensional systems. Every 1-D processing block is simply converted to an N-dimensional processing block. Since this chapter and the next lay the foundations of our later work, this material should be mastered.

2.1 DISCRETE FOURIER TRANSFORM (DFT)

The *discrete Fourier transform (DFT)* was introduced in Chap. 1.8. That discussion will be expanded here. The DFT is the FT of an impulsively sampled signal $f(t)$ as

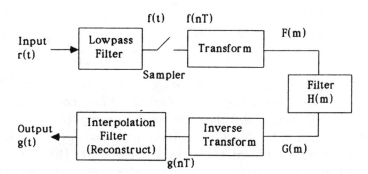

Fig. 2.0.1 Block diagram
of general digital signal
processing system.

$$F_s(j\omega) = \int_{-\infty}^{\infty} \Big[\sum_{n=-\infty}^{\infty} f(t)U_0(t-nT) \Big] e^{-j\omega t} dt = \sum_{n=-\infty}^{\infty} f(nT)e^{-jn\omega T} \qquad (2.1.1)$$

from Eq. 1.8.1 where the sampling frequency is $\omega_s = 2\pi/T$. $F_s(j\omega)$ describes the spectrum of the sampled signal $f(nT)$. As shown in Fig. 1.8.1, the DFT is also the zero-order approximation of the FT of the unsampled signal $f(t)$ where

$$F(j\omega) = \int_{-\infty}^{\infty} f(t)e^{-j\omega t} dt \cong T \sum_{n=-\infty}^{\infty} f(nT)e^{-jn\omega T} = TF_s(j\omega) \qquad (2.1.2)$$

T times the DFT of a signal approximates its FT at frequencies below the Nyquist frequency ($|\omega| < \omega_s/2$). The approximation error approaches zero as the sampling interval $T \to 0$. However there is no frequency aliasing error and therefore zero approximation error if the FT is band-limited to frequencies less than $\omega_s/2$.

DFT's are usually computed at N equally-spaced frequencies between 0 and ω_s. These frequencies are called the *frequency bins*. Since the frequency separation between spectral components equals ω_s/N, the frequency bins equal $\omega_m = m(\omega_s/N)$ for $m = 0, 1, \ldots, N-1$. The *frequency bin number* equals $(m+1)$ and is independent of frequency. The first frequency bin $m = 0$ corresponds to a frequency of dc. The second frequency bin $m = 1$ corresponds to the frequency separation ω_s/N, and so forth. To facilitate writing equations in a shorthand notation, we denote $F(m) \equiv F_s(j\omega_m) \cong F(j\omega_m)/T$ and $f(n) \equiv f(nT)$. Then Eq. 2.1.1 can be rewritten as[1]

$$F(m) = \sum_{n=-\infty}^{\infty} f(n)e^{-j2\pi nm/N} = \sum_{n=-\infty}^{\infty} f(n)W_N^{nm} \qquad (2.1.3)$$

where $W_N = e^{-j2\pi/N}$ is an important constant. It is a unit vector having an angle of $-2\pi/N$ radians $= -360/N$ degrees. In real-time analysis, the signal $f(t)$ is observed over some arbitrary time window such as $0 \le t \le T_w$ (or some other T_w-wide window like $-T_w/2 \le t \le T_w/2$). Only a finite number of samples can be stored for numerical evaluation. Assuming that N equally-spaced samples are collected over $0 \le t \le T_w$, then the DFT equals

$$F(m) = \sum_{n=0}^{N-1} f(n)W_N^{nm} \tag{2.1.4}$$

(see Table 1.8.1). The time interval between samples is $T = T_w/N$. Since only N data points are used, then only N of the $F(m)$ coefficients can be independent. Only N coefficients of $F(m)$ therefore need be computed. For example, there are $N/2$ spectral coefficients (for N even) at frequencies from $m = 0$ to $(N-1)/2$. If $F(m)$ is described in rectangular form, there are $N/2$ real parts (Re $F(m)$) and $N/2$ imaginary parts (Im $F(m)$). If $F(m)$ is described in polar form, there are $N/2$ magnitude parts ($|F(m)|$) and $N/2$ phase parts (arg $F(m)$).

The DFT has several interesting properties. Eq. 2.1.1 can be rewritten as

$$F(m) = \sum_{n=-\infty}^{\infty} f(nT)e^{-jn\omega T} = \sum_{n=-\infty}^{\infty} f(nT)\cos(n\omega T) - j\sum_{n=-\infty}^{\infty} f(nT)\sin(n\omega T) \tag{2.1.5}$$

using Euler's identity where $\omega = m\omega_s/N$. This equation shows that:

1. $F(m)$ is always real and even if $f(nT)$ is even about $n = 0$, i.e.
$f(nT) = f(-nT)$.
2. $F(m)$ is always imaginary and odd if $f(nT)$ is odd about $n = 0$, i.e.
$f(nT) = -f(-nT)$.
3. $F(m)$ and $F(-m)$ are always complex conjugates for real-valued $f(nT)$, i.e.
$F(m) = F^*(-m)$.
4. $F(m)$ is periodic over an N-point frequency interval $\omega_s = 2\pi/T$, i.e.
$F(m) = F(m+N)$.
5. $F(m)$ and $F(N-m)$ are always complex conjugates from properties 3 and 4, i.e.
$F(m) = F^*(N-m)$ or $F(-m) = F(N-m)$.

Although properties 1–3 are true for both DFT's and FT's, properties 4 and 5 only hold for DFT's. They are produced by the sampling operation. Properties 4 and 5 show that although $F(m)$ can be defined arbitrarily inside the interval $0 \le m \le N/2$, its behavior is replicated outside this interval. These conclusions also hold for the finite sum case given by Eq. 2.1.4.

EXAMPLE 2.1.1   Consider a signal

$$f(t) = 3e^{-t}\sin(3t)U_{-1}(t) \tag{2.1.6}$$

Compute its FT and determine an appropriate sampling rate. Then determine the number of samples N that should be used. Finally compute its DFT and compare with its FT.

SOLUTION   From Table 1.5.1 or 1.6.1, the FT equals

$$F(j\omega) = F(s)\Big|_{s=j\omega} = \frac{9}{(s+1)^2 + 3^2}\Big|_{s=j\omega} = \frac{9}{10 + 2j\omega - \omega^2} \tag{2.1.7}$$

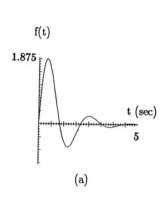

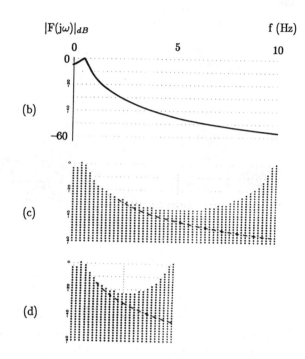

Fig. 2.1.1 (a) Signal $f(t)$, (b) its magnitude spectrum $|F(j\omega)|$, and its DFT magnitude spectrum for (c) 10 Hz and (d) 5 Hz sampling rates in Example 2.1.1.

The magnitude of $F(j\omega)$ equals

$$|F(j\omega)| = \frac{9}{\sqrt{(10 - \omega^2)^2 + (2\omega)^2}} = \frac{9}{\sqrt{\omega^4 - 16\omega^2 + 100}} \tag{2.1.8}$$

which is plotted in Fig. 2.1.1b. In Example 1.3.2, it was shown that the response was down -20 dB from its dc value at a frequency of about 10 rad/sec. Sampling at more than $2(10$ rad/sec$) \cong 3$ Hz reduces the aliasing distortion to 0.5% (see Eq. 1.3.13). Since the DFT frequency separation equals $1/NT$ Hz and $1/NT \leq 0.2$ Hz is needed for good resolution in the spectrum of Fig. 2.1.1b, then $N \geq 3$ Hz/0.2 Hz $= 15$ samples are required. The DFT magnitude spectra are shown in Figs. 2.1.1c–d for sampling frequencies of 5 Hz and 10 Hz, respectively. The frequency bin size is held constant at 0.2 Hz so 25 or 50 samples are taken. We see that little aliasing error is present for $f_s = 10$ Hz but it is quite noticeable at $f_s = 5$ Hz. ∎

The DFT can also be obtained from the z-transform as discussed in Chap. 1.8. Using the z-transform notation of Eq. 1.8.16, $F(m)$ can be expressed as

$$F(m) = F(z)\Big|_{z=W_N^{-m}=e^{j2\pi m/N}} = \sum_{n=0}^{N-1} f(n)z^{-n}\Big|_{z=W_N^{-m}} = \sum_{n=0}^{N-1} f(n)W_N^{nm} \tag{2.1.9}$$

Therefore sampling the z-transform $F(z)$ at equally-spaced points counterclockwise around the unit circle produces the DFT spectral coefficients. This is shown in Fig. 2.1.2 which emphasizes the fact that the $F(m)$ coefficients are periodic with period N.

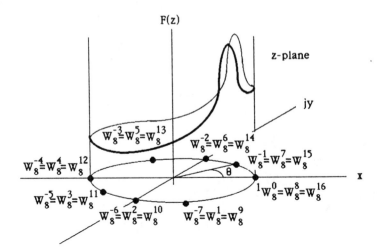

Fig. 2.1.2 Sample points in z-plane to obtain DFT.

2.1.1 CHIRP TRANSFORM (CZT)

The DFT results by evaluating the z-transform around the unit circle in the z-plane. The chirp z-transform, or simply *chirp transform (CZT)*, results by evaluating the z-transform around spiral contours instead. The CZT allows the behavior of a ZT to be analyzed everywhere in the z-plane.[2] The DFT is a special case of the CZT.

The z-transform for windowed data equals

$$F(z) = \sum_{n=0}^{N-1} f(n)z^{-n} \tag{2.1.10}$$

Evaluating $F(z)$ around the unit circle as $z_m = e^{s_m T}$, where $s_m T = j\omega_m T = jm\omega T$, frequency separation $\omega T = 2\pi/N$, and $W_N = e^{-j\frac{2\pi}{N}}$, gives the DFT as

$$F(z_m) = \sum_{n=0}^{N-1} f(n)e^{-jnm\omega T}$$

$$\equiv F(m) = \sum_{n=0}^{N-1} f(n)W_N^{nm}, \qquad m = 0, 1, \ldots, N-1 \tag{2.1.11}$$

The chirp transform results by evaluating $F(z)$ around the spiral $z_m = z_o e^{s_m T}$ having an initial offset $z_o = e^{s_o T}$ as shown in Fig. 2.1.3. This contour corresponds to a straight line in the s-plane having $s_m T = (\sigma_o + j\omega_o)T + m(\sigma + j\omega)T$. Using Eq. 2.1.10, the chirp transform $F_c(z_m)$ equals

$$F_c(z_m) = \sum_{n=0}^{N-1} f(n)e^{-n(\sigma_o T + j\omega_o T)}e^{-nm(\sigma T + j\omega T)}$$

$$\equiv F_c(m) = \sum_{n=0}^{N-1} f(n)A^n W^{nm} \qquad m = 0, 1, \ldots, M-1 \tag{2.1.12}$$

The constants A and W equal

$$A = e^{-s_o T} = e^{-\sigma_o T} e^{-j\omega_o T} = A_o e^{-j\omega_o T}$$
$$W = e^{-sT} = e^{-\sigma T} e^{-j\omega T} = W_o e^{-j\omega T} \qquad (2.1.13)$$

and are used to simplify Eq. 2.1.12. Comparing Eqs. 2.1.11 and 2.1.12, the CZT coefficient $F_c(m)$ and the DFT coefficient $F(m)$ differ by the $A^n W_o^{nm}$ term and have different ωT values.

In the chirp transform, the frequency separation $\omega T \leq 2\pi/M$ is arbitrary. The z_m contour spirals inward at a rate of $-\sigma T$ nepers/point (or outward at σT). It is not necessary for the spiral contour to completely enclose the z-plane origin. The initial starting point z_o can be located anywhere in the z-plane. Since $f(n) \leftrightarrow F(z)$ form transform pairs from the frequency shifting theorem of Table 1.9.2, then $e^{-n(\alpha+j\beta)T} f(n) \leftrightarrow F(ze^{(\alpha+j\beta)T})$. Therefore comparing the ZT and CZT, we see that $F(z)$ and $F_c(z)$ are related as $F_c(z) \leftrightarrow F(ze^{(\sigma_o+m\sigma+j\omega_o)T})$. The z-plane is compressed by $e^{-(\sigma_o+m\sigma)T}$ and rotated by $-\omega_o T$ radians.

The chirp transform $F_c(m)$ can be expressed in the matched filter form of Eq. 1.6.39 by writing $2nm = (m^2 + n^2) - (m - n)^2$. Then Eq. 2.1.12 becomes

$$F_c(m) = \sum_{n=0}^{N-1} f(n) A^n W^{\frac{m^2}{2}} W^{\frac{n^2}{2}} W^{-\frac{(m-n)^2}{2}} = W^{\frac{m^2}{2}} \sum_{n=0}^{N-1} f(n) A^n W^{\frac{n^2}{2}} W^{-\frac{(m-n)^2}{2}}$$

$$= s(m) \sum_{n=0}^{N-1} \frac{f(n) a(n) s(n)}{s(m-n)} = \left[\left(f(n) a(n) s(n) \right) * \frac{1}{s(n)} \right] s(m) \qquad (2.1.14)$$

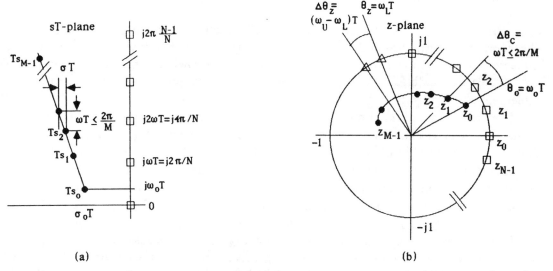

(a)   (b)

Fig. 2.1.3 (a) sT-plane contours and (b) z-plane contours showing sample points for DFT ($\square$), CZT ($\bullet$), and zoom transform ($\triangle$).

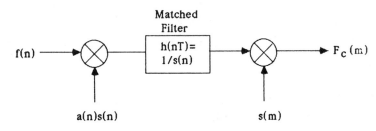

Fig. 2.1.4 Time domain implementation of chirp transform.

$s(n)$ is a sweep modulating signal and $a(n)$ an offset modulating signal where

$$a(n) = A^n = (A_o e^{-j\omega_o T})^n = A_o^n e^{-jn\omega_o T}$$

$$s(n) = W^{\frac{n^2}{2}} = (W_o e^{-j\omega T})^{\frac{n^2}{2}} = W_o^{\frac{n^2}{2}} e^{-\frac{j\omega T n^2}{2}}$$

$$(2.1.15)$$

Taking $A_o = W_o = 1$, $\omega_o T = 0$, and $\omega T = 2\pi/N$ reduces the CZT to the DFT. In this case, $a(n) = 1$, $|s(n)| = 1$, and $\arg s(n) = -\pi n^2/N$. $s(n)$ is a complex exponential with linearly decreasing frequency $\omega_i = d \arg s(n)/dn = -n\omega T = -2\pi n/N$ which is called a *chirp signal* in radar. Thus the name CZT is given to the algorithm of Eq. 2.1.14.

Eq. 2.1.14 is implemented as shown in Fig. 2.1.4. The input $f(n)$ is multiplied by the product $a(n)s(n)$, matched filtered (convolved) with an impulse response $1/s(n)$, and finally multiplied by $s(m)$. This results in the chirp transform coefficient $F_c(m)$. This system is the discrete time equivalent of the continuous time spectrum analyzer shown in Fig. 1.6.6c. It generates discrete Fourier transforms $F(m)$ in discrete time.

2.1.2 ZOOM TRANSFORM (ZmT)

The *zoom transform (ZmT)* is a narrowband DFT implemented using a chirp transform. The zoom transform "zooms" in on a narrow frequency band and expands it. Its implementation is shown in Fig. 2.1.5. It is the narrowband spectrum analyzer of Figs. 1.6.6a or 1.6.6b followed by a regular DFT, or equivalently, a narrowband DFT implemented in software.

The input signal $r(t)$ is multiplied by the complex exponential $\exp(-j\omega_o t)$. This frequency translates the $R(j\omega)$ spectrum. The mixer output $R(j(\omega - \omega_o))$ is lowpass filtered by $H(j\omega)$ to obtain a narrowband spectrum. The lowpass filter bandwidth B is chosen so that $B \ll \omega_o$. The filter output is sampled and processed by a regular DFT. The sampling frequency ω_s is twice the filter bandwidth B. The DFT output therefore equals the $R(j\omega)$ spectrum for ω in the range from $\omega_L = \omega - B$ to $\omega_U = \omega + B$. The frequency spacing is $\omega_s/N = 2B/N$ rad/sec.

From the spectrum analyzer viewpoint of Fig. 2.1.5, suppose that the frequency band to be analyzed lies between ω_L and ω_U with a required resolution of $\Delta\omega$. Then the mixer frequency ω_o, lowpass filter bandwidth B, minimum number of points N, and sampling frequency ω_s used by the DFT equal

$$\omega_o = \frac{\omega_U + \omega_L}{2}, \qquad B = \frac{\omega_U - \omega_L}{2},$$

$$\omega_s = 2B = \omega_U - \omega_L, \qquad N = \frac{\omega_U - \omega_L}{\Delta\omega} = \frac{2B}{\Delta\omega} \qquad (2.1.16)$$

The zoom transform can be implemented using the system equations given by Eqs. 1.6.34 or 1.6.35 and the regular DFT of Eq. 2.1.4, or a narrowband DFT.

For the discrete bandpass filter form of Fig. 2.1.5a, these equations are:

$$r(n) = f(n) * \left[h(n)e^{jnm_o}\right], \qquad R(m) = F(m)H(m - m_o)$$
$$C(m) = R(m) * U_0(m + m_o) = R(m + m_o) = H(m)F(m + m_o)$$
$$c(n) = r(n)e^{-jnm_o} = \left[h(n)e^{jnm_o} * f(n)\right]e^{-jnm_o} \qquad (2.1.17)$$
$$F_z(m) = \text{DFT}\left[c(n)\right]$$

The sampling switch has been moved to the mixer to form output samples. The sampling frequency equals $f_s = 1/T$ and the mixer frequency ω_o falls on DFT bin $m_o = N(f_o/f_s)$.

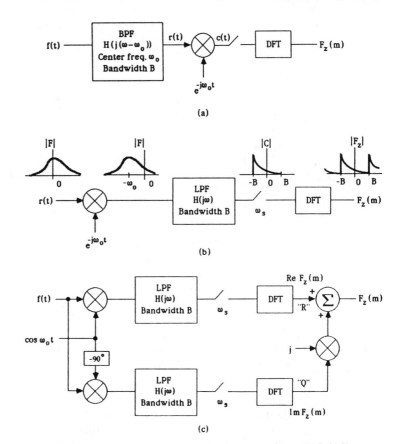

Fig. 2.1.5 Narrowband spectrum analyzer using multipliers and (b) bandpass filter, (b) lowpass filter, and (c) dual-channel lowpass filters.

The discrete lowpass filter form of Fig. 2.1.5b is described by

$$r(n) = f(n)e^{-jnm_o}, \qquad R(m) = F(m) * U_0(m + m_o) = F(m + m_o)$$
$$C(m) = H(m)R(m) = H(m)F(m + m_o), \qquad c(n) = h(n) * \left[f(n)e^{-jnm_o} \right] \qquad (2.1.18)$$
$$F_z(m) = \text{DFT}\left[c(n) \right]$$

This system requires complex number processing because of the $\exp(-jnm_o)$ mixer. Expressing $\exp(-jnm_o) = \cos(nm_o) + j\sin(nm_o)$ and separating this system into the real channel "R" and quadrature channel "Q" gives the parallel system in Fig. 2.1.5c. This parallel system, composed of two Fig. 2.1.5b channels, generates the real and imaginary parts of the zoom transform $F_z(m)$.

The zoom transform can also be evaluated using the chirp transform $(M = N)$ with starting point $z_0 = \exp(j\omega_L T)$, end point $z_{N-1} = \exp(j\omega_U T)$, and frequency separation $\omega T = (\omega_U - \omega_L)T/N$. This is called a *narrowband DFT*.

2.1.3 MATRIX FORM OF THE DFT, CZT, AND ZmT

It is usually easier to represent the DFT, CZT, and ZmT in their matrix forms. They are all special cases of the chirp transform evaluated along the special contours shown in Fig. 2.1.3 and can be written as

$$
\begin{bmatrix}
F(0) \\
F(1) \\
F(2) \\
\vdots \\
F(M-1)
\end{bmatrix}
=
\begin{bmatrix}
z_0^0 & z_0^{-1} & z_0^{-2} & \cdots & z_0^{-(N-1)} \\
z_1^0 & z_1^{-1} & z_1^{-2} & \cdots & z_1^{-(N-1)} \\
z_2^0 & z_2^{-1} & z_2^{-2} & \cdots & z_2^{-(N-1)} \\
\vdots & \vdots & \vdots & \cdots & \vdots \\
z_{M-1}^0 & z_{M-1}^{-1} & z_{M-1}^{-1} & \cdots & z_{M-1}^{-(N-1)}
\end{bmatrix}
\begin{bmatrix}
f(0) \\
f(1) \\
f(2) \\
\vdots \\
f(N-1)
\end{bmatrix}
\qquad (2.1.19)
$$

where the $z_0, \ldots, z_{M-1}$ are M points along these contours. The points equal

$$\text{DFT}: \qquad z_m = e^{j\omega_m T}, \qquad\qquad \omega_m = m\frac{\omega_s}{N}, \qquad M = N$$

$$\text{CZT}: \qquad z_m = z_0 e^{(\sigma_m + j\omega_m)T}, \qquad \omega_m = m\frac{k\omega_s}{M}, \qquad\qquad 0 < k \le 1 \qquad (2.1.20)$$

$$\text{ZmT}: \qquad z_m = e^{j\omega_m T}, \qquad\qquad \omega_m = \omega_L + m\frac{\omega_U - \omega_L}{N}, \qquad M = N$$

and $\omega_s = 2\pi/T$ is the sampling frequency. ω_s must be chosen high enough to reduce the aliasing error to an acceptable level as shown in Examples 1.3.2 and 2.1.1.

EXAMPLE 2.1.2 Suppose that a pulsed-cosine telephone signalling tone in the 697–1633 Hz range must be resolved and detected within 10 Hz. Determine the spectrum analyzer parameters to accomplish this.

SOLUTION A narrowband spectrum analyzer is needed which has

$$f_L = 697 \text{ Hz}, \qquad f_U = 1633 \text{ Hz}, \qquad \Delta f = 10 \text{ Hz} \qquad (2.1.21)$$

From Eq. 2.1.16, the system parameters that must be used in Fig. 2.1.5 equal

$$f_o = 1165 \text{ Hz}, \qquad B = 468 \text{ Hz}, \qquad f_s = 936 \text{ Hz}, \qquad N \geq \frac{2(468 \text{ Hz})}{10 \text{ Hz}} = 94 \text{ pts}$$

$$(2.1.22)$$

A regular 94-point (or more) DFT of Eq. 2.1.20a is needed to analyze spectrum between 697 and 1633 Hz.

If the ZmT (or narrowband DFT) of Eq. 2.1.20c is used instead, then

$$f_L = 697 \text{ Hz}, \qquad f_U = 1633 \text{ Hz}, \qquad \Delta f = 10 \text{ Hz}, \qquad N \geq 94 \text{ pts} \qquad (2.1.23)$$

where the telephone signal is first band-limited by a lowpass filter to 4 KHz. It is then sampled at the Nyquist rate of $f_s = 2(4 \text{ KHz}) = 8 \text{ KHz}$.

If the signal is only band-limited and then analyzed using a regular DFT of Eq. 2.1.20a (between 0 and 8 KHz) and not using the specialized system of Fig. 2.1.5, then $N = f_s/\Delta f = 8 \text{ KHz}/10 \text{ Hz} = 800$ points are required to achieve the same resolution. This is about a nine-fold data increase over the narrowband DFT. ■

2.1.4 NONUNIFORMLY SAMPLED DFT, CZT, AND ZmT

Nonuniform sampling was discussed in Chap. 1.3. We will now extend that discussion using a DFT matrix formulation. Under nonuniform sampling, the zero-order approximation of the short-term FT of an unsampled signal $f(t)$ equals (see Eq. 1.6.33)

$$F(j\omega) = \int_0^{NT} f(t)e^{-j\omega t}dt \cong \sum_{n=0}^{N-1}(t_n - t_{n-1})f(t_n)e^{-j\omega t_n}, \qquad \text{backward difference}$$

$$\text{or} \cong \sum_{n=0}^{N-1}(t_{n+1} - t_n)f(t_n)e^{-j\omega t_n}, \qquad \text{forward difference}$$

$$(2.1.24)$$

Evaluating $F(j\omega)$ at N different frequencies ω_m gives a set of equations which can be expressed in matrix form. The forward difference approximation equals

$$
\begin{bmatrix} F(0) \\ F(1) \\ F(2) \\ \vdots \\ F(N-1) \end{bmatrix} = \exp\left\{-j\begin{bmatrix} \omega_0 t_0 & \omega_0 t_1 & \omega_0 t_2 & \cdots & \omega_0 t_{N-1} \\ \omega_1 t_0 & \omega_1 t_1 & \omega_1 t_2 & \cdots & \omega_1 t_{N-1} \\ \omega_2 t_0 & \omega_2 t_1 & \omega_2 t_2 & \cdots & \omega_2 t_{N-1} \\ \vdots & \vdots & \vdots & \cdots & \vdots \\ \omega_{N-1}t_0 & \omega_{N-1}t_1 & \omega_{N-1}t_2 & \cdots & \omega_{N-1}t_{N-1} \end{bmatrix}\right\}
$$

$$
\times \begin{bmatrix} t_1 - t_0 & 0 & 0 & \cdots & 0 \\ 0 & t_2 - t_1 & 0 & \cdots & 0 \\ 0 & 0 & t_3 - t_2 & \cdots & 0 \\ \vdots & \vdots & \vdots & \ddots & 0 \\ 0 & 0 & 0 & \cdots & t_N - t_{N-1} \end{bmatrix} \begin{bmatrix} f(0) \\ f(1) \\ f(2) \\ \vdots \\ f(N-1) \end{bmatrix}
$$

$$(2.1.25)$$

Eq. 2.1.25 can be expressed as $F = W \Delta t \, f$ in shorthand form. Reducing the time indices by one in the Δt matrix produces the backward difference approximation. Evaluating the $\omega_m t_n$ term of the W matrix gives $e^{-j\omega_m t_n}$. Under the DFT uniform sampling condition, $t_n = nT$ and $\omega_m = m\omega_s/N$. Therefore $\omega_m t_n = mn\omega_s T/N = 2\pi mn/N$ and $t_n - t_{n-1} = T$. This simplifies the W matrix as will be discussed in the next section. Uniform time sampling also permits the $\Delta t \, f$ matrix product to be simplified to a time scaled matrix Tf. Eq. 2.1.25 applies to the nonuniformly sampled ZmT when the ω_m frequencies are selected according to Eq. 2.1.20c. The nonuniformly sampled CZT uses $e^{-[(\sigma_o + \sigma_m) + j(\omega_o + \omega_m)]t_n}$ terms in the first matrix as required by Eq. 2.1.20b.

2.2 FAST FOURIER TRANSFORM (FFT)

The fast Fourier transform (FFT) refers to a collection of algorithms for computing the DFT in a simplified manner.[3] This large variety of FFT methods allow the DFT to be determined more rapidly and with greater accuracy than simply evaluating the DFT directly using Eq. 2.1.4. Many programs are available which implement the FFT in software. FFT hardware modules are also used.

The DFT was given by Eq. 2.1.4 where $W_N = \exp(-j2\pi/N)$. Fig. 2.1.2 shows that W_N^{nm} is periodic or cyclic and repeats itself when nm equals any integer multiple of N. The FFT takes advantage of this redundancy. Eq. 2.1.4 can be expressed in an $N \times N$ matrix form $F = Wf$ as

$$
\begin{bmatrix}
F(0) \\
F(1) \\
F(2) \\
\vdots \\
F(N-1)
\end{bmatrix}
=
\begin{bmatrix}
0 & 0 & 0 & \cdots & 0 \\
0 & 1 & 2 & \cdots & (N-1) \\
0 & 2 & 4 & \cdots & 2(N-1) \\
\vdots & \vdots & \vdots & \cdots & \vdots \\
0 & N-1 & 2(N-1) & \cdots & (N-1)^2
\end{bmatrix}
\begin{bmatrix}
f(0) \\
f(1) \\
f(2) \\
\vdots \\
f(N-1)
\end{bmatrix}
\tag{2.2.1}
$$

where only the power or exponent nm of the W_N^{nm} term appears in the W matrix. This shorthand notation further simplifies the appearance and interpretation of Eq. 2.1.4.

Let us choose $N = 8$ and consider the DFT matrix in more detail. Eq. 2.2.1 becomes

$$
\begin{bmatrix}
F(0) \\
F(1) \\
F(2) \\
F(3) \\
F(4) \\
F(5) \\
F(6) \\
F(7)
\end{bmatrix}
=
\begin{bmatrix}
0 & 0 & 0 & 0 & 0 & 0 & 0 & 0 \\
0 & 1 & 2 & 3 & 4 & 5 & 6 & 7 \\
0 & 2 & 4 & 6 & 8 & 10 & 12 & 14 \\
0 & 3 & 6 & 9 & 12 & 15 & 18 & 21 \\
0 & 4 & 8 & 12 & 16 & 20 & 24 & 28 \\
0 & 5 & 10 & 15 & 20 & 25 & 30 & 35 \\
0 & 6 & 12 & 18 & 24 & 30 & 36 & 42 \\
0 & 7 & 14 & 21 & 28 & 35 & 42 & 49
\end{bmatrix}
\begin{bmatrix}
f(0) \\
f(1) \\
f(2) \\
f(3) \\
f(4) \\
f(5) \\
f(6) \\
f(7)
\end{bmatrix}
\tag{2.2.2}
$$

Several of the W_8^{nm} terms are shown in Fig. 2.1.2. These terms are periodic and repeat when $nm = 8r$ for any integer r. The principle value of W_8^{nm} is given by the modulus $\text{mod}_8(nm)$. Using these moduli values, then Eq. 2.2.2 can be rewritten as

$$
\begin{bmatrix} F(0) \\ F(1) \\ F(2) \\ F(3) \\ F(4) \\ F(5) \\ F(6) \\ F(7) \end{bmatrix}
=
\begin{bmatrix}
0 & 0 & 0 & 0 & 0 & 0 & 0 & 0 \\
0 & 1 & 2 & 3 & 4 & 5 & 6 & 7 \\
0 & 2 & 4 & 6 & 0 & 2 & 4 & 6 \\
0 & 3 & 6 & 1 & 4 & 7 & 2 & 5 \\
0 & 4 & 0 & 4 & 0 & 4 & 0 & 4 \\
0 & 5 & 2 & 7 & 4 & 1 & 6 & 3 \\
0 & 6 & 4 & 2 & 0 & 6 & 4 & 2 \\
0 & 7 & 6 & 5 & 4 & 3 & 2 & 1
\end{bmatrix}
\begin{bmatrix} f(0) \\ f(1) \\ f(2) \\ f(3) \\ f(4) \\ f(5) \\ f(6) \\ f(7) \end{bmatrix}
\qquad (2.2.3)
$$

Notice that the entries in row i are identical to the entries in column i. Also notice that the entries in row i and row $(10 - i)$ total 8 (or column i and column $(10 - i)$). This is due to the characteristics of W_N^{nm} discussed in Eq. 2.1.5 where

$$
W_N^{nm} = (W_N^{-nm})^* = W_N^{n(m+N)} = (W_N^{n(N-m)})^* \qquad (2.2.4)
$$

All FFT's make use of these relations to exploit the redundancies and symmetries in the $N \times N$ matrix of Eq. 2.2.1. This matrix can be decomposed into products of matrices with more zero entries (sparse matrices) than the original matrix.[4]

The DFT given by Eq. 2.1.4 can be easily represented in flow graph form. A directed branch with gain W_N^{nm} connects every $f(n)$ node to every $F(m)$ node. There are N^2 directed branches. The $N = 8$ case is shown in Fig. 2.2.1. For visual simplicity, only the branches involved in the $F(1)$ computation are shown.

The amount and type of computations required in Eq. 2.2.1 determines the execution time for the DFT. Assuming that the W_N^{nm} coefficients have been formed, then each of the $F(m)$ components requires N complex multiplications and $(N - 1)$ complex additions (see Eq. 2.1.4). Since there are N spectral components, the DFT requires N^2 complex multiplications and $N(N - 1)$ complex additions. This is summarized in Table 2.8.2. To compute W_N^{nm} and fill the matrix in Eq. 2.2.1, N^2 complex multiplications are required. However if coefficient symmetry is noted as was shown in Eq. 2.2.4, no more than N complex multiplications are needed. In practice, the complex multiplications require the most time (say T_m) so the DFT computation time is proportional to $T_m N^2$.

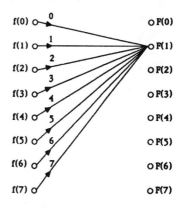

Fig. 2.2.1 8-point DFT in flow graph form. For simplicity, only $F(1)$ of the eight $F(m)$ spectral coefficients is shown.

Another consideration is memory size. The maximum memory required to implement Eq. 2.2.1 requires $2N$ locations for all the $F(m)$ and $f(n)$ data values and N^2 locations for the W_N^{nm} coefficients. This shows a total of $N(N+2)$ words of memory are needed. Since the time and memory size both vary as N^2, computing the DFT by the direct method using Eq. 2.2.1 or Fig. 2.2.1 takes a long time and requires a large memory. For example, a 1024-point DFT using a computer having a complex multiply time of 1 μsec would require no less than $(1\mu\text{sec})(1024^2) \cong 1$ second to complete and would require no less than $1024^2 \cong 1$ Mbyte of memory. DFT cycle times of one second are unacceptable in most real-time applications while 1 Mbyte memories (with their associated read/write times) are rather large. Often DFT cycle times in the μsec to $m$sec range are necessary using small memories less than 8 Kbyte. Obviously, orders of magnitude improvement are necessary to implement practical DFT systems.

FFT techniques reduce the DFT computations to approximately $N \log_2 N$ complex multiplications and memories no larger than $2N$ bytes. In the 1024-point example above, the FFT could be completed in $(1\mu\text{sec})(1024) \log_2(1024) \cong 10$ $m$sec with only a memory of 2048 bytes which is a vast improvement.

The FFT reduces the number of arithmetic operations required by the DFT by decomposing the DFT into successively smaller DFT's. The way in which this is done varies from algorithm to algorithm. However the methods generally fall into two basic classes called *decimation-in-time* and *decimation-in-frequency*. In the time approach, the signal sequence $f(n)$ is decomposed or decimated into smaller subsequences. In the frequency approach, the spectral sequence $F(m)$ is decomposed into smaller subsequences. In either case, roughly $N \log_2 N$ complex multiplies and memories of $2N$ bytes are needed. These FFT methods will now be discussed in detail. The algorithms to be considered require N to be an integer power of 2 (i.e., $N = 2^k$ where k is any integer). FFT algorithms can be developed for any N that is not prime using a more general approach as Harris and others have shown.[4]

2.3 TIME DECIMATION ALGORITHMS

Time decimation algorithms separate the N-sample sequence $f(n)$ into two $(N/2)$-sample subsequences consisting of even-numbered points $f(2n)$ and odd-numbered points $f(2n+1)$. Using this approach, the DFT given by Eq. 2.1.4 can be rewritten as

$$
\begin{aligned}
F(m) &= \sum_{n=0}^{N-1} f(n)W_N^{nm} = \sum_{n=0}^{(N/2)-1} f(2n)W_N^{2nm} + \sum_{n=0}^{(N/2)-1} f(2n+1)W_N^{(2n+1)m} \\
&= \sum_{n=0}^{(N/2)-1} f(2n)(W_N^{nm})^2 + W_N^m \sum_{n=0}^{(N/2)-1} f(2n+1)(W_N^{nm})^2
\end{aligned} \tag{2.3.1}
$$

for $m = 0, 1, \ldots, N - 1$. Notice that $W_N^2 = W_{N/2}$ since doubling the angle of W_N is equivalent to halving its base N (i.e., $W_N^2 = e^{j2(2\pi/N)} = e^{j2\pi/(N/2)} = W_{N/2}$). Using this result in Eq. 2.3.1 gives

$$F(m) = \sum_{n=0}^{(N/2)-1} f(2n)W_{N/2}^{nm} + W_N^m \sum_{n=0}^{(N/2)-1} f(2n+1)W_{N/2}^{nm} = F_2(m) + W_N^m F_3(m) \quad (2.3.2)$$

The N-point DFT sequence $F(m)$ has been *decimated* or broken down into two $(N/2)$-point DFT subsequences $F_2(m)$ and $F_3(m)$. Although $F(m)$ is periodic in m with period N, $F_2(m)$ and $F_3(m)$ are periodic in m with only half the period $N/2$.

The process can be repeated to decimate the two $(N/2)$-point DFT sequences $F_2(m)$ and $F_3(m)$ into four $(N/4)$-point DFT subsequences $F_4(m)$, $F_5(m)$, $F_6(m)$, and $F_7(m)$,

$$F_2(m) = \sum_{n=0}^{(N/2)-1} f(2n)W_{N/2}^{nm} = \sum_{n=0}^{(N/4)-1} f(4n)W_{N/2}^{2nm} + \sum_{n=0}^{(N/4)-1} f(4n+2)W_{N/2}^{(2n+1)m}$$

$$= \sum_{n=0}^{(N/4)-1} f(4n)W_{N/4}^{nm} + W_{N/2}^m \sum_{n=0}^{(N/4)-1} f(4n+2)W_{N/4}^{nm} = F_4(m) + W_{N/2}^m F_5(m)$$

$$F_3(m) = \sum_{n=0}^{(N/4)-1} f(4n+1)W_{N/4}^{nm} + W_{N/2}^m \sum_{n=0}^{(N/4)-1} f(4n+3)W_{N/4}^{nm} = F_6(m) + W_{N/2}^m F_7(m)$$

$$(2.3.3)$$

$F_4(m)$ through $F_7(m)$ are periodic in m with period $N/4$.

The decimation process can be continued until it terminates in a single term. This requires $\log_2 N$ steps. In the next to the last step, 2^{N-2} four-point subsequences are obtained having the form

$$F_x(m) = F_y(m) + W_2^m F_z(m) \quad (2.3.4)$$

They can be decimated to 2^{N-1} two-point subsequences. The first pair is

$$F_y(m) = f(0) + W_2^m f(N/2)$$
$$F_z(m) = f(1) + W_2^m f((N/2)+1) \quad\quad (2.3.5)$$

In the last step, the process terminates in single terms like $f(0), f(N/2), f(1)$, and $f((N/2)+1)$. As a result, 2^N one-point DFT sequences are obtained as

$$f(0), \ f(N/2), \ f(N/4), \ f(3N/4),\ldots, f((N/2)-1), \ f(N-1) \quad (2.3.6)$$

This is simply the input data rearranged in a different sequence. Combining all these subsequences together starting with Eq. 2.3.6 and working backwards to Eq. 2.3.1 yields $F(m)$ as given by Eq. 2.1.4.

These operations can be simply represented in flow graph form. Consider the first decimation stage described by Eqs. 2.3.1 and 2.3.2. The N-point sample sequence $f(n)$ is rearranged into two $(N/2)$-point subsequences consisting of even samples $f(2n)$ and odd samples $f(2n + 1)$. Performing two $(N/2)$-point DFT transforms on these sequences yields $F_2(m)$ and $F_3(m)$. Recognizing that these sequences are periodic in m with period $N/2$, then Eq. 2.3.2 is described by the flow graph of Fig. 2.3.1. This graph generates the $F(m)$ sequence from $F_2(m)$ and $F_3(m)$ subsequences.

Now consider the second decimation stage described by Eq. 2.3.3. This equation is described by Fig. 2.3.2. It generates $F_2(m)$ (or $F_3(m)$) sequences from $F_4(m)$ and $F_5(m)$ (or $F_6(m)$ and $F_7(m)$) subsequences. Fig. 2.3.2 has exactly the same structure as Fig. 2.3.1 except the branch gains are squared. After $(\log_2(N) - 1)$ decimations, we obtain 2^{N-1} two-point equations having the form of Eq. 2.3.5. These are described by the flow graph of Fig. 2.3.3. Since the basic graph resembles a butterfly, these flow graphs are referred to as *butterfly diagrams*. At the last $\log_2(N)$ decimation, 2^N one-point sequences are obtained which are simply the original data rearranged in a different order.

With this general background, let us return to the 8-point DFT given by Eq. 2.2.3. Combining Figs. 2.3.1–2.3.3 results in Fig. 2.3.4. It is easy to verify that this figure describes Eq. 2.2.3 by simply computing $F(0), F(1), \ldots, F(7)$ in terms of $f(0), f(1), \ldots, f(7)$ using signal flow graph theory. Since only forward paths and no feedback paths are present, computation consists of only summing simple forward path gains. For example, $F(1) = f(0) + f(1)W_8^1 + f(2)W_8^2 + \ldots + f(7)W_8^7$ which checks with Eq. 2.2.3. Close inspection of the flow graph verifies that there are only $N \log_2(N) = 24$ complex multiplies and $N \log_2(N) = 24$ complex additions. Direct evaluation of Eq. 2.2.3 would require $N^2 = 64$ complex multiplies and $N(N-1) = 32$ complex additions.

This basic flow graph can be manipulated into other forms by rearranging nodes. As long as branch gains are preserved, the DFT will always be properly computed. However, the order in which the data is accessed and stored will change as well as the sequence of arithmetic operations. This leads to different FFT forms. For example, three extra columns of nodes can be inserted ahead of the existing columns of nodes in Fig. 2.3.4 and connected with unity gain branches. Then all existing branch gains can be moved backward into these new branches leaving only ± 1 gain branches in the original graph. This results in the flow graph of Fig. 2.3.5.

In this form there are only $(N/2) \log_2(N) = 12$ complex multiplies and $N \log_2(N) = 24$ complex additions. This reduces the multiplications needed in Fig. 2.3.4 by half. Also multiply cycles are separated from addition cycles. In both cases, the input sequence $f(n)$ is stored in *bit-reversed* order. If $(n_2\ n_1\ n_0)$ is the binary word representing index n, then $n = (n_0\ n_1\ n_2)$ is the bit-reversed binary word. For example, $(000)_2 = (0)_{10} \rightarrow (000)_2 = (0)_{10}$, $(001)_2 = (1)_{10} \rightarrow (100)_2 = (4)_{10}$, $(010)_2 = (2)_{10} \rightarrow (010)_2 = (2)_{10}$, and $(011)_2 = (3)_{10} \rightarrow (110)_2 = (6)_{10}$, etc. The decimal equivalent (base 10) of the binary word (base 2) is also shown. $f(n)$ is read in sequence as $[f(0), f(1), f(2), \ldots, f(6), f(7)]$. However $f(n)$ is stored in bit-reversed sequence $[f(0), f(4), f(2), \ldots, f(3), f(7)]$. $2N$ registers are needed to store computational data and N registers are needed to store W_N^{nm} coefficients.

The FFT represented in Fig. 2.3.5 can be changed into other forms. By reordering the input to the normal sequence but the output into a bit-reversed sequence, the FFT of Fig. 2.3.6 is obtained. By reordering the output of Fig. 2.3.6 into a normal sequence, the FFT of Fig. 2.3.7 is obtained.

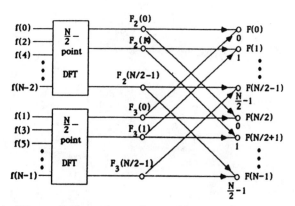

Fig. 2.3.1 Time decimation of one N-point DFT into two $(N/2)$-point DFT's.

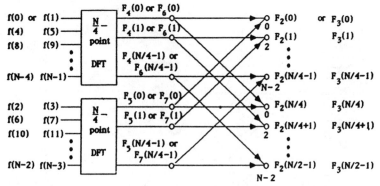

Fig. 2.3.2 Time decimation of one $(N/2)$-point DFT into two $(N/4)$-point DFT's.

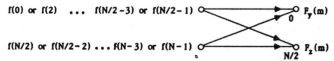

Fig. 2.3.3 Time decimation of one 2-point DFT into two 1-point DFT's.

The last computational stage in the FFT of Fig. 2.3.7 has a peculiar symmetry not seen in the other forms. Fig. 2.3.8 shows an FFT that utilizes this symmetry in every computational stage. The input is in bit-reversed order and the output is in normal order. Here the same computational algorithm can be used in all stages. Only the interstage gains need be changed. This permits sequential data accessing and storing.

FFT's can also be manipulated from a matrix viewpoint rather than a flow graph approach. Using this technique, the FFT matrix is factored into products of sparse matrices having certain symmetries. For example, consider the 8-point FFT given by Eq. 2.2.3 where the entries are the powers of W_8 needed by the transform. The time decomposition with bit-reversed input and normal output shown in Fig. 2.3.5 can be expressed in matrix form as

$$
\begin{bmatrix} F(0) \\ F(1) \\ F(2) \\ F(3) \\ F(4) \\ F(5) \\ F(6) \\ F(7) \end{bmatrix}
=
\begin{bmatrix}
0 & \cdot & \cdot & \cdot & 0 & \cdot & \cdot & \cdot \\
\cdot & 0 & \cdot & \cdot & \cdot & 1 & \cdot & \cdot \\
\cdot & \cdot & 0 & \cdot & \cdot & \cdot & 2 & \cdot \\
\cdot & \cdot & \cdot & 0 & \cdot & \cdot & \cdot & 3 \\
0 & \cdot & \cdot & \cdot & 4 & \cdot & \cdot & \cdot \\
\cdot & 0 & \cdot & \cdot & \cdot & 5 & \cdot & \cdot \\
\cdot & \cdot & 0 & \cdot & \cdot & \cdot & 6 & \cdot \\
\cdot & \cdot & \cdot & 0 & \cdot & \cdot & \cdot & 7
\end{bmatrix}
\begin{bmatrix}
0 & \cdot & 0 & \cdot & \cdot & \cdot & \cdot & \cdot \\
\cdot & 0 & \cdot & 2 & \cdot & \cdot & \cdot & \cdot \\
0 & \cdot & 4 & \cdot & \cdot & \cdot & \cdot & \cdot \\
\cdot & 0 & \cdot & 6 & \cdot & \cdot & \cdot & \cdot \\
\cdot & \cdot & \cdot & \cdot & 0 & \cdot & 0 & \cdot \\
\cdot & \cdot & \cdot & \cdot & \cdot & 0 & \cdot & 2 \\
\cdot & \cdot & \cdot & \cdot & 0 & \cdot & 4 & \cdot \\
\cdot & \cdot & \cdot & \cdot & \cdot & 0 & \cdot & 6
\end{bmatrix}
$$

$$(2.3.7)$$

$$
\times
\begin{bmatrix}
0 & 0 & \cdot & \cdot & \cdot & \cdot & \cdot & \cdot \\
0 & 4 & \cdot & \cdot & \cdot & \cdot & \cdot & \cdot \\
\cdot & \cdot & 0 & 0 & \cdot & \cdot & \cdot & \cdot \\
\cdot & \cdot & 0 & 4 & \cdot & \cdot & \cdot & \cdot \\
\cdot & \cdot & \cdot & \cdot & 0 & 0 & \cdot & \cdot \\
\cdot & \cdot & \cdot & \cdot & 0 & 4 & \cdot & \cdot \\
\cdot & \cdot & \cdot & \cdot & \cdot & \cdot & 0 & 0 \\
\cdot & \cdot & \cdot & \cdot & \cdot & \cdot & 0 & 4
\end{bmatrix}
\begin{bmatrix} f(0) \\ f(4) \\ f(2) \\ f(6) \\ f(1) \\ f(5) \\ f(3) \\ f(7) \end{bmatrix}
$$

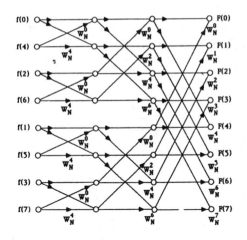

Fig. 2.3.4 FFT for $N = 8$ using time decimation with bit-reversed input and normal output.

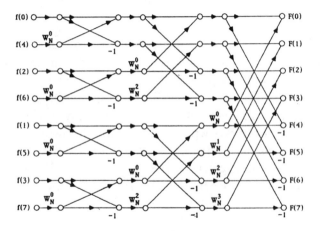

Fig. 2.3.5 FFT for $N = 8$ using time decimation with bit-reversed input and normal output using modified Fig. 2.3.4.

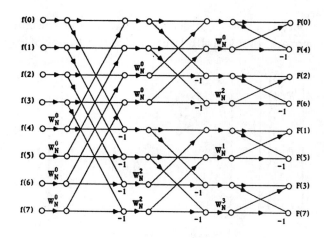

Fig. 2.3.6 FFT for $N = 8$ using time decimation with normal input and bit-reversed output.

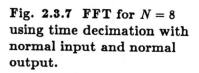

Fig. 2.3.7 FFT for $N = 8$ using time decimation with normal input and normal output.

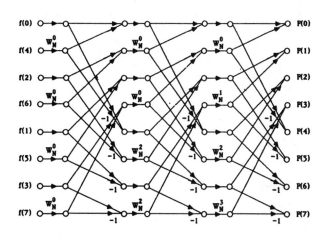

Fig. 2.3.8 FFT for $N = 8$ using time decimation with bit-reversed input, normal output, and identical stages.

Here the dots represent zero values and the numbers represent the powers of W_8. The normal input and bit-reversed output case of Fig. 2.3.6 becomes

$$
\begin{bmatrix} F(0) \\ F(4) \\ F(2) \\ F(6) \\ F(1) \\ F(5) \\ F(3) \\ F(7) \end{bmatrix}
=
\begin{bmatrix}
0 & 0 & \cdot & \cdot & \cdot & \cdot & \cdot & \cdot \\
0 & 4 & \cdot & \cdot & \cdot & \cdot & \cdot & \cdot \\
\cdot & \cdot & 0 & 2 & \cdot & \cdot & \cdot & \cdot \\
\cdot & \cdot & 0 & 6 & \cdot & \cdot & \cdot & \cdot \\
\cdot & \cdot & \cdot & \cdot & 0 & 1 & \cdot & \cdot \\
\cdot & \cdot & \cdot & \cdot & 0 & 5 & \cdot & \cdot \\
\cdot & \cdot & \cdot & \cdot & \cdot & \cdot & 0 & 3 \\
\cdot & \cdot & \cdot & \cdot & \cdot & \cdot & 0 & 7
\end{bmatrix}
\begin{bmatrix}
0 & \cdot & 0 & \cdot & \cdot & \cdot & \cdot & \cdot \\
\cdot & 0 & \cdot & 0 & \cdot & \cdot & \cdot & \cdot \\
0 & \cdot & 4 & \cdot & \cdot & \cdot & \cdot & \cdot \\
\cdot & 0 & \cdot & 4 & \cdot & \cdot & \cdot & \cdot \\
\cdot & \cdot & \cdot & \cdot & 0 & \cdot & 2 & \cdot \\
\cdot & \cdot & \cdot & \cdot & 0 & \cdot & 2 \\
\cdot & \cdot & \cdot & \cdot & 0 & \cdot & 6 & \cdot \\
\cdot & \cdot & \cdot & \cdot & 0 & \cdot & 6
\end{bmatrix}
$$

$$
\times
\begin{bmatrix}
0 & \cdot & \cdot & \cdot & 0 & \cdot & \cdot & \cdot \\
\cdot & 0 & \cdot & \cdot & \cdot & 0 & \cdot & \cdot \\
\cdot & \cdot & 0 & \cdot & \cdot & \cdot & 0 & \cdot \\
\cdot & \cdot & \cdot & 0 & \cdot & \cdot & \cdot & 0 \\
0 & \cdot & \cdot & \cdot & 4 & \cdot & \cdot & \cdot \\
\cdot & 0 & \cdot & \cdot & \cdot & 4 & \cdot & \cdot \\
\cdot & \cdot & 0 & \cdot & \cdot & \cdot & 4 & \cdot \\
\cdot & \cdot & \cdot & 0 & \cdot & \cdot & \cdot & 4
\end{bmatrix}
\begin{bmatrix} f(0) \\ f(1) \\ f(2) \\ f(3) \\ f(4) \\ f(5) \\ f(6) \\ f(7) \end{bmatrix}
$$

$$(2.3.8)$$

The normal input and normal output case of Fig. 2.3.7 is

$$
\begin{bmatrix} F(0) \\ F(1) \\ F(2) \\ F(3) \\ F(4) \\ F(5) \\ F(6) \\ F(7) \end{bmatrix}
=
\begin{bmatrix}
0 & 0 & \cdot & \cdot & \cdot & \cdot & \cdot & \cdot \\
\cdot & \cdot & 0 & 1 & \cdot & \cdot & \cdot & \cdot \\
\cdot & \cdot & \cdot & \cdot & 0 & 2 & \cdot & \cdot \\
\cdot & \cdot & \cdot & \cdot & \cdot & \cdot & 0 & 3 \\
0 & 4 & \cdot & \cdot & \cdot & \cdot & \cdot & \cdot \\
\cdot & \cdot & 0 & 5 & \cdot & \cdot & \cdot & \cdot \\
\cdot & \cdot & \cdot & \cdot & 0 & 6 & \cdot & \cdot \\
\cdot & \cdot & \cdot & \cdot & \cdot & \cdot & 0 & 7
\end{bmatrix}
\begin{bmatrix}
0 & \cdot & 0 & \cdot & \cdot & \cdot & \cdot & \cdot \\
\cdot & 0 & \cdot & 0 & \cdot & \cdot & \cdot & \cdot \\
\cdot & \cdot & \cdot & \cdot & 0 & \cdot & 2 & \cdot \\
\cdot & \cdot & \cdot & \cdot & 0 & \cdot & 2 \\
0 & \cdot & 4 & \cdot & \cdot & \cdot & \cdot & \cdot \\
\cdot & 0 & \cdot & 4 & \cdot & \cdot & \cdot & \cdot \\
\cdot & \cdot & \cdot & \cdot & 0 & \cdot & 6 & \cdot \\
\cdot & \cdot & \cdot & \cdot & 0 & \cdot & 6
\end{bmatrix}
$$

$$
\times
\begin{bmatrix}
0 & \cdot & \cdot & \cdot & 0 & \cdot & \cdot & \cdot \\
\cdot & 0 & \cdot & \cdot & \cdot & 0 & \cdot & \cdot \\
\cdot & \cdot & 0 & \cdot & \cdot & \cdot & 0 & \cdot \\
\cdot & \cdot & \cdot & 0 & \cdot & \cdot & \cdot & 0 \\
0 & \cdot & \cdot & \cdot & 4 & \cdot & \cdot & \cdot \\
\cdot & 0 & \cdot & \cdot & \cdot & 4 & \cdot & \cdot \\
\cdot & \cdot & 0 & \cdot & \cdot & \cdot & 4 & \cdot \\
\cdot & \cdot & \cdot & 0 & \cdot & \cdot & \cdot & 4
\end{bmatrix}
\begin{bmatrix} f(0) \\ f(1) \\ f(2) \\ f(3) \\ f(4) \\ f(5) \\ f(6) \\ f(7) \end{bmatrix}
$$

$$(2.3.9)$$

The bit-reversed input and normal output case of Fig. 2.3.8 having identical stages has

$$
\begin{bmatrix} F(0) \\ F(1) \\ F(2) \\ F(3) \\ F(4) \\ F(5) \\ F(6) \\ F(7) \end{bmatrix}
=
\begin{bmatrix}
0 & 0 & \cdot & \cdot & \cdot & \cdot & \cdot & \cdot \\
\cdot & \cdot & 0 & 1 & \cdot & \cdot & \cdot & \cdot \\
\cdot & \cdot & \cdot & \cdot & 0 & 2 & \cdot & \cdot \\
\cdot & \cdot & \cdot & \cdot & \cdot & \cdot & 0 & 3 \\
0 & 4 & \cdot & \cdot & \cdot & \cdot & \cdot & \cdot \\
\cdot & \cdot & 0 & 5 & \cdot & \cdot & \cdot & \cdot \\
\cdot & \cdot & \cdot & \cdot & 0 & 6 & \cdot & \cdot \\
\cdot & \cdot & \cdot & \cdot & \cdot & \cdot & 0 & 7
\end{bmatrix}
\begin{bmatrix}
0 & 0 & \cdot & \cdot & \cdot & \cdot & \cdot & \cdot \\
\cdot & \cdot & 0 & 0 & \cdot & \cdot & \cdot & \cdot \\
\cdot & \cdot & \cdot & \cdot & 0 & 2 & \cdot & \cdot \\
\cdot & \cdot & \cdot & \cdot & \cdot & \cdot & 0 & 2 \\
0 & 4 & \cdot & \cdot & \cdot & \cdot & \cdot & \cdot \\
\cdot & \cdot & 0 & 4 & \cdot & \cdot & \cdot & \cdot \\
\cdot & \cdot & \cdot & \cdot & 0 & 6 & \cdot & \cdot \\
\cdot & \cdot & \cdot & \cdot & \cdot & \cdot & 0 & 6
\end{bmatrix}
$$

$$
\times
\begin{bmatrix}
0 & 0 & \cdot & \cdot & \cdot & \cdot & \cdot & \cdot \\
\cdot & \cdot & 0 & 0 & \cdot & \cdot & \cdot & \cdot \\
\cdot & \cdot & \cdot & \cdot & 0 & 0 & \cdot & \cdot \\
\cdot & \cdot & \cdot & \cdot & \cdot & \cdot & 0 & 0 \\
0 & 4 & \cdot & \cdot & \cdot & \cdot & \cdot & \cdot \\
\cdot & \cdot & 0 & 4 & \cdot & \cdot & \cdot & \cdot \\
\cdot & \cdot & \cdot & \cdot & 0 & 4 & \cdot & \cdot \\
\cdot & \cdot & \cdot & \cdot & \cdot & \cdot & 0 & 4
\end{bmatrix}
\begin{bmatrix} f(0) \\ f(4) \\ f(2) \\ f(6) \\ f(1) \\ f(5) \\ f(3) \\ f(7) \end{bmatrix}
\tag{2.3.10}
$$

Forming the matrix products from right to left corresponds to performing the FFT flow graph operations from left to right. Each of these factored matrices represents an FFT stage. For example, consider Eq. 2.3.7 and move from right to left across the equation. At the same time, move from left to right across its flow graph description in Fig. 2.3.5. The input data is in bit-reversed sequence. The first FFT stage butterflies adjacent inputs with $[0\ k]$ gains. This corresponds to the $[0\ k]$ block diagram in Fig. 2.3.7. The second FFT stage butterflies every input pair separated by two. This corresponds to the $[0\ .\ k]$ skewed block pattern in Eq. 2.3.7. The last FFT stage butterflies every input separated by four. This corresponds to the $[0\ .\ .\ .\ k]$ skewed pattern in Eq. 2.3.7. Finally the output data is in normal order.

These matrices are 75% filled with zeros which account for the reduced flow graph complexity. Such matrix patterns nicely complement the flow graphs and will be useful when we discuss other transforms.

2.4　FREQUENCY DECIMATION ALGORITHMS

In time decimation FFT algorithms, the input sequence $f(n)$ is broken down into smaller subsequences of even and odd-numbered points. Alternatively, the output sequence $F(m)$ can be broken down into smaller subsequences which is referred to as *frequency decimation*. In this approach, the N samples $f(n)$ are divided in half to form two $(N/2)$-point subsequences. The DFT given by Eq. 2.1.4 can then be rewritten as

$$
F(m) = \sum_{n=0}^{N-1} f(n) W_N^{nm} = \sum_{n=0}^{(N/2)-1} f(n) W_N^{nm} + \sum_{n=N/2}^{N-1} f(n) W_N^{nm}
$$

$$
= \sum_{n=0}^{(N/2)-1} f(n) W_N^{nm} + W_N^{(N/2)m} \sum_{n=0}^{(N/2)-1} f(n + (N/2)) W_N^{nm}
\tag{2.4.1}
$$

for $m = 0, 1, \ldots, N - 1$. Comparing this result with that of Eq. 2.3.2, the summations of Eq. 2.4.1 are not $(N/2)$-point DFT's although they contain $(N/2)$-points. Since $W_N^{N/2} = -1$, Eq. 2.4.1 can be rewritten as

$$F(m) = \sum_{n=0}^{(N/2)-1} \left\{ f(n) + (-1)^m f(n + (N/2)) \right\} W_N^{nm} \qquad (2.4.2)$$

If the N points of $F(m)$ are separated into two $(N/2)$-point subsequences consisting of even-numbered points $F(2m)$ and odd-numbered points $F(2m + 1)$, then Eq. 2.4.2 can be rewritten as

$$F(2m) = F(k)\Big|_{k \text{ even}} = \sum_{n=0}^{(N/2)-1} \left\{ f(n) + f(n + (N/2)) \right\} W_{N/2}^{nm} = \sum_{n=0}^{(N/2)-1} f_1(n) W_{N/2}^{nm}$$

$$F(2m + 1) = F(k)\Big|_{k \text{ odd}} = \sum_{n=0}^{(N/2)-1} W_N^n \left\{ f(n) - f(n + (N/2)) \right\} W_{N/2}^{nm} = \sum_{n=0}^{(N/2)-1} f_2(n) W_{N/2}^{nm}$$

$$(2.4.3)$$

where $m = 0, 1, 2, \ldots, (N/2) - 1$. $F(2m)$ and $F(2m + 1)$ are two $(N/2)$-point DFT subsequences. $F(m)$ is periodic in m with period N while $F(2m)$ and $F(2m + 1)$ are periodic in m and have period $N/2$.

Now the process can be repeated to decimate the two $(N/2)$-point DFT subsequences $F(2m)$ and $F(2m + 1)$ into four $(N/4)$-point DFT subsequences $F(4m)$, $F(4m + 2)$, $F(4m + 1)$, and $F(4m + 3)$ as

$$F(4m) = F(2k)\Big|_{\substack{k \text{ even} \\ k/2 \text{ even}}}$$

$$= \sum_{n=0}^{(N/4)-1} \left\{ f_1(n) + f_1(n + (N/4)) \right\} W_{N/4}^{nm} = \sum_{n=0}^{(N/4)-1} f_3(n) W_{N/4}^{nm}$$

$$F(4m + 2) = F(2k)\Big|_{\substack{k \text{ even} \\ k/2 \text{ odd}}}$$

$$= \sum_{n=0}^{(N/4)-1} W_{N/2}^n \left\{ f_1(n) - f_1(n + (N/4)) \right\} W_{N/4}^{nm} = \sum_{n=0}^{(N/4)-1} f_4(n) W_{N/4}^{nm}$$

$$(2.4.4)$$

$$F(4m + 1) = F(2k + 1)\Big|_{\substack{k \text{ odd} \\ (k+1)/2 \text{ odd}}}$$

$$= \sum_{n=0}^{(N/4)-1} \left\{ f_2(n) + f_2(n + (N/4)) \right\} W_{N/4}^{nm} = \sum_{n=0}^{(N/4)-1} f_5(n) W_{N/4}^{nm}$$

$$F(4m + 3) = F(2k + 1)\Big|_{\substack{k \text{ odd} \\ (k+1)/2 \text{ even}}}$$

$$= \sum_{n=0}^{(N/4)-1} W_{N/2}^n \left\{ f_2(n) - f_2(n + (N/4)) \right\} W_{N/4}^{nm} = \sum_{n=0}^{(N/4)-1} f_6(n) W_{N/4}^{nm}$$

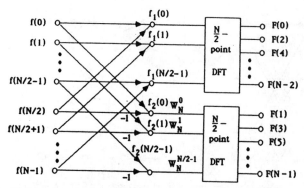

Fig. 2.4.1 Frequency decimation of one N-point DFT into two $(N/2)$-point DFT's.

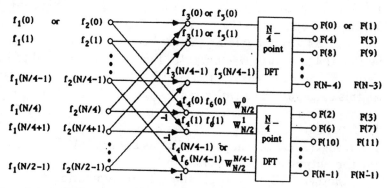

Fig. 2.4.2 Frequency decimation of one $(N/2)$-point DFT into two $(N/4)$-point DFT's.

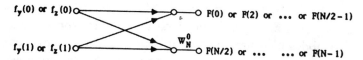

Fig. 2.4.3 Frequency decimation of one 2-point DFT into two 1-point DFT's.

where $m = 0, 1, 2, \ldots, (N/4) - 1$. $F(4m)$ through $F(4m + 3)$ are periodic in m with period $N/4$.

The decimation process continues until it terminates in a single term. In the last step, 2^N one-point DFT sequences are obtained as

$$F(0), \; F(N/2), \; F(N/4), \; F(3N/4), \; \ldots, \; F((N/2) - 1), \; F(N - 1) \qquad (2.4.5)$$

These equations are easily expressed in flow graph form. Eqs. 2.4.1–2.4.3 are shown in Fig. 2.4.1. The N inputs $f(n)$ and $f(n + N/2)$ are summed to form $N/2$ outputs

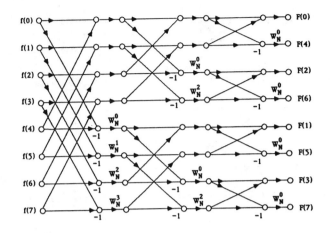

Fig. 2.4.4 FFT for $N = 8$ using frequency decimation with normal input and bit-reversed output.

$f_1(n)$ and differenced to form $N/2$ outputs $f_2(n)$. Weighting $f_2(n)$ by W_N^n, the N total outputs are applied to the next DFT stage as shown in Fig. 2.4.1.

Fig. 2.4.2 represents Eq. 2.4.4. Each of the $N/2$ outputs of Fig. 2.4.1 is processed by Fig. 2.4.2 to form $N/4$ outputs. Fig. 2.4.2 is identical to Fig. 2.4.1 but only half the size. This process repeats itself until only two outputs are left as shown in Fig. 2.4.3. At this stage, all the $F(m)$ given by Eq. 2.4.5 have been generated.

Applying these results to the 8-point DFT given by Eq. 2.2.3, we combine Figs. 2.4.1–2.4.3 to obtain Fig. 2.4.4. As before, only $(N/2)\log_2(N) = 12$ complex multiplies and $N\log_2(N) = 24$ complex additions are needed. The input is in the normal sequence but the output is in bit-reversed order.

Another FFT form is obtained by manipulating Fig. 2.4.4 so that the input is in bit-reversed order but the output is in normal order as shown in Fig. 2.4.5. Rearranging Fig. 2.4.5 so that both inputs and outputs are in normal order results in Fig. 2.4.6. The unusual symmetry in the input stage of Fig. 2.4.6 can be replicated throughout the graph as shown in Fig. 2.4.7.

Every frequency-decimation flow graph can be transformed to an equivalent time-decimation flow graph and vice versa. Transposed graphs are obtained by converting sink nodes to source nodes (or source nodes to sink nodes) and reversing branch directions. Branch gains remain unchanged. Comparing flow graphs shows that Figs. (2.3.5, 2.4.4), (2.3.6, 2.4.5), (2.3.7, 2.4.6), and (2.3.8, 2.4.7) are transpose pairs. In matrix notation, this corresponds to reversing matrix order. Fig. 2.4.4 has (see Eq. 2.3.7)

$$
\begin{bmatrix} F(0) \\ F(4) \\ F(2) \\ F(6) \\ F(1) \\ F(5) \\ F(3) \\ F(7) \end{bmatrix} =
\begin{bmatrix}
0 & 0 & \cdot & \cdot & \cdot & \cdot & \cdot & \cdot \\
0 & 4 & \cdot & \cdot & \cdot & \cdot & \cdot & \cdot \\
\cdot & \cdot & 0 & 0 & \cdot & \cdot & \cdot & \cdot \\
\cdot & \cdot & 0 & 4 & \cdot & \cdot & \cdot & \cdot \\
\cdot & \cdot & \cdot & \cdot & 0 & 0 & \cdot & \cdot \\
\cdot & \cdot & \cdot & \cdot & 0 & 4 & \cdot & \cdot \\
\cdot & \cdot & \cdot & \cdot & \cdot & \cdot & 0 & 0 \\
\cdot & \cdot & \cdot & \cdot & \cdot & \cdot & 0 & 4
\end{bmatrix}
\begin{bmatrix}
0 & \cdot & 0 & \cdot & \cdot & \cdot & \cdot & \cdot \\
\cdot & 0 & \cdot & 0 & \cdot & \cdot & \cdot & \cdot \\
0 & \cdot & 4 & \cdot & \cdot & \cdot & \cdot & \cdot \\
\cdot & 2 & \cdot & 6 & \cdot & \cdot & \cdot & \cdot \\
\cdot & \cdot & \cdot & \cdot & 0 & \cdot & 0 & \cdot \\
\cdot & \cdot & \cdot & \cdot & \cdot & 0 & \cdot & 0 \\
\cdot & \cdot & \cdot & \cdot & 0 & \cdot & 4 & \cdot \\
\cdot & \cdot & \cdot & \cdot & \cdot & 2 & \cdot & 6
\end{bmatrix} \times
$$

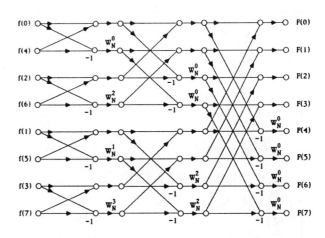

Fig. 2.4.5 FFT for $N = 8$ using frequency decimation with bit-reversed input and normal output.

Fig. 2.4.6 FFT for $N = 8$ using frequency decimation with normal input and normal output.

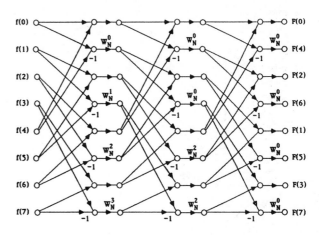

Fig. 2.4.7 FFT for $N = 8$ using frequency decimation with bit-reversed input, normal output, and identical stages.

$$\times \begin{bmatrix} 0 & \cdot & \cdot & \cdot & 0 & \cdot & \cdot & \cdot \\ \cdot & 0 & \cdot & \cdot & \cdot & 0 & \cdot & \cdot \\ \cdot & \cdot & 0 & \cdot & \cdot & \cdot & 0 & \cdot \\ \cdot & \cdot & \cdot & 0 & \cdot & \cdot & \cdot & 0 \\ 0 & \cdot & \cdot & \cdot & 4 & \cdot & \cdot & \cdot \\ \cdot & 1 & \cdot & \cdot & \cdot & 5 & \cdot & \cdot \\ \cdot & \cdot & 2 & \cdot & \cdot & \cdot & 6 & \cdot \\ \cdot & \cdot & \cdot & 3 & \cdot & \cdot & \cdot & 7 \end{bmatrix} \begin{bmatrix} f(0) \\ f(1) \\ f(2) \\ f(3) \\ f(4) \\ f(5) \\ f(6) \\ f(7) \end{bmatrix}$$

(2.4.6)

Fig. 2.4.5 has (see Eq. 2.3.8)

$$\begin{bmatrix} F(0) \\ F(1) \\ F(2) \\ F(3) \\ F(4) \\ F(5) \\ F(6) \\ F(7) \end{bmatrix} = \begin{bmatrix} 0 & \cdot & \cdot & \cdot & 0 & \cdot & \cdot & \cdot \\ \cdot & 0 & \cdot & \cdot & \cdot & 0 & \cdot & \cdot \\ \cdot & \cdot & 0 & \cdot & \cdot & \cdot & 0 & \cdot \\ \cdot & \cdot & \cdot & 0 & \cdot & \cdot & \cdot & 0 \\ 0 & \cdot & \cdot & \cdot & 4 & \cdot & \cdot & \cdot \\ \cdot & 0 & \cdot & \cdot & \cdot & 4 & \cdot & \cdot \\ \cdot & \cdot & 0 & \cdot & \cdot & \cdot & 4 & \cdot \\ \cdot & \cdot & \cdot & 0 & \cdot & \cdot & \cdot & 4 \end{bmatrix} \begin{bmatrix} 0 & \cdot & 0 & \cdot & \cdot & \cdot & \cdot & \cdot \\ \cdot & 0 & \cdot & 0 & \cdot & \cdot & \cdot & \cdot \\ 0 & \cdot & 4 & \cdot & \cdot & \cdot & \cdot & \cdot \\ \cdot & 0 & \cdot & 4 & \cdot & \cdot & \cdot & \cdot \\ \cdot & \cdot & \cdot & \cdot & 0 & \cdot & 0 & \cdot \\ \cdot & \cdot & \cdot & \cdot & \cdot & 0 & \cdot & 0 \\ \cdot & \cdot & \cdot & \cdot & 2 & \cdot & 6 & \cdot \\ \cdot & \cdot & \cdot & \cdot & \cdot & 2 & \cdot & 6 \end{bmatrix}$$

$$\times \begin{bmatrix} 0 & 0 & \cdot & \cdot & \cdot & \cdot & \cdot & \cdot \\ 0 & 4 & \cdot & \cdot & \cdot & \cdot & \cdot & \cdot \\ \cdot & \cdot & 0 & 0 & \cdot & \cdot & \cdot & \cdot \\ \cdot & \cdot & 2 & 6 & \cdot & \cdot & \cdot & \cdot \\ \cdot & \cdot & \cdot & \cdot & 0 & 0 & \cdot & \cdot \\ \cdot & \cdot & \cdot & \cdot & 1 & 5 & \cdot & \cdot \\ \cdot & \cdot & \cdot & \cdot & \cdot & \cdot & 0 & 0 \\ \cdot & \cdot & \cdot & \cdot & \cdot & \cdot & 3 & 7 \end{bmatrix} \begin{bmatrix} f(0) \\ f(4) \\ f(2) \\ f(6) \\ f(1) \\ f(5) \\ f(3) \\ f(7) \end{bmatrix}$$

(2.4.7)

Fig. 2.4.6 has (see Eq. 2.3.9)

$$\begin{bmatrix} F(0) \\ F(1) \\ F(2) \\ F(3) \\ F(4) \\ F(5) \\ F(6) \\ F(7) \end{bmatrix} = \begin{bmatrix} 0 & \cdot & \cdot & \cdot & 0 & \cdot & \cdot & \cdot \\ \cdot & 0 & \cdot & \cdot & \cdot & 0 & \cdot & \cdot \\ \cdot & \cdot & 0 & \cdot & \cdot & \cdot & 0 & \cdot \\ \cdot & \cdot & \cdot & 0 & \cdot & \cdot & \cdot & 0 \\ 0 & \cdot & \cdot & \cdot & 4 & \cdot & \cdot & \cdot \\ \cdot & 0 & \cdot & \cdot & \cdot & 4 & \cdot & \cdot \\ \cdot & \cdot & 0 & \cdot & \cdot & \cdot & 4 & \cdot \\ \cdot & \cdot & \cdot & 0 & \cdot & \cdot & \cdot & 4 \end{bmatrix} \begin{bmatrix} 0 & \cdot & \cdot & \cdot & 0 & \cdot & \cdot & \cdot \\ \cdot & 0 & \cdot & \cdot & \cdot & 0 & \cdot & \cdot \\ 0 & \cdot & \cdot & \cdot & 4 & \cdot & \cdot & \cdot \\ \cdot & 0 & \cdot & \cdot & \cdot & 4 & \cdot & \cdot \\ \cdot & \cdot & 0 & \cdot & \cdot & \cdot & 0 & \cdot \\ \cdot & \cdot & \cdot & 0 & \cdot & \cdot & \cdot & 0 \\ \cdot & \cdot & 2 & \cdot & \cdot & \cdot & 6 & \cdot \\ \cdot & \cdot & \cdot & 2 & \cdot & \cdot & \cdot & 6 \end{bmatrix}$$

$$\times \begin{bmatrix} 0 & \cdot & \cdot & \cdot & 0 & \cdot & \cdot & \cdot \\ 0 & \cdot & \cdot & \cdot & 4 & \cdot & \cdot & \cdot \\ \cdot & 0 & \cdot & \cdot & \cdot & 0 & \cdot & \cdot \\ \cdot & 1 & \cdot & \cdot & \cdot & 5 & \cdot & \cdot \\ \cdot & \cdot & 0 & \cdot & \cdot & \cdot & 0 & \cdot \\ \cdot & \cdot & 2 & \cdot & \cdot & \cdot & 6 & \cdot \\ \cdot & \cdot & \cdot & 0 & \cdot & \cdot & \cdot & 0 \\ \cdot & \cdot & \cdot & 3 & \cdot & \cdot & \cdot & 7 \end{bmatrix} \begin{bmatrix} f(0) \\ f(1) \\ f(2) \\ f(3) \\ f(4) \\ f(5) \\ f(6) \\ f(7) \end{bmatrix}$$

(2.4.8)

and Fig. 2.4.7 has (see Eq. 2.3.10)

$$
\begin{bmatrix} F(0) \\ F(4) \\ F(2) \\ F(6) \\ F(1) \\ F(5) \\ F(3) \\ F(7) \end{bmatrix}
=
\begin{bmatrix}
0 & \cdot & \cdot & \cdot & 0 & \cdot & \cdot & \cdot \\
0 & \cdot & \cdot & \cdot & 4 & \cdot & \cdot & \cdot \\
\cdot & 0 & \cdot & \cdot & \cdot & 0 & \cdot & \cdot \\
\cdot & 0 & \cdot & \cdot & \cdot & 4 & \cdot & \cdot \\
\cdot & \cdot & 0 & \cdot & \cdot & \cdot & 0 & \cdot \\
\cdot & \cdot & 0 & \cdot & \cdot & \cdot & 4 & \cdot \\
\cdot & \cdot & \cdot & 0 & \cdot & \cdot & \cdot & 0 \\
\cdot & \cdot & \cdot & 0 & \cdot & \cdot & \cdot & 4
\end{bmatrix}
\begin{bmatrix}
0 & \cdot & \cdot & \cdot & 0 & \cdot & \cdot & \cdot \\
0 & \cdot & \cdot & \cdot & 4 & \cdot & \cdot & \cdot \\
\cdot & 0 & \cdot & \cdot & \cdot & 0 & \cdot & \cdot \\
\cdot & 0 & \cdot & \cdot & \cdot & 4 & \cdot & \cdot \\
\cdot & \cdot & 0 & \cdot & \cdot & \cdot & 0 & \cdot \\
\cdot & \cdot & 2 & \cdot & \cdot & \cdot & 6 & \cdot \\
\cdot & \cdot & \cdot & 0 & \cdot & \cdot & \cdot & 0 \\
\cdot & \cdot & \cdot & 2 & \cdot & \cdot & \cdot & 6
\end{bmatrix}
$$

$$
\times
\begin{bmatrix}
0 & \cdot & \cdot & \cdot & 0 & \cdot & \cdot & \cdot \\
0 & \cdot & \cdot & \cdot & 4 & \cdot & \cdot & \cdot \\
\cdot & 0 & \cdot & \cdot & \cdot & 0 & \cdot & \cdot \\
\cdot & 1 & \cdot & \cdot & \cdot & 5 & \cdot & \cdot \\
\cdot & \cdot & 0 & \cdot & \cdot & \cdot & 0 & \cdot \\
\cdot & \cdot & 2 & \cdot & \cdot & \cdot & 6 & \cdot \\
\cdot & \cdot & \cdot & 0 & \cdot & \cdot & \cdot & 0 \\
\cdot & \cdot & \cdot & 3 & \cdot & \cdot & \cdot & 7
\end{bmatrix}
\begin{bmatrix} f(0) \\ f(1) \\ f(2) \\ f(3) \\ f(4) \\ f(5) \\ f(6) \\ f(7) \end{bmatrix}
$$

$$(2.4.9)$$

When writing these equations by inspection of the figures, the interstage gains W_N^k are included in the succeeding stage.

The FFT flow graphs of these last two sections form the basis for writing FFT computer programs. Table 2.4.1 shows the FFT program for the flow graph of time decimation with bit-reversed input and normal output in Fig. 2.3.4. Table 2.4.2 shows the FFT program using the flow graph for frequency decimation with normal input and bit-reversed output in Fig. 2.4.4. These programs can perform up to 1024-point FFT's. The actual point size is $N = 2^M$ where M is entered in the subroutine calls.

Under close scrutiny, it becomes clear that the FFT program of Table 2.4.1 describes the flow graph of Fig. 2.3.4. Statements 1 to 2 initializes important constants. Statements 3 to 7 shuffle the normal order data into bit-reversed order. The remaining program performs the actual FFT computation using three nested DO loops. The outer loop (8–20) keeps track of the FFT stage being performed. The middle loop (10–20) keeps track of which butterfly within the stage is being computed. The inner loop (11–15) calculates the power of W_N required by each butterfly as it is computed within the stage. The FFT program in Table 2.4.2 can be similarly analyzed. It is remarkable that such a powerful analysis tool can be coded in such a concise, short, and simple form.

2.5 WINDOWING EFFECT

The DFT produces two spectral distortions in the FT approximation. One is the so-called *windowing effect* which is due to spectral spreading caused by signal truncation. The second is frequency aliasing which is caused by discrete time sampling.

Table 2.4.1 FFT for $N = 1024$ using time decimation in Fig. 2.3.4.

```
            SUBROUTINE FFT(F,N)
            COMPLEX F(1024),U,W,T
      1     M=ALOG2(N)
            NV2=N/2
            NM1=N-1
            PI=3.14159265358979
      2     J=1
      3     DO 7 I=1,NM1
               T=F(J)
               F(J)=F(I)
               F(I)=T
      5        K=NV2
      6        IF (K.GE.J) GO TO 7
               J=J-K
               K=K/2
               GO TO 6
      7        J=J+K
      8     DO 20 L=1,M
               LE=2**L
               LE1=LE/2
               U=(1.0,0.0)
      9        W=CMPLX(COS(PI/FLOAT(LE1)),SIN(PI/FLOAT(LE1)))
     10        DO 20 J=1,LE1
     11           DO 15 I=J,N,LE
                     IP=I+LE
                     T=F(IP)*U
                     F(IP)=F(I)-T
     15              F(I)=F(I)+T
     20           U=U*W
            RETURN
            END
```

Table 2.4.2 FFT for $N = 1024$ using frequency decimation in Fig. 2.4.4.

```
            SUBROUTINE FFT(F,N)
            COMPLEX F(1024),U,W,T
      1     M=ALOG2(N)
      2     PI=3.14159265358979
      3     DO 20 L=1,M
               LE=2**(M+1-L)
               LE1=LE/2
               U=(1.0,0.0)
               W=CMPLX(COS(PI/FLOAT(LE1)),-SIN(PI/FLOAT(LE1)))
               DO 20 J=1,LE1
                  DO 10 I=J,N,LE
                     IP=I+LE1
                     T=F(I)+F(IP)
                     F(IP)=(F(I)-F(IP))*U
     10              F(I)=T
     20           U=U*W
            NV2=N/2
            NM1=N-1
            J=1
            DO 30 I=1,NM1
               IF (I.GE.J) GO TO 25
               T=F(J)
               F(J)=F(I)
               F(I)=T
     25        K=NV2
     26        IF (K.GE.J) GO TO 30
               J=J-K
               K=K/2
               GO TO 26
     30        J=J+K
            RETURN
            END
```

To analyze these effects, consider Fig. 2.5.1. The input signal $f(t)$ is applied to one multiplier input. The other multiplier input is driven by a window function $w(t)$ which is zero everywhere except for $-NT/2 \leq t < NT/2$. The window function *gates on* and continuously samples the signal $f(t)$ for NT seconds. This gated signal is applied to one input of a second multiplier whose other input is driven by an impulse stream $s(t)$ of frequency $1/T$ Hz. The first multiplier truncates the signal $f(t)$ while the second multiplier converts this $f(t)$ into its impulsively sampled equivalent $f_w(t)$. Mathematically,

$$f_w(t) = f(t)w(t)s(t)$$

$$F_w(j\omega) = \frac{1}{(2\pi)^2} F(j\omega) * W(j\omega) * S(j\omega) \qquad (2.5.1)$$

Let us derive the spectral characteristics of $f_w(t)$. The spectrum of $f(t)$ equals $F(j\omega)$. Assume for simplicity that the window function $w(t)$ is a rectangular pulse centered at $t = 0$ with width NT. Denoting its spectrum as $W(j\omega)$ then

$$w(t) = U_{-1}(t + NT/2) - U_{-1}(t - NT/2)$$

$$W(j\omega) = NT \, \frac{\sin(\omega NT/2)}{\omega NT/2} \qquad (2.5.2)$$

(see Eqs. 1.6.19 and 1.6.22). The sampling function $s(t)$ is an impulse stream described by Eq. 1.3.1 whose spectrum is given by Eq. 1.3.3 so that

$$s(t) = \sum_{n=-\infty}^{\infty} U_0(t - nT)$$

$$S(j\omega) = \omega_s \sum_{n=-\infty}^{\infty} U_0(\omega - n\omega_s), \qquad \omega_s = \frac{2\pi}{T} \qquad (2.5.3)$$

Combining Eqs. 2.5.1–2.5.3 gives the windowed output $f_w(t)$ and its spectrum $F_w(j\omega)$,

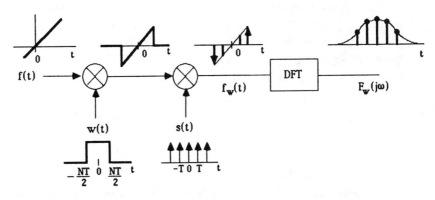

Fig. 2.5.1 Discrete time system explicitly showing windowing, sampling, and discrete transform operations.

$$f_w(t) = \sum_{n=-\infty}^{\infty} f(nT)w(nT)U_0(t - nT)$$

$$F_w(j\omega) = \frac{1}{(2\pi)^2}F(j\omega) * NT\frac{\sin(\omega NT/2)}{\omega NT/2} * \frac{2\pi}{T}\sum_{n=-\infty}^{\infty}U_0(\omega - n\omega_s)$$

(2.5.4)

To obtain a specific result from Eq. 2.5.4, a dc signal $f(t) = 1$ is input to the sampling system in Fig. 2.5.1. From Table 1.6.1, its spectrum $F(j\omega) = 2\pi U_0(\omega)$ is an impulse at $\omega = 0$. Substituting $F(j\omega)$ into Eq. 2.5.4 and manipulating the result gives

$$F_w(j\omega) = NT\frac{\sin(\omega NT/2)}{\omega NT/2} * \frac{1}{T}\sum_{n=-\infty}^{\infty}U_0(\omega - n\omega_s) = N\sum_{n=-\infty}^{\infty}\frac{\sin[(NT/2)(\omega - n\omega_s)]}{(NT/2)(\omega - n\omega_s)}$$

(2.5.5)

This infinite series is not very useful. A closed form equivalent can be obtained by observing that the sampled dc signal $f_w(nT)$ consists of N impulses located at $t = (-(N-1)T/2, \ldots, -T, 0, T, \ldots, (N-1)T/2)$ where N is *odd*. Taking the FT of $f_w(t)$ gives $F_w(j\omega)$ as

$$f_w(t) = \sum_{n=-(N-1)/2}^{(N-1)/2} U_0(t - nT), \qquad N \text{ odd}$$

$$F_w(j\omega) = \sum_{n=-(N-1)/2}^{(N-1)/2} e^{-jn\omega T} = \sum_{n=-(N-1)/2}^{\infty} e^{-jn\omega T} - \sum_{n=(N+1)/2}^{\infty} e^{-jn\omega T}$$

$$= \left[e^{j\omega(N-1)T/2} - e^{-j\omega(N+1)T/2}\right]\sum_{n=0}^{\infty}e^{-jn\omega T} = e^{-j\omega T/2}\frac{e^{j\omega NT/2} - e^{-j\omega NT/2}}{1 - e^{-j\omega T}}$$

$$= \frac{\sin(\omega NT/2)}{\sin(\omega T/2)}$$

(2.5.6)

Eq. 2.5.6 describes the case for an odd number of samples because the sampling function $s(t)$ is synchronized to $t = 0$. An *even* number of samples result when $s(t)$ is shifted a half-clock interval around $t = \pm T/2$. In this case, the sampling function consists of impulses located at $t = (-(N-1)T/2, \ldots, -T/2, T/2, \ldots, (N-1)T/2)$. Then $f_w(t)$ and $F_w(j\omega)$ equal

$$f_w(t) = \sum_{n=0}^{(N/2)-1}\left(U_0[t - (n+1/2)T] + U_0[t + (n+1/2)T]\right), \qquad N \text{ even}$$

$$F_w(j\omega) = \frac{\sin(\omega NT/2)}{\sin(\omega T/2)}$$

(2.5.7)

By the time-shifting theorem of Table 1.6.2, the magnitude spectrums $|F_w(j\omega)|$ are always given by Eqs. 2.5.6b or 2.5.7b regardless of the particular synchronizing used for gathering the N equally-spaced samples of $f(t)$ inside the NT second time window.

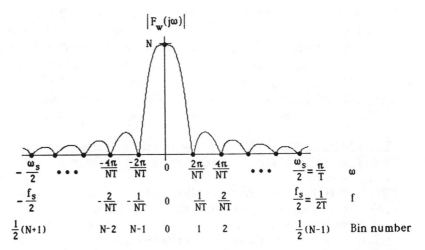

Fig. 2.5.2 **DFT spectrum of dc signal using N samples, sampling frequency f_s, and sampling interval T.**

$|F_w(j\omega)|$ is plotted in Fig. 2.5.2. It is periodic in ω with a period equal to the sampling frequency ($\omega_s = 2\pi/T$) so only a single cycle need be shown. Since $\sin x \cong x$ for $|x| \ll 1$, then $F_w(0) = N$ at dc. The spectrum $F_w(j\omega) = 0$ at the FFT bin frequencies (integer multiples of $\omega_s/N = 2\pi/NT$). The main lobe is two bins wide and the peak side-lobe levels are down -13 dB.

Now consider the distortions produced by windowing and sampling. With continuous sampling over infinite time, the FT correctly establishes that the spectrum of a unit dc signal equals $2\pi U_0(\omega)$. However, by analyzing only T seconds of signal, Eq. 2.5.2 concludes the spectrum has a $(\sin x/x)$ nature. Thus the zero-bandwidth spectrum spreads. Further by impulsively sampling within the window, Eq. 2.5.7 concludes that the spectrum has a $(\sin Nx/\sin x)$ nature. It is the sampling and unavoidable frequency aliasing that converts the $(\sin x/x)$ spectrum into the $(\sin Nx/\sin x)$ spectrum. From an FFT viewpoint, the algorithms yield N spectral components located at the FFT bin frequencies $m\omega_s/N = m2\pi/NT$. These frequencies coincide with the zero-crossing frequencies in Fig. 2.5.2. Therefore the FFT gives $F(0) = N$ and $F(m) = 0$ for $m = 1, 2, \ldots, N-1$ which is the correct spectrum based on apriori knowledge.

Now consider a complex exponential input signal

$$f(t) = e^{j\omega_o t} = \cos(\omega_o t) + j\sin(\omega_o t)$$
$$F(j\omega) = 2\pi U_0(\omega - \omega_o) \tag{2.5.8}$$

It is complex-valued and has only a single spectral component at $\omega = \omega_o$. Repeating the same analysis that was performed on the dc signal in Eqs. 2.5.5–2.5.7, the spectrum of this sampled signal equals

$$F_w(j\omega) = \frac{\sin[(\omega - \omega_o)NT/2]}{\sin[(\omega - \omega_o)T/2]} \tag{2.5.9}$$

$F_w(j\omega)$ is the same spectrum as that shown in Fig. 2.5.2 but the ω-axis must be relabelled as the $(\omega - \omega_o)$-axis. The FFT computes spectral components at frequencies $\omega = m\omega_s/N$ for integer bin values in the range $m = [0, N - 1]$. If the signal frequency ω_o (or signal frequency bin $m_o = N\omega_o/\omega_s = Nf_oT$) falls on an FFT frequency bin m then Eq. 2.5.9 correctly estimates the spectrum of Eq. 2.5.8. Otherwise the spectrum is in error based upon our apriori knowledge of $f(t)$. Maximum error is obtained when ω_o lies halfway between FFT frequency bins as $\omega_o = (m + 0.5)\omega_s/N$.

EXAMPLE 2.5.1   Explain the effects of performing an FFT on the complex-valued signal $\exp(j2\pi f_o t)$ given by Eq. 2.5.8 where f_o is arbitrary. Use a constant width window.

SOLUTION   The effects are illustrated in Fig. 2.5.3. Keeping the window size NT constant fixes the FFT bin size at $1/NT$ Hz. Varying the signal frequency f_o recenters the $(\sin Nx/\sin x)$ characteristic of Fig. 2.5.2 at bin $m_o = Nf_oT$. Only when m_o equals an integer is the characteristic centered on an FFT bin.

Regardless of where the spectral peak of the signal occurs relative to an FFT bin, the FFT always outputs spectra at the bin frequencies. Fig. 2.5.3a shows the case when $m_o = 0$ (or any other integer). All the FFT outputs are zero except at $m = 0$ (or that other integer). Fig. 2.5.3b shows the case when $m_o = 0.25$ (or 25% away from any bin). Now all the FFT outputs are nonzero. Fig. 2.5.3c shows the worst-case situation when $m_o = 0.5$ (or halfway between any two bins). The FFT outputs are largest at the two bins on either side of m_o and decrease to a constant level (see Fig. 2.6.1). ■

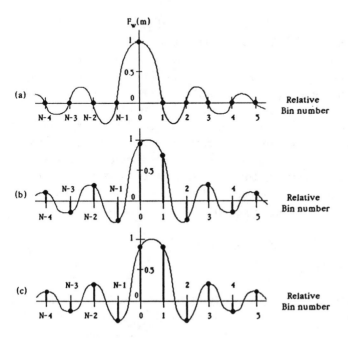

Fig. 2.5.3 FFT spectrum of windowed signal having frequency (a) centered on FFT bin, (b) 25% away from FFT bin, and (c) halfway between FFT bins in Example 2.5.1.

EXAMPLE 2.5.2 Plot the FFT coefficients for a signal $\exp(j2\pi f_o t)$ whose frequency $f_o = 1$ Hz. Sample at a 100 Hz rate. Use sample sizes of 64, 128, 256, 512, and 1024 points.

SOLUTION The signal $f(t)$ is input to the FFT block in Fig. 2.5.1 which outputs the spectra shown in Fig. 2.5.4. Several important points should be observed from these results. First, the FFT frequency bins are separated by $f_s/N = 100$ Hz$/N$. The separation should be small compared with f_o for good spectral resolution, so select $N \gg 100$ Hz$/1$ Hz $= 100$. Therefore good results can only be expected for $N = 1024$ and perhaps 512 points. Using only 64 or 128 points will be totally inadequate. Second, only when $m_o = Nf_o/f_w$ equals an integer will the input frequency coincide with an FFT frequency bin and a perfect spectral line estimate at bin m_o be obtained. The closer the fractional part of m_o is to 0.5, the worse will be the estimate. In this case, $m_o = N/100$ equals (0.64, 1.28, 2.56, 5.12, 10.24). The signal frequency m_o is close to an FFT bin frequency (good estimate) when $N = (128, 512, 1024)$ but almost halfway between FFT bin frequencies (poor estimate) when $N = (64, 256)$. Combining these spectral resolutions and integer cycle sampling requirements, $N = 512$ and 1024 should give reasonably good spectral estimates but the others will give poor estimates as verified by Fig. 2.5.4. ∎

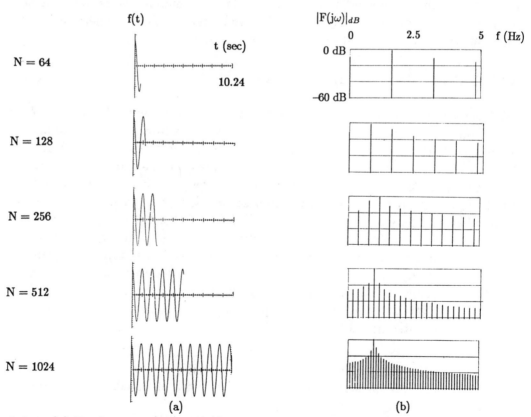

Fig. 2.5.4 (a) Real part of signal $f(t)$ and its (b) **FFT** magnitude spectrum $|F(j\omega)|$ of Example 2.5.2.

Table 2.5.1 FFT problems and remedies.

Distortion due to Spectral Aliasing
  1. Increase sampling frequency $\omega_s = 1/T$. Increases folding frequency (see Chap. 1.3).
  2. Use lowpass filter before sampling. Increases asymptotic roll-off which reduces stop-band spectrum of signal (see Ref. 18, Chap. 4.)

Distortion due to Spectral Leakage
  1. Use window functions with low side-lobes (see Chap. 2.6). Reduces stop-band spectrum of signal.
  2. Increase window width NT. (Increase sampling rate $1/T$ and/or number of samples N.)

Poor Spectral Resolution (Spectral Scalloping or Picket Fence Effect)
  1. Increase number of samples N while holding sampling rate $1/T$ fixed. Increases window width NT and places FFT bins closer together.
  2. Use zero padding. (For given number of samples, add zeros – see Chap. 3.11). Equivalent to extending window width in 1.

These examples show that FFT's can produce poor spectral estimates. They can usually be improved by adjusting the sample rate, number of samples, windowing, and/or prefiltering. Table 2.5.1 summarizes some typical FFT problems and solutions. Windowing will be discussed in the next two sections.

2.6 TIME DOMAIN WINDOWING

We have seen that *windowing* the signal and sampling it for only a finite time NT rather than infinite time produces undesirable spectral spreading. This spreading is called *spectral leakage*. Spectral leakage can be greatly reduced by using window functions.

Window functions smooth the signal at each edge of the sampling window to make the signal and its first- and higher-order derivatives more continuous. This forces the resulting spectrum to be more band-limited which in turn reduces spectral leakage. A number of different time windows are listed in Table 2.6.1 and are shown in Fig. 2.6.1 using $N = 50$. The rectangular window was discussed in the last section. Other well-known and often-used windows include triangular (Bartlett), Hann (raised-cosine or cosine-squared), and Hamming. Many other less well-known but useful windows are also listed. Some of their important frequency domain properties were tabulated and discussed by Harris.[5] They generally depend on N unless N is sufficiently large. These properties are listed in Table 2.6.2 and include:

  1. *3 dB bandwidth:* Frequency where $|W(j\omega)|$ is reduced 3 dB from its dc value $W(0)$.
  2. *Asymptotic roll-off:* High-frequency slope of $|W(j\omega)|$ envelope. If $|W(j\omega)|_{max} \cong 1/\omega^n$ at high frequencies, then the asymptotic roll-off equals $-20n$ dB/dec or $-6n$ dB/oct.
  3. *Peak side-lobe level:* Maximum level attained by $|W(j\omega)|$ in its side-lobes relative to $W(0)$.

4. *6 dB bandwidth:* Frequency where $|W(j\omega)|$ is reduced 6 dB from its dc value $W(0)$.
5. *Equivalent noise bandwidth:* Bandwidth of an ideal lowpass filter (having $|H(j\omega)| = 1, |\omega| < B$ and zero elsewhere) which passes the same energy as the window function.
6. *Coherent signal gain:* DC gain $W(0)$ of the window.
7. *Processing gain:* Ratio of output SNR_o to the input SNR_i.
8. *Scalloping loss:* Ratio of half-bin gain $|W(j\omega_s/2N)|$ to dc gain $W(0)$.
9. *Worst-case processing gain:* Processing gain (in dB) minus scalloping loss squared (in dB).
10. *Overlap correlation:* Autocorrelation of two successive FFT transforms with fractional overlap.

2.6.1 PERFORMANCE PARAMETERS

These properties will now be investigated in detail using the rectangular window as an example. Assume that the window width is NT seconds where T is the sampling interval and N is the number of samples. It will be seen that it is generally simpler to compute the parameters of the continuous time window rather than its sampled version. At sufficiently high sampling rates, the parameters are virtually identical.

The 3 dB bandwidth of a window is a measure of its ability to prevent spectral spreading. Essentially the window imposes an effective nonzero bandwidth on a spectral line. The rectangular window has the sampled spectrum (Eqs. 2.5.6 or 2.5.7)

$$W_s(j\omega) = \frac{\sin(\omega NT/2)}{\sin(\omega T/2)} \tag{2.6.1}$$

The 3 dB bandwidth is found by setting $|W_s(jB)| = W_s(0)/\sqrt{2}$ and solving for B. Since $W_s(0) = N$, this reduces to solving $\sqrt{2}\sin Nx = N\sin x \cong Nx$ ($|x| \ll 1$) for Nx where $x = BT/2$. Solving iteratively yields $Nx = \pm 1.397$. The 3 dB bandwidth of the rectangular window equals

$$B_{3dB} = \frac{2x}{T} = \frac{2.78}{NT} \text{ rad/sec} = \frac{0.443}{NT} \text{ Hz} = 0.443 \text{ bins} \tag{2.6.2}$$

The effective 3 dB bandwidth of a spectral line processed by this window is $2(0.443) = 0.89$ bins (see Fig. 2.5.2 and Table 2.6.2).

The 6 dB bandwidth is a measure of the minimum resolution capability of a window. Two equal-strength signals must be separated in frequency by more than twice the 6 dB window bandwidth if their two spectral peaks are to be resolved. The 6 dB bandwidth is found by setting $|W_s(jB)| = W_s(0)/2$ and solving for B. Solving $2\sin Nx = N\sin x \cong Nx$ iteratively yields $Nx = \pm 1.895$. The 6 dB bandwidth of the rectangular window equals

$$B_{6dB} = \frac{2x}{T} = \frac{3.79}{NT} \text{ rad/sec} = \frac{0.604}{NT} \text{ Hz} = 0.604 \text{ bins} \tag{2.6.3}$$

The effective 6 dB bandwidth of a spectral line processed by this window is $2(0.604) = 1.20$ bins (see Fig. 2.5.2 and Table 2.6.2).

Table 2.6.1 Various window functions $(-N/2 \leq n < N/2)$.[5]

Name	Time Function $w(n)$	Comments
Rectangular	1	Constant weighting [6]
Triangular	$1 - 2|n|/N$	Linear weight (Bartlett, Fejer) [7]
Cosine^p	$\cos^p(n\pi/N)$	pth power of cosine pulse
Cosine, $p = 1$	$\cos(n\pi/N)$	Half-cosine pulse
Hann, $p = 2$	$\cos^2(n\pi/N) = 0.5\left[1 + \cos(2n\pi/N)\right]$	Cosine-squared, raised-cosine [7]
Hamming	$0.54 + 0.46\cos(2n\pi/N)$	First side-lobe cancelled for $(a_0, a_1) = (0.54349, 0.45651)$ [7]
Blackman	$\sum_{m=0}^{k-1} a_m \cos(2mn\pi/N)$ $= a_0 + a_1\cos(2n\pi/N) + a_2\cos(4n\pi/N) + \ldots$	General cosine series expansion [8]

Coefficient	a_0	a_1	a_2	a_3
3-term (-67 dB)	0.42323	0.49755	0.07922	—
3-term (-61 dB)	0.44959	0.49364	0.05677	—
3-term (-58 dB, -18 dB/dec roll-off)	0.42	0.50	0.08	—
3-term (-51 dB, cancels 2nd & 3rd side-lobes)	0.42659	0.49656	0.07685	—
4-term (-92 dB)	0.35875	0.48829	0.14128	0.01168
4-term (-74 dB)	0.40217	0.49703	0.09392	0.00183

Name	Time Function $w(n)$	Comments
Riesz	$1 - (2n/N)^2$	Parabolic [9]
Riemann	$\left[\sin(2n\pi/N)\right]/(2n\pi/N)$	Main lobe of sinc function [10]
de la Valle-Poussin	$1 - 6\left[1 - 2|n|/N\right]\left[2n/N\right]^2, \quad 0 \leq |n| \leq N/4$ $2\left[1 - 2|n|/N\right]^3, \quad N/4 \leq |n| \leq N/2$	Jackson, Parzen, piece-wise cubic [9]
Tukey	$1, \quad 0 \leq |n| \leq \alpha N/2$ $0.5\left[1 + \cos\left\{\left[\pi(n - \alpha N/2)\right]/\left[(1 - \alpha)N\right]\right\}\right],$ $\qquad \alpha N/2 \leq |n| \leq N/2$	Cosine tapered over α% interval [11]
Bohman	$(1 - 2|n|/N)\cos(2\pi|n|/N) + (1/\pi)\sin(2\pi|n|/N)$	Cosine tapered [12]
Poisson	$\exp(-2\alpha|n|/N)$	Exponential [10]
Hann-Poisson	Hann $w(n) \times$ Poisson $w(n)$	Exponentially-weighted Hann
Cauchy	$1/\left[1 + (2\alpha n/N)^2\right]$	Abel, Poisson [13]
Gaussian	$\exp\left[-0.5(2\alpha n/N)^2\right]$	Weierstrass, normal [13]
Dolph-Chebyshev	$(-1)^n \dfrac{\cos\left(N\cos^{-1}[B\cos(\pi n/N)]\right)}{\cosh(N\cosh^{-1} B)}$, where $B = \cosh[(\cosh^{-1} 10^\alpha)/N], \quad 0 \leq |n| \leq N - 1$	Hyperbolic taper [14]
Kaiser-Bessel	$I_0\left(\pi\alpha\sqrt{1 - (2n/N)^2}\,\right)/I_0(\pi\alpha),$ where $I_0(x) = 1 + \sum_{k=1}^{\infty}[(x/2)^k/k!]^2$	Bessel taper [15]
Barcilon-Temes	Prolate-spheroidal wave functions	Minimizes energy outside band of frequencies [16]

Table 2.6.2 Properties of window functions. (Courtesy of F.J. Harris.[5])

WINDOW		HIGHEST SIDE-LOBE LEVEL (dB)	SIDE-LOBE FALL-OFF (dB/OCT)	COHERENT GAIN	EQUIV. NOISE BW (BINS)	3.0-dB BW (BINS)	SCALLOP LOSS (dB)	WORST CASE PROCESS LOSS (dB)	6.0-dB BW (BINS)	OVERLAP CORRELATION (PCNT)	
										75% OL	50% OL
RECTANGLE		−13	−6	1.00	1.00	0.89	3.92	3.92	1.21	75.0	50.0
TRIANGLE		−27	−12	0.50	1.33	1.28	1.82	3.07	1.78	71.9	25.0
COS^a(x)	a = 1.0	−23	−12	0.64	1.23	1.20	2.10	3.01	1.65	75.5	31.8
HANN	a = 2.0	−32	−18	0.50	1.50	1.44	1.42	3.18	2.00	65.9	16.7
	a = 3.0	−39	−24	0.42	1.73	1.66	1.08	3.47	2.32	56.7	8.5
	a = 4.0	−47	−30	0.38	1.94	1.86	0.86	3.75	2.59	48.6	4.3
HAMMING		−43	−6	0.54	1.36	1.30	1.78	3.10	1.81	70.7	23.5
RIESZ		−21	−12	0.67	1.20	1.16	2.22	3.01	1.59	76.5	34.4
RIEMANN		−26	−12	0.59	1.30	1.26	1.89	3.03	1.74	73.4	27.4
DE LA VALLE-POUSSIN		−53	−24	0.38	1.92	1.82	0.90	3.72	2.55	49.3	5.0
TUKEY	a = 0.25	−14	−18	0.88	1.10	1.01	2.96	3.39	1.38	74.1	44.4
	a = 0.50	−15	−18	0.75	1.22	1.15	2.24	3.11	1.57	72.7	36.4
	a = 0.75	−19	−18	0.63	1.36	1.31	1.73	3.07	1.80	70.5	25.1
BOHMAN		−46	−24	0.41	1.79	1.71	1.02	3.54	2.38	54.5	7.4
POISSON	a = 2.0	−19	−6	0.44	1.30	1.21	2.09	3.23	1.69	69.9	27.8
	a = 3.0	−24	−6	0.32	1.65	1.45	1.46	3.64	2.08	54.8	15.1
	a = 4.0	−31	−6	0.25	2;08	1.75	1.03	4.21	2.58	40.4	7.4
HANN	a = 0.5	−35	−18	0.43	1.61	1.54	1.26	3.33	2.14	61.3	12.6
POISSON	a = 1.0	−39	−18	0.38	1.73	1.64	1.11	3.50	2.30	56.0	9.2
	a = 2.0	NONE	−18	0.29	2.02	1.87	0.87	3.94	2.65	44.6	4.7
CAUCHY	a = 3.0	−31	−6	0.42	1.48	1.34	1.71	3.40	1.90	61.6	20.2
	a = 4.0	−35	−6	0.33	1.76	1.50	1.36	3.83	2.20	48.8	13.2
	a = 5.0	−30	−6	0.28	2.06	1.68	1.13	4.28	2.53	38.3	9.0
GAUSSIAN	a = 2.5	−42	−6	0.51	1.39	1.33	1.69	3.14	1.86	67.7	20.0
	a = 3.0	−55	−6	0.43	1.64	1.55	1.25	3.40	2.18	57.5	10.6
	a = 3.5	−69	−6	0.37	1.90	1.79	0.94	3.73	2.52	47.2	4.9
DOLPH-	a = 2.5	−50	0	0.53	1.39	1.33	1.70	3.12	1.85	69.6	22.3
CHEBYSHEV	a = 3.0	−60	0	0.48	1.51	1.44	1.44	3.23	2.01	64.7	16.3
	a = 3.5	−70	0	0.45	1.62	1.55	1.25	3.35	2.17	60.2	11.9
	a = 4.0	−80	0	0.42	1.73	1.65	1.10	3.48	2.31	55.9	8.7
KAISER-	a = 2.0	−46	−6	0.49	1.50	1.43	1.46	3.20	1.99	65.7	16.9
BESSEL	a = 2.5	−57	−6	0.44	1.65	1.57	1.20	3.38	2.20	59.5	11.2
	a = 3.0	−69	−6	0.40	1.80	1.71	1.02	3.56	2.39	53.9	7.4
	a = 3.5	−82	−6	0.37	1.93	1.83	0.89	3.74	2.57	48.8	4.8
BARCILON-	a = 3.0	−53	−6	0.47	1.56	1.49	1.34	3.27	2.07	63.0	14.2
TEMES	a = 3.5	−58	−6	0.43	1.67	1.59	1.18	3.40	2.23	58.6	10.4
	a = 4.0	−68	−6	0.41	1.77	1.69	1.05	3.52	2.36	54.4	7.6
EXACT BLACKMAN		−51	−6	0.46	1.57	1.52	1.33	3.29	2.13	62.7	14.0
BLACKMAN (0.42, 0.50, 0.08)		−58	−18	0.42	1.73	1.68	1.10	3.47	2.35	56.7	9.0
MINIMUM 3-SAMPLE BLACKMAN-HARRIS		−67	−6	0.42	1.71	1.66	1.13	3.45	1.81	57.2	9.6
* MINIMUM 4-SAMPLE BLACKMAN-HARRIS		−92	−6	0.36	2.00	1.90	0.83	3.85	2.72	46.0	3.8
* 61 dB 3-SAMPLE BLACKMAN-HARRIS		−61	−6	0.45	1.61	1.56	1.27	3.34	2.19	61.0	12.6
74 dB 4-SAMPLE BLACKMAN-HARRIS		−74	−6	0.40	1.79	1.74	1.03	3.56	2.44	53.9	7.4
4-SAMPLE KAISER-BESSEL	a = 3.0	−69	−6	0.40	1.80	1.74	1.02	3.56	2.44	53.9	7.4

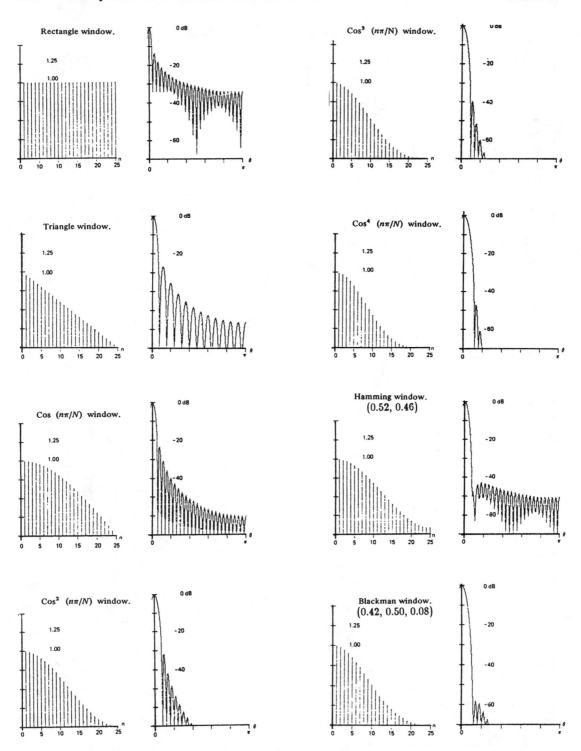

Fig. 2.6.1 Various time domain windows and their magnitude spectra. (Courtesy of F.J. Harris.[5])

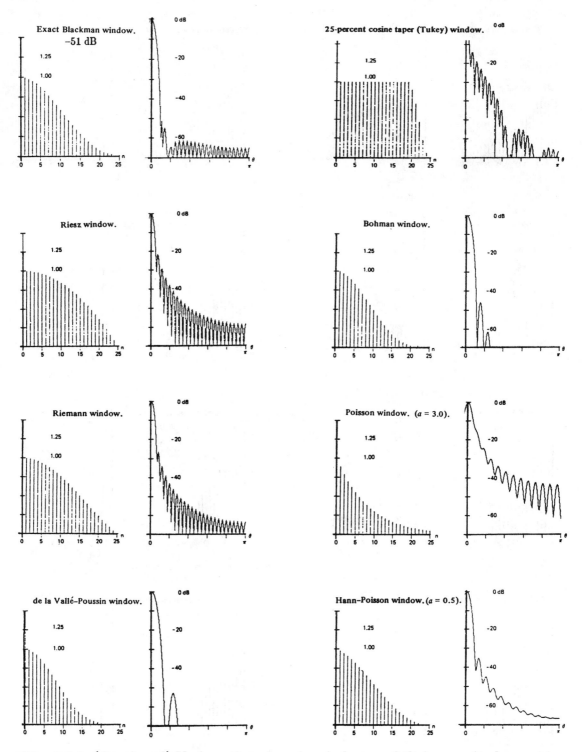

Fig. 2.6.1 (Continued) Various time domain windows and their magnitude spectra. (Courtesy of F.J. Harris.[5])

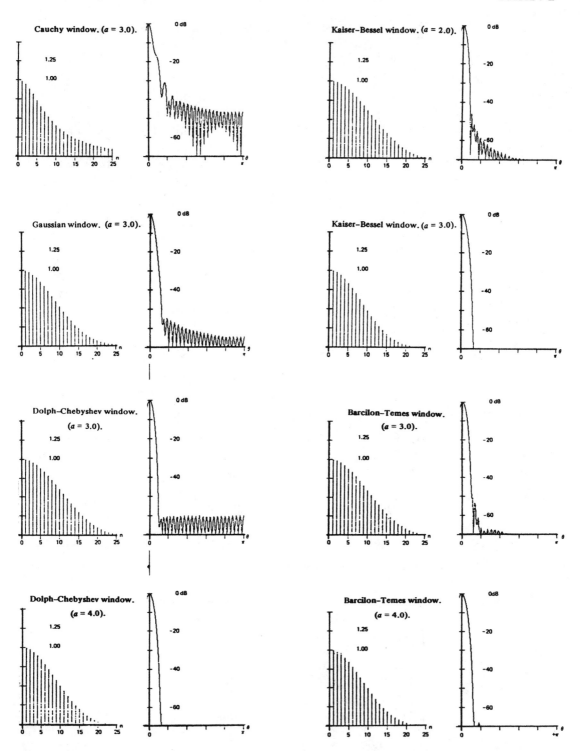

Fig. 2.6.1 (Continued) **Various time domain windows and their magnitude spectra.** (Courtesy of **F.J. Harris.**[5])

Spectral leakage causes a frequency component located at $\omega = \omega_o$ to contribute spectral output at another frequency $\omega = \omega_1$. This contribution is found by evaluating the gain at ω_1 for a window centered at ω_o. Spectral leakage biases or shifts the amplitude and location of spectral estimates and makes it difficult to detect small signals in the presence of large nearby signals or noise. To reduce spectral leakage, the window should exhibit low-amplitude side-lobes, and the transition bandwidth to reach these side-lobes should be narrow. The asymptotic roll-off should be high.

The peak side-lobe levels are found by setting $dW_s(j\omega)/d\omega = 0$ and solving for the frequencies that satisfy the equation. Expressing W_s as $(\sin Nx / \sin x)$ where $x = \omega T/2$, then $dW_s/dx = 0$ yields $N \sin(x) \cos(Nx) - \sin(Nx) \cos(x) = 0$. Rearranging gives $\tan Nx = N \tan x \cong Nx$ for $|x| \ll 1$. Solving iteratively yields $Nx = 0, 3\pi/2, \ldots$ which correspond to $\omega = 0, 3\pi/NT, \ldots$. The first side-lobe occurs at $\omega = 3\pi/NT$. Normalizing $|W_s(j\omega)|$ by $W_s(0)$, the first side-lobe level of the rectangular window is down

$$\frac{|W_s(j3\pi/NT)|}{W_s(0)} = \frac{1}{3\pi/2} = 0.212 = -13 \text{ dB} \tag{2.6.4}$$

The first side-lobe is usually the largest and establishes the peak side-lobe level.

The maximum stop-band attenuation of the envelope usually occurs at the Nyquist frequency $\omega_s/2$ and equals

$$\frac{|W_s(j\omega_s/2)|}{W_s(0)} \cong \frac{2|\text{Re } W(j\omega_s/2)|}{W(0)} \tag{2.6.5}$$

where $W(j\omega)$ is taken to be the envelope of the continuous spectrum. For the rectangular window, $W(0) = NT$ and $|W(j\omega)| \leq 2/\omega$ (see Eq. 2.5.2) so with $N = 50$,

$$\frac{|W_s(j\omega_s/2)|}{W_s(0)} \cong \frac{2(2/\omega)}{NT}\bigg|_{\omega=\pi/T} \leq \frac{4}{N\pi} = \frac{1}{39.3} = -32 \text{ dB} \tag{2.6.6}$$

This 32 dB stop-band rejection is verified in Fig. 2.6.1.

The asymptotic roll-off rate is most easily evaluated using the continuous window. The unsampled window spectrum equals

$$W(j\omega) = NT \frac{\sin(\omega NT/2)}{\omega NT/2} \tag{2.6.7}$$

from Eq. 2.5.2. Since $|W(j\omega)|$ is always bounded by its envelope as $|W(j\omega)| \leq 2/\omega$, the asymptotic roll-off of the rectangular window equals

$$\frac{d \log_{10} |W(j\omega)|}{d \log_{10} \omega}\bigg|_{max} = -1 = -20 \text{ dB/dec} = -6 \text{ dB/oct} \tag{2.6.8}$$

Since the sampled window spectrum $W_s(j\omega)$ is identical to the unsampled window spectrum $W(j\omega)/T$ for frequencies $|\omega| \ll \omega_s/2$, $W_s(j\omega)$ has this same slope over a limited frequency range. However due to frequency aliasing, the asymptotic roll-off of $W_s(j\omega)$

at the Nyquist frequency $\omega_s/2$ always equals zero. This is seen in Fig. 2.5.2 and is true for all windows.

At sufficiently high sampling rates, the asymptotic roll-off of any general (and rational) $W_s(j\omega)$ is $-20n$ dB/dec or $-6n$ dB/oct over limited frequency ranges. n equals the order of the derivative where $d^n w_s(t)/dt^n$ becomes impulsive. For example, the rectangular window has impulsive first-order derivatives so its asymptotic roll-off is -20 dB/dec, but triangular windows have impulsive second-order derivatives so their asymptotic roll-off is -40 dB/dec.

The noise bandwidth NBW of a window is a measure of its ability to reject flat broadband noise. NBW equals the bandwidth of a rectangular lowpass filter (see Eq. 1.3.7) with dc gain $W_s(0)$ that would output the same noise power as the window having a white noise input spectrum. Therefore

$$NBW = \frac{\int_{-\pi/T}^{\pi/T} |W_s(j\omega)|^2 d\omega}{W_s^2(0)} = \frac{N \sum_{n=0}^{N-1} w^2(nT)}{\left[\sum_{n=0}^{N-1} w(nT)\right]^2} \qquad (2.6.9)$$

Since the rectangular window has $w(nT) = 1$, its NBW is unity. Broader main-lobe windows have larger NBW. Generally the NBW of a window should be made as small as possible relative to its 3 dB and 6 dB bandwidths. This tends to minimize the signal-to-noise ratio SNR degradation (i.e., maximize the processing gain $PG = SNR_o/SNR_i$ and maximize the resolution capability of the window).

The following discussion describes coherent and noncoherent signal (H_c and H_n) and power gains, the coherent and noncoherent processing gains (PG_c and PG_n), scalloping loss (SL), and the correlation coefficient $\rho(r)$. The engineer may wish to skip these details on his first reading and now go directly to the high-selectivity window discussion in Chap. 2.6.2 following Eq. 2.6.22.

Windows provide processing gain since they yield a larger spectral output for coherent signals (signals with frequencies that correspond to the FFT frequency bins) than for noncoherent signals (signals with frequencies different from the FFT frequency bins or wide-band noise). The FFT output spectrum is highest when a signal frequency falls on an FFT frequency bin as shown in Fig. 2.5.3.

To describe this mathematically, consider the noisy and complex input signal

$$f(nT) = s(nT) + \eta(nT) = e^{j\omega_o nT} + \eta(nT) = W_N^{-nm_o} + \eta(nT) \qquad (2.6.10)$$

The first term is the complex-valued exponential signal given by Eq. 2.5.8 whose radian frequency equals ω_o. Its FFT frequency equals m_o which is not necessarily an integer ($m_o = N\omega_o/\omega_s$). $\eta(nT)$ is white noise with zero mean and variance σ^2.

The window can be viewed as a processing block with input $f(nT)$, output $g(nT)$, and convolutional gain $W(m)$ where $g(nT) = w(nT)f(nT)$ or $G(m) = W(m)*F(m)/2\pi$. The convolutional block can be replaced by a multiplicative block with gain $H(m)$ having the same input and output where $G(m) = H(m)F(m)$ or $g(nT) = h(nT) * f(nT)$. Equating inputs and outputs of the two blocks and manipulating the result shows that

$H(m) = G(m)/F(m) = [W(m) * F(m)]/2\pi F(m)$. $H(m)$ depends upon the spectral natures of the FFT input spectrum $F(m)$ and weighting spectrum $W(m)$. Evaluating these FFT input and output spectra by substituting Eq. 2.6.10 into Eq. 2.1.4 gives

$$F(m) = \sum_{n=0}^{N-1} W_N^{(m-m_o)n} + \sum_{n=0}^{N-1} \eta(nT) W_N^{mn} = F_s(m) + F_\eta(m) \qquad (2.6.11)$$

$$G(m) = \sum_{n=0}^{N-1} w(nT) W_N^{(m-m_o)n} + \sum_{n=0}^{N-1} w(nT)\eta(nT) W_N^{mn} = G_s(m) + G_\eta(m)$$

Consider the coherent signal case first where the signal frequency falls on an FFT bin frequency so $m = m_o$. At this frequency, the FFT spectral input F_s and output G_s are composed of coherent (matched frequency) components F_s and G_s, and noncoherent (unmatched frequency) components F_η and G_η. Taking the average or expected values of $F(m_o)$ and $G(m_o)$ given by Eq. 2.6.11 yields

$$E\{F(m_o)\} = E\{F_s(m_o)\} + E\{F_\eta(m_o)\} = F_s(m_o) + 0 = N \qquad (2.6.12)$$

$$E\{G(m_o)\} = E\{G_s(m_o)\} + E\{G_\eta(m_o)\} = G_s(m_o) + 0 = \sum_{n=0}^{N-1} w(nT)$$

using Eq. 1.17.4. The noncoherent noise components $E\{F_\eta\}$ and $E\{G_\eta\}$ are zero since $\eta(nT)$ is a zero-mean process having $E\{\eta(nT)\} = 0$. The expected power spectra $E\{|F(m_o)|^2\}$ and $E\{|G(m_o)|^2\}$ equal

$$E\{|F(m_o)|^2\} = E\{|F_s(m_o)|^2\} + E\{|F_\eta(m_o)|^2\} = N^2 + \sigma^2 N \qquad (2.6.13)$$

$$E\{|G(m_o)|^2\} = E\{|G_s(m_o)|^2\} + E\{|G_\eta(m_o)|^2\} = \left[\sum_{n=0}^{N-1} w(nT)\right]^2 + \sigma^2 \sum_{n=0}^{N-1} w^2(nT)$$

using Eqs. 1.17.15 and 2.6.11.

The coherent signal gain $H_n(m_o)$ is defined to equal the ratio of the expected coherent output $E\{G_s(m_o)\}$ and the expected coherent input $E\{F_s(m_o)\}$ so

$$H_c(m_o) = \frac{E\{G_s(m_o)\}}{E\{F_s(m_o)\}} = \frac{\sum_{n=0}^{N-1} w(nT)}{N} = \frac{W_s(0)}{N} \qquad (2.6.14)$$

from Eq. 2.6.12. The coherent signal gain varies as the dc gain $W_s(0)$ of the window. The coherent power gain is defined to equal the square of the coherent signal gain. Both equal unity for the rectangular window but are smaller for improved windows.

The noncoherent power gain $|H_n(m_o)|^2$ is defined to equal the ratio of the average noncoherent power output $E\{|G_\eta(m_o)|^2\}$ and the average noncoherent power input $E\{|F_\eta(m_o)|^2\}$ as

$$|H_n(m_o)|^2 = \frac{E\{|G_\eta(m_o)|^2\}}{E\{|F_\eta(m_o)|^2\}} = \frac{\sum_{n=0}^{N-1} w^2(nT)}{N} \qquad (2.6.15)$$

The noncoherent power gain varies as the sum of the squares of the window terms. The noncoherent signal gain equals the square root of the noncoherent power gain. Both are unity for the rectangular window.

The coherent processing gain PG_c is defined as the ratio of the coherent power gain to the noncoherent power gain, or equivalently, the output SNR_o to the input SNR_i (i.e., signal-to-noise ratios). It measures how well the window passes coherent signals and rejects noncoherent signals and broad-band noise. PG_c should be made as large as possible. Combining Eqs. 2.6.14 and 2.6.15,

$$PG_c = \frac{SNR_o}{SNR_i} = \frac{|H_c(m_o)|^2}{|H_n(m_o)|^2} = \frac{\left[\sum_{n=0}^{N-1} w(nT)\right]^2}{N \sum_{n=0}^{N-1} w^2(nT)} = \frac{1}{NBW} \qquad (2.6.16)$$

PG_c equals the reciprocal of the noise bandwidth NBW from Eq. 2.6.9. Maximizing the coherent processing gain requires minimizing the noise bandwidth. A rectangular window has $PG_c = -10 \log_{10} NBW = 0$ dB. Other windows have smaller PG_c (i.e., larger coherent processing loss) as shown in Table 2.6.2.

Now consider the noncoherent signal case where the signal frequency m_o is not an integer but lies between the FFT frequency bins. This produces the *picket-fence effect* or *spectral scalloping*. It is worst for half-bin frequency shifts as shown in Fig. 2.5.3c and is described by a parameter called *scalloping loss SL*. SL equals the reduction in the coherent signal gain when a signal frequency is halfway between FFT frequency bins. The effects can be readily evaluated from Eq. 2.6.11 by assuming the incoming frequency is $\omega_o + \Delta\omega$ and $\eta = 0$ where $\Delta\omega = (2\pi/NT)/2$, $\Delta f = 1/2NT$, and $m = m_o + 0.5$. Then the expected FFT input and output spectra generated at the bins on either side of the signal frequency m_o are

$$E\{F(m_o + 0.5)\} = \sum_{n=0}^{N-1} W_N^{(m_o+0.5-m_o)n} = \sum_{n=0}^{N-1} W_{2N}^n = W_s\left(\frac{j\omega_s}{2N}\right)\bigg|_{\text{Rect}} \cong 0.637N$$

$$E\{G(m_o + 0.5)\} = \sum_{n=0}^{N-1} w(nT)W_{2N}^n = W_s\left(\frac{j\omega_s}{2N}\right) \qquad (2.6.17)$$

(cf. Eq. 2.6.12). The expected input spectrum $E\{F(m_o + 0.5)\}$ equals the rectangular window gain at the first one-half bin frequency $\omega = \omega_s/2N$ which is $0.637N$. The coherent signal gain is reduced to (cf. Eq. 2.6.14)

$$H_s(m_o + 0.5) = \frac{E\{G(m_o + 0.5)\}}{E\{F(m_o + 0.5)\}} = \frac{W_s(j\omega_s/2N)}{W_s(j\omega_s/2N)|_{\text{Rect}}} \cong \frac{W_s(j\omega_s/2N)}{0.637N} \qquad (2.6.18)$$

Therefore combining Eqs. 2.6.15 and 2.6.18 gives the noncoherent processing gain (or worst-case coherent processing gain) PG_n as

$$PG_n = \frac{|H_c(m_o + 0.5)|^2}{|H_n(m_o)|^2} = \frac{|H_c(m_o + 0.5)|^2}{|H_c(m_o)|^2}\frac{|H_c(m_o)|^2}{|H_n(m_o)|^2} = SL^2 \times PG_c \qquad (2.6.19)$$

It equals the coherent processing gain PG_c reduced by the square of the scalloping loss SL or the sum of the coherent processing gain and the scalloping loss squared both in dB. The worst-case coherent loss $1/PG_n$ is the negative of the PG_n gain in dB.

This parameter measures the FFT processing loss when analyzing a noncoherent signal located halfway between FFT bins. It gives the level of the minimum detectable tone in broadband noise.

Thus taking the square root of PG_c/PG_n (Eqs. 2.6.16 and 2.6.19) yields SL as

$$SL = \frac{|H_c(m_o + 0.5)|}{|H_c(m_o)|} = \frac{\left|\sum_{n=0}^{N-1} w(nT)W_{2N}^n\right|}{\sum_{n=0}^{N-1} w(nT)} = \frac{|W_s(j\omega_s/2N)|}{W_s(0)} \qquad (2.6.20)$$

SL represents the maximum reduction in the coherent signal gain due to processing a noncoherent signal having a one-half bin shift in frequency. The rectangular filter has a scalloping loss and worst-case processing gain of

$$SL = \frac{\sin(\pi/2)/\sin(\pi/2N)}{N} \cong \frac{2}{\pi} = 0.637 = -3.9 \text{ dB}$$

$$PG_n = PG_c \times SL^2 = 10\log_{10}(1.0) + 10\log_{10}(0.637^2) = 0 - 3.9 = -3.9 \text{ dB}$$
$$(2.6.21)$$

by evaluating the window spectrum $W_s(j\omega)$ given by Eq. 2.6.1 at the frequencies required by Eq. 2.6.20. PG_c is given by Eq. 2.6.16.

The FFT is used to analyze data blocks composed of N data points. Data blocks are seldom put end-to-end because of undesirable edge effects. Spectrally-superior data windows usually exhibit small gains near their ends. Significant amounts of end-point data is ignored or missed if nonoverlapping data blocks are used. To avoid losing this data, blocks are often overlapped 50% to 75% as shown in Fig. 2.6.2. Overlapped processing increases work load since overlapped sequences of length N with overlap ratio r have effective lengths of $N/(1-r)$. This will be discussed further with running transforms in Chap. 2.11.

An important consideration in overlapping data is the correlation between successive blocks. Using Eq. 1.17.9, the correlation coefficient $\rho(r)$ equals

$$\rho(r) = \frac{\sum_{n=0}^{rN-1} w(n)w[n + (1-r)N]}{\sum_{n=0}^{N-1} w^2(n)} \qquad (2.6.22)$$

The rectangular window has $\rho(r) = rN/N = r$ using Eq. 2.6.22 so $\rho(r) = 0.5$ for $r = 50\%$ and $\rho(r) = 0.75$ for $r = 75\%$. Since the processing time equals $T_p(r) = T_F/(1-r)$ where $T_F = $ FFT cycle time, then $T_p(0.5) = 2T_F$ and $T_p(0.75) = 4T_F$. A trade-off must be made between the increased computation time T_p and the higher correlation between FFT coefficients.

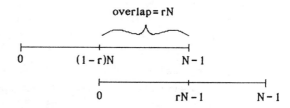

Fig. 2.6.2 Relation between overlapping data blocks.

2.6.2 HIGH-SELECTIVITY WINDOW TYPES [5]

A variety of time windows are shown in Fig. 2.6.1 with their log-magnitude spectrums ($N = 50$). The rectangular window has narrow 3 dB bandwidth but poor stop-band rejection (about -30 dB). The triangular window has greater bandwidth but better stop-band rejection (about -55 dB). $\cos^p(x)$ windows have increasing bandwidths, increasing stop-band rejection, and decreasing first side-lobe levels with increasing p. The Hann window is the special case where $p = 2$. The Hamming window is a Hann window with different weights specially selected to reduce or cancel the first side-lobe level. Using $(a_0, a_1) = (0.54, 0.46)$ reduces this level from -32 dB to -43 dB; $(a_0, a_1) = (0.54349, 0.46651)$ gives complete cancellation. However the Hamming stop-band rejection is only about -50 dB which is not as good as the Hann window. The Blackman window is a Fourier series window where the number and weighting of the various terms are selected to control stop-band performance. Wider transition bands having stop-band rejections of -60 dB and more can be easily obtained using three- and four-term windows. Second and third side-lobes can be cancelled or -60 dB/dec slope achieved using three-term windows.

Other windows have been constructed of sums, products, convolutions, or segments of simple functions or other simple windows. The Riesz window is parabolic and the Riemann window has a sinc pulse shape. They are not too different spectrally from the $\cos(x)$ window. The de la Valle-Poussin window is cubic. It has large 3 dB bandwidth, excellent asymptotic roll-off of -80 dB/dec and one side-lobe about -50 dB down. The Tukey windows are transitional windows having cosine tapers controlled by parameter α where $0 < \alpha < 1$. The rectangular ($\alpha = 1$) and Hann ($\alpha = 0$) windows are limiting cases. Increased bandwidth but greater stop-band rejection is obtained as parameter α is decreased. The Bohman window is somewhat similar spectrally to the de la Valle-Poussin window. The Poisson window is a two-sided exponential. It is a transitional type with fairly narrow 3 dB bandwidth but poor stop-band rejection not too different from that of the triangular window. The Hann-Poisson window is the product of the two individual windows. The Cauchy window is a transitional type having narrow 3 dB bandwidth, wide transition band, and poor stop-band rejection. The Gaussian window has a Gaussian shape. It is another transitional type which is similar to, but gives improved performance over, the Hamming type. The 3 dB bandwidth can be traded off with stop-band rejection. The Dolph-Chebyshev, Kaiser-Bessel, and Barcilon-Temes windows all have fairly narrow bandwidth and a high stop-band rejection level that can be controlled by a transition parameter α. In the Dolph-Chebyshev window, the stop-band rejection is almost constant.

EXAMPLE 2.6.1 Investigate the spectral characteristics of the rectangular window function by attempting to detect a weak spectral line in the presence of a strong nearby spectral line. The composite signal $f(t)$ equals

$$f(t) = \cos[2\pi(10f_e/N)t] + 0.01\cos[2\pi(16f_e/N)t] \qquad (2.6.23)$$
$$= e^{j2\pi(10f_e/N)t} + e^{-j2\pi(10f_e/N)t} + 0.01e^{j2\pi(16f_e/N)t} + 0.01e^{-j2\pi(16f_e/N)t}$$

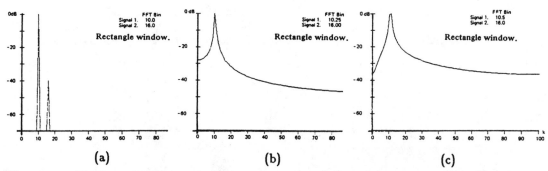

Fig. 2.6.3 FFT spectra for rectangular window and strong signal at (a) $10f_s/N$, (b) $10.25f_s/N$, and (c) $10.5f_s/N$ frequencies in Example 2.6.1. (Courtesy of F.J. Harris.[5])

The signals are located at four frequencies $(\pm 10f_s/N, \pm 16f_s/N)$ and the ratio of signal levels is $1/0.01 = 100 = 40$ dB. Next shift the larger signal $\frac{1}{4}$-bin to $\pm 10.25f_s/N$ and $\frac{1}{2}$-bin to $\pm 10.5f_s/N$. Interpret the FFT spectra of these signals.

SOLUTION The FFT spectra are shown in Fig. 2.6.3. Since the strong line spectrum falls on an FFT frequency in Fig. 2.6.3a, good spectral resolution is obtained. Moving the strong line signal to $10.25f_s/N$ results in the output of Fig. 2.6.3b. Due to the $(\sin Nx/\sin x)$ side-lobe structure of the rectangular window spectrum (see Fig. 2.5.2), the spectrum of the first signal completely masks that of the second signal. Since the side-lobe amplitude of the rectangular window 5.75 bins away from center is only -25 dB down from the peak and the weak signal is down -40 dB at this point, there is no hope of recovering the weak signal. There is also asymmetry about the main lobe centered at the 10.25 bin due to the addition of the two $(\sin Nx/\sin x)$ spectra centered at ± 10.25 bins. The spectral leakage between the positive and negative frequencies is very evident. Moving the larger signal to $10.5f_s/N$ gives the response shown in Fig. 2.6.3c. There is even greater degradation in the out-of-band spectrum (its only about -30 dB down rather than -45 dB as before) with even less hope of observing the signal at $16f_s/N$. ■

EXAMPLE 2.6.2 The rectangular weighting used in Example 2.6.1 failed to provide signal discrimination and allow detection of both line spectra. Now use triangular and $\cos^p x$ windows to attempt signal detection. Use $f(t)$ given by Eq. 2.6.23 and large-signal frequencies of either $\pm 10f_s/N$ or $\pm 10.5f_s/N$.

SOLUTION For some of the windows, poorer resolution occurs when the large signal is at $\pm 10f_s/N$ rather than $\pm 10.5f_s/N$. This is due to spectral enhancement and cancellation (phase effect) when adding the four spectral terms together. Numerical evaluation is needed to determine where the worst-case resolution occurs. The triangular weighting result is shown in Fig. 2.6.4. Since the side-lobe amplitude of the triangular window 5.5 bins from center is down -43 dB and the weak signal is down only -40 dB, the weak signal is barely detectable as verified in Fig. 2.6.4. Also note that the -40 dB/dec asymptotic roll-off allows little leakage between positive and negative frequencies.

Although a $\cos x$ window gives poorer response than the triangular window, $\cos^2 x$, $\cos^3 x$, and $\cos^4 x$ windows give better responses. In the $\cos x$ case, there

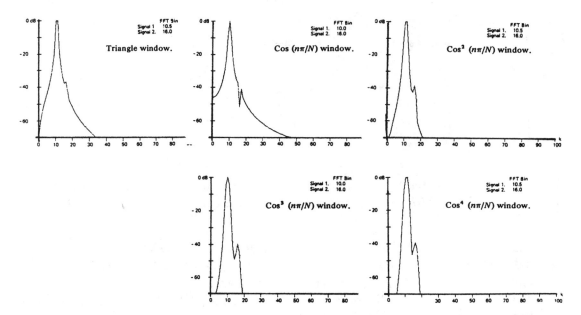

Fig. 2.6.4 **FFT spectra for triangular and $\cos^p(x)$ windows and strong signal at $10f_s/N$ and $10.5f_s/N$ for worst-case resolution in Example 2.6.2. (Courtesy of F.J. Harris.[5])**

is large leakage between the $\pm 10 f_s/N$ frequencies. Also due to phase cancellation, the strong and weak signals cancel at $\pm 16 f_s/N$ so the small second peak is not due to the weak signal. In the $\cos^2 x$ or Hann window case, although the side-lobe amplitude at 5.5 bins from center is down -50 dB, the bandwidth is slightly larger so the weak signal is barely seen. In the $\cos^3 x$ case, the side-lobe level is down more than -60 dB, but the bandwidth is only slightly larger so the weak signal becomes more obvious. Since the $\cos^4 x$ spectrum is very close to the $\cos^3 x$ spectrum in Fig. 2.6.1, we would expect only slightly larger spectral bandwidths in Fig. 2.6.4 which is the case. ∎

EXAMPLE 2.6.3 Investigate the other windows in Fig. 2.6.1. Select the strong tone frequency at $\pm 10 f_s/N$ or $\pm 10.5 f_s/N$ to produce worst-case detection.
SOLUTION The window responses are shown in Fig. 2.6.5. The Hamming window detects the weak signal -35 dB down which is about 3 dB above the side-lobe of the strong signal. Phase cancellation produces the notch around $\pm 13 f_s/N$. The spectral spreading produces about a -50 dB detection floor. The Blackman window has a -17 dB notch between the two signals. The rapid asymptotic roll-off has prevented large spectral smearing. The exact Blackman window has a -24 dB notch between the two signals. The side-lobes of the large signal extends over the entire spectral range giving the -60 dB floor. The Blackman-Harris windows give improved signal lobe definition. The Riesz window detects only the large signal. The -20 dB notch is due to the phase cancellation of the $-10 f_s/N$ side-lobe. The Riemann window response is similar to the Riesz window with a small null at the small signal frequency. The de la Valle-Poussin window isolates the small signal

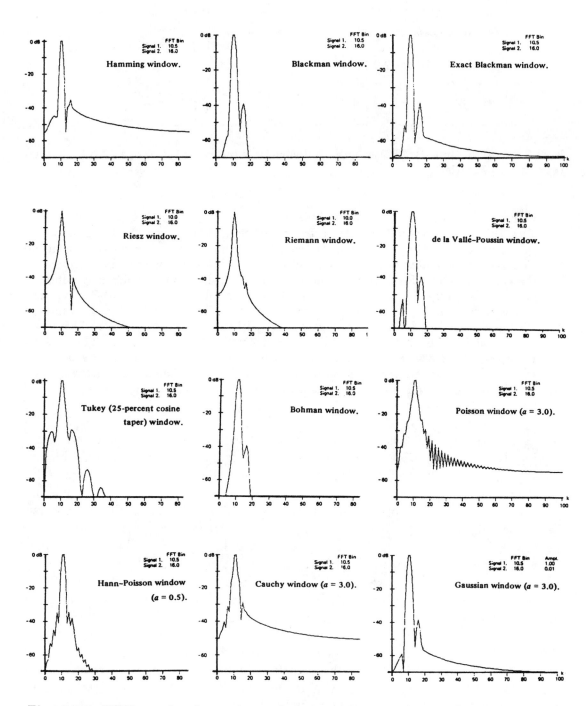

Fig. 2.6.5   FFT spectra for various windows of Example 2.6.3. (Courtesy of F.J. Harris.[5])

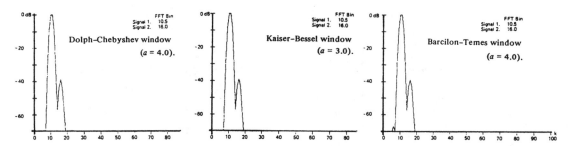

Fig. 2.6.5 (Continued) FFT spectra for various windows of Example 2.6.3. (Courtesy of F.J. Harris.[5])

with a −16 dB notch. The lower side-lobe of the large signal appears at $5f_s/N$. The Tukey windows fail to isolate the small signal due to the high side-lobes of the large signal. Going to larger percentage tapers fail to improve the situation. The Bohman window detects the small signal with a 6 dB notch. It is similar to the $\cos^p x$ windows for larger p. The Poisson, Hann-Poisson, and Cauchy windows are unsatisfactory due to the addition of the large side-lobes at $\pm 10.5 f_s/N$. The Gaussian windows are slightly better for larger α's. The Dolph-Chebyshev windows are unsuitable for smaller α's but give good performance for larger α's. The $\alpha = 4$ window detects both signals with a −18 dB notch between them. The Kaiser-Bessel windows also give good performance with a −22 dB notch between them while holding the leakage to no more than −70 dB over the entire spectrum. The Barcilon-Temes windows are very close to the Kaiser-Bessel windows with a −20 dB notch between signals.   ■

Examples 2.6.1–2.6.3 show the importance of selecting proper windows when analyzing data with the DFT. Different applications (e.g., locating tones embedded in broadband noise or discriminating multiple tones over the frequency band) require different windows.

Consider some of the performance measures listed in Table 2.6.2. Since the worst-case processing loss ranges between 3.0 (good windows) to 3.75 (poor windows), this parameter has little meaning for detecting coherent line spectra in broadband noise. However a good figure of merit in this application is the 3 dB bandwidth/noise bandwidth ratio which ranges between 1.03 (good windows) to 1.22 (poor windows).

Multitone detection requires windows that have a highly concentrated main-lobe and very low side-lobe structure. The transitional windows like Blackman-Harris, Kaiser-Bessel, Dolph-Chebyshev, and Barcilon-Temes perform best in this application. However the Dolph-Chebyshev window is extremely sensitive to coefficient errors as well as having constant level side-lobes which makes it somewhat less attractive. The Kaiser-Bessel and Blackman-Harris windows have coefficients that are easy to generate, and side-lobe levels and time-bandwidth products which are easy to trade off.

2.7 FREQUENCY DOMAIN WINDOWING

Windowing can be applied in the frequency domain as well as the time domain. This is shown in Fig. 2.7.1 which is a rearranged version of Fig. 2.5.1. Multiplication in time corresponds to convolution in frequency. Thus rather than multiplying the sampled data $f_s(t)$ by $w(t)$ and then transforming the product $f_w(t)$, $F_w(j\omega)$ can also be obtained by first transforming $f_s(t)$ and then convolving $F_s(j\omega)$ with the window spectrum $W(j\omega)$. Which method is simpler depends upon the spectral nature of $f_s(t)$ and $w(t)$ since the convolutional block gain is $H_w(j\omega) = F_w(j\omega)/F_s(j\omega) = [F_s(j\omega) * W(j\omega)]/2\pi F_s(j\omega)$ (see discussion leading to Eq. 2.6.11).[17]

Time multiplication of N data points by an N-point window requires N multiplications. Frequency convolution requires N complex multiplications and $N-1$ complex additions per frequency bin for a total of $N(2N-1)$ operations for N bins. If a large number of the N spectral coefficients of $F_s(j\omega)$ or $W(j\omega)$ are zero as they are for narrowband data or high stop-band rejection windows (see Fig. 2.6.1), only a fraction of the $N(2N-1)$ operations need to be performed.

All windows can be expressed as cosine (or Fourier) series expansions of the desired window shape. Since cosine functions are matched to the DFT basis functions (this was introduced in Chap. 1.4 and will be discussed in the next section), all windows are composed of discrete spectra. The simplest windows to convolve directly in the frequency domain are the Blackman types in Table 2.6.1. If a k-term Blackman window is used, then the window spectrum $W(j\omega)$ has only $(2k-1)$ nonzero terms unlike the other types which have an infinite number of terms. Therefore for N-point data, there are $(2k-1)$ complex multiplications and $2(k-1)$ complex additions per bin so only $(4k-3)N$ total operations are required. For $k = 2$ or 3, this is $5N$ or $9N$ which is not an excessive amount of computation.

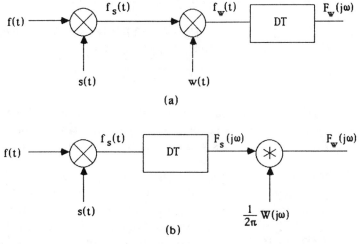

Fig. 2.7.1 Windowing in the (a) time domain and (b) frequency domain.

All windows can be expressed as cosine series expansions where

$$w(n) = \sum_{m=0}^{\infty} a_m \cos\left(\frac{2nm\pi}{N}\right) = 0.5 \sum_{m=0}^{\infty} a_m \left[e^{j2nm\pi/N} + e^{-j2nm\pi/N}\right]$$

$$W(m) = 0.5N \sum_{n=0}^{\infty} a_n \left[U_0(m-n) + U_0(m+n)\right] \tag{2.7.1}$$

The a_n coefficients are computed from Eqs. 1.7.9 or 1.7.12 as described in Chap. 1.7. Convolving the window spectrum $W(m)$ with the sampled signal spectrum $F_s(m)$ gives

$$F_w(m) = \frac{1}{N} F_s(m) * W(m) = 0.5 F_s(m) * \sum_{n=0}^{\infty} a_n \left[U_0(m-n) + U_0(m+n)\right]$$

$$= 0.5 \sum_{n=0}^{\infty} a_n \left[F_s(m-n) + F_s(m+n)\right] \tag{2.7.2}$$

Therefore by simply scaling spectrum F_s by a_n, recentering F_s about different m, and summing the results produces frequency domain windowing. The k-term Blackman window and its special cases are also described by Eqs. 2.7.1 and 2.7.2 but the sums only extend from 0 to $k-1$.

EXAMPLE 2.7.1 Determine the DFT spectrum of a signal $f(t)$ windowed with cosine, Hann, and Hamming windows. The windows are centered about $t = 0$ for $-NT/2 \leq t < NT/2$.
SOLUTION These windows are special cases of the Blackman window. From Table 2.6.1 and Eq. 2.7.1, the cosine window has $a_1 = 1$, the Hann window has $a_0 = a_1 = 0.5$, while the Hamming window has $a_0 = 0.543$ and $a_1 = 0.457$. All the other a_m's equal zero. Therefore, substituting these results into Eq. 2.7.2 gives the frequency domain windowing equations as

$$\text{Cosine:} \quad F_w(m) = 0.5 F_s(m-0.5) + 0.5 F_s(m+0.5)$$

$$\text{Hann:} \quad F_w(m) = 0.25 F_s(m-1) + 0.5 F_s(m) + 0.25 F_s(m+1)$$

$$\text{Hamming:} \quad F_w(m) = 0.229 F_s(m-1) + 0.543 F_s(m) + 0.229 F_s(m+1) \tag{2.7.3}$$

For these windows, it is almost as simple to perform frequency domain windowing by adding scaled spectra as to perform time domain windowing. Remember that Eq. 2.7.3 describes windows centered about $t = 0$ where $-NT/2 \leq t < NT/2$. ∎

EXAMPLE 2.7.2 A 1 Hz signal $f(t) = \cos(2\pi f_o t)$ sampled at 8 Hz beginning at $t = -NT/2$ gives the sample vector $[f_s] = [-1, -0.707, 0, 0.707, 1, 0.707, 0, -0.707]^t$. Write the weighted sample vector using cosine, Hann, and Hamming windows. Find the FFT spectra for the weighted and unweighted data.
SOLUTION The weighted signal vector $[f_w]$ equals $[f_w] = [w][f_s]$ and the weighting functions are given in Table 2.6.1. Sampling the cosine weighting function

$w(n) = \cos(\pi n/N)$ for $N = 8$ and $n = -N/2, \ldots, N/2 - 1$ gives the window matrix $[w] = \text{diag}[0, 0.383, 0.707, 0.924, 1, 0.924, 0.707, 0.383]$. Therefore the cosine-weighted data vector equals

$$[f_w]_{cos} = \text{diag} \begin{bmatrix} 0 \\ 0.383 \\ 0.707 \\ 0.924 \\ 1.000 \\ 0.924 \\ 0.707 \\ 0.383 \end{bmatrix} \begin{bmatrix} -1.000 \\ -0.707 \\ 0 \\ 0.707 \\ 1.000 \\ 0.707 \\ 0 \\ -0.707 \end{bmatrix} = \begin{bmatrix} 0 \\ -0.271 \\ 0 \\ 0.653 \\ 1.000 \\ 0.653 \\ 0 \\ -0.271 \end{bmatrix} \tag{2.7.4}$$

The Hann-weighted data vector is computed using $w(n) = 0.5 + 0.5 \cos(2\pi n/N)$ as

$$[f_w]_{Hann} = \text{diag} \begin{bmatrix} 0 \\ 0.147 \\ 0.500 \\ 0.854 \\ 1.000 \\ 0.854 \\ 0.500 \\ 0.147 \end{bmatrix} \begin{bmatrix} -1.000 \\ -0.707 \\ 0 \\ 0.707 \\ 1.000 \\ 0.707 \\ 0 \\ -0.707 \end{bmatrix} = \begin{bmatrix} 0 \\ -0.104 \\ 0 \\ 0.604 \\ 1.000 \\ 0.604 \\ 0 \\ -0.104 \end{bmatrix} \tag{2.7.5}$$

The Hamming-weighted data vector uses $w(n) = 0.543 + 0.457 \cos(2\pi n/N)$ so

$$[f_w]_{Hamm} = \text{diag} \begin{bmatrix} 0.086 \\ 0.220 \\ 0.543 \\ 0.866 \\ 1.000 \\ 0.866 \\ 0.543 \\ 0.220 \end{bmatrix} \begin{bmatrix} -1.000 \\ -0.707 \\ 0 \\ 0.707 \\ 1.000 \\ 0.707 \\ 0 \\ -0.707 \end{bmatrix} = \begin{bmatrix} -0.086 \\ -0.156 \\ 0 \\ 0.613 \\ 1.000 \\ 0.613 \\ 0 \\ -0.156 \end{bmatrix} \tag{2.7.6}$$

Transforming these weighted data vectors with the DFT given by Eq. 2.2.3 or any one of the FFT algorithms shown in Figs. 2.3.5–2.3.8 or 2.4.4–2.4.7 yields

$$[F_s] = \begin{bmatrix} 0 \\ -4 \\ 0 \\ 0 \\ 0 \\ 0 \\ 0 \\ -4 \end{bmatrix}, \quad [F_w]_{cos} = \begin{bmatrix} -1.765 \\ -2.307 \\ -1.000 \\ 0.307 \\ -0.235 \\ 0.307 \\ -1.000 \\ -2.307 \end{bmatrix}, \quad [F_w]_{Hann} = \begin{bmatrix} -2 \\ -2 \\ -1 \\ 0 \\ 0 \\ 0 \\ -1 \\ -2 \end{bmatrix}, \quad [F_w]_{Hamm} = \begin{bmatrix} -1.828 \\ -2.172 \\ -0.914 \\ 0 \\ 0 \\ 0 \\ -0.914 \\ -2.172 \end{bmatrix}$$

$$\tag{2.7.7} \blacksquare$$

EXAMPLE 2.7.3 Now find the FFT spectra directly using frequency domain windowing rather than time domain windowing.

SOLUTION The weighted data spectra can be found directly from the unweighted data spectrum $[F_s]$ in Eq. 2.7.7a and the algorithms of Eq. 2.7.3. Therefore

Cosine : $F_w(m) = 0.5 F_s(m - 0.5) + 0.5 F_s(m + 0.5)$

$$[F_w]_{Hann} = 0.25 \begin{bmatrix} -4 \\ 0 \\ -4 \\ 0 \\ 0 \\ 0 \\ 0 \\ 0 \end{bmatrix} + 0.5 \begin{bmatrix} 0 \\ -4 \\ 0 \\ 0 \\ 0 \\ 0 \\ 0 \\ -4 \end{bmatrix} + 0.25 \begin{bmatrix} -4 \\ 0 \\ 0 \\ 0 \\ 0 \\ 0 \\ -4 \\ 0 \end{bmatrix} = \begin{bmatrix} -2 \\ -2 \\ -1 \\ 0 \\ 0 \\ 0 \\ -1 \\ -2 \end{bmatrix}$$

$$(2.7.8)$$

$$[F_w]_{Hamm} = 0.229 \begin{bmatrix} -4 \\ 0 \\ -4 \\ 0 \\ 0 \\ 0 \\ 0 \\ 0 \end{bmatrix} + 0.543 \begin{bmatrix} 0 \\ -4 \\ 0 \\ 0 \\ 0 \\ 0 \\ 0 \\ -4 \end{bmatrix} + 0.229 \begin{bmatrix} -4 \\ 0 \\ 0 \\ 0 \\ 0 \\ 0 \\ -4 \\ 0 \end{bmatrix} = \begin{bmatrix} -1.828 \\ -2.172 \\ -0.914 \\ 0 \\ 0 \\ 0 \\ -0.914 \\ -2.174 \end{bmatrix}$$

The Hann and Hamming spectra agree with those in Eq. 2.7.7. The cosine spectra cannot be determined unless the midpoint spectra of F_s are approximated. ∎

EXAMPLE 2.7.4 Sampling a 1 Hz signal $f(t) = \cos(2\pi f_o t)$ at 8 Hz beginning at $t = 0$ gives the sample vector $[f_s] = [1, 0.707, 0, -0.707, -1, -0.707, 0, 0.707]^t$. Find the FFT spectra for unweighted data and for weighted data using cosine, Hann, and Hamming windows.

SOLUTION The time domain windows in Table 2.6.1 extended over $-NT/2 \le t < NT/2$. However in this example, the window covers $0 \le t < NT$. Setting $n = n - N/2$ in Table 2.6.1 produces window functions that are centered about $t = NT/2$ as

Cosine : $w(n) = \cos\left[\pi(n - N/2)/N\right] = \sin(\pi n/N)$

Hann : $w(n) = 0.5 - 0.5\cos(2\pi n/N)$ $(2.7.9)$

Hamming : $w(n) = 0.543 - 0.457\cos(2\pi n/N)$

If the FT transform pair are $w(t) \leftrightarrow W(j\omega)$ then, by the time-shifting theorem of Table 1.6.2, $w(t - NT/2) \leftrightarrow e^{-j\omega NT/2}W(j\omega)$. Therefore the phase spectra $W(m)$ must be increased by $e^{-j\omega NT/2} = e^{-jm\pi}$. Using the Blackman window of Eq. 2.7.1 and the spectrum coefficients of Example 2.7.1, the delayed cosine window has $a_1 = -j0.5$, the delayed Hann window has $a_0 = -a_1 = 0.5$, and the delayed

Hamming window has $a_0 = 0.543$ and $a_1 = -0.457$. All the other a_m's equal zero. Substituting these results into Eq. 2.7.2 gives the frequency domain windowing equations as

Cosine : $\quad F_w(m) = -j0.5F_s(m - 0.5) + j0.5F_s(m + 0.5)$

Hann : $\quad F_w(m) = -0.25F_s(m - 1) + 0.5F_s(m) - 0.25F_s(m + 1)$

Hamming : $\quad F_w(m) = -0.229F_s(m - 1) + 0.543F_s(m) - 0.229F_s(m + 1)$

$$(2.7.10)$$

Eq. 2.7.10 replaces Eq. 2.7.3 for windows covering $0 \leq t < NT$. The FFT spectra for this 1 Hz signal is obtained by multiplying the $F_s(m)$ terms of Eq. 2.7.7a by $\exp(-jm\pi)$. Therefore

$$[F_s] = \text{diag}[e^{-jm\pi}] \, [F'_s] = \text{diag} \begin{bmatrix} 1 \\ -1 \\ 1 \\ -1 \\ 1 \\ -1 \\ 1 \\ -1 \end{bmatrix} \begin{bmatrix} 0 \\ -4 \\ 0 \\ 0 \\ 0 \\ 0 \\ 0 \\ -4 \end{bmatrix} = \begin{bmatrix} 0 \\ 4 \\ 0 \\ 0 \\ 0 \\ 0 \\ 0 \\ 4 \end{bmatrix} \qquad (2.7.11)$$

Substituting $[F_s]$ into Eq. 2.7.10 gives the weighted signal spectra as

Cosine : $\quad F_w(m) = -j0.5F_s(m - 0.5) + j0.5F_s(m + 0.5)$

$$[F_w]_{Hann} = -0.25 \begin{bmatrix} 4 \\ 0 \\ 4 \\ 0 \\ 0 \\ 0 \\ 0 \\ 0 \end{bmatrix} + 0.5 \begin{bmatrix} 0 \\ 4 \\ 0 \\ 0 \\ 0 \\ 0 \\ 0 \\ 4 \end{bmatrix} - 0.25 \begin{bmatrix} 4 \\ 0 \\ 0 \\ 0 \\ 0 \\ 0 \\ 4 \\ 0 \end{bmatrix} = \begin{bmatrix} -2 \\ 2 \\ -1 \\ 0 \\ 0 \\ 0 \\ -1 \\ 2 \end{bmatrix}$$

$$[F_w]_{Hamm} = -0.229 \begin{bmatrix} 4 \\ 0 \\ 4 \\ 0 \\ 0 \\ 0 \\ 0 \\ 0 \end{bmatrix} + 0.543 \begin{bmatrix} 0 \\ 4 \\ 0 \\ 0 \\ 0 \\ 0 \\ 0 \\ 4 \end{bmatrix} - 0.229 \begin{bmatrix} 4 \\ 0 \\ 0 \\ 0 \\ 0 \\ 0 \\ 4 \\ 0 \end{bmatrix} = \begin{bmatrix} -1.828 \\ 2.172 \\ -0.914 \\ 0 \\ 0 \\ 0 \\ -0.914 \\ 2.174 \end{bmatrix}$$

$$(2.7.12)$$

These spectra are identical with those computed in Eq. 2.7.8 except for the phase shift factor $\exp(-jm\pi) = \pm 1$. The cosine case cannot be evaluated since the midpoint spectra are not known. Some form of interpolation must be used. Using simple linear spectral interpolation as $F(m + 0.5) = 0.5[F(m) + F(m + 1)]$ gives

$$[F_w]_{cos} \cong j0.5[-F_s(m) - F_s(m - 1) + F_s(m) + F_s(m + 1)]$$

$$
= -j0.5 \begin{bmatrix} 2 \\ 2 \\ 2 \\ 0 \\ 0 \\ 0 \\ 0 \\ 2 \end{bmatrix} + j0.5 \begin{bmatrix} 2 \\ 2 \\ 0 \\ 0 \\ 0 \\ 0 \\ 2 \\ 2 \end{bmatrix} = j \begin{bmatrix} 0 \\ 0 \\ -1 \\ 0 \\ 0 \\ 0 \\ 1 \\ 0 \end{bmatrix} \tag{2.7.13}
$$

which is a poor approximation to the exact spectra given in Eq. 2.7.7b.   ∎

EXAMPLE 2.7.5   A better approximation for $[F_w]_{cos}$ is obtained by expressing the half-cosine pulse as a Fourier series and truncating this series. Find the cosine window spectra using this approach.

SOLUTION   The half-cosine pulse of Table 2.6.1 has the FS form

$$
w(n) = \cos(n\pi/N) = \frac{2}{\pi}\left[1 + \frac{2}{3}\cos(x) - \frac{2}{15}\cos(2x) + \frac{2}{35}\cos(3x) - \ldots\right] \tag{2.7.14}
$$

where $x = 2n\pi/N$. Comparing this result with the general window in Eq. 2.7.1, $a_0 = 2/\pi = 0.637, a_1 = 4/3\pi = 0.424, a_2 = -4/15\pi = -0.085, a_3 = 4/35\pi = 0.036, \ldots$. Therefore the sampled signal spectrum obtained using half-cosine pulse weighting for $-NT/2 \le t < NT/2$ equals

$$
\begin{aligned}
F_w(m) =& 0.637 F_s(m) + 0.212[F_s(m+1) + F_s(m-1)] \\
& - 0.042[F_s(m+2) + F_s(m-2)] + 0.018[F_s(m+3) + F_s(m-3)] - \ldots
\end{aligned} \tag{2.7.15}
$$

from Eq. 2.7.2. For a window covering on $0 \le t < NT$, $F_w(m)$ instead equals

$$
\begin{aligned}
F_w(m) =& 0.637 F_s(m) - 0.212[F_s(m+1) + F_s(m-1)] \\
& - 0.042[F_s(m+2) + F_s(m-2)] - 0.018[F_s(m+3) + F_s(m-3)] - \ldots
\end{aligned} \tag{2.7.16}
$$

This results by multiplying the individual $F_w(m)$ spectra of Eq. 2.7.15 by $e^{-jm\pi}$. Substituting the signal spectrum $F_s(m)$ from Eq. 2.7.7a into this result gives

$$
[F_w]_{cos} = 0.637 \begin{bmatrix} 0 \\ 4 \\ 0 \\ 0 \\ 0 \\ 0 \\ 0 \\ 4 \end{bmatrix} - 0.212 \begin{bmatrix} 8 \\ 0 \\ 4 \\ 0 \\ 0 \\ 0 \\ 4 \\ 0 \end{bmatrix} - 0.042 \begin{bmatrix} 0 \\ 4 \\ 0 \\ 4 \\ 0 \\ 4 \\ 0 \\ 4 \end{bmatrix} - 0.018 \begin{bmatrix} 0 \\ 0 \\ 4 \\ 0 \\ 8 \\ 0 \\ 4 \\ 0 \end{bmatrix} - \ldots \cong \begin{bmatrix} -1.696 \\ 2.380 \\ -0.920 \\ -0.168 \\ -0.144 \\ -0.168 \\ -0.920 \\ 2.380 \end{bmatrix} \tag{2.7.17}
$$

Truncating the infinite series at the fourth term as shown gives a better approximation to the exact spectrum (Eq. 2.7.7d multiplied by $e^{-jm\pi}$) than the linear interpolation of Eq. 2.7.13. Since the cosine window has an asymptotic roll-off of only -40 dB/dec, more terms must be retained in Eqs. 2.7.15 and 2.7.16 for better accuracy. In windows having faster roll-off, fewer terms would be used.   ∎

2.8 OTHER DISCRETE TRANSFORMS

The discrete Fourier transform is the most widely used transform. However there are a number of other discrete transforms. Some of them are listed in Table 1.4.1. These transforms often offer advantages over the discrete Fourier transform so we wish to investigate them. The mathematical groundwork for discrete transforms shall first be discussed. This discussion will augment Chap. 1.4.

Continuous transforms take the form of

$$F(p) = \int_{t_1}^{t_2} K(p,t)f(t)dt, \qquad p \in R_p$$

$$f(t) = \int_{p_1}^{p_2} K^{-1}(p,t)F(p)dp, \qquad t \in R_t$$

(2.8.1)

Discrete transforms have the form

$$F(m) = \sum_{n=1}^{N} K(m,n)f(n), \qquad m = 1,\ldots,M \quad \text{(or } 0,\ldots,M-1)$$

$$f(n) = \sum_{m=1}^{M} K^{-1}(n,m)F(m), \qquad n = 1,\ldots,N \quad \text{(or } 0,\ldots,N-1)$$

(2.8.2)

where the integrals in Eq. 2.8.1 are replaced by summations. We shall only consider discrete transforms but conclusions to be drawn apply to continuous transforms as well. $TF(m)$ is an approximation of $F(p)$ as discussed in Chap. 1.4 (i.e., $TF(m) \cong F(p)$).

$K(m,n)$ is called the *kernel* and $K^{-1}(n,m)$ is called the *inverse kernel* of the transform. The inverse kernels are the basis functions used to approximate $f(n)$. They describe or *span* the function space of $f(n)$. If the kernel set is *complete,* any $f(n)$ can be expressed as a linear combination of the $K^{-1}(n,m)$ with a total squared-error of zero. The spectral coefficient $F(m)$ equals the amount, level, or contribution of $K(m,n)$ in $f(n)$. Likewise $f(n)$ equals the level of $K^{-1}(n,m)$ in $F(m)$. The spectral frequencies being used are indexed as $1,\ldots,M$. The times being used are indexed as $1,\ldots,N$. These points need not be equally spaced as shown in Chap. 2.1.4. The properties of the transform are determined by the properties of its kernel.

The transform equations of Eq. 2.8.2 are usually expressed in matrix form as

$$\begin{bmatrix} F(1) \\ F(2) \\ \vdots \\ F(M) \end{bmatrix} = \begin{bmatrix} K(1,1) & K(1,2) & \ldots & K(1,N) \\ K(2,1) & K(2,2) & \ldots & K(2,N) \\ \vdots & \vdots & \ldots & \vdots \\ K(M,1) & K(M,2) & \ldots & K(M,N) \end{bmatrix} \begin{bmatrix} f(1) \\ f(2) \\ \vdots \\ f(N) \end{bmatrix}$$

$$\begin{bmatrix} f(1) \\ f(2) \\ \vdots \\ f(N) \end{bmatrix} = \begin{bmatrix} J(1,1) & J(1,2) & \ldots & J(1,M) \\ J(2,1) & J(2,2) & \ldots & J(2,M) \\ \vdots & \vdots & \ldots & \vdots \\ J(N,1) & J(N,2) & \ldots & J(N,M) \end{bmatrix} \begin{bmatrix} F(1) \\ F(2) \\ \vdots \\ F(M) \end{bmatrix}$$

(2.8.3)

where $K^{-1}(n,m)$ has been written as $J(n,m)$ to compact the equations. $K^{-1}(m,n)$ is the transpose of $K^{-1}(n,m)$. The time domain temporal samples $f(n)$ and the frequency domain spectral samples $F(m)$ are expressed in column vectors of dimensions N and M, respectively. The FFT matrices in Eq. 2.2.1–2.2.3 had this form.

The approximation bases $J(n,m)$ are arbitrarily chosen. For maximum simplicity, they should be closely matched to the signal so only a few terms are needed to reconstruct $f(n)$ within some desired accuracy. For high-quality reproduction, many more terms need to be included. The spectral coefficients $F(m)$ are found using some optimization method such as least-squares. Many other optimum criteria are possible but we shall only be concerned with least-squares to show the principle involved.[18]

Suppose that an arbitrarily sampled function $g(n)$ is to be approximated by $f(n)$. The total squared-error E equals

$$E = \int_{-\infty}^{\infty} \Big[g(t) - f(t)\Big]^2 dt$$

$$E = \sum_{n=1}^{N} \Big[g(n) - f(n)\Big]^2 \qquad (2.8.4)$$

Substituting $f(n)$ from Eq. 2.8.2 into the discrete error E equation gives

$$E = \sum_{n=1}^{N} \Big[g(n) - \sum_{m=1}^{M} F(m)J(n,m)\Big]^2 \qquad (2.8.5)$$

Once the time samples $g(n)$ and basis functions $J(n,m)$ are chosen, the error depends upon the spectral coefficients $F(m)$, the number of frequencies M, and the number of sample points N. Fixing N and M, then E depends only upon the $F(m)$.

To minimize the error E, the proper $F(m)$ values must be used. These values are found by setting the partial derivatives of E with respect to $F(m)$ equal to zero for all m. Performing this operation on Eq. 2.8.5 yields

$$\frac{\partial E}{\partial F(k)} = -2\sum_{n=1}^{N} J(n,k)\Big[g(n) - \sum_{m=1}^{M} F(m)J(n,m)\Big] = 0, \qquad k = 1,\ldots,M \qquad (2.8.6)$$

Simplifying Eq. 2.8.6 gives

$$\sum_{n=1}^{N} J(n,k)g(n) = \sum_{m=1}^{M}\sum_{n=1}^{N} F(m)J(n,k)J(n,m), \qquad k = 1,\ldots,M \qquad (2.8.7)$$

which can be expressed in matrix form as

$$
\begin{bmatrix}
\sum J(n,1)g(n) \\
\sum J(n,2)g(n) \\
\vdots \\
\sum J(n,M)g(n)
\end{bmatrix}
=
\begin{bmatrix}
\sum J^2(n,1) & \cdots & \sum J(n,1)J(n,M) \\
\sum J(n,2)J(n,1) & \cdots & \sum J(n,2)J(n,M) \\
\vdots & \ddots & \vdots \\
\sum J(n,M)J(n,M) & \cdots & \sum J^2(n,M)
\end{bmatrix}
\begin{bmatrix}
F(1) \\
F(2) \\
\vdots \\
F(M)
\end{bmatrix}
$$

$$(2.8.8)$$

Each sum goes from $n = 1$ to N. This is a set of M linear equations in the M unknown coefficients $F(1), \ldots, F(M)$. These coefficients are found by inverting the matrix as

$$
\begin{bmatrix} F(1) \\ F(2) \\ \vdots \\ F(M) \end{bmatrix} = \begin{bmatrix} \sum J^2(n,1) & \cdots & \sum J(n,1)J(n,M) \\ \sum J(n,2)J(n,1) & \cdots & \sum J(n,2)J(n,M) \\ \vdots & \ddots & \vdots \\ \sum J(n,M)J(n,M) & \cdots & \sum J^2(n,M) \end{bmatrix}^{-1} \begin{bmatrix} \sum J(n,1)g(n) \\ \sum J(n,2)g(n) \\ \vdots \\ \sum J(n,M)g(n) \end{bmatrix}
$$

$$(2.8.9)$$

Now consider the relation between the number N of sample points $g(n)$ and the number M of spectral coefficients $F(m)$. If $N = M$, there is a unique solution to Eq. 2.8.9 (unless the inverse matrix is not defined). However if $N > M$, the problem is over-constrained and there is generally no solution so statistical averaging techniques are necessary. If $N < M$, the problem is under-constrained and there are an infinite number (family) of solutions.

The smallest or least squared-error is obtained when the spectral coefficients $F(m)$ are selected so they satisfy Eq. 2.8.9. The minimum squared-error is obtained by evaluating Eq. 2.8.5 when the optimum coefficients $F(m)$ are used. This results in

$$
E_{min} = \sum_{n=1}^{N} \left[g^2(n) - 2g(n) \sum_{m=1}^{M} F(m)J(n,m) + \sum_{k=1}^{M} \sum_{m=1}^{M} F(k)F(m)J(n,k)J(n,m) \right]
$$

$$(2.8.10)$$

$$
= \sum_{n=1}^{N} g^2(n) - 2\sum_{n=1}^{N} \left[g(n) \sum_{m=1}^{M} F(m)J(n,m) \right] + \sum_{k=1}^{M} F(k)\left[\sum_{n=1}^{N} \sum_{m=1}^{M} F(m)J(n,k)J(n,m) \right]
$$

where the order of summation has been interchanged in the last term. Substituting Eq. 2.8.7 into this term simplifies Eq. 2.8.10 as

$$
E_{min} = \sum_{n=1}^{N} g(n)\left[g(n) - 2\sum_{m=1}^{M} F(m)J(n,m) + \sum_{k=1}^{M} F(k)J(n,k) \right]
$$

$$(2.8.11)$$

$$
= \sum_{n=1}^{N} g(n)\left[g(n) - \sum_{m=1}^{M} F(m)J(n,m) \right] = \sum_{n=1}^{N} g(n)\left[g(n) - f(n) \right]
$$

The minimum error is the sum of the error between the original signal $g(n)$ and its approximation (or reconstruction) $f(n)$ weighted by the original signal $g(n)$. Alternatively stated, it is the difference between the autocorrelation of the original signal and the cross-correlation between the original and approximate signals.

The spectral coefficients $F(m)$ can be computed with much greater ease if the approximating functions $J(m,n)$ are *orthogonal* over the sampling interval N. Mathematically, this means that the $J(m,n)$ satisfy

$$
\int_{t_1}^{t_2} J(p,t)J(m,t)dt = c_p D_0(p-m)
$$

$$(2.8.12)$$

$$
\sum_{n=1}^{N} J(n,k)J(n,m) = \sum_{n=1}^{N} J^t(k,n)J(n,m) = c_m D_0(k-m)
$$

The columns of the inverse kernel matrix J or K^{-1} (Eq. 2.8.3) are orthogonal with one another. For zero-mean basis functions, this is equivalent to saying that their cross-correlation (or cross-products for $k \neq m$) is zero. Their autocorrelations (for $k = m$) equal c_m. If $c_m = 1$, the approximating functions are normalized to unity and are said to be *orthonormal*. Substituting the orthogonality condition of Eq. 2.8.12b into Eq. 2.8.8 yields a diagonal matrix whose off-diagonal terms are zero. Solving Eq. 2.8.8 for the $F(m)$ coefficients yields

$$F(m) = \frac{\sum_{n=1}^{N} J(n,m)g(n)}{\sum_{n=1}^{N} J^2(n,m)}, \qquad m = 1, \ldots, M \qquad (2.8.13)$$

If the spectral components are orthonormal, the denominator of Eq. 2.8.13 is unity which further simplifies this result.

Another very important and useful observation should be made. Any kernel matrix K which is *symmetric* $(K(n,m) = K(m,n))$ has an inverse kernel matrix J or K^{-1} that is proportional so $K^{-1}(m,n) = cK(n,m) = cK^t(m,n)$ in Eq. 2.8.3. This means that any algorithm used for computing a discrete transform can be used directly to compute the inverse discrete transform but the results must be multiplied by c. Thus, fast transforms having symmetric kernels can be used directly as fast inverse transforms.

Now let us summarize all these ideas. Any sampled signal $g(n)$ can be approximated by a linear combination of basis functions $K^{-1}(n,m)$ (Eq. 2.8.2). The relative weights of the various K^{-1} terms are the spectral coefficients $F(m)$. The optimum $F(m)$ in the least-squares sense satisfy Eq. 2.8.9. If the basis functions are orthogonal, Eq. 2.8.9 reduces to a diagonal matrix and $F(m)$ is given by Eq. 2.8.13. The approximate signal $f(n)$ is a linear combination of the spectral $K^{-1}(n,m)$ terms weighted by the spectral coefficients $F(m)$ (Eq. 2.8.2). The minimum error between the signal $g(n)$ and its approximation $f(n)$ is given by Eq. 2.8.11. The error approaches zero as $N \to \infty$ when the basis set is complete. Furthermore, symmetric kernel matrices have identical inverse kernel matrices multiplied by c. In such cases, the same algorithms can be used for computing both transform and inverse transform coefficients. Having established the foundation for transforms, some well-known transforms will now be reviewed.

A variety of discrete transforms are listed in Table 2.8.1.[19] One convenient classification groups the transforms as statistically optimum (Karhunen-Loéve) and suboptimum (all others). The suboptimum transforms are subgrouped as type 1 and type 2. Type 1 transforms have basis functions (vectors) which lie on the unit circle in the transform domain. The discrete Fourier transform and Walsh-Hadamard transforms are examples of type 1 transforms. Type 2 transforms have basis functions which lie off the unit circle. Most discrete transforms are type 2. The type 2 transforms are further divided into sinusoidal (i.e., sinusoidal basis vectors) or nonsinusoidal (i.e., nonsinusoidal basis vectors). The discrete cosine and discrete sine transforms are examples of sinusoidal type 2 transforms. The Haar and slant transforms are examples of nonsinusoidal type 2 transforms. The basis functions (inverse kernels) are listed in Table 1.4.1.

The spectral coefficients using any of these transforms are found by evaluating the matrix of Eq. 2.8.9. If the transform is orthogonal, Eq. 2.8.13 is evaluated instead.

Table 2.8.1 Some types of discrete transforms.[19]

```
                    ┌──────────────────────────────┐
                    │ Discrete orthogonal transforms│
                    └──────────────────────────────┘
              ┌──────────────┴─────────────────┐
   ┌──────────────────────┐          ┌──────────────────────┐
   │ Optimal transforms   │          │ Suboptimal transforms│
   │ Karhunen-Loéve DKLT  │          └──────────────────────┘
   └──────────────────────┘      ┌───────────┴───────────┐
         ┌────────┐          ┌────────┐
         │ Type 1 │          │ Type 2 │
         └────────┘          └────────┘
```

| Fourier DFT, FFT
Walsh DWT
Cal-Sal DCST
Paley DPT
Hadamard DHT | Type 2-Nonsinusoidal
Haar DHaT
Slant DST
Hadamard-Haar DHHT
Slant-Haar DSHT
Fermat DFNT
Rationalized Haar DRHT | Type 2-Sinusoidal
Discrete Cosine DCosT
Discrete Sine DSinT |

Table 2.8.2 Approximate number of real or complex operations needed to implement fast transforms.[19]

Transform	Real Operations	Complex Operations
DKLT	N^2	
DFT		$N(N-1)$
FFT		$N \log_2 N$
WHT	$N \log_2 N$ additions or subtractions	
DHT	$2(N-1)$	
DST	$N \log_2 N + (2N-4)$	
CHT		$3N-4$
HHT_1	$4(N/2-1)+N = 3N-4$	
HHT_2	$8(N/4-1)+2N = 4N-8$	
HHT_3	$16(N/8-1)+3N = 5N-16$	
HHT_r	$2^{r+1}(N/2^r-1)+rN = \left[(r+2)N - 2^{r+1}\right]$	
SHT_r	$(r+2)N - 2^{r+1}$	
DRT	$N \log_2 N$ additions or subtractions	
DLB	$N \log_2 N$ (integer arithmetic)	
DCosT	$N \log_2 N$	
DSinT		$2N \log_2 2N$

This requires N^2 multiplications and $N(N-1)$ additions. However the fast transform methods can be used to reduce the number of operations required. For example the fast DFT algorithms or FFT's required only $N \log_2 N$ complex multiplications and additions which represents considerable computational savings. The approximate number of real or complex operations (multiply or add) required for the transforms of Table 2.8.1 are listed in Table 2.8.2. Most transforms require about $N \log_2 N$ operations. The Haar transform (HaT), Haar-Hadamard transform (HHT), and slant transform (ST) require fewer operations. The rapid transform (RT) and Walsh-Hadamard transform (WHT) require only additions and subtractions which are usually much faster than multiplications. We shall now discuss the Walsh,[20] Paley, Hadamard,[21] Haar,[22] and slant [23] transforms in detail and introduce the Karhunen-Loéve transform.

2.8.1 WALSH (DWT) AND CAL-SAL (DCST) TRANSFORMS

Walsh functions are sequences of ± 1 pulses. A Walsh function $Wal(p,t)$ of order N is composed of $N/2$ pulses. The eighth-order Walsh functions are shown in Fig. 2.8.1a. These functions are nonperiodic inside the interval N but the entire pulse sequence repeats itself indefinitely. The concept of frequency must be generalized to describe these functions. Consider periodic functions having evenly spaced zero-crossings (zc) like sine waves $\sin(\omega t)$ and square waves $U_{-1}(\sin \omega t)$. *Frequency f* is defined to equal the number of cycles per second (cps or Hertz), or equivalently, one-half the number of zero crossings per second (zps). For nonperiodic functions having unevenly spaced zero-crossings, *generalized frequency p* is defined to equal one-half the average number of zero-crossings per second. Therefore $p = zc/2T$ for even zc and $p = (zc+1)/2T$ for odd zc. *Sequency* equals one-half the number of zero crossings over the same interval.

Walsh functions $Wal(k,t)$ are usually ordered in four ways. This ordering is given different names as:

1. Sequency or Walsh ordering (Fig. 2.8.1a).
2. Cal-Sal ordering (Fig. 2.8.1b).
3. Dyadic or Paley ordering (Fig. 2.8.1c).
4. Natural or Hadamard ordering (Fig. 2.8.1d).

The Cal-Sal ordering groups the Walsh functions $Wal(p,t)$ according to even symmetry $Cal(p/2,t)$ in ascending sequency order and odd symmetry $Sal((p+1)/2,t)$ in descending sequency order. The sequency or Walsh ordering $Wal(p,t)$ groups the Walsh functions in ascending order of sequency, or equivalently, by (Sal, Cal) pairs having the same generalized frequency. The dyadic or Paley ordering $Pal(p,t)$ regroups the sequency ordered Walsh functions $Wal(g(p),t)$ in a Gray code–to–binary converted order. The natural or Hadamard ordering $Had(p,t)$ regroups the sequency ordered Walsh functions $Wal(b(p),t)$ in a bit-reversed and then Gray code–to–binary conversion order. Equivalently, Hadamard ordering regroups Paley in a bit-reversed order.

Since all four transforms contain the same information, only the order distinguishes one transform from another. When spectra should be ordered in increasing sequency, then Walsh ordering is used. When grouping signals by even and odd symmetry, then Cal-Sal ordering is used. When bit-reversed pairs are needed, then Paley and Hadamard ordering is used. We will now investigate these orderings in more detail.

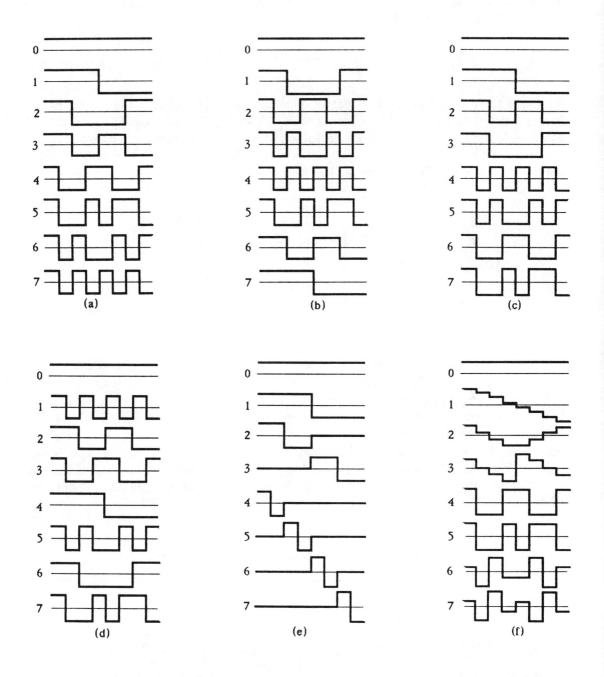

Fig. 2.8.1 Kernels and basis functions for 8-point (a) Walsh, (b) Cal-Sal, (c) Paley, (d) Hadamard, (e) Haar, and (f) slant transforms.

2.8.2 PALEY TRANSFORM (DPT)

The discrete Walsh transform with Paley ordering, or simply the *discrete Paley transform (DPT)*, has a kernel $K(m,n)$ of

$$K(m,n) = \frac{1}{N}Pal(m,n) = \frac{1}{N}(-1)^{\sum_{i=0}^{L-1} b_i(n)b_{L-1-i}(m)}, \qquad L = \log_2 N \qquad (2.8.14)$$

$[Pal_N]$ is the Paley transform of matrix dimension $N \times N$. $b_i(n)$ is the ith bit in the binary representation of n. For example if $N = 8$ so $L = 3$, then the exponent of (-1) equals $b_0(n)b_2(m) + b_1(n)b_1(m) + b_2(n)b_0(m)$. If $n = 6$, then since 6 (in base 10) = 110 (in binary or base 2), it follows that $b_2(6) = 1, b_1(6) = 1$, and $b_0(6) = 0$. The Paley transform is always filled with ± 1's. For $N = 8$, it equals

$$[Pal_N] = N[Pal_N]^{-1} = \begin{bmatrix} + & + & + & + & + & + & + & + \\ + & + & + & + & - & - & - & - \\ + & + & - & - & + & + & - & - \\ + & + & - & - & - & - & + & + \\ + & - & + & - & + & - & + & - \\ + & - & + & - & - & + & - & + \\ + & - & - & + & + & - & - & + \\ + & - & - & + & - & + & + & - \end{bmatrix} \qquad (2.8.15)$$

where only the signs $(\pm)$ of unity are shown. It is a symmetric matrix so the inverse kernel is proportional to the kernel as $[Pal_N]^{-1} = [Pal_N]/N$. Also notice the rows and columns are orthogonal. This is verified from Eq. 2.8.12 since the cross-correlation between any pair of rows or columns is zero. Since the autocorrelation of any row or column equals N, dividing $[Pal_N]$ by $\sqrt{N}$ produces an orthonormal set of Paley functions. The $N = 8$ continuous Paley functions are shown in Fig. 2.8.1c. The $N = 8$ discrete Paley functions of Eq. 2.8.15 are obtained by sampling the continuous functions.

The discrete Paley transform describes the spectrum $F(m)$ of a signal $f(n)$ in terms of discrete Paley functions $Pal(m,n)$ as

$$F(m) = \frac{1}{N}\sum_{n=0}^{N-1} Pal(m,n)f(n), \qquad m = 0,\ldots,N-1$$

$$f(n) = \sum_{m=0}^{N-1} Pal(n,m)F(m), \qquad n = 0,\ldots,N-1 \qquad (2.8.16)$$

EXAMPLE 2.8.1 Consider a parabolic signal $f(t) = t^2$. Determine its Paley transform spectra for $0 \le t < 1$ using 8 terms.

SOLUTION The continuous transform spectra $F(p)$ are obtained by evaluating Eq. 2.8.1a. We will normalize $F(p)$ by $1/T$ so that these coefficients can be directly compared with the discrete spectra $F(m)$. Denoting $F(p)/T$ as $F_n(p)$, then

$$F_n(0) = \frac{1}{T}F(0) = \frac{1}{T}\int_0^T Pal(0,t)t^2\,dt = \int_0^1 t^2\,dt = 1/3 = 0.333,$$

$$F_n(1) = \frac{1}{T}\int_0^T Pal(1,t)t^2\,dt = \int_0^{0.5} t^2\,dt - \int_{0.5}^1 t^2\,dt = -1/4 = -0.25,$$

$$F_n(2) = -1/8 = -0.125, \qquad F_n(3) = 1/16 = 0.0625,$$

$$F_n(4) = -1/16 = -0.0625, \qquad F_n(5) = 1/32 = 0.0313,$$

$$F_n(6) = 1/64 = 0.0156, \qquad F_n(7) = 0$$

(2.8.17)

These are simply t^2 integrals with plus or minus signs evaluated across the $[0,1]$ interval. The discrete transform spectra given by Eq. 2.8.16 are easily evaluated using the Paley matrix of Eq. 2.8.15 and signal $f(n) = (n/8)^2$ as

$$[F]_{Pal} =
\begin{bmatrix}
0.2734 \\
-0.2188 \\
-0.1094 \\
0.0625 \\
-0.0547 \\
0.0313 \\
0.0156 \\
0
\end{bmatrix}
= \frac{1}{512}
\begin{bmatrix}
140 \\
-112 \\
-56 \\
32 \\
-28 \\
16 \\
8 \\
0
\end{bmatrix}
= \frac{1}{8}
\begin{bmatrix}
+ & + & + & + & + & + & + & + \\
+ & + & + & + & - & - & - & - \\
+ & + & - & - & + & + & - & - \\
+ & + & - & - & - & - & + & + \\
+ & - & + & - & + & - & + & - \\
+ & - & + & - & - & + & - & + \\
+ & - & - & + & + & - & - & + \\
+ & - & - & + & - & + & + & -
\end{bmatrix}
\begin{bmatrix}
0/64 \\
1/64 \\
4/64 \\
9/64 \\
16/64 \\
25/64 \\
36/64 \\
49/64
\end{bmatrix}$$

(2.8.18)

The discrete transform spectra are fairly close to the continuous transform spectra. ∎

EXAMPLE 2.8.2 Now consider a 1 Hz cosine signal $f(t) = \cos(2\pi f_o t)$. Find its Paley transform spectra in the interval $0 \le t < 1$ using 8 terms.
SOLUTION The continuous Paley transform spectra equal

$$F_n(0) = \frac{1}{T}\int_0^T Pal(0,t)\cos(2\pi t)\,dt = 0, \qquad F_n(1) = 0,$$

$$F_n(2) = 0, \qquad F_n(3) = 2/\pi = 0.637, \qquad F_n(4) = 0,$$

$$F_n(5) = 0.263, \qquad F_n(6) = 0, \qquad F_n(7) = 0$$

(2.8.19)

The discrete Paley transform spectra are

$$[F]_{Pal} =
\begin{bmatrix}
0 \\
0.2500 \\
0 \\
0.6036 \\
0 \\
0.2500 \\
0 \\
-0.1036
\end{bmatrix}
= \frac{1}{8}
\begin{bmatrix}
+ & + & + & + & + & + & + & + \\
+ & + & + & + & - & - & - & - \\
+ & + & - & - & + & + & - & - \\
+ & + & - & - & - & - & + & + \\
+ & - & + & - & + & - & + & - \\
+ & - & + & - & - & + & - & + \\
+ & - & - & + & + & - & - & + \\
+ & - & - & + & - & + & + & -
\end{bmatrix}
\begin{bmatrix}
1.000 \\
0.707 \\
0 \\
-0.707 \\
-1.000 \\
-0.707 \\
0 \\
0.707
\end{bmatrix}$$

(2.8.20)

∎

The decimation process used to derive FFT's can be used for Paley and other discrete transforms. The signal flow graph of a fast Paley transform (FPT) with bit-reversed input and normal output is shown in Fig. 2.8.2a (cf. Fig. 2.3.6). A FPT subroutine is listed in Table 2.8.3. It is much simpler than the FFT subroutines of Tables 2.4.1 and 2.4.2. Since the basis functions are ± 1 rather than $e^{j\theta}$, no complex multiplications are needed and only real additions and subtractions are required.

2.8.3 HADAMARD TRANSFORM (DHT)

The *discrete Hadamard transform (DHT)* has a kernel $K(m, n)$ of

$$K(m, n) = \frac{1}{N} Had(m, n) = \frac{1}{N}(-1)^{\sum_{i=0}^{L-1} b_i(n) b_i(m)}, \qquad L = \log_2 N \qquad (2.8.21)$$

$[Had_N]$ is the Hadamard transform matrix of dimension $N \times N$. The summation is performed with base 2 arithmetic. $b_i(p)$ is the ith bit in the binary representative of p as with the Paley transform. The Hadamard transform matrix is also always filled with ± 1's. For $N = 8$, it equals

$$[Had_N] = N[Had_N]^{-1} = \begin{bmatrix} + & + & + & + & + & + & + & + \\ + & - & + & - & + & - & + & - \\ + & + & - & - & + & + & - & - \\ + & - & - & + & + & - & - & + \\ + & + & + & + & - & - & - & - \\ + & - & + & - & - & + & - & + \\ + & + & - & - & - & - & + & + \\ + & - & - & + & - & + & + & - \end{bmatrix} \qquad (2.8.22)$$

Since the matrix is symmetric, the kernels and inverse kernels are proportional as $[Had_N]^{-1} = [Had_N]/N$. The kernels and inverse kernels are orthogonal. Dividing $[Had_N]$ by $\sqrt{N}$ produces a set of orthonormal Hadamard functions. The $N = 8$ continuous Hadamard functions are shown in Fig. 2.8.1d. The $N = 8$ discrete Hadamard functions of Eq. 2.8.22 are sampled versions of the continuous functions.

The discrete Hadamard transform describes the spectrum $F(m)$ of a signal $f(n)$ in terms of discrete Hadamard functions $Had(m, n)$ as

$$F(m) = \frac{1}{N} \sum_{n=0}^{N-1} Had(m, n) f(n), \qquad m = 0, \dots, N-1$$

$$f(n) = \sum_{m=0}^{N-1} Had(n, m) F(m), \qquad n = 0, \dots, N-1$$

$$(2.8.23)$$

The Paley and Hadamard matrices (Eqs. 2.8.15 and 2.8.22) are identical except for the ordering of the rows (or columns). Since the WT, CST, PT, and HT are often mixed in the literature, they are collectively referred to as the Walsh-Hadamard transforms.

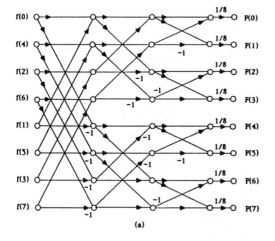

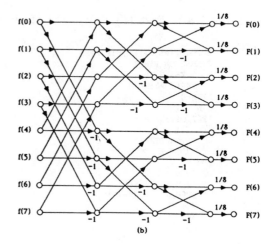

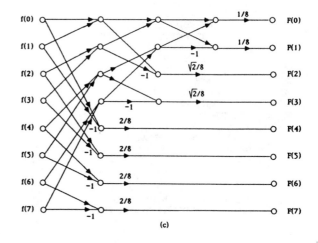

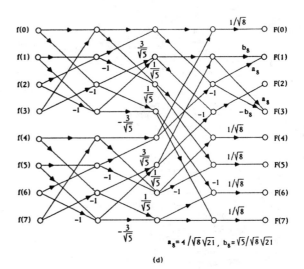

Fig. 2.8.2 Signal flow graphs for 8-point
(a) Paley,
(b) Hadamard,
(c) Haar, and
(d) slant transforms.

Table 2.8.3 Fast (a) Paley, (b) Hadamard, (c) Haar, and (d) slant subroutines.

(a)
```
      SUBROUTINE FPT(N,FTR,FWR)
      DIMENSION FTR(1),FWR(1),Z(1024)
      DO 5 I=1,N
5         FWR(I)=FTR(I)
      ITER=0
      I=N
10    IF(I.GT.1) THEN
          I=I/2
          ITER=ITER+1
          GOTO 10
      ENDIF
      DO 30 M=1,ITER
          IF(M.EQ.1) NP=1
          IF(M.GT.1) NP=NP*2
          MN=N/NP
          MN2=MN/2
          DO 20 MP=1,NP
              IB=(MP-1)*MN
              DO 15 K=1,MN2
                  KK=IB+K
                  KP=KK+MN2
                  J=IB+2*(K-1)+1
                  T1=FWR(J)
                  T2=FWR(J+1)
                  Z(KK)=T1+T2
15                Z(KP)=T1-T2
20        CONTINUE
          DO 25 I=1,N
25            FWR(I)=Z(I)
30    CONTINUE
      DO 35 I=1,N
35        FWR(I)=FWR(I)/FLOAT(N)
      RETURN
      END
```

(b)
```
      SUBROUTINE FHT(N,FTR,FWR)
      DIMENSION FTR(1),FWR(1)
      DO 5 I=1,N
5         FWR(I)=FTR(I)
      ITER=0
      I=N
10    IF(I.GT.1) THEN
          I=I/2
          ITER=ITER+1
          GOTO 10
      ENDIF
      DO 30 M=1,ITER
          IF(M.EQ.1) NP=N/2
          IF(M.GT.1) NP=NP/2
          MN=N/NP
          MN2=MN/2
          DO 20 MP=1,NP
              IB=(MP-1)*MN
              DO 15 K=1,MN2
                  KK=IB+K
                  KP=KK+MN2
                  T1=FWR(KK)
                  T2=FWR(KP)
                  FWR(KK)=T1+T2
                  FWR(KP)=T1-T2
15        CONTINUE
20    CONTINUE
30    CONTINUE
      DO 35 I=1,N
35        FWR(I)=FWR(I)/FLOAT(N)
      RETURN
      END
```

(c)
```
      SUBROUTINE FHAT(N,FTR,FWR)
      DIMENSION FTR(1),FWR(1),Z(1024)
      DO 5 I=1,N
5         FWR(I)=FTR(I)
      ITER=0
      I=N
10    IF(I.GT.1) THEN
          I=I/2
          ITER=ITER+1
          GOTO 10
      ENDIF
      C=SQRT(2.0)
      L=1
      DO 30 M=1,ITER
          IF(M.EQ.1) NP=N/2
          IF(M.GT.1) NP=NP/2
          DO 15 MP=1,NP
              K=(MP-1)*2+1
              MM=MP+NP
              R=C**(ITER-M)
              T1=FWR(K)
              T2=FWR(K+1)
              Z(MP)=T1+T2
15            Z(MM)=(T1-T2)*R/FLOAT(N)
          NN=N/L
          DO 20 I=1,NN
20            FWR(I)=Z(I)
30        L=L*2
      FWR(1)=FWR(1)/FLOAT(N)
      RETURN
      END
```

(d)
```
      SUBROUTINE FST(N,FTR,FWR)
      DIMENSION FTR(1),FWR(1),TR(4)
      ITER=0
      I=N
5     IF(I.GT.1) THEN
          I=I/2
          ITER=ITER+1
          GOTO 5
      ENDIF
      IF (ITER.EQ.2) THEN
      DO 8 I=1,2
          J=5-I
          TR(I)=FTR(I)+FTR(J)
8         TR(I+2)=FTR(I)-FTR(J)
      R=SQRT(5.0)
      FWR(1)=TR(1)+TR(2)
      FWR(2)=(3.0Q0*TR(3)+TR(4))*R/5.0
      FWR(3)=TR(1)-TR(2)
      FWR(4)=(TR(3)-3.0Q0*TR(4))*R/5.0
      ELSE
          DO 10 I=1,N
10            FWR(I)=FTR(I)
      N1=N-1
      MR=0
      DO 20 M=1,N1
      L=N
15    L=L/2
      IF(L.EQ.0) GOTO 20
      IF((MR+L).GT.N1) GOTO 15
```

Table 2.8.3 (Continued) Remainder of fast slant subroutine.

```
      MR=MOD(MR,L)+L                                  35     CONTINUE
      IF(MR.LE.M) GOTO 20                                    ITER=ITER-1
      TR1=FWR(M+1)                                           M=N/2
      FWR(M+1)=FWR(MR+1)                                     N4=1
      FWR(MR+1)=TR1                                          DO 45 I=1,ITER
20    CONTINUE                                                 M=M/2
      DO 35 M=1,ITER                                           N4=N4*2
        IF(M.EQ.1) NP=1                                        X1=FLOAT(N4)**2
        IF(M.GT.1) NP=NP*2                                     AN=SQRT(3*X1/(4*X1-1))
        MN=N/NP                                                BN=SQRT((X1-1)/(4*X1-1))
        MN2=MN/2                                               DO 40 K=1,M
        C=1.0                                                    T1=FWR(M+K)
        DO 30 MP=1,NP                                            T2=FWR(3*M+K)
          IB=(MP-1)*MN                                           FWR(M+K)=AN*T1+T2*BN
          DO 25 K=1,MN2                                 40       FWR(3*M+K)=-T1*BN+T2*AN
            KK=IB+K                                     45     CONTINUE
            KP=KK+MN2                                          ENDIF
            T1=FWR(KK)                                         DO 50 I=1,N
            T2=FWR(KP)                                  50       FWR(I)=FWR(I)/FLOAT(N)
25          FWR(KK)=T1+C*T2                                   RETURN
30          FWR(KP)=T1-C*T2                                   END
          C=-C
```

A fast Hadamard transform (FHT) with normal input and output is shown in signal flow graph form in Fig. 2.8.2b. The FPT subroutine of Table 2.8.3 can be used to obtain the Hadamard spectral coefficients $F(m)$ if the coefficients are reordered. Comparing the DPT matrix (Eq. 2.8.15) and the DHT matrix (Eq. 2.8.22) for example, the Hadamard coefficients are equal to the Paley coefficients reordered in bit-reversed sequence as $(0, 4, 2, 6, 1, 5, 3, 7)$. A FHT subroutine is listed in Table 2.8.3.

EXAMPLE 2.8.3 Determine the 8-point Hadamard transform spectra of the $\cos(2\pi t)$ signal for $0 \leq t < 1$.

SOLUTION Since the Hadamard approximation is equivalent to the Paley approximation in rearranged form, the Paley results of Example 2.8.2 can be used. From Eqs. 2.8.19 and 2.8.20, the cosine signal has continuous and discrete Hadamard transform spectra of

$$
[F_n(p)]_{Had} = \begin{bmatrix} 0 \\ 0 \\ 0 \\ 0 \\ 0 \\ 0.263 \\ 0.637 \\ 0 \end{bmatrix}, \qquad [F(m)]_{Had} = \begin{bmatrix} 0 \\ 0 \\ 0 \\ 0 \\ 0.2500 \\ 0.2500 \\ 0.6036 \\ -0.1036 \end{bmatrix} \qquad (2.8.24)
$$

by reordering coefficients in a $(0, 4, 2, 6, 1, 5, 3, 7)$ bit-reversed sequence. ■

Hadamard transforms have simple recursion relations for generating higher-order kernel matrices. Denoting the Nth order Hadamard kernel matrix as $[Had_N]$, then the

$2N$th order Hadamard kernel matrix equals

$$[Had_{2N}] = \begin{bmatrix} [Had_N] & [Had_N] \\ [Had_N] & -[Had_N] \end{bmatrix} \qquad (2.8.25)$$

Thus, $[Had_{2N}]$ is a block matrix composed of four $\pm[Had_N]$ matrices. For example, the first-, second-, and fourth-order Hadamard matrices equal

$$[Had_1] = [+], \qquad [Had_2] = \begin{bmatrix} + & + \\ + & - \end{bmatrix}, \qquad [Had_4] = \begin{bmatrix} + & + & + & + \\ + & - & + & - \\ + & + & - & - \\ + & - & - & + \end{bmatrix} \qquad (2.8.26)$$

while the eighth-order matrix is given by Eq. 2.8.22.

2.8.4 HAAR TRANSFORM (DHaT)

The *discrete Haar transform (DHaT)* matrices are filled with scaled tri-state signals ± 1 (denoted $+$ or $-$) and 0 (denoted $.$). The lowest order Haar transform matrices equal

$$[Ha_1] = [+], \qquad [Ha_2] = \begin{bmatrix} + & + \\ + & - \end{bmatrix}, \qquad [Ha_4] = \begin{bmatrix} (& + & + & + & + &) & 1 \\ (& + & + & - & - &) & 1 \\ (& + & - & . & . &) & \sqrt{2} \\ (& . & . & + & - &) & \sqrt{2} \end{bmatrix}$$

$$[Ha_8] = \begin{bmatrix} (& + & + & + & + & + & + & + & + &) & 1 \\ (& + & + & + & + & - & - & - & - &) & 1 \\ (& + & + & - & - & . & . & . & . &) & \sqrt{2} \\ (& . & . & . & . & + & + & - & - &) & \sqrt{2} \\ (& + & - & . & . & . & . & . & . &) & 2 \\ (& . & . & + & - & . & . & . & . &) & 2 \\ (& . & . & . & . & + & - & . & . &) & 2 \\ (& . & . & . & . & . & . & + & - &) & 2 \end{bmatrix}$$

$$(2.8.27)$$

The parenthesis inside the matrix means that the entire row must be multiplied by the number outside the parenthesis. Higher-order matrices can be generated using

$$[Ha_{2N}] = \begin{bmatrix} [Ha_N] & \times & [1 & 1] \\ c[I_N] & \times & [1 & -1] \end{bmatrix}, \qquad c = \sqrt{N} \qquad (2.8.28)$$

where $[I_N]$ is the Nth order unit matrix whose off-diagonal terms are zero. Inspecting these matrices and the recursion equation of Eq. 2.8.28, we see the first rows are all $+1$'s and the second rows are half $+1$'s and half -1's. The remaining rows contain positive and negative pulses embedded in 0's. The amplitudes increase in powers of $\sqrt{2}$. Such a transform samples the input signal with finer temporal resolution. This transform is well matched to burst-type signals that tend to be concentrated in localized regions. The $N = 8$ continuous Haar functions are shown in Fig. 2.8.1e. Sampling these functions produce the $N = 8$ discrete transforms of Eq. 2.8.27.

The discrete Haar transform describes the spectrum $F(m)$ of a signal $f(n)$ in terms of discrete Haar functions $Ha(m,n)$ as

$$F(m) = \frac{1}{N} \sum_{n=0}^{N-1} Ha(m,n)f(n), \qquad m = 0,\ldots,N-1$$

$$f(n) = \sum_{m=0}^{N-1} Ha(n,m)F(m), \qquad n = 0,\ldots,N-1$$

(2.8.29)

$[Ha_N]$ is orthogonal but not symmetric as the Paley and Hadamard cases. The Haar matrix satisfies $[Ha_N][Ha_N]^t = N[I_N]$.

EXAMPLE 2.8.4 Determine the 8-point Haar transform spectra of the $\cos(2\pi t)$ signal for $0 \le t < 1$.
SOLUTION The continuous Haar transform spectra equal

$$F_n(0) = \frac{1}{T} \int_0^T Ha(0,t)\cos(2\pi t)dt = 0, \qquad F_n(1) = 0,$$

$$F_n(2) = 0.4502, \qquad F_n(3) = -0.4502, \qquad F_n(4) = 0.1318,$$

$$F_n(5) = 0.1318, \qquad F_n(6) = -0.1318, \qquad F_n(7) = -0.1318$$

(2.8.30)

Using the DHaT given by Eq. 2.8.27d, the discrete Haar spectra equal

$$[F]_{Ha} =$$

$$
\begin{bmatrix}
0 \\
0.2500 \\
0.4268 \\
-0.4268 \\
0.0732 \\
0.1768 \\
-0.0732 \\
-0.1768
\end{bmatrix}
= \frac{1}{8}
\begin{bmatrix}
(\ +\ +\ +\ +\ +\ +\ +\ +\ ) & 1 \\
(\ +\ +\ +\ +\ -\ -\ -\ -\ ) & 1 \\
(\ +\ +\ -\ -\ \cdot\ \cdot\ \cdot\ \cdot\ ) & \sqrt{2} \\
(\ \cdot\ \cdot\ \cdot\ \cdot\ +\ +\ -\ -\ ) & \sqrt{2} \\
(\ +\ -\ \cdot\ \cdot\ \cdot\ \cdot\ \cdot\ \cdot\ ) & 2 \\
(\ \cdot\ \cdot\ +\ -\ \cdot\ \cdot\ \cdot\ \cdot\ ) & 2 \\
(\ \cdot\ \cdot\ \cdot\ \cdot\ +\ -\ \cdot\ \cdot\ ) & 2 \\
(\ \cdot\ \cdot\ \cdot\ \cdot\ \cdot\ \cdot\ +\ -\ ) & 2
\end{bmatrix}
\begin{bmatrix}
1.000 \\
0.707 \\
0 \\
-0.707 \\
-1.000 \\
-0.707 \\
0 \\
0.707
\end{bmatrix}
$$

(2.8.31) ∎

The signal flow graph for a fast Haar transform is shown in Fig. 2.8.2c. Input and output are in normal order. Notice the peculiar upper triangular symmetry produced by the short pulses. A FHaT subroutine is listed in Table 2.8.3.

2.8.5 SLANT TRANSFORM (DST)

The $N = 8$ continuous slant transform uses multi-level digital signals as shown in Fig. 2.8.1f. These signals are quantized triangular, sawtooth, and pulsed waves. Sampling these signals produces the *discrete slant transform (DST)*. The first several slant transform matrices equal

$$[S_2] = \begin{bmatrix} 1 & 1 \\ 1 & 1 \end{bmatrix}, \qquad [S_4] = \begin{bmatrix} (\ 1 & 1 & 1 & 1) \\ (\ 3 & 1 & -1 & -3) & /\sqrt{5} \\ (\ 1 & -1 & -1 & 1) \\ (\ 1 & -3 & 3 & -1) & /\sqrt{5} \end{bmatrix}$$

$$[S_8] = \begin{bmatrix} (\ 1 & 1 & 1 & 1 & 1 & 1 & 1 & 1 &) \\ (\ 7 & 5 & 3 & 1 & -1 & -3 & -5 & -7 &) & /\sqrt{21} \\ (\ 3 & 1 & -1 & -3 & -3 & -1 & 1 & 3 &) & /\sqrt{5} \\ (\ 7 & -1 & -9 & -17 & 17 & 9 & 1 & -7 &) & /\sqrt{105} \\ (\ 1 & -1 & -1 & 1 & 1 & -1 & -1 & 1 &) \\ (\ 1 & -1 & -1 & 1 & -1 & 1 & 1 & -1 &) \\ (\ 1 & -3 & 3 & -1 & -1 & 3 & -3 & 1 &) & /\sqrt{5} \\ (\ 1 & -3 & 3 & -1 & 1 & -3 & 3 & -1 &) & /\sqrt{5} \end{bmatrix} \qquad (2.8.32)$$

These matrices can be recursively generated using

$$[S_{2N}] = \frac{1}{\sqrt{2}} \begin{bmatrix} \begin{array}{cc:c} 1 & 0 & 0 \\ a_N & b_N & \\ \cdots & \cdots & \cdots \\ 0 & & I \\ \hdashline 0 & 1 & 0 \\ -b_N & a_N & \\ \cdots & \cdots & \cdots \\ 0 & & I \end{array} \middle| \begin{array}{cc:c} 1 & 0 & 0 \\ -a_N & b_N & \\ \cdots & \cdots & \cdots \\ 0 & & I \\ \hdashline 0 & 1 & 0 \\ b_N & a_N & \\ \cdots & \cdots & \cdots \\ 0 & & I \end{array} \end{bmatrix} \begin{bmatrix} [S_{N/2}] & [0] \\ \hline [0] & [S_{N/2}] \end{bmatrix}$$

$$(2.8.33)$$

where the coefficients in the block submatrices equal

$$a_2 = 1, \qquad a_{2N} = 2b_{2N}a_N, \qquad b_{2N} = \frac{1}{\sqrt{1 + 4a_N^2}}, \qquad N = 2,4,8,\ldots \qquad (2.8.34)$$

The slant transform well describes signals that tend to be ramp-like with occasional discontinuities.

The discrete slant transform describes the spectrum $F(m)$ of a signal $f(n)$ in terms of discrete slant functions $S(m,n)$ as

$$F(m) = \frac{1}{N} \sum_{n=0}^{N-1} S(m,n)f(n), \qquad m = 0,\ldots,N-1$$

$$(2.8.35)$$

$$f(n) = \sum_{m=0}^{N-1} S(n,m)F(m), \qquad n = 0,\ldots,N-1$$

Like the Haar transform, $[S_N]$ is orthogonal but not symmetric and satisfies $[S_N][S_N]^t = N[I_N]$.

EXAMPLE 2.8.5 Determine the 8-point slant transform spectra of the $\cos(2\pi t)$ signal for $0 \le t < 1$.

SOLUTION The continuous slant transform spectra equal

$$F_n(0) = \frac{1}{T} \int_0^T S(0,t) \cos(2\pi t)\,dt = 0, \qquad F_n(1) = 0,$$

$$F_n(2) = 0.7256, \qquad F_n(3) = 0, \qquad F_n(4) = 0, \qquad (2.8.36)$$

$$F_n(5) = 0, \qquad F_n(6) = -0.0488, \qquad F_n(7) = 0$$

Using the DST of Eq. 2.8.32c, the discrete slant spectra equal

$$[F]_S =$$

$$\begin{bmatrix} 0 \\ 0.2182 \\ 0.6516 \\ -0.1220 \\ 0 \\ -0.1036 \\ -0.0463 \\ 0 \end{bmatrix} = \frac{1}{8} \begin{bmatrix} (\ 1 & 1 & 1 & 1 & 1 & 1 & 1 & 1\ ) & \\ (\ 7 & 5 & 3 & 1 & -1 & -3 & -5 & -7\ ) & /4.583 \\ (\ 3 & 1 & -1 & -3 & -3 & -1 & 1 & 3\ ) & /2.236 \\ (\ 7 & -1 & -9 & -17 & 17 & 9 & 1 & -7\ ) & /10.25 \\ (\ 1 & -1 & -1 & 1 & 1 & -1 & -1 & 1\ ) & \\ (\ 1 & -1 & -1 & 1 & -1 & 1 & 1 & -1\ ) & \\ (\ 1 & -3 & 3 & -1 & -1 & 3 & -3 & 1\ ) & /2.236 \\ (\ 1 & -3 & 3 & -1 & 1 & -3 & 3 & -1\ ) & /2.236 \end{bmatrix} \begin{bmatrix} 1.000 \\ 0.707 \\ 0 \\ -0.707 \\ -1.000 \\ -0.707 \\ 0 \\ 0.707 \end{bmatrix}$$

$$(2.8.37) \qquad \blacksquare$$

A fast slant transform is shown in signal flow graph form in Fig. 2.8.2d. Input and output are in normal order. A FST subroutine is listed in Table 2.8.3.

2.8.6 KARHUNEN-LOÉVE TRANSFORM (DKLT)

The transforms which have been discussed so far are defined or characterized by their basis functions. The transforms are suboptimal from the standpoint that their basis functions are predefined and fixed. They cannot be adjusted to match the signal so the proper transform must be chosen apriori. The *Karhunen-Loéve* or *K-L transform* has basis functions which are not predefined but automatically adjust to optimally describe a signal for a particular application. K-L transforms are fully developed in Chap. 8.6. Since the theory is somewhat involved, we will only show some final results here.

It is shown in Eq. 8.6.31 that the optimum basis functions are determined by solving the matrix equation

$$\begin{bmatrix} R(0,0) & \cdots & R(0,N-1) \\ R(1,0) & \cdots & R(1,N-1) \\ \vdots & \ddots & \vdots \\ R(N-1,0) & \cdots & R(N-1,N-1) \end{bmatrix} \begin{bmatrix} K_{KL}^{-1}(0,m) \\ K_{KL}^{-1}(1,m) \\ \vdots \\ K_{KL}^{-1}(N-1,m) \end{bmatrix} = \lambda_m \begin{bmatrix} K_{KL}^{-1}(0,m) \\ K_{KL}^{-1}(1,m) \\ \vdots \\ K_{KL}^{-1}(N-1,m) \end{bmatrix}$$

$$(2.8.38)$$

where $[R]$ is the autocorrelation matrix of the signal $[f]$. The mth basis function K_{KLm}^{-1} satisfies the eigenvector equation of Eq. 2.8.38 where $RK_{KLm}^{-1} = \lambda_m K_{KLm}^{-1}$ and λ_m is the mth eigenvalue. One set of continuous K-L kernels and inverse kernels is shown in Figs. 2.8.3a–b. The $m = 0$ kernel and inverse kernel are cosines (except for one point) while the other kernels and inverse kernels are unit pulses delayed by the index m. This set is optimum for a high-resolution detection filter with impulse response matrix h_T. This filter outputs a pulse when it detects a cosine signal contaminated with white noise

($SNR_{rms} = -12$ dB). For this situation, it can be shown using the results of Chap. 8.6 that the autocorrelation matrix R equals

$$R = h_T = \begin{bmatrix} 0.25 & 0.177 & 0 & -0.177 & -0.25 & -0.177 & 0 & 0.177 \\ \cdots & \cdots & \cdots & \cdots & \cdots & \cdots & \cdots & \\ 0 & 0 & 0 & 0 & 0 & 0 & 0 & 0 \end{bmatrix} \quad (2.8.39)$$

using 8-points. Only the first row has nonzero elements. The other seven rows contain zeros. The eight eigenvalues of Eq. 2.8.39 are $[\lambda] = [1,0,0,0,0,0,0,0]$. The eight corresponding eigenfunctions K_{KLm}^{-1} are the basis functions. Sampling each basis function produces each column vector of the K-L inverse kernel matrix $[K_{KL}]^{-1}$ which equals

$$[K_{KL}]^{-1} = \begin{bmatrix} 1 & -0.707 & 0 & 0.707 & 1 & 0.707 & 0 & -0.707 \\ 0 & 1 & 0 & 0 & 0 & 0 & 0 & 0 \\ 0 & 0 & 1 & 0 & 0 & 0 & 0 & 0 \\ 0 & 0 & 0 & 1 & 0 & 0 & 0 & 0 \\ 0 & 0 & 0 & 0 & 1 & 0 & 0 & 0 \\ 0 & 0 & 0 & 0 & 0 & 1 & 0 & 0 \\ 0 & 0 & 0 & 0 & 0 & 0 & 1 & 0 \\ 0 & 0 & 0 & 0 & 0 & 0 & 0 & 1 \end{bmatrix} \quad (2.8.40)$$

These basis functions are shown in Fig. 2.8.3c. They are *not orthogonal* because $([K_{KL}]^{-1})^t ([K_{KL}]^{-1}) \neq N[I]$. The K-L kernel matrix $[K_{KL}]$ equals the inverse of Eq. 2.8.40 where

$$[K_{KL}] = \begin{bmatrix} 1 & 0.707 & 0 & -0.707 & -1 & -0.707 & 0 & 0.707 \\ 0 & 1 & 0 & 0 & 0 & 0 & 0 & 0 \\ 0 & 0 & 1 & 0 & 0 & 0 & 0 & 0 \\ 0 & 0 & 0 & 1 & 0 & 0 & 0 & 0 \\ 0 & 0 & 0 & 0 & 1 & 0 & 0 & 0 \\ 0 & 0 & 0 & 0 & 0 & 1 & 0 & 0 \\ 0 & 0 & 0 & 0 & 0 & 0 & 1 & 0 \\ 0 & 0 & 0 & 0 & 0 & 0 & 0 & 1 \end{bmatrix} \quad (2.8.41)$$

Sampling the continuous kernels in Fig. 2.8.3a (inverse kernels in Fig. 2.8.3b) produces each row of Eq. 2.8.41 (Eq. 2.8.40). Once the optimum basis functions K_{KL}^{-1} are established and the transform K_{KL} computed, the K-L transform pair becomes

$$F(m) = \frac{1}{N} \sum_{n=0}^{N-1} K_{KL}(m,n) f(n), \qquad m = 1, \ldots, N$$

$$\qquad (2.8.42)$$

$$f(n) = N \sum_{m=0}^{N-1} K_{KL}^{-1}(n,m) F(m), \qquad n = 1, \ldots, N$$

EXAMPLE 2.8.6 Determine the 8-point K-L transform spectra of the $\cos(2\pi t)$ signal for $0 \leq t < 1$. Use the transform which was derived for the high-resolution detection filter.

SOLUTION Using the continuous kernels of Fig. 2.8.3a, the continuous K-L transform spectra equal

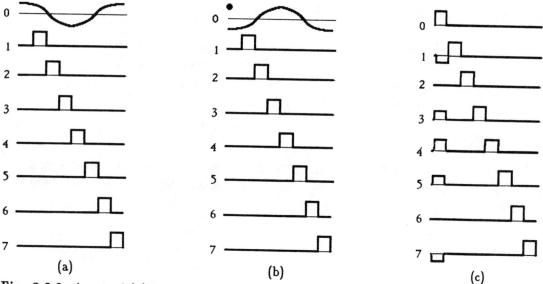

Fig. 2.8.3 A set of (a) kernels, (b) inverse kernels, and (c) sampled basis functions for an 8-point K-L transform.

$$F_n(0) = \frac{1}{T}\int_0^1 K_{KL}(0,t)\cos(2\pi t)dt = \int_0^1 \cos^2(2\pi t)dt = 0.5,$$

$$F_n(1) = \frac{1}{T}\int_{0.125}^{0.25}\cos(2\pi t)dt = \frac{\sin(2\pi t)}{2\pi}\Big|_{t=0.125}^{0.25} = 0.0466, \qquad (2.8.43)$$

$$F_n(2) = -0.0466, \qquad F_n(3) = -0.1125, \qquad F_n(4) = -0.1125,$$

$$F_n(5) = -0.0466, \qquad F_n(6) = 0.0466, \qquad F_n(7) = 0.1125$$

Using the sampled kernels $[K_{KL}]$ of Eq. 2.8.41, the discrete K-L spectra equal

$$[F]_{KL} =$$

$$\begin{bmatrix} 0.5 \\ 0.0884 \\ 0 \\ -0.0884 \\ -0.125 \\ -0.0884 \\ 0 \\ 0.0884 \end{bmatrix} = \frac{1}{8} \begin{bmatrix} 1 & 0.707 & 0 & -0.707 & -1 & -0.707 & 0 & 0.707 \\ 0 & 1 & 0 & 0 & 0 & 0 & 0 & 0 \\ 0 & 0 & 1 & 0 & 0 & 0 & 0 & 0 \\ 0 & 0 & 0 & 1 & 0 & 0 & 0 & 0 \\ 0 & 0 & 0 & 0 & 1 & 0 & 0 & 0 \\ 0 & 0 & 0 & 0 & 0 & 1 & 0 & 0 \\ 0 & 0 & 0 & 0 & 0 & 0 & 1 & 0 \\ 0 & 0 & 0 & 0 & 0 & 0 & 0 & 1 \end{bmatrix} \begin{bmatrix} 1.000 \\ 0.707 \\ 0 \\ -0.707 \\ -1.000 \\ -0.707 \\ 0 \\ 0.707 \end{bmatrix}$$

$$(2.8.44) \qquad \blacksquare$$

A fast K-L transform does not generally exist since the KLT has no fixed, predetermined symmetry. It optimally adjusts to fit the data. The flow graph of the K-L transforms has the form of Fig. 2.2.1. It has only one stage unlike the suboptimum transforms that usually use $\log_2 N$ stages.

2.9 RELATION BETWEEN TRANSFORM COEFFICIENTS

The coefficients of any generalized transform can be related to the coefficients of all the other transforms. For example, the FFT coefficients $F(m)$ can be used to generate the generalized transform coefficients $G(m)$ or vice versa. This is easily shown by equating the time domain response matrices $[f]$ as

$$[f] = [W]^{-1}[F] = [J][G] \tag{2.9.1}$$

from Eqs. 2.8.3b and 3.2.6b. $[F]$ is the FFT spectrum, $[G]$ is the generalized spectrum, $[W]^{-1}$ is the inverse FFT kernel matrix, and $[J]$ is the generalized inverse kernel matrix. Solving for $[G]$ or $[F]$ in Eq. 2.9.1 gives

$$[F] = [W][J][G] = [G_F][G] \quad \text{so} \quad [G_F] = [W][J]$$
$$[G] = [J]^{-1}[W]^{-1}[F] = [F_G][F] \quad \text{so} \quad [F_G] = [J]^{-1}[W]^{-1} = [G_F]^{-1} \tag{2.9.2}$$

Therefore $[G_F]$ is the conversion matrix for converting the $[G]$ spectrum to the $[F]$ spectrum. Conversely $[F_G]$ is the conversion matrix for converting the $[F]$ spectrum to the $[G]$ spectrum. These conversion matrices are the product of $[W]$ and $[J]$ or its inverse. Taking $[W]^{-1}$ to be the inverse kernel matrix for any other generalized transform, Eq. 2.9.2 allows any generalized transform pairs to be interrelated also.

Using these results, we will find the Paley, Hadamard, Haar, and slant transform spectra of the $\cos(2\pi t)$ signal using the 8–point FFT spectra found in Example 2.7.2 (see Eq. 2.7.7a). Consider the Paley transform first. The $N = 8$ FFT kernel matrix is given by Eq. 2.2.3. Its inverse is given by the same matrix but its elements are negative. The Paley inverse kernel matrix is given by Eq. 2.8.15. Therefore from Eq. 2.9.2b,

$$64\,[F_G]_{Pal} =$$

$$
\begin{bmatrix}
+ & + & + & + & + & + & + & + \\
+ & + & + & + & - & - & - & - \\
+ & + & - & - & + & + & - & - \\
+ & + & - & - & - & - & + & + \\
+ & - & + & - & + & - & + & - \\
+ & - & + & - & - & + & - & + \\
+ & - & - & + & + & - & - & + \\
+ & - & - & + & - & + & + & -
\end{bmatrix}
\begin{bmatrix}
0 & 0 & 0 & 0 & 0 & 0 & 0 & 0 \\
0 & -1 & -2 & -3 & -4 & -5 & -6 & -7 \\
0 & -2 & -4 & -6 & 0 & -2 & -4 & -6 \\
0 & -3 & -6 & -1 & -4 & -7 & -2 & -5 \\
0 & -4 & 0 & -4 & 0 & -4 & 0 & -4 \\
0 & -5 & -2 & -7 & -4 & -1 & -6 & -3 \\
0 & -6 & -4 & -2 & 0 & -6 & -4 & -2 \\
0 & -7 & -6 & -5 & -4 & -3 & -2 & -1
\end{bmatrix}
=
$$

$$
\begin{bmatrix}
8 & 0 & 0 & 0 & 0 & 0 & 0 & 0 \\
0 & 2+j4.828 & 0 & 2+j0.828 & 0 & 2-j0.828 & 0 & 2-j4.828 \\
0 & 0 & 4+j4 & 0 & 0 & 0 & 4-j4 & 0 \\
0 & 4.828-j2 & 0 & -0.828+j2 & 0 & -0.828-j2 & 0 & 4.828+j2 \\
0 & 0 & 0 & 0 & 8 & 0 & 0 & 0 \\
0 & 2-j0.828 & 0 & 2-j4.828 & 0 & 2+j4.828 & 0 & 2+j0.828 \\
0 & 0 & 4-j4 & 0 & 0 & 0 & 4+j4 & 0 \\
0 & -0.828-j2 & 0 & 4.828+j2 & 0 & 4.828-j2 & 0 & -0.828+j2
\end{bmatrix}
\tag{2.9.3}
$$

Now substituting the $[F_G]$ matrix just found and the FFT spectra $[F]$ into Eq. 2.9.2b gives the Paley spectra $[G]$ as

$$[G]_{Pal} = \begin{bmatrix} 0 \\ 0.2500 \\ 0 \\ 0.6036 \\ 0 \\ 0.2500 \\ 0 \\ -0.1036 \end{bmatrix} = [F_G] \begin{bmatrix} 0 \\ 4 \\ 0 \\ 0 \\ 0 \\ 0 \\ 0 \\ 4 \end{bmatrix} \qquad (2.9.4)$$

Comparing the $[G]$ found using this method with the Paley spectrum of Eq. 2.8.20, we see they are identical. The FFT–Hadamard conversion matrix is obtained by reordering the rows of Eq. 2.9.3 in a $(0,4,2,6,1,5,3,7)$ sequence. Therefore

$$64\,[F_G]_{Had} =$$

$$\begin{bmatrix} 8 & 0 & 0 & 0 & 0 & 0 & 0 & 0 \\ 0 & 0 & 0 & 0 & 0 & 0 & 0 & 0 \\ 0 & 0 & 4+j4 & 0 & 0 & 0 & 4-j4 & 0 \\ 0 & 0 & 4-j4 & 0 & 0 & 0 & 4+j4 & 0 \\ 0 & 2+j4.828 & 0 & 2+0.828 & 0 & 2-j0.828 & 0 & 2-j4.828 \\ 0 & 2-j0.828 & 0 & 2-j4.828 & 0 & 2+j4.828 & 0 & 2+j0.828 \\ 0 & 4.828-j2 & 0 & -0.828+j2 & 0 & -0.828-j2 & 0 & 4.828+j2 \\ 0 & -0.828-j2 & 0 & 4.828+j2 & 0 & 4.828-j2 & 0 & -0.828+j2 \end{bmatrix}$$

$$(2.9.5)$$

The FFT–Haar and FFT–slant conversion matrices are obtained by combining Eq. 2.9.2b with the kernel matrices of Eq. 2.8.27d and 2.8.32c, respectively, to obtain

$$[F_G]_{Ha} =$$

$$\frac{1}{64} \begin{bmatrix} 8 & 0 & 0 & 0 \\ 0 & 2+j4.828 & 0 & 2+j0.828 \\ 0 & 3.414-j1.414 & 2.828+j2.828 & -0.586+j1.414 \\ 0 & -3.414+j1.414 & 2.828+j2.828 & 0.586-j1.414 \\ 0 & 0.586-j1.414 & 2-j2 & 3.414-j1.414 \\ 0 & 1.414+0.586 & -2+j2 & -1.414-j3.414 \\ 0 & -0.586+j1.414 & 2-j2 & -3.414+j1.414 \\ 0 & -1.414-0.586 & -2+j2 & 1.414+j3.414 \end{bmatrix}$$

$$\begin{bmatrix} 0 & 0 & 0 & 0 \\ 0 & 2-j0.828 & 0 & 2-j4.828 \\ 0 & -0.586-j1.414 & 2.828+j2.828 & 3.414+j1.414 \\ 0 & 0.586+j1.414 & 2.828-j2.828 & -3.414-j1.414 \\ 4 & 3.414+j1.414 & 2+j2 & 0.586+j1.414 \\ 4 & -1.414+j3.414 & -2-j2 & 1.414-j0.586 \\ 4 & -3.414-j1.414 & 2+j2 & -0.586-j1.414 \\ 4 & 1.414-j3.414 & -2-j2 & -1.414+j0.586 \end{bmatrix}$$

$$(2.9.6a)$$

$[F_G]_S =$

$$\frac{1}{64}\begin{bmatrix} 8 & 0 & 0 & 0 \\ 0 & 1.746+j4.215 & 1.746+j1.746 & 1.746+j0.723 \\ 0 & 5.211-j2.159 & 0 & 0.153-j0.370 \\ 0 & -0.976-j2.359 & 3.123+j3.123 & -0.976-j0.404 \\ 0 & 0 & 4-j4 & 0 \\ 0 & -0.828-j2 & 0 & 4.828+j2 \\ 0 & -0.370+j0.153 & 0 & 2.159-j5.212 \\ 0 & 0 & -1.788+j1.788 & 0 \end{bmatrix}$$

$$\begin{bmatrix} 0 & 0 & 0 & 0 \\ 1.746 & 1.746-j0.723 & 1.746-j1.746 & 1.746-j4.125 \\ 0 & 0.153+j0.370 & 0 & 5.211+j2.159 \\ 3.123 & -0.976+j0.404 & 3.123-j3.123 & -0.976+j2.356 \\ 0 & 0 & 4+j4 & 0 \\ 0 & 4.828-j2 & 0 & -0.828+j2 \\ 0 & 2.159+j5.212 & 0 & -0.370-j0.153 \\ 7.154 & 0 & -1.788-j1.788 & 0 \end{bmatrix}$$

$$(2.9.6b)$$

Using these matrices in Eq. 2.9.2b gives the other generalized spectra $[G]$ as

$$[G] = [F_G]\begin{bmatrix} 0 \\ 4 \\ 0 \\ 0 \\ 0 \\ 0 \\ 0 \\ 4 \end{bmatrix} = \begin{bmatrix} 0 \\ 0 \\ 0 \\ 0 \\ 0.2500 \\ 0.2500 \\ 0.6036 \\ -0.1036 \end{bmatrix}_{Had}, \begin{bmatrix} 0 \\ 0.2500 \\ 0.4268 \\ -0.4268 \\ 0.0732 \\ 0.1768 \\ -0.0732 \\ -0.1768 \end{bmatrix}_{Ha}, \begin{bmatrix} 0 \\ 0.2182 \\ 0.6516 \\ -0.1220 \\ 0 \\ -0.1036 \\ -0.0463 \\ 0 \end{bmatrix}_{S} \quad (2.9.7)$$

which agree with the spectra found in Eqs. 2.8.24, 2.8.31, and 2.8.37, respectively.

For later reference, find the conversion matrices $[G_F]$ for converting Paley, Hadamard, Haar, and slant transform spectra into FFT spectra. These matrices result by using Eq. 2.9.2a. For example, the Paley-FFT conversion matrix equals

$[G_F]_{Pal} = [W][J] =$

$$\begin{bmatrix} 0 & 0 & 0 & 0 & 0 & 0 & 0 & 0 \\ 0 & 1 & 2 & 3 & 4 & 5 & 6 & 7 \\ 0 & 2 & 4 & 6 & 0 & 2 & 4 & 6 \\ 0 & 3 & 6 & 1 & 4 & 7 & 2 & 5 \\ 0 & 4 & 0 & 4 & 0 & 4 & 0 & 4 \\ 0 & 5 & 2 & 7 & 4 & 1 & 6 & 3 \\ 0 & 6 & 4 & 2 & 0 & 6 & 4 & 2 \\ 0 & 7 & 6 & 5 & 4 & 3 & 2 & 1 \end{bmatrix}\begin{bmatrix} + & + & + & + & + & + & + & + \\ + & + & + & + & - & - & - & - \\ + & + & - & - & + & + & - & - \\ + & + & - & - & - & - & + & + \\ + & - & + & - & + & - & + & - \\ + & - & + & - & - & + & - & + \\ + & - & - & + & + & - & - & + \\ + & - & - & + & - & + & + & - \end{bmatrix} =$$

$$\begin{bmatrix} 8 & 0 & 0 & 0 & 0 & 0 & 0 & 0 \\ 0 & 2-j4.828 & 0 & 4.828+j2 & 0 & 2+j0.828 & 0 & -0.828+j2 \\ 0 & 0 & 4-j4 & 0 & 0 & 0 & 4+j4 & 0 \\ 0 & 2-j0.828 & 0 & -0.828-j2 & 0 & 2+j4.828 & 0 & 4.828-j2 \\ 0 & 0 & 0 & 0 & 8 & 0 & 0 & 0 \\ 0 & 2+j0.828 & 0 & -0.828+j2 & 0 & 2-j4.828 & 0 & 4.828+j2 \\ 0 & 0 & 4+j4 & 0 & 0 & 0 & 4-j4 & 0 \\ 0 & 2+j4.828 & 0 & 4.828-j2 & 0 & 2-0.828 & 0 & -0.828-j2 \end{bmatrix}$$

$$(2.9.8)$$

which results by combining Eqs. 2.2.3 and 2.8.15. It is the inverse of Eq. 2.9.3. The other matrices are (reorder the columns in bit–reversed sequence for Hadamard)

$$[G_F]_{Had} =$$

$$\begin{bmatrix} 8 & 0 & 0 & 0 & 0 & 0 & 0 & 0 \\ 0 & 0 & 0 & 0 & 2-j4.828 & 2+j0.828 & 4.828+j2 & -0.828+j2 \\ 0 & 0 & 4-j4 & 4+j4 & 0 & 0 & 0 & 0 \\ 0 & 0 & 0 & 0 & 2-j0.828 & 2+j4.828 & -0.828-j2 & 4.828-j2 \\ 0 & 8 & 0 & 0 & 0 & 0 & 0 & 0 \\ 0 & 0 & 0 & 0 & 2+j0.828 & 2-j4.828 & -0.828+j2 & 4.828+j2 \\ 0 & 0 & 4+j4 & 4-j4 & 0 & 0 & 0 & 0 \\ 0 & 0 & 0 & 0 & 2+j4.828 & 2-j0.828 & 4.828-j2 & -0.828-j2 \end{bmatrix}$$

$$[G_F]_{Ha} = \qquad\qquad\qquad\qquad\qquad (2.9.9a)$$

$$\begin{bmatrix} 8 & 0 & 0 & 0 \\ 0 & 2-j4.828 & -3.414-j1.414 & 3.414+j1.414 \\ 0 & 0 & 2.828-j2.828 & 2.828-j2.828 \\ 0 & 2-j0.828 & -0.586-j4.414 & 0.586+j1.414 \\ 0 & 0 & 0 & 0 \\ 0 & 2+j0.828 & -0.586+j4.414 & 0.586-j1.414 \\ 0 & 0 & 2.828+j2.828 & 2.828+j2.828 \\ 0 & 2+j4.828 & 3.414-j1.414 & -3.414+j1.414 \end{bmatrix}$$

$$\begin{bmatrix} 0 & 0 & 0 & 0 \\ 0.586+j1.414 & 1.414-j0.586 & -0.586-j1.414 & -1.414+j0.586 \\ 2+j2 & -2-j2 & 2+j2 & -2-j2 \\ 3.414+j1.414 & -1.414+j3.414 & -3.414-j1.414 & 1.414-j3.414 \\ 4 & 4 & 4 & 4 \\ 3.414-j1.414 & -1.414-j3.414 & -3.414+j1.414 & 1.414+3.41 \\ 2-j2 & -2+j2 & 2-j2 & -2+j2 \\ 0.586-j1.414 & 1.414+j0.586 & -0.586+j1.414 & -1.414_j0.586 \end{bmatrix}$$

$$(2.9.9b)$$

$$[G_F]_S =$$

$$
\begin{bmatrix}
8 & 0 & 0 & 0 & 0 & 0 & 0 & 0 \\
0 & 1.746 - j4.215 & 5.211 + j2.159 & -0.976 + j2.356 & 0 & -0.828 + j2 & -0.370 - j0.153 & 0 \\
0 & 1.746 - j1.746 & 0 & 3.123 - j3.123 & 4 + j4 & 0 & 0 & -1.788 + j1.788 \\
0 & 1.746 + j0.723 & 0.153 + j0.370 & -0.976 - j0.404 & 0 & 4.828 - j2 & 2.159 + j5.212 & 0 \\
0 & 1.746 & 0 & 3.123 & 0 & 0 & 0 & 7.154 \\
0 & 1.746 + j0.723 & 0.153 - j0.370 & -0.976 - j0.404 & 0 & 4.828 - j2 & 2.159 - j5.212 & 0 \\
0 & 1.746 + j1.746 & 0 & 3.123 + j3.123 & 4 - j4 & 0 & 0 & -1.788 - j1.788 \\
0 & 1.746 + j4.215 & 5.211 - j2.159 & -0.976 - j2.356 & 0 & -0.828 - j2 & -0.370 + j0.153 & 0
\end{bmatrix}
$$

$$(2.9.9c)$$

Using these matrices in Eqs. 2.9.2a converts the $[G]$ spectra given by Eq. 2.9.7 into the FFT spectrum $[F] = [0\ 4\ 0\ 0\ 0\ 0\ 0\ 4]^t$.

Let us reconsider frequency domain windowing of Chap. 2.7 in light of these results. The reason that the Blackman and other time domain windows of Eq. 2.7.1 could be easily applied in the FFT frequency domain traces back to the basis functions. Since Fourier series windows have cosine and sine basis functions, they are matched to the FFT basis functions which are complex exponentials (cosines and sines). Thus, the frequency domain spectra $W(n)$ of the window have a pair of discrete frequency components for each coefficient a_m used in $w(n)$.

This suggests that similar results can be obtained using generalized transforms. Since all the basis functions we considered in Fig. 2.8.2 are basically pulses with an infinite number of frequency components, smoothing windows such as those used in the FFT serve only to increase the number of generalized transform coefficients, not reduce them! Thus, smoothing the signal is undesirable. Basically, the window must be composed of one or more basis functions of the transform being considered. For the slant transform composed of pulses, triangular, and sawtooth waves, these would form suitable windows to facilitate frequency domain windowing. However, considering the edge effects and the necessary overlap of data blocks, only the triangular window could be used to eliminate amplitude modulation of the windowed signal.

2.10   TWO-DIMENSIONAL TRANSFORMS

As discussed in Chaps. 1.5 and 1.6, two-dimensional (2-D) transforms are very important since they are used in such areas as seismic analysis and image processing. 2-D transforms convert functions of two variables $f(x, y)$ into transformed functions of two variables $F(u, v)$. The two independent variables (x, y) usually correspond to a pair of time (temporal) variables (t_1, t_2) like delayed signals, spatial variables (x_1, x_2) like images, or a mixture of both (t, x) like distributed signals. In the transform domain, the two independent variables (u, v) correspond to generalized complex frequencies. Higher-dimension transforms (e.g., 3-D for radar) are also used but these are direct generalizations of those to be discussed so the higher-order cases will not be considered.

Two-dimensional continuous transforms have the form [24]

$$F(u, v) = \int_{y_1}^{y_2} \int_{x_1}^{x_2} K(u, v; x, y) f(x, y) \, dx \, dy, \qquad (u, v) \in R_p$$

$$f(x, y) = \int_{v_1}^{v_2} \int_{u_1}^{u_2} J(u, v; x, y) F(u, v) \, du \, dv, \qquad (x, y) \in R_t$$

$$(2.10.1)$$

where $J = K^{-1}$. The analogous discrete transforms take the form

$$F(m_1, m_2) = \sum_{n_2=1}^{N_2} \sum_{n_1=1}^{N_1} K(m_1, m_2; n_1, n_2) f(n_1, n_2), \quad m_1 = 1, ..., M_1, \quad m_2 = 1, ..., M_2$$

$$f(n_1, n_2) = \sum_{m_2=1}^{M_2} \sum_{m_1=1}^{M_1} J(n_1, n_2; m_1, m_2) F(m_1, m_2), \quad n_1 = 1, ..., N_1, \quad n_2 = 1, ..., N_2$$

$$(2.10.2)$$

The integrals of Eq. 2.10.1 are replaced by summations in Eq. 2.10.2. In continuous transforms, $(u, v; x, y)$ can take on any values in the regions of R_p and R_t. However, in discrete transforms, $(m_1, m_2; n_1, n_2)$ denote the integer indices of the sampled (discrete) values in these regions.

Transform manipulation is simplified if the kernel is *separable*. A separable kernel can be factored into a product as

$$K(m_1, m_2; n_1, n_2) = K_1(m_1, n_1) K_2(m_2, n_2)$$

$$J(n_1, n_2; m_1, m_2) = J_1(n_1, m_1) J_2(n_2, m_2)$$

$$(2.10.3)$$

Further simplification results when the kernel is *symmetric*. This means $K_1(m_1, n_1)$ and $K_2(m_2, n_2)$ are equal. Then Eq. 2.10.3 reduces to

$$K(m_1, m_2; n_1, n_2) = K_1(m_1, n_1) K_1(m_2, n_2)$$

$$J(n_1, n_2; m_1, m_2) = J_1(n_1, m_1) J_1(n_2, m_2)$$

$$(2.10.4)$$

Two-dimensional transforms which are separable can be expressed as a product of a pair of one-dimensional transforms like those discussed in Chap. 2.8. Substituting Eq. 2.10.3 into Eq. 2.10.2 and factoring yields

Table 2.10.1 **Table of 2-D discrete transform kernels.**

Discrete Type	$K(m_1, m_2; n_1, n_2)$	$K^{-1}(n_1, n_2; m_1, m_2)$
z–ZT	$z_1^{-n_1} z_2^{-n_2}$	$\dfrac{1}{(j2\pi)^2 T_1 T_2} z_1^{n_1-1} z_2^{n_2-1}$
Fourier–DFT	$e^{-j2\pi(n_1 m_1 + n_2 m_2)/N}$	$\dfrac{1}{N} e^{j2\pi(n_1 m_1 + n_2 m_2)/N}$
Cosine–DCosT	$\cos[2\pi(n_1 m_1 + n_2 m_2)/N]$	Same as K/N
Sine–DSinT	$\sin[2\pi(n_1 m_1 + n_2 m_2)/N]$	Same as K/N
Paley–DPT	$\frac{1}{N}(-1)^{\sum_{i=0}^{L-1}[b_i(n_1)b_{L-1-i}(m_1)+b_i(n_2)b_{L-1-i}(m_2)]}$	Same as NK
Hadamard–DHT	$\frac{1}{N}(-1)^{\sum_{i=0}^{L-1}[b_i(n_1)b_i(m_1)+b_i(n_2)b_i(m_2)]}$	Same as NK
Haar–DHaT	2-D form of Eq. 2.8.28	2-D form of Eq. 3.3.13
Slant–DST	2-D form of Eq. 2.8.33	2-D form of Eq. 3.3.17

$$F(m_1, m_2) = \sum_{n_2=1}^{N_2} K_2(m_2, n_2) \left[\sum_{n_1=1}^{N_1} K_1(m_1, n_1) f(n_1, n_2) \right]$$

$$f(n_1, n_2) = \sum_{m_2=1}^{M_2} J_2(n_2, m_2) \left[\sum_{m_1=1}^{M_1} J_1(n_1, m_1) F(m_1, m_2) \right] \qquad (2.10.5)$$

This result shows that two-dimensional transforms (or inverse transforms) with separable kernels can be computed in two steps. In the first step contained inside the brackets of Eq. 2.10.5, the one-dimensional transform (or inverse transform) is taken along each row of $f(n_1, n_2)$ (or $F(m_1, m_2)$) which yields $F(m_1, n_2)$ (or $f(n_1, m_2)$). In the second step lying outside the brackets of Eq. 2.10.5, the one-dimensional transform (or inverse transform) is taken along each column of $F(m_1, n_2)$ (or $f(n_1, m_2)$) which yields $F(m_1, m_2)$ (or $f(n_1, n_2)$).

Since the 2-D Fourier transform kernel equals

$$K(m_1, m_2; n_1, n_2) = e^{-j2\pi(m_1 n_1 + m_2 n_2)/N} = e^{-j2\pi m_1 n_1/N} \, e^{-j2\pi m_2 n_2/N} \qquad (2.10.6)$$

it is both separable and symmetric. The inverse 2-D Fourier transform has plus rather than minus signs in the exponents and satisfies these same properties. The 2-D Fourier, Paley, Hadamard, and other transforms are listed in Table 2.10.1. Inspection shows that all the entries are separable and all but Haar and slant are symmetric.

When the transforms are symmetric, any algorithm used for computing the transform coefficients $F(m_1, m_2)$ can be used directly to compute the inverse transform coefficients $f(n_1, n_2)$. When the transforms are also separable, 2-D transforms can be obtained by successively applying 1-D transform algorithms.

EXAMPLE 2.10.1 A 2-D rectangular lowpass filter has a frequency response shown in Fig. 2.10.1a where

$$H(z_1, z_2)\Big|_{z_1 = e^{j\omega_1 T_1}, \; z_2 = e^{j\omega_2 T_2}} = 1, \qquad |\omega_1| < a, \quad |\omega_2| < b$$
$$= 0, \qquad \text{otherwise} \tag{2.10.7}$$

a and b are the 3 dB bandwidths of H in the ω_1 and ω_2 directions, respectively. Find the impulse response of the filter.

SOLUTION From a transform viewpoint, $H(z_1, z_2)$ is a Fourier kernel and Eq. 2.10.7 gives its ac steady-state properties. The inverse kernel $h(t_1, t_2)$ equals the impulse response of $H(z_1, z_2)$. Taking the inverse FT yields

$$h(t_1, t_2) = \int_{-\infty}^{\infty} \int_{-\infty}^{\infty} J(j\omega_1, j\omega_2; t_1, t_2) H(j\omega_1, j\omega_2) \, d\omega_1 \, d\omega_2$$
$$= \int_{-a}^{a} e^{j\omega_1 t_1} \, d\omega_1 \int_{-b}^{b} e^{j\omega_2 t_2} \, d\omega_2 = 4ab \frac{\sin(a t_1)}{a t_1} \frac{\sin(b t_2)}{b t_2} \tag{2.10.8}$$

which has a $(\sin x / x)$ characteristic in both the t_1 and t_2 directions. Treating H as discrete and taking the inverse Fourier transform gives

$$h(n_1, n_2) = \sum_{m_2=1}^{M_2} \sum_{m_1=1}^{M_1} J(n_1, n_2; m_1, m_2) H(m_1, m_2)$$

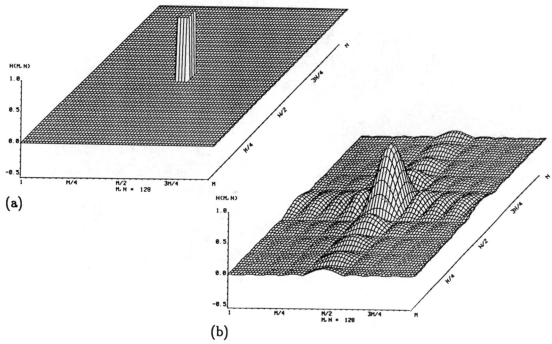

(a)

(b)

Fig. 2.10.1 (a) Magnitude and (b) impulse responses of discrete 2-D rectangular lowpass filter of Example 2.10.1. (Plotted by M. Zhao.)

$$= \sum_{m_1=-A/2}^{(A/2)-1} e^{j2\pi m_1 n_1/N} \sum_{m_2=-B/2}^{(B/2)-1} e^{j2\pi m_2 n_2/N} = \frac{\sin(2\pi n_1 A/N)}{\sin(2\pi n_1/N)} \frac{\sin(2\pi n_2 B/N)}{\sin(2\pi n_2/N)}$$

$$(2.10.9)$$

The discrete case is periodic with a $(\sin Nx/\sin x)$ characteristic. The magnitude and impulse responses of the discrete lowpass filter are shown in Fig. 2.10.1. ∎

EXAMPLE 2.10.2 A 2-D circular lowpass filter has a frequency response

$$H(z_1, z_2)\Big|_{z_1=e^{j\omega_1 T_1},\ z_2=e^{j\omega_2 T_2}} = 1, \qquad \omega_1^2 + \omega_2^2 \leq \omega_c^2$$

$$= 0, \qquad \text{otherwise}$$

$$(2.10.10)$$

This response is shown in Fig. 2.10.2a. It is nonzero in a circular region about the origin rather than a rectangular region as in Fig. 2.10.1a.

SOLUTION The continuous impulse response of the filter equals

$$h(t_1, t_2) =$$

$$\iint_{\omega_1^2+\omega_2^2 \leq \omega_c^2} e^{j(\omega_1 t_1 + \omega_2 t_2)} d\omega_1 d\omega_2 = \int_0^{\omega_c} \int_0^{2\pi} H(r,\theta) d\theta dr = \frac{J_1(\omega_c \sqrt{t_1^2 + t_2^2})}{\omega_c \sqrt{t_1^2 + t_2^2}}$$

$$(2.10.11)$$

where J_1 is a first-order Bessel function. $h(n_1, n_2)$ must be evaluated numerically. The discrete lowpass filter responses are shown in Fig. 2.10.2. ∎

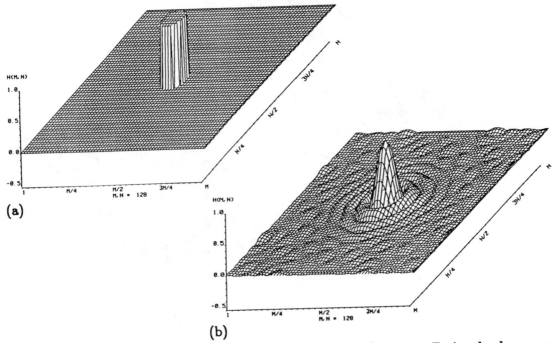

Fig. 2.10.2 (a) Magnitude and (b) impulse responses of discrete 2-D circular lowpass filter of Example 2.10.2. (Plotted by M. Zhao.)

2.11 RUNNING TRANSFORMS

In most applications, deterministic signals have nonperiodic (transient) rather than periodic behavior and stochastic signals have nonstationary statistics which vary with time. Signals are purposely generated, modulated, or coded in these fashions to convey information. To characterize such signals, it is necessary to repetitively perform spectral analysis and to compute short-term spectral coefficients. Such *block processing* requires the use of a sliding sampling window which periodically shifts in time. The resulting transforms which are obtained are called *running, moving,* or *sliding transforms.* A typical running transform is shown in Fig. 2.11.1.

Continuous running transforms have the form

$$F(p; T_o) = K(p,t) * f(t; T_o) = \int_{T_o}^{T_o + NT} K(p, t - T_o) f(t; T_o) dt, \qquad p \in R_p(T_o)$$

$$f(t; T_o) = K^{-1}(p,t) * F(p; T_o) = \int_{p_1}^{p_2} K^{-1}(p, t - T_o) F(p; T_o) dp, \qquad t \in R_t(T_o)$$

$$(2.11.1)$$

where $T_o = k(1 - r)NT$. Discrete running transforms are given by

$$F(m; N_o) = K(m,n) * f(n; N_o) = \sum_{n=N_0}^{N_0 + N - 1} K(m, n - N_o) f(n; N_o), \qquad (2.11.2a)$$

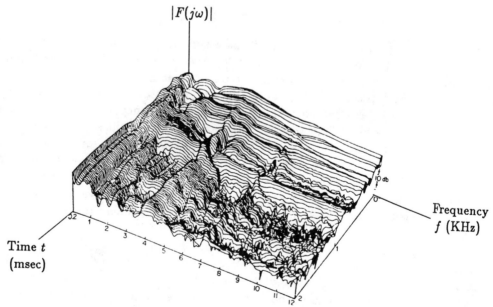

Fig. 2.11.1 Example of magnitude spectra using running transform. (From J. Berman, "Loudspeaker evaluation using digital techniques," Proc. 50th Conv. Audio Engineering Society, London, March 4, 1975.)

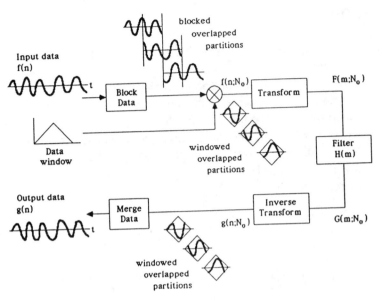

Fig. 2.11.2 Running transform and inverse transform system using windowed data.

$$f(n; N_o) = K^{-1}(n, m) * F(m; N_o) = \sum_{m=N_0}^{N_0+N-1} K^{-1}(n - N_o, m) F(m; N_o),$$ (2.11.2b)

$$m, \ n = 0, \ldots, N - 1$$

where $N_o = k(1 - r)N$. These equations result by generalizing the transforms of Eqs. 2.8.1 and 2.8.2. The window width is $T_w = t_2 - t_1 = NT$ and the window begins sampling at time $T_o = N_oT$ assuming uniform sampling. In practice, the sampling interval T is adjusted to reduce aliasing distortion to acceptable levels. The sampling window width T_w is sized to obtain the desired frequency resolution or bin size. Also T_w must not exceed the short-term stationarity period of the signal where its statistical characteristics are almost constant. The sampling window is moved to integer multiples k of $(1 - r)NT$ to update spectral estimates. The windows overlap by r percent to recover the data lost at the window edges due to edge effects (good FFT windows have small gains at their edges–see Fig. 2.6.1). The percentage overlap r must always be chosen to insure that $1/(1 - r)$ is an integer. $1/(1 - r)$ equals the number of overlapping data blocks.

The general system for processing running transforms is shown in Fig. 2.11.2.[17] The input signal is windowed, transformed, filtered, inverse transformed, and merged to form the output signal. Successively windowing the signal at later times produces the running transforms. If the windows are overlapped and data combined, more spectral resolution can be obtained since this is equivalent to sampling at a higher data rate. Overlapping by r percent increases the effective sampling rate from f_s (no overlap or $r = 0$) to $f_s/(1 - r)$. Letting $r \to 1$ produces continuous sampling.

Now consider the process of merging the running transforms as shown in Fig. 2.11.3. Merging allows transforms of short-term data to be combined to form transforms of long-term data. Merging or adding K windowed data segments, blocks, or files together gives

$$f(t)e(t) = f(t) \sum_{k=0}^{K-1} w[t - k(1-r)NT] \qquad (2.11.3)$$

where the data envelope $e(t)$ equals

$$e(t) = \sum_{k=0}^{K-1} w[t - k(1-r)NT] \qquad (2.11.4)$$

N equals the total number of data points in each window, NT seconds is the length of each temporal window, r is the overlap coefficient (see Eq. 2.6.22), $(1-r)NT$ seconds is the delay between adjacent windows, and K equals the number of merged windows. The overall data length is $[K(1-r)+r]NT$ seconds. For example, Fig. 2.11.3 has an overlap $r = 50\%$, $K = 3$ windows, and an overall window length of $[3(1-0.5)+0.5]N = 2N$.

In order that the signal $f(t)$ not have *envelope distortion* because of amplitude modulation or scalloping, $e(t)$ must be constant over the interval $(1-r)NT \le t \le K(1-r)NT$. This is assured when the window spectrum $W(j\omega)$ contains no harmonics equal to the reciprocal of the shift interval. This can be shown by noting that the envelope $e(t)$ and its spectrum $E(j\omega)$ can be expressed as the sum of $1/(1-r)$ shifted terms as

$$e(t) = \sum_{i=0}^{\frac{1}{1-r}-1} e_i(t) = \sum_{i=0}^{\frac{r}{1-r}} e_1[t - i(1-r)NT]$$

$$E(j\omega) = W_1(j\omega) \sum_{i=0}^{\frac{r}{1-r}} e^{-j\omega i(1-r)NT} \qquad (2.11.5)$$

$$= W_1(j\omega) \frac{1 - e^{-j\omega NT}}{1 - e^{-j\omega(1-r)NT}} = W_1(j\omega) \frac{\sin[\omega NT/2]}{\sin[\omega(1-r)NT/2]} e^{-j\omega rNT/2}$$

$W_1(j\omega)$ is the sampled window (i.e., periodically extended) spectrum with fundamental frequency $2\pi/NT$ rad/sec. Since the window harmonics equal $\omega = m\omega_o = m2\pi/NT$ where m is the harmonic number,

$$E(m) = W_1(m) \frac{\sin[m\pi]}{\sin[m(1-r)\pi]} e^{-jmr\pi} = (-1)^{mr} \frac{1}{1-r} W_1(m), \qquad m = 0 \text{ or } \frac{\pm q}{1-r}$$

$$= 0, \qquad m = \text{any other integer}$$

$$(2.11.6)$$

for any integer q. The envelope spectrum $E(m)$ equals zero at every harmonic m except integer multiples q of $1/(1-r)$. Therefore it is necessary to choose windows whose W_1 spectra have no harmonics equal to or multiples of the number of overlapping blocks.

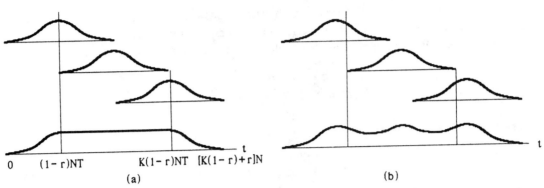

$$0 \qquad (1-r)NT \qquad\qquad K(1-r)NT \quad [K(1-r)+r]N$$

(a)   (b)

Fig. 2.11.3 Envelopes of merged data using (a) proper and (b) improper overlap.

The merging process provides an envelope gain of

$$E(0) = \frac{1}{1-r}W_1(0) \qquad (2.11.7)$$

where $W_1(0)$ is the dc gain of the sampled window. The output level equals $f(t)e(t) = [W_1(0)/(1-r)]f(t)$. Thus merging two overlapping data blocks doubles the level of the output signal, merging three overlapping blocks triples the output level, and so on. Merging blocks end-to-end ($r = 0$) provides no increased envelope gain ($E(0) = 1$).

PROBLEMS

2.1 Find the 4-point DFT of the signal $f(t) = e^{-t}U_{-1}(t)$ using a sampling rate of: (a) 1, (b) 2, (c) 4, (d) 8, (e) 16 Hz.

2.2 Given a signal sequence $[f(n)] = [0, 1, 2, 3]$. (a) Find the DFT of the signal sequence. Use this result to find the DFT of: (b) $f(n)\cos(n\pi)$, (c) $f(n)\sin(n\pi)$.

2.3 Find the 8-point DFT of the signal shown in: (a) Fig. P2.3a for $t \in (-2, 2)$, (b) Fig. P2.3b for $t \in (-2, 2)$, (c) Fig. P2.3c for $t \in (0, 3)$, (d) Fig. P2.3d for $t \in (-2, 2)$.

2.4 Find the 8-point DFT of the signal shown in: (a) Fig. P2.4a for $t \in (0, 2)$, (b) Fig. P2.4b for $t \in (0, 2)$, (c) Fig. P2.4c for $t \in (0, 1)$, (d) Fig. P2.4d for $t \in (0, 5)$.

2.5 Find the 4-point CZT of the signal $f(t) = e^{-t}U_{-1}(t)$ using a sampling rate of: (a) 1, (b) 2, (c) 4, (d) 8, (e) 16 Hz. Set $W_o = 1$, $\omega_o T = 0$, $A_o = 0.5$.

2.6 Given a signal sequence $[f(n)] = [0, 1, 2, 3]$. (a) Find the CZT of the signal sequence. Set $W_o = 1$, $\omega_o T = 0$, $A_o = 0.5$. Use this result to find the CZT of: (b) $f(n)\cos(n\pi)$, (c) $f(n)\sin(n\pi)$.

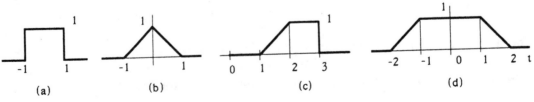

(a)   (b)   (c)   (d)

Fig. P2.3

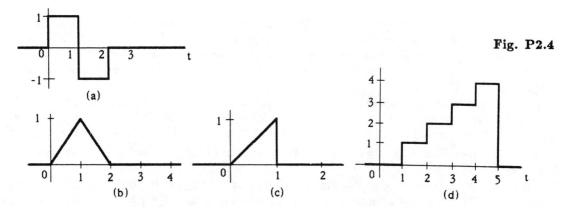

Fig. P2.4

2.7 Find the 8-point CZT of the signal shown in: (a) Fig. P2.3a for $t \in (-2, 2)$, (b) Fig. P2.3b for $t \in (-2, 2)$, (c) Fig. P2.3c for $t \in (0, 3)$, (d) Fig. P2.3d for $t \in (-2, 2)$. Set $W_o = 1$, $\omega_o T = 0$, $A_o = 0.5$.

2.8 Draw the 4-point FFT time-decimation flow graph equivalent of: (a) Fig. 2.3.5, (b) Fig. 2.3.6, (c) Fig. 2.3.7, (d) Fig. 2.3.8. (e) Evaluate the spectrum of the signal sequence $[f(n)] = [0, 1, 2, 3]$.

2.9 Draw the 4-point FFT frequency-decimation flow graph equivalent of: (a) Fig. 2.4.4, (b) Fig. 2.4.5, (c) Fig. 2.4.6, (d) Fig. 2.4.7. (e) Evaluate the spectrum of the signal sequence $[f(n)] = [0, 1, 2, 3]$.

2.10 Draw an 8-point FFT flow graph. Use it to evaluate the spectrum of: (a) Fig. P2.3a for $t \in (-2, 2)$, (b) Fig. P2.3b for $t \in (-2, 2)$, (c) Fig. P2.3c for $t \in (0, 3)$, (d) Fig. P2.3d for $t \in (-2, 2)$. Use normal input and output sequences with time decimation.

2.11 Draw an 8-point FFT flow graph. Use it to evaluate the spectrum of: (a) Fig. P2.4a for $t \in (0, 2)$, (b) Fig. P2.4b for $t \in (0, 2)$, (c) Fig. P2.4c for $t \in (0, 1)$, (d) Fig. P2.4d for $t \in (0, 5)$. Use normal input and output sequences with time decimation.

2.12 Repeat (a) Problem 2.10 and (b) Problem 2.11 but use normal input and output sequences with frequency decimation.

2.13 Evaluate the DFT of Fig. P2.13a using the FFT time decimation flow graphs of: (a) Fig. 2.3.5, (b) Fig. 2.3.6, (c) Fig. 2.3.7, (d) Fig. 2.3.8. (e) Obtain the DFT of Fig. P2.13b using the DFT of Fig. P2.13a.

2.14 Evaluate the DFT of Fig. P2.13a using the FFT frequency decimation flow graphs of: (a) Fig. 2.4.4, (b) Fig. 2.4.5, (c) Fig. 2.4.6, (d) Fig. 2.4.7. (e) Obtain the DFT of Fig. P2.13b using the DFT of Fig. P2.13a.

2.15 The generalized Hamming window has $w(n) = \alpha + (1 - \alpha)\cos(2\pi n/N)$ for $|n| \leq N/2$ and zero elsewhere. $0 \leq \alpha \leq 1$. (a) Compute and (b) sketch the frequency response $W(m)$. (c) How does the width of the main lobe and the side-lobe level vary with α? (d) What α values are used to obtain a Hann and a Hamming window?

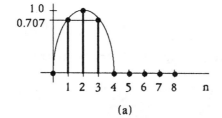

(a)

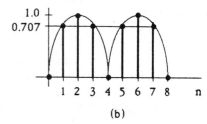

(b)

Fig. P2.13

2.16   Consider a 1 Hz signal $f(t) = \sin(2\pi t)$. Sample at a 2 Hz rate. Write the weighted 8-point data array and compute the corresponding spectrum using: (a) rectangular, (b) Hann, (c) Hamming, (d) cosine windows. (e) Sample at a 4 Hz rate and repeat parts a–d.

2.17   Repeat Problem 2.16 but obtain the spectrum of the weighted data array using a frequency domain approach.

2.18   (a) Write the window equation for triangular, Hann, and Hamming windows using 4 data points. (b) Write the windowed data for the signal $f(t) = e^{-t}U_{-1}(t)$. Use a sampling rate of 2 Hz and begin sampling at $t = 0$. (c) Find the DFT spectra of the various weighted $f(t)$.

2.19   Repeat Problem 2.18 but obtain the spectrum of the weighted data array using a frequency domain approach.

2.20   (a) Write the window equation for triangular, Hann, and Hamming windows using 4 data points. (b) Write the windowed data for the signal sequence $[f(n)] = [0, 1, 2, 3]$. (c) Find the DFT spectra of the various weighted $f(n)$.

2.21   Draw 8-point Fourier, Paley, Hadamard, Haar, and slant transform flow graphs. Use them to evaluate the spectra of: (a) Fig. P2.3a for $t \in (-2, 2)$, (b) Fig. P2.3b for $t \in (-2, 2)$, (c) Fig. P2.3c for $t \in (0, 3)$, (d) Fig. P2.3d for $t \in (-2, 2)$.

2.22   Draw 8-point Fourier, Paley, Hadamard, Haar, and slant transform flow graphs. Use them to evaluate the spectra of: (a) Fig. P2.4a for $t \in (0, 2)$, (b) Fig. P2.4b for $t \in (0, 2)$, (c) Fig. P2.4c for $t \in (0, 1)$, (d) Fig. P2.4d for $t \in (0, 5)$.

2.23   Draw 8-point Fourier, Paley, Hadamard, Haar, and slant transform flow graphs. (a) Use them to evaluate the spectra of Fig. P2.13a. (b) Which spectrum is most band-limited? (c) Find the Paley, Haar, and slant transforms directly from the DFT spectrum. Use the relations between the transform coefficients.

2.24   (a) Generalize the 1-D FFT program of Table 2.4.1 so it can perform a 2-D FFT. (b) Repeat part a for the 1-D FPT program of Table 2.8.3. (c) Repeat part a for the 1-D FHaT program of Table 2.8.5. (d) Repeat part a for the 1-D FST program of Table 2.8.6.

REFERENCES

1. Cooley, J.W. and J.W. Tukey, "An algorithm for the machine computation of complex Fourier series," Math. Computation, vol. 19, pp. 297–301, Apr. 1965.

2. Jong, M.T., *Methods of Discrete Signal and System Analysis,* Chap. 7.9, McGraw-Hill, 1982.

3. Cochran, W.T., et al., "What is the Fast Fourier transform?" IEEE Trans. Audio Electroacoust., vol. AU-15, pp. 45–55, June 1967.
   Rabiner, L.R. and C.M. Rader, ed., *Digital Signal Processing,* IEEE Press, NY, 1972.
   Digital Signal Processing II, IEEE Press, NY, 1976.

4. Cooley, J.W., P.A. Lewis, and P.D. Welch, "Historial notes on the Fast Fourier transform" IEEE Trans. Electroacoust, vol. AU-15, pp. 76–79, June 1967.
   Stearns, S.D., *Digital Signal Analysis,* Chap. 6, Hayden, NY, 1975.
   Harris, F.J., "Fast algorithms for high-speed digital signal processing," Seminar Notes, 209 p., 1986.

5. Harris, F.J., "On the use of windows for harmonic analysis with the discrete Fourier transform," Proc. IEEE, pp. 51–83, Jan. 1978.
   Nuttall, A.H., "Some windows with very good sidelobe behavior," IEEE Trans. ASSP, vol. ASSP-29, pp. 84–91, Feb. 1981.

6. Rice, J.R., *The Approximation of Functions,* vol. 1, Chap. 5.3, pp. 124–131, Addison-Wesley, MA, 1964.

7. Blackman, R.B. and J.W. Tukey, *The Measurement of Power Spectra,* App. B.5, pp. 95–100, Dover, NY, 1958.

8. Rabiner, L.R., B. Gold, and C.A. McGonegal, "An approach to the approximation problem for nonrecursive digital filters," IEEE Trans. Audio Electroacoust., vol. AU-18, pp. 83–106, June 1970.

    Harris, F.J., "High-resolution spectral analysis with arbitrary spectral centers and adjustable spectral resolutions," J. Comput. Elec. Eng., vol. 3, pp. 171–191, 1976.

9. Parzen, E., "Mathematical considerations in the estimation of spectra," Technometrics, vol. 3, no. 2, pp. 167–190, May 1961.

10. Bary, N.K., *A Treatise on Trigonometric Series,* vol. 1, Chap. 1.53, pp. 149–150; Chap. 1.68, pp. 189–192, MacMillan, NY, 1964.

11. Tukey, J.W., "An introduction to the calculations of numerical spectrum analysis," in Spectral Analysis of Time Series, B. Harris, Ed., pp. 25–46, Wiley, NY, 1967.

12. Bohman, H., "Approximate Fourier analysis of distribution functions," Arkiv Foer Matematik, vol. 4, pp. 99–157, 1960.

13. Akhiezer, N.I., *Theory of Approximation,* Chap. 4.64, pp. 118–120, Ungar, NY, 1956.

14. Helms, H.D., "Digital filters with equiripple or minimax responses," IEEE Trans. Audio Electroacoust, vol. AU-19, pp. 87–94, Mar. 1971.

15. Kuo, F.F. and J.F. Kaiser, *System Analysis by Digital Computer,* Chap. 7, pp. 232–238, Wiley, NY, 1966.

16. Barcilon, V. and G. Temes, "Optimum impulse response and the Van der Maas function," IEEE Trans. Circuit Theory, vol. CT-19, pp. 336–342, July 1972.

    Slepian, D. and H. Pollak, "Prolate-spheroidal wave functions, Fourier analysis and uncertainty I," Bell Tel. Syst. J., vol. 40, pp. 43–64, Jan. 1961.

    Landau, H. and H. Pollak, "Prolate-spheroidal wave functions, Fourier analysis and uncertainty II," Bell Tel. Syst. J., vol. 40, pp. 65–84, Jan. 1961.

17. Harris, F.J., "On the use of merged, overlapped, and windowed FFT's to generate synthetic time series data with a specified power spectrum," Rec. 16th Asilomar Conf., pp. 316–321, 1982.

18. Lindquist, C.S., *Active Network Design with Signal Filtering Applications,* Chap. 5.1–5.2, Steward and Sons, CA, 1977.

19. Ahmed, N. and K.R. Rao, *Orthogonal Transforms for Digital Signal Processing,* Springer-Verlag, NY, 1975.

    Elliott, D.F. and K.R. Rao, *Fast Transforms: Algorithms, Analyses, and Applications,* Academic Press, NY, 1983.

20. Walsh, J.L., "A closed set of normal orthogonal functions," Am. J. Math., vol. 55, pp. 5–24, 1923.

21. Hadamard, J., "Resolution d'une question relative aux determinants," Bull. Sci. Math. Ser. 2, vol. 17, pp. 240–246, 1893.

22. Haar, A., "Zur Theorie der orthogonalen Funktionensysteme," Math. Ann., vol. 69, pp. 331–371, 1910.

23. Enomoto, H., and K. Shibata, "Orthogonal transform coding system for television signals," Proc. Symp. Applied Walsh Functions, pp. 11–71, 1971.

    Shibata, K., "Waveform analysis of image signals by orthogonal transformations," Proc. Symp. Applied Walsh Functions, pp. 210–215, 1972.

24. Gonzalez, R.C. and P. Wintz, *Digital Image Processing,* Chap. 3, Addison-Wesley, MA, 1977.

3 TIME DOMAIN ANALYSIS

> *And the serpent said unto the woman, Ye shall not die ...*
> *you shall be as gods ... and (she) did eat ... and he did eat.*
> *And the eyes of them both were opened, and they ...*
> *hid themselves from the presence of the Lord God ...*
> *And the Lord God said unto the serpent ...thou art cursed ...*
> *the Lord God sent him (man) forth from the garden of Eden,*
> *to till the ground from whence he was taken.*
>
> Genesis 3:4–8, 14

This chapter discusses time domain analysis methods. These methods complement the frequency domain analysis techniques presented in the last chapter. From a filter viewpoint, Chap. 3 gives the tools for analyzing the *inverse transform* and *interpolator* blocks (lower two blocks) in the signal processing system of Fig. 2.0.1. The sampler and transform blocks (upper two blocks) were analyzed in Chap. 2. Chap. 4 and the remaining chapters will be concerned with the design of the filter gain block.

This discussion begins by deriving the impulse, step, and sinusoidal responses of digital filters using the inverse *z*-transform. Then the inverse discrete Fourier transform (IDFT) is presented. Both uniform and nonuniform samplings are considered. The fast inverse DFT or IFFT is discussed to reduce the computational time. Other inverse discrete transforms including Paley, Hadamard, Haar, slant, and Karhunen-Loéve are also discussed. They further simplify evaluation and can provide better performance.

Various interpolation filters are investigated which reconstruct continuous time signals from their samples. These include simple interpolators, ideal lowpass filter interpolator, Fourier series interpolator, and generalized series interpolators. This chapter concludes with a discussion of the distortion caused by nonzero sampling duration and signal truncation. The engineer will find that Chaps. 2 and 3 provide a broad and cohesive viewpoint for relating signals, spectra, and transforms.

3.1 INVERSE Z-TRANSFORM (IZT)

The time domain behavior of digital filters is often judged on the basis of their impulse, step, and sinusoidal step responses.[1] Let us now briefly review how these responses are obtained using the inverse *z*-transform introduced in Chap 1.10. This material will expand the earlier discussion. The *z*-transform response $Y(z)$ of a digital filter having a transfer function $H(z)$ equals

$$Y(z) = H(z)X(z) \tag{3.1.1}$$

where $X(z)$ is the z-transform input. The time domain response $y(nT)$ is

$$y(nT) = Z^{-1}\big[H(z)X(z)\big] = h(nT) * x(nT) \qquad (3.1.2)$$

It equals the inverse z-transform of the product of $H(z)$ and $X(z)$ or the convolution of the filter's impulse response $h(nT)$ and its input $x(nT)$.

The *step response* will be considered first. The z-transform of a unit step equals

$$Z\big[U_{-1}(t)\big] = \sum_{n=0}^{\infty} z^{-n} = \frac{z}{z-1}, \qquad |z| > 1 \qquad (3.1.3)$$

using Eq. 1.9.3 or the z-transform in Table 1.9.1.

EXAMPLE 3.1.1 Determine the unit step response for a first-order digital low-pass filter having a transfer function of

$$H(z) = \frac{z}{z-k} = \frac{1}{1 - kz^{-1}}, \qquad |z| > k \qquad (3.1.4)$$

SOLUTION From Eqs. 3.1.1 and 3.1.3, the transformed step response equals ($k \neq 1$)

$$Y(z) = H(z)\frac{z}{z-1} = \frac{z^2}{(z-1)(z-k)} = \frac{K_1 z}{z-1} + \frac{K_2 z}{z-k}, \qquad |z| > \max(1, k) \quad (3.1.5)$$

using the partial fraction expansion of Eq. 1.10.3. Evaluating residues K_1 and K_2,

$$\begin{aligned}
K_1 &= (z-1)\frac{Y(z)}{z}\Big|_{z=1} = \frac{z}{z-k}\Big|_{z=1} = \frac{1}{1-k} \\
K_2 &= (z-k)\frac{Y(z)}{z}\Big|_{z=k} = \frac{z}{z-1}\Big|_{z=k} = \frac{k}{k-1}
\end{aligned} \qquad (3.1.6)$$

Finding the z-transform inverses of $Y(z)$ from Table 1.9.1 gives ($k \neq 1$)

$$y(nT) = \left(\frac{1}{1-k}\right)1^n\, U_{-1}(n) - \left(\frac{k}{1-k}\right)k^n\, U_{-1}(n) \qquad (3.1.7)$$

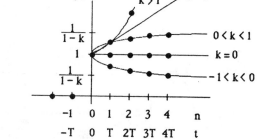

Fig. 3.1.1 Normalized step response of first-order digital lowpass filter in Example 3.1.1.

If $k = 1$, then $y(nT) = U_{-2}(n + 1)$. The step response is plotted in Fig. 3.1.1 for different k values. It is unstable for $k \geq 1$.   ∎

The *impulse response* is also very helpful in assessing digital filter behavior. In analog systems, the unit impulse $U_0(t)$ is defined as

$$U_0(t) = dU_{-1}(t)/dt \qquad (3.1.8)$$

where $U_{-1}(t)$ is the unit step shown in Fig. 1.5.1a. From Table 1.6.1, the FT of the unit impulse equals

$$\mathcal{F}[U_0(t)] = F(j\omega) = 1 \qquad (3.1.9)$$

This signal and its spectrum are shown in Fig. 3.1.2a. Sampling $U_0(t)$ produces conceptual and mathematical problems because the sampled $U_0(t)$ is an impulse $U_0^2(t)$ having area $U_0(0)$. Thus the unit impulse is unsuitable as a digital system input. Its spectrum, evaluated using Eq. 1.3.6, is flat but has infinite height.

However the band-limited version of $U_0(t)$ shown in Fig. 3.1.2b and denoted as $D_0(t)$ is suitable. Using Eq. 3.1.9, $D_0(t)$ has a band-limited spectrum $D_0(j\omega)$ of

$$\mathcal{F}[D_0(t)] = 1, \qquad |\omega| < \omega_s/2 = \omega_N$$
$$= 0, \qquad |\omega| > \omega_s/2 \qquad (3.1.10)$$

Its bandwidth equals the Nyquist frequency ω_N (one-half the sampling frequency ω_s). The signal therefore equals

$$D_0(t) = \frac{1}{T} \frac{\sin(\pi t/T)}{\pi t/T} = \frac{1}{T}, \qquad t = 0$$
$$= 0, \qquad t = nT \qquad (3.1.11)$$

using entry 17 of Table 1.6.1 and the FT symmetry property in Table 1.6.2. It equals $1/T$ at $t = 0$ and is zero at all other sample times $t = nT$. Thus, $D_0(t)$ may be viewed as a single pulse at $t = 0$ with height $1/T$. Its ZT equals

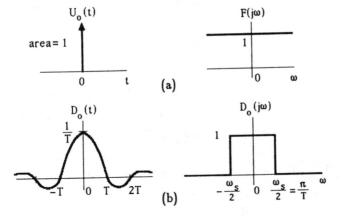

Fig. 3.1.2 (a) Unit analog impulse $U_0(t)$ and (b) unit digital impulse $D_0(t)$.

$$Z\big[D_0(t)\big] = \frac{1}{T}, \qquad \text{all } z \qquad\qquad (3.1.12)$$

$D_0(t)$ is sometimes called the *digital unit pulse* rather than the *digital unit impulse* to emphasize this fact. Sampling $D_0(nT)$ produces $D_0(0)U_0(t) = U_0(t)/T$ which has the flat spectrum $F(j\omega)$ in Fig. 3.1.2a. It is the sum of the band-limited spectra in Fig. 3.1.2b shifted integer multiples of ω_s. For continuous sampling, the sampling interval T approaches zero and Fig. 3.1.2b reduces to Fig. 3.1.2a. We will denote the digital unit pulse as both $U_0(t)$ and $D_0(t)$. Regardless of notation, this signal must always be interpreted as discussed in Eqs. 3.1.10–3.1.12. The application will always make it clear whether $U_0(t)$ denotes a *digital pulse* or an *analog impulse*.

EXAMPLE 3.1.2 Compute the unit impulse response for the first-order lowpass filter of Example 3.1.1. Determine the range of k to insure a causal, stable $H(z)$.
SOLUTION Since $X(z) = 1/T$ from Eq. 3.1.12, the impulse response $Y(z)$ equals

$$Y(z) = \frac{z}{T(z-k)}, \qquad |z| > k \qquad\qquad (3.1.13)$$

From the z-transform pairs in Table 1.9.1, we find

$$y(nT) = \frac{1}{T}k^n\, U_{-1}(n) = \frac{1}{T}\, e^{n\ln k}\, U_{-1}(n) \qquad\qquad (3.1.14)$$

which is plotted in Fig. 3.1.3. Since the region of convergence is $|z| > k$, the system is potentially causal. From the response, $y(nT) = 0$ for $n < 0$ so the system is causal. The system is stable for $|k| < 1$, conditionally stable for $k = 0$, and unstable for $|k| > 1$. ∎

The *sinusoidal* and *cosinusoidal step responses* of digital filters are also important. However, the complex exponential step $e^{j\omega_o t}U_{-1}(t)$ is simpler to describe mathematically and is used more often. It is complex-valued as shown in Fig. 3.1.4. The z-transform of the complex exponential step equals

$$Z\big[e^{j\omega_o t}U_{-1}(t)\big] = \frac{z}{z - e^{j\omega_o T}}, \qquad |z| > 1 \qquad\qquad (3.1.15)$$

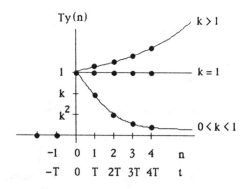

Fig. 3.1.3 Normalized impulse response of first-order digital lowpass filter in Example 3.1.2.

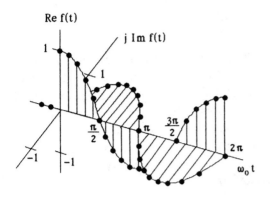

Re f(t)

j Im f(t)

$\frac{\pi}{2}$ π $\frac{3\pi}{2}$ 2π

$\omega_o t$

Fig. 3.1.4 Unit complex exponential step function.

from Eq. 1.9.3 where $\sigma = j\omega_o$. Taking the real and imaginary parts of $e^{j\omega_o t}U_{-1}(t)$ yield the $\cos(\omega_o t)U_{-1}(t)$ cosine step and the $\sin(\omega_o t)U_{-1}(t)$ sine step. Using Euler's identity and manipulating Eq. 3.1.15 yields their transforms as illustrated in Example 1.9.5.

EXAMPLE 3.1.3 Compute the unit complex exponential step response of the first-order lowpass filter of Example 3.1.1. Find the steady-state response for a pole location $0 < k < 1$.

SOLUTION Combining Eqs. 3.1.4 and 3.1.15, the transformed response equals

$$Y(z) = \frac{z}{z-k}\frac{z}{z-e^{j\omega_o T}} = \left(\frac{k}{k-e^{j\omega_o T}}\right)\frac{z}{z-k} + \left(\frac{e^{j\omega_o T}}{e^{j\omega_o T}-k}\right)\frac{z}{z-e^{j\omega_o T}} \quad (3.1.16)$$

Taking the inverse z-transform gives

$$y(nT) = Z^{-1}[Y(z)] = \left[\left(\frac{1}{1-ke^{-j\omega_o T}}\right)e^{jn\omega_o T} + \left(\frac{k}{k-e^{j\omega_o T}}\right)k^n\right]U_{-1}(n) \quad (3.1.17)$$

The second term represents the transient component which goes to zero as $n \to \infty$. The closer k is to zero, the faster it decays. The first term is the steady-state component which can be expressed as

$$y(nT) = |H|\ e^{j(n\omega_o T + \arg H)}U_{-1}(n) \quad (3.1.18)$$

Evaluating the digital filter gain $H(z)$ given by Eq. 3.1.4 at the signal frequency $s = j\omega_o$ or $z = e^{sT} = e^{j\omega_o T}$ gives

$$H(e^{j\omega_o T}) = |H|e^{j\arg H} = \frac{e^{j\omega_o T}}{e^{j\omega_o T}-k} = \frac{1}{1-ke^{-j\omega_o T}} \quad (3.1.19)$$

Analyzing Eq. 3.1.19 shows that the filter has a dc gain of $1/(1-k)$, a Nyquist frequency $(\omega_o T = \pi)$ gain of $1/(1+k)$, and a normalized 3 dB bandwidth $\omega_{3dB}T$ of $\cos^{-1}[(4k-1-k^2)/2k]$. Its phase $(\arg H)$ at a normalized frequency $\Omega = \omega_o T$ equals $-\tan^{-1}[(k\sin\Omega)/(1-k\cos\Omega)]$ which lies between 0 and $-\pi$ radians. When the normalized signal frequency $\omega_o T$ is less than $\omega_{3dB}T$, the signal passes with little relative attenuation and small phase shift. If $\omega_o T$ is greater than $\omega_{3dB}T$ and less than π, it is attenuated and delayed. The greatest relative attenuation is $(1-k)/(1+k)$ and occurs when $k \cong 1$ at the Nyquist frequency $\omega_o T = \pi$. ∎

Since the z-transform and inverse z-transform must be computed at all the points (an infinite number) contained within certain regions, they cannot be numerically evaluated using computer algorithms. However, these transforms can be evaluated at a finite number of points along particular z-plane contours. For example, evaluating the ZT and IZT at equally-spaced points around the unit circle ($|z| = 1$ and $\arg z = -2nm\pi/N$ as shown in Fig. 2.1.3) results in the DFT and IDFT, respectively. Table 1.8.1 summarizes their properties. Implementing fast algorithms result in the FFT and IFFT as discussed in Chaps. 2.3 and 2.4. A large number of other discrete transforms and their fast computer algorithms were presented in Chap. 2.8. The inverse transforms and their fast versions will now be investigated.

3.2 INVERSE DISCRETE FOURIER TRANSFORM (IDFT)

The inverse discrete Fourier transform (IDFT) was introduced in Chap. 1.8 which will now be expanded. The IDFT equals the inverse Fourier transform of a band-limited sampled signal $f_s(t)$. $f_s(t)$ and its spectrum $F_s(j\omega)$ are related to $f(t)$ and $F(j\omega)$ by Eqs. 1.3.2 and 1.3.6. $F(j\omega)$ and $F_s(j\omega)$ are shown in Fig. 3.2.1a.

3.2.1 UNIFORMLY SAMPLED IDFT

The sampled signal $f_s(t)$ is converted back (or reconstructed) into a continuous time signal $f_r(t)$ by passing the signal spectrum $F_s(j\omega)$ through the ideal lowpass filter (Eq. 1.3.7) whose bandwidth $B = \omega_s/2 = \omega_N$ equals the Nyquist frequency. This is shown in Figs. 3.2.1b and 3.2.2. The $F_s(j\omega)$ spectrum can be approximated as a periodic series of impulses were (see Fig. 1.8.2)

$$F_s(j\omega) \cong \frac{\omega_s}{N} \sum_{m=-\infty}^{\infty} F_s\left(\frac{jm\omega_s}{N}\right) U_0\left(\omega - \frac{m\omega_s}{N}\right) \qquad (3.2.1)$$

Lowpass filtering $F_s(j\omega)$ produces the band-limited spectrum $F_r(j\omega)$ where

$$F_r(j\omega) \cong \frac{\omega_s}{N} \sum_{m=-N/2}^{(N/2)-1} F_s\left(\frac{jm\omega_s}{N}\right) U_0\left(\omega - \frac{m\omega_s}{N}\right), \qquad |\omega| \le \omega_s/2$$

$$= 0, \qquad |\omega| > \omega_s/2 \qquad (3.2.2)$$

$F_r(j\omega)$ can be numerically evaluated since it involves only a finite number of terms unlike $F_s(j\omega)$. The upper limit is $((N/2) - 1)$ and not $N/2$ since the extra half-bin below $-N/2$ equals the missing half-bin below $N/2$.

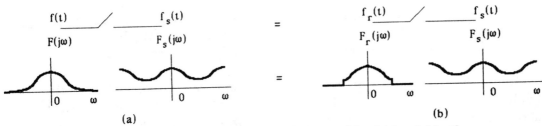

Fig. 3.2.1 (a) Sampling a non-band-limited signal $f(t)$ which yields the same spectrum $F_s(j\omega)$ as (b) sampling a band-limited signal $f_r(t)$.

Taking the IFT of $F_r(j\omega)$ yields the inverse discrete Fourier transform $f_r(nT)$ where

$$f_r(nT) = \frac{1}{2\pi} \int_{-\infty}^{\infty} F_r(j\omega)e^{j\omega nT} d\omega$$

$$= \frac{1}{2\pi} \int_{-\omega_\bullet/2}^{\omega_\bullet/2} \frac{2\pi}{NT} \left[\sum_{m=-N/2}^{(N/2)-1} F_s\left(\frac{jm\omega_s}{N}\right) U_0\left(\omega - \frac{m\omega_s}{N}\right) \right] e^{j\omega nT} d\omega$$

$$= \sum_{m=-N/2}^{(N/2)-1} \frac{1}{NT} \int_{-\omega_\bullet/2}^{\omega_\bullet/2} F_s\left(\frac{jm\omega_s}{N}\right) e^{j\omega nT} U_0\left(\omega - \frac{m\omega_s}{N}\right) d\omega$$

$$= \frac{1}{NT} \sum_{m=-N/2}^{(N/2)-1} F_s\left(\frac{jm\omega_s}{N}\right) e^{jmn\omega_\bullet T/N}$$

(3.2.3)

Using the DFT shorthand notation of Eq. 2.1.3,

$$T f_r(n) = \frac{1}{N} \sum_{m=-N/2}^{(N/2)-1} F_s(m) W_N^{-nm} = \frac{1}{N} \sum_{m=0}^{N-1} F_s(m) W_N^{-nm}$$

(3.2.4)

where $W_N = \exp(-j2\pi/N)$. The summation limits can be shifted because of the periodicity in $F_s(m)$ and W_N (see Eqs. 2.1.5 and 2.2.4).

The discrete Fourier transform and inverse discrete Fourier transform pair are

$$F(m) \equiv F_s(m) = \sum_{n=0}^{N-1} f(n) W_N^{nm}, \qquad m = 0, 1, ..., N-1$$

$$f(n) \equiv T f_r(n) = \frac{1}{N} \sum_{m=0}^{N-1} F(m) W_N^{-nm}, \qquad n = 0, 1, ..., N-1$$

(3.2.5)

An equal number N of samples are used in both time and frequency which results in unique coefficients as was shown in Eq. 2.8.9. The relation between these reconstructed samples $f_r(n)$ and the original samples $f(n)$ will be discussed momentarily. In matrix

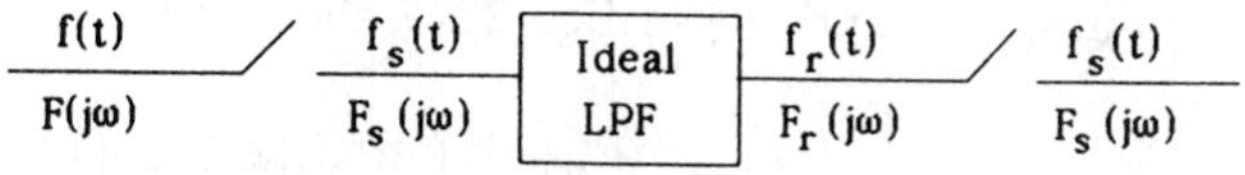

$$\frac{f(t)}{F(j\omega)} \diagup \frac{f_s(t)}{F_s(j\omega)} \boxed{\begin{array}{c}\text{Ideal}\\\text{LPF}\end{array}} \frac{f_r(t)}{F_r(j\omega)} \diagup \frac{f_s(t)}{F_s(j\omega)}$$

Fig. 3.2.2 Relation between the signal $f(t)$, sampled signal $f_s(t)$, and the reconstructed signal $f_r(t)$.

form, Eq. 3.2.5 becomes

$$
\begin{bmatrix} F(0) \\ F(1) \\ F(2) \\ \vdots \\ F(N-1) \end{bmatrix} = \begin{bmatrix} 0 & 0 & 0 & \cdots & 0 \\ 0 & 1 & 2 & \cdots & (N-1) \\ 0 & 2 & 4 & \cdots & 2(N-1) \\ \vdots & \vdots & \vdots & \cdots & \vdots \\ 0 & N-1 & 2(N-1) & \cdots & (N-1)^2 \end{bmatrix} \begin{bmatrix} f(0) \\ f(1) \\ f(2) \\ \vdots \\ f(N-1) \end{bmatrix}
$$
(3.2.6)

$$
\begin{bmatrix} f(0) \\ f(1) \\ f(2) \\ \vdots \\ f(N-1) \end{bmatrix} = \frac{1}{N} \begin{bmatrix} 0 & 0 & 0 & \cdots & 0 \\ 0 & -1 & -2 & \cdots & -(N-1) \\ 0 & -2 & -4 & \cdots & -2(N-1) \\ \vdots & \vdots & \vdots & \vdots & \vdots \\ 0 & -(N-1) & -2(N-1) & \cdots & -(N-1)^2 \end{bmatrix} \begin{bmatrix} F(0) \\ F(1) \\ F(2) \\ \vdots \\ F(N-1) \end{bmatrix}
$$

or simply

$$[F] = [W_N][f]$$
$$[f_r] = \frac{1}{N}[W_N]^*[F] = [W_N]^{-1}[F]$$
(3.2.7)

Only the exponent $\pm nm$ of the $W_N^{\pm nm}$ term appears in the W_N matrix. Notice that the square matrices $[W_N]$ and $N[W_N]^{-1}$ are identical except that their elements have opposite signs. For example, $[W_N]^{-1}$ is given by Eq. 2.2.3 when $N = 8$ using negative elements.

EXAMPLE 3.2.1 Determine the IDFT of the 8-point DFT spectra for the unweighted cosine signal $f(t) = \cos(2\pi t)$ in Example 2.7.4.
SOLUTION The spectrum $[F_s]$ of the unweighted cosine signal was given by Eq. 2.7.11. The IDFT results by substituting this spectrum into Eq. 3.2.6b and evaluating $[f]$ as

$$
[W_N]^{-1} \begin{bmatrix} 0 \\ 4 \\ 0 \\ 0 \\ 0 \\ 0 \\ 0 \\ 4 \end{bmatrix} = \frac{1}{8} \begin{bmatrix} 0 & 0 & 0 & 0 & 0 & 0 & 0 & 0 \\ 0 & -1 & -2 & -3 & -4 & -5 & -6 & -7 \\ 0 & -2 & -4 & -6 & 0 & -2 & -4 & -6 \\ 0 & -3 & -6 & -1 & -4 & -7 & -2 & -5 \\ 0 & -4 & 0 & -4 & 0 & -4 & 0 & -4 \\ 0 & -5 & -2 & -7 & -4 & -1 & -6 & -3 \\ 0 & -6 & -4 & -2 & 0 & -6 & -4 & -2 \\ 0 & -7 & -6 & -5 & -4 & -3 & -2 & -1 \end{bmatrix} \begin{bmatrix} 0 \\ 4 \\ 0 \\ 0 \\ 0 \\ 0 \\ 0 \\ 4 \end{bmatrix} = \begin{bmatrix} 1.000 \\ 0.707 \\ 0 \\ -0.707 \\ -1.000 \\ -0.707 \\ 0 \\ 0.707 \end{bmatrix}
$$
(3.2.8)

The output samples equal the samples of the original signal. This is expected since combining $[f_r]$ and $[F]$ in Eq. 3.2.7 shows that

$$[f_r] = [W_N]^{-1}[F] = [W_N]^{-1}\left([W_N][f]\right) = \left([W_N]^{-1}[W_N]\right)[f] = [I][f] = [f]$$
(3.2.9)

because $[W_N]$ and $[W_N]^{-1}$ are inverses.   ∎

3.2.2 NONUNIFORMLY SAMPLED IDFT [3]

Now we will consider the IDFT. The nonuniformly sampled DFT was discussed in Chap. 2.1.4. A variety of transform matrix pairs can be formulated to approximate the FT and IFT. Some of these are listed in Table 3.2.1. A DFT zero-order approximation was given by Eq. 2.1.24. Using a dual approximation for the IDFT yields the integral approximation transform pair

$$F(m) \equiv F(j\omega_m) = \sum_{n=0}^{N-1} f(t_n)e^{-j\omega_m t_n}\Delta t_n$$
(3.2.10)

$$f(n) \equiv f(t_n) = \frac{1}{2\pi}\sum_{m=0}^{N-1} F(j\omega_m)e^{j\omega_m t_n}\Delta\omega_m$$

The sample times $t = t_n$ and sample frequencies $\omega = \omega_m$. The time interval $\Delta t_n = t_{n+1} - t_n$ and the frequency interval $\Delta\omega_m = \omega_{m+1} - \omega_m$. In matrix form these equations become (see Eq. 2.1.25)

$$
\begin{bmatrix} F(0) \\ F(1) \\ \vdots \\ F(N-1) \end{bmatrix} =
$$

$$
\begin{bmatrix} 1 & 1 & \cdots & 1 \\ 1 & e^{-j\omega_1 t_1} & \cdots & e^{-j\omega_1 t_{N-1}} \\ \vdots & \vdots & \ddots & \vdots \\ 1 & e^{-j\omega_{N-1} t_1} & \cdots & e^{-j\omega_{N-1} t_{N-1}} \end{bmatrix}
\begin{bmatrix} \Delta t_0 & \cdots & & 0 \\ 0 & \Delta t_1 & & 0 \\ \vdots & & \ddots & \vdots \\ 0 & & \cdots & \Delta t_{N-1} \end{bmatrix}
\begin{bmatrix} f(0) \\ f(1) \\ \vdots \\ f(N-1) \end{bmatrix}
$$
(3.2.11)

$$
\begin{bmatrix} f(0) \\ f(1) \\ \vdots \\ f(N-1) \end{bmatrix} =
$$

$$
\frac{1}{2\pi}\begin{bmatrix} 1 & 1 & \cdots & 1 \\ 1 & e^{j\omega_1 t_1} & \cdots & e^{j\omega_{N-1} t_1} \\ \vdots & \vdots & \ddots & \vdots \\ 1 & e^{j\omega_{N-1} t_1} & \cdots & e^{j\omega_{N-1} t_{N-1}} \end{bmatrix}
\begin{bmatrix} \Delta\omega_0 & \cdots & & 0 \\ 0 & \Delta\omega_1 & & 0 \\ \vdots & & \ddots & \vdots \\ 0 & & \cdots & \Delta\omega_{N-1} \end{bmatrix}
\begin{bmatrix} F(0) \\ F(1) \\ \vdots \\ F(N-1) \end{bmatrix}
$$

where we set the initial time $t_0 = 0$ and the initial frequency $\omega_0 = 0$. In short-hand

Table 3.2.1 List of different transform pairs using nonuniform sampling. (From S.R. Yhann, "Iterative transform methods for nonuniformly-sampled data," M.S.E.E. Thesis, Univ. of Miami, 106 p., May 1987.)

Approach	Equation	Eq. No.	Comments
Generalized transform	$[F] = ([W]^h[W])^{-1}[W]^h[f]$ $[f_r] = [W][F]$	3.2.22 3.2.22	Computations large. Does not produce spectral estimate related to continuous time.
Matrix inversion of orthogonal series	$[F] = [W]^{-1}[f]$ $[f_r] = [W][F]$	3.2.18 3.2.18	Matrix inversion ill-conditioned. Signal assumed periodic and band-limited.
Integral approximation	$[F] = [W]^h[\Delta t][f]$ $[f_r] = \frac{1}{2\pi}[W][\Delta\omega][F]$	3.2.12 3.2.12	Spectral estimate good for nonperiodic and non-band-limited signals.
Integral approx.– matrix inversion	$[F] = [W]^h[\Delta t][f]$ $[f_r] = [\Delta t]^{-1}([W]^h)^{-1}[F]$	3.2.24 3.2.24	Matrix inversion ill-conditioned.
Integral approx.– iterative signal recovery	$[F] = [W]^h[\Delta t][f]$ $[\Delta t][f_r]_{k+1}$ $= \frac{1}{\Omega}([W][\Delta\omega][F] - [T][\Delta t][f_r]_k)$	3.2.28 3.2.28	**Signal recovery exact** provided convergence guaranteed.

notation, the *integral approximation* transform pair equal

$$[F] = [W]^h[\Delta t][f]$$
$$[f_r] = \frac{1}{2\pi}[W][\Delta\omega][F] \tag{3.2.12}$$

The $[W]$ matrices containing the $e^{\pm j\omega_m t_n}$ terms are conjugate transposes (i.e., Hermitian where operator $[\ ]^h \equiv [\ ]^{*t}$) of each other. For sampling intervals $\Delta t_n = T$ and $\Delta\omega_m = \omega_s/N$, Eq. 3.2.12 reduces to the uniform sampling case described by Eq. 3.2.7.

Exact reconstruction requires $[f_r] = [f]$. Combining $[f_r]$ and $[F]$ matrices in Eq. 3.2.12 gives

$$[f_r] = \frac{1}{2\pi}[W][\Delta\omega][F] = \frac{1}{2\pi}[W][\Delta\omega][W]^h[\Delta t][f] = [I][f] = [f] \tag{3.2.13}$$

This shows that the sampling interval matrices $[\Delta\omega]$ and $[\Delta t]$ cannot be chosen independently, but must satisfy

$$\frac{1}{2\pi}[W][\Delta\omega][W]^h[\Delta t] = [I] \tag{3.2.14}$$

Solving for $[\Delta\omega]$ gives

$$\frac{1}{2\pi}[\Delta\omega] = \left([W]\right)^{-1}\left([W]^h[\Delta t]\right)^{-1} = \left([W]^h[\Delta t][W]\right)^{-1} \tag{3.2.15}$$

The $[W]$ matrix implicitly involves the $\Delta\omega_m$ and Δt_n terms. Eq. 3.2.15 has no solution for arbitrary $[\Delta t]$. This can be shown by expanding $[\Delta\omega]^{-1}$ as

$$2\pi\begin{bmatrix} \Delta\omega_0 & \cdots & & 0 \\ 0 & \Delta\omega_1 & & 0 \\ \vdots & & \ddots & \vdots \\ 0 & \cdots & & \Delta\omega_{N-1} \end{bmatrix}^{-1} = 2\pi\begin{bmatrix} \Delta\omega_0^{-1} & \cdots & & 0 \\ 0 & \Delta\omega_1^{-1} & & 0 \\ \vdots & & \ddots & \vdots \\ 0 & \cdots & & \Delta\omega_{N-1}^{-1} \end{bmatrix}$$

$$= \begin{bmatrix} e^{-j\omega_0 t_0} & e^{-j\omega_0 t_1} & \cdots & e^{-j\omega_0 t_{N-1}} \\ e^{-j\omega_1 t_0} & e^{-j\omega_1 t_1} & \cdots & e^{-j\omega_1 t_{N-1}} \\ \vdots & \vdots & \ddots & \vdots \\ e^{-j\omega_{N-1} t_0} & e^{-j\omega_{N-1} t_1} & \cdots & e^{-j\omega_{N-1} t_{N-1}} \end{bmatrix}\begin{bmatrix} \Delta t_0 & \cdots & & 0 \\ 0 & \Delta t_1 & & 0 \\ \vdots & & \ddots & \vdots \\ 0 & \cdots & & \Delta t_{N-1} \end{bmatrix}$$

$$\times \begin{bmatrix} e^{j\omega_0 t_0} & e^{j\omega_1 t_0} & \cdots & e^{j\omega_{N-1} t_0} \\ e^{j\omega_0 t_1} & e^{j\omega_1 t_1} & \cdots & e^{j\omega_{N-1} t_1} \\ \vdots & \vdots & \ddots & \vdots \\ e^{j\omega_0 t_{N-1}} & e^{j\omega_1 t_{N-1}} & \cdots & e^{j\omega_{N-1} t_{N-1}} \end{bmatrix} = [X] \tag{3.2.16}$$

where $X(i,k) = \sum_{n=0}^{N-1} \Delta t_n e^{-j(\omega_i - \omega_k)t_n}$. All the diagonal terms of the $[X]$ matrix on the right-hand-side are constant and $X(i,i) = \sum_{n=0}^{N-1} \Delta t_n = T_W$. Then from the left-hand-side, this requires a uniform frequency spacing $\Delta\omega_i = 2\pi/T_W$. However under this condition, the off-diagonal terms on the right-hand-side of the $[X]$ matrix are usually nonzero. Therefore Eq. 3.1.15 has no general solution and the reconstructed signal $[f_r]$ cannot equal but can only approximate the original signal $[f]$.

EXAMPLE 3.2.2   Find the DFT spectrum and IDFT reconstructed signal for a nonuniformly sampled 1 Hz sinusoidal signal of

$$f(t) = \sin 2\pi t \tag{3.2.17}$$

Use the integral approximations of Eq. 3.2.12 with nonuniform time sampling and uniform frequency spacing. Use a $T_W = 4$ second sampling window and zero-order adaptive sampling.
SOLUTION   Zero-order adaptive sampling is described by Eq. 1.3.19. $f(t)$ is sampled whenever the absolute change $|\Delta f_n| = |f(t_n) - f(t_{n-1})|$ exceeds $\pm\epsilon$. The signal is plotted versus time in Fig. 3.2.3a. The sampled signal (linearly interpolated) is plotted versus sample number in Fig. 3.2.3b. Since index is plotted rather than time, the sampled sine wave appears triangular so it is easy to see error in the reconstructed signal.

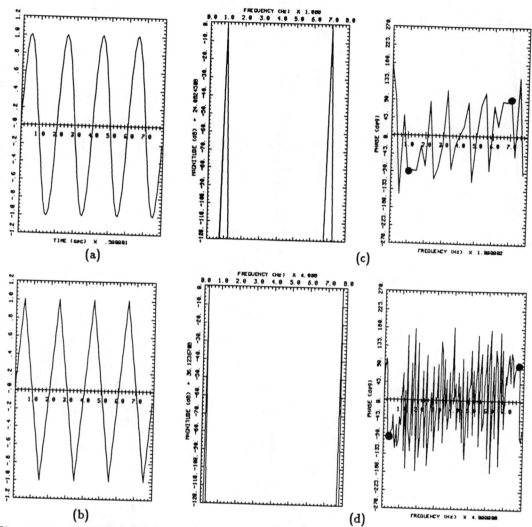

Fig. 3.2.3 (a) Sinusoidal signal and its (b) zero-order adaptively sampled equivalent. Magnitude and phase spectra for (c) 32-points and (d) 128-points in Example 3.2.2. Uniform time spacing, uniform frequency spacing. (From S.R. Yhann, loc. cit.)

Now we apply the integral approximation algorithm of Eq. 3.2.12. The uniform frequency spacing equals $\Delta\omega = 2\pi/T_W = 2\pi/NT$. Using a DFT with uniform time spacing gives the 32-point and 128-point magnitude and phase spectra shown in Figs. 3.2.3c and 3.2.3d, respectively. Using adaptive sampling, the resulting spectra and reconstructed signal are shown in Figs. 3.2.4a and 3.2.4b. Phase spectra are inaccurate at low magnitude levels where they are ignored. Between frequencies of 0 and f_s, the magnitude spectra has two impulses located at 1 Hz and $(f_s - 1)$ Hz. Comparing Figs. 3.2.3c–3.2.4a and Figs. 3.2.3d–3.2.4b, we see that there is significant spectral distortion. This distortion is due to the nonuniform Δt_n pulse widths resulting from zero-order adaptive sampling. Better spectral estimates are obtained for non-band-limited signals. Reconstruction accuracy is dependent

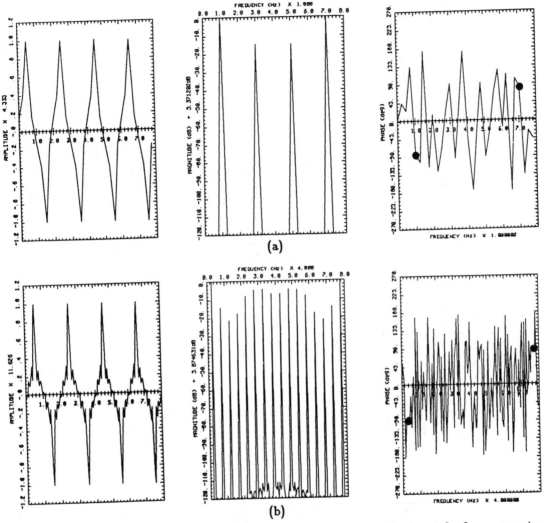

Fig. 3.2.4 Reconstructed sinusoidal signal and its magnitude and phase spectra using integral approximation algorithm in Example 3.2.2 for (a) 32-points and (b) 128-points. (From S.R. Yhann, loc. cit.)

upon the signal and sample size. Figs. 3.2.4a–b show that there is significant reconstructed signal error.

A different approach to approximate FT and IFT transforms utilizes *matrix inversions* of orthogonal series. In this case, the transform pair equal

$$[F] = [W]^{-1}[f]$$
$$[f_r] = [W][F]$$

(3.2.18)

which are short-hand equations for

$$
\begin{bmatrix} F(0) \\ F(1) \\ \vdots \\ F(N-1) \end{bmatrix} = \begin{bmatrix} 1 & 1 & \cdots & 1 \\ 1 & e^{j\omega_1 t_1} & \cdots & e^{j\omega_{N-1} t_1} \\ \vdots & \vdots & \ddots & \vdots \\ 1 & e^{j\omega_1 t_{N-1}} & \cdots & e^{j\omega_{N-1} t_{N-1}} \end{bmatrix}^{-1} \begin{bmatrix} f(0) \\ f(1) \\ \vdots \\ f(N-1) \end{bmatrix}
$$

$$
\begin{bmatrix} f(0) \\ f(1) \\ \vdots \\ f(N-1) \end{bmatrix} = \begin{bmatrix} 1 & 1 & \cdots & 1 \\ 1 & e^{j\omega_1 t_1} & \cdots & e^{j\omega_{N-1} t_1} \\ \vdots & \vdots & \ddots & \vdots \\ 1 & e^{j\omega_1 t_{N-1}} & \cdots & e^{j\omega_{N-1} t_{N-1}} \end{bmatrix} \begin{bmatrix} F(0) \\ F(1) \\ \vdots \\ F(N-1) \end{bmatrix}
\qquad (3.2.19)
$$

where we assume a starting time $t_0 = 0$ and equal frequency spacing $\Delta\omega_m = 2\pi/T_W$ where $\omega_m = m\Delta\omega_m$. This formulation using orthogonal series is

$$
F(m) = \sum_{n=0}^{N-1} f(t_n) w_m^{-1}(t_n)
$$

$$
f_r(t_n) = \sum_{m=0}^{N-1} F(m) w_m(t_n)
\qquad (3.2.20)
$$

where w_m are basis functions evaluated at the sample times t_n. The $F(m)$ coefficients minimize the MSE error. A more general $F(m)$ equation can be used for nonorthogonal w_m. Algebraically the matrix inversion approximation of Eq. 3.2.18 equals the integral approximation of Eq. 3.2.12 when the time and frequency spacing matrices $[\Delta t]$ and $\frac{1}{2\pi}[\Delta\omega]$ are set equal to the identity matrix.

The matrix inversion method requires that $[W]^{-1}$ be defined. Since

$$
[W] = \begin{bmatrix} 1 & 1 & \cdots & 1 \\ 1 & e^{j\omega_1 t_1} & \cdots & e^{j\omega_{N-1} t_1} \\ \vdots & \vdots & \ddots & \vdots \\ 1 & e^{j\omega_1 t_{N-1}} & \cdots & e^{j\omega_{N-1} t_{N-1}} \end{bmatrix} = \begin{bmatrix} 1 & X_0 & X_0^2 & \cdots & X_0^{N-1} \\ 1 & X_1 & X_1^2 & \cdots & X_1^{N-1} \\ \vdots & \vdots & \vdots & \ddots & \vdots \\ 1 & X_{N-1} & X_{N-1}^2 & \cdots & X_{N-1}^{N-1} \end{bmatrix}
\qquad (3.2.21)
$$

where $X_n = e^{j2\pi t_n/T_W}$, the $[W]$ matrix is *always* invertible if $X_i \neq X_k$. This is assured when $t_i \neq t_k$ and $t_i \in (0, T_W)$ where the time window $T_W = \sum_{n=0}^{N-1} t_n$. The $[W]$ matrix has unity magnitude elements and constant angle difference between columns. It also has conjugate symmetry about the column $(1 + N/2)$. The inverse matrix $[W]^{-1}$ has no obvious symmetry.

An element in the $[W]$ matrix equals $W(n, m) = e^{j2\pi m t_n/T_W}$. Its magnitude is unity and its phase equals $2\pi m t_n/T_W$ radians. The phase difference ϕ between two times t_i and t_k is $\phi(i, k) = (2\pi/T_W)(t_i - t_k)$. For large N, the first and last rows in the $[W]$ matrix are almost equal to unity, and are therefore not linearly independent. When two rows are almost equal, the determinant $|W| \cong 0$ and $[W]^{-1}$ evaluated by the computer behaves erratically. Therefore the matrix inversion approach of Eq. 3.2.18 is inherently ill-conditioned for large N and will give poor reconstructions.

EXAMPLE 3.2.3   Find the DFT spectrum and IDFT reconstructed signal for the 1 Hz $\sin(2\pi t)$ signal of Eq. 3.2.17. Use the matrix inversion approximations of Eq. 3.2.18. Adaptively sampling the signal $\sin(2\pi t)$ provides the sampling times t_n.

SOLUTION   The 32-point and 128-point $[W]$ matrices are generated as just illustrated. The resulting spectra and reconstructed signals using Eq. 3.2.18 are shown in Figs. 3.2.5a and 3.2.5b. The magnitude estimates are good (cf. Fig. 3.2.3c–d) for small N. The phase estimates at low magnitude spectra levels are poor. The magnitude and phase spectra do not exhibit symmetry about the Nyquist frequency as obtained with the FFT. The reconstructed signals are good. ∎

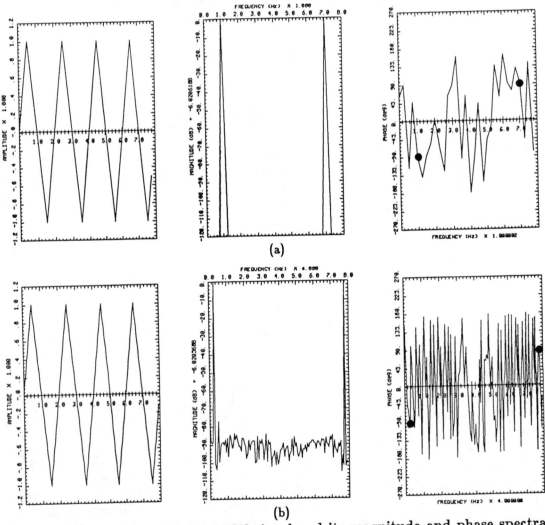

Fig. 3.2.5   Reconstructed sinusoidal signal and its magnitude and phase spectra using orthogonal series approximation algorithm in Example 3.2.3 for (a) 32-points and (b) 128-points. (From S.R. Yhann, loc. cit.)

Another interesting FT-IFT approximation approach uses *generalized transforms*. This transform pair equals

$$[F] = \left([W]^h[W]\right)^{-1}[W]^h[f]$$
$$[f_r] = [W][F]$$

(3.2.22)

$[F]$ results by rewriting Eq. 3.2.18a in a special form as

$$[F] = [W]^{-1}[f] = [W]^{-1}\left([W^h]^{-1}[W^h]\right)[f]$$
$$= \left([W]^{-1}[W^h]^{-1}\right)\left([W]^h[f]\right) = \left([W]^h[W]\right)^{-1}\left([W]^h[f]\right)$$

(3.2.23)

This can be shown to be an MSE approach. The columns of $[W]$ are the basis functions used to decompose $[f]$. $[W]^h[f]$ is the inner product matrix or energy matrix describing the energy in $[f]$ having the form $[W]^h$. The $[W]^h[W]$ product is the energy in the basis function so $([W]^h[W])^{-1}$ is a normalization constant. The basis function need not be orthogonal. A large number of operations are needed to evaluate Eq. 3.2.22 so it is not computationally efficient.

Another approximation results by combining the *integral approximation–matrix inversion* approaches to give the transform pair

$$[F] = [W]^h[\Delta t][f]$$
$$[f_r] = [\Delta t]^{-1}\left([W]^h\right)^{-1}[F]$$

(3.2.24)

This results by incorporating the $[\Delta\omega]$ matrix requirement given by Eq. 3.2.15 into the transform pair of Eq. 3.2.12 as

$$[f_r] = \frac{1}{2\pi}[W][\Delta\omega][F] = [W][W]^{-1}\left([W]^h[\Delta t]\right)^{-1}[F] = [\Delta t]^{-1}\left([W]^h\right)^{-1}[F] \quad (3.2.25)$$

The diagonal matrix $[\Delta t]$ is always invertible since all the $[\Delta t_n]$ elements are nonzero. The $[W]^h$ matrix is always invertible for equal frequency spacings of $2\pi/T_W$. $[W]^h$ is the conjugate transpose of $[W]$. Since determinant $|W| = |W^h|$, $([W]^h)^{-1}$ always exists. However the inverted matrix becomes ill-conditioned for large N as discussed in Eq. 3.2.21. A standard recursive algorithm can be used to evaluate $[f]$ as

$$[f]_{k+1} = [F] - \left([W]^h[\Delta t] - [I]\right)[f]_k$$

(3.2.26)

which results by expressing Eq. 3.2.24a as

$$[F] - [f] = [W]^h[\Delta t][f] - [f] = \left([W]^h[\Delta t] - [I]\right)[f]$$

(3.2.27)

However this algorithm converges only when the norm $||[W]^h[\Delta t] - [I]||$ is less than unity. This condition is not often satisfied.

A better *iterative algorithm* solves the integral approximation of Eq. 3.2.12 as

$$[F] = [W]^h [\Delta t][f]$$

$$[\Delta t][f_r]_{k+1} = \frac{1}{\Omega}\Big([W][\Delta \omega][F] - [T][\Delta t][f_r]_k \Big)$$

(3.2.28)

where $[T] = [W][\Delta \omega][W]^h - \Omega[I]$ and $\Omega = \sum_{m=0}^{N-1} \Delta \omega_m$. The final $[\Delta t][f_r]$ vector is scaled by $[\Delta t]^{-1}$ after convergence to obtain the reconstructed signal $[f_r]$. No assumption is made about $\Delta \omega_m$ or Δt_n. The iterative procedure does not require matrix inversion which introduces large errors for large N. Also the number of time and frequency samples (N and M) need not be equal. This approximation is proved by rewriting Eq. 3.2.10 as

$$f(t_n) = \frac{1}{2\pi}\sum_{m=0}^{N-1} F(m)e^{j\omega_m t_n}\Delta \omega_m = \frac{1}{2\pi}\sum_{m=0}^{N-1}\left[\sum_{k=0}^{N-1} f(t_k)\Delta t_k e^{-j\omega_m t_k}\right]e^{j\omega_m t_n}\Delta \omega_m$$

$$= \frac{1}{2\pi}f(t_n)\Delta t_n \Omega + \frac{1}{2\pi}\sum_{\substack{k=0 \\ k\neq n}}^{N-1} f(t_k)\Delta t_k \left(\sum_{m=0}^{N-1} e^{-j\omega_m(t_k-t_n)}\Delta \omega_m\right)$$

(3.2.29)

This reduction interchanges the order of summation, combines exponential terms, and notes that the product of exponentials is unity when $t_n = t_k$. Solving Eq. 3.2.29 for $f(t_n)\Delta t_n\Omega$ gives

$$f(t_n)\Delta t_n \Omega = \sum_{m=0}^{N-1} F(m)e^{j\omega_m t_n}\Delta \omega_m - \frac{1}{2\pi}\sum_{\substack{k=0 \\ k\neq n}}^{N-1} f(t_k)\Delta t_k \left(\sum_{m=0}^{N-1} e^{-j\omega_m(t_k-t_n)}\Delta \omega_m\right)$$

(3.2.30)

which, in matrix form, equals Eq. 3.2.28. The right-hand term in parenthesis $T(k,n) = \sum_{m=0}^{N-1} e^{-j\omega_m(t_k-t_n)}\Delta \omega_m$ are elements in the $[T]$ matrix which equals

$$[T] = [W][\Delta \omega][W]^h - \Omega[I] = \begin{bmatrix} 0 & X(0,1) & \cdots & X(0,N-1) \\ X(1,0) & 0 & \cdots & X(1,N-1) \\ \vdots & \vdots & \ddots & \vdots \\ X(N-1,0) & X(N-1,1) & \cdots & 0 \end{bmatrix}$$

(3.2.31)

and is used to simplify the recursion equation. It can be shown that $[T]$ controls convergence and convergence is guaranteed when $\|T\| < 1$. This is often satisfied when $\max(\Delta t_i/\Delta t_k)$ is less than 3.

EXAMPLE 3.2.4   Find the DFT spectrum and IDFT reconstructed signal for the 1 Hz $\sin(2\pi t)$ signal of Eq. 3.2.17. Use the iterative algorithm of Eq. 3.2.29.
SOLUTION   The resulting spectra and reconstructed signals using Eq. 3.2.29 are shown in Figs. 3.2.6a and 3.2.6b. The magnitude and phase estimates are the same as those obtained in Fig. 3.2.4 using the integral approximation algorithm. This is expected since the $[F]$ spectral equations are identical (Eqs. 3.2.12 and 3.2.28). However the reconstructed signals are much better and are almost identical to the original signals shown in Figs. 3.2.3a–b. ∎

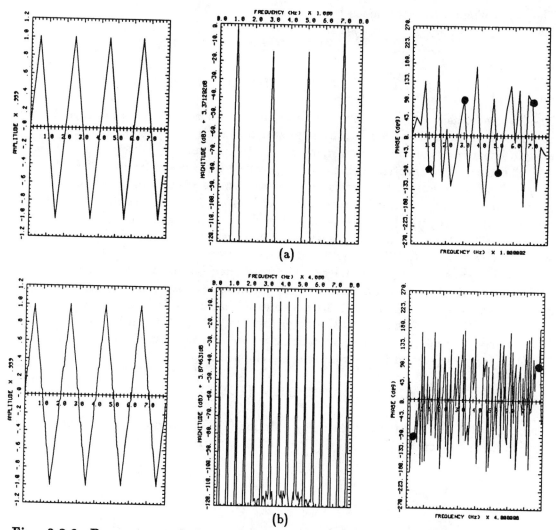

Fig. 3.2.6 Reconstructed sinusoidal signal and its magnitude and phase spectra using iterative algorithm to solve integral approximation in Example 3.2.4 for (a) 32-points and (b) 128-points. (From S.R. Yhann, loc. cit.)

3.3 OTHER INVERSE DISCRETE TRANSFORMS

The discrete Fourier transform is the most widely used transform. However as discussed in Chap. 2.8, there are a number of other very useful discrete transforms.[4] Some of these are listed in Table 2.8.1. Discrete transform pairs have the form (Eq. 2.8.2)

$$F(m) = \sum_{n=0}^{N-1} K(m,n)f(n), \qquad m = 0, \ldots, N-1$$

$$f(n) = \sum_{m=0}^{N-1} K^{-1}(n,m)F(m), \qquad n = 0, \ldots, N-1$$

(3.3.1)

They approximate the continuous transform pairs (Eq. 2.8.1)

$$F(p) = \int_{t_1}^{t_2} K(p,t)f(t)dt, \qquad p \in R_p$$

$$f(t) = \int_{p_1}^{p_2} K^{-1}(p,t)F(p)dp, \qquad t \in R_t$$

(3.3.2)

The inverse kernels $K^{-1}(n,m)$ are the basis functions used to approximate $f(n)$. Any complete kernel set allows $f(n)$ to be approximated with a total squared-error of zero using a linear combination of $K^{-1}(n,m)$ and an infinite number of terms.

The generalized spectral coefficient $F(m)$ equals the amount or level of $K^{-1}(n,m)$ in $f(n)$. The approximation basis $K^{-1}(n,m)$ can be chosen in many ways. For maximum simplicity, minimum bandwidth, and fewest terms, $f(n)$ should be closely matched to $K^{-1}(n,m)$. For high-quality reproduction, a large number of terms must be used. The Paley, Hadamard, Haar, and slant transforms are orthogonal so they satisfy $[K][K]^t = N[I]$. Therefore their inverse matrices $[K]^{-1} = \frac{1}{N}[K]^t$ are simply scaled transposes of the $[K]$ matrices. The Karhunen-Loéve transforms are not usually orthogonal. We shall now discuss these inverse discrete transforms in detail. It should be pointed out when Eq. 3.3.1 is expressed in matrix form as $[F] = [K][f]$ and $[f] = [K]^{-1}[F]$, any of the alternative algorithms in Table 3.2.1 can be used. The $[W]$ matrix must be replaced by the $[K]$ matrix. Sampling can be uniform or nonuniform. Uniform sampling is used in the following development.

3.3.1 INVERSE PALEY TRANSFORM (IDPT)

The discrete Walsh transform with Paley ordering (DPT) has proportional kernels and inverse kernels given by Eq. 2.8.14. The DPT matrix is always filled with ±1's. For $N = 8$, it equals

$$[Pal_N] = N\,[Pal_N]^{-1} = \begin{bmatrix} + & + & + & + & + & + & + & + \\ + & + & + & + & - & - & - & - \\ + & + & - & - & + & + & - & - \\ + & + & - & - & - & - & + & + \\ + & - & + & - & + & - & + & - \\ + & - & + & - & - & + & - & + \\ + & - & - & + & + & - & - & + \\ + & - & - & + & - & + & + & - \end{bmatrix}$$

(3.3.3)

where only the signs $(\pm)$ of the ±1's are shown. The inverse DPT expands a sampled signal $f(n)$ in terms of discrete Paley functions $Pal(m,n)$ using Eq. 3.3.1. In matrix form, the expansions equal

$$[F]_{Pal} = \frac{1}{N}[Pal_N]\,[f]_{Pal}$$

$$[f]_{Pal} = [Pal_N]\,[F]_{Pal}$$

(3.3.4)

The inverse continuous Paley transform (IPT) expands a signal $f(t)$ in terms of the continuous Paley functions $Pal(p,t)$ (see Fig. 2.8.1c) using Eq. 3.3.2. The discrete Paley transforms are obtained by sampling the continuous Paley transforms.

EXAMPLE 3.3.1 Consider a parabolic signal $f(t) = t^2$ for $0 \leq t < 1$. The DPT coefficients were determined in Example 2.8.1 to equal (Eq. 2.8.18)

$$[F]_{Pal} = \begin{bmatrix} 0.2734 \\ -0.2188 \\ -0.1094 \\ 0.0625 \\ -0.0547 \\ 0.0313 \\ 0.0156 \\ 0 \end{bmatrix} \tag{3.3.5}$$

Reconstruct the signal having this discrete spectra.
SOLUTION The IDPT is found using Eq. 3.3.4. Multiplying the Paley matrix by the discrete Paley spectra yields

$$[f]_{Pal} = \begin{bmatrix} 0 \\ 0.0156 \\ 0.0625 \\ 0.1406 \\ 0.2500 \\ 0.3906 \\ 0.5625 \\ 0.7656 \end{bmatrix} = \begin{bmatrix} + & + & + & + & + & + & + & + \\ + & + & + & + & - & - & - & - \\ + & + & - & - & + & + & - & - \\ + & + & - & - & - & - & + & + \\ + & - & + & - & + & - & + & - \\ + & - & + & - & - & + & - & + \\ + & - & - & + & + & - & - & + \\ + & - & - & + & - & + & + & - \end{bmatrix} \begin{bmatrix} 0.2734 \\ -0.2188 \\ -0.1094 \\ 0.0625 \\ -0.0547 \\ 0.0313 \\ 0.0156 \\ 0 \end{bmatrix} \tag{3.3.6}$$

These samples equal $f(n) = (n/8)^2$. This result and those of the other examples will be discussed at length when interpolators are investigated in Chap. 3.8. ∎

EXAMPLE 3.3.2 Now consider a 1 Hz cosine signal $f(t) = \cos(2\pi t)$. Find its samples in the interval $0 \leq t < 1$ using an eighth-order IDPT.
SOLUTION From Example 2.8.2, the discrete Paley spectra of the signal equal

$$[F]_{Pal} = \begin{bmatrix} 0 \\ 0.2500 \\ 0 \\ 0.6036 \\ 0 \\ 0.2500 \\ 0 \\ -0.1036 \end{bmatrix} \tag{3.3.7}$$

(Eq. 2.8.20). Using Eq. 3.3.4, the signal having this DPT spectra equals

$$[f]_{Pal} = \begin{bmatrix} 1.000 \\ 0.707 \\ 0 \\ -0.707 \\ -1.000 \\ -0.707 \\ 0 \\ 0.707 \end{bmatrix} = \begin{bmatrix} + & + & + & + & + & + & + & + \\ + & + & + & + & - & - & - & - \\ + & + & - & - & + & + & - & - \\ + & + & - & - & - & - & + & + \\ + & - & + & - & + & - & + & - \\ + & - & + & - & - & + & - & + \\ + & - & - & + & + & - & - & + \\ + & - & - & + & - & + & + & - \end{bmatrix} \begin{bmatrix} 0 \\ 0.2500 \\ 0 \\ 0.6035 \\ 0 \\ 0.2500 \\ 0 \\ -0.1036 \end{bmatrix} \qquad (3.3.8)$$

■

3.3.2 INVERSE HADAMARD TRANSFORM (IDHT)

The discrete Hadamard transform (DHT) has proportional kernels and inverse kernels given by Eq. 2.8.21. The DHT matrix is filled with $\pm 1's$. For $N = 8$, it equals

$$[Had_N] = N[Had_N]^{-1} = \begin{bmatrix} + & + & + & + & + & + & + & + \\ + & - & + & - & + & - & + & - \\ + & + & - & - & + & + & - & - \\ + & - & - & + & + & - & - & + \\ + & + & + & + & - & - & - & - \\ + & - & + & - & - & + & - & + \\ + & + & - & - & - & - & + & + \\ + & - & - & + & - & + & + & - \end{bmatrix} \qquad (3.3.9)$$

The inverse DHT expands a signal $f(n)$ in terms of discrete Hadamard functions $Had(m,n)$ as (see Fig. 2.8.1d)

$$[F]_{Had} = \frac{1}{N}[Had_N]\,[f]_{Had} \qquad (3.3.10)$$

$$[f]_{Had} = [Had_N]\,[F]_{Had}$$

Analogous results hold for the continuous case using Eq. 3.3.2.

EXAMPLE 3.3.3 Determine the 8-point inverse discrete Hadamard transform for the $\cos(2\pi t)$ signal.

SOLUTION The discrete Hadamard spectra were found in Example 2.8.3 as

$$[F]_{Had} = \begin{bmatrix} 0 \\ 0 \\ 0 \\ 0 \\ 0.2500 \\ 0.2500 \\ 0.6036 \\ -0.1036 \end{bmatrix} \qquad (3.3.11)$$

(Eq. 2.8.24b). Using Eq. 3.3.10, the signal having this DHT spectra is

$$[f]_{Had} = \begin{bmatrix} 1.000 \\ 0.707 \\ 0 \\ -0.707 \\ -1.000 \\ -0.707 \\ 0 \\ 0.707 \end{bmatrix} = \begin{bmatrix} + & + & + & + & + & + & + & + \\ + & - & + & - & + & - & + & - \\ + & + & - & - & + & + & - & - \\ + & - & - & + & + & - & - & + \\ + & + & + & + & - & - & - & - \\ + & - & + & - & - & + & - & + \\ + & + & - & - & - & - & + & + \\ + & - & - & + & - & + & + & - \end{bmatrix} \begin{bmatrix} 0 \\ 0 \\ 0 \\ 0 \\ 0.2500 \\ 0.2500 \\ 0.6036 \\ -0.1036 \end{bmatrix} \qquad (3.3.12)$$

which is the same as the IDPT output of Eq. 3.3.8.   ■

3.3.3 INVERSE HAAR TRANSFORM (IDHaT)

The inverse Haar transform (IDHaT) matrices are transposed Haar transform matrices scaled by $1/N$ where

$$[Ha_N]^{-1} = \frac{1}{N}[Ha_N]^t \qquad (3.3.13)$$

Therefore, interchanging rows and columns in the Haar matrices of Eq. 2.8.28 and multiplying by $1/N$ produce inverse Haar matrices. The inverse DHaT expands a signal $f(n)$ in terms of discrete Haar functions $Ha(m, n)$ as (see Fig. 2.8.1e)

$$[F]_{Ha} = \frac{1}{N}[Ha_N]^t[f]_{Ha}$$
$$[f]_{Ha} = [Ha_N] [F]_{Ha} \qquad (3.3.14)$$

Analogous results hold for the continuous case.

EXAMPLE 3.3.4 Determine the 8-point inverse discrete Haar transform for the $\cos(2\pi t)$ signal.
SOLUTION The discrete Haar spectra were found in Example 2.8.4 to equal

$$[F]_{Ha} = \begin{bmatrix} 0 \\ 0.2500 \\ 0.4268 \\ -0.4268 \\ 0.0732 \\ 0.1768 \\ -0.0732 \\ -0.1768 \end{bmatrix} \qquad (3.3.15)$$

(Eq. 2.8.31). Using Eq. 3.3.14, the signal having this DHaT spectra is

$$[f]_{Ha} = \begin{bmatrix} 1.000 \\ 0.707 \\ 0 \\ -0.707 \\ -1.000 \\ -0.707 \\ 0 \\ 0.707 \end{bmatrix} = \begin{bmatrix} 1 & 1 & \sqrt{2} & \sqrt{2} & 2 & 2 & 2 & 2 \\ + & + & + & \cdot & + & \cdot & \cdot & \cdot \\ + & + & + & \cdot & - & \cdot & \cdot & \cdot \\ + & + & - & \cdot & \cdot & + & \cdot & \cdot \\ + & + & - & \cdot & \cdot & - & \cdot & \cdot \\ + & - & \cdot & + & \cdot & \cdot & + & \cdot \\ + & - & \cdot & + & \cdot & \cdot & - & \cdot \\ + & - & \cdot & - & \cdot & \cdot & \cdot & + \\ + & - & \cdot & - & \cdot & \cdot & \cdot & - \end{bmatrix} \begin{bmatrix} 0 \\ 0.2500 \\ 0.4268 \\ -0.4268 \\ 0.0732 \\ 0.1768 \\ -0.0732 \\ -0.1768 \end{bmatrix} \quad (3.3.16)$$

The numbers above each column indicate the multiplier of that column.   ∎

3.3.4 INVERSE SLANT TRANSFORM (IDST)

The inverse slant transform (IDST) matrices are transposed slant transform matrices scaled by $1/N$ where

$$[S_N]^{-1} = \frac{1}{N}[S_N]^t \tag{3.3.17}$$

Thus, interchanging rows and columns in the slant matrices of Eq. 2.8.33 and multiplying by $1/N$ produce inverse slant matrices. The inverse DST expands a signal $f(n)$ in terms of discrete slant functions $S(m,n)$ as (see Fig. 2.8.1f)

$$[F]_S = \frac{1}{N}[S_N]^t[f]_S$$
$$[f]_S = [S_N][F]_S \tag{3.3.18}$$

Analogous results hold for the continuous case.

EXAMPLE 3.3.5 Determine the 8-point inverse discrete slant transform for the $\cos(2\pi t)$ signal.
SOLUTION The discrete slant spectra were found in Example 2.8.5 to equal

$$[F]_S = \begin{bmatrix} 0 \\ 0.2182 \\ 0.6516 \\ -0.1220 \\ 0 \\ -0.1036 \\ -0.0463 \\ 0 \end{bmatrix} \quad (3.3.19)$$

(Eq. 2.8.37). Using Eq. 3.3.18, the signal having this DST spectra is

$$[f]_s = \begin{bmatrix} 1.000 \\ 0.707 \\ 0 \\ -0.707 \\ -1.000 \\ -0.707 \\ 0 \\ 0.707 \end{bmatrix} = \begin{bmatrix} 1 & \frac{1}{\sqrt{21}} & \frac{1}{\sqrt{5}} & \frac{1}{\sqrt{105}} & 1 & 1 & \frac{1}{\sqrt{5}} & \frac{1}{\sqrt{5}} \\ 1 & 7 & 3 & 7 & 1 & 1 & 1 & 1 \\ 1 & 5 & 1 & -1 & -1 & -1 & -3 & -3 \\ 1 & 3 & -1 & -9 & -1 & -1 & 3 & 3 \\ 1 & 1 & -3 & -17 & 1 & 1 & -1 & -1 \\ 1 & -1 & -3 & 17 & 1 & -1 & -1 & 1 \\ 1 & -3 & -1 & 9 & -1 & 1 & 3 & -3 \\ 1 & -5 & 1 & 1 & -1 & 1 & -3 & 3 \\ 1 & -7 & 3 & -7 & 1 & -1 & 1 & -1 \end{bmatrix} \begin{bmatrix} 0 \\ 0.2182 \\ 0.6516 \\ -0.1220 \\ 0 \\ -0.1036 \\ -0.0463 \\ 0 \end{bmatrix}$$

$$(3.3.20)$$

The multipliers of each column are indicated above the individual columns. ∎

3.3.5 INVERSE KARHUNEN-LOÉVE TRANSFORM (IDKLT)

The inverse K-L transform (IDKLT) matrix must be computed since the K-L transform is not always orthogonal. Therefore the inverse K-L transform does not always equal the conjugate transposed K-L transform matrix scaled by $1/N$ as unitary or orthogonal transforms, i.e.

$$[K_{KL}]^{-1} \neq \frac{1}{N}[K_{KL}]^h \qquad (3.3.21)$$

However $[K_{KL}]$ can be orthogonalized using the Gram-Schmidt process in which case Eq. 3.3.21 becomes an equality. The inverse KLT expands a signal $f(n)$ in terms of discrete K-L functions $K_{KL}(m,n)$ as

$$[F]_{KL} = \frac{1}{N}[K_{KL}][f]_{KL}$$
$$[f]_{KL} = N[K_{KL}]^{-1}[F]_{KL} \qquad (3.3.22)$$

(see Fig. 2.8.3). Analogous results hold for the continuous case.

EXAMPLE 3.3.6 Determine the 8-point inverse discrete K-L transform for the $\cos(2\pi t)$ signal.

SOLUTION The discrete K-L spectra were found in Example 2.8.6 to equal

$$[F]_{KL} = \begin{bmatrix} 0.5000 \\ 0.0884 \\ 0 \\ -0.0884 \\ -0.1250 \\ -0.0884 \\ 0 \\ 0.0884 \end{bmatrix} \qquad (3.3.23)$$

(Eq. 2.8.44). Using Eq. 3.3.22 and $[K_{KL}]^{-1}$ from Eq. 2.8.40, the IKLT gives the signal as

$$[f]_{KL} =$$

$$\begin{bmatrix} 1.000 \\ 0.707 \\ 0 \\ -0.707 \\ -1.000 \\ -0.707 \\ 0 \\ 0.707 \end{bmatrix} = 8 \begin{bmatrix} 1 & -0.707 & 0 & 0.707 & 1 & 0.707 & 0 & -0.707 \\ 0 & 1 & 0 & 0 & 0 & 0 & 0 & 0 \\ 0 & 0 & 1 & 0 & 0 & 0 & 0 & 0 \\ 0 & 0 & 0 & 1 & 0 & 0 & 0 & 0 \\ 0 & 0 & 0 & 0 & 1 & 0 & 0 & 0 \\ 0 & 0 & 0 & 0 & 0 & 1 & 0 & 0 \\ 0 & 0 & 0 & 0 & 0 & 0 & 1 & 0 \\ 0 & 0 & 0 & 0 & 0 & 0 & 0 & 1 \end{bmatrix} \begin{bmatrix} 0.5000 \\ 0.0884 \\ 0 \\ -0.0884 \\ -0.1250 \\ -0.0884 \\ 0 \\ 0.0884 \end{bmatrix}$$

$$(3.3.24) \quad \blacksquare$$

3.4 FAST INVERSE TRANSFORMS

3.4.1 INVERSE FAST FOURIER TRANSFORM (IFFT)

Close inspection of the discrete Fourier transform pair given by Eq. 3.2.5 shows that $F(m)$ and $f(-m)$ are described by *dual* equations. They have the same mathematical form where

$$\sum_{n=0}^{N-1} F(n)W_N^{nm} = Nf(-m) \longleftrightarrow F(m) = \sum_{n=0}^{N-1} f(n)W_N^{nm} \qquad (3.4.1)$$

Since the FFT facilitated fast DFT computation, the same algorithms can be used to compute the IFFT. It is clear from Eq. 3.4.1 that all that is needed is to replace m by $-m$ in the FFT computation and to divide by N.

However, only division by N is required if the $f(-m)$ coefficients generated by the FFT are recognized to be in *reverse sequence* or *inverted-time order*. The $[f]$ data array $[f(0), f(-1), f(-2), \ldots, f(-(N-1))]^t$ must be shuffled into normal forward sequence. Since $f(-m) = f(N-m)$ because of periodicity, the $[f]$ data array also equals $[f(0), f(N-1), f(N-2), \ldots, f(2), f(1)]^t$. The *shuffled* data array $[g]$ having *normal sequence* is formed by setting $g(m) = f(N-m)$. Performing this reverse sequencing operation on the $[f]$ array gives $[g] = [f(0), f(1), \ldots, f(N-1)]^t$ which is the original data in normal order.

The fast inverse FFT for $N = 8$ is performed using any of the time decimation algorithms in Figs. 2.3.5–2.3.8 or the frequency decimation algorithms in Figs. 2.4.4–2.4.7. The FFT programs in Tables 2.4.1 and 2.4.2 can be used for N up to 1024. Eq. 3.4.1 shows that the time response $f(n)$ is obtained by reverse sequencing the FFT outputs and dividing by N.

EXAMPLE 3.4.1 Find the 4-point FFT of a sawtooth signal. Apply the FFT again to the sawtooth spectrum and show the data is in reversed-time order. Shuffle the data into normal forward order.

SOLUTION Taking the sawtooth signal samples to be $[f] = [0, 1, 2, 3]^t$, then the FFT spectrum equals

$$[F] = [W][f] = \begin{bmatrix} 1 & 1 & 1 & 1 \\ 1 & -j & -1 & j \\ 1 & -1 & 1 & -1 \\ 1 & j & -1 & -j \end{bmatrix} \begin{bmatrix} 0 \\ 1 \\ 2 \\ 3 \end{bmatrix} = \begin{bmatrix} 6 \\ -2+j2 \\ -2 \\ -2-j2 \end{bmatrix} \quad (3.4.2)$$

Performing the FFT a second time and dividing by N gives

$$[f'] = \frac{1}{4}[W][F] = \frac{1}{4} \begin{bmatrix} 1 & 1 & 1 & 1 \\ 1 & -j & -1 & j \\ 1 & -1 & 1 & -1 \\ 1 & j & -1 & -j \end{bmatrix} \begin{bmatrix} 6 \\ -2+j2 \\ -2 \\ -2-j2 \end{bmatrix} = \begin{bmatrix} 0 \\ 3 \\ 2 \\ 1 \end{bmatrix} \quad (3.4.3)$$

which is the original data in inverted-time order where $[f'] = [0, 3, 2, 1]^t$. Shuffling the data as $g(n) = f'(4 - n)$ produces $[g] = [0, 1, 2, 3]^t$ which is the original data in normal forward order. ∎

3.4.2 OTHER FAST INVERSE TRANSFORMS

The general discrete transforms given by Eq. 3.3.1 describe $F(m)$ and $f(m)$ as

$$\sum_{n=0}^{N-1} F(n)K^{-1}(m,n) = f(m) \longleftrightarrow F(m) = \sum_{n=0}^{N-1} f(n)K(m,n) \quad (3.4.4)$$

These equations have the same form so are *duals* when $K^{-1}(m,n) = K(m,n)$, i.e. when the kernel and inverse kernel are equal (or proportional). In this case, the transform and inverse transform flow graphs are *identical*. The algorithms used to evaluate fast transforms *can also be used* to evaluate fast inverse transforms. Transforms having symmetrical kernels with matrix elements $K(m,n) = K(n,m)$ satisfy this condition. For example, the Paley and Hadamard transforms have proportional symmetrical kernels and $NK^{-1}(m,n) = K(m,n)$. In the symmetrical but reversed-sign case where the kernels satisfy $K^{-1}(m,n) = K(m,-n)$, time reversal is needed. As just discussed, the DFT falls in this category since $K(m,n) = W_N^{nm}$ and $NK^{-1}(m,n) = W_N^{-nm}$.

In the situation where the kernel and inverse kernel are not equal as $K^{-1}(m,n) \neq K(m,n)$, then the equations of Eq. 3.4.4 are *not duals*. Therefore the fast transform and fast inverse transform flow graphs and algorithms are *different*. In the unitary or orthogonal transform case when kernel and inverse kernel are conjugate transposes as $K^{-1}(m,n) = K^h(m,n)$, their flow graphs are conjugate transposes of one another. Therefore, the fast transform algorithms *cannot be used* as fast inverse transform algorithms but they can be systematically converted. Haar and slant transforms fall in this category.

The signal flow graphs for the fast inverse discrete Paley and Hadamard 8-point transforms are shown in Figs. 3.4.1a–b. They are identical to the FPT and FHT flow graphs in Figs. 2.8.2a–b deleting the $1/N$ multipliers. Therefore the FPT and FHT subroutines of Table 2.8.3 can be used as IFPT and IFHT subroutines. The output data from the subroutines is multiplied by N.

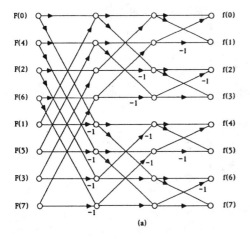

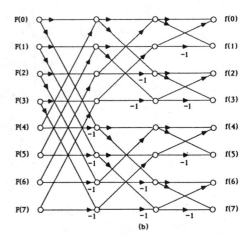

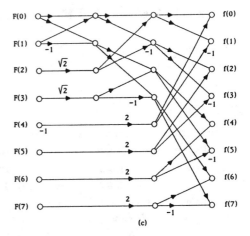

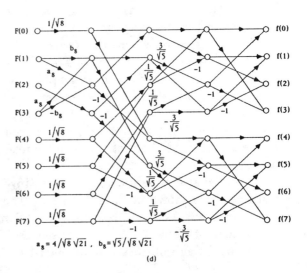

Fig. 3.4.1 Signal flow
graphs for 8-point
(a) Paley,
(b) Hadamard,
(c) Haar, and
(d) slant
inverse transforms.

Table 3.4.1 Fast inverse (a) Haar and (b) slant subroutines.

```
(a)       SUBROUTINE IFHAT(N,FWR,FTR)                        ELSE
          DIMENSION FTR(1),FWR(1),Z(1024)                      DO 10 I=1,N
          DO 5 I=1,N                                             FTR(I)=FWR(I)
             FTR(I)=FWR(I)                          10        ITER=ITER-1
  5          Z(I)=FTR(I)                                      M=1
          ITER=0                                              N4=N/2
          I=N                                                 DO 20 I=1,ITER
  10      IF(I.GT.1) THEN                                        X1=FLOAT(N4)**2
             I=I/2                                               AN=SQRT(3*X1/(4*X1-1))
             ITER=ITER+1                                         BN=SQRT((X1-1)/(4*X1-1))
             GOTO 10                                             DO 15 K=1,M
          ENDIF                                                     T1=FTR(M+K)
          C=SQRT(2.0)                                              T2=FTR(3*M+K)
          L=N/2                                      15            FTR(M+K)=AN*T1-T2*BN
          DO 30 M=1,ITER                                           FTR(3*M+K)=T1*BN+T2*AN
             IF(M.EQ.1) NP=1                                    M=M*2
             IF(M.GT.1) NP=NP*2                     20         N4=N4/2
             DO 15 MP=1,NP                                  N1=N-1
                K=(MP-1)*2+1                                 MR=0
                MM=MP+NP                                     DO 35 M=1,N1
                T1=FTR(MP)                                   L=N
                T2=FTR(MM)                          30       L=L/2
                Z(K)=T1+T2                                   IF(L.EQ.0) GOTO 35
  15            Z(K+1)=T1-T2                                 IF((MR+L).GT.N1) GOTO 30
             NN=N/L                                          MR=MOD(MR,L)+L
             R=C**M                                          IF(MR.LE.M) GOTO 35
             DO 20 I=1,NN                                    TR1=FTR(M+1)
                FTR(I)=Z(I)                                  FTR(M+1)=FTR(MR+1)
                K=NN+I                                        FTR(MR+1)=TR1
  20            IF(M.LT.ITER) FTR(K)=Z(K)*R         35       CONTINUE
  30         L=L/2                                           ITER=ITER+1
          RETURN                                             DO 50 M=1,ITER
          END                                                   IF(M.EQ.1) NP=1
                                                  35            IF(M.GT.1) NP=NP*2
                                                               MN=N/NP
                                                               MN2=MN/2
                                                               C=1.0
(b)       SUBROUTINE IFST(N,FWR,FTR)                           DO 45 MP=1,NP
          DIMENSION FTR(1),FWR(1),TR(4)                           IB=(MP-1)*MN
          ITER=0                                                  DO 40 K=1,MN2
          I=N                                                        KK=IB+K
  5       IF(I.GT.1) THEN                                            KP=KK+MN2
             I=I/2                                                   T1=FTR(KK)
             ITER=ITER+1                                            T2=FTR(KP)
             GOTO 5                                                  FTR(KK)=T1+C*T2
          ENDIF                                                     FTR(KP)=T1-C*T2
          IF (ITER.EQ.2) THEN                                    C=-C
          R=SQRT(5.0)                            50            CONTINUE
          TR(1)=FWR(1)+FWR(3)                                ENDIF
          TR(2)=FWR(1)-FWR(3)                                RETURN
          TR(3)=(3.0*FWR(2)+FWR(4))*R/5.0       40          END
          TR(4)=(FWR(2)-3.0*FWR(4))*R/5.0       45
          DO 8 I=1,2                            50
             J=5-I
             FTR(I)=TR(I)+TR(I+2)
  8          FTR(J)=TR(I)-TR(I+2)
```

The signal flow graphs for the fast inverse discrete Haar and slant 8-point transforms are shown in Figs. 3.4.1c–d. They are the transposes of the FHaT and FST shown in Figs. 2.8.2c–d when the $1/N$ factors are deleted. In other words, they are the same graphs turned end-for-end with the branch directions reversed. The FHaT and FST subroutines of Table 2.8.3 can be modified to be consistent with the transposed flow graph. This results in the IFHaT and IFST subroutines of Table 3.4.1.

In general, there are no fast K-L or inverse K-L transforms because the general K-L transform matrix does not have any fixed, predefined symmetry unlike the other transforms we have discussed. The flow graph of the K-L transform has the form of Fig. 2.2.1. It has only one stage rather than $\log_2 N$ stages like the other transforms. Therefore the transform and its inverse have the same flow graph structure but different branch gains.

EXAMPLE 3.4.2 Find the 8-point fast inverse transforms of the Paley, Hadamard, Haar, and slant spectra for the $\cos(2\pi t)$ signal in the interval $0 \le t < 1$.
SOLUTION These $[F]$ spectra were given by Eqs. 3.3.7, 3.3.11, 3.3.15, and 3.3.19, respectively. The fast 8-point inverse transforms are shown in Figs. 3.4.1a–d. Therefore applying these discrete spectra to the appropriate input nodes of the signal flow graph produces the $f(n)$ sampled signal outputs. The outputs equal those given by the $f(n)$ matrices of Eqs. 3.3.8, 3.3.12, 3.3.16, and 3.3.20, respectively. They are simply the eight samples $[1, 0.707, 0, -0.707, -1, -0.707, 0, 0.707]^t$ of the $\cos(2\pi t)$ signal using an 8 Hz sampling rate starting at time $t = 0$. ∎

3.5 SIMPLE INTERPOLATORS

Now that various inverse transforms have been discussed, the *interpolation* or *reconstruction block* in Fig. 2.0.1 shall be investigated. This block reconstructs the output signal $g(t)$ from its samples $g(nT)$. In the most general sense, reconstruction methods reduce to examining the equation[5]

$$f_r(t) = \sum_{n=-\infty}^{\infty} f(nT)h(t - nT) = f_s(t) * h(t) \qquad (3.5.1)$$

where $h(t)$ is the impulse response of some suitably chosen *reconstruction* or *interpolation filter*. For a causal impulse response of finite length (NT seconds), Eq. 3.5.1 reduces to

$$f_r(t) = \sum_{n=0}^{N-1} f(nT)h(t - nT) = \sum_{n=0}^{N-1} h(nT)f(t - nT) = f_s(t) * h(t) \qquad (3.5.2)$$

$h(t)$ is not easy to select and there are no simple rules to follow.

When all the $f(nT)$ data is used (i.e. infinite filter order), then the ideal lowpass filter gives perfect reconstruction (see Chap. 3.6). The Fourier coefficients give an approximate response that is optimum in the minimum mean squared error (MSE) sense. For finite data length, the Fourier coefficients are still optimum in the MSE sense but are unsuitable practically due to overshoot and sustained oscillation (see Chap. 3.7). Weighted Fourier series using windows (Chap. 10.4) reduce these undesirable characteristics. Heuristic arguments and subjective judgement are important when selecting the interpolator. Let us now discuss the simplest interpolators and then proceed to the more complicated ones.

3.5.1 ZERO-ORDER INTERPOLATOR

The *zero-order interpolator* provides the easiest reconstruction method. This reconstruction filter is often called a sample-and-hold or D-A converter. Its output $f_r(t)$ is shown in Fig. 3.5.1a. $f_r(t)$ holds the sampled signal $f(nT)$ over the entire clock interval until it is updated at the next sample time as

$$f_r(t) = f(nT) = \int_{nT}^{(n+1)T} f_s(t)dt, \qquad nT \le t < (n+1)T \qquad (3.5.3)$$

Setting $f_s(t) = U_0(t)$ or $f(0) = 1/T$ and evaluating Eq. 3.5.3 yields the impulse response of the zero-order hold as

$$
\begin{aligned}
h_0(t) &= 1/T, \qquad 0 \le t < T \\
&= 0, \qquad\quad \text{elsewhere}
\end{aligned}
\qquad (3.5.4)
$$

which is drawn in Fig. 3.5.1b. Rewriting $h_0(t)$ in closed-form and taking its Laplace transform gives the transfer function as

$$
\begin{aligned}
h_0(t) &= \frac{1}{T}\left[U_{-1}(t) - U_{-1}(t - T)\right] \\
H_0(s) &= \frac{1 - e^{-sT}}{sT}, \qquad \text{all } s
\end{aligned}
\qquad (3.5.5)
$$

In ac steady-state, $s = j\omega$ so

$$H_0(j\omega) = e^{-j\omega T/2}\,\frac{\sin(\omega T/2)}{\omega T/2} \qquad (3.5.6)$$

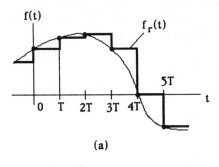

(a)

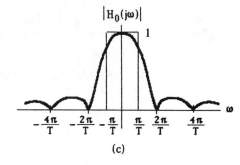

(c)

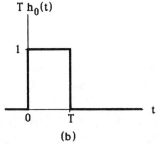

(b)

Fig. 3.5.1 (a) Reconstructed signal, (b) impulse response, and (c) magnitude response of zero-order interpolator.

The magnitude response is shown in Fig. 3.5.1c. It is a poor approximation to the ideal lowpass filter. Nevertheless when T is sufficiently small, $Tf_r(t)$ closely approximates $f(t)$. For further discussion, see Chap. 5.14.

3.5.2 FIRST-ORDER INTERPOLATOR

A *first-order interpolator* connects adjacent samples of $f(t)$ by straight lines as shown in Fig. 3.5.2a. Assuming $f_r(t)$ equals $(a + bt)$, then using the samples $[f(0), f(T)]$ and evaluating $f_r(t)$ at $t = (0, T)$ gives the a and b coefficient values as

$$a = f(0), \qquad b = \frac{1}{T}\Big[f(T) - f(0)\Big] \tag{3.5.7}$$

The reconstructed signal $f_r(t)$ equals

$$f_r(t) = f(nT) + \Big[f((n+1)T) - f(nT)\Big]\frac{t - nT}{T}$$

$$= f(nT)\Big(n + 1 - \frac{t}{T}\Big) + f((n+1)T)\Big(\frac{t}{T} - n\Big), \qquad nT \le t < (n+1)T \tag{3.5.8}$$

Setting $f(0) = 1/T$ and all other samples to zero and evaluating Eq. 3.5.8 gives the impulse response of the first-order interpolator as

$$\begin{aligned}
Th_1(t) &= 1 + t/T, & -T \le t \le 0 \quad (n = -1) \\
&= 1 - t/T, & 0 \le t \le T \quad (n = 0) \\
&= 0, & \text{elsewhere (all other } n)
\end{aligned} \tag{3.5.9}$$

which is shown in Fig 3.5.2b. Although $h_1(t)$ is noncausal, delaying the impulse response one sample makes $h_1(t - T)$ realizable. Expressing $h_1(t)$ in closed form and taking the Laplace transform gives the transfer function of the first-order interpolator as

$$h_1(t) = \frac{1}{T^2}\Big[U_{-2}(t + T) - 2U_{-2}(t) + U_{-2}(t - T)\Big]$$

$$H_1(s) = \frac{e^{sT} - 2 + e^{-sT}}{s^2 T^2} = \Big[\frac{e^{sT} - e^{-sT}}{sT}\Big]^2 \tag{3.5.10}$$

The ac steady-state gain is shown in Fig. 3.5.2c where

$$H_1(j\omega) = T\Big[\frac{\sin(\omega T/2)}{\omega T/2}\Big]^2 \tag{3.5.11}$$

This response is closer to that of the ideal lowpass filter and is an improvement over that of the zero-order hold. Comparing transfer functions (Eqs. 3.5.6 and 3.5.11) shows they are related as $e^{-sT}H_1(s) = H_0^2(s)$ so $h_1(t - T) = h_0(t) * h_0(t)$. $H_1(j\omega)$ has a wider 3 dB bandwidth than $H_0(j\omega)$ but double the asymptotic roll-off (-40 dB/dec). For further discussion, see Chap. 5.15.

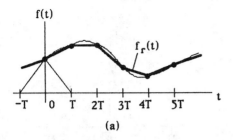

(a)

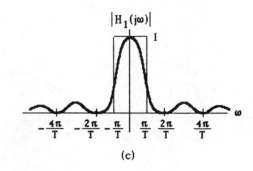

(c)

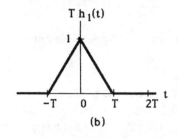

(b)

Fig. 3.5.2 (a) Reconstructed signal, (b) impulse response, and (c) magnitude response of first-order interpolator.

3.5.3 SECOND-ORDER INTERPOLATOR

A *second-order interpolator* connects three adjacent samples of $f(t)$ by a parabola as shown in Fig. 3.5.3a. Equating $f_r(t)$ to $(a + bt + ct^2)$, then evaluating $f_r(t)$ at $t = (-T, 0, T)$ and using samples $[f(-T), f(0), f(T)]$ results in a, b, and c coefficients of

$$a = f(0), \qquad b = \frac{1}{2T}\Big[f(T) - f(-T)\Big], \qquad c = \frac{1}{2T^2}\Big[f(T) - 2f(0) + f(-T)\Big] \quad (3.5.12)$$

The interpolated signal $f_r(t)$ equals

$$
\begin{aligned}
f_r(t) &= f(nT) + \Big\{f((n+1)T) - f((n-1)T)\Big\}\frac{t - nT}{2T} \\
&\quad + \frac{1}{2}\Big\{f((n+1)T) - 2f(nT) + f((n-1)T)\Big\}\Big[\frac{t - nT}{T}\Big]^2 \\
&= \frac{1}{2}f((n-1)T)\Big(n - \frac{t}{T}\Big)\Big(n+1 - \frac{t}{T}\Big) + f(nT)\Big(1 - n + \frac{t}{T}\Big)\Big(n+1 - \frac{t}{T}\Big) \\
&\quad + \frac{1}{2}f((n+1)T)\Big(n - \frac{t}{T}\Big)\Big(n-1 - \frac{t}{T}\Big), \qquad nT \leq t < (n+1)T
\end{aligned}
$$

$$(3.5.13)$$

Setting $f(0) = 1/T$ and all other samples to zero in Eq. 3.5.13 gives the impulse response of a second-order interpolator as

$$
\begin{aligned}
Th_2(t) &= \frac{1}{2}\Big(1 + \frac{t}{T}\Big)\Big(2 + \frac{t}{T}\Big), & -T \leq t \leq 0 \quad (n = -1) \\
&= \Big(1 + \frac{t}{T}\Big)\Big(1 - \frac{t}{T}\Big), & 0 \leq t \leq T \quad (n = 0) \\
&= \frac{1}{2}\Big(1 - \frac{t}{T}\Big)\Big(2 - \frac{t}{T}\Big), & T \leq t \leq 2T \quad (n = 1) \\
&= 0, & \text{elsewhere (all other } n)
\end{aligned}
$$

$$(3.5.14)$$

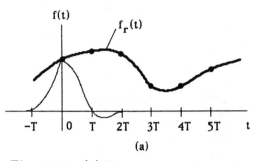

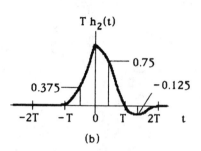

Fig. 3.5.3 (a) Reconstructed signal and (b) impulse response of second-order interpolator.

$h_2(t)$ is plotted in Fig. 3.5.3b. Introducing a delay of one clock period to form $h_2(t-T)$ insures realizability. The interpolation range in Eq. 3.5.13 can instead be taken as $(n-1)T \le t < nT$ (shifted back one period). Then the impulse response is a time inverted version of that shown in Fig. 3.5.3 ($h_2(-t)$ in Eq. 3.5.14).

3.5.4 HIGHER-ORDER LAGRANGE INTERPOLATORS

The preceding interpolators are examples of the lowest-order Lagrange interpolators. *Lagrange interpolators* are obtained by using the Lagrange polynomials[5]

$$q_{N,m}(t) = \prod_{\substack{k=0 \\ k \ne m}}^{N} \frac{t-t_k}{t_m-t_k} = \frac{(t-t_0)(t-t_1)\dots(t-t_N)}{(t_m-t_0)(t_m-t_1)\dots(t_m-t_N)}, \quad m=0,\dots,N \quad (3.5.15)$$

There are $(N+1)$ polynomials each of Nth degree. The mth polynomial equals zero at all sample times except for $t=t_m$ when it equals unity. The $(N+1)$ samples need not be equally spaced. For simplicity, equal spacing will be used here.

The reconstruction filter output for interpolation over a single interval equals

$$f_r(t) = \sum_{m=0}^{N} f(t_m)q_{N,m}(t) = f(t_0)q_{N,0}(t)+f(t_1)q_{N,1}(t)+\dots+f(t_N)q_{N,N}(t),$$
$$t_0 \le nT \le t < (n+1)T \le t_N \qquad (3.5.16)$$

$f_r(t)$ passes through the $(N+1)$ samples from $f(t_0)$ to $f(t_N)$. Only the interpolation from nT to $(n+1)T$ is used. The impulse response of the interpolator is found by letting $f(t) = \frac{1}{T}U_0(t)$. Solving Eq. 3.5.16 for the resulting impulse response $h_N(t)$ yields

$$Th_N(t) = q_{N,N}(t-t_N), \quad -(N-k)T \le t < -(N-k+1)T \quad (n=-N+k)$$
$$\dots \qquad \dots \qquad \dots$$
$$= q_{N,k+1}(t-t_{k+1}), \quad -T \le t < 0 \qquad (n=-1)$$
$$= q_{N,k}(t-t_k), \quad 0 \le t < T \qquad (n=0) \qquad (3.5.17)$$
$$= q_{N,k-1}(t-t_{k-1}), \quad T \le t < 2T \qquad (n=1)$$
$$\dots \qquad \dots \qquad \dots$$
$$= q_{N,0}(t-t_0), \quad kT \le t < (k-1)T \qquad (n=k)$$

where k is the time index where $t_k = nT$. The closed form expression for $h_N(t)$ is

$$Th_N(t) = \sum_{m=0}^{N} q_{N,m}(t - t_m)\Big\{U_{-1}\big[t - (t_m - t_k)\big] - U_{-1}\big[t - (t_m - t_k) + T\big]\Big\},$$
$$-t_N + nT \le t < nT - t_0 + T \qquad (3.5.18)$$

and 0 otherwise. The difference in unit steps forms a single-period selection pulse. This pulse "turns on" the mth Lagrange polynomial $q_{N,m}(t)$ for one period T. Each of the $(N + 1)$ Lagrange polynomials contribute a one-period section to the impulse response which lasts for $(N + 1)T$ seconds. The $h_N(t)$ impulse response begins at $-(t_N - nT)$ and ends at $(nT - t_0 + T)$. Therefore $h_N(t)$ must be delayed at least $(nT - t_N)$ seconds to insure realizability (assuming $t_N > nT$). This Lagrange interpolator has the form shown in Fig. 7.3.1 where $a_0 = q_{N,N}(t), \ldots, a_{N-1} = q_{N,1}(t)$, and $a_N = q_{N,0}(t)$.

Let us verify the $h_N(t)$ impulse response interval $(-t_N + nT, nT - t_0 + T)$ as given by Eq. 3.5.18. The zero-order interpolator has $t_0 = t_N = nT$ from Eq. 3.5.3 so $h_0(t)$ is nonzero for $t = (0, T)$ as verified from Eq. 3.5.4. The first-order interpolator has $t_0 = nT$ and $t_N = (n + 1)T$ from Eq. 3.5.8 so its nonzero $h_1(t)$ interval is $t = (-T, T)$ which checks with Eq. 3.5.9. The second-order interpolator has $t_0 = (n - 1)T$ and $t_N = (n + 1)T$ from Eq. 3.5.13. Then $t = (-T, 2T)$ is the nonzero $h_2(t)$ interval as seen from Eq. 3.5.14.

The third-order interpolator has the impulse response shown in Fig. 3.5.4a. Four samples $[f((n - 1)T), \ldots, f((n + 2)T)]$ are used to determine the interpolation constants and $nT \le t \le (n + 1)T$. The impulse response $Th_3(t)$ is symmetrical about $t = 0$ and has the frequency responses shown in Fig. 3.5.4b.

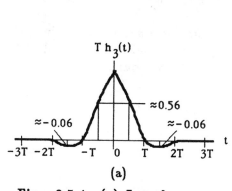

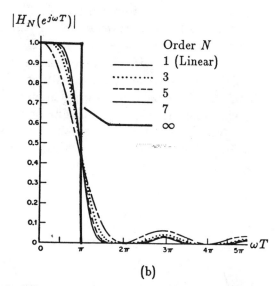

Fig. 3.5.4 (a) Impulse response of third-order interpolator and (b) magnitude responses of third- and higher-order interpolators. ((b) adapted from R.W. Schafer and L.R. Rabiner, "A digital signal processing approach to interpolation," Proc. IEEE, pp. 692–702, June, ©1973, IEEE.)

The frequency responses of higher odd-order Lagrange interpolators are also shown. These responses become more selective as order increases. When $n = \infty$, the filter becomes an ideal lowpass filter having an abrupt cut-off at $1/T$ Hz which is the Nyquist frequency. This case will be discussed next. Other types of interpolators including Newton, Gauss, Stirling, Bessel, and Everett will be investigated in Chaps. 10.6 and 11.5 where nonrecursive and recursive filters are designed.

3.6 IDEAL LOWPASS FILTER INTERPOLATOR

The reconstructed signal $f_r(t)$ of Fig. 3.2.1 can be related to the original signal $f(t)$ in another important way. Taking the IFT of $F_r(j\omega)$ from Eq. 3.2.2, then $f_r(t)$ equals

$$
\begin{aligned}
f_r(t) &= \frac{1}{2\pi} \int_{-\infty}^{\infty} F_r(j\omega) e^{j\omega t} d\omega, \quad \text{all } t \\
&= \frac{1}{2\pi} \int_{-\omega_s/2}^{\omega_s/2} F_s(j\omega) e^{j\omega t} d\omega = \sum_{n=-\infty}^{\infty} f(nT) \frac{1}{2\pi} \int_{-\omega_s/2}^{\omega_s/2} e^{j\omega(t-nT)} d\omega \quad (3.6.1) \\
&= \frac{1}{T} \sum_{n=-\infty}^{\infty} f(nT) \frac{\sin[(\omega_s/2)(t-nT)]}{(\omega_s/2)(t-nT)} = f_s(t) * h_I(t)
\end{aligned}
$$

where $F_s(j\omega)$ is given by Eq. 2.1.1. $f_r(t)$ is the output of the ideal lowpass filter in Fig. 3.2.2 whose impulse response $h_I(t)$ has a $[\sin(\omega_s t/2)/(\omega_s t/2)]$ characteristic. The input to the filter is the impulse sequence $f_s(t)$. Thus an ideal lowpass filter can be used to recover signal $f_r(t)$ from samples $f(nT)$ of the original signal $f(t)$.

Eq. 3.6.1 is important since it is the interpolation equation for numerically reconstructing $f_r(t)$ from the samples $f(nT)$. It is often referred to as the *ideal lowpass filter interpolator*. The $(\sin x/x)$ function is shown in Fig. 3.6.1. It is periodic in t with period T and its zero-crossings occur only at nT. At these times, T times the reconstructed signal $f_r(nT)$ equals the original signal $f(nT)$ which was sampled. This is *always true* whether there is frequency aliasing or not.

However at other times, all the $f(nT)$ samples combine linearly in a $(\sin x/x)$ fashion to form the reconstructed signal $Tf_r(t)$. When $f(t)$ is a band-limited signal with $|B| < \omega_s/2$, there is no aliasing and $Tf_r(t)$ and $f(t)$ are always equal. Unfortunately

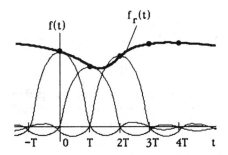

Fig. 3.6.1 Reconstructed signal $f_r(t)$ using an ideal lowpass filter interpolator.

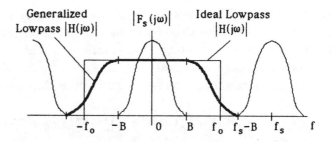

Fig. 3.6.2 Relation between signal bandwidth B, sampling frequency f_s, and non-ideal lowpass reconstruction filter bandwidth f_o.

$f(t)$ is seldom band-limited so there is generally frequency aliasing so $f(t)$ and $Tf_r(t)$ are not equal except when $t = nT$. Nevertheless even with such aliasing, $Tf_r(t)$ produces a useful approximation for $f(t)$. Another important observation is that $f_r(t)$ produced by Eq. 3.6.1 (or the alternative form of Eq. 3.6.5) is nonperiodic unlike the inverse transform methods to be discussed in the next two sections.

A more general form of the ideal lowpass filter reconstruction equation was developed by Papoulis.[8] Consider a signal $f(t)$ having a band-limited spectrum $F(j\omega)$ with $F(j\omega) = 0$ for $|\omega| > B$. Assume that the signal is sampled at a frequency above the Nyquist rate ($\omega_s > 2B$). Then the gain of the ideal lowpass filter (Fig. 3.2.2) used to recover the signal can be more arbitrarily chosen as

$$
\begin{aligned}
H(j\omega) &= 1, & |\omega| &\leq B \\
&= \text{arbitrary}, & B &< |\omega| < \omega_s - B \\
&= 0, & |\omega| &> \omega_s - B
\end{aligned}
\qquad (3.6.2)
$$

This relation is illustrated in Fig. 3.6.2. Due to the higher sampling rate, the lowpass filter no longer needs to be ideal with an abrupt cut-off although it must still have linear passband phase and zero stopband gain. It can have any transition-band shaping (both magnitude and phase) in the frequency range between B and $\omega_s - B$. This is especially convenient since it allows a trade-off between lower interpolation filter orders and higher sampling rates to obtain improved $f_r(t)$ approximations. In this more general filter case, the reconstruction equation becomes

$$
f_r(t) = f_s(t) * h(t) = \frac{1}{T} \sum_{n=-\infty}^{\infty} f(nT)h(t - nT)
\qquad (3.6.3)
$$

$h(t)$ is the impulse response of the *generalized lowpass interpolation filter* whose gain is shown in Fig. 3.6.2.

When numerically evaluating Eq. 3.6.1, only a finite number N of terms can be used. Since $(\sin x / x)$ peaks at $x = 0$ and approaches zero for increasing x, the $N/2$ terms on either side of time t should be used. Evaluating Eq. 3.6.3 in real time for $(k-1)T < t < kT$, then the summation limits should be changed from $n = (-\infty, \infty)$

to $n = (k, k+N-1)$. This equation is then a running or sliding interpolation because the limits slide forward with time (see Chap. 2.11). For data analysis over fixed times the limits are $n = (0, N-1)$ which is the more usual case.

Computational speed is increased using a *Whittaker interpolator*. Eq. 3.6.1 can be rewritten with the sine term outside the sum using the sine identity

$$\sin\left[\pi(t-nT)/T\right] = \sin(\pi t/T)\cos(n\pi) - \cos(\pi t/T)\sin(n\pi) = (-1)^n \sin(\pi t/T) \quad (3.6.4)$$

where $\omega_s/2 = \pi/T$. This produces Whittaker's interpolation equation as

$$f_r(t) = \frac{\sin(\pi t/T)}{\pi} \sum_{n=0}^{N-1} \frac{(-1)^n f(nT)}{t-nT}, \qquad t \neq nT$$

$$= \frac{1}{T}f(t), \qquad t = nT \tag{3.6.5}$$

Here the summation limits are adjusted for windowed data. For continuous sampling, set $n = (-\infty, \infty)$. Although Eq. 3.6.5a reduces to Eq. 3.6.1 when $t = nT$ using L'Hospital's rule, a computer with its finite word length cannot properly compute this limit. Therefore, $f_r(t)$ must be set equal to $f(nT)/T$ at the sample times.

EXAMPLE 3.6.1 Consider the signal $f(t)$ as

$$f(t) = 3e^{-t}\sin(3t)U_{-1}(t) \tag{3.6.6}$$

whose spectrum was determined in Example 2.1.1. Reconstruct $f(t)$ using Whit-

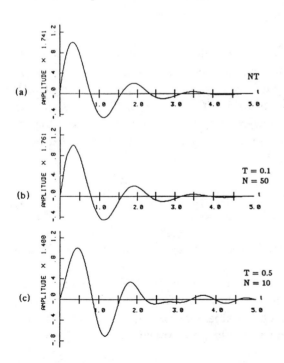

(a)

(b)

(c)

Fig. 3.6.3 (a) Signal $f(t)$ and its Whittaker reconstruction $f_r(t)$ using (b) 50 samples and (c) 10 samples in Example 3.6.1.

taker's approximation.

SOLUTION The two sample sequences are substituted into Eq. 3.6.5. The results are plotted in Fig. 3.6.3. The reconstructed signal $f_r(t)$ is very close to $f(t)$ for $N = 50$ but somewhat in error for $t > 2$ sec for $N = 10$. It was noted in Example 2.1.1 that since the $-20\,\text{dB}$ spectral bandwidth was about 1.5 Hz, sampling at frequencies greater than 3 Hz was needed to maintain low aliasing error. The results in Fig. 3.6.3 confirm this requirement. ∎

It should be noted that Eq. 3.6.1 is the optimum reconstruction equation and the ideal lowpass filter is the perfect interpolator which produces $f(t) = Tf_r(t)$ (for no frequency aliasing). However, the truncated reconstruction of Eq. 3.6.5 (or Eq. 3.6.1 having limits $n = (0, N-1)$) is no longer optimum since the data sequence has finite length and $f(t) \neq Tf_r(t)$ (under any condition) even if an ideal filter is used.

3.7 FOURIER SERIES INTERPOLATOR

The lowpass and Lagrange interpolators of the preceding sections converted discrete time samples into continuous time signals. A different interpolation approach using generalized series is shown in Fig. 3.7.1. The inverse transform and time domain interpolator blocks in Fig. 2.0.1 are combined into what is called a *generalized series interpolator*. This type of interpolator takes discrete generalized frequency spectra samples and directly converts them into continuous time signals. The generalized series interpolator reduces the number of operations required for interpolation.

Let us first consider the *Fourier series interpolator* where

$$f_r(t) = \frac{1}{N}\left[F(0) + 2\sum_{m=1}^{N/2}\left\{\text{Re}[F(m)]\,\cos(m\omega_o t) - \text{Im}[F(m)]\,\sin(m\omega_o t)\right\}\right], \quad 0 \leq t < NT$$

$$(3.7.1)$$

$F(m)$ are the DFT coefficients found using the FFT. For N even, $F(N/2)$ is reduced by half as will be discussed in a moment. To derive this result, consider the Fourier series

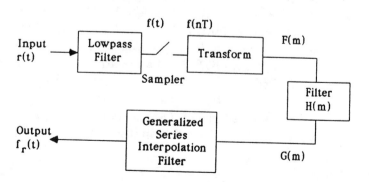

Fig. 3.7.1 Block diagram of general digital signal processing system using a generalized series interpolator.

of a signal $f(t)$ as

$$f(t) = \sum_{m=-\infty}^{\infty} c_m e^{jm\omega_o t} = a_0 + \sum_{m=1}^{\infty} \left[a_m \cos(m\omega_o t) + b_m \sin(m\omega_o t) \right]$$

$$c_m = c_{-m}^* = \frac{1}{2}(a_m - jb_m) = \frac{1}{T_w} \int_0^{T_w} f(t) e^{-jm\omega_o t} dt, \qquad c_0 = a_0$$

(3.7.2)

(see Eqs. 1.7.8 and 1.7.9). The fundamental frequency ω_o equals $2\pi/T_w = 2\pi/NT$ rad/sec or $1/NT$ Hz. Continuous and discrete Fourier series were discussed in Chaps. 1.7 and 1.8 which may now be reviewed if needed.

Truncating the sum and replacing the integral by a sum in Eq. 3.7.2 produces the discrete Fourier series (DFS) (see Eq. 1.8.15). Eq. 3.7.2 then reduces to

$$f_r(t) = \sum_{m=-M}^{M} c_m e^{jm\omega_o t} = a_0 + \sum_{m=1}^{M} \left[a_m \cos(m\omega_o t) + b_m \sin(m\omega_o t) \right]$$

$$c_m = \frac{1}{N} \sum_{n=0}^{N-1} f(nT) e^{-jnm\omega_o T} = \frac{1}{N} \sum_{n=0}^{N-1} f(nT) e^{-j2\pi nm/N}$$

(3.7.3)

Recall that the DFT and IDFT equal (Eq. 3.2.5)

$$f(n) = T f_r(nT) = \frac{1}{N} \sum_{m=0}^{N-1} F(m) W_N^{-nm} = \frac{1}{N} \sum_{m=-(N-1)/2}^{(N-1)/2} F(m) W_N^{-nm}, \qquad N \text{ odd}$$

$$F(m) = F_s(m) = \sum_{n=0}^{N-1} f(n) W_N^{nm}$$

(3.7.4)

where $W_N = e^{-j2\pi/N}$. Comparing Eqs. 3.7.3 and 3.7.4, it is clear that they are identical if $F(m) = N c_m$ (see Eqs. 1.7.11 and 1.7.12). Whenever N is even, the spectral coefficient $F(N/2)$ falls at the Nyquist frequency $\omega_s/2$. In this case, the series is augmented by one so N is replaced by $(N+1)$. Then the single term $F(N/2)$ is replaced by two terms $0.5F(N/2)$ and $0.5F(-N/2)$. This maintains the even symmetry of the spectral coefficients about $\omega = 0$ required by Eqs. 3.7.1 and 3.7.3.

Now let us look into the relation between the original signal $f(t)$ or sequence $f(n)$ and the reconstructed signal $f_r(t)$ or sequence $f_r(n)$. The relation may be found by substituting the DFT of Eq. 3.7.4b into the IDFT of Eq. 3.7.4a to obtain

$$T f_r(n) = \frac{1}{N} \sum_{m=0}^{N-1} W_N^{-nm} \left[\sum_{k=0}^{N-1} f(k) W_N^{km} \right] = \sum_{k=0}^{N-1} f(k) \frac{1}{N} \left[\sum_{m=0}^{N-1} W_N^{m(k-n)} \right]$$

(3.7.5)

Notice in Eq. 3.7.5 that $f(n)$ is a truncated sequence equal to zero for $n < 0$ and $n \geq N$.

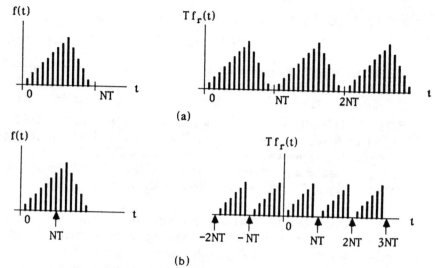

Fig. 3.7.2 (a) Non-truncated and (b) truncated signal $f(t)$ and its Fourier series reconstruction.

Summing all the equally-spaced vectors W_N^{k-n} around the unit circle yields

$$\frac{1}{N}\sum_{m=0}^{N-1} W_N^{m(k-n)} = \sum_{r=-\infty}^{\infty} D_0(k-n-rN) \tag{3.7.6}$$

$1/N$ times this sum equals unity when $k-n = rN$ for $r = 0, \pm 1, \pm 2, \ldots$ and zero otherwise. This results in an infinite length D_0 pulse train. Therefore Eq. 3.7.5 can be reduced to

$$Tf_r(n) = \sum_{r=-\infty}^{\infty} f(n+rN) \tag{3.7.7}$$

Thus, the IDFT generates a periodic sequence $f_r(n)$ with period N. It is the superposition of an infinite number of shifted original IFT truncated sequences $f(n)$. If the original sequence into the DFT is nonzero for a length less than N (or time NT), then each period out of the IDFT is identical to the original sequence. However if the original sequence into the DFT is nonzero for a length greater than or equal to N (or time NT), there will be sequence truncation and signal distortion. This distortion will be computed in Chap 3.10.

The truncation error is illustrated in Fig. 3.7.2a for a time-limited signal $f(t)$. Performing DFT and IDFT on this signal results in the reconstructed signal $f_r(t)$. $f_r(t)$ is periodic in t with period NT. Thus, the DFT and IDFT convert a nonperiodic $f(t)$ into a periodic $f_r(t)$. When $f(t)$ is sampled over its nonzero portions and above its Nyquist rate, then $Tf_r(t)$ is a periodic replica of $f(t)$.

The situation when too few samples are collected is shown in Fig. 3.7.2b. Here $f(t)$ is nonzero outside the sample window so the window is too narrow. Therefore $Tf_r(t)$ is composed of truncated $f(t)$ and there is signal distortion.

EXAMPLE 3.7.1 The DFT and FT of the signal

$$f(t) = 3e^{-t}\sin(3t)U_{-1}(t) \qquad\qquad (3.7.8)$$

was determined in Example 2.1.1. Reconstruct the Fourier series signal $f_r(t)$ for the DFT by computing the IDFT. Use $NT = 5$ seconds and sampling rates of $f_s = 10$ Hz and 2 Hz. Compare with the Whittaker interpolation of Example 3.6.1.

SOLUTION The equivalent signal being input to the DFT is periodic with a period of 5 seconds as shown in Fig. 3.7.3a. Performing the DFT and IDFT produces the signals shown in Figs. 3.7.3b and 3.7.3c. The reconstructed signal in Fig. 3.7.3b is a good approximation of the original signal but the reconstructed signal in Fig. 3.7.3c shows the effect of aliasing. ∎

The reconstructed waveforms in Example 3.7.1 illustrate an important result that was the reason for using window functions as pointed out in Chap. 2.6. The asymptotic roll-off of the spectrum $F(j\omega)$ depends on the smoothness of $f(t)$ (see discussion following Eq. 2.6.8). The smoothness was measured by the number n of time domain derivatives $d^n f(t)/dt^n$ that are continuous. The asymptotic roll-off of $F(j\omega)$ is then $-20(n+1)$ dB/dec $= -6(n+1)$ dB/oct. Larger asymptotic roll-offs in $F(j\omega)$ imply that $F(j\omega)$ is more band-limited, and for adequate sampling rate $f_s > 2B$, that aliasing errors will be smaller. The discontinuity in $df(t)/dt$ at $t = 0$ and 5 seconds in Fig. 3.7.3a implies the spectral roll-off rate of $f(t)$ is only -20 dB/dec. This slow roll-off in conjunction with the 20 dB spectral bandwidth of about 1.5 Hz (compared with the low sampling frequency $f_s = 1/T = 2$ Hz) leads to the distortion of $f_r(t)$ in Fig. 3.7.3c.

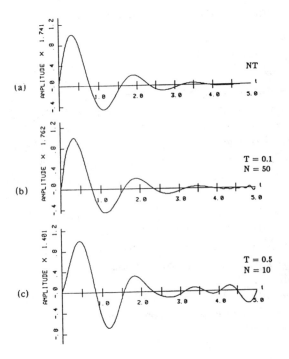

Fig. 3.7.3 (a) Signal $f(t)$ and its Fourier series reconstruction $f_r(t)$ using (b) 50 samples and (c) 10 samples in Example 3.7.1.

Therefore reconstructed signals $f_r(t)$ can often be improved by making the original signals $f(t)$ smoother. One approach is to use the *Campbell smoothing method*.[9] It consists of modifying the original signal as

$$f(t) = \text{arbitrary}, \qquad 0 \le t \le 2NT$$
$$= 0, \qquad \text{elsewhere} \tag{3.7.9}$$

so it has odd symmetry about its midpoint. Algebraically this new signal $f_c(t)$ and its spectrum $F_c(m)$ are defined to equal

$$f_c(t) = f(t) - f(2NT - t), \qquad 0 \le t \le 2NT$$
$$F_c(m) = F(m) - e^{-j2\pi m}F(-m) = F(m) - F(2N - m) \tag{3.7.10}$$

It consists of the original signal added to a time inverted, time shifted, and inverted version of itself. This signal has half-wave odd symmetry where $f_c(t) = -f_c(2NT - t)$. Thus, its Fourier series consists only of sine terms at odd harmonics of $f_o = 1/2NT$ Hz. $f_c(t)$ is usually smoother than the original signal $f(t)$ as will now be demonstrated.

EXAMPLE 3.7.2 Repeat the DFT analysis of Example 3.7.1 but use Campbell's approach for creating a new and smoother signal to analyze.
SOLUTION The original signal $f(t)$ is given by Eq. 3.7.8. Substituting $f(t)$ into Eq. 3.7.10 yields a smoother signal $f_c(t)$ where

$$f_c(t) = 3\left[e^{-t}\sin(3t) - e^{-(10-t)}\sin(3(10-t))\right]\left[U_{-1}(t) - U_{-1}(10-t)\right] \tag{3.7.11}$$

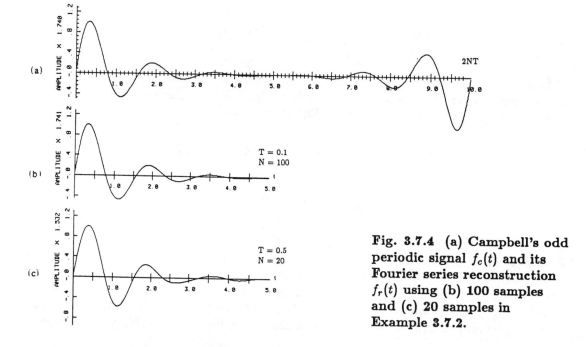

Fig. 3.7.4 (a) Campbell's odd periodic signal $f_c(t)$ and its Fourier series reconstruction $f_r(t)$ using (b) 100 samples and (c) 20 samples in Example 3.7.2.

The original sample interval $NT = 5$ seconds is therefore extended to $2NT = 10$ seconds. The new signal $f_c(t)$ is shown in Fig. 3.7.4a. Performing a DFT and IDFT produces the reconstructed signals $f_r(t)$ in Figs. 3.7.4b and 3.7.4c. The sample size has doubled but the FFT frequency bin remains the same since the sampling rates have not changed. There is noticeable improvement in $f_r(t)$ in Fig. 3.7.4c compared with the Fourier series reconstructed $f_r(t)$ in Fig. 3.7.3c. ∎

When Campbell's correction method is used with Fourier series reconstruction in the signal processing system of Fig. 3.7.1, the output spectrum $G(m)$ from the filter with gain $H(m)$ is corrected. This spectrum is manipulated according to Eq. 3.7.10b to yield $G_c(m) = G(m) - G(2N - m)$. Then $g_c(t)$ is reconstructed using the $G_c(m)$ spectrum and Eq. 3.7.1 for $0 \leq t < NT$.

3.8 GENERALIZED SERIES INTERPOLATOR

The *generalized series interpolator* shown in Fig. 3.7.1 reconstructs $f(t)$ from its spectral samples $F(m)$ as

$$f_r(t) = \sum_{m=0}^{N-1} F(m)K^{-1}(m,t) \tag{3.8.1}$$

The $F(m)$ are the spectral coefficients of $f(t)$ found using the discrete transform. The $K^{-1}(m,t)$ are the continuous basis functions $K^{-1}(p,t)$ of $f(t)$ evaluated at the generalized frequencies $p = m$.

To show this, assume that the spectrum $F(p)$ of $f(t)$ is composed of impulses as

$$F(p) = \frac{p_2 - p_1}{N} \sum_{m=-\infty}^{\infty} F(m)U_0(p - m) \tag{3.8.2}$$

If the spectrum is continuous, then $F(m)$ approximates the area of $F(p)$ over a generalized frequency range between p_2 and p_1 (see Fig. 1.8.2). Substituting Eq. 3.8.2 into the continuous transform pair given by Eq. 3.3.2 yields

$$f(t) = \int_{p_1}^{p_2} F(p)K^{-1}(p,t)dp \cong \frac{p_2 - p_1}{N} \sum_{m=-\infty}^{\infty} F(m)K^{-1}(m,t)$$

$$F(p) = \int_{t_1}^{t_2} f(t)K(p,t)dt \cong \frac{t_2 - t_1}{N} \sum_{n=N_1}^{N_2} f(n)K(p,n) \tag{3.8.3}$$

The $F(p)$ transform is approximated by a sum. The $(p_2 - p_1)(t_2 - t_1)$ product is normalized to some constant like unity. Therefore generalized continuous transforms are reduced to generalized series transforms just as Fourier transforms are reduced to Fourier series (see Chap. 1.4). Evaluating $f(t)$ at discrete times as $f(nT)$, and $F(p)$ at discrete generalized frequencies as $F(m)$ produces the discrete transform pair where

$$f(n) = Tf_r(n) = \frac{1}{N} \sum_{m=0}^{N-1} F(m)K^{-1}(n,m)$$

$$F(m) = \sum_{n=0}^{N-1} f(n)K(m,n)$$

$$\text{(3.8.4)}$$

(setting $t_2 = NT$, $t_1 = 0$, $p_2 = 1/T$, and $p_1 = 0$) which agree with Eq. 3.3.1.

The relation between samples of the interpolated signal $f_r(t)$ and the original signal $f(t)$ that was transformed is found by substituting Eq. 3.8.4b into Eq. 3.8.4a as

$$NTf_r(n) = \sum_{m=0}^{N-1} K^{-1}(n,m)\left[\sum_{k=0}^{N-1} K(m,k)f(k)\right] = \sum_{k=0}^{N-1} f(k)\left[\sum_{m=0}^{N-1} K^{-1}(n,m)K(m,k)\right]$$

$$\text{(3.8.5)}$$

The product of the transform matrix $[K]$ and scaled inverse transform matrix $[K]^{-1}$ equals

$$[K]\,[K]^{-1} = [K]^{-1}\,[K] = N[I]$$

$$\text{(3.8.6)}$$

where $[I]$ is the diagonal unit matrix. Now the matrix elements like $K(m,n)$ are periodic with period N in both the m and n indices. This means that $K(m,n)$ satisfies $K(m,n) = K(m+kN, n+qN)$ for $k,q = 0,\pm1,\pm2,\ldots$. Therefore, summing matrix element products in the kth row of $[K]^{-1}$ and the nth column of $[K]$ yields

$$\frac{1}{N} \sum_{m=0}^{N-1} K^{-1}(n,m)K(m,k) = \sum_{r=-\infty}^{\infty} D_0(k-n-rN)$$

$$\text{(3.8.7)}$$

$1/N$ times this sum equals unity when $k - n = rN$ for $r = 0,\pm1,\pm2,\ldots$ and zero otherwise. Substituting this result into Eq. 3.8.5 yields

$$Tf_r(n) = \sum_{r=-\infty}^{\infty} f(n+rN)$$

$$\text{(3.8.8)}$$

which is the same result as Eq. 3.7.7. Therefore, the DT and IDT combination converts the truncated sample sequence $f(n)$ into a periodic reconstructed sequence. As in the Fourier transform case, signal truncation error results when the sample sequence $f(n)$ has nonzero values outside the sampling window of length N. Using the weighting functions discussed in Chap. 2.6 reduce this error.

Even when the sampled signals are time-limited, time weighting smooths the signals and makes their spectra more band-limited. This results because, from a Fourier series viewpoint, smoothing reduces the discontinuities in the signal and reduces the spectral components at higher frequencies. This happens because the Fourier basis functions (sines and cosines) are smooth and not discontinuous.

This is not the case with general series interpolators however. The Paley, Hadamard, Haar, and slant basis functions were discontinuous. These functions were shown in Fig. 2.8.1. Paley and Hadamard functions were composed of pulses of amplitude ± 1 which extended over the entire sampling interval. Haar functions were composed of these same pulses but they extended over only fractions of the interval. Slant transforms used quantized pulse, triangular, and sawtooth functions over the entire interval. Therefore, signals similar in shape to the basis functions are more band-limited in the generalized frequency sense. They can be described using fewer spectral components. Using smoothing techniques like weighting windows and Campbell's method actually increase the spectral components and accentuate interpolation inaccuracies rather than reduce them. Weighting windows can still be used but they must have the form of the basis functions used by the transform to reduce the number of spectral components.

EXAMPLE 3.8.1 Consider a parabolic signal $f(t) = t^2$ for $0 \le t < 1$ whose Paley spectrum was determined in Example 2.8.1. Determine the eighth-order Paley interpolator.

SOLUTION The continuous and discrete Paley spectra were found to equal

$$[F_n(p)]_{Pal} = \begin{bmatrix} 0.3333 \\ -0.2500 \\ -0.1250 \\ 0.0625 \\ -0.0625 \\ 0.0313 \\ 0.0156 \\ 0 \end{bmatrix}, \quad [F(m)]_{Pal} = \begin{bmatrix} 0.2734 \\ -0.2188 \\ -0.1094 \\ 0.0625 \\ -0.0547 \\ 0.0313 \\ 0.0156 \\ 0 \end{bmatrix} \quad (3.8.9)$$

in Eqs. 2.8.17 and 2.8.18. $F_n(p)$ are the exact coefficients using the PT while $F(m)$ are the approximate coefficients determined using the DPT. Using Eq. 3.8.4a, the reconstructed signal equals

$$f_r(t) \cong 0.2734 \, Pal(0,t) - 0.2188 \, Pal(1,t) - 0.1094 \, Pal(2,t) + 0.0625 \, Pal(3,t)$$
$$- 0.0547 \, Pal(4,t) + 0.0313 \, Pal(5,t) + 0.0156 \, Pal(6,t)$$

$$(3.8.10)$$

Sampling $f_r(t)$ every 0.125 seconds starting at $t = 0$ gives

$$[f]_{Pal} = \begin{bmatrix} 0 \\ 0.0156 \\ 0.0625 \\ 0.1406 \\ 0.2500 \\ 0.3906 \\ 0.5625 \\ 0.7656 \end{bmatrix} = \begin{bmatrix} + & + & + & + & + & + & + & + \\ + & + & + & + & - & - & - & - \\ + & + & - & - & + & + & - & - \\ + & + & - & - & - & - & + & + \\ + & - & + & - & + & - & + & - \\ + & - & + & - & - & + & - & + \\ + & - & - & + & + & - & - & + \\ + & - & - & + & - & + & + & - \end{bmatrix} \begin{bmatrix} 0.2734 \\ -0.2188 \\ -0.1094 \\ 0.0625 \\ -0.0547 \\ 0.0312 \\ 0.0156 \\ 0 \end{bmatrix} \quad (3.8.11)$$

as found in Example 3.3.1 (Eq. 3.3.6). Both the exact ($F(p)$) and approximate

$(F(m))$ results are plotted in Fig. 3.8.1. Since t^2 is a smooth function, we should not expect the Paley approximation (a discontinuous function) to be very accurate unless a large number of terms are used.

∎

EXAMPLE 3.8.2 Now consider a 1 Hz cosine signal $f(t) = \cos(2\pi t)$. Reconstruct it in the interval $0 \le t < 1$ using an eighth-order Paley interpolator.
SOLUTION From Example 2.8.2, the continuous and discrete Paley spectra are

$$[F_n(p)]_{Pal} = \begin{bmatrix} 0 \\ 0 \\ 0 \\ 0.637 \\ 0 \\ 0.263 \\ 0 \\ 0 \end{bmatrix}, \qquad [F(m)]_{Pal} = \begin{bmatrix} 0 \\ 0.2500 \\ 0 \\ 0.6036 \\ 0 \\ 0.2500 \\ 0 \\ -0.1036 \end{bmatrix} \qquad (3.8.12)$$

(Eqs. 2.8.19 and 2.8.20). The reconstructed cosine signal equals

$$f_r(t) \cong 0.25\, Pal(1,t) + 0.6036\, Pal(3,t) + 0.25\, Pal(5,t) - 0.1036\, Pal(7,t) \quad (3.8.13)$$

Its samples were found in Example 3.3.2 and are given by Eq. 3.3.8. These signals are plotted in Fig. 3.8.2.

∎

EXAMPLE 3.8.3 Determine the eighth-order Hadamard interpolator for the $\cos(2\pi t)$ signal.
SOLUTION The continuous and discrete Hadamard spectra were found in Example 2.8.3 to equal

$$[F_n(p)]_{Had} = \begin{bmatrix} 0 \\ 0 \\ 0 \\ 0 \\ 0 \\ 0.263 \\ 0.637 \\ 0 \end{bmatrix}, \qquad [F(m)]_{Had} = \begin{bmatrix} 0 \\ 0 \\ 0 \\ 0 \\ 0.2500 \\ 0.2500 \\ 0.6036 \\ -0.1036 \end{bmatrix} \qquad (3.8.14)$$

(Eq. 2.8.24). The reconstructed cosine signal has

$$f_r(t) \cong 0.25\, Had(4,t) + 0.25\, Had(5,t) + 0.6036\, Had(6,t) - 0.1036\, Had(7,t)$$
$$(3.8.15)$$

and its samples are given by Eq. 3.3.12. The exact and approximate signals are plotted in Fig. 3.8.3.

∎

EXAMPLE 3.8.4 Find the eighth-order Haar interpolator for the $\cos(2\pi t)$ signal.
SOLUTION The continuous and discrete spectra were found in Example 2.8.4 to equal

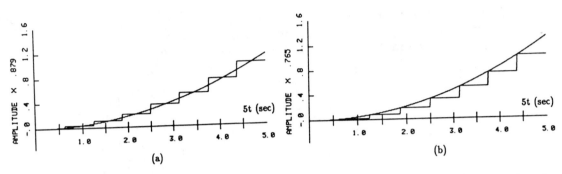

Fig. 3.8.1 Signal and its Paley reconstruction using (a) exact and (b) approximate spectra in Example 3.8.1.

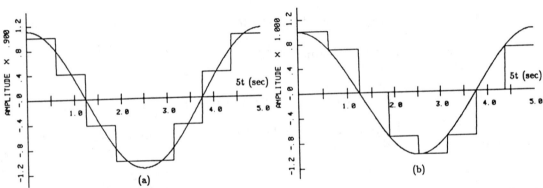

Fig. 3.8.2 Signal and its Paley reconstruction using (a) exact and (b) approximate spectra in Example 3.8.2.

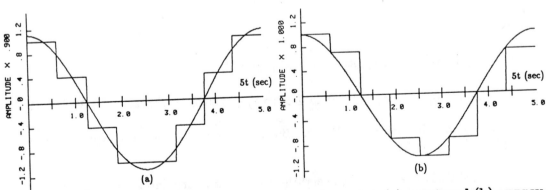

Fig. 3.8.3 Signal and its Hadamard reconstruction using (a) exact and (b) approximate spectra in Example 3.8.3.

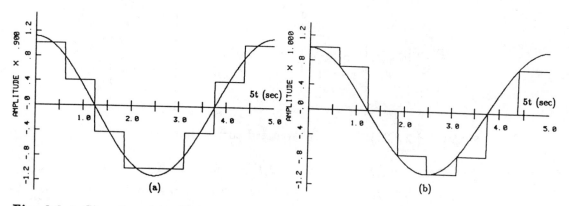

Fig. 3.8.4 Signal and its Haar reconstruction using (a) exact and (b) approximate spectra in Example 3.8.4.

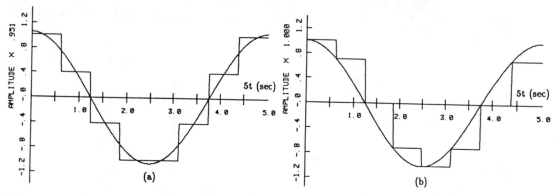

Fig. 3.8.5 Signal and its slant reconstruction using (a) exact and (b) approximate spectra in Example 3.8.5.

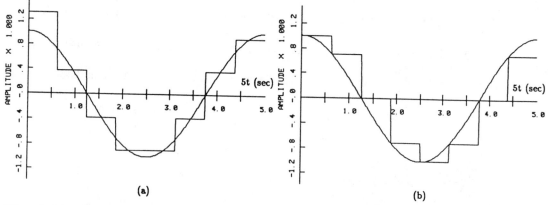

Fig. 3.8.6 Signal and its Karhunen-Loéve reconstruction using (a) exact and (b) approximate spectra in Example 3.8.6.

$$[F_n(p)]_{Ha} = \begin{bmatrix} 0 \\ 0 \\ 0.4502 \\ -0.4502 \\ 0.1318 \\ 0.1318 \\ -0.1318 \\ -0.1318 \end{bmatrix}, \qquad [F(m)]_{Ha} = \begin{bmatrix} 0 \\ 0.2500 \\ 0.4268 \\ -0.4268 \\ 0.0732 \\ 0.1768 \\ -0.0732 \\ -0.1768 \end{bmatrix} \qquad (3.8.16)$$

(Eqs. 2.8.30 and 2.8.31). The reconstructed cosine signal is

$$f_r(t) \cong 0.25\ Ha(1,t) + 0.4268\ Ha(2,t) - 0.4268\ Ha(3,t) + 0.0732\ Ha(4,t)$$
$$+ 0.1768\ Ha(5,t) - 0.0732\ Ha(6,t) - 0.1768\ Ha(7,t) \qquad (3.8.17)$$

Its samples are given by Eq. 3.3.16. These signals are plotted in Fig. 3.8.4. ∎

EXAMPLE 3.8.5 Find the eighth-order slant interpolator for the $\cos(2\pi t)$ signal.
SOLUTION The continuous and discrete spectra were found in Example 2.8.5 to equal

$$[F_n(p)]_S = \begin{bmatrix} 0 \\ 0 \\ 0.7256 \\ 0 \\ 0 \\ 0 \\ -0.0488 \\ 0 \end{bmatrix}, \qquad [F(m)]_S = \begin{bmatrix} 0 \\ 0.2182 \\ 0.6516 \\ -0.1220 \\ 0 \\ -0.1036 \\ -0.0463 \\ 0 \end{bmatrix} \qquad (3.8.18)$$

(Eqs. 2.8.36 and 2.8.37). The reconstructed cosine signal is

$$f_r(t) \cong 0.2182\ S(1,t) + 0.6516\ S(2,t) - 0.1220\ S(3,t) - 0.1036\ S(5,t) - 0.0463\ S(6,t) \qquad (3.8.19)$$

whose samples are given in Eq. 3.3.20. These signals are plotted in Fig. 3.8.5. ∎

EXAMPLE 3.8.6 Determine the eighth-order Karhunen-Loéve interpolator for the $\cos(2\pi t)$ signal.
SOLUTION The continuous and discrete spectra were found in Example 2.8.6 to equal

$$[F_n(p)]_{KL} = \begin{bmatrix} 0.5000 \\ 0.0466 \\ -0.0466 \\ -0.1125 \\ -0.1125 \\ -0.0466 \\ 0.0466 \\ 0.1125 \end{bmatrix}, \qquad [F(m)]_{KL} = \begin{bmatrix} 0.5000 \\ 0.0884 \\ 0 \\ -0.0884 \\ -0.1250 \\ -0.0884 \\ 0 \\ 0.0884 \end{bmatrix} \qquad (3.8.20)$$

(Eqs. 2.8.43 and 2.8.44). The reconstructed cosine signal is

$$f_r(t) \cong 0.5\, K_{KL}^{-1}(0,t) + 0.0884\, K_{KL}^{-1}(1,t) - 0.0884\, K_{KL}^{-1}(3,t)$$
$$- 0.125\, K_{KL}^{-1}(4,t) - 0.0884\, K_{KL}^{-1}(5,t) + 0.0884\, K_{KL}^{-1}(7,t) \qquad (3.8.21)$$

whose samples are given in Eq. 3.3.24. These signals are plotted in Fig. 3.8.6. ■

3.9 DISTORTION DUE TO NONZERO SAMPLE DURATION

Now we shall consider the errors in DFT's and IDFT's that arise from nonzero sample duration, signal truncation, and window size. Up to this point, signals were sampled using the system shown in Fig. 1.3.1 and the ideal impulse streams described by Eq. 1.3.1. Now consider the case shown in Fig. 3.9.1 where pulse streams are used for sampling rather than impulse streams. Impulse streams are impossible to generate in real systems so pulse streams are used.

A single unit pulse $s_D(t)$ of duration D has the spectrum $S_D(j\omega)$ where

$$s_D(t) = U_{-1}(t + D/2) - U_{-1}(t - D/2)$$
$$S_D(j\omega) = D\frac{\sin(\omega D/2)}{\omega D/2} \qquad (3.9.1)$$

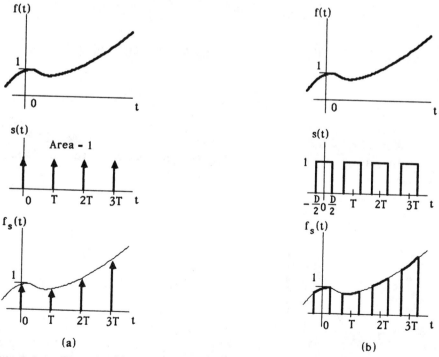

Fig. 3.9.1 Signal $f(t)$, sampling signal $s(t)$, and sampled signal $f_s(t)$ where sampling signal is an (a) impulse stream and (b) pulse stream.

(Eqs. 1.6.19 and 1.6.22). Convolving $s_D(t)$ with a unit impulse stream produces the desired pulse stream of Fig. 3.9.1b. Therefore

$$s(t) = s_D(t) * \sum_{n=-\infty}^{\infty} U_0(t - nT), \qquad \omega_s = \frac{2\pi}{T}$$

$$S(j\omega) = S_D(j\omega)\,\omega_s \sum_{n=-\infty}^{\infty} U_0(\omega - n\omega_s) = 2\pi\,\frac{D}{T}\,\frac{\sin(\omega D/2)}{\omega D/2} \sum_{n=-\infty}^{\infty} U_0(\omega - n\omega_s)$$

$$(3.9.2)$$

The spectrum $S(j\omega)$ of the pulse stream is shown in Fig. 3.9.2. Notice that by increasing the $s_D(t)$ pulse height to $1/D$ and reducing the pulse width D to zero (i.e., unit area sample pulses), the $(\sin x/x)$ distribution flattens to the ideal impulse spectrum. On the other extreme, setting the pulse width D equal to the sampling interval T produces continuous sampling. Here the spectrum $S(j\omega)$ reduces to the single impulse $2\pi U_0(\omega)$. This confirms that Eq. 3.9.2 exhibits the proper limiting behavior.

Now what is the result of sampling $f(t)$ with the pulse stream of Fig. 3.9.1b rather than the impulse stream of Fig. 3.9.1a? Following the procedure used to obtain the sampled data results of Eqs. 1.3.2 and 1.3.3, we can write that

$$f_s(t) = f(t)s(t), \qquad \omega_s = \frac{2\pi}{T}$$

$$(3.9.3)$$

$$F_s(j\omega) = \frac{1}{2\pi}F(j\omega) * S(j\omega) = \frac{D}{T}\,\frac{\sin(\omega D/2)}{\omega D/2} \sum_{n=-\infty}^{\infty} F\left[j(\omega - n\omega_s)\right]$$

When $D \to 0$, then $F_s(j\omega)/D$ given by Eq. 3.9.3b reduces to (Eq. 1.3.6)

$$\frac{1}{D}F_s(j\omega) = \frac{1}{T} \sum_{n=-\infty}^{\infty} F(j(\omega - n\omega_s)) \tag{3.9.4}$$

Therefore, Eq. 3.9.3 shows that the effect of nonzero pulse width or nonzero duration sampling is to cause a $[\sin(\omega D/2)/(\omega D/2)]$ distortion on the envelope of $F_s(j\omega)/D$ given by Eq. 3.9.4.

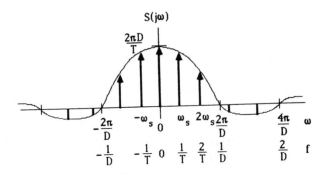

Fig. 3.9.2 Spectrum of sampling pulse stream $s(t)$ in Fig. 3.9.1b.

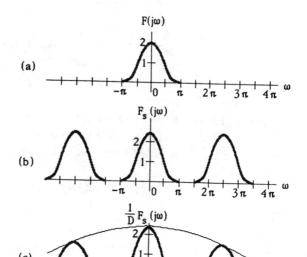

Fig. 3.9.3 Spectra of (a) signal $f(t)$ and sampled signal $f_s(t)$ using (b) impulse sampling and (c) pulse sampling in Example 3.9.1.

A convenient measure of envelope distortion is the *envelope droop* or *sag* at the Nyquist frequency $\omega_s/2$. Expanding $\sin(x)$ in a Taylor series $(x = (D/T)(\pi/2))$ and truncating at the third-order term gives $(\sin x/x) \cong 1 - (x^2/3)$. The envelope droop is therefore approximately $-x^2/3 = 0.82(D/T)^2 = 82(D/T)^2$ percent. For a 10% pulse width $(D/T = 0.1)$, the droop is about 0.8% which is almost negligible.

EXAMPLE 3.9.1 Consider a signal $f(t)$ having a raised-cosine spectrum where

$$F(j\omega) = 1 + \cos\omega, \qquad |\omega| < \pi$$
$$= 0, \qquad\qquad |\omega| \geq \pi \qquad\qquad (3.9.5)$$

Sample the signal at $\omega_s = 2\pi(1.25)$ (or $T = 1/1.25 = 0.8$ second) so that no aliasing is present. Show the effects of using nonzero sample duration $D = 0.4$ second.
SOLUTION The spectrum $F(j\omega)$ is drawn in Fig. 3.9.3a. Substituting Eq. 3.9.5 into Eq. 3.9.3 and solving gives the sampled spectrum $F_s(j\omega)$. Impulsive sampling produces the uniform envelope spectrum shown in Fig. 3.9.3b centered at multiples of $\omega_s = 2.5\pi$ with a dc value $F_s(0)$ of $F(0)/T = 2/0.8 = 2.5$. However, finite duration sampling produces the $F_s(j\omega)/D$ spectrum shown in Fig. 3.9.3c having the $(\sin x/x)$ envelope distortion. The envelope level at ω_s equals $[\sin(\pi D/T)/(\pi D/T)]$ times the dc value of 2.5. In this case where $D/T = 0.5$, the envelope level is reduced 64% to about $0.64(2.5) = 1.6$. Since the sampling frequency ω_s is above the Nyquist rate $2B = 2\pi$, there is no aliasing distortion.   ∎

3.10 DISTORTION DUE TO SIGNAL TRUNCATION

Portions of signals must usually be eliminated or *truncated* to allow fast transform analysis. As discussed in Chap. 2.5, taking N samples separated by a sampling interval T produces a sampling window of width NT seconds. The FT of a signal $f(t)$ equals

$$F(j\omega) = \int_{-\infty}^{\infty} f(t)e^{-j\omega t}dt = \int_{-\infty}^{0} f(t)e^{-j\omega t}dt + \int_{0}^{NT} f(t)e^{-j\omega t}dt + \int_{NT}^{\infty} f(t)e^{-j\omega t}dt$$

$$= F_2(j\omega) + F_1(j\omega) + F_3(j\omega)$$

$$(3.10.1)$$

from Eq. 1.6.1a. $F(j\omega)$ can be divided into the three terms shown where $F(j\omega)$ is the actual spectrum, $F_1(j\omega)$ is the computed DFT spectrum, and $F_2(j\omega)$ and $F_3(j\omega)$ are error spectra. Therefore the spectral error $E(j\omega)$ between the actual and computed spectrum is

$$E(j\omega) = F(j\omega) - F_1(j\omega) = F_2(j\omega) + F_3(j\omega) \qquad (3.10.2)$$

The spectral error is simply the FT of the portion of the signal $f(t)$ outside the time window of $(0, NT)$ seconds. This error was referred to as spectral spreading in Chap. 2.5. Now Eq. 3.10.2 allows us to quantitatively evaluate this error.

EXAMPLE 3.10.1 Compute the spectral error produced by truncating a signal

$$f(t) = e^{-t}U_{-1}(t) \qquad (3.10.3)$$

for an arbitrary window interval NT. Discuss the results.
SOLUTION The spectral error equals

$$F_3(j\omega) = \int_{NT}^{\infty} e^{-(1+j\omega)t}dt = \frac{e^{-(1+j\omega)NT}}{1+j\omega} = \frac{e^{-NT}}{\sqrt{1+\omega^2}}e^{-j(\omega NT + \tan^{-1}\omega)} \qquad (3.10.4)$$

from Eq. 3.10.1 where $F_2(j\omega) = 0$. The actual spectrum equals

$$F(j\omega) = \int_{0}^{\infty} e^{-(1+j\omega)t}dt = \frac{1}{1+j\omega} \qquad (3.10.5)$$

Therefore the normalized spectral error is

$$\frac{F_3(j\omega)}{F(j\omega)} = e^{-(1+j\omega)NT} = e^{-NT}e^{-j\omega NT} \qquad (3.10.6)$$

so the magnitude error equals $\exp(-NT)$ and the phase error equals $-\omega NT$. The normalized magnitude error is less than 5% for a window width of $NT = 3$ seconds and less than 1% for $NT = 5$ seconds. Integral squared error can also be used for analysis as will be discussed in Chap. 4.[10] ∎

EXAMPLE 3.10.2 Examine the spectral error when the signal $f(t)$ of Example 3.6.1 is truncated at $NT = 2, 4$, and 5 seconds. Sample at 20 Hz so there is negligible frequency aliasing error.
SOLUTION The spectral error equals

$$F_3(j\omega) = \int_{NT}^{\infty} 3e^{-t}\sin(3t)e^{-j\omega t}dt \qquad (3.10.7)$$

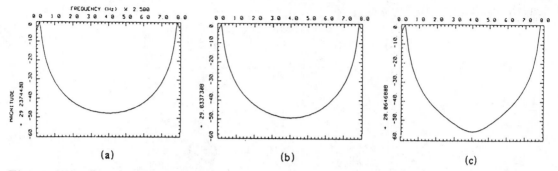

Fig. 3.10.1 Spectra of truncated, sampled signal $f_s(t)$ **for window sizes** NT **of (a) 5, (b) 4, and (c) 2 seconds in Example 3.10.2.**

As was shown in Example 3.6.1, $NT = 5$ is the smallest sampling window that can be used without producing significant spectral error. The magnitude spectra for the three cases are shown in Fig. 3.10.1. Since a sufficiently high sampling rate is used, there is almost no aliasing error. Instead all error is due to truncation. At $NT = 2$, there is significant error while at $NT = 4$, the error is small. ∎

3.11 WINDOW SIZE ADJUSTMENT

It is clear from the discussion of the last section that the window size need cover only the fluctuating portion of the signal (nonzero portion riding on a dc level) to obtain good DFT's. As long as the sampling frequency is above the Nyquist rate and constant, the window size can be increased without changing the signal spectrum $F_s(j\omega)$. The FFT bins are simply placed closer together.

This is illustrated in Fig. 3.11.1a where the sample size increases from N to M. The frequency bin size decreases from ω_s/N to ω_s/M but the folding frequency $\omega_s/2 = \pi/T$ remains constant as shown in Fig. 3.11.1b. The spectral coefficients are evaluated at different frequencies as $F_s(jm\omega_s/M)$ rather than $F_s(jn\omega_s/N)$ but $F_s(j\omega)$ itself remains unchanged. Increasing the number of samples while holding the sampling rate constant improves spectral resolution. It reduces the picket fence effect (see Table 2.5.1).

To illustrate this, consider an N-point DFT $X(m)$ and a $2N$-point DFT $Y(m)$. First consider padding the end of the N sample sequence $x(n)$ with N zeros so its

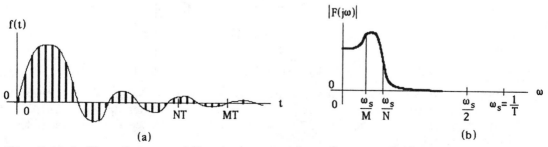

Fig. 3.11.1 Use of zero padding to increase transform resolution.

length is extended to $2N$ as

$$
\begin{aligned}
y(n) &= x(n), & n &= 0, 1, \ldots, N-1 \\
&= 0, & n &= N, N+1, \ldots, 2N-1
\end{aligned}
\tag{3.11.1}
$$

The DFT of $x(n)$ and $y(n)$ equal

$$
X(m) = \sum_{n=0}^{N-1} x(n) W_N^{nm}, \qquad m = 0, 1, \ldots, N-1
$$

$$
Y(m) = \sum_{n=0}^{2N-1} y(n) W_{2N}^{nm} = \sum_{n=0}^{N-1} x(n) W_{2N}^{nm} = \sum_{n=0}^{N-1} x(n) W_N^{nm/2} = X(m/2),
$$
$$
m = 0, 1, \ldots, 2N-1
$$
$$
\tag{3.11.2}
$$

$Y(m)$ has twice as many data points as $X(m)$ in the same frequency range between $\omega = (0, \omega_s)$. Thus this doubles the resolution. $X(m)$ equals every other spectral value of $Y(m)$. Any integer number r of N zeros can be used $(rN$ total zeros$)$. This increases the resolution by r.

EXAMPLE 3.11.1 Double the number of samples for the sawtooth signal $[f]_4 = [0, 1, 2, 3]$ in Example 3.4.1 and show that the original coefficients remain unchanged.
SOLUTION The 4-point FFT spectrum was found using Eq. 3.4.2 as

$$
[F]_4 = \begin{bmatrix} 6 \\ -2+j2 \\ -2 \\ -2-j2 \end{bmatrix}, \qquad [f]_4 = \begin{bmatrix} 0 \\ 1 \\ 2 \\ 3 \end{bmatrix}
\tag{3.11.3}
$$

The 8-point signal equals $[f]_8 = [0, 1, 2, 3, 0, 0, 0, 0]$ using zero padding. The 8-point FFT spectrum equals

$$
[F]_8 = \begin{bmatrix} 6 \\ -1.414 - j4.829 \\ -2+j2 \\ 1.414 - j0.828 \\ -2 \\ 1.414 + j0.828 \\ -2 - j2 \\ -1.414 + j4.829 \end{bmatrix}, \qquad [f]_8 = \begin{bmatrix} 0 \\ 1 \\ 2 \\ 3 \\ 0 \\ 0 \\ 0 \\ 0 \end{bmatrix}
\tag{3.11.4}
$$

There are twice as many frequency bins between 0 and ω_s when $N = 8$ instead of $N = 4$. Therefore every other frequency bin for $N = 8$ lines up with the $N = 4$ frequency bin. Comparing Eq. 3.11.4 with Eq. 3.11.3, we see the spectra at these frequencies are identical. ∎

A dc level or constant k can be added to any signal $x(n)$. The resulting FFT spectrum $Y(m)$ remains unchanged except at $m = 0$ where it increases by kN. This is shown as

$$y(n) = x(n) + k$$

$$Y(m) = \sum_{n=0}^{N-1} \Big[x(n) + k \Big] W_N^{nm} = X(m) + kNU_0(m) \qquad (3.11.5)$$

Alternatively, a complex exponential can be added as $y(n) = x(n) + ke^{m_o n}$. The resulting FFT spectrum $Y(m)$ equals $X(m) + kNU_0(m - m_o)$.

EXAMPLE 3.11.2 Find the 8-point FFT spectrum for the sawtooth signal of Example 3.11.1 which has a quiescent level of three rather than zero.

SOLUTION The 8-point signal equals $[g]_8 = [3, 4, 5, 6, 3, 3, 3, 3]$. The 8-point FFT spectra for a constant level of three is $[G']_8 = [24, 0, 0, 0, 0, 0, 0, 0]$. Superimposing this spectra with that of $[f]_8$ given by Eq. 3.11.4 yields

$$[F]_8 = \begin{bmatrix} 6 \\ -1.414 - j4.829 \\ -2 + j2 \\ 1.414 - j0.828 \\ -2 \\ 1.414 + j0.828 \\ -2 - j2 \\ -1.414 + j4.829 \end{bmatrix} + \begin{bmatrix} 24 \\ 0 \\ 0 \\ 0 \\ 0 \\ 0 \\ 0 \\ 0 \end{bmatrix} = \begin{bmatrix} 30 \\ -1.414 - j4.829 \\ -2 + j2 \\ 1.414 - j0.828 \\ -2 \\ 1.414 + j0.828 \\ -2 - j2 \\ -1.414 + j4.829 \end{bmatrix} \qquad (3.11.6)$$

Alternatively the $[g]_8$ sequence can be FFT'ed directly. This example reemphasizes that any fluctuating signal can be padded using constant values without changing its basic FFT spectrum except at dc $(m = 0)$. ∎

Another padding method pads zeros between samples of $x(n)$ to form $y(n)$ such as

$$y(n) = x(n/2), \qquad n \text{ even}$$
$$= 0, \qquad\qquad n \text{ odd} \qquad (3.11.7)$$

for $n = 0, 1, \ldots, 2N - 1$. The DFT of $y(n)$ equals

$$Y(m) = \sum_{n=0}^{2N-1} y(n) W_{2N}^{nm} = \sum_{n=0}^{N-1} x(n) W_{2N}^{2nm} = \sum_{n=0}^{N-1} x(n) W_N^{nm} = X(m),$$

$$m = 0, 1, \ldots, 2N - 1$$
$$(3.11.8)$$

The spectrum of $Y(m)$ is identical to that of $X(m)$. $Y(m)$ has twice as many data points as $X(m)$ over twice the frequency range from $\omega = (0, 2\omega_s)$. Thus the FFT bin spacing is unchanged. This zero padding algorithm does not make efficient use of zeros since it does not improve the spectral resolution unlike the zero padding algorithm of Eq. 3.11.1.

3.12 CONSTANT DELAY AND CONSTANT MAGNITUDE FILTERS

There are some applications that require transfer functions with linear phase and constant delay, or equivalently, impulse responses with constant time delay. The frequency response of such a filter must satisfy

$$H(z)\Big|_{z=e^{j\omega T}} = H(e^{j\omega T}) = R(e^{j\omega T})e^{-j\omega T_d} \tag{3.12.1}$$

where $R(e^{j\omega T})$ must be real-valued. Its linear phase equals $-\omega T_d$. This insures that the impulse response $h(nT)$ of the system equals $r(nT - T_d)$ where $r(nT)$ satisfies

$$r(nT) = \mathcal{F}^{-1}[R(e^{j\omega T})] = \mathcal{F}^{-1}[R(e^{-j\omega T})] = r(-nT) \tag{3.12.2}$$

and is an even function. Since $R(e^{j\omega T})$ is real-valued, $R(e^{j\omega T}) = R(e^{-j\omega T})$ which implies that $H(z) = H(z^{-1})$ from Table 1.9.2. $R(z)$ is the ratio of two polynomials satisfying $R(z) = R(z^{-1})$. They must have pole pairs (and zero pairs) that are images of each other (i.e., they lie on radial lines at reciprocal distances about the unit circle in the z-plane) and perhaps poles or zeros at the origin.

Therefore *constant delay* or *linear phase shift filters* are characterized by transfer functions having the form

$$H(z) = R(z)z^{-m} = \frac{N(z)N(z^{-1})}{D(z)D(z^{-1})}\, z^{-m} \tag{3.12.3}$$

When $m = 0$, the impulse response of this filter equals

$$\begin{aligned} h(nT) &= Z^{-1}[N(z)/D(z)], && |z| > r_o \quad (r_o \le 1) \\ h(-nT) &= Z^{-1}[N(z^{-1})/D(z^{-1})], && |z| < 1/r_o \end{aligned} \tag{3.12.4}$$

Such $H(z)$ are noncausal but stable since their impulse responses are nonzero for negative time. This was discussed in Chap. 1.12. They can be exactly realized in nonreal time (i.e., off-line) or approximated in real time by choosing $m = T_d/T$ sufficiently large. Notice from Eq. 3.12.4 that if $H(z)$ had a region of convergence $|z| > r_o$, such systems would be unstable or at best conditionally stable for $r_o \ge 1$. Defining the regions of convergence in this way shown insures stability but results in noncausality.

Another filter of interest is called the *allpass filter*. The allpass filter has a constant magnitude response where

$$|H(z)|_{z=e^{j\omega T}} = |H(e^{j\omega T})| = 1 \tag{3.12.5}$$

and an arbitrary phase response. Its transfer function has the form

$$H(z) = \frac{D(z^{-1})}{D(z)}\, z^{-m} \tag{3.12.6}$$

They have pole-zero pairs that are images of each other and perhaps poles or zeros at the origin. Such $H(z)$ can be causal and stable. For $m = 0$, the impulse response equals

$$\begin{aligned} h(nT) &= Z^{-1}[D(z^{-1})/D(z)], && |z| > r_o \\ h(-nT) &= 0 \end{aligned} \tag{3.12.7}$$

PROBLEMS

3.1   Determine the impulse response of the system having the following gain $H(z)$: (a) $1/(1+z^{-1}/2), |z| > 1/2$; (b) $1/(1+z^{-1}/2), |z| < 1/2$; (c) $(1-z^{-1}/2)/(1+z^{-1}/2), |z| > 1/2$; (d) $(1-z^{-1}/2)/(1+3z^{-1}/4+z^{-2}/8), |z| > 1/2$; (e) $(1-az^{-1}/2)/(z^{-1}-a), |z| > 1/|a|$.

3.2   Determine the impulse response of the system having the following gain $H(z)$: (a) $(z^2+3z)/(z-0.5)^3, |z| > 0.5$; (b) $1/[(z+1)^2(z-0.5)], |z| > 1$.

3.3   Determine the step responses of the system having the gain $H(z)$ given in Problem 3.1.

3.4   Determine the step responses of the system having the gain $H(z)$ given in Problem 3.2.

3.5   Determine the complex exponential step responses of the system having the gain $H(z)$ given in Problem 3.1.

3.6   Determine the complex exponential step responses of the system having the gain $H(z)$ given in Problem 3.2.

3.7   Find three different signals all having $X(z)$ as their z-transforms. Specify the region of convergence for each one: (a) $1/[(z-1)(z-2)]$, (b) $z/[(z-0.25)(z-0.5)]$, (c) $z/[(z-2)(z^2+1)]$.

3.8   Use the z-transform and inverse z-transform to find the response of the system described by the following difference equations: (a) $f(n-2) - 1.5f(n-1) + 0.5f(n) = 1$, $n \geq 0$ with initial conditions $f(-2) = f(-1) = 0$; (b) $f(n) - 0.9f(n-1) = 0.5 + 0.9^{n-1}$, $n \geq 0$ where $f(-1) = 5$; (c) $f(n) - 0.5f(n-1) = \sin(n\pi/2)$, $n \geq 0$ where $f(-1) = 0$.

3.9   Given a spectrum $[F(m)] = [0, 1, 2, 1]$. Find the IDFT of: (a) $F(m)$, (b) $F(m)\cos(m\pi)$, (c) $F(m)\sin(m\pi)$, (d) $F(m)e^{-m}\cos(m\pi)$, (e) $F(m)e^{-m}\sin(m\pi)$.

3.10   Find the 4-point IDFT of the spectrum shown in: (a) Fig. P3.10a, (b) Fig. P3.10b, (c) Fig. P3.10c, (d) Fig. P3.10d.

3.11   Find the 4-point IDFT of the spectrum shown in: (a) Fig. P3.11a, (b) Fig. P3.11b, (c) Fig. P3.11c, (d) Fig. P3.11d.

3.12   Find the 8-point IDFT of the spectrum shown in: (a) Fig. P3.10a, (b) Fig. P3.10b, (c) Fig. P3.10c, (d) Fig. P3.10d.

3.13   Find the 8-point IDFT of the spectrum shown in: (a) Fig. P3.11a, (b) Fig. P3.11b, (c) Fig. P3.11c, (d) Fig. P3.11d.

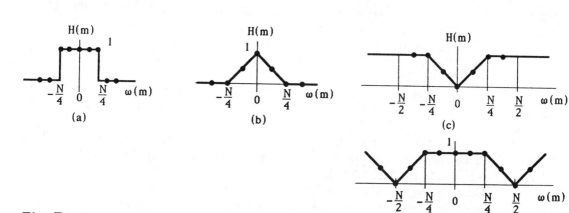

Fig. P3.10

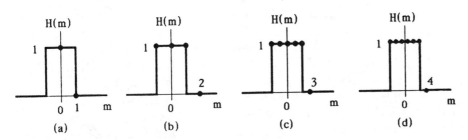

Fig P3.11

3.14 Draw the 4-point IFFT time-decimation flow graph equivalent of: (a) Fig. 2.3.5, (b) Fig. 2.3.6, (c) Fig. 2.3.7, (d) Fig. 2.3.8. (e) Evaluate the signal having the spectral sequence $[F(m)] = [0, 1, 2, 1]$.

3.15 Draw the 4-point IFFT frequency-decimation flow graph equivalent of: (a) Fig. 2.4.4, (b) Fig. 2.4.5, (c) Fig. 2.4.6, (d) Fig. 2.4.7. (e) Evaluate the signal having the spectral sequence $[F(m)] = [0, 1, 2, 1]$.

3.16 Draw an 4-point IFFT flow graph. Use it to evaluate the IDFT of the spectrum shown in: (a) Fig. P3.10a, (b) Fig. P3.10b, (c) Fig. P3.10c, (d) Fig. P3.10d. Use normal input and output sequences with time decimation.

3.17 Draw an 4-point IFFT flow graph. Use it to evaluate the IDFT of: (a) Fig. P3.11a, (b) Fig. P3.11b, (c) Fig. P3.11c, (d) Fig. P3.11d. Use normal input and output sequences with time decimation.

3.18 Draw an 8-point IFFT flow graph. Use it to evaluate the IDFT of the spectrum shown in: (a) Fig. P3.10a, (b) Fig. P3.10b, (c) Fig. P3.10c, (d) Fig. P3.10d. Use normal input and output sequences with time decimation.

3.19 Draw an 8-point IFFT flow graph. Use it to evaluate the IDFT of: (a) Fig. P3.11a, (b) Fig. P3.11b, (c) Fig. P3.11c, (d) Fig. P3.11d. Use normal input and output sequences with time decimation.

3.20 Repeat (a) Problem 3.16 and (b) Problem 3.17 but use normal input and output sequences with frequency decimation.

3.21 Repeat (a) Problem 3.18 and (b) Problem 3.19 but use normal input and output sequences with frequency decimation.

3.22 Use the fast transform flow graphs to find the 4-point inverse Fourier, Paley, Hadamard, Haar, and slant transforms for the spectrum shown in: (a) Fig. P3.10a, (b) Fig. P3.10b, (c) Fig. P3.10c, (d) Fig. P3.10d.

3.23 Use the fast transform flow graphs to find the 4-point inverse Fourier, Paley, Hadamard, Haar, and slant transforms for the spectrum shown in: (a) Fig. P3.11a, (b) Fig. P3.11b, (c) Fig. P3.11c, (d) Fig. P3.11d.

3.24 Use the fast transform flow graphs to find the 8-point inverse Fourier, Paley, Hadamard, Haar, and slant transforms for the spectrum shown in: (a) Fig. P3.10a, (b) Fig. P3.10b, (c) Fig. P3.10c, (d) Fig. P3.10d.

3.25 Use the fast transform flow graphs to find the 8-point inverse Fourier, Paley, Hadamard, Haar, and slant transforms for the spectrum shown in: (a) Fig. P3.11a, (b) Fig. P3.11b, (c) Fig. P3.11c, (d) Fig. P3.11d.

3.26 Consider the interpolator of Fig. 2.0.1. (a) Write the transfer functions for zero-order and first-order interpolators. (b) Sketch their frequency responses. Cascade a lowpass filter after the interpolator to reduce the high-frequency responses. (c) What should their two transfer functions be? (d) If the interpolator-lowpass filter should have an overall response of unity for $|\omega| < \omega_s/2$ and zero elsewhere, what should the two transfer functions be? (e) Why are the high-order Lagrange interpolators of Fig. 3.5.4 useful?

3.27   Suppose that the signal sequence $[f(n)] = [0, 1, 1, 0, 0, 0, 0, 0]$ is input to a Lagrange interpolator. Compute and draw the interpolator output assuming it is: (a) zero-order, (b) first-order, (c) second-order, (d) third-order, (e) infinite order.

3.28   Suppose that the signal sequence $[f(n)] = [0, 1, 2, 3, 0, 0, 0, 0]$ is input to a Lagrange interpolator. Compute and draw the interpolator output assuming it is: (a) zero-order, (b) first-order, (c) second-order, (d) third-order, (e) infinite order.

3.29   (a) Compute the spectral outputs of the Lagrange interpolators of Problem 3.27? (b) How are the signal outputs related to these spectra?

3.30   (a) Compute the spectral outputs of the Lagrange interpolators of Problem 3.28? (b) How are the signal outputs related to these spectra?

3.31   Suppose that the signal sequence $[f(n)] = [0, 1, 1, 0, 0, 0, 0, 0]$ is input to an ideal lowpass filter interpolator. (a) Compute and draw the interpolator output. (b) Compute its spectral output.

3.32   Suppose that the signal sequence $[f(n)] = [0, 1, 2, 3, 0, 0, 0, 0]$ is input to an ideal lowpass filter interpolator. (a) Compute and draw the interpolator output. (b) Compute its spectral output.

3.33   Suppose that the signal $f(t) = e^{-t}U_{-1}(t)$ is input to an ideal lowpass filter interpolator. Use only four samples and a sampling rate of: (a) 1, (b) 2, (c) 4, (d) 8, (e) 16 Hz. Compute and draw the interpolator output. Compute its spectral output.

3.34   Suppose that the signal sequence $[f(n)] = [0, 1, 1, 0, 0, 0, 0, 0]$ is input to Whittaker's reconstruction equation. Compute and draw the interpolator output.

3.35   Suppose that the signal sequence $[f(n)] = [0, 1, 2, 3, 0, 0, 0, 0]$ is input to Whittaker's reconstruction equation. Compute and draw the interpolator output.

3.36   Using Whittaker's reconstruction equation, find the reconstructed signal of $f(t) = e^{-t}U_{-1}(t)$. Use only four samples and a sampling rate of: (a) 1, (b) 2, (c) 4, (d) 8, (e) 16 Hz.

3.37   The frequency domain version of Whittaker's reconstruction equation is computed as follows: Step 1. Sample $f(t)$ as $f(nT) = f_n$; Step 2. Compute $X_p(j\omega) = \sum_{n=-\infty}^{\infty} f_n e^{-jn\omega/\omega_o}$, $\omega_o = \pi/T$; Step 3. Band-limit spectrum to $X(j\omega) = X_p(j\omega), |\omega| < \omega_o$ and zero elsewhere; Step 4. Inverse Fourier transform $X(j\omega)$ to obtain $f_r(t)$. Repeat Example 3.7.1 using these steps and compare the result with the example.

3.38   Suppose that the spectrum of a signal sequence $[f(n)] = [0, 1, 1, 0, 0, 0, 0, 0]$ is input to a Fourier series interpolator. (a) Compute the spectrum of the signal. (b) Compute and draw the interpolator output.

3.39   Suppose that the spectrum of a signal sequence $[f(n)] = [0, 1, 2, 3, 0, 0, 0, 0]$ is input to a Fourier series interpolator. (a) Compute the spectrum of the signal. (b) Compute and draw the interpolator output.

3.40   Using a Fourier series interpolator, find the reconstructed signal of $f(t) = e^{-t}U_{-1}(t)$. Use a 4-point DFT and a sampling rate of: (a) 1, (b) 2, (c) 4, (d) 8, (e) 16 Hz.

3.41   Suppose that the signal sequence $[f(n)] = [0, 1, 1, 0, 0, 0, 0, 0]$ is input to a Campbell smoothing algorithm. (a) Compute the smoothed signal. (b) Compute the spectrum of the smoothed signal.

3.42   Suppose that the signal sequence $[f(n)] = [0, 1, 2, 3, 0, 0, 0, 0]$ is input to a Campbell smoothing algorithm. (a) Compute the smoothed signal. (b) Compute the spectrum of the smoothed signal.

3.43   Use Campbell's algorithm to smooth the signal $f(t) = e^{-t}U_{-1}(t)$. Find the 4-point DFT of the smoothed signal using a sampling rate of: (a) 1, (b) 2, (c) 4, (d) 8, (e) 16 Hz.

3.44   Suppose that the generalized spectra of the signal sequence $[f(n)] = [0, 1, 1, 0, 0, 0, 0, 0]$ is input to Paley, Hadamard, Haar, and slant interpolators. (a) Compute the generalized spectra of this signal for these different transforms. (b) Write the equation for the outputs of the various interpolators.

3.45   Suppose that the generalized spectra of the signal sequence $[f(n)] = [0, 1, 2, 3, 0, 0, 0, 0]$ is input to Paley, Hadamard, Haar, and slant interpolators. (a) Compute the generalized spectra of this signal for these different transforms. (b) Write the equation for the outputs of the various interpolators.

3.46   Using Paley, Hadamard, Haar, and slant interpolators, write the equation for the reconstructed signal of $f(t) = e^{-t}U_{-1}(t)$. Use a sampling rate of: (a) 1, (b) 2, (c) 4, (d) 8, (e) 16 Hz.

3.47   Determine the maximum width pulse for sampling the following signals so the envelope droop does not exceed either 1% or 5%: (a) $e^{-t}U_{-1}(t)$, (b) $e^{-|t|}$, (c) e^{-t^2}.

3.48   Determine the maximum width pulse for sampling the following signals so the envelope droop does not exceed either 1% or 5%: (a) $e^{-t}\cos(t)U_{-1}(t)$, (b) $e^{-t}\sin(t)U_{-1}(t)$.

3.49   Compute the window size needed to insure the spectral error produced by truncating the following signals does not exceed either 1% or 5%: (a) $e^{-t}U_{-1}(t)$, (b) $e^{-|t|}$, (c) e^{-t^2}.

3.50   Compute the window size needed to insure the spectral error produced by truncating the following signals does not exceed either 1% or 5%: (a) $e^{-t}\cos(t)U_{-1}(t)$, (b) $e^{-t}\sin(t)U_{-1}(t)$.

3.51   Consider a signal $f(t) = e^{-t}\sin(t)U_{-1}(t)$. (a) Draw its magnitude spectrum. (b) What is the minimum sampling frequency so that the spectrum is -40 dB at the Nyquist frequency? Compute the aliasing error. (c) How many samples must be collected to resolve the spectrum to 1 rad/sec? (d) What is the minimum data window size? (e) How many additional samples should be taken or zeros padded so the FFT or other fast transform algorithms can be used?

3.52   Repeat Problem 3.51 when the signal $f(t) = e^{-t}\cos(t)U_{-1}(t)$.

3.53   Repeat Problem 3.51 when the signal $f(t)$ equals (a) $e^{-t}U_{-1}(t)$, (b) $e^{-|t|}$, (c) e^{-t^2}.

3.54   A telephone channel has a bandwidth of about 3200 Hz. (a) What is the minimum sampling frequency required? (b) How many samples must be collected to resolve a voice spectrum to 20 Hz? (c) What is the data window size? (d) How many additional samples should be taken or zeros padded so the FFT or other fast algorithms can be used?

3.55   (a) Write the general $H(z)$ transfer functions of first- and second-order allpass filters. (b) Find the impulse response of these filters. (c) Using these results, write the allpass transfer functions which delay an input signal by one clock period. (d) What is the impulse response of these filters?

REFERENCES

1. Lindquist, C.S., *Active Network Design with Signal Filtering Applications*, Chap. 3, Steward and Sons, CA, 1977.
2. Papoulis, A., *Circuits and Systems–A Modern Approach*, Chap. 7, Holt, Rinehart, and Winston, NY, 1980.
3. Lindquist, C.S. and S.R.Yhann, "Transforms for nonuniformly sampled data based on iterative algorithms," Proc. IASTED Intl. Symp. Applied Control, Filtering and Signal Processing, Geneva, June 1987.

Yhann, S.R. and C.S. Lindquist, "Spectral estimation from nonuniformly-sampled data and new matrix inversion methods," Proc. IEEE Miami Technicon, pp. 133-136, 1987.

Yhann, S.R., "Iterative transform methods for nonuniformly-sampled data," M.S.E.E. Thesis, Univ. of Miami, 106 p., May 1987.

4. Elliott, D.F. and K.R. Rao, *Fast Transforms–Algorithms, Analyses, Applications*, Academic Press, NY, 1982.

5. Schafer, R.W. and L.R. Rabiner, "A digital signal approach to interpolation," Proc. IEEE, vol. 61, pp. 692–702, June 1973.

6. Tretter, S.A., *Introduction to Discrete-Time Signal Processing*, Chap. 3, Wiley, NY, 1976. Wylie, C.R., *Advanced Engineering Mathematics*, 4th ed., Chap. 4.2, McGraw-Hill, NY, 1975.

7. Whittaker, E.T., "Expansions of the interpolation-theory," Proc. Roy. Soc. Edinburgh, vol. 35, p. 181, 1915.

8. Papoulis, A., "Error analysis in sampling theory," Proc. IEEE, vol. 54, no. 7, p. 947, July 1966.

9. Campbell, A.B., "A New Sampling Theorem for Causal (Non-Bandlimited) Functions," Ph.D. Dissertation, Univ. of New Mexico, Albuquerque, July 1973.

10. Ref. 1, Chap. 5.

4 IDEAL FILTER RESPONSE

Therefore thus saith the Lord God,
Behold, I lay in Zion for a foundation a stone,
a tried stone, a precious corner stone, a sure foundation:
he that believeth shall not make haste.

Isaiah 28:16

A general digital signal processing system is shown in Fig. 2.0.1. It is composed of six blocks – the lowpass filter, sampler, transform, filter gain, inverse transform, and interpolator. The lowpass filter and sampler was discussed in Chap. 1, the fast transform was considered in Chap. 2, and the fast inverse transform and data interpolator were investigated in Chap. 3. The only remaining block is the *filter block* with transfer function $H(m)$ which will now be considered.

In Chap. 4, transfer functions will be obtained using systematic design procedures. *Classical filters, optimum filters,* and *adaptive filters* will be discussed. Since classical[1] and optimum[2] filters are well known and design data is well tabulated, this chapter will concentrate primarily on the new emerging area of adaptive filters.

The classical, optimum, and adaptive filters to be discussed are usually lowpass types and are described by their analog (continuous) transfer functions. Chap. 5 will present transformations that allow the analog transfer functions of Chap. 4 to be converted into digital (discrete) transfer functions. Chap. 6 will discuss other transformations that convert analog lowpass filters into digital highpass, bandpass, bandstop, and allpass filters.

These next three closely-related chapters will forge indispensible links in the chain of understanding for the engineer. This chapter will give him a wide variety of standard filter types and responses. From these, he may select the one best-suited to meet his particular application. Therefore it is important for the engineer to grasp the philosophical basis of these filter types so they can be critically compared and properly utilized.

4.1 IDEAL LOWPASS FILTERS

Filters are characterized by their transfer functions. A *rational transfer function* $H(s)$ is the ratio of two polynomials

$$H(s) = c \, \frac{\prod_{j=1}^{m}(s+z_j)}{\prod_{i=1}^{n}(s+p_i)} = \frac{\sum_{j=0}^{m} a_j s^j}{\sum_{i=0}^{n} b_i s^i} \qquad (4.1.1)$$

250

where $s = \sigma + j\omega$ is the complex frequency variable. The $-z_j$ are the finite zeros of the filter and the $-p_i$ are the finite poles. There are $(n - m)$ zeros at infinity (or poles if negative). Stable, causal filters have Re $p_i > 0$ for $i = 1, \ldots, n$ and simple (first-order) $j\omega$-axis poles. When the filter impulse response is real, complex poles and zeros exist in complex conjugate pairs.

The *order* of a filter is a measure of its complexity. Order k is equal to the number of finite poles (n) or zeros (m), whichever is greater, so $k = \max(n, m)$. In practical filters, $n > m$ since the response must be zero at infinite frequency so $k = n$. This condition need not be satisfied for ideal filters modelled mathematically. For example, the ideal differentiator has a gain $H(s) = s$.

The *frequency response* of a filter describes its transfer

$$H(j\omega) = H(s)\Big|_{s=j\omega} = |H(j\omega)|\, e^{j\arg H(j\omega)} = \mathrm{Re}H(j\omega) + j\mathrm{Im}H(j\omega) \qquad (4.1.2)$$

$H(j\omega)$ is characterized by its magnitude and phase, or alternatively, its real and imaginary parts over the entire frequency range from zero to infinity. The first form is the polar representation for H while the second is the rectangular representation.

The magnitude response of a general lowpass filter is shown in Fig. 4.1.1. In describing such magnitude responses, the following terminology is often used:

1. *Passband:* Frequencies over which the response is within M_p dB of its maximum value $(\omega \le \omega_p)$.
2. *Stopband:* Frequencies over which the response is attenuated at least M_r dB from its maximum value $(\omega \ge \omega_r)$. Also called *rejection band.*
3. *Transition Band:* Intermediate frequencies lying between passband and stopband $(\omega_p < \omega < \omega_r)$.
4. *Band-edge Frequency (Cut-off Frequency):* Frequency ω_p marking edge of passband.
5. *Stopband Frequency:* Frequency ω_r marking edge of stopband.
6. M_p *Bandwidth:* Equals band-edge frequency.
7. *3 dB Bandwidth (3 dB Frequency):* Maximum frequency at which response is down 3 dB from its maximum value $(\omega = \omega_{3dB} = B)$.

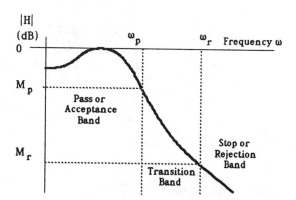

Fig. 4.1.1 Magnitude response of lowpass filter.

An ideal lowpass filter response having unity gain and unity bandwidth is shown in Fig. 1.12.2a. Its gain equals

$$H(j\omega) = e^{-j\omega\tau_o}, \qquad |\omega| \leq 1$$
$$= 0, \qquad |\omega| > 1 \qquad\qquad (4.1.3)$$

It has infinite rejection in its stopband, a transition bandwidth of zero, and a constant gain in its passband. It has linear phase with negative slope τ_o (or constant delay τ_o) through the passband. As shown in Example 1.12.2, the ideal lowpass filter is nonrealizable. No ratio of finite degree polynomials (a rational function) will realize the ideal $H(j\omega)$ exactly. Generally however, the higher the filter order, the better the approximation. This chapter will present various approaches for approximating the ideal lowpass filter.

Suppose that we express the magnitude-squared response of the ideal filter by

$$|H(j\omega)|^2 = \frac{1 + c_1\omega^2 + c_2\omega^4 + \ldots + c_m\omega^{2m}}{1 + d_1\omega^2 + d_2\omega^4 + \ldots + d_n\omega^{2n}} \qquad (4.1.4)$$

Since the magnitude function is always even, the two polynomials must also be even. Since $|H|$ approaches zero as ω approaches infinity, then $n > m$. Classical filters utilize classical polynomials in Eq. 4.1.4. For optimum filters, the c_i and d_i coefficients are selected according to some optimization criterion involving the frequency domain characteristics of $H(j\omega)$, the time domain properties of the impulse response $h(t)$, or both. In adaptive filters the coefficients depend upon the signal and noise the filter processes.

Express the filter transfer function from Eq. 4.1.1 as

$$H(s) = \frac{N(s)}{D(s)} = \frac{1 + a_1 s + a_2 s^2 + \ldots + a_m s^m}{1 + b_1 s + b_2 s^2 + \ldots + b_n s^n} \qquad (4.1.5)$$

The pole and zero locations for $H(s)$ are determined from $H(j\omega)$ using *analytic continuation*. This is a process which recognizes that the magnitude-squared function given by Eq. 4.1.4 can also be expressed as

$$H(s)H(-s) = |H(s)|^2 = |H(j\omega)|^2\Big|_{\omega=s/j=-js} \qquad (4.1.6)$$

Therefore $H(s)$ is found by factoring the product $H(s)H(-s)$. The left-half-plane (LHP) pole-zero pattern is selected for $H(s)$. The right-half-plane (RHP) pattern constitutes $H(-s)$. Combining Eqs. 4.1.4–4.1.6, the LHP and RHP poles and zeros satisfy

$$\text{Poles:} \quad 0 = 1 - d_1 s^2 + d_2 s^4 - \ldots + (-1)^n d_n s^{2n}$$
$$= 1 - (b_1^2 - 2b_2)s^2 + (b_2^2 - 2b_1 b_3)s^4 - (b_3^2 - 2b_2 b_4)s^6 + \ldots$$
$$\text{Zeros:} \quad 0 = 1 - c_1 s^2 + c_2 s^4 - \ldots + (-1)^m c_m s^{2m} \qquad (4.1.7)$$
$$= 1 - (a_1^2 - 2a_2)s^2 + (a_2 - 2a_1 a_3)s^4 - (a_3^2 - 2a_2 a_4)s^6 + \ldots$$

The a_i and b_i filter coefficients and c_i and d_i magnitude-squared coefficients satisfy

$$d_1 = b_1{}^2 - 2b_2 \quad , \qquad c_1 = a_1{}^2 - 2a_2$$
$$d_2 = b_2{}^2 - 2b_1b_3, \qquad c_2 = a_2{}^2 - 2a_1a_3$$
$$d_3 = b_3{}^2 - 2b_2b_4, \qquad c_3 = a_3{}^2 - 2a_2a_4 \tag{4.1.8}$$
$$\cdots \quad , \qquad \cdots$$

Since the $(n+m)$ c_i and d_i coefficients are known, the $(n+m)$ a_i and b_i coefficients can be found by solving this set of simultaneous equations. This method yields the filter transfer function $H(s)$ given by Eq. 4.1.5.

4.2 CLASSICAL FILTERS

Let us first consider *classical filters*. These are filters whose gain expressions are directly related to well-known classical polynomials or curves.[1] A variety of classical filters are listed in Table 4.2.1. Such filters are often grouped into constant magnitude, constant delay or linear phase, and transitional types.

The *constant magnitude* types have magnitudes which closely approximate the brick-wall characteristics of the ideal lowpass filter described by Eq. 4.1.3. The most popular filters of this type are the Butterworth, Chebyshev, and elliptic filters. The *constant delay* types have delays (or phases) which closely approximate the constant passband delay (or linear phase) of the ideal lowpass filter (stopband delay and phase is arbitrary). The Bessel filter is the most popular of this type. The *transitional* types have magnitude and delay characteristics which generally fall between those of the constant magnitude and constant delay types. At least one transitional parameter (like m) generally controls the filter transition from one type to the other.

The characteristics of the major classical filters are compared in Table 4.2.2. These filters are listed in order of decreasing band-edge selectivity. Elliptic filters have the highest selectivity and synchronously-tuned filters the lowest. We shall now briefly describe Butterworth, Chebyshev, elliptic, and Bessel filters. Further details may be found in Ref. 1.

4.2.1 BUTTERWORTH FILTERS

A *Butterworth filter* has a magnitude-squared gain of

$$|H(j\omega)|^2 = \frac{1}{1 + \omega^{2n}} \tag{4.2.1}$$

This magnitude characteristic is plotted in Fig. 4.2.1a for different filter orders n. Notice that all the filters have a 3 dB bandwidth of unity since $|H(j1)| = 1/\sqrt{2} = -3$ dB. Butterworth filters have no finite zeros. All their $j\omega$-axis or transmission zeros are at infinity. The asymptotic roll-off of $|H(j\omega)|$ is $-20n$ dB/dec.

Comparing Eqs. 4.1.4 and 4.2.1, their coefficients equal $c_1 = c_2 = \ldots = c_m = 0$, $d_1 = d_2 = \ldots = d_{n-1} = 0$, and $d_n = 1$. Expanding $|H|^2$ in a Maclaurin series about

Table 4.2.1 Classical filter types and gain functions.
(See pages indicated in Ref. 1.)

Filter Type	$|H(j\omega)|^2$ Gain Function	Comments
Elliptic	$\dfrac{1}{1 + \epsilon^2 Z_n^2(\omega)}$	$Z_n =$ elliptic function (pp. 235–243)
Ultraspherical	$\dfrac{1}{1 + \epsilon^2 [F_n^\alpha(\omega)]^2}$	$F_n^\alpha =$ ultraspherical polynomial with parameter α (pp. 225–228)
Chebyshev	$\dfrac{1}{1 + \epsilon^2 T_n^2(\omega)}$	Poles equally spaced around ellipse, $T_n =$ Chebyshev polynomial (pp. 216–224)
Inverse Chebyshev	$\dfrac{1}{1 + 1/\epsilon^2 T_n^2(1/\omega)}$	Poles are inverses of Chebyshev filter poles (pp. 232–235)
Papoulis	$\dfrac{1}{1 + \epsilon^2 L_n^2(\omega)}$	$L_n =$ Papoulis polynomial (pp. 228–232)
Butterworth	$\dfrac{1}{1 + \epsilon^2 \omega^{2n}}$	Poles equally spaced around circle (pp. 207–216)
Bessel	$\dfrac{P_n^2(0)}{P_n(j\omega) P_n(-j\omega)}$	$P_n(s) = s^n B_n(1/s)$ where $B_n(s) =$ Bessel polynomial (pp. 245–250)
Gaussian	$\cong e^{-0.694\omega^2}$	$|H|$ has Gaussian shape (pp. 251–254)
Equiripple Delay	$\dfrac{\tau_o}{1 + \epsilon^2 T_{2n}(\omega)}$	$T_{2n} =$ Chebyshev polynomial (pp. 250–251)
Parabolic contour	Parabolic curve	Poles equally spaced around parabola (pp. 257–260)
Catenary contour	Hyperbolic curve	Poles equally spaced around hyperbola (pp. 261–264)
Elliptic contour	Elliptic curve	Poles equally spaced around ellipse (pp. 265–268)
Butterworth-Chebyshev	$\dfrac{1}{1 + \omega^{2k} T_{n-k}^2(\omega)}$	Polynomial magnitude interpolation
Butterworth-Thomson Legendre-Thomson	$|s_m| = |s_T|^m |s_B|^{1-m}$ $\arg s_m =$ $\arg s_B - m(\arg s_B - \arg s_T)$	Geometric pole magnitude, linear pole angle interpolations
Chebyshev-Equiripple Delay Papoulis-Equiripple Delay	$|s_m| =$ $|s_C| - m(|s_C| - |s_E|)$ $\arg s_m =$ $\arg s_C - m(\arg s_C - \arg s_E)$	Linear pole magnitude, linear angle interpolations
Gaussian to 6 or 12 dB	—	—
Synchronously-Tuned	$\dfrac{1}{(1 + \omega^2)^n}$	All poles at $s = -1$ (pp. 254–256)

Table 4.2.2 Comparison of classical filters.[1]

Filter Type	In-band Gain	Stopband Gain	Band-edge Selectivity & NBW	Transition Bandwidth	In-band Delay	Pole Q	Step Response Overshoot	Step Response Rise Time	Step Response Delay Time
Elliptic	Equiripple	Equiripple	Highest	Lowest	Ripples & increases	Highest	Highest	2.2	Highest
Modified Inverse Chebyshev	MFM	Equiripple	Higher	Lower	Ripples & increases	Higher	Higher	2.2	Higher
Ultraspherical	Non-equiripple	Mono. dec.	Higher	Lower	Ripples & increases	Higher	Higher	2.2	Higher
Chebyshev	Equiripple	Mono. dec.	High	Low	Ripples & increases	High	Meaium	2.2	High
Inverse Chebyshev	MFM	Equiripple	High	Low	Ripples & increases	High	Medium	2.2	High
Papoulis/ Halpern	Mono. dec.	Mono. dec.	High	Low	Ripples & increases	High	Medium	2.2	Medium
Butterworth	MFM	Mono. dec.	Medium	Medium	No ripple & increases	Medium	Medium	2.2	Medium
Bessel	Mono. dec.	Mono. dec.	Low	High	MFD	Medium	Negligible	2.2	Low
Gaussian	Mono. dec.	Mono. dec.	Low	High	Mono. dec.	Medium	Negligible	2.2	Low
Equiripple Delay	Mono. dec.	Mono. dec.	Low	High	Ripples	Medium	Low	2.2	Low
Parabolic/Catenary/ Elliptic Contour	Mono. dec.	Mono. dec.	Lower	Higher	Ripples & increases	Medium	Low	2.2	Lower
Synchronously-tuned	Mono. dec.	Mono. dec.	Lowest	Highest	Mono. dec.	Lowest	None	2.2	Lowest

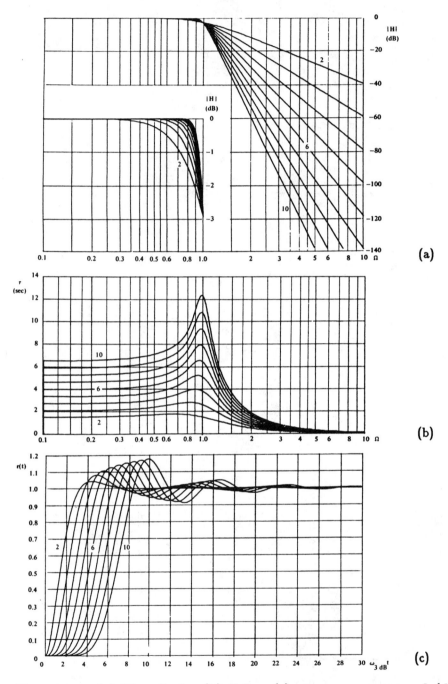

Fig. 4.2.1 (a) Magnitude, (b) delay, (c) step responses, and (d) nomograph of Butterworth filters. ((a)–(c) from A.D. Taylor, "A study of transitional filters," M.S.E.E. Directed Research, Calif. State Univ., Long Beach, Aug. 1975. (d) from J.N. Hallberg, "Filters and nomographs," M.S.E.E. Directed Research, Calif. State Univ., Jan. 1974.)

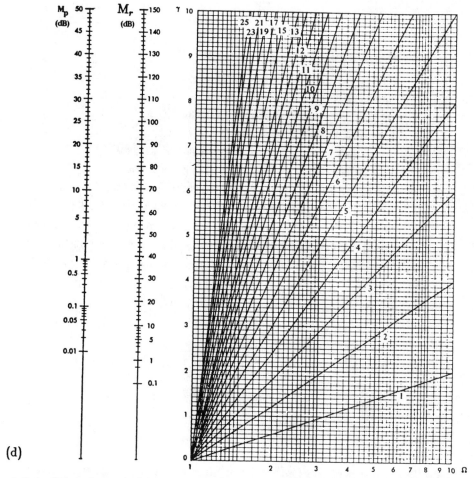

Fig. 4.2.1 (Continued) (a) Magnitude, (b) delay, (c) step responses, and (d) nomograph of Butterworth filters.

$\omega = 0$ gives $|H|^2 = 1 - \omega^{2n} + \omega^{4n} - \dots$. Therefore the first $(2n-1)$ derivatives of $|H(0)|^2$ are zero and the filter is said to be *maximally flat magnitude* or *MFM*. Fig. 4.2.1a shows this MFM behavior very clearly.

As the filter order increases, the magnitude approximation approaches the ideal magnitude response of Fig. 1.12.2a. The ideal response makes an abrupt gain transition at $\omega = 1$ but the approximation has only finite slope. The gain slope at the passband-edge is called the *filter selectivity* or *cut-off rate*. The ideal lowpass filter has a band-edge selectivity of infinity. Butterworth filters have band-edge selectivities of $0.354n$ which increase with filter order. The ideal response has zero gain in its stopband which the approximation never provides.

Substituting the c_i and d_i coefficients into Eq. 4.1.8 and solving for the a_i and b_i coefficients yields the filter polynomials $D_n(s)$ listed in Table 4.2.3. For example, the second-order filter has $n = 2, d_1 = 0$, and $d_2 = 1$ which gives $b_1 = \sqrt{2}, b_2 = 1$ and $a_1 = a_2 = 0$.

Table 4.2.3 Butterworth filter poles.

$D_1 = s + 1$

$D_2 = s^2 + 1.4142s + 1 = (s + 0.7071)^2 + 0.7071^2$

$D_3 = s^3 + 2.0000s^2 + 2.0000s + 1 = (s + 1.0000)\left[(s + 0.5000)^2 + 0.8660^2\right]$

$D_4 = s^4 + 2.6131s^3 + 3.4142s^2 + 2.6131s + 1$

$\quad = \left[(s + 0.3827)^2 + 0.9239^2\right]\left[(s + 0.9239)^2 + 0.3827^2\right]$

$D_5 = s^5 + 3.2361s^4 + 5.2361s^3 + 5.2361s^2 + 3.2361s + 1$

$\quad = (s + 1.0000)\left[(s + 0.3090)^2 + 0.9511^2\right]\left[(s + 0.8090)^2 + 0.5878^2\right]$

$D_6 = s^6 + 3.8637s^5 + 7.4641s^4 + 9.1416s^3 + 7.4641s^2 + 3.8637s + 1$

$\quad = \left[(s + 0.2588)^2 + 0.9659^2\right]\left[(s + 0.7071)^2 + 0.7071^2\right]\left[(s + 0.9659)^2 + 0.2588^2\right]$

$D_7 = s^7 + 4.4940s^6 + 10.0978s^5 + 14.5918s^4 + 14.5918s^3 + 10.0978s^2 + 4.4940s + 1$

$\quad = (s + 1.000)\left[(s + 0.2225)^2 + 0.9749^2\right]\left[(s + 0.6235)^2 + 0.7818^2\right]\left[(s + 0.9010)^2 + 0.4339^2\right]$

$D_8 = s^8 + 5.1258s^7 + 13.1371s^6 + 21.8462s^5 + 25.6884s^4 + 21.8462s^3 + 13.1371s^2 + 5.1258s + 1$

$\quad = \left[(s + 0.1951)^2 + 0.9808^2\right]\left[(s + 0.5556)^2 + 0.8315^2\right]\left[(s + 0.8315)^2 + 0.5556^2\right]\left[(s + 0.9808)^2 + 0.1951^2\right]$

$D_9 = s^9 + 5.7588s^8 + 16.5817s^7 + 31.1634s^6 + 41.9864s^5 + 41.9864s^4 + 31.1634s^3$

$\quad + 16.5817s^2 + 5.7588s + 1$

$\quad = (s + 1.0000)\left[(s + 0.1737)^2 + 0.9848^2\right]\left[(s + 0.5000)^2 + 0.8660^2\right]\left[(s + 0.7660)^2 + 0.6428^2\right]$

$\quad \times \left[(s + 0.9397)^2 + 0.3420^2\right]$

$D_{10} = s^{10} + 6.3925s^9 + 20.4317s^8 + 42.8021s^7 + 64.8824s^6 + 74.2334s^5 + 64.8824s^4 + 42.8021s^3$

$\quad + 20.4317s^2 + 6.3925s + 1$

$\quad = \left[(s + 0.1564)^2 + 0.9877^2\right]\left[(s + 0.4540)^2 + 0.8910^2\right]\left[(s + 0.7071)^2 + 0.7071^2\right]\left[(s + 0.8910)^2 + 0.4540^2\right]$

$\quad \times \left[(s + 0.9877)^2 + 0.1564^2\right]$

The order n of the Butterworth filter is easily determined using the nomograph of Fig. 4.2.1d. The procedure begins by first expressing the required frequency response in the form shown of Fig. 4.2.2a. The pertinent parameters equal:

M_p = maximum passband ripple in dB

M_r = minimum stopband attenuation in dB

Ω_r = normalized stopband frequency

where

$$\Omega_r = \frac{f_r}{f_p} = \frac{\omega_r}{\omega_p} \tag{4.2.2}$$

and

f_p (or ω_p) = band-edge frequency of passband in Hz (or rad/sec)

f_r (or ω_r) = band-edge frequency of stopband in Hz (or rad/sec)

This data is then entered onto the nomograph as shown in Fig. 4.2.2b. A line is drawn through M_p (point P_1) and M_r (point P_2) until it intersects γ at P_3. A horizontal line is then drawn across the graph from P_3. A vertical line is drawn up from the stopband frequency Ω_r at P_4. The horizontal and vertical lines intersect at P_5. The minimum required filter order is read off the first curve lying *above* P_5.

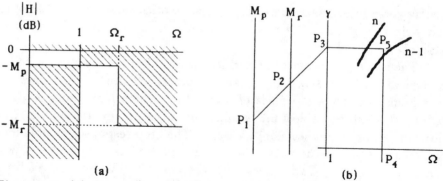

Fig. 4.2.2 (a) General magnitude specification for lowpass filter and (b) entry of data onto nomograph.

EXAMPLE 4.2.1 Determine the order required for a 1 KHz Butterworth filter to meet the specification shown in Fig. 4.2.3a using the Butterworth nomograph. Here 40 dB and 60 dB of rejection is required at frequencies of two and three times the 1.25 dB bandwidth, respectively.

SOLUTION To use the nomograph, we first write that

$$
\begin{aligned}
M_p &= 1.25 \text{ dB (at } f = 1 \text{ KHz)} \\
M_{r1} &= 40 \text{ dB (at } f = 2 \text{ KHz)}, \qquad \Omega_{r1} = 2 \text{ KHz}/1 \text{ KHz} = 2 \\
M_{r2} &= 60 \text{ dB (at } f = 3 \text{ KHz)}, \qquad \Omega_{r2} = 3 \text{ KHz}/1 \text{ KHz} = 3
\end{aligned}
\tag{4.2.3}
$$

Entering this data on the nomograph as shown in Fig. 4.2.3b gives $n_1 \geq 8$ and $n_2 \geq 7$. Thus an eighth-order Butterworth filter is required.   ∎

Filter delay $\tau(j\omega)$ is defined to equal the negative derivative of the phase arg $H(j\omega)$ or $\tau(j\omega) = -d \arg H(j\omega)/d\omega$.[3] Filter phase is the negative integral of its delay. The delay characteristic of the Butterworth filter is shown in Fig. 4.2.1b. Delay expressions are complicated and are evaluated using a computer. The delay variation increases with filter order in the passband. For $n = 3$ the delay variation is about 1 second. For $n = 9$ the delay variation is almost 6 seconds. Even with this amount of delay variation, the

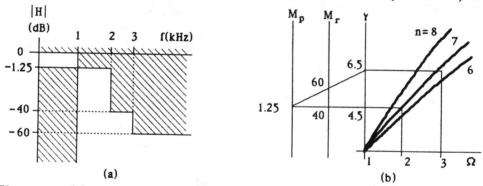

Fig. 4.2.3 (a) Filter specification for Example 4.2.1 and (b) entry of data onto nomograph.

Butterworth filter has better delay characteristics than most of the other filters having fairly narrow transition bands. Usually the delay variation is important but not the absolute delay value.

The unit step response of the Butterworth filter is shown in Fig. 4.2.1c as a function of normalized time $\omega_{3dB}t$ (in radians). Normalized time is used since the step response was calculated using the poles of Table 4.2.3 which have unity ω_{3dB}. From the time scaling theorem of Table 1.6.2, scaling the poles by any nonunity ω_{3dB} scales the time domain responses (in seconds) by ω_{3dB}. The step responses exhibit more overshoot as n increases because the pole damping factors decrease. The $10-90\%$ rise times equal about 2.2 seconds. This agrees with the empirical Valley and Wallman result where $t_r \cong 2.2/\omega_{3dB}$ since $\omega_{3dB} = 1$. The 50% delay times about equal the dc delay values $\tau(0)$ in Fig. 4.2.1b.

EXAMPLE 4.2.2 Determine if a Butterworth filter can be designed to meet the frequency and time requirements shown in Fig. 4.2.4a. Design for minimum filter order and draw the block diagram of the filter if it can be realized.

SOLUTION We begin by determining the required filter order from the frequency response. Since $M_p = 3$ dB, $M_{r1} = 10$ dB at $\Omega_{r1} = 2$, and $M_{r2} = 40$ dB at $\Omega_{r2} = 4$, the Butterworth filter nomograph shows that $n_1 \geq 2$ and $n_2 \geq 4$. Thus at least a fourth-order filter is required.

Now we sketch the step response of the filter noting $n = 4$. Since the bandwidth $\omega_{3dB} = 2\pi(2\ \text{KHz})$, time is denormalized by ω_{3dB} in Fig. 4.2.4a. Transferring Fig. 4.2.4a onto the normalized step response of Fig. 4.2.1c, we see that the time domain requirements are easily met. Since the time differences between the specification

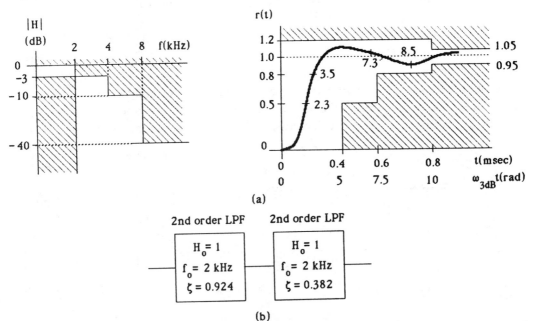

(a)

(b)

Fig. 4.2.4 (a) Filter specifications and (b) block diagram of Butterworth filter of Example 4.2.2.

and actual responses equal

$$\Delta t_1 = 5 - 2.3 = 2.7 \text{ sec} \qquad (r(t) = 0.5)$$
$$\Delta t_2 = 7.5 - 3.5 = 4.0 \text{ sec} \qquad (r(t) = 0.8) \qquad (4.2.4)$$
$$\Delta t_3 = 10 - 7.3 = 2.7 \text{ sec} \qquad (r(t) = 1.05)$$

the time domain specification may be reduced by as much as $\Delta t = \min(2.7, 4.0, 2.7)$ = 2.7 seconds. Alternatively, the bandwidth might be expanded. However, since 2 KHz bandwidth is needed here, this cannot be done.

The transfer function of the filter equals

$$H(s) = \frac{1}{(s_n^2 + 0.765 s_n + 1)(s_n^2 + 1.848 s_n + 1)} \qquad (4.2.5)$$

where $s_n = s/2\pi(2 \text{ KHz})$ using Table 4.2.3. The block diagram realization of the filter is shown in Fig. 4.2.4b.   ■

We shall usually represent nth-order filters as the cascade of $n/2$ (n even) or $(n - 1)/2$ (n odd) second-order stages. Odd-order filters also use a first-order block. Each second-order block is explicitly labelled with its: (1) order and type, (2) dc gain H_o, (3) resonant frequency of pole (ω_n or f_n), and (4) damping factor ς of the pole (omit ς for first-order block). In Chaps. 8–11, such block diagrams are used to summarize filter design parameters. Filter implementations will be discussed in Chap. 12.

We now see the great ease in utilizing classical filter results since we rely on using tabulated transfer functions and curves. The bulk of the work involves determining the acceptable type of filter (e.g., Butterworth) and its required order. The transfer functions can always be easily expressed using the *frequency normalized form* involving s_n. Denormalized forms involving s are more complicated and obscure pertinent information so we will seldom use them.

Now that Butterworth filters have been analyzed in detail, several others will be considered. Since the development of each filter type follows the same procedure as that of the Butterworth filter, our discussions shall be briefer but no less complete.

4.2.2 CHEBYSHEV FILTERS

The *Chebyshev filter*, also called the *equal ripple* or *equiripple filter*, has a magnitude-squared response

$$|H(j\omega)|^2 = \frac{1}{1 + \epsilon^2 T_n^2(\omega)} \qquad (4.2.6)$$

T_n is the nth-order Chebyshev polynomial where

$$T_n(\omega) = \cos(n \cos^{-1} \omega), \qquad |\omega| \le 1 \qquad (4.2.7)$$

The first few Chebyshev polynomials equal

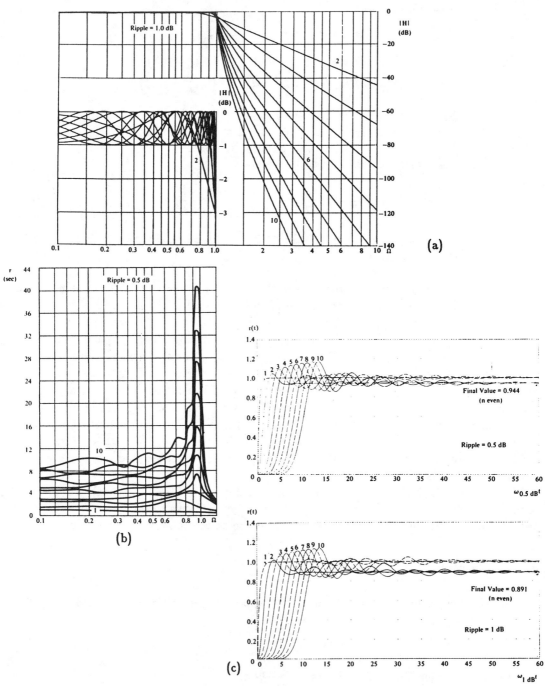

Fig. 4.2.5 (a) Magnitude (1 dB ripple), (b) delay (0.5 dB ripple), (c) step responses (0.5 and 1 dB ripples) and (d) nomograph of Chebyshev filters. ((a) from A.D. Taylor, loc. cit. (c) from K.W. Henderson and W.H. Kautz, "Transient responses of conventional filters," IRE Trans. Circuit Theory, vol. CT-5, pp. 333–347, Dec., ©1958, IEEE. (d) from J.N. Hallberg, loc. cit.)

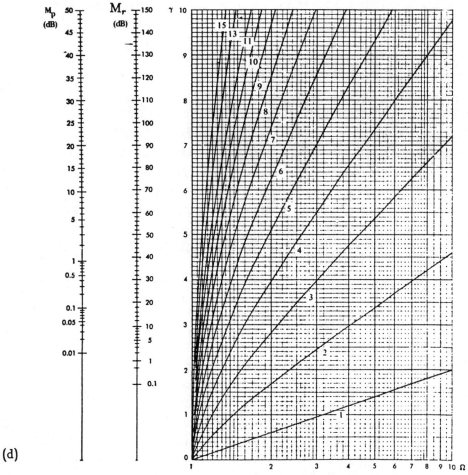

(d)

Fig. 4.2.5 (Continued) (a) Magnitude (1 dB ripple), (b) delay (0.5 dB ripple), (c) step responses (0.5 and 1 dB ripples) and (d) nomograph of Chebyshev filters.

$$
\begin{aligned}
T_1(\omega) &= \omega \\
T_2(\omega) &= 2\omega^2 - 1 \\
T_3(\omega) &= 4\omega^3 - 3\omega \\
T_4(\omega) &= 8\omega^4 - 8\omega^2 + 1 \\
T_5(\omega) &= 16\omega^5 - 20\omega^3 + 5\omega \\
T_6(\omega) &= 32\omega^6 - 48\omega^4 + 18\omega^2 - 1 \\
T_7(\omega) &= 64\omega^7 - 112\omega^5 + 56\omega^3 - 7\omega \\
T_8(\omega) &= 128\omega^8 - 256\omega^6 + 160\omega^4 - 32\omega^2 + 1 \\
T_9(\omega) &= 256\omega^9 - 576\omega^7 + 432\omega^5 - 120\omega^3 + 9\omega \\
T_{10}(\omega) &= 512\omega^{10} - 1280\omega^8 + 1120\omega^6 - 400\omega^4 + 50\omega^2 - 1
\end{aligned}
$$

$$(4.2.8)$$

Table 4.2.4 Chebyshev filter poles.

0.5 dB ripple

$D_1 = s + 2.8628$

$D_2 = (s + 0.7128)^2 + 1.0040^2$

$D_3 = (s + 0.6265)[(s + 0.3132)^2 + 1.0219^2]$

$D_4 = [(s + 0.1754)^2 + 1.0163^2][(s + 0.4233)^2 + 0.4209^2]$

$D_5 = (s + 0.3623)[(s + 0.1120)^2 + 1.0116^2][(s + 0.2931)^2 + 0.6252^2]$

$D_6 = [(s + 0.0777)^2 + 1.0085^2][(s + 0.2121)^2 + 0.7382^2][(s + 0.2898)^2 + 0.2702^2]$

$D_7 = (s + 0.2562)[(s + 0.0570)^2 + 1.0064^2][(s + 0.1597)^2 + 0.8071^2][(s + 0.2308)^2 + 0.4479^2]$

$D_8 = [(s + 0.0436)^2 + 1.0050^2][(s + 0.1242)^2 + 0.8520^2][(s + 0.1859)^2 + 0.5693^2][(s + 0.2193)^2 + 0.1999^2]$

$D_9 = (s + 0.1984)[(s + 0.0345)^2 + 1.0040^2][(s + 0.0992)^2 + 0.8829^2][(s + 0.1520)^2 + 0.6553^2]$
$\qquad \times\, [(s + 0.1864)^2 + 0.3487^2]$

$D_{10} = [(s + 0.0279)^2 + 1.0033^2][(s + 0.0810)^2 + 0.9051^2][(s + 0.1261)^2 + 0.7183^2][(s + 0.1589)^2 + 0.4612^2]$
$\qquad \times\, [(s + 0.1761)^2 + 0.1589^2]$

---

1 dB ripple

$D_1 = s + 1.9652$

$D_2 = (s + 0.5489)^2 + 0.8951^2$

$D_3 = (s + 0.4942)[(s + 0.2471)^2 + 0.9660^2]$

$D_4 = [(s + 0.1395)^2 + 0.9834^2][(s + 0.3369)^2 + 0.4073^2]$

$D_5 = (s + 0.2895)[(s + 0.0895)^2 + 0.9901^2][(s + 0.2342)^2 + 0.6119^2]$

$D_6 = [(s + 0.06218)^2 + 0.9934^2][(s + 0.1699)^2 + 0.7272^2][(s + 0.2321)^2 + 0.2662^2]$

$D_7 = (s + 0.2054)[(s + 0.0457)^2 + 0.9953^2][(s + 0.1281)^2 + 0.7982^2][(s + 0.1851)^2 + 0.4429^2]$

$D_8 = [(s + 0.0350)^2 + 0.9965^2][(s + 0.0997)^2 + 0.8448^2][(s + 0.1492)^2 + 0.5644^2][(s + 0.1760)^2 + 0.1982^2]$

$D_9 = (s + 0.1593)[(s + 0.0277)^2 + 0.9972^2][(s + 0.0797)^2 + 0.8769^2][(s + 0.1221)^2 + 0.6509^2]$
$\qquad \times\, [(s + 0.1497)^2 + 0.3463^2]$

$D_{10} = [(s + 0.0224)^2 + 0.9978^2][(s + 0.1013)^2 + 0.7143^2][(s + 0.0650)^2 + 0.9001^2][(s + 0.1277)^2 + 0.4586^2]$
$\qquad \times\, [(s + 0.1415)^2 + 0.1580^2]$

It is easy to show from Eq. 4.2.7 that these polynomials satisfy the recursion equation that $T_{n+1} = 2\omega T_n - T_{n-1}$ and also that $T_n^2 = \frac{1}{2}(T_{2n} + 1)$. T_n is an even function for n even and an odd function for n odd. The polynomials oscillate between ± 1 in the frequency interval $(-1, 1)$. Beyond this region $|T_n|$ increases monotonically.

ϵ is the parameter which determines the ripple magnitude. As shown in Fig. 4.2.5a, the filter responses ripple throughout the passband between 1 and $1/\sqrt{1 + \epsilon^2}$. Ripples of 0.5, 1, 2, and 3 dB have corresponding ϵ's of 0.35, 0.51, 0.76, and 1, respectively.

The poles of $T(s)T(-s)$ satisfy $[1 + \epsilon^2 T_n^2(s_k/j)] = 0$ from Eq. 4.2.6 using analytic continuation. Solving this equation for the poles $s_k = \alpha_k + j\beta_k$, it can be shown that α_k and β_k are related as $(\alpha_k/\sinh v)^2 + (\beta_k/\cosh v)^2 = 1$ where $v = (1/n)\sinh^{-1}(1/\epsilon)$. This is the standard equation for an ellipse whose semi-major axis has value $a\cosh v$,

Table 4.2.4 (Continued) Chebyshev filter poles.

---

2 dB ripple

$D_1 = s + 1.3076$

$D_2 = (s + 0.4019)^2 + 0.8133^2$

$D_3 = (s + 0.3689)\left[(s + 0.1845)^2 + 0.9231^2\right]$

$D_4 = \left[(s + 0.1049)^2 + 0.9580^2\right]\left[(s + 0.2532)^2 + 0.3968^2\right]$

$D_5 = (s + 0.2183)\left[(s + 0.0675)^2 + 0.9735^2\right]\left[(s + 0.1766)^2 + 0.6016^2\right]$

$D_6 = \left[(s + 0.0470)^2 + 0.9817^2\right]\left[(s + 0.1283)^2 + 0.7187^2\right]\left[(s + 0.1753)^2 + 0.2630^2\right]$

$D_7 = (s + 0.1553)\left[(s + 0.0346)^2 + 0.9866^2\right]\left[(s + 0.0969)^2 + 0.7912^2\right]\left[(s + 0.1400)^2 + 0.4391^2\right]$

$D_8 = \left[(s + 0.0265)^2 + 0.9898^2\right]\left[(s + 0.0754)^2 + 0.8391^2\right]\left[(s + 0.1129)^2 + 0.5607^2\right]\left[(s + 0.1332)^2 + 0.1969^2\right]$

$D_9 = (s + 0.1206)\left[(s + 0.0209)^2 + 0.9919^2\right]\left[(s + 0.0603)^2 + 0.8723^2\right]\left[(s + 0.0924)^2 + 0.6474^2\right]$
$\qquad \times \left[(s + 0.1134)^2 + 0.3444^2\right]$

$D_{10} = \left[(s + 0.0170)^2 + 0.9935^2\right]\left[(s + 0.0767)^2 + 0.7113^2\right]\left[(s + 0.0493)^2 + 0.8962^2\right]\left[(s + 0.0967)^2 + 0.4567^2\right]$
$\qquad \times \left[(s + 0.1072)^2 + 0.1574^2\right]$

---

3 dB ripple

$D_1 = s + 1.0024$

$D_2 = (s + 0.3225)^2 + 0.7772^2$

$D_3 = (s + 0.2986)\left[(s + 0.1493)^2 + 0.9038^2\right]$

$D_4 = \left[(s + 0.0852)^2 + 0.9465^2\right]\left[(s + 0.2056)^2 + 0.3921^2\right]$

$D_5 = (s + 0.1775)\left[(s + 0.0549)^2 + 0.9659^2\right]\left[(s + 0.1436)^2 + 0.5970^2\right]$

$D_6 = \left[(s + 0.0382)^2 + 0.9764^2\right]\left[(s + 0.1044)^2 + 0.7148^2\right]\left[(s + 0.1427)^2 + 0.2616^2\right]$

$D_7 = (s + 0.1265)\left[(s + 0.0281)^2 + 0.9827^2\right]\left[(s + 0.0789)^2 + 0.7881^2\right]\left[(s + 0.1140)^2 + 0.4373^2\right]$

$D_8 = \left[(s + 0.0216)^2 + 0.9868^2\right]\left[(s + 0.0614)^2 + 0.8365^2\right]\left[(s + 0.0920)^2 + 0.5590^2\right]\left[(s + 0.1055)^2 + 0.1963^2\right]$

$D_9 = (s + 0.0983)\left[(s + 0.0171)^2 + 0.9896^2\right]\left[(s + 0.0491)^2 + 0.8702^2\right]\left[(s + 0.0753)^2 + 0.6459^2\right]$
$\qquad \times \left[(s + 0.0923)^2 + 0.3437^2\right]$

$D_{10} = \left[(s + 0.0138)^2 + 0.9915^2\right]\left[(s + 0.0401)^2 + 0.8944^2\right]\left[(s + 0.0625)^2 + 0.7099^2\right]\left[(s + 0.0788)^2 + 0.4558^2\right]$
$\qquad \times \left[(s + 0.0873)^2 + 0.1570^2\right]$

---

the semi-minor axis has a value sinh v, and whose foci are located at $\omega = \pm 1$. Thus, the poles lie on an ellipse whose two axes coincide with the s-plane axes.

In practice, it is more convenient to utilize tables of Chebyshev filter poles rather than to generate them from Eq. 4.2.8. The pole locations for Chebyshev filters having in-band ripples of 0.5, 1, 2, and 3 dB are listed in Table 4.2.4 (for other ripples, see Ref. 4). These poles are normalized so that 1 rad/sec corresponds to the point where the gain is reduced by the *amount of ripple* and not 3 dB as for the Butterworth filters.

Chebyshev filters with 1 dB ripple have the magnitude characteristics shown in Fig. 4.2.5a. The gain has been drawn so it has a 3 dB cut-off frequency of unity. We see that the Chebyshev filter has a more rapid cut-off than the Butterworth filter. Its band-edge selectivity is proportional to n^2 so its transition band is narrower than that of the Butterworth filter. Of course, both filters have the same asymptotic slope for a given filter order. The price paid for this rapid cut-off rate is delay having large variations within the filter passband.

The delay characteristics of 1 dB Chebyshev filters are shown in Fig. 4.2.5b. For small in-band ripple (small ϵ), they do not differ greatly from the Butterworth characteristics. However, for large ϵ and large ripple, the delay peaking and variation do increase greatly. This is because the Chebyshev poles have a smaller damping factor ς than the Butterworth poles. Delay peaking[3] depends on $1/\varsigma$. For only 1 dB ripple, the delay variation is about 2 seconds for $n = 3$. For $n = 9$ the delay variation is 30 seconds which is significantly larger.

The unit step responses of 1, 2, and 3 dB Chebyshev filters are shown in Fig. 4.2.5c and are normalized to unity (not 3 dB) ripple frequency. Due to the higher-Q poles, the delay time and overshoot increase with filter order but not significantly with increasing ripple. Notice that the even-order filters exhibit more percentage overshoot than odd-order filters because even-order filters have a dc gain reduced by the ripple value rather than 0 dB.

The Chebyshev filter nomograph is shown in Fig. 4.2.5d and applies for any ripple. The same lowpass filter specifications and data entry format of Fig. 4.2.2 are used. It should be pointed out that all nomographs in this chapter have been constructed so they have the same M_p, M_r, Ω_r, and γ axes. Thus once the (γ, Ω_r) combinations have been obtained, they can be used on all the nomographs which speeds up their use. For example, the two data points (γ, Ω_r) of Example 4.2.1 are $(4.5, 2)$ and $(6.5, 3)$.

EXAMPLE 4.2.3 Determine the order of the Chebyshev filter required to meet the specifications of Example 4.2.1 where $M_p = 1.25$ dB, $M_{r1} = 40$ dB at $\Omega_{r1} = 2$, and $M_{r2} = 60$ dB at $\Omega_{r2} = 3$.

SOLUTION The minimum order filter results when $M_p = 1.25$ dB rather than some lesser value. Entering this data on Fig. 4.2.5d yields $n_1 \geq 5$ and $n_2 \geq 5$. Thus, a fifth-order Chebyshev filter having an in-band ripple of 1.25 dB can be used. Since we found before that an eighth-order Butterworth filter was required, we can reduce the order by three (almost half) using a Chebyshev filter. ∎

4.2.3 ELLIPTIC FILTERS

The *elliptic filter* has an equal-ripple response in both its passband and its stopband. Permitting stopband ripple allows the transition band to be narrower than that of the Chebyshev filter. Elliptic filters have the narrowest transition band of all commonly used filters. They have finite zeros of transmission. The first zero pair is located just outside of the passband and forces the response to rapidly approach zero as frequency approaches Ω_r. The asymptotic roll-offs are either -20 (n odd) or -40 dB/dec (n even). The magnitude responses of third-order elliptic filters having 0.1 dB in-band ripple and different stopband ripples are shown in Fig. 4.2.6a. The response of a third-order Butterworth filter having a 0.1 dB frequency of unity is also shown. The great improvement in reducing the width of the transition band is obvious.

The elliptic filter has a magnitude-squared function

$$|H(j\omega)|^2 = \frac{1}{1 + \epsilon^2 Z_n^2(\omega)} = \frac{D_n^2(\omega)}{D_n^2(\omega) + \epsilon^2 N_n^2(\omega)} \qquad (4.2.9)$$

where Z_n is the nth-order elliptic function and ϵ is the parameter which determines the attenuation at the cut-off frequency $\omega = 1$. Elliptic functions have equal ripple in both the passband and stopband. Z_n has the property that $Z_n(1/\omega) = 1/Z_n(\omega)$ so that its value at a frequency $1/\omega$ is equal to the reciprocal of its value at a frequency ω. Thus the poles of Z_n are the reciprocals of its zeros.

$Z_n(\omega)$ is a complicated function to deal with. In practice, design tables are used to obtain the pole and zero locations. The poles and zeros of third, fourth, and fifth-order elliptic filters having 0.28 dB passband ripple are listed in Table 4.2.5. A variety of stopband frequencies Ω_r are given. They are tabulated for integer values of θ where $\theta = \sin^{-1}(1/\Omega_r)$. As the edge of the stopband Ω_r approaches unity, θ approaches 90 degrees. Data for 1.25 dB elliptic filters of orders three, four, and five are listed in Tables 4.8.2 and 4.8.3 of Ref. 1. Complete filter data is listed in Zverev.[5]

The delay characteristics of some third-order elliptic filters having 60 dB stopband rejection are shown in Fig. 4.2.6b. Since the poles of elliptic filters almost fall on an ellipse, they are somewhat similar to Chebyshev delay curves. They exhibit less delay variation but more delay peaking. Delay increases with in-band ripple and filter order.

The unit step responses for third-order elliptic filters having 1, 2, and 3 dB in-band ripples (and angles from 10 to 80 degrees) are shown in Fig. 4.2.6c. Note that for constant in-band ripple, the step response depends on angle θ or, equivalently, the stopband frequency Ω_r. Increasing θ corresponds to decreasing Ω_r and to narrowing the transition band. The low-frequency delay and thus the delay time decreases as Ω_r decreases. Overshoot increases with filter order and decreases as Ω_r decreases. The elliptic filter step response approaches that of the corresponding Chebyshev filter as $\theta \to 0$ or $\Omega_r \to \infty$.

The elliptic filter nomograph is shown in Fig. 4.2.6d. It holds for any M_p-M_r combination. The curves are located higher on the nomograph than those for the Chebyshev filter (Fig. 4.2.5d) and much higher than those for the Butterworth filter (Fig. 4.2.1d) for a given filter order. This indicates lower elliptic filter orders.

EXAMPLE 4.2.4 Determine the order of the elliptic filter required to meet the specification of Example 4.2.1 for (γ, Ω_r) combinations of (4.5, 2) and (6.5, 3).
SOLUTION Entering the data on the elliptic filter nomograph, we see $n_1 \geq 4$ and $n_2 \geq 4$. Thus, only a fourth-order elliptic filter is required. Earlier in Examples 4.2.1 and 4.2.3, we found that eighth-order Butterworth and fifth-order Chebyshev filters were needed so this represents a further savings in filter order. It should be noted, however, that a fourth-order elliptic filter requires a pair of transmission zeros unlike the other filters. This increases filter complexity. Therefore, although one filter stage (first-order) is eliminated, there is little savings in the total complexity between the Chebyshev filter and elliptic filter. ∎

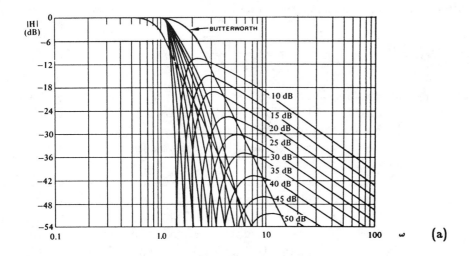

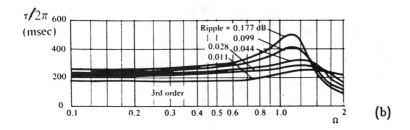

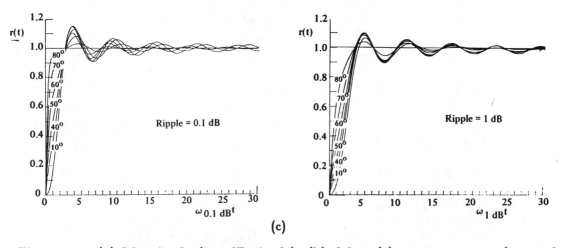

Fig. 4.2.6 (a) Magnitude (0.1 dB ripple), (b) delay, (c) step responses (0.1 and 1 dB ripples) for third-order elliptic filters and (d) general nomograph. ((a) from Feedback, vol. 1, no. 1, p. 8, KTI, Santa Clara, CA, April 1973. (c) from A.G.J. Holt, J.P. Gray, and J.K. Fidler, "Transient response of elliptic function filters," IRE Trans. Circuit Theory, vol. CT-15, pp. 71–73, March, ©1968, IEEE.)

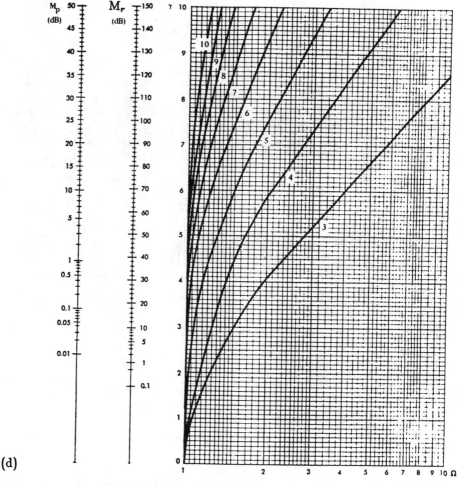

(d)

Fig. 4.2.6 (Continued) (a) Magnitude (0.1 dB ripple), (b) delay, (c) step responses (0.1 and 1 dB ripples) for third-order elliptic filters and (d) general nomograph.

EXAMPLE 4.2.5 Write the transfer function for a fourth-order elliptic lowpass filter having a maximum low-frequency gain of 10, an in-band ripple of 0.28 dB, a corner frequency of 1 KHz, and a stopband rejection of 60 dB. Draw the block diagram for the filter.

SOLUTION The design data for fourth-order filters is listed in Table 4.2.5b. Although θ or Ω_r are not given, the M_r values are listed in the third column. Thus no calculation of θ, Ω_r, or M_r is necessary. Using $M_r = 60.12$ dB, the two pole pairs and the one zero pair equal

$$p_1, \; p_1^* = -0.507 \pm j0.453, \qquad p_2, \; p_2^* = -0.194 \pm j1.05, \qquad z_1, z_1^* = \pm j3.14$$
$$(4.2.10)$$

Thus the transfer function equals

Table 4.2.5 Poles and zeros for (a) third-, (b) fourth-, and (c) fifth-order 0.28 dB elliptic filters. (Adapted from A.I. Zverev, *Handbook of Filter Synthesis*, pp. 178, 198, 220, copyright ©1967, Wiley. Reprinted by permission of John Wiley & Sons, Inc., NY.)

θ	Ω_r	M_r	α_1	α_2	β_1	β_2	β_3
11.0	5.2408	84.39	0.5002303	0.2032345	0.4391769	1.0497315	6.215646
12.0	4.8097	81.36	0.5006536	0.2026531	0.4400273	1.0497638	5.701423
13.0	4.4454	78.56	0.5011144	0.2020203	0.4409542	1.0497985	5.266618
14.0	4.1336	75.98	0.5016129	0.2013362	0.4419584	1.0498356	4.894214
15.0	3.8637	73.56	0.5021492	0.2006004	0.4430405	1.0498750	4.571732
16.0	3.6280	71.31	0.5027235	0.1998127	0.4442013	1.0499166	4.289813
17.0	3.4203	69.18	0.5033361	0.1989731	0.4454417	1.0499602	4.041300
18.0	3.2361	67.18	0.5039872	0.1980811	0.4467627	1.0500058	3.820626
19.0	3.0716	65.28	0.5046770	0.1971367	0.4481651	1.0500531	3.623399
20.0	2.9238	63.48	0.5054058	0.1961394	0.4496500	1.0501020	3.446101
21.0	2.7904	61.76	0.5061739	0.1950891	0.4512185	1.0501525	3.285888
22.0	2.6695	60.12	0.5069816	0.1939855	0.4528719	1.0502041	3.140431
23.0	2.5593	58.56	0.5078293	0.1928282	0.4546113	1.0502569	3.007807
24.0	2.4586	57.05	0.5087193	0.1916169	0.4564382	1.0503105	2.886413
25.0	2.3662	55.61	0.5096459	0.1903513	0.4585539	1.0503648	2.774903
26.0	2.2812	54.22	0.5106156	0.1890311	0.4603599	1.0504194	2.672139
27.0	2.2027	52.88	0.5116269	0.1876558	0.4624579	1.0504742	2.577149
28.0	2.1301	51.59	0.5126801	0.1862251	0.4646494	1.0505287	2.489103
29.0	2.0627	50.34	0.5137758	0.1847387	0.4669364	1.0505828	2.407283
30.0	2.0000	49.13	0.5149145	0.1831960	0.4693206	1.0506361	2.331070
31.0	1.9416	47.95	0.5160967	0.1815967	0.4718041	1.0506882	2.259921
32.0	1.8871	46.82	0.5173230	0.1799404	0.4743888	1.0507386	2.193363
33.0	1.8361	45.71	0.5185941	0.1782266	0.4770771	1.0507871	2.130982
34.0	1.7883	44.64	0.5199106	0.1764548	0.4798712	1.0508331	2.072410
35.0	1.7434	43.59	0.5212732	0.1746246	0.4827735	1.0508761	2.017322
36.0	1.7013	42.57	0.5226827	0.1727354	0.4857867	1.0509157	1.965429
37.0	1.6616	41.58	0.5241399	0.1707868	0.4889135	1.0509512	1.916475
38.0	1.6243	40.61	0.5256457	0.1687783	0.4921566	1.0509820	1.870229
39.0	1.5890	39.66	0.5272009	0.1667093	0.4955192	1.0510075	1.826685
40.0	1.5557	38.73	0.5288066	0.1645794	0.4990045	1.0510269	1.785057
41.0	1.5243	37.83	0.5304639	0.1623879	0.5026156	1.0510396	1.745777
42.0	1.4945	36.94	0.5321738	0.1601344	0.5063503	1.0510446	1.708493
43.0	1.4663	36.07	0.5339377	0.1578183	0.5102303	1.0510410	1.673069
44.0	1.4396	35.22	0.5357567	0.1554390	0.5142541	1.0510280	1.639380
45.0	1.4142	34.38	0.5376323	0.1529960	0.5183939	1.0510044	1.607311
46.0	1.3902	33.56	0.5395661	0.1504888	0.5226922	1.0509693	1.576760
47.0	1.3673	32.76	0.5415597	0.1479168	0.5271410	1.0509213	1.547632
48.0	1.3456	31.96	0.5436148	0.1452795	0.5317451	1.0508592	1.519839
49.0	1.3250	31.18	0.5457334	0.1425764	0.5365099	1.0507816	1.493303
50.0	1.3054	30.42	0.5479175	0.1398070	0.5414409	1.0506869	1.467949
51.0	1.2868	29.66	0.5501695	0.1369709	0.5465439	1.0505737	1.443712
52.0	1.2692	28.92	0.5524917	0.1340676	0.5518253	1.0504401	1.420528
53.0	1.2521	28.18	0.5548869	0.1310967	0.5572917	1.0502844	1.398341
54.0	1.2361	27.46	0.5573580	0.1280579	0.5629501	1.0501044	1.377098
55.0	1.2208	26.75	0.5599082	0.1249510	0.5688081	1.0498980	1.356750
56.0	1.2062	26.04	0.5625410	0.1217757	0.5748737	1.0496628	1.337251
57.0	1.1924	25.35	0.5652603	0.1185321	0.5811554	1.0493964	1.318559
58.0	1.1792	24.66	0.5680703	0.1152201	0.5876625	1.0490961	1.300636
59.0	1.1666	23.98	0.5709758	0.1118399	0.5944047	1.0487589	1.283446
60.0	1.1547	23.30	0.5739819	0.1083918	0.6013925	1.0483817	1.266954

(b) Pole-zero diagram: zeros at $\pm j\beta_3$, $\pm j\beta_2$, $\pm j\beta_1$ on the $j\omega$ axis; poles at $-\alpha_1$, $-\alpha_2$ on the σ axis.

(a) Pole-zero diagram: zeros at $\pm j\beta_2$, $\pm j\beta_1$ on the $j\omega$ axis; poles at $-\alpha_1$, $-\alpha_0$ on the σ axis.

θ	Ω_r	M_r	α_0	α_1	β_1	β_2
6	9.5668	71.10	0.74651	0.36927	1.07981	11.0392
7	8.2055	67.07	0.74767	0.36840	1.08007	9.4661
8	7.1853	63.58	0.74900	0.36740	1.08037	8.2868
9	6.3925	60.50	0.75052	0.36626	1.08071	7.3700
10	5.7588	57.74	0.75223	0.36499	1.08109	6.6370
11	5.2408	55.24	0.75412	0.36359	1.08150	6.0377
12	4.8097	52.96	0.75620	0.36205	1.08195	5.5386
13	4.4454	50.86	0.75848	0.36038	1.08243	5.1166
14	4.1336	48.91	0.76095	0.35858	1.08294	4.7552
15	3.8637	47.09	0.76362	0.35664	1.08348	4.4423
16	3.6280	45.39	0.76650	0.35456	1.08406	4.1688
17	3.4203	43.79	0.76958	0.35235	1.08466	3.9277
18	3.2361	42.28	0.77288	0.35000	1.08528	3.7137
19	3.0716	40.84	0.77640	0.34752	1.08593	3.5224
20	2.9238	39.48	0.78014	0.34490	1.08660	3.3505
21	2.7904	38.18	0.78410	0.34215	1.08730	3.1951
22	2.6695	36.94	0.78830	0.33926	1.08800	3.0541
23	2.5593	35.75	0.79275	0.33624	1.08873	2.9256
24	2.4586	34.61	0.79744	0.33308	1.08947	2.8079
25	2.3662	33.51	0.80238	0.32979	1.09021	2.6999
26	2.2904	32.45	0.80759	0.32635	1.09097	2.6003
27	2.2027	31.43	0.81307	0.32279	1.09172	2.5083
28	2.1301	30.44	0.81884	0.31909	1.09248	2.4231
29	2.0627	29.49	0.82489	0.31525	1.09324	2.3438
30	2.0000	28.57	0.83124	0.31128	1.09399	2.2701
31	1.9416	27.67	0.83791	0.30717	1.09473	2.2012
32	1.8871	26.80	0.84490	0.30293	1.09546	2.1368
33	1.8361	25.95	0.85223	0.29855	1.09617	2.0765
34	1.7883	25.12	0.85991	0.29406	1.09686	2.0199
35	1.7434	24.32	0.86795	0.28941	1.09752	1.9666
36	1.7013	23.54	0.87637	0.28464	1.09816	1.9165
37	1.6616	22.77	0.88519	0.27974	1.09876	1.8692
38	1.6243	22.02	0.89442	0.27471	1.09932	1.8245
39	1.5890	21.29	0.90409	0.26956	1.09984	1.7823
40	1.5557	20.58	0.91421	0.26428	1.10031	1.7423
41	1.5243	19.88	0.92480	0.25887	1.10072	1.7044
42	1.4945	19.19	0.93589	0.25335	1.10107	1.6684
43	1.4663	18.52	0.94750	0.24771	1.10136	1.6343
44	1.4396	17.86	0.95966	0.24195	1.10158	1.6018
45	1.4142	17.22	0.97240	0.23608	1.10173	1.5710
46	1.3902	16.58	0.98574	0.23010	1.10179	1.5415
47	1.3673	15.96	0.99972	0.22402	1.10176	1.5135
48	1.3456	15.35	1.01436	0.21784	1.10164	1.4868
49	1.3250	14.75	1.02972	0.21157	1.10142	1.4613
50	1.3054	14.16	1.04583	0.20520	1.10109	1.4369
51	1.2868	13.58	1.06272	0.19875	1.10065	1.4137
52	1.2692	13.01	1.08045	0.19223	1.10009	1.3914
53	1.2521	12.45	1.09906	0.18563	1.09941	1.3702
54	1.2361	11.90	1.11860	0.17897	1.09860	1.3498
55	1.2208	11.37	1.13914	0.17226	1.09766	1.3303
56	1.2062	10.84	1.16072	0.16551	1.09657	1.3117
57	1.1924	10.32	1.18341	0.15872	1.09534	1.2938
58	1.1792	9.81	1.20728	0.15190	1.09395	1.2767
59	1.1666	9.31	1.23239	0.14508	1.09242	1.2603
60	1.1547	8.83	1.25884	0.13825	1.09073	1.2446

Table 4.2.5 (Continued) Poles and zeros for (a) third-, (b) fourth-, and (c) fifth-order 0.28 dB elliptic filters. (Adapted from A.I. Zverev, loc. cit., ©1967, Wiley.)

θ	Ω_r	M_r	a_0	a_1	a_2	β_1	β_2	β_3	β_4
11.0	5.2408	107.94	0.42916	0.34307	0.12854	0.6441	1.0330	5.5057	8.3625
12.0	4.8097	104.14	0.43006	0.34299	0.12804	0.6452	1.0330	5.0520	8.1241
13.0	4.4454	100.63	0.43104	0.34291	0.12749	0.6463	1.0329	4.6684	7.4993
14.0	4.1336	97.39	0.43210	0.34282	0.12690	0.6476	1.0329	4.3401	6.9638
15.0	3.8637	94.36	0.43325	0.34272	0.12627	0.6490	1.0328	4.0559	6.4997
16.0	3.6280	91.52	0.43448	0.34261	0.12560	0.6504	1.0328	3.8076	6.0936
17.0	3.4203	88.85	0.43580	0.34248	0.12489	0.6520	1.0328	3.5888	5.7353
18.0	3.2361	86.33	0.43720	0.34235	0.12413	0.6536	1.0327	3.3946	5.4168
19.0	3.0716	83.94	0.43870	0.34220	0.12333	0.6554	1.0326	3.2212	5.1318
20.0	2.9238	81.66	0.44029	0.34204	0.12248	0.6572	1.0326	3.0654	4.8753
21.0	2.7904	79.50	0.44197	0.34187	0.12160	0.6592	1.0325	2.9246	4.6433
22.0	2.6695	77.43	0.44375	0.34168	0.12067	0.6613	1.0324	2.7970	4.4323
23.0	2.5593	75.45	0.44562	0.34147	0.11970	0.6634	1.0324	2.6807	4.2397
24.0	2.4586	73.54	0.44759	0.34125	0.11869	0.6657	1.0323	2.5743	4.0631
25.0	2.3662	71.71	0.44967	0.34101	0.11763	0.6681	1.0322	2.4767	3.9007
26.0	2.2812	69.95	0.45185	0.34074	0.11654	0.6706	1.0321	2.3868	3.7507
27.0	2.2027	68.25	0.45413	0.34046	0.11540	0.6732	1.0320	2.3038	3.6119
28.0	2.1301	66.60	0.45653	0.34015	0.11422	0.6759	1.0319	2.2270	3.4829
29.0	2.0627	65.01	0.45904	0.33981	0.11301	0.6787	1.0318	2.1556	3.3629
30.0	2.0000	63.47	0.46166	0.33945	0.11175	0.6817	1.0317	2.0892	3.2508
31.0	1.9416	61.97	0.46440	0.33906	0.11045	0.6847	1.0315	2.0274	3.1460
32.0	1.8871	60.51	0.46727	0.33864	0.10911	0.6879	1.0314	1.9695	3.0476
33.0	1.8361	59.10	0.47027	0.33819	0.10773	0.6912	1.0313	1.9154	2.9553
34.0	1.7883	57.72	0.47339	0.33770	0.10631	0.6946	1.0311	1.8646	2.8683
35.0	1.7434	56.37	0.47666	0.33718	0.10485	0.6981	1.0310	1.8170	2.7864
36.0	1.7013	55.06	0.48006	0.33662	0.10336	0.7018	1.0308	1.7722	2.7089
37.0	1.6616	53.78	0.48361	0.33601	0.10182	0.7056	1.0306	1.7299	2.6356
38.0	1.6243	52.53	0.48731	0.33535	0.10025	0.7095	1.0304	1.6901	2.5662
39.0	1.5890	51.30	0.49117	0.33465	0.09864	0.7135	1.0302	1.6525	2.5003
40.0	1.5557	50.10	0.49519	0.33390	0.09699	0.7177	1.0300	1.6170	2.4377
41.0	1.5243	48.92	0.49939	0.33309	0.09531	0.7220	1.0298	1.5833	2.3781
42.0	1.4945	47.76	0.50376	0.33222	0.09359	0.7265	1.0296	1.5515	2.3213
43.0	1.4663	46.63	0.50832	0.33129	0.09183	0.7311	1.0294	1.5213	2.2672
44.0	1.4396	45.51	0.51307	0.33029	0.09004	0.7358	1.0291	1.4926	2.2154
45.0	1.4142	44.42	0.51803	0.32922	0.08822	0.7407	1.0288	1.4654	2.1660
46.0	1.3902	43.34	0.52320	0.32807	0.08636	0.7457	1.0286	1.4396	2.1187
47.0	1.3673	42.27	0.52859	0.32684	0.08446	0.7509	1.0283	1.4150	2.0733
48.0	1.3456	41.23	0.53422	0.32551	0.08254	0.7562	1.0280	1.3916	2.0299
49.0	1.3250	40.19	0.54010	0.32410	0.08058	0.7617	1.0277	1.3693	1.9881
50.0	1.3054	39.17	0.54624	0.32258	0.07860	0.7673	1.0273	1.3481	1.9480
51.0	1.2868	38.16	0.55266	0.32096	0.07658	0.7731	1.0270	1.3279	1.9095
52.0	1.2690	37.17	0.55937	0.31922	0.07453	0.7791	1.0266	1.3087	1.8724
53.0	1.2521	36.18	0.56639	0.31736	0.07246	0.7852	1.0262	1.2903	1.8366
54.0	1.2361	35.21	0.57374	0.31537	0.07036	0.7915	1.0259	1.2728	1.8021
55.0	1.2208	34.24	0.58144	0.31323	0.06823	0.7980	1.0254	1.2561	1.7689
56.0	1.2062	33.29	0.58952	0.31095	0.06608	0.8046	1.0250	1.2402	1.7368
57.0	1.1924	32.34	0.59799	0.30850	0.06390	0.8114	1.0246	1.2250	1.7057
58.0	1.1792	31.40	0.60690	0.30588	0.06170	0.8184	1.0241	1.2104	1.6757
59.0	1.1666	30.46	0.61626	0.30308	0.05948	0.8255	1.0236	1.1966	1.6467
60.0	1.1547	29.53	0.62612	0.30008	0.05725	0.8328	1.0231	1.1834	1.6185

(c)

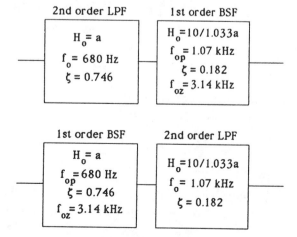

Fig. 4.2.7 Block diagrams of fourth-order elliptic filter of Example 4.2.4.

$$H(s) = \frac{10}{1.033} \frac{[(0.68)(1.07)/3.14]^2(s_n^2 + 3.14^2)}{[(s_n + 0.507)^2 + 0.453^2][(s_n + 0.194)^2 + 1.05^2]} \qquad (4.2.11)$$

where $s_n = s/2\pi(1\text{ KHz})$. Note the normalization constant $(10/1.033)$ is selected so that $H(0) = -0.28\text{ dB} = 1/1.033$. The filter can be realized as the cascade of a lowpass and bandstop filter as shown in Fig. 4.2.7. Two realizations are possible depending upon which pole pair is grouped with the zero pair.   ∎

4.2.4 BESSEL FILTERS

Butterworth filters have fairly good magnitude and delay characteristics. Chebyshev and elliptic filters have more magnitude selectivity but their delay characteristics exhibit increased ripple and peaking. Eq. 4.1.3 shows that the ideal passband delay should be constant or the passband phase should be linear (and zero at dc). Any deviation from constant delay is called *delay distortion* and any deviation from linear phase is called *phase distortion*. Thus Butterworth filters have moderate delay and phase distortion while Chebyshev and elliptic filters have large distortions.

Now filters will be considered that have small delay and phase distortions. Such filters are said to be *linear phase, maximally flat delay*, or *constant delay* types. The transfer function of an ideal one second delay filter equals

$$H(s) = e^{-s\tau_o}\Big|_{\tau_o = 1} = \frac{1}{e^s} = \frac{1}{\sinh s + \cosh s} \qquad (4.2.12)$$

The ac steady-state transfer function of any filter can be expressed as

$$H(j\omega) = \frac{a(\omega^2) + j\omega b(\omega^2)}{c(\omega^2) + j\omega d(\omega^2)} \qquad (4.2.13)$$

where $a, b, c,$ and d are rational polynomials of ω^2. The phase of $H(j\omega)$ equals

$$\begin{aligned}
\theta(j\omega) = \arg H(j\omega) &= \tan^{-1}\left(\frac{\omega b}{a}\right) - \tan^{-1}\left(\frac{\omega d}{c}\right) \\
&= \omega\left[\left(\frac{b}{a}\right) - \left(\frac{d}{c}\right)\right] - \frac{\omega^3}{3}\left[\left(\frac{b}{a}\right)^3 - \left(\frac{d}{c}\right)^3\right] + \frac{\omega^5}{5}\left[\left(\frac{b}{a}\right)^5 - \left(\frac{d}{c}\right)^5\right] - \ldots \quad (4.2.14) \\
&= -f_1\omega - f_3\omega^3 - f_5\omega^5 - \ldots
\end{aligned}$$

where $\tan^{-1}(x)$ is expanded in a Maclaurin series as $(x - x^3/3 + x^5/5 - \ldots)$ and like powers of ω are collected. Therefore the delay equals

$$\tau(j\omega) = -\frac{d\theta(j\omega)}{d\omega} = f_1 + 3f_3\omega^2 + 5f_5\omega^4 + \ldots \qquad (4.2.15)$$

Linear phase or maximally flat delay (MFD) is obtained by setting as many consecutive derivatives of delay equal to zero as possible excluding the first derivative. Thus $f_1 \neq 0$ and $f_3 = f_5 = \ldots = 0$ for a MFD response.

The *Bessel filter* is the class of linear phase (or MFD) filter having all of its zeros at infinity. It is analogous to the Butterworth MFM filter. Its transfer function equals

$$H(s) = \frac{b_0}{b_0 + b_1 s + \ldots + b_n s^n} = \frac{P_n(0)}{P_n(s)} \qquad (4.2.16)$$

The $P_n(s)$ are polynomials which are obtained from Bessel polynomials $B_n(s)$ as $P_n(s) = s^n B_n(1/s)$. The poles of the Bessel filter are listed in Table 4.2.6 (for orders through 31, see Ref. 6). These poles have been frequency normalized to produce unity 3 dB bandwidth. The approximate Bessel filter poles are located on the Butterworth unit circle. Their imaginary parts are separated by $2/n$ of the diameter beginning and ending with half this value.

The magnitude characteristics are shown in Fig. 4.2.8a. The filter band-edge selectivity equals about 0.491 and is independent of order. It is less than that of even a second-order Butterworth filter. Bessel filters have very poor selectivity compared with the other filters that have been discussed.

Although the Bessel filter has poor selectivity, it has excellent delay characteristics (MFD) as seen in Fig. 4.2.8b. The dc delay is approximately

$$\tau(0) \cong \sqrt{(2n-1)\ln 2} \qquad (4.2.17)$$

The unit step response of the Bessel filter is shown in Fig. 4.2.8c. The rise time is about 2.2 seconds but the delay time increases with filter order according to Eq. 4.2.17. The overshoot is less that 1%.

The order of a Bessel filter can be easily determined from the nomograph of Fig. 4.2.8d. The curves are located lower on the nomograph than those of even the Butterworth filter which confirms that the Bessel filter has very poor selectivity.

EXAMPLE 4.2.6 Determine the order of a Bessel filter to meet the specification of Example 4.2.1 where (γ, Ω_r) was $(4.5, 2)$ and $(6.5, 3)$.
SOLUTION Entering the data on the Bessel filter nomograph, we see that it is impossible to meet this specification. ∎

EXAMPLE 4.2.7 A lowpass Bessel filter must be designed to meet the following specifications: (1) low-frequency gain = 20 dB, (2) 3 dB frequency = 1 KHz, (3) 20 dB frequency = 2.7 KHz, (4) 60 dB frequency = 10 KHz, and (5) minimum high-frequency roll-off = −60 dB/dec. Determine the required order of the filter. Write its transfer function and draw it in block diagram form.
SOLUTION From the roll-off requirement, the order must be at least equal to $n \geq -60/-20 = 3$. From the nomograph, we find $n_1 \geq 4$ and $n_2 \geq 4$. The transfer function of the filter must therefore equal

$$H(s) = \frac{10(1.431^2)(1.604^2)}{[s_n^2 + 2(0.958)(1.431)s_n + 1.431^2][s_n^2 + 2(0.621)(1.604)s_n + 1.604^2]} \qquad (4.2.18)$$

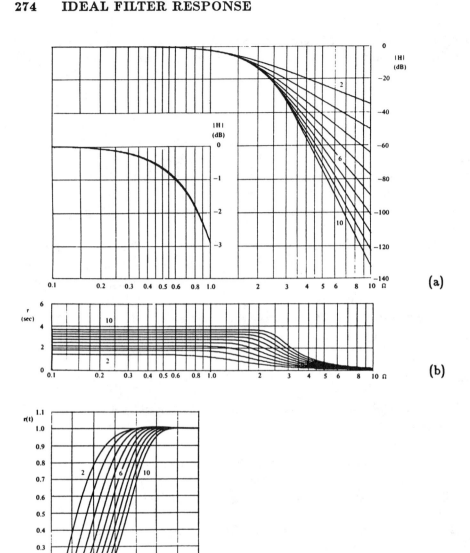

(a)

(b)

(c)

Fig. 4.2.8 (a) Magnitude, (b) delay, (c) step responses, and (d) nomograph of Bessel filter. ((a)–(c) from A.D. Taylor, loc. cit. and (d) from J.N. Hallberg, loc. cit.)

Fig. 4.2.9 Block diagram of Bessel filter of Example 4.2.6.

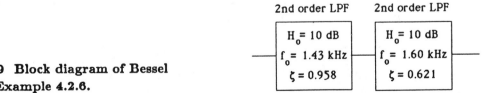

2nd order LPF	2nd order LPF
$H_0 = 10$ dB	$H_0 = 10$ dB
$f_0 = 1.43$ kHz	$f_0 = 1.60$ kHz
$\zeta = 0.958$	$\zeta = 0.621$

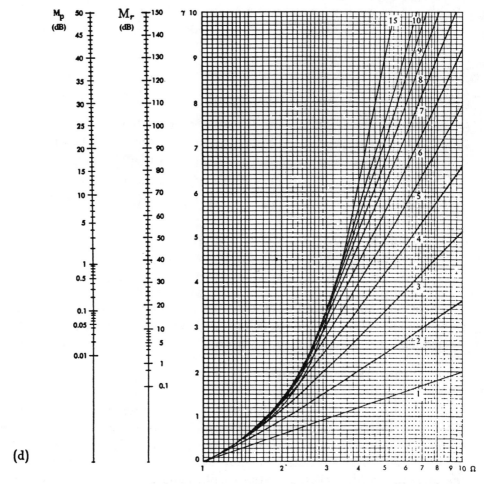

Fig. 4.2.8 (Continued) (a) Magnitude, (b) delay, (c) step responses, and (d) general nomograph of Bessel filter.

where $s_n = s/2\pi(1 \text{ KHz})$ from Table 4.2.6. The block diagram of the filter is easily drawn in Fig. 4.2.9. ∎

4.2.5 CLASSICAL FILTERS IN RETROSPECT

We have seen that there are a variety of classical filter responses which can be utilized in design (for others see Ref. 1). These filters can be loosely grouped as: (1) high selectivity filters (elliptic and Chebyshev), (2) transitional or compromise type filters (Butterworth), and (3) constant delay or linear phase filters (Bessel).

For applications which require high selectivity and are phase insensitive, the Chebyshev and elliptic filters can be utilized. For applications which are phase sensitive but do not require high selectivity, Bessel filters can be utilized. Phase sensitive applications requiring high selectivity require compromises and Butterworth filters are often used.

Table 4.2.6 Bessel filter poles.

$D_1 = s + 1$

$D_2 = (s + 1.1018)^2 + 0.6362^2$

$D_3 = (s + 1.3243)\left[(s + 1.0486)^2 + 1.0006^2\right]$

$D_4 = \left[(s + 1.3709)^2 + 0.4107^2\right]\left[(s + 0.9958)^2 + 1.2578^2\right]$

$D_5 = (s + 1.5040)\left[(s + 1.3825)^2 + 0.7186^2\right]\left[(s + 0.9588)^2 + 1.4727^2\right]$

$D_6 = \left[(s + 1.5734)^2 + 0.3211^2\right]\left[(s + 1.3835)^2 + 0.9725^2\right]\left[(s + 0.9318)^2 + 1.6639^2\right]$

$D_7 = (s + 1.6871)\left[(s + 1.6147)^2 + 0.5903^2\right]\left[(s + 1.3811)^2 + 1.1936^2\right]\left[(s + 0.9113)^2 + 1.8396^2\right]$

$D_8 = \left[(s + 1.7596)^2 + 0.2727^2\right]\left[(s + 1.6389)^2 + 0.8237^2\right]\left[(s + 1.3756)^2 + 1.3900^2\right]\left[(s + 0.8939)^2 + 2.0008^2\right]$

$D_9 = (s + 1.8570)\left[(s + 1.8076)^2 + 0.5123^2\right]\left[(s + 1.6527)^2 + 1.0317^2\right]\left[(s + 1.3679)^2 + 1.5680^2\right]$
$\qquad \times \left[(s + 0.8785)^2 + 2.1503^2\right]$

$D_{10} = \left[(s + 1.9301)^2 + 0.2425^2\right]\left[(s + 1.8446)^2 + 0.7284^2\right]\left[(s + 1.6640)^2 + 1.2229^2\right]\left[(s + 1.3624)^2 + 1.7359^2\right]$
$\qquad \times \left[(s + 0.8670)^2 + 2.2956^2\right]$

The final selection is based upon experience and always involves certain compromises and trade-offs. Usually these trade-offs are between high filter selectivity, low delay distortion, and well-behaved step response. Now we shall briefly review optimum filters.

4.3 OPTIMUM FILTERS

Optimum filters are filters which optimize some criterion under certain constraints.[2] Usually the criterion requires the minimization of a performance or error index E. The *performance index* provides a quantitative measure of the performance of the optimum filter. It is chosen to emphasize the most desirable filter characteristics and deemphasize the undesirable filter characteristics. Optimality is application dependent and chosen largely on qualitative judgements. Performance indices involving integrals are the most universally utilized and have been given standard names.

In time domain optimization, the common performance index integrals are

$$IAE = \int_0^T |e(t)|dt, \qquad ITSE = \int_0^T te^2(t)dt$$

$$ISE = \int_0^T e^2(t)dt, \qquad ISTSE = \int_0^T t^2 e^2(t)dt \qquad (4.3.1)$$

$$ITAE = \int_0^T t|e(t)|dt, \qquad IEXSE = \int_0^T e^{2kt}e^2(t)dt$$

where $e(t)$ is the error between the desired and the actual filter responses and T is some arbitrary time between zero and infinity. These performance indices are the integrals of absolute error (*IAE*), squared-error (*ISE*), time-weighted absolute error (*ITAE*), time-weighted squared-error (*ITSE*), time-squared-weighted squared-error (*ISTSE*), and exponentially-weighted squared-error (*IEXSE*). Any time limits can be used.

The more general performance index, of which the *IAE, ISE, ITAE, ITSE, ISTSE,* and *IEXSE* indices are subclasses, is the least pth approximation having

$$E = \int_a^b w(t)e^p(t)dt \tag{4.3.2}$$

$w(t)$ is some arbitrary weighting function, p is the power to which the error is raised, and (a,b) represents the time interval over which the error is monitored. When absolute error is utilized, $e^p(t)$ is replaced by $|e(t)|^p = \sqrt{e^{2p}(t)}$.

The performance index is a function of the error. The error $e(t)$ is equal to the difference between the desired filter response $d(t)$ and the actual filter response $c(t)$. For example, if a unit step response is desired, then $e(t) = U_{-1}(t) - c(t)$ where $c(t)$ is the step response of the filter. A typical $e(t)$ response is shown in Fig. 4.3.1. The following observations can be made about the various performance indices:

1. *IAE:* Error is equally weighted over the interval from 0 to T. Response errors tend to be distributed over the interval.

2. *ISE:* Although error-squared is equally-weighted over the entire interval, large errors are emphasized (or overrated depending upon the requirements) and small errors tend to be neglected (or underrated). The behavior of the rising edge of the transient essentially determines the optimum filter. Overshoots and ringing tend to be overlooked so that large overshoot and long settling time often occur.

3. *ITAE:* Error is weighted by time over the entire interval. Thus, possibly large initial errors are neglected and the later oscillatory behavior of the response is emphasized. The large initial errors are overlooked so that the optimum filter is relatively independent of the initial response. Large overshoot but short settling time often result.

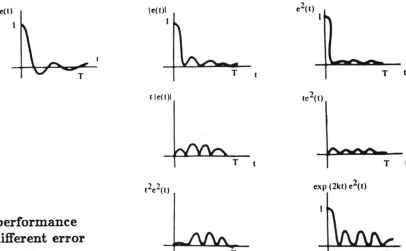

Fig. 4.3.1 Various performance integrals involving different error functions.

4. *ITSE:* Error-squared is weighted by t over the time interval. This tends to further deemphasize the initial response errors and emphasize the latter oscillatory behavior. Large overshoot, short settling time, and low-level ringing often result.

5. *ISTSE:* Error-squared is weighted by t^2 over the time interval. This further deemphasizes the initial response errors and greatly emphasizes the later oscillatory behavior. Large overshoot, low-level ringing, and short settling time often result.

6. *IEXSE:* Error-squared is exponentially-weighted over the time interval. The first two terms in the *IEXSE* series are *ISE* and *ISTSE.* Large errors are emphasized and small errors are neglected over the entire interval. Large overshoots and short settling times often result.

Optimization in the frequency domain is analogous to that in the time domain. The same performance indices are generally used:

$$IAE = \int_0^\Omega f[E(j\omega)]d\omega, \qquad IFSE = \int_0^\Omega \omega f[E^2(j\omega)]d\omega$$

$$ISE = \int_0^\Omega f[E^2(j\omega)]d\omega, \qquad ISFSE = \int_0^\Omega \omega^2 f[E^2(j\omega)]d\omega \qquad (4.3.3)$$

$$IFAE = \int_0^\Omega \omega f[E(j\omega)]d\omega, \qquad IEXSE = \int_0^\Omega e^{2k\omega} f[E^2(j\omega)]d\omega$$

$f[E(j\omega)]$ is some function of the error between the desired and the actual filter frequency responses. Ω is some arbitrary frequency between zero and infinity (any frequency limits can be used). The same error interpretations discussed previously apply here. Since $E(j\omega)$ is generally a complex number having magnitude and phase as $E(j\omega) = |E(j\omega)|exp[j \arg E(j\omega)]$, the various indices must be carefully applied.

IAE, ISE, IFAE, IFSE, ISFSE, and *IEXSE* indices are subclasses of the least *p*th approximation having

$$E = \int_a^b W_1(j\omega)|E(j\omega)|^p d\omega + \int_a^b W_2(j\omega) \arg E(j\omega)d\omega + \int_a^b W_3(j\omega) \frac{d \arg E(j\omega)}{d\omega} d\omega$$

$$(4.3.4)$$

Here $W_1(j\omega)$, $W_2(j\omega)$, and $W_3(j\omega)$ are arbitrary weighting functions chosen to emphasize or deemphasize portions of the magnitude, phase, and delay responses of the filter, respectively.

Filters can be simultaneously optimized in both the frequency and time domains. This requires combining a frequency domain performance index (Eq. 4.3.4) and a time domain performance index (Eq. 4.3.2) such as simple summing. However, because of the increased computational complexity, optimization in both domains is not often used.

Let us illustrate the design method for optimizing an all-pole lowpass filter having a transfer function

$$H_n(s) = \frac{1}{1 + b_1 s + \ldots + b_{n-1} s^{n-1} + s^n} \qquad (4.3.5)$$

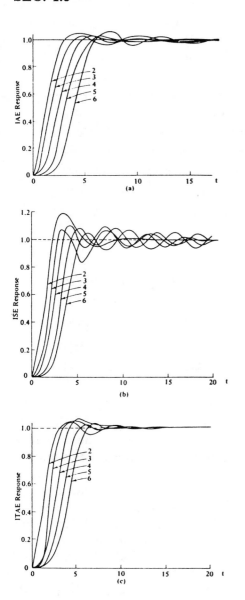

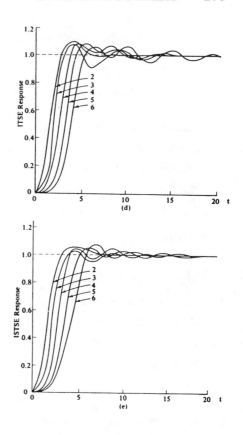

Fig. 4.3.2 Step responses of optimum lowpass filter using (a) IAE, (b) ISE, (c) ITAE, (d) ITSE, and (e) ISTSE performance indices. (From V.W. Eveleigh, *Adaptive Control and Optimization*, Chap. 2, McGraw-Hill, NY, 1967.)

whose dc gain equals unity. Suppose that a unit step signal is input and the desired output also equals $U_{-1}(t)$. The step response of the filter equals

$$c_n(t) = h_n(t) * U_{-1}(t) = \mathcal{L}^{-1}\left[H_n(s)/s\right] \qquad (4.3.6)$$

The resulting error is $e(t) = U_{-1}(t) - c_n(t)$. $e(t)$ can then be substituted into the desired performance index of Eq. 4.3.1. The optimum filter results by minimizing the performance index by properly choosing the b_i coefficients, under the constraint that $H_n(s)$ is realizable, i.e., $h_n(t) = \mathcal{L}^{-1}[H_n(s)] = 0$ for $t < 0$. In general, closed-form solutions do not exist so iterative computer techniques are used. The coefficients so

Table **4.3.1** Optimum filter types and gain functions.
(See pages indicated in Ref. 2.)

Filter Type	Comments
Constant Magnitude Types	
ITAE Step Response	Close to Butterworth filter (pp. 288–289)
ISE Step Response	Close to ideal lowpass filter step response (pp. 289–290)
ISE Gain/Impulse Response	Spurious magnitude response (pp. 290–291)
ISE Gain/Phase Response	Magnitude ripples (pp. 295–297)
Minimum Rise Time Response	Fastest step response (pp. 297–300)
Constant Delay Types	
ISE Delay Response	Good overall response (pp. 292–294)
Minimax Delay Response	Magnitude ripples (pp. 300–303)
Minimax Phase Response	Good overall response (pp. 303–307)
Valand's MFD Response	Excessive delay peaking (pp. 307–309)
Budak's MFD Response	Preringing in step response (pp. 309–310)
Pade's MFD Response	Constant delay (pp. 310–311)

obtained are usually tabulated for ready reference. The optimum filters which result using the *IAE*, *ISE*, *ITAE*, *ITSE*, and *ISTSE* performance indices have the responses shown in Fig. 4.3.2. Comparing these responses with one another verifies our comments concerning the peculiarities of the different indices.

To further interpret these performance indices, consider a second-order lowpass filter having a gain function $H_2(s) = 1/(s^2 + 2\varsigma s + 1)$. Various performance indices are shown in Fig. 4.3.3 for a unit step input which is also the desired output. The optimum filter is the one whose ς minimizes the performance index. The optimum filters which result have different damping factors. $\varsigma \cong 0.7$ for *IAE* and *ITAE*, and $\varsigma \cong 0.6$ for *ISE* and *ITSE*. Also shown are $\varsigma = 0$ for *IE* and $\varsigma \cong 0.5$ for *ITE*.

The minimum values of the various performance indices are not significant. Although the minimum *ITSE* index is less than the other indices, this does not imply that *ITSE* filters are better. Filter comparisons must be made using the same performance index with different $h_n(t)$ or $H_n(s)$. However the selectivities of the various performance indices can be compared. The *selectivity* (the value of $d^2 I/d\varsigma^2$ at $\varsigma = \varsigma_{opt}$) is a measure of the degree of concavity of the performance index. From Fig. 4.3.3, *ITAE* is most selective and *ISE* is the least selective.

Some of the most interesting optimum filters are listed in Table 4.3.1 where they have been grouped into constant magnitude types and constant delay types. Their characteristics are compared in Table 4.3.2.

Caution must be exercised when choosing an optimum filter because of certain peculiarities which these filters usually exhibit. These peculiarities often become evident

Table 4.3.2 Comparison of optimum filters.[2]

Filter Type	In-band Gain	Stopband Gain	Band-edge Selectivity & NBW	Transition Bandwidth	In-band Delay	Pole Q	Step Response Overshoot	Step Response Rise Time	Step Response Delay
ITAE Step Response	No ripple	Mono. dec.	Medium	Medium	No ripple & increases	Medium	Low	2.2	Medium
ISE Step Response	Ripples	Mono. dec.	Medium	High	Almost constant	High	Medium	2.2	Low
ISE Gain/Impulse Response	Dec. with peaking	Mono. dec.	High	High	Ripples with peaking	High	Medium	2.2	Low
ISE Delay Response	Mono. dec.	Mono. dec.	Low	High	Constant with peaking	Low	Low	2.2	Low
ISE Gain/Phase Response	Dec. with peaking	Mono. dec.	Medium	Medium	Ripples with peaking	Low	Low	2.2	Low
Minimum Rise Time Response	Mono. dec.	Equiripple	Medium	Low	Ripples & increases	High	Low	Lowest	Low
Minimax Delay Response	Ripples	Mono. dec.	Low	High	Equiripple (except $\tau(0)$)	Medium	Low	2.2	Low
Minimax Phase Response	Mono. dec.	Mono. dec.	Low	High	Almost equiripple	Medium	Low	2.2	Low
Valand's MFD Response	Mono. dec.	Mono. dec.	Low	High	Ripples with peaking	Medium	Medium	2.2	Medium
Budak's MFD Response	Mono. dec.	Mono. dec.	Low	High	Mono. dec.	Medium	Negligible	2.2	Low
Padé's MFD Response (for LPF)	Mono. dec.	Mono. dec. or constant	Low	High	Mono. dec.	Medium	Low	2.2	Low

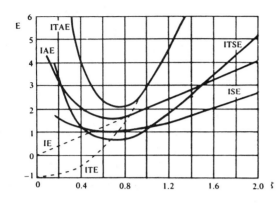

Fig. 4.3.3 Various performance indices for step response of second-order lowpass filter. (Adapted from J.J. D'Azzo and C.H. Houpis, *Control System Analysis and Synthesis,* **Chap. 17, McGraw-Hill, NY, 1960.)**

only after close scrutiny. Of the filters of Table 4.3.1, only the *ITAE* step response, *ISE* delay, and the minimax phase response have good time domain and frequency domain characteristics. The minimax phase response filter is the only widely used optimum filter. Nevertheless, optimum filters can often provide advantages over classical filters. They can also be tailored to meet specialized design requirements.

We will not pursue optimum filters further here but the engineer is encouraged to utilize to Ref. 2 for the details and a number of further examples.

4.4 ADAPTIVE FILTERS

Adaptive filters are optimum filters whose transfer functions adapt to changing filtering requirements. Such adaptation is usually needed because of fluctuating signal and/or noise conditions. Adaptive filters have gain functions or pole/zero patterns that are periodically recomputed according to some algorithm and then readjusted to optimally process incoming signals while discriminating against noise. Unfortunately, both classical and optimum filters have fixed pole/zero positions and are nonadaptive. Their inability to adapt to changing conditions makes such nonadaptive filters unsuitable for many applications.

To vividly illustrate when adaptive filters are most useful, consider the filter shown in Fig. 4.4.1. A signal $r(t)$ is input to a filter having transfer function $H(j\omega)$ to form an output signal $c(t)$. $r(t)$ is expressed as the sum of a signal $s(t)$ and a noise $n(t)$. The filter block $P(j\omega)$ is useful for post-processing as shall be seen later. Assume that the optimum lowpass filter has the transfer function

$$H(j\omega) = \frac{|S(j\omega)|^k}{|S(j\omega)|^k + |N(j\omega)|^k} = \frac{1}{1 + |SNR(j\omega)|^{-k}} \qquad (4.4.1)$$

where $S(j\omega)/N(j\omega)$ is the input spectral signal-to-noise ratio SNR. Eq. 4.4.1 reduces to Eq. 4.1.3 (for delay $\tau_o = 0$) only when the spectral SNR is constant and band-limited to B as

$$|SNR(j\omega)| = \frac{|S(j\omega)|}{|N(j\omega)|} = c_o, \quad |\omega| \le B$$

$$= 0, \quad |\omega| > B \qquad (4.4.2)$$

One such condition is when the signal $s(t)$ has a flat, band-limited spectrum $|S(j\omega)| = S_o$ and the noise is white with a flat spectrum $|N(j\omega)| = N_o$. Then $SNR = 0$ when $|S(j\omega)| = 0$ and $|N(j\omega)| \neq 0$. Under the constant band-limited SNR condition of Eq. 4.4.2, $H(j\omega)$ becomes

$$H(j\omega) = \frac{1}{1 + c_o^{-k}}, \qquad |\omega| \leq B$$

$$= 0, \qquad |\omega| > B \tag{4.4.3}$$

For a large passband $|SNR| = S_o/N_o \equiv c_o$, the passband gain $H(j\omega) \cong 1$ for $|\omega| < B$ which agrees with Eq. 4.1.3. Otherwise the gain is lower. A constant delay τ_o or linear phase $-\omega\tau_o$ can be assigned to $H(j\omega)$ in the passband. From a practical standpoint, processing delay τ_o insures filter realizability (see Chap. 1.12).

Summarizing these ideas, the ideal lowpass filter is only optimal or ideal with respect to Eq. 4.4.1 when both the signal and noise spectra are equal or scaled versions of one another and the signal spectrum is band-limited. Unfortunately this is virtually never the case so the ideal lowpass filter performs suboptimally in most applications. In the remainder of this chapter, filters will be investigated that do give optimal performance by adapting to changing signal and noise environments.

Filters are often designed in two steps. First, the ideal analog filter transfer function is derived based upon some criterion to be discussed in a moment. The ideal digital filter transfer function is then formed using Eq. 1.3.3. Such ideal filters are not necessarily realizable. Therefore second, the ideal digital filter transfer function is approximated by one that is optimum, realizable, and stable. The optimization follows the procedure discussed in Chap. 4.3.

We have chosen to formulate these filter designs in the frequency domain rather than the time domain. This allows us to use simple frequency domain equations involving spectra, to directly use classical and optimum filter results, and to draw on all of the vast analog design information available. The adaptive filters to be described are listed in Table 4.4.1. By contrast, filters will be designed directly in the time domain in Chap. 5. Although the methods used there will be simpler compared with those used in this chapter, they are usually applied to nonadaptive filters. The frequency and time domain designs are shown in Fig. 4.4.2 in block diagram form.

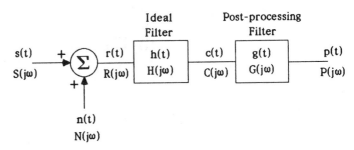

Fig. 4.4.1 Block diagram of basic linear filtering system assuming additive noise.

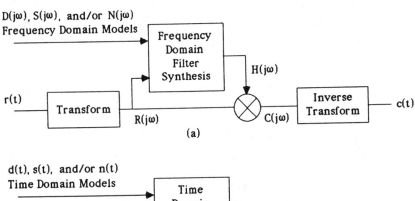

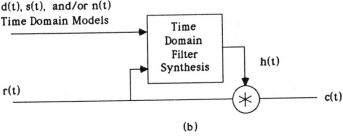

Fig. 4.4.2 Filter implementation in (a) frequency domain and (b) time domain.

4.4.1 DISTORTION BALANCE FILTERS

One particularly satisfying design approach involves balancing the signal and noise distortions.[7] The ideal filter equations are generated by equating the degradation caused by the filter distorting or altering the signal to the distortion caused by the noise leaking through the filter. The signal degradation $E_S(j\omega)$ is the difference between input and output signals. The noise leakage $E_N(j\omega)$ equals the output noise. From Fig. 4.4.1,

$$E_S(j\omega) = S(j\omega) - H(j\omega)S(j\omega) = [1 - H(j\omega)]S(j\omega)$$
$$E_N(j\omega) = H(j\omega)N(j\omega) \tag{4.4.4}$$

where $H(j\omega)$ is the ideal filter gain. For more general results, the signal distortion is weighted by $W_S(j\omega)$ and the noise distortion by $W_N(j\omega)$. Then equating the weighted signal error to the weighted noise error at every frequency ω gives

$$E_S(j\omega)W_S(j\omega) = E_N(j\omega)W_N(j\omega)$$
$$[1 - H(j\omega)]S(j\omega)W_S(j\omega) = H(j\omega)N(j\omega)W_N(j\omega) \tag{4.4.5}$$

This insures that the total error

$$E(j\omega) = W_S(j\omega)E_S(j\omega) + W_N(j\omega)E_N(j\omega) \tag{4.4.6}$$

is composed of equally-weighted signal and equally-weighted noise. Solving Eq. 4.4.5 for the ideal filter transfer function $H(j\omega)$ gives

$$H(j\omega) = \frac{S(j\omega)W_S(j\omega)}{S(j\omega)W_S(j\omega) + N(j\omega)W_N(j\omega)} \qquad (4.4.7)$$

This result has a number of interesting applications. If we wish to form a filter that balances complex signal and noise levels equally, we set $W_S(j\omega) = W_N(j\omega)$ so

$$H(j\omega) = \frac{S(j\omega)}{S(j\omega) + N(j\omega)} \qquad (4.4.8)$$

However, if we wish to form a filter that balances signal and noise power levels, we set $W_S(j\omega) = S^*(j\omega)$ and $W_N(j\omega) = N^*(j\omega)$. $H(j\omega)$ is then given by Eq. 4.4.1 where $k = 2$. A multitude of other possibilities exist. For example, if we wish to ignore high-frequency signal and low-frequency noise, we might choose $W_S(j\omega) = 1/(1 + j\omega)$ and $W_N(j\omega) = j\omega/(1 + j\omega)$. Alternatively, if we wish to balance the derivative of the signal from noise, $W_S(j\omega) = j\omega$ and $W_N(j\omega) = 1$. The possibilities are limitless.

4.4.2 GENERAL ESTIMATION FILTERS

A different approach can be used to derive another ideal filter transfer function.[7] Reconsider Fig. 4.4.1 where the input $r(t)$ and output $c(t)$ equal

$$\begin{aligned} r(t) &= s(t) + n(t) &\text{or}&\quad R(j\omega) = S(j\omega) + N(j\omega) \\ c(t) &= h(t) * r(t) &\text{or}&\quad C(j\omega) = H(j\omega)R(j\omega) \end{aligned} \qquad (4.4.9)$$

Let $d(t)$ equal the desired optimal output. Then define the output error $e(t)$ to equal

$$e(t) = d(t) - c(t) \qquad \text{or} \qquad E(j\omega) = D(j\omega) - C(j\omega) \qquad (4.4.10)$$

A wide variety of optimization criteria can be used to minimize the output error as discussed in Chap. 4.3. One of the most universally accepted is the integral-squared error (ISE) criterion where (see Eq. 4.3.1b)

$$ISE = \int_{-\infty}^{\infty} |e(t)|^2 dt = \int_{-\infty}^{\infty} |d(t) - c(t)|^2 dt \qquad (4.4.11)$$

The ideal filter is the one whose transfer function $H(j\omega)$ minimizes the error index ISE. Using Parseval's theorem given by Eq. 1.6.15, Eq. 4.4.11 can be rewritten in the frequency domain as

$$ISE = \frac{1}{2\pi} \int_{-\infty}^{\infty} E(j\omega)E^*(j\omega)d\omega = \frac{1}{2\pi} \int_{-\infty}^{\infty} \left[D(j\omega) - C(j\omega)\right]\left[D^*(j\omega) - C^*(j\omega)\right]d\omega \qquad (4.4.12)$$

where $E^*(j\omega)$ is the conjugate error spectrum. When $e(t)$ is real-valued, then $E^*(j\omega) = E(-j\omega)$. Since $C(j\omega) = H(j\omega)R(j\omega)$ is a functional of $H(j\omega)$, Eq. 4.4.12 is minimized by setting the derivative of the integral with respect to $H(j\omega)$ equal to zero. The integrand can be rewritten as

$$D(j\omega)D^*(j\omega) - H(j\omega)R(j\omega)D^*(j\omega)$$
$$-H^*(j\omega)R^*(j\omega)D(j\omega) + H(j\omega)H^*(j\omega)R(j\omega)R^*(j\omega) \tag{4.4.13}$$

Differentiating this functional with respect to $H(j\omega)$ and setting the result equal to zero yields (assume $H(j\omega)$ is real)

$$-R(j\omega)D^*(j\omega) - R^*(j\omega)D(j\omega) + H(j\omega)R(j\omega)R^*(j\omega) + H^*(j\omega)R(j\omega)R^*(j\omega) = 0 \tag{4.4.14}$$

Solving Eq. 4.4.14 for $H(j\omega)$ gives (the same result holds for complex $H(j\omega)$)

$$H(j\omega) = \frac{D(j\omega)R^*(j\omega)}{R(j\omega)R^*(j\omega)} = \frac{\text{Desired input output cross correlation spectrum}}{\text{Input autocorrelation spectrum}} \tag{4.4.15}$$

The spectral product RR^* equals the power spectrum $S_{rr}(j\omega)$ of the input autocorrelation $R_{rr}(\tau)$ when $r(t)$ is ergodic (see Eq. 4.4.31). Likewise the cross-spectral product DR^* equals the power spectrum $S_{dr}(j\omega)$ of the cross-correlation $R_{dr}(\tau)$ between the desired output and input. When the input is not ergodic, RR^* and DR^* are *estimates* of the autocorrelation and cross-correlation spectra.

Eq. 4.4.15 can also be expressed as

$$H(j\omega) = \frac{D(j\omega)R^*(j\omega)}{R(j\omega)R^*(j\omega)} = \frac{D(j\omega)}{R(j\omega)} = \frac{\text{Desired output}}{\text{Available input}} \tag{4.4.16}$$

When a deterministic input-output filter viewpoint is taken, Eq. 4.4.16 is used. Eq. 4.4.15 utilizes the stochastic correlation filter viewpoint. These equations are equivalent.

Eq. 4.4.15 is the transfer function for the general estimation filter. This ideal filter gives the best estimate (in the minimum ISE sense) of a desired signal $d(t)$ from an input signal $r(t)$ as the ratio of the two power spectra. When $r(t)$ is composed of a signal $s(t)$ contaminated with additive noise $n(t)$, Eq. 4.4.15 reduces to

$$H(j\omega) = \frac{D(j\omega)[S^*(j\omega) + N^*(j\omega)]}{[S(j\omega) + N(j\omega)][S^*(j\omega) + N^*(j\omega)]} \tag{4.4.17}$$

4.4.3 GENERAL DETECTION FILTERS

Another problem of common interest is the detection of the presence or absence of a signal in noise. The ideal filter detects a desired signal $d(t)$ by producing an output signal pulse $s_o(0)$ in the total output $c(t)$ such that the SRR_o ratio of the peak output signal-to-total rms output is maximized. Its transfer function equals[7]

$$H(j\omega) = \frac{D^*(j\omega)}{R(j\omega)R^*(j\omega)} = \frac{\text{Conjugate spectrum of desired signal to be detected}}{\text{Input autocorrelation spectrum}}$$
$$\tag{4.4.18}$$

This filter produces a narrow output pulse (which reduces temporal or time ambiguity) centered at time $t = 0$ with undefined shape.

The ideal filter which detects $d(t)$ by producing $s_o(0)$ in $c(t)$, such that the peak output signal-to-rms output noise SNR_o is maximized, has

$$H(j\omega) = \frac{D^*(j\omega)}{N(j\omega)N^*(j\omega)} = \frac{\text{Conjugate spectrum of desired signal to be detected}}{\text{Noise autocorrelation spectrum}}$$

(4.4.19)

To prove Eq. 4.4.18, consider the peak output signal (at time zero)-to-total rms output ratio squared (see Eq. 4.4.59)

$$
\begin{aligned}
SRR_o^2 &= \frac{|s_o(0)|^2}{\int_{-\infty}^{\infty} |c(t)|^2 dt} = \frac{\left|(1/2\pi)\int_{-\infty}^{\infty} S(j\omega)H(j\omega)d\omega\right|^2}{(1/2\pi)\int_{-\infty}^{\infty} |H(j\omega)|^2 |R(j\omega)|^2 d\omega} \\
&= \frac{\left|\int_{-\infty}^{\infty} [S(j\omega)/R(j\omega)]H(j\omega)R(j\omega)d\omega\right|^2}{2\pi \int_{-\infty}^{\infty} |H(j\omega)|^2 |R(j\omega)|^2 d\omega} \\
&\leq \frac{\int_{-\infty}^{\infty} \left[|S(j\omega)|^2/|R(j\omega)|^2\right] d\omega \int_{-\infty}^{\infty} |H(j\omega)|^2 |R(j\omega)|^2 d\omega}{2\pi \int_{-\infty}^{\infty} |H(j\omega)|^2 |R(j\omega)|^2 d\omega} \\
&= \frac{1}{2\pi} \int_{-\infty}^{\infty} \frac{|S(j\omega)|^2}{|R(j\omega)|^2} d\omega = \frac{1}{2\pi} \int_{-\infty}^{\infty} |SRR_i(j\omega)|^2 d\omega
\end{aligned}
$$

(4.4.20)

It equals the integral of the spectral signal-to-total input ratio squared. This results by noting that the squared integral of a product never exceeds the product of integrals with squared integrand (Schwartz's inequality). $s_o(0)$ is obtained by evaluating the inverse Fourier transform of $S_o(j\omega)$ at time $t = 0$. The integral of the total output $|c(t)|^2$ is found using Parseval's theorem where $C(j\omega) = H(j\omega)R(j\omega)$. The equality condition in Eq. 4.4.20 results when

$$H(j\omega)R(j\omega) = k\left[\frac{S(j\omega)}{R(j\omega)}\right]^*$$

(4.4.21)

Solving for $H(j\omega)$ gives (let $k = 1$)

$$H(j\omega) = \frac{S^*(j\omega)}{|R(j\omega)|^2}$$

(4.4.22)

This is identical to Eq. 4.4.18 when $S(j\omega)$ is the desired output $D(j\omega)$ to be detected and $R(j\omega)$ is the input spectrum rejected. If the output signal s_o is to peak at time T_p rather than time zero, then Eq. 4.4.22 is multiplied by $\exp(-j\omega T_p)$.

4.4.4 GENERAL ESTIMATION-DETECTION FILTERS

The general estimation filter of Eq. 4.4.15 optimally estimates signals in the presence of additive noise. The general detection filter of Eq. 4.4.18 maximizes the peak output signal relative to the total rms output. Combining these two results as

$$H(j\omega) = H_E(j\omega)H_D(j\omega) \tag{4.4.23}$$

gives us an estimation-detection filter which performs both operations. This filter best estimates the signal and then outputs a narrow pulse when the signal is present.

4.4.5 WIENER ESTIMATION FILTERS

Suppose that the generalized estimation filter is used for estimating signal. Setting the desired signal to be estimated equal to the input signal $s(t)$ as $d(t) = s(t)$, then $D(j\omega) = S(j\omega)$ and Eq. 4.4.17 becomes

$$
\begin{aligned}
H(j\omega) &= \frac{S(j\omega)[S^*(j\omega) + N^*(j\omega)]}{[S(j\omega) + N(j\omega)][S^*(j\omega) + N^*(j\omega)]} \\
&= \frac{|S(j\omega)|^2 + S(j\omega)N^*(j\omega)}{|S(j\omega)|^2 + |N(j\omega)|^2 + S(j\omega)N^*(j\omega) + S^*(j\omega)N(j\omega)}
\end{aligned}
\tag{4.4.24}
$$

This is called the *correlated estimation* or *correlated Wiener* filter. It is coherent in the sense that the phase spectra of the signal and noise are used when computing $H(j\omega)$. The phase of $H(j\omega)$ is generally nonzero. Since SN^* and S^*N are complex conjugates of each other, their sum can also be expressed as $2 \operatorname{Re} SN^*$.

When the signal and noise are orthogonal, or equivalently, are uncorrelated and the noise has zero mean, then $S(j\omega)N^*(j\omega) = 0$ (see discussion for Eq. 4.4.26) and Eq. 4.4.24 reduces to

$$H(j\omega) = \frac{|S(j\omega)|^2}{|S(j\omega)|^2 + |N(j\omega)|^2} \tag{4.4.25}$$

This is the *uncorrelated estimation* or *uncorrelated Wiener* filter. From the distortion balance filter viewpoint of Eq. 4.4.7, this filter results when signal and noise powers are equally weighted. It is noncoherent since it discards the phase spectra of the signal and noise. The phase of $H(j\omega)$ is always zero.

Before proceeding to other filter types, notice that the signal spectrum $S(j\omega)$, conjugate noise spectrum $N^*(j\omega)$, their product $S(j\omega)N^*(j\omega)$, and the cross-correlation spectrum $S_{sn}(j\omega)$ (see Eq. 1.17.15) equal

$$
\begin{aligned}
S(j\omega) &= \mathcal{F}\big[s(t)\big], \qquad N^*(j\omega) = \mathcal{F}\big[n^*(-t)\big] \\
\widehat{S}_{sn}(j\omega) &= \mathcal{F}\big[s(t) * n^*(-t)\big] = S(j\omega)N^*(j\omega) \\
S_{sn}(j\omega) &= \mathcal{F}\big[E\{\mathbf{s}(t+\tau)\mathbf{n}^*(\tau)\}\big] = \mathcal{F}\big[R_{sn}(t)\big]
\end{aligned}
\tag{4.4.26}
$$

$\mathbf{s}$ and $\mathbf{n}$ represent random signal and noise processes, respectively. s and n are the actual signal and noise, respectively. They are particular members of the ensemble of all those possible formed by the $\mathbf{s}$ and $\mathbf{n}$ random processes. S_{sn} is the apriori (assumed) cross-correlation spectrum while SN^* is the aposteriori (actual) cross-correlation spectral estimate $\widehat{S}_{sn}$. Eq. 4.4.26 shows that SN^* equals S_{sn} only when the signal-noise convolutional product $s(t) * n^*(-t)$ equals the cross-correlation function R_{sn}. This occurs only when both $s(t)$ and $n(t)$ are ergodic processes.

The cross-correlation spectrum S_{sn} is zero when the signal and noise are orthogonal, or equivalently, are uncorrelated and the noise is a zero-mean process (see Eqs. 1.17.9 and 1.17.12). In such cases, the cross-correlation estimate SN^* product is *set equal to zero* in Eq. 4.4.24 which gives Eq. 4.4.25. This is done even though the actual SN^* product is nonzero over the frequency ranges where $S \neq 0$ and $N \neq 0$, or equivalently, the nonzero signal and noise spectra are overlapping. Nonoverlapping or disjoint S and N have $SN^* = 0$ and are very seldom encountered in practice. Correlation is discussed further in Chap. 4.4.7.

4.4.6 PULSE-SHAPING ESTIMATION FILTERS

Suppose that we wish to form a new signal $p(t)$ from the output signal $c(t)$ in Fig. 4.4.1. This is done using the post-processing block with gain $G(j\omega)$ where

$$p(t) = g(t) * c(t) \qquad \text{or} \qquad P(j\omega) = G(j\omega)C(j\omega) \qquad (4.4.27)$$

Then the gain $G(j\omega)$ equals

$$G(j\omega) = \frac{P(j\omega)}{C(j\omega)} \qquad (4.4.28)$$

where $P(j\omega)$ is the desired output spectrum of $p(t)$. This post-processing block can be used to generate a variety of useful signals. After $G(j\omega)$ is determined, it is combined with gain $H(j\omega)$ to form a single gain block (i.e., $H(j\omega)G(j\omega) \to H(j\omega)$). The blocks are separated here only for conceptual simplicity.

It might be useful to reshape $c(t)$ to have a different waveform $p(t)$ or to estimate different signals. We could first use the uncorrelated Wiener filter whose transfer function $H(j\omega)$ is given by Eq. 4.4.25 to estimate $s(t)$. Then choosing $G(j\omega) = P(j\omega)/S(j\omega)$ results in the desired output signal $p(t)$. Combining $H(j\omega)$ and $P(j\omega)$ into a single transfer function gives

$$H(j\omega) = \frac{P(j\omega)S^*(j\omega)}{|S(j\omega)|^2 + |N(j\omega)|^2} \qquad (4.4.29)$$

which is the *pulse-shaping (uncorrelated) estimation filter*. $G(j\omega)$ can also be combined with the correlated estimation filter of Eq. 4.4.24 to give the *pulse-shaping (correlated) estimation filter* as

$$H(j\omega) = \frac{P(j\omega)[S^*(j\omega) + N^*(j\omega)]}{|S(j\omega)|^2 + |N(j\omega)|^2 + S(j\omega)N^*(j\omega) + S^*(j\omega)N(j\omega)} \qquad (4.4.30)$$

Pulse-shaping filters can also be viewed as a high-resolution detection filter (see Chap. 4.4.10) followed by a pulse shaper. The impulse $U_0(t)$ from the high-resolution detection filter $H(j\omega)$ excites the impulse response $p(t)$ from the post-processor filter $G(j\omega)$.

Pulse-shaping filters yield the maximum output SNR_o possible under the constraint that the output signal pulse $p(t) = \mathcal{F}^{-1}[P(j\omega)]$ has a specific form.

4.4.7 CORRELATION FILTERS

The estimate $\widehat{R}_{ss}(t)$ of the autocorrelation $R_{ss}(t)$ of an ergodic signal $s(t)$ and the estimate $\widehat{S}_{ss}(j\omega)$ of its power spectrum $S_{ss}(j\omega)$ equal

$$
\begin{aligned}
R_{ss}(t) &= \mathcal{F}^{-1}[S_{ss}(j\omega)] = E\{\mathbf{s}(t+\tau)\mathbf{s}^*(\tau)\} \equiv s(t) * s^*(-t) = \widehat{R}_{ss}(t) \\
S_{ss}(j\omega) &= \mathcal{F}[R_{ss}(t)] = \mathcal{F}[E\{\mathbf{s}(t+\tau)\mathbf{s}^*(\tau)\}] \equiv S(j\omega)S^*(j\omega) = \widehat{S}_{ss}(j\omega)
\end{aligned}
\tag{4.4.31}
$$

from Eqs. 1.17.21 and 1.17.22. The post-processing block $G(j\omega)$ in Fig. 4.4.1 can be used to generate the autocorrelation function. From Eqs. 4.4.27 and 4.4.31, its output $P(j\omega) = S(j\omega)S^*(j\omega)$ while its input $C(j\omega) = S(j\omega)$. Then its transfer function $G(j\omega) = P(j\omega)/C(j\omega) = S^*(j\omega)$.

Therefore combining $P(j\omega)$ with the uncorrelated Wiener filter described by Eq. 4.4.25 gives the ideal autocorrelation estimation filter gain as

$$
H(j\omega) = \frac{S^*(j\omega)|S(j\omega)|^2}{|S(j\omega)|^2 + |N(j\omega)|^2}
\tag{4.4.32}
$$

Another autocorrelation estimation filter results by using the correlated Wiener filter of Eq. 4.4.24. Its gain equals

$$
H(j\omega) = \frac{|S(j\omega)|^2[S^*(j\omega) + N^*(j\omega)]}{|S(j\omega)|^2 + |N(j\omega)|^2 + S(j\omega)N^*(j\omega) + S^*(j\omega)N(j\omega)}
\tag{4.4.33}
$$

The estimate of the cross-correlation $R_{sn}(t)$ between two ergodic signals, $s(t)$ and $n(t)$, and its power spectrum $S_{sn}(j\omega)$ equal

$$
\begin{aligned}
R_{sn}(t) &= \mathcal{F}^{-1}[S_{sn}(j\omega)] = E\{\mathbf{s}(t+\tau)\mathbf{n}^*(\tau)\} \equiv s(t) * n^*(-t) = \widehat{R}_{sn}(t) \\
S_{sn}(j\omega) &= \mathcal{F}[R_{sn}(t)] = \mathcal{F}[E\{\mathbf{s}(t+\tau)\mathbf{n}^*(\tau)\}] \equiv S(j\omega)N^*(j\omega) = \widehat{S}_{sn}(j\omega)
\end{aligned}
\tag{4.4.34}
$$

Proceeding as before, $G(j\omega) = N^*(j\omega)$ which can be combined with the uncorrelated Wiener filter. This gives the ideal cross-correlation estimation filter gain as

$$
H(j\omega) = \frac{N^*(j\omega)|S(j\omega)|^2}{|S(j\omega)|^2 + |N(j\omega)|^2}
\tag{4.4.35}
$$

Using the correlated Wiener filter instead gives

$$
H(j\omega) = \frac{S(j\omega)N^*(j\omega)[S^*(j\omega) + N^*(j\omega)]}{|S(j\omega)|^2 + |N(j\omega)|^2 + S(j\omega)N^*(j\omega) + S^*(j\omega)N(j\omega)}
\tag{4.4.36}
$$

The correlation estimates obtained using the correlated Wiener filters (Eqs. 4.4.33 and 4.4.36) are usually better than those obtained using the uncorrelated Wiener filters (Eqs. 4.4.32 and 4.4.35).[8]

4.4.8 MATCHED DETECTION FILTERS

There is a class of problems which requires the detection of the presence of a signal $s(t)$ embedded in additive noise $n(t)$. In this case, we wish to maximize the peak output SNR_o in order to detect the signal and reject the rms noise. Using Eq. 4.4.18 where $D(j\omega) = S(j\omega)$ and $R(j\omega) = N(j\omega)$, then

$$H(j\omega) = \frac{S^*(j\omega)}{N(j\omega)N^*(j\omega)} = \frac{S^*(j\omega)}{|N(j\omega)|^2} \qquad (4.4.37)$$

This is the *classical detection, matched,* or *prewhitening* filter. $H(j\omega)$ weights each spectral input by the ratio of the conjugate signal spectrum to the noise power spectrum. The filter introduces a phase which is opposite to signal phase. Thus all the output spectral components of a signal similar to the expected signal will be in phase. The output tends to be concentrated into a narrow pulse which occurs when the signal occurs.

Notice from Eq. 4.4.37 that the filter's spectral output $C = HR = H(S + N)$ consists of a signal component $SS^*/|N|^2$ and a noise component $NS^*/|N|^2$. The signal component is the square of the expected input spectral SNR_i at that frequency. The noise component is the conjugate SNR_i. From a correlation viewpoint, the signal component is the signal autocorrelation and the noise component is the signal-noise cross-correlation when the noise is white (i.e., $|N(j\omega)| = 1$). In either case, the noise component is zero if the input signal and noise are orthogonal or are uncorrelated with zero mean noise. Otherwise the noise component is nonzero.

The performance of detection filters is often measured by the SNR gain which is the ratio of the output SNR_o to the input SNR_i. It is invariant to changes in the input signal or noise level and will be discussed in Chap. 4.4.14.

From the distortion balance filter viewpoint, the classical detection filter first discriminates between signal and scaled noise power levels (so $W_S(j\omega) = S^*(j\omega)$ and $W_N(j\omega) = kN^*(j\omega)$). If the constant k is chosen to be much larger than the squared input spectral SNR_i ratio (i.e., $k \gg |S|^2/|N|^2$), then Eq. 4.4.7 reduces to

$$H(j\omega) = \frac{S(j\omega)S^*(j\omega)}{kN(j\omega)N^*(j\omega)} \qquad (4.4.38)$$

Following $H(j\omega)$ by an inverse filter $P(j\omega) = k/S(j\omega)$, the overall gain characteristic becomes that of the classical detection filter given by Eq. 4.4.37. This shows that by relying entirely on noise rejection, the estimation filter becomes a detection filter.

4.4.9 INVERSE DETECTION FILTERS

An *inverse detection filter* outputs an impulse $U_0(t)$ when *only* signal $s(t)$ and no noise is applied. It outputs a narrow pulse when a signal similar to $s(t)$ is applied. Using Eq. 4.4.9b, the transfer function of an inverse detection filter equals

$$H(j\omega) = \frac{\mathcal{F}[U_0(t)]}{\mathcal{F}[s(t)]} = \frac{1}{S(j\omega)} \tag{4.4.39}$$

When signal plus noise is applied to the filter, its input equals $R = S + N$. The output of the filter then equals $C = HR = 1 + (N/S)$ or $c(t) = U_0(t) + \mathcal{F}^{-1}[1/SNR_i]$. The undesired output component due to nonzero input noise is $\mathcal{F}^{-1}[1/SNR_i]$. When the spectral signal-to-noise ratio is large, $|SNR_i| \gg 1$ so $|1/SNR_i| \ll 1$ and then this undesired output component is negligible. However $|SNR_i|$ is seldom large so this component is present which produces output error. This error can be significantly reduced using the high-resolution detection filter.

4.4.10 HIGH-RESOLUTION DETECTION FILTERS

Another conceptually useful filter is the *high-resolution detection filter* which outputs an impulse when *both* the signal $s(t)$ and noise $n(t)$ are applied. It outputs a narrow pulse when a signal similar to $s(t) + n(t)$ is applied. This requires a post-processing gain $G(j\omega)$ after the estimation filter in Fig. 4.4.1 of

$$G(j\omega) = \frac{\mathcal{F}[U_0(t)]}{\mathcal{F}[s(t)]} = \frac{1}{S(j\omega)} \tag{4.4.40}$$

which is an inverse filter. Combining this with the uncorrelated Wiener filter in Eq. 4.4.25 results in the ideal high-resolution detection filter where

$$H(j\omega) = \frac{S^*(j\omega)}{|S(j\omega)|^2 + |N(j\omega)|^2} \tag{4.4.41}$$

Eq. 4.4.41 is identical to the detection filter of Eq. 4.4.22 when $S(j\omega)$ and $N(j\omega)$ are assumed to be orthogonal. The Wiener filter estimates the signal $s(t)$ and the inverse filter converts $s(t)$ into an impulse. Another high-resolution filter results using the correlated Wiener filter of Eq. 4.4.24 as

$$H(j\omega) = \frac{S^*(j\omega) + N^*(j\omega)}{|S(j\omega)|^2 + |N(j\omega)|^2 + S(j\omega)N^*(j\omega) + S^*(j\omega)N(j\omega)} \tag{4.4.42}$$

The high-resolution detection filter of Eq. 4.4.42 functions as a matched detection filter (Eq. 4.4.37) at frequencies where the spectral SNR_i is low. It acts as an inverse or deconvolution filter (Eq. 4.4.39) at high spectral SNR_i. The filter deconvolves $S(j\omega)$ with $1/S(j\omega)$ to produce a single impulse. Rewriting Eq. 4.4.41 as

$$H(j\omega) = \frac{1}{S(j\omega)} \frac{1}{1 + |N(j\omega)|^2/|S(j\omega)|^2} = \frac{1}{S(j\omega)} \frac{1}{1 + |SNR_i|^{-2}} \qquad (4.4.43)$$

$H(j\omega)$ depends upon the input spectral signal-to-noise ratio SNR_i for constant $S(j\omega)$.

From a distortion balance viewpoint, this is a filter which discriminates between signal and noise power levels (so $W_S(j\omega) = S^*(j\omega)$ and $W_N(j\omega) = N^*(j\omega)$) followed by an inverse filter having gain $G(j\omega) = 1/S(j\omega)$.

4.4.11 TRANSITIONAL FILTERS

There are situations which require a filter whose transfer function can make a transition from one type to another type. These are called *transitional filters*. The characteristics of two transfer functions are blended together. The transition is controlled by one parameter (or more) like α. These transfer functions are usually formed heuristically.

For example, suppose that we want to form a transitional detection filter that can make a transition from an inverse type ($\alpha = 0$) to a high-resolution type ($\alpha = 0.5$) to a matched type ($\alpha = 1$). One way combines Eqs. 4.4.37, 4.4.39, and 4.4.41 as

$$H(j\omega) = \frac{S^*(j\omega)}{(1-\alpha)|S(j\omega)|^2 + \alpha|N(j\omega)|^2} \qquad (4.4.44)$$

Choosing some α between 0 and 1 yields a transfer function that combines the attributes of these filter types. As $\alpha \to 0$, the filter functions as an inverse type. As $\alpha \to 0.5$, it becomes a high-resolution type. As $\alpha \to 1$, it performs as a matched type. Another example of a transitional detection filter is

$$H(j\omega) = \frac{1}{S(j\omega)^{2(1-\alpha)}} \qquad (4.4.45)$$

by generalizing Eq. 4.4.39. When $\alpha = 0.5$, it functions as an inverse filter.

An example of a transitional estimation filter is

$$H(j\omega) = \frac{|S(j\omega)|^2 + \alpha S(j\omega)N^*(j\omega)}{|S(j\omega)|^2 + |N(j\omega)|^2 + \alpha\big(S(j\omega)N^*(j\omega) + S^*(j\omega)N(j\omega)\big)} \qquad (4.4.46)$$

As $\alpha \to 0$, the filter becomes uncorrelated Wiener. As $\alpha \to 1$, the filter becomes correlated Wiener. The possibilities for transitional filters are endless. The transitional filter approach gives the designer great freedom in formulating adaptive filters.

4.4.12 STATISTICALLY OPTIMUM FILTERS

A variety of adaptive filters have been derived. In general, their gains are expressed as the ratio of two functions $X(j\omega)$ and $Y(j\omega)$ where

$$H(j\omega) = \frac{Y(j\omega)}{X(j\omega)} \qquad (4.4.47)$$

The X and Y functions involve sums, products, and other operations on the signal spectrum $S(j\omega)$ and noise spectrum $N(j\omega)$. Which S and N spectra should be used to form $H(j\omega)$? The statistically optimum approach requires

$$H(j\omega) = \frac{E\{Y(j\omega)\}}{E\{X(j\omega)\}} \qquad (4.4.48)$$

where $E\{\ \}$ is the expectation operator. Eq. 4.4.48 results by expressing Eq. 4.4.47 as $Y = HX$, treating X and Y as stochastic processes rather than deterministic signals, taking expectations of both sides, and solving for $H(j\omega)$ which is deterministic. The optimum $H(j\omega)$ equals the ratio of the expected (statistically averaged) X and Y which the filter processes. The transfer functions of Chap. 4.4 are therefore tabulated in Table 4.4.1 using expectation signs around their numerators and denominators. Some examples of terms in $E\{X(j\omega)\}$ and $E\{Y(j\omega)\}$ include:

$E\{S(j\omega)\}$, $E\{N(j\omega)\}$ = expected signal and noise spectra
$E\{|S(j\omega)|\}$, $E\{|N(j\omega)|\}$ = expected signal and noise magnitude spectra
$E\{|S(j\omega)|^2\}$, $E\{|N(j\omega)|^2\}$ = expected signal and noise autocorrelation (power) spectra
$E\{S(j\omega)N^*(j\omega)\}$ = expected signal-noise cross-correlation spectrum
$E\{W_S S(j\omega)\}$, $E\{W_N N(j\omega)\}$ = expected weighted signal and noise spectra   (4.4.49)

4.4.13 ADAPTATION AND LEARNING PROCESS

Adaptive filters undergo a *learning process* and perform *adaptation* to form the filter transfer function $H(j\omega)$. The transfer function is updated either occasionally or continuously. The ideal filter transfer functions listed in Table 4.4.1 are formed using the expected spectra $E\{X\}$ and $E\{Y\}$ like those in Eq. 4.4.49. They are either software programmed into the filter or estimated. As will be shown in Chap. 4.5, Class 1 adaptive filters store $E\{X\}$ and $E\{Y\}$ and update them (at most) periodically. Class 2 and 3 adaptive filters continuously form estimates $\widehat{E}\{X\}$ of $E\{X\}$ and $\widehat{E}\{Y\}$ of $E\{Y\}$ as the filter operates.

The expected signal spectrum $E\{S(j\omega)\}$ is sometimes called the *learning signal.* The expected noise spectrum $E\{N(j\omega)\}$ is sometimes called the *learning noise.* Mathematically,

$$E\{S(j\omega)\} = E\{\mathcal{F}[s(t)]\} \neq E\{\mathcal{F}[\mathsf{s}(t)]\}$$
$$E\{N(j\omega)\} = E\{\mathcal{F}[n(t)]\} \neq E\{\mathcal{F}[\mathsf{n}(t)]\} \qquad (4.4.50)$$

Since operators $E\{\ \}$ and $\mathcal{F}[\ ]$ are linear, they can be applied in any order as $E\{\mathcal{F}[\ ]\}$ or $\mathcal{F}[E\{\ \}]$. Therefore the inequalities result because $s(t) \neq \mathsf{s}(t)$ and $n(t) \neq \mathsf{n}(t)$ (a member from an ensemble does not equal the ensemble). The learning signal and learning noise equal the inverse Fourier transforms of their expected spectra in Eq. 4.4.50 as

$$\mathcal{F}^{-1}[E\{S(j\omega)\}] = E\{s(t)\} \neq E\{\mathsf{s}(t)\}$$
$$\mathcal{F}^{-1}[E\{N(j\omega)\}] = E\{n(t)\} \neq E\{\mathsf{n}(t)\} \qquad (4.4.51)$$

The inequalities of Eqs. 4.4.50 and 4.4.51 become equalities when the signal $s(t)$ (or $n(t)$) is either a deterministic process or an ergodic stochastic process. This is the usual case. For further details, see Chap. 1.17.

The noise spectrum can be expressed in polar form as

$$N(j\omega) = |N(j\omega)|e^{j \text{ arg } N(j\omega)} \tag{4.4.52}$$

The expected noise spectrum equals

$$E\{N(j\omega)\} = E\{|N(j\omega)|e^{j \text{ arg } N(j\omega)}\} = E\{|N(j\omega)|\} E\{e^{j \text{ arg } N(j\omega)}\} = E\{\mathcal{F}[n(t)]\} \tag{4.4.53}$$

It can be expressed as the product of expectations because the magnitude and phase are independent (see Eq. 1.17.10). When the phase of $E\{N(j\omega)\}$ is random and uniformly distributed between $(-\pi, \pi)$, then the expected noise phase is zero and $E\{e^{j \text{ arg } N(j\omega)}\}$ is zero. Then the expected noise spectrum is also zero or

$$E\{N(j\omega)\} = 0 \tag{4.4.54}$$

Even though the expected noise spectrum is zero, the expected noise magnitude spectrum is not. It can take on a variety of forms. The most common forms are:

$$\begin{aligned} E\{|N(j\omega)|\} &= 1, && \text{white noise} \\ &= 1/\omega, && 1/f \text{ or integrated white noise} \\ &= 1/\omega^2, && 1/f^2 \text{ or doubly integrated white noise} \\ &= e^{-\omega^2}, && \text{Gaussian noise} \end{aligned} \tag{4.4.55}$$

The expected noise power spectra is the square of the expected noise magnitude spectra in Eq. 4.4.55 so that

$$\begin{aligned} E\{|N(j\omega)|^2\} &= 1, && \text{white noise} \\ &= 1/\omega^2, && 1/f \text{ or integrated white noise} \\ &= 1/\omega^4, && 1/f^2 \text{ or doubly integrated white noise} \\ &= e^{-2\omega^2}, && \text{Gaussian noise} \end{aligned} \tag{4.4.56}$$

The remainder of the book uses the frequency domain viewpoint in formulating adaptive filters and Eqs. 4.4.50–4.4.56 will be used repeatedly. In future terminology, the learning signal and the learning noise will usually be called the expected signal and expected noise, respectively. They are the inverse Fourier transforms of their expected spectra and are computed using Eq. 4.4.51. It must be remembered that we apply the term *expected* to *frequency domain spectra* rather than time domain signals unlike most other books.

4.4.14 PERFORMANCE MEASURES

The signal-to-noise ratio (SNR) can be defined in a variety of ways to best describe a particular problem. These include (in order of increasing computational difficulty) the: (1) spectral SNR, (2) rms SNR or simply the SNR, and the (3) peak SNR.

The spectral SNR is defined as

$$\text{Spectral } SNR = \frac{S(j\omega)}{N(j\omega)} = SNR(j\omega) \qquad (4.4.57)$$

which is a function of frequency. It is the ratio of the signal spectrum to the noise spectrum evaluated at a particular frequency ω. The spectral SNR has a magnitude and an angle. Generally only the magnitude is of interest. The Bode plot of $|SNR(j\omega)|$ shows the frequency ranges over which the signal dominates or the noise dominates.

The rms SNR, which is usually simply called the SNR, equals

$$\text{Rms } SNR = \frac{\sqrt{P_s}}{\sqrt{P_n}} = \sqrt{\frac{\int_{-\infty}^{\infty} |s(t)|^2 \, dt}{\int_{-\infty}^{\infty} |n(t)|^2 \, dt}} = \sqrt{\frac{\int_{-\infty}^{\infty} |S(j\omega)|^2 \, d\omega}{\int_{-\infty}^{\infty} |N(j\omega)|^2 \, d\omega}} \qquad (4.4.58)$$

It is the square root of the ratio of the signal power P_s to the noise power P_n. It can be evaluated in either the time domain (integrate $|s(t)|^2$ and $|n(t)|^2$), or in the frequency domain using Parseval's theorem (integrate $|S(j\omega)|^2$ and $|N(j\omega)|^2$).

The peak SNR is defined as

$$\text{Peak } SNR = \frac{|s(T_p)|}{\sqrt{P_N}} = \frac{|s(T_p)|}{\sqrt{\int_{-\infty}^{\infty} |n(t)|^2 \, dt}} = \frac{\sqrt{2\pi} \, |s(T_p)|}{\sqrt{\int_{-\infty}^{\infty} |N(j\omega)|^2 \, d\omega}} \qquad (4.4.59)$$

It is the ratio of the peak signal $s(T_p)$ which occurs at time T_p (or peak temporal signal) to the square root of the noise power. The noise power can be evaluated in time (integrate $|n(t)|^2$) or in frequency (integrate $|N(j\omega)|^2$). It is usually difficult to find the peak signal value $s(T_p)$ since this requires solving $ds(T_p)/dt = 0$.

Regardless of which performance measure is used, SNR is usually expressed in dB rather than its numeric value as

$$SNR_{dB} = 20 \log_{10} SNR = 10 \log_{10} SNR^2 \qquad (4.4.60)$$

for SNR real. When SNR is the ratio r of signals, $20 \log r$ converts it to dB. When SNR is the ratio r of signal powers, $10 \log r$ converts it to dB.

The signal-to-noise ratio gain G_{SNR} provided by a filter of gain $H(j\omega)$ equals

$$G_{SNR} = \frac{SNR_o}{SNR_i} = SNR_{o\ dB} - SNR_{i\ dB} \qquad (4.4.61)$$

Any of the SNR measures of Eqs. 4.4.57–4.4.59 can be used. The spectral SNR gain equals

$$\text{Spectral } G_{SNR} = \frac{S_o(j\omega)/N_o(j\omega)}{S_i(j\omega)/N_i(j\omega)} = \frac{H(j\omega)S_i(j\omega)/H(j\omega)N_i(j\omega)}{S_i(j\omega)/N_i(j\omega)} = 1 \qquad (4.4.62)$$

using the spectral SNR of Eq. 4.4.57. The output signal spectrum $S_o = HS_i$ while the output noise spectrum $N_o = HN_i$. Since the spectral G_{SNR} equals unity and is independent of the $H(j\omega)$ filter used, the spectral G_{SNR} is of little interest. If the rms SNR of Eq. 4.4.58 is used instead, then the rms SNR gain equals

$$\text{Rms } G_{SNR} = \frac{\int_{-\infty}^{\infty} |S_o(j\omega)|^2 d\omega \Big/ \int_{-\infty}^{\infty} |N_o(j\omega)|^2 d\omega}{\int_{-\infty}^{\infty} |S_i(j\omega)|^2 d\omega \Big/ \int_{-\infty}^{\infty} |N_i(j\omega)|^2 d\omega} \qquad (4.4.63)$$

while the peak SNR gain equals

$$\text{Peak } G_{SNR} = \frac{s_o(t)_{pk} \Big/ \sqrt{\int_{-\infty}^{\infty} |N_o(j\omega)|^2 d\omega}}{s_i(t)_{pk} \Big/ \sqrt{\int_{-\infty}^{\infty} |N_i(j\omega)|^2 d\omega}} \qquad (4.4.64)$$

The filter output equals $c(t) = s_o(t) + n_o(t)$. Therefore, the output error equals $n_o(t) = c(t) - s_o(t)$ which is the output noise. A measure of the distortion in the output is the ratio of the noise energy to the signal energy as

$$\text{Distortion} = \frac{\int_{-\infty}^{\infty} |n_o(t)|^2 dt}{\int_{-\infty}^{\infty} |s_o(t)|^2 dt} = \frac{\int_{-\infty}^{\infty} |N_o(j\omega)|^2 d\omega}{\int_{-\infty}^{\infty} |S_o(j\omega)|^2 d\omega} = \frac{\int_{-\infty}^{\infty} |H(j\omega)|^2 |N_i(j\omega)|^2 d\omega}{\int_{-\infty}^{\infty} |H(j\omega)|^2 |S_i(j\omega)|^2 d\omega}$$

$$= 1/\text{rms } SNR_o^2$$

$$(4.4.65)$$

It is also equal to the reciprocal of the squared rms SNR_o using Eq. 4.4.47. Distortion measures the normalized mean-squared-error in the output signal.

Another performance index is the noise bandwidth NBW. It is defined as

$$NBW = \int_0^{\omega_s} |H(j\omega)|^2 d\omega \Big/ 2|H(j\omega)|^2_{max} \qquad (4.4.66)$$

In digital filters, NBW is often normalized by the Nyquist frequency $\omega_s/2$ to yield

$$NBW_n = \frac{NBW}{\omega_s/2} = \int_0^{\omega_s} |H(j\omega)|^2 d\omega \Big/ \omega_s |H(j\omega)|^2_{max} \qquad (4.4.67)$$

To obtain maximum noise rejection, the NBW and NBW_n of the filter are minimized.

Table 4.4.1 Ideal Class 1 filter transfer functions $H(j\omega)$.

Filter Name	Transfer Function $H(j\omega)$	Eq. Number

Distortion Balance ($W_S(j\omega)$ = Signal weighting, $W_N(j\omega)$ = Noise weighting)

$$\frac{E\{S(j\omega)W_S(j\omega)\}}{E\{S(j\omega)W_S(j\omega)\} + E\{N(j\omega)W_N(j\omega)\}}$$ (4.4.7)

General Estimation ($D(j\omega)$ = Desired output)

$$H_E(j\omega) = \frac{E\{D(j\omega)R^*(j\omega)\}}{E\{R(j\omega)R^*(j\omega)\}}$$ (4.4.15)

General Detection

$$H_D(j\omega) = \frac{E\{D^*(j\omega)\}}{E\{R(j\omega)R^*(j\omega)\}}$$ (4.4.18)

Estimation–Detection

$$H_E(j\omega)H_D(j\omega)$$ (4.4.23)

Uncorrelated Estimation (Uncorrelated Wiener)

$$\frac{E\{|S(j\omega)|^2\}}{E\{|S(j\omega)|^2\} + E\{|N(j\omega)|^2\}}$$ (4.4.25)

Correlated Estimation (Correlated Wiener)

$$\frac{E\{S(j\omega)(S^*(j\omega) + N^*(j\omega))\}}{E\{|S(j\omega) + N(j\omega)|^2\}}$$ (4.4.24)

Transitional 1 Estimation (Uncorrelated Wiener $\alpha = 0$, Correlated Wiener $\alpha = 1$)

$$\frac{E\{|S(j\omega)|^2 + \alpha S(j\omega)N^*(j\omega)\}}{E\{|S(j\omega)|^2 + |N(j\omega)|^2 + \alpha(S(j\omega)N^*(j\omega) + S^*(j\omega)N(j\omega))\}}$$ (4.4.46)

Pulse-Shaping Estimation ($P(j\omega)$ = Desired pulse spectrum, Uncorrelated Wiener)

$$\frac{E\{P(j\omega)S^*(j\omega)\}}{E\{|S(j\omega)|^2\} + E\{|N(j\omega)|^2\}}$$ (4.4.29)

Pulse-Shaping Estimation ($P(j\omega)$ = Desired pulse spectrum, Correlated Wiener)

$$\frac{E\{P(j\omega)(S^*(j\omega) + N^*(j\omega))\}}{E\{|S(j\omega)|^2 + |N(j\omega)|^2 + S(j\omega)N^*(j\omega) + S^*(j\omega)N(j\omega)\}}$$ (4.4.30)

Table **4.4.1** (Continued) Ideal Class 1 filter transfer functions $H(j\omega)$.

Filter Name	Transfer Function $H(j\omega)$	Eq. Number

Classical Detection (Matched)

$$\frac{E\{S^*(j\omega)\}}{E\{|N(j\omega)|^2\}} \tag{4.4.37}$$

High-Resolution Detection (Uncorrelated Wiener)

$$\frac{E\{S^*(j\omega)\}}{E\{|S(j\omega)|^2\} + E\{|N(j\omega)|^2\}} \tag{4.4.41}$$

High-Resolution Detection (Correlated Wiener)

$$\frac{E\{S^*(j\omega) + N^*(j\omega)\}}{E\{|S(j\omega)|^2 + |N(j\omega)|^2 + S(j\omega)N^*(j\omega) + S^*(j\omega)N(j\omega)\}} \tag{4.4.42}$$

Inverse Detection (Deconvolution)

$$\frac{1}{E\{S(j\omega)\}} \tag{4.4.39}$$

Transitional 1 Detection (Inverse $\alpha = 0$, High-resolution $\alpha = 0.5$, Matched $\alpha = 1$)

$$\frac{E\{S^*(j\omega)\}}{(1 - \alpha)E\{|S(j\omega)|^2\} + \alpha E\{|N(j\omega)|^2\}} \tag{4.4.44}$$

Transitional 2 Detection (Inverse $\alpha = 0.5$)

$$\frac{1}{E\{S(j\omega)^{2(1-\alpha)}\}} \tag{4.4.45}$$

Autocorrelation Estimation (Uncorrelated Wiener)

$$\frac{E\{S^*(j\omega)|S(j\omega)|^2\}}{E\{|S(j\omega)|^2\} + E\{|N(j\omega)|^2\}} \tag{4.4.32}$$

Autocorrelation Estimation (Correlated Wiener)

$$\frac{E\{(S^*(j\omega) + N^*(j\omega))|S(j\omega)|^2\}}{E\{|S(j\omega)|^2 + |N(j\omega)|^2 + S(j\omega)N^*(j\omega) + S^*(j\omega)N(j\omega)\}} \tag{4.4.33}$$

Cross-Correlation Estimation (Uncorrelated Wiener)

$$\frac{E\{N^*(j\omega)|S(j\omega)|^2\}}{E\{|S(j\omega)|^2\} + E\{|N(j\omega)|^2\}} \tag{4.4.35}$$

Cross-Correlation Estimation (Correlated Wiener)

$$\frac{E\{S(j\omega)N^*(j\omega)(S^*(j\omega) + N^*(j\omega))\}}{E\{|S(j\omega)|^2 + |N(j\omega)|^2 + S(j\omega)N^*(j\omega) + S^*(j\omega)N(j\omega)\}} \tag{4.4.36}$$

4.5 ADAPTIVE FILTER CLASSIFICATION

It should be noted that the ideal filters in Table 4.4.1 can only be formulated if the expected signal spectrum $E\{S(j\omega)\}$ and expected noise spectrum $E\{N(j\omega)\}$ (and/or their auto- and cross-correlation spectra) are known apriori. Such apriori knowledge is seldom available so the algorithms must be reformulated using less information.

Ideal filter design can be classified in order of increasing difficulty based upon the apriori assumptions about the signal and noise models. The classes are:[9]

Class 1: Models for both the signal and noise expectation spectra (or power spectra) are known apriori with sufficient accuracy to synthesize the ideal filter.

Class 2: Models for either the signal expectation spectrum or the noise expectation spectrum but not both are known apriori. The unknown spectral model must be explicitly estimated aposteriori from the ideal filter input.

Class 3: Models for neither the signal nor the noise expectation spectra are known apriori. They must be estimated aposteriori from the ideal filter input. Of course, feature(s) of either the signal or the noise must be known apriori to allow the signal and noise to be distinguished.

Class 0 and Class 4 filters form limiting classes and are of no interest. In Class 0 filters, both the signal and noise spectra (actual, not expected) are known so no filtering is needed to recover the signal. In Class 4 filters, no distinguishing features between signal and noise are known so no filter can ever be formulated to separate them.

Class 1 filters change their transfer functions only when the expected signal and/or expected noise spectra (apriori estimates) are updated. Class 2 and 3 filters are always adapting their transfer functions since they are continuously estimating the unknown expected signal and/or noise spectra (aposteriori estimates). From this viewpoint, Class 2 and 3 filters are *dynamically adaptive* but Class 1 filters are only *statically adaptive* (i.e., dynamically nonadaptive). However all Class 1, 2, and 3 filters are adaptive since they tailor their transfer functions to the expected signal and noise spectral inputs they process. The classical and optimum filters of Chaps. 4.2–4.3 are nonadaptive. Adaptive filters have greater capability and potential than nonadaptive filters.

The algorithms of Table 4.4.1 apply to Class 1 filters only. They can be extended so that they apply to Class 2 and 3 filters by replacing the expectations $E\{\ \}$ by their estimates $\widehat{E}\{\ \}$. For example, the Class 1 uncorrelated estimation filter of Table 4.4.1 can be converted as

$$\text{Class 1}: \quad \frac{E\{|S(j\omega)|^2\}}{E\{|S(j\omega)|^2\} + E\{|N(j\omega)|^2\}}$$

$$\text{Class 2}: \quad \frac{E\{|S(j\omega)|^2\}}{E\{|S(j\omega)|^2\} + \widehat{E}\{|N(j\omega)|^2\}}$$

$$\text{Class 3}: \quad \frac{\widehat{E}\{|S(j\omega)|^2\}}{\widehat{E}\{|S(j\omega)|^2\} + \widehat{E}\{|N(j\omega)|^2\}}$$

(4.5.1)

4.5.1 SPECTRAL ESTIMATES

The unknown expected spectra must be estimated to formulate the filter. In the ideal filter arrangement of Fig. 4.4.1, the filter input $r(t)$ is the sum of signal $s(t)$ and noise $n(t)$. Therefore, Fourier transforming $r(t)$ and rearranging its spectrum $R(j\omega)$ gives

$$
\begin{aligned}
R(j\omega) &= S(j\omega) + N(j\omega) &\text{so}& \quad E\{R(j\omega)\} = E\{S(j\omega)\} + E\{N(j\omega)\} \\
N(j\omega) &= R(j\omega) - S(j\omega) &\text{so}& \quad E\{N(j\omega)\} = E\{R(j\omega)\} - E\{S(j\omega)\} \quad (4.5.2) \\
S(j\omega) &= R(j\omega) - N(j\omega) &\text{so}& \quad E\{S(j\omega)\} = E\{R(j\omega)\} - E\{N(j\omega)\}
\end{aligned}
$$

Once $S(j\omega)$ and $N(j\omega)$ have been combined as a sum to form $R(j\omega)$, they can only be separated by filtering. $R(j\omega)$ can be observed but $S(j\omega)$ and $N(j\omega)$ cannot be. Since $E\{N(j\omega)\}$ and/or $E\{S(j\omega)\}$ is unknown in Class 2 and 3 problems, $E\{R(j\omega)\}$ is unknown also. However $E\{R(j\omega)\}$ and other useful expectation spectra can be estimated as shall now be shown.

The input spectrum and the magnitude-squared input spectrum equal[10]

$$
\begin{aligned}
R(j\omega) &= S(j\omega) + N(j\omega) \\
|R(j\omega)|^2 &= |S(j\omega)|^2 + |N(j\omega)|^2 + S(j\omega)N^*(j\omega) + S^*(j\omega)N(j\omega)
\end{aligned} \quad (4.5.3)
$$

from Eq. 4.5.2a. Taking the expectation of Eq. 4.5.3 gives

$$
E\{R(j\omega)\} = E\{S(j\omega)\} + E\{N(j\omega)\}
$$
$$
E\{|R(j\omega)|^2\} = E\{|S(j\omega)|^2\} + E\{|N(j\omega)|^2\} + E\{S(j\omega)N^*(j\omega)\} + E\{S^*(j\omega)N(j\omega)\}
$$
$$(4.5.4)$$

When the expected noise spectrum has a zero mean (so $E\{N\} = 0$ from Eq. 1.17.5) and the expected signal and noise spectra are orthogonal (so $E\{SN^*\} = 0$ from Eq. 1.17.11), Eq. 4.5.4 simplifies to

$$
\begin{aligned}
E\{R(j\omega)\} &= E\{S(j\omega)\} \\
E\{|R(j\omega)|^2\} &= E\{|S(j\omega)|^2\} + E\{|N(j\omega)|^2\}
\end{aligned} \quad (4.5.5)
$$

Treating the signal and noise to be apriori orthogonal is equivalent to assuming that the noise spectrum has a zero mean and that the expected signal and noise spectra are uncorrelated (so $E\{SN^*\} = E\{S\}E\{N^*\}$ from Eq. 1.17.12) or independent (so $E\{f(S)g(N^*)\} = E\{f(S)\}E\{g(N^*)\}$ from Eq. 1.17.10).

Estimates of expectations are found by approximating expectations $E\{\ \}$ by averages $\langle\ \rangle$ (i.e., assume ergodicity). Replacing $E\{\ \}$ by $\langle\ \rangle$ on the left-hand-side of Eq. 4.5.5 gives $\hat{E}\{\ \}$ estimates involving the input spectrum as

$$
\begin{aligned}
\hat{E}\{R(j\omega)\} &\equiv \langle R(j\omega)\rangle \cong E\{S(j\omega)\} \\
\hat{E}\{|R(j\omega)|^2\} &\equiv \langle|R(j\omega)|^2\rangle \cong E\{|S(j\omega)|^2\} + E\{|N(j\omega)|^2\}
\end{aligned} \quad (4.5.6)
$$

Averaging or smoothing the squared-input $\langle |R|^2 \rangle = \langle |S|^2 \rangle + \langle |N|^2 \rangle + \langle SN^* + S^*N \rangle$ gives a good estimate of $E\{|S|^2\} + E\{|N|^2\}$. Smoothing reduces the aposteriori cross-correlation $(SN^* + S^*N)$ term as much as possible so that it approaches the apriori cross-correlation $E\{SN^* + S^*N\} \cong \langle SN^* + S^*N \rangle$. Combining Eqs. 4.5.4–4.5.6 in different ways gives a variety of estimates involving signal and noise spectra in terms of input spectra only as

$$\hat{E}\{S(j\omega)\} = \langle R(j\omega) \rangle$$
$$\hat{E}\{N(j\omega)\} = R(j\omega) - \langle R(j\omega) \rangle \quad \text{or} \quad \langle R(j\omega) - \langle R(j\omega) \rangle \rangle$$
$$\hat{E}\{|S(j\omega)|^2\} = |\langle R(j\omega) \rangle|^2 \tag{4.5.7}$$
$$\hat{E}\{|N(j\omega)|^2\} = \langle |R(j\omega)|^2 \rangle - |\langle R(j\omega) \rangle|^2$$
$$\hat{E}\{|S(j\omega)|^2 + |N(j\omega)|^2\} = \langle |R(j\omega)|^2 \rangle$$
$$\hat{E}\{S(j\omega)N^*(j\omega) + S^*(j\omega)N(j\omega)\} = |R(j\omega)|^2 - \langle |R(j\omega)|^2 \rangle$$

4.5.2 RELATION BETWEEN CLASS 1, 2, AND 3 ALGORITHMS

In Class 2 algorithms, the expected noise spectrum $E\{N\}$ is generally unknown and must be estimated. Taking the estimate of both sides of Eq. 4.5.2b gives

$$E\{N(j\omega)\} = E\{R(j\omega)\} - E\{S(j\omega)\}$$
$$\hat{E}\{N(j\omega)\} = \hat{E}\{R(j\omega)\} - \hat{E}\{S(j\omega)\} = R(j\omega) - kE\{S(j\omega)\} \tag{4.5.8}$$

This Class 2 estimate is more accurate than its Class 3 counterpart in Eq. 4.5.7b. The estimate $\hat{E}\{R\}$ of the expected input spectrum $E\{R\}$ is given by Eq. 4.5.6a. The estimate $\hat{E}\{S\}$ is assumed to equal k times the expected signal spectrum $E\{S\}$. In Class 2 filters, the expected signal spectrum $E\{S\}$ is known but the magnitude scaling constant k is unknown apriori. k must be estimated via some aposteriori algorithm.

Therefore Class 2 filters are derived from Class 1 filters by replacing $E\{N\}$ by the $\hat{E}\{N\}$ of Eq. 4.5.8b in all of the Class 1 ideal filters in Table 4.4.1. The resulting Class 2 filter algorithms are listed in Table 4.5.1. Almost all Class 2 transfer functions are more complicated than Class 1 transfer functions since they involve the total input spectrum $R(j\omega)$, some form of spectral smoothing or averaging of $R(j\omega)$ indicated by $\langle R(j\omega) \rangle$, and the expected signal spectrum $E\{S(j\omega)\}$. As long as the Class 2 transfer functions involve these terms but no k's, they can be readily implemented. Those involving k are more involved to realize since k must be estimated.

Class 3 filters are derived from Class 1 filters by using the estimates given by Eq. 4.5.7 in all of the Class 1 filters of Table 4.4.1. The resulting Class 3 filter algorithms are also listed in Table 4.5.1. They are simpler than the Class 1 and 2 types. For example, the general estimation filter has (Eq. 4.4.15)

$$H(j\omega) = \frac{E\{S(j\omega)R^*(j\omega)\}}{E\{|R(j\omega)|^2\}} \tag{4.5.9}$$

Evaluating this equation for the three classes of uncorrelated estimation filters gives

$$
\text{Class 1 :} \qquad \frac{E\{|S(j\omega)|^2\}}{E\{|S(j\omega)|^2\} + E\{|N(j\omega)|^2\}}
$$

$$
\text{Class 2 :} \qquad \frac{E\{|S(j\omega)|^2\}}{E\{|S(j\omega)|^2\} + E\{|R(j\omega) - kS(j\omega)|^2\}} \qquad (4.5.10)
$$

$$
\text{Class 3 :} \qquad \frac{|\langle R(j\omega)\rangle|^2}{\langle |R(j\omega)|^2\rangle}
$$

If k cannot be estimated, a different algorithm not involving k is used. For example, the Class 2 filter denominator containing k, S, and R can be replaced by the Class 3 filter denominator, which involves R only, as

$$
\text{Class 2 :} \qquad \frac{E\{|S(j\omega)|^2\}}{\langle |R(j\omega)|^2\rangle} \qquad (4.5.11)
$$

The numerator remains unchanged. This Class 2 algorithm is simpler to evaluate since k need not be estimated. Other Class 3 filters can be obtained using different estimates. Some of these Class 3 filters are listed in Table 9.10.1. Experience shows that adaptive filtering algorithms may sometimes best be generalized heuristically.

4.5.3 FREQUENCY DOMAIN SMOOTHING

Now let us consider the smoothing or averaging operator $\langle\ \rangle$ required by all Class 2 and 3 algorithms. One smoothing approach is to perform *block-to-block averaging* or *intra-block averaging* of the FFT's. Intra-block averaging is the same as *temporal* or *time domain averaging* where

$$
\langle R(j\omega)\rangle = R(j\omega)B(j\omega) \qquad \text{or} \qquad r_b(t) = r(t) * b(t) \qquad (4.5.12)
$$

Multiplying each spectral component $R(j\omega)$ with a window $B(j\omega)$ is equivalent to convolving $r(t)$ with $b(t)$. For example, the power spectrum $|R(j\omega)|^2$ can be averaged over many blocks M of overlapping data as

$$
\langle |R(j\omega)|^2\rangle \cong \frac{1}{M}\sum_{k=1}^{M} B_k |R_k(j\omega)|^2 \qquad (4.5.13)
$$

This reduces the effects of the undesirable but unavoidable aposteriori cross-correlations as discussed in Eq. 4.5.6b. However this also introduces more computational delay. Using a block weighting constant B_k gives more algorithm flexibility. Eq. 4.5.13 can be reexpressed in the Eq. 4.5.12 form as

$$
\langle |R(j\omega)|^2\rangle \cong \frac{1}{M}|R(j\omega)|^2 \sum_{k=1}^{M} B_k W(j\omega)e^{-j\omega k(1-r)NT}
$$

$$
B(j\omega) = \frac{1}{M}\sum_{k=1}^{M} B_k W(j\omega)e^{-j\omega k(1-r)NT} \qquad (4.5.14)
$$

Table 4.5.1   Ideal Class 2 and 3 filter transfer functions $H(j\omega)$.
(Also see Table 9.10.1.)

Filter Name	Class 2	Class 3

Distortion Balance ($W_S(j\omega)$ = Signal weighting, $W_N(j\omega)$ = Noise weighting)

$$\frac{E\{S(j\omega)W_S(j\omega)\}}{E\{R(j\omega)W_N(j\omega) - kS(j\omega)[W_N(j\omega) - W_S(j\omega)]\}}$$

$$\frac{\langle R(j\omega)W_S(j\omega)\rangle}{\langle R(j\omega)W_S(j\omega)\rangle + (R(j\omega) - \langle R(j\omega)\rangle)W_N(j\omega)}$$

Uncorrelated Estimation (Uncorrelated Wiener)

$$\frac{E\{|S(j\omega)|^2\}}{E\{|S(j\omega)|^2 + |R(j\omega) - kS(j\omega)|^2\}} \qquad \frac{|\langle R(j\omega)\rangle|^2}{\langle |R(j\omega)|^2\rangle}$$

Correlated Estimation (Correlated Wiener)

$$\frac{E\{S(j\omega)(R^*(j\omega) + (1-k)S^*(j\omega))\}}{E\{|R(j\omega) + (1-k)S(j\omega)|^2\}} \qquad \frac{\langle\langle R(j\omega)\rangle\ R^*(j\omega)\rangle}{\langle |R(j\omega)|^2\rangle}$$

Transitional 1 Estimation (Uncorrelated Wiener $\alpha = 0$, Correlated Wiener $\alpha = 1$)

$$\frac{E\{|S(j\omega)|^2 + \alpha S(j\omega)(R^*(j\omega) - kS^*(j\omega))\}}{E\{|S(j\omega)|^2 + |R(j\omega) - kS(j\omega)|^2 + \alpha[S(j\omega)(R^*(j\omega) - kS^*(j\omega)) + S^*(j\omega)(R(j\omega) - kS(j\omega))]\}}$$

$$\frac{|\langle R(j\omega)\rangle|^2 + (1-\alpha)\langle R(j\omega)\rangle(R^*(j\omega) - \langle R^*(j\omega)\rangle)}{(1-\alpha)\langle |R(j\omega)|^2\rangle}$$

Pulse-Shaping Estimation ($P(j\omega)$ = Desired pulse spectrum, Uncorrelated Wiener)

$$\frac{E\{P(j\omega)S^*(j\omega)\}}{E\{|S(j\omega)|^2 + |R(j\omega) - kS(j\omega)|^2\}} \qquad \frac{P(j\omega)\langle R^*(j\omega)\rangle}{\langle |R(j\omega)|^2\rangle}$$

Pulse-Shaping Estimation ($P(j\omega)$ = Desired pulse spectrum, Correlated Wiener)

$$\frac{E\{P(j\omega)(R^*(j\omega) + (1-k)S^*(j\omega))\}}{E\{|R(j\omega) + (1-k)S(j\omega)|^2\}} \qquad \frac{\langle P(j\omega)R^*(j\omega)\rangle}{\langle |R(j\omega)|^2\rangle}$$

Classical Detection (Matched)

$$\frac{E\{S^*(j\omega)\}}{E\{|R(j\omega) - kS(j\omega)|^2\}} \qquad \frac{\langle R^*(j\omega)\rangle}{|R(j\omega)\ - \langle R(j\omega)\rangle|^2}$$

High-Resolution Detection (Uncorrelated Wiener)

$$\frac{E\{S^*(j\omega)\}}{E\{|S(j\omega)|^2 + |R(j\omega) - kS(j\omega)|^2\}} \qquad \frac{\langle R^*(j\omega)\rangle}{\langle |R(j\omega)|^2\rangle}$$

Table 4.5.1   (Continued) Ideal Class 2 and 3 filter transfer functions $H(j\omega)$.

Filter Name	Class 2	Class 3

High-Resolution Detection (Correlated Wiener)

$$\frac{E\{R^*(j\omega) + (1-k)S^*(j\omega)\}}{E\{|R(j\omega) + (1-k)S(j\omega)|^2\}} \qquad\qquad \frac{\langle R^*(j\omega)\rangle}{\langle |R(j\omega)|^2\rangle}$$

Inverse Detection (Deconvolution)

$$\frac{1}{E\{S(j\omega)\}} \qquad\qquad \frac{1}{\langle R(j\omega)\rangle}$$

Transitional 1 Detection (Inverse $\alpha = 0$, High-Resolution $\alpha = 0.5$, Matched $\alpha = 1$)

$$\frac{E\{S^*(j\omega)\}}{E\{(1-\alpha)|S(j\omega)|^2 + \alpha|R(j\omega) - kS(j\omega)|^2\}}$$

$$\frac{\langle R^*(j\omega)\rangle}{(1-\alpha)|\langle R(j\omega)\rangle|^2 + \alpha|R(j\omega) - \langle R(j\omega)\rangle|^2}$$

Transitional 2 Detection (Inverse $\alpha = 0.5$)

$$\frac{1}{E\{S(j\omega)^{2(1-\alpha)}\}} \qquad\qquad \frac{1}{\langle R(j\omega)\rangle^{2(1-\alpha)}}$$

Autocorrelation Estimation (Uncorrelated Wiener)

$$\frac{E\{S^*(j\omega)|S(j\omega)|^2\}}{E\{|S(j\omega)|^2 + |R(j\omega) - kS(j\omega)|^2\}} \qquad \frac{\langle R^*(j\omega)\rangle\,|\langle R(j\omega)\rangle|^2}{\langle |R(j\omega)|^2\rangle}$$

Autocorrelation Estimation (Correlated Wiener)

$$\frac{E\{(R^*(j\omega) + (1-k)S^*(j\omega))|S(j\omega)|^2\}}{E\{|R(j\omega) + (1-k)S(j\omega)|^2\}} \qquad \frac{|\langle R(j\omega)\rangle|^2}{R(j\omega)}$$

Cross-Correlation Estimation (Uncorrelated Wiener)

$$\frac{E\{(R^*(j\omega) - kS^*(j\omega))|S(j\omega)|^2\}}{E\{|S(j\omega)|^2 + |R(j\omega) - kS(j\omega)|^2\}} \qquad \frac{(R^*(j\omega) - \langle R^*(j\omega)\rangle)\,|\langle R(j\omega)\rangle|^2}{\langle |R(j\omega)|^2\rangle}$$

Cross-Correlation Estimation (Correlated Wiener)

$$\frac{E\{S(j\omega)(R^*(j\omega) - kS^*(j\omega))(R^*(j\omega) + (1-k)S^*(j\omega))\}}{E\{|R(j\omega) + (1-k)S(j\omega)|^2\}}$$

$$\frac{(R^*(j\omega) - \langle R^*(j\omega)\rangle)\,\langle R(j\omega)\rangle}{R(j\omega)}$$

$W(j\omega)$ is given by Eq. 2.6.1 and r is the block overlap factor (see Chap. 2.11).

Smoothing within the block or *inner-block averaging* often achieves acceptable results without large delays. Inner-block averaging is the same as *spectral* or *frequency domain averaging* where

$$\langle R(j\omega)\rangle = \frac{1}{2\pi}R(j\omega) * B(j\omega) \qquad \text{or} \qquad r_b(t) = r(t)b(t) \qquad (4.5.15)$$

Averaging each spectral component by $B(j\omega)$ is equivalent to convolving $R(j\omega)$ with $B(j\omega)$. Such spectral averaging is identical to temporal weighting of the input $r(t)$ by $b(t)$. It reduces the spurious cross-correlation between $S(j\omega)$ and $N(j\omega)$ but also lowers the spectral resolution.

A variety of time domain windows $W(j\omega)$ were tabulated in Table 2.6.1. These can be used as frequency domain smoothing windows $B(j\omega)$ also. For example, the rectangular, triangular, Hann, and Gaussian windows equal ($\alpha = 2/p$)

$$\begin{aligned} B_R(j\omega) &= 1, & B_T(j\omega) &= 1 - \alpha\omega/\omega_s \\ B_H(j\omega) &= 0.5[1 + \cos(\alpha\omega/\omega_s)], & B_G(j\omega) &= e^{-2\alpha\omega/\omega_s} \end{aligned} \qquad (4.5.16)$$

for $|\omega| < p\omega_s/2$ and zero elsewhere. The smoothing width is selected to be some percentage p of the sampling frequency $1/T$ Hertz. For the Gaussian case, 100% smoothing corresponds to a width of two standard deviations. Now let us design several Class 2 and 3 adaptive filters.

4.6 CLASS 1 ADAPTIVE FILTER DESIGN

To illustrate the different characteristics of these filters, we present a design example.[7] Suppose that the expected signal $E\{s(t)\}$ is the tri-level pulse shown in Fig. 4.6.1a. The expected signal may be viewed as two 1 *msec* pulses (amplitudes 3 and 10) side-by-side or the sum of overlapping 1 *msec* (amplitude 7) and 2 *msec* (amplitude 3) pulses. The expected spectrum $E\{S(j\omega)\}$ of $E\{s(t)\}$ is shown in Fig. 4.6.1b. It is composed of two $(\sin x/x)$ magnitude characteristics having zero-crossing bandwidths $1/T$ of $1/1$ *msec* $= 1$ KHz and $1/2$ *msec* $= 500$ Hz, respectively.

The expected noise is assumed to be the sum of a $1/f^2$ power spectrum and a flat power spectrum as shown in Fig. 4.6.1c. Their relative amplitudes are chosen so their spectral magnitudes cross at 1 KHz. Their phases are assumed to be random and uniformly distributed between $-\pi$ and π.

The total input $r(t)$ is shown in Fig. 4.6.1d. Here three tri-level signal pulses like the one shown in Fig. 4.6.1a have been combined at times $t = 20$, 40, and 260 *msec*. The levels have been chosen so the peak SNR is about 0 dB and the average SNR is about -20 dB. This input is applied to the filter of Fig. 4.4.1. The following six different filters will be considered:

1. Absolute estimation

$$H_1(j\omega) = \frac{E\{|S(j\omega)|\}}{E\{|S(j\omega)|\} + E\{|N(j\omega)|\}} \qquad (4.6.1)$$

2. Wiener estimation

$$H_2(j\omega) = \frac{E\{|S(j\omega)|^2\}}{E\{|S(j\omega)|^2\} + E\{|N(j\omega)|^2\}} \qquad (4.6.2)$$

3. Squared estimation

$$H_3(j\omega) = H_2^2(j\omega) \qquad (4.6.3)$$

4. Detection with square root noise power

$$H_4(j\omega) = \frac{E\{S^*(j\omega)\}}{E\{|N(j\omega)|\}} \qquad (4.6.4)$$

5. Classical detection

$$H_5(j\omega) = \frac{E\{S^*(j\omega)\}}{E\{|N(j\omega)|^2\}} \qquad (4.6.5)$$

6. High-resolution detection

$$H_6(j\omega) = \frac{E\{S^*(j\omega)\}}{E\{|S(j\omega)|^2\} + E\{|N(j\omega)|^2\}} \qquad (4.6.6)$$

The absolute estimation filter $H_1(j\omega)$ is a type of distortion balance filter of Eq. 4.4.7 having weightings $W_S(j\omega) = [S^*(j\omega)/S(j\omega)]^{1/2}$ and $W_N(j\omega) = [N^*(j\omega)/N(j\omega)]^{1/2}$. This filter discriminates between rms signal and noise levels unlike the filter in Eq. 4.4.8 which discriminates between complex signal and noise levels. The Wiener estimation filter $H_2(j\omega)$ matches Eq. 4.4.25. The squared estimation filter $H_3(j\omega)$ is the square of the Wiener filter $H_2(j\omega)$. These are all special cases of Eq. 4.4.1 where $k = 1, 2,$ and 4, respectively. The detection filter with expected square-root noise power $H_4(j\omega)$ is a special case of the general detection filter given by Eq. 4.4.18. Here the power spectrum of the signal to be rejected is the square root of the expected noise power spectrum. The classical detection filter $H_5(j\omega)$ and high-resolution detection filter $H_6(j\omega)$ match Eqs. 4.4.37 and 4.4.41, respectively.

Substituting the expected signal spectrum $E\{S(j\omega)\}$ of Fig. 4.6.1b and the expected noise spectrum $E\{|N(j\omega)|\}$ of Fig. 4.6.1c into Eqs. 4.6.1–4.6.6 yields the ideal estimation filter responses of Fig. 4.6.2a and the ideal detection filter responses of Fig. 4.6.2b. Since $E\{|S(j\omega)|\} \cong 250$ and $E\{|N(j\omega)|\} \sim 1/\omega$ as $\omega \to 0$, all the filters exhibit zero dc gains. Their asymptotic low-frequency behavior is bounded by

$$|H_1(j\omega)| = \frac{E\{S(j\omega)\}}{E\{|N(j\omega)|\}} = |H_4(j\omega)| \sim \omega, \qquad |H_2(j\omega)| = |H_1(j\omega)|^2 \sim \omega^2,$$

$$|H_3(j\omega)| = |H_1(j\omega)|^4 \sim \omega^4, \qquad |H_5(j\omega)| = |H_6(j\omega)| = \frac{E\{|S(j\omega)|\}}{E\{|N(j\omega)|\}^2} \sim \omega^2 \qquad (4.6.7)$$

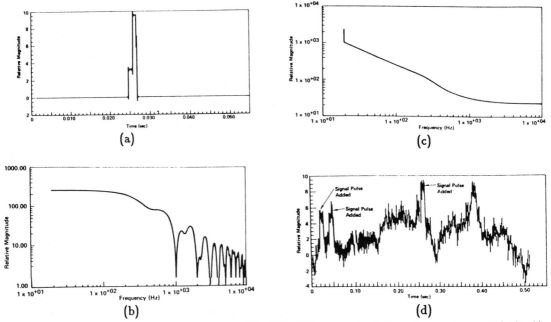

Fig. 4.6.1 (a) Expected signal pulse $E\{s(t)\}$, (b) expected signal spectrum $E\{S(j\omega)\}$, (c) expected noise spectrum $E\{N(j\omega)\}$, and (d) total input $r(t)$. (From **W.H. Haas** and **C.S. Lindquist**, "New classes of ideal linear filters for filter design applications," Rec. 13th Asilomar Conf., pp. 294–297, ©1979, IEEE.)

Since $E\{|N(j\omega)|\} \sim 1$ as $\omega \to \infty$ and $E\{|S(j\omega)|\} = (\sin x/x)$, the high-frequency filter gains vary as $(\sin x/x)$ (for H_1, H_4, H_5, H_6), $(\sin x/x)^2$ (for H_2), or $(\sin x/x)^4$ (for H_3). This high-frequency behavior is clearly seen in Figs. 4.6.2a and 4.6.2b. The ideal filter required is bandpass with a *comb* (like fingers of a comb) characteristic (they are repetitive but not periodic).

The time domain responses of the estimation type filters are shown in Fig. 4.6.3 and the detection type filters are shown in Fig. 4.6.4. From these figures, it becomes clear that the ideal filter transfer functions in Eqs. 4.6.1–4.6.6 have been listed in order of increasing signal distortion. Comparing the filter input $r(t)$ with these filter outputs, the following comments can be made.

The estimation filters recover the signal $s(t)$ from the input $r(t)$ as best they can. The absolute estimation filter $H_1(j\omega)$ response in Fig. 4.6.3a recovers the three pulses fairly well at $t = 20, 40$, and 260 *msec* but allows the spurious or noise pulses at $t = 280$ and 380 *msec* to pass. The Wiener estimation filter $H_2(j\omega)$ response in Fig. 4.6.3b allows less low-frequency signal and noise and high-frequency noise to pass (compare with Fig. 4.6.2a). This $H_2(j\omega)$ response is a bandpass filtered version of the $H_1(j\omega)$ response having 3 dB frequencies of about 100 Hz and 2 KHz. The squared estimation filter $H_3(j\omega)$ response in Fig. 4.6.3c is poorer because it is even more band-limited than $H_2(j\omega)$. The $H_3(j\omega)$ response appears to be a bandpass filtered version of the $H_2(j\omega)$ response and has 3 dB frequencies of about 550 Hz and 750 Hz. From a filter response

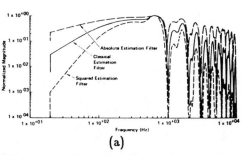 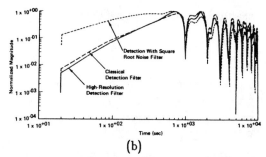

(a)   (b)

Fig. 4.6.2 Magnitude responses of (a) ideal estimation filters and (b) ideal detection filters. (From W.H. Haas and C.S. Lindquist, loc. cit., ©1979, IEEE.)

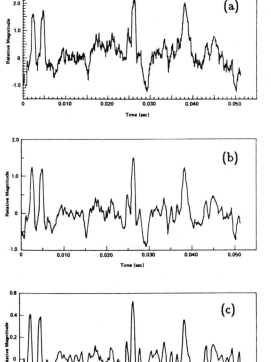 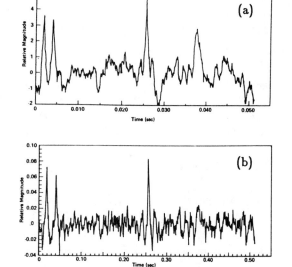

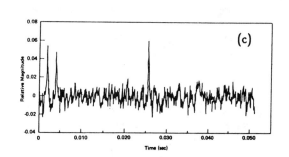

Fig. 4.6.3 Time domain responses of (a) absolute estimation H_1, (b) Wiener estimation H_2, and (c) squared-estimation H_3 filters. (From W.H. Haas and C.S. Lindquist, loc. cit., ©1979, IEEE.)

Fig. 4.6.4 Time domain responses of (a) detection with square root noise H_4, (b) classical detection H_5, and (c) high-resolution detection H_6 filters. (From W.H. Haas and C.S. Lindquist, loc. cit., ©1979, IEEE.)

standpoint, the impulse or pulse response of $H_3(j\omega)$ will tend to equal[11]

$$r_{BP}(t) = r_{LP}(Bt/2)\cos(\omega_o t) \tag{4.6.8}$$

ω_o is the center frequency of the filter where $f_o \cong \sqrt{550(750)} = 650$ Hz, B is the filter bandwidth where $B \cong 750 - 550 = 200$ Hz, and $r_{LP}(t)$ is the step response of a fourth-order synchronously-tuned lowpass filter having $H_{LP}(s) = 1/(s+1)^4$. It is clear from Fig. 4.6.3c that we tend to have a *ringing* response where the period of the oscillation is about 1.5 *msec* (or frequency $f_o \cong 1/1.5$ *msec* $\cong 650$ Hz) and envelope rise time is about 2 *msec* (or $f_{3dB} \cong 0.35/2$ *msec* $\cong 200$ Hz).

The detection filters detect the presence of the signal $s(t)$ in the input $r(t)$ as best they can. The detection filter with square root noise $H_4(j\omega)$ response in Fig. 4.6.4a correctly indicates that signal is present at $t = 20$, 40, and 260 *msec* but incorrectly indicates that signal is present at $t = 280$ and 380 *msec*. The time domain responses of $H_1(j\omega)$ and $H_4(j\omega)$ are not too different as we would expect since Eqs. 4.6.1 and 4.6.4 show their frequency responses are essentially equal in low spectral SNR environments. The ideal detection filter $H_5(j\omega)$ response gives good signal detection at $t = 20$, 40, and 260 *msec*. It surpresses the spurious signal detection pulses at $t = 280$ and 380 *msec* unlike $H_4(j\omega)$. The high-resolution filter $H_6(j\omega)$ response is even more improved where pulses of level 0.05 occurring at $t = 20$, 40, and 260 *msec* rise above the fairly constant noise floor of average peak level 0.01.

4.7 CLASS 2 ADAPTIVE FILTER DESIGN

To illustrate the use and performance of the Class 2 algorithms, we select the high-resolution detection filter of Table 4.5.1 and the filter structure of Fig. 4.6.1a.[12] Next we arbitrarily choose the signal and noise models. The expected signal is selected to be a square pulse of length 3.2 seconds centered about zero where (see Fig. 4.7.1a)

$$E\{s(t)\} = U_{-1}(t+3.2) - U_{-1}(t-3.2) \tag{4.7.1}$$

The expected signal spectrum equals

$$E\{S(j\omega)\} = 6.4\,\frac{\sin 3.2\omega}{3.2\omega} \tag{4.7.2}$$

(see Eqs. 1.6.19 and 1.6.22). The expected noise is assumed to be random and Gaussian with a magnitude spectrum of

$$
\begin{aligned}
E\{|N(j\omega)|\} &= 1,\ |\omega| < B\ \text{(and 0, otherwise)}, &&\text{band} - \text{limited white noise}\\
&= 1/\omega, &&\text{integrated white noise} \quad\quad (4.7.3)\\
&= 1/\omega^2, &&\text{doubly integrated white noise}
\end{aligned}
$$

The expected phase spectrum is assumed to be uniformly distributed between $-180°$ and $180°$. The expected noise is computed from its expected spectrum using Eqs. 4.4.51b and 4.4.55.

Consider the band-limited white noise case. The input signal is delayed 20.2 seconds from its expected value and magnitude scaled by a constant k. It begins at 17 seconds and ends at 23.4 seconds. The total input spectrum is $R(j\omega) = ke^{-j20.2\omega}E\{S(j\omega)\} + E\{N(j\omega)\}$. Inverse Fourier transforming this spectrum gives the input $r(t)$ shown in Fig. 4.7.1b. The peak signal-to-rms noise ratio was adjusted by k to equal approximately -20 dB which is an arbitrary choice. When k is sufficiently small, then noise dominates the input and $R(j\omega) \cong E\{N(j\omega)\}$ which is the situation in Fig. 4.7.1b. When k is sufficiently large, then $R(j\omega) \cong kE\{S(j\omega)\}$. Substituting the expected signal and noise spectra into the Class 1 high-resolution $H(j\omega)$ equation of Table 4.4.1 yields

$$H(j\omega) = \frac{E\{S^*(j\omega)\}}{E\{|S(j\omega)|\}^2 + E\{|N(j\omega)|^2\}} = \frac{6.4(\sin 3.2\omega)/3.2\omega}{1 + [6.4(\sin 3.2\omega)/3.2\omega]^2} \qquad (4.7.4)$$

Since $E\{|N(j\omega)|\} \gg E\{|S(j\omega)|\}$, then $H(j\omega)$ reduces to

$$H(j\omega) \cong \frac{E\{S^*(j\omega)\}}{E\{|N(j\omega)|^2\}} = E\{S^*(j\omega)\} \qquad (4.7.5)$$

The ideal filter response is shown in Fig. 4.7.1d. The filter behaves as a correlation filter which convolves the input with the expected signal (see Chap. 4.4.7). Passing the input $r(t)$ through this filter yields the output $c(t)$ shown in Fig. 4.7.1c where

$$C(j\omega) \cong E\{S^*(j\omega)\}R(j\omega) = kE\{S(j\omega)\}E\{S^*(j\omega)\} + E\{N(j\omega)\}E\{S^*(j\omega)\}$$
$$c(t) \cong kE\{s(t)\} * E\{s^*(-t)\} + E\{n(t)\} * E\{s^*(-t)\} \qquad (4.7.6)$$

which is the sum of the matched (or coherent) component $E\{s(t)\} * E\{s^*(-t)\}$ and the unmatched (or noncoherent) component $E\{n(t)\} * E\{s^*(-t)\}$. The output tends to equal a triangular autocorrelation pulse (matched component) which is twice as wide as the input pulse. The output cross-correlation noise (unmatched component) level is very low. This filter provides an average SNR gain of 14.3.

Another filter is shown in Fig. 4.7.2 for the same signal pulse but noise is assumed to have a $1/\omega^2$ magnitude spectrum where

$$H(j\omega) = \frac{\omega^2[6.4(\sin 3.2\omega)/3.2\omega]}{1 + \omega^2[6.4(\sin 3.2\omega)/3.2\omega]^2} \qquad (4.7.7)$$

SNR_i is adjusted to equal -20 dB. Since $E\{|N(j\omega)|\} \gg E\{|S(j\omega)|\}$, then from Eq. 4.7.5, $H(j\omega)$ approximately equals

$$H(j\omega) \cong \frac{E\{S^*(j\omega)\}}{E\{|N(j\omega)|^2\}} = \omega^4 E\{S^*(j\omega)\} \qquad (4.7.8)$$

The ideal filter gain is ω^4 times the autocorrelation filter gain in Fig. 4.7.1. The filter behaves as an autocorrelator followed by four differentiators. Differentiating the triangular input pulse of Fig. 4.7.1c four times produces a large narrow output pulse

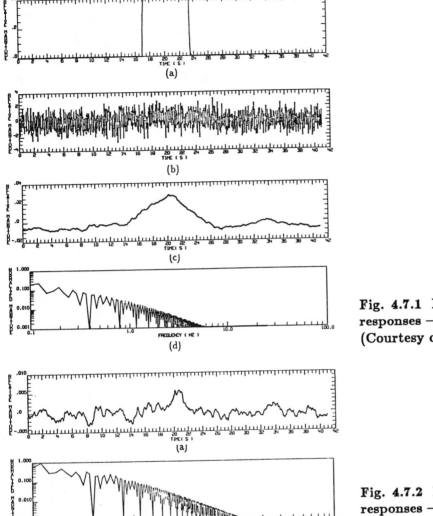

Fig. 4.7.1 Nonadaptive filter
responses – white noise.
(Courtesy of W.H. Haas.[9])

Fig. 4.7.2 Nonadaptive filter
responses – $1/f^2$ noise.
(Courtesy of W.H. Haas.[9])

(high-resolution theory would predict a triplet) located at 20.2 seconds, the peak time of
the triangular pulse. There are spurious half-amplitude negative output pulses located
at 17 and 23.4 seconds, the edges of the input triangular pulse. The SNR gain is 68.1.

Now we turn to the Class 2 adaptive filter. Using the Class 2 high-resolution
detection filter from Table 4.5.1 where

$$H(j\omega) = \frac{E\{S^*(j\omega)\}}{E\{|S(j\omega)|\}^2 + E\{|N(j\omega)|\}^2} \simeq \frac{E\{S^*(j\omega)\}}{\langle |R(j\omega)|^2 \rangle} \qquad (4.7.9)$$

$\langle |R|^2 \rangle$ is an estimate of $E\{|S|\}^2 + E\{|N|\}^2$ which is the input power spectral density
(PSD). It is obtained by smoothing or averaging the power spectrum $|R|^2$ of the filter

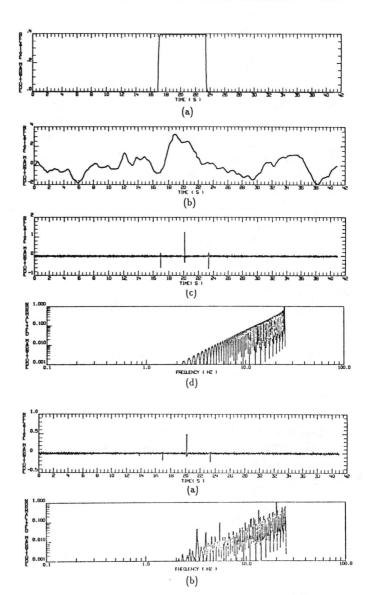

Fig. 4.7.3 Adaptive filter responses – white noise. (Courtesy of W.H. Haas.[9])

Fig. 4.7.4 Adaptive filter responses – $1/f^2$ noise. (Courtesy of W.H. Haas.[9])

input. Smoothing the power spectrum is necessary to reduce the effects of aposteriori cross-correlation between $S(j\omega)$ and $N(j\omega)$. Although the signal and noise spectra are assumed to be uncorrelated apriori (i.e., $E\{SN^*\} = 0$), once specific members of the S and N ensembles are used, $S(j\omega)$ and $N(j\omega)$ will be correlated aposteriori where $S(j\omega)N(j\omega) \neq 0$. Smoothing can be applied temporally (between $R(j\omega)$ obtained from different FFT blocks) or spectrally (between several $R(j\omega)$ within the same block). In this design $\langle |R(j\omega)|^2 \rangle$ is formed by spectrally averaging $R(j\omega)$ with its two neighbors on both sides as

$$\langle R(j\omega) \rangle \cong \frac{R(j\omega_{k-1}) + R(j\omega_k) + R(j\omega_{k+1})}{3}, \qquad 0 \leq k < N \qquad (4.7.10)$$

The same signal pulse and white noise are input to this Class 2 adaptive filter as shown in Fig. 4.7.1a–b. The resulting adaptive filter transfer function (Fig. 4.7.3b) is shown with the filter output in Fig. 4.7.3a. Comparing this response with that shown in Fig. 4.7.1c, we see it is somewhat degraded with a higher noise content. The SNR gain is 9.1 which is below the SNR gain of 14.3 found in Fig. 4.7.1. The Class 2 adaptive filter does not give quite as good a SNR gain performance as the Class 1 adaptive filter. This is to be expected since Class 2 filters are designed using less apriori information than Class 1 filters.

Repeating this process with the same signal pulse but doubly-integrated white noise in Fig. 4.7.2a–b yields the transfer function (Fig. 4.7.4b) and the filter output shown in Fig. 4.7.4a. These responses are very close. Their transfer functions differ only slightly. The SNR gain is 57.5 which is slightly below the SNR gain of 68.1 provided by the Class 1 adaptive filter in Fig. 4.7.2.

This example shows that adaptive filtering algorithms are simple to apply and give impressive results. Here no noise model was assumed but only the spectral shape $E\{S^*(j\omega)\}$ of the expected signal, not its level, was used. The filter adapted its shape to the received input $R(j\omega)$. This particular simulation shows the algorithm works well when the signal is weak–exactly when we want it to.

4.8 CLASS 3 ADAPTIVE FILTER DESIGN

Class 3 filters are the most difficult to design since expectation spectra must be estimated aposteriori from the filter input. Their transfer functions are given in Table 4.5.1. They have surprisingly simple forms–simpler than Class 1 and 2 algorithms. They require no apriori knowledge about the signal and noise that are being processed except that the noise spectrum must have zero mean.

For example, the general estimation filter of Table 4.4.1 equals

$$H(j\omega) = \frac{E\{S(j\omega)R^*(j\omega)\}}{E\{|R(j\omega)|^2\}} \tag{4.8.1}$$

Evaluating this equation for the three classes of correlated estimation filters gives

$$\text{Class 1}: \quad \frac{E\{S(j\omega)(S^*(j\omega) + N^*(j\omega))\}}{E\{|S(j\omega) + N(j\omega)|^2\}}$$

$$\text{Class 2}: \quad \frac{E\{S(j\omega)(R^*(j\omega) + (1-k)S^*(j\omega))\}}{\langle|R(j\omega)|^2\rangle_p} \tag{4.8.2}$$

$$\text{Class 3}: \quad \frac{\langle\langle R(j\omega)\rangle_p R^*(j\omega)\rangle_p}{\langle|R(j\omega)|^2\rangle_{2p}}$$

The uncorrelated estimation filters were described by Eqs. 4.5.10 and 4.5.11. Class 2 and 3 filters require smoothing as indicated by the $\langle \, \rangle$ operator. A variety of smoothing windows are listed in Table 2.6.1. Typical frequency domain smoothing windows are the rectangular, triangular, Hann, and Gaussian as given by Eq. 4.5.16. The subscript p

indicates the percentage width windows used in the smoothing. The smoothing window width is selected to be some percentage p of the sampling frequency $1/T$ Hertz. The smoothing window is generally nonzero for $|\omega| < p\omega_s/2$ and zero elsewhere.

It is important to use the proper smoothing width windows in both the numerator and denominator of Eqs. 4.8.2b and 4.8.2c. The percentage smoothing p of the numerator should be equal to (or less than) the p of the denominator. This is not the case in $H_1(j\omega)$ given by Eq. 4.8.2c unless the denominator window has twice the width $(2p)$ of the two numerator windows (each p).

Some other Class 3 correlated estimation filters are

$$H_2(j\omega) = \frac{\langle\langle R(j\omega)\rangle_p R^*(j\omega)\rangle_p}{\langle|R(j\omega)|^2\rangle_p}$$

$$H_3(j\omega) = \frac{\langle R(j\omega)\rangle_p R^*(j\omega)}{\langle|R(j\omega)|^2\rangle_p} \qquad (4.8.3)$$

$$H_4(j\omega) = \frac{\langle\langle R(j\omega)\rangle_p R^*(j\omega)\rangle_p}{\langle\langle|R(j\omega)|^2\rangle_p\rangle_p}$$

The H_2 algorithm tends to produce large peaking over $p/2$ FFT bins on either side of the passband. H_2 is a smoothed version of H_3. The H_3 algorithm produces better results but is sensitive to noise since only single smoothing is used in the numerator. The H_4 algorithm uses double smoothing in the denominator. It tends to produce moderate peaking over the $p/2$ bins on either side of the passband like H_2. We use the H_1 algorithm given by Eq. 4.8.2c because it tends to have a monotonic gain (rolls off) over the $p/2$ bins on either side of the passband.

Consider the following Class 3 estimation problem. Assume that the expected signal is a continuous cosine at FFT frequency bins (3, 125) and the expected noise is white. 128-point FFT's are used. For comparison purposes, the input rms signal-to-noise ratio SNR_i, output rms SNR_o, rms SNR gain, and rms distortion or error indices are used:

$$SNR_{idB} = 20\log_{10}\left[\frac{s_i(t)_{rms}}{n_i(t)_{rms}}\right], \qquad SNR_{odB} = 20\log_{10}\left[\frac{s_o(t)_{rms}}{n_o(t)_{rms}}\right]$$

$$G_{SNR} = SNR_{odB} - SNR_{idB}, \qquad \text{Distortion} = \sqrt{\frac{\int_0^{NT}|c(t) - s_o(t)|^2 dt}{\int_0^{NT}|s_o(t)|^2 dt}} = \frac{1}{SNR_o}$$

$$(4.8.4)$$

Several typical correlated estimation (CE) filter inputs, outputs, and gains are shown in Fig. 4.8.1 using Eq. 4.8.2c. These responses are for rectangular frequency smoothing for window widths p of 3, 7, 12, and 18% at a constant SNR_i ratio of 9 dB. The uncorrelated estimation (UE) filter responses are shown in Fig. 4.8.2 using a gain from Table 4.5.1 of

$$H(j\omega) = \frac{|\langle R(j\omega)\rangle_p|^2}{\langle|R(j\omega)|^2\rangle_p} \qquad (4.8.5)$$

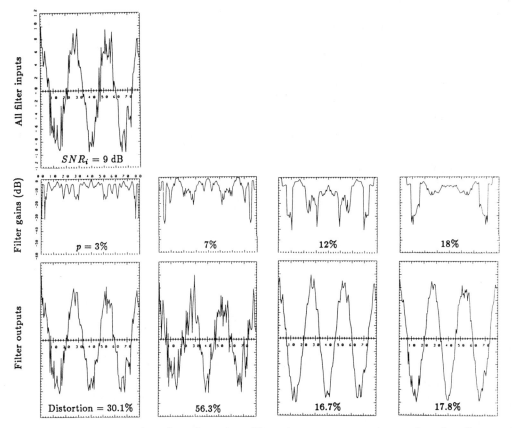

Fig. 4.8.1 Typical correlated estimation filter inputs, outputs, and gains for rectangular frequency smoothing at a constant SNR$_i$ of 9 dB. (From N. Rizq, "Theory and design of Class 3 adaptive filters," M.S.E.E. Thesis, 146 p., Univ. of Miami, Dec. 1986.)

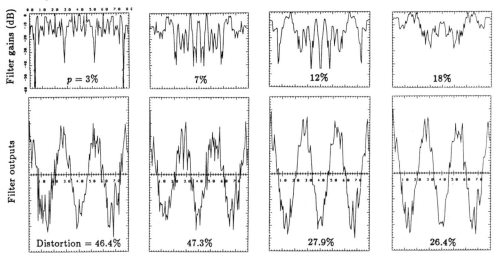

Fig. 4.8.2 Typical uncorrelated estimation filter inputs, outputs, and gains for rectangular frequency smoothing at a constant SNR$_i$ of 9 dB. (From N. Rizq, loc. cit.)

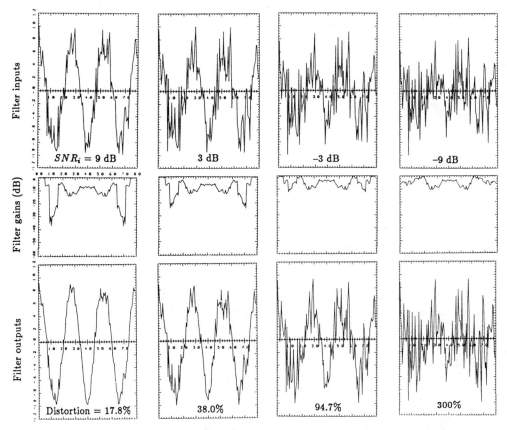

Fig. 4.8.3 Typical correlated estimation filter inputs, outputs, and gains for 18% rectangular smoothing. (From N. Rizq, loc. cit.)

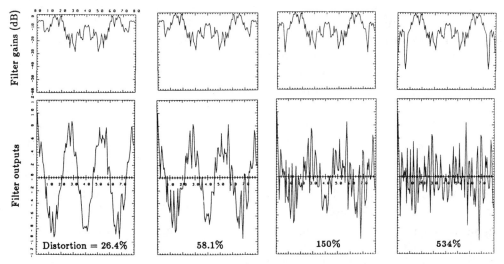

Fig. 4.8.4 Typical uncorrelated estimation filter inputs, outputs, and gains for 18% rectangular smoothing. (From N. Rizq, loc. cit.)

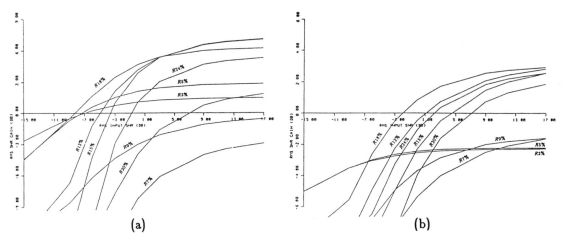

Fig. 4.8.5 SNR gain as a function of input SNR$_i$ and window width p for (a) CE and (b) UE filters. (From N. Rizq, loc. cit.)

The CE filter gains are slightly smoother than the UE filter gains. The CE filters do not provide quite as low an average stopband gain as the UE filters. For example, the CE filter has an average stopband gain of about –10 dB compared to about –15 dB for the UE filter ($p = 7\%$). Thus, the CE filters pass slightly more noise than the UE filters. The CE filters have slightly higher passband gain than the UE filters, so they also pass slightly more signal. For example, the bin 3 gains are about –3 dB for CE and –10 dB for UE ($p = 7\%$). Therefore the passband/stopband gain ratio is about –3 + 10 = 7 dB for CE and –10 + 15 = 5 dB for UE filters. Thus the CE filter provides slightly more processing gain (7 – 5 = 2 dB) than the UE filter. This better performance is expected since the correlated filter uses phase information unlike the uncorrelated filter.

Another set of correlated estimation filter responses are shown in Fig. 4.8.3 for various SNR_i levels using 18% smoothing and variable SNR_i. The corresponding set for uncorrelated estimation filters are shown in Fig. 4.8.4. We see that the CE filters are effective in recovering signal from the noise at SNR_i greater than about 3 dB while the UE filters must have SNR_i greater than 9 dB. The CE filters are superior since they require less SNR_i to operate well.

The SNR gain is plotted as a function of input SNR_i for various window widths p in Fig. 4.8.5. These curves only apply to the sine/cosine signal–white noise case. There is an input SNR_i threshold above which there is SNR gain. Below this threshold, the filter does not recover the signal well. The lowest SNR_i threshold for CE filters is about –8 dB for windows having about 4% and 18% width. The highest SNR gain of 8 dB occurs for about 15% smoothing. The UE filters provide SNR gain for window widths greater than about 10%. The lowest SNR_i threshold for UE filters is about –2 dB for 18% window width. The largest SNR gain is about 3 dB. For Class 3 filters,[10] reducing the signal frequency bin m_o or increasing the window width p so that $pN/2 \geq 2m_o$ or $p \geq 4m_o/N$ causes the filter gain to notch at frequency m_o with corresponding loss in the SNR gain around $p = 4m_o/N$. For CE filters, $p \geq 2m_o/N$ instead and the gain notches at $p = 2m_o/N$. For $N = 128$ and $m_o = 3$, this occurs around $p = 4(3)/128 \cong 9\%$ for

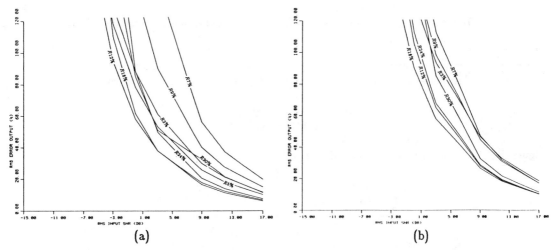

(a)   (b)

Fig. 4.8.6 Output distortion as a function of input SNR$_i$ and window width p for (a) CE and (b) UE filters. (From N. Rizq, loc. cit.)

CE filters and $p = 2(3)/128 \cong 5\%$ for UE filters. Fig. 4.8.5 confirms that there is only SNR loss, not gain, around these smoothing widths.

The rms output distortion is plotted in Fig. 4.8.6. It decreases as the SNR_i increases. Using a smoothing width around 18% gives the minimum distortion for any SNR_i level. The CE and UE curves are almost identical except that they are shifted from one another by about 4 dB. Therefore the CE filters can provide the same distortion performance but at 4 dB lower SNR_i than the UE filters.

4.9 ADAPTIVE FILTERS USING FEEDBACK

The adaptive filter system shown in Fig. 4.4.1 and redrawn in Fig. 4.9.1a is nonfeedback in nature. The gain block is denoted as $T(j\omega)$. In some applications, the unity-gain feedback system of Fig. 4.9.1b is necessary. The nonfeedback algorithms of Tables 4.4.1 and 4.5.1 that describe $T(j\omega)$ in Fig. 4.9.1a are easily converted into a new set of feedback algorithms that describe $G(j\omega)$ in Fig. 4.9.1b as

$$T(j\omega) = \frac{G(j\omega)}{1 + G(j\omega)} \tag{4.9.1}$$

by Mason's gain rule. Solving for $G(j\omega)$ gives

$$G(j\omega) = \frac{T(j\omega)}{1 - T(j\omega)} \tag{4.9.2}$$

Substituting $T(j\omega)$ from Tables 4.4.1 and 4.5.1 into Eq. 4.9.2 and solving for $G(j\omega)$ yields the unity-gain feedback algorithms listed in Tables 4.9.1 and 4.9.2.

The unity-feedforward gain system of Fig. 4.9.1c which describes $H(j\omega)$ is also easily described as[14]

Table 4.9.1 Ideal Class 1 feedback filter transfer functions $G(j\omega)$.

Filter Name	Class 1

Distortion Balance $(W_S(j\omega) = \text{Signal weighting}, W_N(j\omega) = \text{Noise weighting})$

$$\frac{E\{S(j\omega)W_S(j\omega)\}}{E\{N(j\omega)W_N(j\omega)\}}$$

General Estimation $(D(j\omega) = \text{Desired output})$

$$\frac{H_E(j\omega)}{1 - H_E(j\omega)} = \frac{E\{D(j\omega)R^*(j\omega)\}}{E\{(R(j\omega) - D(j\omega))R^*(j\omega)\}}$$

General Detection

$$\frac{H_D(j\omega)}{1 - H_D(j\omega)} = \frac{E\{D^*(j\omega)\}}{E\{R(j\omega)R^*(j\omega) - D^*(j\omega)\}}$$

Estimation–Detection

$$\frac{H_E(j\omega)H_D(j\omega)}{1 - H_E(j\omega)H_D(j\omega)}$$

Uncorrelated Estimation (Uncorrelated Wiener)

$$\frac{E\{|S(j\omega)|^2\}}{E\{|N(j\omega)|^2\}}$$

Correlated Estimation (Correlated Wiener)

$$\frac{E\{|S(j\omega)|^2 + S(j\omega)N^*(j\omega)\}}{E\{|N(j\omega)|^2 + S^*(j\omega)N(j\omega)\}}$$

Transitional 1 Estimation (Uncorrelated Wiener $\alpha = 0$, Correlated Wiener $\alpha = 1$)

$$\frac{E\{|S(j\omega)|^2 + \alpha S(j\omega)N^*(j\omega)\}}{E\{|N(j\omega)|^2 + \alpha S^*(j\omega)N(j\omega)\}}$$

Pulse-Shaping Estimation $(P(j\omega) = \text{Desired pulse spectrum, Uncorrelated Wiener})$

$$\frac{E\{P(j\omega)S^*(j\omega)\}}{E\{|S(j\omega)|^2 + |N(j\omega)|^2 - P(j\omega)S^*(j\omega)\}}$$

Pulse-Shaping Estimation $(P(j\omega) = \text{Desired pulse spectrum, Correlated Wiener})$

$$\frac{E\{P(j\omega)(S^*(j\omega) + N^*(j\omega))\}}{E\{|S(j\omega)|^2 + |N(j\omega)|^2 + (S(j\omega) - P(j\omega))N^*(j\omega)) + S^*(j\omega)(N(j\omega) - P(j\omega))\}}$$

Table 4.9.1 (Continued) Ideal Class 1 feedback filter transfer functions $G(j\omega)$.

Filter Name	Class 1

Classical Detection (Matched)

$$\frac{E\{S^*(j\omega)\}}{E\{|N(j\omega)|^2 - S^*(j\omega)\}}$$

High-Resolution Detection (Uncorrelated Wiener)

$$\frac{E\{S^*(j\omega)\}}{E\{|S(j\omega)|^2 + |N(j\omega)|^2 - S^*(j\omega)\}}$$

High-Resolution Detection (Correlated Wiener)

$$\frac{E\{S^*(j\omega) + N^*(j\omega)\}}{E\{(S(j\omega) + N(j\omega) - 1)(S^*(j\omega) + N^*(j\omega))\}}$$

Inverse Detection (Deconvolution)

$$\frac{1}{E\{S(j\omega) - 1\}}$$

Transitional 1 Detection (Inverse $\alpha = 0$, High-resolution $\alpha = 0.5$, Matched $\alpha = 1$)

$$\frac{E\{S^*(j\omega)\}}{E\{(1 - \alpha)|S(j\omega)|^2 + \alpha|N(j\omega)|^2 - S^*(j\omega)\}}$$

Transitional 2 Detection (Inverse $\alpha = 0.5$)

$$\frac{1}{E\{S(j\omega)^{2(1-\alpha)} - 1\}}$$

Autocorrelation Estimation (Uncorrelated Wiener)

$$\frac{E\{S^*(j\omega)|S(j\omega)|^2\}}{E\{|S(j\omega)|^2(1 - S^*(j\omega)) + |N(j\omega)|^2\}}$$

Autocorrelation Estimation (Correlated Wiener)

$$\frac{E\{(S^*(j\omega) + N^*(j\omega))|S(j\omega)|^2\}}{E\{|S(j\omega)|^2(1 - S^*(j\omega) - N^*(j\omega)) + |N(j\omega)|^2 + S(j\omega)N^*(j\omega) + S^*(j\omega)N(j\omega)\}}$$

Cross-Correlation Estimation (Uncorrelated Wiener)

$$\frac{E\{N^*(j\omega)|S(j\omega)|^2\}}{E\{|S(j\omega)|^2(1 - N^*(j\omega)) + |N(j\omega)|^2\}}$$

Cross-Correlation Estimation (Correlated Wiener)

$$\frac{E\{S(j\omega)N^*(j\omega)(S^*(j\omega) + N^*(j\omega))\}}{E\{|S(j\omega)|^2 + |N(j\omega)|^2 + S(j\omega)N^*(j\omega)(1 - S^*(j\omega) - N^*(j\omega)) + S^*(j\omega)N(j\omega)\}}$$

Table 4.9.2 Ideal Class 2 and 3 feedback filter transfer functions $G(j\omega)$.

Filter Name	Class 2	Class 3

Distortion Balance ($W_S(j\omega)$ = Signal weighting, $W_N(j\omega)$ = Noise weighting)

$$\frac{E\{S(j\omega)W_S(j\omega)\}}{E\{R(j\omega)W_N(j\omega) - S(j\omega)[kW_N(j\omega) - (k-1)W_S(j\omega)]\}}$$

$$\frac{\langle R(j\omega)W_S(j\omega)\rangle}{(R(j\omega) - \langle R(j\omega)\rangle)W_N(j\omega)}$$

Uncorrelated Estimation (Uncorrelated Wiener)

$$\frac{E\{|S(j\omega)|^2\}}{E\{|R(j\omega) - kS(j\omega)|^2\}}$$

$$\frac{|\langle R(j\omega)\rangle|^2}{\langle |R(j\omega)|^2\rangle - |\langle R(j\omega)\rangle|^2}$$

Correlated Estimation (Correlated Wiener)

$$\frac{E\{S(j\omega)(R^*(j\omega) + (1-k)S^*(j\omega))\}}{E\{|R(j\omega) + (1-k)S(j\omega)|^2 - S(j\omega)(R^*(j\omega) + (1-k)S^*(j\omega))\}}$$

$$\frac{\langle\langle R(j\omega)\rangle\, R^*(j\omega)\rangle}{\langle |R(j\omega)|^2\rangle - \langle\langle R(j\omega)\rangle\, R^*(j\omega)\rangle}$$

Transitional 1 Estimation (Uncorrelated Wiener $\alpha = 0$, Correlated Wiener $\alpha = 1$)

$$\frac{E\{|S(j\omega)|^2 + \alpha S(j\omega)(R^*(j\omega) - kS^*(j\omega))\}}{\begin{array}{c}E\{|S(j\omega)|^2 + |R(j\omega) - kS(j\omega)|^2 + \alpha[S(j\omega)(R^*(j\omega) - kS^*(j\omega)) \\ + S^*(j\omega)(R(j\omega) - kS(j\omega))] - |S(j\omega)|^2 - \alpha S(j\omega)(R^*(j\omega) - kS^*(j\omega))\}\end{array}}$$

$$\frac{|\langle R(j\omega)\rangle|^2 + (1-\alpha)\langle R(j\omega)\rangle(R^*(j\omega) - \langle R^*(j\omega)\rangle)}{(1-\alpha)\langle |R(j\omega)|^2\rangle - |\langle R(j\omega)\rangle|^2 - (1-\alpha)\langle R(j\omega)\rangle(R^*(j\omega) - \langle R^*(j\omega)\rangle)}$$

Pulse-Shaping Estimation ($P(j\omega)$ = Desired pulse spectrum, Uncorrelated Wiener)

$$\frac{E\{P(j\omega)S^*(j\omega)\}}{E\{|S(j\omega)|^2 + |R(j\omega) - kS(j\omega)|^2 - P(j\omega)S^*(j\omega)\}}$$

$$\frac{P(j\omega)\langle R^*(j\omega)\rangle}{\langle |R(j\omega)|^2\rangle - P(j\omega)\langle R^*(j\omega)\rangle}$$

Pulse-Shaping Estimation ($P(j\omega)$ = Desired pulse spectrum, Correlated Wiener)

$$\frac{E\{P(j\omega)(R^*(j\omega) + (1-k)S^*(j\omega))\}}{E\{|R(j\omega) + (1-k)S(j\omega)|^2 - P(j\omega)(R^*(j\omega) + (1-k)S^*(j\omega))\}}$$

$$\frac{\langle P(j\omega)R^*(j\omega)\rangle}{\langle |R(j\omega)|^2\rangle - \langle P(j\omega)R^*(j\omega)\rangle}$$

Classical Detection (Matched)

$$\frac{E\{S^*(j\omega)\}}{E\{|R(j\omega) - kS(j\omega)|^2 - S^*(j\omega)\}}$$

$$\frac{\langle R^*(j\omega)\rangle}{|R(j\omega) - \langle R(j\omega)\rangle|^2 - \langle R^*(j\omega)\rangle}$$

High-Resolution Detection (Uncorrelated Wiener)

$$\frac{E\{S^*(j\omega)\}}{E\{|S(j\omega)|^2 + |R(j\omega) - kS(j\omega)|^2 - S^*(j\omega)\}}$$

$$\frac{\langle R^*(j\omega)\rangle}{\langle |R(j\omega)|^2\rangle - \langle R^*(j\omega)\rangle}$$

Table 4.9.2 (Continued) Ideal Class 2 and 3 feedback filter transfer functions $G(j\omega)$.

Filter Name	Class 2	Class 3

High-Resolution Detection (Correlated Wiener)

$$\frac{E\{R^*(j\omega) + (1-k)S^*(j\omega)\}}{E\{|R(j\omega) + (1-k)S(j\omega)|^2 - R^*(j\omega) - (1-k)S^*(j\omega)\}} \qquad \frac{\langle R^*(j\omega)\rangle}{\langle|R(j\omega)|^2\rangle - \langle R^*(j\omega)\rangle}$$

Inverse Detection (Deconvolution)

$$\frac{1}{E\{S(j\omega) - 1\}} \qquad\qquad \frac{1}{\langle R(j\omega)\rangle - 1}$$

Transitional 1 Detection (Inverse $\alpha = 0$, High-Resolution $\alpha = 0.5$, Matched $\alpha = 1$)

$$\frac{E\{S^*(j\omega)\}}{E\{(1-\alpha)|S(j\omega)|^2 + \alpha|R(j\omega) - kS(j\omega)|^2 - S^*(j\omega)\}}$$

$$\frac{\langle R^*(j\omega)\rangle}{(1-\alpha)|\langle R(j\omega)\rangle|^2 + \alpha|R(j\omega) - \langle R(j\omega)\rangle|^2 - \langle R^*(j\omega)\rangle}$$

Transitional 2 Detection (Inverse $\alpha = 0.5$)

$$\frac{1}{E\{S(j\omega)^{2(1-\alpha)} - 1\}} \qquad\qquad \frac{1}{\langle R(j\omega)\rangle^{2(1-\alpha)} - 1}$$

Autocorrelation Estimation (Uncorrelated Wiener)

$$\frac{E\{S^*(j\omega)|S(j\omega)|^2\}}{E\{|S(j\omega)|^2 + |R(j\omega) - kS(j\omega)|^2 - S^*(j\omega)|S(j\omega)|^2\}}$$

$$\frac{\langle R^*(j\omega)\rangle \; |\langle R(j\omega)\rangle|^2}{\langle|R(j\omega)|^2\rangle - \langle R^*(j\omega)\rangle \; |\langle R(j\omega)\rangle|^2}$$

Autocorrelation Estimation (Correlated Wiener)

$$\frac{E\{(R^*(j\omega) + (1-k)S^*(j\omega))|S(j\omega)|^2\}}{E\{|R(j\omega) + (1-k)S(j\omega)|^2 - (R^*(j\omega) + (1-k)S^*(j\omega))|S(j\omega)|^2\}}$$

$$\frac{|\langle R(j\omega)\rangle|^2}{R(j\omega) - |\langle R(j\omega)\rangle|^2}$$

Cross-Correlation Estimation (Uncorrelated Wiener)

$$\frac{E\{(R^*(j\omega) - kS^*(j\omega))|S(j\omega)|^2\}}{E\{|S(j\omega)|^2 + |R(j\omega) - kS(j\omega)|^2 - (R^*(j\omega) - kS^*(j\omega))|S(j\omega)|^2\}}$$

$$\frac{(R^*(j\omega) - \langle R^*(j\omega)\rangle) \; |\langle R(j\omega)\rangle|^2}{\langle|R(j\omega)|^2\rangle - (R^*(j\omega) - \langle R^*(j\omega)\rangle) \; |\langle R(j\omega)\rangle|^2}$$

Cross-Correlation Estimation (Correlated Wiener)

$$\frac{E\{S(j\omega)(R^*(j\omega) - kS^*(j\omega))(R^*(j\omega) + (1-k)S^*(j\omega))\}}{E\{|R(j\omega) + (1-k)S(j\omega)|^2 - S(j\omega)(R^*(j\omega) - kS^*(j\omega))(R^*(j\omega) + (1-k)S^*(j\omega))\}}$$

$$\frac{(R^*(j\omega) - \langle R^*(j\omega)\rangle) \; \langle R(j\omega)\rangle}{R(j\omega) - (R^*(j\omega) - \langle R^*(j\omega)\rangle) \; \langle R(j\omega)\rangle}$$

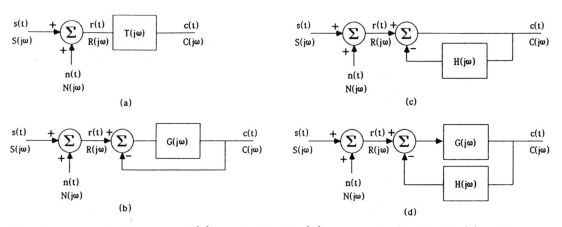

Fig. 4.9.1 Block diagram of (a) nonfeedback, (b) unity-gain feedback, (c) unity-gain feedforward, and (d) general feedback systems.

$$T(j\omega) = \frac{1}{1 + H(j\omega)}, \qquad H(j\omega) = \frac{1 - T(j\omega)}{T(j\omega)} = \frac{1}{G(j\omega)} \qquad (4.9.3)$$

Comparing Eqs. 4.9.2 and 4.9.3, $H(j\omega)$ in Fig. 4.9.1c has a transfer function which is the reciprocal of $G(j\omega)$ listed in Tables 4.9.1 and 4.9.2. Now let us briefly describe some of these results.

The general nonfeedback estimation filter of Table 4.4.1 has

$$T(j\omega) = \frac{E\{D(j\omega)R^*(j\omega)\}}{E\{R(j\omega)R^*(j\omega)\}} \qquad (4.9.4)$$

The general unity-gain feedback estimation filter of Table 4.9.1 has

$$G(j\omega) = \frac{E\{D(j\omega)R^*(j\omega)\}}{E\{R^*(j\omega)(R(j\omega) - D(j\omega))\}} \qquad (4.9.5)$$

which is more complicated. At frequencies where the input power spectrum $E\{RR^*\}$ is much greater than the desired input-output spectral product $E\{DR^*\}$, the loop gain is much less than unity. Then $G(j\omega)$ reduces to

$$G(j\omega) = \frac{E\{D(j\omega)R^*(j\omega)\}}{E\{R(j\omega)R^*(j\omega)\}}, \qquad E\{R(j\omega)R^*(j\omega)\} \gg E\{D(j\omega)R^*(j\omega)\} \qquad (4.9.6)$$

and equals the general estimation nonfeedback $T(j\omega)$ in Table 4.4.1.

The general unity-gain feedback detection filter of Table 4.9.1 has

$$G(j\omega) = \frac{E\{D^*(j\omega)\}}{E\{R(j\omega)R^*(j\omega)\} - E\{D^*(j\omega)\}} \qquad (4.9.7)$$

At frequencies where the input power spectrum $E\{RR^*\}$ is much greater than the desired spectrum $E\{D^*\}$, the loop-gain is much less than unity. $G(j\omega)$ equals

$$G(j\omega) = \frac{E\{D^*(j\omega)\}}{E\{R(j\omega)R^*(j\omega)\}}, \qquad E\{R(j\omega)R^*(j\omega)\} \gg E\{D^*(j\omega)\} \qquad (4.9.8)$$

which is the general detection nonfeedback $T(j\omega)$ in Table 4.4.1.

The only feedback filter that has a simpler transfer function than its corresponding nonfeedback filter is the uncorrelated Wiener estimation filter of Table 4.9.1 where

$$G(j\omega) = \frac{E\{|S(j\omega)|^2\}}{E\{|N(j\omega)|^2\}} \qquad (4.9.9)$$

$T(j\omega)$ reduces to this result when $E\{|N(j\omega)|\} \gg E\{|S(j\omega)|\}$ as shown in Table 4.4.1.

One further comment should be made. Since general feedback systems having nonunity $G(j\omega)$ and $H(j\omega)$ have overall gain

$$T(j\omega) = \frac{G(j\omega)}{1 + G(j\omega)H(j\omega)} \qquad (4.9.10)$$

either $G(j\omega)$ or $H(j\omega)$ can be determined when the other is known. Solving Eq. 4.9.10 for $G(j\omega)$ and $H(j\omega)$ gives

$$G(j\omega) = \frac{T(j\omega)}{1 - T(j\omega)H(j\omega)} = \left[\frac{1}{T(j\omega)} - H(j\omega)\right]^{-1}$$

$$H(j\omega) = \frac{1}{T(j\omega)} - \frac{1}{G(j\omega)} \qquad (4.9.11)$$

These results show that the Class 2 and 3 adaptive filter forms are viable alternatives to the Class 1 adaptive filters. Algebraically it makes little difference whether nonfeedback systems (Fig. 4.9.1a and Tables 4.4.1 and 4.5.1) or feedback systems (Figs. 4.9.1b–4.9.1c and Tables 4.9.1–4.9.2) are used when implementing these algorithms. However the presence of k and the need to estimate it makes the feedback formulations more complicated.

4.10 HOMOMORPHIC FILTERS

Homographic filters are nonlinear filters that separate signal and noise which have been combined using some operation other than addition.[15] Homomorphic filters satisfy generalized superposition. If $T[\ ]$ is a homomorphic transformation and $\otimes$ represents the operation used to combine the signal and noise, then

$$T\big[k_1 s(t) \otimes k_2 n(t)\big] = k_1 T\big[s(t)\big] + k_2 T\big[n(t)\big] = k_1 s_T(t) + k_2 n_T(t) \qquad (4.10.1)$$

The transformed signal $T[s(t)]$ is denoted as $s_T(t)$ while the transformed noise $T[n(t)]$ is denoted as $n_T(t)$. The inverse homomorphic transformation $T^{-1}[\ ]$ is used to recover the signal. Since Eq. 4.10.1 has the form of additive signal $s_T(t)$ and noise $n_T(t)$, all the algorithms of this chapter such as those tabulated in Tables 4.4.1, 4.5.1, and 4.9.1–4.9.2 can be used to separate them. A general homomorphic filter is shown in Fig. 4.10.1a. Let us now consider two common cases – multiplicative and convolutional filters.

4.10.1 MULTIPLICATIVE FILTERS

We have been considering systems whose inputs $r(t)$ were composed of additive signal $s(t)$ plus noise $n(t)$ as (Fig. 4.4.1)

$$r(t) = s(t) + n(t)$$
$$R(j\omega) = S(j\omega) + N(j\omega)$$

(4.10.2)

Now consider systems having multiplicative signal and noise input as

$$r(t) = s(t)n(t)$$
$$R(j\omega) = \frac{1}{2\pi} S(j\omega) * N(j\omega)$$

(4.10.3)

The processing algorithms which have been discussed in this chapter described additive inputs like those of Eq. 4.10.2. They do not apply to multiplicative inputs as those in Eq. 4.10.3. Multiplicative inputs arise in such areas as amplitude modulation, range compression, and image processing.

Taking the logarithm of Eq. 4.10.3 yields

$$\log r(t) = \log s(t) + \log n(t)$$
$$r_L(t) = \log r(t) = s_L(t) + n_L(t)$$
$$R_L(j\omega) = S_L(j\omega) + N_L(j\omega)$$

(4.10.4)

The logarithm operation converts products into sums. Since Eq. 4.10.4 now is an additive form like Eq. 4.10.2, all the algorithms of Tables 4.4.1, 4.5.1, and 4.9.1–4.9.2 can be used. After filtering, the antilogarithm or exponentiation must be taken to remove the logarithmic distortion.

This is the basic concept of multiplicative filtering. Let us now summarize the fundamental filtering equations. Additive input systems use (Fig. 4.10.1b):

$$r(t) = s(t) + n(t)$$
$$R(j\omega) = \mathcal{F}\big[r(t)\big] = S(j\omega) + N(j\omega)$$
$$C(j\omega) = H(j\omega)R(j\omega)$$
$$c(t) = \mathcal{F}^{-1}\big[C(j\omega)\big]$$

(4.10.5)

Multiplicative systems instead use (Fig. 4.10.1c):

$$r(t) = s(t)n(t)$$
$$r_L(t) = \log\ r(t) = \log\ s(t) + \log\ n(t) = s_L(t) + n_L(t)$$

(4.10.6a–b)

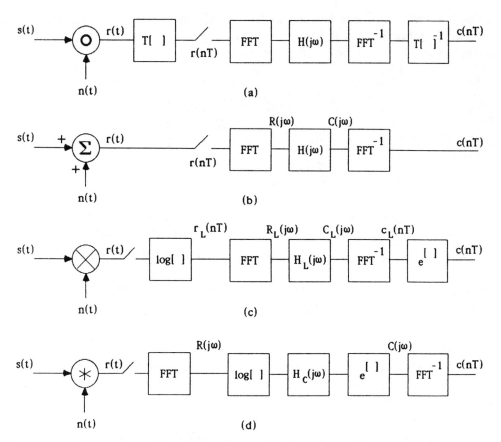

Fig. 4.10.1 (a) General homomorphic, (b) additive, (c) multiplicative, and (d) convolutional filter systems.

$$R_L(j\omega) = S_L(j\omega) + N_L(j\omega)$$
$$C_L(j\omega) = H_L(j\omega)R_L(j\omega)$$
$$c_L(t) = \mathcal{F}^{-1}[C_L(j\omega)]$$
$$c(t) = \exp c_L(t)$$

$$(4.10.6c\text{–}f)$$

This is the set of equations which describe and implement multiplicative filters.

To show how the filters of Tables 4.4.1 and 4.5.1 which describe Fig. 4.4.1 are converted, consider the uncorrelated Wiener filter where

$$\text{Class 1}: \quad \frac{E\{|S(j\omega)|^2\}}{E\{|S(j\omega)|^2\} + E\{|N(j\omega)|^2\}}$$

$$\text{Class 2}: \quad \frac{E\{|S(j\omega)|^2\}}{E\{|S(j\omega)|^2\} + E\{|R(j\omega) - kS(j\omega)|^2\}}$$

$$(4.10.7)$$

$$\text{Class 3}: \quad \frac{|\langle R(j\omega)\rangle|^2}{\langle |R(j\omega)|^2\rangle}$$

These are converted into multiplicative transfer functions using $R_L(j\omega)$, $S_L(j\omega)$, and $N_L(j\omega)$ in place of $R(j\omega)$, $S(j\omega)$, and $N(j\omega)$, respectively. Making these substitutions using Eq. 4.10.4, then Eq. 4.10.7 becomes

$$
\begin{aligned}
&\text{Class 1}: && \frac{E\{|S_L(j\omega)|^2\}}{E\{|S_L(j\omega)|^2\} + E\{|N_L(j\omega)|^2\}} \\[2mm]
&\text{Class 2}: && \frac{E\{|S_L(j\omega)|^2\}}{E\{|S_L(j\omega)|^2\} + E\{|R_L(j\omega) - S_L(j\omega) - \log k|^2\}} \\[2mm]
&\text{Class 3}: && \frac{|\langle R_L(j\omega)\rangle|^2}{\langle|R_L(j\omega)|^2\rangle}
\end{aligned}
\tag{4.10.8}
$$

These equations describe three classes of multiplicative filters in Fig. 4.10.2c. This is an extremely simple conversion process. Taking the Fourier transforms of the logarithms of the expected $r(t)$, $s(t)$, and $n(t)$ and manipulating the results gives

$$
\begin{aligned}
E\{R_L(j\omega)\} &= \mathcal{F}\big[E\{r_L(t)\}\big] = \mathcal{F}\big[\log E\{r(t)\}\big] \\
E\{S_L(j\omega)\} &= \mathcal{F}\big[E\{s_L(t)\}\big] = \mathcal{F}\big[\log E\{s(t)\}\big] \\
E\{N_L(j\omega)\} &= \mathcal{F}\big[E\{n_L(t)\}\big] = \mathcal{F}\big[\log E\{n(t)\}\big]
\end{aligned}
\tag{4.10.9}
$$

give the needed expected spectra to use in Tables 4.4.1, 4.5.1, and 4.9.1–4.9.2.

It should be remembered that $r(t)$, $s(t)$, and $n(t)$ are treated as complex numbers in Eq. 4.10.9. For example

$$
\begin{aligned}
s(t) &= \exp s_L(t) = |s(t)|e^{j\,\arg\,s(t)} = |s(t)|e^{-j\pi\,\mathrm{sgn}\,\,s(t)} \\
s_L(t) &= \log s(t) = \log|s(t)| + j\,\arg s(t) = |s_L(t)|e^{j\,\arg\,s_L(t)}
\end{aligned}
\tag{4.10.10}
$$

The first 3 terms in $s(t)$ assume that $s(t)$ is complex-valued. The last term assumes that $s(t)$ is real-valued so it is positive $(\arg s(t) = 0°)$, negative $(\arg s(t) = -180°)$, or zero $(\arg s(t) = \text{arbitrary})$. Therefore, the expected spectra of complex- or real-valued $r(t)$, $s(t)$, and $n(t)$ are evaluated in the same way.

4.10.2 CONVOLUTIONAL FILTERS

Suppose the signal and noise are convolved or correlated to form the input $r(t)$ as

$$
\begin{aligned}
r(t) &= s(t) * n(t) \\
R(j\omega) &= S(j\omega)N(j\omega)
\end{aligned}
\tag{4.10.11}
$$

Such inputs arise in multi-path communications, speech processing, correlation studies, and probability density function filtering. Comparing Eq. 4.10.11 with Eq. 4.10.3, $r(t)$ and $R(j\omega)$ are duals so the logarithmic operation can be applied again. Then

$$r(t) = s(t) * n(t)$$
$$\log R(j\omega) = \log \left[S(j\omega)N(j\omega)\right] = \log S(j\omega) + \log N(j\omega) \qquad (4.10.12)$$
$$R_C(j\omega) = S_C(j\omega) + N_C(j\omega)$$

In this case, taking the log of the spectrum (not the input itself as in Eq. 4.10.4) produces the generalized superposition required by Eq. 4.10.1.

Therefore convolutional systems use (Fig. 4.10.1d)

$$r(t) = s(t) * n(t)$$
$$R(j\omega) = S(j\omega)N(j\omega)$$
$$R_C(j\omega) = \log R(j\omega) = \log S(j\omega) + \log N(j\omega) = S_C(j\omega) + N_C(j\omega)$$
$$C_C(j\omega) = H_C(j\omega)R_C(j\omega) \qquad\qquad (4.10.13)$$
$$C(j\omega) = \exp C_C(j\omega)$$
$$c(t) = \mathcal{F}^{-1}\left[C(j\omega)\right]$$

The optimum uncorrelated estimation filters given by Eq. 4.10.7 can be applied to convolutional filters by replacing $R(j\omega)$, $S(j\omega)$, and $N(j\omega)$ with $R_C(j\omega)$, $S_C(j\omega)$, and $N_C(j\omega)$, respectively. Making these changes in Eq. 4.10.7 yields

$$\text{Class 1:} \quad \frac{E\{|\log S(j\omega)|^2\}}{E\{|\log S(j\omega)|^2\} + E\{|\log N(j\omega)|^2\}}$$

$$\text{Class 2:} \quad \frac{E\{|\log S(j\omega)|^2\}}{E\{|\log S(j\omega)|^2\} + E\{|\log(R(j\omega) - kS(j\omega)|^2\}} \qquad (4.10.14)$$

$$\text{Class 3:} \quad \frac{|\langle \log R(j\omega)\rangle|^2}{\langle|\log R(j\omega)|^2\rangle}$$

It is important to notice that these convolutional filters are different from the multiplicative filters given by Eq. 4.10.8 since, for example,

$$S_C(j\omega) = \log S(j\omega) = \log\{\mathcal{F}\left[s(t)\right]\} \neq \mathcal{F}\{\log[s(t)]\} = S_L(j\omega) \qquad (4.10.15)$$

Remember that in convolutional filters, the logarithm is used to process spectra as

$$S(j\omega) = \exp S_C(j\omega) = |S(j\omega)|e^{j \arg S(j\omega)}$$
$$S_C(j\omega) = \log S(j\omega) = \log |S(j\omega)| + j \arg S(j\omega) \qquad (4.10.16)$$

$S_C(j\omega)$ is called the *complex cepstrum* of the spectrum $S(j\omega)$. The real part of $S_C(j\omega)$ equals $\log |S(j\omega)|$ and is called the *cepstrum*. Therefore the cepstrum of a signal $s(t)$ is simply a scaled version of its magnitude spectrum plotted in dB. Thus the cepstrum logarithmically compresses a magnitude spectrum.

4.10.3 OTHER FILTERS

We have described how homomorphic filters are formed for multiplicative inputs and convolutional inputs. These homomorphic transformations to process these inputs are listed in Table 4.10.1. More complicated processing may be required using additional transforms such as those listed in Table 4.10.2.

An example of such more complicated processing is provided by considering a complex signal $s(t)$ which is encoded by its phase $\theta_s(t)$ where

$$s(t) = s_r(t) + js_i(t), \qquad \theta_s(t) = \tan^{-1} \frac{s_i(t)}{s_r(t)} \qquad (4.10.17)$$

Comparing Eqs. 4.10.17 and 4.10.1, the $\otimes$ and T homomorphic operators equal

$$[\,]\otimes(\,) = \tan^{-1} \frac{(\,)}{[\,]}, \qquad T[\,] = \log\left[\tan[\,]\right] \qquad (4.10.18)$$

Combining T and $\otimes$ shows that

$$T[s_r \otimes s_i] = \log\left[\tan\left[\tan^{-1}\frac{s_i}{s_r}\right]\right] = \log\frac{s_i}{s_r} = \log s_i - \log s_r = s_{iT} + s_{rT} \qquad (4.10.19)$$

Table 4.10.1 Basic homomorphic T operations to
convert nonlinear inputs into linear inputs.

Input	Input	$T[\,]$	$T^{-1}[\,]$	Output
Additive	$s(t) + n(t)$	1	1	$s(t) + n(t)$
Multiplicative	$s(t)n(t)$	$\ln[\,]$	$\exp[\,]$	$\ln s(t) + \ln n(t)$
Convolutional	$s(t) * n(t)$ or $S(j\omega)N(j\omega)$	$\ln[\,]$	$\exp[\,]$	$\ln S(j\omega) + \ln N(j\omega)$

Table 4.10.2 Other homomorphic T operations to convert
nonlinear input functions into linear input functions.

Name	$T[\,]$	$T^{-1}[\,]$
Linear	1	1
Power	$[\,]^n$	$[\,]^{1/n}$
Logarithmic	$\ln[\,]$	$\exp[\,]$
	$\exp[\,]$	$\ln[\,]$
Trigonometric and Hyperbolic	$\sin[\,]$ or $\sin^{-1}[\,]$	$\sin^{-1}[\,]$ or $\sin[\,]$
(Change $\sin[\,]$ to $\sinh[\,]$, etc.)	$\cos[\,]$ or $\cos^{-1}[\,]$	$\cos^{-1}[\,]$ or $\cos[\,]$
	$\tan[\,]$ or $\tan^{-1}[\,]$	$\tan^{-1}[\,]$ or $\tan[\,]$

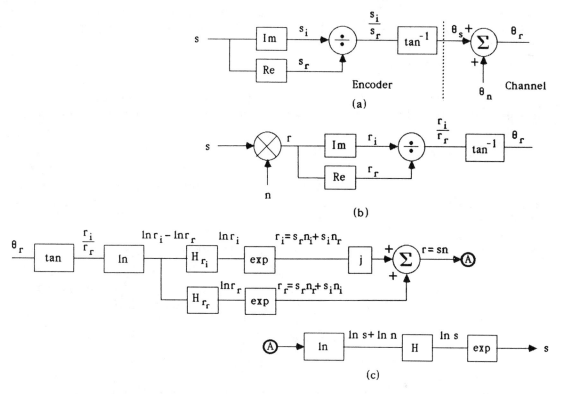

Fig. 4.10.2 (a) Encoder and channel, (b) equivalent encoder and channel, and (c) decoder for phase modulated system.

The encoder, channel, and decoder models are shown in Fig. 4.10.2. Two estimation filters with gains H_r and H_i are used to separate $\log s_r$ and $\log s_i$. The two log signals are passed through exp blocks to recover estimates of s_r and s_i. Forming $\hat{s} = \hat{s}_r + j\hat{s}_i$ yields an estimate of the original complex signal s.

Now assume that the phase θ_s is contaminated by additive channel noise which we choose to express as

$$\theta_n(t) = \tan^{-1}\frac{n_i(t)}{n_r(t)} \tag{4.10.20}$$

Combining Eqs. 4.10.17 and 4.10.20, then the total input $\theta_r(t)$ equals

$$\theta_r(t) = \theta_s(t) + \theta_n(t) = \tan^{-1}\frac{s_i(t)}{s_r(t)} + \tan^{-1}\frac{n_i(t)}{n_r(t)}$$

$$= \tan^{-1}\frac{s_r(t)n_i(t) + s_i(t)n_r(t)}{s_r(t)n_r(t) - s_i(t)n_i(t)} = \tan^{-1}\frac{r_i(t)}{r_r(t)} \tag{4.10.21}$$

An equivalent complex input $r(t)$ can be defined from the numerator and denominator of Eq. 4.10.21 to equal

$$r(t) = r_r(t) + jr_i(t) = \left[s_r(t) + js_i(t)\right]\left[n_r(t) + jn_i(t)\right] \tag{4.10.22}$$

This equation shows that adding noise to the phase encoded signal is equivalent to phase encoding the product of the signal and the noise. The decoder in Fig. 4.10.2c estimates $s(t)$ from the $\theta_r(t)$ input.

One common problem in homomorphic filters is that the T transform is usually single-valued but the T^{-1} inverse transform is not. T^{-1} is often multivalued so that some algorithm must be developed for selecting just one value from the many multiple values available. This insures single-valuedness which is required for realization.

4.11 HOMOMORPHIC FILTER DESIGN

The basic concepts of homomorphic filters have been introduced. Now we shall present design examples for Class 1 multiplicative and convolutional estimation filters. This will illustrate how to separate signals combined in some nonlinear fashion.

4.11.1 MULTIPLICATIVE ESTIMATION FILTER DESIGN

The multiplicative filters of Fig. 4.10.1c are described by Eq. 4.10.6. Consider the multiplicative input $r(t)$ where signal $s(t)$ and noise $n(t)$ are combined as

$$r(t) = s(t)n(t)$$
$$\log r(t) = \log s(t) + \log n(t) \tag{4.11.1}$$

For convenience, but not necessity, let us assume that $s(t)$ and $n(t)$ are real-valued. They can be represented in polar form as

$$s(t) = |s(t)|e^{j \arg s(t)} = |s(t)|e^{-j\pi[1-\text{sgn } s(t)]/2}$$
$$n(t) = |n(t)|e^{j \arg n(t)} = |n(t)|e^{-j\pi[1-\text{sgn } n(t)]/2} \tag{4.11.2}$$
$$r(t) = |r(t)|e^{j \arg r(t)} = |r(t)|e^{-j\pi[1-\text{sgn } r(t)]/2}$$

The phases of $s(t)$, $n(t)$, and $r(t)$ are either 0 or $-\pi$ radians. Taking the logarithm of Eq. 4.11.2 gives

$$\log s(t) = \log|s(t)| + j \arg s(t) = \log|s(t)| - j\pi[1 - \text{sgn } s(t)]/2$$
$$\log n(t) = \log|n(t)| + j \arg n(t) = \log|n(t)| - j\pi[1 - \text{sgn } n(t)]/2 \tag{4.11.3}$$
$$\log r(t) = \log|s(t)| + \log|n(t)| - j\pi[2 - \text{sgn } s(t) - \text{sgn } n(t)]/2$$

Eq. 4.11.3 shows that although $s(t)$, $n(t)$, and $r(t)$ are real-valued, their logarithms are complex-valued. The real part of the logarithm equals the log of the magnitude function (e.g., $\log|s(t)|$). The imaginary part of the logarithm equals the phase of the function which is $0°$ or $-180°$ (e.g., $\arg s(t)$).

The input $r(t)$ can be processed by a single complex filter (see Fig. 4.11.1a) or by two real filters (see Fig. 4.11.1b). These filters have $H(s)$ transfer functions with complex or real coefficients, respectively. For simplicity, we consider the two-filter case.

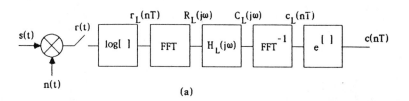

(a)

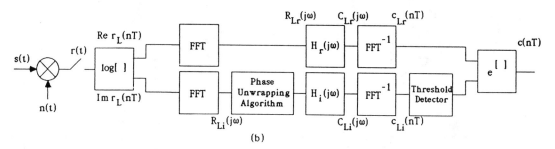

(b)

Fig. 4.11.1 Multiplicative filter using (a) one complex filter and (b) two real filters.

Suppose that the cosine signal $s(t)$ and the white noise $n(t)$ shown in Figs. 4.11.2a and 4.11.2b, respectively, are multiplicatively combined to form $r(t)$ as shown in Fig. 4.11.2c. The logarithms of $s(t)$ and $n(t)$ are also shown in the figures. When $|s(t)| \leq 1$, then $\log s(t)$ is the bounded between 0 (where $s(t) = \pm 1$) and $-\infty$ (where $s(t) = 0$). Numerical overflow problems are reduced by limiting $s(t)$ to some small finite value as $\log |s(t)| \equiv \epsilon$ for $|s(t)| \leq \delta$. The imaginary part of $\log s(t)$ equals $0°$ for $s(t) > 0$, $-180°$ for $s(t) < 0$, and $0°$ or $-180°$ for $s(t) = 0$. The white noise exhibits similar properties.

The magnitude filter with gain $H_r(j\omega)$ in Fig. 4.11.1b estimates the $\log |s(t)|$ and rejects $\log |n(t)|$. The phase filter with gain $H_i(j\omega)$ estimates arg $s(t)$ and rejects arg $n(t)$. From the plots of Figs. 4.11.2a and 4.11.2b, it can be seen that $\log |s(t)|$ and $\log |n(t)|$ are dissimilar enough that they can be separated by a Class 1 uncorrelated magnitude estimation filter. However, arg $s(t)$ and arg $n(t)$ are very similar (only $0°$ and $-180°$). Since they make step transitions, they have magnitude spectra with only -20 dB/dec asymptotic roll-off and are difficult to separate. Their phase spectra are unlike (periodic versus random) which should be used when formulating the phase filter. For simplicity in this example, a Class 1 uncorrelated phase estimation filter will be used.

$$S_L(j\omega) = S_{Lr}(j\omega) + S_{Li}(j\omega) = \mathcal{F}\big[\log|s(t)|\big] + j\mathcal{F}\big[\arg\ s(t)\big]$$
$$N_L(j\omega) = N_{Lr}(j\omega) + N_{Li}(j\omega) = \mathcal{F}\big[\log|n(t)|\big] + j\mathcal{F}\big[\arg\ n(t)\big] \qquad (4.11.4)$$
$$R_L(j\omega) = R_{Lr}(j\omega) + R_{Li}(j\omega) = \mathcal{F}\big[\log|r(t)|\big] + j\mathcal{F}\big[\arg\ r(t)\big]$$

then the Class 1 magnitude and phase filters have gains

$$H_r(j\omega) = \frac{E\{|S_{Lr}(j\omega)|^2\}}{E\{|S_{Lr}(j\omega)|^2\} + E\{|N_{Lr}(j\omega)|^2\}}$$

$$H_i(j\omega) = \frac{E\{|S_{Li}(j\omega)|^2\}}{E\{|S_{Li}(j\omega)|^2\} + E\{|N_{Li}(j\omega)|^2\}} \qquad (4.11.5)$$

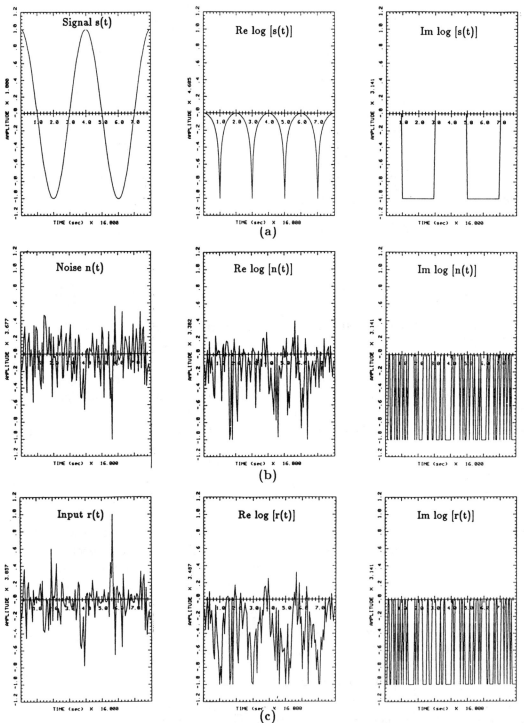

Fig. 4.11.2 (a) Signal $s(t)$ and $\log[s(t)]$, (b) noise $n(t)$ and $\log[n(t)]$, (c) total input $r(t)$ and $\log[r(t)]$, (d) gains of magnitude and phase filters, (e) magnitude and phase filter outputs, (f) multiplicative filter input and output signals. (From L.E. McLaren, Special project, Univ. of Miami, 1986.)

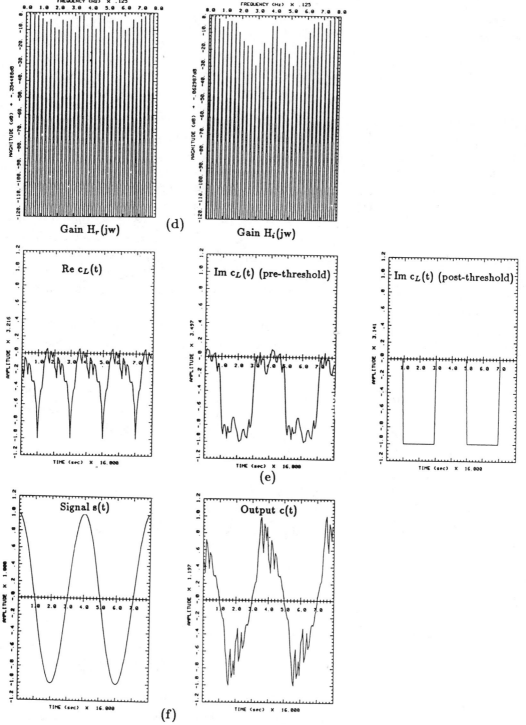

Fig. 4.11.2 (Continued) (a) Signal $s(t)$ and $\log[s(t)]$, (b) noise $n(t)$ and $\log[n(t)]$, (c) total input $r(t)$ and $\log[r(t)]$, (d) gains of magnitude and phase filters, (e) magnitude and phase filter outputs, (f) multiplicative filter input and output signals.

There is another problem related to phase. The phase angles in Eq. 4.11.2 and elsewhere equal

$$\arg s(t) = \text{ARG } s(t) \pm 2\pi k_s$$
$$\arg n(t) = \text{ARG } n(t) \pm 2\pi k_n \qquad\qquad (4.11.6)$$
$$\arg r(t) = \text{ARG } r(t) \pm 2\pi k_r$$

$\arg[\ ]$ is the actual phase and $\text{ARG}[\ ]$ is the principal value phase computed by the FFT. $\arg[\ ]$ can have any value but $\text{ARG}[\ ]$ can only have a value in the range from $-180°$ to $180°$. k_s, k_n, and k_r are integers. Unless $|\arg[\ ]| \leq 180°$, the phase computed by the FFT does not equal the actual phase. $\arg[\ ] \neq \text{ARG}[\ ]$ unless the proper $\pm 2\pi k$ factor is added. This adjustment is accomplished by using a *phase unwrapping* algorithm.[16] For example, $\arg r(t) = 0°$ or $-180°$ in Fig. 4.11.2c when computed directly from $r(t)$. Since $\arg s(t)$ and $\arg n(t) = 0$ or $-180°$, then computing phase indirectly using $\arg r(t) = \arg s(t) + \arg n(t)$ gives $\arg r(t) = 0°$ or $-360°$, and $-180°$. When $\arg s(t)$ and $\arg n(t)$ have similar signs, the phase assigned to $\arg r(t)$ can only be determined by observing $s(t)$ and $n(t)$ or by using a phase unwrapping algorithm. The phase unwrapping block has been included in Fig. 4.11.1b. The magnitude and phase filter gains are shown in Fig. 4.11.2d.

Passing the real part (or the imaginary part) of $r_L(t)$ through filter $H_r(j\omega)$ (or $H_i(j\omega)$) gives the outputs in Fig. 4.11.2e. Comparing $c_L(t)$ and $s_L(t)$ in Figs. 4.11.2a and 4.11.2e, we see that the signal magnitude is fairly well recovered and that the signal phase is very well recovered.

Since the phase filter output is continuous between $0°$ and $-360°$ or $0°$ to $-180°$ for wrapped phase, it must be threshold detected to limit it to $0°$ or $-180°$ so the resulting $c(t)$ output will be real-valued. A variety of threshold detection algorithms could be used. For example,

$$\text{Minimax threshold} = \frac{1}{2}\Big[c_{Li\ max} + c_{Li\ min}\Big]$$
$$\text{Average threshold} = \frac{1}{N}\sum_{n=0}^{N-1} c_{Li}(nT) \qquad\qquad (4.11.7)$$
$$\text{Median threshold} = c_{Li\ median}$$

When $c_{Li}(t) >$ threshold, then $c_{Li}(t)$ is set to $0°$. When $c_{Li}(t) <$ threshold, then $c_{Li}(t)$ is set to $-180°$. When $c_{Li}(t) =$ threshold, then $c_{Li}(t)$ is randomly set to $0°$ or $-180°$.

The output of the exponential block in Fig. 4.11.1b is

$$\exp c_L(t) = e^{\log |c(t)| + j\arg\ c(t)} = e^{\log |c(t)|} e^{j\arg\ c(t)} = |c(t)| e^{j\arg\ c(t)} \qquad (4.11.8)$$

which is shown in Fig. 4.11.2f. Comparing the input and output signals in Fig. 4.11.2f, we see that the signal is adequately recovered. This is actually a rather remarkable result since the signal and noise were combined nonlinearly. They have been separated using linear filters and nonlinear homographic transformations.

4.11.2 CONVOLUTIONAL ESTIMATION FILTER DESIGN

The convolutional filter of Fig. 4.10.1d is described by Eq. 4.10.13. Let us now illustrate the design of a single-channel estimation convolutional filter. The output of the homomorphic logarithm block is the sum of the complex signal spectra $S_C(j\omega) = \log|S(j\omega)| + j\arg S(j\omega)$ and noise spectra $N_C(j\omega) = \log|N(j\omega)| + j\arg N(j\omega)$. Therefore from Eq. 4.10.13, the output of the log block equals

$$R_C(j\omega) = \log|R(j\omega)| + j\arg R(j\omega)$$
$$= \log|S(j\omega)| + \log|N(j\omega)| + j\big[\arg S(j\omega) + \arg N(j\omega)\big] \qquad (4.11.9)$$

The Class 1 estimation filter gain is given by Eq. 4.10.14 as

$$H_C(j\omega) = \frac{E\{|\log S(j\omega)|^2\}}{E\{|\log S(j\omega)|^2 + |\log N(j\omega)|^2\}}$$

$$= \frac{E\{|\log|S(j\omega)||^2 + |\arg S(j\omega)|^2\}}{E\{|\log|S(j\omega)||^2 + |\log|N(j\omega)||^2 + |\arg S(j\omega)|^2 + |\arg N(j\omega)|^2\}} \qquad (4.11.10)$$

Both the real and imaginary parts of $\log S(j\omega)$ and $\log N(j\omega)$ are used to formulate the gain $H_C(j\omega)$ of the single-channel convolutional filter unlike the dual-channel multiplicative filter in Fig. 4.11.2. The Class 1 estimation filter described by Eq. 4.11.10 effectively separates $S_C(j\omega)$ and $N_C(j\omega)$ spectra that are dissimilar and where $|\log|S(j\omega)|| \cong |\arg S(j\omega)|$ and $|\log|N(j\omega)|| \cong |\arg N(j\omega)|$. When $|\log|S(j\omega)|| \gg |\arg S(j\omega)|$ and $|\log|N(j\omega)|| \gg |\arg N(j\omega)|$, the filter gain $H_C(j\omega)|$ depends only upon the spectral magnitudes of $S_C(j\omega)$ and $N_C(j\omega)$ (i.e., their real parts). Reversing the inequality signs, $H_C(j\omega)$ depends only upon their spectral phases (i.e., their imaginary parts).

Consider the sine wave $s(t)$ of Fig. 4.11.3a and the square pulse $n(t)$ of Fig. 4.11.3b. The convolutional product $r(t) = s(t) * n(t)$ is shown in Fig. 4.11.3c. Their magnitude and phase spectra are also shown where $|R(j\omega)| = |S(j\omega)|\,|N(j\omega)|$ and $\arg R(j\omega) = \arg S(j\omega) + \arg N(j\omega)$. They are somewhat dissimilar. From a filter viewpoint, $n(t)$ is the impulse response $h(t)$ of a lowpass filter. From the spectra, $s(t)$ falls just outside the passband of $n(t)$ which is evident from the shape of $r(t)$.

The $H_C(j\omega)$ filter gain utilizes $\log S(j\omega)$ and $\log N(j\omega)$ which are shown in Figs. 4.11.3d and 4.11.3e, respectively. The noise spectrum has a finer structure than the signal spectrum. The filter gain $H_C(j\omega)$ is shown in Fig. 4.11.3g. It tends to behave as a wideband bandpass filter. Comparing the magnitude and phase spectra of $\log S(j\omega)$ and $\log N(j\omega)$ in Fig. 4.11.3d–e (both in dB), we see that the phase spectra are 20–30 dB larger than the magnitude spectra. Therefore $H_C(j\omega)$ is determined primarily by phase so we anticipate good phase recovery but possibly poor magnitude recovery. The filter input $\log R(j\omega)$ is shown in Fig. 4.11.3f and the filter output $\log C(j\omega)$ is shown in Figs. 4.11.3h–4.11.3i. Comparing these figures with $\log S(j\omega)$ in Figs. 4.11.3a and 4.11.3d, we see that the phase of $\log S(j\omega)$ is better estimated than its magnitude. The signal is fairly will recovered by comparing these figures with Fig. 4.11.3d. Applying $\log C(j\omega)$ to the exp block in Eq. 4.10.13 yields $C(j\omega)$. Taking the inverse FFT yields $c(t)$ which is shown in Fig. 4.11.3j. The signal is fairly well reconstructed.

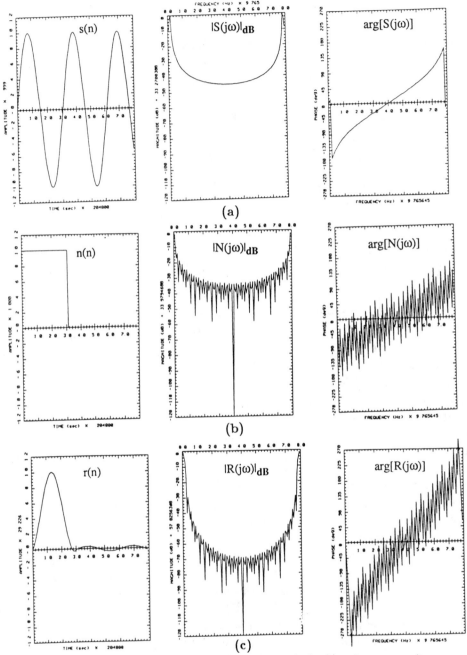

Fig. 4.11.3 (a) Signal $s(t)$, magnitude spectrum $|S(j\omega)|$, phase spectrum $\arg S(j\omega)$; (b) noise $n(t)$, magnitude spectrum $|N(j\omega)|$, phase spectrum $\arg N(j\omega)$; (c) total input $r(t)$, magnitude spectrum $|R(j\omega)|$, phase spectrum $\arg R(j\omega)$; real and imaginary parts of (d) $\log S(j\omega)$, (e) $\log N(j\omega)$, (f) $\log R(j\omega)$; (g) estimation filter gain; (h) real output spectrum $\log |C(j\omega)|$, (i) imaginary output spectrum $\arg C(j\omega)$, (j) convolved filter output signal. (From **S.W. Cox**, "Homomorphic estimation and detection of convolved signals," M.S.E.E. Thesis, 231 pp., Florida Atlantic Univ., Aug. 1988.)

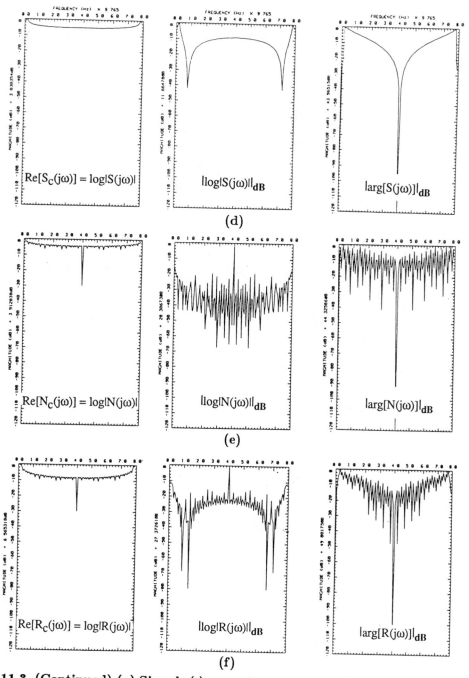

Fig. 4.11.3 (Continued) (a) Signal $s(t)$, magnitude spectrum $|S(j\omega)|$, phase spectrum $\arg S(j\omega)$; (b) noise $n(t)$, magnitude spectrum $|N(j\omega)|$, phase spectrum $\arg N(j\omega)$; (c) total input $r(t)$, magnitude spectrum $|R(j\omega)|$, phase spectrum $\arg R(j\omega)$; real and imaginary parts of (d) $\log S(j\omega)$, (e) $\log N(j\omega)$, (f) $\log R(j\omega)$; (g) estimation filter gain; (h) real output spectrum $\log|C(j\omega)|$, (i) imaginary output spectrum $\arg C(j\omega)$, (j) convolved filter output signal.

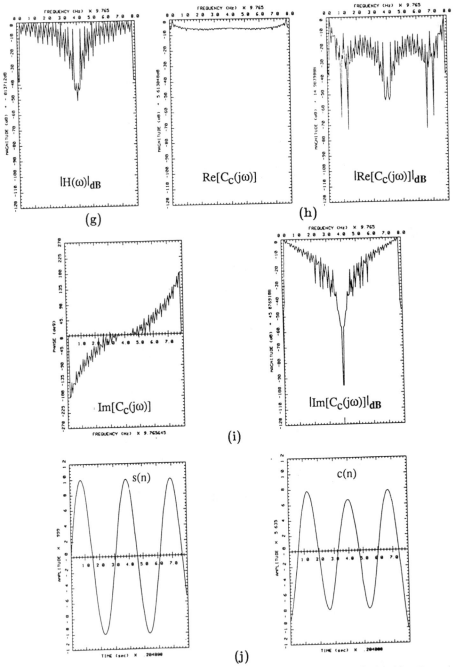

Fig. 4.11.3 (Continued) (a) Signal $s(t)$, magnitude spectrum $|S(j\omega)|$, phase spectrum $\arg S(j\omega)$; (b) noise $n(t)$, magnitude spectrum $|N(j\omega)|$, phase spectrum $\arg N(j\omega)$; (c) total input $r(t)$, magnitude spectrum $|R(j\omega)|$, phase spectrum $\arg R(j\omega)$; real and imaginary parts of (d) $\log S(j\omega)$, (e) $\log N(j\omega)$, (f) $\log R(j\omega)$; (g) estimation filter gain; (h) real output spectrum $\log |C(j\omega)|$, (i) imaginary output spectrum $\arg C(j\omega)$, (j) convolved filter output signal.

4.12 AMPLITUDE DOMAIN FILTERS

A signal is generally a function of both amplitude and time. Although we usually treat *time* as the independent variable and *amplitude* as the dependent variable, they need not be uniquely related. A broader viewpoint of time domain signals treats them as functions of two independent variables, *time and amplitude*, where $x(t) = g(x,t)$ or

$$\text{Signal} = \text{Function}(\text{Time}, \text{Amplitude}) \tag{4.12.1}$$

Signals can be encoded directly into the amplitude domain such that they are independent of short-term time ordering. Filters that process such signals are called *amplitude domain filters* or *pdf filters*.[17] Such filters process sequences of numbers whose short-term ordering is irrelevant. The information encoded into the amplitude domain is not localized with respect to time but is statistically spread out over some finite interval. This behavior is somewhat analogous to the frequency spectrum of a signal. One cannot associate a particular frequency with some point in time. Frequency is a property of the time interval itself.

Amplitude domain filters utilize the following probability results. A signal $x(t)$ is characterized by its probability density function (pdf) which is denoted as $f_X(x)$. If x and y are independent random variables, then defining z to be their sum gives

$$z = x + y \tag{4.12.2}$$

The pdf of z, written as $f_Z(z)$, equals

$$f_Z(z) = f_X(x) * f_Y(z - x) \tag{4.12.3}$$

$f_X(x)$ and $f_Y(y)$ are the pdf's of x and y, respectively. The pdf's of x and y are convolved to find the pdf of z. The characteristic functions Φ of x, y, and z equal

$$\Phi_X(j\omega) = E\{e^{j\omega x}\} = \int_{-\infty}^{\infty} f_X(x)e^{j\omega x}dx, \qquad \Phi_Y(j\omega) = E\{e^{j\omega y}\} = \int_{-\infty}^{\infty} f_Y(y)e^{j\omega y}dy$$

$$\Phi_Z(j\omega) = E\{e^{j\omega z}\} = \int_{-\infty}^{\infty} f_Z(z)e^{j\omega z}dz \tag{4.12.4}$$

It can be shown using Eq. 4.12.3 that the Φ's are related as

$$\Phi_Z(j\omega) = \Phi_X(j\omega)\Phi_Y(j\omega) \tag{4.12.5}$$

Close inspection of Eq. 4.12.4 shows that the characteristic functions are also Fourier transforms where

$$F_Z(j\omega) = \mathcal{F}\big[f_Z(z)\big] = \Phi_Z(-j\omega) \tag{4.12.6}$$

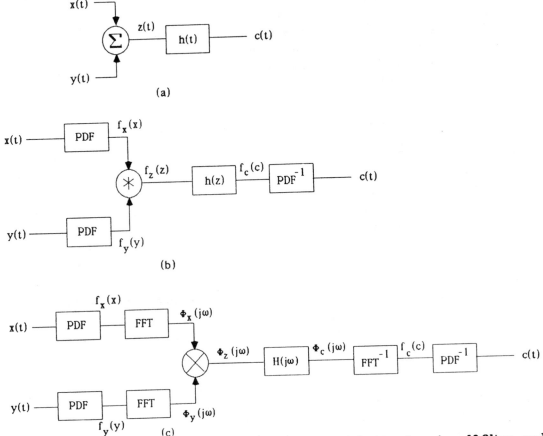

Fig. 4.12.1 Block diagram of (a) additive filter and (b) time domain pdf filter, and (c) frequency domain pdf filter.

Fourier transforms (via FFT's) process the pdf functions, which describe amplitudes instead of the usual time functions, and produce characteristic function spectra. The convolutional filtering described in Chap. 4.10.2 can be used to separate $f_X(x)$ and $f_Y(y)$ in Eq. 4.12.3. Alternatively, the multiplicative filtering of Chap. 4.10.1 can separate $\Phi_X(j\omega)$ and $\Phi_Y(j\omega)$ in Eq. 4.12.5.

These concepts are summarized in the following figures. Fig. 4.12.1a shows an additive filter with impulse response $h(t)$. It processes z given by Eq. 4.12.2 to form the output c where

$$c(t) = h(t) * z(t) \qquad (4.12.7)$$

Very often, $h(t)$ is an estimation filter used to form $\hat{x}(t)$. Fig. 4.12.1b shows a pdf filter. It is the pdf dual of Fig. 4.12.1a. Here x and y are input to pdf processing blocks which histogram the data to form the pdf estimates of $f_X(x)$ and $f_Y(y)$, respectively. These estimates are convolved together (see Eq. 4.12.3) to form an estimate of $f_Z(z)$. Then $f_Z(z)$ is filtered by $h(t)$ as given in Eq. 4.12.7. An estimation type filter yields $\hat{f}_X(x)$. The frequency domain equivalent of the pdf filter of Fig. 4.12.1b is shown in

Fig. 4.12.1c. It results by adding FFT blocks after the pdf blocks. The FFT blocks convert pdf's to characteristic functions. Φ_X and Φ_Y are multiplied together to form Φ_Z as given by Eq. 4.12.5. Then Φ_Z is filtered by $H(j\omega)$ to form

$$\Phi_C(j\omega) = H(j\omega)\Phi_Z(j\omega) \tag{4.12.8}$$

which is the frequency domain equivalent of Eq. 4.12.7. Taking the inverse FFT of $\Phi_C(j\omega)$ produces the pdf $f_C(c)$.

Now let us consider a different problem. Suppose that the inputs x and y are switched or selected to form z as shown in Fig. 4.12.2a rather than summed. The switching function $d(t)$ is a binary signal which selects x (for $d(t) = 1$) or y (for $d(t) = 0$). If p is the probability that $d(t) = 1$, then the pdf and characteristic function of z equal

$$f_Z(z) = pf_X(x) + (1 - p)f_Y(y)$$
$$\Phi_Z(j\omega) = p\Phi_X(j\omega) + (1 - p)\Phi_Y(j\omega) \tag{4.12.9}$$

Notice that $f_Z(z)$ is the weighted sum of $f_X(x)$ and $f_Y(y)$ in the switched case rather than the convolutional product in the additive case described by Eq. 4.12.3. This is a fortunate situation as shown in Figs. 4.12.2b and 4.12.2c because $f_X(x)$ and $f_Y(y)$

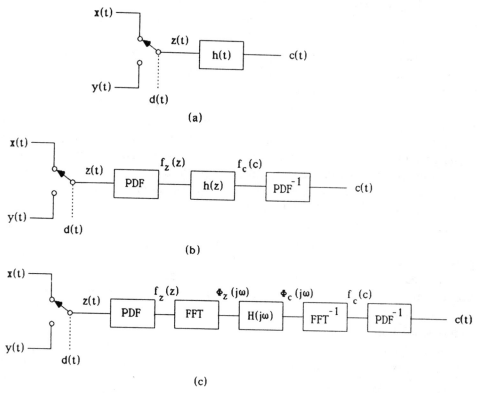

Fig. 4.12.2 Block diagram of (a) switched filter, (b) time domain pdf filter, and (c) frequency domain pdf filter.

can be separated by linear filtering as described in Chap. 4.4 rather than the more complicated convolutional (Eq. 4.12.3) or multiplicative (Eq. 4.12.4) filtering in Chap. 4.10. It may appear rather surprising that this nonlinear time switching problem reduces to a linear pdf filtering problem!

Now that the basic theory of amplitude domain filters has been discussed, let us review the encoding and decoding of binary signals. One suitable method for encoding data $d(t)$ of duration T seconds beginning at time t is the following. When $d(t) = 1$, the pdf signal $x(t)$, determined by sampling $F_X(x; \text{true})$, will be selected as $z(t)$. When $d(t) = 0$ pdf signal $x(t)$, determined by sampling $F_X(x; \text{false})$, will be selected as $z(t)$. A sequence of fixed length L of random numbers are generated. These L random values between $(0,1)$ are used to sample the appropriate inverse cumulative distribution functions (cdf), $F_X(x; \text{true})$ or $F_X(x; \text{false})$. This encodes $d(t)$ at time t into a stream of length T seconds of L pulses having amplitudes x which are distributed as $f_X(x; \text{true})$ or $f_X(x; \text{false})$. The amplitude domain information is recovered by forming a histogram of the pulse stream to estimate the pdf for the specified encoding time interval.

An example of this encoding procedure is shown in Fig. 4.12.3. We have arbitrarily chosen two disjoint pdf's to represent $d(t)$ over the data interval T. These pdf's, denoted as $f_X(x; \text{true})$ and $f_X(x; \text{false})$, are shown in Fig. 4.12.3a. Their corresponding cdf's are shown in Fig. 4.12.3b. The inverse cdf's, shown in Fig. 4.12.3c, are obtained by inverting the cdf's. Assume that the data $d(t) = (1, 0, 1)$ is to be encoded using 32 pulses of appropriate amplitude for each bit. Generating three sets of 32 random numbers between $(0,1)$ and using these to sample the inverse cdf's in Fig. 4.12.3c produces the encoded data $z(t)$ of Fig. 4.12.3d. The expected power spectrum of $z(t)$ is also shown. It is wide-band and flat within ± 3 dB. It was formed by averaging 64 power spectra of independent sequences of 2048 events.

These encoded signals have inherently wide bandwidth. In practical systems, bandwidth is very important and should be minimized. One way to reduce the wide bandwidth is to order the transmission sequence for minimum amplitude changes between adjacent events of the sequences. For example, the modulated sequence for bit i might be ordered according to ascending amplitude while the modulated sequence for bit $i+1$ might be ordered according to descending amplitude, and so on. This strategy also allows the demodulator to detect channel errors which are not consistent with this monotonic ordering.

An example of such amplitude sorting and ordering is given in Fig. 4.12.3e. The power spectrum of this encoded signal is significantly reduced. The power spectrum for the ordered signal is strongly periodic at odd harmonics of the sequence length. The broad-band noise is 30 dB below the dominant peak and high frequency noise falls off at -20 dB/dec. We are completely free to rearrange the time domain ordering of the events since they are all jointly independent.

Amplitude domain filters provide another means of processing in specialized applications. This approach enables us to effortlessly combine and apply probability theory (Chap. 1.17), statistical estimation/detection theory (Chap. 4.4), and classical linear filtering theory to amplitude domain filters.

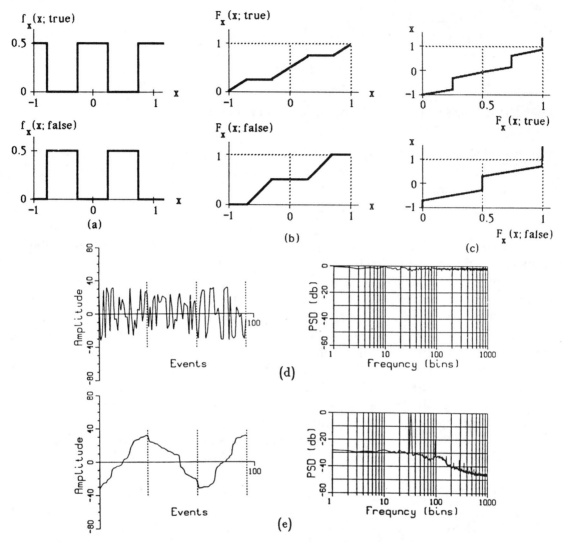

Fig. 4.12.3 Examples of (a) pdf's $f_X(x; true)$ and $f_X(x; false)$, (b) cdf's $F_X(x; true)$ and $F_X(x; false)$, (c) inverse cdf's, (d) unordered z for $(1,0,1)$ data pattern and its power spectrum, and (e) ordered z and its power spectrum. (From J.J. Reis, "Amplitude domain filtering," M.S.E.E. Thesis, Calif. State Univ., Long Beach, May 1985.)

4.13   AMPLITUDE DOMAIN FILTER DESIGN

We will now apply the concepts of amplitude domain filtering to the problem of transmitting binary data through a randomly connected channel.[18] The process to be considered is shown in Fig. 4.13.1. Binary data is modulated prior to its application to the transmission channel. The channel introduces a severe nonlinear distortion by randomly replacing elements of the modulated signal with Gaussian noise. This distorted signal is fed to the demodulator to recover the original signal. The random replacement of

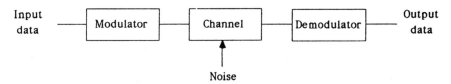

Fig. 4.13.1 Block diagram of communication system.

segments of the transmitted signal with Gaussian noise is equivalent to the transmission of data through the randomly connected channel of Fig. 4.12.2. Each of the blocks in Fig. 4.13.1 will now be described in detail.

4.13.1 BIPHASE MODULATOR

The biphase modulator performs pdf encoding of the binary data. This encoding is designed to match the characteristics of the signal to be transmitted with the characteristics of the transmission channel. If the channel is known to be noisy, then an encoding process is selected to minimize the effects of noise.

Assume that the data rate is f_d samples/sec so the data period is $T = 1/f_d$ seconds. Each binary bit can be in either the "true" (1) or the "false" (0) state. By associating one pdf with the true state and a different pdf with the false state, we generate an information carrying or modulated signal which can be sent over the randomly connected channel. Random sequences of length L are generated and transmitted for each bit.

Sequences for the true state will be composed of random variables (r.v.) drawn from some appropriate $f_X(x; \text{true})$ pdf. Sequences for the false state will be composed of random variables drawn from a complementary or a phase-shifted version of the same pdf $f_X(x; \text{false})$. For example, we could use locally periodic pdf's of

$$f_X(x; \text{true}) = \frac{1}{2A} \frac{1 + \cos(\omega_o x)}{1 + \text{sinc}(\omega_o A)} \left[U_{-1}(x + A) - U_{-1}(x - A) \right]$$

$$f_X(x; \text{false}) = \frac{1}{2A} \frac{1 + \cos(\omega_o x + \pi)}{1 + \text{sinc}(\omega_o A)} \left[U_{-1}(x + A) - U_{-1}(x - A) \right]$$

(4.13.1)

The pdf's are 180° out of phase with each other resulting in a form of biphase modulation. $f_X(x; \text{true})$ and $f_X(x; \text{false})$ are plotted in Fig. 4.13.2a for $\omega_o = 2\pi$ and $A = 32$ which are the parameters used in the simulation to be described later.

Depending upon the state of the data bit, the transmitted sequence is composed of either L events from the r.v. generator $f_X(x; \text{true})$ or L events from the r.v. generator $f_X(x; \text{false})$. For very short sequence lengths, it will be somewhat difficult to recover the encoded data since the pdf's cannot be well estimated.

Typical waveforms for the time domain signals associated with true data, false data, and noise for a sequence length of 32 events are shown in Figs. 4.13.2b, 4.13.2c, and 4.13.2d, respectively. The Gaussian noise simulates channel interference during drop-out. Also shown are the histograms based upon these samples. These raw pdf estimates are filtered by simple Gaussian lowpass filters whose cut-off frequencies are about 1/4 that of the pdf periodicity. This filtering operation, known as *Parzen density estimation*, will be discussed in a moment.

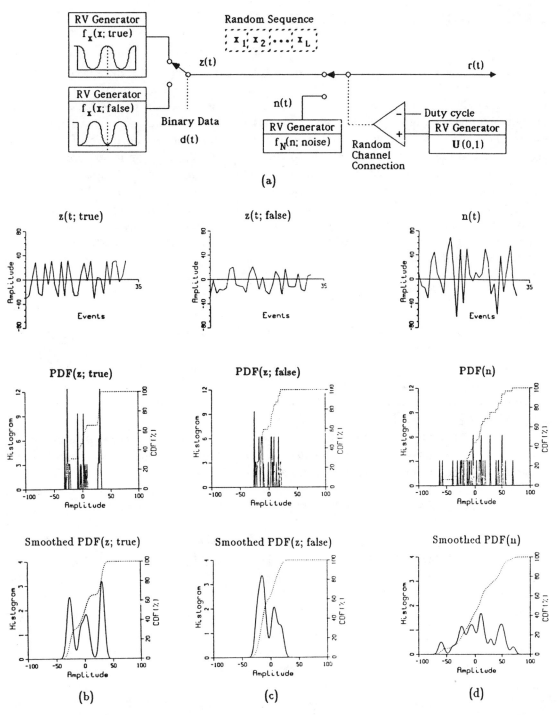

Fig. 4.13.2 (a) Biphase modulator and randomly connected channel model. Signal, pdf, and smoothed pdf for (b) true data, (c) false data, and (d) noise. (From J.J. Reis, loc. cit.)

4.13.2 RANDOMLY CONNECTED CHANNEL

The randomly connected channel is a type of channel distortion where the channel undergoes instantaneous dropout. This situation occurs when digital data is transmitted over poorly maintained voice-grade telephone circuits where the connections are noisy.

The model for the randomly connected channel is shown in Fig. 4.12.2. The pdf output signal is made up of random segments of the pdf input signal and Gaussian noise. The pdf of the output signal is a linear combination of the pdf's of the input signal and the Gaussian noise. As shown in Chap. 4.12, this nonlinear time domain channel distortion is mapped into a linear filtering problem in the pdf domain.

Two typical kinds of channel distortion reduce the performance of pdf filters – gain and offset, and linear filtering. Variations in gain a and offset b are not considered distortions in linear networks since their output is similarly changed. However in the pdf domain, gain and offset cause distortions which scale and shift the amplitude-axis in time domain signals as

$$w(t) = ax(t) + b$$

$$f_W(w) = \frac{1}{|a|} f_X\left(\frac{w-b}{a}\right) \tag{4.13.2}$$

A practical modem (modulator/demodulator) needs to be sophisticated enough to adapt to such gain and offset conditions.

If the channel subjects the transmission signal to any form of linear time domain filtering (other than $h(t) = U_0(t)$ or $H(j\omega) = 1$), then the transmitted events will no longer be independent and severe amplitude distortion may result. Even a simple lowpass filter (especially if it is not of the linear phase type) could distort the transmitted signal beyond recognition. However amplitude and phase equalizers having magnitude and delay characteristics inverse to those of the channel can be cascaded before the demodulator for compensation and eliminate this type of error.

4.13.3 BIPHASE DEMODULATOR

The overall block diagram for the biphase demodulator is shown in Fig. 4.13.3. It demodulates the biphase carrier signal but is not especially sophisticated or optimal. However, it illustrates the methods of pdf filtering. Temporal information from the channel is converted into a pdf by the pdf operator. The resulting pdf estimate will converge to the true pdf with probability one as the sample size approaches infinity provided the channel is ergodic. The ergodic nature of the channel depends wholly upon the nature of the channel noise, since the biphase modulating signals are ergodic by design. The channel noise is also ergodic when it is Gaussian with pdf $N(\mu = 0, \sigma = A)$ and autocorrelation $R_{nn}(\tau) = U_0(\tau)$.

This pdf estimate is processed by two matched filters. Each filter is tuned to its expected pdf function with gain

$$H(j\omega; \text{true}) = \Phi_X^*(j\omega; \text{true}) \quad \text{or} \quad h(x; \text{true}) = f_X(-x; \text{true})$$

$$H(j\omega; \text{false}) = \Phi_X^*(j\omega; \text{false}) \quad \text{or} \quad h(x; \text{false}) = f_X(-x; \text{false}) \tag{4.13.3}$$

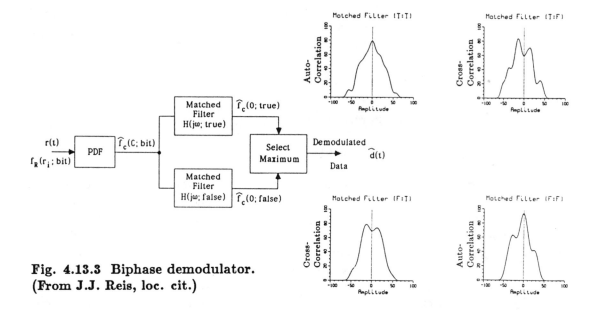

Fig. 4.13.3 Biphase demodulator.
(From J.J. Reis, loc. cit.)

Such a selection is equivalent to cross-correlating the channel pdf against the reference modulating pdf's. In the case of zero channel noise, the output from the filter which matches (or mismatches) the modulating pdf will actually be the autocorrelation function (or cross-correlation function) for that pdf, since

$$\widehat{R}_{XX}(x;\text{bit}) = f_X(x;\text{bit}) * f_X(-x;\text{bit}) = \text{Autocorrelation of } f_X(x;\text{bit})$$

$$\widehat{R}_{XY}(x;\text{bit}) = f_X(x;\text{bit}) * f_X(-x;\overline{\text{bit}}) = \text{Cross-correlation of } f_X(x;\text{bit}) \ \& \ f_X(-x;\overline{\text{bit}})$$

$$\widehat{S}_{XX}(j\omega;\text{bit}) = \Phi_X(j\omega;\text{bit})\Phi_X^*(j\omega;\text{bit})$$

$$\widehat{S}_{XY}(j\omega;\text{bit}) = \Phi_X(j\omega;\text{bit})\Phi_X^*(j\omega;\overline{\text{bit}})$$

$$(4.13.4)$$

where bit = true or bit = false. Since the autocorrelation functions $\widehat{R}_{XX}(x;\text{bit})$ and $\widehat{R}_{XX}(x;\overline{\text{bit}})$ are maximum at $x = 0$, we can assign the received bit decision to that channel which is maximum as

$$\widehat{R}_{XX}(0;\text{true}) \geq \widehat{R}_{XX}(0;\text{false}), \quad \text{Mark the received bit } \widehat{d}(t) \text{ as true}$$
$$\text{Otherwise,} \quad \text{Mark the received bit } \widehat{d}(t) \text{ as false}$$

$$(4.13.5)$$

When the sequence length associated with a data bit is very short, the probability density function estimated by the pdf operator will be very impulsive. Clearly, the estimated pdf will not closely match the pdf of the transmitted random variable. We can estimate those portions of the pdf which are poorly estimated by smoothing. This consists of convolving a suitable kernel with the pdf, or multiplying their respective characteristic functions, as

$$f_X(x; \text{smooth}) = f_X(x; \text{bit}) * h(-x; \text{smooth})$$
$$\Phi_X(j\omega; \text{smooth}) = \Phi_X(j\omega; \text{bit}) H_X^*(j\omega; \text{smooth})$$

(4.13.6)

Parzen density estimation uses Gaussian smoothing as

$$h(x; \text{Parzen}) = \frac{1}{\sqrt{2\pi}\sigma_P} \exp\left(\frac{-x^2}{2\sigma_P^2}\right)$$

(4.13.7)

$$H(j\omega; \text{Parzen}) = \exp(-\omega^2/2\sigma_P^2)$$

Since the biphase modulated waveforms are amplitude-limited to $\pm A$, the channel signal is clipped to $\pm A$ before filtering. This clipping operation is suboptimal and leads to a biased estimate in favor of $\widehat{f}_X(x; \text{true})$. We know apriori that valid channel signals cannot exceed $\pm A$ in amplitude. Therefore, when the channel presents the demodulator with a signal greater than $\pm A$, the demodulator "knows" unambiguously that the signal event is due totally to channel noise. The optimal strategy is to discard these high amplitude noise signals, which reduces the effective length of the encoding sequence. However, we have chosen the simpler (although suboptimal) implementation of keeping the sequence length constant.

This simplification is justified provided the channel is not too noisy. For example, for Gaussian noise with a normal pdf $N(0, A)$, the total probability that the channel noise exceeds $\pm A$ in amplitude equals $\Pr(x \geq |A|) = 2[0.5 - \text{erf}(A/\sigma)] = 0.31732$. Furthermore, if the channel is randomly connected to the noise source $100Q$ % of the time, then the expected number of large amplitude channel errors in the channel sequence $z(t)$ will be, $N = \Pr(x \geq |A|)QL$. For $Q = 0.25$ (25% of the received signal is replaced by Gaussian noise), only 8% of the received signal will be clipped. The clipping action will place probability impulses into the channel signal, which in turn will bias the demodulator decision in favor of $\widehat{f}_X(x; \text{true})$ over $\widehat{f}_X(x; \text{false})$

4.13.4 SIMULATION RESULTS

A simulation was performed where pdf-encoded binary data was transmitted over a randomly connected channel. The modulated signals had sequence lengths of 1, 2, 4, 8, 16, and 32 events. The duty cycles for the random disconnection of the channel were 0%, 25%, 50%, 75%, and 100%. The amplitude of the modulated signals was ± 32 units. The standard deviation of the Gaussian replacement noise was also ± 32 units. A 100,000 event Monte Carlo process was performed for each of the above combinations of sequence lengths and duty cycles. The Monte Carlo process was terminated prior to the 100,000 event total if the number of transmission errors reached 500.

The modulator consists of two pdf random variable generators using Eq. 4.13.1 where $\omega_o = 2\pi$ and $A = 32$. A uniform random number (*u.r.n.*) generator was used to generate the random bit stream $d(t)$ for transmission over the channel as follows: If *u.r.n.* ≥ 0.5, then transmit bit $d(t) = \text{true}$; otherwise transmit bit $d(t) = \text{false}$. For each bit, the modulator generates a sequence of length L of samples from the appropriate $f_X(x; \text{bit})$. This sequence $z(t)$ is then applied to the channel as shown in Fig. 4.13.2.

The channel is modeled by a data selector which randomly replaces elements of the transmitted sequence with Gaussian noise whose pdf is given by Eq. 4.13.7 where $\sigma_P = 32$ units. The amount of noise randomly inserted by the channel is controlled by another uniform random number generator. Each element of the transmitted sequence $r(t)$ is subject for replacement according to the following rule: If $u.r.n. \geq$ duty cycle, then leave the channel connected (do not add noise) with samples from $f_Z(z; \text{bit})$; otherwise replace the transmitted value with samples from $f_N(n; \text{noise})$.

The receiver histograms the channel data, clips it to $\pm A$ units in amplitude, and normalizes the result to form the estimated pdf for the raw received channel signal, $\widehat{f}_R(r_i; \text{channel})$. This pdf is filtered in the frequency domain using FFT's. The transfer functions for the matched filters and Parzen smoother are multiplied together so that both filtering operations can be performed in a single step. Combining Eqs. 4.13.3 and 4.13.6, the composite transfer functions are

$$
\begin{aligned}
H(j\omega; \text{true}) &= H^*(j\omega; \text{Parzen})\Phi_X^*(j\omega; \text{true}) \\
H(j\omega; \text{false}) &= H^*(j\omega; \text{Parzen})\Phi_X^*(j\omega; \text{false})
\end{aligned}
\tag{4.13.8}
$$

The Parzen smoother has $\sigma_P = A/8f_o = 32/8(1) = 4$ amplitude-units and $\omega_{3\text{dB}} = \sqrt{\ln 2}\,\sigma_P/2\pi = 0.26$ cycles/amplitude-unit. The tails of the Parzen function are limited to $\pm 3\sigma_P$.

After the data is filtered in the frequency domain, an inverse FFT is taken to obtain the correlation signals. These amplitude domain signals are not necessarily real-valued so magnitudes are taken to obtain only positive densities. The resulting correlation signals are normalized to unit area. Typical demodulator outputs from the true and false matched filters are shown in Fig. 4.13.3. The correlation peak at zero is largest from the filter that is matched to the transmitted bit. When the transmitted bit is true, the output from the true matched filter is larger than the output from the false matched filter, and vice versa. The received bit is determined according to the output signal which has the largest value at $x = 0$ following Eq. 4.13.5.

A standard likelihood function is also computed on the raw (unfiltered) channel pdf $\widehat{f}_X(x_i; \text{bit})$. Since all events from the modulated signals and the replacement Gaussian noise are independent, the likelihood ratio should produce a lower bound on the receiver bit error performance. The likelihood ratio R is given by

$$
R = \prod_{i=1}^{L} \frac{f_X(x_i; \text{true})^{\widehat{f}_X(x_i)}}{f_X(x_i; \text{false})^{\widehat{f}_X(x_i)}}
\tag{4.13.9}
$$

To avoid numeric problems due to the computer's limited dynamic range, the simulation uses the logarithmic form as

$$
\lambda = \log R = \sum_{i=1}^{L} \left[\widehat{f}_X(x_i) \log f_X(x_i; \text{true}) - \widehat{f}_X(x_i) \log f_X(x_i; \text{false}) \right]
\tag{4.13.10}
$$

The maximum likelihood decision becomes,

$$\text{If } \lambda \geq 0, \qquad \text{Mark the received bit } \widehat{d}(t) \text{ as true}$$
$$\text{Otherwise,} \qquad \text{Mark the received bit } \widehat{d}(t) \text{ as false} \tag{4.13.11}$$

4.13.5 BER AND INFORMATION CHARACTERISTICS

The bit error rate BER is plotted in Fig. 4.13.4a as a function of the sequence length L. A maximum sequence length of 32 events was used due to the extensive simulation time. The BER decreases exponentially with increasing channel length and can be made arbitrarily small using sufficiently large modulating sequences. This holds for all levels of channel noise except the 100% duty cycle case where the BER bit error rate is about 50% independent of sequence length. This limiting condition is equivalent to random guessing.

The BER is not zero for zero channel noise (duty cycle = 0%) since the modulating pdf's overlap significantly. We anticipated that the demodulator would produce a biased estimate of the channel signal. This biased behavior is easily observed in the performance curves for sequence length $L = 1$ and zero channel noise. An unbiased estimator would show a BER of 0.5. The BER performance of the matched filter demodulator is never more than a few percent worse than the performance of the optimal maximum likelihood estimator.[16]

The BER data is plotted in Fig. 4.13.4a with respect to the modulation sequence length and the duty cycle of the channel noise. BER is not plotted as a function of signal-to-noise ratio since SNR is a constant and independent of the sequence length. SNR depends only upon the duty cycle of the channel replacement noise. However, obviously more information about the transmitted bit is being sent as the sequence length increases. Information and sequence length can be related as follows.

The information I contained in a sequence of independent symbols equals

$$I = \sum_{i=1}^{L} f_i \log_2(1/f_i), \qquad f_i = \text{probability for symbol } i \tag{4.13.12}$$

The information contained in the transmitted channel signal is the sum of the information contained in both the modulated signal and channel noise, since the modulated signal and channel noise are disjoint. Hence,

$$I_{\text{Channel}} = I_{\text{Signal}} + I_{\text{Noise}} \tag{4.13.13}$$

We define a performance index as

$$I_{\text{Decision}} = \left(I_{S|T} + I_{N|T} \right) - \left(I_{S|F} + I_{N|F} \right)$$
$$S|T = \text{Signal assuming true bit,} \qquad S|F = \text{Signal assuming false bit}$$
$$N|T = \text{Noise assuming true bit,} \qquad N|F = \text{Noise assuming false bit}$$
$$\tag{4.13.14}$$

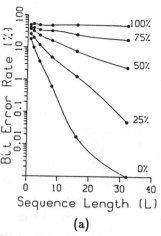

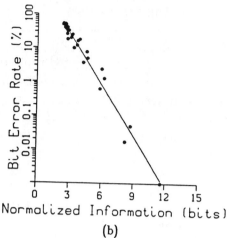

(a)

(b)

Fig. 4.13.4 Bit error rate as a function of (a) sequence length and (b) channel signal net information. (From J.J. Reis, loc. cit.)

The first (second) bracketed term corresponds to the information contained in the channel signal assuming the data bit is true (false). Their difference represents the net amount of information *not* common to both signals. Assuming that the information associated with the channel noise is equally represented in both the true and the false pdf's, $I_{N|T} = I_{N|F}$. Then the performance index reduces to

$$I_D = I_{S|T} - I_{S|F} = \text{net signal information}$$

$$= \sum_i \widehat{f_i}(\text{true}) \log_2 \left[1/f_i(\text{true}) \right] + \sum_i \widehat{f_i}(\text{false}) \log_2 \left[1/f_i(\text{false}) \right]$$

$$= \sum_i \left(\widehat{f_i}(\text{true}) \log_2 \left[1/f_i(\text{true}) \right] - \widehat{f_i}(\text{false}) \log_2 \left[1/f_i(\text{false}) \right] \right)$$

$$(4.13.15)$$

This expression is analogous to that for the log likelihood ratio given in Eq. 4.13.10. The log likelihood ratio (performance index) can therefore be interpreted as the net signal information with respect to the apriori (aposteriori or estimated) modulating pdf's. The expected value of I_D, scaled for sequence length, is given by

$$\sqrt{L} \, E\{I_D\} = \frac{\sum |\lambda(\text{true})| + \sum |\lambda(\text{false})|}{\sum \text{bits(true)} + \sum \text{bits(false)}} \, \sqrt{L} \qquad (4.13.16)$$

using λ from Eq. 4.13.10. A plot of the BER verses $\sqrt{L} \, E\{I_D\}$ is given in Fig. 4.13.4b. The resulting data is well-behaved over the simulation range of 0% to 100% duty cycle and sequence lengths from 1 to 32 events. The BER decreases exponentially with increasing channel information. This behavior is exactly analogous to that found in conventional phase-modulated (PM) systems. For a channel length of 32 events and a BER of 10^{-5}, our system is equivalent to the ideal PM encoder operating in an additive Gaussian noise environment at a SNR of 10 dB.

Selecting the pdf's given by Eq. 4.13.1 are clearly suboptimal in view of the apriori knowledge that the channel noise was Gaussian. Both the Gaussian pdf and the $f_X(x; \text{true})$ peak in the same place at $x = 0$. Furthermore, the channel signal was clipped at $\pm A$ which inserted probability impulses in the $f_X(x; \text{true})$ pdf. Hence one of the modulating carriers was highly correlated with the channel noise yielding biased decisions. In retrospect, the modulating pdf's should have been selected such that the pdf of the channel noise overlapped both pdf's equally. Of course we could have eliminated this problem by choosing the channel noise to be uniformly distributed instead of Gaussian distributed. However even with suboptimal modulating pdf's, system performance was still excellent.

The matched filters used in the demodulator ignored all knowledge of the channel noise. A more sophisticated demodulator would have incorporated an estimation filter as well as a high-resolution filter for the channel noise. The output of the channel noise filter can be used to tag those demodulated bits which are likely in error.

PROBLEMS

4.1 A band-limiting filter has its 0.28 dB frequency = 300 Hz and its 40 dB frequency = 400 Hz. Determine the orders of the Butterworth, Chebyshev, and elliptic filters which meet this specification. Plot the magnitude response of the minimum order elliptic filter and draw its block diagram realization.

4.2 Determine the order of the Bessel, Butterworth, Chebyshev, and elliptic filters which meet the following gain specifications: (1) 3 dB frequency = 3200 Hz, (2) 20 dB frequency = 4000 Hz, and (3) 40 dB frequency = 5000 Hz.

4.3 A phase detector requires 60 dB rejection at 4 Hz and select a 1.25 dB bandwidth of 1 Hz. (a) Determine the orders of the Butterworth, Chebyshev, and elliptic filters which meet this specification. (b) If a third-order filter is desired and M_p, M_r, and ω_r must remain fixed, what bandwidth ω_p should be specified?

4.4 (a) Determine the orders of the Butterworth, Chebyshev, and elliptic filters which meet the following specification: (1) maximum low-frequency gain = 0 dB, (2) 0.28 dB frequency = 2 KHz, (3) 40 dB frequency = 3 KHz, and (4) 60 dB frequency = 4 KHz. (b) Plot the magnitude response of the elliptic filter. How much rejection can be obtained?

4.5 Determine the orders of the Butterworth, Chebyshev, and elliptic filters which meet the magnitude response shown in Fig. P4.5.

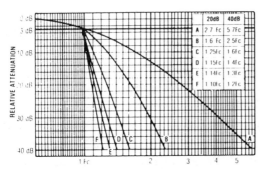

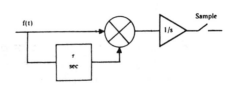

Fig. P4.10

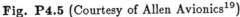
Fig. P4.5 (Courtesy of Allen Avionics[19])

4.6   Determine the transfer function of a fourth-order elliptic filter having a maximum low-frequency gain of 10, an in-band ripple of 0.28 dB, a corner frequency of 10 KHz, and a minimum stopband rejection of 40 dB. Draw its block diagram realization.

4.7   (a) Draw the block diagram of a third-order elliptic filter having $f_{0.28dB} = 3.2$ KHz and $M_s \geq 40$ dB. Label completely. (b) Draw the magnitude and step responses for this filter.

4.8   (a) The ideal delay filter has a transfer function of e^{-s}. Plot its step response. (b) Plot the unit step response of a fourth-order Bessel filter (assume a delay time of 1 sec). Compare it with the response of part (a). (c) Draw the block diagram of a fourth-order Bessel filter which will yield a delay of 1 $msec$.

4.9   (a) An ideal delay filter has a transfer function of e^{-s}. Plot its magnitude and delay responses. Classify the filter (e.g., lowpass). (b) The Bessel filter has a MFD characteristic. Therefore, a good gain approximation for constant delay is $H_n(s) = P_n(-s)/P_n(s)$ where P_n is the nth-order Bessel filter polynomial listed in Table 4.2.6. Plot the magnitude and delay responses of this filter. (c) Suppose a third-order filter of part (b) is used to obtain 1 $msec$ of delay. Design the filter in block diagram form.

4.10   A correlation detector consists of a delay element, multiplier, and integrator as shown in Fig. P4.10. Algebraically, it performs the operation $R(\tau) = \int_{-\infty}^{\infty} f(\tau)f(t - \tau)dt$. Suppose that a Bessel filter is used for the delay element. (a) Plot the magnitude and step responses of an ideal delay filter and (b) the Bessel filter. (c) Suppose that a third-order Bessel filter is used to obtain 1 $msec$ of delay. Design the filter in block diagram form. (d) Describe what filter parameters to vary to obtain delays from 0 to 1 $msec$.

Problems 4.11–4.18 use material in Ref. 1.

4.11*   (a) The McBride/Schaefgen/Steiglitz optimum ISE filter approximates the ideal step response shown in Fig. 5.4.1.* What is the maximum overshoot of the ideal filter? (b) How does increasing filter order appear to improve the response? (c) The ideal filter has unity 3 dB bandwidth. Plot the magnitude response of the optimum filters. Is unity bandwidth obtained (note that unity time delay was obtained)?

4.12*   (a) Write the transfer function of the McBride/Schaefgen/Steiglitz optimum third-order filter. Is it a minimum phase filter? (b) Show how an allpass function can be used in conjunction with a minimum phase function to obtain $H(s)$. Rewrite $H(s)$ in this form. (c) Draw a block diagram realization of the filter. Is unity bandwidth obtained (note that unity time delay was obtained)?

4.13*   (a) How does increasing filter order improve the response of the Pottle/Wong optimum ISE gain filter? (b) Draw the block diagram realization of the optimum fifth-order filter.

4.14*   Design a fourth-order Ariga/Sato optimum ISE filter in block diagram form. Use the type that has the narrowest transition band with the least delay peaking. The 3 dB bandwidth must equal 1 KHz.

4.15*   The step response of the Ariga/Sato fifth-order filter is shown in Fig. 5.6.2.* The step responses of fifth-order Butterworth and Bessel filters are also shown. However, the bandwidths of all three filters are different. Normalize their 3 dB bandwidths to unity. Then replot and compare their step responses.

4.16*   (a) Explain the difference in approaches used for determining the filter polynomials of Valand's linear phase filter and the Bessel filter. (b) Compare the phase responses of the two filters.

4.17*   (a) Design a 1 $msec$ Pade delay filter in block diagram form. Use first and second-order lowpass and allpass filter approximations. (b) Compare their magnitude responses and bandwidths.

4.18* An ideal delay filter has a transfer function of e^{-s} as discussed in Problem 4.8. Repeat that problem and show that $H_n(s) = P_n(s)/P_{-n}(s)$ is the gain of a Pade filter having equal indices.

4.19 Design a Class 1 adaptive filter to process a step signal having $E\{s(t)\} = U_{-1}(t)$ which is contaminated by uncorrelated white noise having $E\{|N(j\omega)|\} = 1$. (a) Write the transfer functions $H(j\omega)$ of the uncorrelated Wiener estimation, matched detection, and high-resolution detection filters. (b) Plot their magnitude responses. (c) Write their $H(s)$ transfer functions. (d) Which filters are causal? Which filters are stable? (e) Using a flow chart, explain how to implement these filters in software using FFT's.

4.20 Design a Class 1 adaptive filter to process a signum signal having $E\{s(t)\} = \text{sgn } t$ which is contaminated by uncorrelated 60 Hz sinusoidal noise (peak-to-peak amplitude = 1 volt, frequency = 60 Hz, phase is random). Repeat parts (a)–(e) of Problem 4.19.

4.21 Design a Class 1 adaptive filter to process an exponential step signal having $E\{s(t)\} = e^{-t}U_{-1}(t)$ which is contaminated by uncorrelated white noise having $E\{|N(j\omega)|\} = 1$. Repeat parts (a)–(e) of Problem 4.19.

4.22 Design a Class 1 adaptive filter to process a signal pulse having $E\{s(t)\} = U_{-1}(t + T/2) - U_{-1}(t - T/2)$ (unit amplitude, length T, centered at $t = 0$) which is contaminated by uncorrelated white noise having $E\{|N(j\omega)|\} = 1$. Repeat parts (a)–(e) of Problem 4.19.

4.23 Design a Class 1 adaptive filter to process a pulsed-cosine signal having $E\{s(t)\} = \cos(2\pi t)[U_{-1}(t + 0.1) - U_{-1}(t - 0.1)]$ (peak amplitude = 1, frequency = 1 Hz, pulse duration = 0.2 sec) which is contaminated by uncorrelated white noise having $E\{|N(j\omega)|\} = 1$. Repeat parts (a)–(e) of Problem 4.19.

4.24 Design Class 2 and 3 adaptive filters to process a step signal having $E\{s(t)\} = U_{-1}(t)$ which is contaminated by uncorrelated white noise having $E\{|N(j\omega)|\} = 1$ and random phase arg $N(j\omega)$ between $(-\pi, \pi)$. (a) Smooth (inner block) the expected signal and noise power spectra with rectangular, triangular, and Hann windows. Let $T = 1$ and smoothing width $p = 10\%$. Assume $R(j\omega) = E\{S(j\omega)\} + E\{N(j\omega)\}$. (b) Write the transfer functions of Class 2 uncorrelated Wiener estimation, matched detection, and high-resolution detection filters. (c) Repeat part b for Class 3 filters. (d) Plot their magnitude responses. (e) How do they compare with their Class 1 filter counterparts.

4.25 Design Class 2 and 3 adaptive filters to process a signum signal having $E\{s(t)\} = \text{sgn } t$ which is contaminated by uncorrelated 60 Hz sinusoidal noise (peak-to-peak amplitude = 1 volt, frequency = 60 Hz, phase is random). Repeat parts (a)–(e) of Problem 4.24.

4.26 Design Class 2 and 3 adaptive filters to process an exponential step signal having $E\{s(t)\} = e^{-t}U_{-1}(t)$ which is contaminated by uncorrelated white noise having $E\{|N(j\omega)|\} = 1$ and random phase arg $N(j\omega)$ between $(-\pi, \pi)$. Repeat parts (a)–(e) of Problem 4.24.

4.27 Design Class 2 and 3 adaptive filters to process a signal pulse having $E\{s(t)\} = U_{-1}(t + T/2) - U_{-1}(t - T/2)$ (unit amplitude, length T, centered at $t = 0$) which is contaminated by uncorrelated white noise having $E\{|N(j\omega)|\} = 1$ and random phase arg $N(j\omega)$ between $(-\pi, \pi)$. Repeat parts (a)–(e) of Problem 4.24.

4.28 Design Class 2 and 3 adaptive filters to process a pulsed-cosine signal having $E\{s(t)\} = \cos(2\pi t)[U_{-1}(t + 0.1) - U_{-1}(t - 0.1)]$ (peak amplitude = 1, frequency = 1 Hz, pulse duration = 0.2 sec) which is contaminated by uncorrelated white noise having $E\{|N(j\omega)|\} = 1$ and random phase arg $N(j\omega)$ between $(-\pi, \pi)$. Repeat parts (a)–(e) of Problem 4.24.

4.29 Redesign the Class 1 adaptive filter of Problem 4.19 in the feedback form of Fig. 4.9.1. The filter processes a step signal having $E\{s(t)\} = U_{-1}(t)$ which is contaminated by uncorrelated white noise having $E\{|N(j\omega)|\} = 1$. Uncorrelated Wiener estimation, matched detection, and high-resolution detection filters must be designed. (a) Write the forward transfer function $G(j\omega)$ using unity feedback. (b) Write the feedback transfer function $H(j\omega)$ using unity feedforward gain. (c) Write the feedback transfer function $H(j\omega)$ assuming the forward gain $G(s) = 1/(s+1)$.

4.30 Redesign the Class 1 adaptive filter of Problem 4.20 to have the feedback form of Fig. 4.9.1. The filter processes a signum signal having $E\{s(t)\} = \text{sgn } t$ which is contaminated by uncorrelated 60 Hz sinusoidal noise (peak-to-peak amplitude = 1 volt, frequency = 60 Hz, phase is random). Repeat parts (a)–(c) of Problem 4.29.

4.31 Redesign the Class 1 adaptive filter of Problem 4.21 to have the feedback form of Fig. 4.9.1. The filter processes an exponential step signal having $E\{s(t)\} = e^{-t}U_{-1}(t)$ which is contaminated by uncorrelated white noise having $E\{|N(j\omega)|\} = 1$. Repeat parts (a)–(c) of Problem 4.29.

4.32 Redesign the Class 1 adaptive filter of Problem 4.22 to have the feedback form of Fig. 4.9.1. The filter processes a signal pulse having $E\{s(t)\} = U_{-1}(t+T/2) - U_{-1}(t-T/2)$ (unit amplitude, length T, centered at $t = 0$) which is contaminated by uncorrelated white noise having $E\{|N(j\omega)|\} = 1$. Repeat parts (a)–(c) of Problem 4.29.

4.33 Redesign the Class 1 adaptive filter of Problem 4.23 to have the feedback form of Fig. 4.9.1. The filter processes a pulsed-cosine signal having $E\{s(t)\} = \cos(2\pi t)[U_{-1}(t+0.1) - U_{-1}(t-0.1)]$ (peak amplitude = 1, frequency = 1 Hz, pulse duration = 0.2 sec) which is contaminated by uncorrelated white noise having $E\{|N(j\omega)|\} = 1$. Repeat parts (a)–(c) of Problem 4.29.

4.34 Design a Class 1 multiplicative filter to separate a step signal having $E\{s(t)\} = U_{-1}(t)$ which is mixed with uncorrelated white noise having $E\{|N(j\omega)|\} = 1$. (a) Write the time domain signals that must be processed. (b) Write the transfer functions $H(j\omega)$ of the uncorrelated Wiener estimation, matched detection, and high-resolution detection filters. (c) Plot their magnitude responses. (d) Using a flow chart, explain how to implement these filters in software using FFT's.

4.35 Design a Class 1 multiplicative filter to separate a signum signal having $E\{s(t)\} = \text{sgn } t$ which is mixed with uncorrelated 60 Hz sinusoidal noise (peak-to-peak amplitude = 1 volt, frequency = 60 Hz, phase is random). Repeat parts (a)–(d) of Problem 4.34.

4.36 Design a Class 2 multiplicative filter to separate an exponential step signal having $E\{s(t)\} = e^{-t}U_{-1}(t)$ which is mixed with uncorrelated white noise having $E\{|N(j\omega)|\} = 1$. Repeat parts (a)–(d) of Problem 4.34.

4.37 Design a Class 2 multiplicative filter to separate a signal pulse having $E\{s(t)\} = U_{-1}(t+T/2) - U_{-1}(t-T/2)$ (unit amplitude, length T, centered at $t = 0$) which is mixed with uncorrelated white noise having $E\{|N(j\omega)|\} = 1$. Repeat parts (a)–(d) of Problem 4.34.

4.38 Design a Class 3 multiplicative filter to separate a pulsed-cosine signal having $E\{s(t)\} = \cos(2\pi t)[U_{-1}(t+0.1) - U_{-1}(t-0.1)]$ (peak amplitude = 1, frequency = 1 Hz, pulse duration = 0.2 sec) which is mixed with uncorrelated white noise having $E\{|N(j\omega)|\} = 1$. Repeat parts (a)–(d) of Problem 4.34.

4.39 Design a Class 1 convolutional filter to separate a step signal having $E\{s(t)\} = U_{-1}(t)$ which is convolved with uncorrelated white noise having $E\{|N(j\omega)|\} = 1$. Repeat parts (a)–(d) of Problem 4.34.

4.40 Design a Class 1 convolutional filter to separate a signum signal having $E\{s(t)\} = \text{sgn } t$ which is convolved with uncorrelated 60 Hz sinusoidal noise (peak-to-peak amplitude = 1 volt, frequency = 60 Hz, phase is random). Repeat parts (a)–(d) of Problem 4.34.

4.41   Design a Class 2 convolutional filter to separate an exponential step signal having $E\{s(t)\} = e^{-t}U_{-1}(t)$ which is convolved with uncorrelated white noise having $E\{|N(j\omega)|\} = 1$. Repeat parts (a)–(d) of Problem 4.34.

4.42   Design a Class 2 convolutional filter to separate a signal pulse having $E\{s(t)\} = U_{-1}(t + T/2) - U_{-1}(t - T/2)$ (unit amplitude, length T, centered at $t = 0$) which is convolved with uncorrelated white noise having $E\{|N(j\omega)|\} = 1$. Repeat parts (a)–(d) of Problem 4.34.

4.43   Design a Class 3 convolutional filter to separate a pulsed-cosine signal having $E\{s(t)\} = \cos(2\pi t)[U_{-1}(t + 0.1) - U_{-1}(t - 0.1)]$ (peak amplitude = 1, frequency = 1 Hz, pulse duration = 0.2 sec) which is convolved with uncorrelated white noise having $E\{|N(j\omega)|\} = 1$. Repeat parts (a)–(d) of Problem 4.34.

REFERENCES

1. Lindquist, C.S., *Active Network Design with Signal Filtering Applications*, Chap. 4, Steward and Sons, CA, 1977.

2. Ref. 1, Chap. 5.

3. Ref. 1, Chap. 2.10.

4. Christian, E. and E. Eisenmann, *Filter Design Tables and Graphs*, Wiley, NY, 1966.

5. Zverev, A.I., *Handbook of Filter Synthesis*, Chap. 5, Wiley, NY, 1967.

6. Orchard, H.J., "Roots of the maximally flat-delay polynomials," IEEE Trans. Circuit Theory, vol. CT-12, pp. 452–454, Sept. 1965.

7. Haas, W.H. and C.S. Lindquist, "New classes of ideal linear filters for filter design applications," Rec. 13th Asilomar Conf. Circuits, Systems, and Computers, pp. 294–297, 1979.

8. Lindquist, C.S. and W.H. Haas, "Some new adaptive correlation filters," Proc. 28th Midwest Symp. Circuits and Systems, pp. 479–488, 1985.

9. Haas, W.H. and C.S. Lindquist, "Linear detection filtering in the context of a least-squares estimator for signal processing applications," Proc. IEEE Intl. Conf. ASSP, pp. 651–654, 1978.

10. Lindquist, C.S., "Class 3 adaptive frequency domain filters," Proc. 28th Midwest Symp. Circuits and Systems, pp. 472–478, 1985.

11. Ref. 1, Chap. 6.

12. Haas, W.H. and C.S. Lindquist, "An adaptive FFT-based detection filter," Proc. 26th Midwest Symp. Circuits and Systems, pp. 265–269, 1983.

13. Lindquist, C.S., "Frequency domain adaptive algorithms for Class 3 correlated estimation filters," Rec. 19th Asilomar Conf. Circuits, Systems, and Computers, pp. 411–415, 1985.

14. Lindquist, C.S., H.B. Rozran, and W.H. Haas, "A new set of adaptive optimum filters," Rec. 17th Asilomar Conf. Circuits, Systems, and Computers, pp. 383–387, 1983.

15. Gold, B. and C.M. Rader, *Digital Processing of Signals*, Chap. 8, McGraw-Hill, NY, 1969. Rabiner, L.R. and B. Gold, *Theory and Application of Digital Signal Processing*, Chap. 12, Prentice-Hall, NY, 1975.

16. Tribolet, J.M., "A new phase unwrapping algorithm," IEEE Trans. ASSP, vol. ASSP-25, pp. 170–177, April 1977.

17. Reis, James J., "Amplitude domain filtering," M.S.E.E. Thesis, Calif. State Univ., Long Beach, 125 p., May 1985.

18. Reis, J.J. and C.S. Lindquist, "Amplitude domain filtering of biphase signals in a randomly connected channel," Proc. IEEE Intl. Symp. Circuits and Systems, pp. 341–344, 1986.

19. Precision LC Filters, Catalog 22F, 35 p., Allen Avionics, Inc., Mineola, NY, 1987.

5 DIGITAL TRANSFORMATIONS

*A new heart also will I give you, and
a new spirit will I put within you: and
I will take away the stony heart out of your flesh ...
And I will put my spirit within you and ...
ye shall be my people, and I will be your God.*

Ezekiel 36:26–28

Classical, optimum, and adaptive filters were investigated in the last chapter. This chapter will present analog-to-digital transformations which convert these continuous-time analog filters into discrete-time digital filters.

Historically, two approaches have been used for obtaining these transformations. One is carried out in the frequency domain while the other is carried out in the time domain. The frequency domain approach uses numerical integration approximations. The time domain approach approximates the analog filter's response to an impulse, step, or higher-order input. A variety of transformations result using these two approaches. Their time and frequency domain characteristics will be carefully discussed and interrelated. Since the transformation must be properly selected by the engineer to insure good design and system performance, it is important that he thoroughly digest this chapter.

5.1 S-DOMAIN TO Z-DOMAIN TRANSFORMATIONS

Digital transformations directly convert *unsampled* analog transfer functions $H(s)$ into digital transfer functions $H(z)$ and have the form

$$H(z) = H(s)\Big|_{s=g(z)} \quad \text{or} \quad H(s) = H(z)\Big|_{z=g^{-1}(s)} \tag{5.1.1}$$

5.1.1 Z-TRANSFORM

One digital transformation, the z-transform, was introduced in Chap. 1.9. It converted the *sampled* continuous-time transfer function $H_s(s)$ into a discrete-time transfer function $H(z)$ as

$$H(z) = H_s(s)\Big|_{sT=\ln z} \quad \text{or} \quad H_s(s) = H(z)\Big|_{z=e^{sT}} \tag{5.1.2}$$

359

using Eq. 1.9.1. The mapping properties of $z = e^{sT}$ between the s-plane and z-plane were discussed in relation to Figs. 1.9.6 and 1.9.7. The time domain and frequency domain characteristics of $H(z)$ were analyzed in Chaps. 2 and 3, respectively. This substitution produces a rational function $H(z)$ in $(\ln z)$ but not z. Irrational $H(z)$ have an infinite number of poles and/or zeros and are difficult to design.

5.1.2 MATCHED Z-TRANSFORM

Now assume that $z = e^{sT}$ is used to transform the *unsampled* analog transfer function $H(s)$ in Eq. 5.1.1 rather than the sampled transfer function $H_s(s)$ in Eq. 5.1.2. This is called the *matched z-transform*. It maps the s-plane poles $-p_k$ and zeros $-z_k$ into z-plane poles $e^{-p_k T}$ and zeros $e^{-z_k T}$, respectively. $H(z)$ is formed from $H(s)$ as

$$H(s) = a\,\frac{\prod_{k=1}^{m}(s+z_k)}{\prod_{k=1}^{n}(s+p_k)}, \qquad H(z) = b\,\frac{\prod_{k=1}^{m}(z-e^{-z_k T})}{\prod_{k=1}^{n}(z-e^{-p_k T})} \qquad (5.1.3)$$

where $H(s)$ is the (unsampled) analog transfer function. This method is easy to apply, as will be shown in Chap. 11.4, but it does not preserve the magnitude and phase response characteristics of $H(j\omega)$. It does preserve the time domain impulse response $h(t)$. In a moment, we will present a general method for developing digital transforms that will give better results.

Let us find the matched z-transform of an analog integrator. Consider a noncausal integrator with gain $H(s) = 1/s$ (for Re $s = 0$). Its impulse response equals

$$
\begin{aligned}
h(t) = \mathcal{L}^{-1}[H(s)] = 0.5 \ \mathrm{sgn}\, t &= -0.5, & t &< 0 \\
&= 0, & t &= 0 \qquad (5.1.4)\\
&= 0.5, & t &> 0
\end{aligned}
$$

(see Table 1.5.1). The causal integrator has $H(s) = (1/s) + \pi U_0(-js)$ (for Re $s \geq 0$) (see Eq. 1.6.9) so its impulse response equals

$$
\begin{aligned}
h(t) = \mathcal{L}^{-1}[H(s)] = 0.5(1 + \mathrm{sgn}\, t) &= 0, & t &< 0 \\
&= 0.5, & t &= 0 \qquad (5.1.5)\\
&= 1, & t &> 0
\end{aligned}
$$

Since $sT = \ln z$, the digital integrator has gain $H(z) = (T/\ln z)$ (for all z except along negative real axis) and an impulse response of (see Eq. 1.10.1)

$$h(nT) = Z^{-1}\left[\frac{T}{\ln z}\right] = \frac{1}{j2\pi T}\oint_C H(z)z^{n-1}dz = \frac{1}{j2\pi T}\oint_C \frac{z^{n-1}}{\ln z}dz \qquad (5.1.6)$$

To be well-defined $H(z)$ has a branch cut along the negative real axis. The contour is shown in Fig. 5.1.1. Manipulating Eq. 5.1.6 gives

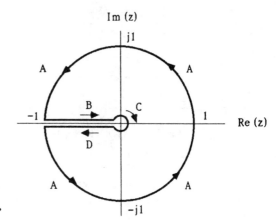

Fig. 5.1.1 Integration contour for matched-z transform of analog integrator.

$$\left[\oint_A + \oint_B + \oint_C + \oint_D \right] \left[\frac{1}{j2\pi T} \frac{z^{n-1}}{\ln z} \right] dz = \frac{1}{\pi} \int_\theta^{n\pi} \frac{\sin\theta}{\theta} d\theta + (-1)^n \int_0^\infty \frac{r^{n-1}}{\ln^2 r + \pi^2} dr \tag{5.1.7}$$

The integral along contour C ($\oint_C$) equals zero, while the integrals along contours B and D ($\oint_B$ and $\oint_D$) combine into the second integral in Eq. 5.1.7. $\oint_A$ is given by the first integral. These integrals must be evaluated numerically which gives

$$\begin{aligned} h(nT) &= -0.5, & n &= \ldots, -2, -1 \\ &= 0, & n &= 0 \\ &= 0.5, & n &= 1, 2, \ldots \end{aligned} \tag{5.1.8}$$

which matches the impulse response of the noncausal integrator of Eq. 5.1.4. The causal digital integrator has $H(z) = (T/\ln z) + \pi U_0(-jT/\ln z)$ (for $|z| > 1$) and an impulse response given by Eq. 5.1.5.

5.1.3 GENERAL TRANSFORMS

Combining Eqs. 5.1.1 and 5.1.2 allows a sampled $H_s(s)$ to be obtained from an unsampled $H(s)$, now to be denoted as $H(p)$ where $H(s) \to H(p)$, or vice versa as

$$\begin{aligned} H(p)\Big|_{p=g(z)} &\to H(z)\Big|_{z=e^{sT}} \to H_s(s) \\ H_s(s)\Big|_{sT=\ln z} &\to H(z)\Big|_{z=g^{-1}(p)} \to H(p) \end{aligned} \tag{5.1.9}$$

Transforming a convenient and/or desirable analog transfer function $H(p)$ using $p = g(z)$ generates $H(z)$. Then transforming $H(z)$ using $z = e^{sT}$ yields $H_s(s)$ which is the resulting sampled analog transfer function. Although p and s are both complex analog frequency variables used to describe their respective continuous transfer functions, they are seldom equal so different symbols must be used to distinguish between them.

What are some possible $p = g(z)$ transforms? Setting $z = e^{pT}$ or $pT = \ln z$, several transforms can be established by expanding $(\ln z)$ in a Taylor series as

$$pT = (z - 1) - \frac{1}{2}(z - 1)^2 + \frac{1}{3}(z - 1)^3 - \cdots, \qquad 0 < |z| \le 2$$

$$pT = \frac{z - 1}{z} + \frac{1}{2}\left(\frac{z - 1}{z}\right)^2 + \frac{1}{3}\left(\frac{z - 1}{z}\right)^3 + \cdots, \qquad \text{Re } z > 1/2 \qquad (5.1.10)$$

$$pT = 2\left[\frac{z - 1}{z + 1} + \frac{1}{3}\left(\frac{z - 1}{z + 1}\right)^3 + \frac{1}{5}\left(\frac{z - 1}{z + 1}\right)^5 + \cdots\right], \qquad \text{Re } z \ge 0$$

When z is close to unity, the error $(z - 1)$ is small so these series converge rapidly. They can be truncated after the first term to yield

$$H_{111} : \quad pT = z - 1$$

$$H_{001} : \quad pT = \frac{z - 1}{z} = 1 - z^{-1} \qquad (5.1.11)$$

$$H_{011} : \quad pT = 2\frac{z - 1}{z + 1}$$

These transforms are called the forward Euler (H_{111}), backward Euler (H_{001}), and bilinear (H_{011}), respectively. Each produce a unique transfer function $H(z)$ from $H(p)$ and will be discussed in detail in later sections.

Higher-order transforms can be obtained by truncating these series at higher-order terms. For example, truncating the series after the second term gives

$$pT = (z - 1) - \frac{1}{2}(z - 1)^2 = -\frac{1}{2}(z^2 - 4z + 3)$$

$$pT = \frac{z - 1}{z} + \frac{1}{2}\left(\frac{z - 1}{z}\right)^2 = \frac{1}{2}\frac{3z^2 - 4z + 1}{z^2} \qquad (H_{002}) \qquad (5.1.12)$$

$$pT = 2\left[\frac{z - 1}{z + 1} + \frac{1}{3}\left(\frac{z - 1}{z + 1}\right)^3\right] = \frac{8}{3}\frac{z^3 - 1}{z^3 + 3z^2 + 3z + 1}$$

This is a clumsy derivation approach. A general approach will now be presented to methodically derive such transforms.

5.2 NUMERICAL INTEGRATION TRANSFORMS H_{dmn}

One especially simple way of obtaining digital transformations is to approximate continuous integrators by discrete integrators as was done in Example 1.9.7. Equating the integrator transfer function $H(p) = 1/p$ to the discrete time equivalent $H(z) = 1/g(z)$ given by Eq. 5.1.9 results in $p = g(z)$ which is the digital transform.[1] The discrete integrators generated by this method should have the following properties.

First, they should be suitable for use as general p-to-z transformations. Second, they should convert rational $H(p)$ into rational $H(z)$. Third, stable and causal $H(p)$ should convert into stable and causal $H(z)$. Fourth, the low-frequency magnitude responses of $H(p)$ and $H(z)$ should coincide. Fifth, the magnitude error should be small over a large frequency range. Finally, the phase error should also be small over a large frequency range. The rationality condition permits $H(z)$ to be realized (assuming $H(z)$ is causal) using only delay elements z^{-1}, multipliers k_i, and adders. The causality condition insures realizability while stability insures that only stable $H(z)$ will be obtained. Of course, the magnitude and phase approximation errors should always be as small as possible. This is insured when the $j\omega$-axis maps into the unit circle (i.e., $p = j\omega \rightarrow |z| = 1$). Matching of the responses at low frequencies provides proper magnitude normalization.

To begin, the transfer function $H_{dmn}(z)$ of the discrete integrator is chosen to equal

$$H_{dmn}(z) = \frac{z^{-d}\sum_{j=0}^{m-d} a_j z^{-j}}{\sum_{i=0}^{n} b_i z^{-i}}, \qquad b_0 = 1 \qquad (5.2.1)$$

d is the number of zeros at the origin, m is the number of finite zeros including those at the origin, and n is the number of finite poles in the (z^{-1})-plane. The a_j and b_i coefficients are selected so that $H_{dmn}(z)$ well approximates the transfer function of an analog integrator $H(p) = 1/p$ as

$$H_{dmn}(z)\Big|_{z=e^{sT}} = \frac{1}{s + E(s)} = \frac{1}{p} \qquad (5.2.2)$$

The error term $E(s)$ provides an error measure for choosing the a_j and b_i coefficients. The coefficients could be chosen to set the $(n+m+1)$ lowest-order terms to zero in the Maclaurin or power series expansion of $E(s)$. Alternatively, they could be chosen to set $E(j\omega)$ to zero at $(n+m+1)$ frequencies, or even to minimize the integral of $E(j\omega)$ over some frequency range.

The first method of zeroing the lowest-order terms is especially convenient because it guarantees zero error at zero frequency. It is simple to apply using the power series for e^{sT} and leads to a solution without numerical iteration or integration.

The a_j and b_i coefficients are easily determined as follows: (1) choose n (the number of finite poles), m (the number of finite zeros), and d (the number of zeros at the origin); (2) replace $z = e^{sT} = 1 + sT + (sT)^2/2! + \ldots$ by this power series expansion; (3) use long division to express $H_{dmn}(z)$ in the form of Eq. 5.2.2; (4) set the $(n+m+1)$ lowest-order terms in $E(s)$ to zero and solve these equations for the a_j and b_i coefficients in $H_{dmn}(z)$.

It can be shown that the coefficients are found by inverting the matrix[1]

$$
\begin{bmatrix} -1 \\ 0 \\ 0 \\ \vdots \\ 0 \\ 0 \\ \vdots \\ 0 \end{bmatrix}
=
\begin{bmatrix}
0 & 0 & \cdots & 0 & 1 & \cdots & 1 \\
1 & \dfrac{1^0}{0!} & \cdots & \dfrac{m^0}{0!} & \dfrac{1^1}{1!} & \cdots & \dfrac{n^1}{1!} \\
0 & \dfrac{1^1}{1!} & \cdots & \dfrac{m^1}{1!} & \dfrac{1^2}{2!} & \cdots & \dfrac{n^2}{2!} \\
\vdots & \vdots & \cdots & \vdots & \vdots & \cdots & \vdots \\
0 & \dfrac{1^{k-1}}{(k-1)!} & \cdots & \dfrac{m^{k-1}}{(k-1)!} & \dfrac{1^k}{k!} & \cdots & \dfrac{n^k}{k!} \\
\vdots & \vdots & \cdots & \vdots & \vdots & \cdots & \vdots \\
0 & \dfrac{1^{n+m-1}}{(n+m-1)!} & \cdots & \dfrac{m^{n+m-1}}{(n+m-1)!} & \dfrac{1^{n+m}}{(n+m)!} & \cdots & \dfrac{n^{n+m}}{(n+m)!}
\end{bmatrix}
\begin{bmatrix} a_0 \\ a_1 \\ a_2 \\ \vdots \\ a_m \\ b_1 \\ \vdots \\ b_n \end{bmatrix}
\quad (5.2.3)
$$

for the H_{0mn} case, and

$$
[-1\ 0\ 0\ \ldots\ 0\ 0\ \ldots\ 0]^t =
$$

$$
\begin{bmatrix}
0 & 0 & \cdots & 0 & 1 & \cdots & 1 \\
\dfrac{1^0}{0!} & \dfrac{2^0}{0!} & \cdots & \dfrac{m^0}{0!} & \dfrac{1^1}{1!} & \cdots & \dfrac{n^1}{1!} \\
\dfrac{1^1}{1!} & \dfrac{2^1}{1!} & \cdots & \dfrac{m^1}{1!} & \dfrac{1^2}{2!} & \cdots & \dfrac{n^2}{2!} \\
\vdots & \vdots & \cdots & \vdots & \vdots & \cdots & \vdots \\
\dfrac{1^{k-1}}{(k-1)!} & \dfrac{2^{k-1}}{(k-1)!} & \cdots & \dfrac{m^{k-1}}{(k-1)!} & \dfrac{1^k}{k!} & \cdots & \dfrac{n^k}{k!} \\
\vdots & \vdots & \cdots & \vdots & \vdots & \cdots & \vdots \\
\dfrac{1^{n+m-2}}{(n+m-2)!} & \dfrac{2^{n+m-2}}{(n+m-2)!} & \cdots & \dfrac{m^{n+m-2}}{(n+m-2)!} & \dfrac{1^{n+m-1}}{(n+m-1)!} & \cdots & \dfrac{n^{n+m-1}}{(n+m-1)!}
\end{bmatrix}
\begin{bmatrix} a_1 \\ a_2 \\ a_3 \\ \vdots \\ a_m \\ b_1 \\ \vdots \\ b_n \end{bmatrix}
$$

$$(5.2.4)$$

for the H_{1mn} case. To illustrate this method, consider the following example.

EXAMPLE 5.2.1   Find the digital filter transfer function $H_{011}(z)$ that best approximates the analog integrator having $H(p) = 1/p$.

SOLUTION   Following the procedure just outlined, $H_{011}(e^{sT})$ equals

$$
H_{011}(e^{sT}) = \frac{a_0 + a_1 z^{-1}}{1 + b_1 z^{-1}}\bigg|_{z=e^{sT}} = \frac{a_0 + a_1(1 - sT + (sT)^2/2! - \ldots)}{1 + b_1(1 - sT + (sT)^2/2! - \ldots)}
$$
$$
= \frac{1}{c_1 s + (c_0 + c_2 s^2 + c_3 s^3 + \ldots)} = \frac{1}{s + E(s)} = \frac{1}{p}
\qquad (5.2.5)
$$

Equating coefficients gives

$$
c_0 = \frac{1 + b_1}{a_0 + a_1}, \qquad c_1 = \frac{T(a_1 - a_0 b_1)}{(a_0 + a_1)^2}, \qquad c_2 = \frac{T^2(a_0 b_1 - a_1)(a_0 - a_1)}{(a_0 + a_1)^3}, \qquad \ldots
$$

$$(5.2.6)$$

The best discrete integrator results when $c_1 = 1$ and $c_i = 0$ for $i = 0, 2, 3, \ldots$.

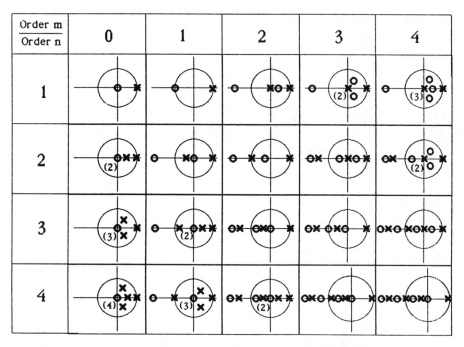

Fig. 5.2.1 Pole and zero locations for optimum digital integrators with zero delay of Table 5.2.1. (From J.S. Graham, "Discrete integrator transformations for digital filter synthesis," M.S.E.E. Directed Research, Calif. State Univ., Long Beach, Oct. 1984.)

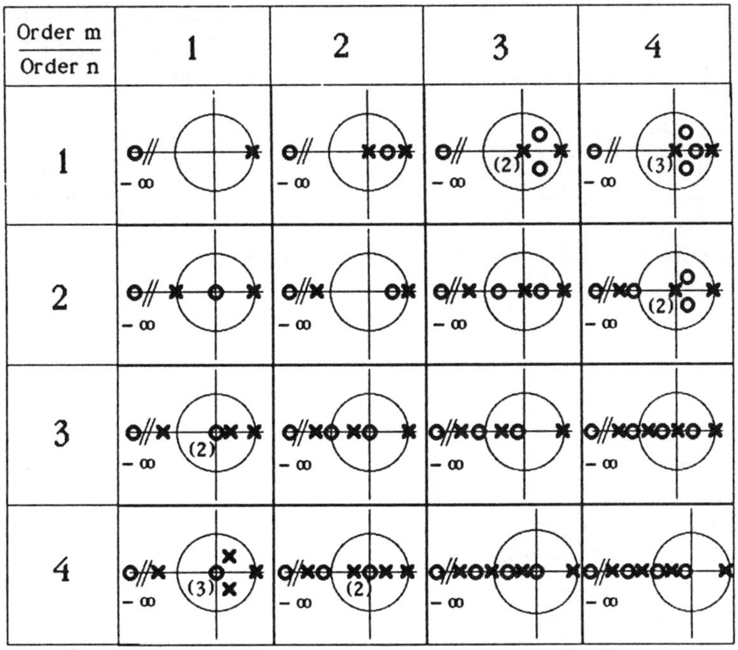

Fig. 5.2.2 Pole and zero locations for optimum digital integrators with unit delay of Table 5.2.2. (From J.S. Graham, loc. cit.)

Table 5.2.1 Optimum digital integrator $H_{0mn}(z)$ with zero delay.[1]

$\dfrac{m}{n}$	0	1
1	$*\quad \dfrac{T}{1-z^{-1}}$	$*\quad \dfrac{T}{2}\dfrac{1+z^{-1}}{1-z^{-1}}$
2	$*\quad \dfrac{2T}{3-4z^{-1}+z^{-2}}$	$*\quad \dfrac{2T(1+2z^{-1})}{5-4z^{-1}-z^{-2}}$
3	$*\quad \dfrac{6T}{11-18z^{-1}+9z^{-2}-2z^{-3}}$	$*\quad \dfrac{6T(1+3z^{-1})}{17-9z^{-1}-9z^{-2}+z^{-3}}$
4	$*\quad \dfrac{12T}{25-48z^{-1}+36z^{-2}-16z^{-3}+3z^{-4}}$	$*\quad \dfrac{12T(1+4z^{-1})}{37-8z^{-1}-36z^{-2}+8z^{-3}-z^{-4}}$

$\dfrac{m}{n}$	2	3
1	$*\quad \dfrac{T}{12}\dfrac{5+8z^{-1}-z^{-2}}{1-z^{-1}}$	$*\quad \dfrac{T}{24}\dfrac{9+19z^{-1}-5z^{-2}+z^{-3}}{1-z^{-1}}$
2	$*\quad \dfrac{T}{3}\dfrac{1+4z^{-1}+z^{-2}}{1-z^{-2}}$	$\dfrac{T(10+57z^{-1}+24z^{-2}-z^{-3})}{33+24z^{-1}-57z^{-2}}$
3	$\dfrac{3T(1+6z^{-1}+3z^{-2})}{10+9z^{-1}-18z^{-2}-z^{-3}}$	$\dfrac{3T(1+9z^{-1}+9z^{-2}+z^{-3})}{11+27z^{-1}-27z^{-2}}$
4	$\dfrac{12T(1+8z^{-1}+6z^{-2})}{43+80z^{-1}-108z^{-2}-16z^{-3}+z^{-4}}$	$\dfrac{12T(1+12z^{-1}+18z^{-2}+4z^{-3})}{47+192z^{-1}-108z^{-2}-128z^{-3}-3z^{-4}}$

$\dfrac{m}{n}$	4
1	$*\quad \dfrac{T}{720}\dfrac{251+646z^{-1}-246z^{-2}+106z^{-3}-19z^{-4}}{1-z^{-1}}$
2	$\dfrac{T}{90}\dfrac{281+2056z^{-1}+1176z^{-2}+104z^{-3}+11z^{-4}}{11+16z^{-1}-27z^{-2}}$
3	$\dfrac{3T}{10}\dfrac{47+552z^{-1}+192z^{-2}+152z^{-3}-3z^{-4}}{55+216z^{-1}+135z^{-2}-136z^{-3}}$
4	$\dfrac{6T}{5}\dfrac{1+16z^{-1}+36z^{-2}+16z^{-3}+z^{-4}}{5+32z^{-1}-32z^{-3}-5z^{-4}}$

Table 5.2.2 Optimum digital integrator $H_{1mn}(z)$ with unit delay.[1]

$\dfrac{m}{n}$	1	2
1	* $\dfrac{Tz^{-1}}{1-z^{-1}}$	* $\dfrac{T}{2}\dfrac{3z^{-1}-z^{-2}}{1-z^{-1}}$
2	* $\dfrac{2Tz^{-1}}{1-z^{-2}}$	$\dfrac{2T(2z^{-1}+z^{-2})}{1+4z^{-1}-5z^{-2}}$
3	$\dfrac{6Tz^{-1}}{2+3z^{-1}-6z^{-2}+z^{-3}}$	$\dfrac{6T(z^{-1}+z^{-2})}{1+9z^{-1}-9z^{-2}-z^{-3}}$
4	$\dfrac{12Tz^{-1}}{3+10z^{-1}-18z^{-2}+6z^{-3}-z^{-4}}$	$\dfrac{12T(2z^{-1}+3z^{-2})}{3+44z^{-1}-36z^{-2}-12z^{-3}+z^{-4}}$

$\dfrac{m}{n}$	3	4
1	* $\dfrac{T}{12}\dfrac{23z^{-1}-16z^{-2}+5z^{-3}}{1-z^{-1}}$	* $\dfrac{T}{24}\dfrac{55z^{-1}+59z^{-2}+37z^{-3}-9z^{-4}}{1-z^{-1}}$
2	$\dfrac{T}{3}\dfrac{17z^{-1}+14z^{-2}-z^{-3}}{1+8z^{-1}-9z^{-2}}$	$\dfrac{T}{3}\dfrac{413z^{-1}+436z^{-2}-69z^{-3}+10z^{-4})}{19+232z^{-1}-251z^{-2}}$
3	$\dfrac{3T(3z^{-1}+6z^{-2}+z^{-3})}{1+18z^{-1}-9z^{-2}-10z^{-3}}$	$\dfrac{3T(43z^{-1}+123z^{-2}+35z^{-3}-z^{-4})}{11+297z^{-1}-27z^{-2}-281z^{-3}}$
4	$\dfrac{12T(z^{-1}+3z^{-2}+z^{-3})}{1+28z^{-1}-28z^{-3}-z^{-4}}$	$\dfrac{12T(4z^{-1}+18z^{-2}+12z^{-3}+z^{-4})}{3+128z^{-1}+108z^{-2}-192z^{-3}-47z^{-4}}$

Three degrees of freedom are available (a_0, a_1, b_1) so we can only set $c_1 = 1$ and $c_0 = c_2 = 0$. Higher-order c_i will not in general equal zero. Substituting these three coefficient conditions into Eq. 5.2.6 yields $b_1 = -1$ and $a_0 = a_1 = T/2$. Then Eq. 5.2.5 reduces to

$$H_{011}(z) = \frac{T}{2}\frac{1+z^{-1}}{1-z^{-1}}\bigg|_{z=e^{\bullet T}} = 1\bigg/ s\left[1 - \frac{(sT)^2}{12} - \frac{(sT)^4}{120} - \cdots\right] \qquad (5.2.7)$$

The resulting approximation has error terms beginning with s^3 and progressing to higher order. $H_{011}(z)$ is called the *bilinear discrete integrator*. The s-to-z transform thus found equals

$$p = \frac{2}{T}\frac{1-z^{-1}}{1+z^{-1}} \qquad \text{or} \qquad z^{-1} = \frac{1-(pT/2)}{1+(pT/2)} \qquad (5.2.8)$$

and is called the *bilinear transformation*. It is one of the most widely used transforms due to its desirable mapping properties as shall soon be seen.   ■

This procedure can be used to generate transforms of any order. Two sets of discrete integrator transfer functions $H_{dmn}(z)$ are listed in Tables 5.2.1 (zero delay $d = 0$) and 5.2.2 (unit delay $d = 1$). The p-to-z transforms are simply $p = 1/H_{dmn}(z)$. The tables list thirty-six transfer functions up to fourth-order. Although they can all be used as general transforms, they cannot all be used as causal discrete integrators since some are not stable. However, seventeen of them have poles (or zeros) lying inside the unit circle making them usable as discrete integrators (or discrete differentiators) as shown in Figs. 5.2.1 and 5.2.2. The suitable transforms for discrete integrators are marked with an asterisk (*) in each table. Now let us discuss some of these transformations in more detail. Integrators using different approaches will be discussed in Chap. 11.5.

5.3 FORWARD EULER TRANSFORM H_{111}

$H_{111}(z)$ is called the *forward Euler discrete integrator* (FEDI) transfer function. Since

$$H_{111}(z) = \frac{Tz^{-1}}{1 - z^{-1}} = \frac{Y(z)}{X(z)} \tag{5.3.1}$$

from Table 5.2.2, the input-output difference equation is

$$(1 - z^{-1})Y(z) = Tz^{-1}X(z)$$
$$y(n) - y(n - 1) = Tx(n - 1) \tag{5.3.2}$$

From a numerical integration standpoint, the integral of $x(t)$ over the interval $[(n - 1)T, nT]$ is approximated by the area of a rectangle of height $Tx(n - 1)$ and width T. This corresponds to the shaded area under the $x(t)$ curve as shown in Table 5.3.1.

The ac steady-state FEDI gain is found by substituting $z = e^{j\omega T}$ into $H_{111}(z)$ to obtain

$$H_{111}(e^{j\omega T}) = \left.\frac{T}{z - 1}\right|_{z = e^{j\omega T}} = \frac{\omega T/2}{\sin(\omega T/2)} \frac{e^{-j\omega T/2}}{j\omega} \tag{5.3.3}$$

The magnitude and phase of H_{111} are plotted in Fig. 5.3.1 along with that of the analog integrator $H(j\omega) = 1/j\omega$.

The fractional magnitude error between $|H_{111}|$ and $|H|$ is found by dividing the two magnitudes and subtracting unity as

$$e_m(\omega) = \frac{|H_{111}|}{|H|} - 1 = \frac{\omega T/2}{\sin(\omega T/2)} - 1 = (\omega T/2)\csc(\omega T/2) - 1 \tag{5.3.4}$$

e_m is simply the reciprocal of the $(\sin x/x)$ function plotted in Fig. 1.6.4 minus one. The low-frequency error equals

$$e_m(\omega) \cong \frac{(\omega T/2)^2}{6}, \qquad \left|\frac{\omega T}{2}\right| \ll 1 \tag{5.3.5}$$

Table 5.3.1 Some useful two-point transforms and their characteristics.[2]

Type	Transform	Numerical Integration	Realization
H_{111}	$s \rightarrow \dfrac{z-1}{T}$	$(n-1)T \qquad nT$	
H_{001}	$s \rightarrow \dfrac{1}{T}\dfrac{z-1}{z}$	$(n-1)T \qquad nT$	
H_{011}	$s \rightarrow \dfrac{2}{T}\dfrac{z-1}{z+1}$	$(n-1)T \qquad nT$	
MBDI	$s \rightarrow \dfrac{2}{T}\dfrac{z(z-1)}{z+1}$	$(n-2)T \; (n-1)T \; nT$	
H_{121}	$s \rightarrow \dfrac{2}{T}\dfrac{z-1}{z(3z-1)}$	$(n-2)T \; (n-1)T \; nT$	
H_{112}	$s \rightarrow \dfrac{1}{2T}\dfrac{z^2-1}{z}$	$(n-2)T \; (n-1)T \; nT$	
LDI	$s \rightarrow \dfrac{1}{T}\dfrac{z-1}{z^{\frac{1}{2}}}$	$(n-1)T \; (n-\frac{1}{2})T \; nT$	
ODI	$s \rightarrow \dfrac{2}{T}\dfrac{z-1}{z^{\frac{1}{4}}+z^{\frac{3}{4}}}$	$(n-1)T \; (n-\frac{1}{4})T$ $(n-\frac{3}{4})T \quad nT$	

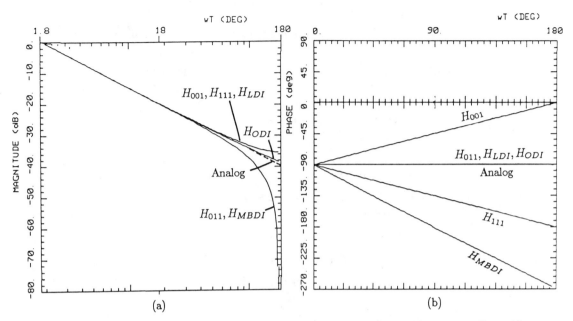

Fig. 5.3.1 (a) Magnitude and (b) phase responses of some optimum and continuous integrators.

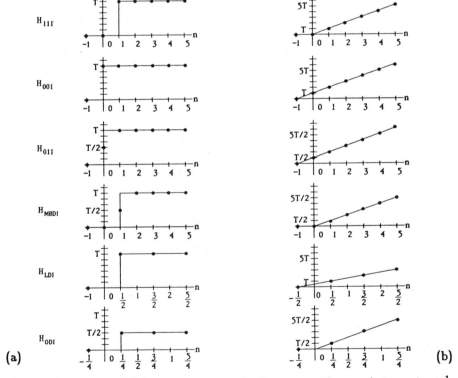

Fig. 5.3.2 (a) Impulse and (b) step responses of some optimum integrators.[1]

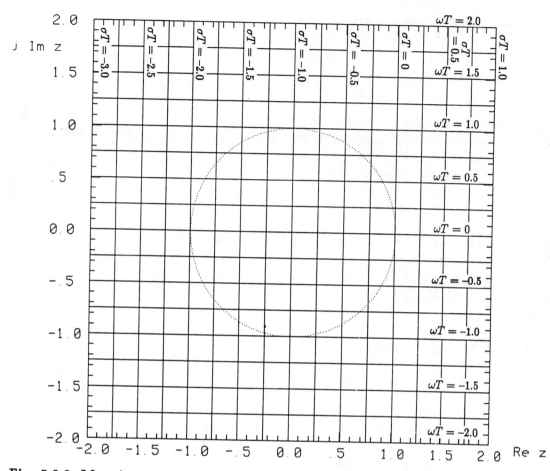

Fig. 5.3.3 Mapping properties of forward Euler transform. (Plotted by M. Zhao.)

which is found by expanding $\sin x = x - x^3/3! + \ldots$ in Eq. 5.3.3 and simplifying the result. The dc error $e_m(0)$ equals zero. The maximum error equals 57% from Eq. 5.3.4 and occurs at the Nyquist frequency $\omega_N = \pi/T$. The phase error e_p between arg H_{111} and arg H is found by subtracting these terms to form

$$e_p(\omega) = (-\omega T/2 - \pi/2) - (-\pi/2) = -(\omega T/2) \qquad (5.3.6)$$

e_p equals zero at dc and decreases linearly for increasing frequency. There is good agreement between the low-frequency FEDI and analog magnitude responses in Fig. 5.3.1a but the error increases monotonically to 57% at $\omega T = \pi$. The phase error magnitude decreases linearly with frequency as seen in Fig. 5.3.1b. e_p equals $-90°$ when $\omega T = \pi$.

The impulse response of $H_{111}(z)$ is also of interest. Since $H_{111}(z)$ can be expanded in a power series as

$$H_{111}(z) = \frac{Tz^{-1}}{1 - z^{-1}} = Tz^{-1} \sum_{n=0}^{\infty} z^{-n} \qquad (5.3.7)$$

then the impulse response equals

$$h_{111}(nT) = \frac{1}{T}Z^{-1}\big[H_{111}(z)\big] = 0, \qquad n = \ldots, -2, -1, 0 \tag{5.3.8}$$
$$= 1, \qquad n = 1, 2, 3, \ldots$$

h_{111} consists of a unit step centered at $t = 0$. This is shown in Fig. 5.3.2. Since the analog integrator impulse response equals $h(t) = U_{-1}(t)$ as given by Eq. 5.1.5, $h_{111}(nT)$ is in error only at $n = 0$.

From a mapping standpoint, equating Eq. 5.3.1 to $1/p$ and solving for p yields

$$p_n = pT = z - 1 \qquad \text{or} \qquad z = p_n + 1 \tag{5.3.9}$$

The polar mapping from the normalized p_n-plane to the z-plane is shown in Fig. 5.3.3. The z-plane is identical to the p_n-plane but is translated one unit to the left (see Fig. 1.9.6). Since portions of the left-half p-plane map outside the z-plane unit circle, stable and causal analog $H(p)$ filters may transform into unstable and causal digital $H(z)$ filters using H_{111}.

5.4 BACKWARD EULER TRANSFORM H_{001}

$H_{001}(z)$ is called the *backward Euler discrete integrator* (BEDI) transfer function. Since

$$H_{001}(z) = \frac{T}{1 - z^{-1}} = \frac{Y(z)}{X(z)} \tag{5.4.1}$$

from Table 5.2.1, the input-output difference equation is

$$(1 - z^{-1})Y(z) = TX(z) \tag{5.4.2}$$
$$y(n) - y(n-1) = Tx(n)$$

The BEDI approximates the area of $x(t)$ between $[(n-1)T, nT]$ by the area of a rectangle of height $Tx(nT)$ and width T. As shown in Table 5.3.1, the BEDI approximates the $x(t)$ integral by extending $x(nT)$ backward rather than forward as the FEDI.

The ac steady-state gain of the BEDI equals

$$H_{001}(e^{j\omega T}) = \frac{\omega T/2}{\sin(\omega T/2)}\frac{e^{j\omega T/2}}{j\omega} \tag{5.4.3}$$

Comparing Eqs. 5.4.3 and 5.3.3, we see that the BEDI has the same magnitude as the FEDI. Therefore the BEDI magnitude error is also given by Eq. 5.3.4. However the phase error $e_p(\omega) = \omega T/2$ is opposite in sign from that for the FEDI given by Eq. 5.3.6. The magnitude and phase plots of the BEDI are shown in Fig. 5.3.1. This integrator was also analyzed in Examples 1.9.7 and 1.9.8.

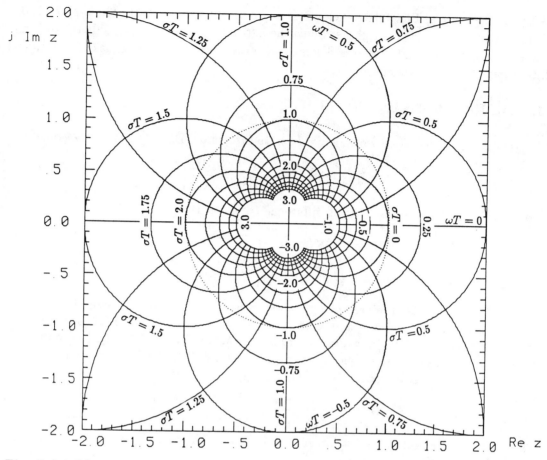

Fig. 5.4.1 Mapping properties of backward Euler transform. (Plotted by M. Zhao.)

The impulse response of $H_{001}(z)$ is

$$h_{001}(nT) = \frac{1}{T}Z^{-1}\left[\frac{T}{1-z^{-1}}\right] = 0, \qquad n = \ldots, -2, -1$$

$$= 1, \qquad n = 0, 1, 2, \ldots$$

(5.4.4)

which consists of a unit step located at $t = 0$. This is shown in Fig. 5.3.2. The BEDI step response begins one clock interval before the FEDI step response. The BEDI response coincides with the unit impulse response of the analog integrator given by Eq. 5.1.5 except at $n = 0$.

The p-to-z transform equals

$$p_n = pT = 1 - z^{-1} \qquad \text{or} \qquad z = (1 - p_n)^{-1}$$

(5.4.5)

The polar mapping from the p_n-plane to the z-plane is shown in Fig. 5.4.1. The (z^{-1})-plane is identical to the p_n-plane rotated by 180 degrees and translated one unit to the left (it is the same as Fig. 5.3.3 when ωT is augmented by π radians). The z-plane is obtained by inverting the (z^{-1})-plane. Since the left-half p_n-plane maps inside the unit circle, stable and causal $H(p)$ analog filters always produce stable and causal $H(z)$ digital filters using the H_{001} transform.

5.5 BILINEAR TRANSFORM H_{011}

$H_{011}(z)$ is called the *bilinear* or *Tustin discrete integrator* (BDI) transfer function. Since

$$H_{011}(z) = \frac{T}{2}\frac{1+z^{-1}}{1-z^{-1}} = \frac{Y(z)}{X(z)} \tag{5.5.1}$$

from Table 5.2.1, the input-output difference equation is

$$(1-z^{-1})Y(z) = \frac{T}{2}(1+z^{-1})X(z)$$

$$y(n) - y(n-1) = \frac{T}{2}\big[x(n) + x(n-1)\big] \tag{5.5.2}$$

The BDI approximates the area of $x(t)$ between $[(n-1)T, nT]$ by the area of a trapezoid whose bases are $x(n-1)$ and $x(n)$ and width T as shown in Table 5.3.1. Alternatively viewed, it is the area of a rectangle with average height $0.5[x(n-1) + x(n)]$ and width T. The BDI transfer function is the average of the FEDI and BEDI transfer functions.

The ac steady-state gain of the BDI equals

$$H_{011}(e^{j\omega T}) = \frac{T}{j2}\frac{\cos(\omega T/2)}{\sin(\omega T/2)} = \frac{\omega T}{2}\frac{\cot(\omega T/2)}{j\omega} \tag{5.5.3}$$

whose magnitude and phase is plotted in Fig. 5.3.1. The magnitude and phase error of H_{011} relative to $H = 1/j\omega$ equal

$$e_m(\omega) = (\omega T/2)\cot(\omega T/2) - 1, \qquad e_p(\omega) = (-\pi/2) - (-\pi/2) = 0 \tag{5.5.4}$$

The low-frequency magnitude error equals

$$e_m(\omega) \cong -\frac{(\omega T/2)^2}{3}, \qquad \left|\frac{\omega T}{2}\right| \ll 1 \tag{5.5.5}$$

which is found by expanding $\tan x = x + x^3/3 + \dots$ in Eq. 5.5.4a and simplifying the results. The BDI low-frequency magnitude error is twice that of the FEDI and BEDI given by Eq. 5.3.5. Also notice that the BDI gain equals zero at the Nyquist frequency which is a very desirable property in many applications as we will discuss later. The bilinear transform is used very often due to its good frequency domain behavior. The impulse response of $H_{011}(z)$ equals

$$h_{011}(nT) = \frac{1}{T}Z^{-1}\left[\frac{T}{2}\frac{1+z^{-1}}{1-z^{-1}}\right] = 0, \qquad n = \dots, -2, -1$$

$$= 0.5, \qquad n = 0 \tag{5.5.6}$$

$$= 1, \qquad n = 1, 2, 3, \dots$$

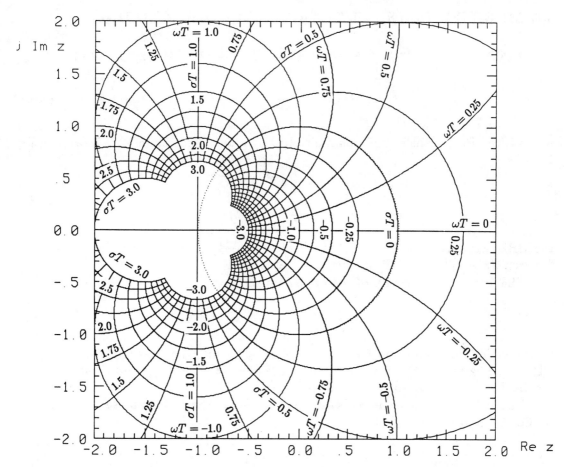

Fig. 5.5.1 Mapping properties of bilinear transform. (Plotted by M. Zhao.)

by simply combining Eqs. 5.3.8 and 5.4.4. The BDI impulse response matches that of the analog integrator exactly.

The mapping from the p_n-plane to the z-plane is shown in Fig. 5.5.1 (it is the same as the Smith chart used for transmission line analysis) where

$$p_n = \frac{pT}{2} = \frac{1 - z^{-1}}{1 + z^{-1}} \qquad \text{or} \qquad z = \frac{1 + p_n}{1 - p_n} \qquad (5.5.7)$$

The H_{011} transform maps the p_n-plane imaginary axis into the z-plane unit circle and the left-half p_n-plane maps inside the unit circle. Therefore stable and causal $H(p)$ analog filters always produce stable and causal $H(z)$ filters using the bilinear transform.

5.6   MODIFIED BILINEAR TRANSFORM H_{MBDI} AND H_{121}

The *modified bilinear discrete integrator* (MBDI) transfer function is given by[3]

$$H_{MBDI}(z) = \frac{T}{2}\frac{z^{-1}(1+z^{-1})}{1-z^{-1}} = \frac{Y(z)}{X(z)} \tag{5.6.1}$$

which is the H_{011} transfer function multiplied by a unit delay term z^{-1}. Rewriting Eq. 5.6.1 as

$$(1-z^{-1})Y(z) = \frac{T}{2}z^{-1}(1+z^{-1})X(z)$$

$$y(n) - y(n-1) = \frac{T}{2}\big[x(n-1) + x(n-2)\big] \tag{5.6.2}$$

the MBDI approximates the area of $x(t)$ between $[(n-1)T, nT]$ by the area of a trapezoid between $[(n-2)T, (n-1)T]$ as shown in Table 5.3.1.

The ac steady-state gain of the MBDI equals

$$H_{MBDI}(e^{j\omega T}) = e^{-j\omega T}H_{011}(e^{j\omega T}) \tag{5.6.3}$$

by combining Eqs. 5.6.1 and 5.5.3. The magnitude error e_m is identical to that for the BDI case as given by Eq. 5.5.4a. However the phase error $e_p(\omega) = -\omega T$ rather than 0 for the H_{011} case as given by Eq. 5.5.4b.

The MBDI impulse response is a delayed version of the H_{011} impulse response given by Eq. 5.5.6 so

$$\begin{aligned}
h_{MBDI}(nT) = h_{011}((n-1)T) &= 0, & n &= \ldots, -2, -1, 0 \\
&= 0.5, & n &= 1 \\
&= 1, & n &= 2, 3, \ldots
\end{aligned} \tag{5.6.4}$$

as shown in Fig. 5.3.2.

However the MBDI is suboptimum with respect to our error criterion established in Eq. 5.2.2. From Table 5.2.2, the optimum MBDI transfer function is

$$H_{121}(z) = \frac{3T}{2}\frac{z^{-1}(1-z^{-1}/3)}{1-z^{-1}} = \frac{Y(z)}{X(z)} \tag{5.6.5}$$

The input-output difference equation is

$$(1-z^{-1})Y(z) = \frac{3T}{2}z^{-1}\Big(1 - \frac{1}{3}z^{-1}\Big)X(z)$$

$$y(n) - y(n-1) = \frac{3T}{2}\Big[x(n-1) - \frac{1}{3}x(n-2)\Big] \tag{5.6.6}$$

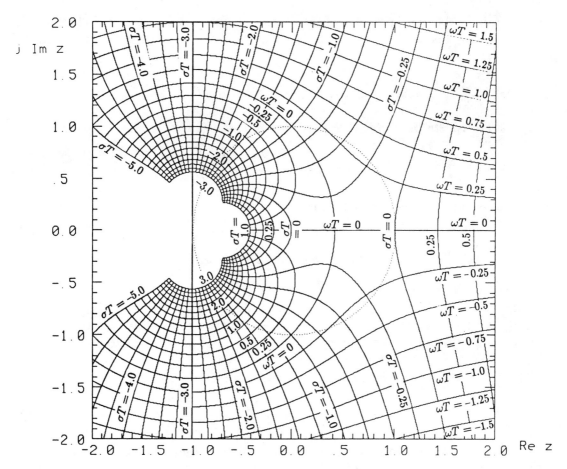

Fig. 5.6.1 **Mapping properties of modified bilinear transform. (Plotted by M. Zhao.)**

$H_{121}(z)$ approximates the area under $x(t)$ between $[(n-1)T, nT]$ by the area of a two rectangles having heights $x(n-1)$ and $0.5[x(n-1) - x(n-2)]$ and width T.

The ac steady-state gain equals

$$H_{121}(e^{j\omega T}) = \frac{3}{2}\left(1 - \frac{1}{3}e^{-j\omega T}\right) H_{111}(e^{j\omega T}) \qquad (5.6.7)$$

by combining Eqs. 5.6.5 and 5.3.3. The magnitude error e_m and phase error e_p are complicated. They are smaller than the MBDI errors of Eq. 5.6.3 as will be shown in Chap. 5.9.

The p-to-z mapping for the MBDI equals

$$p_n = \frac{pT}{2} = \frac{z(z-1)}{1+z} \qquad \text{or} \qquad z = \frac{1}{2}(1+p_n) \pm \frac{1}{2}\sqrt{p_n^2 + 6p_n + 1} \qquad (5.6.8)$$

which is shown in Fig. 5.6.1. The mapping is multiple-valued and two z values result for every p_n value. The imaginary axis of the p_n-plane maps outside the unit circle so stable and causal $H(p)$ analog filters may not transform into stable and causal $H(z)$ digital filters.

5.7 LOSSLESS TRANSFORM H_{LDI} AND H_{112}

The *lossless discrete integrator* (LDI) transfer function equals[3]

$$H_{LDI}(z) = \frac{Tz^{-0.5}}{1 - z^{-1}} = \frac{Y(z)}{X(z)} \tag{5.7.1}$$

The input-output difference equation is

$$(1 - z^{-1})Y(z) = Tz^{-0.5}X(z)$$
$$y(n) - y(n-1) = Tx(n - 0.5) \tag{5.7.2}$$

The LDI approximates the area of $x(t)$ between $[(n-1)T, nT]$ by the area of a rectangle of height $x(n - 0.5)$ and width T as shown in Table 5.3.1. The LDI differs from the BDI in that it uses the mid-point sample $x(n - 0.5)$ rather than the average of the two end-point samples $0.5[x(n-1) + x(n)]$. This requires a bi-phase clock to obtain samples at half-intervals as well as full intervals.

From Table 5.2.1, the optimum H_{112} transfer function is

$$H_{112}(z) = \frac{2Tz^{-1}}{1 - z^{-2}} \tag{5.7.3}$$

H_{112} is identical to H_{LDI} except that the sampling frequency is only half as much (i.e., $z^{-0.5} = e^{-sT} \rightarrow z^{-1}$ and $T \rightarrow 2T$).

The ac steady-state gain of the LDI equals

$$H_{LDI}(e^{j\omega T}) = \frac{1}{j\omega} \frac{\omega T/2}{\sin(\omega T/2)} \tag{5.7.4}$$

whose magnitude and phase are plotted in Fig. 5.3.1. We see that the LDI has the same magnitude as the FEDI and BEDI, and the same phase as the BDI. The magnitude error e_m is given by Eq. 5.3.4. The phase error e_p equals zero.

The LDI impulse response is

$$h_{LDI}(nT) = \frac{1}{T} Z^{-1} \left[\frac{Tz^{-0.5}}{1 - z^{-1}} \right] = 0, \qquad n - 0.5 = \dots, -2, -1 \tag{5.7.5}$$
$$= 1, \qquad n - 0.5 = 0, 1, 2, \dots$$

which is shown in Fig. 5.3.2. It is a step delayed by a one-half clock period.

The *p*-to-*z* mapping for the LDI transform equals

$$p_n = Tp = z^{0.5} - z^{-0.5} \qquad \text{or} \qquad z = \left[\left(\frac{p_n}{2} \right) \pm \sqrt{\left(\frac{p_n}{2} \right)^2 + 1} \right]^2 \tag{5.7.6}$$

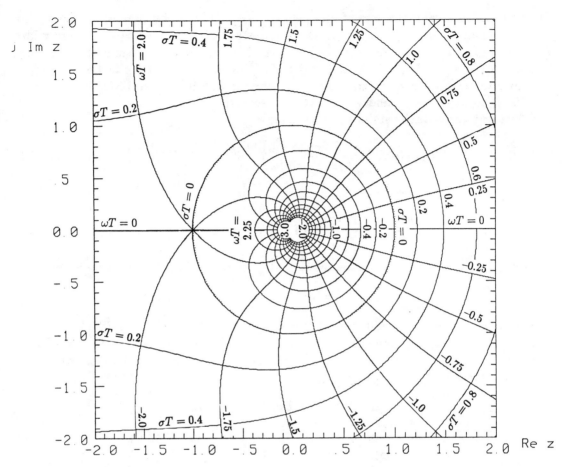

Fig. 5.7.1 Mapping properties of lossless transform. (Plotted by M. Zhao.)

which is plotted in Fig. 5.7.1. It maps the p_n-plane imaginary axis into the z-plane unit circle for only small values of p_n. Stable and causal $H(p)$ analog filters may not map into stable and causal $H(z)$ digital filters.

5.8 OPTIMUM TRANSFORM H_{ODI} AND H_{134}

The *optimum discrete integrator* (ODI) transfer function equals[2]

$$H_{ODI}(z) = \frac{T}{2} \frac{z^{-a} + z^{-b}}{1 - z^{-1}} = \frac{Y(z)}{X(z)} \tag{5.8.1}$$

The input-output difference equation is

$$(1 - z^{-1})Y(z) = \frac{T}{2}(z^{-a} + z^{-b})X(z)$$

$$y(n) - y(n-1) = \frac{T}{2}\big[x(n-a) + x(n-b)\big] \tag{5.8.2}$$

The ODI approximates the area of $x(t)$ between $[(n-1)T, nT]$ by sampling twice within the interval and forming a trapezoid whose bases are $x(n-a)$ and $x(n-b)$ and width T as shown in Fig. 5.3.1. Alternatively viewed, it is the area of a rectangle of average height $0.5[x(n-a) + x(n-b)]$ and width T. Constants a and b can be selected to minimize the fractional magnitude error e_m while maintaining zero phase error $e_p = 0$. For H_{ODI} to be causal, a and b must be nonnegative.

The ac steady-state gain of the ODI equals

$$H_{ODI}(e^{j\omega T}) = \frac{T}{j2} \frac{\cos[(b-a)(\omega T/2)]}{\sin(\omega T/2)} e^{j(1-a-b)\omega T/2} \tag{5.8.3}$$

The magnitude and phase error equal

$$e_m = \frac{\omega T}{2} \frac{\cos[(b-a)(\omega T/2)]}{\sin(\omega T/2)} - 1, \qquad e_p = (1-a-b)(\omega T/2) \tag{5.8.4}$$

from Eq. 5.8.3. For zero errors, we set $e_m = 0$ and $e_p = 0$ in Eq. 5.8.4 which gives

$$x \cos[(b-a)\omega T/2] = \sin(\omega T/2) \qquad \text{and} \qquad a+b=1 \tag{5.8.5}$$

Solving these equations simultaneously yields $a = 0.211325$ and $b = 0.788675$. Since a and b represent fractional time shifts when sampling, they should be chosen conveniently. For example, using a four-phase clock gives timing references at 0, 25, 50, and 75% of the time interval T. Thus, selecting

$$a = 0.25, \qquad b = 1 - a = 0.75 \tag{5.8.6}$$

results in practical timing values that still insure zero phase shift (Eq. 5.8.5b) and small magnitude error (Eq. 5.8.5a).

It is useful to note that the H_{011} and H_{LDI} are special cases of the ODI when (a, b) equals $(0, 1)$ and $(0.5, 0.5)$, respectively. Therefore the ODI is a fairly general case. The ODI impulse response is

$$h_{ODI}(nT) = \frac{1}{T} Z^{-1} \left[\frac{T}{2} \frac{z^{-a} + z^{-b}}{1 - z^{-1}} \right] = 0.5, \qquad n-a, \ n-b = 1, 2, 3, \ldots$$
$$= 0, \qquad \text{elsewhere} \tag{5.8.7}$$

which is shown in Fig. 5.3.2 for the special case where $a = 0.25$ and $b = 0.75$. It is a delayed one-half unit step with two unit samples at the 25% and 75% points in every interval T.

The p-to-z mapping for the ODI in this case is

$$p_n = \frac{pT}{2} = \frac{1 - z^{-1}}{z^{-1/4} + z^{-3/4}} = \frac{1 - z^{-0.5}}{z^{-0.25}} \qquad \text{or} \qquad z = \left[\left(\frac{p_n}{2} \right) \pm \sqrt{ \left(\frac{p_n}{2} \right)^2 + 1 } \right]^4 \tag{5.8.8}$$

which is plotted in Fig. 5.8.1. The ODI transform maps the p_n-plane imaginary axis

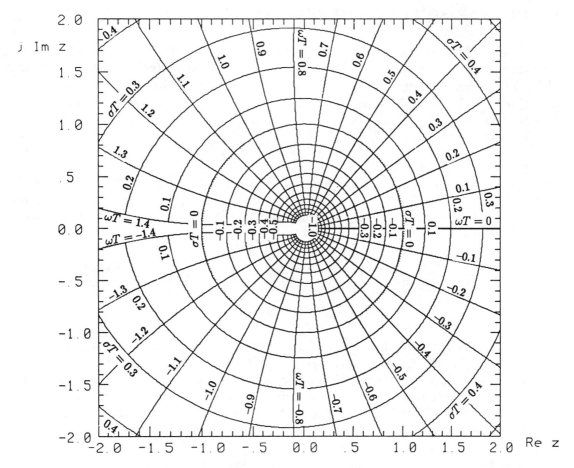

Fig. 5.8.1 Mapping properties of optimum transform. (Plotted by M. Zhao.)

into the z-plane unit circle. Stable and causal $H(p)$ analog filters always map into stable and causal $H(z)$ digital filters.

Increasing the clock frequency by four $(1/T \rightarrow 4/T$ so $z = e^{sT} \rightarrow z^4)$, then H_{ODI} becomes

$$H_{ODI}(z^4) = 2Tz^{-1}\frac{1+z^{-2}}{1-z^{-4}} = 2T\frac{z^{-1}}{1-z^{-2}} = H_{112}(z) \qquad (5.8.9)$$

which is identical to H_{112} given by Eq. 5.7.1. Pole-zero cancellation at $z = \pm j1$ reduces H_{ODI} to H_{112}. However the four-pole H_{ODI} is suboptimum by the error criterion of Eq. 5.2.2. From Table 5.2.2, the optimum ODI transfer function is

$$H_{134}(z) = 12Tz^{-1}\frac{1+3z^{-1}+z^{-2}}{1+28z^{-1}-28z^{-3}-z^{-4}} \qquad (5.8.10)$$

which is more involved.

5.9 COMPARISON OF NUMERICAL INTEGRATION TRANSFORMS

For greater understanding, let us further compare and discuss the numerical integration transforms described in the preceding sections. We choose the integral of absolute error (IAE) to measure accumulated error. The IAE between the impulse response of the digital filter and that of the ideal analog integrator (a unit step) over the time interval $(-5T, 5T)$ equals

$$IAE(t) = \int_{-5T}^{t} |h(t) - U_{-1}(t)| dt \qquad (5.9.1)$$

from Eq. 4.3.1. The impulse responses of the lowest-order filters of Tables 5.2.1 and 5.2.2 are shown in Fig. 5.9.1 along with their $IAE(t)$ errors. These continuous responses are obtained from the discrete output values using the Whittaker reconstruction method of Eq. 3.6.5 where

$$h(t) = Th_r(t) = \sum_{n=0}^{N-1} h(nT) \frac{\sin(\pi(t-nT)/T)}{\pi(t-nT)/T} = \frac{\sin(\pi t/T)}{\pi} \sum_{n=0}^{N-1} \frac{(-1)^n h(nT)}{\pi(t-nT)} \qquad (5.9.2)$$

for $t \neq nT$ and $h(nT)$ otherwise.

Most of the impulse responses are highly under-damped and oscillatory like H_{013}, H_{031}, H_{041}, H_{121}, H_{131}, and H_{141}. H_{014}, H_{022}, H_{044}, H_{112}, and H_{134} each have a pole at $z = -1$ (see Figs. 5.2.1 and 5.2.2) so are conditionally stable and have responses which never settle to unity. Poor time domain response suggests these transforms may produce poor digital filters in Eq. 5.1.1. The best responses (in order of increasing overshoot) are H_{002}, H_{003}, H_{MBDI}, H_{011}, and H_{004}. The remaining responses H_{001}, H_{021}, H_{012}, and H_{111} are poorer.

In terms of the $IAE(t)$ index, the best responses (in order of increasing $ISE(5)$ value) are H_{011}, H_{021}, H_{012}, H_{031}, H_{003}, H_{002}, and H_{004}. These responses are monotonic and have increasing rise times. The worst responses (in order of decreasing $ISE(5)$ value) are H_{112}, H_{141}, H_{131}, H_{014}, H_{022}, and H_{121}. These time domain results (overshoot and $ISE(5)$ value) suggest that the best transform is H_{011} (BDI) followed by H_{003} and H_{002}.

The IAE between complex gain $H(e^{j\omega T})$ and that of the ideal analog integrator $(1/j\omega)$ over the frequency interval $(0, 0.5)$ equals

$$IAE(j\omega) = \int_{0}^{\omega} |H(e^{j\omega T}) - 1/j\omega| d\omega \qquad (5.9.3)$$

from Eq. 4.3.2. The $IAE(j\omega)$ errors are plotted versus frequency in Fig. 5.9.2. The sampling period $T = 2\pi$ so the maximum frequency considered is $\omega_N = \pi/T = 0.5$. The frequency response errors can be quite different at at low frequency and the Nyquist frequency. Higher-order discrete integrators have superior performance at low frequencies. H_{014}, H_{022}, H_{044}, H_{112}, and H_{134} gains approach infinity near the Nyquist frequency since they each have a pole at $z = -1$ (see Figs. 5.2.1 and 5.2.2).

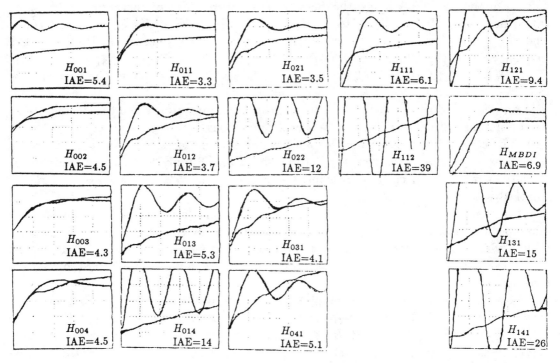

Fig. 5.9.1 Impulse responses of optimum digital integrators with IAE indices. (From J.S. Graham, loc. cit.)

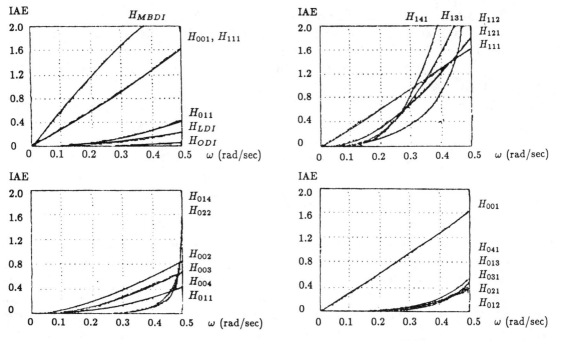

Fig. 5.9.2 IAE indices of gain error $H(e^{j\omega T}) - 1/j\omega$. (From J.S. Graham, loc. cit.)

From the $IAE(j\omega)$ index the best responses (in order of increasing $IAE(j0.5)$ value) are H_{ODI}, H_{LDI}, H_{012}, H_{021}, H_{031}, H_{011}, H_{013}, H_{041}, H_{004}, H_{003}, and H_{002}. The H_{001}, H_{111}, H_{121}, M_{BDI}, and H_{141} are significantly poorer. In the restricted frequency range $(0, 0.2)$, the best $IAE(j\omega)$ filters are H_{022}, H_{014}, H_{013}, H_{ODI}, H_{041}, H_{031}, H_{012}, H_{021}, and H_{LDI}. Therefore from the $IAE(j\omega)$ viewpoint, the best filter is the H_{ODI} followed by H_{031}, H_{012}, H_{021}, and H_{011}.

Another goodness index is the implementation complexity which can be measured by the number of delay, multiply, and add operations required for H_{dmn} (see Chap. 1.2). These are determined from Fig. 5.9.3 which shows the various discrete integrator structures. This is an expanded version of Table 5.3.1. Simply counting the number of operations required for each discrete integrator gives Table 5.9.1. The H_{LDI} and H_{ODI} (marked with asterisks) require fractional delays.

For example, H_{001} requires one delay, one multiply (scaler), and one (two-input) adder which total 3; H_{002} requires two delays, three multiplies, and two (two-input) adders which total 7. We assume that two-input adders or sequential addition is used rather than parallel addition (i.e., microprocessor versus high-level software implementation).

Table 5.9.1 shows the least complex integrators are H_{001} and H_{111} (3 operations total). These are followed by H_{011}, H_{112}, H_{LDI} (4); H_{MBDI} (5); H_{121}, H_{ODI} (6); H_{002} (7); H_{021}, H_{022} (8); and H_{012}, H_{131} (9). The most complex are H_{041} (14); H_{004}, H_{014} (13); H_{013}, H_{141} (12); H_{031} (11); and H_{003} (10). Naturally the higher-order H_{dmn} structures are more complex and computationally less efficient (order $= \max(m,n)$). The most efficient structures are the H_{001} and H_{111} followed by H_{011}, H_{112}, and H_{MBDI}.

We conclude from these time domain, frequency domain, and complexity results that H_{011}, H_{012}, and H_{021} are the best all-round zero-delay discrete integrators or p-to-z transforms. H_{011} has the best impulse response and least complexity, H_{012} has the smallest frequency response error, while H_{021} falls between H_{011} and H_{012}. The p_n-to-z transforms are:

$$H_{011}: \quad p_n = \frac{pT}{2} = \frac{1 - z^{-1}}{1 + z^{-1}}$$

$$H_{012}: \quad p_n = \frac{pT}{2} = \frac{5 - 4z^{-1} - z^{-2}}{1 + 2z^{-1}} \qquad (5.9.4)$$

$$H_{021}: \quad p_n = \frac{pT}{12} = \frac{1 - z^{-1}}{5 + 8z^{-1} - z^{-2}}$$

The best unit delay integrators are H_{111}, H_{121}, H_{131}, and H_{141}. Of these, H_{111} has the smallest impulse response error and least complexity; H_{141} has the smallest frequency response error at low frequencies but the greatest complexity. H_{121} and H_{131} are compromises between H_{111} and H_{141}. In general, the overall performance of H_{0mn} discrete integrators is superior to that of the H_{1mn} types. The p_n-to-z transforms are:

$$H_{111}: \quad p_n = pT = \frac{1 - z^{-1}}{z^{-1}} = z - 1 \qquad (5.9.5a)$$

Table 5.9.1 Implementation complexity of
discrete integrators.[1] (* = fractional delay.)

Transform	Add	Multiply	Delay	Total
H_{001}	1	1	1	3
H_{002}	2	3	2	7
H_{003}	3	4	3	10
H_{004}	4	5	4	13
H_{011}	2	1	1	4
H_{012}	3	4	2	9
H_{013}	4	5	3	12
H_{014}	4	5	4	13
H_{021}	3	3	2	8
H_{022}	3	2	3	8
H_{031}	4	4	3	11
H_{041}	5	5	4	14
H_{111}	1	1	1	3
H_{112}	1	1	2	4
H_{121}	2	2	2	6
H_{131}	3	3	3	9
H_{141}	4	4	4	12
H_{MBDI}	2	1	2	5
H_{LDI}	1	1	2*	4*
H_{ODI}	2	1	3*	6*

$$H_{131}: \quad p_n = \frac{pT}{12} = \frac{1 - z^{-1}}{23z^{-1} - 16z^{-2} + 5z^{-3}}$$

$$H_{141}: \quad p_n = \frac{pT}{24} = \frac{1 - z^{-1}}{55z^{-1} + 59z^{-2} + 37z^{-3} - 9z^{-4}}$$

(5.9.5b–c)

Another important question is stability. From a mapping standpoint, the jv-axis of the p_n-plane is mapped into different contours by each of these transforms. Causal and stable analog systems should map into causal and stable digital systems (poles inside unit circle). As will be shown in a moment, only H_{001}, H_{002}, and H_{011} map the p_n-plane jv-axis inside or onto the z-plane unit circle. Therefore H_{001}, H_{002}, and H_{011} can be used with no worry of producing unstable digital filters from stable analog filters. However all the other transforms are capable of mapping poles outside the unit circle so stability is not guaranteed. These other transforms must be used with caution.

Finding the mapping properties from the p_n-plane to the z-plane usually requires computerized techniques as illustrated by the transforms that have been considered.

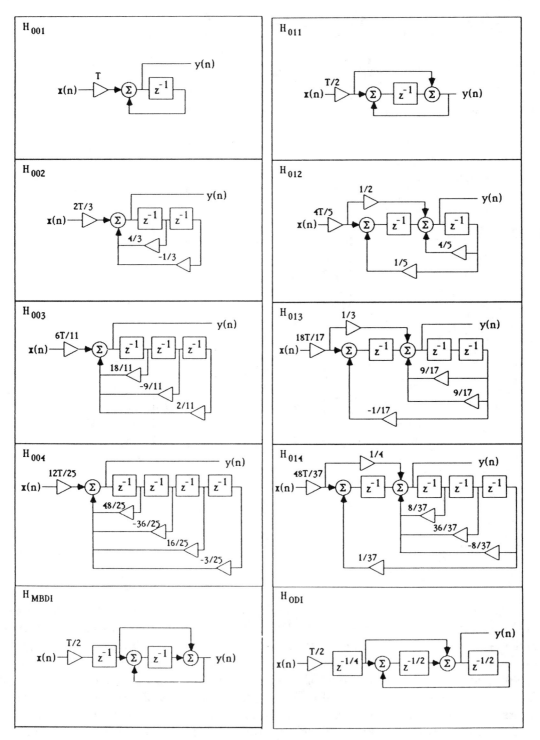

Fig. 5.9.3 Optimum digital integrator implementations. (From J.S. Graham, loc. cit.)

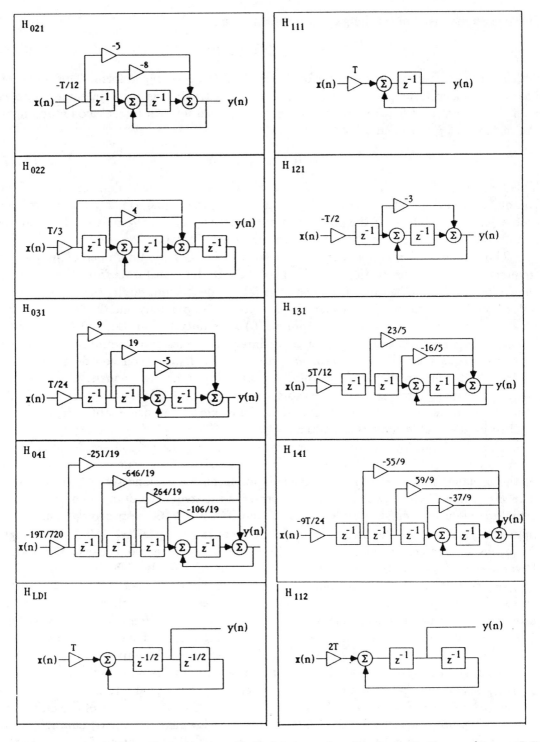

Fig. 5.9.3 (Continued) Optimum digital integrator implementations. (From J.S. Graham, loc. cit.)

However some insights can be gained by considering

$$1 - pH_{dmn}(z) = 0 \qquad (5.9.6)$$

which is the general integrator transform (Eq. 5.2.2) in rearranged form. This equation has a root locus form (see Chap. 1.16) where p is treated as an adjustable complex gain. The usual root locus method can only be applied for real p (real-axis mappings). We instead set $p = jv$ and solve

$$1 - jvH_{dmn}(z) = 0 \qquad (5.9.7)$$

This maps the left-half-plane and its jv-axis boundary into some z-plane contour. If the contour falls entirely inside (outside) the unit circle, the transform H_{dmn} will produce stable (unstable) filters. If the contour is only partially inside the unit circle, stability is not assured and must be determined directly from $H_{dmn}(z)$.

The various jv-axis mappings are shown in Fig. 5.9.4. A number of important frequency mapping observations can be made about the transforms from the root loci shown in Fig. 5.9.4. H_{001} and H_{002} contours fall inside the unit circle, the H_{011} contour falls on the unit circle, while all other H_{0mn} contours fall partially outside the unit circle. Since the analog frequency v varies from dc (0) to infinity (∞) in Eq. 5.9.7, the digital integrator has root loci that originate at the integrator poles and terminate on its zeros. Since the H_{0mn} filters have no zeros at infinity unlike H_{1mn}, their contours terminate on their finite zeros rather than at infinity. Increasing sampling frequency in H_{003} and H_{004} make these integrators more stable (poles closer to origin) but this makes the other integrators less stable (poles further from origin) or even unstable. H_{1mn} filters tend to be less stable than H_{0mn} filters.

The H_{011} transform exactly preserves the frequency domain characteristics (with frequency compression) of the analog lowpass filter since H_{011} maps the entire jv-axis of the p-plane ($0 \le v < \infty$) onto the unit circle of the z-plane ($0° \le \omega T < 180°$). The H_{LDI} and H_{ODI} transforms exactly preserve the frequency domain characteristics over low-frequency ranges ($0 \le v \le \sqrt{2}/C$ and $0 \le v \le 2/C$, respectively) since they map the jv-axis near the origin onto the unit circle ($0° \le \omega T \le 180°$). The high-frequency characteristics of the analog filter cannot be obtained because the jv-axis beyond $v = \sqrt{2}/C$ or $2/C$ is not mapped onto the unit circle. The H_{001}, H_{111}, and H_{MBDI} transforms distort the frequency domain behavior since they do not map the jv-axis of the p-plane onto the unit circle at all. However because they map the jv-axis around $v = 0$ to be tangent to the unit circle at $z = 1$, the low-frequency characteristics are somewhat maintained at sufficiently high sampling rates. Of these three transforms, H_{001} gives the least distortion since its root locus is closest the unit circle. In addition, H_{MBDI} maps the jv-axis around $v = \infty$ to be tangent to the unit circle at $z = -1$ so H_{MBDI} also somewhat maintains the high-frequency filter characteristics. The H_{112} transform maps the jv-axis near the origin ($0 \le v \le 2/C$) onto two arcs of the unit circle ($0° \le \omega T \le 90°$ and $180° \ge \omega T \ge 90°$). Therefore lowpass (highpass) characteristics become bandpass (bandstop) characteristics so that H_{112} is not useful as a analog lowpass-to-digital lowpass filter transform.

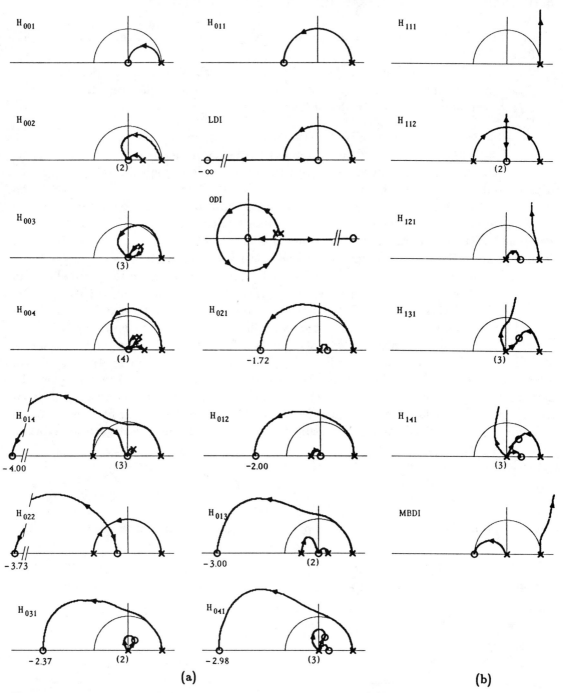

Fig. 5.9.4 Left-hand p-plane mapping into z-plane for (a) H_{0mn} and (b) H_{1mn} digital integrators. (From J.S. Graham, loc. cit.)

5.10 NUMERICAL INTEGRATION FILTER ORDERS

Digital filters generally have frequency domain specifications like that drawn in Fig. 5.10.1a (cf. Fig. 4.2.2) where

f_p = maximum passband frequency (Hz) for M_p (dB) maximum ripple

f_r = minimum stopband frequency (Hz) for M_r (dB) minimum rejection

f_s = sampling frequency (Hz) $= 1/T$ seconds

and

$\omega_p T = 360^\circ f_p/f_s$ = normalized passband frequency in degrees

$\omega_r T = 360^\circ f_r/f_s$ = normalized stopband frequency in degrees

$\Omega_d = \omega_r T/\omega_p T$ = passband/stopband frequency ratio of digital filter

The normalized stopband frequency Ω_d of the digital filter equals $\omega_r T/\omega_p T = f_r/f_p$. From such (M_p, M_r, Ω_d) specifications, the digital filter order must be determined. To do this, the classical, optimum, and adaptive filter transfer functions of Chap. 4 must be combined with the digital transformations of Chap. 5 which have just been discussed. This can be easily done using magnitude response nomographs.[4]

The nomographs for analog filters are well-established and have the form shown in Fig. 5.10.2. The frequency domain gain requirements for the analog filter (M_p, M_r, Ω_a) shown in Fig. 5.10.1b are transferred onto the nomograph. The filter order n is determined by reading the order of the first curve falling above the data point (γ, Ω_a).

The analog filter nomograph can also be used for digital filters by relating the digital filter frequency Ω_d and analog filter frequency Ω_a. This can be easily done as will be shown in a moment. Then by plotting this relation on a graph directly below the nomograph as shown in Fig. 5.10.2, the digital filter frequency Ω_d is graphically converted into the equivalent analog filter frequency Ω_a. Then the analog filter nomograph is used as usual to obtain the required filter order n. Now let us investigate the relation between Ω_d and Ω_a.

Consider the bilinear transform H_{011} given by Eq. 5.9.4a where

$$p = \frac{2}{T}\frac{1 - z^{-1}}{1 + z^{-1}} \tag{5.10.1}$$

Setting the analog frequency variable $p = u + jv$ and the digital frequency variable

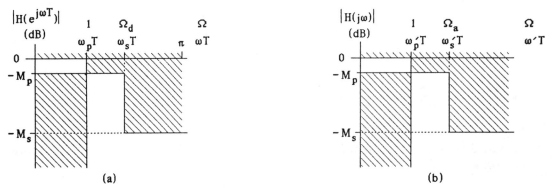

Fig. 5.10.1 Magnitude response specification of (a) digital and (b) analog filters.

$z = e^{j\omega T}$, then

$$jv = j\frac{\tan(\omega T/2)}{T/2} = j\omega\frac{\tan(\omega T/2)}{\omega T/2} \tag{5.10.2}$$

using the reciprocal of Eq. 5.5.3. This relates the digital filter frequency ω (or normalized frequency ωT) to the analog filter frequency v (or normalized frequency vT). Since the digital filter specification requires a normalized passband frequency of $\omega_p T$ and a normalized stopband frequency of $\omega_r T$, then the equivalent analog filter frequencies are

$$v_p T = 2\tan(\omega_p T/2) \text{ rad} \qquad \text{or} \qquad v_p = \omega_p\frac{\tan(\omega_p T/2)}{\omega_p T/2} \text{ rad/sec}$$

$$v_r T = 2\tan(\omega_r T/2) \text{ rad} \qquad \text{or} \qquad v_r = \omega_r\frac{\tan(\omega_r T/2)}{\omega_r T/2} \text{ rad/sec} \tag{5.10.3}$$

in radians/second or

$$2\pi(f_p'/f_s) = 2\tan(\pi f_p/f_s) \text{ rad} \qquad \text{or} \qquad f_p' = f_p\frac{\tan(\pi f_p/f_s)}{\pi f_p/f_s} \text{ Hz}$$

$$2\pi(f_r'/f_s) = 2\tan(\pi f_r/f_s) \text{ rad} \qquad \text{or} \qquad f_r' = f_r\frac{\tan(\pi f_r/f_s)}{\pi f_r/f_s} \text{ Hz} \tag{5.10.4}$$

in Hertz. f_i is the digital filter frequency and f_i' is the equivalent analog filter frequency. The analog frequency ratio Ω_a equals

$$\Omega_a = \frac{f_r'}{f_p'} = \frac{360° f_r'/f_s}{360° f_p'/f_s} = \frac{\tan(180° f_r/f_s)}{\tan(180° f_p/f_s)} = \frac{\tan(\omega_r T/2)}{\tan(\omega_p T/2)} \tag{5.10.5}$$

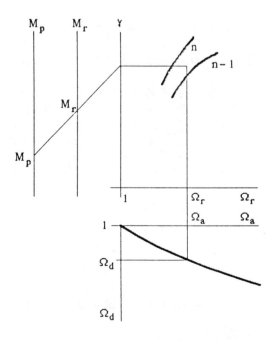

Fig. 5.10.2 Nomograph for computing digital filter order.

which is a function of f_p, f_r, and f_s. It is not a function of simply $\Omega_d = f_r/f_p$ as it was for analog filters where $\Omega_a = f_r'/f_p'$ in Fig. 5.10.1b. Ω_a can be related to Ω_d and f_s/f_r as shown in Fig. 5.10.3. Here Ω_a is plotted versus Ω_d for different f_s/f_r (sampling frequency/rejection frequency) ratios. To interpret these curves, expand $\tan x = x + x^3/3 + \ldots$ and assume $|x| \ll 1$. Then Eq. 5.10.5 reduces to

$$\Omega_a \cong \frac{(\omega_r T/2)[1+(\omega_r T/2)^2/3]}{(\omega_p T/2)[1+(\omega_p T/2)^2/3]} \cong \Omega_d\left[1+\frac{(\omega_r T)^2-(\omega_p T)^2}{12}\right]$$

$$= \Omega_d\left\{1+3.29\left[(f_r/f_s)^2-(f_p/f_s)^2\right]\right\} \cong \Omega_d \qquad (5.10.6)$$

This approximation shows that when the sampling frequency is sufficiently high (say $f_s/f_r > 10$), the normalized analog frequency Ω_a is equal to the digital filter frequency Ω_d corrected by a second-order frequency-squared difference term $3.29[(f_r/f_s)^2 - (f_p/f_s)^2]$. This behavior is clearly seen in Fig. 5.10.3.

Now consider the modified bilinear transform H_{MBDI} whose gain is given by Eq. 5.6.3. Combining Eqs. 5.5.3 and 5.6.3 and proceeding as before setting $p = u + jv$ and $z = e^{j\omega T}$ yields

$$u + jv = j(2/T)\tan(\omega T/2)e^{-j\omega T} = (2/T)\left[\sin(\omega T)+j\cos(\omega T)\right]\tan(\omega T/2) \quad (5.10.7)$$

At frequencies ω much smaller than the sampling frequency $2\pi/T$, $\sin(\omega T) \cong \omega T$, $\cos(\omega T) \cong 1$, and $\tan(\omega T/2) \cong \omega T/2$ so Eq. 5.10.7 reduces to

$$u + jv \cong \omega^2 T + j\frac{\tan(\omega T/2)}{T/2} \cong 0 + j\omega\frac{\tan(\omega T/2)}{\omega T/2} \qquad (5.10.8)$$

which is the same as the H_{011} frequencies in Eq. 5.10.2. Thus the modified bilinear transform is also described by Eqs. 5.10.2–5.10.6. Since $\Omega_a \geq \Omega_d$ using the H_{011} and H_{MBDI} transforms, the filter order is *reduced* based on Ω_a rather than Ω_d.

EXAMPLE 5.10.1 Determine the order of a digital Butterworth filter needed to satisfy the following specifications: (1) sampling frequency $f_s = 10$ KHz, (2) dc gain $= 0$ dB, (3) passband ripple ≤ 0.28 dB for frequencies ≤ 1 KHz, (4) stopband rejection ≥ 40 dB for frequencies ≥ 2 KHz. Use the bilinear transform.
SOLUTION The normalized passband and stopband frequencies for the digital filter and the digital stopband frequency ratio equal

$$\omega_p T = 2\pi(f_p/f_s) = 360°(1\text{ KHz}/10\text{ KHz}) = 36° = 0.628\text{ rad}$$

$$\omega_r T = 2\pi(f_r/f_s) = 360°(2\text{ KHz}/10\text{ KHz}) = 72° = 1.257\text{ rad} \qquad (5.10.9)$$

$$\Omega_d = \frac{f_r}{f_p} = \frac{72°}{36°} = 2$$

Then the equivalent analog filter frequencies are

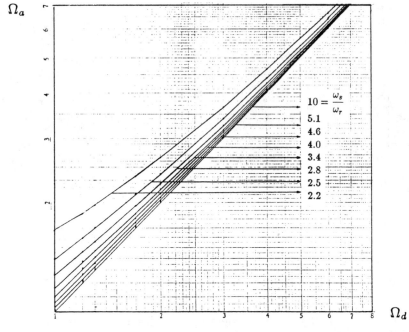

Fig. 5.10.3 Relation between Ω_a and Ω_d for various f_s/f_r ratios in Eq. 5.10.5. (From E. Chiang, "Digital filter design using analog filter nomographs and digital filter transformations," M.S.E.E. Directed Research, Calif. State Univ., Long Beach, Dec. 1983.)

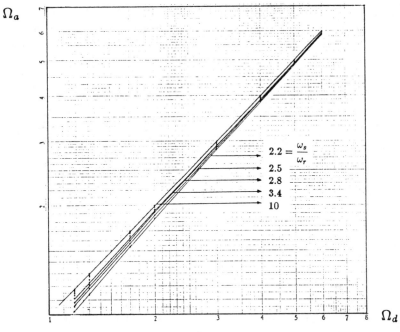

Fig. 5.10.4 Relation between Ω_a and Ω_d for various f_s/f_r ratios in Eq. 5.10.13. (From E. Chiang, loc. cit.)

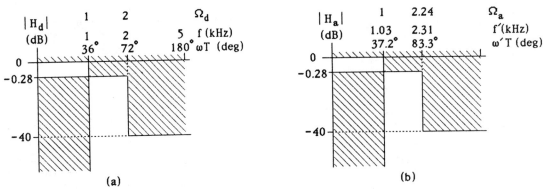

Fig. 5.10.5 **(a) Digital and (b) equivalent analog filter specification of Example 5.10.1.**

$$2\pi(f'_p/f_s) = 2\tan(36^\circ/2) = 37.2^\circ \quad \text{or} \quad f'_p = 1 \text{ KHz} \frac{\tan(0.628/2)}{0.628/2} = 1.034 \text{ KHz}$$

$$2\pi(f'_r/f_s) = 2\tan(72^\circ/2) = 83.3^\circ \quad \text{or} \quad f'_r = 2 \text{ KHz} \frac{\tan(1.257/2)}{1.257/2} = 2.313 \text{ KHz}$$

$$(5.10.10)$$

from Eq. 5.10.4. The analog stopband frequency ratio equals 2.2 since

$$\Omega_a = \frac{83.3^\circ}{37.2^\circ} = \frac{\tan(72^\circ/2)}{\tan(36^\circ/2)} = 2.24 \cong 2.20 \tag{5.10.11}$$

from Eq. 5.10.5. This is close to its 2.20 approximation of Eq. 5.10.6. These filter parameters are summarized in Fig. 5.10.5. Entering the data $(M_p, M_r, \gamma, \Omega_a) = (0.28, 40, 5.2, 2.24)$ onto the Butterworth filter nomograph shows $n \geq 8$. ∎

EXAMPLE 5.10.2 Determine the minimum orders of digital Chebyshev, elliptic, and Bessel filters needed to satisfy the specifications of Example 5.10.1.
SOLUTION By simply entering the point $(\gamma, \Omega_a) = (5.2, 2.24)$ determined in Example 5.10.1 onto these filter nomographs, the filter orders can be read off directly. Summarizing the results: Chebyshev $n = 5$, elliptic $n = 4$, Bessel impossible. ∎

Next consider the forward Euler transform H_{111} whose gain is given by Eq. 5.3.3. Setting $p = u + jv$ and $z = e^{j\omega T}$ yields

$$u + jv = j\omega \frac{\sin(\omega T/2)}{\omega T/2} e^{j\omega T/2} = -\omega \frac{\sin^2(\omega T/2)}{\omega T/2} + j\omega \frac{\sin(\omega T/2)\cos(\omega T/2)}{\omega T/2} \tag{5.10.12}$$

Since $\sin(x)\cos(x) = 0.5\sin(2x)$, the imaginary term can be rewritten as $v = (\sin\omega T)/T$. The backward Euler transform H_{001} (Eq. 5.4.3) is described by the same equation but u is negative. Eq. 5.10.12, like the root loci contours of Fig. 5.9.4, shows that the z-plane unit circle (frequency ω) does not map into the jv-axis of the p-plane except for sufficiently small ωT. Then $\omega T \cong 0$ and $z = e^{j\omega T} \cong 1 + j\omega T$ so that Eq. 5.10.12 reduces to

$$u + jv \cong -\frac{\omega^2 T}{2} + j\omega \frac{\sin(\omega T)}{\omega T} \cong 0 + j\omega \frac{\sin(\omega T)}{\omega T} \qquad (5.10.13)$$

The equivalent analog filter frequencies become

$$2\pi(f'_p/f_s) = \sin(2\pi f_p/f_s) \text{ rad} \qquad \text{or} \qquad f'_p = f_p \frac{\sin(2\pi f_p/f_s)}{2\pi f_p/f_s} \text{ Hz}$$

$$2\pi(f'_r/f_s) = \sin(2\pi f_r/f_s) \text{ rad} \qquad \text{or} \qquad f'_r = f_r \frac{\sin(2\pi f_r/f_s)}{2\pi f_r/f_s} \text{ Hz} \qquad (5.10.14)$$

so the analog frequency ratio Ω_a equals

$$\Omega_a = \frac{f'_r}{f'_p} = \frac{360° f'_r/f_s}{360° f'_p/f_s} = \frac{\sin(360° f_r/f_s)}{\sin(360° f_p/f_s)} \qquad (5.10.15)$$

which is plotted in Fig. 5.10.4. Expanding $\sin x = x - x^3/3! + \ldots$ and assuming $|x| \ll 1$, then Eq. 5.10.15 reduces to

$$\Omega_a \cong \frac{(\omega_r T)[1 - (\omega_r T)^2/6]}{(\omega_p T)[1 - (\omega_p T)^2/6]} \cong \Omega_d \left[1 - \frac{(\omega_r T)^2 - (\omega_p T)^2}{6} \right]$$

$$= \Omega_d \left\{ 1 - 6.6[(f_r/f_s)^2 - (f_p/f_s)^2] \right\} \cong \Omega_d \qquad (5.10.16)$$

which is identical to Eq. 5.10.6 except the correction term is twice as large and has opposite sign. The H_{112} transform has $u + jv = 0 + j\sin(\omega T)$ exactly so $vT = \sin(\omega T)$ (manipulate Eq. 5.7.3 like Eq. 5.7.4). Therefore Eqs. 5.10.14–5.10.16 hold for H_{112}.

The lossless transform (LDI) has

$$jv = j\omega \frac{\sin(\omega T/2)}{\omega T/2} \qquad (5.10.17)$$

from Eq. 5.7.4 so the equivalent analog filter frequencies equal

$$2\pi(f'_p/f_s) = 2\sin(\pi f_p/f_s) \text{ rad} \qquad \text{or} \qquad f'_p = f_p \frac{\sin(\pi f_p/f_s)}{(\pi f_p/f_s)} \text{ Hz}$$

$$2\pi(f'_r/f_s) = 2\sin(\pi f_r/f_s) \text{ rad} \qquad \text{or} \qquad f'_r = f_p \frac{\sin(\pi f_r/f_s)}{(\pi f_r/f_s)} \text{ Hz} \qquad (5.10.18)$$

The analog frequency ratio Ω_a is

$$\Omega_a = \frac{f'_r}{f'_p} = \frac{180° f'_r/f_s}{180° f'_p/f_s} = \frac{\sin(180° f_r/f_s)}{\sin(180° f_p/f_s)} \qquad (5.10.19)$$

which reduces to Eq. 5.10.15 (with one-half argument) for small $|\omega T/2| \ll 1$ products.

Table 5.10.1 Normalized analog stopband
frequency ratio Ω_a of some discrete transforms.

Transform	Analog Frequency Ratio Ω_a
$H_{001}, H_{111}, H_{112}$	$\dfrac{\sin(\omega_r T)}{\sin(\omega_p T)}$
H_{011}	$\dfrac{\tan(\omega_r T/2)}{\tan(\omega_p T/2)}$
H_{MBDI}	$\dfrac{\tan(\omega_r T/2)\cos(\omega_r T)}{\tan(\omega_p T/2)\cos(\omega_p T)}$
H_{LDI}	$\dfrac{\sin(\omega_r T/2)}{\sin(\omega_p T/2)}$
H_{ODI}	$\dfrac{\sin(\omega_r T/4)}{\sin(\omega_p T/4)}$

The optimum transform (ODI) has

$$jv = j\frac{2}{T}\frac{\sin(\omega T/2)}{\cos(\omega T/4)} = j\omega\frac{\sin(\omega T/4)}{\omega T/4} \qquad (5.10.20)$$

from Eq. 5.8.3. Therefore the equivalent analog filter frequencies and the analog frequency ratio are also described by Eqs. 5.10.14 and 5.10.15 using one-quarter arguments. Thus, the H_{LDI} and H_{ODI} filter order approaches that of H_{001}, H_{111}, and H_{112} filters. Since $\Omega_a \leq \Omega_d$ using the H_{001}, H_{111}, H_{112}, H_{LDI}, and H_{ODI} transforms, the filter order is *increased* based on Ω_a rather than Ω_d. Comparing Eq. 5.10.5 (Fig. 5.10.3) and Eq. 5.10.15 (Fig. 5.10.4), the digital filter orders are lower using the H_{011} and H_{MBDI} transforms but higher using the H_{001}, H_{111}, H_{112}, H_{LDI}, and H_{ODI} transforms. All the Ω_a ratios for the various transforms are summarized in Table 5.10.1.

It is generally assumed that the frequency responses of the digital filter obtained using these discrete transforms are close to the frequency responses of the analog filters at high sampling rates. But how high a sampling rate is sufficient? To find the sampling frequency limits, we specify the percentage frequency error which is allowed. For example, frequency errors could be 1%, 5%, 10%, or even 20%. Let us now determine the permissible sampling frequencies within these percentage errors.

First, we consider the backward Euler H_{001}, forward Euler H_{111}, and H_{112} discrete transforms. The frequency transformation is given by Eq. 5.10.13 as

$$vT = \sin(\omega T) \qquad (5.10.21)$$

The frequency range $0° \leq \omega T \leq 90°$ maps into $0 \leq vT \leq 1$. Expanding the sine as $\sin(\omega T) = (\omega T)[1 - (\omega T)^2/3! + \ldots]$, then the fractional error $vT/\omega T$ never exceeds $(\omega T)^2/3!$. If the percentage error is required to be within 1%, then

$$(\omega T)^2/6 \leq 0.01, \qquad \omega T \leq 0.245 = 14° \quad \text{or} \quad f_s = 1/T \geq 26f \qquad (5.10.22)$$

Table 5.10.2 Maximum normalized frequency to reduce
percentage error for some discrete transforms.

Transform	Frequency mapping vT	$\leq 1\%$	$\leq 5\%$	$\leq 10\%$	$\leq 20\%$
$H_{001}, H_{111}, H_{112}$	$\sin(\omega T)$	$14°$	$32°$	$45°$	$65°$
H_{011}	$2\tan(\omega T/2)$	$20°$	$43°$	$59°$	$80°$
H_{MBDI}	$2\tan(\omega T/2)\cos(\omega T)$	$9°$	$20°$	$28°$	$40°$
H_{LDI}	$2\sin(\omega T/2)$	$28°$	$63°$	$90°$	$129°$
H_{ODI}	$4\sin(\omega T/4)$	$56°$	$126°$	$180°$	$259°$

This means that when designing digital filters using H_{001}, H_{111}, or H_{112} transforms, the frequency mapping error is less than 1% for normalized frequencies less than $14°$. Alternatively stated, the sampling frequency must be 26 times greater than the maximum specified filter frequency to have less than 1% mapping error. The normalized frequency reduces to about $32°$, $45°$, $65°$ (or about $11f$, $8f$, and $6f$) for percentage errors of 5%, 10%, and 20%, respectively. These results are listed in Table 5.10.2.

Next, we consider the bilinear transform H_{011} whose frequency transformation is given by Eq. 5.10.2 as

$$vT/2 = \tan(\omega T/2) \qquad (5.10.23)$$

The frequency range $0° \leq \omega T < 180°$ maps into $0 \leq vT < \infty$. Following the same procedure, we expand $\tan(\omega T/2) = (\omega T/2)[1 + (\omega T/2)^2/3 + \ldots]$. The fractional error is the order of $(\omega T)^2/12$ which is half that incurred using H_{001} and H_{111}. For a $vT/\omega T$ percentage error of 1%, then

$$(\omega T)^2/12 \leq 0.01, \qquad \omega T \leq 0.346 = 20° \quad \text{or} \quad f_s \geq 18f \qquad (5.10.24)$$

A larger frequency range or a lower sampling frequency is possible when the H_{011} transform is used to design a digital filter rather than H_{001} and H_{111}.

Now we consider the modified bilinear transform H_{MBDI} whose frequency transformation is given by Eq. 5.10.7 where

$$vT/2 = \tan(\omega T/2)\cos(\omega T) \qquad (5.10.25)$$

The frequency range $0° \leq \omega T \leq 45°$ maps into $0 \leq vT \leq 0.59$. Expanding $\tan(\omega T/2) = (\omega T/2)[1+(\omega T/2)^2/3+\ldots]$ and $\cos(\omega T) = [1-(\omega T)^2/2!+\ldots]$, then Eq. 5.10.25 becomes

$$\frac{vT}{2} = \frac{\omega T}{2}\left[1 - \frac{5(\omega T)^2}{12} + \frac{(\omega T)^4}{120} + \ldots\right] \qquad (5.10.26)$$

The fractional error is the order of $5(\omega T)^2/12$. Using the percentage error of 1%, then

$$5(\omega T)^2/12 \leq 0.01, \qquad \omega T \leq 0.155 = 8.9° \quad \text{or} \quad f_s \geq 41f \qquad (5.10.27)$$

The sampling frequency is larger than those required for the other transforms.

Next consider the lossless discrete transform H_{LDI} whose frequency transformation is also described by Eq. 5.10.17 as

$$vT/2 = \sin(\omega T/2) \tag{5.10.28}$$

The frequency range $0° \leq \omega T \leq 180°$ maps into $0 \leq vT \leq 2$. Using the sine expansion of Eq. 5.10.21 with a one-half argument, the fractional error never exceeds $(\omega T/2)^2/3!$. For a 1% percentage error in $vT/\omega T$,

$$(\omega T)^2/24 \leq 0.01, \qquad \omega T \leq 0.49 = 28° \quad \text{or} \quad f_s = 1/T \geq 13f \tag{5.10.29}$$

The percentage errors are about half of those obtained using H_{001}, H_{111}, and H_{112} and is the smallest compared to all the other transforms in Table 5.10.2.

The optimum discrete transform H_{ODI} has the frequency transformation as

$$vT/2 = 2\sin(\omega T/4) \tag{5.10.30}$$

from Eq. 5.10.20. The frequency range $0° \leq \omega T \leq 360°$ maps into $0 \leq vT \leq 4$. Using the sine and cosine series given earlier, Eq. 5.10.30 becomes

$$\frac{vT}{2} = \frac{\omega T}{2}\left[1 - \frac{(\omega T)^2}{96} + \cdots\right] \tag{5.10.31}$$

The fractional error of frequency transformation using the H_{ODI} is bounded by $(\omega T)^2/96$. Using a $vT/\omega T$ percentage error of 1%, then

$$(\omega T)^2/96 \leq 0.01, \qquad \omega T \leq 0.98 = 56° \quad \text{or} \quad f_s \geq 6.4f \tag{5.10.32}$$

Recall that H_{ODI} has twice the period of H_{LDI} while H_{112} has only half the period. Therefore they have the same error (Eq. 5.10.29) based upon a 2π interval. However only the H_{LDI} transform maintains the filter type.

From the frequency mapping viewpoint of Table 5.10.2, we anticipate that the transforms can be ordered (best to worst) as H_{ODI} (56°), H_{LDI} (28°), H_{011} (20°); H_{001}, H_{111}, and H_{112} (14°); and H_{MBDI} (9°). For a fixed sampling frequency f_s, H_{LDI} gives the smallest mapping error while H_{MBDI} gives the largest error. The design examples of the next section confirm that this is indeed the case.

5.11 NUMERICAL INTEGRATION FILTER DESIGN EXAMPLES

The design procedure for digital filters using discrete transforms is straightforward. It consists of the following steps:

1a. Select a suitable analog filter type (e.g., Butterworth, Chebyshev, elliptic, etc.) and
1b. Choose a particular p_n-to-z transform from Tables 5.2.1 and 5.2.2 based upon the discussion of Chap. 5.9.

2. Determine the required analog filter order as described in Chap. 5.10.
3. Write the analog transfer function $H(p_n)$ having unity bandwidth and the desired magnitude from the appropriate table of Chap. 4.2.
4. Compute the discrete transformation from Table 5.11.1 or compute using Eq. 5.2.2. The analog frequency normalization constants C equal

$$
\begin{aligned}
H_{001}, H_{111}: \quad & C = v_p T && = \sin(\omega_p T) \\
H_{011}: \quad & C = v_p T/2 = \tan(\omega_p T/2) \\
H_{MDBI}: \quad & C = v_p T/2 = \tan(\omega_p T/2)\cos(\omega_p T) \\
H_{112}: \quad & C = 2v_p T && = 2\sin(\omega_p T) \\
H_{LDI}: \quad & C = v_p T && = 2\sin(\omega_p T/2) \\
H_{ODI}: \quad & C = v_p T/2 = 2\sin(\omega_p T/4)
\end{aligned}
\tag{5.11.1}
$$

where $\omega_p T$ radians $= 360° f_p/f_s$ degrees and $p_n = p/v_p = pT/v_p T$.

5. Compute the digital transfer function $H(z)$ by substituting the p_n-to-z transformation found in Step 4 into $H(p_n)$ found in Step 3.
6. Implement the transfer function using one of the techniques discussed in Chap. 12.

This procedure is easily carried out. *Steps 1-3* – Consider a first-order analog Butterworth lowpass filter having $H(p_n) = 1/(p_n + 1)$ from Chap. 4.2 where $v_p = 1$. Design the corresponding digital filter having $\omega_p = 1$ rad/sec with a sampling frequency f_s of 1 Hz. *Step 4* – Since $\omega_p T = 360° f_p/f_s = 57.3°$, the frequency constants are

$$
\begin{aligned}
H_{001}, H_{111}: \quad & C = \sin(57.3°) = (48.2°) = 0.8415 \\
H_{011}: \quad & C = \tan(57.3°/2) = (62.6°/2) = 0.5463 \\
H_{MBDI}: \quad & C = \tan(57.3°/2)\cos(57.3°) = (33.8°/2) = 0.2952 \\
H_{LDI}: \quad & C = 2\sin(57.3°/2) = 2(54.9°/2) = 0.9589
\end{aligned}
\tag{5.11.2}
$$

from Eq. 5.11.1. *Step 5* – Substituting the proper transforms of Table 5.11.1 into $H(p_n) = 1/(p_n + 1)$ gives the transfer functions of the digital Butterworth filters as

$$
H_{001}: \quad H(z) = 1 \bigg/ \left[\frac{1}{C}\left(1 - z^{-1}\right) + 1\right] = \frac{0.457}{1 - 0.5430z^{-1}}
$$

$$
H_{111}: \quad H(z) = 1 \bigg/ \left[\frac{1}{C}\left(\frac{1 - z^{-1}}{z^{-1}}\right) + 1\right] = \frac{0.8415z^{-1}}{1 - 0.1585z^{-1}}
$$

$$
H_{011}: \quad H(z) = 1 \bigg/ \left[\frac{1}{C}\left(\frac{1 - z^{-1}}{1 + z^{-1}}\right) + 1\right] = \frac{0.3534(1 + z^{-1})}{1 - 0.2932z^{-1}}
\tag{5.11.3}
$$

$$
H_{MBDI}: \quad H(z) = 1 \bigg/ \left[\frac{1}{C}\left(\frac{1 - z^{-1}}{z^{-1}(1 + z^{-1})}\right) + 1\right] = \frac{0.2952z^{-1}(1 + z^{-1})}{1 - 0.7049z^{-1} + 0.2952z^{-2}}
$$

$$
H_{LDI}: \quad H(z) = 1 \bigg/ \left[\frac{1}{C}\left(\frac{1 - z^{-1}}{z^{-0.5}}\right) + 1\right] = \frac{0.9589z^{-0.5}}{1 + 0.9589z^{-0.5} - z^{-1}}
$$

Increasing the sampling frequency ω_s to 10 rad/sec but maintaining $\omega_p = 1$ in the digital filter gives $\omega_p T = 360° f_p/f_s = 36°$. Eq. 5.11.1 requires that the transform constants be changed to

$$H_{001}, H_{111} : \quad C = \sin(36°) = (33.7°) = 0.5878$$
$$H_{011} : \quad C = \tan(36°/2) = (37.2°/2) = 0.3249$$
$$H_{MBDI} : \quad C = \tan(36°/2)\cos(36°) = (30.1°/2) = 0.2629 \tag{5.11.4}$$
$$H_{LDI} : \quad C = 2\sin(36°/2) = 2(35.4°/2) = 0.6180$$

The digital Butterworth transfer functions become

$$H_{001} : \quad H(z) = \frac{0.3702}{1 - 0.6298z^{-1}}$$

$$H_{111} : \quad H(z) = \frac{0.5876z^{-1}}{1 - 0.4124z^{-1}}$$

$$H_{011} : \quad H(z) = \frac{0.2451(1 + z^{-1})}{1 - 0.5098z^{-1}} \tag{5.11.5}$$

$$H_{MBDI} : \quad H(z) = \frac{0.2627z^{-1}(1 + z^{-1})}{1 - 0.7373z^{-1} + 0.2627z^{-2}}$$

$$H_{LDI} : \quad H(z) = \frac{0.6180z^{-0.5}}{1 + 0.6180z^{-0.5} - z^{-1}}$$

Increasing the sampling rate ω_s still further to 100 rad/sec with ω_p still unity gives $\omega_p T = 360° f_p/f_s = 3.6°$. Due to the high sampling rate relative to the corner frequency, the normalized corner frequency $\omega_p T$ of the digital filter is small ($\omega_p T \ll 1$ rad or $360° f_p/f_s \ll 57.3°$). In such cases, Eq. 5.11.1 shows that $v_p T \cong \omega_p T$ (or $v_p \cong \omega_p$) regardless of which transform is used. The transform constants equal

$$H_{001}, H_{111} : \quad C = \sin(3.6°) = (3.60°) = 0.06279$$
$$H_{011} : \quad C = \tan(3.6°/2) = (3.60°/2) = 0.03143$$
$$H_{MBDI} : \quad C = \tan(3.6°/2)\cos(3.6°) = (3.59°/2) = 0.03136 \tag{5.11.6}$$
$$H_{LDI} : \quad C = 2\sin(3.6°/2) = 2(3.60°/2) = 0.06282$$

The digital Butterworth filter transfer functions become

$$H_{001} : \quad H(z) = \frac{0.05908}{1 - 0.9409z^{-1}}$$

$$H_{111} : \quad H(z) = \frac{0.06279z^{-1}}{1 - 0.9372z^{-1}}$$

$$H_{011} : \quad H(z) = \frac{0.03047(1 + z^{-1})}{1 - 0.9391z^{-1}} \tag{5.11.7}$$

$$H_{MBDI} : \quad H(z) = \frac{0.03135z^{-1}(1 + z^{-1})}{1 - 0.96865z^{-1} + 0.03135z^{-2}}$$

$$H_{LDI} : \quad H(z) = \frac{0.06282z^{-0.5}}{1 + 0.0.06282z^{-0.5} - z^{-1}}$$

Table 5.11.1 Some analog lowpass-to-digital lowpass transformations.

H_{dmn}	Discrete transform	Transformation $p_n = g(z)$
H_{001}	Backward Euler	$p_n = \dfrac{1}{C}\left(1 - z^{-1}\right)$ $C = \sin(BT)$
H_{111}	Forward Euler	$p_n = \dfrac{1}{C}\left(\dfrac{1 - z^{-1}}{z^{-1}}\right)$ $C = \sin(BT)$
H_{011}	Bilinear	$p_n = \dfrac{1}{C}\left(\dfrac{1 - z^{-1}}{1 + z^{-1}}\right)$ $C = \tan(BT/2)$
H_{MBDI}	Modified Bilinear	$p_n = \dfrac{1}{C}\left(\dfrac{1 - z^{-1}}{z^{-1}(1 + z^{-1})}\right)$ $C = \tan(BT/2)\cos(BT)$
H_{112}	Discrete	$p_n = \dfrac{1}{C}\left(\dfrac{1 - z^{-2}}{z^{-1}}\right)$ $C = 2\sin(BT)$
H_{LDI}	Lossless	$p_n = \dfrac{1}{C}\left(\dfrac{1 - z^{-1}}{z^{-0.5}}\right)$ $C = 2\sin(BT/2)$
H_{ODI}	Optimum	$p_n = \dfrac{1}{C}\left(\dfrac{1 - z^{-0.5}}{z^{-0.25}}\right)$ $C = 2\sin(BT/4)$

The frequency responses of $H(p_n)$ and the various $H(z)$ given by Eqs. 5.11.3, 5.11.5, and 5.11.7 are shown in Figs. 5.11.1–5.11.3, respectively. At high sampling rates like $\omega_{3dB}T = 3.6°$, Fig. 5.11.3 shows that all the transforms give similar magnitude results for normalized frequencies up to about 90°. Then the H_{001}, H_{111}, and H_{LDI} responses continue on a -20 dB/dec asymptote. The H_{011} and H_{MBDI} responses drop to zero. Lowering the sampling frequency so $\omega_{3dB}T = 36°$, Fig. 5.11.2 shows that the H_{001}, H_{111}, and H_{LDI} magnitude responses start to separate. At still lower sampling rates of $\omega_{3dB}T = 57°$, only the H_{011} and H_{MBDI} magnitude responses maintain their shape. Only the H_{011} phase tends to match the analog filter phase. Obviously phase is much more sensitive than magnitude when comparing transforms. This demonstrates

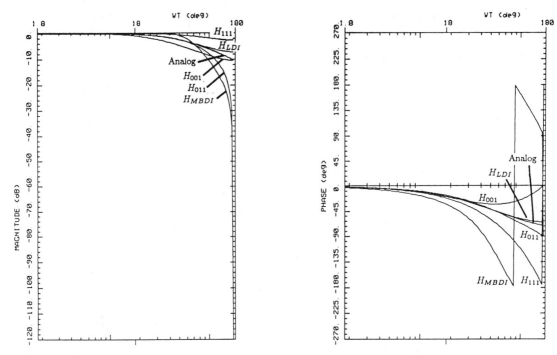

Fig. 5.11.1 Frequency responses of first-order digital Butterworth filters having $\omega_{3dB} = 1$ and $f_s = 1$.

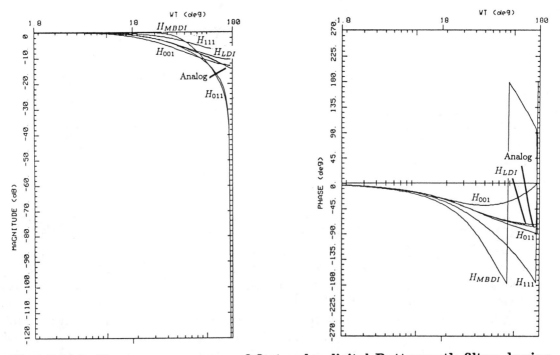

Fig. 5.11.2 Frequency responses of first-order digital Butterworth filters having $\omega_{3dB} = 1$ and $\omega_s = 10$.

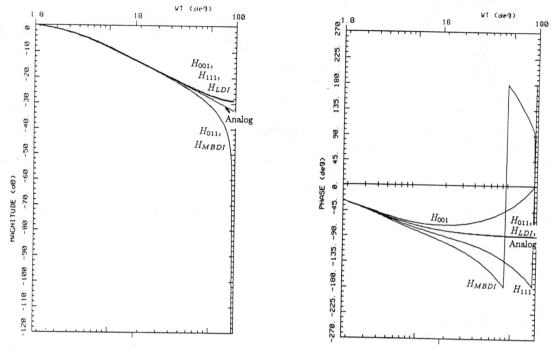

Fig. 5.11.3 Frequency responses of first-order digital Butterworth filters having $\omega_{3dB} = 1$ and $\omega_s = 100$.

that the transform must be carefully selected to maintain responses for normalized frequencies above the 14–28° range as listed in Table 5.10.2.

Analog filter poles p'_k and zeros z'_k can be directly converted into digital filter poles p_k and zeros z_k using the inverse p_n-to-z transform where $z = g^{-1}(p_n)$. Therefore the analog filter transfer function $H(p_n)$ is indirectly transformed into a digital transfer function $H(z)$ as[4]

$$H(p_n) = a\,\frac{\prod_{k=1}^{m}(p + z'_k)}{\prod_{k=1}^{n}(p + p'_k)}, \qquad H(z) = H(p_n)\Big|_{p_n = g(z)} = b\,\frac{\prod_{k=1}^{m}(z + z_k)}{\prod_{k=1}^{n}(z + p_k)} \qquad (5.11.8)$$

A design example is now presented to illustrate this procedure.

EXAMPLE 5.11.1 Determine the transfer function of a third-order Butterworth lowpass filter having a dc gain of 0 dB, a 3 dB bandwidth of 1 KHz, and a 10 KHz sampling rate. Use the bilinear transform. Find the poles and zeros directly.
SOLUTION The Butterworth analog filter transfer function $H(p_n)$ equals

$$H(p_n) = \frac{1}{(p_n + 1)(p_n{}^2 + p_n + 1)}, \qquad p_n = pT/v_p T \qquad (5.11.9)$$

from Chap. 4.2. Since the digital filter has a normalized passband of $\omega_p T = 360° f_p/f_s = 36°$, then the normalized frequency constant $C = \tan(\omega_p T/2) = \tan(18°) = 0.3249$. Therefore from Table 5.11.1, the bilinear transform p_n equals

$$p_n = \frac{1}{C}\frac{1-z^{-1}}{1+z^{-1}} = 3.078\frac{1-z^{-1}}{1+z^{-1}} \quad \text{or} \quad z = \frac{1+Cp_n}{1-Cp_n} = \frac{1+0.3249p_n}{1-0.3249p_n} \quad (5.11.10)$$

Solving the p_n equation for z yields the inverse bilinear transform as shown. The digital transfer function $H(z)$ is found by substituting Eq. 5.11.10 into the analog transfer function $H(p_n)$ of Eq. 5.11.9 which gives

$$H(z) = 1\bigg/\left[3.078\left(\frac{1-z^{-1}}{1+z^{-1}}\right)+1\right]\left[3.078^2\left(\frac{1-z^{-1}}{1+z^{-1}}\right)^2 + 3.078\left(\frac{1-z^{-1}}{1+z^{-1}}\right)+1\right]$$

$$= \frac{0.01807(1+z^{-1})^3}{(1-0.5095z^{-1})(1-1.2510z^{-1}+0.5457z^{-2})}$$

$$(5.11.11)$$

The constant 0.01807 normalizes the dc gain $H(1)$ of the digital filter to unity.

The analog filter poles and zeros in Eq. 5.11.9 are directly converted into digital filter poles and zeros using Eq. 5.11.10 as

$$p_{n1} = -1 \quad \text{so} \quad z_1 = \frac{1+0.3249(-1)}{1-0.3249(-1)} = 0.5095$$

$$p_{n2}, p_{n2}^* = (-0.5 \pm j0.866) \quad \text{so}$$

$$z_2, z_2^* = \frac{1+0.3249(-0.5 \pm j0.866)}{1-0.3249(-0.5 \pm j0.866)} = 0.6254 \pm j0.3934 = 0.7388e^{\pm j32.17^\circ}$$

$$|z_{n3}| = \infty \quad \text{so} \quad z_3 = \frac{1+0.3249(\infty)}{1-0.3249(\infty)} = -1$$

$$(5.11.12)$$

These poles and zeros agree with those in Eq. 5.11.11. ∎

The behavior of a third-order digital Butterworth lowpass filter having sampling rates $\omega_s = 4\omega_{3dB}$ and $20\omega_{3dB}$ using H_{011}, $H_{012}, H_{021}, H_{111}$, and H_{121} transforms is shown in Fig. 5.11.4. The actual and ideal responses are plotted along with the IAE errors (Eq. 4.3.2) between them. The higher the sampling rate, the closer are the actual responses to the ideal response and the smaller are the IAE errors. H_{021} has the smallest total error followed by $H_{011}, H_{121}, H_{111}$, and H_{012}. H_{111} has a smaller total error than H_{012} because its phase error is much smaller than its magnitude error. The step responses are shown in Fig. 5.11.5. They are all somewhat in error at the lower sampling rate but approach the ideal Butterworth response at the higher rate.

To illustrate how these transforms control a more band-limited transfer function, consider a third-order 1.25 dB ripple digital elliptic lowpass filter. Its gain is shown in Fig. 5.11.6 for $\omega_s = 4\omega_{1.25dB}$ and $20\omega_{1.25dB}$. The H_{011} transform always maintains the equiripple characteristic. The other responses do not unless the sampling rate is sufficiently high. At the higher $20\omega_{1.25dB}$ rate, H_{121} is the best followed by $H_{021}, H_{011}, H_{111}$, and H_{012}. H_{111} is still not equiripple at this higher rate. The step responses are shown in Fig. 5.11.7. They approach the ideal response at higher sampling rates.

This frequency transform design technique will be used to design recursive filters in Chap. 11.2 which the engineer may wish to peruse. Now let us review some time domain design methods.

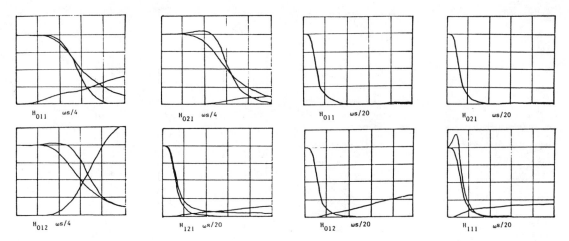

Fig. 5.11.4 Frequency responses of third-order digital Butterworth filters having $\omega_s = 4\omega_{3dB}$ and $20\omega_{3dB}$. (From J.S. Graham, loc. cit.)

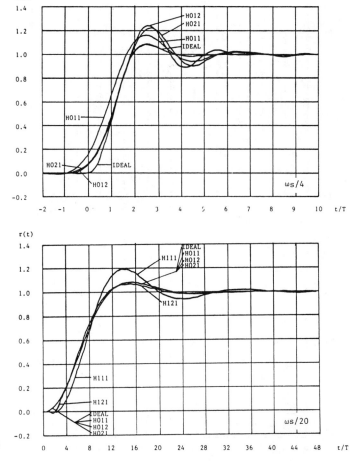

Fig. 5.11.5 Step responses of third-order digital Butterworth filters having $\omega_s = 4\omega_{3dB}$ and $20\omega_{3dB}$. (From J.S. Graham, loc. cit.)

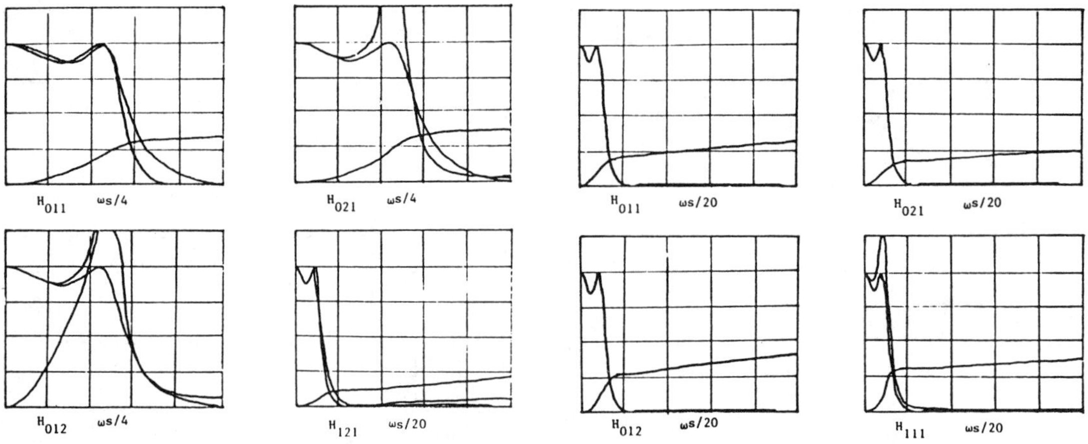

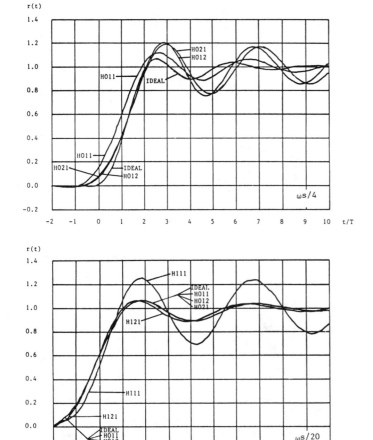

Fig. 5.11.6 Frequency responses of third-order digital elliptic filters having $\omega_s = 4\omega_p$ and $20\omega_p$. (From J.S. Graham, loc. cit.)

Fig. 5.11.7 Step responses of third-order digital elliptic filters having $\omega_s = 4\omega_p$ and $20\omega_p$. (From J.S. Graham, loc. cit.)

5.12 INPUT-INVARIANT TRANSFORMS

An *input-invariant transform* produces a digital filter transfer function $H(z)$ whose output at the sample times $t = nT$ is identical to (or simulates) the output of an analog filter having the same input. This is shown in Fig. 5.12.1 where $H(s)$ is the analog transfer function, $x(t)$ is the input, and $y(t)$ is the desired response. The error $e(t)$ will be zero at $t = nT$ if $H(z)$ satisfies

$$H(z) = \frac{Y(z)}{X(z)} = \frac{Z[y(n)]}{Z[x(n)]} = \frac{Z\left[\mathcal{L}^{-1}\{H(s)X(s)\}_s\right]}{Z\left[\mathcal{L}^{-1}\{X(s)\}_s\right]} \qquad (5.12.1)$$

The s subscript indicates the sampled version of the Laplace transform. The sampled-data equivalent of Fig. 5.12.1 using an interpolator with gain $H_I(s)$ is shown in Fig. 5.12.2. Its overall gain is analogous to Eq. 5.12.1 and equals

$$\frac{Y(s)}{X_s(s)} = \frac{H(s)X(s)}{X_s(s)} = H_I(s)H(s) \qquad (5.12.2)$$

where

$$H_I(s) = \frac{X(s)}{X_s(s)} = \frac{X(s)}{X(z)\big|_{z=e^{sT}}} = \frac{\mathcal{L}[x(t)]}{Z[x(nT)]\big|_{z=e^{sT}}} \qquad (5.12.3)$$

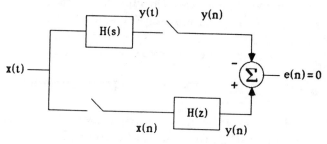

Fig. 5.12.1 Input-invariant transform model.

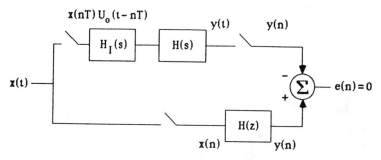

Fig. 5.12.2 Sampled-data input-invariant transform model.

Fig. 5.12.3 Transforming *s*-plane transfer functions into *z*-plane transfer functions.

$H_I(s)$ is an interpolation block which shapes the signal that is filtered by $H(s)$. Interpolators were discussed in Chaps. 3.5–3.8. We will expand that discussion in the following sections.

Generally, the *z*-transform of the product in Eq. 5.12.1 does not equal the product of *z*-transforms or

$$Z\left[\mathcal{L}^{-1}\{H(s)X(s)\}_s\right] \neq Z\left[\mathcal{L}^{-1}\{H(s)\}_s\right] Z\left[\mathcal{L}^{-1}\{X(s)\}_s\right] \tag{5.12.4}$$

$$Z\left[\{h(t) * x(t)\}_s\right] \neq Z\left[h_s(t)\right] Z\left[x_s(t)\right]$$

as shown in Fig. 5.12.3. The *z*-transform of a product commutes to the product of *z*-transforms *only* for blocks that are separated by sampling switches. Using this result in Eq. 5.12.1 shows that generally, $H(z) \neq Z\left[h_s(t)\right]$. The digital filter gain does *not equal* the *z*-transform of the impulse response of the analog filter. $H(z)$ depends upon both the analog filter gain $H(s)$ and the $X(s)$ input it processes (see Example 5.18.1).

The input-invariant method preserves the time domain response parameters of $y(t)$ formed by passing $x(t)$ through a desirable filter. $x(t)$ might be an impulse, step, ramp, sine step, cosine step, or some other useful input. In applications where such preservation is necessary like video or radar, this is an excellent transform method. Time domain performance is maintained whether there is frequency aliasing or not.

5.13 IMPULSE-INVARIANT TRANSFORM

The *impulse-invariant transform* preserves the impulse response $h(t)$ of the analog filter. From Eq. 5.12.1, the transfer function of the digital filter required to do this equals[6]

$$H_0(z) = \frac{Y(z)}{1/T} = TZ\left[y(n)\right] = TZ\left[h(n)\right] = TZ\left[\mathcal{L}^{-1}\{H(s)\}_s\right]$$

$$= Z\left[\sum_{n=-\infty}^{\infty} \mathcal{L}^{-1}\{H(s - jn\omega_s)\}\right] \tag{5.13.1}$$

$H_s(s)$ is given by Eq. 1.3.3, $X(s) = \mathcal{L}[U_0(t)] = 1$, and $X(z) = Z[D_0(nT)] = 1/T$ as found in Eq. 3.1.12. The impulse-invariant transform is also called the *zero-order approximation* since $H_0(z)$ is a DFT (see Eq. 1.8.16) which zero-order approximates the Fourier transform $H(j\omega)$ of $h(t)$.

If the analog filter transfer function is band-limited so $H(j\omega) = 0$ for $|\omega| > \omega_s/2$, then there is no aliasing and Eq. 5.13.1 reduces to

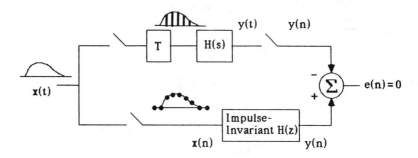

Fig. 5.13.1 Impulse-invariant transform model.

$$H_0(e^{j\omega T}) = H(j\omega), \qquad |\omega| \le \omega_s/2 \qquad\qquad (5.13.2)$$

The digital filter has exactly the same frequency response as the analog filter. When the analog filter response is not exactly band-limited, the digital filter response will only approximate the analog filter response because of aliasing. The system model of the impulse-invariant transform is shown in Fig. 5.13.1. Only scaling and no interpolation is needed since $H_I(s) = 1/(1/T) = T$ from Eq. 5.12.3.

EXAMPLE 5.13.1 Determine the impulse-invariant transfer function to approximate a first-order lowpass filter having $H(s) = 1/(s+1)$.
SOLUTION Since the analog filter impulse response $h(t) = e^{-t}U_{-1}(t)$, the impulse response of the digital filter equals

$$h_0(n) = e^{-nT}U_{-1}(n) = Z^{-1}\big[H_0(z)/T\big] \qquad\qquad (5.13.3)$$

The digital and analog filter impulses are identical at the sampling times as shown in Fig. 5.13.2b. Then the impulse-invariant transfer function equals

$$H_0(z) = TZ\big[e^{-nT}U_{-1}(n)\big] = \frac{T}{1 - e^{-T}z^{-1}} \qquad\qquad (5.13.4)$$

from Table 1.9.1. The magnitude responses for the filters are shown Fig. 5.13.2a for $T = 0$ and 0.5. Since $H(j\omega)$ is not band-limited, significant frequency aliasing is present if the sampling interval T is not sufficiently small. In this case, we must choose $\omega_s = 2\pi/T \gg B_{3dB} = 1$ or $T \ll 2\pi$ based on the 3 dB bandwidth of $H(j\omega)$. Since this is not the case here, Fig. 5.13.2a shows there is appreciable frequency aliasing error. Setting $Y(z) = H_0(z)X(z)$ and taking the inverse z-transform yields the time-domain processing algorithm as

$$(1 - e^{-T}z^{-1})Y(z) = TX(z)$$
$$y(n) = e^{-T}y(n-1) + Tx(n) \qquad\qquad (5.13.5)$$

∎

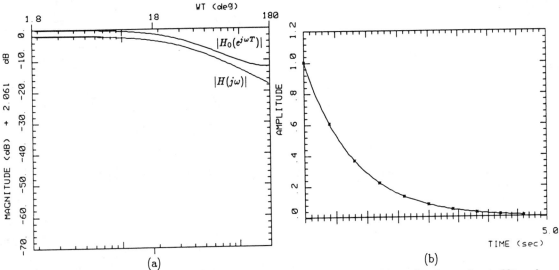

Fig. 5.13.2 (a) Magnitude and (b) impulse responses of impulse-invariant filter in Example 5.13.1.

EXAMPLE 5.13.2 Evaluate the step response of the impulse-invariant transfer function found in Example 5.13.1.

SOLUTION The transformed step response equals

$$R(z) = \frac{z}{z-1} H_0(z) = \frac{Tz^2}{(z-1)(z-e^{-T})}, \qquad |z| > 1 \qquad (5.13.6)$$

from Eq. 5.13.4 and $Z[U_{-1}(t)] = z/(z-1)$. Then the sampled step response equals

$$r(n) = Z^{-1}[R(z)] = Z^{-1}\left[\frac{T}{1-e^{-T}}\left(\frac{1}{z-1} - \frac{e^{-T}}{z-e^{-T}}\right)\right] = T\frac{1-e^{-(n+1)T}}{1-e^{-T}} U_{-1}(n)$$
$$(5.13.7)$$

This response is plotted in Fig. 5.13.3 for $T = 0$ and 0.5 seconds. From Eq. 5.13.7, the initial and final values of $r(nT)$ equal

$$r(0) = T = R(z)\Big|_{|z|=\infty} = H_0(\infty), \qquad r(\infty) = \frac{T}{1-e^{-T}} = (z-1)R(z)\Big|_{z=1} = H_0(1)$$
$$(5.13.8)$$

using the theorems of Table 1.9.2.   ∎

It is clear from Fig. 5.13.3 that there is a steady-state error between the digital filter and analog filter step responses. This error equals $[r(\infty) - 1]$ which approaches $T/2$ for $T \gg 1$. Since the error is due to the dc digital filter gain $H_0(1)$ equalling $T/(1-e^{-T})$ rather than unity, it can be totally eliminated by magnitude normalizing $H_0(z)$ by $(1-e^{-T})/T$.

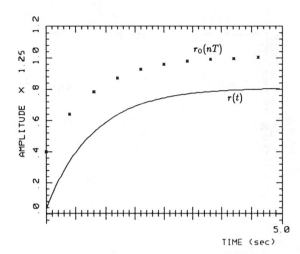

Fig. 5.13.3 Step responses
of impulse-invariant filter in
Example 5.13.2.

This dc gain correction was suggested by Fowler.[7] It is usually called the *dc-adjusted impulse-invariant method* but can be applied to any $H(z)$. Correcting the impulse-invariant transfer function $H_0(z)$ of Eq. 5.13.1 yields the dc-adjusted version $H_{0A}(z)$,

$$H_{0A}(z) = \frac{H_0(z)}{H_0(1)} \qquad (5.13.9)$$

Although this is a general method that can be applied to correct $H_0(z)$ for any input like steps and ramps, a higher-order transform more closely matched to a particular input is usually used instead of $H_{0A}(z)$. These higher-order transforms will be discussed in the next sections.

EXAMPLE 5.13.3 Determine the dc-adjusted version of the impulse-invariant filter $H(s) = 1/(s+1)$ found in Example 5.13.1.
SOLUTION Evaluating the dc gain $H_0(1)$ of $H_0(z)$ found in Eq. 5.13.4 yields $H_0(1) = T/(1 - e^{-T})$. Therefore, the dc-adjusted transfer function becomes

$$H_{0A}(z) = \frac{H_0(z)}{H_0(1)} = \frac{1 - e^{-T}}{1 - e^{-T}z^{-1}} \qquad (5.13.10)$$

Converting into time-domain algorithms,

$$(1 - e^{-T}z^{-1})Y(z) = (1 - e^{-T})X(z)$$
$$y(n) = e^{-T}y(n-1) + (1 - e^{-T})x(n) \qquad (5.13.11)$$

The magnitude response $|H_{0A}(e^{j\omega T})|$ and step response $y(n)$ are plotted in Fig. 5.13.4 along with the equivalent analog filter responses for $T = 0$ and 0.5. The dc gains and steady-state step responses are now equal unlike those in Figs. 5.13.2a and 5.13.3, respectively. ∎

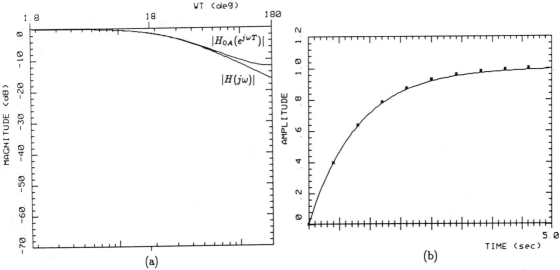

Fig. 5.13.4 (a) Magnitude and (b) step responses of dc-adjusted impulse-invariant filter in Example 5.13.3.

5.14 STEP-INVARIANT TRANSFORM

The *step-invariant transform* preserves the step response $r(t)$ of the analog filter. The transfer function of the digital filter required to do this is

$$H_1(z) = \frac{Y(z)}{z/(z-1)} = \frac{z-1}{z} Z\left[y(n)\right] = \frac{z-1}{z} Z\left[r(n)\right] = \frac{z-1}{z} Z\left[\mathcal{L}^{-1}\left\{\frac{H(s)}{s}\right\}_s\right]$$

(5.14.1)

where $X(s) = \mathcal{L}[U_{-1}(t)] = 1/s$ and $X(z) = Z[U_{-1}(nT)] = z/(z-1)$. The system model of the transform is shown in Fig. 5.14.1. The equivalent sampled data transfer function equals

$$H_1(s) = \frac{z-1}{z}\bigg|_{z=e^{sT}} \frac{H(s)}{s} = \frac{1-e^{-sT}}{s} H(s)$$

(5.14.2)

The $H_I(s)$ interpolator associated with the step-invariant transform is

$$H_I(s) = \frac{H_1(s)}{H(s)} = \frac{X(s)}{X(z)\big|_{z=e^{sT}}} = \frac{1/s}{z/(z-1)\big|_{z=e^{sT}}} = \frac{1-e^{-sT}}{s}$$

(5.14.3)

using Eq. 5.12.3. This is the transfer function of the *zero-order hold* in Chap. 3.5.1. Its impulse response equals $h_I(t) = \mathcal{L}^{-1}[H_1(s)/H(s)] = U_{-1}(t) - U_{-1}(t-T)$. In the sampled-data system shown in Fig. 5.14.1, zero-order interpolation is used between adjacent samples (see Fig. 3.5.1). Therefore the signal $x(t)$ is constructed of steps between samples.

EXAMPLE 5.14.1 Determine the step-invariant transfer function which approximates $H(s) = 1/(s+1)$.

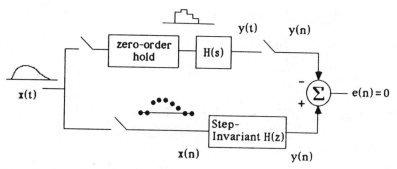

Fig. 5.14.1 Step-invariant transform model.

SOLUTION Since the analog filter step response $r(t) = \mathcal{L}^{-1}[1/s(s+1)] = (1 - e^{-t})U_{-1}(t)$, then the z-transform of the sampled step response equals

$$R(z) = Z\left[r(nT)\right] = \frac{z}{z-1} - \frac{z}{z-e^{-T}}, \qquad |z| > e^{-T} \qquad (5.14.4)$$

Substituting this result into Eq. 5.14.1 yields the step-invariant transfer function

$$H_1(z) = \frac{z-1}{z}\;\frac{z(1-e^{-T})}{(z-1)(z-e^{-T})} = \frac{1-e^{-T}}{z-e^{-T}} \qquad (5.14.5)$$

Converting this into the time-domain algorithm yields

$$(1 - e^{-T}z^{-1})Y(z) = z^{-1}(1-e^{-T})X(z)$$
$$y(n) = e^{-T}y(n-1) + (1-e^{-T})x(n-1) \qquad (5.14.6)$$

It differs from the impulse-invariant algorithm only in that $x(n-1)$ is used rather than $x(n)$ (cf. Eq. 5.13.5). Since $H_1(z) = z^{-1}H_{0A}(z)$ (cf. Eqs. 5.13.10 and 5.14.7), the magnitude responses are identical (Fig. 5.13.4a) while the step response of $H_1(z)$ is delayed one unit from that of $H_{0A}(z)$ (Fig. 5.13.4b). It is identical to the sampled analog filter step response $r(nT)$. ∎

5.15 RAMP-INVARIANT TRANSFORM

The *ramp-invariant transform* preserves the ramp response of the analog filter. The digital filter transfer function required to do this is

$$H_2(z) = \frac{Y(z)}{Tz/(z-1)^2} = \frac{(z-1)^2}{Tz}Z\left[y(n)\right] = \frac{(z-1)^2}{Tz}Z\left[\mathcal{L}^{-1}\left\{\frac{H(s)}{s^2}\right\}_s\right] \qquad (5.15.1)$$

where $\mathcal{L}[U_{-2}(t)] = 1/s^2$ and $Z[U_{-2}(nT)] = Tz/(z-1)^2$. The system model of this transform is shown in Fig. 5.15.1. The equivalent sampled-data transfer function $H_2(s)$ for the ramp-invariant transform is

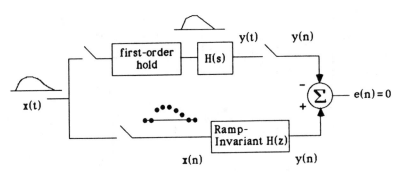

Fig. 5.15.1 Ramp-invariant transform model.

$$H_2(s) = \frac{(z-1)^2}{Tz}\bigg|_{z=e^{sT}} \frac{H(s)}{s^2} = \frac{e^{sT} - 2 + e^{-sT}}{Ts^2} H(s) \qquad (5.15.2)$$

The ramp-invariant interpolator has a gain $H_I(s)$ of

$$H_I(s) = \frac{H_2(s)}{H(s)} = \frac{1/s^2}{Tz/(z-1)^2\big|_{z=e^{sT}}} = \frac{e^{sT} - 2 + e^{-sT}}{Ts^2} \qquad (5.15.3)$$

using Eq. 5.12.3. This is the transfer function for a *first-order hold* system having impulse response $h_I(t) = \mathcal{L}^{-1}[H_2(s)/H(s)] = [U_{-2}(t+T) - 2U_{-2}(t) + U_{-2}(t-T)]/T$. As discussed in Chap. 3.5.2, this system takes a signal $x(t)$ and reexpresses it using first-order interpolation between adjacent samples (see Fig. 3.5.2a). Therefore $x(t)$ is constructed of ramps between samples as shown in Fig. 5.15.1.

EXAMPLE 5.15.1 Determine the ramp-invariant approximation of $H(s) = 1/(s+1)$.

SOLUTION The analog filter ramp response equals

$$y(t) = \mathcal{L}^{-1}\left[\frac{H(s)}{s^2}\right] = \mathcal{L}^{-1}\left[\frac{1}{(s+1)s^2}\right] = \mathcal{L}^{-1}\left[\frac{1}{s^2} - \frac{1}{s} + \frac{1}{s+1}\right] \qquad (5.15.4)$$
$$= (t - 1 + e^{-t})U_{-1}(t)$$

Then from Table 1.9.1, the sampled version of $y(t)$ has the z-transform

$$Y(z) = Z[y(n)] = \frac{Tz}{(z-1)^2} - \frac{z}{z-1} + \frac{z}{z-e^{-T}} \qquad (5.15.5)$$

Therefore using Eq. 5.15.1, the transfer function of the ramp-invariant filter equals

$$H_2(z) = \frac{(z-1)^2}{Tz}Y(z) = 1 - \frac{z-1}{T} + \frac{(z-1)^2}{T(z-e^{-T})}$$
$$= \frac{(T - 1 + e^{-T})z - (Te^{-T} - 1 + e^{-T})}{T(z - e^{-T})} \qquad (5.15.6)$$

The time-domain algorithm then becomes

$$T(1 - e^{-T}z^{-1})Y(z) = \left[(T - 1 + e^{-T}) - (Te^{-T} - 1 + e^{-T})z^{-1}\right]X(z)$$

$$y(n) = e^{-T}y(n-1) + \frac{1}{T}(T - 1 + e^{-T})x(n) - \frac{1}{T}(Te^{-T} - 1 + e^{-T})x(n-1)$$

$$(5.15.7)$$

This ramp-invariant filter is more complex than the step-invariant filter given by Eq. 5.14.6 since it uses both terms $x(n)$ and $x(n-1)$. ∎

5.16 SINE, COSINE, AND COMPLEX EXPONENTIAL STEP-INVARIANT TRANSFORMS

The *sine step-invariant transform* preserves the sine step response of the analog filter. The digital filter transfer function is[8]

$$
\begin{aligned}
H_{1s}(z) &= \frac{Y(z)}{z\sin b/(z^2 - 2z\cos b + 1)} = \frac{z^2 - 2z\cos b + 1}{z\sin b}Z\left[y(n)\right] \\
&= \frac{z^2 - 2z\cos b + 1}{z\sin b}Z\left[\mathcal{L}^{-1}\left\{\frac{\omega_o}{s^2 + \omega_o^2}H(s)\right\}_s\right]
\end{aligned}
$$

$$(5.16.1)$$

where $\mathcal{L}[\sin(\omega_o t)U_{-1}(t)] = \omega_o/(s^2 + \omega_o^2)$, $Z[\sin(n\omega_o T)U_{-1}(n)] = z\sin b/(z^2 - 2z\cos b + 1)$, and $b = \omega_o T$. The equivalent sampled-data transfer function $H_{1s}(s)$ for the sine step-invariant transform is

$$H_{1s}(s) = \frac{e^{sT} - 2\cos b + e^{-sT}}{\sin b}\frac{\omega_o}{s^2 + \omega_o^2}H(s)$$

$$(5.16.2)$$

In ac steady-state the digital filter gain equals

$$H_{1s}(e^{j\omega T}) = \frac{2\omega_o(\cos\omega T - \cos\omega_o T)}{(\omega_o^2 - \omega^2)\sin\omega_o T}H(j\omega)$$

$$(5.16.3)$$

from Eq. 5.16.2. We can express $\omega = \omega_o + \Delta\omega$. For frequencies ω close to the signal frequency ω_o, then $|\Delta\omega| \ll \omega_o$. Eq. 5.16.3 can then be rewritten as

$$H_{1s}(e^{j\omega T}) \cong T\frac{\sin(T\Delta\omega/2)}{T\Delta\omega/2}H(j\omega)$$

$$(5.16.4)$$

since $\cos(\omega T) - \cos(\omega_o T) \cong -2\sin(\omega_o T)\sin(T\Delta\omega/2)$. In addition, when $|T\Delta\omega/2| \ll 1$ or $|\Delta\omega| \ll 2f_s$ then

$$H_{1s}(e^{j\omega T}) \cong TH(j\omega), \qquad |\omega - \omega_o| \ll \min(\omega_o, 2f_s)$$

$$(5.16.5)$$

The analog filter response is maintained in the vicinity of the signal frequency ω_o.

The interpolator associated with the sine step-invariant transform has a gain from Eq. 5.16.2 of

$$H_I(s) = \frac{H_{1s}(s)}{H(s)} = \frac{e^{sT} - 2\cos b + e^{-sT}}{\sin b}\frac{\omega_o}{s^2 + \omega_o^2}$$

$$(5.16.6)$$

Its impulse response $h_I(t)$ equals $\mathcal{L}^{-1}[H_{1s}(s)/H(s)]$ so

$$h_I(t) = \frac{\sin\omega_o(t+T)\, U_{-1}(t+T) + \cos\omega_o T\,\sin\omega_o t\, U_{-1}(t) + \sin\omega_o(t-T)\, U_{-1}(t-T)}{\sin\omega_o T}$$

$$= \frac{\sin\omega_o(t+T)}{\sin\omega_o T}, \qquad -T \le t < 0 \qquad\qquad (5.16.7)$$

$$= \frac{\sin\omega_o(T-t)}{\sin\omega_o T}, \qquad 0 \le t < T$$

and zero elsewhere. Therefore this transform uses a sine interpolation of arc length $\omega_o T$ radians between sample points. For sufficiently small $\omega_o T \ll 1$, then $\sin\omega_o T \cong \omega_o T$ and the sine interpolator reduces to a ramp interpolator or first-order hold.

EXAMPLE 5.16.1   Determine the sine step-invariant transfer function which approximates $H(s) = 1/(s+1)$.
SOLUTION   Since the analog filter sine step response equals

$$r(t) = \mathcal{L}^{-1}\left[\frac{\omega_o}{(s+1)(s^2+\omega_o^2)}\right]$$

$$= \left[\frac{\omega_o}{1+\omega_o^2}e^{-t} + \frac{1}{j2(1+j\omega_o)}e^{j\omega_o t} - \frac{1}{j2(1-j\omega_o)}e^{-j\omega_o t}\right]U_{-1}(t) \qquad (5.16.8)$$

then the z-transform of the sampled sine step response equals

$$R(z) = \left[\frac{\omega_o}{1+\omega_o^2}\frac{z}{z-e^{-T}} + \frac{1}{j2(1+j\omega_o)}\frac{z}{z-e^{j\omega_o T}} - \frac{1}{j2(1-j\omega_o)}\frac{z}{z-e^{-j\omega_o T}}\right]$$

$$= \frac{\omega_o}{1+\omega_o^2}\left[\frac{z}{z-e^{-T}} - \frac{z-\cos\omega_o T - (1/\omega_o)\sin\omega_o T}{z^2 - 2z\cos\omega_o T + 1}\right]$$

$$(5.16.9)$$

for $|z| > e^{-T}$. Substituting this result into Eq. 5.16.1 yields the sine step-invariant transfer function

$$H_{1s}(z) = \frac{Y(z)}{X(z)} = \frac{\omega_o}{1+\omega_o^2} \times$$

$$\frac{z[e^{-T} - 3\cos\omega_o T + (1/\omega_o)\sin\omega_o T] + [1 + e^{-T}(\cos\omega_o T - (1/\omega_o)\sin\omega_o T)]}{z(z-e^T)}$$

$$(5.16.10)$$

Converting this into a time-domain algorithm gives

$$y(n) = e^{-T}y(n-1) + \frac{\omega_o}{1+\omega_o^2}\left\{[e^{-T} - 3\cos\omega_o T + (1/\omega_o)\sin\omega_o T]x(n-1)\right.$$

$$\left. + [1 + e^{-T}(\cos\omega_o T - (1/\omega_o)\sin\omega_o T)]x(n-2)\right\}$$

$$(5.16.11)$$

It differs from the step-invariant algorithm since $x(n-2)$ is used in addition to $x(n-1)$ (cf. Eq. 5.14.6). ∎

The *cosine step-invariant transform* preserves the cosine step response of the analog filter. The digital filter transfer function is

$$H_{1c}(z) = \frac{Y(z)}{z(z - \cos b)/(z^2 - 2z\cos b + 1)} = \frac{z^2 - 2z\cos b + 1}{z(z - \cos b)} Z[y(n)]$$

$$= \frac{z^2 - 2z\cos b + 1}{z(z - \cos b)} Z\left[\mathcal{L}^{-1}\left\{\frac{s}{s^2 + \omega_o^2} H(s)\right\}_s\right]$$

(5.16.12)

where $\mathcal{L}[\cos(\omega_o t)U_{-1}(t)] = s/(s^2 + \omega_o^2)$, $Z[\cos(n\omega_o T)U_{-1}(n)] = z(z - \cos b)/(z^2 - 2z\cos b + 1)$, and $b = \omega_o T$. The equivalent sampled-data transfer function $H_{1c}(s)$ for the cosine step-invariant transform is

$$H_{1c}(s) = \frac{e^{sT} - 2\cos b + e^{-sT}}{e^{sT} - \cos b} \frac{s}{s^2 + \omega_o^2} H(s)$$

(5.16.13)

In ac steady-state the digital filter gain equals

$$H_{1c}(e^{j\omega T}) = \frac{j2\omega}{\omega_o^2 - \omega^2} \frac{\cos\omega T - \cos\omega_o T}{(\cos\omega T - \cos\omega_o T) + j\sin\omega T} H(j\omega)$$

(5.16.14)

from Eq. 5.16.13. When $\omega \cong \omega_o$, this result can be simplified as Eq. 5.16.3 to equal

$$H_{1c}(e^{j\omega T}) \cong \frac{-j}{\Delta\omega} \frac{1}{1 - j\sin(\omega_o T)/2\sin(\omega_o T)\sin(T\Delta\omega/2)} H(j\omega)$$

(5.16.15)

Further when $|T\Delta\omega/2| \ll 1$ or $|\Delta\omega| \ll 2f_s$, then

$$H_{1c}(e^{j\omega T}) \cong \frac{j}{\Delta\omega} \frac{2}{j} \sin(T\Delta\omega/2) H(j\omega) \cong TH(j\omega), \qquad |\omega - \omega_o| \ll \min(\omega_o, 2f_s)$$

(5.16.16)

The digital and analog filter responses are almost equal in the vicinity of ω_o.

The cosine step-invariant transform has an interpolator gain of

$$H_I(s) = \frac{H_{1c}(s)}{H(s)} = \frac{1 - 2e^{-sT}\cos b + e^{-2sT}}{1 - e^{-sT}\cos b} \frac{s}{s^2 + \omega_o^2}$$

(5.16.17)

from Eq. 5.16.12. Its impulse response $h_I(t)$ equals $\mathcal{L}^{-1}[H_{1c}(s)/H(s)]$ which can be expressed as

$$h_I(t) = \mathcal{L}^{-1}\left[\frac{(1 - e^{-sT}\cos b)^2 + (1 - \cos^2 b)e^{-2sT}}{1 - e^{-sT}\cos b} \frac{s}{s^2 + \omega_o^2}\right]$$

$$= \mathcal{L}^{-1}\left\{\left[1 - e^{-sT}\cos b + e^{-2sT}\sin^2 b \sum_{n=0}^{\infty}(-\cos b)^n e^{-nsT}\right] \frac{s}{s^2 + \omega_o^2}\right\}$$

$$= \cos \omega_o T \; U_{-1}(t) - \cos \omega_o T \; \cos \omega_o(t - T) \; U_{-1}(t - T)$$

$$+ \sin^2 \omega_o T \sum_{n=0}^{\infty} (-1)^n \cos^n \omega_o T \cos \omega_o(t - (n+2)T) \; U_{-1}(t - (n+2)T)$$

$$(5.16.18)$$

The first line results by completing the square of the numerator in Eq. 5.16.17 while the second line expands $1/(1 - e^{-sT} \cos b)$ in the binomial series $\sum_{n=0}^{\infty}(-\cos b)e^{-nsT}$. Since $\sin^2 x \cong x^2 \to 0$ when $x \to 0$, using small $\omega_o T$ reduces Eq. 5.16.18 to

$$
\begin{aligned}
h_{1c}(t) &= \cos \omega_o T \; U_{-1}(t) - \cos \omega_o T \; \cos \omega_o(t - T) \; U_{-1}(t - T) \\
&= \quad \cos \omega_o T, \qquad 0 \leq t < T \\
&= -\sin \omega_o T \sin \omega_o(t - T) \cong -\omega_o T \sin \omega_o(t - T), \qquad t \geq 2T
\end{aligned}
$$

$$(5.16.19)$$

The approximate $h_{1c}(t)$ is a cosine pulse of arc length $\omega_o T$ radians followed by low level ringing. As $\omega_o T \to 0$, this interpolator becomes a zero-order hold.

The *complex exponential step-invariant transform* preserves the complex exponential step response of the analog filter as shown in Fig. 5.16.1. The digital filter transfer function is

$$H_{1e}(z) = \frac{Y(z)}{z/(z - e^{j\omega_o T})} = \frac{z - e^{j\omega_o T}}{z} Z\big[y(n)\big] = \frac{z - e^{j\omega_o T}}{z} Z\left[\mathcal{L}^{-1}\left\{\frac{1}{s - j\omega_o} H(s)\right\}_s\right]$$

$$(5.16.20)$$

where $\mathcal{L}[e^{j\omega_o t}U_{-1}(t)] = 1/(s - j\omega_o)$, $Z[e^{jn\omega_o T}U_{-1}(n)] = z/(z - e^{j\omega_o T})$, and $b = \omega_o T$. The equivalent sampled-data transfer function $H_{1e}(s)$ for the complex exponential step-invariant transform is

$$H_{1e}(s) = \frac{e^{sT} - e^{j\omega_o T}}{e^{sT}} \; \frac{1}{s - j\omega_o} \; H(s)$$

$$(5.16.21)$$

In ac steady-state the digital filter gain equals

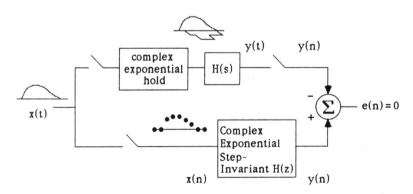

Fig. 5.16.1 Complex exponential step-invariant transform model.

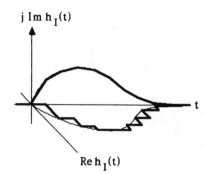

Fig. 5.16.2 Impulse response of complex exponential hold.

$$H_{1e}(e^{j\omega T}) = \frac{1 - e^{j(\omega_o - \omega)T}}{j(\omega - \omega_o)} \, H(j\omega) = \frac{1 - e^{-jT\Delta\omega}}{j\Delta\omega} \, H(j\omega) \qquad (5.16.22)$$

For $|T\Delta\omega| \ll 1$ or $|\Delta\omega| \ll 1/T = f_s$, Eq. 5.16.22 becomes

$$H_{1e}(e^{j\omega T}) \cong TH(j\omega), \qquad |\omega - \omega_o| \ll \min(\omega_o, f_s) \qquad (5.16.23)$$

since $e^{-jT\Delta\omega} = 1 - jT\Delta\omega + \dots$. The filter responses are identical in the vicinity of ω_o. The interpolator used by the complex exponential step-invariant transform has a gain

$$H_I(s) = \frac{H_{1e}(s)}{H(s)} = \frac{1 - e^{j\omega_o T}e^{-sT}}{s - j\omega_o} \qquad (5.16.24)$$

Its impulse response $h_I(t)$ equals $\mathcal{L}^{-1}[H_{1e}(s)/H(s)]$ so

$$h_I(t) = e^{j\omega_o t}U_{-1}(t) - e^{j\omega_o T}e^{j\omega_o(t-T)}U_{-1}(t) = e^{j\omega_o t}, \qquad 0 \le t < T \qquad (5.16.25)$$

and zero elsewhere. This interpolator is called a complex exponential zero-order hold. It outputs a $e^{j\omega_o t}$ complex exponential pulse for each impulsive input as shown in Fig. 5.16.2. Since $e^{j\omega_o t} = \cos\omega_o t + j\sin\omega_o t$, then $e^{j\omega_o t} = 1 + j\omega_o t \to 1 + j0$ when $\omega_o T \to 0$. For small $\omega_o T$, the real part of $h_I(t)$ functions as a zero-order hold while the imaginary part functions as a first-order hold.

5.17 COMPARISON OF TRANSFORMS

The numerical integration transforms were compared in Chap. 5.9. Comparisons are somewhat arbitrary due to the fact that the accuracy of the digital filter response compared to that of the analog filter response depends upon (1) the form of the input signal, (2) the filter transfer function being simulated, (3) the digital transform used, and (4) the sampling interval T.

Nevertheless, we shall now compare several others — namely the step-invariant or ramp-invariant, dc-adjusted impulse-invariant, and bilinear transforms. For a first-order lowpass filter having $H(s) = 1/(s+1)$, the transfer functions are

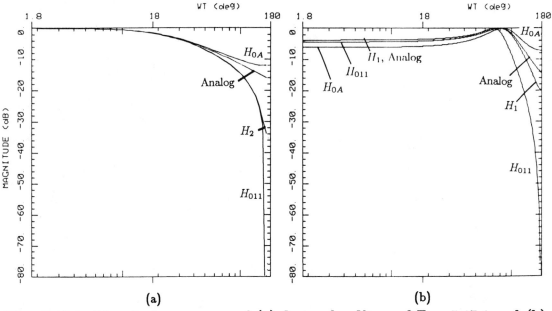

Fig. 5.17.1 Magnitude responses of (a) first-order filters of Eq. 5.17.1 and (b) second-order systems of Eq. 5.17.2.

$$H(s) = \frac{1}{(1/T)sT + 1}$$

$$H_2(s) = \frac{(T - 1 + e^{-T})z - (Te^{-T} - 1 + e^{-T})}{T(z - e^{-T})}$$

$$H_{0A}(z) = \frac{(1 - e^{-T})z}{z - e^{-T}} \qquad\qquad (5.17.1)$$

$$H_{011}(z) = \frac{T(z + 1)}{(T + 2)z + (T - 2)}$$

The ac steady-state gain magnitudes where $s = j\omega$ and $z = e^{j\omega T}$ are plotted in Fig. 5.17.1a for $T = 0.5$. H_{0A} is the closest to H. H_2 and H_{011} have somewhat larger errors.

Now consider the case for a second-order lowpass filter having $H(s) = 10/(s^2 + 2s + 10)$ for $T = 0.5$. Then the transfer functions become

$$H(s) = \frac{10}{(1/T^2)(sT)^2 + (2/T)(sT) + 10}$$

$$H_1(z) = \frac{(1 - e^{-T}[\cos(3T) + \frac{1}{3}\sin(3T)])z + e^{-T}[e^{-T} - \cos(3T) + \frac{1}{3}\sin(3T)]}{z^2 - 2e^{-T}\cos(3T)z + e^{-2T}}$$

$$H_{0A}(z) = \frac{(1 - 2e^{-T}\cos(3T) + e^{-2T})z}{z^2 - 2e^{-T}\cos(3T)z + e^{-2T}} \qquad\qquad (5.17.2)$$

$$H_{011}(z) = \frac{10T^2(z + 1)^2}{(10T^2 + 4T + 4)z^2 + (20T^2 - 8)z + (10T^2 - 4T + 4)}$$

The gain magnitudes are plotted in Fig. 5.17.1b. In this case, H_1 is closest to H while H_{0A} and H_{011} exhibit larger errors.

5.18 ANALOG-TO-DIGITAL SYSTEM CONVERSION

So far in this chapter, we have discussed various techniques for transforming an analog transfer function into a digital transfer function. Another often-encountered design problem requires implementing an entire analog system in digital form which shall now be discussed. Very often, the system is represented by block diagrams with transfer functions assigned to each block. The blocks may be linear or nonlinear. Nonlinear blocks like $y = x^2$, $|x|$, $\tan(x)$, and $\log(x)$ are often very simple to implement digitally even though their analog forms may be very difficult to realize.

The system will generally involve feedback of various signals. Feedback systems are often more difficult to design to operate exactly as their analog counterpart because signals must be computed sequentially around feedback loops. Such signals can only be computed in terms of previous values but not present values (which are not known) so delay blocks must be added to insure realizability. Another problem is that the simulation of products $[H_1(s)H_2(s)]$ is usually different from the product of simulations $[H_1(s)][H_2(s)]$ (see Fig. 5.12.2). It depends upon the particular s-to-z transform used.

EXAMPLE 5.18.1 Investigate the differences in simulating

$$H(s) = H_1(s)H_2(s) = \frac{1}{s}\frac{1}{s+1} = \frac{Y(z)}{X(z)} \tag{5.18.1}$$

by simulating H_1 and H_2 individually versus simulating a single transfer function H. Use the impulse-invariant transform.

SOLUTION The individual impulse-invariant transforms equal

$$H_1(z) = TZ\left[\mathcal{L}^{-1}\left\{\frac{1}{s}\right\}_s\right] = \frac{Tz}{z-1}$$

$$H_2(z) = TZ\left[\mathcal{L}^{-1}\left\{\frac{1}{s+1}\right\}_s\right] = \frac{Tz}{z-e^{-T}} \tag{5.18.2}$$

from Eq. 5.14.1. Then the time-domain algorithm is

$$(1 - z^{-1})(1 - e^{-T}z^{-1})Y(z) = T^2X(z)$$

$$y(n) = (1 + e^{-T})y(n-1) - e^{-T}y(n-2) + T^2x(n) \tag{5.18.3}$$

This is the recursion equation when the individual gains H_1 and H_2 are simulated. Alternatively, the entire product H_1H_2 is transformed by

$$H(z) = TZ\left[\mathcal{L}^{-1}\left\{\frac{1}{s(s+1)}\right\}_s\right] = TZ\left[(1 - e^{-t})U_{-1}(t)\right]$$

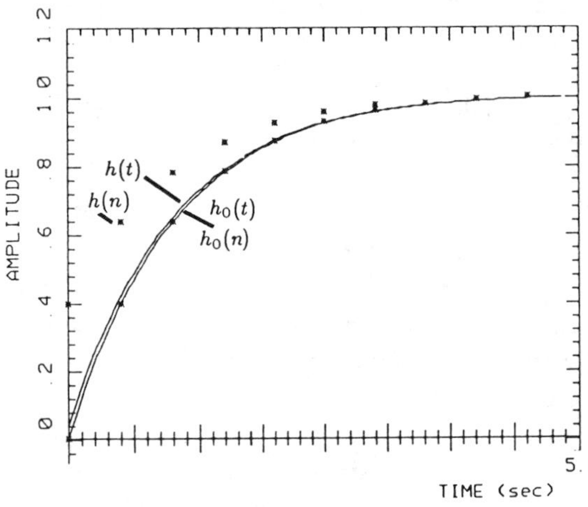

Fig. 5.18.1 Impulse-invariant filter response of Example 5.18.1.

$$= T\left(\frac{z}{z-1} - \frac{z}{z-e^{-T}}\right) = \frac{T(1-e^{-T})z}{(z-1)(z-e^{-T})} \tag{5.18.4}$$

Then the time-domain algorithm becomes

$$(1 - z^{-1})(1 - e^{-T}z^{-1})Y(z) = T(1 - e^{-T})z^{-1}X(z)$$

$$y(n) = (1 + e^{-T})y(n-1) - e^{-T}y(n-2) + T(1-e^{-T})x(n-1) \tag{5.18.5}$$

Comparing Eqs. 5.18.2 and 5.18.4, we see that the transform of the product does not equal the product of transforms so $H(z) \neq H_1(z)H_2(z)$. The transforms have the same poles but H has only one zero at the origin unlike $H_1 H_2$ which has two. This causes the $x(n-1)$ term to appear in Eq. 5.18.5 which does not appear in Eq. 5.18.3. The impulse responses for these two digital filters are plotted in Fig. 5.18.1 for $T = 0$ and 0.5. h_0 given by Eq. 5.18.5 is almost exact while h of Eq. 5.18.3 is in error. ∎

A procedure for implementing an analog system in digital form is the following:[8]

1. Temporarily replace all nonlinear elements with their linearized (small-signal) equivalents.
2. Simulate each block of the system using a suitable s-to-z transformation described in the previous sections.
3. Add delay inside all nonrealizable closed loops to make their simulations realizable.
4. Insert constant gain blocks inside feedback loops to make the overall system poles of the simulated system agree with those of the analog system.
5. Add a compensation block at the system input to make the overall $H(z)$ equal the desired simulation of $H(s)$.
6. Replace all nonlinear elements with their simulation equivalents.

To illustrate this procedure, consider the analog phase-locked loop (PLL) shown in Fig. 5.18.2a and its linearized model in Fig. 5.18.2b. $H_1(s)$ is a first-order lowpass filter. $H_2(s)$ is an integrator that forces the steady-state output $x_6(t)$ to track the input $x_1(t)$. Under this condition, the error $x_3(t)$ equals zero. a is an amplifier-limiter of gain a whose output cannot exceed ± 1. The overall PLL transfer function equals

$$H(s) = \frac{H_1(s)H_2(s)}{1 + H_1(s)H_2(s)} = \frac{a}{s^2 + s + a} \qquad (5.18.6)$$

which is lowpass. For reasonable system performance, we select $a = 1$. The phase-locked loop can be converted into digital form following the procedure just outlined as will now be described.

Step 1 (Linearize nonlinear elements). The amplifier-limiter has unit gain until its input reaches ± 1. Then its gain drops to zero. Therefore, this nonlinear element is replaced by a unit gain block.

Step 2 (z-transform blocks). Next the analog blocks are converted into digital form using some s-to-z transform. Choosing the impulse-invariant transform, the transfer functions $H_1(z)$ and $H_2(z)$ become

$$H_1(z) = TZ\left[\mathcal{L}^{-1}\left\{\frac{1}{s}\right\}_s\right] = \frac{Tz}{z - 1}$$

$$H_2(z) = TZ\left[\mathcal{L}^{-1}\left\{\frac{1}{s+1}\right\}_s\right] = \frac{Tz}{z - e^{-T}} \qquad (5.18.7)$$

(see Eq. 5.18.2). The equivalent circuit is shown in Fig. 5.18.3. Both $H_1(z)$ and $H_2(z)$ are causal and stable.

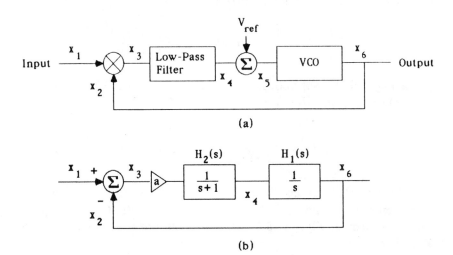

(a)

(b)

Fig. 5.18.2 (a) Analog phase-locked loop and its (b) linearized model.

Step 3 (Make realizable). Since the feedback path has unity gain, the present value of the feedback signal $x_2(n)$ depends on the present value of the output signal $x_6(n)$. The loop is nonrealizable in real-time so delay z^{-1} must be added to make it realizable.

Step 4 (Match poles). The impulse-invariant transform $H(z)$ of the overall gain $H(s)$ given by Eq. 5.18.6 equals

$$H(z) = TZ\left[\mathcal{L}^{-1}\{H(s)\}_s\right] = TZ\left[\mathcal{L}^{-1}\left\{\frac{1}{s^2 + s + 1}\right\}_s\right]$$

$$= TZ\left[\left\{\frac{1}{0.866}e^{-0.5t}\sin(0.866t)\,U_{-1}(t)\right\}_s\right] = \frac{1}{0.866}\frac{z\,e^{-0.5T}\sin(0.866T)}{z^2 - 2ze^{-0.5T}\cos(0.866T) + e^{-T}}$$

$$(5.18.8)$$

In terms of $H_1(z)$ and $H_2(z)$ found in Eq. 5.18.7, the overall gain equals

$$H(z) = \frac{H_1(z)H_2(z)}{1 + Kz^{-1}H_1(z)H_2(z)} = \frac{T^2z^2}{(z-1)(z-e^{-T}) + KT^2z}$$

$$= \frac{T^2z^2}{z^2 - (1 + e^{-T} - KT^2)z + e^{-T}}$$

$$(5.18.9)$$

where an arbitrary gain-delay block Kz^{-1} has been added in the feedback loop. The poles of Eqs. 5.18.8 and 5.18.9 can be matched by equating the coefficients of z and solving for K as

$$1 + e^{-T} - KT^2 = 2e^{-0.5T}\cos(0.866T)$$

$$K = \frac{1 - 2e^{-0.5T}\cos(0.866T) + e^{-T}}{T^2}$$

$$(5.18.10)$$

Step 5 (Match overall gain). Adding an input compensation block $H_3(z)$ to Eq. 5.18.9 yields an overall gain expression

$$H(z) = \frac{H_1(z)H_2(z)H_3(z)}{1 + H_1(z)H_2(z)}$$

$$(5.18.11)$$

Equating Eqs. 5.18.8 and 5.18.9, substituting Eqs. 5.18.7 and 5.18.8 into Eq. 5.18.11, and solving for the $H_3(z)$ yields

$$H_3(z) = \left[1 + \frac{1}{H_1(z)H_2(z)}\right]H(z) = \frac{e^{-0.5T}\sin(0.866T)}{0.866T^2z}$$

$$(5.18.12)$$

Step 6 (Replace nonlinear elements). The limiter which was linearized in Step 1 is returned to its normal state. The complete digital filter system is shown in block diagram form in Fig. 5.18.3a. The blocks are expressed in their algorithmic form in Fig. 5.18.3b.

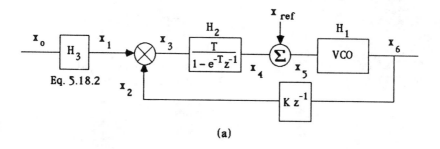

$$x_1(n) = \frac{e^{-0.5T}\sin(0.866T)}{0.866T^2}\, x_0(n-1)$$

$$x_3(n) = x_1(n)x_2(n)$$

$$x_4(n) = e^{-T}x_4(n-1) + Tx_3(n)$$

$$x_5(n) = x_4(n) + x_{ref}$$

$$x_6(n) = \sin[x_5(n) * nT]$$

$$x_2(n) = \frac{1 - 2e^{-0.5T}\cos(0.866T) + e^{-T}}{T^2}\, x_6(n-1)$$

(b)

Fig. 5.18.3 (a) **Block diagram for impulse-invariant digital phase-locked loop and its (b) simulation equations.**

To obtain a different set of equations, let us now convert the analog system of Fig. 5.18.2 into digital form using the dc-adjusted impulse-invariant transform.

Step 1 (Linearize nonlinear blocks). Same as before.

Step 2 (z-transform blocks). Using the dc-adjusted impulse-invariant transform of Eq. 5.13.8 yields

$$H_{1A}(z) = \frac{H_1(z)}{H_1(1)} = \text{undefined} \quad \rightarrow \quad H_1(z) = \frac{Tz}{z-1}$$

$$H_{2A}(z) = \frac{H_2(z)}{H_2(1)} = \frac{1-e^{-T}}{T}\,\frac{Tz}{z-e^{-T}} = \frac{(1-e^{-T})z}{z-e^{-T}} \tag{5.18.13}$$

Since H_1 is an integrator with infinite gain at dc, H_{1A} is undefined so H_1 is used instead.

Step 3 (Make realizable). Since the present value of the loop signal $x(n)$ depends on the present value of another loop signal $g(n)$, the feedback path is nonrealizable in real-time. Delay z^{-1} must be added to make it realizable as before.

Step 4 (Match poles). The dc-adjusted impulse-invariant transform of the overall gain $H_A(z)$ equals

$$H_A(z) = \frac{H(z)}{H(1)} = \frac{1 - 2e^{-0.5T}\cos(0.866T) + e^{-T}}{e^{-0.5T}\sin(0.866T)}\,\frac{z\,e^{-0.5T}\sin(0.866T)}{z^2 - 2ze^{-0.5T}\cos(0.866T) + e^{-T}}$$

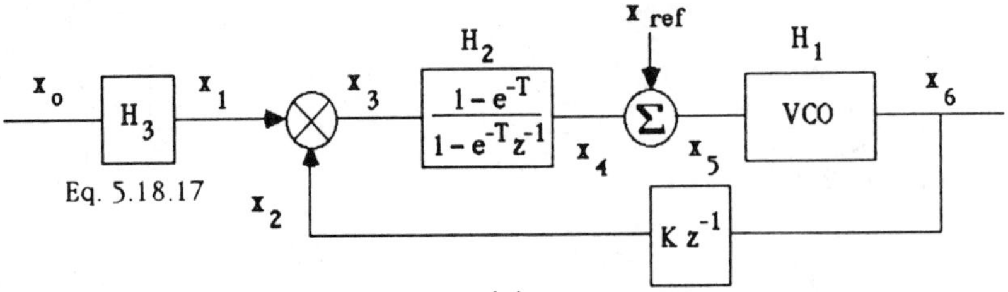

(a)

$$x_1(n) = \frac{1 - 2e^{-0.5T}\cos(0.866T) + e^{-T}}{T(1 - e^{-T})} x_0(n-1)$$

$$x_3(n) = x_1(n)x_2(n)$$

$$x_4(n) = e^{-T}x_4(n-1) + (1 - e^{-T})x_3(n)$$

$$x_5(n) = x_4(n) + x_{ref}$$

$$x_6(n) = \sin[x_5(n) * nT]$$

$$x_2(n) = \frac{1 - 2e^{-0.5T}\cos(0.866T) + e^{-T}}{T(1 - e^{-T})} x_6(n-1)$$

(b)

Fig. 5.18.4 (a) Block diagram for dc-adjusted impulse-invariant digital phase-locked loop and its (b) simulation equations.

$$= \frac{z(1 - 2e^{-0.5T}\cos(0.866T) + e^{-T})}{z^2 - 2ze^{-0.5T}\cos(0.866T) + e^{-T}}$$

$$(5.18.14)$$

using $H(z)$ from Eq. 5.18.8. From $H_1(z)$ and $H_{2A}(z)$ of Eq. 5.18.13, $H_A(z)$ equals

$$H_A(z) = \frac{H_1(z)H_{2A}(z)}{1 + Kz^{-1}H_1(z)H_{2A}(z)} = \frac{Tz^2(1 - e^{-T})}{(z-1)(z-e^{-T}) + zKT(1 - e^{-T})}$$

$$= \frac{Tz^2(1 - e^{-T})}{z^2 - z[(1 + e^{-T}) - KT(1 - e^{-T})] + e^{-T}}$$

$$(5.18.15)$$

The gain K can be adjusted to equate the poles in Eqs. 5.18.14 and 5.18.15. This is accomplished by matching the coefficients of the z terms in these equations which yields

$$2e^{-0.5T}\cos(0.866T) = (1 + e^{-T}) - KT(1 - e^{-T})$$

$$K = \frac{1 - 2e^{-0.5T}\cos(0.866T) + e^{-T}}{T(1 - e^{-T})}$$

$$(5.18.16)$$

Step 5 (Match overall gain). Now add an input gain block $H_3(z)$ having gain (see Eq. 5.18.11)

$$H_3(z) = \left[1 + \frac{1}{H_{1A}(z)H_{2A}(z)}\right]H_A(z) = \frac{1 - 2e^{-0.5T}\cos(0.866T) + e^{-T}}{T(1 - e^{-T})z} \qquad (5.18.17)$$

which results by substituting Eqs. 5.18.13 and 5.18.14 into Eq. 5.18.17. The complete digital system is shown in Fig. 5.18.4a and the time-domain algorithms are listed in Fig. 5.18.4b.

Now consider the PLL responses. The input signal $x_0(n)$ is shown in Fig. 5.18.5a. Its frequency f_o is stepped slightly higher at time $f = 0$. The sampling frequency $f_s \cong 10 f_o$. The VCO output signal $x_6(n)$ is shown in Fig. 5.18.5b. The error signal $x_3(n)$ from the multiplier or phase detector is shown in Fig. 5.18.5c. The lowpass filter extracts the low-frequency spectra of $x_3(n)$ to drive the PLL into lock. The dc-adjusted impulse-invariant LPF output is shown in Fig. 5.18.5d and the impulse-invariant LPF output is shown in Fig. 5.18.5e. This $x_4(n)$ output is added to a bias x_{ref} and their

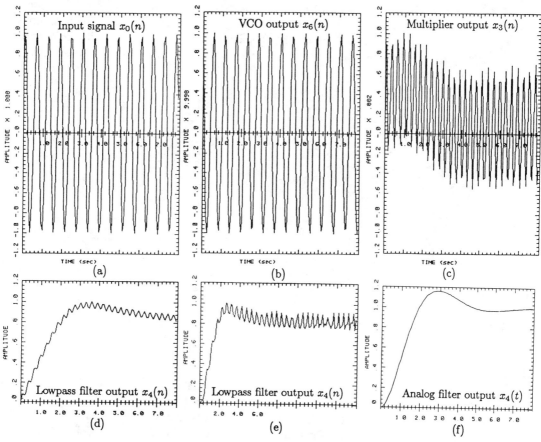

Fig. 5.18.5 (a) Input signal $x_0(n)$, (b) VCO output $x_6(n)$, (c) multiplier output $x_3(n)$, output $x_4(n)$ of (d) dc-adjusted impulse-invariant lowpass filter and (e) impulse-invariant lowpass filter, and (f) output $x_4(t)$ of analog lowpass filter. (From S. Yhann and A.R.H. Taib, Special project, Univ. of Miami, May 1986.)

sum is applied to the VCO. This causes the VCO output frequency to increase until the PLL acquires (1) frequency lock and then (2) phase lock. Since the input frequency variation was a step (or the input phase variation was a ramp), the output frequency is an oscillatory step. The step response of the analog PLL equals

$$r(t) = \mathcal{L}^{-1}\left[\frac{1}{s(s^2 + s + 1)}\right] = \left[1 - \frac{1}{0.866}e^{-0.5t}\sin(0.866t + 60°)\right]U_{-1}(t) \quad (5.18.18)$$

which is shown in Fig. 5.18.5f. It has about 15% overshoot. The digital filter responses exhibit about this amount of overshoot. The $\omega_n t$ time scales differ because various ω_n were used in the simulations. Thus, the digital PLL performs almost identically to the analog PLL as should be the case.

PROBLEMS

5.1 The magnitude responses of a variety of medium cut-off analog lowpass filters are shown in Fig. P5.1. (a) Determine what type of classical analog filters (Butterworth, Chebyshev, elliptic, Bessel) can be used and their orders. (b) Determine the corresponding frequencies of the equivalent digital lowpass filter using the bilinear transform and other transforms in Table 5.11.1. Assume that $\omega_s = 10$.

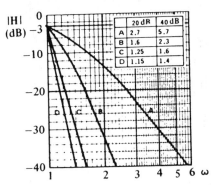

Fig. P5.1 (Courtesy of Allen Avionics[10])

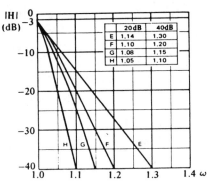

Fig. P5.2 (Courtesy of Allen Avionics[10])

5.2 The magnitude responses of a variety of sharp cut-off digital lowpass filters are shown in Fig. P5.2. Determine what type of classical filters (Butterworth, Chebyshev, elliptic, Bessel) can be used and their orders. Use the bilinear transform and other transforms in Table 5.11.1. Assume that $\omega_s = 15$.

5.3 The magnitude responses of a variety of linear phase digital lowpass filters are shown in Fig. P5.3. (a) Determine what type of classical filters (Butterworth, Chebyshev, elliptic, Bessel) can be used and their orders. Use the bilinear transform and other transforms in Table 5.11.1. Assume that $\omega_s = 20$. (b) Determine the corresponding frequencies of the equivalent analog lowpass filter.

5.4 A digital lowpass filter has the following requirements: maximum in-band gain = 0 dB, 3 dB passband $f_p \le 941$ Hz, 40 dB stopband $f_r \ge 1209$ Hz. (a) Determine the minimum orders of the Butterworth, Chebyshev, elliptic, and Bessel filters. Use the bilinear transform

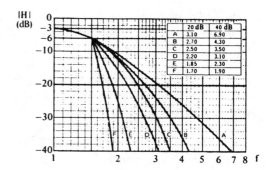

Fig. P5.3 (Courtesy of Allen Avionics[10])

and other transforms in Table 5.11.1. Assume that $f_s = 8f_p$. (b) Determine the corresponding frequencies of the equivalent analog lowpass filter.

5.5 Consider a digital lowpass filter having the following requirements: maximum in-band gain = 0 dB, $M_p = 0.28$ dB for $f_p \leq 941$ Hz, $M_r = 40$ dB for $f_r \geq 1209$ Hz. Determine the minimum orders of the Butterworth, Chebyshev, elliptic, and Bessel filters. Use the bilinear transform and other transforms in Table 5.11.1. Assume that $f_s = 16f_p$.

5.6 A digital lowpass filter has the following requirements: maximum in-band gain = 0 dB, 3 dB passband $f \leq 3.2$ KHz, 40 dB stopband $f \geq 4.8$ KHz. (a) Determine the minimum orders of the Butterworth, Chebyshev, elliptic, and Bessel filters. Use the bilinear transform and other transforms in Table 5.11.1. Assume that $f_s = 19.2$ KHz. (b) Determine the corresponding frequencies of the equivalent analog lowpass filter.

5.7 Consider a digital lowpass filter having the following requirements: maximum in-band gain = 0 dB, $M_p = 0.28$ dB for $f_p \leq 3.2$ KHz, $M_r = 40$ dB for $f_r \geq 4.8$ KHz. Determine the minimum orders of the Butterworth, Chebyshev, elliptic, and Bessel filters. Use the bilinear transform and other transforms in Table 5.11.1. Assume that $f_s = 38.4$ KHz.

5.8 Consider a digital lowpass filter having the following requirements: dc gain = 0 dB, 3 dB passband $f_p \leq 9.6$ KHz, 40 dB stopband $f_r \geq 19.2$ KHz. Determine the minimum orders of the Butterworth, Chebyshev, elliptic, and Bessel filters. Use the bilinear transform and other transforms in Table 5.11.1. Assume that $f_s = 8f_p$.

5.9 Consider a digital lowpass filter having the following requirements: dc gain = 0 dB, 3 dB passband $f_p \leq 9.6$ KHz, 60 dB stopband $f_r \geq 19.2$ KHz. Determine the minimum orders of the Butterworth, Chebyshev, elliptic, and Bessel filters. Use the bilinear transform and other transforms in Table 5.11.1. Assume that $f_s = 16f_p$.

5.10 Given an analog differentiator $H(s) = s$ and a sampling rate of 100 rad/sec. Find the digital differentiator $H(z)$ using the: (a) forward Euler, (b) backward Euler, (c) bilinear, (d) modified bilinear, and (e) lossless transforms.

5.11 Given an analog lowpass filter $H(s) = 1/(s+1)$ and a sampling rate of 100 rad/sec. Find the digital lowpass filter $H(z)$ using the: (a) forward Euler, (b) backward Euler, (c) bilinear, (d) modified bilinear, and (e) lossless transforms.

5.12 Given an analog lowpass filter $H(s) = 1/(s^2 + \sqrt{2}s + 1)$ and a sampling rate of 10 rad/sec. Find the digital lowpass filter $H(z)$ using the: (a) forward Euler, (b) backward Euler, (c) bilinear, (d) modified bilinear, and (e) lossless transforms.

5.13 Consider an analog lowpass filter having $H(s) = (s+1)/[\sqrt{2}s^2 + s + 1]$ and a sampling rate of 100 rad/sec. Find the digital lowpass filter $H(z)$ using the: (a) forward Euler, (b) backward Euler, (c) bilinear, (d) modified bilinear, and (e) lossless transforms.

5.14 Given an analog integrator $H(s) = 1/s$ and a variable sampling rate $1/T$ Hz. Plot the root locus of the poles of the digital integrator $H(z)$ using the: (a) forward Euler, (b) backward

Euler, (c) bilinear, (d) modified bilinear, and (e) lossless transforms.

5.15   Given an analog lowpass filter $H(s) = 1/(s + 1)$ and a variable sampling rate $1/T$ Hz. Plot the root locus of the poles of the digital lowpass filter $H(z)$ using the: (a) forward Euler, (b) backward Euler, (c) bilinear, (d) modified bilinear, and (e) lossless transforms.

5.16   Given an analog lowpass filter $H(s) = 1/(s^2 + \sqrt{2}s + 1)$ and a variable sampling rate $1/T$ Hz. Plot the root locus of the poles of the digital lowpass filter $H(z)$ using the: (a) forward Euler, (b) backward Euler, (c) bilinear, (d) modified bilinear, and (e) lossless transforms.

Solve only one of the parts in Problems 5.17–5.19.

5.17*   Design a digital lowpass filter to satisfy the following Touch-Tone telephone filter specifications: maximum in-band gain = 0 dB, $f_p = 941$ Hz, and $f_s = 20f_p$. Find its $H(z)$ using the bilinear transform. Draw its block diagram realization as the cascade of first- and second-order blocks. Realize this digital lowpass filter as: (a) 4th order Butterworth; (b) 4th order Chebyshev with in-band ripple of 0.5 dB, (c) 1 dB, (d) 2 dB, (e) 3 dB; (f) 3rd order elliptic with in-band ripple of 0.28 dB and stopband rejection of 40 dB, (g) 60 dB; (h) 4th order Bessel.

5.18*   Design a digital lowpass filter to satisfy the following voice filter specifications: maximum in-band gain = 0 dB, $f_p = 3200$ Hz, and $f_s = 15f_p$. Find its $H(z)$ using the bilinear transform. Draw its block diagram realization as the cascade of first- and second-order blocks. Realize this digital lowpass filter as: (a) 4th order Butterworth; (b) 4th order Chebyshev with in-band ripple of 0.5 dB, (c) 1 dB, (d) 2 dB, (e) 3 dB; (f) 3rd order elliptic with in-band ripple of 0.28 dB and stopband rejection of 40 dB, (g) 60 dB; (h) 4th order Bessel.

5.19*   Design a digital lowpass filter to satisfy the following data modem noise filter specifications: maximum in-band gain = 0 dB, $f_p = 9600$ Hz, and $f_s = 10f_p$. Find its $H(z)$ using the bilinear transform. Draw its block diagram realization as the cascade of first- and second-order blocks. Realize this digital lowpass filter as: (a) 4th order Butterworth; (b) 4th order Chebyshev with in-band ripple of 0.5 dB, (c) 1 dB, (d) 2 dB, (e) 3 dB; (f) 3rd order elliptic with in-band ripple of 0.28 dB and stopband rejection of 40 dB, (g) 60 dB; (h) 4th order Bessel.

5.20   Given an analog differentiator $H(s) = s$. (a) Design a digital integrator using the impulse-, step-, and ramp-invariant transforms. (b) Draw their pole-zero patterns and frequency responses. (c) Draw their impulse responses.

5.21   Given an analog lowpass filter $H(s) = 1/(s + 1)$ and a sampling rate of 100 rad/sec. Find the digital lowpass filter $H(z)$ using the: (a) impulse-invariant, (b) dc-adjusted impulse-invariant, (c) step-invariant transform.

5.22   Given an analog lowpass filter $H(s) = 1/(s^2 + \sqrt{2}s + 1)$ and a sampling rate of 100 rad/sec. Find the digital lowpass filter $H(z)$ using the: (a) impulse-invariant, (b) dc-adjusted impulse-invariant, (c) step-invariant transforms.

5.23   Consider an analog lowpass filter having $H(s) = (s+1)/[\sqrt{2}s^2 + s + 1]$ and a sampling rate of 100 rad/sec. Find the digital lowpass filter $H(z)$ using the: (a) impulse-invariant, (b) dc-adjusted impulse-invariant, and (c) step-invariant transforms.

5.24   Given an analog integrator $H(s) = s$ and a variable sampling rate $1/T$ Hz. Plot the root locus of the poles of the digital integrator $H(z)$ using the: (a) impulse-invariant, (b) dc-adjusted impulse-invariant, and (c) step-invariant transforms.

5.25   Given an analog lowpass filter $H(s) = 1/(s+1)$ and a variable sampling rate $1/T$ Hz. Plot the root locus of the poles of digital lowpass filter $H(z)$ using the: (a) impulse-invariant, (b) dc-adjusted impulse-invariant, and (c) step-invariant transforms.

5.26   Given an analog lowpass filter $H(s) = 1/(s^2 + \sqrt{2}s + 1)$ and a variable sampling rate $1/T$ Hz. Plot the root locus of the poles of digital lowpass filter $H(z)$ using the: (a) impulse-invariant, (b) dc-adjusted impulse-invariant, and (c) step-invariant transforms.

5.27   Given an analog lowpass filter $H(s) = (s+1)/[\sqrt{2}s^2 + s + 1]$ and a variable sampling rate $1/T$ Hz. Plot the root locus of the poles of digital lowpass filter $H(z)$ using the: (a) impulse-invariant, (b) dc-adjusted impulse-invariant, and (c) step-invariant transforms.

Solve only one of the parts in Problems 5.28–5.30.

5.28*   Design a digital lowpass filter to satisfy the following touch-tone telephone filter specifications: maximum in-band gain = 0 dB, $f_p = 941$ Hz, and $f_s = 20f_p$. Find its $H(z)$ using the impulse-invariant transform. Draw its block diagram realization as the cascade of first- and second-order blocks. Realize this digital lowpass filter as: (a) 4th order Butterworth; (b) 4th order Chebyshev with in-band ripple of 0.5 dB, (c) 1 dB, (d) 2 dB, (e) 3 dB; (f) 3rd order elliptic with in-band ripple of 0.28 dB and stopband rejection of 40 dB, (g) 60 dB; (h) 4th order Bessel.

5.29*   Design a digital lowpass filter to satisfy the following voice filter specifications: maximum in-band gain = 0 dB, $f_p = 3200$ Hz, and $f_s = 15f_p$. Find its $H(z)$ using the dc-adjusted impulse-invariant transform. Draw its block diagram realization as the cascade of first- and second-order blocks. Realize this digital lowpass filter using: (a) 4th order Butterworth; (b) 4th order Chebyshev with in-band ripple of 0.5 dB, (c) 1 dB, (d) 2 dB, (e) 3 dB; (f) 3rd order elliptic with in-band ripple of 0.28 dB and stopband rejection of 40 dB, (g) 60 dB; (h) 4th order Bessel.

5.30*   Design a digital lowpass filter to satisfy the following data modem noise filter specifications: maximum in-band gain = 0 dB, $f_p = 9600$ Hz, and $f_s = 10f_p$. Find its $H(z)$ using the step-invariant transform. Draw its block diagram realization as the cascade of first- and second-order blocks. Realize this digital lowpass filter using: (a) 4th order Butterworth; (b) 4th order Chebyshev with in-band ripple of 0.5 dB, (c) 1 dB, (d) 2 dB, (e) 3 dB; (f) 3rd order elliptic with in-band ripple of 0.28 dB and stopband rejection of 40 dB, (g) 60 dB; (h) 4th order Bessel.

5.31   Suppose $H_1(z) = 0.457z/(z - 0.523)$, $H_2(z) = 0.3534(z+1)/(z - 0.2932)$, $H_3(z) = 0.8415/(z - 0.1585)$ where $H_i(z) = Y_i(z)/X_i(z)$. Write the software equations for implement the lowpass $H_i(z)$.

5.32   Write the software equations to implement a 4th order digital lowpass $H(z)$

$$H(z) = \frac{Y(z)}{X(z)} = \frac{1 + a_1 z^{-1} + a_2 z^{-2} + a_3 z^{-3} + a_4 z^{-4}}{1 + b_1 z^{-1} + b_2 z^{-2} + b_3 z^{-3} + b_4 z^{-4}}$$

The $H(z)$ filter coefficients $(a_0, a_1, a_2, a_3, a_4)$ and (b_1, b_2, b_3, b_4) equal:
(a) Bessel (0.0947, 0.0379, 0.568, 0.379, 0.0947) and (0.228, 0.269, 0.0136, 0.00523),
(b) Butterworth (0.469, 1.86, 2.79, 1.86, 0.466) and (−0.782, 0.680, −0.183, 0.0301),
(c) 0.5 dB Chebyshev (0.0308, 0.123, 0.185, 0.123, 0.0308) and (−1.38, 1.48, −0.802, 0.229),
(d) 2 dB Chebyshev (0.0187, 0.0749, 0.112, 0.0747, 0.0187) and (−1.81, 2.06, −1.28, 0.411),
(e) 3 dB Chebyshev (0.0156, 0.0622, 0.0934, 0.0622, 0.0156) and (−1.94, 2.26, −1.46, 0.485),
(f) 0.28 dB elliptic (0.124, 0.307, 0.307, 1, 0) and (0.612, −0.622, 0.147, 0).

5.33   Design the digital equivalent of the delta modulation system shown in Fig. P5.33. (a) Write the digital transfer function for an integrator $1/s$ using a BEDI. (b) Write the set of equations that describe the digital system. (c) Implement this system in software using Fortran type statements. Draw a flow chart to describe the program.

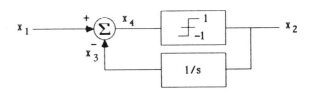

Fig. P5.33

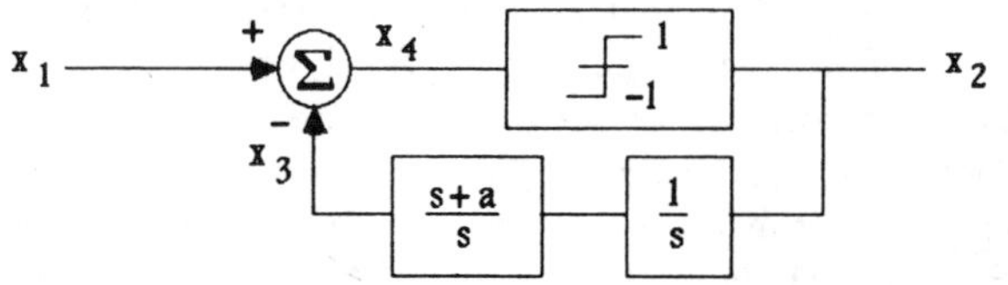

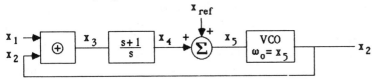

Fig. P5.34

5.34 Design the digital equivalent of the delta modulation system shown in Fig. P5.34. (a) Write the digital transfer function for the $(s + a)/s^2$ feedback block using a bilinear transform. (b) Write the set of equations that describe the digital system. (c) Implement this system in software using Fortran type statements. Draw a flow chart to describe the program.

5.35 A phase-locked loop (PLL) model is shown in Fig. P5.35. The output x_3 of an exclusive-or phase detector is lowpass filtered. The LPF output x_4 is summed with x_{ref} to form x_5. This signal drives a voltage-controlled oscillator having unity gain. The PLL is to be digitized using a sampling frequency $f_s = 2$ Hz. (a) Digitally simulate the system using the procedure described in Chap. 5.17. (b) Write the set of equations that describe the digital system. (c) Implement this system in software using Fortran type statements. Draw a flow chart to describe the program.

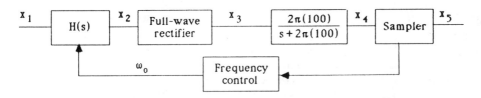

Fig. P5.35

5.36 The block diagram of a constant bandwidth B, variable center frequency ω_o bandpass filter (BPF) having $H(s) = Bs/(s^2 + Bs + \omega_o^2)$ is shown in Fig. P5.36. $B = 200$ Hz and $f_o = 1200$ Hz to 2400 Hz. This system determines the average rectified signal level in 200 Hz bands over the f_o frequency range. The BPF is to be digitized using a sampling frequency $f_s = 4800$ Hz. (a) Digitally simulate the system using the procedure described in Chap. 5.18. (b) Write the set of equations that describe the digital system. (c) Implement this system in software using Fortran type statements. Draw a flow chart to describe the program.

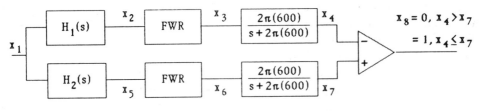

Fig. P5.36

(additional block diagram)

$x_8 = 0, \ x_4 > x_7$

$= 1, \ x_4 \leq x_7$

$B = 2\pi(600 \text{ Hz}), \quad \omega_1 = 2\pi(1200 \text{ Hz}) \quad H_i(s) = \dfrac{Bs}{s^2 + Bs + \omega_i^2}$

$\omega_2 = 2\pi(2400 \text{ Hz})$

Fig. P5.37

5.37 A frequency-shift keying (FSK) modem decodes a 1200 Hz signal as a "0" and 2400 Hz as a "1". The block diagram of the modem includes bandpass filters, full-wave rectifiers, and lowpass filters as shown in Fig. P5.37. The modem is to be digitized using a sampling frequency $f_s = 9600$ Hz. (a) Digitally simulate the system using the procedure described in Chap. 5.17. (b) Write the set of equations that describe the digital system. (c) Implement this system in software using Fortran type statements. Draw a flow chart to describe the program.

REFERENCES

1. Graham, J.S. and C.S. Lindquist, "A study of optimum discrete integrators in digital filter design," Proc. IEEE Intl. Symp. Circuits and Systems, pp. 1332–1335, 1983.
   Graham, J.S., "Discrete integrator transformations for digital filter synthesis," M.S.E.E. Directed Research, Calif. State Univ., Long Beach, 94 p., Oct. 1984.
2. El-Masry, E.I. and A.A. Sakla, "Low-sensitivity digital ladder filters," Rec. 13th Asilomar Conf. Circuits, Systems, and Computers, pp. 273–278, 1979.
3. Bruton, L.T., "Low-sensitivity digital ladder filters," IEEE Trans. Circuits and Systems, vol. CAS-22, pp. 168–176, March 1975.
   —, "Topological equivalence of inductorless ladder structures using integrators," IEEE Trans. Circuit Theory, vol. CT-20, pp. 434–437, July 1973.
4. Lindquist, C.S., *Active Network Design with Signal Filtering Applications*, Chaps. 4–5, Steward and Sons, CA, 1977.
5. Chiang, E., "Digital filter design using analog filter nomographs and digital transformations," M.S.E.E. Directed Research, Calif. State Univ., Long Beach, 80 p., Dec. 1983.
6. Kaiser, J.F. and F.F. Kuo, *System Analysis by Digital Computer*, Chap. 7 (Digital Filters), Wiley, NY, 1966.
7. Fowler, M.E., "A new numerical method for simulation," Simulation Magazine, p. 324, May 1965.
8. Antoniou, A., "Invariant-sinusoid approximation method for recursive digital filters," Electronics Letters, vol. 9, pp. 498–500, Oct. 1973.
9. Stearns, S.D., *Digital Signal Analysis*, Chap. 11, Hayden, NJ, 1975.
   Smith, J.M., *Mathematical Modeling and Digital Simulation for Engineers and Scientists*, Chap. 4, Wiley-Interscience, NY, 1977.
10. Precision LC Filters, Catalog 22F, 35 p., Allen Avionics, Inc., Mineola, NY, 1987. (These are analog filter responses. In Problems 5.2 and 5.3, they are treated as digital filter responses.)

6 FREQUENCY TRANSFORMATIONS

And, behold, thou shalt conceive in thy womb ...
and bring forth a son, and shalt call his name Jesus.
He shall be great, and shall be called the Son of the Highest;
And he shall reign over the house of Jacob for ever;
and of his kingdom there shall be no end.

Luke 1:31–33

A variety of classical, optimum, and adaptive lowpass filters were investigated in Chap. 4. A collection of *s-to-z* transforms were presented in Chap. 5 which convert analog filters into digital filters. In this chapter, frequency transformations will be discussed which allow highpass, bandpass, bandstop, and allpass digital filters to be derived directly from lowpass analog filters. This important topic will be explored in detail for it will allow all the lowpass filter information in previous chapters to be used for designing these other filter types. The relations between frequency and time domain will be carefully investigated. In general, these relationships are easily expressed mathematically but they require computer evaluation. Other important but less known transforms will be introduced for use in certain applications.

Filters are classified by the form of their magnitude characteristics $|H(j\omega)|$. The filter classifications are:

1. *Lowpass filters*: Pass low frequencies and reject (attenuate) high frequencies.
2. *Highpass filters*: Pass high frequencies and reject (attenuate) low frequencies.
3. *Bandpass filters*: Pass a prescribed band (range) of frequencies and reject all other frequencies.
4. *Bandstop (band-reject, band-elimination, or notch) filters*: Reject a prescribed band of frequencies and pass all other frequencies.
5. *Allpass filters*: Pass all frequencies equally but modify their phases.

The ideal filter magnitude characteristics are shown in Fig. 6.0.1 with some typical time domain responses. These classifications are very useful in generically describing the type of filter being discussed. Many examples of these filters will be presented later in the chapter.

A variety of transformations are used in filter design. Some are summarized in Fig. 6.0.2. The classical analog-to-analog (A-A) transforms $p = p(s)$ convert analog lowpass filters into the other analog filter types (i.e., highpass, bandpass, bandstop, allpass). These transforms are fully discussed in Ref. 1. The analog-to-digital (A-D) discrete transforms $p = g(z)$ of Chap. 5.2 convert analog filters into digital filters. These transforms maintain the filter type. Constantinides introduced digital-to-digital

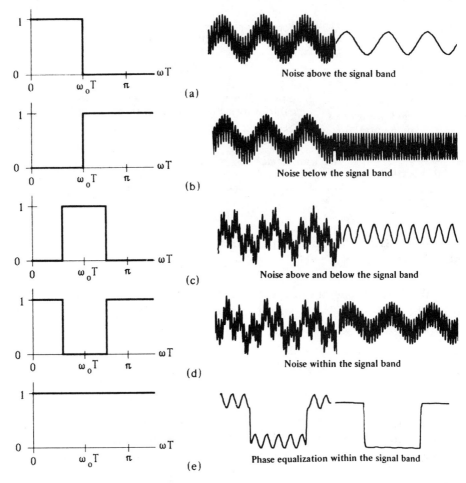

Fig. 6.0.1 Magnitude characteristics and time domain responses of (a) lowpass, (b) highpass, (c) bandpass, (d) bandstop, and (e) allpass filters.

(D-D) allpass transforms that converted digital lowpass filters into the other digital filter types. We use a new method that directly transforms analog lowpass filters into the other digital filter types. Our direct method combines the A-A transforms of Ref. 1 and the A-D transforms of Chap. 5.2 into a single equivalent transform as shown in Fig. 6.0.2. We will now briefly illustrate these transforms.

The A-A transforms are listed in Table 6.0.1.[1] Constantinides D-D transforms are also listed.[2] They have the general form

$$q(z) = \prod_{k=1}^{n} \frac{z^{-1} - \alpha_k}{1 - \alpha_k^* z^{-1}} \tag{6.0.1}$$

where $|\alpha_k| < 1$ for stability. Choosing $n = 1$ gives lowpass and highpass transforms while $n = 2$ gives bandpass and bandstop transforms. Larger n gives more general transforms.

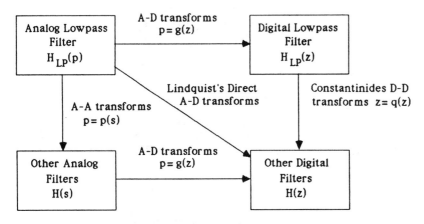

Fig. 6.0.2 Digital filter highpass, bandpass, bandstop, and allpass transforms.

To illustrate the use of these D-D transforms, let us obtain a digital highpass filter from the first-order Butterworth analog lowpass filter having the normalized transfer function as (see Table 4.2.3)

$$H_{LP}(p_n) = \frac{1}{1 + p_n} \qquad \text{where} \qquad p_n = p/v_p \tag{6.0.2}$$

This corresponds to the top-left block in Fig. 6.0.2. The corner frequency ω_p of both the digital lowpass and highpass filter is specified to be 1 radian/second. Using a sampling frequency ω_s of 100 radians/second, the normalized digital frequency βT equals 0.02π rad or $3.6°$. Selecting the bilinear transform from Table 5.11.1 as the A-D transform

$$p_n = \frac{1}{0.03143}\left(\frac{1 - z^{-1}}{1 + z^{-1}}\right) \tag{6.0.3}$$

and substituting p_n into Eq. 6.0.2 gives the digital lowpass transfer function as (cf. Eq. 5.11.7)

$$H_{LP}(z) = \frac{0.03047(1 + z^{-1})}{1 - 0.9391z^{-1}} \tag{6.0.4}$$

This is the top-right block in Fig. 6.0.2. Since the desired digital highpass filter has the same normalized digital frequency $\omega_p T$, the α value from Table 6.0.1 equals

$$\alpha = \frac{\cos[(3.6° - 3.6°)/2]}{\cos[(3.6° + 3.6°)/2]} = 1.002 \tag{6.0.5}$$

Then using the Constantinides lowpass-highpass D-D transform in Table 6.0.1, we obtain

$$H_{HP}(z) = H_{LP}(q(z))\Big|_{q(z) = -\frac{z^{-1} - 1.002}{1 - 1.002z^{-1}}} = \frac{0.9695(1 - z^{-1})}{1 - 0.9391z^{-1}} \tag{6.0.6}$$

Table 6.0.1 Some analog-to-analog and digital-to-digital transformations.[1,2]

Transform	Analog-Analog	Digital-Digital (Constantinides)
LP–LP	$p = s_n$	$q(z) = \dfrac{z^{-1} - \alpha}{1 - \alpha z^{-1}}$ $\alpha = \dfrac{\sin[(\beta - \omega_p)T/2]}{\sin[(\beta + \omega_p)T/2]}$
LP–HP	$p = \dfrac{1}{s_n}$	$q(z) = -\dfrac{z^{-1} - \alpha}{1 - \alpha z^{-1}}$ $\alpha = \dfrac{\cos[(\beta - \omega_p)T/2]}{\cos[(\beta + \omega_p)T/2]}$
LP–BP	$p = Q\left(\dfrac{1}{s_n} + s_n\right)$	$q(z) = -\dfrac{z^{-2} - \dfrac{2\alpha k}{k+1}z^{-1} + \dfrac{k-1}{k+1}}{1 - \dfrac{2\alpha k}{k+1}z^{-1} + \dfrac{k-1}{k+1}z^{-2}}$ $\alpha = \dfrac{\cos[(\omega_U + \omega_L)T/2]}{\cos[(\omega_U - \omega_L)T/2]}$ $k = \cot[(\omega_U - \omega_L)T/2]\tan(\beta T/2)$
LP–BS	$p = \dfrac{1}{Q\left(\dfrac{1}{s_n} + s_n\right)}$	$q(z) = \dfrac{z^{-2} - \dfrac{2\alpha}{1+k}z^{-1} + \dfrac{1-k}{1+k}}{1 - \dfrac{2\alpha}{1+k}z^{-1} + \dfrac{1-k}{1+k}z^{-2}}$ $\alpha = \dfrac{\cos[(\omega_U + \omega_L)T/2]}{\cos[(\omega_U - \omega_L)T/2]}$ $k = \tan[(\omega_U - \omega_L)T/2]\tan(\beta T/2)$

ω_p = band-edge frequency of digital lowpass and highpass filters.
ω_U, ω_L = upper and lower band-edge frequencies of digital bandpass and bandstop filters.
β = bandwidth of transformed digital lowpass filter.

This method is time consuming because we must transform the analog lowpass filter to a digital lowpass filter before obtaining the desired digital highpass filter through Constantinides' digital allpass transformations.

Using our direct D-D transform, we convert the analog lowpass filter into a digital highpass filter directly. Selecting the bilinear discrete transform and using our direct D-D transform of Chap. 6.1 gives

$$H_{HP}(z) = H_{LP}(p_n)\bigg|_{p_n = 0.03143\left(\frac{1 + z^{-1}}{1 - z^{-1}}\right)} = \frac{0.9695(1 - z^{-1})}{1 - 0.9391 z^{-1}} \qquad (6.0.7)$$

Eq. 6.0.6 agrees with Eq. 6.0.7. This example shows that the major advantage of our method over the Constantinides' method is that it saves time. It is unnecessary to transform the analog lowpass filter into a digital digital filter and then convert it into a digital bandpass, bandstop, or allpass filter.

6.1 LOWPASS–TO–HIGHPASS TRANSFORMATIONS [3]

The *analog lowpass-to-analog highpass transformation (p-to-s)* converts analog lowpass filter transfer functions into analog highpass filter transfer functions as

$$H_{HP}(s) = H_{LP}(p)\Big|_{p = \frac{\omega_o}{s}} \tag{6.1.1}$$

In analog highpass filters, p is the analog lowpass frequency variable and s is the analog highpass frequency variable. ω_o is the *corner frequency* or *stopband bandwidth* of the analog highpass filter. If the lowpass filter has gain $H_{LP}(p)$, then the corresponding highpass filter has gain $H_{LP}(s)$ where p is replaced by $\omega_o/s = 1/s_n$. Thus, $H_{HP}(s)$ is easily determined from $H_{LP}(p)$.

Expressing the transformation of Eq. 6.1.1 as

$$s = |s|e^{j \arg s}, \qquad p = |p|e^{j \arg p} = \frac{\omega_o}{|s|}e^{-j \arg s} \tag{6.1.2}$$

then

$$|p| = \frac{\omega_o}{|s|}, \qquad \arg s = -\arg p \tag{6.1.3}$$

For an analog lowpass filter having any pole-zero distribution in the p-plane, the pole-zero distribution of the analog highpass filter is obtained by simply *inverting* the p-plane around ω_o. Thus, analog lowpass poles or zeros at $p = 0, \omega_n e^{j\theta}$, and ∞ become analog highpass poles or zeros at $s = \infty, \omega_o e^{-j\theta}/\omega_n$, and 0. Thus, the critical frequencies are located at reciprocal distances from the origin. Their angles are negatives of each other. Denoting the complex frequency variables p and s as

$$p = u + jv, \qquad s = \sigma + j\omega \tag{6.1.4}$$

then in ac steady-state, $p = jv$ and $s = j\omega$. Since $j\omega = -j/v$, the frequency axis is inverted such that $p = j0, j1$, and $j\infty$ map into $s = -j\infty, -j\omega_o$, and $j0$, respectively.

6.1.1 HIGHPASS FILTER MAGNITUDE, PHASE, AND DELAY RESPONSES

Magnitude, phase, and delay responses are easily determined from Eq. 6.1.1. The magnitude responses of analog highpass filters are related to those of analog lowpass filters as

$$|H_{HP}(j\omega/\omega_o)| = |H_{LP}(-j/\omega)| = |H_{LP}(j/\omega)| \tag{6.1.5}$$

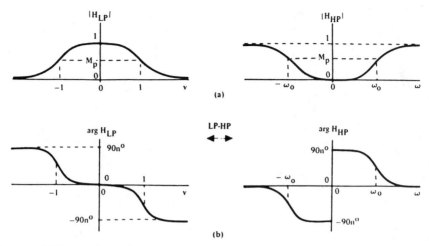

Fig. 6.1.1 Effect of analog lowpass-to-analog highpass transformation upon (a) magnitude and (b) phase responses of filters.

(recalling $|H|$ is usually an even function). The lowpass filter behavior at low frequencies becomes the exact highpass filter behavior at high frequencies and vice versa as shown in Fig. 6.1.1a. This figure vividly shows the process of frequency inversion using the lowpass-to-highpass transformation.

The phase responses of analog highpass filters can also be easily expressed as

$$\arg H_{HP}(j\omega/\omega_o) = \arg H_{LP}(-j/\omega) = -\arg H_{LP}(j/\omega) \qquad (6.1.6)$$

(recalling $\arg H$ is usually an odd function). Thus, the highpass filter phase is obtained by simply taking the negative of the lowpass filter phase after it is inverted about $\omega = 1$ and translated to $\omega = \omega_o$. This is illustrated in Fig. 6.1.1b.

The delay responses (phase derivatives) of analog highpass filters can also be easily written as

$$\tau_{HP}(j\omega) = (1/\omega^2)\tau_{LP}(j/\omega) \qquad (6.1.7)$$

Thus, to obtain the delay of the highpass filter, the lowpass filter delay characteristic is inverted about $\omega = 1$ and multiplied by $1/\omega^2$. In general, analog highpass and lowpass filters have quite different delay characteristics.

EXAMPLE 6.1.1 Determine the magnitude and phase responses of Butterworth analog highpass filters from the magnitude and phase responses of unity-bandwidth Butterworth analog lowpass filters shown in Fig. 6.1.2a.

SOLUTION The magnitude characteristic H_n of the analog highpass filter is obtained by rotating the magnitude characteristic H_n of the analog lowpass filter about $\omega = 1$. To obtain the phase characteristic $\arg H_n$, we rotate the phase characteristic $\arg H_n$ of the analog lowpass filter about $\omega = 1$ and change its sign. These characteristics are shown in Fig. 6.1.2b. ∎

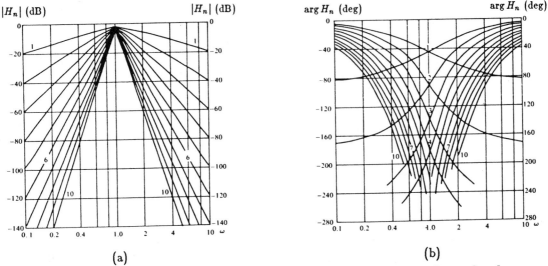

Fig. 6.1.2 (a) Magnitude and (b) phase responses of Butterworth analog lowpass and highpass filters. (Reprinted from EDN, March 15, 1971, ©1988 CAHNERS PUBLISHING COMPANY, a Division of Reed Publishing USA.)

6.1.2 HIGHPASS FILTER IMPULSE AND STEP RESPONSES

The step and impulse responses of analog highpass filters can be expressed in terms of their analogous lowpass filter responses. If the analog lowpass filter has step response $r_{LP}(t)$ and impulse response $h_{LP}(t)$, then it can be shown that the analogous highpass filter responses equals[3]

$$h_{HP}(t) = \mathcal{L}^{-1}\left[H_{LP}\left(\frac{1}{s}\right)\right] = \frac{-1}{t}\int_0^\infty \sqrt{t\tau}J_1(2\sqrt{t\tau})h_{LP}(\tau)d\tau \; U_{-1}(t) + r_{LP}(\infty) \, U_0(t)$$

$$r_{HP}(t) = \int_{-\infty}^t h_{HP}(\tau)d\tau = \int_0^\infty J_0(2\sqrt{t\tau})h_{LP}(\tau)d\tau \; U_{-1}(t)$$

$$(6.1.8)$$

These are very useful results. They show that the step (or impulse) response of a highpass filter is the integral of the product of the lowpass filter impulse response times a zero-order Bessel function of the first kind J_0 (or J_0' plus a U_0 term).

Although h_{LP} may be a monotonic response, J_0 or J_0' can cause h_{HP} or r_{HP} to be oscillatory since J_0 and J_0' are themselves oscillatory. This is illustrated in Fig. 6.1.3a by the step response of a Butterworth highpass filter. It exhibits greater oscillatory behavior than the step response of the Butterworth lowpass filter in Fig. 4.2.1c from which it was derived.

Notice the second term in $h(t)$ given by Eq. 6.1.8a is an impulse of area $r_{LP}(\infty)$ which occurs at $t = 0$. This impulse allows r_{HP} to make an instantaneous transition from zero to $r_{LP}(\infty)$ at time $t = 0$. Although these results are easily written mathematically and are useful conceptually, they require computer evaluation.

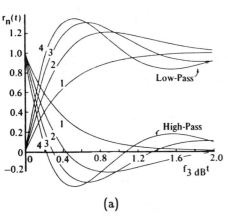

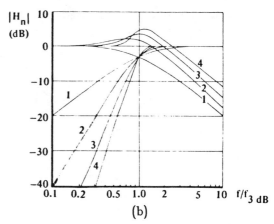

(a)   (b)

Fig. 6.1.3   (a) Step and (b) magnitude responses of Butterworth analog highpass filters and their complementary analog lowpass filters. (From J.R. Ashley, "Butterworth filters as loudspeaker frequency-dividing networks," Proc. IEEE, vol. 58, pp. 959–960, June, ©1970, IEEE.)

6.1.3 DIRECT ANALOG LOWPASS–TO–DIGITAL HIGHPASS TRANSFORMATIONS [4]

A number of analog-to-digital (p-to-z) transforms were derived in Chap. 5.2 where

$$H(z) = H(p)\bigg|_{p \,=\, g(z) \,=\, \dfrac{1}{H_{dmn}}} \qquad (6.1.9)$$

They were summarized as discrete integrators in Tables 5.2.1 and 5.2.2. Their transfer functions $H_{dmn} = 1/p = 1/g(z)$ are given by Eq. 5.2.2. In digital highpass filters, p is the analog filter frequency variable and s is the digital highpass filter frequency variable where $z = e^{sT}$.

Combining the analog lowpass-to-analog highpass transformation of Eq. 6.1.1 with the analog-to-digital transform of Eq. 6.1.9 gives the direct *analog lowpass-to-digital highpass transformation* as

$$H_{HP}(z) = H_{HP}(p_{HP})\bigg|_{p_{HP} \,=\, \frac{1}{H_{dmn}(z)}} = H_{LP}(p_{LP})\bigg|_{p_{LP} \,=\, \frac{v_o}{p_{HP}}\big|_{p_{HP} \,=\, \frac{1}{H_{dmn}(z)}}} \qquad (6.1.10)$$

The analog lowpass frequency variable p_{LP} can be expressed as

$$p_{LP} = \frac{v_o}{p_{HP}}\bigg|_{p_{HP} \,=\, \frac{1}{H_{dmn}(z)}} = v_o H_{dmn}(z) \qquad (6.1.11)$$

Therefore analog lowpass filters can be converted directly into digital highpass filters using Eq. 6.1.10 and the desired H_{dmn} from Tables 5.2.1 and 5.2.2. Because of the frequency inversion, these tables can be viewed as tabulating analog lowpass-to-digital highpass transforms as well as discrete integrator transfer functions.

Since the H_{dmn} are all different, each one of them produces its own unique analog lowpass-to-digital highpass transformation. Some of the major discrete transforms H_{dmn} from Tables 5.2.1 and 5.2.2 and their frequency mappings of Eq. 5.11.1 are

$$
\begin{aligned}
H_{001}: &\quad p_{HP}T = 1 - z^{-1} &&\text{and}&& v_{HP}T = \sin(\omega_{HP}T) \\[4pt]
H_{111}: &\quad p_{HP}T = \frac{1 - z^{-1}}{z^{-1}} &&\text{and}&& v_{HP}T = \sin(\omega_{HP}T) \\[4pt]
H_{011}: &\quad \frac{p_{HP}T}{2} = \frac{1 - z^{-1}}{1 + z^{-1}} &&\text{and}&& v_{HP}T/2 = \tan(\omega_{HP}T/2) \\[4pt]
H_{MBDI}: &\quad \frac{p_{HP}T}{2} = \frac{1 - z^{-1}}{z^{-1}(1 + z^{-1})} &&\text{and}&& v_{HP}T/2 = \tan(\omega_{HP}T/2)\cos(\omega_{HP}T) \\[4pt]
H_{112}: &\quad 2p_{HP}T = \frac{1 - z^{-2}}{z^{-1}} &&\text{and}&& v_{HP}T = \sin(\omega_{HP}T) \\[4pt]
H_{LDI}: &\quad p_{HP}T = \frac{1 - z^{-1}}{z^{-0.5}} &&\text{and}&& v_{HP}T/2 = \sin(\omega_{HP}T/2) \\[4pt]
H_{ODI}: &\quad \frac{p_{HP}T}{2} = \frac{1 - z^{-0.5}}{z^{-0.25}} &&\text{and}&& v_{HP}T/4 = \sin(\omega_{HP}T/4)
\end{aligned}
$$

$$(6.1.12)$$

Direct analog lowpass-to-digital highpass transformations can be easily determined from these equations. For example, consider the bilinear transform H_{011} and its frequency mapping function given by Eqs. 5.5.1 and 5.5.3 where

$$
p_{HP} = \frac{1}{H_{011}(z)} = \frac{2}{T}\frac{1 - z^{-1}}{1 + z^{-1}} \quad\text{and}\quad \frac{v_{HP}T}{2} = \tan\left(\frac{\omega_{HP}T}{2}\right) \tag{6.1.13}
$$

Substituting Eq. 6.1.13 into Eq. 6.1.11 and simplifying the result gives

$$
p_{LP} = \left.\frac{v_o}{p_{HP}}\right|_{p_{HP} = \frac{1}{H_{dmn}(z)}} = \frac{v_oT}{2}\frac{1 + z^{-1}}{1 - z^{-1}} = C\left(\frac{1 + z^{-1}}{1 - z^{-1}}\right) \tag{6.1.14}
$$

where $C = v_oT/2 = bT/2 = \tan(BT/2)$. B and b are the digital and analog highpass filter stopband bandwidths, respectively. An infinite of analog lowpass-to-digital highpass transformations can be derived following this procedure. Several such transformations are listed in Table 6.1.1.

The frequency mapping characteristics of these highpass transformations can be determined from their root loci. The highpass root loci can be obtained from the lowpass root loci in Fig. 5.9.4. Since the analog lowpass-to-digital highpass transforms are reciprocals of the analog lowpass-to-digital lowpass transforms (i.e., $p_{HP} = 1/p_{LP} = H_{dmn}(z)$), the highpass root loci are identical to the lowpass root loci with only two notational changes. First the poles and zeros are interchanged because $p_{HP} = 1/p_{LP}$ and secondly therefore, the branch directions of the root loci are reversed. Thus we can make the following observations about the highpass transforms after modifying the root loci shown in Fig. 5.9.4.

Table 6.1.1 Some analog lowpass-to-digital highpass transformations.

H_{dmn}	Discrete transform	Transformation $p_{LP} = g(z)$
H_{001}	Backward Euler	$p_{LP} = C\left(\dfrac{1}{1-z^{-1}}\right)$ $$C = \sin(BT)$$
H_{111}	Forward Euler	$p_{LP} = C\left(\dfrac{z^{-1}}{1-z^{-1}}\right)$ $$C = \sin(BT)$$
H_{011}	Bilinear	$p_{LP} = C\left(\dfrac{1+z^{-1}}{1-z^{-1}}\right)$ $$C = \tan(BT/2)$$
H_{MBDI}	Modified Bilinear	$p_{LP} = C\left(\dfrac{z^{-1}(1+z^{-1})}{1-z^{-1}}\right)$ $$C = \tan(BT/2)\cos(BT)$$
H_{112}	Discrete	$p_{LP} = C\left(\dfrac{z^{-1}}{1-z^{-2}}\right)$ $$C = 2\sin(BT)$$
H_{LDI}	Lossless	$p_{LP} = C\left(\dfrac{z^{-0.5}}{1-z^{-1}}\right)$ $$C = 2\sin(BT/2)$$
H_{ODI}	Optimum	$p_{LP} = C\left(\dfrac{z^{-0.25}}{1-z^{-0.5}}\right)$ $$C = 2\sin(BT/4)$$

The H_{011} transform exactly preserves the frequency domain characteristics (with frequency compression) of the analog lowpass filter since H_{011} maps the entire jv-axis of the p-plane ($\infty > v \geq 0$) onto the unit circle of the z-plane ($0° < \omega T \leq 180°$). The H_{LDI} and H_{ODI} transforms exactly preserve the frequency domain characteristics over high-frequency ranges ($\infty > z \geq \sqrt{2}/C$ and $\infty > z \geq 2/C$, respectively) since they map the jv-axis near infinity onto the unit circle ($0° < \omega T \leq 180°$). The low-frequency characteristics of the analog filter cannot be obtained because the jv-axis below $v = \sqrt{2}/C$ or $2/C$ is not mapped onto the unit circle. The H_{001}, H_{111}, and H_{MBDI} transforms distort the frequency domain behavior since they do not map the jv-axis of the p-plane onto the unit circle at all. However because they map the jv-axis around $v = \infty$ to be tangent to the unit circle at $z = 1$, the high-frequency characteristics are somewhat maintained at sufficiently high sampling rates. Of these three transforms, H_{001} gives the least distortion since its root locus is closest the unit circle. In addition, H_{MBDI} maps the jv-axis around $v = 0$ to be tangent to the unit circle at $z = -1$ so H_{MBDI} also somewhat maintains the low-frequency filter characteristics. The H_{112} transform maps the jv-axis near infinity ($\infty > v \geq 2/C$) onto two arcs of the unit circle ($0° < \omega T \leq 90°$ and $180° > \omega T \geq 90°$). Therefore highpass (lowpass) characteristics become bandpass (bandstop) characteristics so that H_{112} is not useful as a analog lowpass-to-digital highpass filter transform.

The ac steady-state relation between the digital highpass frequency ω_{HP}, the analog highpass frequency v_{HP}, and the analog lowpass frequency v_{LP} is found by setting $z = e^{j\omega T}$, $p_{HP} = jv_{HP}$, and $p_{LP} = jv_{LP}$ in Eq. 6.1.14 as

$$v_{LP} = -\frac{v_o T/2}{v_{HP} T/2} = -\frac{C}{\tan(\omega T/2)} \qquad (6.1.15)$$

To interpret this frequency mapping, assume that the digital frequency ωT is small where $|\omega T| \ll 1$. Then $\tan(\omega T/2) \cong \omega T/2$ and Eq. 6.1.15 reduces to

$$v_{LP} \cong -\frac{C}{\omega T/2} = -\frac{B}{\omega} \qquad (6.1.16)$$

This relation is identical to the analog lowpass-to-analog highpass mapping of Eq. 6.1.4. This approximation shows that when the sampling frequency is sufficiently high so that the normalized digital frequency ωT is small, then the analog lowpass-to-digital highpass transformation using H_{011} behaves like the analog lowpass-to-analog highpass transformation shown in Fig. 6.1.1. All the H_{dmn} exhibit this same limiting behavior.

To demonstrate the usefulness of these transformations, we design a first-order Butterworth digital highpass filter by directly transforming a first-order Butterworth analog lowpass filter having $H_{LP}(p_n) = 1/(p_n+1)$. The corresponding first-order digital highpass filters have transfer functions which equal

$$H_{001}: \quad H_{HP}(z) = 1\Big/\Big[C\frac{1}{1-z^{-1}}+1\Big] = \frac{1-z^{-1}}{(1+C)-z^{-1}}$$

$$H_{111}: \quad H_{HP}(z) = 1\Big/\Big[C\frac{z^{-1}}{1-z^{-1}}+1\Big] = \frac{1-z^{-1}}{1-(1-C)z^{-1}}$$

$$H_{011}: \quad H_{HP}(z) = 1\Big/\Big[C\frac{1+z^{-1}}{1-z^{-1}}+1\Big] = \frac{1-z^{-1}}{(1+C)-(1-C)z^{-1}} \qquad (6.1.17)$$

$$H_{MBDI}: \quad H_{HP}(z) = 1\Big/\Big[C\frac{z^{-1}(1+z^{-1})}{1-z^{-1}}+1\Big] = \frac{1-z^{-1}}{1-(1-C)z^{-1}+Cz^{-2}}$$

$$H_{LDI}: \quad H_{HP}(z) = 1\Big/\Big[C\frac{z^{-0.5}}{1-z^{-1}}+1\Big] = \frac{1-z^{-1}}{1+Cz^{-0.5}-z^{-1}}$$

These are analogous to the first-order digital lowpass filters listed in Eq. 5.11.3. The digital highpass filter is specified to have a 3 dB stopband bandwidth $\omega_{HP}T = BT$ which maps into $v_{HP}T = bT$ for the analog highpass filter using Eq. 6.1.12. BT also determines C as shown in Table 6.1.1.

6.1.4 HIGHPASS FILTER ORDER

Highpass filters are specified in the frequency domain, the time domain, or in both domains simultaneously. Frequency domain specifications are especially convenient since the classical and optimum filters developed in Chap. 4.2 can be utilized. The nomographs of Chap. 4.2 are especially valuable in determining the required filter order.

Generally the frequency domain specifications of analog highpass filters have the form shown in Fig. 6.1.4a. The maximum passband ripple is M_p for frequencies greater than f_o. The minimum stopband rejections are $M_{r1}, M_{r2}, \ldots$ for $f_1, f_2, \ldots$. To utilize the design information of Chap. 4.2, this data is converted into equivalent data for an analog lowpass filter. The conversion consists of simply normalizing frequencies in the analog highpass filter by f_o. Then the reciprocal normalized frequency is formed which describes frequency in the analog lowpass filter. Finally nomographs are used to determine analog filter order as shown in the following example.

> **EXAMPLE 6.1.2** Consider two analog band-splitting filters to separate frequencies below 941 Hz and frequencies above 1209 Hz with gains as shown in Fig. 6.1.4. Determine the minimum order Butterworth, Chebyshev, and elliptic lowpass and highpass analog filters to obtain at least 40 dB of stopband rejection with no more than 3 dB in-band ripple.
>
> **SOLUTION**   The maximum frequency from the low group (941 Hz) and the minimum frequency from the high group (1209 Hz) form the band-edges of the passband and stopband. Normalizing these frequencies by one another as $\Omega_a = 1209/941 = 1.285$ shows the lowpass and the highpass analog filters have identical order. Entering this data $(M_p, M_r, \gamma, \Omega_a) = (3, 40, 4, 1.285)$ onto the nomographs of Chap. 4.2 shows that the minimum orders are: Butterworth $n = 19$, Chebyshev $n = 8$, and elliptic $n = 5$. ∎

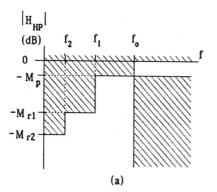

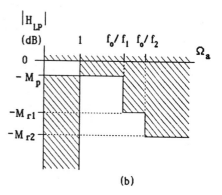

Fig. 6.1.4 (a) Analog highpass filter specification and its (b) equivalent analog lowpass filter specification.

Digital highpass filters have the magnitude response specifications shown in Fig. 6.1.5a. The frequencies involved equal:

$$\omega_p T = 2\pi\omega_p/\omega_s = \text{normalized passband frequency in radians}$$
$$= 360° f_p/f_s = \text{normalized passband frequency in degrees}$$
$$\omega_r T = 2\pi\omega_r/\omega_s = \text{normalized stopband frequency in radians}$$
$$= 360° f_r/f_s = \text{normalized stopband frequency in degrees} \qquad (6.1.18)$$
$$\Omega_d = \frac{\omega_p T}{\omega_r T} = \frac{360° f_p/f_s}{360° f_r/f_s} = \frac{\text{digital passband frequency}}{\text{digital stopband frequency}}$$

This digital highpass filter specification can be converted into its equivalent analog lowpass filter specification as shown in Fig. 6.1.5b. The normalized analog lowpass filter frequency $v_{LP}T$ depends upon: (1) the sampling frequency $\omega_s = 2\pi/T$, (2) the digital filter frequency ω_{HP}, and (3) the p_n-to-z transform of Table 6.1.1.

To compute the required digital highpass filter order, we use the normalized digital highpass frequency ratio Ω_d of Eq. 6.1.18 to find the equivalent analog lowpass frequency ratio Ω_a. Alternatively, we can also use the Ω_a-Ω_d mapping curves of Figs. 5.10.3 and 5.10.4. Then we transfer the $(M_p, M_r, \gamma, \Omega_a)$ data onto the analog filter nomograph and read off the order n. Using the normalized digital highpass frequencies $\omega_p T$ and $\omega_r T$, the equivalent normalized analog lowpass frequency ratio $\Omega_a = v_r T/v_p T$ equals

$$H_{001}, H_{111}, H_{112}: \quad \Omega_a = \frac{\sin(\omega_p T)}{\sin(\omega_r T)} = \frac{\sin(360° f_p/f_s)}{\sin(360° f_r/f_s)}$$

$$H_{011}: \quad \Omega_a = \frac{\tan(\omega_p T/2)}{\tan(\omega_r T/2)} = \frac{\tan(180° f_p/f_s)}{\tan(180° f_r/f_s)}$$

$$H_{MBDI}: \quad \Omega_a = \frac{\tan(\omega_p T/2)\cos(\omega_p T)}{\tan(\omega_r T/2)\cos(\omega_r T)} \longrightarrow \frac{\tan(180° f_p/f_s)}{\tan(180° f_r/f_s)} \qquad (6.1.19)$$

$$H_{LDI}: \quad \Omega_a = \frac{\sin(\omega_p T/2)}{\sin(\omega_r T/2)} = \frac{\sin(180° f_p/f_s)}{\sin(180° f_r/f_s)}$$

$$H_{ODI}: \quad \Omega_a = \frac{\sin(\omega_p T/4)}{\sin(\omega_r T/4)} = \frac{\sin(90° f_p/f_s)}{\sin(90° f_r/f_s)}$$

combining Eqs. 6.1.12 and 6.1.18. These ratios are the reciprocal of those listed in Table 5.10.1. The H_{LDI} and H_{ODI} frequency ratios reduce to the H_{001}, H_{111}, and H_{112} frequency ratios for small ωT values; likewise the H_{MBDI} frequency ratio reduces to the H_{011} frequency ratio. When $\omega_{HP}T$ is sufficiently small and the sampling rate $1/T$ is sufficiently high, Eq. 6.1.19 shows that $\Omega_d \to \Omega_a$ independent of which transform is used. Eqs. 6.1.18–6.1.19 allow digital highpass filter specifications to be easily converted into analog lowpass filter specifications.

EXAMPLE 6.1.3 Determine the order of a digital elliptic highpass filter having 1.25 dB in-band ripple and 40 dB minimum stopband attenuation. The ripple

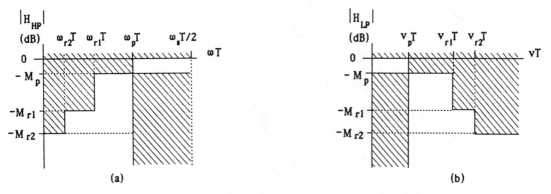

Fig. 6.1.5 (a) Digital highpass filter specification and its (b) equivalent analog lowpass filter specification.

frequency is 1 KHz and the stopband frequency is 370 Hz. Use sampling frequencies of 10 KHz and 100 KHz. Consider different transforms.

SOLUTION The normalized passband and stopband digital frequencies $\omega_{HP}T = 360° f_{HP}/f_s$ are $36°$ and $13.3°$ using a sampling frequency of 10 KHz. At 100 KHz sampling, the normalized passband and stopband digital frequencies are $3.6°$ and $1.33°$, respectively. Since different transforms may be used, the ratios given by Eq. 6.1.19 will be considered. Therefore at a 10 KHz sampling rate, the equivalent normalized analog lowpass frequency ratios Ω_a equal

$$H_{001}, H_{111} : \quad \Omega_a = \frac{\sin(36°)}{\sin(13.3°)} = 2.56 \quad \text{(or 2.70)}$$

$$H_{011} : \quad \Omega_a = \frac{\tan(36°/2)}{\tan(13.3°/2)} = 2.79 \quad \text{(or 2.71)}$$

$$H_{MBDI} : \quad \Omega_a = \frac{\tan(36°/2)\cos(36°)}{\tan(13.3°/2)\cos(13.3°)} = 2.31 \quad \text{(or 2.70)}$$
(6.1.20)

$$H_{LDI} : \quad \Omega_a = \frac{\sin(36°/2)}{\sin(13.3°/2)} = 2.69 \quad \text{(or 2.71)}$$

The ratios that result from sampling at 100 KHz are shown in parenthesis. Notice that the Ω_a's are almost equal regardless of which transform is used especially at high sampling rates. When $\Omega_a \cong 2.7$, then $(M_p, M_r, \gamma, \Omega_a) = (1.25, 40, 4.5, 2.7)$. We determine the order of the digital filter by using the elliptic filter nomograph of Chap. 4.2. The nomograph shows that $n \geq 3$ so a third-order elliptic filter with 1.25 dB in-band ripple meets this specification using any of these transforms. ■

EXAMPLE 6.1.4 Determine the orders of digital Butterworth, Chebyshev, and elliptic highpass filters needed to satisfy the following specifications: (1) sampling frequency $f_s = 20$ KHz, (2) passband ripple $M_p \leq 1$ dB for frequencies $f_p \geq 3.5$ KHz, (3) stopband rejection $M_r \geq 45$ dB for frequencies $f_r \geq 1.5$ KHz. The particular p_n-to-z transform to be used is not specified.

SOLUTION Since $\omega_p T = 360° f_p/f_s = 63°$ and $\omega_r T = 27°$, the equivalent analog lowpass filter has a stopband/passband frequency ratio from Eq. 6.1.19 of

$$H_{001}, H_{111}: \quad \Omega_a = \frac{\sin(63°)}{\sin(27°)} = 1.96$$

$$H_{011}: \quad \Omega_a = \frac{\tan(63°/2)}{\tan(27°/2)} = 2.55$$

$$H_{MBDI}: \quad \Omega_a = \frac{\tan(63°/2)\cos(63°)}{\tan(27°/2)\cos(27°)} = 1.30 \qquad (6.1.21)$$

$$H_{LDI}: \quad \Omega_a = \frac{\sin(63°/2)}{\sin(27°/2)} = 2.24$$

Entering $M_p = 1$ dB, $M_r = 45$ dB, and $\Omega_a = 2.55$ onto the filter nomographs of Chap. 4.2 yields: Butterworth $n \geq 7$, Chebyshev $n \geq 5$, elliptic $n \geq 4$. However when $\Omega_a = 1.96$, the filter orders increase to: Butterworth $n \geq 9$, Chebyshev $n \geq 5$, elliptic $n \geq 4$. Filter order is unreasonably high for $\Omega_a = 1.30$. It is instructive to determine the analog highpass filter frequencies analogous to the digital highpass frequencies in Fig. 6.1.5. These frequencies are

$$H_{001}, H_{111}: \quad v_p T = \sin(63°) = 51.1°, \qquad v_r T = \sin(27°) = 26.0°$$

$$H_{011}: \quad v_p T = 2\tan(63°/2) = 70.2°, \quad v_r T = 2\tan(27°/2) = 27.5°$$

$$H_{MBDI}: \quad v_p T = 2\tan(63°/2)\cos(63°) = 31.9°,$$

$$v_r T = 2\tan(27°/2)\cos(27°) = 24.5°$$

$$H_{LDI}: \quad v_p T = 2\sin(63°/2) = 62.2°, \quad v_r T = 2\sin(27°/2) = 26.8°$$

$$(6.1.22)$$

from Eq. 6.1.12. The 3 dB frequencies of the fifth-order Butterworth filters are $v_{1dB}T/v_{3dB}T = 0.8736$ from Chap. 4.2. ∎

6.1.5 HIGHPASS FILTER DESIGN PROCEDURE

The design procedure for digital highpass filters using direct transformations is straightforward. It consists of the following steps:

1a. Select a suitable analog filter type (e.g., Butterworth, Chebyshev, elliptic, etc.) and
1b. Choose a particular discrete transform from Tables 5.2.1 and 5.2.2 based upon the discussion of Chap. 5.9.
2. Determine the required analog lowpass filter order as described in Chap. 6.1.4.
3. Write the transfer function $H_{LP}(p_n)$ of the analog lowpass filter having unity bandwidth and the desired magnitude from the appropriate table of Chap. 4.2.
4. Compute the direct analog lowpass-to-digital highpass transformation from Table 6.1.1 or compute using Eq. 6.1.11.
5. Compute the transfer function $H_{HP}(z)$ of the digital highpass filter by substituting the p_n-to-z transformation found in Step 4 into $H_{LP}(p_n)$ found in Step 3.
6. Implement this digital highpass transfer function using one of the techniques discussed in Chap. 12.

Let us illustrate this procedure by designing a third-order Butterworth digital high-pass filter. Its 3 dB corner frequency f_p is specified to be 1 KHz with sampling frequencies f_s of 10 KHz and 100 KHz. *Steps 1-3* – From Chap. 4.2, the third-order Butterworth analog lowpass filter has a normalized transfer function

$$H_{LP}(p_n) = \frac{1}{p_n^3 + 2p_n^2 + 2p_n + 1} \tag{6.1.23}$$

where $p_n = p_{LP}T/v_{LP}T$. *Step 4* – The normalized digital passband frequency $\omega_{HP}T = 360° f_p/f_s = 36°$ for $f_s = 10$ KHz and $3.6°$ for $f_s = 100$ KHz. The equivalent analog highpass frequencies $v_{HP}T$ are found from Eq. 6.1.12. To reduce distortion, we use the H_{011} and H_{LDI} transforms for the wide bandwidth case and the H_{001}, H_{111}, H_{MBDI} transforms for the narrow bandwidth case (see Table 5.10.2). The H_{011} and H_{LDI} transforms require analog frequency constants of

$$\begin{aligned} H_{011} : \quad & C = v_{HP}T/2 = \tan(\omega_{HP}T/2) = \tan(18°) = 0.32492 \\ H_{LDI} : \quad & C = v_{HP}T = 2\sin(\omega_{HP}T/2) = 2\sin(18°) = 0.61803 \end{aligned} \tag{6.1.24}$$

for a digital passband of $36°$. Using H_{001}, H_{111}, and H_{MBDI} transforms for a passband of $3.6°$, the analog frequency constants equal

$$\begin{aligned} H_{001}, H_{111} : \quad & C = v_{HP}T = \sin(\omega_{HP}T) = \sin(3.6°) = 0.06279 \\ H_{MBDI} : \quad & C = v_{HP}T/2 = \tan(\omega_{HP}T/2)\cos(\omega_{HP}T) = \tan(18°)\cos(3.6°) = 0.03136 \end{aligned} \tag{6.1.25}$$

The p_n-to-z transforms are determined using Table 6.1.1 with the normalized frequency constants of Eqs. 6.1.24 and 6.1.25. This gives $p_n = p_{HP}T/v_{HP}T$ as

$$H_{011} : \quad p_n = 0.32492\left(\frac{1 + z^{-1}}{1 - z^{-1}}\right) \qquad H_{001} : \quad p_n = 0.062791\left(\frac{1}{1 - z^{-1}}\right)$$

$$H_{LDI} : \quad p_n = 0.61803\left(\frac{z^{-0.5}}{1 - z^{-1}}\right) \qquad H_{111} : \quad p_n = 0.062791\left(\frac{z^{-1}}{1 - z^{-1}}\right)$$

$$H_{MBDI} : \quad p_n = 0.031364\left(\frac{z^{-1}(1 + z^{-1})}{1 - z^{-1}}\right) \tag{6.1.26}$$

Step 5 – Now substitute each p_n-to-z mapping function of Eq. 6.1.26 into the normalized lowpass filter transfer function of Eq. 6.1.23. This gives the transfer functions $H_{HP}(z)$ of the third-order Butterworth digital highpass filter as

$$H_{011} : \quad H_{HP}(z) = \frac{0.52762(1 - 3z^{-1} + 3z^{-2} - z^{-3})}{1 - 1.76004z^{-1} + 1.18289z^{-2} - 0.27806z^{-3}}$$

$$H_{LDI} : \quad H_{HP}(z) = \frac{1 - 3z^{-1} + 3z^{-2} - z^{-3}}{1 + 1.23607z^{-0.5} - 2.23607z^{-1} - 2.23607z^{-1.5} + 2.23607z^{-2}}$$
$$+ 1.23607^{-2.5} - z^{-3}$$

$$\tag{6.1.27}$$

at a sampling frequency of 10 KHz (digital passband of $36°$) and

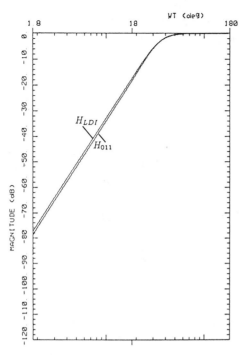
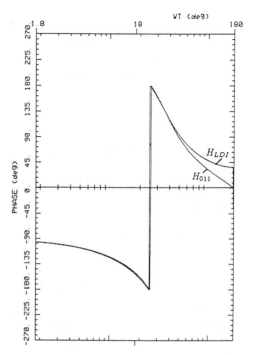

Fig. 6.1.6 Magnitude and phase responses of third-order Butterworth digital high-pass filter using H_{011} and H_{LDI} transforms (see Eq. 6.1.27). $f_{3dB} = 1$ KHz, $f_s = 10$ KHz, and $BT = 36°$. (Plotted by C. Powell. Based upon T. Wongterarit, "Design of digital filters from analog lowpass filters using frequency transformations," M.S.E.E. Thesis, Univ. of Miami, 116 p., May 1986.)

$$H_{001}: \quad H_{HP}(z) = \frac{0.88206(1 - 3z^{-1} + 3z^{-2} - z^{-3})}{1 - 2.87467z^{-1} + 2.75694z^{-2} - 0.88206z^{-3}}$$

$$H_{111}: \quad H_{HP}(z) = \frac{1 - 3z^{-1} + 3z^{-2} - z^{-3}}{1 - 2.87442z^{-1} + 2.75672z^{-2} - 0.88206z^{-3}} \qquad (6.1.28)$$

$$H_{MBDI}: \quad H_{HP}(z) = \frac{1 - 3z^{-1} + 3z^{-2} - z^{-3}}{1 - 2.93727z^{-1} + 2.93924z^{-2} - 1.06073z^{-3} + 0.06085z^{-4}}$$
$$- 0.001875z^{-5} + 0.00003085z^{-6}$$

at a sampling frequency of 100 KHz (digital passband of 3.6°). The magnitude and phase responses of Eqs. 6.1.27 and 6.1.28 are shown in Figs. 6.1.6 and 6.1.7, respectively. They have flat Butterworth magnitude responses. Their Nyquist frequency gains are unity and their dc gains are zero. Their 3 dB corner frequencies are about 1 KHz. All the transforms preserve the analog Butterworth filter responses fairly well.

Now let us design a third-order elliptic digital highpass filter with 1.25 dB in-band ripple and a minimum stop-band rejection of 40 dB. Its passband ripple frequency is 1 KHz and its stopband is less than 370 Hz. We will use the same sampling frequencies as were used for the third-order Butterworth filter. *Steps 1-3* – The normalized transfer function of this elliptic analog lowpass filter is[5]

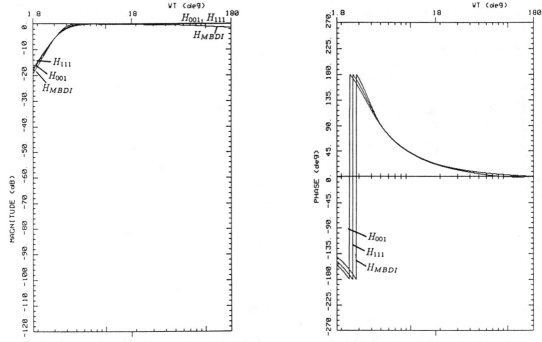

Fig. 6.1.7 Magnitude and phase responses of third-order Butterworth digital high-pass filter using H_{001}, H_{111}, and H_{MBDI} transforms (see Eq. 6.1.28). $f_{3dB} = 1$ KHz, $f_s = 100$ KHz, and $BT = 3.6°$. (Plotted by C. Powell. Based upon T. Wongterarit, loc. cit.)

$$H_{LP}(p_n) = \frac{K(p_n^2 + 2.6999^2)}{(p_n + 0.48069)[(p_n + 0.20841)^2 + 0.96178^2]}$$

$$= \frac{0.06386(p_n^2 + 7.2895)}{(p_n + 0.48069)(p_n^2 + 0.41682p_n + 0.96846)}$$

$$(6.1.29)$$

where $p_n = p_{LP}T/v_{LP}T$. *Step 4* – Since we use the same sampling frequencies (10 KHz and 100 KHz) and the same corner frequency (1 KHz) as before, the normalized digital passband filter frequencies $\omega_{HP}T$ and the analog lowpass frequencies $v_{LP}T$ are the same as those given by Eqs. 6.1.24 and 6.1.25. Therefore, we can use the p_n-to-z transforms given by Eq. 6.1.26. *Step 5* – Substituting these transforms into Eq. 6.1.29 gives the transfer functions of the third-order elliptic digital highpass filter as

$$H_{011}: \quad H_{HP}(z) = \frac{0.48470(1 - 2.94290z^{-1} + 2.94290z^{-2} - z^{-3})}{1 - 1.62025z^{-1} + 1.05194z^{-2} - 0.15005z^{-3}}$$

$$H_{LDI}: \quad H_{HP}(z) = \frac{1 - 2.94760z^{-1} + 2.94760z^{-2} - z^{-3}}{1 + 1.55172z^{-0.5} - 2.26359z^{-1} - 2.59635z^{-1.5} + 2.26359z^{-2}}$$
$$+ 1.55172z^{-2.5} - z^{-3}$$

$$(6.1.30)$$

for a 10 KHz sampling frequency (digital passband of 36°) and

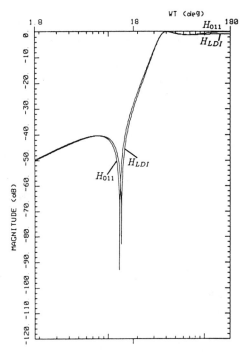
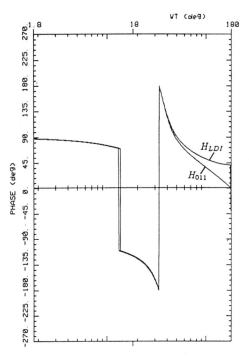

Fig. 6.1.8 Magnitude and phase responses of third-order elliptic digital highpass filter using H_{011} and H_{LDI} transforms (see Eq. 6.1.30). $f_{1.25dB} = 1$ KHz, $f_s = 10$ KHz, and $BT = 36°$. (Plotted by C. Powell. Based upon T. Wongterarit, loc. cit.)

$$H_{001}: \quad H_{HP}(z) = \frac{0.85826(1 - 2.99892z^{-1} + 2.99838z^{-2} - 0.99946z^{-3})}{1 - 2.85036z^{-1} + 2.70861z^{-2} - 0.85779z^{-3}}$$

$$H_{111}: \quad H_{HP}(z) = \frac{1 - 3z^{-1} + 3.000541z^{-2} - 1.000541z^{-3}}{1 - 2.84235z^{-1} + 2.69230z^{-2} - 0.84942z^{-3}} \qquad (6.1.31)$$

$$H_{MBDI}: \quad H_{HP}(z) = 1 - 3z^{-1} + 3.00013z^{-2} - 0.99987z^{-3} - 0.0001350z^{-4}$$
$$-0.0001350z^{-5}$$
$$\overline{ 1 - 2.92125z^{-1} + 2.92315z^{-2} - 1.07679z^{-3} + 0.07705z^{-4}}$$
$$-0.001698z^{-5} + 0.00006628z^{-6}$$

for a 100 KHz sampling frequency (digital passband of 3.6°). The magnitude and phase responses of Eqs. 6.1.30 and 6.1.31 are plotted in Figs. 6.1.8 and 6.1.9, respectively. The magnitude responses of Fig. 6.1.8 have equiripple in both the passband and stopband as required of elliptic filters. The passband ripple is 1.25 dB and the passband corner frequency is 1 KHz. The minimum stopband rejection is 40 dB below 370 Hz. The magnitude responses of Fig. 6.1.9 are not quite as good with H_{MBDI} being the poorest.

6.1.6 COMPLEMENTARY TRANSFORMATION

In general, the analog lowpass-to-digital highpass transformation preserves the magnitude characteristic of a filter at high sampling rates but modifies its step response.

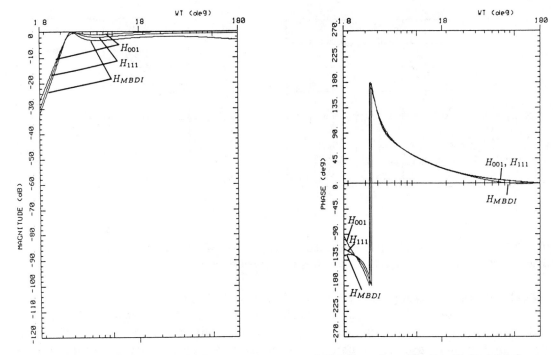

Fig. 6.1.9 Magnitude and phase responses of third-order elliptic digital highpass filter using H_{001}, H_{111}, and H_{MBDI} transforms (see Eq. 6.1.31). $f_{1.25dB} = 1$ KHz, $f_s = 100$ KHz, and $BT = 3.6°$. (Plotted by C. Powell. Based upon T. Wongterarit, loc. cit.)

There is a different transformation which preserves the step response but modifies the magnitude characteristic which is desirable in some applications. This transformation is called the complementary or low-transient lowpass-to-highpass transformation.[6]

The step response of a Butterworth lowpass filter is shown in Fig. 6.1.10a. To maintain the step response characteristics in terms of overshoot (undershoot), rise time (fall time), delay time (storage time), settling time, etc., the complementary highpass filter step and impulse responses must equal

$$r_{HP}(nT) = U_{-1}(nT) - r_{LP}(nT)$$
$$h_{HP}(nT) = U_0(nT) - h_{LP}(nT)$$

(6.1.32)

This condition forces the time domain responses to be compliments of one another. To determine the relation between the complementary filter transfer functions, take the z-transform of the impulse responses in Eq. 6.1.32. This yields the *complementary digital lowpass-to-digital highpass transformation* where

$$H_{HP}(z) = 1 - H_{LP}(z)$$

(6.1.33)

$H_{LP}(z)$ is magnitude normalized so that $H_{LP}(1) = 1$. This insures that the dc gain

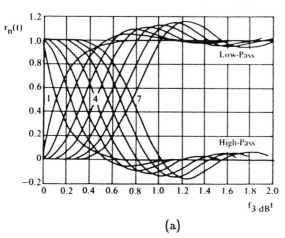
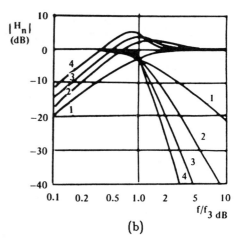

Fig. 6.1.10 (a) Step and (b) magnitude responses of Butterworth analog lowpass filters and their complementary analog highpass filters.

$H_{HP}(1)$ of the digital highpass filter equals zero. If the lowpass filter $H_{LP}(z)$ has n poles and m finite zeros, then Eq. 6.1.33 shows that $H_{HP}(z)$ has the same n poles but that the original zeros are modified and $(n - m)$ additional zeros are introduced. Thus, stable digital lowpass filters remain stable under this transformation, but minimum-phase filters may become nonminimum phase. All-pole digital lowpass filters convert into digital highpass filters having n finite zeros where one of these zeros is located at $z = 1$. Such highpass filters have low-frequency asymptotic roll-offs of only 20 dB/dec as shown in Fig. 6.1.10b. Performing the lowpass-to-highpass transformation on the filters in Fig. 6.1.10 produce the filters in Fig. 6.1.3. The -20 dB/dec roll-off is evident for the lowpass filter. Again we see that out-of-band rejection is not increased by increasing filter order and the only advantage of increasing the highpass filter order is to obtain the desired step response. Therefore the complementary or low-transient transformation is not often utilized due to the poor out-of-band rejection obtained. Nevertheless, it is a useful transform because it gives the ability to maintain time domain behavior.

EXAMPLE 6.1.5 A third-order Butterworth lowpass filter was discussed in Example 5.11.1. Find the highpass filter that has a complementary step response.
SOLUTION The lowpass filter transfer function equals

$$H_{LP}(z) = H_{LP}(p_{LP}) \Big|_{p_{LP} \, = \, \frac{1}{H_{011}(z)}} = \frac{1}{(p_n + 1)(p_n^2 + p_n + 1)} \Big|_{p_n \, = \, 3.078 \frac{1 - z^{-1}}{1 + z^{-1}}}$$

$$= \frac{0.01807(1 + z^{-1})^3}{(1 - 0.5095z^{-1})(1 - 1.2510z^{-1} + 0.5457z^{-2})} \tag{6.1.34}$$

from Eq. 5.11.11. Since $H_{LP}(1) = 1$, we can use Eq. 6.1.33 to find the complementary highpass filter transfer function as

$$H_{CHP}(z) = \frac{0.9819 - 1.8145z^{-1} + 1.1289z^{-2} - 0.2761z^{-3}}{1 - 1.7605z^{-1} + 1.1831z^{-2} - 0.2780z^{-3}} \tag{6.1.35}$$

■

6.2 LOWPASS-TO-BANDPASS TRANSFORMATIONS [7]

The *analog lowpass-to-analog bandpass transformation* (p-to-s) converts analog lowpass filter transfer functions into analog bandpass filter transfer functions as

$$H_{BP}(s) = H_{LP}(p)\Big|_{p = \frac{\omega_o}{B}\left(\frac{s}{\omega_o} + \frac{\omega_o}{s}\right)} = \frac{s^2 + \omega_o^2}{Bs} \tag{6.2.1}$$

In analog bandpass filters, p is the analog lowpass frequency variable and s is the analog bandpass frequency variable. ω_o is the *center frequency* of the bandpass filter and B is the *bandwidth* of its passband. Denoting the upper and lower band-edge frequencies as ω_U and ω_L, respectively, then

$$\omega_o = \sqrt{\omega_U \omega_L}, \qquad B = \omega_U - \omega_L \tag{6.2.2}$$

For simplicity, the frequency normalized transformation is often used where $s_n = s/\omega_o$ and $p = Q(s_n + 1/s_n)$. Then Eq. 6.2.1 reduces to

$$H_{BP}(s) = H_{LP}(p)\Big|_{p = Q\left(s_n + \frac{1}{s_n}\right)} \tag{6.2.3}$$

Q is called the *quality factor* of the bandpass filter and equals

$$Q = \frac{\omega_o}{B} \tag{6.2.4}$$

Q is defined in terms of the M_p-bandwidth B. The passband ripple M_p can have any value such as 0.01, 0.5, or 3 dB.

To investigate the effect this transformation has upon the poles and zeros of the analog lowpass filter, rewrite Eqs. 6.2.1 and 6.2.3 as

$$s^2 - pBs + \omega_o^2 = 0, \qquad s_n^2 - \frac{p}{Q}s_n + 1 = 0 \tag{6.2.5}$$

Solving for s (or s_n) using the biquadratic equation gives

$$s = \frac{pB}{2} \pm \sqrt{\left(\frac{pB}{2}\right)^2 - \omega_o^2}, \qquad s_n = \frac{s}{\omega_o} = \frac{p}{2Q} \pm \sqrt{\left(\frac{p}{2Q}\right)^2 - 1} \tag{6.2.6}$$

Every pole and zero p of the analog lowpass filter transforms into a pair of poles and zeros s_n in the analog bandpass filter. The location of these critical frequencies depends upon the p and Q of the transformation. In the *narrowband* or *high-Q* case where $|p| \ll 2Q$, then s (or s_n) equals

$$s = \frac{pB}{2} \pm j\omega_o\sqrt{1 - \left(\frac{pB}{2\omega_o}\right)^2} \cong \frac{pB}{2} \pm j\omega_o, \qquad s_n = \frac{s}{\omega_o} \cong \frac{p}{2Q} \pm j1 \tag{6.2.7}$$

In the *wideband* or *low-Q* case where $|p| \gg 2Q$, then s (or s_n) instead equals

$$s = \frac{pB}{2}\left[1 \pm \sqrt{1-\left(\frac{2\omega_o}{pB}\right)^2}\right] \cong \frac{pB}{2}\left(1 \pm \left[1 - 0.5\left(\frac{2\omega_o}{pB}\right)^2\right]\right) \cong pB, \quad \frac{Q\omega_o}{p} \qquad (6.2.8)$$

$$s_n = \frac{s}{\omega_o} \cong \frac{p}{Q}, \quad \frac{Q}{p}$$

Therefore when $|p| \ll 2Q$, the analog bandpass filter poles and zeros are obtained from the analog lowpass filter poles and zeros by scaling p by half the bandwidth $B/2$ and then translating vertically to $\pm j\omega_o$. However when $|p| \gg 2Q$, then one analog bandpass pole-zero is obtained by scaling p by B. The second pole-zero is obtained by inverting the first pole-zero and scaling by ω_o^2. For intermediate values of p, the pole-zero locations must be calculated using Eq. 6.2.6. These results show that real lowpass poles-zeros remain real under low-Q transformations but become complex under high-Q transformations. The relation given by Eq. 6.2.6 between the analog bandpass filter poles and zeros and those of the analog lowpass filter is plotted in Fig. 6.2.1.[8] Scaling p by the filter Q as

$$\frac{p}{Q} = x + jy \qquad (6.2.9)$$

then s_n is simply read off Fig. 6.2.1 as

$$s_n = \frac{s}{\omega_o} = \sigma + j\omega \qquad (6.2.10)$$

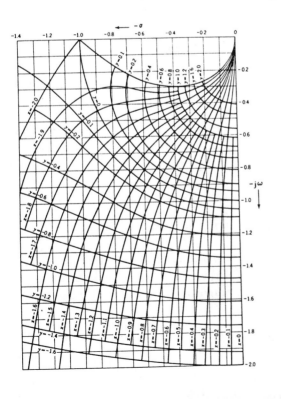

Fig. 6.2.1 Mapping properties of analog lowpass-to-analog bandpass transformation. (From M.S. Ghausi, *Principles and Design of Linear Active Circuits*, p. 437, Fig. 15-6, McGraw-Hill, NY, 1965.)

6.2.1 BANDPASS FILTER MAGNITUDE, PHASE, AND DELAY RESPONSES

The ac steady-state responses of analog bandpass filters are easily determined from Eq. 6.2.1. Setting $p = jv$ and $s = j\omega$ as was done in Eq. 6.1.4, then

$$v = \frac{\omega_o}{B}\left(\frac{\omega}{\omega_o} - \frac{\omega_o}{\omega}\right) = \frac{1}{B}\left(\omega - \frac{\omega_o^2}{\omega}\right) = \frac{\Delta\omega(\omega + \omega_o)}{B\omega} \tag{6.2.11}$$

The frequency variation from center frequency ω_o is expressed as $\Delta\omega = \omega - \omega_o$. When the frequency derivation is small so that $\Delta\omega \cong 0$, then $\omega \cong \omega_o$ and $v \cong 2\Delta\omega/B$. Substituting Eq. 6.2.11 into Eq. 6.2.1 shows that the magnitude, phase, and delay responses of analog bandpass filters are related to those of analog lowpass filters as

$$|H_{BP}(j\omega)| = |H_{LP}(jv)|\Big|_{v = Q\left(\frac{\omega}{\omega_o} - \frac{\omega_o}{\omega}\right)}$$

$$\arg H_{BP}(j\omega) = \arg H_{LP}(jv)\Big|_{v = Q\left(\frac{\omega}{\omega_o} - \frac{\omega_o}{\omega}\right)} \tag{6.2.12}$$

$$\tau_{BP}(j\omega) = -\frac{d \arg H_{BP}(j\omega)}{d\omega} = \frac{1 + (\omega_o/\omega)^2}{B}\,\tau_{LP}(jv)\Big|_{v = Q\left(\frac{\omega}{\omega_o} - \frac{\omega_o}{\omega}\right)}$$

The magnitude characteristic of the analog bandpass filter has geometric symmetry about its center frequency ω_o. Thus, its gain at frequency ω is identical to that at frequency ω_o/ω. Therefore, the shape of the gain characteristic is preserved for transformations of any Q (i.e., low, intermediate, or high) as shown in Fig. 6.2.2a. The phase characteristic also has geometric symmetry about ω_o. It is identical to the phase

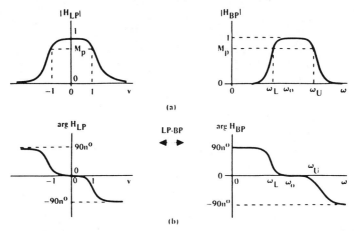

Fig. 6.2.2 Effect of analog lowpass-to-analog bandpass transformation upon (a) magnitude and (b) phase responses of filter.

of the analog lowpass filter after phase is recentered from 0 to ω_o. This is shown in Fig. 6.2.2b. The delay of a low-Q and high-Q analog bandpass filter is shown in Fig. 6.2.3. The delay becomes more asymmetrical with decreasing Q.

The ac steady-state passband characteristics of analog bandpass filters can be easily found by setting $v \cong 2\Delta\omega/B$ in Eq. 6.2.12. This shows that for small frequency deviations around its center frequency, the analog bandpass filter exhibits the same responses as its equivalent analog lowpass filter exhibits around dc (scale τ_{LP} by $2/B$).

6.2.2 BANDPASS FILTER IMPULSE AND COSINE STEP RESPONSES

Lowpass filters are characterized by their step responses. Bandpass filters are characterized by their responses to a cosine step input $\cos(\omega_o t)U_{-1}(t)$ where ω_o is the center frequency of the filter. Note that this analog bandpass filter input corresponds to the analog lowpass filter step input $1/p$, since using the analog lowpass-to-analog bandpass transformation,

$$\mathcal{L}^{-1}\left[\frac{1}{p}\,\Bigg|_{p=\frac{s^2+\omega_o^2}{Bs}}\right] = \mathcal{L}^{-1}\left[\frac{Bs}{s^2+\omega_o^2}\right] = B\cos(\omega_o t)\,U_{-1}(t) \tag{6.2.13}$$

from Table 1.5.1. When the impulse response of the analog lowpass filter is known, then the impulse response of the equivalent analog bandpass filter can be found but it results in a complicated expression.[7] This expression reduces to the simple form

$$h_{BP}(t) = \mathcal{L}^{-1}\left[H_{BP}(s)\right] = Bh_{LP}(Bt/2)\cos(\omega_o t)\,U_{-1}(t) \tag{6.2.14}$$

for narrowband analog bandpass filters whose roots are described by Eq. 6.2.7. The response consists of a cosine step whose envelope equals $Bh_{LP}(Bt/2)$ where h_{LP} is the impulse response of the analog lowpass filter. The cosine step response equals

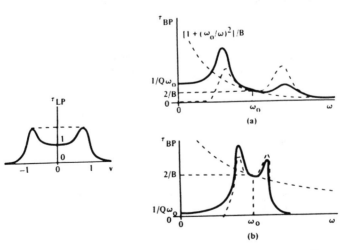

Fig. 6.2.3 Effect of analog lowpass-to-analog bandpass transformation upon delay response of filters under (a) low-Q and (b) high-Q conditions.

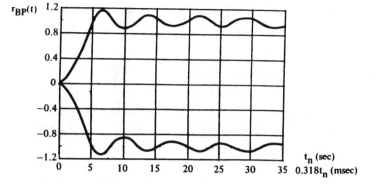

Fig. 6.2.4 Envelope of cosine step response of third-order elliptic analog bandpass filter of Example 6.2.1.

$$r_{BP}(t) = \mathcal{L}^{-1}\big[R_{BP}(s)\big] = r_{LP}(Bt/2)\cos(\omega_o t)\, U_{-1}(t) \qquad (6.2.15)$$

Therefore, narrowband analog bandpass filters have impulse and cosine step responses which can be drawn by inspection using the impulse and step responses of their equivalent analog lowpass filters from Chap. 4.2.

EXAMPLE 6.2.1   A third-order elliptic analog bandpass filter has a center frequency f_o of 10 KHz, a bandwidth B of 1 KHz, and a Q of 10. It has 1.25 dB in-band ripple. Sketch its cosine step response assuming unity passband gain.

SOLUTION   Treating this as a high-Q filter, its cosine step response is easily sketched using its analogous lowpass filter step response which is shown in Fig. 4.2.6c. Using $M_p = 1.25 \cong 1$ dB, $\Omega_r = 12° \cong 10°$, and $M_r = 60$ dB,[5] we first sketch the envelope of the bandpass filter response $r_{LP}(t_n)$ as shown in Fig. 6.2.4. The time scale is denormalized as

$$t = \frac{t_n}{B/2} = \frac{t_n}{2\pi(1\ \text{KHz})/2} = 0.318 t_n \ (msec) \qquad (6.2.16)$$

This envelope is filled in with the cosine signal $\cos[2\pi(10\ \text{KHz})t]\, U_{-1}(t)$. Because $f_o \gg B$, this signal simply "darkens" the envelope of Fig. 6.2.4. ∎

6.2.3 DIRECT ANALOG LOWPASS–TO–DIGITAL BANDPASS TRANSFORMATIONS [4]

The analog lowpass-to-analog bandpass transformation of Eq. 6.2.1 can be combined with the analog-to-digital transforms of Eq. 6.1.9 to give the direct *analog lowpass-to-digital bandpass transformation* as

$$
\begin{aligned}
H_{BP}(z) &= H_{BP}(p_{BP})\Big|_{p_{BP}\,=\,\dfrac{1}{H_{dmn}(z)}} \\[2mm]
&= H_{LP}(p_{LP})\Big|_{p_{LP}\,=\,Q\left(\dfrac{p_{BP}}{v_o} + \dfrac{v_o}{p_{BP}}\right)\Big|_{p_{BP}\,=\,\dfrac{1}{H_{dmn}(z)}}}
\end{aligned}
\qquad (6.2.17)
$$

In digital bandpass filters, p is the analog frequency variable and s is the digital bandpass frequency variable where $z = e^{sT}$. The analog lowpass frequency variable p_{LP} can be expressed as

$$p_{LP} = Q\left(\frac{p_{BP}}{v_o} + \frac{v_o}{p_{BP}}\right)\Bigg|_{p_{BP} = \frac{1}{H_{dmn}(z)}} = Q\left[\frac{1}{v_o H_{dmn}(z)} + v_o H_{dmn}(z)\right] \quad (6.2.18)$$

Since the H_{dmn} are all different, each one of them produces its own unique analog lowpass-to-digital bandpass transformation.

These direct analog lowpass-to-digital bandpass transformations can be easily determined. For example, consider the bilinear transform H_{011} and its frequency mapping function given by Eqs. 5.5.1 and 5.5.3 where

$$p_{BP} = \frac{2}{T}\frac{1 - z^{-1}}{1 + z^{-1}} \quad \text{and} \quad \frac{v_{BP}T}{2} = \tan\left(\frac{\omega_{BP}T}{2}\right) \quad (6.2.19)$$

Substituting Eq. 6.2.19 into Eq. 6.2.18 and simplifying the result gives

$$p_{LP} = \frac{1}{C}\left(\frac{1 - Dz^{-1} + z^{-2}}{1 - z^{-2}}\right) \quad (6.2.20)$$

where $C = v_o T/Q = bT/2 = \tan(BT/2)$ and $D = 2\cos(\omega_o T)$. ω_o and v_o (B and b) are the digital and analog bandpass filter center frequencies (bandwidths), respectively. An infinite number of analog lowpass-to-digital bandpass transformations can be derived following this procedure. Several such transforms are listed in Table 6.2.1.

The various jv-axis mappings are shown in Fig. 6.2.5. A number of important frequency mapping observations can be made about the bandpass transforms from these root loci. The H_{001} contour falls inside the unit circle and the H_{011} contour falls on the unit circle. These transforms always produce stable filters. The other H_{dmn} contours fall partially outside the unit circle so they produce filters that may be unstable. The H_{011} transform exactly preserves the frequency domain characteristics (with frequency compression) of the analog lowpass filter since H_{011} maps the entire jv-axis of the p-plane ($-\infty < v < \infty$) onto the unit circle of the z-plane ($0° < \omega T < 180°$). The H_{LDI} transform also exactly preserves the frequency domain characteristics over low-frequency ranges ($-\infty < v \leq (2 + D)/2C$) since it maps the jv-axis near the origin onto the unit circle ($0° < \omega T \leq 180°$). The high-frequency characteristics of the analog filter cannot be obtained because the jv-axis beyond $v = (2 + D)/2C$ is not mapped onto the unit circle. The H_{001}, H_{111}, and H_{MBDI} transforms distort the frequency domain behavior since they do not map the jv-axis onto the unit circle at all. However because they map the jv-axis around $v = -\infty$ to be tangent to the unit circle at $z = 1$, the high-frequency characteristics are somewhat maintained at sufficiently high sampling rates. Of these three transforms, H_{001} usually gives the least distortion at low frequencies since its root locus is closest the unit circle. In addition, H_{MBDI} maps the jv-axis around $v = \infty$ to be tangent to the unit circle at $z = -1$ so H_{MBDI} also compresses and maintains the high-frequency filter characteristics.

Table 6.2.1 Some analog lowpass-to-digital bandpass transformations.[4]

H_{dmn}	Discrete transform	Transformation $p_{LP} = g(z)$
H_{001}	Backward Euler	$p_{LP} = \dfrac{1}{C}\left(\dfrac{D - 2z^{-1} + z^{-2}}{1 - z^{-1}}\right)$
		$C = \sin(BT)$
		$D = 1 + \sin^2(\omega_o T)$
H_{111}	Forward Euler	$p_{LP} = \dfrac{1}{C}\left(\dfrac{1 - 2z^{-1} + Dz^{-2}}{z^{-1}(1 - z^{-1})}\right)$
		$C = \sin(BT)$
		$D = 1 + \sin^2(\omega_o T)$
H_{011}	Bilinear	$p_{LP} = \dfrac{1}{C}\left(\dfrac{1 - Dz^{-1} + z^{-2}}{1 - z^{-2}}\right)$
		$C = \tan(BT/2)$
		$D = 2\cos(\omega_o T)$
H_{MBDI}	Modified Bilinear	$p_{LP} = \dfrac{1}{C}\left(\dfrac{1 - 2z^{-1} + Dz^{-2} + Ez^{-3} + Fz^{-4}}{z^{-1}(1 - z^{-2})}\right)$
		$C = \tan(BT/2)\cos(BT)$
		$D = 1 + F = 1 + \tan^2(\omega_o T/2)\cos^2(\omega_o T)$
		$E = 2F = 2\tan^2(\omega_o T/2)\cos^2(\omega_o T)$
		$F = \tan^2(\omega_o T/2)\cos^2(\omega_o T)$
H_{112}	Discrete	$p_{LP} = \dfrac{1}{C}\left(\dfrac{1 - Dz^{-2} + z^{-4}}{z^{-1}(1 - z^{-2})}\right)$
		$C = 2\sin(BT)$
		$D = 2\cos(2\omega_o T)$
H_{LDI}	Lossless	$p_{LP} = \dfrac{1}{C}\left(\dfrac{1 - Dz^{-1} + z^{-2}}{z^{-0.5}(1 - z^{-1})}\right)$
		$C = 2\sin(BT/2)$
		$D = 2\cos(\omega_o T)$
H_{ODI}	Optimum	$p_{LP} = \dfrac{1}{C}\left(\dfrac{1 - Dz^{-0.5} + z^{-1}}{z^{-0.25}(1 - z^{-0.5})}\right)$
		$C = 2\sin(BT/4)$
		$D = 2\cos(\omega_o T/2)$

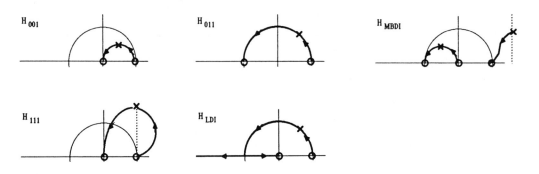

Fig. 6.2.5 Mapping of jv-axis of p_n-plane into z-plane using some bandpass transforms of Table 6.2.1.

The ac steady-state relation between the digital bandpass frequency $\omega_{BP}T$, the analog bandpass frequency $v_{BP}T$, and the analog lowpass frequency $v_{LP}T$ is found by setting $z = e^{j\omega T}$, $p_{BP} = jv_{BP}$, and $p_{LP} = jv_{LP}$ in Eq. 6.2.20 as

$$v_{LP} = \frac{1}{C}\left[\frac{\cos(\omega_o T) - \cos(\omega T)}{\sin(\omega T)}\right] \tag{6.2.21}$$

To interpret this frequency mapping, assume that the digital frequency ωT is small where $|\omega T| \ll 1$. Then $\sin \omega T \cong \omega T$, $\cos(\omega T/2) \cong 1 - (\omega T/2)^2/2$, and Eq. 6.2.21 reduces to

$$v_{LP} = \frac{1}{C}\left[\frac{(\omega T)^2 - (\omega_o T)^2 + (\omega_o T)^2(\omega T)^2/8}{\omega T}\right] \cong \frac{1}{C}\left[\frac{(\omega T)^2 - (\omega_o T)^2}{\omega T}\right] = \frac{\Delta\omega(\omega + \omega_o)}{B\omega} \tag{6.2.22}$$

This relation is identical to the analog lowpass-to-analog bandpass mapping of Eq. 6.2.11. This approximation shows that when the sampling frequency is sufficiently high so that the normalized digital frequency ωT is small, then the analog lowpass-to-digital bandpass transformation using H_{011} behaves like the analog lowpass-to-analog bandpass transformation shown in Fig. 6.2.2. All the H_{dmn} exhibit this limiting behavior.

To demonstrate the usefulness of these transformations, we design a first-order Butterworth digital bandpass filter by directly transforming a first-order Butterworth analog lowpass filter. The digital bandpass filter is specified to have a center frequency and a 3 dB bandwidth of 1 rad/sec. Since $Q = \omega_o/B$, the Q of this bandpass filter equals 1 so it is relatively wideband. Because of sampling, the magnitude response of the digital bandpass filter is periodic beyond the Nyquist frequency $\omega_s/2$. When the analog bandpass filter has insufficient stopband rejection at dc and $\omega_s/2$ relative to its center frequency, significant aliasing distortion occurs because the shifted magnitude responses overlap as shown in Fig. 1.3.3. For this reason, since the digital bandpass filter is relatively wideband, the sampling frequency must be selected to be large relative to its center frequency $\omega_o = 1$ rad/sec. Thus we choose a sampling frequency of 100 rad/sec. Then the digital bandpass filter has a center frequency $\omega_o T$ and 3 dB bandwidth BT both equal to $3.6°$. From Table 6.2.1, we compute some analog lowpass-to-digital bandpass transformations as

$$H_{001}: \quad p_n = 15.92597\left(\frac{1.00394 - 2z^{-1} + z^{-2}}{1 - z^{-1}}\right)$$

$$H_{111}: \quad p_n = 15.92597\left(\frac{1 - 2z^{-1} + 1.00394z^{-2}}{z^{-1}(1 - z^{-1})}\right)$$

$$H_{011}: \quad p_n = 31.82052\left(\frac{1 - 1.99605z^{-1} + z^{-2}}{1 - z^{-2}}\right) \tag{6.2.23}$$

$$H_{MBDI}: \quad p_n = 31.88343\left(\frac{1 - 2z^{-1} + 1.00098z^{-2} + 0.001967z^{-3} + 0.0009837z^{-4}}{z^{-1}(1 - z^{-2})}\right)$$

$$H_{LDI}: \quad p_n = 15.91811\left(\frac{1 - 1.99605z^{-1} + z^{-2}}{z^{-0.5}(1 - z^{-1})}\right)$$

Substituting these p_n-to-z transforms into the first-order Butterworth analog lowpass filter transfer function $H_{LP}(p_n) = 1/(p_n + 1)$ yields

$$H_{001}: \quad H_{BP}(z) = \frac{0.05886(1 - z^{-1})}{1 - 1.93375z^{-1} + 0.93744z^{-2}}$$

$$H_{111}: \quad H_{BP}(z) = \frac{0.06279z^{-1}(1 - z^{-1})}{1 - 1.93721z^{-1} + 0.94115z^{-2}}$$

$$H_{011}: \quad H_{BP}(z) = \frac{0.03047(1 - z^{-2})}{1 - 1.93524z^{-1} + 0.93906z^{-2}} \tag{6.2.24}$$

$$H_{MBDI}: \quad H_{BP}(z) = \frac{0.03136z^{-1}(1 - z^{-2})}{1 - 1.96864z^{-1} + 1.00098z^{-2} - 0.02940z^{-3} + 0.0009837z^{-4}}$$

$$H_{LDI}: \quad H_{BP}(z) = \frac{0.06282z^{-0.5}(1 - z^{-1})}{1 + 0.06282z^{-0.5} - 1.99605z^{-1} - 0.06282z^{-1.5} + z^{-2}}$$

The magnitude and phase responses of these first-order Butterworth digital bandpass filters are plotted in Fig. 6.2.6. The magnitude responses have ± 20 dB/dec asymptotes about $3.6°$. The H_{011} and H_{MBDI} responses have zero gain at the Nyquist frequency.

Now the Q of the digital bandpass filter is increased to 10 so that it becomes narrowband. Since the digital bandwidth $B = 0.1$ rad/sec is ten times smaller than the previous design, the sampling frequency can be decreased without introducing appreciable aliasing distortion. Let us choose a sampling frequency of 10 rad/sec. The digital center frequency $\omega_o T = 36°$ and the 3 dB bandwidth frequency $BT = 3.6°$. The p_n-to-z transforms then become

$$H_{001}: \quad p_n = 15.92597\left(\frac{1.34549 - 2z^{-1} + z^{-2}}{1 - z^{-1}}\right)$$

$$H_{111}: \quad p_n = 15.92597\left(\frac{1 - 2z^{-1} + 1.34549z^{-2}}{z^{-1}(1 - z^{-1})}\right) \tag{6.2.25a–c}$$

$$H_{011}: \quad p_n = 31.82052\left(\frac{1 - 1.61803z^{-1} + z^{-2}}{1 - z^{-2}}\right)$$

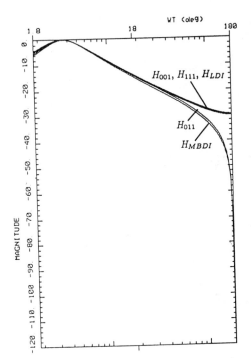 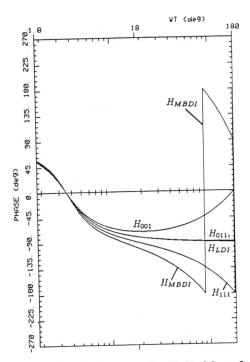

Fig. 6.2.6 Magnitude and phase responses of first-order Butterworth digital band-pass filter using H_{001}, H_{111}, H_{011}, H_{MBDI}, and H_{LDI} transforms (see Eq. 6.2.24). $\omega_o = 1$ rad/sec, $\omega_s = 100$ rad/sec, $\omega_o T = 3.6°$, and $Q = 1$. (Plotted by C. Powell. Based upon T. Wongterarit, loc. cit.)

$$H_{MBDI}: \quad p_n = 31.88343\left(\frac{1 - 2z^{-1} + 1.06910z^{-2} + 0.13820z^{-3} + 0.06910z^{-4}}{z^{-1}(1 - z^{-2})}\right)$$

$$H_{LDI}: \quad p_n = 15.91811\left(\frac{1 - 1.61803z^{-1} + z^{-2}}{z^{-0.5}(1 - z^{-1})}\right) \qquad (6.2.25\text{d-e})$$

The resulting first-order Butterworth digital bandpass filter transfer functions become

$$H_{001}: \quad H_{BP}(z) = \frac{0.04459(1 - z^{-1})}{1 - 1.46476z^{-1} + 0.71009z^{-2}}$$

$$H_{111}: \quad H_{BP}(z) = \frac{0.06279z^{-1}(1 - z^{-1})}{1 - 1.93721z^{-1} + 1.28270z^{-2}}$$

$$H_{011}: \quad H_{BP}(z) = \frac{0.03047(1 - z^{-2})}{1 - 1.56874z^{-1} + 0.93906z^{-2}} \qquad (6.2.26)$$

$$H_{MBDI}: \quad H_{BP}(z) = \frac{0.03136z^{-1}(1 - z^{-2})}{1 - 1.96864z^{-1} + 1.06910z^{-2} + 0.10683z^{-3} + 0.06910z^{-4}}$$

$$H_{LDI}: \quad H_{BP}(z) = \frac{0.06282z^{-0.5}(1 - z^{-1})}{1 + 0.06282z^{-0.5} - 1.61803z^{-1} - 0.06282z^{-1.5} + z^{-2}}$$

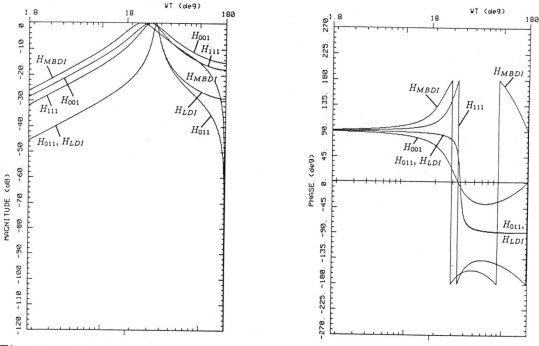

Fig. 6.2.7 Magnitude and phase responses of first-order Butterworth digital bandpass filter using H_{001}, H_{111}, H_{011}, H_{MBDI}, and H_{LDI} transforms (see Eq. 6.2.26). $\omega_o = 1$ rad/sec, $\omega_s = 10$ rad/sec, $\omega_o T = 36°$, and $Q = 10$. (Plotted by C. Powell. Based upon T. Wongterarit, loc. cit.)

Their magnitude and phase responses are plotted in Fig. 6.2.7. They have ± 20 dB/dec asymptotes similar to the bandpass responses in Fig. 6.2.6. The center frequencies of the H_{001}, H_{111}, and H_{MBDI} are shifted down from 36° because of frequency mapping errors (due to the lower sampling rate) as discussed in Table 5.10.2.

Now let us design a third-order Butterworth digital bandpass filter by transforming a third-order Butterworth analog lowpass filter. This digital bandpass filter is specified to have a center frequency of 1 KHz, a 3 dB bandwidth of 100 Hz, and a $Q = 10$. The upper and lower 3 dB frequencies are 1051 Hz and 951 Hz. Choosing a sampling frequency of 10 KHz, the normalized center frequency and bandwidth is 36° and 3.6°, respectively. Therefore, the p_n-to-z transforms of Eq. 6.2.25 apply to this design. Substituting them into the third-order analog lowpass filter transfer function

$$H(p_n) = \frac{1}{p_n^3 + 2p_n^2 + 2p_n + 1} \tag{6.2.27}$$

gives the third-order Butterworth digital bandpass filter transfer functions of

$$H_{001}: \quad H_{BP}(z) = \frac{0.00009258(1 - 3z^{-1} + 3z^{-2} - z^{-3})}{1 - 4.41398z^{-1} + 8.58897z^{-2} - 9.41387z^{-3}} \tag{6.2.28a}$$
$$+ 6.13952z^{-4} - 2.26730z^{-5} + 0.37397z^{-6}$$

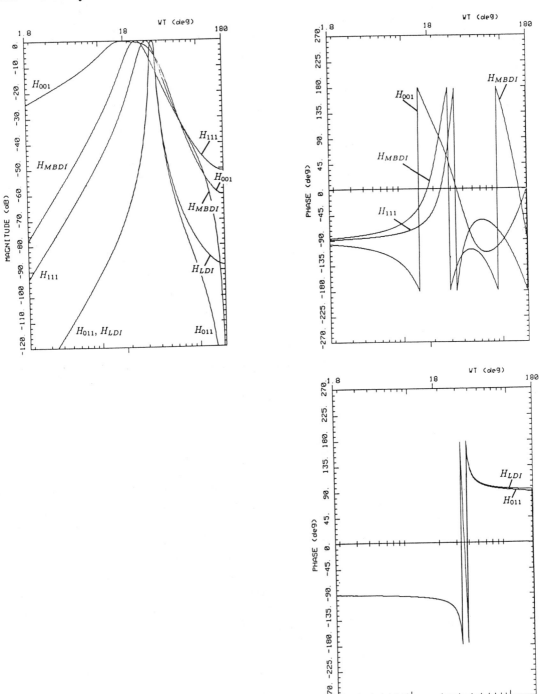

Fig. 6.2.8 Magnitude and phase responses of third-order Butterworth digital band-pass filter using H_{001}, H_{111}, H_{011}, H_{MBDI}, and H_{LDI} transforms (see Eq. 6.2.28). f_o = 1 KHz, f_s = 10 KHz, $\omega_o T = 36^o$, and $Q = 10$. (Plotted by C. Powell. Based upon T. Wongterarit, loc. cit.)

$$H_{111}: \quad H_{BP}(z) = \frac{0.0002476z^{-3}(1 - 3z^{-1} + 3z^{-2} - z^{-3})}{1 - 5.87442z^{-1} + 15.41646z^{-2} - 22.83461z^{-3}}$$
$$+ 20.10616z^{-4} - 9.98748z^{-5} + 2.21377z^{-6}$$

$$H_{011}: \quad H_{BP}(z) = \frac{0.00002915(1 - 3z^{-2} + 3z^{-4} - z^{-6})}{1 - 4.75245z^{-1} + 10.40461z^{-2} - 13.08895z^{-3}}$$
$$+ 9.97768z^{-4} - 4.37044z^{-5} + 0.88189z^{-6}$$

$$H_{MBDI}: \quad H_{BP}(z) = \frac{0.00003085z^{-3}(1 - 3z^{-2} + 3z^{-4} - z^{-6})}{1 - 5.93727z^{-1} + 14.95835z^{-2} - 20.09618z^{-3}}$$
$$+ 14.80520z^{-4} - 5.47347z^{-5} + 1.02872z^{-6}$$
$$- 0.50952z^{-7} + 0.19498z^{-8} + 0.03471z^{-9}$$
$$+ 0.01311z^{-10} + 0.001170z^{-11} + .000000325z^{-12}$$

$$\text{(6.2.28b–e)}$$

$$H_{LDI}: \quad H_{BP}(z) = \frac{0.002479z^{-1.5}(1 - 3z^{-1} + 3z^{-2} - z^{-3})}{1 + 0.12564z^{-0.5} - 4.84621z^{-1} - 0.53198z^{-1.5}}$$
$$+ 10.82555z^{-2} + 0.98607z^{-2.5} - 13.90294z^{-3}$$
$$- 0.98607z^{-3.5} + 10.82555z^{-4} + 0.53198z^{-4.5}$$
$$- 4.84621z^{-5} - 0.12564z^{-5.5} + z^{-6}$$

Their magnitude and phase responses are plotted in Fig. 6.2.8. They exhibit the Butterworth maximally flat magnitude response. They have ± 60 dB/dec asymptotic roll-off except for H_{MBDI} which is very distorted. Only the H_{011} and H_{LDI} responses are very good as expected since the H_{011} and H_{LDI} transforms exactly preserve the frequency domain characteristics.

6.2.4 BANDPASS FILTER ORDER

Bandpass filters are specified in the frequency domain, the time domain, or in both domains simultaneously. In the time domain, narrowband bandpass filter approximations given by Eqs. 6.2.14–6.2.15 are used. In the frequency domain, the responses described by Eq. 6.2.12 and the nomographs of Chap. 4.2 are utilized. In general, the frequency domain specifications of analog bandpass filters have the form shown in Fig. 6.2.9a. f_L and f_U are the required corner frequencies for less than M_p in-band gain variation. Frequencies $(f_1, f_2), (f_3, f_4), \ldots$ are the frequencies for stopband rejections of $M_{r1}, M_{r2}, \ldots$. In many applications, the gain characteristic will have geometric symmetry about the center frequency f_o so that $f_o^2 = f_L f_U = f_1 f_2 = f_3 f_4 = \ldots$. To utilize the design information of Chap. 4.2, this analog bandpass filter data is converted into equivalent data for an analog lowpass filter using

$$\Omega_i = Q \left| \frac{f_i}{f_o} - \frac{f_o}{f_i} \right| \tag{6.2.29}$$

We normalize each stopband frequency f_i by f_o, form its reciprocal, take the absolute

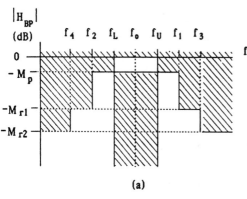

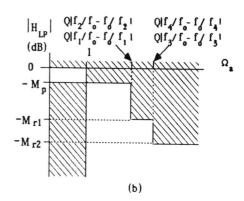

Fig. 6.2.9 (a) Analog bandpass filter specification and its (b) equivalent analog lowpass filter specification.

value of the difference, and multiply by Q. This gives each normalized stopband frequency Ω_i of the equivalent analog lowpass filter. Since each pair of upper and lower frequencies are assumed to have geometric symmetry, their (Ω_i, Ω_{i+1}) are equal so that $\Omega_i = \Omega_{i+1}$. The equivalent analog lowpass filter specification is shown in Fig. 6.2.9b. f_o corresponds to $\Omega_o = 0$, the band-edge frequencies f_U and f_L correspond to $\Omega_U = \Omega_L = 1$, and the Ω_i correspond to the various cut-off frequencies. The analog lowpass filter orders are easily determined using this specification and the nomographs.

Another useful result which simplifies analog filter order determination is use of the shaping factor $S_{M_p}^{M_r} = B_r/B_p$. M_p and M_r are arbitrary attenuations; B_p and B_r are the corresponding bandwidths of the passband and stopband. It is easily shown that the shaping factor is independent of Q and is invariant under the analog lowpass-to-analog bandpass transformation.[7] Thus, the shaping factors can be calculated directly from the bandwidth ratio for the analog bandpass filter whose frequency responses have geometric symmetry. Then setting $\Omega_a = S_{M_p}^{M_r}$ for the analog lowpass filter allows the order to be determined directly from the nomograph. The shaping factor approach greatly reduces the work required to find bandpass filter orders.

EXAMPLE 6.2.2 Determine the minimum order Butterworth, Chebyshev, and elliptic analog bandpass filters required to meet the following magnitude characteristics (refer to Fig. 6.2.9a). The 1.25 dB passband lies between 500–2000 Hz. The 40 dB stopband lies below 250 Hz and above 4000 Hz.

SOLUTION To determine the filter orders, we first determine the center frequency f_o, bandwidth B, and Q of the analog bandpass filter as

$$f_o = \sqrt{500(2000)} = 1000 \text{ Hz}, \quad B = 2000 - 500 = 1500 \text{ Hz}, \quad Q = \frac{1000}{1500} = 0.667$$

$$(6.2.30)$$

from Eqs. 6.2.2 and 6.2.4. Therefore using Eq. 6.2.29, the normalized lowpass stopband frequencies are

$$\Omega_1 = 0.667 \left| \frac{4000}{1000} - \frac{1000}{4000} \right| = 2.5, \quad \Omega_2 = 0.667 \left| \frac{250}{1000} - \frac{1000}{250} \right| = 2.5 \quad (6.2.31)$$

which are equal since they have geometric symmetry about f_o. Using the shaping factor approach, $B_{1.25dB} = 2000 - 500 = 1500$ Hz and $B_{40dB} = 4000 - 250 = 3750$ Hz so the 1.25-40 dB shaping factor equals $S_{1.25}^{40} = 3750/1500 = 2.5$. Therefore the stopband frequency $\Omega_a = S_{1.25}^{40} = 2.5$ directly. This eliminates the need to evaluate Eqs. 6.2.30 and 6.2.31. The analog bandpass filter requirements are now expressed in terms of the equivalent analog lowpass filter specification as shown in Fig. 6.2.9b. Entering $(M_p, M_r, \gamma, \Omega_a) = (1.25, 40, 4.5, 2.5)$ onto the nomographs of Chap. 4.2, we find: Butterworth $n = 6$, Chebyshev $n = 4$, elliptic $n = 3$. ∎

Some bandpass filter applications specify magnitude responses which have arithmetic symmetry rather than geometric symmetry. In these situations, the nomographs can still be utilized assuming that bandpass filters having geometric symmetry are acceptable. The technique requires that the most stringent stopband frequency be selected as $\Omega_a = \min(\Omega_1, \Omega_2)$ at a given attenuation M_r. Further details of arithmetically symmetrical bandpass filters are discussed in Ref. 7.

EXAMPLE 6.2.3 Determine the minimum order Butterworth, Chebyshev, and elliptic analog bandpass filters required to meet the following magnitude characteristics having arithmetic symmetry. The frequency pairs (500, 1000) Hz form the 1.25 dB passband, (250, 1250) Hz mark the 40 dB stopband, and (125, 1375) Hz mark the 60 dB stopband in Fig. 6.2.9a.

SOLUTION In this case, the analog bandpass filter parameters equal

$$f_o = \sqrt{500(1000)} = 707 \text{ Hz}, \quad B = 1000 - 500 = 500 \text{ Hz}, \quad Q = \frac{707}{500} = 1.41$$

$$(6.2.32)$$

Computing the upper and lower stopband frequency pairs of the equivalent analog lowpass filter gives $(\Omega_1, \Omega_2) = (1.70, 3.50)$ and $(\Omega_3, \Omega_4) = (2.02, 7.75)$ using

$$\Omega_i = 1.41 \left| \frac{f_i}{707} - \frac{707}{f_i} \right|$$

$$(6.2.33)$$

The equivalent analog lowpass filter response is drawn in Fig. 6.2.9b. The required stopband frequencies Ω_i are unequal and so that the smallest frequency of each frequency pair is chosen as $\Omega_{ai} = \min(\Omega_i, \Omega_{i+1})$. This gives the most stringent gain requirement. In this case, $(M_p, M_r, \gamma, \Omega_a)$ equals (1.25, 40, 4.5, 1.70) and (1.25, 60, 6.5, 2.02). From the nomographs of Chap. 4.2, we find: Butterworth $n = 10, 11$ (use $n = 11$); Chebyshev $n = 6, 6$ (use $n = 6$); elliptic $n = 4, 5$ (use $n = 5$). ∎

Digital bandpass filters have magnitude response specifications of the form shown in Fig. 6.2.10a. The frequencies involved equal:

$$\omega_U T = 2\pi\omega_U/\omega_s = \text{normalized upper passband frequency in radians}$$
$$= 360° f_U/f_s = \text{normalized upper passband frequency in degrees}$$
$$\omega_L T = 2\pi\omega_L/\omega_s = \text{normalized lower passband frequency in radians}$$
$$= 360° f_L/f_s = \text{normalized lower passband frequency in degrees}$$

$$(6.2.34a\text{--}b)$$

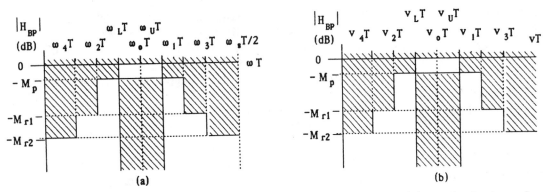

Fig. 6.2.10   (a) Digital bandpass filter specification and its (b) equivalent analog lowpass filter specification.

$$\omega_i T = 2\pi\omega_i/\omega_s = \text{normalized stopband frequency in radians}$$
$$= 360° f_i/f_s = \text{normalized stopband frequency in degrees} \qquad (6.2.34\text{c–d})$$
$$\Omega_{di} = \frac{(\omega_{i+1} - \omega_i)T}{(\omega_U - \omega_L)T} = \frac{360°(f_{i+1} - f_i)/f_s}{360°(f_U - f_L)/f_s} = \frac{\text{digital stopband bandwidth}}{\text{digital passband bandwidth}}$$

This digital bandpass filter specification can be converted into the equivalent analog bandpass filter specification using the following procedure. The result is shown in Fig. 6.2.10b. The normalized digital bandpass frequencies ωT are converted into the equivalent analog bandpass frequencies $v_{BP}T$ using Eq. 5.11.1. For example, the H_{011} transform uses $v_i T = 2\tan(\omega_i T/2)$. Then the center frequency $v_o T$, bandwidth bT, and quality factor q of the equivalent analog bandpass filter is determined as

$$v_o T = \sqrt{v_U T \; v_L T}, \qquad bT = v_U T - v_L T, \qquad q = \frac{v_o T}{bT} \qquad (6.2.35)$$

Using Eq. 6.2.29, the normalized digital bandpass frequencies $v_i T$ are converted into the equivalent normalized analog lowpass frequency ratio Ω_i as

$$\Omega_i = q\left| \frac{v_i T}{v_o T} - \frac{v_o T}{v_i T} \right| = \left| \frac{(v_i T)^2 - (v_o T)^2}{v_i T \; bT} \right| \qquad (6.2.36)$$

Evaluating Eq. 6.2.36 for the transforms in Table 6.2.1 gives

$$H_{001}, H_{111}, H_{112}: \quad \Omega_a = \left| \frac{\sin^2(\omega_r T) - \sin^2(\omega_o T)}{\sin(\omega_r T)\sin(BT)} \right|$$

$$H_{011}: \quad \Omega_a = \left| \frac{\tan^2(\omega_r T/2) - \tan^2(\omega_o T/2)}{\tan(\omega_r T/2)\tan(BT/2)} \right|$$

$$H_{MBDI}: \quad \Omega_a = \left| \frac{\tan^2(\omega_r T/2)\cos^2(\omega_r T) - \tan^2(\omega_o T/2)\cos^2(\omega_o T)}{\tan(\omega_r T/2)\tan(BT/2)\cos(\omega_r T)\cos(BT)} \right|$$

$$H_{LDI}: \quad \Omega_a = \left| \frac{\sin^2(\omega_r T/2) - \sin^2(\omega_o T/2)}{\sin(\omega_r T/2)\sin(BT/2)} \right| \qquad (6.2.37\text{a–d})$$

$$H_{ODI}: \quad \Omega_a = \left| \frac{\sin^2(\omega_r T/4) - \sin^2(\omega_o T/4)}{\sin(\omega_r T/4)\sin(BT/4)} \right| \tag{6.2.37e}$$

When the digital bandpass filter frequency pairs are distributed so that the equivalent analog bandpass filter frequency pairs have geometric symmetry about f_o, then $\Omega_{ai} = \Omega_i$. Usually however, this is not the case so $\Omega_{ai} = \min(\Omega_i, \Omega_{i+1})$. Both situations are illustrated in Examples 6.2.4–6.2.5.

EXAMPLE 6.2.4 Determine the minimum order Butterworth, Chebyshev, and elliptic digital bandpass filter required to satisfy the magnitude characteristics shown in Fig. 6.2.11a. The sampling rate is 12 KHz. Use a bilinear transform.

SOLUTION The digital bandpass filter frequencies are first normalized by the 12 KHz sampling frequency using $\omega_i T = 2\pi f_i/f_s$. This converts $(0.8, 0.9, 1.1, 1.2, 6.0)$ KHz into $(24°, 27°, 33°, 36°, 180°)$ as shown in Fig. 6.2.11a. Then the equivalent analog bandpass filter frequencies are computed using $v_i T = 2\tan(\omega_i T/2)$. These frequencies are $(24.4°, 27.5°, 33.9°, 37.2°, \infty)$ as shown in Fig. 6.2.11b. Therefore from Eq. 6.2.35, the analog bandpass filter parameters $v_o T$, bT, and q are

$$v_o T = \sqrt{(27.5°)(33.9°)} = 30.6°, \quad bT = 33.9° - 27.5° = 6.4°, \quad q = \frac{30.6°}{6.4°} = 4.75 \tag{6.2.38}$$

The corresponding digital bandpass filter parameters $\omega_o T$, BT, and Q are

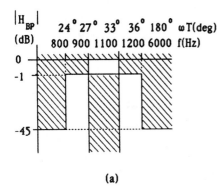

(a)

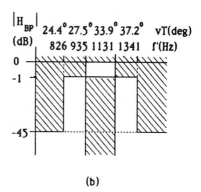

(b)

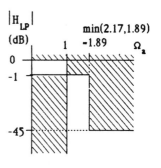

(c)

Fig. 6.2.11 (a) Digital bandpass filter specification and its equivalent (b) analog bandpass filter and (c) analog lowpass filter specifications of Example 6.2.4.

$$\omega_o T = 2\tan^{-1}(v_o T/2) = 2\tan^{-1}\sqrt{\tan(27^\circ/2)\tan(33^\circ/2)} = 29.9^\circ$$

$$\cong \sqrt{\omega_U T\ \omega_L T} = \sqrt{(27^\circ)(33^\circ)} = 29.8^\circ \tag{6.2.39}$$

$$BT = \omega_U T - \omega_L T = 33^\circ - 27^\circ = 6^\circ, \quad Q = \frac{\omega_o T}{BT} = \frac{29.9^\circ}{6^\circ} = 4.98$$

The equivalent analog lowpass filter frequencies are computed using Eq. 6.2.36 as

$$\Omega_i = 4.75 \left| \frac{v_i T}{30.6^\circ} - \frac{30.6^\circ}{v_i T} \right| = \left| \frac{\tan^2(\omega_i T/2) - \tan^2(29.9^\circ/2)}{\tan(\omega_i T/2)\tan(6^\circ/2)} \right| \tag{6.2.40}$$

The upper and lower 40 dB stopband frequencies of the analog bandpass filter are found to equal $\Omega_1 = 2.17$ and $\Omega_2 = 1.89$ as shown in Fig. 6.2.11c. Using the nomographs of Chap. 4.2 where $(M_p, M_r, \gamma, \Omega_a) = (1, 40, 4.6, 1.89)$, we find: Butterworth $n = 9$, Chebyshev $n = 6$, and elliptic $n = 4$. ∎

EXAMPLE 6.2.5 Determine the order of an elliptic digital bandpass filter to meet the following specifications: (1) maximum mid-band gain = 0 dB, (2) passband ripple $M_p \le 1.25$ dB for frequencies $f_p \le 951$ Hz and $f_p \ge 1051$ Hz, (3) stopband rejection $M_r \ge 40$ dB for frequencies $f_r \le 851$ Hz and $f_r \ge 1151$ Hz, and (4) sampling frequency $f_s = 10$ KHz. Consider different transforms.
SOLUTION We find the order of the elliptic analog lowpass filter to meet these requirements. From Eq. 6.2.39, the digital bandpass filter parameters are

$$\omega_o T \cong 360^\circ \frac{\sqrt{0.951(1.051)}}{10} = 36^\circ, \quad BT = 360^\circ \frac{1.051 - 0.951}{10} = 3.6^\circ,$$

$$Q \cong \frac{36^\circ}{3.6^\circ} = 10; \quad \omega_1 T = 360^\circ \frac{1.151}{10} = 41.4^\circ, \quad \omega_2 T = 360^\circ \frac{0.851}{10} = 30.6^\circ \tag{6.2.41}$$

Using Eq. 6.2.37, the equivalent analog stopband frequency ratios Ω_i equal

$$H_{001}, H_{111} : \quad \Omega_1 = \left| \frac{\sin^2(41.4^\circ) - \sin^2(36^\circ)}{\sin(41.4^\circ)\sin(3.6^\circ)} \right| = 2.76$$

$$\Omega_2 = \left| \frac{\sin^2(30.6^\circ) - \sin^2(36^\circ)}{\sin(30.6^\circ)\sin(3.6^\circ)} \right| = 3.30$$

$$H_{011} : \quad \Omega_1 = \left| \frac{\tan^2(20.7^\circ) - \tan^2(18^\circ)}{\tan(20.7^\circ)\tan(1.8^\circ)} \right| = 2.85 \tag{6.2.42a–c}$$

$$\Omega_2 = \left| \frac{\tan^2(15.3^\circ) - \tan^2(18^\circ)}{\tan(15.3^\circ)\tan(1.8^\circ)} \right| = 3.21$$

$$H_{MBDI} : \quad \Omega_1 = \left| \frac{\tan^2(20.7^\circ)\cos^2(41.4^\circ) - \tan^2(18^\circ)\cos^2(36^\circ)}{\tan(20.7^\circ)\tan(1.8^\circ)\cos(41.4^\circ)\cos(3.6^\circ)} \right| = 2.52$$

$$\Omega_2 = \left| \frac{\tan^2(15.3^\circ)\cos^2(30.6^\circ) - \tan^2(18^\circ)\cos^2(36^\circ)}{\tan(15.3^\circ)\tan(1.8^\circ)\cos(30.6^\circ)\cos(3.6^\circ)} \right| = 3.53$$

$$H_{LDI}: \quad \Omega_1 = \left| \frac{\sin^2(20.7^\circ) - \sin^2(18^\circ)}{\sin(20.7^\circ)\sin(1.8^\circ)} \right| = 2.81$$

$$\Omega_2 = \left| \frac{\sin^2(15.3^\circ) - \sin^2(18^\circ)}{\sin(15.3^\circ)\sin(1.8^\circ)} \right| = 3.25$$

(6.2.42d)

The $(M_p, M_r, \gamma, \Omega_a)$ data is entered onto the elliptic filter nomograph as $(1.25, 40, 4.5, \Omega_a)$ where $\Omega_a = \min(\Omega_1, \Omega_2)$. The nomograph shows that the filter order is three regardless of which discrete transform is used (for any $\Omega_i \geq 2.37$). ∎

6.2.5 BANDPASS FILTER DESIGN PROCEDURE

The design procedure for digital bandpass filters using direct transformations is straightforward. It consists of the following steps:

1a. Select a suitable analog filter type (e.g., Butterworth, Chebyshev, elliptic, etc.) and
1b. Choose a particular discrete transform from Tables 5.2.1 and 5.2.2 based upon the discussion of Chap. 5.9.
2. Determine the required analog lowpass filter order as described in Chap. 6.2.4.
3. Write the transfer function $H_{LP}(p_n)$ of the analog lowpass filter having unity bandwidth and the desired magnitude from the appropriate table of Chap. 4.2.
4. Evaluate the direct analog lowpass-to-digital bandpass transformation from Table 6.2.1 or compute using Eq. 6.2.18.
5. Compute the transfer function $H_{BP}(z)$ of the digital bandpass filter by substituting the p_n-to-z transformation found in Step 4 into $H_{LP}(p_n)$ found in Step 3.
6. Implement this digital bandpass transfer function using one of the techniques discussed in Chap. 12.

Let us illustrate this procedure by designing an elliptic digital bandpass filter which meets the specifications of Example 6.2.5. *Steps 1-2* – From Example 6.2.5, a third-order filter is required regardless of which transform is used. *Step 3* – The transfer function of this third-order elliptic analog lowpass filter is[5]

$$H(p_n) = \frac{0.06386(p_n^2 + 7.2895)}{(p_n + 0.48069)(p_n^2 + 0.41682p_n + 0.96846)}$$

(6.2.43)

where $p_n = p_{LP}T/v_{LP}T$. *Step 4* – From the specification, the center frequency and bandwidth of the digital bandpass filter is 1 KHz and 100 Hz, respectively, so $Q = 10$. Using a sampling frequency of 10 KHz, the normalized center frequency and bandwidth is 36° and 3.6°, respectively. Therefore, we can utilize the p_n-to-z transforms of Eq. 6.2.25. *Step 5* – Substituting these transforms into Eq. 6.2.43 gives the transfer functions of the third-order elliptic digital bandpass filters as

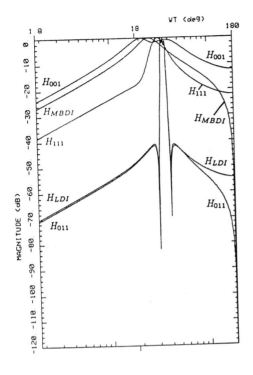

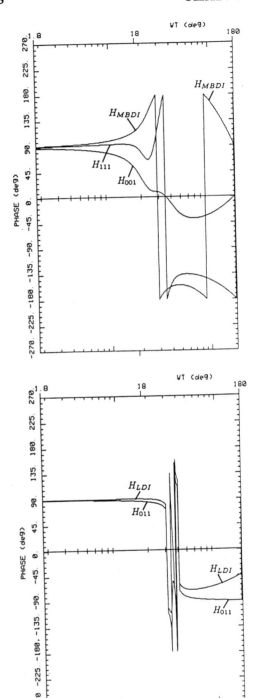

Fig. 6.2.12 Magnitude and phase responses of third-order elliptic digital bandpass filter using H_{001}, H_{111}, H_{011}, H_{MBDI}, and H_{LDI} transforms (see Eq. **6.2.44**). $\omega_o = 1$ rad/sec, $\omega_s = 10$ rad/sec, $\omega_o T = 36°$, and $Q = 10$. (Plotted by C. Powell. Based upon T. Wongterarit, loc. cit.)

$H_{001}:$ $H(z) = 0.002899(1 - 3.95769z^{-1} + 6.61152z^{-2}$
$$\dfrac{-5.82883z^{-3} + 2.71874z^{-4} - 0.54375z^{-5})}{\begin{array}{l}1 - 4.43739z^{-1} + 8.73177z^{-2} - 9.67895z^{-3} \\ + 6.37439z^{-4} - 2.37022z^{-5} + 0.39306z^{-6}\end{array}}$$

$H_{111}:$ $H(z) = 0.004010z^{-1}(1 - 5z^{-1} + 10.71972z^{-2}$
$$\dfrac{-12.15917z^{-4} + 7.24980z^{-5} - 1.81035z^{-6})}{\begin{array}{l}1 - 5.94365z^{-1} + 15.75931z^{-2} - 23.56173z^{-3} \\ + 20.92166z^{-4} - 10.47052z^{-5} + 2.33485z^{-6}\end{array}}$$

$H_{011}:$ $H(z) = 0.001964(1 - 3.21294z^{-1} + 3.57073z^{-2}$
$$\dfrac{-3.57073z^{-4} + 3.21294z^{-5} - z^{-6})}{\begin{array}{l}1 - 4.80607z^{-1} + 10.64234z^{-2} - 13.54273z^{-3} \\ + 10.44415z^{-4} - 4.62873z^{-5} + 0.94517z^{-6}\end{array}}$$ (6.2.44)

$H_{MBDI}:$ $H(z) = 0.002003z^{-1}(1 - 4z^{-1} + 5.14537z^{-2} + 0.000001907z^{-3}$
$$- 5.43133z^{-4} + 4.01910z^{-5} - 0.54002z^{-6} - 0.000002004z^{-7}$$
$$\dfrac{-0.17401z^{-8} - 0.01910z^{-9} - 0.000004697z^{-10})}{1 - 5.97185z^{-1} + 15.09585z^{-2} - 20.27223z^{-3}}$$
$$+ 14.80595z^{-4} - 5.28973z^{-5} + 0.89061z^{-6}$$
$$- 0.48828z^{-7} + 0.19422z^{-8} + 0.04030z^{-9}$$
$$+ 0.01372z^{-10} + 0.001258z^{-11} + .000000325z^{-12}$$

$H_{LDI}:$ $H(z) = 0.008020z^{-0.5}(1 - 4.20729z^{-1} + 7.76778z^{-2}$
$$\dfrac{-7.76778z^{-3} + 4.20729z^{-4} - z^{-5})}{1 + 0.05638z^{-0.5} - 4.84948z^{-1} - 0.23873z^{-1.5}}$$
$$+ 10.83737z^{-2} + 0.44249z^{-2.5} - 13.92006z^{-3}$$
$$- 0.44249z^{-3.5} + 10.83737z^{-4} + 0.23873z^{-4.5}$$
$$- 4.84948z^{-5} - 0.05638z^{-5.5} + z^{-6}$$

The magnitude and phase responses are shown in Fig. 6.2.12. Only the H_{011} and H_{LDI} filters behave as third-order elliptic bandpass filters. Their center frequency is 1 KHz and their 1.25 dB bandwidth is 100 Hz. They have equal ripple in both the passband and stopband. The stopband rejection is about −40 dB at frequencies of 890 Hz and 1120 Hz (or 32.0° and 40.5°). The H_{001}, H_{111} and H_{MBDI} filter responses exhibit great distortion as discussed in Fig. 6.2.5.

6.2.6 FREQUENCY DISCRIMINATOR TRANSFORMATIONS

Discriminator transformations are used to design frequency discriminators for FM applications. These are bandpass filters whose gains are proportional to input frequency (i.e., they are band-limited differentiators). A lowpass frequency discriminator with unit bandwidth has a linear magnitude response from $\omega T = 0$ to 1 and then falls off for

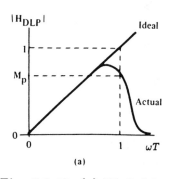
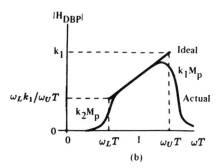

Fig. 6.2.13 (a) Digital lowpass and (b) digital low-Q bandpass frequency discriminators.

$1 < \omega T < \pi$. A bandpass frequency discriminator has a linear magnitude response over some passband $0 < \omega_L T < \omega T < \omega_U T < \pi$ where $\omega_U T - \omega_L T = BT$. Both types are shown in Fig. 6.2.13. *The lowpass discriminator* has a transfer function

$$H_{DLP}(z) = \frac{H_{LP}(z)}{H_{dmn}(z)} \qquad (6.2.45)$$

where H_{LP} is a lowpass filter gain chosen to obtain the desired stopband characteristics. For a maximally flat passband characteristic, for example, H_{LP} is selected to be Butterworth where $H_{DLP}(z) = p_{LP}/B_n(p_{LP})$ for $p_{LP} = 1/H_{dmn}(z)$ and B_n is the nth-order Butterworth polynomial.[9]

If a *low-Q bandpass discriminator* is desired, then it has a transfer function[10]

$$H_{DBP}(z) = \frac{H_{BP}(z)}{H_{dmn}(z)} \qquad (6.2.46)$$

This transformation yields gain linearity with respect to $\log \omega T$. If linearity with respect to ωT is required instead, then use the narrowband transformation in Eq. 6.2.7. These transformations are suitable for wide-bandwidth or low-Q discriminators. However, for narrow bandwidths, they are not too useful because of the small change in gain. In these

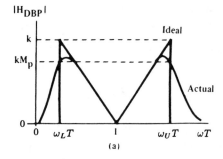
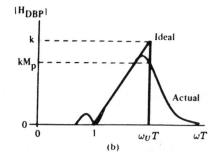

Fig. 6.2.14 Digital high-Q bandpass frequency discriminators using (a) H_{BP} and (b) modified H_{BP}.

cases, a *high-Q bandpass discriminator* is required whose transfer function equals[11]

$$H_{DBP}(z) = Q\left[\frac{1}{v_o H_{dmn}(z)} + v_o H_{dmn}(z)\right] H_{BP}(z) \qquad (6.2.47)$$

Its response is shown in Fig. 6.2.14a. By modifying the parameters used in only the H_{BP} portion of Eq. 6.2.47 so that $\omega_o'T = \sqrt{\omega_o T \, \omega_U T}$ and $B'T = \omega_U T - \omega_o T$, a bandpass filter results which eliminates the lower side of the gain characteristic as shown in Fig. 6.2.14b.[12] Again, the gain exhibits linearity with respect to $(\omega T - 1/\omega T)$. The narrow-band transformation may be used to obtain linearity with respect to $(\omega T - 1)$.

The unit step response of lowpass discriminators equal

$$r_{DLP}(nT) = h_{LP}(nT) \qquad (6.2.48)$$

The unit step response of the low-Q bandpass discriminator and the unit cosine step response of the high-Q bandpass discriminator equal

$$r_{DBP}(nT) = h_{BP}(nT) \qquad (6.2.49)$$

Thus, these various discriminator responses can be drawn directly from those of their analogous lowpass and bandpass filters.

6.3 LOWPASS-TO-BANDSTOP TRANSFORMATIONS [13]

The *analog lowpass-to-analog bandstop transformation* (*p*-to-*s*) converts analog lowpass filter transfer functions into analog bandstop filter transfer functions as

$$H_{BS}(s) = H_{LP}(p)\Big|_{\frac{1}{p} = \frac{\omega_o}{B}\left(\frac{s}{\omega_o} + \frac{\omega_o}{s}\right)} = H_{HP}(p)\Big|_{p = \frac{\omega_o}{B}\left(\frac{s}{\omega_o} + \frac{\omega_o}{s}\right)} \qquad (6.3.1)$$

In analog bandstop filters, p is the analog lowpass frequency variable and s is the analog bandstop frequency variable. ω_o is the *center* or *notch frequency* of the bandstop filter and B is the *bandwidth* of its stopband. The *quality factor* of the bandstop filter is $Q = \omega_o/B$. These parameters are given by Eqs. 6.2.2 and 6.2.4. For simplicity, the frequency normalized transformation is often used where $s_n = s/\omega_o$ and $1/p = Q(s_n + 1/s_n)$. Then Eq. 6.3.1 reduces to

$$H_{BS}(s) = H_{LP}(p)\Big|_{\frac{1}{p} = Q\left(s_n + \frac{1}{s_n}\right)} = H_{HP}(p)\Big|_{p = Q\left(s_n + \frac{1}{s_n}\right)} \qquad (6.3.2)$$

This transformation is the reciprocal of the analog lowpass-to-analog bandpass transformation given by Eq. 6.2.3. Thus it is interesting and very useful to observe that analog bandstop filters can also be obtained by using the analog lowpass-to-analog bandpass transformation on analog highpass filter transfer functions.

The poles and zeros of the analog bandstop filter can be easily related to those of the analog lowpass filter. Rewriting Eqs. 6.3.1 and 6.3.2 as

$$s^2 - \frac{B}{p}s + \omega_o^2 = 0, \qquad s_n^2 - \frac{s_n}{pQ} + 1 = 0 \qquad (6.3.3)$$

Eq. 6.3.3 is identical to Eq. 6.2.5 where p is replaced by $1/p$. Therefore, the analog bandpass filter results can be used by interchanging p with $1/p$. Solving Eq. 6.3.3 for s (or s_n) using the biquadratic equation gives

$$s = \frac{B}{2p} \pm \sqrt{\left(\frac{B}{2p}\right)^2 - \omega_o^2}, \qquad s_n = \frac{s}{\omega_o} = \frac{1}{2Qp} \pm \sqrt{\left(\frac{1}{2Qp}\right)^2 - 1} \qquad (6.3.4)$$

(cf. Eq. 6.2.6). As with analog bandpass filters, every pole and zero p of the analog lowpass filter transforms into a pair of poles and zeros s_n in the analog bandstop filter. The location of the critical frequencies depends upon the p and Q of the transformation. In the *narrowband* or *high-Q* case where $|1/p| \ll 2Q$, then from Eq. 6.2.7, s (or s_n) equals

$$s \cong \frac{B}{2p} \pm j\omega_o, \qquad s_n = \frac{s}{\omega_o} \cong \frac{1}{2Qp} \pm j1 \qquad (6.3.5)$$

In the *wideband* or *low-Q* case when $|1/p| \gg 2Q$, then from Eq. 6.2.8, s (or s_n) instead equals

$$s = \frac{B}{p}, \quad pQ\omega_o; \qquad s_n = \frac{s}{\omega_o} \cong \frac{1}{pQ}, \quad pQ \qquad (6.3.6)$$

Therefore when $|1/p| \ll 2Q$, the analog bandstop filter poles and zeros are obtained from the analog lowpass filter poles and zeros by inverting p, scaling $1/p$ by half the bandwidth $B/2$, and then translating vertically to $\pm j\omega_o$. However when $|1/p| \gg 2Q$, then one analog bandstop filter pole-zero is obtained by scaling $1/p$ by B. The second pole-zero is obtained by inverting the first pole-zero and scaling by ω_o^2. For intermediate values of p, the pole-zero locations must be calculated using Eq. 6.3.4. These results show that real lowpass filter poles-zeros remain real under low-Q transformations but become complex under high-Q transformations. The roots may also be obtained directly from Fig. 6.2.1 by letting

$$\frac{1}{pQ} = x + jy \qquad (6.3.7)$$

and simply reading off s_n as

$$s_n = \frac{s}{\omega_o} = \sigma + j\omega \qquad (6.3.8)$$

6.3.1 BANDSTOP FILTER MAGNITUDE, PHASE, AND DELAY RESPONSES

The ac steady-state responses of analog bandstop filters are easily determined from Eq. 6.3.1. Setting $p = jv$ and $s = j\omega$, then

$$-\frac{1}{v} = \frac{\omega_o}{B}\left(\frac{\omega}{\omega_o} - \frac{\omega_o}{\omega}\right) = \frac{1}{B}\left(\omega - \frac{\omega_o^2}{\omega}\right) = \frac{\Delta\omega(\omega + \omega_o)}{B\omega} \qquad (6.3.9)$$

The frequency variation from center frequency ω_o is expressed as $\Delta\omega = \omega - \omega_o$. When the frequency deviation is small so that $\Delta\omega \cong 0$, then $\omega = \omega_o$ and $-1/v = 2\Delta\omega/B$. Substituting Eq. 6.3.9 into Eq. 6.3.1 shows that the magnitude, phase, and delay responses of analog bandstop filters are related to those of analog highpass filters as

$$|H_{BS}(j\omega)| = |H_{HP}(jv)|\Big|_{v = Q\left(\frac{\omega}{\omega_o} - \frac{\omega_o}{\omega}\right)}$$

$$\arg H_{BS}(j\omega) = \arg H_{HP}(jv)\Big|_{v = Q\left(\frac{\omega}{\omega_o} - \frac{\omega_o}{\omega}\right)} \qquad (6.3.10)$$

$$\tau_{BS}(j\omega) = -\frac{d\,\arg H_{BS}(j\omega)}{d\omega} = \frac{(B/\omega)^2[1 + (\omega_o/\omega)^2]}{B[1 - (\omega_o/\omega)^2]^2}\,\tau_{LP}(jv)\Big|_{v = Q\left(\frac{\omega}{\omega_o} - \frac{\omega_o}{\omega}\right)}$$

The magnitude characteristic of the analog bandstop filter has geometric symmetry about its center frequency ω_o. Thus, its gain at frequency ω is identical to that at frequency ω_o/ω. Therefore, the shape of the gain characteristic is preserved for transformations of any Q as shown in Fig. 6.3.1a. The phase characteristic also has geometric symmetry about ω_o. It is identical to the phase of the analog highpass filter after phase is recentered from 0 to ω_o. This is shown in Fig. 6.3.1b. It is important to observe the phase change of $180(n - m)$ degrees at ω_o. The delay of an analog bandstop filter can be determined from the analog lowpass filter delay. The delay of a low-Q and high-Q analog bandstop filter is shown in Fig. 6.3.2. The delay expression is more complicated than that for bandpass filters (cf. Eq. 6.2.12). The delay becomes more asymmetrical with decreasing Q.

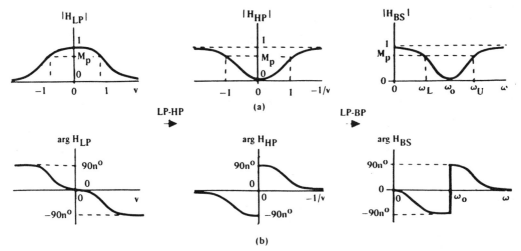

Fig. 6.3.1 Effect of analog lowpass-to-analog bandstop transformation upon (a) magnitude and (b) phase responses of filter.

The ac steady-state passband characteristics of analog bandstop filters can be easily found by setting $-1/v \cong 2\Delta\omega/B$ in Eq. 6.3.10. This shows that for small frequency deviations around its center frequency, the analog bandstop filter exhibits the same responses as its equivalent analog highpass filter exhibits around dc (scale τ_{HP} by $2/B$).

6.3.2 BANDSTOP FILTER IMPULSE AND COSINE STEP RESPONSES

Bandstop filters are characterized by their responses to a cosine step input $\cos(\omega_o t)U_{-1}(t)$ where ω_o is the center frequency of the filter. When the impulse response of the analog highpass filter is known, then the impulse response of the analog bandstop filter can be found but it results in a complicated expression.[13] This expression reduces to a simple form for narrowband analog bandstop filters whose roots are described by Eq. 6.3.5. The impulse and cosine step responses of these analog bandstop filters equal

$$h_{BS}(t) = B h_{HP}(Bt/2) \cos(\omega_o t) \, U_{-1}(t)$$
$$r_{BS}(t) = r_{HP}(Bt/2) \cos(\omega_o t) \, U_{-1}(t) \tag{6.3.11}$$

by analogy with Eqs. 6.2.14 and 6.2.15. Therefore, narrowband analog bandstop filters have impulse and cosine step responses which can be drawn by inspection using the impulse and step responses of their equivalent analog highpass filters from Chap. 6.1.

> **EXAMPLE 6.3.1** A fourth-order Butterworth analog bandstop filter has a center frequency f_o of 10 KHz, a 3 dB bandwidth B of 1 KHz, and unity passband gain. Sketch its cosine step response. Also sketch the response of a complementary fourth-order Butterworth analog bandstop filter having the same parameters.
> **SOLUTION** Since $Q = 10 \text{ KHz}/1 \text{ KHz} = 10$, this is a high-$Q$ filter. Its cosine step response is easily sketched using its analogous highpass filter step response

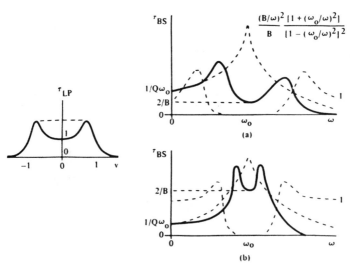

Fig. 6.3.2 Effect of analog lowpass-to-analog bandstop transformation upon delay response of filter under (a) low-Q and (b) high-Q conditions.

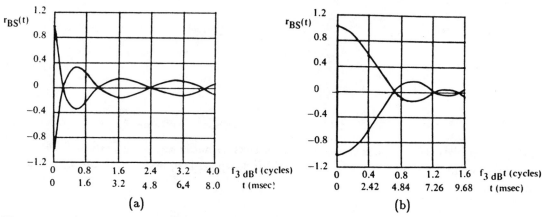

Fig. 6.3.3 Envelope of cosine step response of (a) fourth-order Butterworth analog bandstop filter and (b) its complementary bandpass filter of Example 6.3.1.

which is shown in Fig. 6.1.4. Using Eq. 6.3.11b, the envelope response $r_{HP}(t_n)$ of the bandstop filter is shown in Fig. 6.3.3a. The time scale is denormalized as

$$t = \frac{t_n}{B/2} = \frac{t_n}{1 \text{ KHz}/2} = 2t_n \ (msec) \tag{6.3.12}$$

This envelope is filled in with the cosine signal $\cos[2\pi(10 \text{ KHz})t] \ U_{-1}(t)$. Because $f_o \gg B$, this signal simply "darkens" the envelope of Fig. 6.3.3a.

The cosine step response of the complementary analog bandstop filter is readily drawn using the analog highpass filter step response in Fig. 6.1.10. This response gives the envelope response $r_{HP}(t_n)$ of the bandstop filter shown in Fig. 6.3.3b. The time scale is denormalized as

$$t = \frac{t_n}{0.33B/2} = 6.06t_n \ (msec) \tag{6.3.13}$$

Remembering that the complementary fourth-order Butterworth analog highpass filter has a 3 dB bandwidth $f_{3dB} = 0.33$ from Fig. 6.1.10, f_{3dB} must be denormalized accordingly. The envelope is filled in with the cosine signal $\cos[2\pi(10 \text{ KHz})t] \times U_{-1}(t)$. Comparing this response with that of Fig. 6.3.3a shows that there is reduced overshoot, reduced ringing, and increased storage time. The penalty paid for maintaining the time domain response is reduced asymptotic roll-off in the stopband of the magnitude response of the complementary filter. ∎

6.3.3 DIRECT ANALOG LOWPASS–TO–DIGITAL BANDSTOP TRANSFORMATIONS [4]

The analog lowpass-to-analog bandstop transformation of Eq. 6.3.1 can be combined with the analog-to-digital transforms of Eq. 6.1.9 to give the direct *analog lowpass-to-digital bandstop transformation* as

$$H_{BS}(z) = H_{BS}(p_{BS})\Big|_{p_{BS}\,=\,\dfrac{1}{H_{dmn}(z)}} = H_{BP}(p_{BP})\Big|_{p_{BP}\,=\,H_{dmn}(z)}$$

$$= H_{LP}(p_{LP})\Big|_{\dfrac{1}{p_{LP}}\,=\,Q\left(\dfrac{p_{BS}}{v_o}+\dfrac{v_o}{p_{BS}}\right)\Big|_{p_{BS}\,=\,\dfrac{1}{H_{dmn}(z)}}} \tag{6.3.14}$$

In digital bandstop filters, p is the analog frequency variable and s is the digital bandstop frequency variable where $z = e^{sT}$. The analog lowpass frequency variable p_{LP} can be expressed as

$$\frac{1}{p_{LP}} = Q\left(\frac{p_{BS}}{v_o}+\frac{v_o}{p_{BS}}\right)\Big|_{p_{BS}\,=\,\dfrac{1}{H_{dmn}(z)}} = Q\left[\frac{1}{v_o H_{dmn}(z)}+v_o H_{dmn}(z)\right] \tag{6.3.15}$$

Since the H_{dmn} are all different, each one of them produces its own unique analog lowpass-to-digital bandstop transformation.

These direct analog lowpass-to-digital bandstop transformations can be easily determined. For example, consider the bilinear transform H_{011} and its frequency mapping function given by Eqs. 5.5.1 and 5.5.3 where

$$p_{BS} = \frac{2}{T}\frac{1-z^{-1}}{1+z^{-1}} \quad\text{and}\quad \frac{v_{BS}T}{2} = \tan\left(\frac{\omega_{BS}T}{2}\right) \tag{6.3.16}$$

Substituting Eq. 6.3.16 into Eq. 6.3.15 and simplifying the result gives

$$\frac{1}{p_{LP}} = Q\left(\frac{2}{v_o T}\frac{1-z^{-1}}{1+z^{-1}}+\frac{v_o T}{2}\frac{1+z^{-1}}{1-z^{-1}}\right) = \frac{1}{C}\left(\frac{1-Dz^{-1}+z^{-2}}{1-z^{-2}}\right) \tag{6.3.17}$$

where $C = \tan(BT/2)$ and $D = 2\cos(\omega_o T)$. Solving for p_{LP} gives

$$p_{LP} = C\left(\frac{1-z^{-2}}{1-Dz^{-1}+z^{-2}}\right) \tag{6.3.18}$$

An infinite number of analog lowpass-to-digital bandstop transformations can be derived following this procedure. Several such transforms are listed in Table 6.3.1. It is useful to observe that the bandstop transforms of Table 6.3.1 are the reciprocal of the bandpass transforms of Table 6.2.1. By letting the center frequency $\omega_o T = 0$, the bandstop transforms of Table 6.3.1 reduce to the highpass transforms of Table 6.1.1.

The frequency mapping characteristics of these bandstop transformations can be determined from their root loci. The bandstop root loci can be obtained from the bandpass root loci in Fig. 6.2.5. Since the analog lowpass-to-digital bandstop transforms are reciprocals of the analog lowpass-to-digital bandpass transforms (i.e., $p_{BS} = 1/p_{BP}$), the bandstop root loci are identical to the bandpass root loci with only two notational

Table 6.3.1 Some analog lowpass-to-digital bandstop transformations.[4]

H_{dmn}	Discrete transform	Transformation $p_{LP} = g(z)$
H_{001}	Backward Euler	$p_{LP} = C\left(\dfrac{1 - z^{-1}}{D - 2z^{-1} + z^{-2}}\right)$ $C = \sin(BT)$ $D = 1 + \sin^2(\omega_o T)$
H_{111}	Forward Euler	$p_{LP} = C\left(\dfrac{z^{-1}(1 - z^{-1})}{1 - 2z^{-1} + Dz^{-2}}\right)$ $C = \sin(BT)$ $D = 1 + \sin^2(\omega_o T)$
H_{011}	Bilinear	$p_{LP} = C\left(\dfrac{1 - z^{-2}}{1 - Dz^{-1} + z^{-2}}\right)$ $C = \tan(BT/2)$ $D = 2\cos(\omega_o T)$
H_{MBDI}	Modified Bilinear	$p_{LP} = C\left(\dfrac{z^{-1}(1 - z^{-2})}{1 - 2z^{-1} + Dz^{-2} + Ez^{-3} + Fz^{-4}}\right)$ $C = \tan(BT/2)\cos(BT)$ $D = 1 + F = 1 + \tan^2(\omega_o T/2)\cos^2(\omega_o T)$ $E = 2F = 2\tan^2(\omega_o T/2)\cos^2(\omega_o T)$ $F = \tan^2(\omega_o T/2)\cos^2(\omega_o T)$
H_{112}	Discrete	$p_{LP} = C\left(\dfrac{z^{-1}(1 - z^{-2})}{1 - Dz^{-2} + z^{-4}}\right)$ $C = 2\sin(BT)$ $D = 2\cos(2\omega_o T)$
H_{LDI}	Lossless	$p_{LP} = C\left(\dfrac{z^{-0.5}(1 - z^{-1})}{1 - Dz^{-1} + z^{-2}}\right)$ $C = 2\sin(BT/2)$ $D = 2\cos(\omega_o T)$
H_{ODI}	Optimum	$p_{LP} = C\left(\dfrac{z^{-0.25}(1 - z^{-0.5})}{1 - Dz^{-0.5} + z^{-1}}\right)$ $C = 2\sin(BT/4)$ $D = 2\cos(\omega_o T/2)$

changes – interchange poles and zeros and reverse branch directions. Thus we can make the following observations about the bandstop transforms after modifying the root loci shown in Fig. 6.2.5.

The H_{001} contour falls inside the unit circle and the H_{011} contour falls on the unit circle. These transforms always produce stable filters. The other H_{dmn} contours fall partially outside the unit circle so they produce filters that may be unstable. The H_{011} transform exactly preserves the frequency domain characteristics (with frequency compression) of the analog lowpass filter since H_{011} maps the entire jv-axis of the p-plane $(-\infty < v < \infty)$ onto the unit circle of the z-plane $(0° < \omega T < 180°)$. The H_{LDI} transform also exactly preserves the frequency domain characteristics over high-frequency ranges $((2+D)/2C \le v < \infty)$ since it maps the jv-axis near infinity onto the unit circle $(0° < \omega T \le 180°)$. The low-frequency characteristics of the analog filter cannot be obtained because the jv-axis below $|v| = (2+D)/2C$ is not mapped onto the unit circle. The H_{001}, H_{111}, and H_{MBDI} transforms distort the frequency domain behavior since they do not map the jv-axis onto the unit circle at all. However because they map the jv-axis around $v = 0$ to be tangent to the unit circle at $z = 1$, the low-frequency characteristics are somewhat maintained at sufficiently high sampling rates. Of these three transforms, H_{001} usually gives the least distortion at low frequencies and the deepest notch since its root locus is closest the unit circle. In addition, H_{MBDI} maps the jv-axis around $v = 0$ to be tangent to the unit circle at $z = -1$. H_{MBDI} usually gives the least rejection at the notch frequency.

The ac steady-state relation between the digital bandstop frequency $\omega_{BS}T$, the analog bandstop frequency $v_{BS}T$, and the analog lowpass frequency $v_{LP}T$ can be easily determined by setting $z = e^{j\omega T}$, $p_{HP} = jv_{HP}$, and $p_{LP} = jv_{LP}$ in Eq. 6.3.18 as

$$\frac{1}{v_{LP}} = -\frac{1}{C}\left[\frac{\cos(\omega_o T) - \cos(\omega T)}{\sin(\omega T)}\right], \qquad v_{LP} = -C\left[\frac{\sin(\omega T)}{\cos(\omega_o T) - \cos(\omega T)}\right] \qquad (6.3.19)$$

When $|\omega T| \ll 1$, then $\sin \omega T \cong \omega T$, $\cos(\omega T) \cong 1 - (\omega T)^2/2!$, and Eq. 6.3.19 reduces to Eq. 6.3.9 where

$$v_{LP} \cong -C\left[\frac{\omega T}{(\omega T)^2 - (\omega_o T)^2}\right] = -\frac{B\omega}{\Delta\omega(\omega + \omega_o)} \qquad (6.3.20)$$

This relationship is identical to the analog lowpass-to-analog bandpass mapping of Eq. 6.2.22. This approximation shows that when the sampling frequency is sufficiently high so that the normalized digital frequency ωT is small, then the analog lowpass-to-digital bandstop transformation using H_{011} behaves like the analog lowpass-to-analog bandstop transformation shown in Fig. 6.3.1. All the H_{dmn} exhibit this same limiting behavior.

To demonstrate the usefulness of these transformations, we design a first-order Butterworth digital bandstop filter by directly transforming a first-order Butterworth analog lowpass filter. The digital bandstop filter is specified to have a center frequency and a 3 dB bandwidth of 1 rad/sec. Since $Q = \omega_o/B$, then Q of this bandstop filter equals 1 so it is relatively wideband. Because the bandstop filter magnitude response is not band-limited, significant aliasing distortion occurs because the shifted magnitude

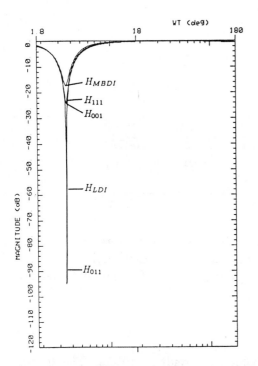
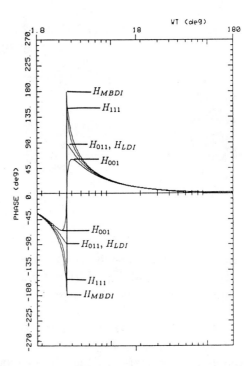

Fig. 6.3.4 Magnitude and phase responses of first-order Butterworth digital bandstop filter using H_{001}, H_{111}, H_{011}, H_{MBDI}, and H_{LDI} transforms (see Eq. 6.3.22).
$\omega_o = 1$ rad/sec, $\omega_s = 100$ rad/sec, $\omega_o T = 3.6°$, and $Q = 1$. (Plotted by C. Powell.
Based upon T. Wongterarit, loc. cit.)

responses overlap as shown in Fig. 1.3.3. We choose a sampling frequency of 100 rad/sec.
Then the digital bandstop filter has a center frequency $\omega_o T$ and a 3 dB bandwidth BT
where both equal 3.6°. From Table 6.3.1, we compute some analog lowpass-to-digital
bandstop transformations as

$$H_{001}: \quad p_n = 0.062791\left(\frac{1 - z^{-1}}{1.00394 - 2z^{-1} + z^{-2}}\right)$$

$$H_{111}: \quad p_n = 0.062791\left(\frac{z^{-1}(1 - z^{-1})}{1 - 2z^{-1} + 1.00394z^{-2}}\right)$$

$$H_{011}: \quad p_n = 0.031426\left(\frac{1 - z^{-2}}{1 - 1.99606z^{-1} + z^{-2}}\right) \qquad (6.3.21)$$

$$H_{MBDI}: \quad p_n = 0.031364\left(\frac{z^{-1}(1 - z^{-2})}{1 - 2z^{-1} + 1.00098z^{-2} + 0.001967z^{-3} + 0.0009837z^{-4}}\right)$$

$$H_{LDI}: \quad p_n = 0.062822\left(\frac{z^{-0.5}(1 - z^{-1})}{1 - 1.99605z^{-1} + z^{-2}}\right)$$

These transformations are reciprocals of those computed for the analogous bandpass filter in Eq. 6.2.23. Substituting these p_n-to-z transforms into the first-order Butterworth
analog lowpass filter transfer function $H_{LP}(p_n) = 1/(p_n + 1)$ yields

$$H_{001}: \quad H(z) = \frac{1.00394 - 2z^{-1} + z^{-2}}{1.06673 - 2.06279z^{-1} + z^{-2}}$$

$$H_{111}: \quad H(z) = \frac{1 - 2z^{-1} + 1.00394z^{-2}}{1 - 1.93721z^{-1} + 0.94115z^{-2}}$$

$$H_{011}: \quad H(z) = \frac{1 - 1.99606z^{-1} + z^{-2}}{1.03143 - 1.99605z^{-1} + 0.96857z^{-2}} \qquad (6.3.22)$$

$$H_{MBDI}: \quad H(z) = \frac{1 - 2z^{-1} + 1.00098z^{-2} + 0.001967z^{-3} + 0.0009837z^{-4}}{1 - 1.96864z^{-1} + 1.00098z^{-2} - 0.02940z^{-3} + 0.0009837z^{-4}}$$

$$H_{LDI}: \quad H(z) = \frac{1 - 1.99605z^{-1} + z^{-2}}{1 + 0.06282z^{-0.5} - 1.99605z^{-1} - 0.06282z^{-1.5} + z^{-2}}$$

The magnitude and phase responses of these first-order Butterworth digital bandstop filters are plotted in Fig. 6.3.4. All these filters exhibit bandstop characteristics. However, only the H_{011} and H_{LDI} transforms provide infinite rejection at the center frequency. H_{MBDI} provides the least rejection.

6.3.4 BANDSTOP FILTER ORDER

Bandstop filters are specified in an analogous fashion to bandpass filters. The specifications may be in the frequency domain, the time domain, or in both domains simultaneously. In the time domain, narrowband bandstop filter approximations given by Eq. 6.3.11 are used. In the frequency domain, the responses described by Eq. 6.3.10 and the nomographs of Chap. 4 are utilized. In general, the frequency domain specifications of analog bandstop filters have the form shown in Fig. 6.3.5a. f_L and f_U are the required corner frequencies for less than M_p in-band gain variation. Frequencies $(f_1, f_2), (f_3, f_4), \ldots$ are the frequencies for stopband rejections of $M_{r1}, M_{r2}, \ldots$. In many applications, the gain characteristic will have geometric symmetry about the center frequency f_o so that $f_o^2 = f_L f_U = f_1 f_2 = f_3 f_4 = \ldots$. To utilize the design information of Chap. 4.2, the analog bandstop filter data is converted into equivalent data for an analog lowpass filter using

$$\Omega_i = \frac{1}{Q \left| \dfrac{f_i}{f_o} - \dfrac{f_o}{f_i} \right|} \qquad (6.3.23)$$

We normalize each stopband frequency f_i by f_o, form its reciprocal, take the absolute value of the difference, multiply by Q, and reciprocate. Alternatively, in analog bandstop filters having geometric symmetry, we simply determine the shaping factor $S_{M_p}^{M_r} = B_p/B_r$ (note the inverted bandwidth ratio compared to the bandpass filter case in Eq. 6.2.29). M_p and M_r are arbitrary attenuations; B_p and B_r are the corresponding bandwidths of the passband and stopband. The equivalent analog lowpass filter specification is then drawn as shown in Fig. 6.3.5b. f_o corresponds to $\Omega_o = \infty$, the band-edge frequencies f_U and f_L correspond to $\Omega_U = \Omega_L = 1$, and the Ω_i correspond

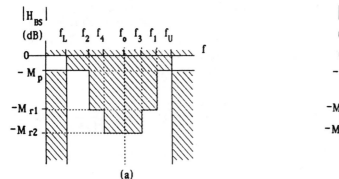

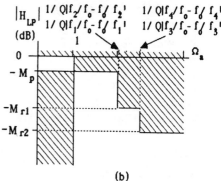

Fig. 6.3.5 (a) Analog bandstop filter specification and its (b) equivalent analog lowpass filter specification.

to the various cut-off frequencies. The analog lowpass filter orders are easily determined using this specification and the nomographs.

Some bandstop filter applications specify magnitude responses which have arithmetic symmetry rather than geometric symmetry. In these situations, the nomographs can still be utilized assuming that bandstop filters having geometric symmetry are acceptable. The techniques requires that the most stringent stopband frequency be selected as $\Omega_{ai} = \min(\Omega_i, \Omega_{i+1})$ at a given attenuation M_r. Further details of the arithmetical bandstop filters are discussed in Ref. 13.

EXAMPLE 6.3.2 Determine the minimum order Butterworth, Chebyshev, and elliptic analog bandstop filters required to meet the following magnitude characteristics (refer to Fig. 6.3.5a). The 1.25 dB passband lies below 250 Hz and above 4000 Hz. The 40 dB stopband lies between 500–2000 Hz.

SOLUTION To determine the filter orders, we first determine the center frequency f_o, bandwidth B, and Q of the analog bandstop filter as

$$f_o = \sqrt{250(4000)} = 1000 \text{ Hz}, \quad B = 4000 - 250 = 3750 \text{ Hz}, \quad Q = \frac{1000}{3750} = 0.267$$
$$(6.3.24)$$

Therefore using Eq. 6.3.23, the normalized lowpass stopband frequencies are

$$\Omega_1 = 1 \Big/ \frac{1}{0.267} \left| \frac{2000}{1000} - \frac{1000}{2000} \right| = 2.5, \quad \Omega_2 = 1 \Big/ \frac{1}{0.267} \left| \frac{500}{1000} - \frac{1000}{500} \right| = 2.5$$
$$(6.3.25)$$

which are equal since the frequencies have geometric symmetry about f_o. Using the shaping factor approach, $B_{1.25dB} = 4000 - 250 = 3750$ Hz and $B_{40dB} = 2000 - 500 = 1500$ Hz so the 1.25-40 dB shaping factor $S_{1.25}^{40} = 3750/1500 = 2.5$. Therefore the stopband frequency equals $\Omega_a = 2.5$ directly. This eliminates the need to evaluate Eqs. 6.3.24 and 6.3.25. The equivalent lowpass filter specification is drawn in Fig. 6.3.5b. Entering $(M_p, M_r, \gamma, \Omega_a) = (1.25, 40, 4.5, 2.5)$ onto the nomographs of Chap. 4.2, we find: Butterworth $n = 6$, Chebyshev $n = 4$, elliptic $n = 3$. ∎

EXAMPLE 6.3.3 Determine the minimum order Butterworth, Chebyshev, and elliptic bandstop filters required to meet the following magnitude characteristics having arithmetic symmetry. The frequency pairs (125, 1375) Hz mark the edges of the 1.25 dB passband, (250, 1250) Hz are the corners of the 40 dB stopband, and (500, 100) Hz are the corners of the 60 dB stopband in Fig. 6.3.5a.

SOLUTION In this case, the analog bandstop filter parameters equal

$$f_o = \sqrt{125(1375)} = 415 \text{ Hz}, \quad B = 1375 - 125 = 1250 \text{ Hz}, \quad Q = \frac{415}{1250} = 0.332$$

$$(6.3.26)$$

Computing the upper and lower stopband frequency pairs of the equivalent analog lowpass filter gives $(\Omega_1, \Omega_2) = (1.12, 2.85)$ and $(\Omega_3, \Omega_4) = (1.51, 8.04)$ using

$$\Omega_i = 1 \left/ \frac{1}{0.332} \left| \frac{f_i}{415} - \frac{415}{f_i} \right| \right. \tag{6.3.27}$$

The equivalent analog lowpass filter is drawn in Fig. 6.3.5b. The required stopband frequencies Ω_i are unequal and so that the smallest of each frequency pair is chosen as $\Omega_{ai} = \min(\Omega_i, \Omega_{i+1})$. This gives the most stringent gain requirement. In this case, $(M_p, M_r, \gamma, \Omega_a)$ equals $(1.25, 40, 4.5, 1.12)$ and $(1.25, 60, 6.5, 1.51)$. From the nomographs of Chap. 4.2, we find: Butterworth impractical; Chebyshev $n = 12, 9$ (use $n = 12$); elliptic $n = 6, 6$ (use $n = 6$).   ∎

Digital bandstop filters have the magnitude response specifications of the form shown in Fig. 6.3.6a with frequencies given by Eq. 6.2.34. It is analogous to the digital bandpass filter shown in Fig. 6.2.10a. The digital bandstop filter specification can be converted into the equivalent analog highpass filter specification following the same procedure used for digital bandpass filters. The center frequency ω_o, bandwidth B, and quality factor Q of the equivalent analog bandstop filter are described by Eq. 6.2.35. The equivalent analog highpass frequency ratio Ω_i is given by Eq. 6.2.36 and its reciprocal gives the equivalent lowpass frequency ratio. Summarizing the analog bandstop filter equations gives

$$v_o T = \sqrt{v_U T \, v_L T}, \qquad bT = v_U T - v_L T, \qquad q = \frac{v_o T}{bT}$$

$$\Omega_i = 1 \left/ q \left| \frac{v_i T}{v_o T} - \frac{v_o T}{v_i T} \right| \right. = \left| \frac{v_i T \, bT}{(v_i T)^2 - (v_o T)^2} \right| \tag{6.3.28}$$

The normalized digital bandstop frequencies ωT are converted into the equivalent analog bandstop frequencies $v_{BS} T$ using Eq. 5.11.1. For example, the H_{011} transform uses $v_i T = 2 \tan(\omega_i T/2)$. Evaluating Eq. 6.3.28 for the transforms in Table 6.3.1 gives

$$H_{001}, H_{111}, H_{112}: \quad \Omega_a = \left| \frac{\sin(\omega_r T) \sin(BT)}{\sin^2(\omega_r T) - \sin^2(\omega_o T)} \right|$$

$$(6.3.29a\text{--}b)$$

$$H_{011}: \quad \Omega_a = \left| \frac{\tan(\omega_r T/2) \tan(BT/2)}{\tan^2(\omega_r T/2) - \tan^2(\omega_o T/2)} \right|$$

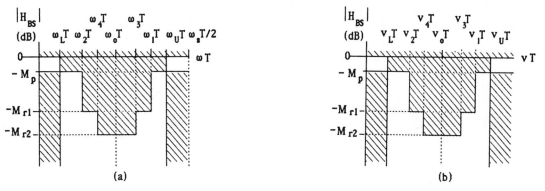

Fig. 6.3.6 (a) Digital bandstop filter specification and its (b) equivalent analog bandstop filter specification.

$$H_{MBDI}: \quad \Omega_a = \left| \frac{\tan(\omega_r T/2)\tan(BT/2)\cos(\omega_r T)\cos(BT)}{\tan^2(\omega_r T/2)\cos^2(\omega_r T) - \tan^2(\omega_o T/2)\cos^2(\omega_o T)} \right|$$

$$H_{LDI}: \quad \Omega_a = \left| \frac{\sin(\omega_r T/2)\sin(BT/2)}{\sin^2(\omega_r T/2) - \sin^2(\omega_o T/2)} \right| \qquad (6.3.29c\text{–}e)$$

$$H_{ODI}: \quad \Omega_a = \left| \frac{\sin(\omega_r T/4)\sin(BT/4)}{\sin^2(\omega_r T/4) - \sin^2(\omega_o T/4)} \right|$$

EXAMPLE 6.3.4 Determine the minimum order Butterworth, Chebyshev, and elliptic digital bandstop filters required to meet the magnitude characteristics shown in Fig. 6.3.7a. The sampling frequency is 3000 Hz. Use a bilinear transform.

SOLUTION The digital frequencies are first normalized by the 3 KHz sampling frequency using $\omega_i T = 2\pi f_i/f_s$. This converts (350, 430, 600, 700, 1500) Hz into $(42°, 51.6°, 72°, 84°, 180°)$ Hz as shown in Fig. 6.3.7a. Then the equivalent analog bandstop filter frequencies are computed using $v_i T = 2\tan(\omega_i T/2)$. These frequencies are $(44°, 55.4°, 83.3°, 103°, \infty)$ as shown in Fig. 6.3.7b. Next the analog bandstop filter parameters $v_o T, bT$, and q are computed as

$$v_o T = \sqrt{(44°)(103°)} = 67.3°, \quad bT = 103° - 44° = 59°, \quad q = \frac{67.3°}{59°} = 1.14$$

$$(6.3.30)$$

The equivalent analog highpass filter frequencies are computed using Eq. 6.2.29 as

$$\Omega_i = 1.14 \left| \frac{v_i T}{67.3°} - \frac{67.3°}{v_i T} \right| = \left| \frac{\tan^2(\omega_i T/2) - \tan^2(\omega_o T/2)}{\tan(\omega_i T/2)\tan(BT/2)} \right| \qquad (6.3.31)$$

The upper and lower 40 dB stopband frequencies of the analog highpass filter are found to equal $\Omega_1 = 0.490$ and $\Omega_2 = 0.446$ as shown in Fig. 6.3.7c. Following the method illustrated in Fig. 6.1.4, inverting frequency converts the analog highpass filter into an analog lowpass filter as shown in Fig. 6.3.7d where $\Omega_1 = 2.04$ and $\Omega_2 = 2.24$. Choosing $\Omega_a = \min(2.04, 2.24) = 2.04$ and entering $(M_p, M_r, \gamma, \Omega_a) = (0.5, 40, 4.9, 2.04)$ onto the nomographs of Chap. 4.2, we find: Butterworth $n = 8$, Chebyshev $n = 5$, and elliptic $n = 4$. ∎

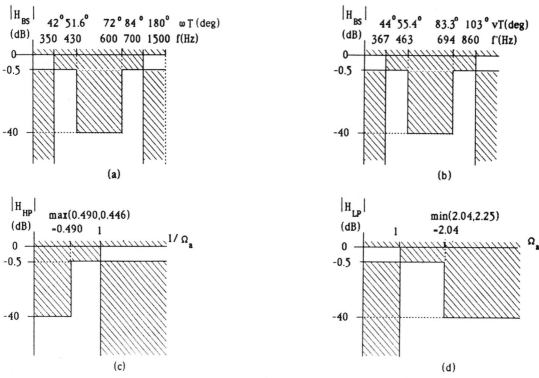

Fig. 6.3.7 (a) Digital bandstop filter specification and its equivalent (b) analog bandstop filter, (c) analog highpass filter, and (c) analog lowpass filter specifications of Example 6.3.4.

6.3.5 BANDSTOP FILTER DESIGN PROCEDURE

The design procedure for digital bandstop filters using direct transformations is straightforward. It consists of the following steps:

1a. Select a suitable analog filter type (e.g., Butterworth, Chebyshev, elliptic, etc.) and
1b. Choose a particular discrete transform from Tables 5.2.1 and 5.2.2 based upon the discussion of Chap. 5.9.
2. Determine the required analog lowpass filter order as described in Chap. 6.3.4.
3. Write the transfer function $H_{LP}(p_n)$ of the analog lowpass filter having unity bandwidth and the desired magnitude from the appropriate table of Chap. 4.2.
4. Evaluate the direct analog lowpass-to-digital bandstop transformation from Table 6.3.1 or compute using Eq. 6.3.15.
5. Compute the transfer function $H_{BS}(z)$ of the digital bandstop filter by substituting the p_n-to-z transformation found in Step 4 into $H_{LP}(p_n)$ found in Step 3.
6. Implement this digital bandstop transfer function using one of the techniques discussed in Chap. 12.

Let us illustrate this procedure by designing a third-order Butterworth digital bandstop filter using the transformations given in Table 6.3.1. *Steps 1-2* – Not necessary since the filter type and order have been specified. *Step 3* – The transfer function of the third-order Butterworth analog lowpass filter is

$$H(p_n) = \frac{1}{p_n^3 + 2p_n^2 + p_n + 1} \tag{6.3.32}$$

where $p_n = p_{LP}T/v_{LP}T$. *Step 4* – The digital bandstop filter is specified to have a center frequency of 1 KHz, a 3 dB bandwidth of 100 Hz, and $Q = 10$. The upper and lower 3 dB frequencies are 1051 Hz and 951 Hz, respectively. Choosing a sampling frequency of 10 KHz, the digital center frequency and bandwidth are $36°$ and $3.6°$, respectively. From Table 6.3.1, we compute some analog lowpass-to-digital bandstop transforms as

$$H_{001} : \quad p_n = 0.062791\left(\frac{1 - z^{-1}}{1.34549 - 2z^{-1} + z^{-2}}\right)$$

$$H_{111} : \quad p_n = 0.062791\left(\frac{z^{-1}(1 - z^{-1})}{1 - 2z^{-1} + 1.34549z^{-2}}\right)$$

$$H_{011} : \quad p_n = 0.031426\left(\frac{1 - z^{-2}}{1 - 1.61803z^{-1} + z^{-2}}\right) \tag{6.3.33}$$

$$H_{MBDI} : \quad p_n = 0.031364\left(\frac{z^{-1}(1 - z^{-2})}{1 - 2z^{-1} + 1.06910z^{-2} + 0.13820z^{-3} + 0.06910z^{-4}}\right)$$

$$H_{LDI} : \quad p_n = 0.062822\left(\frac{z^{-0.5}(1 - z^{-1})}{1 - 1.61803z^{-1} + z^{-2}}\right)$$

These transformations are reciprocals of those computed for the analogous bandpass filter in Eq. 6.2.25. Substituting these transforms into Eq. 6.3.32 gives the transfer functions of the third-order Butterworth digital bandstop filters as

$$H_{001} : \quad H(z) = 0.91092(1 - 4.45934z^{-1} + 8.85823z^{-2} - 9.91289z^{-3}$$
$$\frac{+6.58364z^{-4} - 2.46325z^{-5} + 0.37397z^{-6})}{1 - 4.41398z^{-1} + 8.65511z^{-2} - 9.54382z^{-3}}$$
$$+ 6.23493z^{-4} - 2.29079z^{-5} + 0.37397z^{-6}$$

$$H_{111} : \quad H(z) = 1 - 6z^{-1} + 16.03647z^{-2} - 24.14590z^{-3}$$
$$\frac{+21.57694z^{-4} - 10.86208z^{-5} + 2.43581z^{-6}}{1 - 5.87442z^{-1} + 15.41646z^{-2} - 22.83461z^{-3}} \tag{6.3.34a–c}$$
$$+ 20.10616z^{-4} - 9.98748z^{-5} + 2.21377z^{-6}$$

$$H_{011} : \quad H(z) = 0.93909(1 - 4.85410z^{-1} + 10.85410z^{-2} - 13.94427z^{-3}$$
$$\frac{+10.85410z^{-4} - 4.85410z^{-5} + z^{-6})}{1 - 4.75245z^{-1} + 10.40461z^{-2} - 13.08895z^{-3}}$$
$$+ 9.97768z^{-4} - 4.37044z^{-5} + 0.88189z^{-6}$$

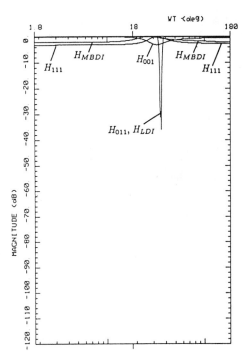 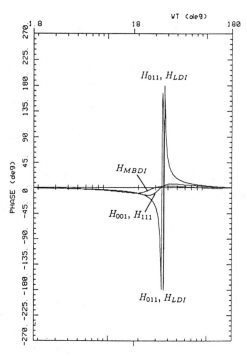

Fig. 6.3.8 Magnitude and phase responses of third-order Butterworth digital band-stop filter using H_{001}, H_{111}, H_{011}, H_{MBDI}, and H_{LDI} transforms (see Eq. 6.3.34). $f_o = 1$ KHz, $f_s = 10$ KHz, $\omega_o T = 36°$, and $Q = 10$. (Plotted by C. Powell. Based upon T. Wongterarit, loc. cit.)

$H_{MBDI}:$

$$H(z) = \frac{\begin{aligned}&1 - 6z^{-1} + 15.20730z^{-2} - 20.41459z^{-3}\\ &+ 14.80703z^{-4} - 5.14217z^{-5} + 0.77871z^{-6}\\ &- 0.46991z^{-7} + 0.19792z^{-8} + 0.03525z^{-9}\\ &+0.01927z^{-10} + 0.001979z^{-11} + 0.0003299z^{-12}\end{aligned}}{\begin{aligned}&1 - 5.93727z^{-1} + 14.95835z^{-2} - 20.09618z^{-3}\\ &+ 14.80520z^{-4} - 5.47347z^{-5} + 1.02872z^{-6}\\ &- 0.50952z^{-7} + 0.19975z^{-8} + 0.02532z^{-9}\\ &+ 0.01821z^{-10} + 0.001680z^{-11} + .0000003299z^{-12}\end{aligned}}$$

$(6.3.34d\text{--}e)$

$H_{LDI}:$

$$H(z) = \frac{\begin{aligned}&1 - 4.85410z^{-1} + 10.85410z^{-2} - 13.94427z^{-3}\\ &+10.85410z^{-4} - 4.85410z^{-5} + z^{-6}\end{aligned}}{\begin{aligned}&1 + 0.12564z^{-0.5} - 4.84621z^{-1} - 0.53198z^{-1.5}\\ &+ 10.82555z^{-2} + 0.98607z^{-2.5} - 13.90294z^{-3}\\ &- 0.98607z^{-3.5} + 10.82555z^{-4} + 0.53198z^{-4.5}\\ &- 4.84621z^{-5} - 0.12564z^{-5.5} + z^{-6}\end{aligned}}$$

The magnitude and phase responses are plotted in Fig. 6.3.8. We see that only the digital bandstop filters designed using the H_{001} and H_{LDI} transforms give good responses. The other transforms do not.

Now we will design a third-order elliptic digital bandstop filter meeting the same specifications. *Steps 1-3* – The transfer function is given by Eq. 6.2.43. *Step 4* – The center frequency and bandwidth are $36°$ and $3.6°$, respectively. Therefore we utilize the p_n-to-z transforms of Eq. 6.3.33. *Step 5* – Substituting these transforms into Eq. 6.2.43 gives the transfer functions of the third-order elliptic digital bandstop filters as

H_{001} :

$$H(z) = 13.96521 \frac{\begin{aligned} &(1 - 4.45905z^{-1} + 8.85704z^{-2} - 9.91082z^{-3} \\ &+ 6.58190z^{-4} - 2.46251z^{-5} + 0.41042z^{-6}) \end{aligned}}{\begin{aligned} &1 - 4.40459z^{-1} + 8.61287z^{-2} - 9.46653z^{-3} \\ &+ 6.16125z^{-4} - 2.25392z^{-5} + 0.36604z^{-6} \end{aligned}}$$

H_{111} :

$$H(z) = 15.05851 \frac{\begin{aligned} &(1 - 6z^{-1} + 16.03702z^{-2} - 24.14806z^{-3} \\ &+ 21.58037z^{-4} - 9.00482z^{-5} + 1.81088z^{-6} \end{aligned}}{\begin{aligned} &1 - 5.84235z^{-1} + 15.25582z^{-2} - 22.49033z^{-3} \\ &+ 19.72027z^{-4} - 9.76227z^{-5} + 2.16010z^{-6} \end{aligned}}$$

H_{011} :

$$H(z) = 14.48887 \frac{\begin{aligned} &(1 - 4.85366z^{-1} + 10.85250z^{-2} - 13.94195z^{-3} \\ &+ 10.85250z^{-4} - 4.85366z^{-5} + z^{-6}) \end{aligned}}{\begin{aligned} &1 - 4.80607z^{-1} + 10.64234z^{-2} - 13.54273z^{-3} \\ &+ 10.44415z^{-4} - 4.62873z^{-5} + 0.94517z^{-6} \end{aligned}}$$

H_{MBDI} :

$$H(z) = 15.65851 \frac{\begin{aligned} &(1 - 6z^{-1} + 15.20743z^{-2} - 20.41486z^{-3} \\ &+ 14.80691z^{-4} - 5.14162z^{-5} + 0.77857z^{-6} \\ &- 0.47021z^{-7} + 0.19805z^{-8} + 0.03526z^{-9} \\ &+ 0.01928z^{-10} + 0.001980z^{-11} + 0.0003299z^{-9}) \end{aligned}}{\begin{aligned} &1 - 5.92125z^{-1} + 14.89420z^{-2} - 20.01370z^{-3} \\ &+ 14.80526z^{-4} - 5.56053z^{-5} + 1.09318z^{-6} \\ &- 0.51824z^{-7} + 0.19969z^{-8} + 0.02268z^{-9} \\ &+ 0.01790z^{-10} + 0.001604z^{-11} + 0.0000003299z^{-12} \end{aligned}}$$

(6.3.35)

H_{LDI} :

$$H(z) = 15.65851 \frac{\begin{aligned} &(1 - 4.85355z^{-1} + 10.85210z^{-2} \\ &- 13.94824z^{-3} + 10.85210z^{-4} - 4.85355z^{-5} + z^{-6}) \end{aligned}}{\begin{aligned} &1 + 0.15773z^{-0.5} - 4.84648z^{-1} - 0.66761z^{-1.5} \\ &+ 10.82653z^{-2} + 1.23721z^{-2.5} - 13.90438z^{-3} \\ &- 1.23721z^{-3.5} + 10.82653z^{-4} + 0.66761z^{-4.5} \\ &- 4.84648z^{-5} - 0.15773z^{-5.5} + z^{-6} \end{aligned}}$$

The magnitude and phase responses are shown in Fig. 6.3.9. Again only the bandstop filters designed using the bilinear transform and lossless transforms give good responses.

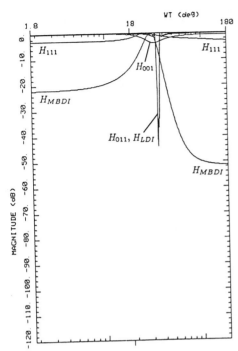
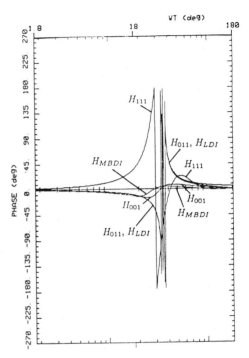

Fig. 6.3.9 Magnitude and phase responses of third-order elliptic digital bandstop filter using H_{001}, H_{111}, H_{011}, H_{MBDI}, and H_{LDI} transforms (see Eq. 6.3.35). $\omega_o = 1$ rad/sec, $\omega_s = 100$ rad/sec, $\omega_o T = 36°$, and $Q = 10$. (Plotted by C. Powell. Based upon T. Wongterarit, loc. cit.)

6.3.6 COMPLEMENTARY TRANSFORMATION

Complementary digital bandstop filters can be obtained from digital bandpass filters by utilizing the complementary digital lowpass-to-digital highpass transformation of Eq. 6.1.32. This gives the flexibility to design digital bandstop filters having desirable time domain performance.

The complementary transformation relates the cosine step and impulse responses of digital bandpass and bandstop filters as

$$r_{BS}(nT) = U_{-1}(nT) - r_{BP}(nT)$$
$$h_{BS}(nT) = U_0(nT) - h_{BP}(nT)$$

$$(6.3.36)$$

This condition forces the time domain responses to be compliments of one another. This maintains the cosine step response characteristics in terms of overshoot (undershoot), rise time (fall time), delay time (storage time), settling time, etc. To determine the relation between the complementary filter transfer functions, take the z-transform of the impulse responses in Eq. 6.3.36. This yields the *complementary digital bandpass-to-digital bandstop transformation* where

$$H_{BS}(z) = 1 - H_{BP}(z)$$

$$(6.3.37)$$

$H_{BP}(z)$ is magnitude normalized so that $H_{BP}(e^{j\omega_o T}) = 1$. This insures that the center frequency gain $H_{BS}(e^{j\omega_o T})$ of the bandstop filter equals zero. If the bandpass filter $H_{BP}(z)$ has n poles and m finite zeros, then Eq. 6.3.37 shows that $H_{BS}(z)$ has the same n poles but that the original zeros are modified and $(n-m)$ additional zeros are introduced. Thus, stable digital bandpass filters remain stable under this transformation, but minimum-phase filters may become nonminimum phase. All-pole digital bandpass filters convert into all-pole digital bandstop filters having n finite zeros where one of these zeros is located at $z = e^{j\omega_o T}$. Such bandstop filters have mid-band roll-offs (relative to $\omega - \omega_o$) of only ± 20 dB/dec. Stopband rejection is not increased by increasing filter order. Nevertheless, the complementary transformation is useful because it gives us the ability to maintain time domain behavior.

EXAMPLE 6.3.5 The transfer functions of first-order Butterworth digital bandpass filters having a center frequency of $36°$ and a Q of 10 are listed in Eq. 6.2.26. Find the digital bandstop filter for the bilinear transform case that has a complementary cosine step response.

SOLUTION   The bandpass filter transfer function using the bilinear transform equals

$$H_{BP}(z) = \frac{0.030469(1 - z^{-2})}{1 - 1.56874z^{-1} + 0.93906z^{-2}} \tag{6.3.38}$$

from Eq. 6.2.26. Since $H_{BP}(e^{j36°}) = 1$, we can use Eq. 6.3.37 directly to find the complementary bandstop filter transfer function as

$$H_{CBS}(z) = \frac{0.96953 - 1.56874z^{-1} + 0.96953z^{-2}}{1 - 1.56874z^{-1} + 0.93906z^{-2}} \tag{6.3.39}$$

∎

6.4 LOWPASS-TO-ALLPASS TRANSFORMATION [14]

Analog lowpass filters are transformed into digital allpass filters using the direct *analog lowpass-to-digital allpass transformation* where

$$H_{AP}(z) = H_{AP}(p_{AP})\bigg|_{p_{AP} = \frac{1}{H_{dmn}(z)}} = \frac{H_{LP}(p_{LP})}{H_{LP}(-p_{LP})}\bigg|_{p_{LP} = \frac{1}{H_{dmn}(z)}} \tag{6.4.1}$$

The equivalent but simpler equation of Eq. 3.12.6 directly converts digital lowpass filters into digital allpass filters using

$$H_{AP}(z) = \frac{H_{LP}(z)}{H_{LP}(z^{-1})} = H_{LP}(z)H_{LP}^{-1}(z^{-1}) \tag{6.4.1}$$

This transformation produces zeros which are z^{-1} reciprocal images of the poles, and vice versa. A lowpass root which equals $|z|e^{\arg z}$ transforms into an allpass image root which equals $|z|^{-1}e^{-\arg z}$ (reciprocal magnitude, negative angle). Stable, causal allpass filters (the usual case of interest) only have poles inside the unit circle and zeros outside the unit circle. They have no poles or zeros on the unit circle since pole-zero cancellation occurs. Thus, only lowpass filters having no finite zeros, except perhaps additional zeros on the unit circle, are used in Eq. 6.4.1. Allpass filters have gains, phase, and delay responses which satisfy

$$|H_{AP}(e^{j\omega T})| = 1, \quad \arg H_{AP}(e^{j\omega T}) = 2 \arg H_{LP}(e^{j\omega T}), \quad \tau_{AP}(e^{j\omega T}) = 2\tau_{LP}(e^{j\omega T})$$
$$(6.4.3)$$

The phase characteristic of the allpass filter is twice that of the lowpass filter and must always be monotonically nonincreasing. Allpass filters can only introduce positive delay. The delay is twice that of its analogous lowpass filter.[15]

In most cases, the lowpass filter selected is an nth order all-pole filter having $H_{LP}(z) = D_n(0)/D_n(z)$. Then the analogous allpass filter has a transfer function

$$H_{AP}(z) = \frac{D_n(z^{-1})}{D_n(z)} \tag{6.4.4}$$

Usually, the lowpass filter is selected so that it provides the desired delay response. For example if H_{AP} is used to delay compensate some delay characteristic τ_c, then τ_{AP} and τ_{LP} must satisfy

$$\tau_{AP} = 2\tau_{LP} = \text{constant} - \tau_c \tag{6.4.5}$$

In general, the optimization methods of Chap. 4.3 are used to determine the parameters of the required H_{AP} delay equalizer.

Although the magnitude, phase, and delay responses of allpass filters are easy to determine from their transfer functions, their step responses are not. Unfortunately, there are no convenient transformation theorems known which directly relate lowpass and allpass filter step responses. They must be calculated directly from their transfer functions using z-transform methods.

EXAMPLE 6.4.1   The delay of the third-order Butterworth lowpass filter of Example 5.11.1 is used for delay compensation. Write the transfer function of the resulting third-order Butterworth allpass filter.

SOLUTION   The transfer functions of the Butterworth analog lowpass and allpass filters equal

$$H_{LP}(p_n) = \frac{1}{(1+p_n)(1+p_n+p_n^2)}, \quad H_{AP}(p_n) = \frac{(1-p_n)(1-p_n+p_n^2)}{(1+p_n)(1+p_n+p_n^2)}$$
$$(6.4.6)$$

from Eq. 5.11.9. The corresponding Butterworth digital lowpass filter has

$$H_{LP}(z) = \frac{0.01807(1+z^{-1})^3}{(1-0.5095z^{-1})(1-1.2510z^{-1}+0.5457z^{-2})} \tag{6.4.7}$$

from Eq. 5.11.11. Since this lowpass filter has no finite zeros except on the unit circle, it may be used to obtain an allpass filter. From Eq. 6.4.2, the transfer function of the allpass filter equals

$$H_{AP}(z) = \frac{H_{LP}(z)}{H_{LP}(z^{-1})} = \frac{(z^{-1} - 0.5095)(z^{-2} - 1.2510z^{-1} + 0.5457)}{(1 - 0.5095z^{-1})(1 - 1.2510z^{-1} + 0.5457z^{-2})} \quad (6.4.8)$$

There is pole-zero cancellation at $z = -1$. The zeros of $H_{AP}(z)$ are reciprocal images of the poles as required of allpass filters. ∎

PROBLEMS

6.1 The magnitude responses of a variety of medium cut-off analog highpass filters are shown in Fig. P6.1. (a) Determine what type of classical analog filters (Butterworth, Chebyshev, elliptic, Bessel) can be used and their orders. (b) Determine the corresponding frequencies of the equivalent digital highpass filter using the bilinear transform and other transforms in Table 6.1.1. Assume that $\omega_s = 10$.

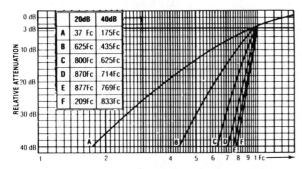

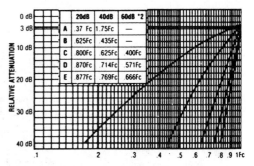

Fig. P6.1 (Courtesy of Allen Avionics[16])   Fig. P6.2 (Courtesy of Allen Avionics[16])

6.2 The magnitude responses of a variety of sharp cut-off digital highpass filters are shown in Fig. P6.2. Determine what type of classical filters (Butterworth, Chebyshev, elliptic, Bessel) can be used and their orders. Use the bilinear transform and other transforms in Table 6.1.1. Assume that $\omega_s = 20$.

6.3 A digital highpass filter has the following requirements: maximum in-band gain = 0 dB, 3 dB passband $f_p \geq 1209$ Hz, 40 dB stopband $f_r \leq 941$ Hz. (a) Determine the minimum orders of the Butterworth, Chebyshev, elliptic, and Bessel filters. Use the bilinear transform and other transforms in Table 6.1.1. Assume that $f_s = 8f_p$. (b) Determine the corresponding frequencies of the equivalent analog highpass filter.

6.4 Consider a digital highpass filter having the following requirements: maximum in-band gain = 0 dB, $M_p = 0.28$ dB for $f_p \geq 1209$ Hz, $M_r = 40$ dB for $f_r \leq 941$ Hz. Determine the minimum orders of the Butterworth, Chebyshev, elliptic, and Bessel filters. Use the bilinear transform and other transforms in Table 6.1.1. Assume that $f_s = 16f_p$.

6.5 A digital highpass filter has the following requirements: maximum in-band gain = 0 dB, 3 dB passband $f_p \geq 3.2$ KHz, 40 dB stopband $f_r \leq 2.1$ KHz. (a) Determine the minimum orders of the Butterworth, Chebyshev, elliptic, and Bessel filters. Use the bilinear transform and other transforms in Table 6.1.1. Assume that $f_s = 19.2$ KHz. (b) Determine the corresponding frequencies of the equivalent analog highpass filter.

6.6 Consider a digital highpass filter having the following requirements: maximum in-band gain = 0 dB, $M_p = 0.28$ dB for $f_p \geq 3.2$ KHz, $M_r = 40$ dB for $f_r \leq 2.1$ KHz. Determine the minimum orders of the Butterworth, Chebyshev, elliptic, and Bessel filters. Use the bilinear transform and other transforms in Table 6.1.1. Assume that $f_s = 38.4$ KHz.

6.7 Consider a digital highpass filter having the following requirements: dc gain = 0 dB, 3 dB passband $f_p \geq 9.6$ KHz, 30 dB stopband $f_r \leq 4.8$ KHz. Determine the minimum orders of the Butterworth, Chebyshev, elliptic, and Bessel filters. Use the bilinear transform and other transforms in Table 6.1.1. Assume that $f_s = 8f_p$.

6.8 Consider a digital highpass filter having the following requirements: dc gain = 0 dB, 3 dB passband $f_p \geq 9.6$ KHz, 40 dB stopband $f_r \leq 4.8$ KHz. Determine the minimum orders of the Butterworth, Chebyshev, elliptic, and Bessel filters. Use the bilinear transform and other transforms in Table 6.1.1. Assume that $f_s = 16f_p$.

6.9 Consider an analog highpass filter $H(s) = s/(s+1)$ and a sampling rate of 50 rad/sec. Find the digital highpass filter $H(z)$ using the: (a) forward Euler, (b) backward Euler, (c) bilinear, (d) modified bilinear, and (e) lossless transforms.

6.10 Consider an analog lowpass filter $H(s) = 1/(s+1)$ and its corresponding highpass filter $H(s) = s/(s+1)$ and a sampling rate of 100 rad/sec. Find the digital highpass filter $H(z)$ using the: (a) forward Euler, (b) backward Euler, (c) bilinear, (d) modified bilinear, and (e) lossless transforms.

6.11 Given an analog lowpass filter $H(s) = 1/(s^2 + \sqrt{2}s + 1)$ and a sampling rate of 10 rad/sec. Find the digital highpass filter $H(z)$ using the: (a) forward Euler, (b) backward Euler, (c) bilinear, (d) modified bilinear, and (e) lossless transforms.

Solve only one of the parts in Problems 6.12–6.13.

6.12* Design a digital highpass filter to satisfy the following Touch-Tone telephone filter specifications: maximum in-band gain = 0 dB, $f_p = 1209$ Hz, and $f_s = 20f_p$. Find its $H(z)$ using the bilinear transform. Draw its block diagram realization as the cascade of first- and second-order blocks. Realize this digital highpass filter as: (a) 4th order Butterworth; (b) 4th order Chebyshev with in-band ripple of 0.5 dB, (c) 1 dB, (d) 2 dB, (e) 3 dB; (f) 3rd order elliptic with in-band ripple of 0.28 dB and stopband rejection of 40 dB, (g) 60 dB; (h) 4th order Bessel.

6.13* Design a digital highpass filter to satisfy the following voice filter specifications: maximum in-band gain = 0 dB, $f_p = 3200$ Hz, and $f_s = 15f_p$. Find its $H(z)$ using the bilinear transform. Draw its block diagram realization as the cascade of first- and second-order blocks. Realize this digital highpass filter as: (a) 4th order Butterworth; (b) 4th order Chebyshev with in-band ripple of 0.5 dB, (c) 1 dB, (d) 2 dB, (e) 3 dB; (f) 3rd order elliptic with in-band ripple of 0.28 dB and stopband rejection of 40 dB, (g) 60 dB; (h) 4th order Bessel.

6.14 Consider an analog highpass filter $H(s) = s/(s+1)$ and a variable sampling rate $1/T$ Hz. Plot the root locus of the poles of the digital highpass filter $H(z)$ using the: (a) forward Euler, (b) backward Euler, (c) bilinear, (d) modified bilinear, and (e) lossless transforms.

6.15 Consider an analog lowpass filter $H(s) = 1/(s+1)$ and a variable sampling rate $1/T$ Hz. Plot the root locus of the poles of the digital highpass filter $H(z)$ using the: (a) forward Euler, (b) backward Euler, (c) bilinear, (d) modified bilinear, and (e) lossless transforms.

6.16 Given an analog lowpass filter $H(s) = 1/(s^2 + \sqrt{2}s + 1)$ and a variable sampling rate $1/T$ Hz. Plot the root locus of the poles of the digital highpass filter $H(z)$ using the: (a) forward Euler, (b) backward Euler, (c) bilinear, (d) modified bilinear, and (e) lossless transforms.

6.17 Given an analog lowpass filter $H(s) = 1/[(s+1)(s^2 + s + 1)]$ and a variable sampling rate $1/T$ Hz. Plot the root locus of the poles of the digital highpass filter $H(z)$ using the: (a) forward Euler, (b) backward Euler, (c) bilinear, (d) modified bilinear, and (e) lossless transforms.

6.18   Suppose $H_1(z) = 0.457(z-1)/(z+0.523)$, $H_2(z) = 0.3534(z-1)/(z+0.2932)$, and $H_3(z) = 0.8415(z-1)/(z+0.1585)$ where $H_i(z) = Y_i(z)/X_i(z)$. Write the software equations for implement these highpass $H_i(z)$.

6.19   Write the software equations to implement these highpass filters: (a) third-order Butterworth of Eq. 6.1.27 and (b) Eq. 6.1.28; (c) third-order elliptic of Eq. 6.1.30 and (d) Eq. 6.1.31.

6.20   The magnitude response of a digital bandpass filter having various shaping factors is shown in Fig. P6.20. (a) Can the bandpass filter be obtained from a lowpass filter using the LP-BP transformation? (b) Determine what type of classical filters (Butterworth, Chebyshev, elliptic, Bessel) can be used and their orders. Use the bilinear transform and other transforms in Table 6.2.1. Assume that $f_s = 10f_o$.

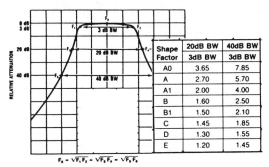

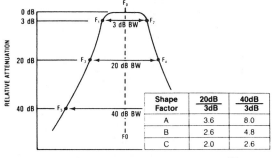

Fig. P6.20 (Courtesy of Allen Avionics[16])   Fig. P6.21 (Courtesy of Allen Avionics[16])

6.21   The magnitude responses of a variety of digital bandpass filters are shown in Fig. P6.21. Assume that $f_s = 20f_o$. Repeat Problem 6.20.

6.22   A 300 Baud acoustical data coupler uses two digital bandpass filters with the following specifications. Filter 1: $f_o = 1170$ Hz, $B_{0.28dB} = 400$ Hz, and $B_{40dB} = 3200$ Hz. Filter 2: $f_o = 2125$ Hz, $B_{0.28dB} = 400$ Hz, and $B_{40dB} = 3000$ Hz. Determine what type of classical filters (Butterworth, Chebyshev, elliptic, Bessel) can be used and their orders. Use the bilinear transform and other transforms in Table 6.2.1. Assume that $f_s = 10f_o$.

6.23   3400/3750 Hz bandpass filters are used to recover telephone signalling tones. Assume that the 3 dB frequencies of the filter are 3325 and 3625 Hz. Suppose the filter must provide 40 dB rejection at 3000 and 3950 Hz. (a) Determine what type of classical filters (Butterworth, Chebyshev, elliptic, Bessel) can be used and their orders. Use the bilinear transform and other transforms in Table 6.2.1. Assume that $f_s = 10f_o$. (b) To minimize order, a third-order elliptic bandpass shall be used. How much rejection can be obtained at 3000 and 3950 Hz? (c) What is the filter Q?

6.24   Touch-tone systems use bandpass filters to decode two frequencies (697, 770, 852, 941, 1209, 1339, 1477, 1633 Hz). Assume that the adjacent filter responses intersect at their 20 dB frequencies and that their Q (based on their ripple bandwidths) equals 25. Determine what type of classical filters (Butterworth, Chebyshev, elliptic, Bessel) can be used and their orders. Use the bilinear transform and other transforms in Table 6.2.1. Assume that $f_s = 10f_o$.

Solve only one of the parts in Problem 6.25.

6.25*   Design the following digital bandpass filters having unity passband gain, a center frequency of 2400 Hz, and a passband of 150 Hz in block diagram form: (a) 2nd order Butterworth, (b) 3rd order Butterworth; (c) 3rd order Chebyshev with in-band ripple of 0.5 dB, (d) 1 dB, (e) 2 dB, (f) 3 dB; (g) 3rd order elliptic with in-band ripple of 0.28 dB and stopband rejection of 40 dB, (h) 60 dB; (i) 3rd order Bessel. Use the bilinear transform and $f_s = 10f_o$.

6.26   Suppose $H_1(z) = 0.457(z-1)(z+1)/(z+0.523)$, $H_2(z) = 0.3534(z-1)(z+1)/(z+0.2932)$, and $H_3(z) = 0.8415(z-1)(z+1)/(z+0.1585)$ where $H_i(z) = Y_i(z)/X_i(z)$. Write the software equations for implement these bandpass $H_i(z)$.

6.27   Write the software equations to implement these bandpass filters: (a) first-order Butterworth of Eq. 6.2.24 and (b) Eq. 6.2.26; (c) third-order Butterworth of Eq. 6.2.28, (d) third-order elliptic of Eq. 6.2.44.

6.28   The magnitude responses of some digital bandstop filters are shown in Fig. P6.28. (a) Can the bandstop filter be obtained from a lowpass filter using the LP-BS transformation? (b) Determine what type of classical filters (Butterworth, Chebyshev, elliptic, Bessel) can be used and their orders. Use the bilinear transform and other transforms in Table 6.3.1. Assume that $f_s = 10f_o$.

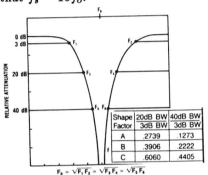

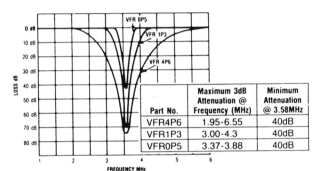

Fig. P6.28 (Courtesy of Allen Avionics[16])   Fig. P6.29 (Courtesy of Allen Avionics[16])

6.29   The magnitude responses of some digital bandstop filters are shown in Fig. P6.29. Assume that $f_s = 20f_o$. Repeat Problem 6.28.

6.30   The telephone company uses tone frequencies within the voice band (200–3200 Hz) to transmit signalling, supervisory, and dialing information between carrier terminals. Single-frequency (SF) signalling systems use a 2600 Hz tone which must often be notched out by a bandstop filter. Assume that that it has a 3 dB bandwidth of 60 Hz and that it provides 40 dB of rejection over a 30 Hz bandwidth. (a) Determine what type of classical digital filters (Butterworth, Chebyshev, elliptic, Bessel) can be used and their orders. Use the bilinear transform and other transforms in Table 6.3.1. Assume that $f_s = 10f_o$. (b) What is the filter Q?

6.31   The telephone company uses 2805 Hz notch filters. Suppose they have the following gain specifications: –0.5 dB frequencies = 2100, 3500 Hz; –3 dB frequencies = 2650, 2960 Hz; –50 dB frequencies = 2795, 2815 Hz. Determine what type of classical digital filters (Butterworth, Chebyshev, elliptic, Bessel) can be used and their orders. Use the bilinear transform and other transforms in Table 6.3.1. Assume that $f_s = 10f_o$. (b) What is the filter Q?

Solve only one of the parts in Problem 6.32.

6.32*   Design the following digital bandstop filters having unity passband gain in block diagram form: (a) 2nd order Butterworth filter with a center frequency of 1 KHz and a 3 dB bandwidth of 100 Hz. (b) 2nd order Butterworth filter with a center frequency of 2600 Hz and a 3 dB bandwidth of 60 Hz. (c) 2nd order Butterworth filter with a center frequency of 1010 Hz and a 0.5 dB bandwidth of 1220 Hz. (d) 4th order Chebyshev filter with a center frequency of 1000 Hz and a 1 dB bandwidth of 375 Hz. (e) 3rd order elliptic filter with a center frequency of 1000 Hz and a 0.28 dB bandwidth of 375 Hz. Use the bilinear transform and $f_s = 10f_o$.

6.33   Suppose $H_1(z) = 0.457(z^2+1)/(z+0.523)$, $H_2(z) = 0.3534(z^2+1)/(z+0.2932)$, and $H_3(z) = 0.8415(z^2+1)/(z+0.1585)$ where $H_i(z) = Y_i(z)/X_i(z)$. Write the software equations for implement these bandstop $H_i(z)$.

6.34   Write the software equations to implement these bandstop filters: (a) first-order Butterworth of Eq. 6.3.22, (b) third-order Butterworth of Eq. 6.3.34, (c) third-order elliptic of Eq. 6.3.35.

6.35   A 1 $msec$ Pade analog allpass filter has the following parameters: (a) first-order has $(H_o, f_o) = (1, 318 \text{ Hz})$; (b) second-order has $(H_o, f_o, \varsigma) = (1, 540 \text{ Hz}, 0.87)$. Find the gain $H(z)$ of the digital allpass filter. Use the bilinear transform and other transforms in Table 5.11.1. Assume that $f_s = 10f_o$.

6.36   A second-order allpass filter is cascaded with a fifth-order Butterworth lowpass filter to reduce its in-band delay variation. The analog allpass filter has parameters $(H_o, f_o, \varsigma) = (1, 730 \text{ Hz}, 0.87)$. Find the gain $H(z)$ of the digital allpass filter. Use the bilinear transform and other transforms in Table 5.11.1. Assume that $f_s = 10f_o$.

6.37   A second-order analog allpass filter has parameters $(H_o, f_o, \varsigma) = (1, 110 \text{ Hz}, 0.87)$. Find the gain $H(z)$ of the digital allpass filter. Use the bilinear transform and other transforms in Table 5.11.1. Assume that $f_s = 15f_o$.

REFERENCES

1. Lindquist, C.S., *Active Network Design with Signal Filtering Applications*, Chap. 6, Steward and Sons, CA, 1977.
2. Constantinides, A.G., "Spectral transformations for digital filters," Proc. IEE, vol. 117, pp. 1585–1590, Aug. 1970.
   Bogner, R.E. and A.G. Constantinides, eds., *Introduction to Digital Filtering*, Chap. 5, Wiley, NY, 1975.
3. Ref. 1, Chap. 6, Sec. 3.
4. Wongterarit, T., "Design of digital filters from analog lowpass filters using frequency transformations," M.S.E.E. Thesis, Univ. of Miami, 116 p., May 1986.
5. Ref. 1, Chap. 4, Sec. 8.
6. Blinchikoff, H.J., "Low-transient highpass filters," IEEE Trans. Circuit Theory, vol. CT-17, pp. 663–667, Nov. 1970.
   Ref. 1, Chap. 6, Sec. 3.
7. Ref. 1, Chap. 6, Sec. 4.
8. Ghausi, M.S., *Principles and Design of Linear Active Circuits*, Chap. 15.1–15.2, McGraw-Hill, NY, 1965.
   Jones, H.E., " On the roots of an undamped quadratic after a lowpass to bandpass transformation," Proc. IEEE, vol. 57, p. 1451, Aug. 1969.
9. Budak, A. and K.E. Waltz, " Linear amplitude Butterworth active filter," Proc. IEEE, vol. 58, pp. 274–275, Feb. 1970.
10. Joseph, R.D., " Maximally linear bandpass frequency discriminator," Proc. IEEE, vol. 59, pp. 1712–1713, Dec. 1971.
11. Budak, A. and K.E. Waltz, " An active Butterworth active filter," Proc. IEEE, vol. 58, pp. 795–796, May 1970.
12. Dangl, A., " A study of band-pass discriminators," M.S.E.E. Directed Research, California State University, Long Beach, June 1976.
13. Ref. 1, Chap 6, Sec. 5.
14. Ref. 1, Chap 6, Sec. 6.
15. Ref. 1, Chap 2, Sec. 11.
16. Precision LC Filters, Catalog 22F, 35 p., Allen Avionics, Inc., Mineola, NY, 1987. (These are analog filter responses. In Problems 6.2, 6.20, 6.21, 6.28, and 6.29, they are treated as digital filter responses.)

7 DIGITAL FILTER CLASSIFICATION

*The Spirit of the Lord is upon me (Jesus), because
he hath anointed me to preach the gospel to the poor;
he hath sent me to heal the broken-hearted,
to preach deliverance to the captives,
and recovering of sight to the blind,
to set at liberty them that are bruised,
to preach the acceptable year of the Lord.*

Luke 4:18–19

The previous chapters have been concerned with the analysis and design of the digital filter system shown in Fig. 2.0.1. General concepts and analysis tools were presented in Chap. 1. Discrete transforms (frequency domain analysis) and discrete inverse transforms (time domain analysis) were described in Chaps. 2 and 3, respectively. Systematic procedures for synthesizing classical, optimum, and adaptive analog filters were presented in Chap. 4. The various digital transformations of Chap. 5 converted these analog filters into digital filters. Chap. 6 presented frequency transformations which allowed digital highpass, bandpass, bandstop, and allpass filters to be obtained directly from analog lowpass filters.

Now $H(z)$ or $H(p)$ must be implemented in hardware or software. Chap. 7 overviews and interrelates hardware and software implementation methods in general terms. Their advantages, disadvantages, and trade-offs will be discussed. The design details of each method will be fully explored in the following four chapters (Chaps. 8–11) which the engineer may wish to peruse. A variety of implementation structures will be presented in Chap. 12.

7.1 DIGITAL FILTER CLASSIFICATION

There are three basic classes of digital filters:

1. Fast transform filters (Chaps. 8 and 9)
2. Nonrecursive filters (Chap. 10)
3. Recursive filters (Chap. 11)

Fast transform filters implement the generalized frequency response $H(p)$ of the digital filter using fast transform algorithms. The best known of these are the FFT filters. The ac steady-state frequency response $H(e^{j\omega T})$ of the digital filter gain $H(z)$ is synthesized using the FFT. These algorithms are performed in software or by specialized hardware modules.

Recursive and nonrecursive filters realize $H(z)$ in hardware using delays (shift registers), scalers (multipliers), and summers (adders), or in customized microprocessor hardware. These filters are also realized in software using sets of recursion equations, by performing time domain convolution, or by FFT and IFFT evaluation.

FFT filters can be realized in recursive or nonrecursive form but this requires that a transfer function $H(z)$ be formulated which adequately approximates the FFT gain response $H(e^{j\omega T})$. The resulting $H(z)$ filters can be of low order compared with the FFT dimension but the methods required to determine $H(z)$ involve optimization and are tedious to perform. Therefore there is little incentive to convert FFT filters into $H(z)$ filters. Nonrecursive filters are often realized using FFT's because of simplicity.

Let us now discuss each of these filter implementations. The details will be presented in later chapters so this discussion is intended only as an overview.

7.2 FAST TRANSFORM FILTERS

Fast transform filters are based on any convenient transform like Fourier, Walsh, Paley, Hadamard, Haar, slant, or Karhunen-Loéve (K-L). Any transform having orthogonal basis functions can be used. Nonorthogonal basis functions can also be used but computations are more involved and lengthy. FFT filters are the best known of the fast transform filters.

Fast transform filters operate in either the time domain or in the generalized frequency domain. In the generalized frequency domain, these filters operate by evaluating the spectral product

$$c = T^{-1}C = T^{-1}(H_T R) = T^{-1}(H_T(Tr)) \qquad (7.2.1)$$

as shown in Fig. 7.2.1a. r is the input vector, c is the output vector, and H_T is the filter gain matrix. The fast transform T converts the sample vector r into the generalized spectrum vector R. The gain matrix H_T is synthesized directly using generalized algorithms from Chap. 4. The spectral output vector C is computed by multiplying H_T and R. The time domain output vector c is obtained by taking the fast inverse transform T^{-1} of C.

The time domain fast transform filter operates by evaluating the temporal product

$$c = h_T r \qquad (7.2.2)$$

as shown in Fig. 7.2.1b. Although the filter gain matrix H_T depends upon the transform that is used, the filter impulse response matrix h_T does not. The time domain filter can operate faster than its frequency domain equivalent since the T and T^{-1} transforms do not need to be computed. H_T and h_T are fixed in Class 1 filters and are continuously updated in Class 2 and 3 filters. Filter algorithm types such as estimation, detection, and correlation can be changed with ease.

FFT filters are implemented using the fast Fourier transform and fast inverse Fourier transform. The frequency domain FFT filter operates using

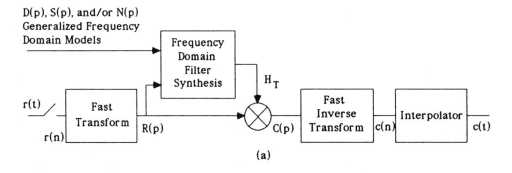

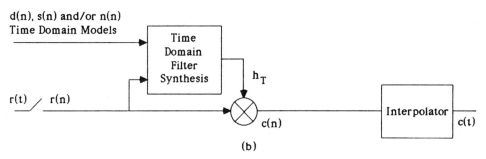

Fig. 7.2.1 Implementation of fast transform digital filter in (a) frequency domain and (b) time domain.

$$c = F^{-1}C = F^{-1}(H_F R) = F^{-1}(H_F(Fr)) \qquad (7.2.3)$$

The time domain FFT filter operates by evaluating

$$c = h_F r \qquad (7.2.4)$$

Frequency domain fast transform filters are implemented in software at the present time because generalized transform T and T^{-1} hardware modules are not available. These transforms must be programmed in software which imposes a higher processing overhead and reduced sampling rate. This is not the case for frequency domain FFT filters because FFT hardware modules are available. However, since generalized transforms rarely utilize computationally difficult basis functions (like complex exponentials used by the FFT) but rather simple functions (like a pulse stream of ±1's for Walsh, Paley, and Hadamard transforms), the generalized transforms can be simply generated.

Time domain fast transform filters do not have this problem since they do not require that T or T^{-1} be evaluated or even known. Notice that only multiplication is required in Eqs. 7.2.2 and 7.2.4. Low-speed fast transform filters use software multiplication while high-speed filters use high-speed hardware multipliers.

The advantage of fast transform filters is the simplicity and variety of algorithms which can be used. Class 1, 2, and 3 algorithms produce filters which are inherently adaptive. In addition to being adaptive, these filters are extremely flexible in meeting

Table 7.2.1 Advantages and disadvantages of fast transform filters.

Table 7.2.1 Advantages and disadvantages of fast transform filters.

Advantages:
  1. Simplicity of generalized frequency domain approach.
  2. Easy to convert frequency domain algorithms into time domain algorithms.
  3. Ease of implementing adaptive systems.
  4. Flexibility in updating system to meet changing needs.
  5. Matching the transform to the application requires fewer samples which reduces memory sizes and increases through-put.
  6. Time domain systems are often faster than frequency domain systems.

Disadvantages:
  1. Low filter through-put unless specialized FFT or generalized transform processing chips are used. Generalized transforms potentially have higher through-put than the FFT.

changing requirements since the algorithms can be updated with great ease. Fast transform filters can be implemented in either the frequency domain or the time domain depending upon preference. The gain matrix H_T or impulse response matrix h_T are derived using apriori information about the statistical nature of signal s and noise n which make up the input r (Class 1) and/or the r input itself (Class 2 and 3). Noncausal transfer functions can be implemented (off-line or not in real time) with as much ease as causal types. This is unlike most on-line real-time hardware systems.

The disadvantage of fast transform filters is filter reduced through-put. If all the T, T^{-1}, and H_T modules or the h_T module are implemented in software, they present a high processing overhead for the system. However if specialized hardware fast transform modules or fast multipliers are used to perform these functions, the amount of data processed can be significantly increased (from 10 to 1000) (see Chap. 9.11).

7.3 NONRECURSIVE FILTERS

The z-transform transfer function of any digital filter equals

$$H(z) = \frac{C(z)}{R(z)} = \frac{N(z)}{D(z)} = \frac{\sum_{k=-\infty}^{\infty} a_k z^{-k}}{\sum_{k=-\infty}^{\infty} b_k z^{-k}} \qquad (7.3.1)$$

Rearranging this equation as $C(z) = H(z)R(z)$, taking the inverse z-transforms of both sides, and solving for $c(n)$ gives

$$c(n) = \sum_{k=-\infty}^{\infty} a_k r(n-k) - \sum_{\substack{k=-\infty \\ k \neq 0}}^{\infty} b_k c(n-k) \qquad (7.3.2)$$

Based on the form of Eqs. 7.3.1 and 7.3.2, such digital filters are classified in two broad classes—*nonrecursive filters* and *recursive filters*.

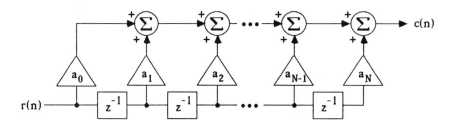

Fig. 7.3.1 General nonrecursive filter.

Nonrecursive filters are filters which have $D(z) = b_0 = 1$ in Eq. 7.3.1. They are all-zero filters with no finite poles except possibly at the origin $(z = 0)$ which act as delay elements. From Eq. 7.3.2, they have an output c which depends only upon the inputs r. If the nth output $c(n)$ depends upon previous inputs up to and including the nth input $r(n)$, then the filter is causal. Otherwise the filter is noncausal. Nonrecursive filters are also called *finite-impulse-response (FIR)* filters. For nonrecursive filters having a finite number of zeros, say $2N + 1$, then $H(z)$ of Eq. 7.3.1 reduces to

$$H(z) = \frac{C(z)}{R(z)} = \sum_{k=-N}^{N} a_k z^{-k}$$

$$c(n) = \sum_{k=-N}^{N} a_k r(n - k)$$

(7.3.3)

assuming the impulse response is centered about $n = 0$. If the response $h(t)$ is centered about $n = N$ rather than zero, then $H(z)$ is multiplied by a delay z^{-N} to recenter the output. A general nonrecursive filter is shown in Fig. 7.3.1 which matches Eq. 7.3.3 for $k = 0$ to N.

Nonrecursive filters have a number of desirable properties. These are listed in Table 7.3.1. They are always stable and can always be made causal. Nonrecursive filters can be designed to have exact linear phase which is very useful in many applications. They can be implemented via the FFT. This frequency domain approach eliminates the longer convolutional time domain approach. The design methods are easy and well-defined. Unfortunately there are no exact closed-form expressions for the responses so most designs require several iterations to obtain suitable corner frequencies, ripples, etc.

Nonrecursive filters have several undesirable characteristics. Unless FFT methods are used which easily implement high-order FIR filters, they generally have low selectivity. Nonrecursive filters require large storage and a great number of computations. Accordingly, their outputs may be delayed large amounts.

Nonrecursive filters can be implemented using a variety of techniques. The most popular methods include: (1) Fourier series (Chap. 10.3), (2) weighted Fourier series (Chap. 10.4), (3) frequency sampling (Chap. 10.5), and (4) numerical analysis (Chap. 10.6) as listed in Table 7.3.2. Also many of the approaches used to design recursive filters can also be used to design nonrecursive filters.

Table 7.3.1 Advantages and disadvantages of nonrecursive and recursive filters.

Nonrecursive Filters	Recursive Filters
1. Always stable.	Can be unstable. Stable only if poles inside unit circle.
2. Can be exactly linear phase.	Linear phase only with phase equalization.
3. FFT implementation possible permitting fast convolution.	Unnecessary.
4. Design methods easy and well-defined.	Design methods more difficult but well-defined. Easy if frequency transforms are used.
5. No exact closed-form equations so most designs require iteration.	Some designs are exact with closed-form equations.
6. Cannot usually use these tabulated filters of Chap. 4.	Easy to incorporate classical, optimum, and adaptive filter tabulated results of Chap. 4 into design.
7. Low selectivity unless high order (orders 4-10 times as much).	Low-order filters give sharp cut-off or high selectivity.
8. More storage and greater number of arithmetic computations. Long delays may be necessary.	Small storage, short delays, and small number of arithmetic computations.

These nonrecursive filter design methods are carried out in the frequency domain. The transfer function $H(z)$ given by Eq. 7.3.3 is evaluated in ac steady-state as

$$H(e^{j\omega T}) = \sum_{k=-N}^{N} a_k e^{-jk\omega T} \tag{7.3.4}$$

where $z = e^{j\omega T}$. In the Fourier series approach, the desired response $H_d(j\omega)$ is approximated by a Fourier series. This series takes the form of Eq. 7.3.4 except that the summation is extended to include an infinite number of terms ($k = \pm\infty$). Truncating this series to $(2N + 1)$ terms yields a nonrecursive filter whose gain response approximates the desired response. The approximate response is optimum in the integral least-squares (ISE) sense.

Unfortunately this approximate response exhibits ringing and oscillation (Gibb's phenomenon) around frequencies where the gain makes abrupt transitions. The effects of Gibb's phenomenon can be reduced and virtually eliminated by using the time domain windows discussed in Chap. 2.5. This leads to a suboptimal response in the ISE sense but a better response in practical applications.

The frequency sampling approach simply sets the filter gain $H(e^{j\omega T})$ equal to the desired gain $H_d(j\omega)$ at $(2N+1)$ frequencies. This allows the $(2N+1)$ coefficients (a_k's) to be determined algebraically by solving a set of $(2N+1)$ simultaneous equations. If the

Table 7.3.2 Summary of nonrecursive and recursive filter design approaches.

Design Method	Nonrecursive Filters	Recursive Filters
FREQUENCY DOMAIN		
1. Fourier Series	Optimum ISE gain response; ringing oscillation due to Gibb's effect.	Same but suboptimal.
2. Weighted Fourier Series	Suboptimal ISE but improved gain response; reduced Gibb's effect.	Same but suboptimal.
3. Frequency Sampling	Exact gain response at $2N+1$ frequencies; ringing and oscillation present.	Same as nonrecursive.
4. Frequency Transforms		
a. General s-to-z	Same as recursive.	Most popular approach; easy to apply, warps frequency response.
b. Bilinear	Not applicable	Most widely used; alias free, warps frequency response.
c. Matched-z	Same as recursive.	Converts analog roots to digital roots directly.
d. Numerical Interpolation/ Extrapolation	Easy to apply; alternative s-to-z transform.	Same as nonrecursive.
e. Numerical Differentiation	Easy to apply; alternative s-to-z transform.	Same as nonrecursive.
f. Numerical Integration	Not applicable.	Easy to apply; alternative s-to-z transform. Almost preserves time domain response.
TIME DOMAIN		
1. Invariant Response	Same as recursive.	Preserves time domain response.
a. Impulse Invariant (z-transform)	Same as recursive.	Preserves impulse response, aliased gain response.
b. Step-Invariant	Same as recursive.	Preserves step response, aliased and distorted gain response.
c. Modified-Impulse Invariant	Same as recursive	Approximately preserves impulse and gain responses.
2. Matched z-Transform	Same as recursive	Does not preserve impulse and gain responses.

frequencies are chosen to be equally spaced between dc and the sampling frequency ω_s (a frequency increment of ω_s/N), then Eq. 7.3.4 takes the form of an FFT which may be solved by fast transform methods. Large oscillations in the gain response usually occur between the specified frequencies. This often necessitates an iterative design procedure and judicious selection of the desired gain values to control the response.

A different approach utilizes a number of classical numerical analysis algorithms for interpolation and extrapolation (see Table 10.6.1). These include Newton, Gauss, Stirling, Bessel, and Everett formulas. They can be applied to a variety of problems such as midpoint and missing data interpolators that cannot be solved by the methods just discussed. These numerical formulas can be manipulated to give a set of differentiation algorithms (see Table 10.6.2). The resulting filters can be used as differentiators. Alternatively, their transfer functions $H(z)$ can be set equal to s to yield a set of general s-to-z frequency transformations. These results augment the transforms listed in Tables 5.2.1 and 5.2.2. The numerical analysis design approach usually requires many iterations so it is not as popular.

Adaptive nonrecursive filters can be designed using a variety of algorithms such as Newton, steepest descent, and LMS (Chaps. 10.7–8). These adaptive filters still use Eq. 7.3.3 but all the filter weights $a_k(n)$ are updated with every sample.

Nonlinear nonrecursive filters are also useful. Median filters will be discussed in Chap. 10.9 to illustrate how nonlinear processing can provide some unique signal processing advantages.

7.4 RECURSIVE FILTERS

Recursive filters are filters whose $H(z)$ transfer functions have the form of Eq. 7.3.1. They have finite poles as well as zeros. From Eq. 7.3.2, the nth output $c(n)$ depends both upon the inputs $r(m)$ (all m) and other outputs $c(m)$ (all $m \neq n$). If $m \leq n$, then the filter is causal. Otherwise the filter is noncausal. Recursive filters are also called *infinite-impulse-response (IIR)* filters. Since recursive filters can have impulse responses which are of finite length, this term is a misnomer. A general recursive filter is shown in Fig. 7.4.1.

Recursive filters have many desirable properties as listed in Table 7.3.1. Their design is easy using classical frequency transformations like those in Tables 5.2.1 and 5.2.2. With these transforms, the well-known classical and optimum filters of Chap. 4 can be utilized. Low-order recursive filters can have gains with sharp cut-off and high selectivity. These filters can be implemented using little storage, short delays, and a small number of arithmetic computations.

Recursive filters have several undesirable properties. They are (absolutely) stable only if their poles lie inside the unit circle (assuming causality). They can never have linear phase like FIR filters. Unless frequency transforms are used, recursive filters are generally more difficult to design. Nevertheless, their desirable characteristics generally far outweigh these undesirable properties so recursive filters are widely used.

Recursive filters can be implemented using a variety of techniques. The most popular methods include: (1) frequency transforms (Chap. 11.2), (2) invariant time

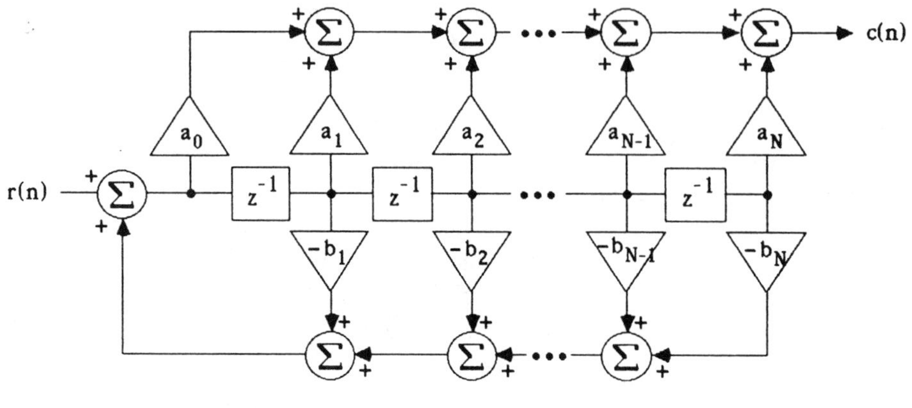

Fig. 7.4.1 General recursive filter.

domain response design (Chap. 11.3), (3) matched z-transform design (Chap. 11.4), and (4) numerical integration algorithms (Chap. 11.5) as listed in Table 7.3.2. Some of the nonrecursive filter design approaches can also be used to design recursive filters.

In the frequency transform approach, an analog filter transfer function $H(p)$ is directly converted into a digital filter transfer function $H(z)$ using

$$H(z) = H(p)\Big|_{p = g(z)} \qquad (7.4.1)$$

$g(z)$ is the desired digital transform from Chap. 5.2. The resulting $H(z)$ generally has frequency domain and time domain responses that differ considerably from the analog filters from which they were derived. Only p-to-z mappings that transform the jv-axis of the p-plane onto the unit circle of the z-plane preserve certain frequency domain characteristics like passband and stopband magnitude response ripples and phase variations. Delay and time domain responses are distorted. Of the many transforms available, the bilinear transform is the most widely used where

$$H(z) = H(p)\Big|_{p = \dfrac{2}{T}\dfrac{z-1}{z+1}} \qquad (7.4.2)$$

This transform has the desirable mapping property just mentioned.

The invariant time domain response approach preserves a particular temporal response of the analog filter used to form the digital filter. For example, the time domain impulse-invariant transform uses

$$H_0(z) = T Z\big[h(nT)\big] \qquad (7.4.3)$$

where $h(nT)$ is the sampled impulse response of the analog filter. Equivalently, $H(z)$ is the z-transform of $h(nT)$. When the sampling frequency $f_s = 1/T$ Hz is sufficiently high compared with the stopband frequency of the analog filter, the frequency response characteristics are preserved. Otherwise there is noticeable aliasing distortion.

The matched z-transform approach is a pole-zero mapping algorithm which directly converts the poles and zeros of the analog filter into equivalent poles and zeros for the digital filter using the z-transform $z_k = e^{-p_k T}$. The results obtained using this approach are related to those using the impulse-invariant method.

In problems requiring integration, the classical numerical analysis algorithms for interpolation and extrapolation (see Table 10.6.1) can be manipulated to give a set of integration algorithms (see Table 11.5.1). Also the p-to-z transforms listed in Tables 5.2.1 and 5.2.2 can be used.

Adaptive recursive filters can also be defined (Chap. 11.6). They use the same algorithms utilized by adaptive nonrecursive filters. They are more difficult to design and stability is usually a problem. Kalman filters (Chap. 11.7) are a type of recursive filter which incorporate a signal model as part of the filter and are extremely useful.

7.5 SELECTION OF IMPLEMENTATION FORM

As we see, there are a variety of basic approaches available for designing digital filters. Most depend upon utilizing well-known classical and optimum analog filters. All the frequency and time domain methods in Table 7.3.2 excluding the numerical analysis approach rely on analog filter data. Nonlinear filters like median filters lie in new research territory and have unique properties.

Considering the design approaches listed in Table 7.3.2, time domain responses are preserved exactly only by the invariant response method. If the sampling frequency f_s is much greater than the filter and input signal bandwidths, then the frequency domain characteristics of the digital filter are approximately the same.

No approach maintains the frequency domain response exactly. Magnitude and phase responses like ripple can be maintained exactly using only special p-to-z transforms like the bilinear. The delay response is distorted. When the sampling frequency f_s is greater than twice the filter's stopband frequency (for rather large rejections like -60 dB), the time domain responses are approximately the same.

The Fourier series, weighted Fourier's series, and frequency sampling design approaches approximate frequency responses fairly well for high-order filters. However the time domain responses are not well maintained. The matched z-transform maintains neither the time domain nor the frequency domain responses very well. However, a variation of this transform can be used to give good results in both domains.

Finally the numerical differentiation and integration algorithms lead to filters that have frequency responses that closely approximate their analog counterpart (for $\omega < \omega_s/2$). The time domain response of discrete integrators often do not match the ideal response of a step but rather have large overshoot and long ringing time. The time domain response of the analog differentiator is an impulse which the resulting digital filters cannot exhibit. Digital differentiators tend to oscillate and ring like the digital integrators.

In summary, only high-sampling rates relative to analog filter bandwidth help preserve step responses in the frequency domain approaches. For low-order, the special p-to-z transforms like the bilinear are the most popular. For high-order filters, FFT implementations of the filter are excellent for their convenience and performance. In Chaps. 8–11, we will review these design techniques in great detail.

8 FAST TRANSFORM FILTER DESIGN

> *They crucified him, and two other with him,*
> *on either side one, and Jesus in the midst ...*
> *After this, Jesus knowing that all things were now*
> *accomplished, that the scripture might be fulfilled, saith ...*
> *It is finished: and he bowed his head, and give up the ghost.*
>
> John 19:18, 28, 30

Digital filters were classified in Chap. 7. Each of these filter classes is described in the next four chapters. Chap. 8 is concerned with the design of generalized fast transform filters. Chap. 9 will examine FFT filters. Fast transform filters were briefly introduced in Chap. 7.2. These filters are usually designed in software since fast transform hardware modules are not generally available. Fast transform filters operate at higher speeds and have higher through-put than FFT filters because the transforms usually have simpler basis functions.

This chapter will develop fast transform filter algorithms using five different approaches. These include: (1) FFT scalar filters, (2) fast transform filters derived from FFT scalar filters, (3) fast transform vector filters, (4) fast transform scalar filters, and (5) Karhunen-Loéve optimum scalar filters. The filtering requirements and hardware capabilities determine which algorithm is best suited to the application. Since the adaptive filters of Chaps. 4.4–4.9 will be used so extensively, the engineer should briefly review that material before progressing deeply into this chapter.

8.1 FAST TRANSFORM ALGORITHMS

Fast transform filters are implemented in the frequency domain as shown in Fig. 8.1.1a. These filters utilize forward and inverse generalized transforms. Efficient fast transform algorithms are used for rapid computation. The block diagram is labelled using the following notation. A discrete sequence of N time or frequency samples is treated as a column vector of length N. Lower case letters correspond to time domain signals and upper case letters to their generalized frequency domain spectra. s is the signal vector, n is the noise vector, and $r = s + n$ is the input vector. The generalized spectrum of s using any orthogonal transform is $S = Ts$. R and C are vectors of the input and output spectra. Orthogonal transform T and its inverse T^{-1} are square $N \times N$ matrices. When transform T is also unitary, then $(1/N)TT^h = I = TT^{-1}$ and $T^h = NT^{-1}$. The inverse transform equals $1/N$ times the transpose of the conjugate transform. The filter gain

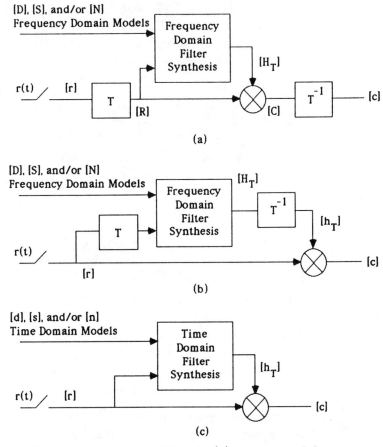

Fig. 8.1.1 Implementation of filter in (a) generalized frequency domain, (b) time domain using transforms, and (c) time domain without using transforms.

H_T based upon transform T is an $N \times N$ matrix. H_T depends upon the transform used (e.g. Fourier, Paley, Haar, and slant).

Fast transform filters, implemented in the frequency domain as shown in Fig. 8.1.1a, are described by the matrix equations

$$R = Tr$$
$$C = H_T R \qquad \qquad (8.1.1)$$
$$c = T^{-1}C$$

(same as Eq. 7.2.1). Fast transform filters use suboptimum transforms (type 1 and 2) and optimum transforms as shown in Fig. 2.8.1. Several such transforms including Walsh, Paley, Hadamard, Haar, slant, and Karhunen-Loéve were discussed in Chaps. 2.8 and 3.3. Fast Fourier transform filters implemented in the frequency domain use

Table 8.1.1 Different fast transform filter
types described in sections of Chap. 8 based
upon transform and gain matrix used.

Transform		Scalar B	Vector $\bar{B}$
FFT	A	8.2	8.4
Hybrid (A$\bar{A}$)			8.3
Other transforms	$\bar{A}$	8.5	8.4
K-L optimum transforms	$\bar{\bar{A}}$	8.6	$\times$

$$R = Fr$$
$$C = H_F R \qquad\qquad (8.1.2)$$
$$c = F^{-1}C$$

(see Eq. 7.2.3). F and F^{-1} represent the fast Fourier and inverse fast Fourier transforms, respectively.

Fast transform filters can also be implemented in the time domain as shown in Figs. 8.1.1b and 8.1.1c. They are described by the single matrix equation

$$c = h_T r = T^{-1}H_T T r \qquad\qquad (8.1.3)$$

which is obtained by manipulating Eq. 8.1.1. h_T is the impulse response of the fast transform filter. Filter gain H_T and filter impulse response h_T matrices are related as

$$H_T = T h_T T^{-1}, \qquad h_T = T^{-1}H_T T \qquad\qquad (8.1.4)$$

The frequency domain approach in Eq. 8.1.1 requires N^2 operations (complex multiplications and additions) to compute C and $N \log_2 N$ operations to compute R and c using two fast transforms. This is a total of $(N + \log_2 N)N \cong N^2$ operations which reduces to only $(1+\log_2 N)N \cong N \log_2 N$ when H_T is diagonal. The time domain approach requires N^2 operations and no fast transforms in Eq. 8.1.3. This is slightly less than the frequency domain approach when H_T is nondiagonal (vector filters) but much more when H_T is diagonal (scalar filters). Converting H_T to h_T requires $2N^3$ operations which becomes excessive when H_T is continuously updated as is the case for Class 2 and 3 filters. However as discussed in Chap. 8.6, h_T can be generated directly so Eq. 8.1.4 need not be used. Therefore the time domain approach usually provides faster processing.

The question now is: How is H_T determined? The answer depends upon the method used to formulate the filter. Fast transform filters fall into two general categories – scalar filters and vector filters. *Scalar filters* have gain matrices H_T that are diagonal. *Vector filters* have gain matrices H_T that are nondiagonal. FFT, other fast transforms, or K-L optimum transforms can be used. We consider 3 transform types and 2 matrix types which form $3 \times 2 = 6$ different combinations. From a Karnaugh map standpoint, transform variable A has three states: A (FFT), $\bar{A}$ (other transforms), and $\bar{\bar{A}}$ (K-L optimum transforms). Matrix B has two states: B (diagonal gain matrices, scalar filters) and $\bar{B}$ (nondiagonal gain matrices, vector filters). Table 8.1.1 shows these six combinations. One does not exist theoretically ($\bar{\bar{A}}\bar{B}$ marked with "$\times$"). The remaining five filter possibilities and one special hybrid case $A\bar{A}$ are described in the different sections (Chaps. 8.2–8.6) of Chap. 8 as indicated in Table 8.1.1.

8.2 FAST FOURIER TRANSFORM (FFT) SCALAR FILTERS

The transfer functions of adaptive analog filters were formulated in Chaps. 4.4 and 4.5. In this section, these filters are reformulated in the FFT frequency domain which allows us to utilize all of the vast analog design information available in Chap. 4. This is easily accomplished by making appropriate changes in the adaptive filter equations derived in Chaps. 4.4 and 4.5.[1] The method consists of simply replacing $t \to n$, $j\omega \to m$, and integrals by sums as we will now discuss. The filters to be described are listed in Tables 8.2.1 and 8.2.2. Their formulations are based upon the following assumptions (case AB in Table 8.1.1):
(a) The FFT is used, so
(b) The basis functions are complex exponentials.
(c) The gain matrix is diagonal.

The block diagram of an adaptive digital filter is shown in Fig. 8.2.1. It is a sampled or digitized version of Fig. 4.4.2. A discrete signal $s(n)$ and noise $n(n)$ are added together to form input $r(n)$. The input is applied to a filter having transfer function $H(m)$ to form the output signal $c(n)$ as

$$r(n) = s(n) + n(n) \quad \text{or} \quad R(m) = \text{FFT}[r(n)] = S(m) + N(m)$$

$$c(n) = \text{FFT}^{-1}[C(m)] = h(n) * r(n) \quad \text{or} \quad C(m) = H(m)R(m) \tag{8.2.1}$$

8.2.1 DISTORTION BALANCE FFT SCALAR FILTERS

Signal and noise distortions are balanced in distortion balance filters. The degradation caused by the filter distorting or altering the signal to the distortion is equated to the noise leaking through the filter. The signal degradation $E_S(m)$ is the difference

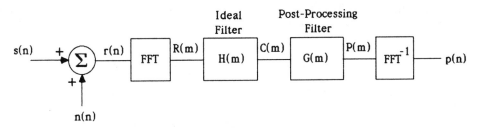

Fig. 8.2.1 Block diagram of FFT filter assuming additive signal and noise.

between input and output signal spectra. The noise leakage $E_N(m)$ equals the output noise spectrum. Therefore

$$E_S(m) = S(m) - H(m)S(m) = [1 - H(m)]S(m)$$
$$E_N(m) = H(m)N(m) \tag{8.2.2}$$

where $H(m)$ is the ideal filter gain. For more general results, the signal distortion is weighted by $W_S(m)$ and the noise distortion by $W_N(m)$. Then equating the weighted signal error to the weighted noise error at every frequency m gives

$$E_S(m)W_S(m) = E_N(m)W_N(m)$$
$$[1 - H(m)]S(m)W_S(m) = H(m)N(m)W_N(m) \tag{8.2.3}$$

This insures that the total error

$$E(m) = E_S(m)W_S(m) + E_N(m)W_N(m) \tag{8.2.4}$$

is composed of equally-weighted signal and equally-weighted noise. Solving Eq. 8.2.3 for the ideal filter transfer function $H(m)$ gives

$$H(m) = \frac{S(m)W_S(m)}{S(m)W_S(m) + N(m)W_N(m)} \tag{8.2.5}$$

This result has a number of interesting applications. If we wish to form a filter that balances complex signal and noise levels equally, we set $W_S(m) = W_N(m)$ so

$$H(m) = \frac{S(m)}{S(m) + N(m)} \tag{8.2.6}$$

However, if we wish to form a filter that balances signal and noise power levels, we set $W_S(m) = S^*(m)$ and $W_N(m) = N^*(m)$. A multitude of other possibilities exist. For example, if we wish to ignore high-frequency signal and low-frequency noise, we might choose $W_S(m) = 1/[1 + H_{dmn}^{-1}(m)]$ and $W_N(m) = 1/[1 + H_{dmn}(m)]$. Alternatively, if we wish to balance the derivative of the signal from noise, $W_S(m) = H_{dmn}^{-1}(m)$ and $W_N(m) = 1$. In these cases, we use the frequency transforms of Chap. 5 where $j\omega = H_{dmn}^{-1}(m)$ (cf. Chap. 4.4.1). The possibilities are limitless.

8.2.2 GENERAL ESTIMATION FFT SCALAR FILTERS

General estimation filters estimate the desired optimal output $d(n)$ from the input $r(n)$. The output error $e(n)$ equals

$$e(n) = d(n) - c(n) \qquad \text{or} \qquad E(m) = D(m) - C(m) \tag{8.2.7}$$

One optimization criteria that can be used is to minimize the mean-squared-error (MSE) where

$$MSE = \|e\| = \sum_{n=0}^{N-1} |e(n)|^2 = \sum_{n=0}^{N-1} |d(n) - c(n)|^2 \tag{8.2.8}$$

The ideal filter is the one whose transfer function $H(m)$ minimizes the error index MSE. Using Parseval's theorem given in Table 1.8.1, Eq. 8.2.8 can be rewritten in the frequency domain as

$$MSE = \frac{1}{N} \sum_{m=0}^{N-1} E(m)E^*(m) = \frac{1}{N} \sum_{m=0}^{N-1} [D(m) - C(m)][D^*(m) - C^*(m)] \quad (8.2.9)$$

where $E^*(m)$ is the conjugate error spectrum. When $e(n)$ is real-valued, then $E^*(m) = E(-m)$. Since $C(m) = H(m)R(m)$ is a functional of $H(m)$, Eq. 8.2.9 is minimized by setting the derivative of the sum with respect to $H(m)$ equal to zero. The summation term can be rewritten as

$$D(m)D^*(m) - H(m)R(m)D^*(m) - H^*(m)R^*(m)D(m) + H(m)H^*(m)R(m)R^*(m)$$
$$(8.2.10)$$

Differentiating MSE with respect to $H(m)$ and setting the result equal to zero yields (assume $H(m)$ is real)

$$\frac{d\,MSE}{dH(m)} = -R(m)D^*(m) - R^*(m)D(m) + H(m)R(m)R^*(m) + H^*(m)R(m)R^*(m) = 0$$
$$(8.2.11)$$

Solving Eq. 8.2.11 for $H(m)$ gives (the same result holds for complex $H(m)$)

$$H(m) = \frac{D(m)R^*(m)}{R(m)R^*(m)} = \frac{\text{Desired input output cross correlation spectrum}}{\text{Input autocorrelation spectrum}} \quad (8.2.12)$$

When the input is ergodic, RR^* and DR^* equal the autocorrelation and cross-correlation spectra. When the input is not ergodic, they are *estimates* of the correlation spectra. Eq. 8.2.12 can also be expressed as

$$H(m) = \frac{D(m)R^*(m)}{R(m)R^*(m)} = \frac{D(m)}{R(m)} = \frac{\text{Desired output}}{\text{Available input}} \quad (8.2.13)$$

When a deterministic input-output filter viewpoint is taken, Eq. 8.2.13 is used. Eq. 8.2.12 utilizes the stochastic correlation filter viewpoint. These equations are equivalent.

Eq. 8.2.12 is the transfer function for the general estimation filter. This ideal filter gives the best estimate (in the minimum MSE sense) of a desired signal $d(n)$ from an input signal $r(n)$ as the ratio of the two power spectra. When $r(n)$ is composed of a signal $s(n)$ contaminated with additive noise $n(n)$, Eq. 8.2.12 reduces to

$$H(m) = \frac{D(m)[S^*(m) + N^*(m)]}{[S(m) + N(m)][S^*(m) + N^*(m)]} \quad (8.2.14)$$

8.2.3 GENERAL DETECTION FFT SCALAR FILTERS

General detection filters detect the presence or absence of a signal in noise. The ideal filter detects a desired signal $d(n)$ by producing an output signal pulse $s_o(0)$ in the total output $c(n)$ such that the SRR_o ratio of the peak output signal-to-total rms output is maximized. Its transfer function equals

$$H(m) = \frac{D^*(m)}{R(m)R^*(m)} = \frac{\text{Conjugate spectrum of desired signal to be detected}}{\text{Input autocorrelation spectrum}}$$

(8.2.15)

This filter produces a narrow output pulse (which reduces temporal or time ambiguity) centered at time $t = 0$ with undefined shape.

The ideal filter which detects $d(n)$ by producing $s_o(0)$ in $c(n)$, such that the peak output signal-to-rms output noise SNR_o is maximized, has

$$H(m) = \frac{D^*(m)}{N(m)N^*(m)} = \frac{\text{Conjugate spectrum of desired signal to be detected}}{\text{Noise autocorrelation spectrum}}$$

(8.2.16)

To prove Eq. 8.2.15, consider the peak output signal (at time zero)-to-total rms output ratio squared (see Eq. 8.2.56)

$$
\begin{aligned}
SRR_o^2 &= \frac{|s_o(0)|^2}{\sum_{n=0}^{N-1}|c(n)|^2} = \frac{\left|(1/N)\sum_{n=0}^{N-1}S(m)H(m)\right|^2}{(1/N)\sum_{n=0}^{N-1}|H(m)|^2|R(m)|^2} \\
&= \frac{\left|\sum_{n=0}^{N-1}[S(m)/R(m)]H(m)R(m)\right|^2}{N\sum_{n=0}^{N-1}|H(m)|^2|R(m)|^2} \\
&\leq \frac{\sum_{n=0}^{N-1}\left[|S(m)|^2/|R(m)|^2\right]\sum_{n=0}^{N-1}|H(m)|^2|R(m)|^2}{N\sum_{n=0}^{N-1}|H(m)|^2|R(m)|^2} \\
&= \frac{1}{N}\sum_{n=0}^{N-1}\frac{|S(m)|^2}{|R(m)|^2} = \frac{1}{N}\sum_{n=0}^{N-1}|SRR_i(m)|^2
\end{aligned}
$$

(8.2.17)

It equals the sum of the spectral signal-to-total input ratio squared. This results by noting that the squared sum of a product never exceeds the product of sums with squared terms (Cauchy's inequality). $s_o(0)$ is obtained by evaluating the discrete inverse Fourier transform of $S_o(m)$ at time $t = 0$. The sum of the total output $c^2(n)$ is found using Parseval's theorem where $C(m) = H(m)R(m)$. The equality condition in Eq. 8.2.17 results when

$$H(m)R(m) = k\left[\frac{S(m)}{R(m)}\right]^*$$

(8.2.18)

Solving for $H(m)$ gives (let $k = 1$)

$$H(m) = \frac{S^*(m)}{|R(m)|^2}$$

(8.2.19)

This is identical to Eq. 8.2.15 when $S(m)$ is the desired output $D(m)$ to be detected and $R(m)$ is the input spectrum rejected. If the output signal s_o is to peak at time T_p rather than time zero, then Eq. 8.2.19 is multiplied by $\exp(-mT_p)$.

8.2.4 GENERAL ESTIMATION-DETECTION FFT SCALAR FILTERS

The general estimation filter of Eq. 8.2.12 optimally estimates signals in the presence of additive noise. The general detection filter of Eq. 8.2.15 maximizes the peak output signal relative to the total rms output. Combining these two results as

$$H(m) = H_E(m)H_D(m) \qquad (8.2.20)$$

gives us an estimation-detection filter which performs both operations. This filter best estimates the signal and then outputs a narrow pulse when the signal is present.

8.2.5 WIENER ESTIMATION FFT SCALAR FILTERS

Suppose that the generalized estimation filter is used for estimating signal. Setting the desired signal to be estimated equal to the input signal $s(n)$ as $d(n) = s(n)$, then $D(m) = S(m)$ and Eq. 8.2.14 becomes

$$
\begin{aligned}
H(m) &= \frac{S(m)[S^*(m) + N^*(m)]}{[S(m) + N(m)][S^*(m) + N^*(m)]} \\
&= \frac{|S(m)|^2 + S(m)N^*(m)}{|S(m)|^2 + |N(m)|^2 + S(m)N^*(m) + S^*(m)N(m)}
\end{aligned}
\qquad (8.2.21)
$$

This is called the *correlated estimation* or *correlated Wiener* filter. It is coherent in the sense that the phase spectra of the signal and noise are used when computing $H(m)$. The phase of $H(m)$ is generally nonzero. Since SN^* and S^*N are complex conjugates of each other, their sum can also be expressed as $2 \operatorname{Re} SN^*$.

When the signal and noise are orthogonal, or equivalently, are uncorrelated and the noise has zero mean, then $S(m)N^*(m) = 0$ (see discussion for Eq. 8.2.23) and Eq. 8.2.21 reduces to

$$H(m) = \frac{|S(m)|^2}{|S(m)|^2 + |N(m)|^2} \qquad (8.2.22)$$

This is the *uncorrelated estimation* or *uncorrelated Wiener* filter. From the distortion balance filter viewpoint of Eq. 8.2.5, this filter results when signal and noise powers are equally weighted. It is noncoherent since it discards the phase spectra of the signal and noise. The phase of $H(m)$ is always zero.

Before proceeding to other filter types, notice that the signal spectrum $S(m)$, conjugate noise spectrum $N^*(m)$, their product $S(m)N^*(m)$, and the cross-correlation spectrum $S_{sn}(m)$ (see Eq. 1.17.15) equal

$$S(m) = \text{FFT}[s(n)], \qquad N^*(m) = \text{FFT}[n^*(-n)]$$

$$\widehat{S}_{sn}(m) = \text{FFT}[s(n) * n^*(-n)] = S(m)N^*(m) \qquad (8.2.23)$$

$$S_{sn}(m) = \text{FFT}[E\{s(n+k)n^*(k)\}] = \text{FFT}[R_{sn}(n)]$$

s and n represent random signal and noise processes, respectively. s and n are the actual signal and noise, respectively. They are particular members of the ensemble of all those possible formed by the s and n random processes. S_{sn} is the apriori (assumed) cross-correlation spectrum while SN^* is the aposteriori (actual) cross-correlation spectral estimate $\widehat{S}_{sn}$. Eq. 8.2.23 shows that SN^* equals S_{sn} only when the signal-noise convolutional product $s(n) * n^*(-n)$ equals the cross-correlation function R_{sn}. This occurs only when both s(n) and n(n) are ergodic processes.

The cross-correlation spectrum S_{sn} is zero when the signal and noise are orthogonal, or equivalently, are uncorrelated and the noise is a zero-mean process (see Eqs. 1.17.9 and 1.17.12). In such cases, the cross-correlation estimate SN^* product is *set equal to zero* in Eq. 8.2.21 which gives Eq. 8.2.22. This is done even though the actual SN^* product is nonzero over the frequency ranges where $S \neq 0$ and $N \neq 0$, or equivalently, the nonzero signal and noise spectra are overlapping. Nonoverlapping or disjoint S and N have $SN^* = 0$ and are very seldom encountered in practice. Correlation is discussed further in Chap. 8.2.7.

8.2.6 PULSE-SHAPING ESTIMATION FFT SCALAR FILTERS

Suppose that we wish to form a new signal $p(n)$ from the output signal $c(n)$ in Fig. 8.2.1. This is done using the post-processing block with gain $G(m)$ where

$$p(n) = g(n) * c(n) \qquad \text{or} \qquad P(m) = G(m)C(m) \qquad (8.2.24)$$

Then the gain $G(m)$ equals

$$G(m) = \frac{P(m)}{C(m)} \qquad (8.2.25)$$

where $P(m)$ is the desired output spectrum of $p(n)$. This $G(m)$ post-processing block can be used to generate a variety of useful signals. After $G(m)$ is determined, it is combined with gain $H(m)$ to form a single gain block (i.e., $H(m)G(m) \rightarrow H(m)$). The blocks are separated here only for conceptual simplicity.

It might be useful to reshape $c(n)$ to have a different waveform $p(n)$ or to estimate different signals. We could first use the uncorrelated Wiener filter whose transfer function $H(m)$ is given by Eq. 8.2.22 to estimate $s(n)$. Then choosing $G(m) = P(m)/S(m)$ results in the desired output signal $p(n)$. Combining $H(m)$ and $P(m)$ into a single transfer function gives

$$H(m) = \frac{P(m)S^*(m)}{|S(m)|^2 + |N(m)|^2} \qquad (8.2.26)$$

which is the *pulse-shaping (uncorrelated) estimation filter*. $G(m)$ can also be combined with the correlated estimation filter of Eq. 8.2.21 to give the *pulse-shaping (correlated)* *estimation filter* as

$$H(m) = \frac{P(m)[S^*(m) + N^*(m)]}{|S(m)|^2 + |N(m)|^2 + S(m)N^*(m) + S^*(m)N(m)} \qquad (8.2.27)$$

Pulse-shaping filters can also be viewed as a high-resolution detection filter (see Chap. 8.2.10) followed by a pulse shaper. The impulse $D_0(n)$ from the high-resolution detection filter $H(m)$ excites the impulse response $p(n)$ from the post-processor filter $G(m)$. Pulse-shaping filters yield the maximum output SNR_o possible under the constraint that the output signal pulse $p(n) = \text{FFT}^{-1}[P(m)]$ has a specific form.

8.2.7 CORRELATION FFT SCALAR FILTERS

The estimate $\widehat{R}_{ss}(n)$ of the autocorrelation $R_{ss}(n)$ of an ergodic signal $s(n)$ and the estimate $\widehat{S}_{ss}(m)$ of its power spectrum $S_{ss}(m)$ equal

$$R_{ss}(n) = \text{FFT}^{-1}[S_{ss}(m)] = E\{s(n+k)s^*(k)\} \equiv s(n) * s^*(-n) = \widehat{R}_{ss}(n)$$
$$S_{ss}(m) = \text{FFT}[R_{ss}(n)] = \text{FFT}[E\{s(n+k)s^*(k)\}] \equiv S(m)S^*(m) = \widehat{S}_{ss}(m)$$
$$(8.2.28)$$

from Eqs. 1.17.21 and 1.17.22. The post-processing block $G(m)$ in Fig. 8.2.1 can be used to generate the autocorrelation function. From Eqs. 8.2.24 and 8.2.28, its output $P(m) = S(m)S^*(m)$ while its input $C(m) = S(m)$. Then its transfer function $G(m) = P(m)/C(m) = S^*(m)$.

Therefore combining $P(m)$ with the uncorrelated Wiener filter described by Eq. 8.2.22 gives the ideal autocorrelation estimation filter gain as

$$H(m) = \frac{S^*(m)|S(m)|^2}{|S(m)|^2 + |N(m)|^2} \qquad (8.2.29)$$

Another autocorrelation estimation filter results by using the correlated Wiener filter of Eq. 8.2.21. Its gain equals

$$H(m) = \frac{|S(m)|^2[S^*(m) + N^*(m)]}{|S(m)|^2 + |N(m)|^2 + S(m)N^*(m) + S^*(m)N(m)} \qquad (8.2.30)$$

The estimate of the cross-correlation $R_{sn}(n)$ between two ergodic signals, $s(n)$ and $n(n)$, and its power spectrum $S_{sn}(m)$ equal

$$R_{sn}(n) = \text{FFT}^{-1}[S_{sn}(m)] = E\{s(n+k)n^*(k)\} \equiv s(n) * n^*(-n) = \widehat{R}_{sn}(n)$$
$$S_{sn}(m) = \text{FFT}[R_{sn}(n)] = \text{FFT}[E\{s(n+k)n^*(k)\}] \equiv S(m)N^*(m) = \widehat{S}_{sn}(m)$$
$$(8.2.31)$$

Proceeding as before, $G(m) = N^*(m)$ which can be combined with the uncorrelated Wiener filter. This gives the ideal cross-correlation estimation filter gain as

$$H(m) = \frac{N^*(m)|S(m)|^2}{|S(m)|^2 + |N(m)|^2} \tag{8.2.32}$$

Using the correlated Wiener filter instead gives

$$H(m) = \frac{S(m)N^*(m)[S^*(m) + N^*(m)]}{|S(m)|^2 + |N(m)|^2 + S(m)N^*(m) + S^*(m)N(m)} \tag{8.2.33}$$

The correlation estimates obtained using the correlated Wiener filters (Eqs. 8.2.30 and 8.2.33) are usually better than those obtained using the uncorrelated Wiener filters (Eqs. 8.2.29 and 8.2.32).

8.2.8 MATCHED DETECTION FFT SCALAR FILTERS

The matched detection filter detects the presence of a signal $s(n)$ embedded in additive noise $n(n)$. In this case, we wish to maximize the peak output SNR_o in order to detect the signal and reject the rms noise. Using Eq. 8.2.15 where $D(m) = S(m)$ and $R(m) = N(m)$, then

$$H(m) = \frac{S^*(m)}{N(m)N^*(m)} = \frac{S^*(m)}{|N(m)|^2} \tag{8.2.34}$$

This is the *classical detection, matched,* or *prewhitening* filter. $H(m)$ weights each spectral input by the ratio of the conjugate signal spectrum to the noise power spectrum. The filter introduces a phase which is opposite to signal phase. Thus all the output spectral components of a signal similar to the expected signal will be in phase. The output tends to be concentrated into a narrow pulse which occurs when the signal occurs.

Notice from Eq. 8.2.34 that the filter's spectral output $C = HR = H(S + N)$ consists of a signal component $SS^*/|N|^2$ and a noise component $NS^*/|N|^2$. The signal component is the square of the input spectral SNR_i at that frequency. The noise component is the conjugate SNR_i. From a correlation viewpoint, the signal component is the signal autocorrelation and the noise component is the signal-noise cross-correlation when the noise is white (i.e., $|N(m)| = 1$). In either case, the noise component is zero if the input signal and noise are orthogonal or are uncorrelated with zero mean noise. Otherwise the noise component is nonzero.

The performance of detection filters is often measured by the SNR gain which is the ratio of the output SNR_o to the input SNR_i. It is invariant to changes in the input signal or noise level and will be discussed in Chap. 8.2.14.

From the distortion balance filter viewpoint, the classical detection filter first discriminates between signal and scaled noise power levels (so $W_S(m) = S^*(m)$ and $W_N(m) = kN^*(m)$). If the constant k is chosen to be much larger than the squared input spectral SNR_i ratio (i.e., $k \gg |S|^2/|N|^2$), then Eq. 8.2.5 reduces to

$$H(m) = \frac{S(m)S^*(m)}{kN(m)N^*(m)} \tag{8.2.35}$$

Following $H(m)$ by an inverse filter $P(m) = k/S(m)$, the overall gain characteristic becomes that of the classical detection filter given by Eq. 8.2.34. This shows that by relying entirely on noise rejection, the estimation filter becomes a detection filter.

8.2.9 INVERSE DETECTION FFT SCALAR FILTERS

An *inverse detection filter* outputs an impulse $D_0(n)$ when *only* signal $s(n)$ and no noise is applied. It outputs a narrow pulse when a signal similar to $s(n)$ is applied. Using Eq. 8.2.13, the transfer function of an inverse detection filter equals

$$H(m) = \frac{\text{FFT}\big[D_0(n)\big]}{\text{FFT}\big[s(n)\big]} = \frac{1}{S(m)} \qquad (8.2.36)$$

When signal pulse noise is applied to the filter, its input equals $R = S+N$. The output of the filter then equals $C = HR = 1+(N/S)$ or $c(n) = D_0(n)+\text{FFT}^{-1}[1/SNR_i]$ from Eq. 8.2.54. The undesired output component due to nonzero input noise is $\text{FFT}^{-1}[1/SNR_i]$. When the spectral signal-to-noise ratio is large, $|SNR_i| \gg 1$ so $|1/SNR_i| \ll 1$ and this undesired output component is negligible. However $|SNR_i|$ is seldom large so then this component is present which produces output error. This error can be significantly reduced using the high-resolution detection filter.

8.2.10 HIGH-RESOLUTION DETECTION FFT SCALAR FILTERS

Another conceptually useful filter is the *high-resolution detection filter* which outputs an impulse when *both* the signal $s(n)$ and noise are applied. It outputs a narrow pulse when a signal similar to $s(n) + n(n)$ is applied. This requires a post-processing gain $G(m)$ after the estimation filter in Fig. 8.2.1 of

$$G(m) = \frac{\text{FFT}\big[D_0(n)\big]}{\text{FFT}\big[s(n)\big]} = \frac{1}{S(m)} \qquad (8.2.37)$$

which is an inverse filter. Combining this with the uncorrelated Wiener filter in Eq. 8.2.22 results in the ideal high-resolution detection filter where

$$H(m) = \frac{S^*(m)}{|S(m)|^2 + |N(m)|^2} \qquad (8.2.38)$$

Eq. 8.2.38 is identical to the detection filter of Eq. 8.2.19 when $S(m)$ and $N(m)$ are assumed to be orthogonal. The Wiener filter estimates the signal $s(n)$ and the inverse filter converts $s(n)$ into an impulse. Another high-resolution filter results using the correlated Wiener filter of Eq. 8.2.21 as

$$H(m) = \frac{S^*(m) + N^*(m)}{|S(m)|^2 + |N(m)|^2 + S(m)N^*(m) + S^*(m)N(m)} \qquad (8.2.39)$$

The high-resolution detection filter of Eq. 8.2.39 functions as a matched detection filter (Eq. 8.2.34) at frequencies where the spectral SNR_i is low. It acts as an inverse or deconvolution filter (Eq. 8.2.36) at high spectral SNR_i. The filter deconvolves $S(m)$ with $1/S(m)$ to produce a single impulse. Rewriting Eq. 8.2.38 as

$$H(m) = \frac{1}{S(m)} \; \frac{1}{1 + |N(m)|^2/|S(m)|^2} = \frac{1}{S(m)} \; \frac{1}{1 + |SNR_i|^{-2}} \qquad (8.2.40)$$

$H(m)$ depends upon the input spectral signal-to-noise ratio SNR_i for constant $S(m)$.

From a distortion balance viewpoint, this is a filter which discriminates between signal and noise power levels (so $W_S(m) = S^*(m)$ and $W_N(m) = N^*(m)$) followed by an inverse filter having gain $G(m) = 1/S(m)$.

8.2.11 TRANSITIONAL FFT SCALAR FILTERS

There are situations which require a filter whose transfer function can make a transition from one type to another type. These are called *transitional filters*. The characteristics of two transfer functions are blended together. The transition is controlled by one parameter (or more) like α. These transfer functions are usually formed heuristically.

For example, suppose that we want to form a transitional detection filter that can make a transition from an inverse type ($\alpha = 0$) to a high-resolution type ($\alpha = 0.5$) to a matched type ($\alpha = 1$). One way combines Eqs. 8.2.34, 8.2.36, and 8.2.38 as

$$H(m) = \frac{S^*(m)}{(1 - \alpha)|S(m)|^2 + \alpha|N(m)|^2} \qquad (8.2.41)$$

Choosing some α between 0 and 1 yields a transfer function that combines the attributes of these filter types. As $\alpha \to 0$, the filter functions as an inverse type. As $\alpha \to 0.5$, it becomes a high-resolution type. As $\alpha \to 1$, it performs as a matched type. Another example of a transitional detection filter is

$$H(m) = \frac{1}{S(m)^{2(1-\alpha)}} \qquad (8.2.42)$$

by generalizing Eq. 8.2.36. When $\alpha = 0.5$, it functions as an inverse filter.

An example of a transitional estimation filter is

$$H(m) = \frac{|S(m)|^2 + \alpha S(m)N^*(m)}{|S(m)|^2 + |N(m)|^2 + \alpha\big(S(m)N^*(m) + S^*(m)N(m)\big)} \qquad (8.2.43)$$

As $\alpha \to 0$, the filter becomes uncorrelated Wiener. As $\alpha \to 1$, the filter becomes correlated Wiener. The possibilities for transitional filters are endless. The transitional filter approach gives the designer great freedom in formulating adaptive filters.

8.2.12 STATISTICALLY OPTIMUM FFT SCALAR FILTERS

A variety of adaptive filters have been derived. In general, their gains are expressed as the ratio of two functions $X(m)$ and $Y(m)$ where

$$H(m) = \frac{Y(m)}{X(m)} \tag{8.2.44}$$

The X and Y functions involve sums, products, and other operations on the signal spectrum $S(m)$ and noise spectrum $N(m)$. Which S and N spectra should be used to form $H(m)$? The statistically optimum approach requires

$$H(m) = \frac{E\{Y(m)\}}{E\{X(m)\}} \tag{8.2.45}$$

where $E\{\ \}$ is the expectation operator. Eq. 8.2.45 results by expressing Eq. 8.2.44 as $Y = HX$, treating X and Y as stochastic processes rather than as deterministic signals, taking expectations of both sides, and solving for $H(m)$ which is deterministic. The optimum $H(m)$ equals the ratio of the expected (statistically averaged) X and Y which the filter processes. The transfer functions of Chap. 8.2 are therefore tabulated in Table 8.2.1 using expectation signs around their numerators and denominators. Some examples of terms in $E\{X(m)\}$ and $E\{Y(m)\}$ include:

$E\{S(m)\},\ E\{N(m)\}$ = expected signal and noise spectra
$E\{|S(m)|\},\ E\{|N(m)|\}$ = expected signal and noise magnitude spectra
$E\{|S(m)|^2\},\ E\{|N(m)|^2\}$ = expected signal and noise autocorrelation (power) spectra
$E\{S(m)N^*(m)\}$ = expected signal-noise cross-correlation spectrum
$E\{S(m)W_S\},\ E\{N(m)W_N\}$ = expected weighted signal and noise spectra

$$\tag{8.2.46}$$

8.2.13 ADAPTATION AND LEARNING PROCESS

Adaptive filters undergo a *learning process* and perform *adaptation* to form the filter transfer function $H(m)$. The transfer function is updated either occasionally or continuously. The ideal filter transfer functions listed in Tables 8.2.1 and 8.2.2 are formed using the expected spectra $E\{X\}$ and $E\{Y\}$ like those in Eq. 8.2.46. They are either software programmed into the filter or estimated. Class 1 adaptive filters store $E\{X\}$ and $E\{Y\}$ and update them (at most) periodically. Class 2 and 3 adaptive filters continuously form estimates $\widehat{E}\{X\}$ of $E\{X\}$ and $\widehat{E}\{Y\}$ of $E\{Y\}$ as the filter operates.

The expected signal spectrum $E\{S(m)\}$ is sometimes called the *learning signal*. The expected noise spectrum $E\{N(m)\}$ is sometimes called the *learning noise*. Mathematically,

$$E\{S(m)\} = E\{\text{FFT}[s(n)]\} = \text{FFT}[E\{s(n)\}] \neq E\{\text{FFT}[\mathbf{s}(n)]\}$$
$$E\{N(m)\} = E\{\text{FFT}[n(n)]\} = \text{FFT}[E\{n(n)\}] \neq E\{\text{FFT}[\mathbf{n}(n)]\} \tag{8.2.47}$$

Since operators $E\{\ \}$ and $\mathrm{FFT}[\ ]$ are linear, they can be applied in any order as $E\{\mathrm{FFT}[\ ]\}$ and $\mathrm{FFT}[E\{\ \}]$. Therefore the inequalities result because $s(t) \neq \mathbf{s}(t)$ and $n(t) \neq \mathbf{n}(t)$ (a member from an ensemble does not equal the ensemble). The learning signal and learning noise equal the inverse FFT's of their expected spectra in Eq. 8.2.47,

$$\begin{aligned}
\mathrm{FFT}^{-1}\big[E\{S(m)\}\big] &= E\{s(n)\} \neq E\{\mathbf{s}(n)\} \\
\mathrm{FFT}^{-1}\big[E\{N(m)\}\big] &= E\{n(n)\} \neq E\{\mathbf{n}(n)\}
\end{aligned} \tag{8.2.48}$$

The inequalities of Eqs. 8.2.47 and 8.2.48 become equalities when the signal $\mathbf{s}(n)$ (or noise $\mathbf{n}(n)$) is either a deterministic process or an ergodic stochastic process. This is the usual case. For further details, see Chap. 1.17.

The noise spectrum can be expressed in polar form as

$$N(m) = |N(m)|e^{j\,\arg N(m)} \tag{8.2.49}$$

The expected noise spectrum equals

$$E\{N(m)\} = E\{|N(m)|e^{j\,\arg N(m)}\} = E\{|N(m)|\}\,E\{e^{j\,\arg N(m)}\} = E\{\mathrm{FFT}[n(n)]\} \tag{8.2.50}$$

It can be expressed as the product of expectations because the magnitude and phase are independent (see Eq. 1.17.10). When the phase of $E\{N(m)\}$ is random and uniformly distributed between $(-\pi, \pi)$, then the expected noise phase $E\{\arg N(m)\}$ is zero and $E\{e^{j\,\arg N(m)}\}$ is zero. Then the expected noise spectrum is also zero or

$$E\{N(m)\} = 0 \tag{8.2.51}$$

Even though the expected noise spectrum is zero, the expected noise magnitude spectrum is not. It can take on a variety of forms. Expected magnitude noise spectra $E\{|N(j\omega)|\}$ were defined in the continuous frequency domain by their Fourier transforms of Eqs. 4.4.55 and 4.4.56. The magnitude and phase spectra are combined and converted into the discrete FFT frequency domain by using the sampling theorem of Eq. 1.3.6. The expected discrete noise spectrum equal

$$\begin{aligned}
E\{N_s(j\omega)\} &= \frac{\omega_s}{2\pi} \sum_{k=-\infty}^{\infty} E\{N(j(\omega - k\omega_s))\} \\
&= \frac{\omega_s}{2\pi} \sum_{k=-\infty}^{\infty} E\{|N(j(\omega - k\omega_s))|\}\,E\{e^{j(\omega - k\omega_s)}\}, \qquad |\omega| \le \omega_s/2
\end{aligned} \tag{8.2.52}$$

$$E\{N(m)\} = E\{N_s(j\omega)\big|_{\omega = m\omega_s/N}\}, \qquad |\omega| \le \omega_s/2$$

$E\{N(m)\}$ is periodic about multiples of the Nyquist frequency N. The magnitude of $E\{N(m)\}$ can be bounded by Schwartz's inequality (Eq. 1.3.10) for $|m| \le N/2$ as

$$E\{|N(m)|\} = E\{|N_s(j\omega)|_{\omega = m\omega_s/N}\} \le \frac{\omega_s}{2\pi} \sum_{k=-\infty}^{\infty} E\{|N(j(\omega - k\omega_s))|_{\omega = m\omega_s/N}\}$$

$$= \frac{\omega_s}{2\pi} , \qquad \text{white noise}$$

$$= \frac{N}{2\pi} \sum_{k=-\infty}^{\infty} \frac{1}{m - kN} , \qquad \text{integrated white noise}$$

$$\text{(8.2.53)}$$

$$= \frac{N^2}{2\pi\omega_s} \sum_{k=-\infty}^{\infty} \frac{1}{(m - kN)^2} , \qquad \text{doubly integrated white noise}$$

$$= \frac{\omega_s}{2\pi} \sum_{k=-\infty}^{\infty} e^{-(\omega_s/N)^2 (m-kN)^2} , \qquad \text{Gaussian noise}$$

8.2.14 PERFORMANCE MEASURES

The signal-to-noise ratio (SNR) can be defined in a variety of ways to best describe a particular problem. These include (in order of increasing computational difficulty) the: (1) spectral SNR, (2) rms SNR or simply the SNR, and the (3) peak SNR.

The spectral SNR is defined as

$$\text{Spectral } SNR = SNR(m) = \frac{S(m)}{N(m)} \qquad \text{(8.2.54)}$$

which is a function of frequency. It is the ratio of the signal spectrum to the noise spectrum evaluated at a particular frequency m. The spectral SNR has a magnitude and an angle. Generally only the magnitude is of interest. The Bode plot of $|SNR(m)|$ shows the frequency ranges over which the signal dominates or the noise dominates.

The rms SNR, which is usually simply called the SNR, equals

$$\text{Rms } SNR = \frac{\sqrt{P_s}}{\sqrt{P_n}} = \sqrt{\frac{\sum_{n=0}^{N-1} |s(n)|^2}{\sum_{n=0}^{N-1} |n(n)|^2}} = \sqrt{\frac{\sum_{m=0}^{N-1} |S(m)|^2}{\sum_{m=0}^{N-1} |N(m)|^2}} \qquad \text{(8.2.55)}$$

It is the square root of the ratio of the signal power P_s to the noise power P_n. It can be evaluated in either the time domain (sum $|s(n)|^2$ and $|n(n)|^2$), or in the frequency domain using Parseval's theorem (sum $|S(m)|^2$ and $|N(m)|^2$).

The peak SNR is defined as

$$\text{Peak } SNR = \frac{|s(T_p)|}{\sqrt{P_n}} = \frac{|s(T_p)|}{\sqrt{\frac{1}{N} \sum_{n=0}^{N-1} |n(n)|^2}} = \frac{|s(T_p)|}{\sqrt{\sum_{m=0}^{N-1} |N(m)|^2}} \qquad \text{(8.2.56)}$$

It is the ratio of the peak signal $s(T_p)$ which occurs at time T_p (or peak temporal signal) to the rms noise or the square root of the noise power. The noise power can be evaluated in time (sum $|n(n)|^2$) or in frequency (sum $|N(m)|^2$). It is usually difficult to find the peak signal value $s(T_p)$ since this requires solving $ds(t)/dt|_{t=n_pT} = 0$.

Regardless of which performance measure is used, SNR is usually expressed in dB rather than its numeric value as

$$SNR_{dB} = 20 \log_{10} SNR = 10 \log_{10} SNR^2 \qquad (8.2.57)$$

for SNR real. When SNR is the ratio r of signals, $20 \log r$ converts it to dB. When SNR is the ratio r of signal powers, $10 \log r$ converts it to dB.

The signal-to-noise ratio gain G_{SNR} provided by a filter of gain $H(m)$ equals

$$G_{SNR} = \frac{SNR_o}{SNR_i} = SNR_{o\ dB} - SNR_{i\ dB} \qquad (8.2.58)$$

Any of the SNR measures of Eqs. 8.2.54–8.2.56 can be used. The spectral SNR gain equals

$$\text{Spectral } G_{SNR} = \frac{S_o(m)/N_o(m)}{S_i(m)/N_i(m)} = \frac{H(m)S_i(m)/H(m)N_i(m)}{S_i(m)/N_i(m)} = 1 \qquad (8.2.59)$$

using the spectral SNR of Eq. 8.2.54. The output signal spectrum $S_o = HS_i$ while the output noise spectrum $N_o = HN_i$. Since the spectral G_{SNR} equals unity and is independent of the $H(m)$ filter used, the spectral G_{SNR} is of little interest. If the rms SNR of Eq. 8.2.55 is used instead, then the rms SNR gain equals

$$\text{Rms } G_{SNR} = \frac{\sqrt{\sum_{m=0}^{N-1}|S_o(m)|^2 \Big/ \sum_{m=0}^{N-1}|N_o(m)|^2}}{\sqrt{\sum_{m=0}^{N-1}|S_i(m)|^2 \Big/ \sum_{m=0}^{N-1}|N_i(m)|^2}} \qquad (8.2.60)$$

while the peak SNR gain equals

$$\text{Peak } G_{SNR} = \frac{s_o(n)_{pk} \Big/ \sqrt{\sum_{n=0}^{N-1}|n_o(n)|^2}}{s_i(n)_{pk} \Big/ \sqrt{\sum_{n=0}^{N-1}|n_i(n)|^2}} \qquad (8.2.61)$$

The filter output equals $c(n) = s_o(n) + n_o(n)$. Therefore, the output error equals $n_o(n) = c(n) - s_o(n)$ which is the output noise. A measure of the distortion in the output is the ratio of the noise energy to the signal energy as

$$\text{Distortion} = \frac{\sum_{n=0}^{N-1}|n_o(n)|^2}{\sum_{n=0}^{N-1}|s_o(n)|^2} = \frac{\sum_{m=0}^{N-1}|N_o(m)|^2}{\sum_{m=0}^{N-1}|S_o(m)|^2} = \frac{\sum_{m=0}^{N-1}|H(m)|^2|N_i(m)|^2}{\sum_{m=0}^{N-1}|H(m)|^2|S_i(m)|^2}$$

$$= 1/\text{rms } SNR_o^2$$

$$(8.2.62)$$

It is also equal to the reciprocal of the squared rms SNR_o. Distortion measures the normalized mean-squared-error in the output signal.

Another performance index is the noise bandwidth NBW. It is defined as

$$NBW = \sum_{m=0}^{N-1} |H(m)|^2 \bigg/ 2|H(m)|_{max}^2 \qquad (8.2.63)$$

In digital filters, NBW is often normalized by the Nyquist frequency $N/2$ to yield

$$NBW_n = \frac{NBW}{N/2} = \sum_{m=0}^{N-1} |H(m)|^2 \bigg/ N|H(m)|_{max}^2 \qquad (8.2.64)$$

To obtain maximum noise rejection, the NBW and NBW_n of the filter are minimized.

8.2.15　GAIN MATRICES OF FFT SCALAR FILTERS

Let us now formulate the gain matrices of some Class 1 estimation, detection, and correlation FFT filters for $N = 4$ to show these computations. We use the sawtooth signal $s = [0, 1, 2, 3]^t$ and zero-mean random noise $n = [1, -1, 1, -1]^t$. The input r equals $s + n = [1, 0, 3, 2]^t$. The FFT signal and noise spectra are $S = Fs$ and $N = Fn$ so

$$S = \begin{bmatrix} 1 & 1 & 1 & 1 \\ 1 & -j & -1 & j \\ 1 & -1 & 1 & -1 \\ 1 & j & -1 & -j \end{bmatrix} \begin{bmatrix} 0 \\ 1 \\ 2 \\ 3 \end{bmatrix} = \begin{bmatrix} 6 \\ -2+j2 \\ -2 \\ -2-j2 \end{bmatrix}, \quad N = \begin{bmatrix} 1 & 1 & 1 & 1 \\ 1 & -j & -1 & j \\ 1 & -1 & 1 & -1 \\ 1 & j & -1 & -j \end{bmatrix} \begin{bmatrix} 1 \\ -1 \\ 1 \\ -1 \end{bmatrix} = \begin{bmatrix} 0 \\ 0 \\ 4 \\ 0 \end{bmatrix}$$

$$(8.2.65)$$

The spectrum of the random noise changes from data block to data block. To achieve statistical averaging, we smooth the N given by Eq. 8.2.65b. Assuming that the random noise has a white spectrum, then the expected noise magnitude spectrum equals $E\{|N|\} = [2, 2, 2, 2]^t$ which is level-adjusted to have the same energy as N where $MSE = (1/4) \sum_{m=0}^{3} |N(m)|^2 = 4$. The expected signal $E\{S\} = S$ since S is deterministic. S and N are assumed to be independent so that $E\{SN^*\} = E\{S\}E\{N^*\} = 0$. With this spectral information, we can now formulate some of the filters of Table 8.2.1.

The uncorrelated Wiener estimation filter gain $H_{F\ UE}$ is $E\{|S(m)|^2\}/E\{|S(m)|^2 + |N(m)|^2\}$. Substituting in the appropriate spectra from Eq. 8.2.65 gives

$$H_{F\ UE} = \text{diag}\left[\frac{6^2}{6^2 + 2^2}, \frac{|-2+j2|^2}{|-2+j2|^2 + 2^2}, \frac{|-2|^2}{|-2|^2 + 2^2}, \frac{|-2-j2|^2}{|-2-j2|^2 + 2^2}\right]$$

$$= \text{diag}[0.9,\ 0.667,\ 0.5,\ 0.667] \qquad (8.2.66)$$

The correlated Wiener estimation filter gain $H_{F\ CE}$ equals $E\{SS^* + SN^*\}/E\{|S|^2 + |N|^2 + SN^* + S^*N\}$ which reduces to $H_{F\ UE}$ since the signal and noise are independent and $E\{SN^*\} = 0$.

The inverse detection filter gain $H_{F\ I}$ is $1/E\{S\}$ where

$$H_{F\ I} = \text{diag}[0.167,\ -0.25 - j0.25,\ -0.5,\ -0.25 + j0.25] \qquad (8.2.67)$$

The high-resolution detection filter gain $H_{F\ UHR}$ using an uncorrelated Wiener approach equals $E\{S^*\}/E\{|S|^2 + |N|^2\}$ so

$$\text{Table 8.2.1   Ideal Class 1 scalar filter transfer functions } H(m).$$

Filter Name	Transfer Function $H(m)$	Eq. Number

Distortion Balance ($W_S(m)$ = Signal weighting, $W_N(m)$ = Noise weighting)

$$\frac{E\{S(m)W_S(m)\}}{E\{S(m)W_S(m)\} + E\{N(m)W_N(m)\}}$$ (8.2.5)

General Estimation ($D(m)$ = Desired output)

$$H_E(m) = \frac{E\{D(m)R^*(m)\}}{E\{R(m)R^*(m)\}}$$ (8.2.12)

General Detection

$$H_D(m) = \frac{E\{D^*(m)\}}{E\{R(m)R^*(m)\}}$$ (8.2.15)

Estimation–Detection

$$H_E(m)H_D(m)$$ (8.2.20)

Uncorrelated Estimation (Uncorrelated Wiener)

$$\frac{E\{|S(m)|^2\}}{E\{|S(m)|^2\} + E\{|N(m)|^2\}}$$ (8.2.22)

Correlated Estimation (Correlated Wiener)

$$\frac{E\{S(m)(S^*(m) + N^*(m))\}}{E\{|S(m) + N(m)|^2\}}$$ (8.2.21)

Transitional 1 Estimation (Uncorrelated Wiener $\alpha = 0$, Correlated Wiener $\alpha = 1$)

$$\frac{E\{|S(m)|^2 + \alpha S(m)N^*(m)\}}{E\{|S(m)|^2 + |N(m)|^2 + \alpha(S(m)N^*(m) + S^*(m)N(m))\}}$$ (8.2.43)

Pulse-Shaping Estimation ($P(m)$ = Desired pulse spectrum, Uncorrelated Wiener)

$$\frac{E\{P(m)S^*(m)\}}{E\{|S(m)|^2\} + E\{|N(m)|^2\}}$$ (8.2.26)

Pulse-Shaping Estimation ($P(m)$ = Desired pulse spectrum, Correlated Wiener)

$$\frac{E\{P(m)(S^*(m) + N^*(m))\}}{E\{|S(m)|^2 + |N(m)|^2 + S(m)N^*(m) + S^*(m)N(m)\}}$$ (8.2.27)

Table 8.2.1 (Continued) Ideal Class 1 scalar filter transfer functions $H(m)$.

Filter Name	Transfer Function $H(m)$	Eq. Number

Classical Detection (Matched)

$$\frac{E\{S^*(m)\}}{E\{|N(m)|^2\}} \tag{8.2.34}$$

High-Resolution Detection (Uncorrelated Wiener)

$$\frac{E\{S^*(m)\}}{E\{|S(m)|^2\} + E\{|N(m)|^2\}} \tag{8.2.38}$$

High-Resolution Detection (Correlated Wiener)

$$\frac{E\{S^*(m) + N^*(m)\}}{E\{|S(m)|^2 + |N(m)|^2 + S(m)N^*(m) + S^*(m)N(m)\}} \tag{8.2.39}$$

Inverse Detection (Deconvolution)

$$\frac{1}{E\{S(m)\}} \tag{8.2.36}$$

Transitional 1 Detection (Inverse $\alpha = 0$, High-resolution $\alpha = 0.5$, Matched $\alpha = 1$)

$$\frac{E\{S^*(m)\}}{(1-\alpha)E\{|S(m)|^2\} + \alpha E\{|N(m)|^2\}} \tag{8.2.41}$$

Transitional 2 Detection (Inverse $\alpha = 0.5$)

$$\frac{1}{E\{S(m)^{2(1-\alpha)}\}} \tag{8.2.42}$$

Autocorrelation Estimation (Uncorrelated Wiener)

$$\frac{E\{S^*(m)|S(m)|^2\}}{E\{|S(m)|^2\} + E\{|N(m)|^2\}} \tag{8.2.29}$$

Autocorrelation Estimation (Correlated Wiener)

$$\frac{E\{(S^*(m) + N^*(m))|S(m)|^2\}}{E\{|S(m)|^2 + |N(m)|^2 + S(m)N^*(m) + S^*(m)N(m)\}} \tag{8.2.30}$$

Cross-Correlation Estimation (Uncorrelated Wiener)

$$\frac{E\{N^*(m)|S(m)|^2\}}{E\{|S(m)|^2\} + E\{|N(m)|^2\}} \tag{8.2.32}$$

Cross-Correlation Estimation (Correlated Wiener)

$$\frac{E\{S(m)N^*(m)(S^*(m) + N^*(m))\}}{E\{|S(m)|^2 + |N(m)|^2 + S(m)N^*(m) + S^*(m)N(m)\}} \tag{8.2.33}$$

Table 8.2.2   Ideal Class 2 and 3 scalar filter transfer functions $H(m)$.
(Also see Table 9.10.1.)

Filter Name	Class 2	Class 3

Distortion Balance ($W_S(m)$ = Signal weighting, $W_N(m)$ = Noise weighting)

$$\frac{E\{S(m)W_S(m)\}}{E\{R(m)W_N(m) - kS(m)[W_N(m) - W_S(m)]\}}$$

$$\frac{\langle R(m)W_S(m)\rangle}{\langle R(m)W_S(m)\rangle + (R(m) - \langle R(m)\rangle)W_N(m)}$$

Uncorrelated Estimation (Uncorrelated Wiener)

$$\frac{E\{|S(m)|^2\}}{E\{|S(m)|^2 + |R(m) - kS(m)|^2\}}$$

$$\frac{|\langle R(m)\rangle|^2}{\langle |R(m)|^2\rangle}$$

Correlated Estimation (Correlated Wiener)

$$\frac{E\{S(m)(R^*(m) + (1-k)S^*(m))\}}{E\{|R(m) + (1-k)S(m)|^2\}}$$

$$\frac{\langle\langle R(m)\rangle R^*(m)\rangle}{\langle |R(m)|^2\rangle}$$

Transitional 1 Estimation (Uncorrelated Wiener $\alpha = 0$, Correlated Wiener $\alpha = 1$)

$$\frac{E\{|S(m)|^2 + \alpha S(m)(R^*(m) - kS^*(m))\}}{\begin{array}{c}E\{|S(m)|^2 + |R(m) - kS(m)|^2\\ +\alpha[S(m)(R^*(m) - kS^*(m)) + S^*(m)(R(m) - kS(m))]\}\end{array}}$$

$$\frac{|\langle R(m)\rangle|^2 + (1-\alpha)\langle R(m)\rangle(R^*(m) - \langle R^*(m)\rangle)}{(1-\alpha)\langle |R(m)|^2\rangle}$$

Pulse-Shaping Estimation ($P(m)$ = Desired pulse spectrum, Uncorrelated Wiener)

$$\frac{E\{P(m)S^*(m)\}}{E\{|S(m)|^2 + |R(m) - kS(m)|^2\}}$$

$$\frac{P(m)\langle R^*(m)\rangle}{\langle |R(m)|^2\rangle}$$

Pulse-Shaping Estimation ($P(m)$ = Desired pulse spectrum, Correlated Wiener)

$$\frac{E\{P(m)(R^*(m) + (1-k)S^*(m))\}}{E\{|R(m) + (1-k)S(m)|^2\}}$$

$$\frac{\langle P(m)R^*(m)\rangle}{\langle |R(m)|^2\rangle}$$

Classical Detection (Matched)

$$\frac{E\{S^*(m)\}}{E\{|R(m) - kS(m)|^2\}}$$

$$\frac{\langle R^*(m)\rangle}{|R(m) - \langle R(m)\rangle|^2}$$

High-Resolution Detection (Uncorrelated Wiener)

$$\frac{E\{S^*(m)\}}{E\{|S(m)|^2 + |R(m) - kS(m)|^2\}}$$

$$\frac{\langle R^*(m)\rangle}{\langle |R(m)|^2\rangle}$$

Table 8.2.2   (Continued) Ideal Class 2 and 3 scalar filter transfer functions $H(m)$.

Filter Name	Class 2	Class 3

High-Resolution Detection (Correlated Wiener)

$$\frac{E\{R^*(m) + (1-k)S^*(m)\}}{E\{|R(m) + (1-k)S(m)|^2\}} \qquad\qquad \frac{\langle R^*(m)\rangle}{\langle |R(m)|^2\rangle}$$

Inverse Detection (Deconvolution)

$$\frac{1}{E\{S(m)\}} \qquad\qquad \frac{1}{\langle R(m)\rangle}$$

Transitional 1 Detection (Inverse $\alpha = 0$, High-Resolution $\alpha = 0.5$, Matched $\alpha = 1$)

$$\frac{E\{S^*(m)\}}{E\{(1-\alpha)|S(m)|^2 + \alpha|R(m) - kS(m)|^2\}}$$

$$\frac{\langle R^*(m)\rangle}{(1-\alpha)|\langle R(m)\rangle|^2 + \alpha|R(m) - \langle R(m)\rangle|^2}$$

Transitional 2 Detection (Inverse $\alpha = 0.5$)

$$\frac{1}{E\{S(m)^{2(1-\alpha)}\}} \qquad\qquad \frac{1}{\langle R(m)\rangle^{2(1-\alpha)}}$$

Autocorrelation Estimation (Uncorrelated Wiener)

$$\frac{E\{S^*(m)|S(m)|^2\}}{E\{|S(m)|^2 + |R(m) - kS(m)|^2\}} \qquad\qquad \frac{\langle R^*(m)\rangle\,|\langle R(m)\rangle|^2}{\langle |R(m)|^2\rangle}$$

Autocorrelation Estimation (Correlated Wiener)

$$\frac{E\{(R^*(m) + (1-k)S^*(m))|S(m)|^2\}}{E\{|R(m) + (1-k)S(m)|^2\}} \qquad\qquad \frac{|\langle R(m)\rangle|^2}{R(m)}$$

Cross-Correlation Estimation (Uncorrelated Wiener)

$$\frac{E\{(R^*(m) - kS^*(m))|S(m)|^2\}}{E\{|S(m)|^2 + |R(m) - kS(m)|^2\}} \qquad\qquad \frac{(R^*(m) - \langle R^*(m)\rangle)\,|\langle R(m)\rangle|^2}{\langle |R(m)|^2\rangle}$$

Cross-Correlation Estimation (Correlated Wiener)

$$\frac{E\{S(m)(R^*(m) - kS^*(m))(R^*(m) + (1-k)S^*(m))\}}{E\{|R(m) + (1-k)S(m)|^2\}}$$

$$\frac{(R^*(m) - \langle R^*(m)\rangle)\,\langle R(m)\rangle}{R(m)}$$

$$H_{F\ UHR} = \text{diag}[0.15,\ -0.167 - j0.167,\ -0.25,\ -0.167 + j0.167] \qquad (8.2.68)$$

The high-resolution detection filter gain $H_{F\ CHR}$ using a correlated Wiener approach is $E\{S^* + N^*\}/E\{|S|^2 + |N|^2 + SN^* + S^*N\}$ which reduces to $H_{F\ UHR}$. The matched detection filter has a gain $H_{F\ M}$ of $E\{S^*\}/E\{|N|^2\}$ where

$$H_{F\ M} = \text{diag}[1.5,\ -0.5 - j0.5,\ -0.5,\ -0.5 + j0.5] \qquad (8.2.69)$$

The signal autocorrelation filter gain $H_{F\ USS}$ based on the uncorrelated Wiener filter is $E\{S^*|S|^2\}/E\{|S|^2 + |N|^2\}$ so that

$$H_{F\ USS} = \text{diag}[5.4,\ -1.333 - j1.333,\ -1,\ -1.333 + j1.333] \qquad (8.2.70)$$

The signal-noise cross-correlation filter gain $H_{F\ USN}$ based on the uncorrelated Wiener filter is $E\{N^*|S|^2\}/E\{|S|^2 + |N|^2\}$. When the signal and noise are independent and the noise has zero mean, then $E\{N^*|S|^2\} = E\{N^*\}E\{|S|^2\} = 0$ so

$$H_{F\ USN} = \text{diag}[0,\ 0,\ 0,\ 0] \qquad (8.2.71)$$

The transitional 1 detection FFT filter gain $H_{F\ TD1}$ equals $E\{S^*\}/E\{(1-\alpha)|S|^2 + \alpha|N|^2\}$ where

$$H_{F\ TD1} = \text{diag}\left[\frac{1}{6(1-\alpha)},\ \frac{1}{(-2+j2)(1-\alpha)},\ \frac{-1}{2+6\alpha},\ \frac{1}{(-2-j2)(1-\alpha)}\right] \qquad (8.2.72)$$

The transitional 2 detection FFT filter gain $H_{F\ TD2}$ equals

$$H_{F\ TD2} = \text{diag}\left[6^{-2(1-\alpha)},\ (-2+j2)^{-2(1-\alpha)},\ (-2)^{-2(1-\alpha)},\ (-2-j2)^{-2(1-\alpha)}\right] \qquad (8.2.73)$$

The transitional estimation FFT filter gain $H_{F\ TE1}$ equals $E\{|S|^2 + \alpha SN^*\}/E\{|S|^2 + |N|^2 + \alpha(SN^* + S^*N)\}$ which reduces to the non-transitional uncorrelated estimation FFT filter of Eq. 8.2.66 since $E\{SN^*\} = 0$.

8.3 FFT-BASED VECTOR FILTERS

The block diagram of all fast transform vector or scalar filters is shown in Fig. 8.3.1. It is a vector relabelled equivalent of Fig. 8.2.1 when the post-processing block is eliminated. The filter gain H_T depends upon the transform used (e.g. Fourier, Paley, Haar, slant, and K-L) as we shall now show. These filters are based on the following assumptions (case $A\bar{A}\bar{B}$ in Table 8.1.1):
(a) The FFT and any other fast transform are used, so
(b) One set of basis functions are complex exponentials; the other is arbitrarily chosen.
(c) The gain matrix is nondiagonal.

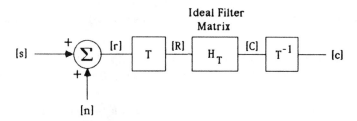

Fig. 8.3.1 Block diagram of fast transform filter assuming additive signal and noise.

8.3.1 THEORY OF FFT-BASED VECTOR FILTERS

A variety of FFT scalar filter gains H_F are listed in Tables 8.2.1 and 8.2.2. Regardless of the particular estimation, detection, or correlation algorithm used, H_F can always be expressed as a diagonal matrix as

$$H_F = \text{diag}\left[H(0),\ldots,H(m),\ldots,H(N-1)\right]$$
$$= \begin{bmatrix} H(0) & 0 & \cdots & 0 \\ 0 & H(1) & \cdots & 0 \\ \vdots & \vdots & \ddots & \vdots \\ 0 & 0 & \cdots & H(N-1) \end{bmatrix} \tag{8.3.1}$$

$H(m)$ is the mth diagonal entry in the H_F gain matrix. The $H(m)$ entries in the gain matrix are computed using the desired $H(m)$ equation in Tables 8.2.1 and 8.2.2. The diagonal Fourier filter matrix H_F can be used to derive the corresponding filter matrix H_T for any specified transform T. Although the filter matrices will be different using different transforms, the filters will have identical outputs.

From the system block diagram of Fig. 8.3.1, we can write that

$$R = Tr, \qquad C = H_T R, \qquad c = T^{-1}C \tag{8.3.2}$$

as given by Eq. 8.1.1. Combining these equations gives

$$c_T = T^{-1}H_T Tr \tag{8.3.3}$$

In the Fourier transform case, F and F^{-1} are the Fourier transform and inverse Fourier transform matrices, respectively. In this case Eq. 8.3.3 equals

$$c_F = F^{-1}H_F Fr \tag{8.3.4}$$

Equating the two filter outputs c_F and c_T yields

$$c_F = F^{-1}H_F Fr = T^{-1}H_T Tr = c_T \tag{8.3.5}$$

This requires that the gain matrices be related as

$$F^{-1}H_F Fr = T^{-1}H_T Tr \qquad (8.3.6)$$

Multiplying each side of Eq. 8.3.6 by T on the left and T^{-1} on the right, and noting that $TT^{-1} = I$ gives the gain matrix H_T as

$$H_T = T F^{-1} H_F F T^{-1} = (FT^{-1})^{-1} H_F (FT^{-1}) \qquad (8.3.7)$$

I is the $N \times N$ identity matrix filled with 1's on the diagonal. Fig. 8.3.2a shows the block diagram of an FFT filter which has TT^{-1} and $T^{-1}T$ identity matrix blocks inserted at both ends. Fig. 8.3.2b shows how the block diagram can be repartitioned to form Eq. 8.3.7.

There are five $N \times N$ matrices in Eq. 8.3.7 so $4N^3$ multiplications (at most) are required to directly convert the Fourier gain matrix H_F to another transform gain matrix H_T. Although matrix H_T may appear to be impractical to compute, it can be obtained using fast Fourier transforms and fast T transforms as shall now be described. Let u_i be the $1 \times N$ unit selection vector having 1 as the ith entry and the rest 0's for $i = 1, 2, \ldots, N$. Then H_T can be generated using the following procedure:

1. Perform N fast T^{-1} transforms on the u_i's from u_1 to u_N. This indirectly generates (the columns of) the T^{-1} matrix.
2. Perform N fast Fourier transforms on the resulting column vectors from Step 1. This generates the (FT^{-1}) matrix.
3. Multiply the ith row of the matrix produced in Step 2 by the ith diagonal entry of H_F. This generates the $(H_T FT^{-1})$ matrix. Since H_F is diagonal, this requires N^2 matrix multiplications.
4. Perform N fast F^{-1} transforms on the resulting column vectors from Step 3. This generates the $(F^{-1} H_T FT^{-1})$ matrix.
5. Finally perform N fast T transforms on the resulting column vectors from Step 4 which produces the H_T matrix.

$4N$ fast transforms are performed following this procedure ($2N$ times on the right side of H_F and $2N$ times on the left). Performing the fast F^{-1} transforms and then the fast T transforms only requires $2N^2 \log_2 N$ operations (most of which are additions and subtractions). Therefore a total of $(1 + 2 \log_2 N)N^2 \cong 2N^2 \log_2 N$ operations are required to evaluate Eq. 8.3.7.

To simplify the computation of Eq. 8.3.7, we can compute

$$G = FT^{-1} \qquad (8.3.8)$$

Then Eq. 8.3.7 can be rewritten as

$$H_T = G^{-1} H_F G \qquad (8.3.9)$$

Once the T transform has been selected, G and G^{-1} can be computed using Eq. 8.3.8 in Step 2 and stored. Later when H_F is updated, we start at Step 3. To reduce storage

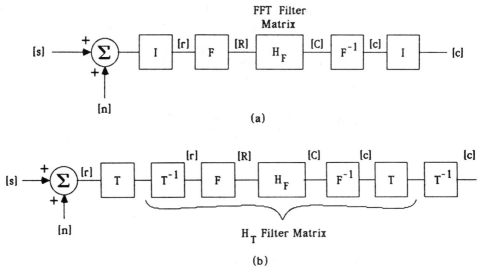

Fig. 8.3.2 Block diagram of (a) FFT filter assuming additive signal and noise and (b) rearranged form to generate FFT-based fast transform filter.

requirements at the expense of increased computation time, only G need be stored since G^{-1} can be computed. Some 8×8 G matrices were listed in Chap. 2.9. Multiplying H_F and G requires N^2 operations; multiplying G^{-1} and $(H_F G)$ requires N^3 operations. Therefore, the simpler Eq. 8.3.9 approach requires $(1+N)N^2 \cong N^3$ operations which is computationally more expensive than the Eq. 8.3.7 approach.

Although H_F is diagonal (a scalar filter), H_T is usually nondiagonal (a vector filter). Although H_T is much slower to compute than H_F for Fourier filters, the actual process of filtering using $C = H_T R$ with fast transform T can be considerably faster. When N gets large, the H_T matrix usually becomes more and more sparse (but not diagonal) and is mostly filled with zeros. Therefore, computation time can be further reduced from $2N^2 \log_2 N$ by *performing only nonzero multiplications/additions* using a simple IF statement in the filter program. General fast transform filtering is not as computationally intensive as it might first appear.

8.3.2 GAIN MATRICES OF FFT-BASED VECTOR FILTERS

To illustrate the matrix manipulations required for FFT-based fast transform filters, we now compute the gain matrices for some Class 1 estimation, detection, and correlation FFT-based Paley filters using $N = 4$. Consider again the sawtooth signal $s = [0,1,2,3]^t$ and zero-mean random noise $n = [1,-1,1,-1]^t$. The noisy input r equals $s + n = [1,0,3,2]^t$. As found in Chap. 8.2.15, the Fourier signal and noise spectra are $S = Fs$ and $N = Fn$ so

$$S = \begin{bmatrix} 6 \\ -2+j2 \\ -2 \\ -2-j2 \end{bmatrix}, \quad N = \begin{bmatrix} 0 \\ 0 \\ 4 \\ 0 \end{bmatrix}, \quad \hat{N} = \begin{bmatrix} 2 \\ 2 \\ 2 \\ 2 \end{bmatrix} \qquad (8.3.10)$$

$\widehat{N}$ is a smoothed version of N to achieve statistical averaging. $\widehat{N}$ has been level-adjusted to have the same energy as N ($MSE = 4$). The expected signal spectrum $E\{S\} = S$ since S is deterministic. The expected noise spectrum $E\{N\} = 0$ since N is a zero-mean process. Therefore the cross-correlation spectrum $E\{SN^*\} = E\{S\}E\{N^*\} = 0$ when S and N are assumed to be independent. A less stringent condition which insures that $E\{SN^*\} = 0$ is that S and N are orthogonal.

The gains of a variety of Class 1 FFT scalar filters were computed in Eqs. 8.2.66–8.2.73. The gain of the uncorrelated estimation FFT scalar filter listed in Table 8.2.1 is $E\{|S(m)|^2\}/E\{|S(m)|^2 + |N(m)|^2\}$ where

$$H_{F\ UE} = \text{diag}\left[\frac{6^2}{6^2 + 2^2}, \frac{|-2+j2|^2}{|-2+j2|^2 + 2^2}, \frac{|-2|^2}{|-2|^2 + 2^2}, \frac{|-2-j2|^2}{|-2-j2|^2 + 2^2}\right] \quad (8.3.11)$$

$$= \text{diag}[0.9,\ 0.667,\ 0.5,\ 0.667]$$

Since the gain $H_{F\ CE}$ of the correlated estimation FFT scalar filter equals $E\{|S|^2 + SN^*\}/E\{|S|^2+|N|^2+SN^*+S^*N\}$ and $E\{SN^*\} = 0$ assuming S and N are orthogonal, then $H_{F\ CE}$ equals $H_{F\ UE}$. The Fourier-inverse Paley transform product $G = FT^{-1} = (1/N)FT^h$ equals

$$G = \frac{1}{4}\begin{bmatrix} 1 & 1 & 1 & 1 \\ 1 & -j & -1 & j \\ 1 & -1 & 1 & -1 \\ 1 & j & -1 & -j \end{bmatrix}\begin{bmatrix} 1 & 1 & 1 & 1 \\ 1 & 1 & -1 & -1 \\ 1 & -1 & 1 & -1 \\ 1 & -1 & -1 & 1 \end{bmatrix} = \frac{1}{2}\begin{bmatrix} 2 & 0 & 0 & 0 \\ 0 & 1-j1 & 0 & 1+j1 \\ 0 & 0 & 2 & 0 \\ 0 & 1+j1 & 0 & 1-j1 \end{bmatrix} \quad (8.3.12)$$

The 8-point G is given by Eq. 2.9.8. Now $G^{-1} = (FT^{-1})^{-1} = (FT^{-1})^h = G^h$ so G^{-1} is the transposed conjugate of G since both F and T are assumed to be unitary transforms. Therefore the equivalent gain $G^{-1}H_F G$ of the uncorrelated and correlated estimation FFT-based Paley filters using Eq. 8.3.9 equals

$$H_{Pal\ UE} = H_{Pal\ CE} = G^{-1}H_{F\ UE}G = G^{-1}\text{diag}[0.9,\ 0.667,\ 0.5,\ 0.667]G$$

$$= \frac{1}{4}\begin{bmatrix} 2 & 0 & 0 & 0 \\ 0 & 1+j1 & 0 & 1-j1 \\ 0 & 0 & 2 & 0 \\ 0 & 1-j1 & 0 & 1+j1 \end{bmatrix}\begin{bmatrix} 0.9 & 0 & 0 & 0 \\ 0 & 0.667 & 0 & 0 \\ 0 & 0 & 0.5 & 0 \\ 0 & 0 & 0 & 0.667 \end{bmatrix}\begin{bmatrix} 2 & 0 & 0 & 0 \\ 0 & 1-j1 & 0 & 1+j1 \\ 0 & 0 & 2 & 0 \\ 0 & 1+j1 & 0 & 1-j1 \end{bmatrix} \quad (8.3.13)$$

which is the same as the FFT filter gain $H_{F\ UE}$. The gains are very seldom preserved. Now let us convert some detection and correlation FFT scalar filters into Paley filters.

It can be shown that any 4×4 diagonal FFT filter gain matrix $H_F = \text{diag}[a, b, c, d]$ converts into the FFT-based Paley gain matrix H_{Pal} as

$$H_{Pal} = \begin{bmatrix} a & 0 & 0 & 0 \\ 0 & \frac{1}{2}(b+d) & 0 & \frac{j}{2}(b-d) \\ 0 & 0 & c & 0 \\ 0 & -\frac{j}{2}(b-d) & 0 & \frac{1}{2}(b+d) \end{bmatrix} \quad (8.3.14)$$

When $d = b^*$, then $(b + d)/2 = \text{Re } b$ and $j(b - d)/2 = -\text{Im } b$. Using this result, the FFT filter gains can be directly converted without using matrix multiplication as in Eq. 8.3.13. The inverse detection FFT scalar filter gain is $1/E\{S\}$ where

$$H_{F\ I} = \text{diag}\left[0.167,\ -0.25 - j0.25,\ -0.5,\ -0.25 + j0.25\right] \tag{8.3.15}$$

The FFT-based Paley filter gain equals

$$H_{Pal\ I} = \begin{bmatrix} 0.167 & 0 & 0 & 0 \\ 0 & -0.25 & 0 & 0.25 \\ 0 & 0 & -0.5 & 0 \\ 0 & -0.25 & 0 & -0.25 \end{bmatrix} \tag{8.3.16}$$

The high-resolution detection FFT scalar filter gain $H_{F\ CHR}$ using an uncorrelated Wiener approach is $E\{S^*\}/E\{|S|^2 + |N|^2\}$ so

$$H_{F\ UHR} = \text{diag}\left[0.15,\ -0.167 - j0.167,\ -0.25,\ -0.167 + j0.167\right] \tag{8.3.17}$$

The high-resolution detection FFT scalar filter gain $H_{F\ CHR}$ using the correlated Wiener approach is $E\{S^* + N^*\}/E\{|S|^2 + |N|^2 + SN^* + S^*N\}$ which equals $H_{F\ UHR}$ because $E\{N\} = 0$ and $E\{SN^*\} = 0$. Converting $H_{F\ UHR}$ to the Paley filter gain gives

$$H_{Pal\ UHR} = \begin{bmatrix} 0.15 & 0 & 0 & 0 \\ 0 & -0.167 & 0 & 0.167 \\ 0 & 0 & -0.25 & 0 \\ 0 & -0.167 & 0 & -0.167 \end{bmatrix} \tag{8.3.18}$$

The FFT matched detection scalar filter has a gain of $E\{S^*\}/E\{|N|^2\}$ where

$$H_{F\ M} = \text{diag}\left[1.5,\ -0.5 - j0.5,\ -0.5,\ -0.5 + j0.5\right] \tag{8.3.19}$$

The FFT-based matched detection Paley filter has a gain of

$$H_{Pal\ M} = \begin{bmatrix} 1.5 & 0 & 0 & 0 \\ 0 & -0.5 & 0 & 0.5 \\ 0 & 0 & -0.5 & 0 \\ 0 & -0.5 & 0 & -0.5 \end{bmatrix} \tag{8.3.20}$$

The signal autocorrelation FFT scalar filter gain using the uncorrelated Wiener approach is $E\{S^*|S|^2\}\ /E\{|S|^2 + |N|^2\}$ so that

$$H_{F\ USS} = \text{diag}\left[5.4,\ -1.333 - j1.333,\ -1,\ -1.333 + j1.333\right] \tag{8.3.21}$$

In the correlated Wiener approach, $H_{F\ CSS}$ equals $E\{|S|^2(S^* + N^*)\}/\ E\{|S|^2 + |N|^2 + SN^* + S^*N\}$ which reduces to $H_{F\ USS}$ when $E\{N\} = 0$ and $E\{SN^*\} = 0$. Converting these FFT scalar filter gains gives

$$H_{Pal\ USS} = \begin{bmatrix} 5.4 & 0 & 0 & 0 \\ 0 & -1.333 & 0 & 1.333 \\ 0 & 0 & -1 & 0 \\ 0 & -1.333 & 0 & -1.333 \end{bmatrix} \qquad (8.3.22)$$

The signal-noise cross-correlation scalar filter gain based upon the uncorrelated Wiener approach is $E\{N^*|S|^2\}/E\{|S|^2 + |N|^2\}$. Since $E\{N^*|S|^2\} = 0$ when S and N are independent and the noise has zero mean, then

$$H_{F\ USN} = \text{diag}[0,\ 0,\ 0,\ 0] \qquad (8.3.23)$$

In this case, $H_{F\ CSN}$ using the correlated Wiener approach is $E\{SN^*(S^*+N^*)\}/E\{|S|^2 + |N|^2 + SN^* + S^*N\}$ reduces to $H_{F\ USN}$. The Paley filter gain $H_{Pal\ CSN}$ equals $H_{F\ CSN}$.

An example of a Paley transitional filter is the transitional 2 detection filter where (see Eq. 8.2.73)

$$H_{Pal\ TD2} = \begin{bmatrix} 6^{-2(1-\alpha)} & 0 & 0 & 0 \\ 0 & \text{Re}\ (-2+j2)^{-2(1-\alpha)} & 0 & \text{Im}\ (-2+j2)^{-2(1-\alpha)} \\ 0 & 0 & (-2)^{-2(1-\alpha)} & 0 \\ 0 & -\text{Im}\ (-2+j2)^{-2(1-\alpha)} & 0 & \text{Re}\ (-2+j2)^{-2(1-\alpha)} \end{bmatrix}$$
$$(8.3.24)$$

8.3.3 FFT-BASED VECTOR FILTER DESIGN EXAMPLES

We shall now design several estimation and detection FFT-based fast transform filters having the form of Fig. 8.3.1 where $N = 128$. The FFT scalar estimation filter using the uncorrelated Wiener approach is given by Eq. 8.2.22. The signal s is assumed to be a cosine pulse as shown in Fig. 8.3.3a. The noise n is assumed to be white with random phase. The signal and noise are added together to form the input r as shown in Fig. 8.3.3b. The peak signal-to-rms noise ratio is selected to equal about 10 dB. The FFT spectra of s, n, and r are computed using Eq. 8.2.1a. Substituting them into Eq. 8.2.22 yields the Fourier transfer function magnitude shown in Fig. 8.3.3d. Multiplying this gain by the input spectrum gives the output spectrum. Taking the inverse transform of C using Eq. 8.2.1b gives the output c shown in Fig. 8.3.3c. The signal is recovered but the spurious noise is high.

The H_T gains are now computed for the Paley, Haar, and slant filters using Eq. 8.3.7. These H_T gains are nondiagonal using generalized transforms. Both the gain matrix and its main diagonal are plotted for the Paley filter in Fig. 8.3.3e, Haar filter in Fig. 8.3.3f, and slant filter in Fig. 8.3.3g. The FFT gain has period $N/2$ while the other gains have period N. All these estimation filters have identical outputs.

A detection example is shown in Fig. 8.3.4. The signal $s(t)$ is assumed to be a triangular pulse as shown in Fig. 8.3.4a. The noise $n(t)$ is assumed to be white with random phase. The signal and noise are added together to form the input $r(t)$ as shown

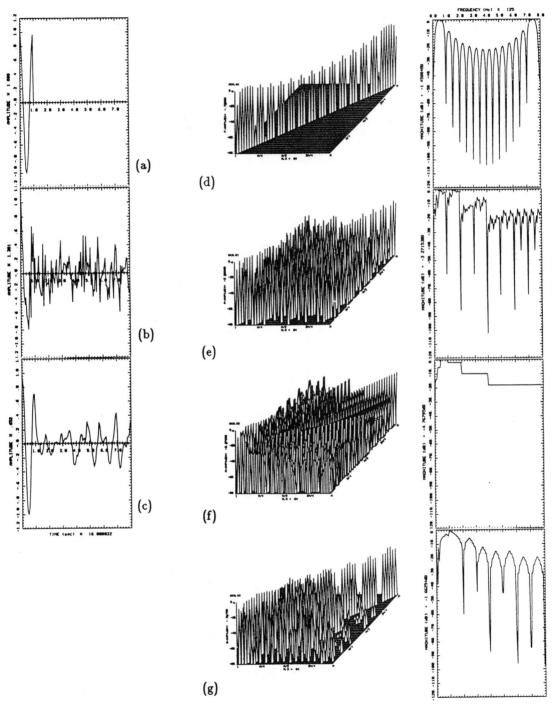

Fig. 8.3.3 (a) Signal, (b) total input, (c) output, and gain (H_T gain matrix and its main diagonal) of (d) Fourier, (e) Paley, (f) Haar, and (g) slant estimation FFT-based fast transform filters. (Plotted by C. Powell and based upon H.E. Lin, "On digital filter design by orthogonal transforms," Special project, Univ. of Miami, 1986.)

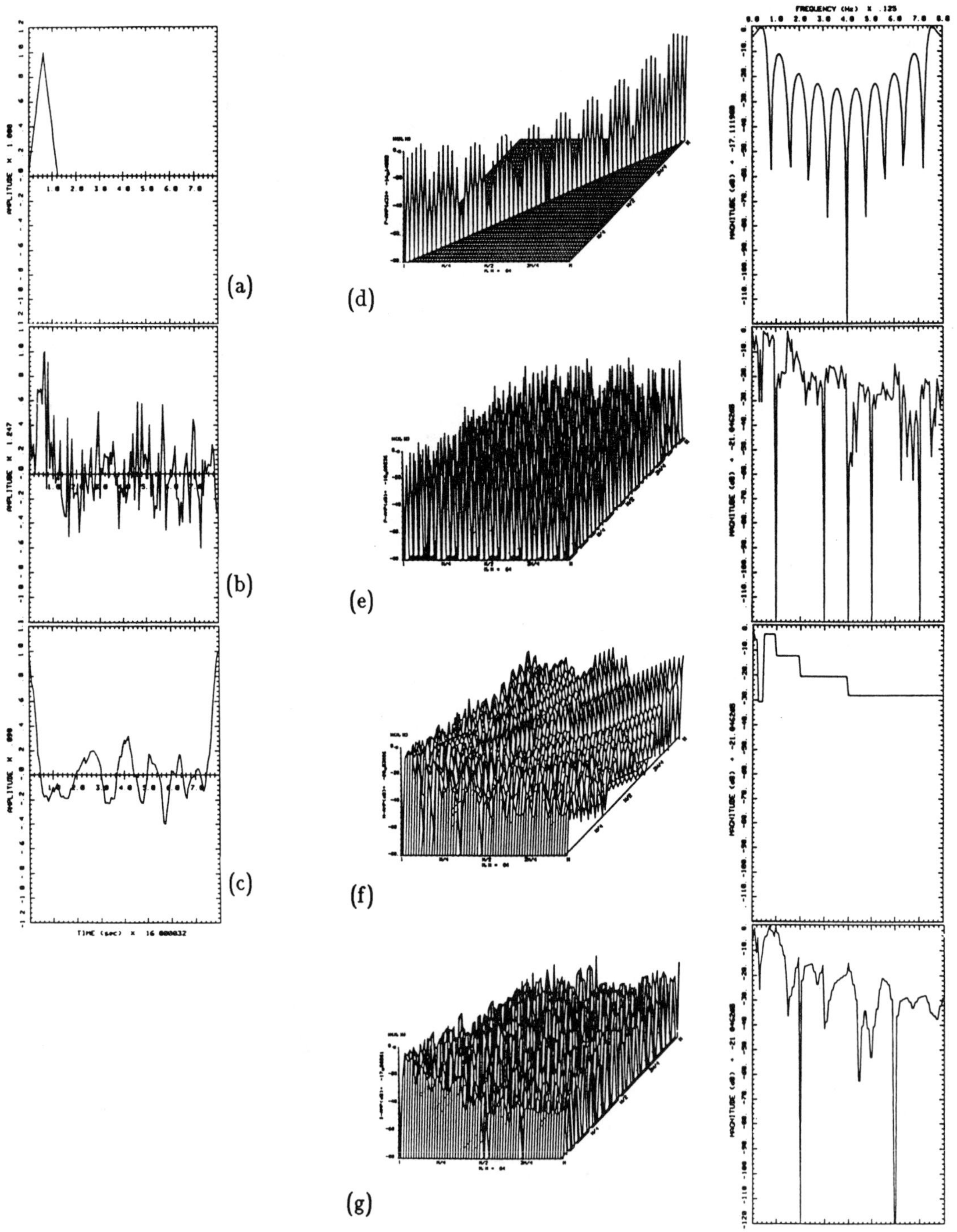

Fig. 8.3.4 (a) Signal, (b) total input, (c) output, and gain (H_T gain matrix and its main diagonal) of (d) Fourier, (e) Paley, (f) Haar, and (g) slant detection FFT-based fast transform filters. (Plotted by C. Powell and based upon H.E. Lin, loc. cit.)

in Fig. 8.3.4b. The peak signal-to-noise ratio is again selected to equal about 10 dB. Computing the FFT spectra of $s(t), n(t)$, and $r(t)$ and substituting them into Eq. 8.2.34 yields the Fourier transfer function of the matched detection scalar filter shown in Fig. 8.3.4d. The output $c(t)$ is shown in Fig. 8.3.4c. The signal is detected but the output pulse width is wide and oscillatory. This process is repeated for the Paley, Haar, and slant transforms. These gains are presented for Paley in Fig. 8.3.4e, Haar in Fig. 8.3.4f, and slant in Fig. 8.3.4g. All filter outputs are identical.

These FFT-based filters have square gain matrices with nonzero elements. These filters are optimum in the FFT sense but not in the generalized transform sense. We shall now derive a set of nondiagonal square gain matrices that are optimum.

8.4 FAST TRANSFORM VECTOR FILTERS

In Chap. 8.2, the analog filters of Chaps. 4.4–4.5 were converted into digital FFT scalar filters. In Chap. 8.3, these scalar filters were converted into fast transform vector filters having the *same outputs* as the FFT filters. In this section, we derive a new set of fast transform filters that have *improved outputs*. The assumptions used to derive this set of filters are (case $\bar{A}\bar{B}$ in Table 8.1.1):

(a) Any fast transform is used, so

(b) The basis functions are arbitrarily chosen.

(c) The gain matrix is nondiagonal.

These filters still use the Fourier, Paley, Haar, slant, and other generalized transforms but are derived directly following the optimization methods of Chap. 8.2. These optimizations are carried out in the generalized frequency domain rather than the FFT frequency domain. This produces a better set of filters than those obtained by simply converting FFT filters as was done in Chap. 8.3. The gain matrices of these filters usually are still nondiagonal so they are vector filters. Vector filters are designed in the generalized frequency domain or in the time domain using the generalized filter of Fig. 8.3.1 where

$$r = s + n \quad \text{or} \quad R = Tr = S + N$$
$$c = T^{-1}C = h_T r \quad \text{or} \quad C = H_T R \tag{8.4.1}$$

Both the filter gain matrix $H_T = T h_T T^{-1}$ and the impulse response matrix or transmission matrix $h_T = T^{-1} H_T T$ are square. H_T is generally nondiagonal and the gain is time-varying.

8.4.1 DISTORTION BALANCE VECTOR FILTERS

Signal and noise distortions are balanced in distortion balance filters. The degradation caused by the filter distorting or altering the signal to the distortion is equated to the noise leaking through the filter. The signal degradation E_S is the difference between input and output signal spectra. The noise leakage E_N equals the output noise spectrum.

Therefore

$$E_S = S - H_T S = (I - H_T)S$$
$$E_N = H_T N \qquad (8.4.2)$$

where H_T is the ideal filter gain and I is the identity matrix whose diagonal is filled with 1's and 0's elsewhere. For more general results, the signal distortion is weighted by W_S and the noise distortion by W_N. Then equating the weighted signal error to the weighted noise error at every frequency p gives

$$E_S W_S^t = E_N W_N^t$$
$$(I - H_T)S W_S^t = H_T N W_N^t \qquad (8.4.3)$$

"t" indicates the transpose operation. This insures that the total error

$$E = E_S W_S^t + E_N W_N^t \qquad (8.4.4)$$

is composed of equally-weighted signal and equally-weighted noise. Solving Eq. 8.4.3 for the ideal filter transfer function H_T gives (superscript "-1" indicates the inverse operation)

$$H_T = S W_S^t \left(S W_S^t + N W_N^t \right)^{-1} \qquad (8.4.5)$$

This result has a number of interesting applications. If we wish to form a filter that balances complex signal and noise levels equally, we set $W_S = W_N = W$ so

$$H_T = S W^t \left(S W^t + N W^t \right)^{-1} \qquad (8.4.6)$$

However, if we wish to form a filter that balances signal and noise power levels, we set $W_S = S^*$ and $W_N = N^*$. A multitude of other possibilities exist. For example, if we wish to ignore high-frequency signal and low-frequency noise, we might choose $W_S = T[\text{FFT}^{-1}[(1 + H_{dmn}^{-1})^{-1}]]$ and $W_N = T[\text{FFT}^{-1}[(1 + H_{dmn})^{-1}]]$. Alternatively, if we wish to balance the derivative of the signal from noise, $W_S = T[\text{FFT}^{-1}[H_{dmn}^{-1}]]$ and $W_N = 1$. In these cases, we use the frequency transforms of Chap. 5 where $p = T[\text{FFT}^{-1}[m]] = T[\text{FFT}^{-1}[H_{dmn}^{-1}]]$ (cf. Chap. 4.4.1). The possibilities are limitless.

8.4.2 GENERAL ESTIMATION VECTOR FILTERS

General estimation filters estimate the desired optimal output d from the input r. The output error e equals

$$e = d - c \qquad \text{or} \qquad E = D - C \qquad (8.4.7)$$

One optimization criteria that can be used is to minimize the mean-squared-error (MSE) where

$$MSE = ||e|| = e^h e = (d^h - c^h)(d - c) \qquad (8.4.8)$$

"h" indicates the Hermitian or conjugate-transpose operation which satisfies $A^h = (A^*)^t = (A^t)^*$. The matrix A^h equals the transpose t or the conjugate $*$ of the transposed matrix. If an element of matrix A equals $a_{ik} = u_{ik} + jv_{ik}$, then the corresponding element of A^h equals $a_{ik}^h = (u_{ik} + jv_{ik})^* = u_{ki} - jv_{ki}$. The ideal filter is the one whose transfer function H_T minimizes the error index MSE. Using Parseval's theorem where $a^h b = (1/N)A^h B$ since $a^h b = (T^{-1}A)^h(T^{-1}B) = A^h(T^{-1})^h T^{-1} B = A^h(T^h T)^{-1} B = A^h(NI)^{-1}B$, Eq. 8.2.8 can be rewritten in the frequency domain as

$$MSE = ||E|| = E^h E = (D^h - C^h)(D - C) \qquad (8.4.9)$$

The $1/N$ constant has been dropped for convenience so that Eqs. 8.4.8 and 8.4.9 are exact duals. $E^h = (E^*)^t$ is the transposed conjugate error spectrum. When e is real-valued, then $E^h = E^t$ for real-valued T transforms. Since $C = H_T R$ is a functional of H_T, Eq. 8.4.9 is minimized by setting the derivative of the sum with respect to H_T equal to zero. MSE can be rewritten as

$$MSE = D^h D - R^h H_T^h D - D^h H_T R + R^h H_T^h H_T R \qquad (8.4.10)$$

using $(AB)^h = B^h A^h$. Differentiating MSE with respect to H_T and setting the result equal to zero yields (assume H_T is real)

$$\frac{d\ MSE}{dH_T} = -R^h D - D^h R + R^h(H_T + H_T^h)R = 0 \qquad (8.4.11)$$

since $d(ABC)/dB = d(AB^h C)/dB = AC$ and $d(AB^h BC)/dB = A(B + B^h)C$. Rearranging Eq. 8.4.11 gives

$$H_T + H_T^h = DR^{-1} + (DR^{-1})^h \qquad (8.4.12)$$

Solving Eq. 8.4.12 for H_T by equating like terms gives (the same result holds for complex H_T)

$$H_T = DR^{-1} = DR^h(R^h)^{-1}R^{-1}$$
$$= (DR^h)\left(RR^h\right)^{-1} = \frac{\text{Desired input output cross correlation spectrum}}{\text{Input autocorrelation spectrum}}$$

$$(8.4.13)$$

This results by introducing an identity matrix $I = (R^h)(R^h)^{-1}$ between D and R^{-1} and then using $(AB)^{-1} = B^{-1}A^{-1}$. When the input is ergodic, RR^h and DR^h equal the autocorrelation and cross-correlation spectra. When the input is not ergodic, they are *estimates* of the correlation spectra. Eq. 8.4.13 cannot be expressed in the deterministic form $H_T = DR^{-1}$ = desired output/available input because H_T reduces to a scalar filter. A deterministic input-output filter is still formulated as a stochastic correlation filter using Eq. 8.4.13 unlike the FFT filter of Eq. 8.2.13.

Eq. 8.4.13 is the transfer function for the general estimation filter. This ideal filter gives the best estimate (in the minimum MSE sense) of a desired signal d from an input signal r as the ratio of the two power spectra. When r is composed of a signal s contaminated with additive noise n, Eq. 8.4.13 reduces to

$$H_T = D(S^h + N^h)\left[(S + N)(S^h + N^h)\right]^{-1} \qquad (8.4.14)$$

8.4.3 GENERAL DETECTION VECTOR FILTERS

General detection filters detect the presence or absence of a signal in noise. The ideal filter detects a desired signal d by producing an output signal pulse $s_o(0)$ in the total output c such that the SRR_o ratio of the peak output signal-to-total rms output is maximized. Its transfer function equals

$$H_T = \left(Tu_0D^h\right)\left(RR^h\right)^{-1} = \frac{\text{Conjugate spectrum of desired signal to be detected}}{\text{Input autocorrelation spectrum}} \qquad (8.4.15)$$

This filter produces a narrow output pulse (which reduces temporal or time ambiguity) centered at time $t = 0$ with undefined shape.

The ideal filter which detects d by producing $s_o(0)$ in c, such that the peak output signal-to-rms output noise SNR_o is maximized, has

$$H_T = \left(Tu_0D^h\right)\left(NN^h\right)^{-1} = \frac{\text{Conjugate spectrum of desired signal to be detected}}{\text{Noise autocorrelation spectrum}} \qquad (8.4.16)$$

To prove Eq. 8.4.15, consider the peak output signal (at time zero)-to-total rms output ratio squared (see Eq. 8.4.56)

$$\begin{aligned}
SRR_o^2 &= \frac{|s_o(0)|^2}{\sum_{n=0}^{N-1}|c(n)|^2} = \frac{\left|\text{tr } u_0^t T^{-1}S_o\right|^2}{\text{tr }(T^{-1}C)(T^{-1}C)^h} \\
&= \frac{\left|\text{tr } T^{-1}H_TSu_0^t\right|^2}{\text{tr }(T^{-1}H_TR)(T^{-1}H_TR)^h} = \frac{\left|\text{tr } T^{-1}(H_TR)(R^{-1}S)u_0^t\right|^2}{\text{tr }(T^{-1}H_TR)(T^{-1}H_TR)^h} \\
&\leq \frac{\text{tr }(T^{-1}H_TR)(T^{-1}H_TR)^h \text{ tr }(R^{-1}Su_0^t)(R^{-1}Su_0^t)^h}{\text{tr }(T^{-1}H_TR)(T^{-1}H_TR)^h} \\
&= \text{tr }(R^{-1}Su_0^t)(R^{-1}Su_0^t)^h = \sum_{n=0}^{N-1}|SRR_i|^2
\end{aligned} \qquad (8.4.17)$$

It equals the sum of the spectral signal-to-total input ratio squared. The first line results by using $s_o = T^{-1}S_o$, $c = T^{-1}C$, $S_o = H_TS$, $C = H_TR$, and incorporating the trace definition where tr $A = \sum a_{ii}$ (sum of diagonal terms). u_i is the ith unit selection vector having 1 as the ith entry and 0's elsewhere. The second line follows from tr (AB)

$= \mathrm{tr}\ (BA)$ and by inserting an $I = RR^{-1}$ identity matrix between H_T and S in the numerator. Line three follows using Cauchy's inequality where $|\mathrm{tr}\ AB|^2 \leq \mathrm{tr}\ |A|^2\ \mathrm{tr}\ |B|^2$ in the numerator. Equality holds when $B = kA^h$. Applying this equality condition in Eq. 8.4.17 gives

$$T^{-1}H_T R = k(R^{-1}Su_o^t)^h = ku_o S^h (R^{-1})^h \qquad (8.4.18)$$

where $(ABC)^h = C^h B^h A^h$ and $u_0^t = u_0^h$. Solving for H_T gives (let $k = 1$)

$$H_T = (Tu_o S^h)\left(RR^h\right)^{-1} \qquad (8.4.19)$$

after multiplying with T on the left, R^{-1} on the right, and simplifying the result where $(R^{-1})^h (R^{-1}) = (RR^h)^{-1}$. This is identical to Eq. 8.4.15 when S is the desired output D to be detected and R is the input spectrum rejected. If the output signal s_o is to peak at time p rather than time zero, then Eq. 8.4.19 uses u_p rather than u_o.

8.4.4 GENERAL ESTIMATION-DETECTION VECTOR FILTERS

The general estimation filter of Eq. 8.4.13 optimally estimates signals in the presence of additive noise. The general detection filter of Eq. 8.4.15 maximizes the peak output signal relative to the total rms output. Combining these two results as

$$H_T = H_E H_D \qquad (8.4.20)$$

gives us an estimation-detection filter which performs both operations. This filter best estimates the signal and then outputs a narrow pulse when the signal is present.

8.4.5 WIENER ESTIMATION VECTOR FILTERS

Suppose that the generalized estimation filter is used for estimating signal. Setting the desired signal to be estimated equal to the input signal s as $d = s$, then $D = S$ and Eq. 8.4.14 becomes

$$\begin{aligned} H_T &= S(S^h + N^h)\left[(S + N)(S^h + N^h)\right]^{-1} \\ &= (SS^h + SN^h)\left(SS^h + NN^h + SN^h + NS^h\right)^{-1} \end{aligned} \qquad (8.4.21)$$

This is called the *correlated estimation* or *correlated Wiener* filter. It is coherent in the sense that the phase spectra of the signal and noise are used when computing H_T. The phase of H_T is generally nonzero. Since $NS^h = (SN^h)^h$, then SN^h and NS^h are complex conjugate transposes of each other, and their sum can be expressed as $2\ \mathrm{Re}\ SN^h$.

When the signal and noise are orthogonal, or equivalently, are uncorrelated and the noise has zero mean, then $SN^h = 0$ (see discussion for Eq. 8.4.23) and Eq. 8.4.21

reduces to

$$H_T = SS^h \left(SS^h + NN^h \right)^{-1} \tag{8.4.22}$$

This is the *uncorrelated estimation* or *uncorrelated Wiener* filter. From the distortion balance filter viewpoint of Eq. 8.4.5, this filter results when signal and noise powers are equally weighted. It is noncoherent since it discards the phase spectra of the signal and noise. The phase of H_T is always zero.

Before proceeding to other filter types, notice that the signal spectrum S, conjugate noise spectrum N^h, their product SN^h, and the cross-correlation spectrum S_{sn} (see Eq. 1.17.15) equal

$$S = Ts, \qquad N^h = n^h T^h = n^h T^{-1}$$
$$\widehat{S}_{sn} \equiv T\left[sn^h\right]T^{-1} = SN^h = T\widehat{R}_{sn}T^{-1} \tag{8.4.23}$$
$$S_{sn} = TE\{sn^h\}T^{-1} = TR_{sn}T^{-1}$$

s and n represent random signal and noise processes, respectively. s and n are the actual signal and noise, respectively. They are particular members of the ensemble of all those possible formed by the s and n random processes. S_{sn} is the apriori (assumed) cross-correlation spectrum while SN^h is the aposteriori (actual) cross-correlation spectral estimate $\widehat{S}_{sn}$. Eq. 8.4.23 shows that SN^h equals S_{sn} only when the signal-noise matrix product sn^h equals the cross-correlation function R_{sn}. This occurs only when both s and n are ergodic processes.

The cross-correlation spectrum S_{sn} is zero when the signal and noise are orthogonal, or equivalently, are uncorrelated and the noise is a zero-mean process (see Eqs. 1.17.9 and 1.17.12). In such cases, the cross-correlation estimate SN^h product is *set equal to zero* in Eq. 8.4.21 which gives Eq. 8.4.22. This is done even though the actual SN^h product is nonzero over the generalized frequency ranges where $S \neq 0$ and $N \neq 0$, or equivalently, the nonzero signal and noise spectra are overlapping. Nonoverlapping or disjoint S and N have $SN^h = 0$ and are very seldom encountered in practice. Correlation is discussed further in Chap. 8.4.7.

8.4.6 PULSE-SHAPING ESTIMATION VECTOR FILTERS

Suppose that we wish to form a new signal p from the output signal c in Fig. 8.3.1. This is done using the post-processing block with gain G_T where

$$p = g_T c \qquad \text{or} \qquad P = G_T C \tag{8.4.24}$$

and $G_T = Tg_T T^{-1}$. Then the gain G_T equals

$$G_T = P_d C_d^{-1} = \left(\text{diag } P\right)\left(\text{diag } C\right)^{-1} \tag{8.4.25}$$

where P is the desired output spectrum of p. Since matrix G_T generally contains N^2 nonzero elements but P and C each contain only N elements, the general G_T solution is underconstrained. $N^2 - N$ elements of G_T may be arbitrarily chosen. The simplest solution is to constrain G_T to be diagonal (i.e., a scalar filter) so it has at most N nonzero elements. Column vectors P and C are converted into diagonal matrices as $P_d = \text{diag } P$ and $C_d = \text{diag } C$. This G_T post-processing block can be used to generate a variety of useful signals. After G_T is determined, it is combined with gain H_T to form a single gain block (i.e., $H_T G_T \to H_T$). The blocks are separated here only for conceptual simplicity.

It might be useful to reshape c to have a different waveform p or to estimate different signals. We could first use the uncorrelated Wiener filter whose transfer function H_T is given by Eq. 8.4.22 to estimate s. Then choosing $G_T = P_d S_d^{-1}$ results in the desired output signal p. Combining H_T and P into a single transfer function gives

$$H_T = P_d S_d^{-1} S S^h \left(S S^h + N N^h \right)^{-1} = P_d U S^h \left(S S^h + N N^h \right)^{-1} \qquad (8.4.26)$$

which is the *pulse-shaping (uncorrelated) estimation filter*. $U = S_d^{-1} S$ where U is the unit column vector of all 1's. The identity matrix I and unit vector U are related as $I = \text{diag } U$. G_T can also be combined with the correlated estimation filter of Eq. 8.4.21 to give the *pulse-shaping (correlated) estimation filter* as

$$H_T = P_d U \left(S^h + N^h \right) \left(S S^h + N N^h + S N^h + N S^h \right)^{-1} \qquad (8.4.27)$$

Pulse-shaping filters can also be viewed as a high-resolution detection filter (see Chap. 8.4.10) followed by a pulse shaper. The impulse D_0 from the high-resolution detection filter H_T excites the impulse response p from the post-processor filter G_T. Pulse-shaping filters yield the maximum output SNR_o possible under the constraint that the output signal pulse $p = T^{-1} P$ has a specific form.

8.4.7 CORRELATION VECTOR FILTERS

The estimate $\widehat{R}_{ss}$ of the autocorrelation R_{ss} of an ergodic signal s and the estimate $\widehat{S}_{ss}$ of its power spectrum S_{ss} equal

$$R_{ss} = T^{-1} S_{ss} T = E\{ss^h\} \equiv ss^h = \widehat{R}_{ss}$$
$$S_{ss} = T R_{ss} T^{-1} = T E\{ss^h\} T^{-1} \equiv S S^h = \widehat{S}_{ss} \qquad (8.4.28)$$

from Eqs. 1.17.21 and 1.17.22. The post-processing block G_T in Fig. 8.2.1 can be used to generate the autocorrelation matrix. From Eqs. 8.4.24 and 8.4.28, this block outputs matrix $P = S S^h$ in response to input vector $C = S$. Then its transfer function $G_T = P C^{-1} = S S^h S^{-1} = (S^h S)^{-1} S S^h S^h$. The $S^h S$ normalization term equals the spectral energy in the signal.

Therefore combining P with the uncorrelated Wiener filter described by Eq. 8.4.22 and setting $S^h S = 1$ for simplicity gives the ideal autocorrelation estimation filter gain as

$$H_T = S S^h S^h \left(S S^h + N N^h \right)^{-1} \qquad (8.4.29)$$

Another autocorrelation estimation filter results by using the correlated Wiener filter of Eq. 8.4.21. Its gain equals

$$H_T = SS^h(S^h + N^h)\left(SS^h + NN^h + SN^h + NS^h\right)^{-1} \qquad (8.4.30)$$

The estimate of the cross-correlation R_{sn} between two ergodic signals, **s** and **n**, and its power spectrum S_{sn} equal

$$
\begin{aligned}
R_{sn} &= T^{-1}S_{sn}T = E\{\mathbf{sn}^h\} \equiv sn^h = \widehat{R}_{sn} \\
S_{sn} &= TR_{sn}T^{-1} = TE\{\mathbf{sn}^h\}T^{-1} \equiv SN^h = \widehat{S}_{sn}
\end{aligned}
\qquad (8.4.31)
$$

Proceeding as before, $G_T = SN^hS^h$ which can be combined with the uncorrelated Wiener filter. This gives the ideal cross-correlation estimation filter gain as

$$H_T = SN^hS^h\left(SS^h + NN^h\right)^{-1} \qquad (8.4.32)$$

Using the correlated Wiener filter instead gives

$$H_T = SN^h(S^h + N^h)\left(SS^h + NN^h + SN^h + NS^h\right)^{-1} \qquad (8.4.33)$$

The correlation estimates obtained using the correlated Wiener filters (Eqs. 8.4.30 and 8.4.33) are usually better than those obtained using the uncorrelated Wiener filters (Eqs. 8.4.29 and 8.4.32).

8.4.8 MATCHED DETECTION VECTOR FILTERS

The matched detection filter detects the presence of a signal s embedded in additive noise n. In this case, we wish to maximize the peak output SNR_o in order to detect the signal and reject the rms noise. Using Eq. 8.4.15 where $D = S$ and $R = N$, then

$$H_T = Tu_0S^h\left(NN^h\right)^{-1} \qquad (8.4.34)$$

This is the *classical detection, matched,* or *prewhitening* filter. H_T weights each spectral input by the ratio of the conjugate signal spectrum to the noise power spectrum. The filter introduces a phase which is opposite to signal phase. Thus all the output spectral components of a signal similar to the expected signal will be in phase. The output tends to be concentrated into a narrow pulse which is outputted when the signal occurs.

Notice from Eq. 8.4.34 that the filter's spectral output $C = HR = H(S + N)$ consists of a signal component $Tu_0(N^{-1}S)^h(N^{-1}S)$ and a noise component $Tu_0(N^{-1}S)^h$. The signal component is the square of the input spectral SNR_i vector. The noise component is the Hermitian of the SNR_i. From a correlation viewpoint, the signal component is the signal autocorrelation and the noise component is the signal-noise cross-correlation when the noise is white (i.e., $|N| = I$). In either case, the noise component is zero if the input signal and noise are orthogonal or are uncorrelated with zero mean noise. Otherwise the noise component is nonzero.

The performance of detection filters is often measured by the SNR gain which is the ratio of the output SNR_o to the input SNR_i. It is invariant to changes in the input signal or noise level and will be discussed in Chap. 8.4.14.

From the distortion balance filter viewpoint, the classical detection filter first discriminates between signal and scaled noise power levels (so $W_S = S^h$ and $W_N = kN^h$). If the constant k is chosen to be much larger than the squared input spectral SNR_i ratio (i.e., $k \gg (SS^h)(NN^h)^{-1}$), then Eq. 8.4.5 reduces to

$$H_T = SS^h \left(kNN^h \right)^{-1} \tag{8.4.35}$$

Following H_T by an inverse filter $P = kS_d^{-1}$, the overall gain characteristic becomes that of the classical detection filter given by Eq. 8.4.34. This shows that by relying entirely on noise rejection, the estimation filter becomes a detection filter.

8.4.9 INVERSE DETECTION VECTOR FILTERS

An *inverse detection filter* outputs an impulse $D_0 = u_0$ when *only* signal s and no noise is applied. It outputs a narrow pulse when a signal similar to s is applied. Using Eq. 8.4.13 the transfer function of an inverse detection filter equals

$$H_T = (\text{diag } Tu_0) \left(\text{diag } Ts \right)^{-1} = (Tu_0)_d \, S_d^{-1} \tag{8.4.36}$$

When signal pulse noise is applied to the filter, its input equals $R = S + N$. The output of the filter then equals $C = HR = I + S^{-1}N$ or $c = D_0 + u_0 SNR_i^{-1}$ from Eq. 8.4.53. The undesired output component due to nonzero input noise is $u_o SNR_i^{-1}$. When the spectral signal-to-noise ratio is large, $|SNR_i| \gg I$ so $|SNR_i|^{-1} \ll I$ and this undesired output component is negligible. However $|SNR_i|$ is seldom large so then this component is present which produces output error. This error can be eliminated using the high-resolution detection filter.

8.4.10 HIGH-RESOLUTION DETECTION VECTOR FILTERS

Another conceptually useful filter is the *high-resolution detection filter* which outputs an impulse when *both* the signal s and noise are applied. It outputs a narrow pulse when a signal similar to $s + n$ is applied. This requires a post-processing gain G_T after the estimation filter in Fig. 8.4.1 of

$$G_T = (\text{diag } Tu_0) \left(\text{diag } Ts \right)^{-1} = (Tu_0)_d \, S_d^{-1} \tag{8.4.37}$$

which is an inverse filter. Combining this with the uncorrelated Wiener filter in Eq. 8.4.22 results in the ideal high-resolution detection filter where

$$H_T = Tu_0 S^h \left(SS^h + NN^h \right)^{-1} \tag{8.4.38}$$

Eq. 8.4.38 is identical to the detection filter of Eq. 8.4.19 when S and N are assumed to be orthogonal. The Wiener filter estimates the signal s and the inverse filter converts s into an impulse. Another high-resolution filter results using the correlated Wiener filter of Eq. 8.4.21 as

$$H_T = Tu_0\left(S^h + N^h\right)\left(SS^h + NN^h + SN^h + NS^h\right)^{-1} \tag{8.4.39}$$

The high-resolution detection filter of Eq. 8.4.39 functions as a matched detection filter (Eq. 8.4.34) at frequencies where the spectral SNR_i is low. It acts as an inverse or deconvolution filter (Eq. 8.4.36) at high spectral SNR_i. The filter deconvolves S with S^{-1} to produce a single impulse. Rewriting Eq. 8.4.38 as

$$H_T = Tu_0 S^{-1}\left(I + (NN^h)(SS^h)^{-1}\right)^{-1} = Tu_0 S^{-1}\left(I + |SNR_i|^{-2}\right)^{-1} \tag{8.4.40}$$

H_T depends upon the input spectral signal-to-noise ratio SNR_i for constant S. From a distortion balance viewpoint, this is a filter which discriminates between signal and noise power levels (so $W_S = S^*$ and $W_N = N^*$) followed by an inverse filter having gain $G_T = S_d^{-1}$.

8.4.11 TRANSITIONAL VECTOR FILTERS

There are situations which require a filter whose transfer function can make a transition from one type to another type. These are called *transitional filters*. The characteristics of two transfer functions are blended together. The transition is controlled by one parameter (or more) like α. These transfer functions are usually formed heuristically.

For example, suppose that we want to form a transitional detection filter that can make a transition from an inverse type ($\alpha = 0$) to a high-resolution type ($\alpha = 0.5$) to a matched type ($\alpha = 1$). One way combines Eqs. 8.4.34, 8.4.36, and 8.4.38 as

$$H_T = US^h\left((1 - \alpha)SS^h + \alpha NN^h\right)^{-1} \tag{8.4.41}$$

Choosing some α between 0 and 1 yields a transfer function that combines the attributes of these filter types. As $\alpha \to 0$, the filter functions as an inverse type. As $\alpha \to 0.5$, it becomes a high-resolution type. As $\alpha \to 1$, it performs as a matched type. Another example of a transitional detection filter is

$$H_T = (Tu_0)_d \left(S_d^{2(1-\alpha)}\right)^{-1} \tag{8.4.42}$$

by generalizing Eq. 8.4.36. When $\alpha = 0.5$, it functions as an inverse filter.

An example of a transitional estimation filter is

$$H_T = \left(SS^h + \alpha SN^h\right)\left(SS^h + NN^h + \alpha(SN^h + NS^h)\right)^{-1} \tag{8.4.43}$$

As $\alpha \to 0$, the filter becomes uncorrelated Wiener. As $\alpha \to 1$, the filter becomes correlated Wiener. The possibilities for transitional filters are endless. The transitional filter approach gives the designer great freedom in formulating adaptive filters.

8.4.12 STATISTICALLY OPTIMUM VECTOR FILTERS

A variety of adaptive filters have been derived. In general, their gains are expressed as the ratio of two functions X and Y where

$$H_T = Y X^{-1} \qquad\qquad (8.4.44)$$

The X and Y functions involve sums, products, and other operations on the signal spectrum S and noise spectrum N. Which S and N spectra should be used to form H_T? The statistically optimum approach requires

$$H_T = E\{Y\} \left(E\{X\} \right)^{-1} \qquad\qquad (8.4.45)$$

where $E\{\ \}$ is the expectation operator. Eq. 8.4.45 results by expressing Eq. 8.4.44 as $Y = H_T X$, treating X and Y as stochastic processes rather than as deterministic signals, taking expectations of both sides, and solving for H_T which is deterministic. The optimum H_T equals the ratio of the expected (statistically averaged) X and Y which the filter processes. The transfer functions of Chap. 8.4 are therefore tabulated in Table 8.4.1 using expectation signs around their numerators and denominators. Some examples of terms in $E\{X\}$ and $E\{Y\}$ include:

$E\{S\}$, $E\{N\}$ = expected signal and noise spectra
$E\{|S|\}$, $E\{|N|\}$ = expected signal and noise magnitude spectra
$E\{SS^h\}$, $E\{NN^h\}$ = expected signal and noise autocorrelation (power) spectra
$E\{SN^h\}$ = expected signal-noise cross-correlation spectrum
$E\{SW_S^t\}$, $E\{NW_N^t\}$ = expected weighted signal and noise spectra   (8.4.46)

8.4.13 ADAPTATION AND LEARNING PROCESS

Adaptive filters undergo a *learning process* and perform *adaptation* to form the filter transfer function H_T. The transfer function is updated either occasionally or continuously. The ideal filter transfer functions listed in Tables 8.4.1 and 8.4.2 are formed using the expected spectra $E\{X\}$ and $E\{Y\}$ like those in Eq. 8.4.46. They are either software programmed into the filter or estimated. Class 1 adaptive filters store $E\{X\}$ and $E\{Y\}$ and update them (at most) periodically. Class 2 and 3 adaptive filters continuously form estimates $\widehat{E}\{X\}$ of $E\{Y\}$ and $\widehat{E}\{X\}$ of $E\{Y\}$ as the filter operates.

Estimates $\widehat{E}\{W\}$ of $E\{W\}$ can be formed in a variety of ways. Most estimates involve some form of diagonal smoothing of the entire W matrix, or localized smoothing and/or threshold detecting within the W matrix. Unfortunately no comprehensive theory now exists to guide forming vector filter estimates so they are formed empirically.

The expected signal spectrum $E\{S\}$ is sometimes called the *learning signal*. The expected noise spectrum $E\{N\}$ is sometimes called the *learning noise*. Mathematically,

$$
\begin{aligned}
E\{S\} &= E\{Ts\} = TE\{s\} \neq E\{T\mathbf{s}\} \\
E\{N\} &= E\{Tn\} = TE\{n\} \neq E\{T\mathbf{n}\}
\end{aligned}
\tag{8.4.47}
$$

Since operators $E\{\ \}$ and $T[\ ]$ are linear, they can be applied in any order as $E\{T[\ ]\}$ and $T[E\{\ \}]$. Therefore the inequalities result because $s(t) \neq \mathbf{s}(t)$ and $n(t) \neq \mathbf{n}(t)$ (a member from an ensemble does not equal the ensemble). The learning signal and learning noise equal the inverse transforms of their expected spectra as

$$
\begin{aligned}
T^{-1}E\{S\} &= E\{s\} \neq E\{\mathbf{s}\} \\
T^{-1}E\{N\} &= E\{n\} \neq E\{\mathbf{n}\}
\end{aligned}
\tag{8.4.48}
$$

The inequalities of Eqs. 8.4.47 and 8.4.48 become equalities when the signal s (or noise n) is either a deterministic process or an ergodic stochastic process. This is the usual case. For further details, see Chap. 1.17.

The noise spectrum can be expressed in polar form as

$$
N = |N|e^{j \arg N}
\tag{8.4.49}
$$

The expected noise spectrum equals

$$
E\{N\} = E\{|N|e^{j \arg N}\} = E\{|N|\}E\{e^{j \arg N}\} = E\{Tn\} = TE\{n\}
\tag{8.4.50}
$$

It can be expressed as the product of expectations because the magnitude and phase are independent (see Eq. 1.17.10). For zero mean noise, $E\{n\} = 0$ so Eq. 8.4.50 shows that the expected noise spectrum is also zero or

$$
E\{N\} = 0
\tag{8.4.51}
$$

From Eq. 8.4.50, this requires that $E\{e^{j \arg N}\}$ equal zero since the expected noise magnitude spectrum cannot be zero. For real-valued transforms and noise, the phase of N equals $\arg N = 0$ and $\pm\pi$. Therefore, the phase of $E\{N\}$ is equally distributed with probability $p(\pm180°) = q$ and $p(0°) = 1 - 2q$. The expected noise phase $E\{\arg N\} = 0$. The expected noise magnitude spectrum $E\{|N(j\omega)|\}$ can take on a variety of forms and is defined in the continuous frequency domain by its Fourier transform of Eqs. 4.4.55 and 4.4.56. The expected noise phase spectrum $E\{\arg N(j\omega)\}$ is random between $(-\pi, \pi)$. These magnitude and phase spectra are combined and converted into the discrete FFT noise spectrum $E\{N(m)\}$ using Eqs. 8.2.52 and 8.2.53. To convert the FFT noise spectrum into the generalized noise spectrum $E\{N(p)\}$ associated with transform T, we use

$$
E\{N(p)\} = TE\{n\} = T \text{ FFT}^{-1}\big[E\{N(m)\}\big] = G^{-1}E\{N(m)\}
\tag{8.4.52}
$$

where $G^{-1} = TF^{-1}$ is given by Eqs. 2.9.2a or 8.3.8. In the white noise case, the magnitude spectrum $E\{|N|\}$ is constant regardless of the transform used.

8.4.14 PERFORMANCE MEASURES

The signal-to-noise ratio (SNR) can be defined in a variety of ways to best describe a particular problem. These include (in order of increasing computational difficulty) the: (1) spectral SNR, (2) rms SNR or simply the SNR, and the (3) peak SNR.

The spectral SNR is defined as

$$\text{Spectral } SNR = SNR(p) = \frac{S(p)}{N(p)} = u_p^t N_d^{-1} S \qquad (8.4.53)$$

which is a function of generalized frequency. $N_d^{-1} S$ is the spectral SNR column vector whose pth entry is $S(p)/N(p)$. In this case the noise vector is diagonalized as $N_d^{-1} = \text{diag}[N^{-1}(0), \ldots, N^{-1}(N-1)]$. The spectral SNR is the ratio of the signal spectrum to the noise spectrum evaluated at a particular frequency p. It has a magnitude and an angle of $0°$ or $\pm 180°$ (for real-valued T). Generally only the magnitude is of interest. The Bode plot of $|SNR(p)|$ shows the frequency ranges over which the signal dominates or the noise dominates.

The rms SNR, which is usually simply called the SNR, equals

$$\text{Rms } SNR = \frac{\sqrt{P_s}}{\sqrt{P_n}} = \sqrt{\frac{s^h s}{n^h n}} = \sqrt{\frac{S^h S}{N^h N}} = \sqrt{(S^h S)(N^h N)^{-1}} \qquad (8.4.54)$$

It is square root of the ratio of the signal power P_s to the noise power P_n. It can be evaluated in either the time domain (sum $|s(n)|^2$ and $|n(n)|^2$), or in the frequency domain using Parseval's theorem (sum $|S(p)|^2$ and $|N(p)|^2$).

The peak SNR is defined as

$$\text{Peak } SNR = \frac{|s(n_p)|}{\sqrt{P_n}} = \frac{|s(n_p)|}{\sqrt{n^h n}/\sqrt{N}} = \frac{|s(n_p)|}{\sqrt{N^h N}} \qquad (8.4.55)$$

where $\sqrt{N}$ is the square root of the number of points. It is the ratio of the peak signal $s(n_p)$ which occurs at time n_p (or peak temporal signal) to the rms noise or the square root of the noise power. The noise power can be evaluated in time (sum $|n(n)|^2$) or in frequency (sum $|N(p)|^2$). It is usually difficult to find the peak signal value $s(n_p)$ since this requires solving $ds(t)/dt|_{t=n_p T} = 0$.

Regardless of which performance measure is used, SNR is usually expressed in dB rather than its numeric value as

$$SNR_{dB} = 20 \log_{10} SNR = 10 \log_{10} SNR^2 \qquad (8.4.56)$$

for SNR real. When SNR is the ratio r of signals, $20 \log r$ converts it to dB. When SNR is the ratio r of signal powers, $10 \log r$ converts it to dB.

The signal-to-noise ratio gain G_{SNR} provided by a filter of gain H_T equals

$$G_{SNR} = \frac{SNR_o}{SNR_i} = SNR_{o\ dB} - SNR_{i\ dB} \qquad (8.4.57)$$

Any of the SNR measures of Eqs. 8.4.53–8.4.55 can be used. The spectral SNR gain equals

$$\text{Spectral } G_{SNR} = \frac{S_o(p)/N_o(p)}{S_i(p)/N_i(p)} = \left(u_p^t N_o^{-1} S_o\right)\left(u_p^t N_i^{-1} S_i\right)^{-1}$$

$$= u_p^t (H_T N_i)^{-1}(H_T S_i) S_i^{-1} N_i (u_p^t)^{-1} = u_p^t(N_i^{-1}(H_T^{-1}H_T)(S_i S_i^{-1})N_i)(u_p^t)^{-1} = 1$$

$$(8.4.58)$$

using the spectral SNR of Eq. 8.4.53. The output signal spectrum $S_o = H_T S_i$ while the output noise spectrum $N_o = H_T N_i$. Since the spectral G_{SNR} equals unity and is independent of the H_T filter used, the spectral G_{SNR} is of little interest. If the rms SNR of Eq. 8.4.54 is used instead, then the rms SNR gain equals

$$\text{Rms } G_{SNR} = \sqrt{\frac{S_o^h S_o/N_o^h N_o}{S_i^h S_i/N_i^h N_i}} = \sqrt{\frac{\left(S_i^h H_T^h H_T S_i\right)\left(N_i^h H_T^h H_T N_i\right)^{-1}}{\left(S_i^h S_i\right)\left(N_i^h N_i\right)^{-1}}}$$

$$= \sqrt{\left(S_i^h S_i S_i^h H_T^h H_T S_i\right)\left(N_i^h H_T^h H_T N_i N_i^h N_i\right)^{-1}}$$

$$= \sqrt{\left(S_i^h(H_T^h H_T)S_i\right)\left[(N_i^h(H_T^h H_T)N_i)(S_i^h S_i)\right]^{-1}(N_i^h N_i)}$$

$$(8.4.59)$$

while the peak SNR gain equals

$$\text{Peak } G_{SNR} = \frac{s_{o\ pk}/\sqrt{n_o^h n_o}}{s_{i\ pk}/\sqrt{n_i^h n_i}} = \frac{s_{o\ pk}/\sqrt{N_o^h N_o}}{s_{i\ pk}/\sqrt{N_i^h N_i}}$$

$$(8.4.60)$$

The filter output equals $c = s_o + n_o$. Therefore, the output error equals $n_o = c - s_o$ which is the output noise. A measure of the distortion in the output is the ratio of the noise energy to the signal energy as

$$\text{Distortion} = \frac{n_o^h n_o}{s_o^h s_o} = \frac{N_o^h N_o}{S_o^h S_o} = \frac{N_i^h H_T^h H_T N_i}{S_i^h H_T^h H_T S_i} = \frac{1}{\text{rms } SNR_o^2}$$

$$(8.4.61)$$

It is also equal to the reciprocal of the squared rms SNR_o. Distortion measures the normalized mean-squared-error in the output signal.

Another performance index is the noise bandwidth NBW. It is defined as

$$NBW = \frac{\sum_{m=0}^{N-1}\sum_{n=0}^{N-1}|H_T(m,n)|^2}{2\left|H_T(m,n)\right|_{max}^2}$$

$$(8.4.62)$$

In digital filters, NBW is often normalized by the Nyquist frequency $N/2$ to yield

$$NBW_n = \frac{NBW}{N/2} = \frac{\sum_{m=0}^{N-1}\sum_{n=0}^{N-1}|H_T(m,n)|^2}{N\left|H_T(m,n)\right|_{max}^2}$$

$$(8.4.63)$$

To obtain maximum noise rejection, the NBW and NBW_n of the filter are minimized.

8.4.15 GAIN MATRICES OF VECTOR FILTERS

Let us now calculate the gain matrices of some Class 1 estimation, detection, and correlation Paley vector filters in Table 8.4.1 using $N = 4$ to illustrate these computations. We again use the sawtooth signal $s = [0, 1, 2, 3]^t$ and zero-mean random noise $n = [1, -1, 1, -1]^t$ of Chap. 8.3.2. The noisy input r equals $s + n = [1, 0, 3, 2]^t$. The Paley signal and noise spectra are $S = Ts$ and $N = Tn$ so

$$S = Ts = \frac{1}{4} \begin{bmatrix} 1 & 1 & 1 & 1 \\ 1 & 1 & -1 & -1 \\ 1 & -1 & 1 & -1 \\ 1 & -1 & -1 & 1 \end{bmatrix} \begin{bmatrix} 0 \\ 1 \\ 2 \\ 3 \end{bmatrix} = \begin{bmatrix} 1.5 \\ -1 \\ -0.5 \\ 0 \end{bmatrix}, \quad N = Tn = T \begin{bmatrix} 1 \\ -1 \\ 1 \\ -1 \end{bmatrix} = \begin{bmatrix} 0 \\ 0 \\ 1 \\ 0 \end{bmatrix}$$

$$(8.4.64)$$

From Eqs. 8.4.53–8.4.57, the various input SNR's equal

$$\text{Rms } SNR_i = \sqrt{s^h s / n^h n} = \sqrt{14/4} = 1.87 = 5 \text{ dB}$$

$$\text{Peak } SNR_i = \sqrt{N} \, |s(n_p)| / \sqrt{n^h n} = \sqrt{4} \, |3| / \sqrt{4} = 3 = 8.5 \text{ dB}$$

$$(8.4.65)$$

$$\text{Spectral } SNR_i = N_d^{-1} S = \text{diag}[\infty, \infty, 1, \infty] \begin{bmatrix} 1.5 \\ -1 \\ -0.5 \\ 0 \end{bmatrix} = \begin{bmatrix} \infty \\ \infty \\ -0.5 \\ \infty \end{bmatrix}$$

The energy in the signal and noise are $s^h s = \text{tr } ss^h$ and $n^h n = \text{tr } nn^h$, and the energy in their product is $n^h s = \text{tr } ns^h = \text{tr } sn^h$, so

$$s^h s = [0 \ \ 1 \ \ 2 \ \ 3] \begin{bmatrix} 0 \\ 1 \\ 2 \\ 3 \end{bmatrix} = 14 = 4 S^h S = 4 [1.5 \ \ -1 \ \ -0.5 \ \ 0] \begin{bmatrix} 1.5 \\ -1 \\ -0.5 \\ 0 \end{bmatrix}$$

$$n^h n = [1 \ \ -1 \ \ 1 \ \ -1] \begin{bmatrix} 1 \\ -1 \\ 1 \\ -1 \end{bmatrix} = 4 = 4 N^h N = 4 [0 \ \ 0 \ \ 1 \ \ 0] \begin{bmatrix} 0 \\ 0 \\ 1 \\ 0 \end{bmatrix} \qquad (8.4.66)$$

$$n^h s = [1 \ \ -1 \ \ 1 \ \ -1] \begin{bmatrix} 0 \\ 1 \\ 2 \\ 3 \end{bmatrix} = -2 = 4 N^h S = 4 [0 \ \ 0 \ \ 1 \ \ 0] \begin{bmatrix} 1.5 \\ -1 \\ -0.5 \\ 0 \end{bmatrix}$$

The aposteriori or actual signal and noise autocorrelations equal ss^h and nn^h so

$$ss^h = \begin{bmatrix} 0 \\ 1 \\ 2 \\ 3 \end{bmatrix} [0 \ \ 1 \ \ 2 \ \ 3] = \begin{bmatrix} 0 & 0 & 0 & 0 \\ 0 & 1 & 2 & 3 \\ 0 & 2 & 4 & 6 \\ 0 & 3 & 6 & 9 \end{bmatrix} \qquad (8.4.67a)$$

$$nn^h = \begin{bmatrix} 1 \\ -1 \\ 1 \\ -1 \end{bmatrix} [1 \quad -1 \quad 1 \quad -1] = \begin{bmatrix} 1 & -1 & 1 & -1 \\ -1 & 1 & -1 & 1 \\ 1 & -1 & 1 & -1 \\ -1 & 1 & -1 & 1 \end{bmatrix} \qquad (8.4.67b)$$

The aposteriori cross-correlation between the signal and noise equals sn^h so

$$sn^h = \begin{bmatrix} 0 \\ 1 \\ 2 \\ 3 \end{bmatrix} [1 \quad -1 \quad 1 \quad -1] = \begin{bmatrix} 0 & 0 & 0 & 0 \\ 1 & -1 & 1 & -1 \\ 2 & -2 & 2 & -2 \\ 3 & -3 & 3 & -3 \end{bmatrix} \qquad (8.4.68)$$

The aposteriori cross-correlation ns^h between the noise and signal is the transpose of that given by Eq. 8.4.68. The smoothed estimates of the correlation functions, formed by averaging each diagonal (use periodic extension), equal

$$\langle ss^h \rangle = \frac{1}{4} \begin{bmatrix} 14 & 8 & 6 & 8 \\ 8 & 14 & 8 & 6 \\ 6 & 8 & 14 & 8 \\ 8 & 6 & 8 & 14 \end{bmatrix}, \qquad \langle nn^h \rangle = \frac{1}{4} \begin{bmatrix} 4 & -4 & 4 & -4 \\ -4 & 4 & -4 & 4 \\ 4 & -4 & 4 & -4 \\ -4 & 4 & -4 & 4 \end{bmatrix}$$

$$\langle sn^h \rangle = \frac{1}{4} \begin{bmatrix} -2 & 2 & -2 & 2 \\ 2 & -2 & 2 & -2 \\ -2 & 2 & -2 & 2 \\ 2 & -2 & 2 & -2 \end{bmatrix}, \qquad \langle ns^h \rangle = \frac{1}{4} \begin{bmatrix} -2 & 2 & -2 & 2 \\ 2 & -2 & 2 & -2 \\ -2 & 2 & -2 & 2 \\ 2 & -2 & 2 & -2 \end{bmatrix} \qquad (8.4.69)$$

The aposteriori power spectra SS^h of the signal and NN^h of the noise and the cross-power spectrum SN^h between the signal and noise equal

$$SS^h = \begin{bmatrix} 1.5 \\ -1 \\ -0.5 \\ 0 \end{bmatrix} [1.5 \quad -1 \quad -0.5 \quad 0] = \begin{bmatrix} 2.25 & -1.5 & -0.75 & 0 \\ -1.5 & 1 & 0.5 & 0 \\ -0.75 & 0.5 & 0.25 & 0 \\ 0 & 0 & 0 & 0 \end{bmatrix}$$

$$NN^h = \begin{bmatrix} 0 \\ 0 \\ 1 \\ 0 \end{bmatrix} [0 \quad 0 \quad 1 \quad 0] = \begin{bmatrix} 0 & 0 & 0 & 0 \\ 0 & 0 & 0 & 0 \\ 0 & 0 & 1 & 0 \\ 0 & 0 & 0 & 0 \end{bmatrix} \qquad (8.4.70)$$

$$SN^h = \begin{bmatrix} 1.5 \\ -1 \\ -0.5 \\ 0 \end{bmatrix} [0 \quad 0 \quad 1 \quad 0] = \begin{bmatrix} 0 & 0 & 1.5 & 0 \\ 0 & 0 & -1 & 0 \\ 0 & 0 & -0.5 & 0 \\ 0 & 0 & 0 & 0 \end{bmatrix}$$

which is the transpose of the NS^h aposteriori cross-power spectrum. The smoothed aposteriori power spectra, formed by averaging each diagonal (use periodic extension), are

$$\langle SS^h \rangle = \frac{1}{4} \begin{bmatrix} 3.5 & -1 & -1.5 & -1 \\ -1 & 3.5 & -1 & -1.5 \\ -1.5 & -1 & 3.5 & -1 \\ -1 & -1.5 & -1 & 3.5 \end{bmatrix}, \quad \langle NN^h \rangle = \frac{1}{4} \begin{bmatrix} 1 & 0 & 0 & 0 \\ 0 & 1 & 0 & 0 \\ 0 & 0 & 1 & 0 \\ 0 & 0 & 0 & 1 \end{bmatrix}$$

$$\langle SN^h \rangle = \frac{1}{4} \begin{bmatrix} -1.5 & -1 & 1.5 & 0 \\ 0 & -1.5 & -1 & 1.5 \\ 1.5 & 0 & -1.5 & -1 \\ -1 & 1.5 & 0 & -1.5 \end{bmatrix}, \quad \langle NS^h \rangle = \frac{1}{4} \begin{bmatrix} -1.5 & 0 & 1.5 & -1 \\ -1 & -1.5 & 0 & 1.5 \\ 1.5 & -1 & -1.5 & 0 \\ 0 & 1.5 & -1 & -1.5 \end{bmatrix}$$

$$(8.4.71)$$

$\langle NS^h \rangle$ is the transpose of $\langle SN^h \rangle$. Filter transfer functions involve such estimates of expected spectra. Since the expected power spectra of the deterministic signal is $E\{SS^h\} = SS^h$, the noise is uncorrelated so $E\{NN^h\} = kI$, and the signal and noise are assumed to be orthogonal so $E\{SN^h\} = 0$, one set of estimates is (see Chap. 8.4.13)

$$\widehat{E}\{SS^h\} = \begin{bmatrix} 2.25 & -1.5 & -0.75 & 0 \\ -1.5 & 1 & 0.5 & 0 \\ -0.75 & 0.5 & 0.25 & 0 \\ 0 & 0 & 0 & 0 \end{bmatrix}, \quad \widehat{E}\{NN^h\} = \begin{bmatrix} 0.25 & 0 & 0 & 0 \\ 0 & 0.25 & 0 & 0 \\ 0 & 0 & 0.25 & 0 \\ 0 & 0 & 0 & 0.25 \end{bmatrix}$$

$$\widehat{E}\{SN^h\} = \widehat{E}\{NS^h\} = \begin{bmatrix} 0 & 0 & 0 & 0 \\ 0 & 0 & 0 & 0 \\ 0 & 0 & 0 & 0 \\ 0 & 0 & 0 & 0 \end{bmatrix} \qquad (8.4.72)$$

The time and Paley frequency domain characteristics of the signal and noise are summarized by Eqs. 8.4.65–8.4.72. Now let us use the spectral information to design some Class 1 estimation, detection, and correlation Paley filters from Table 8.4.1. The uncorrelated Wiener estimation filter gain $H_{Pal\ UE}$ equals $E\{SS^h\}(E\{SS^h + NN^h\})^{-1}$. Substituting in the estimated spectra of Eq. 8.4.72 gives

$$H_{Pal\ UE} = \begin{bmatrix} 2.25 & -1.5 & -0.75 & 0 \\ -1.5 & 1 & 0.5 & 0 \\ -0.75 & 0.5 & 0.25 & 0 \\ 0 & 0 & 0 & 0 \end{bmatrix} \begin{bmatrix} 2.5 & -1.5 & -0.75 & 0 \\ -1.5 & 1.25 & 0.5 & 0 \\ -0.75 & 0.5 & 0.5 & 0 \\ 0 & 0 & 0 & 0.25 \end{bmatrix}^{-1}$$

$$= \begin{bmatrix} 0.6 & -0.4 & -0.2 & 0 \\ -0.4 & 0.267 & 0.133 & 0 \\ -0.2 & 0.133 & 0.067 & 0 \\ 0 & 0 & 0 & 0 \end{bmatrix} \qquad (8.4.73)$$

The correlated Wiener estimation filter gain $H_{Pal\ CE}$ equals $E\{SS^h + SN^h\}(E\{SS^h + NN^h + SN^h + NS^h\})^{-1}$ which reduces to $H_{Pal\ CE}$ when the signal and noise are orthogonal $(E\{SN^*\} = 0)$.

The inverse detection filter gain $H_{Pal\ I}$ is $(Tu_0)_d(E\{S_d\})^{-1}$ so

$$H_{Pal\ I} = \frac{1}{4} \begin{bmatrix} 1 & 0 & 0 & 0 \\ 0 & 1 & 0 & 0 \\ 0 & 0 & 1 & 0 \\ 0 & 0 & 0 & 1 \end{bmatrix} \begin{bmatrix} 1.5 & 0 & 0 & 0 \\ 0 & -1 & 0 & 0 \\ 0 & 0 & -0.5 & 0 \\ 0 & 0 & 0 & 0 \end{bmatrix}^{-1} = \begin{bmatrix} 0.167 & 0 & 0 & 0 \\ 0 & -0.25 & 0 & 0 \\ 0 & 0 & -0.5 & 0 \\ 0 & 0 & 0 & \infty \end{bmatrix}$$

(8.4.74)

where $T u_0 = (1/4)[1,1,1,1]^t$. The high-resolution detection filter gain $H_{Pal\ UHR}$ using an uncorrelated Wiener approach equals $E\{T u_0 S^h\}(E\{SS^h + NN^h\})^{-1}$ so

$$H_{Pal\ UHR} = \begin{bmatrix} 1.5 & -1 & -0.5 & 0 \\ 1.5 & -1 & -0.5 & 0 \\ 1.5 & -1 & -0.5 & 0 \\ 1.5 & -1 & -0.5 & 0 \end{bmatrix} \begin{bmatrix} 2.5 & -1.5 & -0.75 & 0 \\ -1.5 & 1.25 & 0.5 & 0 \\ -0.75 & 0.5 & 0.5 & 0 \\ 0 & 0 & 0 & 0.25 \end{bmatrix}^{-1}$$

$$= \begin{bmatrix} 0.1 & -0.0667 & -0.0333 & 0 \\ 0.1 & -0.0667 & -0.0333 & 0 \\ 0.1 & -0.0667 & -0.0333 & 0 \\ 0.1 & -0.0667 & -0.0333 & 0 \end{bmatrix}$$

(8.4.75)

The high-resolution detection filter gain $H_{Pal\ CHR}$ using a correlated Wiener approach is $E\{T u_0 S^h + N^h\}(E\{SS^h + NN^h + SN^h + NS^h\})^{-1}$ which reduces to $H_{Pal\ CHR}$. The matched detection filter has a gain $H_{Pal\ M}$ of $E\{T u_0 S^h\}(E\{NN^h\})^{-1}$ where

$$H_{Pal\ M} = \begin{bmatrix} 1.5 & -1 & -0.5 & 0 \\ 1.5 & -1 & -0.5 & 0 \\ 1.5 & -1 & -0.5 & 0 \\ 1.5 & -1 & -0.5 & 0 \end{bmatrix} \begin{bmatrix} 0.25 & 0 & 0 & 0 \\ 0 & 0.25 & 0 & 0 \\ 0 & 0 & 0.25 & 0 \\ 0 & 0 & 0 & 0.25 \end{bmatrix}^{-1} = \begin{bmatrix} 1.5 & -1 & -0.5 & 0 \\ 1.5 & -1 & -0.5 & 0 \\ 1.5 & -1 & -0.5 & 0 \\ 1.5 & -1 & -0.5 & 0 \end{bmatrix}$$

(8.4.76)

The signal autocorrelation filter gain $H_{Pal\ USS}$ based on the uncorrelated Wiener filter is $E\{SS^h US^h\}(E\{SS^h + NN^h\})^{-1}$ so that

$$H_{Pal\ USS} = \begin{bmatrix} 0 & 0 & 0 & 0 \\ 0 & 0 & 0 & 0 \\ 0 & 0 & 0 & 0 \\ 0 & 0 & 0 & 0 \end{bmatrix}$$

$$= \begin{bmatrix} 2.25 & -1.5 & -0.75 & 0 \\ -1.5 & 1 & 0.5 & 0 \\ -0.75 & 0.5 & 0.25 & 0 \\ 0 & 0 & 0 & 0 \end{bmatrix} \begin{bmatrix} 1.5 & -1 & -0.5 & 0 \\ 1.5 & -1 & -0.5 & 0 \\ 1.5 & -1 & -0.5 & 0 \\ 1.5 & -1 & -0.5 & 0 \end{bmatrix} \begin{bmatrix} 2.5 & -1.5 & -0.75 & 0 \\ -1.5 & 1.25 & 0.5 & 0 \\ -0.75 & 0.5 & 0.5 & 0 \\ 0 & 0 & 0 & 0.25 \end{bmatrix}^{-1}$$

(8.4.77)

The signal-noise cross-correlation filter gain $H_{Pal\ USN}$ based on the uncorrelated Wiener filter equals $E\{SN^h US^h\}(E\{SS^h + NN^h\})^{-1}$. When S and N are independent,

$$H_{Pal\ USN} = \begin{bmatrix} 0 & 0 & 0 & 0 \\ 0 & 0 & 0 & 0 \\ 0 & 0 & 0 & 0 \\ 0 & 0 & 0 & 0 \end{bmatrix}$$

(8.4.78)

8.4.16 GAIN MATRICES OF FFT VECTOR FILTERS

The gains of a variety of Class 1 FFT scalar filters were computed in Eqs. 8.2.66–8.2.73. Let us now compute the gains of the analogous Class 1 FFT vector filters using again the sawtooth signal $s = [0, 1, 2, 3]^t$ and zero-mean random noise $n = [1, -1, 1, -1]^t$. As found in Chap. 8.2.15, the Fourier signal and noise spectra are $S = Fs$ and $N = Fn$ so

$$S = \begin{bmatrix} 6 \\ -2 + j2 \\ -2 \\ -2 - j2 \end{bmatrix}, \quad N = \begin{bmatrix} 0 \\ 0 \\ 4 \\ 0 \end{bmatrix}, \quad \widehat{N} = \begin{bmatrix} 2 \\ 2 \\ 2 \\ 2 \end{bmatrix} \qquad (8.4.79)$$

$\widehat{N}$ is a smoothed (constant energy) version of N to achieve statistical averaging. The cross-correlation spectrum $E\{SN^h\} = 0$ when S and N are independent or orthogonal.

The gain $H_{F\ UE}$ of the uncorrelated estimation FFT vector filter listed in Table 8.4.1 is $E\{SS^h\}(E\{SS^h + NN^h\})^{-1}$ where

$$H_{F\ UE} = \begin{bmatrix} 0.6 & -0.2 - j0.2 & -0.2 & -0.2 + j0.2 \\ -0.2 + j0.2 & 0.133 & 0.0667 - j0.0667 & -j0.133 \\ -0.2 & 0.0667 + j0.0667 & 0.0667 & 0.0667 - j0.0667 \\ -0.2 - j0.2 & j0.133 & 0.0667 + j0.0667 & 0.133 \end{bmatrix}$$

$$(8.4.80)$$

(cf. Eq. 8.2.66). The correlated Wiener estimation FFT vector filter gain $H_{F\ CE}$ equals $E\{SS^h + SN^h\}(E\{SS^h + NN^h + SN^h + NS^h\})^{-1}$ which reduces to $H_{F\ UE}$ when the signal and noise are orthogonal $(E\{SN^h\} = 0)$.

Now let us consider some detection filters. They all include the term $(Tu_p)_d$ in their gains. This term determines the time n_p that the output pulse $s_o(n_p)$ occurs. T corresponds to the discrete transform that is used and the unit selection vector u_p has a "1" in its pth row. The Fourier transform has

$$T = \begin{bmatrix} 1 & 1 & 1 & 1 \\ 1 & -j & -1 & j \\ 1 & -1 & 1 & -1 \\ 1 & j & -1 & -j \end{bmatrix} \qquad (8.4.81)$$

Therefore when the output signal is to occur at time $n = 0$, then

$$(Tu_0)_d = \left(\begin{bmatrix} 1 & 1 & 1 & 1 \\ 1 & -j & -1 & j \\ 1 & -1 & 1 & -1 \\ 1 & j & -1 & -j \end{bmatrix} \begin{bmatrix} 1 \\ 0 \\ 0 \\ 0 \end{bmatrix} \right)_d = \left(\begin{bmatrix} 1 \\ 1 \\ 1 \\ 1 \end{bmatrix} \right)_d = \begin{bmatrix} 1 & 0 & 0 & 0 \\ 0 & 1 & 0 & 0 \\ 0 & 0 & 1 & 0 \\ 0 & 0 & 0 & 1 \end{bmatrix} = I \ (8.4.82)$$

which is the identity matrix I. The inverse detection FFT vector filter gain is $(Tu_0)_d \times (E\{S_d\})^{-1}$ where

$$H_{F\ I} = \begin{bmatrix} 0.167 & 0 & 0 & 0 \\ 0 & -0.25 - j0.25 & 0 & 0 \\ 0 & 0 & -0.5 & 0 \\ 0 & 0 & 0 & -0.25 + j0.25 \end{bmatrix} \qquad (8.4.83)$$

(cf. Eq. 8.2.67). The high-resolution detection FFT vector filter gain $H_{F\ UHR}$ using an uncorrelated Wiener approach is $E\{Tu_0S^h\}(E\{SS^h + NN^h\})^{-1}$ so

$$H_{F\ UHR} = \begin{bmatrix} 0.1 & -0.0333 - j0.0333 & -0.033 & -0.033 + j0.0333 \\ 0.1 & -0.0333 - j0.0333 & -0.033 & -0.033 + j0.0333 \\ 0.1 & -0.0333 - j0.0333 & -0.033 & -0.033 + j0.0333 \\ 0.1 & -0.0333 - j0.0333 & -0.033 & -0.033 + j0.0333 \end{bmatrix} \qquad (8.4.84)$$

(cf. Eq. 8.2.68). The high-resolution detection FFT vector filter gain $H_{F\ CHR}$ using the correlated Wiener approach is $E\{Tu_0(S^h + N^h)\}(E\{SS^h + NN^h + SN^h + NS^h\}^{-1}$ which equals $H_{F\ UHR}$ because $E\{SN^*\} = 0$. The FFT matched detection filter has a gain of $E\{Tu_0S^h\}(E\{NN^h\})^{-1}$ where (cf. Eq. 8.2.69).

$$H_{F\ M} = \begin{bmatrix} 1.5 & -0.5 - j0.5 & -0.5 & -0.5 + j0.5 \\ 1.5 & -0.5 - j0.5 & -0.5 & -0.5 + j0.5 \\ 1.5 & -0.5 - j0.5 & -0.5 & -0.5 + j0.5 \\ 1.5 & -0.5 - j0.5 & -0.5 & -0.5 + j0.5 \end{bmatrix} \qquad (8.4.85)$$

The signal autocorrelation FFT vector filter gain $H_{F\ USS}$ using the uncorrelated Wiener approach is $E\{SS^hS^h\}(E\{SS^h + NN^h\})^{-1}$ so that

$$H_{F\ USS} = 0.0167 \begin{bmatrix} 36 & -12 - j12 & -12 & -12 + j12 \\ -12 + j12 & 8 & 4 - j4 & -j8 \\ -12 & 4 + j4 & 4 & 4 - j4 \\ -12 - j12 & j8 & 4 + j4 & 8 \end{bmatrix} \begin{bmatrix} 6 \\ -2 - j2 \\ -2 \\ -2 + j2 \end{bmatrix}^t \qquad (8.4.86)$$

(cf. Eq. 8.2.70). In the correlated Wiener approach, $H_{F\ CSS}$ equals $E\{SS^h(S^h + N^h)\}(E\{SS^h + NN^h + SN^h + NS^h\})^{-1}$ which reduces to $H_{F\ USS}$ when $E\{SN^h\} = 0$. The signal-noise cross-correlation filter gain $H_{F\ USN}$ based upon the uncorrelated Wiener approach is $E\{SN^hS^h\}\ (E\{SS^h + NN^h\})^{-1}$ so

$$H_{F\ USN} = 0.0167(4) \begin{bmatrix} 0 & 0 & 6 & 0 \\ 0 & 0 & -2 + j2 & 0 \\ 0 & 0 & -2 & 0 \\ 0 & 0 & -2 - j2 & 0 \end{bmatrix} \begin{bmatrix} 6 \\ -2 - j2 \\ -2 \\ -2 + j2 \end{bmatrix}^t \qquad (8.4.87)$$

(cf. Eq. 8.2.71). Since $E\{SN^hS^h\} = E\{S\}E\{N^h\}E\{S^h\} = 0$ when S and N are independent, then $H_{F\ CSN}$ using the correlated Wiener approach is $E\{SN^h(S^h + N^h)\}(E\{SS^h + NN^h + SN^h + NS^h\})^{-1}$ which reduces to $H_{F\ USN}$.

Let us now compute the filter outputs using some uncorrelated estimation filter gains. The signal equals $s = [0, 1, 2, 3]^t$ and the noise equals $n = [1, -1, 1, -1]^t$. The input equals $r = [1, 0, 3, 2]^t$ and its spectrum equals $R = Fr = [6, -2 + j2, 2, -2 - j2]^t$. The FFT scalar filter has a diagonal gain matrix given by Eq. 8.2.66. The output spectrum $C = HR = [5.4, -1.33 + j1.33, 1, -1.33 - j1.33]^t$ and the output signal $c = T^{-1}C$. Energy normalizing c gives $1.23c = [1.14, 0.531, 2.78, 2.17]^t$ ($1.23c$ has the same energy as s). The MSE error between the desired output $d = s$ and $1.23c$ is $e^he = 2.83$. The FFT vector filter has a nondiagonal gain matrix given by Eq. 8.4.80.

Smoothing gain by averaging each diagonal row produces conjugate diagonal symmetry. The first row of $\langle H_{F\ UE} \rangle$ equals $[0.233, -0.0666 - j0.133, -0.1, -0.0666 + j0.133]^t$. The output spectrum is $[2, -0.8 + j1.2, -0.4, -0.8 - j1.2]^t$ and the energy normalized output signal is $[0, 0, 2.07, 3.11]^t$. The MSE error equals 1.02 which is much smaller than 2.83. Repeating this process using the Paley vector filter gain of Eq. 8.4.73 gives an energy normalized output signal of $[0, 0, 2.88, 2.39]^t$ and a reduced MSE error of 2.13.

This demonstrates that vector filters can provide better processing than scalar filters. Since vector filters have square gain matrices, N^2 multiplications are needed to determine filter outputs. Approximating these matrices by diagonal matrices reduces computational complexity and time as will be described in Chap. 8.5.

8.4.17 VECTOR FILTER DESIGN EXAMPLES

Now let us design several estimation and detection vector filters having the form of Fig. 8.3.1 where $N = 128$.[2] The estimation vector filter using the uncorrelated Wiener approach is given by Eq. 8.4.22. The signal s is assumed to be the cosine pulse shown in Fig. 8.3.3a. The noise n is assumed to be uncorrelated and white with random phase. The signal and noise are added together to form the input r shown in Fig. 8.3.3b. The peak signal-to-rms noise ratio is about 10 dB. Computing the FFT spectra of s, n, and r using Eq. 8.4.1a and substituting them into Eq. 8.4.22 (without $E\{\ \}$ operators) yields the Fourier transfer function shown in Fig. 8.4.1a. The diagonally-smoothed gain is also shown. Multiplying this smoothed gain by the input spectrum gives the output spectrum. Taking the inverse transform using Eq. 8.4.1b gives the output c shown in Fig. 8.4.1a. The signal is recovered and the spurious noise is low.

This process is repeated for the Paley, Haar, and slant transforms. The input r is transformed, the appropriate filter gain is computed using Eq. 8.4.22, and the output response computed. These gains, smoothed gains, and responses are presented for the Paley filter in Fig. 8.4.1b, Haar filter in Fig. 8.4.1c, and slant filter in Fig. 8.4.1d. The Haar filter recovers the signal fairly well but with high spurious noise. The Paley and slant filters do not recover the signal very well but they do reduce the noise.

The detection example is shown in Fig. 8.4.2. The signal $s(t)$ is assumed to be the triangular pulse shown in Fig. 8.3.4a. The noise $n(t)$ is assumed to be white with random phase. The signal and noise are added together to form the input $r(t)$ as shown in Fig. 8.3.4b. The peak signal-to-noise ratio is again selected to equal about 10 dB. The matched detection filter matrix is given by Eq. 8.4.34. Computing the FFT spectra of $s(t), n(t)$, and $r(t)$ and substituting them into Eq. 8.4.34 (without $E\{\ \}$ operators) yields the Fourier transfer function shown in Fig. 8.4.2a. The diagonally-smoothed gain is also shown. Multiplying this smoothed gain by the input spectra gives the output spectrum. Taking the inverse transform gives the output $c(t)$ shown in Fig. 8.4.2a. The signal is detected and the (negative) output pulse width is narrow and fairly noise free.

This process is repeated for the Paley, Haar, and slant transforms. These gains, smoothed gains, and responses are presented for Paley in Fig. 8.4.2b, Haar in Fig. 8.4.2c, and slant in Fig. 8.4.2d. The Paley has a narrow output pulse and low noise. The Haar also has a narrow output pulse but higher noise. The slant pulse is oscillatory with a low noise content.

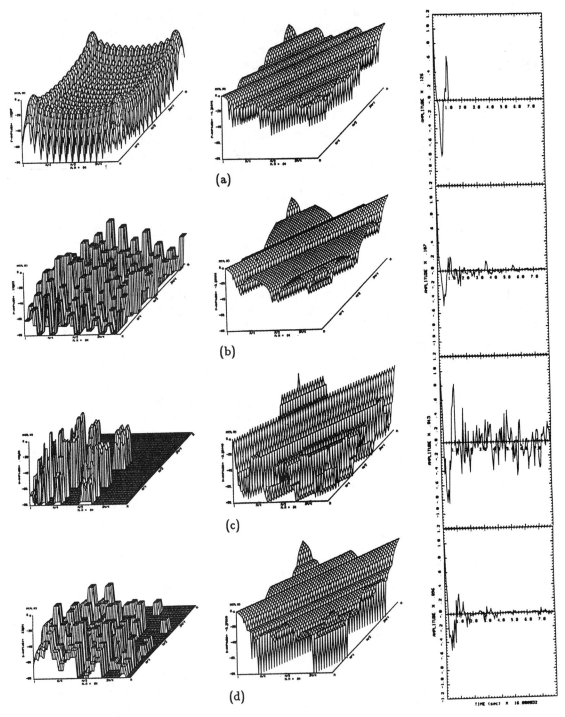

Fig. 8.4.1 Gain, smoothed gain, and output of (a) Fourier, (b) Paley, (c) Haar, and (d) slant estimation vector filters. For signal and total input, see Figs. 8.5.1a–b. (Plotted by A. Ravitz.)

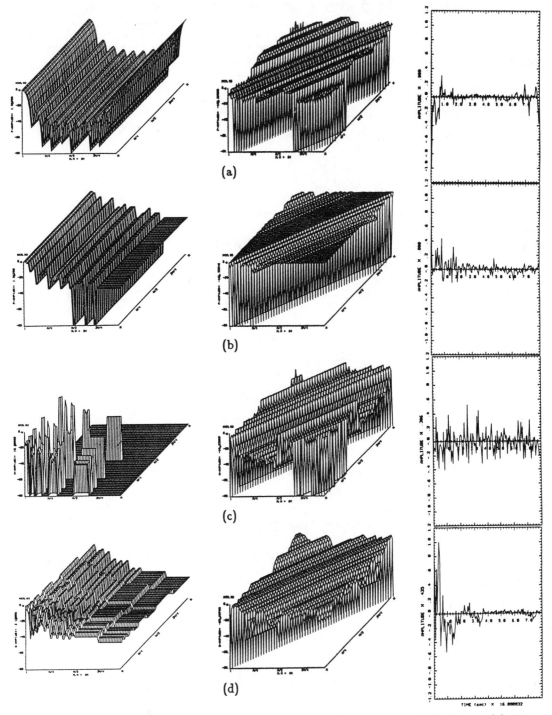

Fig. 8.4.2 Gain, smoothed gain, and output of (a) Fourier, (b) Paley, (c) Haar, and (d) slant detection vector filters. For signal and total input, see Figs. 8.5.2a–b. (Plotted by A. Ravitz.)

Table 8.4.1 Ideal Class 1 vector filter transfer functions H_T.

Filter Name	Transfer Function H_T	Eq. Number

Distortion Balance (W_S = Signal weighting, W_N = Noise weighting)

$$E\{SW_S^t\} \left(E\{SW_S^t + NW_N^t\} \right)^{-1}$$

(8.4.5)

General Estimation (D = Desired output)

$$H_E = E\{DR^h\} \left(E\{RR^h\} \right)^{-1}$$

(8.4.13)

General Detection

$$H_D = E\{Tu_0 D^h\} \left(E\{RR^h\} \right)^{-1}$$

(8.4.15)

Estimation–Detection

$$H_E H_D$$

(8.4.20)

Uncorrelated Estimation (Uncorrelated Wiener)

$$E\{SS^h\} \left(E\{SS^h + NN^h\} \right)^{-1}$$

(8.4.22)

Correlated Estimation (Correlated Wiener)

$$E\{SS^h + SN^h\} \left(E\{SS^h + NN^h + SN^h + NS^h\} \right)^{-1}$$

(8.4.21)

Transitional 1 Estimation (Uncorrelated Wiener $\alpha = 0$, Correlated Wiener $\alpha = 1$)

$$E\{SS^h + \alpha SN^h\} \left(E\{SS^h + NN^h + \alpha(SN^h + NS^h)\} \right)^{-1}$$

(8.4.43)

Pulse-Shaping Estimation (P = Desired pulse spectrum, Uncorrelated Wiener)

$$E\{P_d US^h\} \left(E\{SS^h + NN^h\} \right)^{-1}$$

(8.4.26)

Pulse-Shaping Estimation (P = Desired pulse spectrum, Correlated Wiener)

$$E\{P_d U(S^h + N^h)\} \left(E\{SS^h + NN^h + SN^h + NS^h\} \right)^{-1}$$

(8.4.27)

Table 8.4.1 (Continued) Ideal Class 1 vector filter transfer functions H_T.

Filter Name	Transfer Function H_T	Eq. Number

Classical Detection (Matched)

$$E\{Tu_0 S^h\} \left(E\{NN^h\}\right)^{-1} \qquad (8.4.34)$$

High-Resolution Detection (Uncorrelated Wiener)

$$E\{Tu_0 S^h\} \left(E\{SS^h + NN^h\}\right)^{-1} \qquad (8.4.38)$$

High-Resolution Detection (Correlated Wiener)

$$E\{Tu_0(S^h + N^h)\} \left(E\{SS^h + NN^h + SN^h + NS^h\}\right)^{-1} \qquad (8.4.39)$$

Inverse Detection (Deconvolution)

$$(Tu_0)_d \left(E\{S_d\}\right)^{-1} \qquad (8.4.36)$$

Transitional 1 Detection (Inverse $\alpha = 0$, High-resolution $\alpha = 0.5$, Matched $\alpha = 1$)

$$E\{US^h\} \left(E\{(1-\alpha)SS^h + \alpha NN^h\}\right)^{-1} \qquad (8.4.41)$$

Transitional 2 Detection (Inverse $\alpha = 0.5$)

$$(Tu_0)_d \left(E\{S_d^{2(1-\alpha)}\}\right)^{-1} \qquad (8.4.42)$$

Autocorrelation Estimation (Uncorrelated Wiener)

$$E\{SS^h S^h\} \left(E\{SS^h + NN^h\}\right)^{-1} \qquad (8.4.29)$$

Autocorrelation Estimation (Correlated Wiener)

$$E\{SS^h(S^h + N^h)\} \left(E\{SS^h + NN^h + SN^h + NS^h\}\right)^{-1} \qquad (8.4.30)$$

Cross-Correlation Estimation (Uncorrelated Wiener)

$$E\{SN^h S^h\} \left(E\{SS^h + NN^h\}\right)^{-1} \qquad (8.4.32)$$

Cross-Correlation Estimation (Correlated Wiener)

$$E\{SN^h(S^h + N^h)\} \left(E\{SS^h + NN^h + SN^h + NS^h\}\right)^{-1} \qquad (8.4.33)$$

Table 8.4.2   Ideal Class 2 and 3 vector filter transfer functions H_T.

Filter Name	Class 2	Class 3

Distortion Balance (W_S = Signal weighting, W_N = Noise weighting)

$$E\{SW_S^t\}\left(E\{RW_N^t - kS(W_N^t - W_S^t)\}\right)^{-1} \qquad \langle RW_S^t\rangle\left(\langle RW_S^t\rangle + (R - \langle R\rangle)W_N^t\right)^{-1}$$

Uncorrelated Estimation (Uncorrelated Wiener)

$$E\{SS^h\}\left(E\{SS^h + (R - kS)(R^h - kS^h)\}\right)^{-1} \qquad \langle R\rangle\langle R^h\rangle\left(\langle RR^h\rangle\right)^{-1}$$

Correlated Estimation (Correlated Wiener)

$$E\{S(R^h + (1-k)S^h)\}$$
$$\times\left(E\{(R + (1-k)S)(R^h + (1-k)S^h)\}\right)^{-1} \qquad \langle\langle R\rangle R^h\rangle\left(\langle RR^h\rangle\right)^{-1}$$

Transitional 1 Estimation (Uncorrelated Wiener $\alpha = 0$, Correlated Wiener $\alpha = 1$)

$$E\{SS^h + \alpha S(R^h - kS^h)\}$$
$$\times\Big(E\{SS^h + (R - kS)(R^h - kS^h)$$
$$\quad + \alpha\big(S(R^h - kS^h) + (R - kS)S^h\big)\}\Big)^{-1}$$
$$\left(\langle R\rangle\langle R^h\rangle + (1-\alpha)\langle R\rangle(R^h - \langle R^h\rangle)\right)\left((1-\alpha)\langle RR^h\rangle\right)^{-1}$$

Pulse-Shaping Estimation (P = Desired pulse spectrum, Uncorrelated Wiener)

$$E\{P_dUS^h\}\left(E\{SS^h + (R - kS)(R^h - kS^h)\}\right)^{-1} \qquad P_dU\langle R^h\rangle\left(\langle RR^h\rangle\right)^{-1}$$

Pulse-Shaping Estimation (P = Desired pulse spectrum, Correlated Wiener)

$$E\{P_dU(R^h + (1-k)S^h)\}$$
$$\times\left(E\{(R + (1-k)S)(R^h + (1-k)S^h)\}\right)^{-1} \qquad \langle P_dUR^h\rangle\left(RR^h\right)^{-1}$$

Classical Detection (Matched)

$$E\{Tu_0S^h\}\left(E\{(R - kS)(R^h - kS^h)\}\right)^{-1} \qquad Tu_0\langle R^h\rangle\left((R - \langle R\rangle)(R^h - \langle R^h\rangle)\right)^{-1}$$

High-Resolution Detection (Uncorrelated Wiener)

$$E\{Tu_0S^h\}\left(E\{SS^h + (R - kS)(R^h - kS^h)\}\right)^{-1} \qquad Tu_0\langle R^h\rangle\left(\langle RR^h\rangle\right)^{-1}$$

Table 8.4.2 (Continued) Ideal Class 2 and 3 vector filter transfer functions H_T.

Filter Name	Class 2	Class 3

High-Resolution Detection (Correlated Wiener)

$$E\{Tu_0(R^h + (1-k)S^h)\}$$
$$\times \left(E\{(R + (1-k)S)(R^h + (1-k)S^h)\}\right)^{-1} \qquad Tu_0\langle R^h\rangle \left(\langle RR^h\rangle\right)^{-1}$$

Inverse Detection (Deconvolution)

$$(Tu_0)_d \left(E\{S_d\}\right)^{-1} \qquad\qquad (Tu_0)_d \left(\langle R_d\rangle\right)^{-1}$$

Transitional 1 Detection (Inverse $\alpha = 0$, High-Resolution $\alpha = 0.5$, Matched $\alpha = 1$)

$$E\{US^h\} \left(E\{(1-\alpha)SS^h + \alpha((R - kS)(R^h - kS^h))\}\right)^{-1}$$
$$\langle UR^h\rangle \left((1-\alpha)\langle R\rangle\langle R^h\rangle + \alpha(R - \langle R\rangle)(R^h - \langle R^h\rangle)\right)^{-1}$$

Transitional 2 Detection (Inverse $\alpha = 0.5$)

$$(Tu_0)_d \left(E\{S_d^{2(1-\alpha)}\}\right)^{-1} \qquad\qquad (Tu_0)_d \left(\langle R_d\rangle^{2(1-\alpha)}\right)^{-1}$$

Autocorrelation Estimation (Uncorrelated Wiener)

$$E\{SS^hS^h\}$$
$$\times \left(E\{SS^h + (R - kS)(R^h - kS^h)\}\right)^{-1} \qquad \langle R\rangle\langle R^h\rangle\langle R^h\rangle \left(\langle RR^h\rangle\right)^{-1}$$

Autocorrelation Estimation (Correlated Wiener)

$$E\{SS^h(R^h + (1-k)S^h)\}$$
$$\times \left(E\{(R + (1-k)S)(R^h + (1-k)S^h)\}\right)^{-1} \qquad \langle R\rangle\langle R^h\rangle\left(R_d\right)^{-1}$$

Cross-Correlation Estimation (Uncorrelated Wiener)

$$E\{S(R^h - kS^h)S^h\}$$
$$\times \left(E\{SS^h + (R - kS)(R^h - kS^h)\}\right)^{-1} \qquad \langle R\rangle\langle R^h\rangle(R^h - \langle R^h\rangle) \left(\langle RR^h\rangle\right)^{-1}$$

Cross-Correlation Estimation (Correlated Wiener)

$$E\{S(R^h - kS^h)(R^h + (1-k)S^h)\}$$
$$\times \left(E\{(R + (1-k)S)(R^h + (1-k)S^h)\}\right)^{-1} \qquad \langle R\rangle(R^h - \langle R^h\rangle) \left(R_d\right)^{-1}$$

8.5 FAST TRANSFORM SCALAR FILTERS

In Chap. 8.4, we derived the gain matrices for a variety of fast transform vector filters following the optimization methods used for the FFT scalar filters of Chap. 8.2. These square gain matrices are optimum in different ways but are not computationally efficient because they are nondiagonal rather than diagonal. Therefore to increase efficiency, we shall now design scalar filter matrices gain matrices subject to the following restrictions (case $\bar{A}B$ in Table 8.1.1):
(a) Any fast transform is used, so
(b) The basis functions are arbitrarily chosen.
(c) The gain matrix is diagonal.

8.5.1 THEORY OF SCALAR FILTERS

Fast transform scalar filters can be generated by combining the FFT scalar filters of Chap. 8.2 and vector filters of Chap. 8.4. Vector filters utilize Eq. 8.4.1 where

$$r = s + n \quad \text{or} \quad R = Tr = S + N$$
$$c = T^{-1}C = h_T r \quad \text{or} \quad C = H_T R \tag{8.5.1}$$

Scalar filters utilize Eq. 8.2.1 where (changing transform F to T),

$$r(n) = s(n) + n(n) \quad \text{or} \quad R(m) = T[r(n)] = S(m) + N(m)$$
$$c(n) = T^{-1}[C(m)] \quad \text{or} \quad C(m) = H_T(m)R(m) \tag{8.5.2}$$

Therefore scalar filters utilize the same equation as vector filters but with the additional constraint that the gain matrix H_T be diagonal as $H_T = \text{diag}[H(0), \ldots, H(m), \ldots, H(N-1)]$. Therefore the $C = H_T R$ equation could be rewritten as $C(m) = H_T(m)R(m)$ to emphasize this fact. The other equations in Eq. 8.5.1 remain unchanged. This diagonal constraint requires that the spectral output $R(m)$ at generalized frequency bin m is due only to the spectral input $C(m)$ at the same frequency bin. The frequency components $C(p)$ at other bins ($m \neq p$) do not contribute to the $C(m)$ output. The FFT scalar filters described by Eq. 8.5.2 satisfy this condition. Vector filters are sometimes mistakenly said to be nonlinear because the output at a particular frequency depends upon all the other frequencies (i.e., frequency mixing). However, vector filters are linear as will be discussed in a moment.

Since scalar filters differ from vector filters only in that the gain matrix H_T is diagonal, then making the unnecessary but useful notational change $C = H_T R \longrightarrow C(m) = H_T(m)R(m)$ explicitly converts vector filters to scalar filters. Under this diagonal H_T constraint, the vector filter gain matrices of Tables 8.4.1 and 8.4.2 reduce to the scalar filter matrices of Tables 8.2.1 and 8.2.2. Comparing each filter type in Table 8.2.1 with that in Table 8.4.1 (or Table 8.2.2 with Table 8.4.2), we see they are identical making the following equivalences:

$$S \longrightarrow S(m), \quad S^h \longrightarrow S^*(m), \qquad\qquad SS^h \longrightarrow |S(m)|^2$$
$$N \longrightarrow N(m), \quad N^h \longrightarrow N^*(m), \qquad\qquad NN^h \longrightarrow |N(m)|^2 \qquad (8.5.3)$$
$$P \longrightarrow P(m), \quad Tu_pS^h \longrightarrow T(p,m)S^*(m), \quad SN^h \longrightarrow S(m)N^*(m)$$

and so on. *Therefore, Tables 8.2.1 and 8.2.2 can be used for all fast transform scalar filters.* For T transforms that have 1's in their first row and column which is the usual case, $Tu_0S^h \longrightarrow S^*(m)$ for the detection filters. In the more general Tu_pS^h case, the detection filter gains are modified as $Tu_pS^h \longrightarrow T(p,m)S^*(m)$.

The reader will find it instructive to directly rederive the gain expressions directly in Chap. 8.2 beginning with Eq. 8.2.1 using the slight transform notation change that $R(m) = T[r(n)] = S(m) + N(m)$ and $c(n) = T^{-1}[C(m)]$. Progressing through the equations will show that there are no changes whatever (try it!). Therefore to eliminate redundancy, we choose not to repeat these derivations in this section. *Chap. 8.2.2 will be now treated as describing all fast transform scalar filters – not just FFT scalar filters.*

The fast transform filter impulse response equals

$$h_T = T^{-1}H_TT \quad \text{or} \quad Th_T = H_TT \qquad (8.5.4)$$

by combining the results of Eq. 8.5.1. Eq. 8.5.4 is called a *similarity transformation* since Th_T is similar to H_TT. The diagonal symmetry in the H_T matrix of scalar filters results when the h_T impulse response matrix has special symmetries. These special symmetries depend upon the transform used as we shall now briefly discuss. Linear filters utilize linear T transforms. In the time domain, linearity is described by the convolution operation where $c(t) = h_T(t) * r(t)$ for continuous time systems. For discrete time systems, $c = h_Tr$. Eq. 1.11.7 describes h_T for the time-varying filter and Eq. 1.11.9 describes h_T for the time-invariant filter. The h_T matrices of time-invariant filters have diagonal symmetry where each diagonal contains identical elements. Time-invariant discrete filter matrices also have circular symmetry where $h_T(n,m) = h_T(\text{mod}_N(n+k), \text{mod}_N(m+k))$ for $k = 1, 2, \ldots, N-1$. These discrete filters have matrices whose upper and lower diagonals, displaced from the main diagonal by k and $k - N$ are equal. Causal time-invariant continuous filters have lower-triangular h_T matrices. Anti-causal time-invariant continuous filters have upper-triangular matrices.

It can be shown that time-invariant discrete filters have a diagonal or scalar gain matrix H_F in the Fourier transform case. Therefore a nondiagonal or vector gain matrix H_F indicates that Fourier vector filters are time-varying. However other transforms usually produce vector gain matrices H_T even when the discrete filters are time-invariant. In general, vector filters have improved outputs which are obtained at the expense of increased complexity and slower processing time. We note in passing that since the impulse response h_T is an $N \times N$ matrix, the output depends at most on only the last N inputs so the filter is a nonrecursive or FIR type. The more general recursive or IIR type is described by Eq. 10.1.8.

8.5.2 GAIN MATRICES OF SCALAR FILTERS

We shall now derive the gain matrices of some Paley scalar filters for $N = 4$. We again use the sawtooth signal $s = [0, 1, 2, 3]^t$ and zero-mean random noise $n = [1, -1, 1, -1]^t$.

The Paley signal and noise spectra are $S = Ts$ and $N = Tn$ so

$$S = Ts = \frac{1}{4}\begin{bmatrix} 1 & 1 & 1 & 1 \\ 1 & 1 & -1 & -1 \\ 1 & -1 & 1 & -1 \\ 1 & -1 & -1 & 1 \end{bmatrix}\begin{bmatrix} 0 \\ 1 \\ 2 \\ 3 \end{bmatrix} = \begin{bmatrix} 1.5 \\ -1 \\ -0.5 \\ 0 \end{bmatrix}, \quad N = Tn = T\begin{bmatrix} 1 \\ -1 \\ 1 \\ -1 \end{bmatrix} = \begin{bmatrix} 0 \\ 0 \\ 1 \\ 0 \end{bmatrix}$$

$$(8.5.5)$$

We assume that the random noise has a white spectrum so that the expected noise magnitude spectrum $E\{|N|\} = 0.5[1,1,1,1]^t$. $E\{|N|\}$ has the same MSE energy as N where $MSE = 4N^hN = 4$. The expected signal $E\{S\} = S$ since S is deterministic. S and N are assumed to be orthogonal so that $E\{SN^*\} = 0$. The more stringent condition of independence where $E\{f(S)g(N^*)\} = E\{f(S)\}E\{g(N^*)\}$ can be invoked when needed.

Some Class 1 estimation and detection filters will now be designed using Table 8.2.1. The uncorrelated Wiener estimation filter gain $H_{Pal\ UE}$ is $E\{|S(m)|^2\}/E\{|S(m)|^2 + |N(m)|^2\}$. Substituting in the appropriate spectra gives

$$H_{Pal\ UE} = \text{diag}\left[\frac{1.5^2}{1.5^2 + 0.5^2}, \frac{(-1)^2}{(-1)^2 + 0.5^2}, \frac{(-0.5)^2}{(-0.5)^2 + 0.5^2}, \frac{0^2}{0^2 + 0.5^2}\right]$$

$$= \text{diag}[0.9, 0.8, 0.5, 0] \quad (8.5.6)$$

The correlated estimation filter gain $H_{Pal\ CE}$ is $E\{|S|^2 + SN^*\}/E\{|S|^2 + |N|^2 + SN^* + S^*N\}$ which equals $H_{Pal\ UE}$ when S and N are orthogonal. The inverse detection filter gain $H_{Pal\ I}$ is $1/E\{S\}$ where

$$H_{Pal\ I} = \text{diag}[2.667, -4, -8, \infty] \quad (8.5.7)$$

The high-resolution detection filter $H_{Pal\ UHR}$ gain is $E\{S^*\}/E\{|S|^2 + |N|^2\}$ using an uncorrelated Wiener approach so

$$H_{Pal\ UHR} = \text{diag}[0.6, -0.8, -1, 0] \quad (8.5.8)$$

The high-resolution detection filter gain $H_{Pal\ CHR}$ is $E\{S^* + N^*\}/E\{|S|^2 + |N|^2 + SN^* + S^*N\}$ using a correlated Wiener approach which equals $H_{Pal\ UHR}$ because $E\{N\} = 0$ and $E\{SN^*\} = 0$. The matched detection filter has a gain $H_{Pal\ M}$ of $E\{S^*\}/E\{|N|^2\}$ where

$$H_{Pal\ M} = \text{diag}[6, -4, -2, 0] \quad (8.5.9)$$

The signal autocorrelation filter gain $H_{Pal\ USS}$ is $E\{S^*|S|^2\}/E\{|S|^2\} + E\{|N|^2\}$ using the uncorrelated Wiener approach so that

$$H_{Pal\ USS} = \text{diag}[1.35, -0.8, -0.25, 0] \quad (8.5.10)$$

The signal-noise cross-correlation filter gain $H_{Pal\ USN}$ is $E\{N^*|S|^2\}/E\{|S|^2\} + E\{|N|^2\}$ based on the uncorrelated Wiener filter. When the signal and noise are independent and the noise has zero mean, then $E\{N^*|S|^2\} = E\{N^*\}E\{|S|^2\} = 0$ so

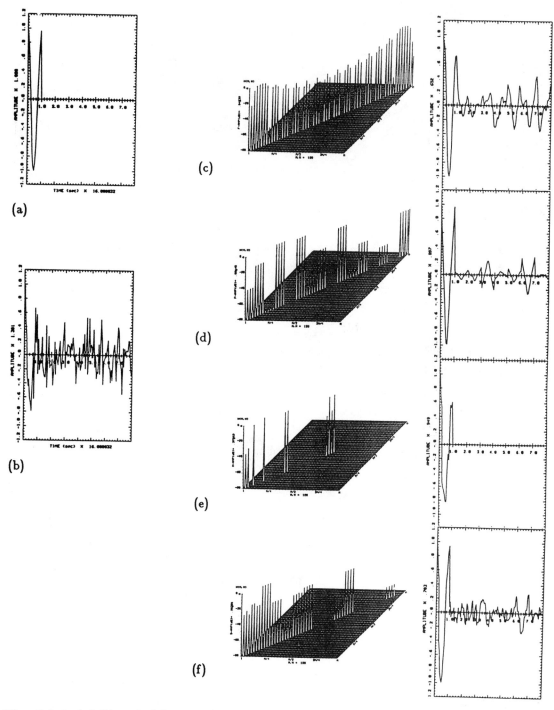

Fig. 8.5.1 (a) Signal, (b) total input, (c) output and gain of Fourier, (d) Paley, (e) Haar, and (f) slant estimation scalar filters. (Plotted by A. Ravitz and based upon C.S. Lindquist and H.E. Lin, "Design of Class 1 estimation filters using generalized transforms," Proc. 29th Midwest Symp. Circuits and Systems, ©1986, IEEE.)

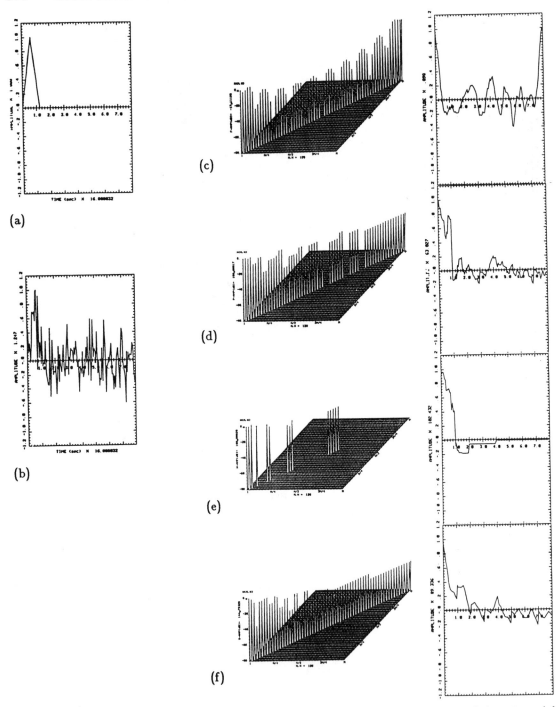

Fig. 8.5.2 (a) Signal, (b) total input, (c) output and gain of Fourier, (d) Paley, (e) Haar, and (f) slant detection scalar filters. (Plotted by A. Ravitz and based upon C.S. Lindquist and H.E. Lin, "Design of Class 1 detection filters using generalized transforms," Rec. 20th Asilomar Conf. Circuits and Systems, pp. 596–599, ©1986, IEEE.)

$$H_{Pal\ USN} = \text{diag}[0,0,0,0] \qquad (8.5.11)$$

Notice that these Paley scalar filters have different gains that the FFT-based Paley vector filters derived in Chap. 8.3. This is verified by comparing the equation pairs for estimation (Eqs. 8.3.13, 8.5.6), inverse (Eqs. 8.3.16, 8.5.7), high-resolution (Eqs. 8.3.18, 8.5.8), matched (Eqs. 8.3.20, 8.5.9), autocorrelation (Eqs. 8.3.22, 8.5.10), and cross-correlation (Eqs. 8.3.23, 8.5.11). The scalar filters of Chap. 8.5 usually provide better performance than the FFT-based vector filters of Chap. 8.3.

8.5.3 SCALAR FILTER DESIGN EXAMPLES

Several estimation and detection correlation scalar filters are now designed for $N = 128$.[3] The uncorrelated Wiener estimation scalar filter has the gain $H(m) = S(m)S^*(m)/[S(m)S^*(m) + N(m)N^*(m)]$ from Table 8.2.1. As before, the signal s is assumed to be a cosine pulse and the noise n is uncorrelated and white with random phase. The signal and total input are shown in Figs. 8.5.1a and 8.5.1b, respectively. The signal-to-noise ratio is about 10 dB. Computing the Fourier, Paley, Haar, and slant spectra and substituting in the $H(m)$ equation gives the various gain functions shown in Figs. 8.5.1c–f. The Fourier filter output has the highest noise content, the Haar filter output is virtually noise free, and the Paley and slant filters recover the signal with the lowest distortion but with significant noise background.

The matched detection scalar filter has a gain $H(m) = T(p,m)S^*(m)/N(m)N^*(m)$ from Table 8.2.1. $T(0,m) = 1$ when the output signal peak is to occur at time $p = 0$. The signal is assumed to be triangular with the same noise as before. The signal and total input are shown in Figs. 8.5.1a and 8.5.1b. The Fourier, Paley, Haar, and slant filter gains and outputs are shown in Figs. 8.5.2c–f, respectively. The Fourier filter output is highly oscillatory and centered about $t = 0$. The other filter outputs begin at $t = 0$. The Haar filter output has the lowest noise content. The Haar and slant filter outputs have the narrowest width and provide the highest resolution.

The best filter performance is generally obtained when the signal and noise spectra are disjoint or nonoverlapping. Good performance is usually obtained when the signal and noise spectra are highly dissimilar. For signals which are not matched to the basis functions of the transforms, our simulations show that slant seems to be the best in recovering these signals while Haar is the best at eliminating the white noise. Paley always performs well with any signal and noise.

8.6 KARHUNEN-LOÉVE SCALAR FILTERS

In Chap 8.5, we derived a set of diagonalized gain matrices for fast transform scalar filters. In this section, we present the *Karhunen-Loéve* or *K-L* optimum scalar filters which also use diagonal matrices. The K-L transform is sometimes called the *Hotelling* or *eigenvector* transform.[3] These filters utilize optimum basis functions which are formed from signal and noise correlation matrices. They differ from the suboptimum scalar filters of Chap 8.5 because the basis functions are *derived* rather than *defined* apriori. The drawback of K-L filters is that they require extensive computations to form these optimum basis functions. Therefore the basic assumptions are (case $\bar{\bar{A}}B$ in Table 8.1.1):

(a) An optimized fast transform is used, and

(b) The basis functions are optimized based upon the input statistics.

(c) The gain matrix is diagonal.

8.6.1 THEORY OF KARHUNEN-LOÉVE SCALAR FILTERS

The transform size (related to generalized spectral bandwidth) and therefore its efficiency in describing the signal depends on how well-matched the basis functions of the orthogonal transform are to the signal. Given a particular signal and noise, it is very desirable to choose the most efficient transform having small $N \times N$ size as well as to perform computations using fast transforms. Reducing size is especially important in some applications like image processing where a large number of samples are stored and processed.

Basically filtering reduces to separating signal (wanted) and noise (unwanted) spectra. The easiest filter problem involves separating signal and noise spectra that are disjoint or nonoverlapping. That is, the signal and noise lie at different generalized frequency bins. A more difficult problem is to separate overlapping spectra where both phases are known, or one phase is deterministic and the other phase is random. Therefore, the best transform to choose is the one where some of its basis functions coincide with the signal and the rest of its basis functions (some or all) coincide with the noise.

This is the motivation for K-L filters. The optimum basis functions are actually generated for each signal and noise problem rather than selected from those basis functions that are already defined (like FFT, etc.). K-L basis functions are optimum in two ways: the approximation error is minimized and the expansion (i.e, spectral) coefficients are uncorrelated. The Karhunen-Loéve analysis method in continuous time will now be described. Then we shall convert the results into discrete time.

A signal $x(t)$ is to be approximated over the time interval $(0, NT)$ in the form

$$x(t) = \sum_{k=1}^{\infty} \alpha_k \phi_k(t), \qquad 0 \le t < NT \tag{8.6.1}$$

In the general case, $x(t)$ and α_k are random and $\phi_k(t)$ is deterministic. The $\phi_k(t)$ are the basis functions used to approximate $x(t)$. The α_k expansion coefficients are chosen such that the expected integral squared error (ISE) is minimized where

$$ISE = \min E\left\{ \int_0^{NT} \left| x(t) - \sum_{k=1}^{\infty} \alpha_k \phi_k(t) \right|^2 dt \right\} \tag{8.6.2}$$

The set of basis functions is said to be complete if $ISE = 0$ for all $x(t)$. Setting the derivative of the ISE error with respect to the α_j coefficient equal to zero gives

$$\frac{\partial ISE}{\partial \alpha_j} = E\left\{ \int_0^{NT} 2\left[x(t) - \sum_{k=1}^{\infty} \alpha_k \phi_k(t)\right] \phi_j^*(t) \, dt \right\} = 0 \qquad (8.6.3)$$

Rearranging Eq. 8.6.3 and moving the expectation operation inside the integral gives

$$\int_0^{NT} E\{x(t)\phi_j^*(t)\} \, dt = \sum_{k=1}^{\infty} \alpha_k \int_0^{NT} \phi_k(t)\phi_j^*(t) \, dt \qquad (8.6.4)$$

where $E\{\phi_k \phi_j^*\} = \phi_k \phi_j^*$ since ϕ_k is deterministic. When the $\phi_k(t)$ basis functions are orthogonal over the $(0, NT)$ interval so

$$\int_0^{NT} \phi_k(t)\phi_j^*(t) \, dt = \begin{cases} c_k, & \text{if } j = k \\ 0, & \text{otherwise} \end{cases} \qquad (8.6.5)$$

then the right-hand side of Eq. 8.6.4 reduces to a single α_k term. Solving for α_k gives

$$\alpha_k = \frac{\int_0^{NT} E\{x(t)\phi_k^*(t)\} \, dt}{\int_0^{NT} \phi_k(t)\phi_k^*(t) \, dt} \longrightarrow \int_0^{NT} E\{x(t)\phi_k^*(t)\} \, dt \qquad (8.6.6)$$

The denominator equals unity when the basis functions are orthonormal ($c_k = 1$).

Eq. 8.6.6 shows that the optimum α_k coefficient depends upon: (1) the basis functions $\phi_k(t)$ and (2) the input $x(t)$. The α_k's are generally correlated with one another. However the basis functions can be properly chosen so that the α_k's are *uncorrelated*. When the α_k are uncorrelated, fewer coefficients are needed in Eq. 8.6.2 to achieve some given ISE level or a lower ISE level is obtained using a fixed number of terms. This observation is the motivation for the K-L expansion which generates the basis functions $\phi_k(t)$ such that the α_k are uncorrelated. When $E\{x(t)\} = 0$ then $E\{\alpha_k\} = 0$. In this zero-mean input case, the α_k coefficients are uncorrelated as well as orthogonal. It shall now be shown that this uncorrelated condition is satisfied when the $\phi_k(t)$ basis functions satisfy the integral equation

$$\int_0^{NT} R_{xx}(t, \tau)\phi_k(\tau)d\tau = \lambda_k \phi_k(t) \qquad (8.6.7)$$

where $E\{\alpha_k \alpha_k^*\} = \lambda_k$. Consider the signal $x(t)$ of Eq. 8.6.1 evaluated at two different times, t and τ, where

$$x(t) = \sum_{k=1}^{\infty} \alpha_k \phi_k(t), \qquad x(\tau) = \sum_{j=1}^{\infty} \alpha_j \phi_j(\tau) \qquad (8.6.8)$$

The autocorrelation $R_{xx}(t, \tau)$ of $x(t)$ is defined as

$$R_{xx}(t, \tau) = E\{x(t)x^*(\tau)\} = E\left\{ \sum_{k=1}^{\infty} \alpha_k \phi_k(t) \sum_{j=1}^{\infty} \alpha_j^* \phi_j^*(\tau) \right\}$$

$$= E\left\{ \sum_{k=1}^{\infty}\sum_{j=1}^{\infty} \alpha_k \alpha_j^* \phi_k(t)\phi_j^*(\tau) \right\} = \sum_{k=1}^{\infty}\sum_{j=1}^{\infty} \phi_k(t)\phi_j^*(\tau)E\{\alpha_k \alpha_j^*\} \qquad (8.6.9)$$

Multiplying both sides of this equation by $\phi_l(\tau)$ and integrating over the interval $(0, NT)$ with respect to τ gives

$$
\int_0^{NT} R_{xx}(t,\tau)\phi_l(\tau)\ d\tau = \int_0^{NT} \phi_l(\tau)\left[\sum_{k=1}^{\infty}\sum_{j=1}^{\infty}\phi_k(t)\phi_j^*(\tau)E\{\alpha_k\alpha_j^*\}\right]\ d\tau
$$

$$
= \sum_{k=1}^{\infty}\sum_{j=1}^{\infty}\phi_k(t)\left[\int_0^{NT}\phi_l(\tau)\phi_j^*(\tau)\ d\tau\right]E\{\alpha_k\alpha_j^*\} = \sum_{k=1}^{\infty}\phi_k(t)E\{\alpha_k\alpha_l^*\}
$$

(8.6.10)

using the orthonormal condition of Eq. 8.6.5. The α_k coefficients are uncorrelated when

$$
E\{\alpha_k\alpha_j^*\} = \begin{cases} \lambda_k, & \text{if } j = k \\ 0, & \text{otherwise} \end{cases}
$$

(8.6.11)

Under this condition, Eq. 8.6.10 reduces to

$$
\int_0^{NT} R_{xx}(t,\tau)\phi_k(\tau)\ d\tau = \phi_k(t)E\{|\alpha_k|^2\} = \lambda_k\phi_k(t)
$$

(8.6.12)

Using Eq. 8.6.6, the correlation between the coefficients is indeed zero since

$$
E\{\alpha_k\alpha_l^*\} = E\left\{\int_0^{NT}\int_0^{NT} x(t)x^*(\tau)\phi_k^*(t)\phi_l(\tau)\ dtd\tau\right\}
$$

$$
= \int_0^{NT}\left[\int_0^{NT} R_{xx}(t,\tau)\phi_l(\tau)d\tau\right]\phi_k^*(t)\ dt = \int_0^{NT}\lambda_l\phi_l(t)\phi_k^*(t)\ dt = \lambda_k U_0(k-l)
$$

(8.6.13)

Thus, the basis functions satisfy the general integral equation of Eq. 8.6.12 as

$$
\int_0^{NT} R_{xx}(t,\tau)\phi(\tau)\ d\tau = \lambda\phi(t)
$$

(8.6.14)

The kernel of the equation is the correlation function $R_{xx}(t,\tau)$. The λ_k values for which the equation is satisfied are the *eigenvalues* of the equation. The $\phi_k(t)$ basis functions are the corresponding *eigenfunctions*. The theory of linear integral equations gives the following properties of this expansion: There is at least one real number $\lambda \neq 0$ and a function $\phi(t)$ satisfying Eq. 8.6.14 (i.e., the solution is non-trivial). Since the kernel $R_{xx}(t,\tau)$ is nonnegative definite, it can be uniformly expanded over the interval $(0, NT)$ in the series

$$
R_{xx}(t,\tau) = \sum_{k=1}^{\infty}\lambda_k\phi_k(t)\phi_k^*(\tau)
$$

(8.6.15)

If $R_{xx}(t,\tau)$ is positive definite, the eigenfunctions form a complete orthonormal set. If $R_{xx}(t,\tau)$ is not positive definite, the eigenfunctions do not form a complete set. In this case, the basis functions can be augmented with enough additional functions to form a complete set. Finally, the expected energy in $x(t)$ equals

$$
E\left\{\int_0^{NT}|x(t)|^2dt\right\} = \int_0^{NT} R_{xx}(t,t)\ dt = \sum_{k=1}^{\infty}\lambda_k
$$

(8.6.16)

With this choice of basis functions, the K-L series of Eq. 8.6.1 converges since

$$E\left\{\left|x(t) - \sum_{k=1}^{\infty} \alpha_k \phi_k(t)\right|^2\right\} = R_{xx}(t,t) - \sum_{k=1}^{\infty} \lambda_k |\phi_k(t)|^2 =$$

$$R_{xx}(t,t) - \sum_{k=1}^{\infty} E\{x(t)\alpha_k^*\}\phi_k^*(t) - \sum_{j=1}^{\infty} E\{x^*(t)\alpha_j\}\phi_j(t) + \sum_{k=1}^{\infty}\sum_{j=1}^{\infty} E\{\alpha_k\alpha_j^*\}\phi_k(t)\phi_j^*(t)$$

(8.6.17)

where the last step follows by substituting α_k and α_l from Eq. 8.6.6 and simplifying.

The K-L transform in continuous time can be extended to discrete time as follows. We approximate a discrete signal $x(n)$ over the time interval $(0, NT)$ in the form

$$x(n) = \sum_{k=1}^{\infty} \alpha_k \phi_k(n), \qquad n = 0, 1, \ldots, N-1 \tag{8.6.18}$$

In the general case, $x(n)$ and α_k are random and $\phi_k(n)$ is deterministic. The $\phi_k(n)$ are the basis functions used to approximate $x(n)$. The α_k expansion coefficients are chosen such that the expected total squared error (MSE) is minimized where

$$MSE = \min E\left\{ \sum_{n=0}^{N-1} \left| x(n) - \sum_{k=1}^{\infty} \alpha_k \phi_k(n) \right|^2 \right\} \tag{8.6.19}$$

The set of basis functions is said to be complete if $MSE = 0$ for all $x(n)$. Setting the derivative of the MSE error with respect to the α_j coefficient equal to zero gives

$$\frac{\partial MSE}{\partial \alpha_j} = E\left\{ \sum_{n=0}^{N-1} 2\left[x(n) - \sum_{k=1}^{\infty} \alpha_k \phi_k(n) \right] \phi_j^*(n) \right\} = 0 \tag{8.6.20}$$

Rearranging Eq. 8.6.20 and moving the expectation operation inside the integral gives

$$\sum_{n=0}^{N-1} E\{x(n)\phi_j^*(n)\} = \sum_{k=1}^{\infty} \alpha_k \sum_{n=0}^{N-1} \phi_k(n)\phi_j^*(n) \tag{8.6.21}$$

where $E\{\phi_k\phi_j^*\} = \phi_k\phi_j^*$ since ϕ_k is deterministic. When the $\phi_k(n)$ basis functions are orthogonal over the $(0, NT)$ interval so

$$\sum_{n=0}^{N-1} \phi_k(n)\phi_j^*(n) = \begin{cases} c_k, & \text{if } j = k \\ 0, & \text{otherwise} \end{cases} \tag{8.6.22}$$

then the right-hand side of Eq. 8.6.21 reduces to a single α_k term. When the basis functions are scaled to be orthonormal, $c_k = 1$. Solving Eq. 8.6.22 in this orthonormal case for α_k gives

$$\alpha_k = \frac{\sum_{n=0}^{N-1} E\{x(n)\phi_k^*(n)\}}{\sum_{n=0}^{N-1} \phi_k(n)\phi_k^*(n)} \longrightarrow \sum_{n=0}^{N-1} E\{x(n)\phi_k^*(n)\} \tag{8.6.23}$$

The α_k's are generally correlated with one another. However the basis functions can be properly chosen so that they are uncorrelated. It shall now be shown that this condition is satisfied when the $\phi_k(n)$ basis functions satisfy the equation

$$\sum_{m=0}^{N-1} R_{xx}(n,m)\phi_k(m) = \lambda_k\phi_k(n) \tag{8.6.24}$$

where $E\{\alpha_k\alpha_k^*\} = \lambda_k$. Consider the signal $x(n)$ of Eq. 8.6.1 evaluated at two different times, n and m, where

$$x(n) = \sum_{k=1}^{\infty} \alpha_k\phi_k(n), \qquad x(m) = \sum_{j=1}^{\infty} \alpha_j\phi_j(m) \tag{8.6.25}$$

The autocorrelation $R_{xx}(n,m)$ of $x(n)$ is defined as

$$R_{xx}(n,m) = E\{x(n)x^*(m)\} = E\left\{\sum_{k=1}^{\infty}\alpha_k\phi_k(n)\sum_{j=1}^{\infty}\alpha_j^*\phi_j^*(m)\right\}$$

$$= E\left\{\sum_{k=1}^{\infty}\sum_{j=1}^{\infty}\alpha_k\alpha_j^*\phi_k(n)\phi_j^*(m)\right\} = \sum_{k=1}^{\infty}\sum_{j=1}^{\infty}\phi_k(n)\phi_j^*(m)E\{\alpha_k\alpha_j^*\} \tag{8.6.26}$$

Multiplying both sides of this equation by $\phi_l(m)$ and summing over the interval $(0, NT)$ with respect to m gives

$$\sum_{m=0}^{N-1} R_{xx}(n,m)\phi_l(m) = \sum_{m=0}^{N-1} \phi_l(m)\left[\sum_{k=1}^{\infty}\sum_{j=1}^{\infty}\phi_k(n)\phi_j^*(m)E\{\alpha_k\alpha_j^*\}\right]$$

$$= \sum_{k=1}^{\infty}\sum_{j=1}^{\infty}\phi_k(n)\left[\sum_{m=0}^{N-1}\phi_l(m)\phi_j^*(m)\right]E\{\alpha_k\alpha_j^*\} = \sum_{k=1}^{\infty}\phi_k(n)E\{\alpha_k\alpha_l^*\} \tag{8.6.27}$$

using the orthonormal condition of Eq. 8.6.22. The α_k coefficients are uncorrelated when

$$E\{\alpha_k\alpha_j^*\} = \begin{cases} \lambda_k, & \text{if } j = k \\ 0, & \text{otherwise} \end{cases} \tag{8.6.28}$$

Under this condition, Eq. 8.6.27 reduces to

$$\sum_{m=0}^{N-1} R_{xx}(n,m)\phi_k(m) = \phi_k(n)E\{|\alpha_k|^2\} = \lambda_k\phi_k(n) \tag{8.6.29}$$

where using Eq. 8.6.23, the correlation between the coefficients is indeed zero since

$$E\{\alpha_k\alpha_l^*\} = E\left\{\sum_{n=0}^{N-1}\sum_{m=0}^{N-1} x(n)x^*(m)\phi_k^*(n)\phi_l(m)\right\}$$

$$= \sum_{n=0}^{N-1}\left[\sum_{n=0}^{N-1} R_{xx}(n,m)\phi_l(m)dm\right]\phi_k^*(n) = \sum_{n=0}^{N-1}\lambda_l\phi_l(n)\phi_k^*(n) = \lambda_k U_0(k-l) \tag{8.6.30}$$

The K-L basis functions satisfy Eq. 8.6.24 which can be written in matrix form as $R\phi_k = \lambda_k\phi_k$ or $[R\phi_k - \lambda_k]\phi_k = 0$ where

$$
\begin{bmatrix}
R(0,0) & R(0,1) & \cdots & R(0,N-1) \\
R(1,0) & R(1,1) & \cdots & R(1,N-1) \\
\vdots & \vdots & \ddots & \vdots \\
R(N-1,0) & R(N-1,1) & \cdots & R(N-1,N-1)
\end{bmatrix}
\begin{bmatrix}
\phi_k(0) \\
\phi_k(1) \\
\vdots \\
\phi_k(N-1)
\end{bmatrix}
= \lambda_k
\begin{bmatrix}
\phi_k(0) \\
\phi_k(1) \\
\vdots \\
\phi_k(N-1)
\end{bmatrix}
$$

$$(8.6.31)$$

Thus, the ϕ_k basis functions satisfy the matrix equation of Eq. 8.6.31. The kernel of the equation is the correlation function $R_{xx}(n,m)$. The $\{\lambda_k\}$ values for which the equation is satisfied are the *eigenvalues* of the equation. The $\{\phi_k\}$ basis functions are the corresponding *eigenvectors*. The theory of linear integral equations gives the following properties of the expansion: There is at least one real number $\lambda \neq 0$ and a function $\phi(n)$ satisfying Eq. 8.6.31 (i.e., the solution is non-trivial). Since the kernel $R_{xx}(n,m)$ is nonnegative definite, it can be uniformly expanded over the interval $(0, NT)$ in the series

$$R_{xx}(n,m) = \sum_{k=1}^{\infty} \lambda_k\phi_k(n)\phi_k^*(m) \tag{8.6.32}$$

If $R_{xx}(n,m)$ is positive definite, the eigenfunctions form a complete orthonormal set. If $R_{xx}(n,m)$ is not positive definite, the eigenfunctions do not form a complete set. In this case, the basis functions can be augmented with enough additional functions to form a complete set. Finally, the expected energy in $x(n)$ equals

$$E\left\{\sum_{n=0}^{N-1}|x(n)|^2\right\} = \sum_{n=0}^{N-1}R_{xx}(n,n) = \sum_{k=1}^{\infty}\lambda_k \tag{8.6.33}$$

With this choice of basis functions, the K-L series of Eq. 8.6.18 converges since

$$E\left\{\left|x(n) - \sum_{k=1}^{N}\alpha_k\phi_k(n)\right|^2\right\} = R_{xx}(n,n) - \sum_{k=1}^{N}\lambda_k|\phi_k(n)|^2 =$$

$$R_{xx}(n,n) - \sum_{k=1}^{N}E\{x(n)\alpha_k^*\}\phi_k^*(n) - \sum_{j=1}^{N}E\{x^*(n)\alpha_j\}\phi_j(n) + \sum_{k=1}^{N}\sum_{j=1}^{N}E\{\alpha_k\alpha_j^*\}\phi_k(n)\phi_j^*(n)$$

$$(8.6.34)$$

where the last step follows by substituting α_k and α_l from Eq. 8.6.23 and simplifying.

8.6.2 GAIN MATRICES OF KARHUNEN-LOÉVE SCALAR FILTERS

Let us design a K-L estimation filter to estimate a signal in the presence of random noise. We again use the sawtooth signal $s = [0, 1, 2, 3]^t$ and zero-mean $n = [1, -1, 1, -1]^t$. We computed the estimate of the signal correlation matrix in Eq. 8.4.67a as

$$R_{ss} = \begin{bmatrix} 0 & 0 & 0 & 0 \\ 0 & 1 & 2 & 3 \\ 0 & 2 & 4 & 6 \\ 0 & 3 & 6 & 9 \end{bmatrix} \tag{8.6.35}$$

The noise is assumed to be white and to have uncorrelated samples so the noise correlation matrix equals

$$R_{nn} = \begin{bmatrix} 1 & 0 & 0 & 0 \\ 0 & 1 & 0 & 0 \\ 0 & 0 & 1 & 0 \\ 0 & 0 & 0 & 1 \end{bmatrix} \tag{8.6.36}$$

This apriori estimate is used rather than the aposteriori estimate given by Eq. 8.4.67b. Both have the same energy $tr(R_{nn}) = 4$. The impulse response matrix h_T of the uncorrelated Wiener filter equals

$$h_T = R_{ss}(R_{ss} + R_{nn})^{-1} \tag{8.6.37}$$

from Table 8.7.1. Substituting Eqs. 8.6.35 and 8.6.36 into Eq. 8.6.37 gives

$$h_T = \begin{bmatrix} 0 & 0 & 0 & 0 \\ 0 & 1 & 2 & 3 \\ 0 & 2 & 4 & 6 \\ 0 & 3 & 6 & 9 \end{bmatrix} \begin{bmatrix} 1 & 0 & 0 & 0 \\ 0 & 2 & 2 & 3 \\ 0 & 2 & 5 & 6 \\ 0 & 3 & 6 & 10 \end{bmatrix}^{-1} = \begin{bmatrix} 0 & 0 & 0 & 0 \\ 0 & 0.0667 & 0.133 & 0.2 \\ 0 & 0.133 & 0.267 & 0.4 \\ 0 & 0.2 & 0.4 & 0.6 \end{bmatrix} \tag{8.6.38}$$

The eigenvalues of h_T are obtained by evaluating

$$|h_T - \lambda I| = \det \left| \begin{bmatrix} 0 & 0 & 0 & 0 \\ 0 & 0.0667 & 0.133 & 0.2 \\ 0 & 0.133 & 0.267 & 0.4 \\ 0 & 0.2 & 0.4 & 0.6 \end{bmatrix} - \lambda \begin{bmatrix} 1 & 0 & 0 & 0 \\ 0 & 1 & 0 & 0 \\ 0 & 0 & 1 & 0 \\ 0 & 0 & 0 & 1 \end{bmatrix} \right|$$

$$= \det \begin{vmatrix} -\lambda & 0 & 0 & 0 \\ 0 & 0.0667 - \lambda & 0.133 & 0.2 \\ 0 & 0.133 & 0.267 - \lambda & 0.4 \\ 0 & 0.2 & 0.4 & 0.6 - \lambda \end{vmatrix} = 0 \tag{8.6.39}$$

Solving for the eigenvalues gives

$$\lambda_1 = 0.933, \quad \lambda_2 = 0, \quad \lambda_3 = 0, \quad \lambda_4 = 0 \tag{8.6.40}$$

The eigenvectors corresponding to these eigenvalues are found by solving $h_T \phi_k = \lambda_k \phi_k$ using some iterative approach such as

$$\phi_{k,j+1} = \lambda_k h_T^{-1} \phi_{k,j} \tag{8.6.41}$$

$\phi_{k,j}$ is the jth estimate of the kth eigenvector ϕ_k. This gives eigenvectors of

$$\phi_1 = \begin{bmatrix} 0 \\ 0.267 \\ 0.535 \\ 0.802 \end{bmatrix}, \quad \phi_2 = \begin{bmatrix} 0 \\ -0.559 \\ -0.778 \\ 0.705 \end{bmatrix}, \quad \phi_3 = \begin{bmatrix} 0 \\ 0.962 \\ -0.096 \\ -0.256 \end{bmatrix}, \quad \phi_4 = \begin{bmatrix} 1 \\ 0 \\ 0 \\ 0 \end{bmatrix} \quad (8.6.42)$$

The K-L inverse transform T^{-1} is the matrix of basis functions whose columns are the eigenvectors of h_T so

$$T^{-1} = [\phi_1, \phi_2, \phi_3, \phi_4] = \begin{bmatrix} 0 & 0 & 0 & 1 \\ 0.267 & -0.559 & 0.962 & 0 \\ 0.535 & -0.778 & -0.096 & 0 \\ 0.802 & 0.705 & -0.256 & 0 \end{bmatrix} \quad (8.6.43)$$

T^{-1} is not orthogonal since $(T^{-1})^h(T^{-1})$ is not diagonal. However it can be orthonormalized so $(T_o^{-1})^h(T_o^{-1}) = I$ using the Gram-Schmidt orthogonalization process as[5]

$$T_o^{-1} = [\phi_1, \phi_2, \phi_3, \phi_4] = \begin{bmatrix} 0 & 0 & 0 & 1 \\ 0.267 & -0.060 & -0.962 & 0 \\ 0.535 & 0.840 & 0.096 & 0 \\ 0.802 & -0.540 & 0.256 & 0 \end{bmatrix} \quad (8.6.44)$$

The K-L transform of kernels equals $T_o = (T_o^{-1})^h$ which is the transpose of Eq. 8.6.44. Then the corresponding optimum filter gain matrix is given by Eq. 8.6.35 as

$$H_T = T_o h_T T_o^{-1} = \mathrm{diag}[0.933, 0, 0, 0] \quad (8.6.45)$$

which is a diagonal matrix. Using the K-L transform of Eq. 8.6.44, the various spectra of interest, such as $S = T_o s$, equal

$$S = \begin{bmatrix} 3.742 \\ 0 \\ 0 \\ 0 \end{bmatrix}, \quad N = \begin{bmatrix} -0.535 \\ 1.439 \\ 0.802 \\ 1 \end{bmatrix}, \quad R = \begin{bmatrix} 3.207 \\ 1.439 \\ 0.802 \\ 1 \end{bmatrix}, \quad H_T = \mathrm{diag} \begin{bmatrix} 0.933 \\ 0 \\ 0 \\ 0 \end{bmatrix}^t, \quad C = \begin{bmatrix} 2.993 \\ 0 \\ 0 \\ 0 \end{bmatrix}$$

$$(8.6.46)$$

The output signal equals $c = h_T r = T_o^{-1} C = [0, 0.8, 1.6, 2.4]^t$. The MSE error equals $e^h e = 0.56$ where $e = s - c$. This is the same output and error as the Fourier and Paley vector filters having matrix gains given by Eqs. 8.4.80 and 8.4.73, respectively. The output error is reduced using a larger number of points.

　　Since the K-L transform depends on the specified correlation matrices, no general fast algorithm exists to compute T or its inverse. This is unimportant when the time domain filtering approach of Fig. 8.1.1c is used since T is not used or computed.

8.6.3　KARHUNEN-LOÉVE SCALAR FILTER DESIGN EXAMPLES

Now we shall design several estimation and detection K-L filters in the time domain having the form of Fig. 8.1.1c where $N = 128$.[4] The K-L uncorrelated Wiener estimation filter has an impulse response given by Eq. 8.6.37. The signal s is assumed to be a cosine pulse as shown in Fig. 8.6.1a. The noise n is assumed to be white and uncorrelated with

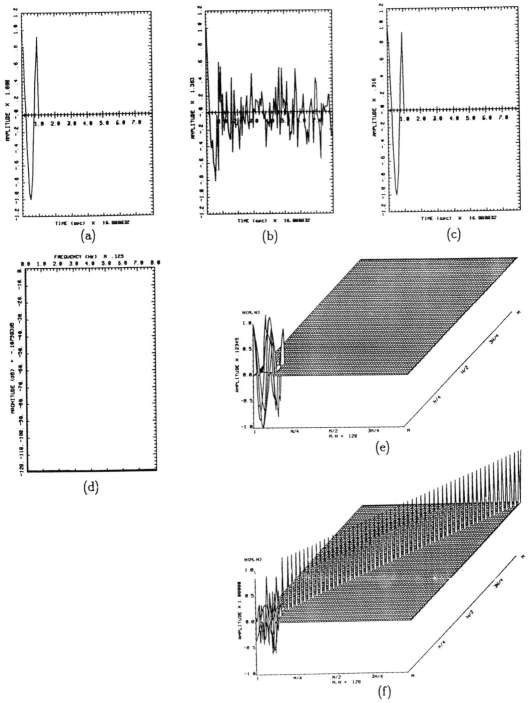

Fig. 8.6.1 (a) Signal, (b) total input, (c) output, (d) filter gain, (e) filter impulse response, and (f) basis functions of K-L estimation filter. (From M.K. Almeer, "Theory and design of digital filters using Karhunen-Loéve transform," M.S.E.E. Thesis, Univ. of Miami, 118 p., June 1987.)

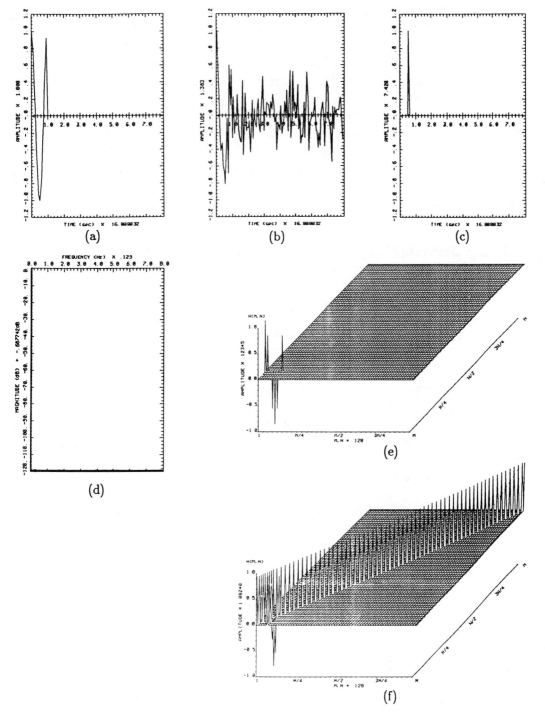

Fig. 8.6.2 (a) Signal, (b) total input, (c) output, (d) filter gain, (e) filter impulse response, and (f) basis functions of K-L detection filter. (From M.K. Almeer, loc. cit.)

random phase. The signal and noise are added together to form the input r as shown in Fig. 9.6.1b. The peak signal-to-noise ratio is about 10 dB. The signal autocorrelation matrix is computed as ss^h using Eq. 8.4.67a. The noise autocorrelation matrix is set equal to $R_{nn} = kI$. The noise constant $k = n^h n/N$ equally distributes the energy into N frequency bins. The signal and noise are assumed to be orthogonal so we set the cross-correlation matrix $R_{sn} = 0$.

Substituting these correlation matrices into Eq. 8.6.38 yields the impulse response matrix h_T of the K-L estimation filter shown in Fig. 8.6.1e. Multiplying h_T by the input r gives the filter output c shown in Fig. 8.6.1c. The signal appears to be perfectly recovered with no noise. This K-L scalar filter performs much better than the Fourier, Paley, Haar, and slant filters shown in Figs. 8.4.1 and 8.5.1. The K-L transform matrix T is plotted in Fig. 8.6.1f. The rows are the basis functions or eigenfunctions of h_T from Eq. 8.6.41. The K-L filter gain is shown in Fig. 8.6.1d and is a diagonal matrix whose diagonal elements are the eigenvalues of h_T.

The K-L detection filter having the same signal and noise input is shown in Fig. 8.6.2. Its gain is given by Eq. 8.6.39 and it has an impulse response given by Eq. 8.6.40b. Substituting the signal and noise autocorrelation matrices discussed earlier yields the impulse response h_T shown in Fig. 8.6.2e. The filter output in Fig. 8.6.2c was adjusted to occur at time $p = 10$ (i.e., use u_{10} rather than u_0 in the h_T equation). The K-L basis functions for this detection filter are shown in Fig. 8.6.2f.

These results are very interesting and illustrate several important points:

1. The K-L basis functions are single noise pulses delayed in time from $n = 0$ to 127 (or diagonal terms of unity in the T matrix). This set is complete because any input can be expressed as a linear combination of these pulses. One (or more) of these pulses is replaced with the signal (Fig. 8.6.2f shows this clearly). Thus, one (or more) of the K-L basis functions is matched to the signal while the remainder are matched to the noise.

2. The h_T impulse response matrices equal

$$h_{T\ UE} = R_{ss}(R_{ss} + R_{nn})^{-1} \cong \frac{1}{k} R_{ss} I = \frac{1}{k} R_{ss}$$

$$h_{T\ UHR} = u_p s^h (R_{ss} + R_{nn})^{-1} \cong \frac{1}{k} u_p s^h$$

$$(8.6.47)$$

The inverse matrix approximately reduces to the identity matrix I scaled by $(1/k)$. Therefore h_T of the estimation filter has the same shape as the signal autocorrelation $R_{ss} = ss^h$ as shown in Fig. 8.6.1e. The detection filter has an h_T of all zeros except for the pth column which equals the conjugate signal s^*. This is shown in Fig. 8.6.2e.

3. The H_T gain matrices are diagonal with only one (or more) nonzero entry. Therefore the frequency domain filter is bandpass with only a one-bin (or more) bandwidth. Its center frequency occurs at the bin corresponding to the index or number of the signal basis function – in these cases very close to zero. This location is arbitrary since the eigenvalues or gains can be ordered in any fashion. Once the order is selected however, the basis or eigenfunctions are ordered to match their corresponding eigenvalues.

4. The signal spectrum is only one-bin (or more) wide because it is matched to the signal basis function in the K-L transform. The signal components occurs at the filter's center frequency. The noise spectrum is wideband and flat because it is matched to the set of delayed pulses (except at the signal frequency). There is a spectral component of the noise at the signal frequency.

5. The signal spectral component produces the impulse response of the filter which has the desired shape – the signal for the estimation filter and the pulse for the detection filter. Both responses appear to be perfect. However this is not the case because the noise spectral component also produces the same responses but at a much lower level. In this case where the input $SNR_i \cong 10$ dB, the rms noise is about three times below the peak signal. Assuming that the noise energy is equally spread over the 128 frequency bins, then the output $SNR_o \cong 52$ dB. This is a SNR gain of about $128 \cong 42$ dB which is excellent.

6. There is low-level noise in the filter output but noise is not recognizable because it produces the same response as signal – not the low-level noise appearing over the entire NT time window as in all the scalar and vector filter outputs discussed previously in this chapter. This is easily seen in the simple four-point example of Chap. 8.6.3. Following Eq. 8.6.46, we saw that the filter output $c = s_o + n_o = [0, 0.8, 1.6, 2.4]^t$. The output signal s_o and noise n_o components both equal $s_o = n_o = [0, 0.4, 0.8, 1.2]^t$. This masking of noise which appears as signal is unique to K-L filters. By incorporating smoothing into the algorithms used to form the autocorrelations $R_{ss} = ss^h$ and $R_{nn} = kI$ as was done in Eq. 8.4.67–8.4.69, less dramatic and unusual results are obtained.

We see that K-L filters have greatly expanded our frequency and time domain understanding. They open new avenues for viewing and developing transforms and for signal and noise analysis.

8.7 TIME DOMAIN VECTOR AND SCALAR FILTERS

All vector and scalar filters can be implemented in the frequency domain (see Fig. 8.1.1a) or in the time domain (see Figs. 8.1.1b or 8.1.1c). The filters have a frequency response gain H_T and a time domain impulse response h_T that are related by

$$h_T = T^{-1}H_T T, \qquad H_T = Th_T T^{-1} \qquad (8.7.1)$$

from Eq. 8.5.4. The H_T gain matrices are listed in Tables 8.4.1 and 8.4.2 for a variety of Class 1, 2, and 3 filter types. When properly interpreted, the matrices apply to both vector and scalar filters. They can be converted to h_T impulse responses using Eq. 8.7.1 when T is known. However h_T can be determined directly from time domain information in the following manner. The resulting h_T impulse responses for vector and scalar filters are listed in Tables 8.7.1 and 8.7.2.

Consider the Class 1 uncorrelated Wiener estimation vector filter having a gain of

$$H_T = E\{SS^h\} \left(E\{SS^h + NN^h\}\right)^{-1} \qquad (8.7.2)$$

Since $S = Ts$, $S^h = s^h T^h$, $N = Tn$, and $N^h = n^h T^h$, we can rewrite Eq. 8.7.2 as

Table 8.7.1 Ideal Class 1 vector and scalar filter impulse responses h_T.

Filter Name	Impulse Response h_T	Analogous Eq. Number

Distortion Balance (w_s = Signal weighting, w_n = Noise weighting)

$$E\{sw_s^t\}\left(E\{sw_s^t + nw_n^t\}\right)^{-1}$$

(8.4.5)

General Estimation (d = Desired output)

$$h_E = E\{dr^h\}\left(E\{rr^h\}\right)^{-1}$$

(8.4.13)

General Detection

$$h_D = E\{u_0 d^h\}\left(E\{rr^h\}\right)^{-1}$$

(8.4.15)

Estimation–Detection

$$h_E h_D$$

(8.4.20)

Uncorrelated Estimation (Uncorrelated Wiener)

$$E\{ss^h\}\left(E\{ss^h + nn^h\}\right)^{-1}$$

(8.4.22)

Correlated Estimation (Correlated Wiener)

$$E\{ss^h + sn^h\}\left(E\{ss^h + nn^h + sn^h + ns^h\}\right)^{-1}$$

(8.4.21)

Transitional 1 Estimation (Uncorrelated Wiener $\alpha = 0$, Correlated Wiener $\alpha = 1$)

$$E\{ss^h + \alpha sn^h\}\left(E\{ss^h + nn^h + \alpha(sn^h + ns^h)\}\right)^{-1}$$

(8.4.43)

Pulse-Shaping Estimation (p = Desired pulse shape, Uncorrelated Wiener)

$$E\{p_d Us^h\}\left(E\{ss^h + nn^h\}\right)^{-1}$$

(8.4.26)

Pulse-Shaping Estimation (p = Desired pulse shape, Correlated Wiener)

$$E\{p_d U(s^h + n^h)\}\left(E\{ss^h + nn^h + sn^h + ns^h\}\right)^{-1}$$

(8.4.27)

Table 8.7.1 (Continued) Ideal Class 1 vector and scalar filter impulse responses h_T.

Filter Name	Impulse Response h_T	Analogous Eq. Number

Classical Detection (Matched)

$$E\{u_0 s^h\} \left(E\{nn^h\}\right)^{-1} \qquad (8.4.34)$$

High-resolution Detection (Uncorrelated Wiener)

$$E\{u_0 s^h\} \left(E\{ss^h + nn^h\}\right)^{-1} \qquad (8.4.38)$$

High-resolution Detection (Correlated Wiener)

$$E\{u_0(s^h + n^h)\} \left(E\{ss^h + nn^h + sn^h + ns^h\}\right)^{-1} \qquad (8.4.39)$$

Inverse Detection (Deconvolution)

$$(u_0)_d \left(E\{s_d\}\right)^{-1} \qquad (8.4.36)$$

Transitional 1 Detection (Inverse $\alpha = 0$, High-resolution $\alpha = 0.5$, Matched $\alpha = 1$)

$$E\{Us^h\} \left(E\{(1 - \alpha)ss^h + \alpha nn^h\}\right)^{-1} \qquad (8.4.41)$$

Transitional 2 Detection (Inverse $\alpha = 0.5$)

$$(u_0)_d \left(E\{s_d^{2(1-\alpha)}\}\right)^{-1} \qquad (8.4.42)$$

Autocorrelation Estimation (Uncorrelated Wiener)

$$E\{ss^h Us^h\} \left(E\{ss^h + nn^h\}\right)^{-1} \qquad (8.4.29)$$

Autocorrelation Estimation (Correlated Wiener)

$$E\{ss^h U(s^h + n^h)\} \left(E\{ss^h + nn^h + sn^h + ns^h\}\right)^{-1} \qquad (8.4.30)$$

Cross-Correlation Estimation (Uncorrelated Wiener)

$$E\{sn^h Us^h\} \left(E\{ss^h + nn^h\}\right)^{-1} \qquad (8.4.32)$$

Cross-Correlation Estimation (Correlated Wiener)

$$E\{sn^h U(s^h + n^h)\} \left(E\{ss^h + nn^h + sn^h + ns^h\}\right)^{-1} \qquad (8.4.33)$$

Table 8.7.2   Ideal Class 2 and 3 vector and scalar filter impulse responses h_T.

Filter Name	Class 2	Class 3

Distortion Balance (w_s = Signal weighting, w_N = Noise weighting)

$$E\{sw_s^t\}\left(E\{rw_n^t - ks(w_n^t - w_s^t)\}\right)^{-1} \qquad \langle rw_s^t\rangle\left(\langle rw_s^t\rangle + (r - \langle r\rangle)w_n^t\right)^{-1}$$

Uncorrelated Estimation (Uncorrelated Wiener)

$$E\{ss^h\}\left(E\{ss^h + (r - ks)(r^h - ks^h)\}\right)^{-1} \qquad \langle r\rangle\langle r^h\rangle\left(\langle rr^h\rangle\right)^{-1}$$

Correlated Estimation (Correlated Wiener)

$$E\{s(r^h + (1-k)s^h)\}$$
$$\times\left(E\{(r + (1-k)s)(r^h + (1-k)s^h)\}\right)^{-1} \qquad \langle\langle r\rangle r^h\rangle\left(\langle rr^h\rangle\right)^{-1}$$

Transitional 1 Estimation (Uncorrelated Wiener $\alpha = 0$, Correlated Wiener $\alpha = 1$)

$$E\{ss^h + \alpha s(r^h - ks^h)\}$$
$$\times\Big(E\{ss^h + (r - ks)(r^h - ks^h)$$
$$\quad + \alpha\big(s(r^h - ks^h) + (r - ks)s^h\big)\}\Big)^{-1}$$
$$\left(\langle r\rangle\langle r^h\rangle + (1-\alpha)\langle r\rangle(r^h - \langle r^h\rangle)\right)\left((1-\alpha)\langle rr^h\rangle\right)^{-1}$$

Pulse-shaping Estimation (p = Desired pulse spectrum, Uncorrelated Wiener)

$$E\{p_d U s^h\}\left(E\{ss^h + (r - ks)(r^h - ks^h)\}\right)^{-1} \qquad p_d U\langle r^h\rangle\left(\langle rr^h\rangle\right)^{-1}$$

Pulse-shaping Estimation (p = Desired pulse spectrum, Correlated Wiener)

$$E\{p_d U(r^h + (1-k)s^h)\}$$
$$\times\left(E\{(r + (1-k)s)(r^h + (1-k)s^h)\}\right)^{-1} \qquad \langle p_d U r^h\rangle\left(rr^h\right)^{-1}$$

Classical Detection (Matched)

$$E\{u_0 s^h\}\left(E\{(r - ks)(r^h - ks^h)\}\right)^{-1} \qquad u_0\langle r^h\rangle\left((r - \langle r\rangle)(r^h - \langle r^h\rangle)\right)^{-1}$$

High-resolution Detection (Uncorrelated Wiener)

$$E\{u_0 s^h\}\left(E\{ss^h + (r - ks)(r^h - ks^h)\}\right)^{-1} \qquad u_0\langle r^h\rangle\left(\langle rr^h\rangle\right)^{-1}$$

Table 8.7.2   (Continued) Ideal Class 2 and 3 vector filter impulse responses h_T.

Filter Name	Class 2	Class 3

High-resolution Detection (Correlated Wiener)

$$E\{u_0(r^h + (1-k)s^h)\}$$
$$\times \left(E\{(r + (1-k)s)(r^h + (1-k)s^h)\} \right)^{-1} \qquad\qquad u_0\langle r^h\rangle \left(\langle rr^h\rangle \right)^{-1}$$

Inverse Detection (Deconvolution)

$$(u_0)_d \left(E\{s_d\} \right)^{-1} \qquad\qquad\qquad (u_0)_d \left(\langle r_d\rangle \right)^{-1}$$

Transitional 1 Detection (Inverse $\alpha = 0$, High-resolution $\alpha = 0.5$, Matched $\alpha = 1$)

$$E\{Us^h\} \left(E\{(1-\alpha)ss^h + \alpha((r-ks)(r^h - ks^h))\} \right)^{-1}$$
$$\langle Ur^h\rangle \left((1-\alpha)\langle r\rangle\langle r^h\rangle + \alpha(r - \langle r\rangle)(r^h - \langle r^h\rangle) \right)^{-1}$$

Transitional 2 Detection (Inverse $\alpha = 0.5$)

$$(u_0)_d \left(E\{s_d^{2(1-\alpha)}\} \right)^{-1} \qquad\qquad (u_0)_d \left(\langle r_d\rangle^{2(1-\alpha)} \right)^{-1}$$

Autocorrelation Estimation (Uncorrelated Wiener)

$$E\{ss^h Us^h\}$$
$$\times \left(E\{ss^h + (r-ks)(r^h - ks^h)\} \right)^{-1} \qquad\qquad \langle r\rangle\langle r^h\rangle U\langle r^h\rangle \left(\langle rr^h\rangle \right)^{-1}$$

Autocorrelation Estimation (Correlated Wiener)

$$E\{ss^h U(r^h + (1-k)s^h)\}$$
$$\times \left(E\{(r + (1-k)s)(r^h + (1-k)s^h)\} \right)^{-1} \qquad\qquad \langle r\rangle\langle r^h\rangle \left(r_d \right)^{-1}$$

Cross-Correlation Estimation (Uncorrelated Wiener)

$$E\{s(r^h - ks^h)Us^h\}$$
$$\times \left(E\{ss^h + (r-ks)(r^h - ks^h)\} \right)^{-1} \qquad\qquad \langle r\rangle\langle r^h\rangle U(r^h - \langle r^h\rangle) \left(\langle rr^h\rangle \right)^{-1}$$

Cross-Correlation Estimation (Correlated Wiener)

$$E\{s(r^h - ks^h)U(r^h + (1-k)s^h)\}$$
$$\times \left(E\{(r + (1-k)s)(r^h + (1-k)s^h)\} \right)^{-1} \qquad\qquad \langle r\rangle(r^h - \langle r^h\rangle) \left(r_d \right)^{-1}$$

$$H_T = E\{Tss^hT^h\} \left(E\{Tss^hT^h + Tnn^hT^h\}\right)^{-1}$$

$$= TE\{ss^h\}T^h \left(TE\{ss^h + nn^h\}T^h\right)^{-1}$$

$$= TE\{ss^h\} \, T^h(T^h)^{-1} \left(E\{ss^h + nn^h\}\right)^{-1}T^{-1} \tag{8.7.3}$$

$$= TE\{ss^h\} \left(E\{ss^h + nn^h\}\right)^{-1}T^{-1}$$

Converting H_T to h_T using Eq. 8.7.1 gives the impulse response of the Class 1 uncorrelated Wiener estimation filter as

$$h_T = T^{-1}H_TT = T^{-1}T \, E\{ss^h\} \left(E\{ss^h + nn^h\}\right)^{-1} T^{-1}T$$

$$= E\{ss^h\} \left(E\{ss^h + nn^h\}\right)^{-1} \tag{8.7.4}$$

which is independent of any T transform.

Now consider the Class 1 high-resolution detection filter having

$$H_T = E\{Tu_0S^h\} \left(E\{SS^h + NN^h\}\right)^{-1} \tag{8.7.5}$$

Proceeding as before gives

$$H_T = E\{Tu_0s^hT^h\} \left(E\{T(ss^h + nn^h)T^h\}\right)^{-1} = TE\{u_0s^h\} \left(E\{ss^h + nn^h\}\right)^{-1}T^{-1}$$

$$h_T = T^{-1}H_TT = E\{u_0s^h\} \left(E\{ss^h + nn^h\}\right)^{-1} \tag{8.7.6}$$

Following this procedure converts the gain matrices of Tables 8.4.1 and 8.4.2 into the impulse response matrices of Tables 8.7.1 and 8.7.2. The time domain filter of Fig. 8.1.1c requires N^2 operations (multiplications and additions) to obtain output $c(n)$ and no fast transforms. The frequency domain filter of Fig. 8.1.1a also requires N^2 operations plus $N\log_2 N$ operations to perform the T and T^{-1} transforms. Therefore the time domain approach is often faster than the frequency domain approach.

8.8 FAST TRANSFORM SPECTRA

A number of fast transform spectra have been used to compute filter gains in this chapter. Therefore to increase understanding of the gains obtained, we present a collection of fast transform spectra in Table 8.8.1. The transforms considered are Fourier, Paley, Haar, and slant. This table shows that spectra become narrowband when the signal shape is close to the basis functions of the transform. The spectra become impulsive with a one-bin (or more) bandwidth when the signal matches one (or more) of the basis functions. Thus, frequency domain algorithms (in Tables 8.2.1, 8.2.2, 8.4.1, and 8.4.2) must be properly matched to signal and noise spectra. Likewise, time domain algorithms (Tables 8.7.1 and 8.7.2) must be properly matched to signal and noise functions.

EXAMPLE 8.6.1 Consider step, exponential, pulse, pulsed-cosine, and sawtooth signals. Discuss their spectra using Fourier, Paley, Haar, and slant transforms.

SOLUTION Assume that the sampling window covers $0 \leq t < NT$. The sampled signal is replicated on both sides of this window so that it becomes periodic over $-\infty < t < \infty$. Sampling a step signal and replicating the result produces a dc signal. Using the Paley, Haar, and slant basis functions shown in Fig. 2.8.1 and sine-cosine functions for Fourier, a dc signal corresponds to the first basis function. Therefore the spectrum consists of an impulse at bin 1 and zero elsewhere regardless of which transform is used. Thus the step signal has only a one-bin bandwidth.

The exponential signal is shown as entry 6 in Table 8.8.1. Since it is most closely approximated by the sawtooth basis function of the slant transform in Fig. 2.8.1, the slant transform spectrum has the highest asymptotic roll-off.

The pulse signal is shown in entry 2. For short pulse duration, the pulse-type basis functions of the Haar transform in Fig. 2.8.1 give the best approximation. Therefore the Haar spectrum has the fewest components. The Fourier transform has the largest spectral bandwidth. These comments also apply to the pulsed-cosine signal of entry 8.

The sawtooth signal is shown in entry 5. Since this is one of the slant basis functions, the slant spectrum has only a one-bin bandwidth. The Haar spectrum consists of only a few discrete components.   ∎

PROBLEMS

8.1   Design a Class 1 FFT scalar filter to process a step signal having $E\{s(t)\} = U_{-1}(t)$ which is contaminated by uncorrelated white noise having $E\{|N(j\omega)|\} = 1$. Use sample time $T = 0.5$ and $N = 4$ points. Write the gain matrices H of the (a) Wiener estimation filters, (b) matched, high-resolution, and inverse detection filters, (c) autocorrelation and cross-correlation filters. (d) Plot their magnitude responses. (e) Find the filter outputs assuming $n(t) = [1, -1, 1, -1]^t$.

8.2   Design a Class 1 FFT scalar filter to process an exponential step signal having $E\{s(t)\} = e^{-t} U_{-1}(t)$ which is contaminated by uncorrelated white noise having $E\{|N(j\omega)|\} = 1$. Repeat parts (a)–(e) of Problem 8.1.

8.3   Design a Class 1 FFT scalar filter to process a signal pulse having $E\{s(t)\} = U_{-1}(t + NT/4) - U_{-1}(t - NT/4)$ (unit amplitude, length $NT/2$, centered at $t = 0$) which is contaminated by uncorrelated white noise having $E\{|N(j\omega)|\} = 1$. Repeat parts (a)–(e) of Problem 8.1.

8.4   Design a Class 1 FFT scalar filter to process a pulsed-cosine signal having $E\{s(t)\} = \cos(\omega_s t/3)[U_{-1}(t + NT/4) - U_{-1}(t - NT/4)]$ (peak amplitude = 1, frequency = $f_s/3$ Hz, pulse duration = $NT/2$ sec) which is contaminated by uncorrelated white noise having $E\{|N(j\omega)|\} = 1$. Repeat parts (a)–(e) of Problem 8.1.

8.5   Design a Class 1 FFT scalar filter to process a sawtooth signal having $E\{s(t)\} = [1 - (t/NT)]U_{-1}(t)$ which is contaminated by uncorrelated white noise having $E\{|N(j\omega)|\} = 1$. Repeat parts (a)–(e) of Problem 8.1.

8.6   Design Class 3 FFT scalar filters to process a step signal having $E\{s(t)\} = U_{-1}(t)$ which is contaminated by uncorrelated white noise having $E\{|N(j\omega)|\} = 1$ and random phase arg $N(j\omega)$ between $(-\pi, \pi)$. (a) Smooth (inner block) the expected signal and noise power

Table 8.8.1 Table of Fourier, Paley, Haar, and slant fast transform spectra.

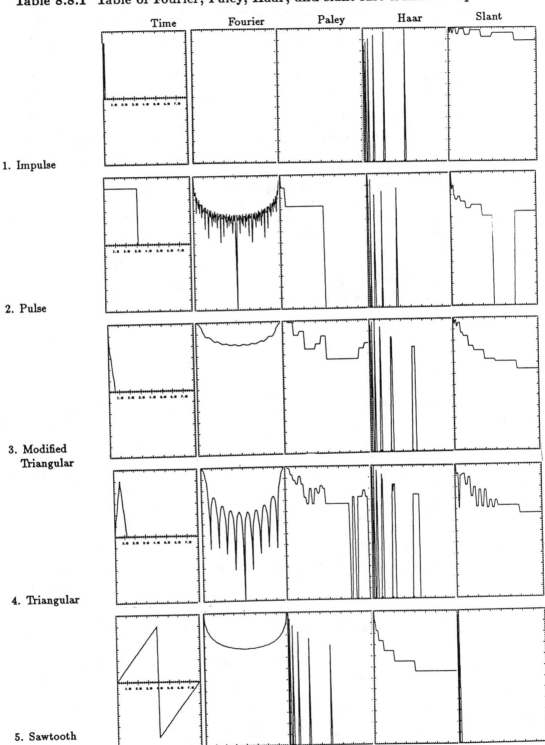

Table 8.8.1 (Continued) Table of Fourier, Paley, Haar, and slant fast transform spectra.

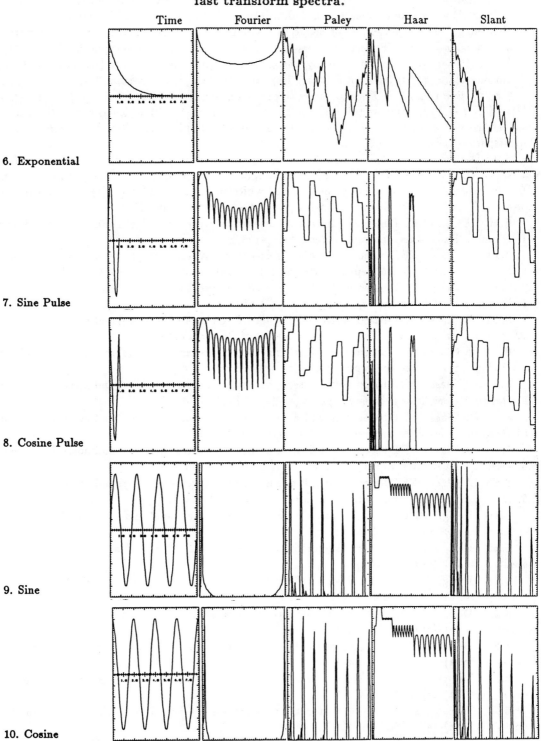

spectra with rectangular, triangular, and Hann windows. Let $T = 1$ and smoothing width $p = 10\%$. Assume $R(j\omega) = E\{S(j\omega)\}$ $+E\{N(j\omega)\}$. (b) Write the gain matrices of Class 2 uncorrelated Wiener estimation, matched detection, and high-resolution detection filters. (c) Plot their magnitude responses. (d) How do they compare with their Class 1 filter counterparts?

8.7   Design Class 3 FFT scalar filters to process an exponential step signal having $E\{s(t)\}$ $= e^{-t}$ $U_{-1}(t)$ which is contaminated by uncorrelated white noise having $E\{|N(j\omega)|\} = 1$ and random phase arg $N(j\omega)$ between $(-\pi, \pi)$. Repeat parts (a)–(d) of Problem 8.6.

8.8   Design Class 3 FFT scalar filters to process a signal pulse having $E\{s(t)\}$ $= U_{-1}(t + T/2) - U_{-1}(t - T/2)$ (unit amplitude, length T, centered at $t = 0$) which is contaminated by uncorrelated white noise having $E\{|N(j\omega)|\} = 1$ and random phase arg $N(j\omega)$ between $(-\pi, \pi)$. Repeat parts (a)–(d) of Problem 8.6.

8.9   Design Class 3 FFT scalar filters to process a pulsed-cosine signal having $E\{s(t)\}$ $= \cos(2\pi t)[U_{-1}(t + 0.1) - U_{-1}(t - 0.1)]$ (peak amplitude $= 1$, frequency $= 1$ Hz, pulse duration $= 0.2$ sec) which is contaminated by uncorrelated white noise having $E\{|N(j\omega)|\} = 1$ and random phase arg $N(j\omega)$ between $(-\pi, \pi)$. Repeat parts (a)–(e) of Problem 8.6.

8.10   A Class 1 FFT-based vector filter is needed to process the signal and noise in Problem 8.1. Design a Paley filter and repeat parts (a)–(e) of Problem 8.1.

8.11   A Class 1 FFT-based vector filter is needed to process the signal and noise in Problem 8.1. Design a Haar filter and repeat parts (a)–(e) of Problem 8.1.

8.12   A Class 1 FFT-based vector filter is needed to process the signal and noise in Problem 8.1. Design a slant filter and repeat parts (a)–(e) of Problem 8.1.

8.13   A Class 1 FFT-based vector filter is needed to process the signal and noise in Problem 8.2. Design a Paley filter and repeat parts (a)–(e) of Problem 8.1.

8.14   A Class 1 FFT-based vector filter is needed to process the signal and noise in Problem 8.2. Design a Haar filter and repeat parts (a)–(e) of Problem 8.1.

8.15   A Class 1 FFT-based vector filter is needed to process the signal and noise in Problem 8.2. Design a slant filter and repeat parts (a)–(e) of Problem 8.1.

8.16   A Class 1 FFT-based vector filter is needed to process the signal and noise in Problem 8.5. Design a Paley filter and repeat parts (a)–(e) of Problem 8.1.

8.17   A Class 1 FFT-based vector filter is needed to process the signal and noise in Problem 8.5. Design a Haar filter and repeat parts (a)–(e) of Problem 8.1.

8.18   A Class 1 FFT-based vector filter is needed to process the signal and noise in Problem 8.5. Design a slant filter and repeat parts (a)–(e) of Problem 8.1.

8.19   Design a Paley Class 1 vector filter to process the signal and noise in Problem 8.1. Repeat parts (a)–(e) of Problem 8.1.

8.20   Design a Haar Class 1 vector filter to process the signal and noise in Problem 8.1. Repeat parts (a)–(e) of Problem 8.1.

8.21   Design a slant Class 1 vector filter to process the signal and noise in Problem 8.1. Repeat parts (a)–(e) of Problem 8.1.

8.22   Design a Fourier Class 1 vector filter to process the signal and noise in Problem 8.1. Repeat parts (a)–(e) of Problem 8.1.

8.23   Design a Paley Class 1 vector filter to process the signal and noise in Problem 8.2. Repeat parts (a)–(e) of Problem 8.1.

8.24   Design a Haar Class 1 vector filter to process the signal and noise in Problem 8.2. Repeat parts (a)–(e) of Problem 8.1.

8.25   Design a slant Class 1 vector filter to process the signal and noise in Problem 8.2. Repeat parts (a)–(e) of Problem 8.1.

8.26 Design a Fourier Class 1 vector filter to process the signal and noise in Problem 8.2. Repeat parts (a)–(e) of Problem 8.1.

8.27 Design a Paley Class 1 vector filter to process the signal and noise in Problem 8.5. Repeat parts (a)–(e) of Problem 8.1.

8.28 Design a Haar Class 1 vector filter to process the signal and noise in Problem 8.5. Repeat parts (a)–(e) of Problem 8.1.

8.29 Design a slant Class 1 vector filter to process the signal and noise in Problem 8.5. Repeat parts (a)–(e) of Problem 8.1.

8.30 Design a Fourier Class 1 vector filter to process the signal and noise in Problem 8.5. Repeat parts (a)–(e) of Problem 8.1.

8.31 Design a Paley Class 3 vector filter to process the signal and noise in Problem 8.6. Repeat parts (a)–(d) of Problem 8.6.

8.32 Design a Haar Class 3 vector filter to process the signal and noise in Problem 8.6. Repeat parts (a)–(d) of Problem 8.6.

8.33 Design a slant Class 3 vector filter to process the signal and noise in Problem 8.6. Repeat parts (a)–(d) of Problem 8.6.

8.34 Design a Fourier Class 3 vector filter to process the signal and noise in Problem 8.6. Repeat parts (a)–(d) of Problem 8.6.

8.35 Design a Paley Class 3 vector filter to process the signal and noise in Problem 8.7. Repeat parts (a)–(d) of Problem 8.6.

8.36 Design a Haar Class 3 vector filter to process the signal and noise in Problem 8.7. Repeat parts (a)–(d) of Problem 8.6.

8.37 Design a slant Class 3 vector filter to process the signal and noise in Problem 8.7. Repeat parts (a)–(d) of Problem 8.6.

8.38 Design a Fourier Class 3 vector filter to process the signal and noise in Problem 8.7. Repeat parts (a)–(d) of Problem 8.6.

8.39 A Paley Class 1 scalar filter is needed to process the signal and noise in Problem 8.1. Repeat parts (a)–(e) of Problem 8.1.

8.40 A Haar Class 1 scalar filter is needed to process the signal and noise in Problem 8.1. Repeat parts (a)–(e) of Problem 8.1.

8.41 A slant Class 1 scalar filter is needed to process the signal and noise in Problem 8.1. Repeat parts (a)–(e) of Problem 8.1.

8.42 A Paley Class 1 scalar filter is needed to process the signal and noise in Problem 8.2. Repeat parts (a)–(e) of Problem 8.1.

8.43 A Haar Class 1 scalar filter is needed to process the signal and noise in Problem 8.2. Repeat parts (a)–(e) of Problem 8.1.

8.44 A slant Class 1 scalar filter is needed to process the signal and noise in Problem 8.2. Repeat parts (a)–(e) of Problem 8.1.

8.45 A Paley Class 1 scalar filter is needed to process the signal and noise in Problem 8.5. Repeat parts (a)–(e) of Problem 8.1.

8.46 A Haar Class 1 scalar filter is needed to process the signal and noise in Problem 8.5. Repeat parts (a)–(e) of Problem 8.1.

8.47 A slant Class 1 scalar filter is needed to process the signal and noise in Problem 8.5. Repeat parts (a)–(e) of Problem 8.1.

8.48 A Paley Class 3 scalar filter is needed to process the signal and noise in Problem 8.6. Repeat parts (a)–(d) of Problem 8.6.

8.49 A Haar Class 3 scalar filter is needed to process the signal and noise in Problem 8.6. Repeat parts (a)–(d) of Problem 8.6.

8.50   A slant Class 3 scalar filter is needed to process the signal and noise in Problem 8.6. Repeat parts (a)–(d) of Problem 8.6.

8.51   A Paley Class 3 scalar filter is needed to process the signal and noise in Problem 8.7. Repeat parts (a)–(d) of Problem 8.6.

8.52   A Haar Class 3 scalar filter is needed to process the signal and noise in Problem 8.7. Repeat parts (a)–(d) of Problem 8.6.

8.53   A slant Class 3 scalar filter is needed to process the signal and noise in Problem 8.7. Repeat parts (a)–(d) of Problem 8.6.

8.54   Design a Class 1 K-L scalar filter to process the signal and noise in Problem 8.1. Repeat parts (a)–(e) of Problem 8.1.

8.55   Design a Class 1 K-L scalar filter to process the signal and noise in Problem 8.2. Repeat parts (a)–(e) of Problem 8.1.

8.56   Design a Class 1 K-L scalar filter to process the signal and noise in Problem 8.3. Repeat parts (a)–(e) of Problem 8.1.

8.57   Design a Class 1 K-L scalar filter to process the signal and noise in Problem 8.4. Repeat parts (a)–(e) of Problem 8.1.

8.58   Design a Class 1 K-L scalar filter to process the signal and noise in Problem 8.5. Repeat parts (a)–(e) of Problem 8.1.

8.59   Design a Class 3 K-L scalar filter to process the signal and noise in Problem 8.6. Repeat parts (a)–(d) of Problem 8.6.

8.60   Design a Class 3 K-L scalar filter to process the signal and noise in Problem 8.7. Repeat parts (a)–(d) of Problem 8.6.

8.61   Design a Class 3 K-L scalar filter to process the signal and noise in Problem 8.8. Repeat parts (a)–(d) of Problem 8.6.

8.62   Design a Class 3 K-L scalar filter to process the signal and noise in Problem 8.9. Repeat parts (a)–(d) of Problem 8.6.

REFERENCES

1. Haas, W.H. & C.S. Lindquist, "An approach to the frequency domain design of FIR sampled-data filters," Proc. IEEE Intl. Symp. Circuits and Systems, pp. 21–23, 1979.

2. Lindquist, C.S. & H.E. Lin, "Design of Class 1 estimation filters using generalized transforms," Proc. 29th Midwest Symp. Circuits and Systems, 1986.
   —, "Design of Class 1 detection filters using generalized transforms," Rec. 20th Asilomar Conf. Circuits, Systems, and Computers, pp. 596–599, 1986.
   —, "Improved Class 1 detection filters using generalized transforms," Proc. IEEE Intl. Symp. Circuits and Circuits, pp. 660–663, 1987.

3. Hotelling, H., "Analysis of a complex function of statistical variables into principal components," J. Educ. Psychol., vol. 24, pp. 417–441; 498–520, 1933.
   Karhunen, K., "Uber lineare methoden in der wahrscheinlichkeitsrechnung," Ann. Acad. Sci. Fennicae, serial A137, 1947.
   Loéve, M., "Fonctions aléatoires de second ordre," in P. Lévy, it Processus Stochastiques et Mouvement Brownien, Hermann, Paris, France, 1948.

4. Almeer, Meer K., "Theory and design of digital filters using Karhunen-Loeve transform," M.S.E.E. Thesis, Univ. of Miami, 118 p., June 1987.

5. Gonzalez, R.C. and P. Wintz, *Digital Image Processing*, 2nd ed., pp. 122–125, Addison-Wesley, Reading, MA, 1987.

9 FFT FILTER DESIGN

And he (Jesus) said unto them ... But ye shall receive power,
after that the Holy Ghost is come upon you;
and ye shall be witnesses unto me ...
unto the uttermost part of the earth.
And when he had spoken these things, while they beheld
he was taken up; and a cloud received him out of their sight.

Acts 1:7–9

The general theory of fast transform filters was developed in the last chapter. This chapter applies these concepts to the design of FFT scalar filters. Our objective is to illustrate how a variety of specialized scalar filter systems can be implemented using the FFT. We are not concerned with designing the most efficient or fastest implementations nor with designing the other fast transform filter types in Table 8.1.1. The FFT scalar filters which are presented can easily be converted into these other forms.

This chapter begins with a discussion of the general structure of FFT scalar filters. Then a variety of design examples using different FFT algorithms will be presented. These will be adaptive filters using many of the algorithms introduced in Tables 8.2.1 and 8.2.2. Finally the speed and through-put rate of some FFT scalar filters will be discussed. Since the adaptive filters of Chaps. 4.4–4.9 and 8.2 will be used so extensively, the engineer should review that material before progressing deeply into this chapter.

FFT filters are usually implemented using computer software. However, new customized FFT hardware modules can be used to speed up FFT computations allowing increased sampling rate. FFT filters generally operate at lower speeds than other fast transform filters which use simpler basis functions. Of course, all the filters of this chapter can be reformulated into the time domain forms of Tables 8.6.1 and 8.6.2 to eliminate transform computations altogether and to possibly achieve increased speed.

9.1 FFT FILTER STRUCTURES

FFT digital filters take the two basic forms shown in Fig. 9.1.1. They require synthesis of an FFT filter transfer function $H(m)$ or impulse response $h(n)$. In nonadaptive systems, $H(m)$ might be a classical filter like Butterworth, Chebyshev, elliptic, or Bessel as described in Chap. 4.2. $H(m)$ might also be an optimum filter obtained using one of the optimization methods discussed in Chap. 4.3. The classical and optimum transfer functions are converted from analog to digital using the transforms of Chaps. 5 and 6. In adaptive systems however, $H(m)$ is synthesized using an ideal filter transfer function

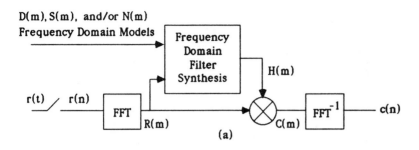

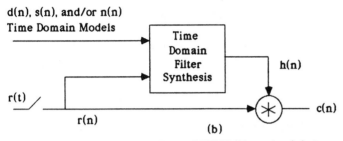

Fig. 9.1.1 Implementation of FFT filter in (a) frequency domain and (b) time domain. (Nonfeedback forms.)

listed in Tables 8.2.1 and 8.2.2. Depending upon apriori assumptions about the signal and noise being processed, these transfer functions take on different forms.

FFT filters designed in the frequency domain have the form shown in Fig. 9.1.1a. These filters operate by evaluating a spectral product as

$$c(n) = \text{FFT}^{-1}\big[C(m)\big] = \text{FFT}^{-1}\big[H(m)R(m)\big] \tag{9.1.1}$$

The total input $r(n)$ is converted to an input spectrum $R(m)$ by the FFT block. This spectrum in conjunction with apriori information about the signal spectrum $S(m)$ and noise spectrum $N(m)$ are processed to form an equivalent transfer function $H(m)$. Then the input spectrum $R(m)$ is multiplied by $H(m)$ to form the output spectrum $C(m)$. Passing $C(m)$ through an inverse FFT block (denoted as FFT^{-1}) produces the time domain output $c(n)$. The filter transfer function may be updated with every input sample or only as often as needed. This depends upon the stationarity characteristics of the input signal $r(n)$. Signals having slowly varying or long stationarity periods need be updated much less frequently than those having short periods.

FFT filters designed in the time domain have the form shown in Fig. 9.1.1b. These filters operate by evaluating a convolutional product as

$$c(n) = r(n) * h(n) \tag{9.1.2}$$

This equation results by replacing frequency multiplication by time convolution in Eq. 9.1.1. The time domain approach is no more complex than the frequency domain approach because the impulse response $h(n)$ can be evaluated directly using Tables

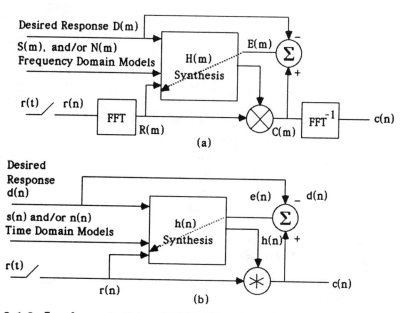

Fig. 9.1.2 Implementation of FFT filter in (a) frequency domain
and (b) time domain. (Feedback forms.)

8.6.1 and 8.6.2 with apriori information about the signal $s(n)$ and noise $n(n)$. The
convolutional product can be performed in software or in hardware using nonrecursive
or recursive filters to be discussed in Chaps. 10 and 11, respectively. Eq. 9.1.2 can also
be computed directly using matrix forms (see Eq. 8.1.4).

In the realizations of Fig. 9.1.1, the FFT filters are periodically resynthesized or
updated to be optimum for the measured inputs. Alternatively the FFT filters might be
connected in a feedback loop to seek the optimum as shown in Fig. 9.1.2. These filters
function the same as those in Fig. 9.1.1 except that $H(m)$ or $h(n)$ is synthesized according to some algorithm related to the spectral error $E(m)$ or time domain error $e(n)$. The
error $e(n)$ is defined to equal the difference between the system output $c(n)$ and some desired output $d(n)$ or their spectra.

9.2 FFT ESTIMATION FILTERS [1]

One major class of problems involves estimating a signal embedded in additive noise
using the general adaptive linear filter in Fig. 8.2.1. As discussed in Chap. 8.2.2, the
general estimation transfer function $H(m)$ equals

$$H(m) = \frac{E\{D(m)R^*(m)\}}{E\{R(m)R^*(m)\}}$$

$$= \frac{\text{Desired input output cross correlation spectrum}}{\text{Input autocorrelation spectrum}}$$

(9.2.1)

from Eq. 8.2.13. Assuming that signal is corrupted by additive noise and that signal
must be estimated, then

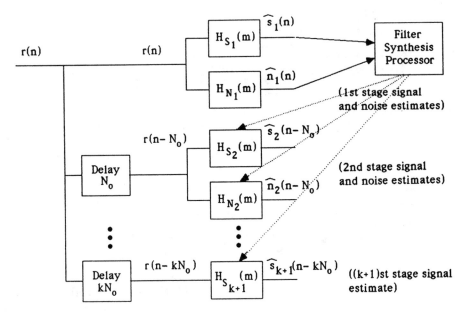

Fig. 9.2.1 Block diagram of an aposteriori estimation filter.

$$R(m) = S(m) + N(m)$$
$$D(m) = S(m)$$

(9.2.2)

where $S(m)$ and $N(m)$ are the signal and noise spectra, respectively. Substituting Eq. 9.2.2 into Eq. 9.2.1 yields a correlated Wiener filter having the gain

$$H(m) = \frac{E\{|S(m)|^2\} + E\{S(m)N^*(m)\}}{E\{|S(m)|^2\} + E\{|N(m)|^2\} + E\{2\,\mathrm{Re}[S(m)N^*(m)]\}}$$

(9.2.3)

(Eq. 8.2.21). If the signal and noise (zero mean) are assumed to be uncorrelated apriori, then their expected spectral product $E\{S(m)N^*(m)\}$ is equal to zero. In this case, the optimum filter reduces to the uncorrelated Wiener filter of Eq. 8.2.22 where

$$H(m) = \frac{E\{|S(m)|^2\}}{E\{|S(m)|^2\} + E\{|N(m)|^2\}}$$

(9.2.4)

The behavior of this filter and others was shown using an illustrative example in Chap. 4.6. This FFT estimation filter example should be reviewed before proceeding further.

In deriving Eq. 9.2.4, the assumption that signal and noise are uncorrelated is made simply because the relative phases between the signal and noise are unknown apriori. However once the signal and noise is estimated for some specific time interval, then the cross-spectral density terms can be estimated aposteriori to further improve signal estimates. One possible adaptive configuration is shown in Fig. 9.2.1.

This configuration consists of a series of estimation filter stages in parallel each delayed from the preceding stage by delay N_o. By stage-wise delaying the input signal $r(n)$, cross-correlation terms can be evaluated and used to reformulate the optimum transfer function for the next stage. Successive refiltering of the input allows the signal estimates to be improved.

To apply Eq. 9.2.3 in succeeding stages, both signal $S(m)$ and noise $N(m)$ must be estimated. Two uncorrelated Class 1 Wiener filters can be used in the first stage to estimate $S(m)$ and $N(m)$, respectively, using apriori expected spectra information as

$$H_{S_1}(m) = \frac{E\{|S(m)|^2\}}{E\{|S(m)|^2\} + E\{|N(m)|^2\}}$$

$$H_{N_1}(m) = \frac{E\{|N(m)|^2\}}{E\{|S(m)|^2\} + E\{|N(m)|^2\}}$$

(9.2.5)

The first-stage spectral estimates of the signal $s_1(n)$ and noise $n_1(n)$ are

$$S_1(m) = H_{S_1}(m) R(m)$$
$$N_1(m) = H_{N_1}(m) R(m)$$

(9.2.6)

These outputs can next be used to estimate the second-stage transfer functions. Assuming now that correlated Wiener filters are used, these transfer functions equal

$$H_{S_2}(m) = \frac{E\{|S(m)|^2\} + E\{S(m)\}N_1^*(m)}{E\{|S(m)|^2\} + |N_1(m)|^2 + 2\,\text{Re}[E\{S(m)\}N_1^*(m)]}$$

$$H_{N_2}(m) = \frac{|N_1(m)|^2 + E\{S^*(m)\}N_1(m)}{E\{|S(m)|^2\} + |N_1(m)|^2 + 2\,\text{Re}[E\{S(m)\}N_1^*(m)]}$$

(9.2.7)

Notice that the actual noise N_1 spectrum is used in Eq. 9.2.7 rather than the expected spectrum as was done in Eq. 9.2.3. The second-stage spectral estimates of the signal $s_2(n)$ and noise $n_2(n)$ are

$$S_2(m) = H_{S_2}(m) R(m)$$
$$N_2(m) = H_{N_2}(m) R(m)$$

(9.2.8)

This process can be carried out indefinitely. However little improvement results using additional stages as we shall see in a moment.

The filter shown in Fig. 9.2.1 is fairly general. Other algorithms can be used in the blocks. Detection and estimation filters can be combined with nonlinear logic to further improve the signal estimate in each stage. It should be pointed out that $s(n - kN_o)$ and $n(n - kN_o)$ is applied to the stage of this system – not $s(n)$ and $n(n)$. This maintains the synchronization between inputs and filters so the cross-correlation is properly estimated and maintained from stage-to-stage.

To illustrate the results obtained using this approach, consider the signal and noise shown in Fig. 9.2.2. The signal consists of a pulse of length 0.3 seconds beginning at $t = 2.4$ seconds. The noise has a $1/f^2$ power spectrum at low frequencies and a flat spectrum at high frequencies. The total input is shown in Fig. 9.2.2c where the abrupt signal edges make the signal somewhat apparent. The signal and noise spectra are similiar to those shown in Figs. 4.6.1b (use only a single $(\sin x)/x$ term) and 4.6.1c,

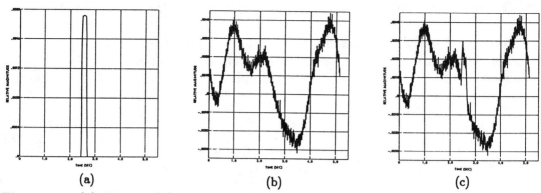

Fig. 9.2.2 (a) Signal, (b) noise, and (c) total input. (From W.H. Haas and C.S. Lindquist, "Least-squares estimation filtering using an adaptive technique based on aposteriori cross-spectral density functions," Rec. 12th Asilomar Conf. Circuits, Systems, and Computers, pp. 273-274, ©1978, IEEE.)

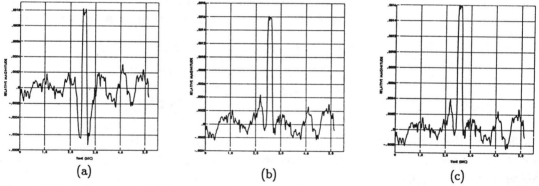

Fig. 9.2.3 Output of (a) first stage, (b) second stage, and (c) fourth stage of aposteriori estimation filter. (From W.H. Haas and C.S. Lindquist, loc. cit., ©1978, IEEE.)

respectively. The peak SNR is arbitrarily selected to equal about $1 \cong 0$ dB. The rms SNR is around -20 dB. Performing an FFT on the apriori signal $s(n)$ and noise $n(n)$ generates the $E\{S(m)\}$ and $E\{N(m)\}$ spectra to be substituted into Eq. 9.2.5. This produces optimum transfer functions H_{S_1} and H_{N_1} whose spectral outputs are computed using Eq. 9.2.6. The output $s_1(n)$ of H_{S_1} is shown in Fig. 9.2.3a and results by taking the inverse FFT of $S_1(m)$. Fig. 9.2.3a shows that a large amount of the noise in the total input of Fig. 9.2.2c has been filtered out by H_{S_1}. The peak SNR is about 9 dB so the peak SNR gain equals 9 dB and considerable improvement has been obtained.

The N_1 spectrum is substituted into Eq. 9.2.7 to produce the second-stage transfer functions H_{S_2} and H_{N_2}. Their spectral outputs S_2 and N_2 are given by Eq. 9.2.8. Taking the inverse FFT of $S_2(m)$ yields $s_2(n)$ shown in Fig. 9.2.3b. The noise has been further reduced and the peak SNR is about 12 dB so a little further improvement has been obtained.

This process is repeated until the fourth-stage transfer functions are obtained and the output $s_4(n)$ computed which is shown in Fig. 9.2.3c. There is negligible improvement over $s_2(n)$ shown in Fig. 9.2.3b. Clearly the additional stages were unnecessary in this example.

9.3 FFT ESTIMATION FILTERS USING ADAPTIVE LINE ENCHANCERS (ALE) AND NOISE CANCELLERS (ANC)[2]

Now we shall present other estimation filters called the adaptive line enhancer (ALE) and the adaptive noise canceller (ANC) which extract and smooth signals from noisy sources. These systems were originally described using time domain estimation algorithms developed by Widrow and others.[3] Here we present generalized ALE and ANC frequency domain algorithms.

The ALE and single-input ANC systems are shown in Fig. 9.3.1. The ALE functions best when estimating a narrowband signal from an input which consists primarily of broadband noise. The ANC is a system which estimates the signal by estimating the noise. It then subtracts it from the total input to form the signal estimate.

Compare Fig. 9.3.1 with the general signal processing system shown in Fig. 8.2.1 whose estimation and detection algorithms are given in Tables 8.2.1 and 8.2.2. Since Figs. 8.2.1 and 9.3.1 are identical, we see that the ALE gain equals

$$H_{ALE}(m) = H(m) \tag{9.3.1}$$

while the ANC gain equals

$$H_{ANC}(m) = 1 - H(m) = 1 - H_{ALE}(m) \tag{9.3.2}$$

Eq. 9.3.1 shows that the ALE transfer functions are identical to those already discussed in Tables 8.2.1 and 8.2.2. Eq. 9.3.2 shows that the ANC transfer functions differ from the ALE transfer function by unity. In other words, H_{ALE} and H_{ANC} are complementary. The H_{ANC} transfer functions are given in Table 9.3.1. For example the Class 1 uncorrelated Wiener filters have

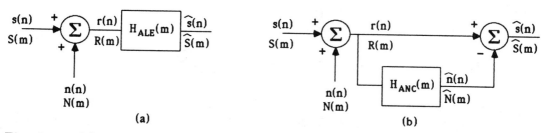

Fig. 9.3.1 (a) ALE and (b) one-channel ANC filters.

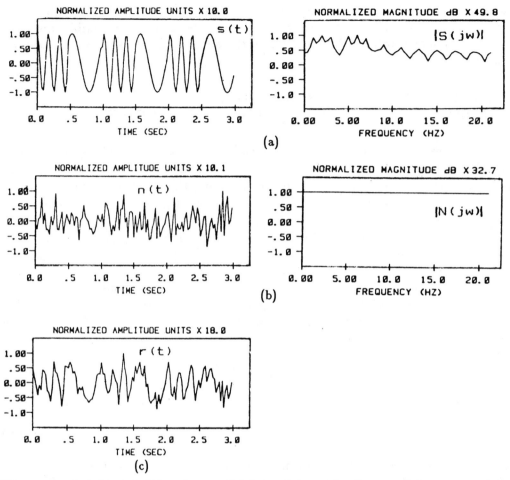

Fig. 9.3.2 (a) Signal, (b) noise, and (c) total input in time and frequency domains. (From C.S. Lindquist, K.C. Severence, and H.B. Rozran, "Frequency domain algorithms for adaptive line enhancer and adaptive noise canceller systems," Rec. 18th Asilomar Conf. Circuits, Systems, and Computers, pp. 233–237, ©1984, IEEE.)

$$H_{ALE}(m) = \frac{E\{|S(m)|^2\}}{E\{|S(m)|^2\} + E\{|N(m)|^2\}}$$

$$H_{ANC}(m) = 1 - H_{ALE}(m) = \frac{E\{|N(m)|^2\}}{E\{|S(m)|^2\} + E\{|N(m)|^2\}}$$

$$(9.3.3)$$

To illustrate how the ALE and ANC systems function, consider the following example described in Figs. 9.3.2–9.3.3. A signal composed of 1.9 and 5.9 Hz frequencies and its spectrum is shown in Fig. 9.3.2a. Noise is assumed to be white as shown in Fig. 9.3.2b. The total input is shown in Fig. 9.3.2c where the peak SNR has been adjusted to about 0 dB.

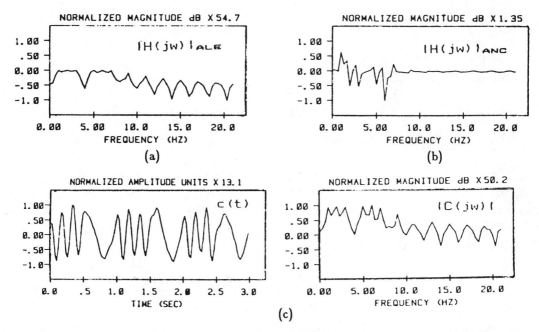

Fig. 9.3.3 Magnitude responses of (a) ALE and (b) ANC and (c) filter outputs. (From C.S. Lindquist, K.C. Severence, and H.B. Rozran, loc. cit., ©1984, IEEE.)

The ALE and ANC transfer functions are computed using Eq. 9.3.3 and are shown in Figs. 9.3.3a and 9.3.3b, respectively. In this white noise case where $E\{|N(m)|^2\} = 1$,

$$H_{ALE} \cong 1 \ \text{ or } \ H_{ANC} \cong \frac{1}{E\{|S(m)|^2\}}, \quad \text{for} \ \ E\{|S(m)|\} \gg E\{|N(m)|\}$$

$$H_{ALE} \cong E\{|S(m)|^2\} \ \text{ or } \ H_{ANC} \cong 1, \quad \text{for} \ \ E\{|S(m)|\} \ll E\{|N(m)|\}$$

(9.3.4)

We see that H_{ALE} is unity where the spectral SNR is high (around 1.9 and 5.9 Hz) and equal to $E\{|S(m)|^2\}$ at frequencies where the spectral SNR is low (everywhere else). H_{ANC} is unity where the spectral SNR is low (everywhere but around 1.9 and 5.9 Hz) and proportional to $1/E\{|S(m)|^2\}$ where the spectral SNR is high (around 1.9 and 5.9 Hz). Both filter outputs are identical and are shown in Fig. 9.3.3c.

The more general ANC system shown in Fig 9.3.4 has two inputs. One input is the total input $r(n)$ composed of signal $s(n)$ plus noise $n(n)$. The other input is a reference noise $n_o(n)$. This system estimates the signal $s(n)$ by first estimating the noise $\widehat{n(n)}$ from the reference noise $n_o(n)$. It then subtracts it from the total input $r(n) = s(n) + n(n)$ to form $\widehat{s(n)} = s(n) + [n(n) - \widehat{n(n)}]$. To be effective, the reference noise must be correlated with the input noise and must be uncorrelated with the desired signal. The more the noise is uncorrelated with the signal $s(n)$, the better is the signal estimate $\widehat{s(n)}$.

Suppose that $n(n)$ and $n_o(n)$ are related as $n_o(n) = h_o(n) * n(n)$ or $N_o(m) = H_o(m)N(m)$ where $H_o(m)$ is some unknown transfer function. Fig. 9.3.4 shows that

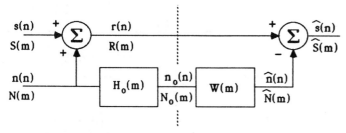

Fig. 9.3.4 Two-channel ANC filter.

$\widehat{n(n)} = n(n)$ gives perfect noise cancellation. This requires $H_o(m)W(m) = 1$ so $W(m) = 1/H_o(m)$, but $H_o(m)$ is unknown. To determine $W(m)$, we use the general estimation filter having $H_E(m) = E\{D(m)R^*(m)\}/E\{R(m)R^*(m)\}$ from Table 8.2.1. Applying this to $W(m)$, $N_o(m)$ becomes the filter input $R(m)$ and $N(m)$ the desired output $D(m)$. Then the noise filter gain equals

$$W(m) = \frac{E\{N(m)N_o^*(m)\}}{E\{|N_o(m)|^2\}} \simeq \frac{N_o^*(m)E\{N(m)\}}{|N_o(m)|^2} \qquad (9.3.5)$$

which is approximately a Class 2 filter having one unknown $E\{N(m)\}$. Since $N(m) = R(m) - S(m)$, then $W(m)$ can be expressed as

$$W(m) = \frac{E\{R(m)N_o^*(m)\}}{E\{|N_o(m)|^2\}} - \frac{E\{S(m)N_o^*(m)\}}{E\{|N_o(m)|^2\}} \qquad (9.3.6)$$

If we then assume that $S(m)$ and $N_o(m)$ are apriori uncorrelated, then $W(m)$ becomes

$$W(m) = \frac{E\{R(m)N_o^*(m)\}}{E\{|N_o(m)|^2\}} \simeq \frac{R(m)N_o^*(m)}{|N_o(m)|^2} \qquad (9.3.7)$$

which is approximately Class 1 since both $R(m)$ and $N_o(m)$ are known apriori.

Eq. 9.3.7 allows $W(m)$ to be computed as $W_1(m)$ with no apriori information about $S(m)$ and $N(m)$ where

$$W_1(m) = \frac{E\{R(m)N_o^*(m)\}}{E\{|N_o(m)|^2\}} \simeq \frac{R(m)N_o^*(m)}{|N_o(m)|^2}$$
$$N_1(m) = W_1(m)N_o(m) \qquad (9.3.8)$$

To improve this gain estimate, we include the spurious cross-correlation term $E\{S(m)N_o^*(m)\}$ in Eq. 9.3.6. $S(m)$ can be aposteriori estimated as

$$S_1(m) = R(m) - N_1(m) \qquad (9.3.9)$$

Then the updated noise filter gain $W_2(m)$ and its output $N_2(m)$ become

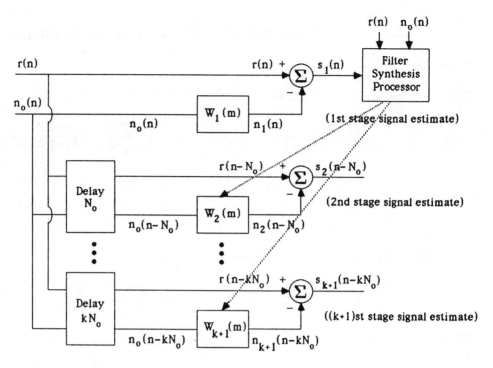

Fig. 9.3.5 Block diagram of a two-channel ANC aposteriori estimation filter.

$$W_2(m) = W_1(m) - \frac{S_1(m)N_o^*(m)}{|N_o(m)|^2}$$

$$N_2(m) = W_2(m)N_o(m)$$

(9.3.10)

This process can be continued until there is negligible improvement in $N_i(m)$ or an acceptably small deviation in $|W_i(m) - W_{i-1}(m)|$. These operations are summarized in Fig. 9.3.5 and are similar to those discussed previously in Fig. 9.2.1.

9.4 FFT DETECTION FILTERS [4,5]

Another class of problems involve detection of the presence or absence of a signal embedded in additive noise. General detection filters take the form of Fig. 9.4.1. As discussed in Chap. 8.2.3, the general detection transfer function $H(m)$ yields

$$H(m) = \frac{E\{D^*(m)\}}{E\{R(m)R^*(m)\}}$$

$$= \frac{\text{Conjugate spectrum of desired signal to be detected}}{\text{Autocorrelation spectrum of signal to be rejected}}$$

(9.4.1)

from Eq. 8.2.15. This filter maximizes the ratio of the peak output signal to total

Table 9.3.1 One-channel ANC filter transfer functions $H_{ANC}(m)$.

Filter Name	Class 1	Class 3

Distortion Balance ($W_S(m)$ = Signal weighting, $W_N(m)$ = Noise weighting)

$$\frac{E\{N(m)W_N(m)\}}{E\{S(m)W_S(m) + N(m)W_N(m)\}}$$

$$\frac{(R(m) - \langle R(m)\rangle)W_N(m)}{\langle R(m)W_S(m)\rangle + (R(m) - \langle R(m)\rangle)W_N(m)}$$

Uncorrelated Estimation (Uncorrelated Wiener)

$$\frac{E\{|N(m)|^2\}}{E\{|S(m)|^2\} + E\{|N(m)|^2\}}$$

$$\frac{\langle |R(m)|^2\rangle - |\langle R(m)\rangle|^2}{\langle |R(m)|^2\rangle}$$

Correlated Estimation (Correlated Wiener)

$$\frac{E\{N(m)(S^*(m) + N^*(m))\}}{E\{|S(m) + N(m)|^2\}}$$

$$\frac{\langle |R(m)|^2\rangle - \langle\langle R(m)\rangle \, R^*(m)\rangle}{\langle |R(m)|^2\rangle}$$

Transitional 1 Estimation (Uncorrelated Wiener $\alpha = 0$, Correlated Wiener $\alpha = 1$)

$$\frac{E\{|N(m)|^2 + \alpha S^*(m)N(m)\}}{E\{|S(m)|^2 + |N(m)|^2 + \alpha(S(m)N^*(m) + S^*(m)N(m))\}}$$

$$\frac{(1-\alpha)\langle |R(m)|^2\rangle - |\langle R(m)\rangle|^2 - (1-\alpha)\langle R(m)\rangle(R^*(m) - \langle R^*(m)\rangle)}{(1-\alpha)\langle |R(m)|^2\rangle}$$

Pulse-Shaping Estimation ($P(m)$ = Desired pulse spectrum, Uncorrelated Wiener)

$$\frac{E\{(S(m) - P(m))S^*(m) + |N(m)|^2\}}{E\{|S(m)|^2\} + E\{|N(m)|^2\}}$$

$$\frac{\langle |R(m)|^2\rangle - P(m)\langle R^*(m)\rangle}{\langle |R(m)|^2\rangle}$$

Pulse-Shaping Estimation ($P(m)$ = Desired pulse spectrum, Correlated Wiener)

$$\frac{E\{(S(m) + N(m) - P(m))(S^*(m) + N^*(m))\}}{E\{|S(m) + N(m)|^2\}}$$

$$\frac{\langle |R(m)|^2\rangle - \langle P(m)R^*(m)\rangle}{\langle |R(m)|^2\rangle}$$

Classical Detection (Matched)

$$\frac{E\{|N(m)|^2 - S^*(m)\}}{E\{|N(m)|^2\}}$$

$$\frac{|R(m) - \langle R(m)\rangle|^2 - \langle R^*(m)\rangle}{|R(m) - \langle R(m)\rangle|^2}$$

High-Resolution Detection (Uncorrelated Wiener)

$$\frac{E\{|S(m)|^2 + |N(m)|^2 - S^*(m)\}}{E\{|S(m)|^2\} + E\{|N(m)|^2\}}$$

$$\frac{\langle |R(m)|^2\rangle - \langle R^*(m)\rangle}{\langle |R(m)|^2\rangle}$$

Table 9.3.1 (Continued) One-channel ANC filter transfer functions $H_{ANC}(m)$.

Filter Name	Class 1	Class 3

High-Resolution Detection (Correlated Wiener)

$$\frac{E\{(S(m) + N(m) - 1)(S^*(m) + N^*(m))\}}{E\{|S(m) + N(m)|^2\}} \qquad\qquad \frac{\langle |R(m)|^2\rangle - \langle R^*(m)\rangle}{\langle |R(m)|^2\rangle}$$

Inverse Detection (Deconvolution)

$$\frac{E\{S(m) - 1\}}{E\{S(m)\}} \qquad\qquad \frac{\langle R(m)\rangle - 1}{\langle R(m)\rangle}$$

Transitional 1 Detection (Inverse $\alpha = 0$, High-Resolution $\alpha = 0.5$, Matched $\alpha = 1$)

$$\frac{E\{(1 - \alpha)|S(m)|^2 + \alpha|N(m)|^2 - S^*(m)\}}{E\{(1 - \alpha)|S(m)|^2 + \alpha|N(m)|^2\}}$$

$$\frac{(1 - \alpha)|\langle R(m)\rangle|^2 + \alpha|R(m) - \langle R(m)\rangle|^2 - \langle R^*(m)\rangle}{(1 - \alpha)|\langle R(m)\rangle|^2 + \alpha|R(m) - \langle R(m)\rangle|^2}$$

Transitional 2 Detection (Inverse $\alpha = 0.5$)

$$\frac{E\{S(m)^{2(1-\alpha)} - 1\}}{E\{S(m)^{2(1-\alpha)}\}} \qquad\qquad \frac{\langle R(m)\rangle^{2(1-\alpha)} - 1}{\langle R(m)\rangle^{2(1-\alpha)}}$$

Autocorrelation Estimation (Uncorrelated Wiener)

$$\frac{E\{(1 - S^*(m))|S(m)|^2 + |N(m)|^2\}}{E\{|S(m)|^2\} + E\{|N(m)|^2\}} \qquad\qquad \frac{\langle |R(m)|^2\rangle - \langle R^*(m)\rangle\,|\langle R(m)\rangle|^2}{\langle |R(m)|^2\rangle}$$

Autocorrelation Estimation (Correlated Wiener)

$$\frac{E\{(S(m) + N(m) - |S(m)|^2)(S^*(m) + N^*(m))\}}{E\{|S(m) + N(m)|^2\}} \qquad\qquad \frac{R(m) - |\langle R(m)\rangle|^2}{R(m)}$$

Cross-Correlation Estimation (Uncorrelated Wiener)

$$\frac{E\{(1 - N^*(m))|S(m)|^2 + |N(m)|^2\}}{E\{|S(m)|^2\} + E\{|N(m)|^2\}} \qquad \frac{\langle |R(m)|^2\rangle - (R^*(m) - \langle R^*(m)\rangle)\,|\langle R(m)\rangle|^2}{\langle |R(m)|^2\rangle}$$

Cross-Correlation Estimation (Correlated Wiener)

$$\frac{E\{(S(m) + N(m) - S(m)N^*(m))(S^*(m) + N^*(m))\}}{E\{|S(m) + N(m)|^2\}}$$

$$\frac{R(m) - (R^*(m) - \langle R^*(m)\rangle)\,\langle R(m)\rangle}{R(m)}$$

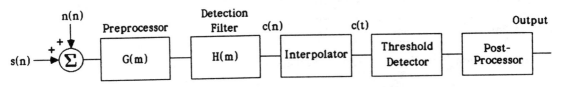

Fig. 9.4.1 Block diagram of a detection filter.

rms output without regard to the exact output detection pulse shape, time resolution of the detection pulse, probability of false detections from correlated noise, and other such factors. From a slightly different viewpoint, $H(m)$ transforms the input $R(m)$ into a measurement of the relative likelihood that a signal $D(m)$ is present in $R(m)$. The detection decision is performed by applying a threshold to the resulting likelihood measurements and post-processing the result.

The classical detection filter results when signal spectrum $S(m)$ must be detected and the noise spectrum $N(m)$ must be rejected so that $D(m) = S(m)$ and $R(m) = N(m)$. Eq. 9.4.1 reduces to

$$H(m) = \frac{E\{S^*(m)\}}{E\{N(m)N^*(m)\}} \tag{9.4.2}$$

which agrees with Eq. 8.2.34. The classical detection filter output equals

$$D(m) = H(m)R(m) = \frac{[S(m) + N(m)]E\{S^*(m)\}}{E\{N(m)N^*(m)\}} \tag{9.4.3}$$

In situations where the cross-spectral density $N(m)E\{S^*(m)\}$ is large, considerable false detections may result from the noise that is correlated to the signal.

One way to alleviate this problem is to set this cross-spectral density term equal to zero in the output given by Eq. 9.4.3 which yields

$$D(m) = \frac{E\{S(m)S^*(m)\}}{E\{N(m)N^*(m)\}} \tag{9.4.4}$$

Substituting this $D(m)$ into $H(m)$ given by Eq. 9.2.1 shows that the resulting optimal estimation transfer function required to give this output $D(m)$ equals

$$H(m) = \frac{E\{D(m)R^*(m)\}}{E\{R(m)R^*(m)\}} = \frac{E\{S(m)R^*(m)\}}{E\{R(m)R^*(m)\}} \frac{E\{S^*(m)\}}{E\{N(m)N^*(m)\}} \tag{9.4.5}$$

This equation can be partitioned as

$$H(m) = H_E(m)H_D(m) \tag{9.4.6}$$

where $H_E(m)$ is a classical estimation filter described by Eq. 9.2.1 and $H_D(m)$ is a classical detection filter described by Eq. 9.4.2. Eq. 9.4.6 is identical to Eq. 8.2.20.

Assuming that the signal $S(m)$ and noise $N(m)$ are uncorrelated apriori, then Eq. 9.4.5 reduces to

$$H(m) = \frac{E\{|S(m)|^2\}}{E\{|S(m)|^2\} + E\{|N(m)|^2\}} \frac{E\{S^*(m)\}}{E\{N(m)N^*(m)\}} \qquad (9.4.7)$$

Another useful filter is the high-resolution detection type described by Eq. 8.2.38. This filter maximizes the peak output signal to total rms output ratio. In this case, $D(m) = S(m)$ and $R(m) = S(m) + N(m)$ so that Eq. 9.4.1 reduces to

$$H(m) = \frac{E\{S^*(m)\}}{E\{|R(m)|^2\}} = \frac{E\{S^*(m)\}}{E\{|S(m) + N(m)|^2\}} \qquad (9.4.8)$$

The high-resolution detection filter outputs an impulse when it detects $E\{R(m)\}$ at its input. At frequencies where the spectral SNR ratio is low ($E\{|S(m)|\}/E\{|N(m)|\} \ll 1$), this filter acts as the classical detection filter of Eq. 9.4.2. At frequencies where the SNR ratio is high ($E\{|S(m)|\}/E\{|N(m)|\} \gg 1$), it acts as an inverse filter (Eq. 8.2.36). Alternatively, this filter may be viewed as an optimum estimation filter (a correlated Wiener type given by Eq. 8.2.21) followed by a filter with gain $E\{S^*(m)\}/E\{S^*(m) + N^*(m)\}$ (Eq. 8.2.15) and an inverse filter having gain $1/E\{S(m)\}$ (Eq. 8.2.36).

To illustrate a typical detection problem, consider Fig. 9.4.2. The signal $s(n)$ is chosen to be a 0.5 second pulse having a $\sin x/x$ spectrum as shown in Fig. 9.4.2a. The noise $n(n)$ is chosen to have a Gaussian power spectrum at low frequencies and a white (constant) spectrum at high frequencies. These spectrum intersect at 2 Hz as shown in Fig. 9.4.2b. Phase is assumed to be random and uniformly distributed between $(-\pi, \pi)$. The total input $r(n)$ is shown in Fig. 9.4.2c. The signal pulse occurring at $t = 0$ and $t = 10$ is barely discernable. The peak SNR is about 0 dB while the rms SNR is about -20 dB.

The classical detection filter was described by Eq. 9.4.2. Substituting $E\{S(m)\}$ and $E\{|N(m)|^2\}$ from Figs. 9.4.2a and 9.4.2b into that equation yields the transfer function and output shown in Fig. 9.4.3a. The output clearly detected the presence of the signal pulse at $t = 0$ and 10 but the ringing type response gives large temporal ambiguity and poor resolution leading to possibly false detections. The rms noise level however is very low. The peak SNR is about 15 dB.

The estimation-detection filter combination was described by Eq. 9.4.7. Its transfer function and output are shown in Fig. 9.4.3b. Although the transfer function is somewhat different than that of the classical detection filter in Fig. 9.4.3a (it has larger bandwidth), the output responses are very similar.

The high-resolution detection filter was described by Eq. 9.4.8. Its transfer function and output are shown in Fig. 9.4.3c. It has a peculiar appearing gain characteristic. At low spectral SNR, gain $H_{HR} \cong H_D$ and it behaves as the classical detection filter in Fig. 9.4.3a. At high spectral SNR, $H_{HR} \cong 1/E\{S(m)\}$ where $E\{S(m)\}$ is the $\sin x/x$ spectrum of Fig. 9.4.2a. Inverting this signal spectrum (e.g., the -20 dB/dec asymptote becomes a $+20$ dB/dec asymptote) yields the approximate H_{HR} under this high SNR condition. Combining these two results yields the high-resolution filter gain in Fig. 9.4.3c. It is close to the classical detection filter gain in Fig. 9.4.3a at low

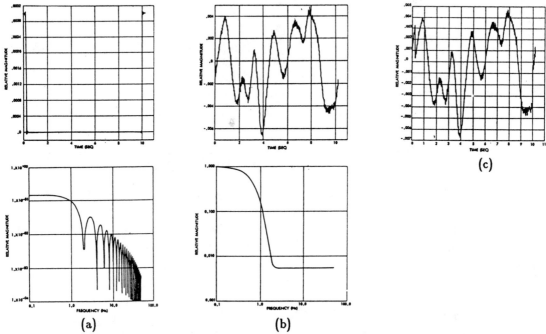

Fig. 9.4.2 (a) Signal and its spectrum, (b) noise and its spectrum, and (c) total input. (From W.H. Haas and C.S. Lindquist, "Linear detection filtering in the context of a least-squares estimator for signal processing applications," Proc. IEEE Intl. Conf. ASSP, pp. 651–654, ©1978, IEEE.)

spectral SNR but has the inverted $\sin x/x$ gain peaks at $\omega = 0, 3, 5, 7, 9, \ldots$ rad/sec in regions of high spectral SNR. The high-resolution detection filter clearly has superior detection performance. It produces sharp pulses (virtually impulses) at $t = 0$ and 10 giving excellent temporal resolution. The output noise level is somewhat larger than that of the other two filters however. The peak SNR is about 12 dB.

These three filters clearly show the trade offs in output pulse shape, time resolution, and output noise floor. Since threshold detectors are used at these filter outputs to establish the presence or absence of a signal, the filters of Figs. 9.4.3a and 9.4.3b would exhibit higher probability of false detections from signal (due to ringing) but a slightly lower probability to noise (due to lower rms noise level).

A more general adaptive FFT detection system is shown in Fig. 9.4.4. It consists of a bank of FFT detection filters each processing overlapping blocks of input data $R(m)$. Every filter with gain $H_i(m)$ is designed to detect a particular signal $s_i(n)$ in the input $r(n)$. The input $r(n)$, signal $s(n)$, and noise $n(n)$ are related to $s_i(n)$ as

$$r(n) = s(n) + n(n) = s_i(n) + n_i(n) \tag{9.4.9}$$

so that

$$n_i(n) = r(n) - s(n) = n(n) + [s(n) - s_i(n)] \tag{9.4.10}$$

Eq. 9.4.9 shows that given a total input $r(n)$, every particular signal $s_i(n)$ being detected

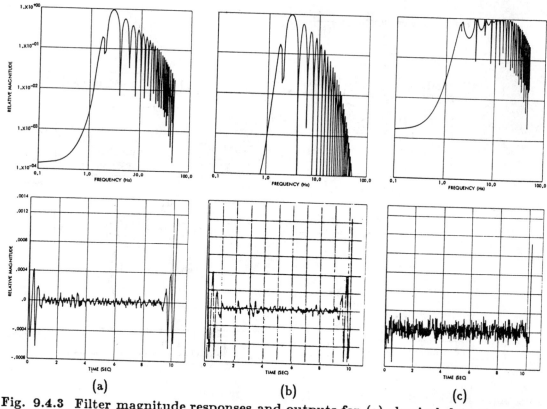

Fig. 9.4.3 Filter magnitude responses and outputs for (a) classical detection filter (Eq. 9.4.2), (b) estimation-detection filter (Eq. 9.4.7), and (c) high-resolution detection filter (Eq. 9.4.8). (From **W.H. Haas and C.S. Lindquist**, loc. cit., ©1978, **IEEE.**)

has an associated particular noise $n_i(n)$. Eq. 9.4.10 shows that every particular noise $n_i(n)$ is composed of the noise $n(n)$ augmented by a signal error $[s(n) - s_i(n)]$.

Consider a Class 2 detection problem where the expected signal spectra $E\{S_i(m)\}$ are known but the expected noise spectrum $E\{N_i(m)\}$ is not known apriori. Then the optimum detection filter has a transfer function

$$H_i(m) = \frac{E\{S_i^*(m)\}}{E\{R(m)R^*(m)\}} = \frac{E\{S_i^*(m)\}}{E\{|R(m)|^2\}}$$

$$= \frac{E\{S_i^*(m)\}}{E\{|S_i(m)|^2\} + E\{|N_i(m)|^2\} + 2E\{\mathrm{Re}[S_i(m)N_i^*(m)]\}} \qquad (9.4.11)$$

from Eq. 9.4.1 where $D(m) = S_i(m)$ and $R(m) = S_i(m) + N_i(m)$. This filter is identical to the high-resolution detection filter of Eq. 9.4.8. $E\{S_i(m)N_i^*(m)\}$ is an estimate of the cross-correlation spectrum between the ith signal $s_i(n)$ and ith noise $n_i(n)$. However in a Class 2 problem, it is assumed that $R(m)$ cannot be separated into its signal and noise components. When signal and noise are assumed to be uncorrelated apriori, then the average of $S_i(m)N_i^*(m)$ from data block to data block should be close to zero. Therefore by averaging (denoted by $\langle \ \rangle$) the power spectrum $|R(m)|^2$, also called the

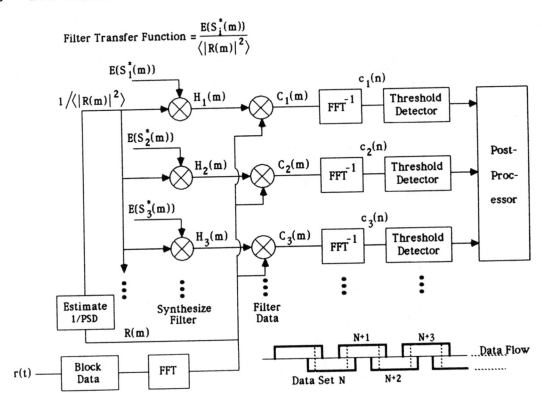

Fig. 9.4.4 Bank of detection filters for detecting different signals S_i in the presence of noise N.[5]

power density spectrum PSD as (see Chap. 4.5.3)

$$\langle |R(m)|^2 \rangle \ \cong \frac{1}{B} \sum_{k=1}^{B} |R_k(m)|^2 \cong E\{|S_i(m)|^2\} + E\{|N_i(m)|^2\} \qquad (9.4.12)$$

over many blocks of overlapping data as shown in Fig. 9.4.4, the effects of the undesirable cross-correlations will be greatly reduced. Then the Class 1 high-resolution $H_i(m)$ given by Eq. 9.4.11 becomes Class 2 where (see Table 4.5.1)

$$H_i(m) = \frac{E\{S_i^*(m)\}}{\langle |R(m)|^2 \rangle} \cong \frac{E\{S_i^*(m)\}}{E\{|S_i(m)|^2\} + E\{|N_i(m)|^2\}} \qquad (9.4.13)$$

It approaches the transfer function of the uncorrelated high-resolution filter given by Eq. 4.4.41. A different averaging approach will be discussed in the next section (see Eq. 9.5.10).

Notice in Fig. 9.4.4 that $H_i(m)$ is obtained by complex multiplication of the expected signal spectrum $E\{S_i^*(m)\}$ and the reciprocal of the PSD estimate $\langle |R(m)|^2 \rangle$ given by Eq. 9.4.12. The spectral outputs $C_i(m)$ of the detection filter are obtained by multiplying $H_i(m)$ by the input spectrum $R_i(m)$ corresponding to data block $r_i(n)$. The temporal outputs $c_i(n)$ are obtained by taking the inverse FFT of $C_i(m)$. Each $c_i(n)$ is threshold detected and post-processed to obtain the desired detection information.

To illustrate the behavior of the system in Fig. 9.4.4, consider the detection of a 1.6 second pulse centered at t = 20.2 seconds in the presence of three different types of noise as shown in Fig. 9.4.5a. The noise inputs have white, $1/f^2$, and $1/f^4$ power spectra. The peak SNR are adjusted to different levels. We will assume that 20 $msec$ block lengths (NT) are used for FFT analysis. This corresponds to rectangular window weighting. The equivalent window or preprocessor gain $G(m)$ in Fig. 9.4.1 is the digitized version of $G(j\omega) = NT\sin(\omega NT/2)/(\omega NT/2)$ from Eq. 2.5.2.

The optimum detection filters were derived from Eqs. 9.4.11 and 9.4.13 for the signal $s(n)$ in Fig. 9.4.5a and the three noises in Figs. 9.4.5b–9.4.5d. The peak SNR's were about $0.5 = -3$ dB but the signal cannot be seen in the noise since the rms SNR is much lower. The data blocks covered 2 sec of input and were overlapped 25%. The overlap reduces the transients or edge effects due to windowing the input $r(n)$. $\langle |R(m)|^2 \rangle$ was found by averaging 42 seconds of $r(n)$, or equivalently, 42 sec/2 sec(1 − 0.25) = 28 blocks of $R_k(m)$ data using Eq. 9.4.12. The resulting $H_i(m)$ transfer functions obtained from Eq. 9.4.13 for the $1/f^2$ and $1/f^4$ noise cases are shown in Figs. 9.4.6a and 9.4.6b, respectively.

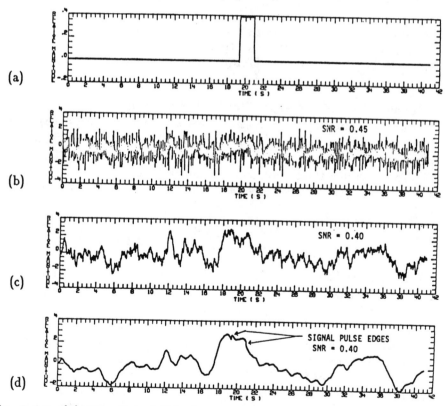

Fig. 9.4.5 (a) Signal and (b) noise having power spectrum that is white, (c) $1/f^2$, and (d) $1/f^4$, respectively. (From W.H. Haas and C.S. Lindquist, "An adaptive FFT-based detection filter," Proc. 26th Midwest Symp. Circuits and Systems, pp. 265–269, ©1983, IEEE.)

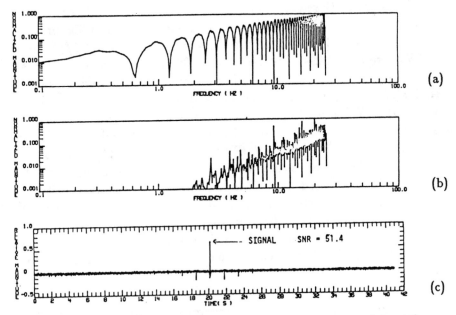

Fig. 9.4.6 Magnitude responses of detection filter for (a) $1/f^2$ noise, (b) $1/f^4$ noise, and (c) output of detection filter in (b). (From **W.H. Haas** and **C.S. Lindquist**, loc. cit., ©**1983, IEEE.**)

The output of the $1/f^4$ filter is shown in Fig. 9.4.6c. It exhibits a strong pulse (virtually an impulse) at $t = 20.2$ seconds with spurious responses on both sides. Clearly the detection filter in Fig. 9.4.6c is locating the signal buried in the noise of Fig. 9.4.5d very well.

To determine the improvement quantitatively, consider the peak SNR gain which is defined to equal (see Eq. 4.4.64)

$$\text{Peak } G_{SNR} = \frac{\text{Peak output } SNR_o}{\text{Peak input } SNR_i} \tag{9.4.14}$$

SNR gains are listed in Table 9.4.1 for Class 1 (dynamically nonadaptive) and Class 2 (dynamically adaptive) versions of these three filters. The Class 1 filter had its gain computed from Eq. 9.4.11 using $E\{S(m)\}$ and $E\{|N(m)|^2\}$. The Class 2 filter had its gain computed according to the procedure outlined in Fig. 9.4.4 and described by Eqs. 9.4.12 and 9.4.13. It is clear that the SNR gain increases with increasing SNR level for both Class 1 and Class 2 filters. For low SNR's the Class 2 filter performance is very close to the Class 1 filter. At high SNR's the Class 2 filter performance is significantly less than that of the Class 1 filter. However the output signal pulse of the Class 2 filter is narrower than that produced by the Class 1 filter. Therefore it can better resolve closely-spaced signal pulses.

Table 9.4.1　SNR gain of FFT detection filters.
(From W.H. Haas and C.S. Lindquist, loc. cit., ©1983, IEEE.)

Background Noise PSD	Filter Type	Pulse Width T_p						
		0.1	0.4	0.8	1.6	3.2	6.4	12.8
		SNR Preliminary [1]			Gain Performance [2]			
1	Ideal	1.9	3.7	5.3	7.5	10.5	14.3	18.3
	Adaptive	1.9	3.7	5.1	6.6	8.5	9.1	9.1
$\frac{1}{f^2}$	Ideal	6.7	6.2	5.7	4.7	3.7	3.1	2.8
	Adaptive	6.7	6.2	5.7	4.7	3.7	3.0	2.7
$\frac{1}{f^4}$	Ideal	520.8	374.6	308.3	211.9	125.5	68.1	35.7
	Adaptive [3]	63.5	161.9	157.5	129.2	88.9	57.7	34.3

[1] SNR = peak signal/rms noise (mean removed)
[2] SNR gain = output SNR/input SNR
[3] For the $1/f^4$ noise case, the output SNR was very large which affects the comparative SNR gain performance of the adaptive FFT filter.

9.5　FFT CORRELATION FILTERS [6,7]

Correlation filters are useful signal processing blocks. They process inputs such as signal plus noise to form an output which is some correlation function $R(n)$ as shown in Fig. 9.5.1. This output may be an autocorrelation estimate for either the signal $s(n)$ as $R_{ss}(n)$ or the noise $n(n)$ as $R_{nn}(n)$. The filter may also form the cross-correlation between the signal and noise as $R_{sn}(n)$ or $R_{ns}(n)$. These spectral correlation outputs can also be combined in a ratio to form the coherence function as shall be discussed in a moment.[8]

The correlation estimates between two signals $s(n)$ and $n(n)$ are defined in the time and frequency domains as

$$\widehat{R}_{ss}(n) = s(n) * s^*(-n) \quad \text{or} \quad \widehat{S}_{ss}(m) = |S(m)|^2$$
$$\widehat{R}_{nn}(n) = n(n) * n^*(-n) \quad \text{or} \quad \widehat{S}_{nn}(m) = |N(m)|^2$$
$$\widehat{R}_{sn}(n) = s(n) * n^*(-n) \quad \text{or} \quad \widehat{S}_{sn}(m) = S(m)N^*(m)$$
$$\widehat{R}_{ns}(n) = s^*(-n) * n(n) = \widehat{R}_{ss}(-n) \quad \text{or} \quad \widehat{S}_{ns}(m) = S^*(m)N(m) = S^*_{sn}(m)$$

(9.5.1)

where $*$ represents the convolution operation (see Eq. 1.17.21). The Fourier transform of a correlation function $R_{xy}(n)$ is the power spectral density (PSD) denoted as $S_{xy}(m)$. We will generally treat the correlation function $R_{xy}(n)$ (Eqs. 1.17.6–1.17.7) and its estimate $\widehat{R}_{xy}(n)$ (Eq. 1.17.21) as being equivalent. For further discussion, see Chap. 4.4.7.

Correlation estimates are made in real-time using only finite amounts of data. The estimates of Eq. 9.5.1 use an infinite amount of data. The approximate correlation estimate $\widehat{\widehat{R}}_{sn}(n)$ and its spectrum $\widehat{\widehat{S}}_{sn}(m)$ are

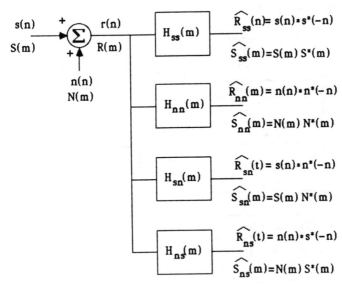

Fig. 9.5.1 Correlation filters to estimate (a) signal autocorrelation, (b) noise autocorrelation, (c) and (d) signal-noise cross-correlations.[6]

$$\widehat{\widehat{R}}_{sn}(n) = \int_0^T s(x)n^*(n+x)dx = [s(n)w(n)] * n^*(-n) = s(n) * [w^*(-n)n^*(-n)]$$

$$\widehat{\widehat{S}}_{sn}(m) = [S(m) * W(m)]N^*(m) = S(m)[N^*(m) * W^*(m)]$$

$$(9.5.2)$$

where $W(m)$ is the digitized version of $W(j\omega)$ as (see Chap. 2.5)

$$W(j\omega) = \frac{1 - e^{-j\omega T}}{j\omega} = T\,\frac{\sin(\omega T/2)}{\omega T/2}e^{-j\omega T} \qquad (9.5.3)$$

This is equivalent to temporal rectangular weighting of the signal $s(n)$ or noise $n(n)$. Thus, the correlator spectral estimate is in error by

$$\text{Spectral error} = \frac{\widehat{\widehat{S}}_{sn}(m)}{\widehat{S}_{sn}(m)} - 1 = \frac{S(m)N^*(m) * [W^*(m) - U_0^*(m)]}{S(m)N^*(m)} \qquad (9.5.4)$$

The finite width sampling window causes one of the input spectra be convolved with a $\sin x/x$ spectrum which introduces spectral distortion. This holds for any autocorrelation (S_{ss} or S_{nn}) and cross-correlation (S_{sn} or S_{ns}) estimates. The spectral distortion cannot later be removed but it can be reduced using other windows.

The coherence function estimate $\widehat{\Gamma}(m)$ measures the similarity between $s(n)$ and $n(n)$ over the window interval. It equals (Eq. 1.17.20)

$$\widehat{\Gamma}(m) = \sqrt{\frac{\widehat{S}_{sn}(m)\widehat{S}_{ns}(m)}{\widehat{S}_{ss}(m)\widehat{S}_{nn}(m)}} \qquad (9.5.5)$$

so the magnitude-squared coherence function estimate is

$$|\widehat{\Gamma}(m)|^2 = \frac{\widehat{S}_{sn}(m)\widehat{S}_{ns}(m)}{\widehat{S}_{ss}(m)\widehat{S}_{nn}(m)} \tag{9.5.6}$$

The spectra used in Eqs. 9.5.5 and 9.5.6 are either estimated correlation spectra (infinite width window) given by Eq. 9.5.1 or their approximations (finite width window) given by Eq. 9.5.2.

$\widehat{\Gamma}(m) = 1$ for infinite width windows. For finite width windows, the coherence function magnitude never exceeds unity so $0 \leq |\Gamma(m)| \leq 1$. The $|\widehat{\Gamma}(m)|$ estimate of $|\Gamma(m)|$ may exceed unity. It measures the ratio of the cross-correlation power to the autocorrelation power between $s(n)$ and $n(n)$ at every frequency between $0 \leq m < \infty$. Signals that are scaled temporal versions of one another have $\Gamma(m) = \pm 1$. Otherwise they are more uncorrelated so $|\Gamma(m)| < 1$. Our interest lies in either estimating the correlation functions of Eq. 9.5.1 and the coherence function of Eq. 9.5.5, or detecting when particular estimated correlation or coherence functions occur.

The correlation filters of Eq. 9.5.1 can be conceptually viewed as some type of signal (noise) estimation filter followed by a pulse-shaping filter whose impulse response is the conjugate expected signal spectrum $E\{S^*\}$ (noise spectrum $E\{N^*\}$). The signal-noise (noise-signal) cross-correlation filter is the same signal (noise) estimation filter followed by a conjugate expected noise spectrum $E\{N^*\}$ (signal spectrum $E\{S^*\}$) pulse-shaping filter. This is shown in Fig. 9.5.2. The coherence function $\Gamma(m)$ equals the ratio of the spectral outputs of the two cross-correlation filters divided by the two autocorrelation filter outputs.

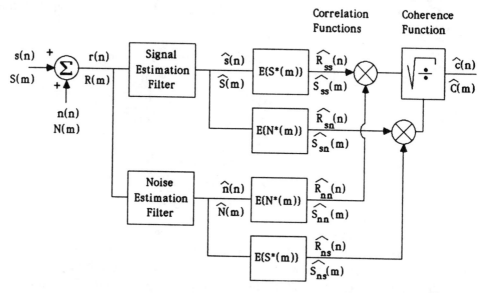

Fig. 9.5.2 Signal and noise estimation filters followed by signal and noise pulse-shaping filters to obtain correlation filters.[6]

Now consider the implementation of the estimation filters in Fig. 9.5.2.[6] A variety of adaptive estimation filters are listed in Tables 8.2.1 and 8.2.2. The distortion balance, general estimation, and Wiener filters are all possibilities. Assume for example that uncorrelated Wiener filters are used. Then the resulting overall correlation filter transfer functions in Fig. 9.5.1 are listed in Table 9.5.1. For example, the Class 1 autocorrelation and cross-correlation filters have (Eqs. 8.2.29 and 8.2.32)

$$H_{ss}(m) = \frac{E\{S^*(m)|S(m)|^2\}}{E\{|S(m)|^2\} + E\{|N(m)|^2\}}$$

$$H_{nn}(m) = \frac{E\{N^*(m)|N(m)|^2\}}{E\{|S(m)|^2\} + E\{|N(m)|^2\}}$$

$$H_{sn}(m) = \frac{E\{N^*(m)|S(m)|^2\}}{E\{|S(m)|^2\} + E\{|N(m)|^2\}}$$

$$H_{ns}(m) = \frac{E\{S^*(m)|N(m)|^2\}}{E\{|S(m)|^2\} + E\{|N(m)|^2\}}$$

(9.5.7)

All the filter inputs are $R(m)$ while the filter outputs are the correlation functions. The constant k in the Class 2 filters estimates the level of $E\{|S|\}$ in $|R|$. The coherence function estimate $\widehat{\Gamma}(m)$ of Eq. 9.5.5 equals the ratio of the spectral products of Eq. 9.5.7 as

$$|\widehat{\Gamma}(m)|^2 = \frac{\widehat{S}_{sn}(m)\widehat{S}_{ns}(m)}{\widehat{S}_{ss}(m)\widehat{S}_{nn}(m)} = \frac{H_{sn}(m)H_{ns}(m)R(m)^2}{H_{ss}(m)H_{nn}(m)R(m)^2} = \frac{H_{sn}(m)H_{ns}(m)}{H_{ss}(m)H_{nn}(m)}$$

(9.5.8)

Now let us implement the correlation filters using the correlated Wiener filters. The correlation transfer functions are also listed in Table 9.5.1 (see Eqs. 8.2.30 and 8.2.33). For Class 1, they are:

$$H_{ss}(m) = \frac{E\{|S(m)|^2(S^*(m) + N^*(m))\}}{E\{|S(m) + N(m)|^2\}}$$

$$H_{nn}(m) = \frac{E\{|N(m)|^2(S^*(m) + N^*(m))\}}{E\{|S(m) + N(m)|^2\}}$$

$$H_{sn}(m) = \frac{E\{S(m)N^*(m)(S^*(m) + N^*(m))\}}{E\{|S(m) + N(m)|^2\}}$$

$$H_{ns}(m) = \frac{E\{S^*(m)N(m)(S^*(m) + N^*(m))\}}{E\{|S(m) + N(m)|^2\}}$$

(9.5.9)

To illustrate the usefulness of these correlation algorithms, consider the signal $s(n)$ and noise $n(n)$ shown in Figs. 9.5.3a and 9.5.3b with their respective spectrum. The noise consists of two components – one is Gaussian with random phase and the other is a scaled and delayed (by 4.25 seconds) version of the signal. Therefore the signal and noise are partially correlated. Their autocorrelation and cross-correlation are shown in Figs. 9.5.3c–e. Passing the signal plus noise through the correlation filter bank of Fig.

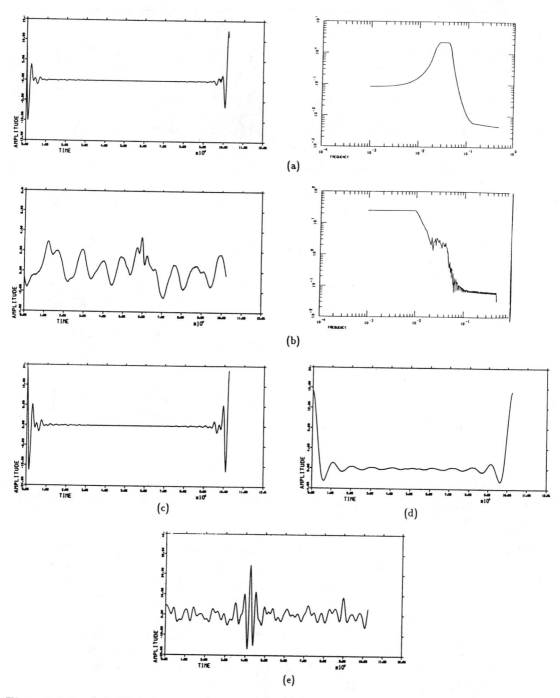

Fig. 9.5.3 (a) Signal and its spectrum, (b) noise and its spectrum, (c) signal autocorrelation, (d) noise autocorrelation, and (e) signal-noise cross-correlation. (From C.S. Lindquist and W.H. Haas, "Some new set adaptive correlation filters," Proc. 28th Midwest Symp. Circuits and Systems, pp. 479–488, ©1985, IEEE.)

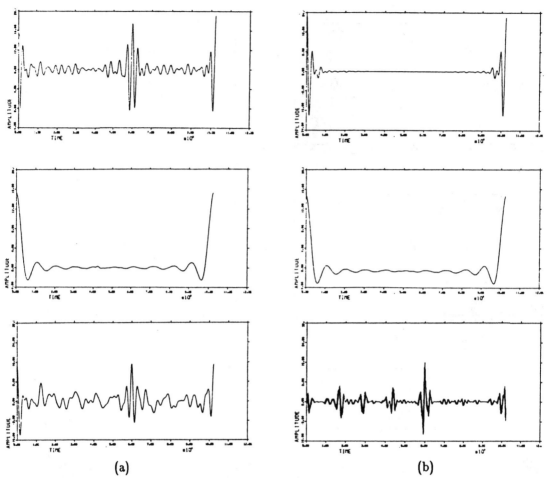

Fig. 9.5.4 Correlation estimates using (a) uncorrelated Wiener filters of Eq. 9.5.7 and (b) correlated Wiener filters of Eq. 9.5.9. (From C.S. Lindquist and W.H. Haas, loc. cit., ©1985, IEEE.)

9.5.1 gives the outputs shown in Fig. 9.5.4. The correlation estimates in Fig. 9.5.4a result using the uncorrelated Wiener estimation filters of Eq. 9.5.7. The R_{ss} estimate is poor, the R_{nn} is good, and the R_{sn} estimate is adequate. The correlation estimates in Fig. 9.5.4b result using the correlated Wiener estimation filters of Eq. 9.5.9. The R_{ss} and R_{nn} estimates are very good while the R_{sn} estimate is again adequate.

These simulations show that set of algorithms given by Eq. 9.5.9 are better than those of Eq. 9.5.7. The reason is that the Eq. 9.5.7 algorithms irretrievably discard the phase information of $E\{S(m)\}$ and $E\{N(m)\}$. However the algorithms given by Eq. 9.5.9 utilize the phase information.

The Class 2 correlation filters of Table 9.5.1 require smoothing of the input where $\langle |R|^2 \rangle = \langle |S|^2 + |N|^2 + SN^* + S^*N \rangle$. This reduces the aposteriori cross-correlation terms $(SN^* + S^*N)$ as much as possible to obtain good estimates of $E\{|S|^2\} + E\{|N|^2\} \cong \langle |S|^2 + |N|^2 \rangle$. One smoothing approach is to perform block-to-block averaging (i.e., intra-block) of the FFT's as was done in Eq. 9.4.12 of the last section. However this introduces more computational delay. Smoothing within the block (i.e., inner-block) also achieves good results without large delays. Averaging each spectral component with a spectral window $B(m)$ is equivalent to convolving $R(m)$ with $B(m)$ as (see Chap. 4.5.3)

$$R_b(m) = R(m) * B(m) \qquad \text{or} \qquad r_b(n) = r(n)b(n) \qquad (9.5.10)$$

Such spectral averaging is equivalent to temporal weighting of the input $r(n)$ itself by $b(n)$. It reduces the spurious cross-correlation between $s(n)$ and $n(n)$ but also lowers the spectral resolution.

This averaging was performed on a Class 2 high-resolution detection filter of Eq. 9.4.13. The results are summarized in Figs. 9.5.5 and 9.5.6. The signal $s(n)$ is a 5 Hz cosine pulse having 2.5 second duration centered at 10 seconds. The noise spectrum $|N(m)|$ was assumed to be Gaussian. In addition, an unwanted 12.5 Hz cosine pulse with 2 second duration was added at 2 seconds. These signals and spectra are shown in Fig. 9.5.5. The peak SNR was adjusted to 2. The filter gain was computed using Eq. 9.5.11. The $B(m)$ window selected was Gaussian so that the filter gain $H(m)$ equals

$$B(m) = e^{-m^2/2\sigma^2}, \qquad H(m) = \frac{S^*(m)}{\langle |R_b(m)|^2 \rangle} = \frac{S^*(m)}{|R(m) * B(m)|^2} \qquad (9.5.11)$$

σ was arbitrarily chosen as a fraction of the FFT bin spacings ($\sigma = 0.01, 1, 10, 50$).

The estimated spectral output $C(m)$, filter magnitude response $|H(m)|$, and temporal output $c(n)$ are shown in Fig. 9.5.6a for $\sigma = 0.01$ which essentially means no smoothing or frequency weighting. Then σ was adjusted to 1, 10, and 50 using the same input. The resulting responses are shown in Figs. 9.5.6b–d, respectively.

Comparing the results we make the following conclusions. Averaging the estimated input spectrum over 1, 10, and 50 bin spacings broadens the spectra and greatly reduces the noise level. The magnitude response should pass 5 Hz, notch 12.5 Hz, and drop to the white noise floor. The magnitude responses show the effects of unwanted cross-correlation. Fig. 9.5.6b is somewhat improved, Fig. 9.5.6c is the best, and Fig. 9.5.6d is somewhat degraded. The 5 second triangular output tends to be the autocorrelation of the 2.5 second input pulse that modulates the cosine signal. All the outputs peak at $t = 10$ seconds. Fig. 9.5.6b has the narrowest pulse width but the output is fairly noisy due to the cross-correlations in $H(m)$. There is also a small spurious response at 2 seconds. The background noise level drops as we progress from Fig. 9.5.6a–d. However the pulse width widens. These results illustrate that time windows perform more than time-limiting the input signal. They also cause spectral window averaging with the input spectrum. This reduces the undesirable cross-correlations in Class 2 adaptive filters.

Consider the two-channel system shown in Fig. 9.5.7. The inputs $r_1(n)$ and $r_2(n)$ are composed of signals $s_1(n)$ and $s_2(n)$ plus noises $n_1(n)$ and $n_2(n)$, respectively. The noises are assumed to be apriori uncorrelated to each other and to each of the signals. The signals are assumed to be delayed and scaled versions of one another as

$s_2(n) = cs_1(n - D)$. We further assume that the signal is a modulated pulse of known maximum duration P but unknown shape. The spectral models of both the signals and noises are assumed to be unknown. This is a Class 3 problem. Suppose that $s_1(n)$ must be estimated.

A Class 3 Wiener estimation filter can be formulated by noting that a Class 1 uncorrelated Wiener estimation filter has

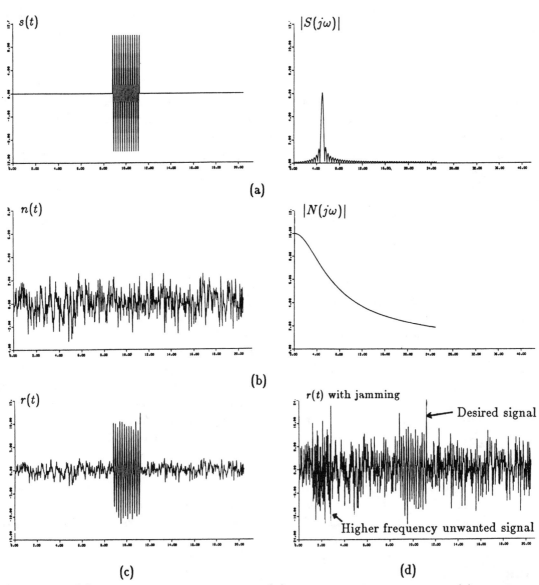

Fig. 9.5.5 (a) Signal and its spectrum, (b) noise and its spectrum, (c) total input, and (d) total input with jamming. (From W.H. Haas and C.S. Lindquist, "Frequency domain algorithms for adaptive filters," Proc. 1984 IEEE Intl. Symp. Circuits and Systems, pp. 503–506, ©1984, IEEE.)

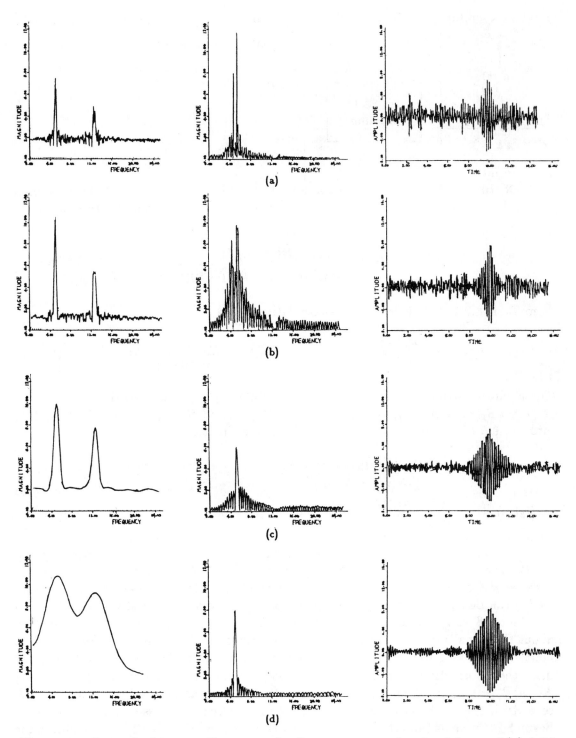

Fig. 9.5.6 Smoothed input, filter magnitude response, and output for σ of (a) 0.01, (b) 1, (c) 10, and (d) 50. (From W.H. Haas and C.S. Lindquist, loc. cit., ©1984, IEEE.)

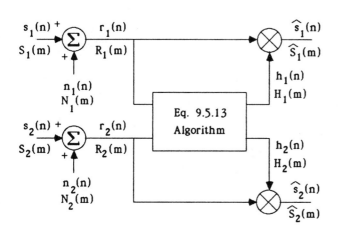

Fig. 9.5.7 Two-channel estimation filter.[6]

$$H(m) = \frac{E\{|S(m)|^2\}}{E\{|S(m)|^2\} + E\{|N(m)|^2\}} \tag{9.5.12}$$

from Table 8.2.1. Consider the cross-correlation filters having

$$H_1(m) = \frac{\langle |R_1(m)R_2^*(m)| \rangle}{\langle |R_1(m)|^2 \rangle}, \qquad H_2(m) = \frac{\langle |R_1^*(m)R_2(m)| \rangle}{\langle |R_2(m)|^2 \rangle} \tag{9.5.13}$$

These filters cross-correlate the two total inputs $R_1(m)$ and $R_2(m)$. Substituting $R_1(m) = S_1(m) + N_1(m)$, $R_2(m) = S_2(m) + N_2(m)$, and $S_2(m) = ce^{-j2\pi mD/NT}S_1(m)$ into Eq. 9.5.13 and simplifying gives the Class 3 estimation filter as

$$H_1(m) = \frac{\langle |[S_1(m) + N_1(m)][S_2^*(m) + N_2^*(m)]| \rangle}{\langle |S_1(m) + N_1(m)|^2 \rangle} = \frac{c\langle |S_1(m)|^2 \rangle}{\langle |S_1(m)|^2 + |N_1(m)|^2 \rangle}$$
$$H_2(m) = \frac{\langle |S_2(m)|^2 \rangle / c}{\langle |S_2(m)|^2 + |N_2(m)|^2 \rangle} \tag{9.5.14}$$

These are approximations to the Class 1 estimation filter whose gain is given by Eq. 9.5.12. The cross-correlation terms $N_1 N_2^*$, $S_1 N_2^*$, and $S_2^* N_1$ all equal zero in Eq. 9.5.14 because only S_1 and S_2 were assumed to be apriori correlated. The averaging indicated by $\langle\ \rangle$ can be carried out using the inner block spectral averaging of Eq. 9.5.10.

More effective averaging is carried out in Eq. 9.5.14 using the apriori information about the signal's maximum pulse duration P. Since the signal autocorrelation can be nonzero for no longer than $2P$ seconds, the numerators can be $2P[\sin(\omega P)/\omega P]$ weighted using this information. This can also be accomplished in the time domain by computing $R_1(m)R_2^*(m)$, inverse transforming to find $r_1(n) * r_2(-n)$, and centering a rectangular P second window at the cross-correlation peak. This zeros the cross-correlation estimate beyond $\pm P$ seconds from the center. This corresponds to $2P[\sin(\omega P)/\omega P]$ weighting the $R_1(m)R_2^*(m)$ spectrum. For improved results, $r_1(n) * r_2(-n)$ can also be smoothed towards the ends using some other arbitrarily chosen window and level-shifted so its mean is zero. This procedure was followed to obtain the smoothed numerator $\langle |R_1(m)R_2^*(m)| \rangle$ in Eq. 9.5.13 while inner block averaging (Eq. 9.5.10) was used to obtain the smoothed denominator $\langle |R_1(m)|^2 \rangle$ for the following example.

A raised-cosine signal pulse not exceeding 1.5 seconds was combined with low-frequency Gaussian noise to form the first input $r_1(n)$ shown in Fig. 9.5.8a. A delayed

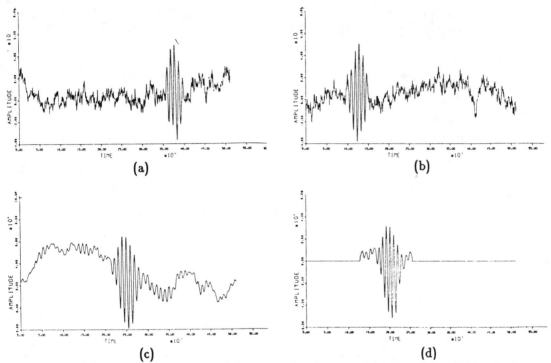

Fig. 9.5.8 (a) Channel 1 input, (b) channel 2 input, (c) cross-correlation between inputs, and (d) truncated and dc-corrected version. (From C.S. Lindquist and W.H. Haas, "Adaptive correlation filters defined in the frequency domain," Rec. 18th Asilomar Conf. Circuits, Systems, and Computers, pp. 327–331, ©1984, IEEE.)

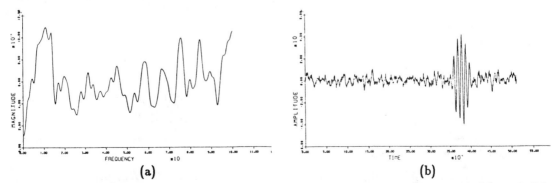

Fig. 9.5.9 (a) Magnitude response of signal estimation filter (Eq. 9.5.14b) and (b) signal estimate $\hat{s}_2(t)$. (From C.S. Lindquist and W.H. Haas, loc. cit., ©1984, IEEE.)

Table 9.5.1 A collection of correlation filters based upon
(a) uncorrelated Wiener estimation filters (Eqs. 8.2.29, 8.2.32)
and (b) correlated Wiener estimation filters (Eqs. 8.2.30, 8.2.33).

Filter Name Class 1	Class 2

Signal Autocorrelation $H_{ss}(m)$ (a)

$$\frac{E\{S^*(m)|S(m)|^2\}}{E\{|S(m)|^2\} + E\{|N(m)|^2\}}$$
$$\frac{E\{S^*(m)|S(m)|^2\}}{\langle|R(m)|^2\rangle}$$

Noise Autocorrelation $H_{nn}(m)$

$$\frac{E\{N^*(m)|N(m)|^2\}}{E\{|S(m)|^2\} + E\{|N(m)|^2\}}$$
$$\frac{E\{(R^*(m) - kS^*(m))|R(m) - kS(m)|^2\}}{\langle|R(m)|^2\rangle}$$

Signal-Noise Cross-Correlation $H_{sn}(m)$

$$\frac{E\{N^*(m)|S(m)|^2\}}{E\{|S(m)|^2\} + E\{|N(m)|^2\}}$$
$$\frac{E\{(R^*(m) - kS^*(m))|S(m)|^2\}}{\langle|R(m)|^2\rangle}$$

Noise-Signal Cross-Correlation $H_{ns}(m)$

$$\frac{E\{S^*(m)|N(m)|^2\}}{E\{|S(m)|^2\} + E\{|N(m)|^2\}}$$
$$\frac{E\{S^*(m)|R(m) - kS(m)|^2\}}{\langle|R(m)|^2\rangle}$$

Signal Autocorrelation $H_{ss}(m)$ (b)

$$\frac{E\{(S^*(m) + N^*(m))|S(m)|^2\}}{E\{|S(m) + N(m)|^2\}}$$
$$\frac{E\{(R^*(m) + (1 - k)S^*(m))|S(m)|^2\}}{|R(m)|^2}$$

Noise Autocorrelation $H_{nn}(m)$

$$\frac{E\{(S^*(m) + N^*(m))|N(m)|^2\}}{E\{|S(m) + N(m)|^2\}}$$
$$\frac{E\{(R^*(m) + (1 - k)S^*(m))|R(m) - kS(m)|^2\}}{|R(m)|^2}$$

Signal-Noise Cross-Correlation $H_{sn}(m)$

$$\frac{E\{(S^*(m) + N^*(m))S(m)N^*(m)\}}{E\{|S(m) + N(m)|^2\}}$$

$$\frac{E\{(R^*(m) + (1 - k)S^*(m))S(m)(R^*(m) - kS^*(m))\}}{|R(m)|^2}$$

Noise-Signal Cross-Correlation $H_{ns}(m)$

$$\frac{E\{(S^*(m) + N^*(m))S^*(m)N(m)\}}{E\{|S(m) + N(m)|^2\}}$$

$$\frac{E\{(R^*(m) + (1 - k)S^*(m))S^*(m)(R(m) - kS(m))\}}{|R(m)|^2}$$

Coherence Function Estimate $|\hat{\Gamma}(m)|^2$

$$\frac{H_{sn}(m)H_{ns}(m)}{H_{ss}(m)H_{nn}(m)}$$
$$\frac{H_{sn}(m)H_{ns}(m)}{H_{ss}(m)H_{nn}(m)}$$

signal pulse was combined with other Gaussian noise to form the second input $r_2(n)$ in Fig. 9.5.8b. The peak SNR's were selected to equal about 10 dB. The cross-correlation $R_{12}(n)$ between $r_1(n)$ and $r_2(n)$ is shown in Fig. 9.5.8c. The correlation peak occurs at 2.5 seconds. Truncating $R_{12}(n)$ for 1.5 seconds centered at 2.5 seconds and adjusting the dc level to zero produces $\langle R_{12}(n) \rangle$ shown in Fig. 9.5.8d. Finally computing $|R_2(m)|^2$ and smoothing using the Gaussian window of Eq. 9.5.11 forms $\langle |R_2(m)|^2 \rangle$. Forming the ratio of $\langle |R_{12}(m)|^2 \rangle$ and $\langle |R_2(m)|^2 \rangle$ in Eq. 9.5.14 gives $H_2(m)$ which is plotted in Fig. 9.5.9a. Filtering $r_2(n)$ with $H_2(m)$ produces the signal estimate $\hat{s}_2(n)$ shown in Fig. 9.5.9b. This is an improved estimate of the noisy $s_1(n)$ input of Fig. 9.5.8a. The noise level has been reduced. Since this is a Class 3 filter where no signal and noise spectral information has been used, the results are somewhat impressive. They show how correlation estimates can be used to make signal estimates.

9.6 FFT DELAY ESTIMATION FILTERS [9,10]

Another class of problems involves delay estimation. In these systems, noisy multiple signals are received at different times and the time differences or delays between signals must be determined. A general two-channel delay estimation system is shown in Fig. 9.6.1a.

The signals $s_1(n)$ and $s_2(n)$ are related as $s_2(n) = s_1(n - D)$ where D is the delay to be estimated by $\hat{D}$. The noises $n_1(n)$ and $n_2(n)$ are assumed to be uncorrelated to each other and to the signals. The noise processes are also assumed to be stationary from FFT data block to data block (i.e., constant statistics over each observation and possibly between intervals).

The total spectral inputs $R_1(m)$ and $R_2(m)$ are applied to the pre-correlation filters $H_1(m)$ and $H_2(m)$. These filters boost the signals, reduce the noise, and shape the resulting output for later processing. Their spectral outputs $C_1(m)$ and $C_2(m)$ are next cross-correlated as

$$c_3(n) = \int_{-\infty}^{\infty} c_1(x)c_2^*(n+x)dx = c_1(n) * c_2^*(-n)$$
$$C_3(m) = C_1(m)C_2^*(m) \tag{9.6.1}$$

using the conjugate multiplication block.

The cross-correlator output $C_3(m)$ is filtered by the post-correlation filter $H_3(m)$ to reduce temporal error in the delay estimate. This filter increases the SNR and reduces the pulse width of the cross-correlation filter output. Its spectral output $C_4(m)$ is converted into a time domain signal $c_4(n)$ by the inverse FFT and interpolator blocks. $c_4(n)$ is applied to a threshold detector. The delay estimate $\hat{D}$ is the time between $n = 0$ and the peak in the output $c_4(n)$. Alternatively viewed, it is the time difference between the peaks in $c_1(n)$ and $c_2(n)$.

System simplification results by noting that the post-correlator spectral output $C_4(m)$ can be related to the spectral inputs $R_1(m)$ and $R_2(m)$ in Fig. 9.6.1a as

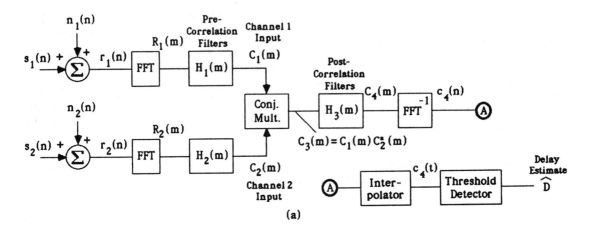

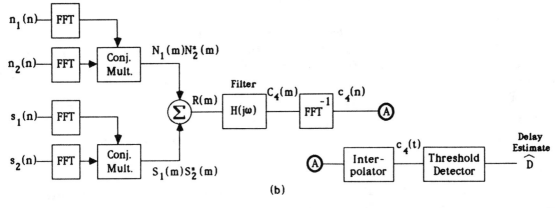

Fig. 9.6.1 (a) Two-channel delay estimation filter and (b) its one-channel equivalent.[9]

$$C_4(m) = H_3(m)C_3(m) = H_3(m)C_1(m)C_2^*(m)$$
$$= H_1(m)H_2^*(m)H_3(m)R_1(m)R_2^*(m) \qquad (9.6.2)$$

An effective overall transfer function $H(m)$ can therefore be defined to equal

$$H(m) = H_1(m)H_2^*(m)H_3(m) \qquad (9.6.3)$$

This shows the pre-correlation filters H_1 and H_2 are redundant. The post-correlation filter H_3 can be replaced by H and the pre-correlation filters removed without changing the system. The filter described by $H(m)$ is driven by an equivalent input spectrum $R(m)$ where

$$R(m) = R_1(m)R_2^*(m) = [S_1(m) + N_1(m)][S_2^*(m) + N_2^*(m)]$$
$$= S_1(m)S_2^*(m) + N_1(m)N_2^*(m) + S_1(m)N_2^*(m) + S_2^*(m)N_1(m) \qquad (9.6.4)$$

which is obtained by expressing $R_1(m)$ and $R_2(m)$ in terms of their respective signals and noises. $R(m)$ is composed of four cross-correlation estimates. If $s_1(n)$ and $n_2(n)$ are assumed to be uncorrelated apriori as well as $s_2(n)$ and $n_1(n)$, then their corresponding cross-spectral terms can be eliminated in Eq. 9.6.4. $R(m)$ can then be reexpressed as

$$R(m) = S(m) + N(m) \tag{9.6.5}$$

where

$$S(m) = S_1(m)S_2^*(m), \qquad N(m) = N_1(m)N_2^*(m) \tag{9.6.6}$$

Using the viewpoint that $R(m)$ is input to transfer function $H(m)$, then the equivalent system becomes the FFT detection filter shown in Fig. 9.6.1b (cf. Fig. 9.4.1).

Several delay estimation systems can now easily be determined. The classical detection system given by Eq. 9.4.2 maximizes the peak SNR where

$$H(m) = \frac{E\{S_1^*(m)S_2(m)\}}{E\{|N_1(m)N_2^*(m)|^2\}} \tag{9.6.7}$$

The high-resolution detection filter given by Eq. 9.4.8 maximizes the peak signal to total rms output ratio where

$$H(m) = \frac{E\{S_1^*(m)S_2(m)\}}{E\{|S_1(m)S_2^*(m)|^2\} + E\{|N_1(m)N_2^*(m)|^2\}} \tag{9.6.8}$$

The filter that minimizes the temporal ambiguity caused by nonzero pulse width in $c_4(n)$ is the inverse or deconvolution filter where

$$H(m) = \frac{1}{E\{S_1(m)S_2^*(m)\}} \tag{9.6.9}$$

All three equations can be described by a single transitional delay estimation filter having

$$H(m) = \frac{E\{S_1^*(m)S_2(m)\}}{(1-\alpha)E\{|S_1(m)S_2^*(m)|^2\} + \alpha E\{|N_1(m)N_2^*(m)|^2\}} \tag{9.6.10}$$

with parameter α. α is the relative weight used to trade off SNR enhancement against temporal ambiguity reduction. Choosing $\alpha = 0$ (inverse filter) minimizes temporal uncertainty, $\alpha = 0.5$ (high resolution) maximizes SRR, while $\alpha = 1$ (matched filter) maximizes peak SNR. Eq. 9.6.10 describes Class 1 delay estimation problems where both the expected signal $E\{S(m)\}$ and noise $E\{N(m)\}$ characterizations of Eq. 9.6.6 are known. Eq. 9.6.10 can be generalized to describe Class 2 problems only when the expected signal $E\{S_1^*(m)S_2(m)\}$ is known as

$$H(m) = \frac{E\{S_1^*(m)S_2(m)\}}{(1-\alpha)E\{|S_1(m)S_2^*(m)|^2\} + \alpha|\langle R_1(m)R_2^*(m)\rangle - E\{S_1(m)S_2^*(m)\}|^2}$$
$$(9.6.11)$$

When the expected signal $E\{S_1^*(m)S_2(m)\}$ shape is known but its level is unknown, Eq. 9.6.11 is modified to

$$H(m) = \frac{E\{S_1^*(m)S_2(m)\}}{\langle|R_1(m)R_2^*(m)|^2\rangle} \qquad (9.6.12)$$

$E\{|R_1(m)R_2^*(m)|^2\}$ is approximated by smoothing or averaging $|R_1(m)R_2^*(m)|^2$ terms over blocks of input data (see Eq. 9.4.12) which generally gives improved results.

In Class 3 problems, both the signal $S(m)$ and noise $N(m)$ must be estimated aposteriori. Since signal should be more correlated than noise, $\langle|R_1(m)R_2^*(m)|^2\rangle$ is an estimate of the expected cross-spectral power $E\{|R_1(m)R_2^*(m)|^2\}$. Thus one possible Class 3 delay estimation filter has

$$H(m) = \frac{1}{\langle|R_1(m)R_2^*(m)|^{2(1-\alpha)}\rangle} \qquad (9.6.13)$$

Setting $\alpha = 0.5$ yields

$$H(m) = \frac{1}{\langle|R_1(m)R_2^*(m)|\rangle} \qquad (9.6.14)$$

which approximates an inverse filter. Setting $\alpha = 0$ gives

$$H(m) = \frac{1}{\langle|R_1(m)R_2^*(m)|^2\rangle} \qquad (9.6.15)$$

which approximates an autocorrelation filter. Although the rationale for making such choices is tenuous, remember that so is the solution to any problem where virtually no models for expected signal and noise can be made. These delay estimation filters are tabulated in Table 9.6.1.[11]

Now let us illustrate these results with two delay examples. The first example is described in Figs. 9.6.2–9.6.4 where the system is required to estimate the delay between two 13 $msec$ cosine pulses with a known separation of 26 $msec$.[9] The expected signals are assumed to have identical magnitude spectra as shown in Fig. 9.6.2a. Expected noises are assumed to have a Gaussian power spectrum at low frequencies and a white spectrum at high frequencies as shown in Fig. 9.6.2b. Their phases are random and uniformly distributed between $(-\pi, \pi)$. The total inputs for channels 1 and 2 are shown in Fig. 9.6.2c. The peak SNR has been adjusted to be fairly large at about 13 dB. FFT's of 1024 points and sampling frequencies of 10 KHz were chosen.

The Class 1 transfer functions shown in Fig. 9.6.3a were calculated using Eq. 9.6.10 for different α's. The Class 2 transfer functions shown in Fig. 9.6.3b were calculated using Eq. 9.6.11. Finally Class 3 transfer functions are shown in Fig. 9.6.3c using Eq. 9.6.13 for $\alpha = 0$. The Class 2 responses have a somewhat similar form to the Class 1 responses but narrower bandwidth. The Class 3 responses equal $1/|S_1(j0)S_2^*(j0)|^2$ at low frequencies (less than 100 Hz where signal dominates) and $1/|N_1(j\infty)N_2^*(j\infty)|^2$ at high frequencies (greater than 1000 Hz where noise dominates).

The output $c_4(n)$ of the post-correlation filter is shown in Fig. 9.6.4 for several of these filters. Examining these responses shows that as α is decreased from unity (maximum SNR) towards zero (minimum temporal ambiguity), the output SNR decreases,

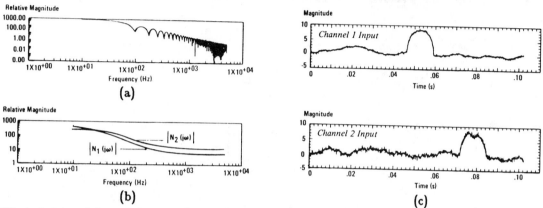

Fig. 9.6.2 (a) Magnitude spectra of signals, (b) noises, and (c) total inputs to channels 1 and 2. (From W.H. Haas and C.S. Lindquist, "Transitional FFT-based filters for delay estimation," Rec. 14th Asilomar Conf. Circuits, Systems, and Computers, pp. 342–344, ©1980, IEEE.)

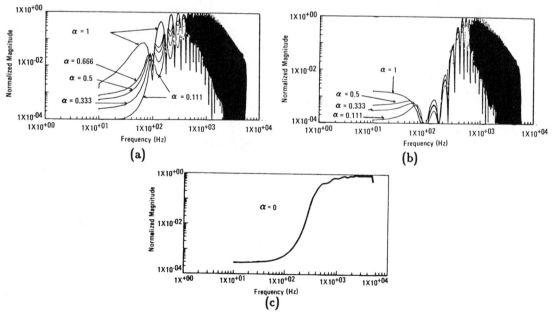

Fig. 9.6.3 Magnitude responses of (a) Class 1 (Eq. 9.6.10), (b) Class 2 (Eq. 9.6.11), and (c) Class 3 (Eq. 9.6.13) delay estimation filters. (From W.H. Haas and C.S. Lindquist, loc. cit., ©1980, IEEE.)

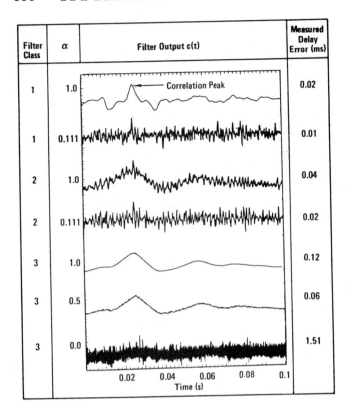

Filter Class	α	Filter Output c(t)	Measured Delay Error (ms)
1	1.0	Correlation Peak	0.02
1	0.111		0.01
2	1.0		0.04
2	0.111		0.02
3	1.0		0.12
3	0.5		0.06
3	0.0		1.51

Time (s)

Fig. 9.6.4 Output c_4 of cross-correlation filters in Fig. 9.6.3. (From W.H. Haas and C.S. Lindquist, loc. cit., ©1980, IEEE.)

the output correlation pulse narrows, and the delay error decreases as anticipated. Minimum delay error is obtained by decreasing α until the cross-correlation peak is still clearly visible (around $SNR = 4$ dB). Reducing α further causes the peak to drop below the noise floor and the delay error to increase rapidly.

Now consider a different problem described in Figs. 9.6.5–9.6.7.[10] Assume the system is required to estimate the delay between Gaussian signal pulses having

$$s_1(nT) = e^{-0.5(nT/3.18 \ msec)^2}, \qquad s_2(nT) = s_1(nT - 0.322346 \ sec) \qquad (9.6.16)$$

The expected noises are assumed to be Gaussian with $1/f^2$ power spectra at low frequencies and white power spectra at high frequencies. Phases are assumed to be random and uniformly distributed between $(-\pi, \pi)$. The noises are shown in Fig. 9.6.5a. Adding the signals given by Eq. 9.6.16 to these noises produces the total inputs shown in Fig. 9.6.5b. The peak input SNR ratio was chosen to equal 8 (i.e., 18 dB). The cross-correlation between $R_1(n)$ and $R_2(n)$ is plotted in Fig. 9.6.5c. The peak occurring at $t = 0.32$ second is clearly evident.

The transfer functions of Class 1 filters are shown in Fig. 9.6.6a while those for Class 2 and 3 filters are shown in Fig. 9.6.6b. The outputs of several of these filters are shown in Fig. 9.6.7. Examining Figs. 9.6.7a–c in succession shows a narrowing of the output pulse accompanied by a lowering of the peak SNR. This requires a trade off between the reduction in temporal uncertainty due to pulse narrowing with the rise in false detections due to a lower SNR ratio. Further delay resolution is obtained from the inverse FFT samples using a sinc function interpolator. Interpolating the Class 1 high-resolution detection filter output gives $t = 0.3223$ which is almost exact. In the Class 3 filter, the smoothed input $\langle |R_1(m)R_2^*(m)| \rangle$ was obtained by convolving the input $|R_1(m)R_2^*(m)|$ with a Gaussian window $W(m)$ where

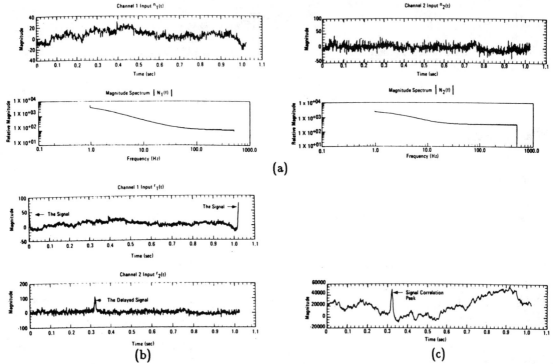

Fig. 9.6.5 (a) Noise inputs n_1 and n_2 and their spectra, (b) total inputs r_1 and r_2 to channels 1 and 2, respectively, and (c) cross-correlation between r_1 and r_2. (From W.H. Haas and C.S. Lindquist, "A synthesis of frequency domain filters for time delay estimation," IEEE Trans. ASSP, pp. 540–548, June, ©1981, IEEE.)

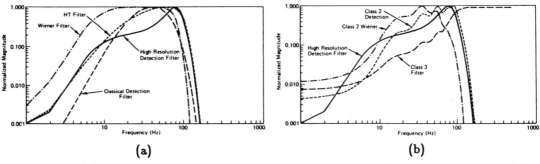

Fig. 9.6.6 (a) Magnitude responses of (a) Class 1 and (b) Class 2 and 3 delay estimation filters. (From C.S. Lindquist and W.H. Haas, loc. cit., ©1981, IEEE.)

Table 9.6.1 A collection of delay estimation filters.[10,11]

Filter Name	Transfer Function $H(m)$	Reference	Eq. Number
Class 1 Classical Detection		Eckart	

$$\frac{E\{S_1^*(m)S_2(m)\}}{E\{|N_1(m)N_2(m)|^2\}} \tag{9.6.7}$$

Class 1 High-Resolution Detection   Haas/Lindquist

$$\frac{E\{S_1^*(m)S_2(m)\}}{E\{|S_1(m)S_2^*(m)|^2\} + E\{|N_1(m)N_2^*(m)|^2\}} \tag{9.6.8}$$

Class 1 Inverse

$$\frac{1}{E\{S_1(m)S_2^*(m)\}} \tag{9.6.9}$$

Class 1 Wiener Uncorrelated Estimation   Hassab/Boucher

$$\frac{E\{|S_1^*(m)S_2(m)|^2\}}{E\{|S_1(m)S_2^*(m)|^2\} + E\{|N_1(m)N_2^*(m)|^2\}}$$

Class 1 Transitional   Haas/Lindquist

$$\frac{E\{S_1^*(m)S_2(m)\}}{(1-\alpha)E\{|S_1(m)S_2^*(m)|^2\} + \alpha E\{|N_1(m)N_2^*(m)|^2\}} \tag{9.6.10}$$

Class 1 Pulse-Shaping Transitional ($P(m) =$ Desired pulse shape)   Haas/Lindquist

$$\frac{E\{S_1^*(m)S_2(m)|P(m)|\}}{(1-\alpha)E\{|S_1(m)S_2^*(m)|^2\} + \alpha E\{|N_1(m)N_2^*(m)|^2\}}$$

Class 1 Hahn-Tretter   Hahn/Tretter

$$\frac{E\{S_1^*(m)S_2(m)\}}{2E\{|S_1(m)S_2^*(m)||N_1(m)N_2^*(m)|\} + E\{|N_1(m)N_2^*(m)|^2\}}$$

Table 9.6.1 (Continued) A collection of delay estimation filters.[10,11]

Filter Name	Transfer Function $H(m)$	Reference	Eq. Number

Class 2 Transitional   Haas/Lindquist

$$\frac{E\{S_1^*(m)S_2(m)\}}{(1-\alpha)E\{|S_1(m)S_2^*(m)|^2\} + \alpha|\langle R_1(m)R_2^*(m)\rangle - E\{S_1(m)S_2^*(m)\}|^2} \qquad (9.6.11)$$

Class 2 Pulse-Shaping Transitional   Haas/Lindquist

$$\frac{E\{S_1^*(m)S_2(m)|P(m)|\}}{(1-\alpha)E\{|S_1(m)S_2^*(m)|^2\} + \alpha|\langle R_1(m)R_2^*(m)\rangle - E\{S_1(m)S_2^*(m)\}|^2}$$

Class 2 High-Resolution Detection   Haas/Lindquist

$$\frac{E\{S_1^*(m)S_2(m)\}}{\langle|R_1(m)R_2^*(m)|^2\rangle} \qquad (9.6.12)$$

Class 2 Wiener Estimation   Hassab/Boucher

$$\frac{E\{|S_1^*(m)S_2(m)|^2\}}{\langle|R_1(m)R_2^*(m)|^2\rangle}$$

Class 3 Transitional   Haas/Lindquist

$$\frac{1}{\langle|R_1(m)R_2^*(m)|^{2(1-\alpha)}\rangle} \qquad (9.6.13)$$

Class 3 Pulse-Shaping Transitional   Haas/Lindquist

$$\frac{|P(m)|}{\langle|R_1(m)R_2^*(m)|^{2(1-\alpha)}\rangle}$$

Class 3 Classical Detection   Roth; Carter, et al

$$\frac{1}{\langle|R_1(m)R_2^*(m)|\rangle} \qquad (9.6.14)$$

Class 3 Autocorrelation   Haas/Lindquist

$$\frac{1}{\langle|R_1(m)R_2^*(m)|^2\rangle} \qquad (9.6.15)$$

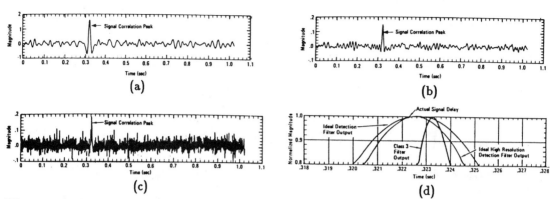

Fig. 9.6.7 Output c_4 of cross-correlation filters in Fig. 9.6.6 for (a) classical detection, (b) high-resolution detection, and (c) Class 3 detection filters with (d) ten-fold resolution increase using sinc function interpolator. (From C.S. Lindquist and W.H. Haas, loc. cit., ©1981, IEEE.)

$$W(m) = e^{-(m/7.07)^2} \qquad (9.6.17)$$

$W(m)$ has a 3 dB smoothing bandwidth of about ±4 bins. It was arbitrarily chosen and yields satisfactory results.

9.7 FFT IMAGE REGISTRATION FILTERS [12]

Image registration requires a precise estimation of the relative displacement or translational offset between two images. Many other image misregistration phenomena like scale, skew, and angle of rotation can be modelled as forms of nonuniform translation. Image registration is therefore a basic operation in image processing. It can also be combined with other fundamental image processing functions such as deblurring (spatial filtering) and object pattern recognition (detection) which will be discussed in the next section. Here we shall be concerned with the estimation of translational offsets with special emphasis on very accurate offset determination.

Image registration is really a spatial form of two-dimensional delay estimation. A general FFT image registration system is shown in Fig. 9.7.1. It is a generalized version of the FFT delay estimation system in Fig. 9.6.1a. A pair of two-dimensional images are input to the system. Windowing is first performed in the lower channel to assure the translated frame image is completely contained in the reference frame image. The window size is inversely proportional to the maximum allowable translational offset to be estimated. The two-dimensional spatial data of the image is converted into two-dimensional spectral data by taking a two-dimensional FFT.

Next pre-correlation filtering of the images is performed to attenuate noise, enhance signal, and shape the image spectra. The filtered images are cross-correlated. Pre-correlation filtering decreases the width of the cross-correlation pulse and reduces noise. It is one of the keys in making accurate estimates by reducing the position error.

The cross-correlation spectrum is next converted into two-dimensional spatial data by performing an inverse FFT.

For precise spatial resolution of the peak location in the cross-correlation function, this spatial data is applied to a sinc function interpolator. The two maximum interpolator outputs $\hat{D}_x$ and $\hat{D}_y$ locate the cross-correlation peak and are estimates of the spatial displacement of one image frame relative to the other. Image registration accuracies to better than $1/100th$ of a *pixel* (i.e., distance between two adjacent *picture elements*) can be obtained using sinc function interpolation but simpler interpolators can be used to reduce computation.

With this general background, let us now discuss the pre-correlator filters in more detail. A large number of filter algorithms can be obtained by simply generalizing the one-dimensional temporal delay filters of Table 9.6.1 into their equivalent two-dimensional spatial image filters. The transfer functions for the three classes of filters which we will consider are:[12]

$$\text{Class 1}: \quad H(m_1, m_2) = \frac{E\{S^*(m_1, m_2)\}\left(\alpha + (1-\alpha)\sqrt{|P(m_1, m_2)|}\right)}{(1-\alpha)E\{|S(m_1, m_2)|^2\} + \alpha E\{|N(m_1, m_2)|^2\}} \qquad (9.7.1)$$

$$\text{Class 2}: \quad H(m_1, m_2) = \frac{E\{S^*(m_1, m_2)\}\left(\alpha + (1-\alpha)\sqrt{|P(m_1, m_2)|}\right)}{(1-\alpha)E\{|S(m_1, m_2)|^2\} + \alpha |R(m_1, m_2)|^2} \qquad (9.7.2)$$

$$\text{Class 3}: \quad H(m_1, m_2) = \frac{\alpha + (1-\alpha)\sqrt{|P(m_1, m_2)|}}{\langle |R(m_1, m_2)|^{1-\alpha}\rangle} \qquad (9.7.3)$$

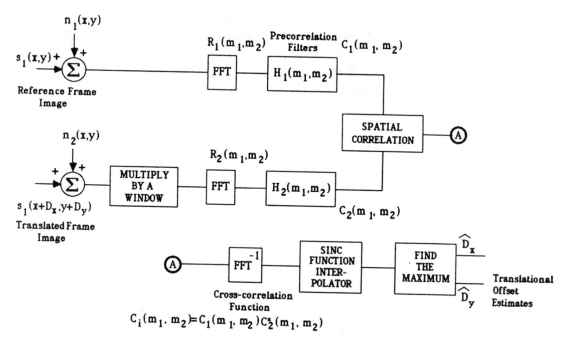

$$C_i(m_1, m_2) = C_1(m_1, m_2)C_2^*(m_1, m_2)$$

Fig. 9.7.1 (a) Image registration filter.[12]

m_1 and m_2 are spatial frequency variables where m_1 represents frequency in the x-direction and m_2 represents frequency in the y-direction. Varying weighting parameter α produces transitional filters which range between those that maximize spatial ambiguity in estimating the position of the cross-correlation peak ($\alpha = 0$) to those that maximize SNR enhancement ($\alpha = 1$).

Spectra $S(m_1, m_2)$ and $N(m_1, m_2)$ are the two-dimensional Fourier transforms of the signal $s(x, y)$ and noise $n(x, y)$ images, respectively. $\langle R(m_1, m_2) \rangle$ is the smoothed input image spectrum of $R(m)$ given by Eq. 9.6.4. Smoothing of $R(m_1, m_2)$ is required to lessen the effects of the spurious cross-correlation term $S(m_1, m_2)N^*(m_1, m_2)$ in $R(m_1, m_2)$.

$P(m_1, m_2)$ is the Fourier transform of the desired cross-correlation function between image signals $s_1(x, y)$ and $s_2(x, y)$. It is essentially a pulse-shaping filter having a gain $P(m_1, m_2)$ given by Eq. 8.2.25 which is matched to the interpolation formula used. $P(m_1, m_2)$ equals

$$P(m_1, m_2) = \mathcal{F}[p(x, y)] \tag{9.7.4}$$

where $p(x, y)$ is the impulse response of the cross-correlation pulse-shaping filter. $p(x, y)$ is computed from the interpolation equation. The impulse is generated by cross-correlating $s_1(x, y)$ and $s_2(x, y)$. Since we assume that $s_2(x, y) = s_1(x + D_x, y + D_y)$, then the cross-correlation estimate equals $R_{s_1 s_2}(x, y) = U_0(x + D_x, y + D_y)$ from Eq. 1.17.7. The spectrum $C(m_1, m_2)$ of $R_{s_1 s_2}$ equals $2\pi e^{-j2\pi D_x / X_o N} e^{-j2\pi D_y / Y_o N}$ where D_x and D_y equal the translational offsets in the x and y directions between the two images $s_1(x, y)$ and $s_2(x, y)$, respectively.

Some examples of interpolation equations are given in Table 9.7.1. These interpolators were discussed in Chaps. 3.5–3.6. For instance, suppose a quadratic interpolator is employed which processes only three adjacent data points. Assume the interpixel spacing (i.e., distance between data points) is X_o. For a one-dimensional problem where the samples are $g(x - X_o), g(x),$ and $g(x + X_o)$ with $g(x)$ the largest, then the peak of $g(x)$ equals $g(x + E\{D_x\})$. The peak occurs where $g'(x) = 0$. Performing this operation on $g(x)$ and solving for $(x + E\{D_x\})$ results in

$$E\{D_x\} = x + \frac{g(x + X_o) - g(x - X_o)}{2\{2g(x) - [g(x + X_o) + g(x - X_o)]\}} \tag{9.7.5}$$

which agrees with Table 9.7.1. The impulse response of the quadratic interpolation equation equals

$$p(x) = 1 - \left(\frac{x - E\{D_x\}}{2X_o}\right)^2, \quad |x - E\{D_x\}| \leq 1 \tag{9.7.6}$$

$$= 0, \quad \text{elsewhere}$$

where $E\{D_x\}$ is the peak location of $p(x)$. The pulse-shaping filter has a gain $P(m)$ which is the Fourier transform of $p(x)$.

Table 9.7.1 Some one-dimensional interpolation equations where
$$0 \le x = \tfrac{t}{T} - n < N - 1.$$

No. Pts. (Pixels)	Interpolator Name	Eq. Number / Spatial Eq. $g(x)$	Location of Peak $(x_p - x)$
1	Zero-order (Constant)	Eq. 3.5.3 g_n	Indeterminant
2	First-order (Linear)	Eq. 3.5.8 $g_n + x(g_{n+1} - g_n)$	$0, \quad g_{n+1} < g_n$ $1, \quad g_{n+1} > g_n$ Indeterminant, $g_{n+1} = g_n$
3	Second-order (quadratic)	Eq. 3.5.13 $g_{n+1} + \tfrac{x}{2}(g_{n+2} - g_n)$ $+ \tfrac{x^2}{2}(g_{n+2} + g_n - 2g_{n+1})$	$\dfrac{g_{n+2} - g_n}{2[2g_{n+1} - (g_{n+2} + g_n)]}$
N	Nth-order (Lagrange)	Eq. 3.5.16 $g_n q_{N,0} + g_{n+1} q_{N,1} + \ldots + g_{n+N} q_{N,N}$	—
N	Nth-order (Fourier series)	Eq. 3.7.1 $g_n q_0 + g_{n+1} q_1 + \ldots + g_{n+N} q_N$	—
N	Nth-order (Newton, Gauss, Stirling, Bessel, Everett)	Table 10.6.1 see Chap. 10.6	—
∞	Sinc $(\sin x/x)$ (Ideal lowpass filter)	Eq. 3.6.1 $\displaystyle\sum_{n=-\infty}^{\infty} g_n \frac{\sin[\pi(x - nT)/T]}{\pi(x - nT)/T}$	—

Alternatively, all samples are used in the sinc function interpolator. It has an impulse response

$$p(x) = \frac{\sin[\pi(x - E\{D_x\})/X_o]}{\pi(x - E\{D_x\})} \tag{9.7.7}$$

$E\{D_x\}$ is determined by finding the maximum value of $g(x)$.

To illustrate use of the image registration algorithms in Eqs. 9.7.1–9.7.3, consider the 256×256 pixel Landsat image shown in Fig. 9.7.2. Two identical images are translated by $D_x = 26.643$ pixels and $D_y = -16.265$ pixels and degraded by additive uncorrelated white Gaussian noise. The rms noise level was set at two image intensity quantization levels. The second frame is windowed to insure it is contained within the first frame. The Class 3 filter algorithm given by Eq. 9.7.3 will be used with $\alpha = 0.5$ and $P(m_1, m_2) = 1$. Therefore the pre-correlation filters have transfer functions $H_i(m_1, m_2) = 1/\sqrt{R_i(m_1, m_2)}$. A sinc function interpolator will be used.

The spatial outputs of the correlator and the sinc interpolator are shown in Figs. 9.7.3 and 9.7.4. Those shown in Fig. 9.7.3 result using no pre-correlation filtering (i.e.,

LOW
INTENSITY ◀— COLOR SCALE —▶ INTENSITY

HIGH

$i_1 (x,y)$

$i_2 (x,y)$

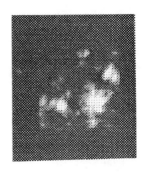

(a)   (b)

Fig. 9.7.2 Landsat images of (a) reference frame and (b) windowed reference frame translated $D_x = 26.643$ pixels and $D_y = -16.265$ pixels. (From **W.H. Haas and C.S. Lindquist**, "The precision registration of digital images – a frequency domain approach," Rec. 15th Asilomar Conf. Circuits, Systems, and Computers, pp. 212–216, ©1981, IEEE. Image digitized by **G. Seetharaman and S. Yhann**.)

$H_i(m_1, m_2) = 1$). Those shown in Fig. 9.7.4 result using the Class 3 pre-correlation filters just described. Due to the large SNR, the unfiltered contour plot of Fig. 9.7.3a exhibits less noise effects than the Class 3 filtered plot of Fig. 9.7.4a. Both plots exhibit cross-correlation peaks in the vicinity of $x = 27$ pixels and $y = 16$ pixels. The tenfold expanded contour plots of Figs. 9.7.3b and 9.7.4b show greater detail.

The sinc function interpolator outputs are shown in Figs. 9.7.3c and 9.7.4c. The cross-correlation peaks are estimated by finding the maximum sinc function levels. For the unfiltered case in Fig. 9.7.3c, $\widehat{D}_x = 26.62$ pixels and $\widehat{D}_y = -16.29$ pixels while for the Class 3 filter in Fig. 9.7.4c, $\widehat{D}_x = 26.64$ pixels and $\widehat{D}_y = -16.26$ pixels. Both estimated locations are almost exact but the Class 3 filter is closer to the actual location of $D_x = 26.643$ pixels and $D_x = -16.265$ pixels with an error $E_x = 0.003$ pixel and $E_y = 0.005$ pixel. Without the interpolator, the 256×256 pixel image registration resolution could be no better than one pixel. With interpolation, better than $1/100th$ of a pixel resolution has been obtained.

Although the FFT approach to image registration is functionally simple, it requires a large number of operations which makes it impractical for many applications. A number of simplifications might be employed.[13]

First, combine the two pre-correlation filters $H_1(m_1, m_2)$ and $H_2(m_1, m_2)$ and the cross-correlation operation into a single filter $H(m)$ as was done for delay estimation in Fig. 9.6.1. Also simpler Class 1 and 2 fixed pre-correlation filters could be used rather than the Class 3 adaptive FFT filter used here.

Second, compute only parts of the cross-correlation function when searching for its maximum. Figs. 9.7.3a and 9.7.4a show only a few contours need to be calculated to confirm the peak exists in the lower left corner of the plot.

Third, after pre-correlation filtering, select only the most prominent image feature for further processing. For example, sharp edges can often be extracted by simple thresholding.

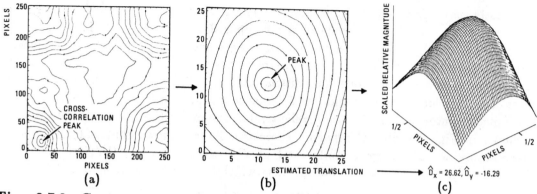

Fig. 9.7.3 Contour plots of (a) cross-correlator output, (b) ten-fold expansion around peak, and (c) sinc function interpolator output expanding 100-fold. No precorrelation filter used. (From **W.H. Haas and C.S. Lindquist**, loc. cit., ©1981, **IEEE.**)

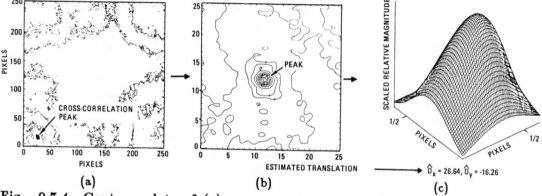

Fig. 9.7.4 Contour plots of (a) cross-correlator output, (b) ten-fold expansion around peak, and (c) sinc function interpolator output expanding 100-fold. Class **3** precorrelation filter used. (From **W.H. Haas and C.S. Lindquist**, loc. cit., ©1981, **IEEE.**)

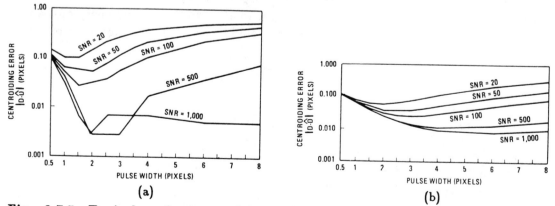

Fig. 9.7.5 Typical performance characteristics of centroiding error versus pulse width for (a) sinc function and (b) quadratic interpolators. (From **W.H. Haas and C.S. Lindquist**, loc. cit., ©1981, IEEE.)

Lastly, use a lower-order interpolator. The performance of a quadratic interpolator and a sinc function interpolator are compared in Fig. 9.7.5.[12] These were determined empirically from a specific pair of images and not analytically for a general pair of images. They are useful to approximate the image registration error that can be obtained using the methods just discussed. Fig. 9.7.5a shows the trade-off between the peak location error and the 50% pulse width as a function of SNR for the sinc function interpolator. The equivalent information for the quadratic interpolator is shown in Fig. 9.7.5b. Comparing these two figures shows that: (1) High-order interpolators are advantageous only for high SNR (in this case, $SNR > 100 = 40$ dB) and (2) produce cross-correlation pulse widths typically between 2 and 3 pixels.

9.8 FFT IMAGE ENHANCEMENT FILTERS [15]

Another image processing problem involves deblurring which requires reducing the *fuzziness* of a blurred image. This is related to pattern recognition, feature extraction, and *edge detection* where the outlines of objects are to be determined.

An FFT edge enhancement system is shown in Fig. 9.8.1. An image having spectrum $R(m_1, m_2)$ is expressed as the sum of an *edge structure* or signal $s(x, y)$ and a *non-edge structure* or noise $n(x, y)$. The image spectrum $R(m_1, m_2)$ is passed through a system blurring transfer function $G(m_1, m_2)$ to produce a total image $I(m_1, m_2)$. The problem is to filter this total image with $H(m_1, m_2)$ to produce an output image $C(m_1, m_2)$ having certain properties.

For the edge enhancement problem we shall consider, assume that the blur transfer function $G(m_1, m_2)$, the expected edge model $E\{E(m_1, m_2)\}$, and the total filter input $R(m_1, m_2)$ are known apriori. However the specific edge structure $S(m_1, m_2)$ and non-edge structure $N(m_1, m_2)$ of the image is unknown. An optimum edge detection filter $H(m_1, m_2)$ must be determined which transforms the expected edge features modeled by $E\{E(m_1, m_2)\}$ into certain desirable features modelled by $F(m_1, m_2)$ and attenuate all other features. The one-dimensional algorithms in Tables 8.2.1 and 8.2.2 are useful when generalized to two-dimensions. If the purpose of $H(m)$ is to maximize the peak absolute intensity of the edge structure output relative to the rms intensity of the non-edge structure, a Class 2 pulse-shaping filter should be used. Its transfer function equals

$$H(m) = \frac{E\{E^*(m)P(m)\}}{\langle|R(m)|^2\rangle G(m)} \tag{9.8.1}$$

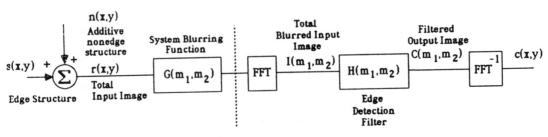

Fig. 9.8.1 Image edge enhancement filter.[15]

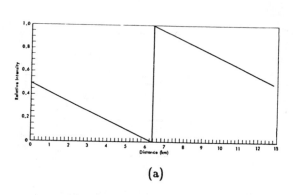

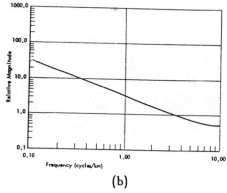

(a)

(b)

Fig. 9.8.2 Model of an image edge $E(x)$ in (a) spatial domain with its (b) magnitude and (c) phase spectra. (From W.H. Haas and C.S. Lindquist, "A frequency domain approach to synthesizing near optimum edge detection filters," Proc. IEEE Intl. Conf. ASSP, pp. 745–748, ©1980, IEEE.)

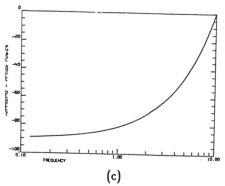

(c)

Again averaging or smoothing a number of inputs $|R_i(m)|^2$ or FFT blocks, as was done in Eq. 9.4.12, forms an estimate of $E\{|S(m)|^2\} + E\{|N(m)|^2\}$ where the cross-spectral term $\langle S(m)N^*(m) + S^*(m)N(m)\rangle$ approaches zero. Inner-block averaging (Eq. 9.5.10) can also be used.

To illustrate these results, consider the image being processed has the edge model shown in Fig. 9.8.2a. Essentially the image around an edge is assumed to make a 13 Km linear transition in relative intensity. An instantaneous white-black transition occurs at the edge itself. The edge magnitude and phase spectra are shown in Figs. 9.8.2b and 9.8.2c, respectively. For simplicity, assume that no blurring is involved so $G(m) = 1$ and that no particular features are to be enhanced so $P(m) = 1$.

The Landsat image being processed is shown in Fig. 9.8.3a. To determine the optimum filter $H(m)$ given by Eq. 9.8.1, the average or smoothed input power spectrum $\langle |R(m)|^2 \rangle$ must be determined. This was accomplished by evaluating the FFT power spectra of each x-direction row of image pixels and averaging them. The result is shown in Fig. 9.8.3b. It has asymptotic roll-offs of $1/f$ below a frequency of 1 Hz/Km and $1/f^2$ above.

Substituting the edge spectrum shown in Figs. 9.8.2b and 9.8.2c and the average image power spectrum shown in Fig. 9.8.3b yields the transfer function $H(m)$ of Eq. 9.8.1. However $H(m)$ is one-dimensional. To convert it into a two-dimensional filter, the $H(m)$ response is rotated so that

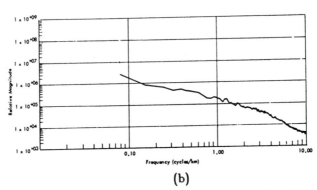

(a)   (b)

Fig. 9.8.3   (a) Landsat image and its (b) expected power spectrum obtained by averaging the image power spectrum of each x-direction row. (Based upon W.H. Haas and C.S. Lindquist, loc. cit., ©1980, IEEE. Image digitized by G. Seetharaman and S. Yhann.)

$$H(m_1, m_2) = H(m)\big|_{m=\sqrt{m_1^2+m_2^2}} \tag{9.8.2}$$

Thus the gain $H(m_1, m_2)$ is assumed to be constant around circles of constant radius $m = \sqrt{m_1^2 + m_2^2}$ from the origin (see Eq. 2.10.10) and Fig. 2.10.2. The magnitude of $H(m)$ is shown in Fig. 9.8.4a.

The output image spectrum $C(m_1, m_2)$ is obtained by multiplying the input image spectrum $R(m_1, m_2)$ shown in Fig. 9.8.3a by the transfer function $H(m_1, m_2)$ of Eq. 9.8.2. The output image $c(x, y)$ is obtained by taking the inverse FFT of $C(m_1, m_2)$. In $c(x, y)$, light shades locate positive-step edges and dark shades locate negative-step edges. The absolute value image $|c(x, y)|$ is shown in Fig. 9.8.4b. $|c(x, y)|$ is used for edge detection where edge polarity is unimportant. In $|c(x, y)|$, light shades indicate edges and dark shades indicate non-edges. The log scaled dB version or cepstrum of this image is shown in Fig. 9.8.4c (see Eq. 4.10.16).

To recover some of the non-edge features in Figs. 9.8.4b and 9.8.4c, two transitional filters were formulated using nth-root or power law versions of $H(m)$ as

$$H_1(m) = \sqrt{H(m)}, \qquad H_2(m) = \sqrt{H_1(m)} = \sqrt{\sqrt{H(m)}} \tag{9.8.3}$$

(i.e., square root and fourth-root filters). The magnitude responses of these mild edge enhancement filters and the resulting images are shown in Fig. 9.8.5. The non-edge image features are slightly maintained in the $H_1(m)$ filter of Fig. 9.8.5a and more maintained in the $H_2(m)$ filter of Fig. 9.8.5b. These mild edge enhancement filters behave differently from the strong edge enhancement filter $H(m)$ of Fig. 9.8.4 which detects sharp intensity variations and rejects non-edge backgrounds. These images demonstrate how transitional filters can be used to obtain different degrees of edge enhancement.

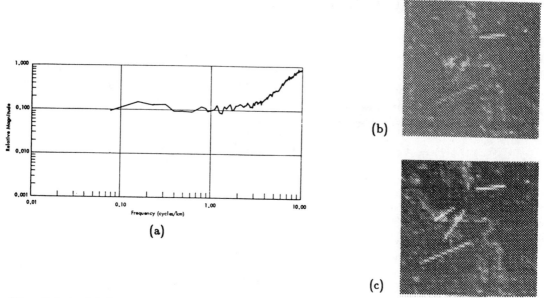

Fig. 9.8.4 (a) Magnitude response of edge enhancement filter, and output images (b) $|c|$ and (c) $\log_{10}|c|^2$ using Eq. 9.8.1. (Based upon W.H. Haas and C.S. Lindquist, loc. cit., ©1980, IEEE. Image digitized by G. Seetharaman and S. Yhann.)

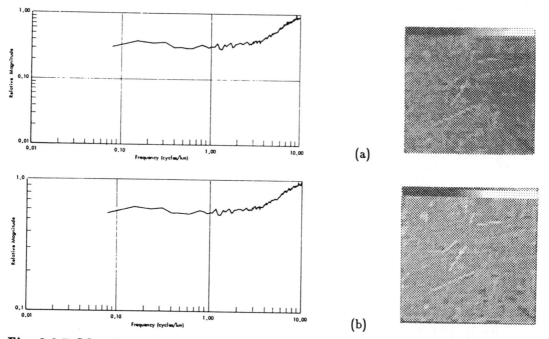

Fig. 9.8.5 Magnitude response of edge enhancement filter and output image for (a) Eq. 9.8.3a and (b) Eq. 9.8.3b. (Based upon W.H. Haas and C.S. Lindquist, loc. cit., ©1980, IEEE. Image digitized by G. Seetharaman and S. Yhann.)

9.9 FFT COMMUNICATION FILTERS [17]

Binary data is transmitted by communication systems using a variety of modulation techniques including on-off keying (OOK) or amplitude-shift keying (ASK), frequency-shift keying (FSK), and phase-shift keying (PSK). The modulated signals obtained using these techniques are shown in Fig. 9.9.1. The corresponding classical amplitude, frequency, and phase demodulation systems are shown in Fig. 9.9.2.

ASK signals are described by

$$s(t) = A\cos(\omega_c t), \quad \text{for binary 1} \atop = 0, \qquad\qquad \text{for binary 0} \Bigg\}, \quad nT \le t < (n+1)T \qquad (9.9.1)$$

$s(t)$ is a pulsed-cosine signal with frequency ω_c and duration T. It can be demodulated using the noncoherent receiver system shown in Fig. 9.9.2a. It consists of a bandpass filter, envelope detector, and threshold detector.

FSK signals are described by

$$s(t) = A\cos(\omega_1 t), \quad \text{for binary 1} \atop = A\cos(\omega_2 t), \quad \text{for binary 0} \Bigg\}, \quad nT \le t < (n+1)T \qquad (9.9.2)$$

$s(t)$ is a pulsed-cosine signal with a frequency of ω_1 or ω_2 and duration T. It can be demodulated as shown in Fig. 9.9.2b. The FSK demodulator consists of two ASK demodulators which operate in parallel whose outputs are subtracted.

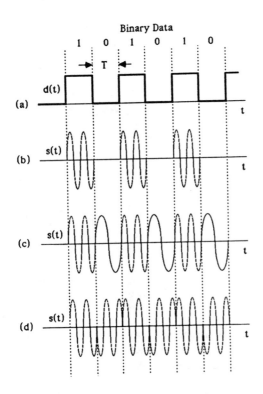

Fig. 9.9.1 (a) Encoded digital data using (b) ASK, (c) FSK, and (d) PSK modulators.[17]

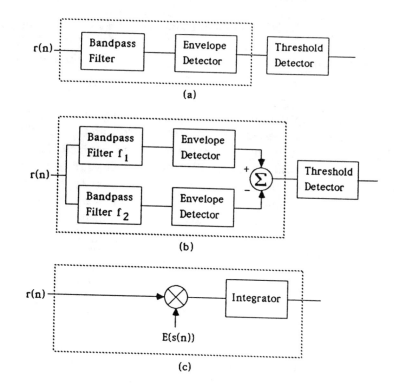

Fig. 9.9.2
(a) Noncoherent ASK,
(b) noncoherent FSK,
and (c) coherent PSK
demodulators.[17]

PSK signals are described by

$$s(t) = A\cos(\omega_c t), \qquad \text{for binary 1}$$
$$= -A\cos(\omega_c t), \qquad \text{for binary 0} \Big\}, \quad nT \le t < (n+1)T \qquad (9.9.3)$$

$s(t)$ is a pulsed-cosine signal of frequency ω_c with an amplitude of A or $-A$. A coherent PSK demodulator is shown in Fig. 9.9.2c.

Using this apriori information about the signal $s(t)$ for each modulation method, we can formulate appropriate adaptive filters using the frequency domain algorithms of Tables 8.2.1 and 8.2.2. These algorithms describe the signal processing system in Fig. 8.2.1. The total input applied to the demodulator equals

$$r(t) = s(t) + n(t) \qquad \text{or} \qquad R(j\omega) = S(j\omega) + N(j\omega) \qquad (9.9.4)$$

If the expected values of both $S(j\omega)$ and $N(j\omega)$ are known apriori, we use Class 1 filters. However, if only one is known apriori, Class 2 filters are used instead. We shall consider both cases.

The expected signal used by all these systems has the basic form (see Eqs. 1.6.25 and 1.6.27)

$$E\{s(t)\} = A\cos(\omega_c t)\big[U_{-1}(t + T/2) - U_{-1}(t - T/2)\big]$$
$$E\{S(j\omega)\} = \frac{A}{2}\big[U_0(\omega + \omega_c) + U_0(\omega - \omega_c)\big] * T\frac{\sin(\omega T/2)}{\omega T/2}$$
$$= \frac{AT}{2}\left[\frac{\sin[(\omega + \omega_c)T/2]}{(\omega + \omega_c)T/2} + \frac{\sin[(\omega - \omega_c)T/2]}{(\omega - \omega_c)T/2}\right] \qquad (9.9.5)$$

Very often, $n(t)$ is a Gaussian random process with zero mean and variance σ. Its phase distribution is assumed to be uniform between $(-\pi, \pi)$. The expected noise spectrum $E\{|N(j\omega)|\}$ is arbitrary. The magnitude spectrum is often taken to be white where

$$E\{|N(j\omega)|\} = N_o \tag{9.9.6}$$

The adaptive filters are obtained by substituting the digitized expected signal and noise into the desired filter of Tables 8.2.1 and 8.2.2. These adaptive filters are very flexible. Estimation types can be used to replace the bandpass filters in Fig. 9.9.2. Alternatively, the bandpass filter and envelope detector combination shown inside the dotted lines can be replaced with a detection type filter. This reduces the hardware needed. Detection filters output narrow pulses when the expected signal is received. We may wish to output pulses rather than impulses. Conceptually this involves following the high-resolution filter with a pulse-shaping filter having an impulse response of an impulse response of (see Eqs. 1.6.19 and 1.6.22)

$$p(t) = U_{-1}(t) - U_{-1}(t - T)$$
$$P(j\omega) = T\,\frac{\sin(\omega T/2)}{\omega T/2}e^{j\omega T/2} \tag{9.9.7}$$

In practice they are combined into one filter called the pulse-shaping estimation filter.

The ASK case is illustrated in Fig. 9.9.3 with encoded data of $(0,0,1,0,1,0)$ using two 5 Hz cosine pulses of 0.5 second duration. Fig. 9.9.3a shows them centered about 1.25 and 2.25 seconds. The expected spectrum is that of a single cosine pulse given by Eq. 9.9.5 and is shown in Fig. 9.9.3b. The noise is assumed to be white as given by Eq. 9.9.6 and shown in Fig. 9.9.3c. The total input and its spectrum are shown in Fig. 9.9.3d for an rms SNR of 0 dB.

Consider a Class 1 uncorrelated high-resolution ASK detection filter that has a gain (see Table 8.2.1)

$$H(m)_{ASK} = \frac{E\{S^*(m)\}}{E\{|S(m)|^2\} + E\{|N(m)|^2\}} \tag{9.9.8}$$

Over frequency ranges where the spectral SNR is large or $E\{|S|\}/E\{|N|\} \gg 1$, $H = 1/E\{S\}$ so it acts as an inverse filter. In ranges where $E\{|S|\}/E\{|N|\} \ll 1$, $H = E\{S^*\}/E\{|N|^2\}$ so it acts as a classical detection filter. Thus, this filter acts in an optimal manner and maximizes the peak output signal-to-total rms output (signal plus noise) ratio. $c(n)$ and $|C(m)|$ are shown in Figs. 9.9.4b and 9.9.4c, respectively.

If a Class 2 algorithm was used instead, the high-resolution ASK detection filter becomes (see Table 8.2.2)

$$H(m)_{ASK} = \frac{E\{S^*(m)\}}{\langle|R(m)|^2\rangle} \tag{9.9.9}$$

Here only the apriori signal spectrum of Eq. 9.9.5 is used and the noise is arbitrary. $R(m)$ is the spectrum of the total received signal which is averaged in some way (Eqs. 9.4.12 or 9.5.10) to reduce undesirable cross-correlation between $E\{S(m)\}$ and $E\{N(m)\}$. The filter output would not be as good as that in Fig. 9.9.4b.

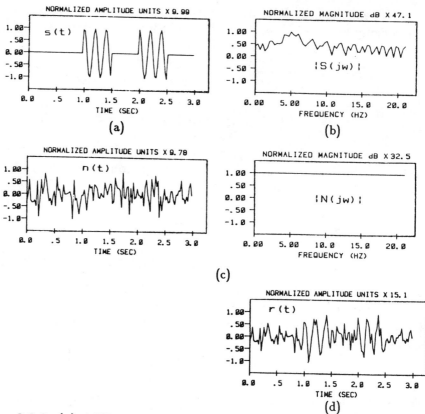

Fig. 9.9.3 (a) ASK signal and its (b) spectrum, (c) noise and its spectrum, and (d) total input. (From C.S. Lindquist and H.B. Rozran, "Amplitude, frequency, and phase demodulation systems using frequency domain adaptive filters," Rec. 18th Asilomar Conf. Circuits, Systems, and Computers, pp. 332–336, ©1984, IEEE.)

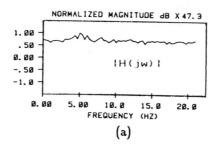

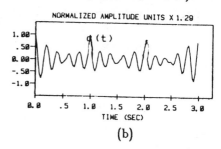

Fig. 9.9.4 Class 1 ASK high-resolution filter showing (a) magnitude response, (b) filter output, and (c) its spectrum. (From C.S. Lindquist and H.B. Rozran, loc. cit., ©1984, IEEE.)

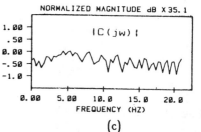

Next consider the FSK demodulator. Now two different adaptive filters are used. Generalizing Eq. 9.9.8 gives the high-resolution FSK filters as

$$H_1(m)_{FSK} = \frac{E\{S_1^*(m)\}}{E\{|S_1(m)|^2\} + E\{|S_2(m)|^2\} + E\{|N(m)|^2\}}$$

$$H_2(m)_{FSK} = \frac{E\{S_2^*(m)\}}{E\{|S_1(m)|^2\} + E\{|S_2(m)|^2\} + E\{|N(m)|^2\}} \tag{9.9.10}$$

where $E\{S_1(m)\}$ and $E\{S_2(m)\}$ are given by Eq. 9.9.5 using $m_c = m_1$ and m_2, respectively. H_1 and H_2 can be combined into a single equivalent uncorrelated high-resolution FSK transfer function as

$$
\begin{aligned}
H(m)_{FSK} &= H_1(m)_{FSK} - H_2(m)_{FSK} \\
&= \frac{E\{S_1^*(m)\} - E\{S_2^*(m)\}}{E\{|S_1(m)|^2\} + E\{|S_2(m)|^2\} + E\{|N(m)|^2\}}
\end{aligned} \tag{9.9.11}
$$

since the outputs are subtracted in Fig. 9.9.2b. This is the gain between the FSK demodulator input and threshold detector input. It has been assumed that both signals are uncorrelated to each other and uncorrelated to the noise. The Class 2 form of Eq. 9.9.11 is

$$H(m)_{FSK} = \frac{E\{S_1^*(m)\} - E\{S_2^*(m)\}}{\langle |R(m)|^2 \rangle} \tag{9.9.12}$$

This shows that ASK can be viewed as a special case of FSK where $m_2 = 0$ (see Fig. 9.9.2b).

To illustrate FSK, assume that $f_1 = 5.9$ Hz and $f_2 = 1.9$ Hz with encoded data of $(1,0,1,0,1,0)$ where $f_1 = $ "1" and $f_2 = $ "0". Both f_1 and f_2 pulses are 0.5 seconds wide. The assumed composite signal shown in Fig. 9.9.5a has f_1 pulses centered about 0.25, 1.25, and 2.25 seconds and f_2 pulses centered at 0.75, 1.75, and 2.75 seconds. The expected spectrum magnitude $|E\{S_1\} - E\{S_2\}|$ is shown in Fig. 9.9.5b. The noise is assumed to be white as shown in Fig. 9.9.3c. The filter input $r(n)$ with a SNR of 0 dB is shown in Fig. 9.9.5c. The FSK filter amplitude response computed from Eq. 9.9.11 is shown in Fig. 9.9.6a. The outputs $c(n)$ and $|C(m)|$ are shown in Figs. 9.9.6b and 9.9.6c, respectively. As with the ASK the results would not be as good if a Class 2 algorithm were used instead.

The PSK case can also be viewed as a special case of FSK where $E\{S_2(m)\} = -E\{S_1(m)\} = -E\{S(m)\}$ in Eqs. 9.9.11 or 9.9.12. Therefore, the high-resolution PSK transfer functions become

$$\text{Class 1}: \quad H(m)_{PSK} = \frac{2E\{S^*(m)\}}{2E\{|S(m)|^2\} + E\{|N(m)|^2\}}$$

$$\text{Class 2}: \quad H(m)_{PSK} = \frac{2E\{S^*(m)\}}{\langle |R(m)|^2 \rangle} \tag{9.9.13}$$

The PSK responses are very close to the FSK responses so Eq. 9.9.13 will not be discussed further.

It is interesting to see how easily the frequency domain algorithms of Tables 8.2.1 and 8.2.2 can be applied to ASK, FSK, and PSK communication systems. Since these systems are adaptive, they give improved performance over a wide range of noise and interference conditions.

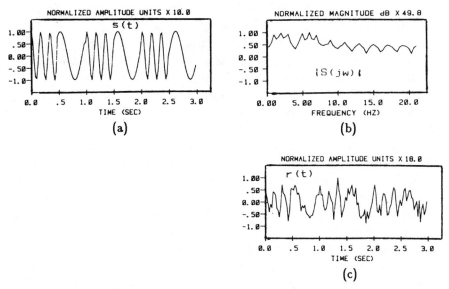

Fig. 9.9.5 (a) FSK signal and its (b) spectrum, and (c) total input. (From C.S. Lindquist and H.B. Rozran, loc. cit., ©1984, IEEE.)

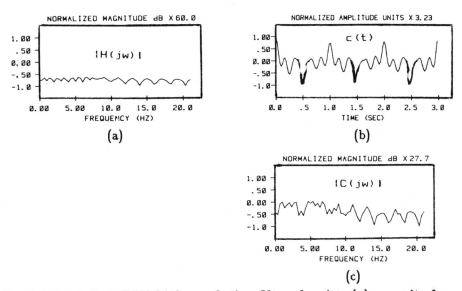

Fig. 9.9.6 Class 1 FSK high-resolution filter showing (a) magnitude response, (b) filter output, and (c) its spectrum. (From C.S. Lindquist and H.B. Rozran, loc. cit., ©1984, IEEE.)

9.10   FFT RADAR FILTERS [18]

Radar processors use three basic approaches: (1) time domain,[19] (2) frequency domain,[20] and (3) spectral estimation.[21] *Doppler radar* measures the return velocities of moving targets by finding doppler shift frequencies. Frequency analysis techniques are needed to quickly process the return spectrum. Only our frequency domain approach will be described here.

A typical radar return is shown in Fig. 9.10.1. Assuming that the radar transmitter-receiver is stationary, then it consists of signal (in this case the jet plane) and noise (clutter). *Clutter* arises from stationary objects like buildings and clouds (dc clutter), and moving objects like other planes, birds, etc. (ac clutter). In Fig. 9.10.1, there are stationary objects giving dc clutter, the plane at the FFT doppler frequency bin b_1, the jet at the FFT doppler frequency bin b_2, and broad-band noise. The target signal return is often 80 dB below the clutter return, the signal has nonzero and variable doppler shift, and clutter return can appear across the frequency band. Therefore the radar filter must accommodate the signal fading and low SNR environment.

Radar returns are coherent in nature. This usually makes algorithms which ignore the phase information ineffective. Algorithms utilizing phase should be used. In addition, neither the expected returned signal nor noise are known apriori. Hence this is the most difficult Class 3 filter problem but the frequency domain algorithms can handle them. Finally, since most radar systems perform FFT's, they already have FFT hardware in place so further hardware is minimized.

Now let us consider the adaptive radar filters themselves. A variety of Class 1–3 algorithms are listed in Tables 8.2.1 and 8.2.2. They involve the expected signal spectrum $E\{S(m)\}$ (or power spectrum $E\{|S(m)|^2\}$), the expected noise power spectrum $E\{|N(m)|^2\}$, and the input spectrum $R(m)$. In radar, $E\{|S(m)|^2\}$ and $E\{|N(m)|^2\}$

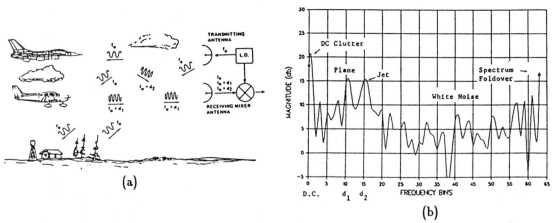

(a)

(b)

Fig. 9.10.1   (a) Radar environment and (b) typical radar return spectrum. (From C.S. Lindquist and L.W. Conboy, "Design of radar systems using frequency domain adaptive filters," Rec. 19th Asilomar Conf. Circuits, Systems, and Computers, pp. 411–415, ©1985, IEEE.)

are not known but must be estimated aposteriori as $\widehat{E}\{|S(m)|^2\}$ and $\widehat{E}\{|N(m)|^2\}$, respectively.

The Class 3 filters of Table 8.2.2 could be used. However let us use a different approach to obtain another set of Class 3 filters shown in Table 9.10.1. We do this by simply replacing the expectations $E\{\ \}$ by their estimates $\widehat{E}\{\ \}$ in the Class 1 filters of Table 8.2.1. For example, the uncorrelated Wiener filters have

$$\text{Class 1}: \quad H(m) = \frac{E\{|S(m)|^2\}}{E\{|S(m)|^2\} + E\{|N(m)|^2\}}$$

$$\text{Class 2}: \quad H(m) = \frac{E\{|S(m)|^2\}}{E\{|S(m)|^2\} + \widehat{E}\{|N(m)|^2\}} \qquad (9.10.1)$$

$$\text{Class 3}: \quad H(m) = \frac{\widehat{E}\{|S(m)|^2\}}{\widehat{E}\{|S(m)|^2\} + \widehat{E}\{|N(m)|^2\}}$$

Several special radar filters were developed. These are shown in Table 9.10.2. They include the autocorrelation, uncorrelated Wiener, Wiener squared, averaged Wiener, and Wiener with noncoherent integration. These filters must be used somewhat differently than those of the preceding sections because of the SNR environments and fading inherent in radar return signals.

The radar system in which these filters are used are shown in Fig. 9.10.2. It is described by the equations

$$C(m) = H_S(m)R(m)$$
$$\widehat{N}(m) = H_N(m)R(m) \quad \text{if} \quad \text{CFAR} = 0$$
$$\qquad\qquad = R(m) - S(m) \quad \text{if} \quad \text{CFAR} = 1 \qquad (9.10.2)$$
$$\widehat{S}(m) = R(m) - \widehat{NN}(m)_k$$
$$\widehat{NN}(m)_k = \widehat{NN}(m)_{k-1} + \mu\widehat{N}(m)_k$$

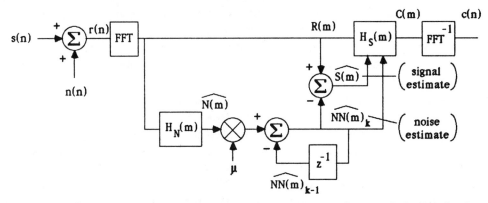

Fig. 9.10.2 Class 3 adaptive radar filter implementation. (From C.S. Lindquist and L.W. Conboy, loc. cit., ©1985, IEEE.)

The pertinent equations are reevaluated with each new kth data block. Signal estimation filter $H_S(m)$ corresponds to one of the transfer functions in Table 9.10.2. The noise estimation filter $H_N(m)$ is also obtained from Table 9.10.2 where the $\widehat{S}(m)$ terms are replaced by $\widehat{N}(m)$. A convergence factor μ is also sometimes used where $0 < \mu < 1$. Then $\widehat{NN}(m)$ is used in place of $\widehat{N}(m)$. Typically μ is in the range from 0.05 to 0.25. μ keeps the system responsive and yet somewhat immune to fades. A smaller μ requires more noise iterations to build up the noise estimate. Larger μ's make the system more sensitive to fading. Now let us make several important comments about these equations.

Radar systems often have long and drastic target signal fades. The target SNR is poor (small) due to large clutter returns which are coherent in nature. Thus a technique was developed to build up a good noise estimate when target signal is not present. This depends on using the constant false alarm rate (CFAR) output of radar systems. When CFAR exceeds a variable threshold, the radar begins its signal tracking. Thus in Eq. 9.10.2, $\widehat{N}(m) = H_N(m)R(m)$ until CFAR switches high and then $\widehat{N}(m) = R(m) - S(m)$. The noise estimate is continually updated even after target tracking has begun. Unfortunately, correlated signal components in the noise estimates reduce the input $r(t)$ levels.

For the radar simulations, the expected signal was taken to be a pulsed complex exponential where

$$E\{s(t)\} = Ae^{m_o t}\left[U_{-1}(t - T_1) - U_{-1}(t - T_2)\right]$$

$$E\{|S(m)|\} = 2\pi A(T_2 - T_1)\frac{\sin[m_o(T_2 - T_1)/2]}{m_o(T_2 - T_1)/2} \qquad (9.10.3)$$

The real part is processed by the "I" or in-phase radar channel; the imaginary part is processed by the "Q" or quadrature-phase radar channel. The signal is gated on at T_1 and off at T_2. Its spectrum is centered at m_o rad/sec and it has a $(\sin x/x)$ shape with a zero-crossing bandwidth of $1/(T_2 - T_1)$ Hz. The signal spectrum can be *dithered* or modulated by $m(t)$ between FFT blocks as $S_m(m;t) = m(t)S(m;t)$. We take $m(t)$ to be a random number uniformly distributed between $(0, M)$ where M is the desired percentage modulation. The magnitude of $m(t)$ scales the spectrum to simulate the peaks and fades caused by target motion, cross-sectional target variation, etc.

Several background noise magnitude spectra are considered which include white, $1/f$, and $1/f^2$ types (i.e., moving clutter). Phase is taken to be random in the range $(-\pi, \pi)$. In addition, a dc clutter spike is also added to simulate stationary clutter. The noises can also be dithered like the signal. All dithers are independent of one another. The rms signal-to-rms noise ratio is set to any desired level.

A radar spectrum return was generated like that shown in Fig. 9.10.1b. The signal $s(t)$, noise $n(t)$, and input $r(t)$ are shown in Fig. 9.10.3. The jet signal was placed at frequency bin 15.3. The plane frequency was placed at bin 10.3. The trees, buildings, and noise give the large dc return spike. White noise was also added. Digitized $(\sin x/x)$ spectral shaping is present due to the finite length data. 64-point FFT's were used with a 50% duty cycle (i.e., 32 nonzero points and a ± 0.5 bin zero-crossing bandwidth).

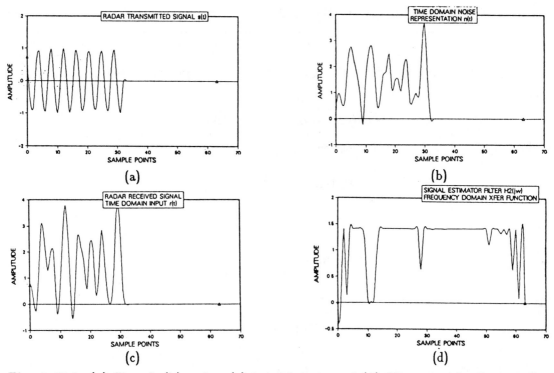

Fig. 9.10.3 (a) Signal, (b) noise, (c) total input, and (d) filter magnitude response with no convergence or modulation. (From C.S. Lindquist and L.W. Conboy, loc. cit., ©1985, IEEE.)

The filtered radar outputs are shown in Fig. 9.10.4 for four different types of filters. Five iterations were used to form noise estimate $\widehat{N}(m)$ before signal insertion from $r(t) = s(t) + n(t)$. Five more iterations were used for building up the signal estimate $\widehat{S}(m)$. Class 1 uncorrelated Wiener filters were used initially for H_S and H_N.

Neither convergence nor dither modulation were used at first. The resulting signal filter H_S is shown in Fig. 9.10.4a. Note that H_S notches at bins 0 and 10.3 while passing the signal at bin 15.3. The autocorrelation filter gives the largest peak at 15.3 and greatest suppression of other spectral data (24.8 dB). Adding signal dither of 50% gives the results in Fig. 9.10.4b. The autocorrelation filter is still good but the spurious response at bin 10.3 has raised in level (rejection is only 10.0 dB). Adding $\mu = 0.1$ convergence for 50% modulation gives the filter outputs shown in Fig. 9.10.4c. All filter performances improved by an average of 9.5 dB. The autocorrelation filter has 13.2 dB more reduction of the spurious spectra than that using the convergence technique.

The autocorrelation filter proved the most effective for locating the target frequency bin. A convergence factor is necessary to counteract signal fading. The filters perform well for input SNR as low as 0.5.

Table 9.10.1 Ideal Class 3 filter transfer functions $H(m)$.

Filter Name	Transfer Function $H(m)$

Distortion Balance ($W_S(m)$ = Signal weighting, $W_N(m)$ = Noise weighting)

$$\frac{\widehat{E}\{S(m)W_S(m)\}}{\widehat{E}\{S(m)W_S(m)\} + \widehat{E}\{N(m)W_N(m)\}}$$

General Estimation ($D(m)$ = Desired output)

$$H_E(m) = \frac{\widehat{E}\{D(m)R^*(m)\}}{\widehat{E}\{R(m)R^*(m)\}}$$

General Detection

$$H_D(m) = \frac{\widehat{E}\{D^*(m)\}}{\widehat{E}\{R(m)R^*(m)\}}$$

Estimation–Detection

$$H_E(m)H_D(m)$$

Uncorrelated Estimation (Uncorrelated Wiener)

$$\frac{\widehat{E}\{|S(m)|^2\}}{\widehat{E}\{|S(m)|^2\} + \widehat{E}\{|N(m)|^2\}}$$

Correlated Estimation (Correlated Wiener)

$$\frac{\widehat{E}\{S(m)(S^*(m) + N^*(m))\}}{\widehat{E}\{|S(m) + N(m)|^2\}}$$

Transitional 1 Estimation (Uncorrelated Wiener $\alpha = 0$, Correlated Wiener $\alpha = 1$)

$$\frac{\widehat{E}\{|S(m)|^2 + \alpha S(m)N^*(m)\}}{\widehat{E}\{|S(m)|^2 + |N(m)|^2 + \alpha(S(m)N^*(m) + S^*(m)N(m))\}}$$

Pulse-Shaping Estimation ($P(m)$ = Desired pulse spectrum, Uncorrelated Wiener)

$$\frac{\widehat{E}\{P(m)S^*(m)\}}{\widehat{E}\{|S(m)|^2\} + \widehat{E}\{|N(m)|^2\}}$$

Pulse-Shaping Estimation ($P(m)$ = Desired pulse spectrum, Correlated Wiener)

$$\frac{\widehat{E}\{P(m)(S^*(m) + N^*(m))\}}{\widehat{E}\{|S(m)|^2 + |N(m)|^2 + S(m)N^*(m) + S^*(m)N(m)\}}$$

Table 9.10.1 (Continued) Ideal Class 3 filter transfer functions $H(m)$.

Filter Name	Transfer Function $H(m)$

Classical Detection (Matched)

$$\frac{\widehat{E}\{S^*(m)\}}{\widehat{E}\{|N(m)|^2\}}$$

High-Resolution Detection (Uncorrelated Wiener)

$$\frac{\widehat{E}\{S^*(m)\}}{\widehat{E}\{|S(m)|^2\} + \widehat{E}\{|N(m)|^2\}}$$

High-Resolution Detection (Correlated Wiener)

$$\frac{\widehat{E}\{S^*(m) + N^*(m)\}}{\widehat{E}\{|S(m)|^2 + |N(m)|^2 + S(m)N^*(m) + S^*(m)N(m)\}}$$

Inverse Detection (Deconvolution)

$$\frac{1}{\widehat{E}\{S(m)\}}$$

Transitional 1 Detection (Inverse $\alpha = 0$, High-resolution $\alpha = 0.5$, Matched $\alpha = 1$)

$$\frac{\widehat{E}\{S^*(m)\}}{(1-\alpha)\widehat{E}\{|S(m)|^2\} + \alpha\widehat{E}\{|N(m)|^2\}}$$

Transitional 2 Detection (Inverse $\alpha = 0.5$)

$$\frac{1}{\widehat{E}\{S(m)^{2(1-\alpha)}\}}$$

Autocorrelation Estimation (Uncorrelated Wiener)

$$\frac{\widehat{E}\{S^*(m)|S(m)|^2\}}{\widehat{E}\{|S(m)|^2\} + \widehat{E}\{|N(m)|^2\}}$$

Autocorrelation Estimation (Correlated Wiener)

$$\frac{\widehat{E}\{(S^*(m) + N^*(m))|S(m)|^2\}}{\widehat{E}\{|S(m)|^2 + |N(m)|^2 + S(m)N^*(m) + S^*(m)N(m)\}}$$

Cross-Correlation Estimation (Uncorrelated Wiener)

$$\frac{\widehat{E}\{N^*(m)|S(m)|^2\}}{\widehat{E}\{|S(m)|^2\} + \widehat{E}\{|N(m)|^2\}}$$

Cross-Correlation Estimation (Correlated Wiener)

$$\frac{\widehat{E}\{S(m)N^*(m)(S^*(m) + N^*(m))\}}{\widehat{E}\{|S(m)|^2 + |N(m)|^2 + S(m)N^*(m) + S^*(m)N(m)\}}$$

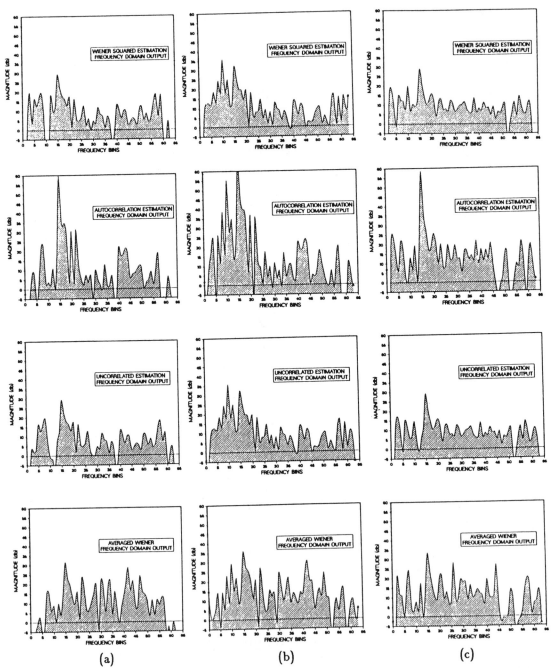

Fig. 9.10.4 Radar outputs: (a) Case 1 – no convergence, no modulation; (b) Case 2 – no convergence, 50% modulation; (c) Case 3 – $\mu = 0.1$ convergence, 50% modulation. (From C.S. Lindquist and L.W. Conboy, loc. cit., ©1985, IEEE.)

Table 9.10.2 Some Class 3 radar filter transfer functions.

Filter Name	Transfer Function $H(m)$

Uncorrelated Estimation (Uncorrelated Wiener)

$$\frac{\widehat{E}\{|S(m)|^2\}}{\widehat{E}\{|S(m)|^2\} + \widehat{E}\{|N(m)|^2\}}$$

Squared Uncorrelated Wiener

$$\left(\frac{\widehat{E}\{|S^*(m)|^2\}}{\widehat{E}\{|S(m)|^2\} + \widehat{E}\{|N(m)|^2\}}\right)^2$$

Averaged Uncorrelated Wiener

$$\frac{\widehat{E}\{|S(m)|^2\}}{\langle R(m)\rangle}$$

Uncorrelated Wiener with Noncoherent Integration

$$\frac{\langle R(m)^2\rangle\ \widehat{E}\{|S(m|^2)\}}{R(m)^2\ \left(\widehat{E}\{|S(m)|^2\} + \widehat{E}\{|N(m)|^2\}\right)}$$

Autocorrelation Estimation (Uncorrelated Wiener)

$$\frac{\widehat{E}\{S^*(m)|S(m)|^2\}}{\widehat{E}\{|S(m)|^2\} + \widehat{E}\{|N(m)|^2\}}$$

9.11 FFT BIOMEDICAL FILTERS [22]

Many biomedical problems such as EKG and EEM analysis require filtering. The electrocardiogram (EKG or ECG) records the electrical activity of the heart. The electroencephalogram (EEG) records the electrical activity of the brain. Both involve a series of waves or patterns that must be analyzed to diagnose illness. We will design some EKG filters used in cardiac pacemakers.

The EKG consists of a series of signals which are customarily denoted as the P-wave, R-wave (also called the QRS-wave), T-wave, and U-wave as shown in Fig. 9.11.1. The P-wave results from depolarization of the heart atria. The QRS-wave represents the electrical depolarization of the heart ventricles. The T-wave is the ventricular recovery or repolarization. The U-wave sometimes occurs after premature QRS-waves and is affected by many factors. In addition, other signals such as EMG muscle noise and interference picked up from sources outside the body are added to the EKG which

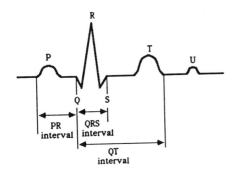

P wave	40–60 msec.
QRS wave	40–100 msec.
T wave	60–240 msec.
PR interval	120–200 msec.
ST interval	130–150 msec.
QT interval	310–410 msec.

(rate of 70 beats/min.)

Fig. 9.11.1 Model of EKG waveform.[22]

further complicates its appearance. A malfunctioning heart exhibits *dysrhythmia* and is analyzed using many EKG parameters or artifacts like the shape of the P-wave and R-wave, and the PR-interval.[23]

The idealized EKG assumes that the R-wave is triangular and that the P-wave and T-wave have raised-cosine shapes. The U-wave will be ignored. The spectral characteristics of these waves are:

$$\text{P} - \text{wave}: \quad S_P(j\omega) = \frac{V_P T_P}{2}\left(\frac{\sin(\omega T_P/2)}{(\omega T_P/2)\left[1 - (\omega T_P/2\pi)^2\right]}\right)e^{-j\omega T_1}$$

$$\text{R} - \text{wave}: \quad S_R(j\omega) = \frac{V_R T_R}{2}\left(\frac{\sin(\omega T_R/4)}{\omega T_R/4}\right)^2 e^{-j\omega T_2} \qquad (9.11.1)$$

$$\text{T} - \text{wave}: \quad S_T(j\omega) = \frac{V_T T_T}{2}\left(\frac{\sin(\omega T_T/2)}{(\omega T_T/2)\left[1 - (\omega T_T/2\pi)^2\right]}\right)e^{-j\omega T_3}$$

These waveforms and their parameter are shown in Fig. 9.11.1. Since the EKG is generated by a nonstationary random process, the V_i and T_j parameters are random variables and have nonstationary statistics. They have a large range of values which can vary greatly between patients and for the same patient at different times.

Typical magnitude and phase spectra of S_P, S_R, and S_T are shown in Figs. 9.11.2a–c. The P-wave has a 3 dB bandwidth of $0.7/T_P$ Hertz, an asymptotic roll-off of -60 dB/dec, and spectral zero-crossings at $2/T_P, 3/T_P, 4/T_P, \ldots$ Hertz. Similar comments can be made for the T-wave. The R-wave has a 3 dB bandwidth of $0.625/T_R$ Hertz, an asymptotic roll-off of -40 dB/dec, and spectral zero-crossings at $2/T_R, 4/T_R, 6/T_R,$ $\ldots$ Hertz. Since the T-wave is usually much longer in duration than the R-wave and P-wave, its bandwidth is much smaller. Because the P-wave and T-wave have a different temporal shape than the R-wave, their magnitude and phase spectra are very dissimilar. This is fortunate because filters can be designed which use this phase information to better separate and identify the components in the EKG.

EKG's are often obscured by various kinds of random additive noise. Although noise may be generated by a variety of sources, EMG muscle noise often predominates. Experimental data shows that it can be treated as an uncorrelated zero-mean Gaussian band-pass process having a spectrum S_{EMG} with magnitude

$$|S_{EMG}(j\omega)| = \frac{N_o}{\sqrt{2\pi}} \exp\left(\frac{-(\omega - \omega_o)^2}{2\sigma^2}\right) \qquad (9.11.2)$$

and random phase $|\arg S_{EMG}| \leq 180°$. Typical parameter values are center frequency $\omega = 2\pi(70\ \text{Hz})$ and standard deviation $\sigma_o = 2\pi(50\ \text{Hz})$. N_0 is adjusted to produce any desired noise level. The magnitude spectrum $|S_{EMG}|$ is drawn in Fig. 9.11.2d. Low-level broad-band white noise S_N is also usually present.

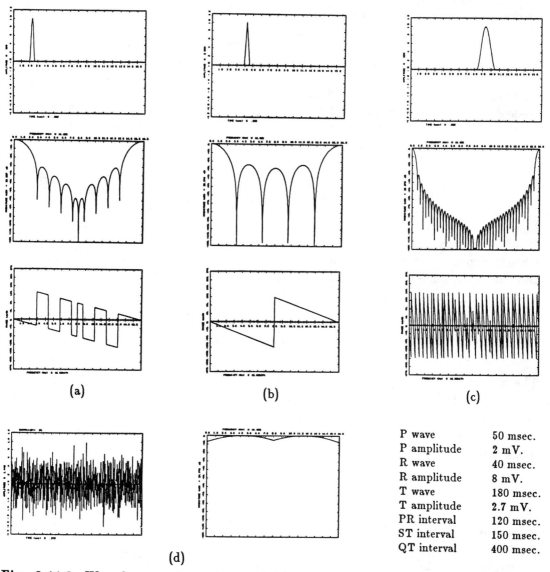

(a)   (b)   (c)

(d)

P wave	50 msec.
P amplitude	2 mV.
R wave	40 msec.
R amplitude	8 mV.
T wave	180 msec.
T amplitude	2.7 mV.
PR interval	120 msec.
ST interval	150 msec.
QT interval	400 msec.

Fig. 9.11.2 Waveform and spectrum of (a) P-wave, (b) R-wave, (c) T-wave, and (d) EMG muscle noise. (From C.S. Lindquist, N. Rizq, and M. Zhao, "Optimum detection of R-waves in EKG signals," Proc. IEEE Miami Technicon, pp. 23–26, ©1987, IEEE.)

A demand-type cardiac pacemaker is made up of a detection filter followed by decision logic and a current pulse generator. The detection filter identifies and detects only a particular component of the EKG waveform but ignores and rejects the remaining components. It is difficult to design since EKG waveforms have such variability in waveshapes, widths, and lengths. We will consider an R-wave demand pacemaker. A high-resolution detection filter will be designed which outputs a narrow pulse every time an R-wave is detected.

From Table 8.2.1, the transfer function of an uncorrelated Wiener type of high-resolution detection filter equals

$$H(m) = \frac{E\{S^*(m)\}}{E\{|S(m)|^2 + |N(m)|^2\}} \tag{9.11.3}$$

For the EKG shown in Fig. 9.11.1, we treat the R-wave as the signal s, and the remaining P-wave, T-wave, EMG muscle noise, and general white background noise as the total noise n. Their respective spectrum were written in Eqs. 9.11.1 and 9.11.2. Using these spectra, Eq. 9.11.3 can be rewritten as

$$H(m) = \frac{E\{S_R^*(m)\}}{E\{|S_R(m)|^2 + |S_P(m) + S_T(m) + S_{EMG}(m) + S_N(m)|^2\}} \tag{9.11.4}$$

The P-wave, R-wave and T-wave, U-wave, and noises are independent of one another. The T-wave cannot occur if the R-wave does not occur so they are not independent. Since the expectation of the product of independent terms is the product of expectations, and the expectation of a sum is the sum of expectations, Eq. 9.11.4 can be rewritten as

$$H(m) = \frac{E\{S_R^*(m)\}}{\begin{array}{c}E\{|S_R(m)|^2 + |S_P(m)|^2 + |S_T(m)|^2 + |S_{EMG}(m)|^2 + |S_N(m)|^2 \\ + S_P(m)S_T^*(m) + S_P^*(m)S_T(m)\}\end{array}} \tag{9.11.5}$$

The magnitude spectra of S_R, S_P, S_T, and S_{EMG} were shown in Fig. 9.11.2. Choosing the signal-to-total rms noise ratio to equal 26 dB, $H(m)$ is evaluated using Eq. 9.11.5 and its magnitude and phase responses are plotted in Fig. 9.11.3a.

A different high-resolution detection filter is the uncorrelated Wiener type which has a gain from Table 8.2.1 of

$$H(m) = \frac{E\{S^*(m) + N^*(m)\}}{E\{|S(m) + N(m)|^2\}} \tag{9.11.6}$$

Evaluating Eq. 9.11.6 using the EKG spectra gives

$$H(m) = \frac{E\{S_R^*(m) + S_P^*(m) + S_T^*(m) + S_{EMG}^*(m) + S_N^*(m)\}}{\begin{array}{c}E\{|S_R(m)|^2 + |S_P(m)|^2 + |S_T(m)|^2 + |S_{EMG}(m)|^2 + |S_N(m)|^2 \\ + S_R(m)S_P^*(m) + S_R^*(m)S_P(m) + S_R(m)S_T^*(m) \\ + S_R^*(m)S_T(m) + S_P(m)S_T^*(m) + S_P^*(m)S_T(m)\}\end{array}} \tag{9.11.7}$$

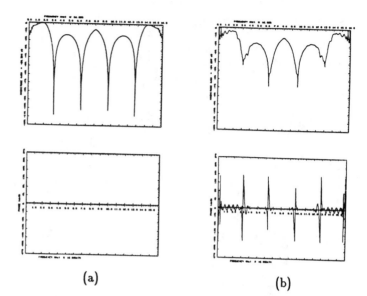

Fig. 9.11.3 Magnitude
and phase response of
(a) uncorrelated Wiener
and (b) correlated Wiener
EKG high-resolution filters.
(From C.S. Lindquist, N.
Rizq, and M. Zhao, loc.
cit., ©1987, IEEE.)

(a)   (b)

The magnitude and phase responses of this filter are shown in Fig. 9.11.3b. The primary difference between the uncorrelated and correlated detection filters are their phase characteristics. The uncorrelated type has zero phase and cannot therefore provide any phase compensation. The correlated type has a nonzero phase which properly compensates the phase of the EKG and provides better R-wave detection.

Now let us process three different EKG patterns using these two detection filters. The filter inputs and outputs are plotted in Fig. 9.11.4. All inputs are contaminated with EMG and white noise. The first EKG pattern is shown in Fig. 9.11.4a which consists of four consecutive R-waves. The two detection filter outputs show that the four R-waves were detected by both filters. The output detection pulses are narrow and occur when the R-waves occur. The noise background is reasonably low. The uncorrelated filter does not reject the P-wave and T-wave very well. However the correlated filter does an excellent job of detecting only the R-wave and rejecting the P-wave and T-wave.

In the second EKG pattern shown in Fig. 9.11.4b, one heartbeat has been skipped. Both filter outputs correctly suppress the second detection pulse which indicates that the second R-wave was missing. In this case, the detection logic of the demand pacemaker would instruct the current pulse generator to output a pulse(s) to the heart to replace the missing R-wave (P-wave) and produce a second heartbeat.

In the third EKG pattern of Fig. 9.11.4c, one heartbeat has been skipped and the next heartbeat signal has only one-half amplitude. The filter outputs indicate that the second R-wave is missing and that the third R-wave has decreased (by one-half) amplitude. The current pulse generator would output a pulse(s) to replace the missing R-wave (P-wave) and produce a second heartbeat. Depending upon where the threshold of the detection logic was set, the demand pacemaker would make an appropriate decision whether or not to output a current pulse(s) to the heart to augment the one-half amplitude R-wave (P-wave).

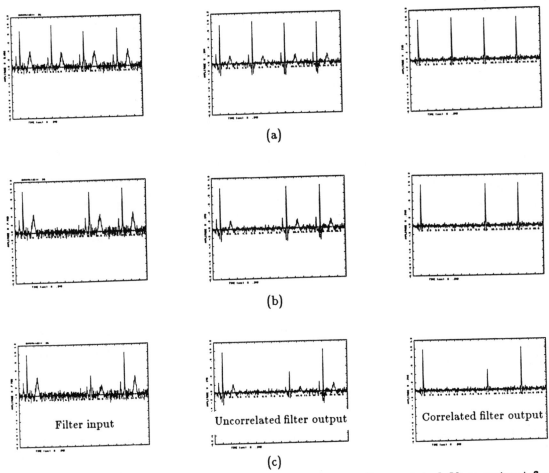

Fig. 9.11.4 Filter input, uncorrelated filter output, and correlated filter output for three EKG waveforms: (a) repetitive, (b) missed second R-wave, (c) missed second R-wave and reduced-amplitude third R-wave. (From C.S. Lindquist, N. Rizq, and M. Zhao, loc. cit., ©1987, IEEE.)

The correlated high-resolution detection filter type performs better than the uncorrelated type. The output detection pulse is very narrow and has a large amplitude relative to the nominal output level. This detection filter can accommodate modest changes in the amplitudes, widths, and intervals of the P-wave, R-wave, T-wave, and changes in the EMG muscle noise and background noise. Our simulations have shown that the correlated detection filter of Eq. 9.11.7 still provides good detection for ±10% and ±20% changes in P-wave, R-wave, and T-wave widths and intervals. Therefore the correlated detection filters provide good performance and are fairly robust. We have used a Class 1 detection filter since the spectra were all assumed to be known apriori. The more practical situation requires us to switch to a Class 2 filter where only spectral information about the R-wave is needed and no assumptions are made about the P-wave, T-wave, EMG muscle noise, or general noise background.

9.12 FFT FILTER IMPLEMENTATION [24]

This chapter has described a variety of FFT filters. Now their implementation using microprocessors will be considered. The basic architectures are discussed and quantitative estimates of processing speeds are made. Since standard microprocessor-based signal processing modules like the FFT are available, such considerations are very important in designing systems.

FFT filter implementations require trade-offs in speed, complexity, power consumption, and cost. When the response time of the filter is reduced so that the sample rate can be increased, then the filter must be more complex, consume more power, and be more expensive. We will investigate systems that are as inexpensive as possible (but unfortunately slow), medium cost systems, and systems where cost is no object (and very fast). Typical FFT filter systems are shown in Figs. 9.12.1a–c, respectively.

The inexpensive, slow speed systems of Fig. 9.12.1a use standard 8-bit microprocessors. To increase system throughput, several microprocessors can be used to give flexibility in speed and cost. The system is divided into blocks of equal complexity (since each processor has the same capability) and independent tasks (since 8-bit processors cannot communicate efficiently with each other except at macroscopic levels). If too many processors are used, the addition of another processor will slow the system down rather than speed it up.

The medium scale system of Fig. 9.12.1b uses slice processors and parallel pipelining to increase speed. This system is often 100 times faster than the 8-bit processor systems of Fig. 9.12.1a for several reasons. First, it typically uses bipolar logic rather than MOS logic. Second, it can do 32-bit parallel arithmetic instead of computing only 8-bits at a time. Third, multiple processors can be used very effectively since instruction fetching is isolated from data fetching. This eliminates much of the bottleneck of having multiple processors share a common bus.

The high speed systems of Fig. 9.12.1c are designed by simply customizing the hardware to exactly implement the signal processing algorithm. Highest speed is achieved by using individual multiply chips for each multiply to be done and individual adder chips for each addition to be done. This is extremely expensive. Control ROM's divide the workload such that all processing elements are always busy. It is the ability of the system to accept as many processing elements as needed that makes this the highest speed architecture.

All three systems can be analyzed in the same way. A set of parameters can be defined which allow the total time to perform a specific algorithm to be computed:

T_m = time for 32-bit, complex, floating-point multiply using a single processing element

T_a = time for 32-bit, complex, floating-point addition using a single processing element

T_s = time to swap data from one processing element to another

N = number of points used in the FFT

M = effective number of processors used simultaneously in calculations

Now recall from Chap. 2.2 that the total number of complex multiplications and additions in an FFT is $0.5N \log_2(N)$. Then the time required to perform an FFT and inverse FFT in any one of the hardware configurations of Fig. 9.12.1 is

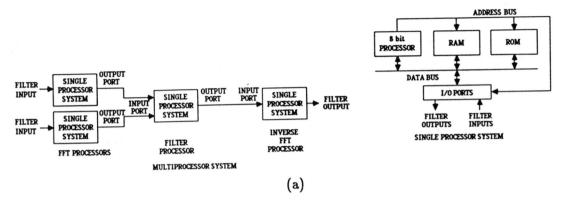

(a)

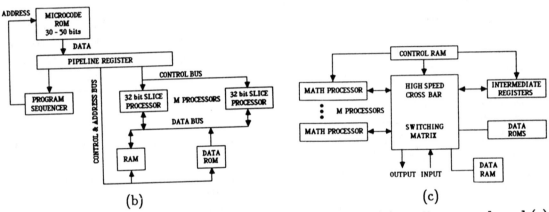

(b)   (c)

Fig. 9.12.1 **FFT** filters implemented with (a) low speed, (b) medium speed, and (c) high-speed microprocessor systems.[24]

$$T_{FFT} = 0.5N \log_2(N) \frac{T_m + T_a}{M} + 2MT_s \qquad (9.12.1)$$

Consider the uncorrelated Wiener estimation filter $H(m)$ of Table 8.2.1 as an example. Since

$$H(m) = \frac{E\{|S(m)|^2\}}{E\{|S(m)|^2\} + E\{|N(m)|^2\}} \qquad (9.12.2)$$

Computing $H(m)$ requires 2 FFT's, 4 swaps, 1 add, and 3 multiplies. Then the system equations evaluated at each new sample

$$\begin{aligned}
R(m) &= \text{FFT}\big[r(n)\big] \\
C(m) &= H(m)R(m) \\
c(n) &= \text{FFT}^{-1}\big[C(m)\big]
\end{aligned} \qquad (9.12.3)$$

require 2 FFT's, 2 swaps, no adds, and 1 multiply.

Therefore the total time T to process a single data point using Eqs. 9.12.2 and 9.12.3 is given by

$$T = 4T_{FFT} + 6MT_s + T_a + 6T_m \qquad (9.12.4)$$

since there are 4 FFT's, 6 swaps, 1 add, and 4 multiplies total. In a Class l filter, $H(m)$ is only computed once or updated infrequently. Then only Eq. 9.12.3 need be evaluated until the update so that the total time T is reduced to

$$T = 2T_{FFT} + 2MT_s + 2T_m \qquad (9.12.5)$$

Typical values of the parameters are:

Name	Low Speed	Medium Speed	High Speed
T_m	100 μsec	10 μsec	2 μsec
T_a	100 μsec	0.5 μsec	0.2 μsec
T_s	10 μsec	0.5 μsec	0.05 μsec

Substituting them into Eq. 9.12.2 assuming a 256-point FFT, the total processing time T is determined as a function of the effective number of processors M. This relation is plotted in Fig. 9.12.2. The reciprocal of T is the maximum sampling rate or throughput of the system.

It is clear that the bulk of the processing time is required to perform the FFT. Using dedicated FFT modules significantly reduces processing time and increases throughput of the FFT system.

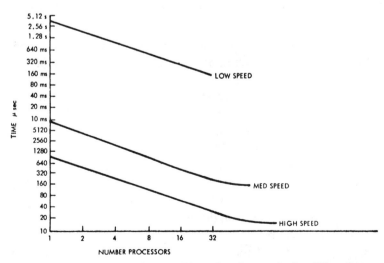

Fig. 9.12.2 Total processing time for 256-point FFT filter implemented with microprocessor systems. (From **W.H. Haas, C.S. Lindquist, and E.N. Evans,** "Microprocessor-based optimum control using adaptive frequency domain design," **Proc. 21st IEEE Conf. Decision and Control, pp. 813–817, ©1982, IEEE.)**

9.13 FAST TRANSFORM FILTER DESIGN IN RETROSPECT

Signal processing and filtering has been described in Chaps. 4, 8, and 9 using a fast transform filter design viewpoint. We will summarize these results before proceeding to the next chapters on more conventional FIR and IIR filter design.

Fast transform filters are simple, high-level, and modularized. These chapters have shown that signal processing can be conceptualized in highly modularized forms. Basic building blocks include estimation, detection, correlation, interpolation, transforms, multiplication, division, convolution, conjugation, nonlinearities, square root, samplers, integration, differentiation, histograms (pdf estimators), number generators (pdf generators), etc. These blocks can be combined to generate large, complex, and sophisticated systems like radars, CAT-scan imagers, and homomorphic processors.

Matrix equations maximize algorithmic simplicity. Block processing of data vectors using scalar and vector filters in matrix form emphasizes high-level algorithms rather than low-level hardware. The compact matrix representations give clear insight and understanding of concepts. New algorithms are more easily derived. (Review Chaps. 8.2 and 8.4.)

Fast transforms can be selected to improve performance. The fast Fourier transform blocks of Chap. 9 can be replaced by any fast transform block like Walsh, Paley, Hadamard, Haar, slant, or K-L from Chap. 8. Fast transform filters can perform better than FFT filters when the transform is matched to the signal and noise being processed. (Review Chaps. 8.6.4 and 8.7.)

Time domain implementation eliminates transforms completely. The fast Fourier transform blocks of Chap. 9 can be eliminated altogether by replacing transfer function H blocks by impulse response h blocks (Tables 8.6.1 and 8.6.2) and making other minor system changes. Time domain implementations sometimes reduce the total number of computations and the processing time. (Review Chaps. 8.1 and 8.6.2.)

Vector filters provide more processing gain than scalar filters. Time-varying systems can be formed using vector filters rather than time-invariant scalar filters. This can further improve system performance at the expense of increased computation. (Review Chap. 8.5.1.)

Filter design emphasizes software rather than hardware. With the advent of the microprocessor, signal processing was revolutionized. Every year, the cost of hardware decreases and its capability, complexity, and speed increases. Therefore, the designer derives maximum benefit by customizing software and using standard hardware.

System modification and up-grade is simple. Since the system is implemented in software rather than hardware, routine algorithm modifications, system reconfiguration, etc. can usually be made with relative ease. System flexibility is of upmost importance.

These comments highlight the important philosophical foundation of Chaps. 4, 8, and 9. The primary disadvantage of the approaches described in these chapters is larger computing times. Because these systems are implemented in high-level software rather than low-level hardware, speed is sacrificed. The more conventional FIR and IIR filter design approaches to be discussed in Chaps. 10 and 11 coupled with the hardware implementation forms of Chap. 12 often produce higher-speed systems. However the design approaches and system capabilities are often less sophisticated.

PROBLEMS

Specify an appropriate sampling rate, window size, and the FFT size (number of points) in each of the following problems. Your instructor may have you use software to evaluate the results.

9.1 Design a Class 1 estimation filter to estimate a step signal having $E\{s(t)\} = U_{-1}(t)$ which is contaminated by uncorrelated white noise having $E\{|N(j\omega)|\} = 1$. (a) Write the transfer functions of the uncorrelated and correlated Wiener signal estimation filters. (b) Write the transfer functions of the uncorrelated and correlated Wiener noise estimation filters. (c) Plot their magnitude responses. (d) What are the transfer functions of the first and second stages of the aposteriori estimation filters in Fig. 9.2.1? (e) Using a flow chart, explain how to implement these filters in software using FFT's.

9.2 Design a Class 1 estimation filter to estimate a signum signal having $E\{s(t)\} = \text{sgn } t$ which is contaminated by uncorrelated 60 Hz sinusoidal noise (peak-to-peak amplitude = 1 volt, frequency = 60 Hz, phase is random). Repeat parts (a)–(e) of Problem 9.1.

9.3 Design a Class 1 estimation filter to estimate an exponential step signal having $E\{s(t)\} = e^{-t}U_{-1}(t)$ which is contaminated by uncorrelated white noise having $E\{|N(j\omega)|\} = 1$. Repeat parts (a)–(e) of Problem 9.1.

9.4 Design a Class 1 estimation filter to estimate a signal pulse having $E\{s(t)\} = U_{-1}(t + T/2) - U_{-1}(t - T/2)$ (unit amplitude, length T, centered at $t = 0$) which is contaminated by uncorrelated white noise having $E\{|N(j\omega)|\} = 1$. Repeat parts (a)–(e) of Problem 9.1.

9.5 Design a Class 1 estimation filter to estimate a pulsed-cosine signal having $E\{s(t)\} = \cos(2\pi t)[U_{-1}(t + 0.1) - U_{-1}(t - 0.1)]$ (peak amplitude = 1, frequency = 1 Hz, pulse duration = 0.2 sec) which is contaminated by uncorrelated white noise having $E\{|N(j\omega)|\} = 1$. Repeat parts (a)–(e) of Problem 9.1.

9.6 Determine the transfer function of a Class 1 adaptive noise cancellor (single channel) to estimate the signal in Problem (a) 9.1, (b) 9.2, (c) 9.3, (d) 9.4, (e) 9.5.

9.7 Design a Class 1 ANC (two channel) to estimate a step signal having $E\{s(t)\} = U_{-1}(t)$ which is contaminated by uncorrelated white noise having $E\{|N(j\omega)|\} = 1$. Assume that the noise reference $n_o(t) = 0.5n(t - 1)$ in Fig. 9.3.5. (a) Write the transfer functions of the uncorrelated and correlated Wiener noise estimation filters. (b) Plot their magnitude responses. (c) What are the transfer functions of the first and second stages of the aposteriori ANC filters in Fig. 9.3.5? (d) Using a flow chart, explain how to implement these filters in software using FFT's.

9.8 Design a Class 1 ANC filter to estimate a signum signal having $E\{s(t)\} = \text{sgn } t$ which is contaminated by uncorrelated 60 Hz sinusoidal noise (peak-to-peak amplitude = 1 volt, frequency = 60 Hz, phase is random). Assume that the noise reference $n_o(t) = n(t - 1)$ in Fig. 9.3.5. Write the Repeat parts (a)–(d) of Problem 9.7.

9.9 Design a Class 1 ANC filter to estimate an exponential step signal having $E\{s(t)\} = e^{-t}U_{-1}(t)$ which is contaminated by uncorrelated white noise having $E\{|N(j\omega)|\} = 1$. Assume that the noise reference $n_o(t) = n(t) * e^{-t}U_{-1}(t)$ in Fig. 9.3.5. Repeat parts (a)–(d) of Problem 9.7.

9.10 Design a Class 1 ANC filter to estimate a signal pulse having $E\{s(t)\} = U_{-1}(t + T/2) - U_{-1}(t - T/2)$ (unit amplitude, length T, centered at $t = 0$) which is contaminated by uncorrelated white noise having $E\{|N(j\omega)|\} = 1$. Assume that the noise reference $n_o(t) = n(t) * U_{-1}(t)$ in Fig. 9.3.5. Repeat parts (a)–(d) of Problem 9.7.

9.11 Design a Class 1 ANC filter to estimate a pulsed-cosine signal having $E\{s(t)\} = \cos(2\pi t)[U_{-1}(t + 0.1) - U_{-1}(t - 0.1)]$ (peak amplitude = 1, frequency = 1 Hz, pulse duration

= 0.2 sec) which is contaminated by uncorrelated white noise having $E\{|N(j\omega)|\} = 1$. Assume that the noise reference $n_o(t) = n(t) * U_{-2}(t)$ in Fig. 9.3.5. Repeat parts (a)–(d) of Problem 9.7.

9.12 Design a Class 1 detection filter to detect a step signal having $E\{s(t)\} = U_{-1}(t)$ which is contaminated by uncorrelated white noise having $E\{|N(j\omega)|\} = 1$. (a) Write the transfer functions of the matched, high-resolution, and inverse signal detection filters which detect the expected signals. (b) Plot their magnitude responses. (c) Write their $H(s)$ transfer functions assuming they are stable but not necessarily causal. (d) Using a flow chart, explain how to implement these filters in software using FFT's.

9.13 Design a Class 1 detection filter to detect a signum signal having $E\{s(t)\} = \text{sgn } t$ which is contaminated by uncorrelated 60 Hz sinusoidal noise (peak-to-peak amplitude = 1 volt, frequency = 60 Hz, phase is random). Repeat parts (a)–(d) of Problem 9.12.

9.14 Design a Class 1 detection filter to detect an exponential step signal having $E\{s(t)\}$ $= e^{-t}U_{-1}(t)$ which is contaminated by uncorrelated white noise having $E\{|N(j\omega)|\} = 1$. Repeat parts (a)–(d) of Problem 9.12.

9.15 Design a Class 1 detection filter to detect a signal pulse having $E\{s(t)\} = U_{-1}(t + T/2) - U_{-1}(t - T/2)$ (unit amplitude, length T, centered at $t = 0$) which is contaminated by uncorrelated white noise having $E\{|N(j\omega)|\} = 1$. Repeat parts (a)–(d) of Problem 9.12.

9.16 Design a Class 1 detection filter to detect a pulsed-cosine signal having $E\{s(t)\} = \cos(2\pi t)[U_{-1}(t + 0.1) - U_{-1}(t - 0.1)]$ (peak amplitude = 1, frequency = 1 Hz, pulse duration = 0.2 sec) which is contaminated by uncorrelated white noise having $E\{|N(j\omega)|\} = 1$. Repeat parts (a)–(d) of Problem 9.12.

9.17 (a) Determine the transfer functions of a bank of Class 1 detection filters like that in Fig. 9.4.4 to detect when the signals described in Problems 9.12, 9.14, 9.15, and 9.16 occur. (b) Using a flow chart, explain how to implement these filters in software using FFT's.

9.18 Design a Class 2 detection filter to detect a step signal having $E\{s(t)\} = U_{-1}(t)$ which is contaminated by uncorrelated white noise having $E\{|N(j\omega)|\} = 1$. (a) Write the transfer functions of the matched, high-resolution, and inverse filters which detect the expected signal. Smooth (inner block) the input spectra with rectangular, triangular, and Hann windows. Let $T = 1$ and smoothing width $p = 10\%$. (b) Using a flow chart, explain how to implement these filters in software using FFT's.

9.19 Design a Class 2 detection filter to detect a signum signal having $E\{s(t)\} = \text{sgn } t$ which is contaminated by uncorrelated 60 Hz sinusoidal noise (peak-to-peak amplitude = 1 volt, frequency = 60 Hz, phase is random). Repeat parts (a)–(b) of Problem 9.18.

9.20 Design a Class 2 detection filter to detect an exponential step signal having $E\{s(t)\}$ $= e^{-t}U_{-1}(t)$ which is contaminated by uncorrelated white noise having $E\{|N(j\omega)|\} = 1$. Repeat parts (a)–(b) of Problem 9.18.

9.21 Design a Class 2 detection filter to detect a signal pulse having $E\{s(t)\} = U_{-1}(t + T/2) - U_{-1}(t - T/2)$ (unit amplitude, length T, centered at $t = 0$) which is contaminated by uncorrelated white noise having $E\{|N(j\omega)|\} = 1$. Repeat parts (a)–(b) of Problem 9.18.

9.22 Design a Class 2 detection filter to detect a pulsed-cosine signal having $E\{s(t)\} = \cos(2\pi t)[U_{-1}(t + 0.1) - U_{-1}(t - 0.1)]$ (peak amplitude = 1, frequency = 1 Hz, pulse duration = 0.2 sec) which is contaminated by uncorrelated white noise having $E\{|N(j\omega)|\} = 1$. Repeat parts (a)–(b) of Problem 9.18.

9.23 (a) Determine the transfer functions of a bank of Class 2 detection filters like that in Fig. 9.4.4 to detect when the signals described in Problems 9.18, 9.20, 9.21, and 9.22 occur. (b) Using a flow chart, explain how to implement these filters in software using FFT's.

9.24 Design Class 1 autocorrelation and cross-correlation filters to correlate a step signal having $E\{s(t)\} = U_{-1}(t)$ which is contaminated by uncorrelated white noise having $E\{|N(j\omega)|\}$

= 1. (a) Write the transfer functions of the correlation filters based upon the uncorrelated Wiener estimation filters (Table 9.5.1a). (b) Write the correlation transfer functions based upon the correlated Wiener estimation filters (Table 9.5.1b). (c) Plot the magnitude responses of the filters in parts (a) and (b). (d) Write the autocorrelation and cross-correlation estimates. (e) Using a flow chart, explain how to implement these filters in software using FFT's.

9.25 Design Class 1 autocorrelation and cross-correlation estimation filters to correlate a signum signal having $E\{s(t)\} = \text{sgn } t$ which is contaminated by uncorrelated 60 Hz sinusoidal noise (peak-to-peak amplitude = 1 volt, frequency = 60 Hz, phase is random). Repeat parts (a)-(e) of Problem 9.24.

9.26 Design Class 1 autocorrelation and cross-correlation estimation filters to correlate a exponential step signal having $E\{s(t)\} = e^{-t}U_{-1}(t)$ which is contaminated by uncorrelated white noise having $E\{|N(j\omega)|\} = 1$. Repeat parts (a)-(e) of Problem 9.24.

9.27 Design Class 1 autocorrelation and cross-correlation estimation filter to correlate a signal pulse having $E\{s(t)\} = U_{-1}(t + T/2) - U_{-1}(t - T/2)$ (unit amplitude, length T, centered at $t = 0$) which is contaminated by uncorrelated white noise having $E\{|N(j\omega)|\} = 1$. Repeat parts (a)-(e) of Problem 9.24.

9.28 Design Class 1 autocorrelation and cross-correlation estimation filters to correlate a pulsed-cosine signal having $E\{s(t)\} = \cos(2\pi t)[U_{-1}(t + 0.1) - U_{-1}(t - 0.1)]$ (peak amplitude = 1, frequency = 1 Hz, pulse duration = 0.2 sec) which is contaminated by uncorrelated white noise having $E\{|N(j\omega)|\} = 1$. Repeat parts (a)-(e) of Problem 9.24.

9.29 Determine the transfer functions of Class 2 autocorrelation and cross-correlation filters to correlate the signal and noise specified in Problem (a) 9.24, (b) 9.25, (c) 9.26, (d) 9.27, (e) 9.28.

9.30 Design a Class 1 delay estimation filter to process step signals having $E\{s_1(t)\} = U_{-1}(t)$ and $s_2(t) = 0.5s_1(t - 1)$ which are contaminated by uncorrelated white noise having $E\{|N(j\omega)|\} = 1$. (a) Write the transfer functions of the classical, high-resolution, and inverse detection filters. (b) Plot their magnitude responses. (c) Write their $H(s)$ transfer functions assuming they are stable but not necessarily causal. (d) Using a flow chart, explain how to implement these filters in software using FFT's.

9.31 Design a Class 1 delay estimation filter to process signum signals having $E\{s_1(t)\} = \text{sgn } t$ and $s_2(t) = s_1(t - 1)$ which are contaminated by uncorrelated 60 Hz sinusoidal noise (peak-to-peak amplitude = 1 volt, frequency = 60 Hz, phase is random). Write the Repeat parts (a)-(d) of Problem 9.30.

9.32 Design a Class 1 delay estimation filter to process exponential step signals having $E\{s_1(t)\} = e^{-t}U_{-1}(t)$ and $s_2(t) = s_1(t) * e^{-t}U_{-1}(t)$ which are contaminated by uncorrelated white noise having $E\{|N(j\omega)|\} = 1$. Repeat parts (a)-(d) of Problem 9.30.

9.33 Design a Class 1 delay estimation filter to process signal pulses having $E\{s_1(t)\} = U_{-1}(t + T/2) - U_{-1}(t - T/2)$ (unit amplitude, length T, centered at $t = 0$) and $s_2(t) = s_1(t) * e^{-t}U_{-1}(t)$ which are contaminated by uncorrelated white noise having $E\{|N(j\omega)|\} = 1$. Repeat parts (a)-(d) of Problem 9.30.

9.34 Design a Class 1 delay estimation filter to process pulsed-cosine signals having $E\{s_1(t)\} = \cos(2\pi t)[U_{-1}(t + 0.1) - U_{-1}(t - 0.1)]$ (peak amplitude = 1, frequency = 1 Hz, pulse duration = 0.2 sec) and $s_2(t) = s_1(t) * U_{-2}(t)$ which are contaminated by uncorrelated white noise having $E\{|N(j\omega)|\} = 1$. Repeat parts (a)-(d) of Problem 9.30.

9.35 Design Class 2 high-resolution delay estimation filters to process the signals and noises specified in Problem (a) 9.30, (b) 9.31, (c) 9.32, (d) 9.33, (e) 9.34.

9.36 Design Class 2 Wiener delay estimation (Hassab/Boucher) filters to process the signals and noises specified in Problem (a) 9.30, (b) 9.31, (c) 9.32, (d) 9.33, (e) 9.34.

9.37   Design Class 3 transitional delay estimation filters to process the signals and noises specified in Problem (a) 9.30, (b) 9.31, (c) 9.32, (d) 9.33, (e) 9.34.

9.38   Design Class 3 classical delay estimation filters to process the signals and noises specified in Problem (a) 9.30, (b) 9.31, (c) 9.32, (d) 9.33, (e) 9.34.

9.39   Design Class 3 autocorrelation delay estimation filters to process the signals and noises specified in Problem (a) 9.30, (b) 9.31, (c) 9.32, (d) 9.33, (e) 9.34.

9.40   Design a Class 1 image registration filter to registrate step image signals having $E\{s_1(x,y)\} = U_{-1}(x,y)$ and $s_2(x,y) = 0.5s_1(x-1,y-1)$ which are contaminated by uncorrelated white noise having $E\{|N(j\omega_1,j\omega_2)|\} = 1$. (a) Write the transfer functions of the classical, high-resolution, and inverse detection filters. (b) Plot their magnitude responses. (c) Write their $H(s)$ transfer functions assuming they are stable but not necessarily causal. (d) Using a flow chart, explain how to implement these filters in software using FFT's.

9.41   Design a Class 1 image registration filter to registrate signum image signals having $E\{s_1(x,y)\} = \mathrm{sgn}\,xy$ and $s_2(x,y) = s_1(x-1,y-1)$ which are contaminated by uncorrelated 60 Hz sinusoidal noise (peak-to-peak amplitude = 1 volt, frequency = 60 Hz, phase is random). Write the Repeat parts (a)–(d) of Problem 9.40.

9.42   Design a Class 1 image registration filter to registrate exponential step image signals having $E\{s_1(x,y)\} = e^{-x-y}U_{-1}(x)U_{-1}(y)$ and $s_2(x,y) = s_1(x,y) * e^{-x-y}U_{-1}(x)U_{-1}(y)$ which are contaminated by uncorrelated white noise having $E\{|N(j\omega_1,j\omega_2)|\} = 1$. Repeat parts (a)–(d) of Problem 9.40.

9.43   Design a Class 1 image registration filter to registrate image signal pulses having $E\{s_1(x,y)\} = [U_{-1}(x+T/2) - U_{-1}(x-T/2)][U_{-1}(x+T/2) - U_{-1}(x-T/2)]$ (unit amplitude, length X_o, Y_o, centered at $x = 0, y = 0$) and $s_2(x,y) = s_1(x,y) * e^{-x-y}U_{-1}(x)U_{-1}(y)$ which are contaminated by uncorrelated white noise having $E\{|N(j\omega_1,j\omega_2)|\} = 1$. Repeat parts (a)–(d) of Problem 9.40.

9.44   Design a Class 1 image registration filter to registrate pulsed-cosine image signals having $E\{s_1(x,y)\} = \cos(2\pi x)[U_{-1}(x+0.1) - U_{-1}(x-0.1)][U_{-1}(x+0.1) - U_{-1}(x-0.1)]$ (peak amplitude = 1, frequency = 1 Hz, pulse duration = 0.2 sec) and $s_2(x,y) = s_1(x,y) * U_{-2}(x,y)$ which are contaminated by uncorrelated white noise having $E\{|N(j\omega_1,j\omega_2)|\} = 1$. Repeat parts (a)–(d) of Problem 9.40.

9.45   Design a Class 1 image enhancement filter to estimate a step image signal having $E\{s(x,y)\} = U_{-1}(x,y)$ which is contaminated by uncorrelated white noise with $E\{|N(j\omega_1,j\omega_2)|\} = 1$. (a) Write the transfer functions of the uncorrelated and correlated Wiener filters. (b) Plot their magnitude responses. (c) Write their $H(s)$ transfer functions assuming they are stable but not necessarily causal. (d) Using a flow chart, explain how to implement these filters in software using FFT's.

9.46   Design a Class 1 image enhancement filter to estimate a signum image signal having $E\{s(x,y)\} = \mathrm{sgn}\,xy$ which is contaminated by uncorrelated 60 Hz sinusoidal noise (peak-to-peak amplitude = 1 volt, frequency = 60 Hz, phase is random). Write the Repeat parts (a)–(d) of Problem 9.45.

9.47   Design a Class 1 image enhancement filter to estimate an exponential step image signal having $E\{s(x,y)\} = e^{-x-y}U_{-1}(x)U_{-1}(y)$ which is contaminated by uncorrelated white noise having $E\{|N(j\omega_1,j\omega_2)|\} = 1$. Repeat parts (a)–(d) of Problem 9.45.

9.48   Design a Class 1 image enhancement filter to estimate an image signal pulse having $E\{s(x,y)\} = [U_{-1}(x+T/2) - U_{-1}(x-T/2)][U_{-1}(x+T/2) - U_{-1}(x-T/2)]$ (unit amplitude, length X_o, Y_o, centered at $x = 0, y = 0$) which is contaminated by uncorrelated white noise having $E\{|N(j\omega_1,j\omega_2)|\} = 1$. Repeat parts (a)–(d) of Problem 9.45.

9.49   Design a Class 1 image enhancement filter to estimate a pulsed-cosine image signal having $E\{s(x,y)\} = \cos(2\pi x)[U_{-1}(x+0.1) - U_{-1}(x-0.1)][U_{-1}(x+0.1) - U_{-1}(x-0.1)]$ (peak amplitude = 1, frequency = 1 Hz, pulse duration = 0.2 sec) which is contaminated by uncorrelated white noise having $E\{|N(j\omega_1, j\omega_2)|\} = 1$. Repeat parts (a)–(d) of Problem 9.45.

9.50   Design a Class 1 image identification filter to detect a step image signal having $E\{s(x,y)\} = U_{-1}(x,y)$ which is contaminated by uncorrelated white noise with $E\{|N(j\omega_1, j\omega_2)|\} = 1$. (a) Write the transfer functions of the classical, high-resolution, and inverse detection filters. (b) Plot their magnitude responses. (c) Write their $H(s)$ transfer functions assuming they are stable but not necessarily causal. (d) Using a flow chart, explain how to implement these filters in software using FFT's.

9.51   Design a Class 1 image identification filter to detect a signum image signal having $E\{s(x,y)\} = \operatorname{sgn} xy$ which is contaminated by uncorrelated 60 Hz sinusoidal noise (peak-to-peak amplitude = 1 volt, frequency = 60 Hz, phase is random). Write the Repeat parts (a)–(d) of Problem 9.50.

9.52   Design a Class 1 image identification filter to detect an exponential step image signal having $E\{s(x,y)\} = e^{-x-y}U_{-1}(x)U_{-1}(y)$ which is contaminated by uncorrelated white noise having $E\{|N(j\omega_1, j\omega_2)|\} = 1$. Repeat parts (a)–(d) of Problem 9.50.

9.53   Design a Class 1 image identification filter to detect an image signal pulse having $E\{s(x,y)\} = [U_{-1}(x+T/2) - U_{-1}(x-T/2)][U_{-1}(x+T/2) - U_{-1}(x-T/2)]$ (unit amplitude, length X_o, Y_o, centered at $x = 0, y = 0$) which is contaminated by uncorrelated white noise having $E\{|N(j\omega_1, j\omega_2)|\} = 1$. Repeat parts (a)–(d) of Problem 9.50.

9.54   Design a Class 1 image identification filter to detect a pulsed-cosine image signal having $E\{s(x,y)\} = \cos(2\pi x)[U_{-1}(x+0.1) - U_{-1}(x-0.1)][U_{-1}(x+0.1) - U_{-1}(x-0.1)]$ (peak amplitude = 1, frequency = 1 Hz, pulse duration = 0.2 sec) which is contaminated by uncorrelated white noise having $E\{|N(j\omega_1, j\omega_2)|\} = 1$. Repeat parts (a)–(d) of Problem 9.50.

9.55   Design Class 1 ASK, FSK, and PSK communication filters to detect exponential step signals having $E\{s(t)\} = \pm e^{-t}U_{-1}(t)$ which are contaminated by uncorrelated white noise having $E\{|N(j\omega)|\} = 1$. (a) Write the transfer functions of the matched, high-resolution, and inverse signal detection filters which detect the expected signals. (b) Plot their magnitude responses. (c) Write their $H(s)$ transfer functions assuming they are stable but not necessarily causal. (d) Using a flow chart, explain how to implement these filters in software using FFT's.

9.56   Design Class 1 ASK, FSK, and PSK communication filters to detect signal pulses having $E\{s(t)\} = \pm[U_{-1}(t+T/2) - U_{-1}(t-T/2)]$ (unit amplitude, length T, centered at $t = 0$) which is contaminated by uncorrelated white noise having $E\{|N(j\omega)|\} = 1$. Repeat parts (a)–(d) of Problem 9.55.

9.57   Design Class 1 ASK, FSK, and PSK communication filters to detect a pulsed-cosine signal having $E\{s(t)\} = \cos(2\pi\omega_o t)[U_{-1}(t+0.1) - U_{-1}(t-0.1)]$ (peak amplitude = 1, frequency = 1 or 2 Hz, pulse duration = 0.2 sec) which is contaminated by uncorrelated white noise having $E\{|N(j\omega)|\} = 1$. Repeat parts (a)–(d) of Problem 9.55.

9.58   Design Class 2 ASK, FSK, and PSK communication filters to detect exponential step signals having $E\{s(t)\} = \pm e^{-t}U_{-1}(t)$ which are contaminated by uncorrelated white noise having $E\{|N(j\omega)|\} = 1$. (a) Write the transfer functions of the matched, high-resolution, and inverse signal detection filters which detect the expected signals. Smooth (inner block) the input spectra with rectangular, triangular, and Hann windows. Let $T = 1$ and smoothing width $p = 10\%$. (b) Using a flow chart, explain how to implement these filters in software using FFT's.

9.59   Design Class 2 ASK, FSK, and PSK communication filters to detect signal pulses having $E\{s(t)\} = \pm[U_{-1}(t+T/2) - U_{-1}(t-T/2)]$ (unit amplitude, length T, centered at $t = 0$) which is contaminated by uncorrelated white noise having $E\{|N(j\omega)|\} = 1$. Repeat parts (a)–(b) of Problem 9.58.

9.60   Design Class 2 ASK, FSK, and PSK communication filters to detect a pulsed-cosine signal having $E\{s(t)\} = \cos(2\pi\omega_o t)[U_{-1}(t+0.1) - U_{-1}(t-0.1)]$ (peak amplitude = 1, frequency = 1 or 2 Hz, pulse duration = 0.2 sec) which is contaminated by uncorrelated white noise having $E\{|N(j\omega)|\} = 1$. Repeat parts (a)–(b) of Problem 9.58.

9.61   Design a Class 1 radar filter to detect a pulsed-cosine signal having $E\{s(t)\} = \cos(2\pi t)[U_{-1}(t+0.1) - U_{-1}(t-0.1)]$ (peak amplitude = 1, frequency = 1 Hz, pulse duration = 0.2 sec) which is contaminated by uncorrelated white noise having $E\{|N(j\omega)|\} = 1$. Consider Fig. 9.10.2. (a) Write the transfer functions of the uncorrelated Wiener, correlated Wiener, and signal autocorrelation estimation filters in Table 9.10.1. (b) Write the transfer functions of the uncorrelated and correlated Wiener noise estimation filters of Table 9.10.1. (c) What is the criterion for detection? (d) Why is the μ factor used? (e) Using a flow chart, explain how to implement these filters in software using FFT's.

9.62   Design a Class 1 radar filter to detect a pulsed-cosine signal having $E\{s(t)\} = \cos(2\pi t)[U_{-1}(t+0.1) - U_{-1}(t-0.1)]$ (peak amplitude = 1, frequency = 1 Hz, pulse duration = 0.2 sec) which is contaminated by uncorrelated white noise having $E\{|N(j\omega)|\} = 1$. Consider Fig. 9.10.2. (a) Write the transfer functions of the classical, high-resolution, and inverse signal estimation filters in Table 9.10.1. (b) Write the transfer functions of the uncorrelated and correlated Wiener noise estimation filters of Table 9.10.1. (c) What is the criterion for detection? (d) Why is the μ factor used? (e) Using a flow chart, explain how to implement these filters in software using FFT's.

9.63   Consider the FFT implementation of the Class 1 estimation filters in Table 8.2.1. Use the low-, medium-, and high-speed microprocessor parameters of Eq. 9.12.3. Compute the processing times for the: (a) uncorrelated Wiener, (b) correlated Wiener, (c) pulse-shaping estimation filter algorithms.

9.64   Consider the FFT implementation of the Class 1 detection filters in Table 8.2.1. Use the low-, medium-, and high-speed microprocessor parameters of Eq. 9.12.3. Compute the processing times for the: (a) matched, (b) high-resolution, (c) inverse filter algorithms.

9.65   Consider the FFT implementation of the Class 1 correlation filters in Table 8.2.1. Use the low-, medium-, and high-speed microprocessor parameters of Eq. 9.12.3. Compute the processing times for the: (a) autocorrelation, (b) cross-correlation filter algorithms.

9.66   Consider the FFT implementation of the Class 2 estimation filters in Table 8.2.2. Use the low-, medium-, and high-speed microprocessor parameters of Eq. 9.12.3. Compute the processing times for the: (a) uncorrelated Wiener, (b) correlated Wiener, (c) pulse-shaping estimation filter algorithms. Smooth (inner block) the input spectra with rectangular, triangular, and Hann windows.

9.67   Consider the FFT implementation of the Class 2 detection filters in Table 8.2.2. Use the low-, medium-, and high-speed microprocessor parameters of Eq. 9.12.3. Compute the processing times for the: (a) matched, (b) high-resolution, (c) inverse filter algorithms. Smooth (inner block) the input spectra with rectangular, triangular, and Hann windows.

9.68   Consider the FFT implementation of the Class 2 correlation filters in Table 8.2.2. Use the low-, medium-, and high-speed microprocessor parameters of Eq. 9.12.3. Compute the processing times for the: (a) autocorrelation, (b) cross-correlation filter algorithms. Smooth (inner block) the input spectra with rectangular, triangular, and Hann windows.

9.69   Consider the FFT implementation of the Class 3 estimation filters in Table 8.2.2. Use the low-, medium-, and high-speed microprocessor parameters of Eq. 9.12.3. Compute the processing times for the: (a) uncorrelated Wiener, (b) correlated Wiener, (c) pulse-shaping estimation filter algorithms. Smooth (inner block) the input spectra with rectangular, triangular, and Hann windows.

9.70   Consider the FFT implementation of the Class 3 detection filters in Table 8.2.2. Use the low-, medium-, and high-speed microprocessor parameters of Eq. 9.12.3. Compute the processing times for the: (a) matched, (b) high-resolution, (c) inverse filter algorithms. Smooth (inner block) the input spectra with rectangular, triangular, and Hann windows.

9.71   Consider the FFT implementation of the Class 3 correlation filters in Table 8.2.2. Use the low-, medium-, and high-speed microprocessor parameters of Eq. 9.12.3. Compute the processing times for the: (a) autocorrelation, (b) cross-correlation filter algorithms. Smooth (inner block) the input spectra with rectangular, triangular, and Hann windows.

REFERENCES

1. Haas, W.H. and C.S. Lindquist, "Least-squares estimation filtering using an adaptive technique based on aposteriori cross-spectral density functions," Rec. 12th Asilomar Conf. Circuits, Systems, and Computers, pp. 273–274, 1978.

2. Lindquist, C.S., K.C. Severence, and H.B. Rozran, "Frequency domain algorithms for adaptive line enhancer and adaptive noise canceller systems," Rec. 18th Asilomar Conf. Circuits, Systems, and Computers, pp. 233–237, 1984.

3. Widrow, B., et al, "Adaptive noise cancelling: principles and applications," Proc. IEEE, vol. 63, pp. 1692–1716, Dec. 1975.

   —, "Stationary and nonstationary learning characteristics of the LMS adaptive filter," Proc. IEEE, vol. 64, pp. 1151–1162, Aug. 1976.

   —, "Adaptive design of digital filters," Proc. IEEE Intl. Conf. ASSP, pp. 243–246, March 1981.

   Widrow, B., "On the speed of convergence of adaptive algorithms," Rec. 14th Asilomar Conf. Circuits, Systems, and Computers, pp. 239–242, 1980.

   Treichler, J.R., "Response of adaptive line enhancer to chirped and doppler-shifted sinusoids," IEEE Trans. ASSP, vol. ASSP-28, pp. 343–348, June 1980.

   Anderson, C.M., E.H. Satorius, and J.R. Zeidler, "Adaptive enhancement of finite bandwidth signals in white Gaussian noise," IEEE Trans. ASSP, vol. ASSP-31, pp. 17–27, Feb. 1983.

   Reed, F.A. and Feintuch, P.L., "A comparison of LMS adaptive cancellers implemented in the frequency domain and the time domain," IEEE Trans. CAS, vol. CAS-28, pp. 610–615, June 1981.

   Friedlander,B., "System identification techniques for adaptive signal processing," IEEE Trans. ASSP, vol. ASSP-30, pp. 240–246, April 1982.

4. Haas, W.H. and C.S. Lindquist, "Linear detection filtering in the context of a least-squares estimator for signal processing applications," Proc. IEEE Intl. Conf. ASSP, pp. 651–654, 1978.

5. —, "An adaptive FFT-based detection filter," Proc. 26th Midwest Symp. Circuits and Systems, pp. 265–269, 1983.

6. Lindquist, C.S. and W.H. Haas, "Adaptive correlation filters defined in the frequency domain," Rec. 18th Asilomar Conf. Circuits, Systems, and Computers, pp. 327–331, 1984.

   —, "Some new adaptive correlation filters," Proc. 28th Midwest Symp. Circuits and Systems, pp. 479–488, 1985.

   —, "Another set of Class 1 correlation filters defined in the frequency domain," Proc. IEEE Intl. Symp. Circuits and Systems, pp. 1075–1078, 1986.

7. Haas, W.H. and C.S. Lindquist, "Frequency domain algorithms for adaptive filters," Proc. IEEE Intl. Symp. Circuits and Systems, pp. 503–506, 1984.

8. Stearns, S.D. and O.M. Solomon, Jr., "On adaptive coherence estimation," Proc. IEEE Intl. Symp. Circuits and Systems, pp. 511–514, 1984.

Chan, Y.T. and R.K. Miskowicz, "Estimation of coherence and time delay with ARMA models," IEEE Trans. ASSP, pp. 295–303, April 1984.

De Figueiredo, R.J.P. and A. Gerber, "Separation of superimposed signals by a cross-correlation method," IEEE Trans. ASSP, pp. 1084–1089, Oct. 1983.

Haas, W.H. and C.S. Lindquist, "Synthesis of frequency domain filters for time delay estimation," IEEE Trans. ASSP, pp. 540–548, June 1981.

Knapp, C.H. and G.C. Carter, "The generalized correlation method for estimation of time delay," IEEE Trans. ASSP, pp. 320–327, Aug. 1976.

Bertran-Salvans, M., "A generalized window approach to spectral estimation," IEEE Trans. ASSP, pp. 7–19, Feb. 1984.

9. Haas, W.H. and C.S. Lindquist, "Transitional FFT-based filters for delay estimation," Rec. 14th Asilomar Conf. Circuits, Systems, and Computers, pp. 342–344, 1980.

10. —, "Synthesis of frequency domain filters for time delay estimation," IEEE Trans. ASSP, vol. ASSP-29, pp. 540–548, June 1981.

11. Carter, G.C., "An overview of time delay estimation research for sonar systems," Rec. 13th Asilomar Conf. Circuits, Systems, and Computers, pp. 349–353, 1979.

Special Delay Estimation Issue, IEEE Trans. ASSP, vol. ASSP–29, June 1981.

Eckart, C., "Optimal rectifier systems for the detection of steady signals," Univ. of Calif., Scripps Inst. Ocean., San Diego, CA, S10 Ref. 52–11, 1952.

Roth, P.R., "Effective measurements using digital signal analysis," IEEE Spectrum, vol. 8, pp. 62–70, April 1971.

Hahn, W.R. and S.A. Tretter, "Optimum processing for delay-vector estimation in passive signal arrays," IEEE Trans. IT, vol. IT-F, pp. 608–614, Sept. 1973.

Carter, G.C., A.H. Nuttall, and P.G. Cable, "The smoothed coherence transform," Proc. IEEE, vol. 61, pp. 1197–1198, Oct. 1971.

Knapp, C.H. and G.C. Carter, "A generalized correlation method for estimation of time delay," IEEE Trans. ASSP, vol. ASSP-24, no. 4, pp. 320–327, Aug. 1976.

Hassab, J.C. and R.F. Boucher, "Optimum estimation of time delay by a generalized correlator," IEEE Trans. ASSP, vol. ASSP-27, no. 4, pp. 373–380, Aug. 1979.

12. Haas, W.H. and C.S. Lindquist, "The precision registration of digital images – a frequency domain approach," Rec. 15th Asilomar Conf. Circuits, Systems, and Computers, pp. 212–216, 1981.

13. Barnea, D.G. and H.F. Silverman, "A class of algorithms for fast digital image registration," IEEE Trans. Comp., vol. C-21, pp. 179–186, Feb. 1972.

Hall, E.L., D.L. Davies, and M.E. Lasey, "The selection of critical subsets for signal, image, and scene matching," IEEE Trans. PAMI, vol. PAMI-2, no. 4, pp. 313–322, July 1980.

14. Westheimer, N. and W.H. Haas, "Autonomous star cataloging for space surveillance missions," Proc. SPIE, vol. 280, April 1981.

15. Haas, W.H. and C.S. Lindquist, "A frequency domain approach to synthesizing near optimum edge detection filters," Proc. IEEE Intl. Conf. ASSP, pp. 745–748, 1980.

16. Twogood, R.E. and R.J. Sherwood, "An overview of digital image processing," Proc. 26th Midwest Symp. Circuits and Systems, pp. 384–388, 1983.

Haas, W.H., G.S. Dixon, and N.Y. Soon, "Performance of mosaic sensor dim target detection algorithms," Proc. SPIE, vol. 178, pp. 25–32, April 1979.

Haas, W.H. and V. Kumar, "The synthesis of FFT-based image enhancement filters," Proc. Workshop on Imaging Trackers and Autonomous Acquisition Applications for Missile Guidance, pp. 94–103, GACIAC-PR-80-01, Nov. 19-20, Redstone Arsenal, AL, 1979.

Shanmugam, K.S., F.M. Dickey, and J.A. Green, "An optimal frequency domain filter for edge detection in digital pictures," IEEE Trans. PMI, vol. PAMI-1, no. 1, pp. 31–49, Jan. 1979.

17. Lindquist, C.S. and H.B. Rozran, "Amplitude, frequency, and phase demodulation systems using frequency domain adaptive filters," Rec. 18th Asilomar Conf. Circuits, Systems, and Computers, pp. 332–336, 1984.

18. Lindquist, C.S. and L.W. Conboy, "Design of radar systems using frequency domain adaptive filters," Rec. 19th Asilomar Conf. Circuits, Systems, and Computers, pp. 411-415, 1985.

Conboy, L.W. "New frequency domain adaptive filters for use in radar systems," M.S.E.E. Thesis, Calif. State Univ., Long Beach, May 1985.

19. Spafford, L., "Optimum radar signal processing in clutter," IEEE Trans. IT, pp. 734–743, Sept. 1968.

Hansen, V., et al., "Adaptive digital MTI processing," Proc. Electronics and Aerospace Conf. and Exposition, pp. 170–176, 1973.

20. Mohan, M. and V. Rao, "An adaptive nonrecursive linear phase MIT filter based on nonuniform spectral resolution of input," IEEE Trans. ASSP, pp. 426–428, Aug. 1979.

Klem, R., "Adaptive clutter suppression in step scan radars," IEEE Trans. AES, pp. 685–688, July 1978.

Cole, E. and E. Harrison, "ASP: Kilotarget tracking at the radar," Rec. Asilomar Conf. Circuits, Systems, and Computers, pp. 594–598, 1979.

Abatzoglou, T., "Maximum likelihood estimates for frequencies of sinusoids of unknown scaling and phase," IEEE Natl. Aerospace and Electr. Conf., pp. 1036–1040, 1981.

21. Bowyer, D., P. Rajasakaran, and W. Gebhart, "Adaptive clutter filtering using autoregressive spectral estimation," IEEE Trans. AES, pp. 538–546, June 1979.

Sawyers, J., "Adaptive pulse-doppler radar signal processing using the maximum entropy method," IEEE Electr. and Aerospace Systems Conf. and Exposition, pp. 454–461, 1980.

22. Lindquist, C.S., N. Rizq, and M. Zhao, "Optimum detection of R-waves in EKG signals," Proc. IEEE Miami Technicon, pp. 23–26, 1987.

23. *Advanced Cardiac Life Support,* Chaps. 6 and 14, American Heart Assoc., National Center, Dallas, TX, 1983.

Advances in Pacemaker Technology, Springer-Verlag, New York, 1975.

24. Haas, W.H., C.S. Lindquist, and E.N. Evans, "Microprocessor-based optimum control using adaptive frequency domain design," Proc. 21st IEEE Conf. Decision and Control, pp. 813–817, 1982.

10 NONRECURSIVE FILTER DESIGN

> *And when the day of Pentecost was fully come,*
> *they were all with one accord in one place.*
> *And suddenly there came a sound from heaven as of a*
> *rushing mighty wind, and it filled all the house where they*
> *were sitting and there appeared unto them cloven tongues*
> *like as of fire, and it sat upon each of them.*
> *And they were all filled with the Holy Ghost ...*
>
> Acts 2:1–4

In the last two chapters, digital filters were designed using fast transforms. Adaptive transfer functions $H(j\omega)$ were determined using a variety of optimization criteria first described in Chap. 4. In this chapter and the next, different methods of approximating $H(j\omega)$ by a causal and stable digital filter transfer function $H(z)$ will be presented. Setting $z = e^{j\omega T}$ in $H(z)$ produces an ac steady-state response which is close to the required $H(j\omega)$ in some optimal manner (see Chap. 4.3).

Based on the form of $H(z)$, two broad classes of digital filters can be defined – *recursive* and *nonrecursive*. A nonrecursive filter has an output value which depends only upon input values. However a recursive filter has an output value which depends upon both input values and other output values. This chapter will discuss nonrecursive filters while the next chapter will discuss recursive filters. Their hardware and software implementations will be discussed in Chap. 12. The advantages, disadvantages, and trade-offs between recursive and nonrecursive filters were discussed in Chaps. 7.3–7.5 which the engineer may want to review before going further into this chapter.

A variety of techniques can be used to design recursive and nonrecursive filters. These are listed in Table 7.3.2. The Fourier series, weighted Fourier series, frequency sampling, and numerical analysis methods will be discussed in this chapter since they are most often applied to nonrecursive filters. However other techniques usually applied to recursive filters and found in the next chapter can also be used.

10.1 NONRECURSIVE FILTERS

The transfer function of any digital filter with a rational transfer function $H(z)$ equals

$$H(z) = \frac{Y(z)}{X(z)} = \frac{\sum_{k=-M}^{M} a_k z^{-k}}{\sum_{k=-N}^{N} b_k z^{-k}} = \frac{c \prod_{k=1}^{2M}(z - z_k)}{\prod_{k=1}^{2N}(z - p_k)} = \sum_{n=-\infty}^{\infty} h(n) z^{-n} \qquad (10.1.1)$$

The filter has $2N$ poles located at $z = p_k$ and $2M$ zeros located at $z = z_k$. Since the output $Y(z)$ equals $H(z)$ times the input $X(z)$, taking the inverse z-transform of both sides yields

$$Y(z) \sum_{k=-N}^{N} b_k z^{-k} = X(z) \sum_{k=-M}^{M} a_k z^{-k}$$

$$\sum_{k=-N}^{N} b_k y(n-k) = \sum_{k=-M}^{M} a_k x(n-k) \tag{10.1.2}$$

Setting $b_0 = 1$ for simplicity and solving for $y(n)$ gives

$$y(n) = \sum_{k=-M}^{M} a_k x(n-k) - \sum_{\substack{k=-N \\ k \neq 0}}^{N} b_k y(n-k) \tag{10.1.3}$$

which is the general time domain algorithm for a *recursive digital filter*. The impulse response of such a filter can be (but not must be) nonzero for all n. Recursive filters are also called *infinite-duration impulse response (IIR) filters*, or less frequently, *ladder, lattice, wave digital, autoregression moving average (ARMA)*, and *autoregression integrated moving average (ARIMA) filters*. Recursive filters can have finite impulse response lengths so the IIR term is a misnomer and must be used with caution.

Nonrecursive filters have $b_0 = 1$ and the other $b_k = 0$ which reduces Eq. 10.1.3 to

$$y(n) = \sum_{k=-M}^{M} a_k x(n-k) \tag{10.1.4}$$

Since the impulse response of these filters can be nonzero at no more than $(2M+1)$ sample times and are therefore of finite length, nonrecursive filters are also called *finite-duration impulse response (FIR), moving average, tapped delay line*, and *transversal filters*. From Eq. 10.1.1, FIR filters are all-zero filters with no finite poles except possibly at the origin. These are different than IIR filters which have nonzero poles. Since nonzero poles arise from feedback in the filter, such filters are described as being recursive in nature. Nonfeedback filters have no finite poles (except possibly at zero) so are said to be nonrecursive in nature.

Another important observation should be made about Eq. 10.1.3. Since the impulse response $h(n) = Z^{-1}[H(z)]$ exists for negative time, the general digital filter described by Eq. 10.1.1 is noncausal. To obtain causal filters that can be implemented in real-time, the limits in Eq. 10.1.1 must be changed so that

$$H(z) = \frac{Y(z)}{X(z)} = \frac{\sum_{k=0}^{M} a_k z^{-k}}{\sum_{k=0}^{N} b_k z^{-k}} = \sum_{n=0}^{\infty} h(n) z^{-n} \tag{10.1.5}$$

Equivalently, the a_k and b_k are set equal to zero for negative k in Eq. 10.1.1. Such causal filters have time domain algorithms equalling

$$y(n) = \sum_{k=0}^{M} a_k x(n-k) - \sum_{k=1}^{N} b_k y(n-k) \tag{10.1.6}$$

rather than those given by Eq. 10.1.3. Here the present value of the output $y(n)$ depends only upon the present and past values of the input $x(n)$ and past values of the output $y(n)$. This is not the case for the noncausal filter described by Eq. 10.1.3 whose $y(n)$ depends upon future values of the input $x(n)$ and output $y(n)$ as well.

Although noncausal systems cannot be implemented by *on-line* systems in *real-time*, they can be implemented by *off-line* or *nonreal-time* systems. Here the input values to be processed are stored in real time by an on-line system. Then they are reprocessed in nonreal-time by the off-line system using noncausal algorithms. Going off-line introduces processing delay which makes a noncausal algorithm causal. Therefore, which form we use is rather arbitrary.

Since there is no loss in generality by setting $M = N$ in Eq. 10.1.1 (set a_k's to zero where needed), we will assume that the general $2N$th-order digital filter is described by

$$H(z) = \frac{Y(z)}{X(z)} = \frac{\sum_{k=-N}^{N} a_k z^{-k}}{\sum_{k=-N}^{N} b_k z^{-k}} = \sum_{n=-\infty}^{\infty} h(n) z^{-n}$$

$$y(n) = \sum_{k=-N}^{N} \left[a_k x(n-k) - b_k y(n-k) \right], \qquad \text{set } b_0 = 0$$

(10.1.7)

This fully symmetrical form usually simplifies analysis. As just discussed, delay can always be added to the noncausal filter of Eq. 10.1.7 so it can be reexpressed as a causal filter in Eq. 10.1.6. N is replaced by $2N$ to maintain the same $(2N+1)$ number of points.

The matrix form of the causal $y(n)$ of Eq. 10.1.6 is (cf. Eq. 1.11.9)

$$
\begin{bmatrix}
\vdots \\
y(-1) \\
y(0) \\
y(1) \\
y(2) \\
\vdots \\
y(N) \\
y(N+1) \\
\vdots
\end{bmatrix}
=
\begin{bmatrix}
\vdots & \vdots & \vdots & \vdots & \vdots & \vdots & \vdots & \vdots \\
\cdots & 0 & 0 & 0 & 0 & 0 & 0 & \cdots \\
\cdots & 0 & a_0 & 0 & 0 & 0 & 0 & \cdots \\
\cdots & 0 & a_1 & a_0 & 0 & 0 & 0 & \cdots \\
\cdots & 0 & a_2 & a_1 & a_0 & 0 & 0 & \cdots \\
\vdots & \vdots & \vdots & \vdots & \vdots & \vdots & \vdots & \vdots \\
\cdots & 0 & a_N & \cdots & a_0 & 0 & 0 & \cdots \\
\cdots & 0 & 0 & a_N & \cdots & a_0 & 0 & \cdots \\
\vdots & \vdots & \vdots & \vdots & \vdots & \vdots & \vdots & \vdots
\end{bmatrix}
\begin{bmatrix}
\vdots \\
x(-1) \\
x(0) \\
x(1) \\
x(2) \\
\vdots \\
x(N) \\
x(N+1) \\
\vdots
\end{bmatrix}
$$

$$
-
\begin{bmatrix}
\vdots & \vdots & \vdots & \vdots & \vdots & \vdots & \vdots & \vdots & \vdots \\
\cdots & 0 & 0 & 0 & 0 & 0 & 0 & & \cdots \\
\cdots & 0 & 0 & 0 & 0 & 0 & 0 & 0 & \cdots \\
\cdots & 0 & b_1 & 0 & 0 & 0 & 0 & 0 & \cdots \\
\cdots & 0 & b_2 & b_1 & 0 & 0 & 0 & 0 & \cdots \\
\vdots & \vdots & \vdots & \vdots & \vdots & \vdots & \vdots & \vdots & \vdots \\
\cdots & 0 & b_N & \cdots & b_1 & 0 & 0 & 0 & \cdots \\
\cdots & 0 & 0 & b_N & \cdots & b_1 & 0 & 0 & \cdots \\
\vdots & \vdots & \vdots & \vdots & \vdots & \vdots & \vdots & \vdots & \vdots
\end{bmatrix}
\begin{bmatrix}
\vdots \\
y(-1) \\
y(0) \\
y(1) \\
y(2) \\
\vdots \\
y(N) \\
y(N+1) \\
\vdots
\end{bmatrix}
$$

(10.1.8)

Nonrecursive systems are described by this same equation but without the second term involving the $[b_k]$ matrix.[1]

10.2 NONRECURSIVE FILTER PROPERTIES

A nonrecursive digital filter of $2N$th-order is characterized by its transfer function

$$H(z) = \sum_{n=-N}^{N} h(n) z^{-n} \tag{10.2.1}$$

from Eq. 10.1.7a. Although $H(z)$ is noncausal, it can be cascaded with a z^{-N} delay block to produce a new causal $H(z)$. Alternatively the sum can be rewritten from $n = 0$ to N which permits both even and odd Nth-order causal filters to be formulated. We choose to use the $n = -N$ to N form of Eq. 10.2.1 to simplify equations. These filters use only an odd number of points and are always even-order. The results are easily extended to odd-order filters which process an even number of points.

An interesting and useful alternative $H(z)$ form results by noticing that $h(n)$ can be written as the sum of even (i.e., symmetric) and odd (i.e., antisymmetric) terms as[2]

$$h(n) = \frac{h(n) + h(-n)}{2} + \frac{h(n) - h(-n)}{2} = h_e(n) + h_o(n) \tag{10.2.2}$$

Substituting Eq. 10.2.2 into Eq. 10.2.1 yields

$$\begin{aligned} H(z) &= \sum_{n=-N}^{N} h_e(n) z^{-n} + \sum_{n=-N}^{N} h_o(n) z^{-n} \\ &= \left[h_e(0) + \sum_{n=1}^{N} h_e(n)(z^{-n} + z^{n}) \right] + \left[\sum_{n=1}^{N} h_o(n)(z^{-n} - z^{n}) \right] \\ &= H_e(z) + H_o(z) \end{aligned} \tag{10.2.3}$$

From the numerical analysis standpoint of Chap. 5, we saw that summing z^k type terms tended to average or smooth $h(n)$ while differencing tended to differentiate $h(n)$. Therefore Eq. 10.2.3 shows that any recursive or nonrecursive filter $H(z)$ can be expressed as the sum of an averaging or smoothing filter $H_e(z)$ and a differencing or differentiating filter $H_o(z)$.

The ac steady-state gain of any nonrecursive digital filter equals

$$\begin{aligned} H(e^{j\omega T}) = H(z) \Big|_{z=e^{j\omega T}} &= \sum_{n=-N}^{N} h(n) e^{-j\omega n T} \\ &= h_e(0) + 2 \sum_{n=1}^{N} h_e(n) \cos(n\omega T) - j2 \sum_{n=1}^{N} h_o(n) \sin(n\omega T) \end{aligned} \tag{10.2.4}$$

which results by setting $z = e^{j\omega T}$ in Eqs. 10.2.1 and 10.2.3. Nonrecursive filters have $N = \infty$ in Eq. 10.2.4. The gain $H(e^{j\omega T})$ has the properties outlined in Chap. 2.1 (see Eq. 2.1.5). When the impulse response $h(n)$ is even or symmetric about $n = 0$ so $h(n) = h_e(n)$, then $H(e^{j\omega T})$ is real, even, and composed only of $\cos(n\omega T)$ terms as

$$H(e^{j\omega T}) = H_e(e^{j\omega T}) = h_e(0) + 2\sum_{n=1}^{N} h_e(n)\cos(n\omega T) \qquad (10.2.5)$$

However if the impulse response $h(n)$ is odd or antisymmetric about $n = 0$ so $h(n) = h_o(n)$, then $H(e^{j\omega T})$ is imaginary, odd, and composed only of $\sin(n\omega T)$ terms as

$$H(e^{j\omega T}) = H_o(e^{j\omega T}) = -j2\sum_{n=1}^{N} h_o(n)\sin(n\omega T) \qquad (10.2.6)$$

Notice though that filters with such symmetric or antisymmetric impulse responses about $n = 0$ are noncausal. Eq. 10.2.1 shows that multiplying their $H(z)$ by z^{-N} destroys this symmetry and makes them both causal. Typical causal impulse responses having even symmetry and odd symmetry are shown in Figs. 10.2.1a and 10.2.1b, respectively. Both even $(2N)$ and odd $(2N-1)$ numbers of data points are considered.

    In ac steady-state, causality requires adding a linear phase of $\arg(e^{-j\omega NT}) = -\omega NT$ radians to the phase of $H(e^{j\omega T})$ in Eq. 10.2.4. In the time domain, this corresponds to delaying the output signal by $-d\arg H/d\omega = NT$ seconds as shown in Fig. 10.2.1. Such causal filters have output signals with either even or odd symmetry about $t = NT$ for any N where

$$\begin{aligned} h(n) &= h(2N-n), &&\text{even symmetry} \\ h(n) &= -h(2N-n), &&\text{odd symmetry} \end{aligned} \qquad (10.2.7)$$

and $0 \le n \le 2N$. Therefore only nonrecursive filters with this very special symmetry have constant delay or linear phase over the entire frequency range. Causal recursive filters having infinite length impulse responses can never have linear phase.

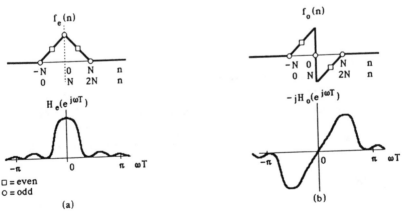

(a)   (b)

Fig. 10.2.1 Typical impulse and gain responses having (a) even symmetry and (b) odd symmetry. (Both even $(2N)$ and odd $(2N+1)$ number of points are shown.)

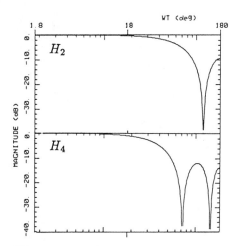 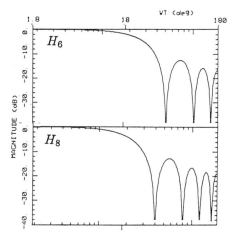

Fig. 10.2.2 Magnitude responses of smoothing filters in Example 10.2.1.

EXAMPLE 10.2.1 Data is to be smoothed using simple averaging between $2N+1$ adjacent data points. Determine the frequency response of the resulting $2N$th-order FIR digital filter transfer function.

SOLUTION Since the output $y(n)$ equals

$$y(n) = \frac{1}{2N+1}\Big[x(n+N)+\ldots+x(n+1)+x(n)+x(n-1)+\ldots+x(n-N)\Big] \quad (10.2.8)$$

taking the z-transform yields the filter transfer function $H(z)$ as

$$H(z) = \frac{Y(z)}{X(z)} = \frac{1}{2N+1}\Big[z^N + \ldots + z^1 + 1 + z^{-1} + \ldots + z^{-N}\Big] \quad (10.2.9)$$

Setting $z = e^{j\omega T}$ yields

$$\begin{aligned}
H(e^{j\omega T}) &= \frac{1}{2N+1}\Big[1 + (e^{j\omega T} + e^{-j\omega T}) + \ldots + (e^{jN\omega T} + e^{-jN\omega T})\Big] \\
&= \frac{1}{2N+1}\Big[1 + 2\cos(\omega T) + \ldots + 2\cos(N\omega T)\Big] \quad (10.2.10) \\
&= \frac{1}{2N+1}\frac{\sin\big[(2N+1)\omega T/2\big]}{\sin(\omega T/2)}
\end{aligned}$$

The magnitude response is plotted in Fig. 10.2.2 for 3, 5, 7, and 9 data points (equivalent to filter orders 2, 4, 6, 8). This averaging scheme acts as a very poor lowpass filter. For example, the second-order (three-data-point) filter has a 3 dB bandwidth of 56° (or $0.16f_s$) and a minimum stopband rejection of only $1/3 = -9.5$ dB at frequencies above 150° (or $0.42f_s$). The transition bandwidth equals $150° - 56° = 94°$ (or $0.26f_s$). ∎

EXAMPLE 10.2.2 The first-order derivative $y(t)$ of $x(t)$ is to be approximated using a four-point difference equation as

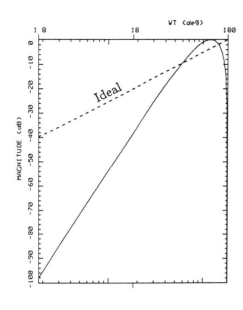

Fig. 10.2.3 Magnitude response of first-order differentiation filter in Example 10.2.2.

$$y(n) = \frac{dx(t)}{dt}\Big|_{t=nT} \cong \frac{1}{T}\left[\frac{1}{2}x(n+2) - x(n+1) + x(n-1) - \frac{1}{2}x(n-2)\right] \quad (10.2.11)$$

Determine the frequency response of this fourth-order FIR differentiator.
SOLUTION Taking the z-transform of both sides and gives

$$H(z) = \frac{Y(z)}{X(z)} = \frac{1}{T}\left[\frac{1}{2}z^2 - z^1 + z^{-1} - \frac{1}{2}z^{-2}\right] \quad (10.2.12)$$

Setting $z = e^{j\omega T}$ yields

$$\begin{aligned}
H(e^{j\omega T}) &= \frac{1}{T}\left(-e^{j\omega T} + e^{-j\omega T}\right) + \frac{1}{2T}\left(e^{j\omega 2T} - e^{-j\omega 2T}\right) \\
&= \frac{j}{T}\left[-2\sin(\omega T) + \sin(2\omega T)\right]
\end{aligned} \quad (10.2.13)$$

which is plotted in Fig. 10.2.3. It has a slope of 60 dB/dec rather than 20 dB/dec and is therefore a poor differentiation algorithm. ∎

10.3 FOURIER SERIES DESIGN (INTEGRAL SQUARED-ERROR ALGORITHM)

The impulse response coefficients $h(n)$ of the transfer function $H(z)$ given by Eq. 10.2.1 can be determined using the Fourier series method discussed in Chap. 1.7. With this technique, a desired frequency response $H_d(j\omega)$ is approximated by $H(e^{j\omega T})$ as

$$H_d(j\omega) \cong H(z)\Big|_{z=e^{j\omega T}} = \sum_{n=-\infty}^{\infty} h(n)z^{-n}\Big|_{z=e^{j\omega T}} \qquad (10.3.1)$$

(see Eq. 1.7.8). The optimization criterion requires minimizing the complex magnitude-squared error E between $H_d(j\omega)$ and $H(e^{j\omega T})$ where

$$E = \min\left[\int_{-\omega_e/2}^{\omega_e/2} |H_d(j\omega) - H(e^{j\omega T})|^2 d\omega\right] = \int_{-\omega_e/2}^{\omega_e/2} |H_d(j\omega)|^2 d\omega - \sum_{n=-N}^{N} h^2(n)$$
$$(10.3.2)$$

The impulse response coefficients $h(n)$ of $H(z)$ which minimize this ISE error equal[3]

$$h(n) = \frac{1}{\omega_s}\int_{-\omega_e/2}^{\omega_e/2} H_d(j\omega)e^{jn\omega T}d\omega, \qquad T = 2\pi/\omega_s \qquad (10.3.3)$$

by analogy with the Fourier series coefficient formula of Eq. 1.7.9. If $H_d(j\omega)$ is even and real, then Eq. 10.3.3 reduces to

$$h_e(n) = \frac{2}{\omega_s}\int_0^{\omega_e/2} H_d(j\omega)\cos(n\omega T)d\omega = h_e(-n) \qquad (10.3.4)$$

so the $h(n)$ coefficients have even symmetry. If $H_d(j\omega)$ is odd and imaginary, then

$$h_o(n) = j\frac{2}{\omega_s}\int_0^{\omega_e/2} H_d(j\omega)\sin(n\omega T)d\omega = -h_o(-n) \qquad (10.3.5)$$

so the $h(n)$ coefficients have odd symmetry.

Eq. 10.3.3 therefore serves as the optimum least-squares filter synthesis equation for Fourier series design. Since the $e^{jn\omega T}$, $\cos(n\omega T)$, and $\sin(n\omega T)$ terms are orthogonal over any frequency interval which is an integer multiple of ω_s wide, the coefficient values $h(n)$ do not change with the number of terms used. Increasing the number $(2N+1)$ of terms in $H(z)$ to approximate $H_d(j\omega)$ improves the approximation and reduces the ISE value in Eq. 10.3.2. From a digital filter standpoint, the IIR filter coefficients (infinite in number) given by Eq. 10.3.3 are truncated to $(2N+1)$ terms to obtain an $2N$th-order FIR filter realization.

EXAMPLE 10.3.1 Determine the transfer function of the optimum ISE digital lowpass filter.
SOLUTION The ideal digital lowpass filter has

$$\begin{aligned} H_d(j\omega) &= 1, & |\omega| \leq \omega_c \\ &= 0, & \omega_c < |\omega| \leq \omega_s/2 \end{aligned} \qquad (10.3.6)$$

Thus, the optimum $h(n)$ coefficients are determined from Eqs. 10.3.3 or 10.3.4 as

$$h(n) = \frac{1}{\omega_s} \int_{-\omega_c}^{\omega_c} e^{j\omega nT} d\omega = \frac{1}{n\pi} \sin(n\omega_c T) \qquad (10.3.7)$$

The $h(n)$ are independent of the total number $(2N+1)$ of data points used. Once the normalized corner frequency $\omega_c T = 2\pi f_c/f_s = 360° f_c/f_s$ has been specified, the coefficients are easily determined.

Suppose we choose $f_c = f_s/4$ so that $\omega_c T = 360°/4 = 90°$. Evaluating Eq. 10.3.7 yields the impulse response coefficients

$$h(0) = \frac{1}{2}, \qquad h(-1) = h(1) = \frac{1}{\pi}, \qquad h(-2) = h(2) = 0,$$

$$h(-3) = h(3) = -\frac{1}{3\pi}, \qquad \cdots \qquad (10.3.8)$$

$h(n)$ is even since $H_d(j\omega)$ is even (Eq. 10.3.4). The filter algorithm and transfer function therefore equal

$$y(n) = \frac{1}{2}x(n) + \frac{1}{\pi}\Big[x(n+1) + x(n-1)\Big] - \frac{1}{3\pi}\Big[x(n+3) + x(n-3)\Big] + \cdots$$

$$H(z) = 0.5 + \frac{1}{\pi}\Big[\cdots - \frac{1}{3}z^3 + z^1 + z^{-1} - \frac{1}{3}z^{-3} + \cdots\Big] \qquad (10.3.9)$$

The IIR filter response of Eq. 10.3.9 can be truncated to any number of terms to produce the desired FIR filter. In ac steady-state where $z = e^{j\omega T}$,

$$H(e^{j\omega T}) = 0.5 + \frac{1}{\pi}\Big[(e^{j\omega T} + e^{-j\omega T}) - \frac{1}{3}(e^{j3\omega T} + e^{-j3\omega T}) + \cdots\Big]$$

$$= 0.5 + \frac{2}{\pi}\Big[\cos(\omega T) - \frac{1}{3}\cos(3\omega T) + \cdots\Big] = 0.5 + \frac{2}{\pi}\sum_{\substack{n=1 \\ n \text{ odd}}}^{\infty} \frac{(-1)^n}{n}\cos(n\omega T)$$

$$(10.3.10)$$

$H(e^{j\omega T})$ is even, real, and composed only of cosine terms since $h(n)$ is even (Eq.

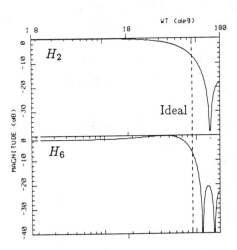
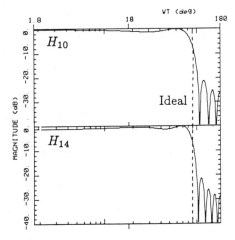

Fig. 10.3.1 Magnitude responses of lowpass filters in Example 10.3.1.

10.2.5). The optimum ISE lowpass filter magnitude responses are plotted in Fig. 10.3.1. They are different from those found in Example 10.2.1. They have a flatter gain characteristic over a wider frequency range and about 1 dB in-band ripple. They provide about twice the minimum stopband rejection for a given order than the lowpass filters in Fig. 10.2.2. They also have a narrower transition band. The three-point second-order filter has a 3 dB bandwidth of 62° (or $0.17f_s$) and a minimum stopband rejection of $0.12 = -18.4$ dB at frequencies above 147° (or $0.41f_s$). This filter is clearly an improvement over that found in Fig. 10.2.2 which has only -9.5 dB rejection and a transition band of 94°.   ∎

EXAMPLE 10.3.2 Design the optimum ISE digital bandpass filter having

$$H_d(j\omega) = 1, \qquad \omega_L \le |\omega| \le \omega_U$$
$$= 0, \qquad 0 \le |\omega| < \omega_L, \quad \omega_U < |\omega| \le \omega_s/2 \tag{10.3.11}$$

SOLUTION From Eq. 10.3.4, the impulse response coefficients equal

$$h(n) = \frac{2}{\omega_s} \int_{\omega_L}^{\omega_U} \cos(2\pi n\omega/\omega_s)d\omega = \frac{1}{n\pi}\left\{ \sin\left[2\pi(n\omega_U/\omega_s)\right] - \sin\left[2\pi(n\omega_L/\omega_s)\right] \right\}$$

$$= \frac{2}{n\pi}\sin\left[n\pi(\omega_U - \omega_L)/\omega_s\right]\, \cos\left[n\pi(\omega_U + \omega_L)/\omega_s\right] \tag{10.3.12}$$

where the arithmetic center frequency $\omega_o = 0.5(\omega_U + \omega_L)$ and the bandwidth $B = \omega_U - \omega_L$. The filter has $Q = \omega_o/B$.

Suppose that $\omega_o = 0.25\omega_s$ and $B = 0.1\omega_s$ so that $\omega_U = 0.3\omega_s$ and $\omega_L = 0.2\omega_s$. Equivalently, the center frequency $\omega_o T = 90°$ while the upper and lower frequencies equal $\omega_U T = 108°$ and $\omega_L T = 72°$. The bandwidth $BT = 36°$. Then Eq. 10.3.12 reduces to

$$h(n) = \frac{2}{n\pi}\sin(0.1n\pi)\cos(0.5n\pi) = \frac{2}{n\pi}\sin(18n°)\cos(90n°) \tag{10.3.13}$$

The odd coefficients are zero while the even coefficients equal

$$h(0) = 0.2, \qquad h(-2) = h(2) = -0.1871, \qquad h(-4) = h(4) = 0.1514,$$
$$h(-6) = h(6) = -0.1009, \qquad h(-8) = h(8) = 0.04677, \tag{10.3.14}$$
$$h(-10) = h(10) = 0, \qquad \cdots$$

They are even since $H_d(j\omega)$ is even. The filter algorithm and transfer function then equal

$$y(n) = 0.2x(n) - 0.1871\left[x(n+2) + x(n-2)\right] + 0.1514\left[x(n+4) + x(n-4)\right] + \ldots$$
$$H(z) = 0.2 - 0.1871\left[z^2 + z^{-2}\right] + 0.1514\left[z^4 + z^{-4}\right] + \ldots$$
$$\tag{10.3.15}$$

The ac steady-state gain equals

$$H(e^{j\omega T}) = 0.2 - 0.3742\cos(2\omega T) + 0.3028\cos(4\omega T) + \ldots \tag{10.3.16}$$

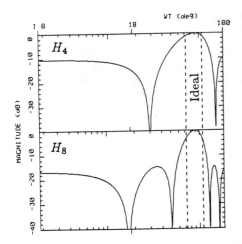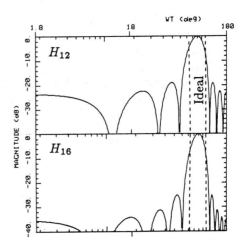

Fig. 10.3.2 Magnitude responses of bandpass filters in Example 10.3.2.

The magnitude responses are shown in Fig. 10.3.2. The stopband rejection and selectivity increase with filter order while the bandwidth decreases. ■

EXAMPLE 10.3.3 Design a digital lowpass differentiator having a transfer function

$$H_d(j\omega) = j2\omega/\omega_s, \qquad |\omega| < \omega_s/2 \qquad (10.3.17)$$
$$= 0, \qquad \text{elsewhere}$$

This filter has a gain of unity at the Nyquist frequency $\omega_s/2$.
SOLUTION For the optimum ISE gain response, we select

$$h(n) = j\frac{2}{\omega_s} \int_{-\omega_s/2}^{\omega_s/2} \frac{\omega}{\omega_s} e^{j2n\pi(\omega/\omega_s)} d\omega = 0, \qquad n = 0$$

$$= \frac{1}{n\pi}\cos(n\pi), \qquad n \neq 0 \qquad (10.3.18)$$

The $h(n)$ are odd since $H_d(j\omega)$ is odd (Eq. 10.3.5). The filter algorithm and transfer function equal

$$y(n) = \frac{1}{\pi}\left[\ldots - \frac{1}{2}x(n+2) + x(n+1) - x(n-1) + \frac{1}{2}x(n-2) - \ldots\right]$$

$$H(z) = \frac{1}{\pi}\left[\ldots + \frac{1}{3}z^3 - \frac{1}{2}z^2 + z^1 - z^{-1} + \frac{1}{2}z^{-2} - \frac{1}{3}z^{-3} + \ldots\right] \qquad (10.3.19)$$

The filter responses equal

$$H(e^{j\omega T}) = j\frac{2}{\pi}\sum_{n=1}^{\infty}\frac{(-1)^n}{n}\sin(n\omega T) \qquad (10.3.20)$$

They are odd, imaginary, and composed only of sine terms since $h(n)$ is odd (Eq. 10.2.6). Their magnitudes are plotted in Fig. 10.3.3. Increasing filter order serves to move the corner frequency of the differentiator towards the Nyquist frequency. This makes the gain approach the more ideal triangular shape. $H_2, H_6, H_{10}, \ldots$ have a 20 dB/dec low-frequency slope. $H_4, H_8, H_{12}, \ldots$ have a 60 dB/dec low-frequency slope and a mid-band slope of 20 dB/dec. Therefore based on Bode plot slope and not ISE error, H_k is better matched over a wider frequency range when $k/2$ is odd rather than even. ∎

EXAMPLE 10.3.4 Design a bandpass differentiator having

$$H_d(j\omega) = j2\omega/\omega_s, \qquad \omega_L \le |\omega| \le \omega_U$$
$$= 0, \qquad\qquad \text{elsewhere} \qquad\qquad (10.3.21)$$

This filter has a gain that increases linearly through the passband from ω_L to ω_U. Its bandwidth $\omega_U - \omega_L$ is less than the Nyquist frequency.

SOLUTION From Eq. 10.3.5, the $h(n)$ coefficients equal

$$h(n) = j\frac{2}{\omega_s} \int_{\omega_L}^{\omega_U} \frac{j2\omega}{\omega_s} \sin(2\pi n\omega/\omega_s)d\omega = \frac{-1}{(\pi n)^2}\Big[\sin(2\pi n\omega_U/\omega_s) - \sin(2\pi n\omega_L/\omega_s)\Big]$$

$$+\frac{2}{\pi n}\Big[(\omega_U/\omega_s)\cos(2\pi n\omega_U/\omega_s) - (\omega_L/\omega_s)\cos(2\pi n\omega_L/\omega_s)\Big] \qquad (10.3.22)$$

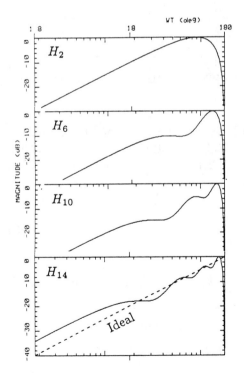
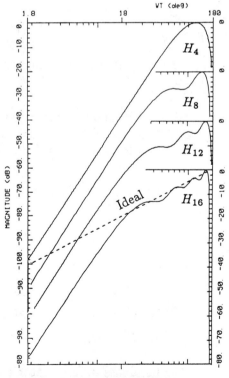

Fig. 10.3.3 Magnitude responses of differentiating filters in Example 10.3.3.

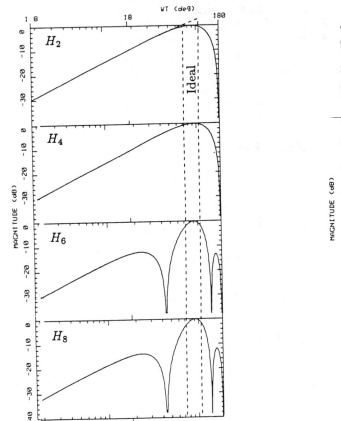

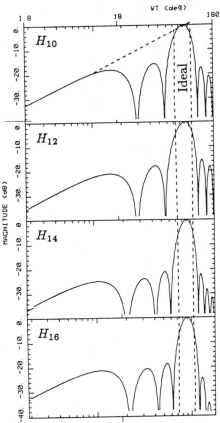

Fig. 10.3.4 Magnitude responses of bandpass differentiating filters in Example 10.3.4.

When the bandpass differentiator becomes a lowpass differentiator because $\omega_U = \omega_s/2$ and $\omega_L = 0$, Eq. 10.3.22 reduces to Eq. 10.3.18. Suppose that $\omega_o = 0.25\omega_s$ and $B = 0.1\omega_s$ so $\omega_U = 0.3\omega_s$ and $\omega_L = 0.2\omega_s$. Then from Eq. 10.3.22,

$$h(n) = \frac{1}{(n\pi)^2}\left[\sin(72n^\circ) - \sin(108n^\circ)\right] + \frac{2}{n\pi}\left[0.3\cos(108n^\circ) - 0.2\cos(72n^\circ)\right]$$
(10.3.23)

and the first few coefficients equal

$$h(0) = 0, \qquad h(-1) = -h(1) = 0.09836, \qquad h(-2) = -h(2) = -0.004026,$$
$$h(-3) = -h(3) = -0.08584, \qquad h(-4) = -h(4) = 0.007127,$$
$$h(-5) = -h(5) = 0.06366, \qquad h(-6) = -h(6) = -0.008632,$$
$$h(-7) = -h(7) = -0.03679, \qquad h(-8) = -h(8) = -0.0083, \qquad \cdots$$
(10.3.24)

The bandpass differentiator algorithm and transfer functions then equal

$$y(n) = 0.09836\big[x(n+1) - x(n-1)\big] - 0.004026\big[x(n+2) - x(n-2)\big]$$
$$- 0.08584\big[x(n+3) - x(n-3)\big] + \dots$$
$$H(z) = 0.09836\big[z^1 - z^{-1}\big] - 0.004026\big[z^2 - z^{-2}\big] - 0.08584\big[z^3 - z^{-3}\big] + \dots$$
$$(10.3.25)$$

In ac steady-state,

$$H(e^{j\omega T}) = -0.1967\sin(\omega T) + 0.008052\sin(2\omega T) + 0.1717\sin(3\omega T) + \dots \quad (10.3.26)$$

The magnitude responses are shown in Fig. 10.3.4. The low-frequency asymptotic slope is 20 dB/dec. The stopband rejection increases with filter order and the bandwidth decreases.   ■

EXAMPLE 10.3.5 Design a wide-band $-90°$ phase-shift digital filter (also called a Hilbert transform filter) with gain

$$H_d(j\omega) = -j\,\mathrm{sgn}\,\omega, \qquad |\omega| < \omega_s/2 \qquad (10.3.27)$$

SOLUTION The optimum ISE coefficients equal

$$h(n) = -j\frac{2}{\omega_s}\int_{-\omega_s/2}^{\omega_s/2}(\mathrm{sgn}\,\omega)e^{j2\pi n(\omega/\omega_s)}d\omega = 0, \qquad n\ \text{even}$$
$$\qquad\qquad (10.3.28)$$
$$= 2/n\pi, \qquad n\ \text{odd}$$

which has odd symmetry since $H_d(j\omega)$ is odd and imaginary. Therefore the filter algorithm and transfer function of the filter equal

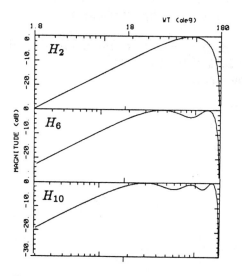
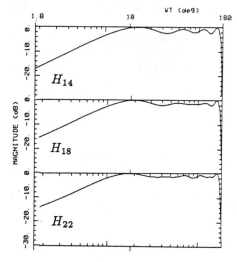

Fig. 10.3.5 Magnitude responses of $-90°$ phase-shifting filters in Example 10.3.5.

$$y(n) = \frac{2}{\pi}\left[\ldots + \frac{1}{3}x(n+3) + x(n+1) - x(n-1) - \frac{1}{3}x(n-3) - \ldots\right]$$

$$H(z) = \frac{2}{\pi}\left[\ldots + \frac{1}{5}z^5 + \frac{1}{3}z^3 + z - z^{-1} - \frac{1}{3}z^{-3} - \frac{1}{5}z^{-5} - \ldots\right] \qquad (10.3.29)$$

where

$$H(e^{j\omega T}) = -j\frac{4}{\pi}\sum_{\substack{n=1 \\ n \text{ odd}}}^{\infty} \frac{1}{n}\sin(n\omega T) \qquad (10.3.30)$$

The magnitude responses are shown in Fig. 10.3.5. The phase responses equal $-90°$ between dc and the Nyquist frequency. The gain exhibits about 1 dB ripple. Its bandwidth widens as the filter order increases. A 10th-order FIR phase shifter provides almost constant phase over a decade frequency range. ∎

10.4 WEIGHTED FOURIER SERIES DESIGN (WINDOWING ALGORITHMS)

It was observed in Chap. 4.3 that an ISE optimization criterion often produces filter characteristics that exhibit large errors (like peaking) over small intervals or small errors (like oscillations or ringing) over large intervals. This is always the case with Fourier series since they are derived using an ISE criterion. The maximum peaking occurs near discontinuities and remains virtually fixed at ±9% or 1.5 dB independent of the approximation order. Increasing filter order increases the oscillation frequency. The oscillations tend to grow in amplitude until the peak values are reached on either side of the discontinuity. This behavior is evident in the filter responses of Figs. 10.3.1–10.3.5. Such performance is usually unacceptable in practice. The oscillation level is reduced by eliminating discontinuities in the magnitude response and making it smoother.

One way of doing this is to smooth the designed gain response $H(e^{j\omega T})$ by allowing a nonzero transition band between the passband and stopband rather than requiring an instantaneous transition. This complicates the calculations of the filter coefficients.

EXAMPLE 10.4.1 Determine the transfer function of the optimum ISE digital lowpass filter having a transition band that is 10% of the passband.

SOLUTION The ideal filter transfer function equals

$$\begin{aligned} H_d(j\omega) &= 1, & |\omega| &\le \omega_c \\ &= 1 - 10(|\omega|/\omega_c - 1), & \omega_c &< |\omega| \le 1.1\omega_c \qquad (10.4.1) \\ &= 0, & 1.1\omega_c &< |\omega| \le \omega_s/2 \end{aligned}$$

The optimum coefficients therefore equal

$$\begin{aligned} h(n) &= \frac{2}{\omega_s}\left[\int_0^{\omega_c}\cos(n\omega T)d\omega + \int_{\omega_c}^{1.1\omega_c}(11 - 10\omega/\omega_c)\cos(n\omega T)d\omega\right] \\ &= -\frac{5}{(n\pi)^2}\frac{\omega_s}{\omega_c}\left[\cos(1.1n\omega_c T) - \cos(n\omega_c T)\right] \qquad (10.4.2) \\ &= \frac{10}{(n\pi)^2}\frac{\omega_s}{\omega_c}\sin(1.05n\omega_c T)\sin(0.05n\omega_c T) \end{aligned}$$

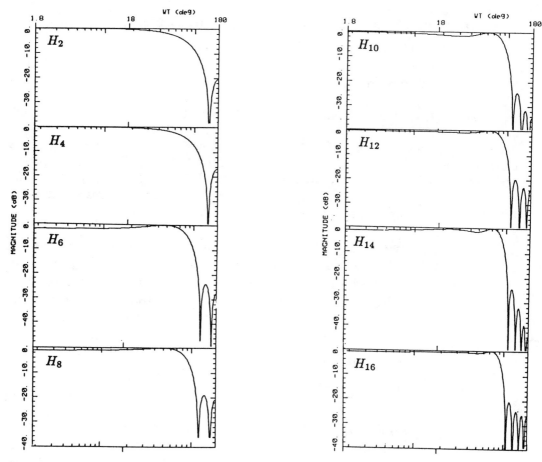

Fig. 10.4.1 Magnitude responses of lowpass filters in Example 10.4.1.

Assume that $\omega_c = 0.25\omega_s$ so $\omega_c T = 90°$. Then

$$h(n) = \frac{40}{(n\pi)^2} \sin(0.525n\pi) \sin(0.025n\pi) \qquad (10.4.3)$$

For small n, Eq. 10.4.3 reduces to $h(n) = (1/n\pi)\sin(1.05n\omega_c T)$ with an error $\leq 10\%$ for $n \leq 5$. Comparing this result with the zero transition band lowpass filter of Eq. 10.3.7, we see that the coefficients are slightly different. The first few filter coefficients equal

$$h(0) = 0.525, \qquad h(-1) = h(1) = 0.3170, \qquad h(-2) = h(2) = -0.02480,$$
$$h(-3) = h(3) = -0.1022, \qquad h(-4) = h(4) = 0.02419,$$
$$h(-5) = h(5) = 0.05732, \qquad h(-6) = h(6) = -0.02320,$$
$$h(-7) = h(7) = -0.03685, \qquad h(-8) = h(8) = 0.02188, \qquad \cdots$$

$$(10.4.4)$$

The filter algorithm and transfer function equal

$$y(n) = 0.525x(n) + 0.3170\big[x(n+1) + x(n-1)\big] - 0.02480\big[x(n+2) + x(n-2)\big]\cdots$$
$$H(z) = 0.525 + 0.3170\big[z^1 + z^{-1}\big] - 0.02480\big[z^2 + z^{-2}\big] + \cdots$$

$$(10.4.5)$$

The ac steady-state gain equals

$$H(e^{j\omega T}) = 0.525 + 0.6340\cos(\omega T) - 0.04960\cos(2\omega T) + \cdots \tag{10.4.6}$$

We see the gain coefficients of $(0.525, 0.634, -0.0496, -0.204, 0.0484, 0.115)$ are close to those of Eq. 10.3.10 of $(0.5, 0.637, 0 , 0.212, 0, 0.127)$. The filter responses are plotted in Fig. 10.4.1. Comparing these with those of Fig. 10.3.1, we see the in-band gain ripple is decreased, the stopband rejection is increased, and the transition band widens. Thus, compromises must be made. ∎

Another way of smoothing $H(e^{j\omega T})$ is by using the window functions discussed in Chaps. 2.5–2.7. With this technique, the Fourier series time sequence $h(n)$ given by Eq. 10.3.3 is multiplied (see Eq. 2.5.1) by a suitable weighting sequence $w(n)$ from Table 2.6.1. Alternatively, the Fourier series coefficients $H(m)$ are convolved with the spectral coefficients $W(m)$ of the window (see Eq. 2.7.2). Notationally, $H(m) \equiv H(e^{j2\pi m/N})$. Summarizing these operations mathematically,

$$h_w(n) = h(n)w(n)$$
$$H_w(e^{j\omega T}) = \frac{1}{2\pi} H(e^{j\omega T}) * W(e^{j\omega T}) \quad \text{or} \tag{10.4.7}$$
$$H_w(m) = \frac{1}{N} H(m) * W(m) = \frac{1}{N}\sum_{k=0}^{N-1} H(k)W(m-k)$$

These are the optimum filter synthesis equations using weighted Fourier series design. Each of these three equations is shown in block diagram form in Fig. 10.4.2. Rather than weighting before filtering as shown in Fig. 10.4.2a, we can convolve spectrums after filtering as shown in Fig. 10.4.2b, or simply filter with the new transfer function H_w as shown in Fig. 10.4.2c.

Let us now review this process using several representative windows from Table 2.6.1 as examples (reread Chaps. 2.6–2.7 if you need to refresh your memory). These windows are centered about $t = 0$.

Hann (raised-cosine) window:

$$h_w(n) = \big[0.5 + 0.5\cos(2n\pi/N)\big]h(n)$$
$$H_w(m) = 0.5H(m) + 0.25H(m-1) + 0.25H(m+1) \tag{10.4.8}$$

Hamming window:

$$h_w(n) = \big[0.543 + 0.457\cos(2n\pi/N)\big]h(n)$$
$$H_w(m) = 0.543H(m) + 0.229H(m-1) + 0.229H(m+1) \tag{10.4.9}$$

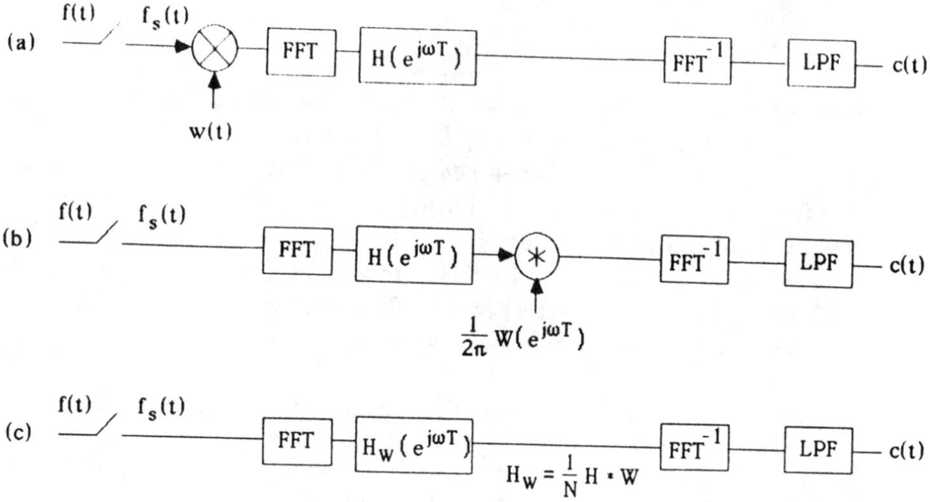

(a) $f(t)$ $f_s(t)$ → ⊗ → FFT → $H(e^{j\omega T})$ — FFT^{-1} → LPF — $c(t)$

$w(t)$

(b) $f(t)$ $f_s(t)$ → FFT → $H(e^{j\omega T})$ → ⊛ → FFT^{-1} → LPF — $c(t)$

$\dfrac{1}{2\pi}W(e^{j\omega T})$

(c) $f(t)$ $f_s(t)$ → FFT → $H_W(e^{j\omega T})$ → FFT^{-1} → LPF — $c(t)$

$H_W = \dfrac{1}{N}H \cdot W$

Fig. 10.4.2 Application of window functions explicitly in (a) time domain, (b) frequency domain, and (c) implicitly in frequency domain.

Blackman window (3-term):

$$h_w(n) = \left[0.423 + 0.498\cos(2n\pi/N) + 0.0792\cos(4n\pi/N)\right]h(n)$$
$$H_w(m) = 0.423H(m) + 0.249H(m-1) + 0.249H(m+1)$$
$$\hspace{3cm} + 0.0396H(m-2) + 0.0396H(m+2) \hspace{2cm} (10.4.10)$$

Since these windows are composed of discrete frequency components which are multiples of the FFT bin frequency ω_s/N, the frequency convolution form $H_w(m)$ is especially convenient. $H(m)$ can be scaled, shifted integer multiples of the FFT bin frequency, and added to form $H_w(m)$ as illustrated in Eqs. 10.4.8–10.4.10.

From this viewpoint, the improvement in the Fourier series transfer function $H(e^{j\omega T})$ depends strictly upon the characteristics of the window gain $W(e^{j\omega T})$. It should have: (1) narrow main-lobe bandwidth ($B \ll \omega_s/N$ if possible), (2) small side-lobe levels (side-lobe gain $= 0$ if possible), and (3) high asymptotic roll-off (as ωT approaches π). These requirements insure that the passband and stopband ripples present in $H(e^{j\omega T})$ will be greatly reduced.

One major design difficulty is that the edge of the passband cannot be determined exactly since the window smears or widens the frequency response around the transition band (see Fig. 4.1.1). This often necessitates several design iterations to obtain the desired corner frequency. The approximate transition bandwidth at a gain discontinuity is greater than the main lobe width of $W(e^{j\omega T})$. Therefore the transition bandwidth $(f_r - f_p)$ of the gain and the main lobe bandwidth kf_s/N of the window satisfy[4]

$$f_r - f_p > kf_s/N \hspace{3cm} (10.4.11)$$

Solving for N, then the number of samples should be no less than

$$N > kf_s/(f_r - f_p) \qquad (10.4.12)$$

The constant k can be obtained from Fig. 2.6.1 by inspection for any desired window. For example, $k = 2$ for rectangular, $k = 4$ for Hann, Hamming, and Bartlett, and $k = 6$ for Blackman windows. When the passband frequency f_p, stopband frequency f_r, sampling frequency f_s, and type of window are specified, Eq. 10.4.12 gives a lower limit on the number of points that must be used in approximating $H_d(j\omega)$. For example, if $f_p = 1$ KHz, $f_r = 2$ KHz, $f_s = 10$ KHz, and a Hamming window is to be used, then $N > 4(10 \text{ KHz})/(2 \text{ KHz} - 1 \text{ KHz}) = 40$. Therefore no less than about 40 samples are necessary based on the transition bandwidth.

EXAMPLE 10.4.2 A lowpass differentiator was designed in Example 10.3.3 using the Fourier series technique. There was excessive ripple in the frequency response shown in Fig. 10.3.3. Reduce the ripple by using Hann and Hamming windows.
SOLUTION The optimum ISE filter coefficients were determined in Eq. 10.3.18 and the optimum ISE gain function was given by Eq. 10.3.19. Combining these results with the Hann and Hamming time weighting functions given by Eqs. 10.4.8a and 10.4.9a gives the weighted Fourier series filter coefficients and transfer functions. For Hann weighting, they equal

$$h_w(n) = \frac{1}{n\pi}\left[0.5 + 0.5\cos(2n\pi/N)\right]\cos(n\pi), \qquad n \neq 0$$
$$= 0, \qquad n = 0 \qquad (10.4.13)$$

$$H_w(z) = \frac{1}{\pi}\sum_{n=1}^{(N-1)/2}(-1)^n\frac{\cos^2(\pi n/N)}{n}\left[z^n - z^{-n}\right]$$

while for Hamming weighting, they equal

$$h_w(n) = \frac{1}{n\pi}\left[0.543 + 0.457\cos(2n\pi/N)\right]\cos(n\pi), \qquad n \neq 0$$
$$= 0, \qquad n = 0 \qquad (10.4.14)$$

$$H_w(z) = \frac{1}{\pi}\sum_{n=1}^{(N-1)/2}(-1)^n\frac{0.543 + 0.457\cos(2n\pi/N)}{n}\left[z^n - z^{-n}\right]$$

N is the number of points and is odd. Evaluating the first few transfer functions $H_w(z)$ of the FIR Hann-weighted lowpass differentiators gives (Eq. 10.4.13b)

$$H_2(z) = \frac{1}{\pi}\left[-0.25(z^1 - z^{-1})\right]$$

$$H_4(z) = \frac{1}{\pi}\left[-0.6545(z^1 - z^{-1}) + 0.09549(z^2 - z^{-2})\right]$$

$$H_6(z) = \frac{1}{\pi}\left[-0.8117(z^1 - z^{-1}) + 0.1944(z^2 - z^{-2}) - 0.0165(z^3 - z^{-3})\right]$$

$$H_8(z) = \frac{1}{\pi}\left[-0.8830(z^1 - z^{-1}) + 0.2934(z^2 - z^{-2}) - 0.0833(z^3 - z^{-3})\right.$$
$$\left. + 0.007538(z^4 - z^{-4})\right]$$
$$(10.4.15)$$

The FIR Hamming-weighted lowpass differentiators have (Eq. 10.4.14b)

$$H_2(z) = \frac{1}{\pi}\Big[-0.3145(z^1 - z^{-1})\Big]$$

$$H_4(z) = \frac{1}{\pi}\Big[-0.6842(z^1 - z^{-1}) + 0.08664(z^2 - z^{-2})\Big]$$

$$H_6(z) = \frac{1}{\pi}\Big[-0.8279(z^1 - z^{-1}) + 0.2207(z^2 - z^{-2}) - 0.04375(z^3 - z^{-3})\Big]$$

$$H_8(z) = \frac{1}{\pi}\Big[-0.8931(z^1 - z^{-1}) + 0.3112(z^2 - z^{-2}) - 0.1048(z^3 - z^{-3})$$
$$+ 0.02839(z^4 - z^{-4})\Big]$$

$$(10.4.16)$$

Alternatively, frequency weighting functions can be used rather than time weighting. Since the Fourier series lowpass differentiator has a transfer function of (Eq. 10.3.20)

$$H(m) = H(e^{j\omega T})\Big|_{\omega = m\omega_\bullet/N} = j\frac{2}{\pi}\sum_{n=1}^{\infty}\frac{(-1)^n}{n}\sin(2mn\pi/N) \qquad (10.4.17)$$

then using Eq. 10.4.8b, the Hann-weighted lowpass differentiator gain equals

$$H_w(m) = j\frac{2}{\pi}\sum_{n=1}^{\infty}\frac{(-1)^n}{n}$$

$$\times \Big[0.5\sin(2mn\pi/N) + 0.25\sin(2n(m-1)\pi/N) + 0.25\sin(2n(m+1)\pi/N)\Big]$$

$$= j\frac{2}{\pi}\sum_{n=1}^{\infty}(-1)^n\frac{0.5}{n}\sin(2mn\pi/N)\big[1 + \cos(2n\pi/N)\big]$$

$$= j\frac{2}{\pi}\sum_{n=1}^{\infty}\frac{(-1)^n}{n}\sin(2mn\pi/N)\cos^2(n\pi/N) \qquad (10.4.18)$$

and the Hamming-weighted gain equals (Eq. 10.4.9b)

$$H_w(m) = \frac{j2}{\pi}\sum_{n=1}^{\infty}\frac{(-1)^n}{n}\sin(2mn\pi/N)\Big[0.543 + 0.457\cos(2n\pi/N)\Big] \qquad (10.4.19)$$

The magnitude responses of these weighted differentiators are plotted in Figs. 10.4.3a and 10.4.3b, respectively. They are almost identical. Comparing these responses with the unweighted differentiator responses of Fig. 10.3.3, we see the in-band ripple is virtually eliminated but the transition band has widened. ∎

Let us summarize what this important weighting procedure has accomplished. An old filter with easily determined Fourier coefficients $h(n)$ and $H(m)$ but undesirable response is easily converted into a new filter with new coefficients $h_w(n)$ and a more desirable response. Once $H_w(n)$ has been determined, it can be used directly as a digital filter as shown in Fig. 10.4.2c without any (explicit) weighting.

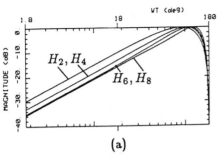
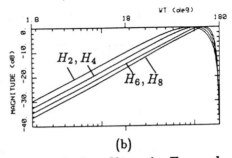

Fig. **10.4.3** Magnitude responses of windowed differentiating filters in Example 10.4.1 using (a) Hann and (b) Hamming weighting.

10.5 FREQUENCY SAMPLING DESIGN (ZERO POINT-ERROR ALGORITHM)

Another method of $(N-1)$st-order nonrecursive filter design is to specify the desired frequency response $H_d(j\omega)$ at N frequencies along the frequency axis $s = j\omega$. The N filter coefficients are then chosen so that the filter gain $H(e^{j\omega T})$ exactly equals these values.[5] This produces zero point-error at these frequencies. By analogy with Eq. 10.3.2, the zero error at every point produces zero total error since

$$E = \sum_{N \text{ values}} |H_d(j\omega) - H(e^{j\omega T})|^2 = 0 \qquad (10.5.1)$$

This is called the *frequency sampling* design method.

For convenience, but not necessity, these N points are taken at intervals of $1/NT = f_s/N$ Hz from dc to the sampling frequency f_s or between the Nyquist frequencies $-f_s/2$ and $f_s/2$. Mathematically, the FIR filter response at frequency ω equals

$$H(e^{j\omega T}) = H(z)\Big|_{z=e^{j\omega T}} = \sum_{n=0}^{N-1} h(n)z^{-n}\Big|_{z=e^{j\omega T}} \qquad (10.5.2)$$

from Eq. 10.2.4. Choosing the frequency ω so that $\omega = m\omega_s/N$, then Eq. 10.5.2 can be rewritten as

$$H_d(m) = H(e^{j\omega_m sT/N}) = H(z)\Big|_{z=e^{j2\pi m/N}} = \sum_{n=0}^{N-1} h(n)z^{-n}\Big|_{z=e^{j2\pi m/N}}$$

$$= \sum_{n=0}^{N-1} h(n)W_N^{nm}, \qquad 0 \le m \le N-1 \qquad (10.5.3)$$

where $W_N = e^{-j2\pi/N}$. Comparing Eq. 10.5.3 with the discrete Fourier transform (DFT) given by Eq. 2.1.9, we see they have identical form. Therefore $H_d(m)$ are the DFT coefficients of the impulse response $h(n)$. Conversely $h(n)$ are the IDFT coefficients of $H_d(m) = |H_d(m)| \exp(j \arg H_d(m))$ where

$$h(n) = \frac{1}{N} \sum_{m=0}^{N-1} H_d(m) W_N^{-nm} = \frac{1}{N} \sum_{m=0}^{N-1} H_d(m) e^{j2\pi nm/N}, \qquad 0 \le n \le N-1 \quad (10.5.4)$$

from Eq. 3.2.5. This is the optimum filter synthesis equation using frequency sampling design. Thus the FFT and IFFT algorithms can used to quickly determine the filter coefficients $h(n)$ from the desired frequency samples $H_d(m)$ (assuming $N = 2^k$). Unequal frequency spacing between samples requires slower DFT algorithms be used rather than the fast algorithms. It should be noted before proceeding that although N values of $H(e^{j\omega T})$ are specified from $0 \le \omega < \omega_s$, the gain magnitude (phase) always has even (odd) symmetry about the Nyquist frequency $\omega_s/2$. This was discussed in relation to Eq. 2.1.5. Therefore N values of $|H|$ and $\arg H$, or alternatively, Re H and Im H, must be specified for $H_d(j\omega)$ between $0 \le \omega < \omega_s/2$. Although there is zero error between the actual gain $H(e^{j\omega T})$ and the desired gain $H_d(j\omega)$ at the sampled frequencies $\omega = m\omega_s/N$, the error is generally nonzero at all other frequencies. To determine the form of the actual gain response between samples, let us derive the filter transfer function $H(z)$. The actual transfer function can be expressed in terms of the gain samples $H_d(m)$ by substituting $h(n)$ given by Eq. 10.5.4 into Eq. 10.5.2, interchanging summation order, and simplifying the summation with respect to n as

$$H(z) = \sum_{n=0}^{N-1} z^{-n} \left[\frac{1}{N} \sum_{m=0}^{N-1} H_d(m) e^{j2\pi nm/N} \right] = \frac{1}{N} \sum_{m=0}^{N-1} H_d(m) \left[\sum_{n=0}^{N-1} \left(e^{j2\pi m/N} z^{-1} \right)^n \right]$$

$$= \frac{1}{N} \sum_{m=0}^{N-1} H_d(m) \frac{1 - z^{-N}}{1 - e^{j2\pi m/N} z^{-1}}$$

$$(10.5.5)$$

In ac steady state where $z = e^{j\omega T}$, Eq. 10.5.5 reduces to

$$H(e^{j\omega T}) = \frac{1}{N} \sum_{m=0}^{N-1} H_d(m) \frac{1 - e^{-jN\omega T}}{1 - e^{-j(\omega T - 2\pi m/N)}} = \frac{1}{N} \sum_{m=0}^{N-1} H_d(m) \frac{1 - e^{-jN(\omega T - 2\pi m/N)}}{1 - e^{-j(\omega T - 2\pi m/N)}}$$

$$= \frac{1}{N} e^{-j\omega(N-1)T/2} \sum_{m=0}^{N-1} H_d(m) e^{j\pi m(N-1)/N} \frac{\sin[N(\omega T - 2m\pi/N)/2]}{\sin[(\omega T - 2m\pi/N)/2]}$$

$$(10.5.6)$$

Eq. 10.5.6 shows that each desired gain value $H_d(m)$ specified at a frequency $\omega = m\omega_s/N$ (or an angle of $2\pi m/N$ radians on the unit circle in the z-plane) provides a $[\sin(N\omega T/2)/\sin(\omega T/2)]$ frequency response interpolation at all other frequencies. This interpolation fills in the gaps of $H(e^{j\omega T})$ between the sample frequencies and fixes its functional form. Notice in Eq. 10.5.6 that if the desired gain samples $H_d(m)$ have phases which satisfy $\arg H_d(m) = -\pi m(N-1)/N$, then the terms being summed are real. The resulting filter transfer function will have linear phase as $\arg H(e^{j\omega T}) = -\omega T(N-1)/2$ and the impulse response $h(n)$ will be delayed by $(N-1)/2$ clock periods or $(N-1)T/2$ seconds (i.e., half the assumed impulse response duration NT seconds in Eq. 10.5.4).

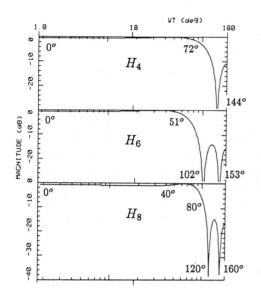

Fig. 10.5.1 Magnitude responses of frequency sampling lowpass filters in Example 10.5.1.

EXAMPLE 10.5.1 Redesign the digital lowpass filter of Example 10.3.1 using the frequency sampling approach.

SOLUTION The lowpass gain function given by Eq. 10.3.6 can be expressed as

$$H_d(m) = 1, \qquad |m| \le N_o = \text{integer}(N\omega_c/\omega_s)$$
$$= 0, \qquad N_o < |m| \le N/2 \tag{10.5.7}$$

to have the same form as Eq. 10.5.3. N_o equals the number of equally-spaced samples of $H_d(j\omega)$ falling in its passband. Although $\arg[H_d(m)]$ equals zero in this example, it can also be selected to meet the linear phase requirement just discussed. Then the frequency sampling filter coefficients equal

$$h(n) = \frac{1}{N} \sum_{m=0}^{N-1} H_d(m)e^{j2\pi nm/N} = \frac{1}{N} \sum_{m=-N_o}^{N_o} e^{j2\pi nm/N}$$

$$= \frac{1}{N}\left[1 + 2\sum_{m=1}^{N_o} \cos(2\pi nm/N)\right], \qquad 0 \le n \le N-1 \tag{10.5.8}$$

from Eq. 10.5.4. The $h(n)$ coefficients depend upon the number of points specified.

Suppose we select $N_o = N/4$ which corresponds to $\omega_c T = 90°$. Evaluating Eq. 10.5.8 for various odd N yields the filter coefficients and substituting them into Eq. 10.2.1 produces the transfer function. The first few transfer functions are

$$H_2(z) = 1 \qquad \text{(not useful lowpass)}$$
$$H_4(z) = 0.6000 + 0.3236\left[z^1 + z^{-1}\right] - 0.1236\left[z^2 + z^{-2}\right]$$
$$H_6(z) = 0.4286 + 0.3210\left[z^1 + z^{-1}\right] + 0.07928\left[z^2 + z^{-2}\right] - 0.1146\left[z^3 + z^{-3}\right]$$
$$H_8(z) = 0.5556 + 0.3199\left[z^1 + z^{-1}\right] - 0.05912\left[z^2 + z^{-2}\right] - 0.1111\left[z^3 + z^{-3}\right]$$
$$+ 0.0725\left[z^4 + z^{-4}\right] \tag{10.5.9}$$

The H_2, H_4, H_6, and H_8 filters use 3, 5, 7, and 9 gain samples at equal angle spacings of $120°$, $72°$, $51.4°$, and $40°$, respectively. The ac steady-state gains obtained by setting $z = e^{j\omega T}$ in Eq. 10.5.9 are plotted in Fig. 10.5.1. Comparing them with those obtained using the Fourier series technique in Fig. 10.3.1, we see these frequency sampling filters exhibit much smaller passband ripple but less stopband rejection than the Fourier series filters. ∎

EXAMPLE 10.5.2 Redesign the digital lowpass filter of Example 10.5.1 using a nonzero transition band as was done in Example 10.4.1.

SOLUTION Denote $B_p = \omega_c/\omega_s$ as the fractional filter bandwidth and $T_p = \omega_t/\omega_c$ as the fractional transition bandwidth of the lowpass filter. Then $B_p T_p = \omega_t/\omega_s$ and the number of samples N_t falling the transition band equals $N_t = B_p T_p N$. Using only integer values of N_t (i.e., 1, 2, 3, ...), then the smallest sample size equals

$$N_{min} = 1/B_p T_p \qquad (10.5.10)$$

From Example 10.4.1, the transition bandwidth $\omega_t = 0.1\omega_c$ (Eq. 10.4.1) and the corner frequency $\omega_c = 0.25\omega_s$. Therefore $T_p = 0.1$, $B_p = 0.25$, and $N_{min} = 1/(0.25)(0.1) = 40$ samples. Thus, large numbers of points are needed to implement

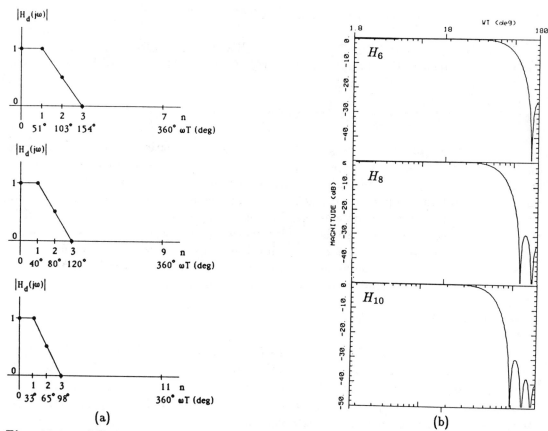

(a)   (b)

Fig. 10.5.2 Magnitude responses of (a) ideal and (b) frequency sampling lowpass filters in Example 10.5.3.

this filter. The filter coefficients are evaluated using Eq. 10.5.4 in matrix form where $[h] = [W_N]^{-1}[H_d]$ as given by Eq. 3.2.7b. The filter coefficients $[h]$ are the inverse DFT of the desired gain coefficients $[H_d]$. ∎

EXAMPLE 10.5.3 Design a lowpass filter having a passband and transition band that is FFT frequency bin wide (f_s/N) using 7, 9, and 11 samples.
SOLUTION Specify $H_d(0) = H_d(-1) = H_d(1) = 1$ and $H_d(-2) = H_d(2) = 0.5$ as shown in Fig. 10.5.2a. The other H_d are zero. Then from Eq. 10.5.4, the filter coefficients equal

$$h(n) = h(-n) = \frac{1}{N}\Big[1 + 2\cos(2n\pi/N) + \cos(4n\pi/N)\Big] \qquad (10.5.11)$$

Evaluating Eq. 10.5.11 and computing the $H_4(z)$ transfer functions gives

$$H_6(z) = 0.5714 + 0.2892\big[z^1 + z^{-1}\big] - 0.04943\big[z^2 + z^{-2}\big] - 0.02549\big[z^3 + z^{-3}\big]$$

$$H_8(z) = 0.4444 + 0.3006\big[z^1 + z^{-1}\big] + 0.04529\big[z^2 + z^{-2}\big] - 0.02549\big[z^3 + z^{-3}\big]$$
$$- 0.01259\big[z^4 + z^{-4}\big]$$

$$H_{10}(z) = 0.3636 + 0.2816\big[z^1 + z^{-1}\big] + 0.1069\big[z^2 + z^{-2}\big] - 0.05556\big[z^3 + z^{-3}\big]$$
$$- 0.04109\big[z^4 + z^{-4}\big] - 0.007067\big[z^5 + z^{-5}\big]$$

$$(10.5.12)$$

Their gains are plotted in Fig. 10.5.2b. Allowing these filters to have a nonzero transition band achieves more stopband rejection than the filters of Example 10.5.1 having a zero transition band. ∎

10.6 NUMERICAL ANALYSIS ALGORITHMS

Another approach for determining digital filter transfer functions is to use numerical analysis formulas. Some basic operations in numerical analysis involve interpolation, extrapolation, integration, and differentiation. A number of formulas have been developed to perform these operations which can be implemented using both recursive and nonrecursive filters. Integration can only be done using recursive filters. Numerical integration was used in Chap. 5 to derive a variety of analog-to-digital transforms which are listed in Tables 5.2.1 and 5.2.2. Applying the reciprocals of these transforms provides numerical differentiation algorithms. Several interpolation formulas were already listed in Table 8.7.1. We shall now review interpolation in more detail.

10.6.1 INTERPOLATION ALGORITHMS

Seven classical interpolation formulas are listed in Table 10.6.1. These include Newton, Gauss, Stirling, Bessel, and Everett. Given the data values $[x(n), x(n+1), \ldots, x(m-1), x(m)]$, these equations allow data $x(t)$ to be estimated at times t other than the sample times $[nT, (n+1)T, \ldots, (m-1)T, mT]$. If $nT \leq t \leq mT$, the process is called *interpolation*. Otherwise it is called *extrapolation*.

Suppose that the data $x(n)$ can be adequately approximated by the series expansions of Table 10.6.1. For example, the Newton forward-difference approximation equals

$$x(p) = x(0) + C(p,1)\Delta x(0) + C(p,2)\Delta^2 x(0) + C(p,3)\Delta^3 x(0) + \ldots \qquad (10.6.1)$$

Evaluating the first few terms yields

$$x(n+p)_1 = x(n)$$

$$x(n+p)_2 = \big[1 - C(p,1)\big]x(n) + C(p,1)x(n+1)$$

$$x(n+p)_3 = \big[1 - C(p,1) + C(p,2)\big]x(n) + \big[C(p,1) - 2C(p,2)\big]x(n+1)$$
$$\qquad + C(p,2)x(n+2)$$

$$x(n+p)_4 = \big[1 - C(p,1) + C(p,2) - C(p,3)\big]x(n)$$
$$\qquad + \big[C(p,1) - 2C(p,2) + 3C(p,3)\big]x(n+1)$$
$$\qquad + \big[C(p,2) - 3C(p,3)\big]x(n+2) + C(p,3)x(n+3)$$

$$x(n+p)_5 = \big[1 - C(p,1) + C(p,2) - C(p,3) + C(p,4)\big]x(n) \qquad (10.6.2)$$
$$\qquad + \big[C(p,1) - 2C(p,2) + 3C(p,3) - 4C(p,4)\big]x(n+1)$$
$$\qquad + \big[C(p,2) - 3C(p,3) + 6C(p,4)\big]x(n+2)$$
$$\qquad + \big[C(p,3) - C(p,4)\big]x(n+3) + C(p,4)x(n+4)$$

$$x(n+p)_6 = \big[1 - C(p,1) + C(p,2) - C(p,3) + C(p,4) - C(p,5)\big]x(n)$$
$$\qquad + \big[C(p,1) - 2C(p,2) + 3C(p,3) - 4C(p,4) + 5C(p,5)\big]x(n+1)$$
$$\qquad + \big[C(p,2) - 3C(p,3) + 6C(p,4) - 10C(p,5)\big]x(n+2)$$
$$\qquad + \big[C(p,3) - C(p,4) + 10C(p,5)\big]x(n+3)$$
$$\qquad + \big[C(p,4) - 5C(p,5)\big]x(n+4) + C(p,5)x(n+5)$$

These equations show how to approximate $x(n+p)$ in terms of the samples $x(n)$, $x(n+1)$, $x(n+2)$, ... Any number N of samples may be used and any value of relative deviation p (possibly noninteger) may be evaluated.

Suppose that we wish to interpolate a series of data to find the midpoint value between samples. Midpoint algorithms are often used to estimate derivatives between samples. They provide more accurate results than those obtained using estimates at the sample points.[6] Since only an even number N of samples can be interpolated to find their midpoint value, only the even numbered equations of Eq. 10.6.2 will be used. Evaluating these equations with the proper p value (i.e., $p = (N-1)/2$) yields

$$y(n+1/2)_2 = \frac{1}{2}\Big[x(n) + x(n+1)\Big]$$

$$y(n+3/2)_4 = \frac{1}{16}\Big[-x(n) + 9x(n+1) + 9x(n+2) - x(n+3)\Big]$$

$$y(n+5/2)_6 = \frac{1}{256}\Big[3x(n) - 25x(n+1) + 150x(n+2) + 150x(n+3) - 25x(n+4)$$
$$\qquad + 3x(n+5)\Big] \qquad (10.6.3)$$

$$y(n+7/2)_8 = \frac{1}{2048}\Big[-5x(n) + 49x(n+1) - 245x(n+2) + 1225x(n+3)$$
$$\qquad + 1225x(n+4) - 245x(n+5) + 49x(n+6) - 5x(n+7)\Big]$$

For convenience, these interpolations should be normalized so that the midpoint occurs at the origin. Translating the samples by $-p/2$ seconds, then the transfer functions $H_N(z)$ of the Newton midpoint interpolation filters equal

$$H_2(z) = \frac{1}{2}\left[z^{1/2} + z^{-1/2}\right]$$

$$H_4(z) = \frac{1}{16}\left[-z^{3/2} + 9z^{1/2} + 9z^{-1/2} - z^{-3/2}\right]$$

$$H_6(z) = \frac{1}{256}\left[3z^{5/2} - 25z^{3/2} + 150z^{1/2} + 150z^{-1/2} - 25z^{-3/2} + 3z^{-5/2}\right]$$

$$H_8(z) = \frac{1}{2048}\left[-5z^{7/2} + 49z^{5/2} - 245z^{3/2} + 1225z^{1/2} + 1225z^{-1/2} - 245z^{-3/2}\right.$$
$$\left. + 49z^{-5/2} - 5z^{-7/2}\right]$$

$$(10.6.4)$$

Setting $z = e^{j\omega T}$, the ac steady-state gains of these filters equal

$$H_2(e^{j\omega T}) = \cos(\omega T/2)$$

$$H_4(e^{j\omega T}) = \frac{1}{8}\left[9\cos(\omega T/2) - \cos(3\omega T/2)\right]$$

$$H_6(e^{j\omega T}) = \frac{1}{128}\left[150\cos(\omega T/2) - 25\cos(3\omega T/2) + 3\cos(5\omega T/2)\right]$$

$$H_8(e^{j\omega T}) = \frac{1}{1024}\left[1225\cos(\omega T/2) - 245\cos(3\omega T/2) + 49\cos(5\omega T/2) - 5\cos(7\omega T/2)\right]$$

$$(10.6.5)$$

which are plotted in Fig. 10.6.1.

These filters have a lowpass characteristic. As the number of samples approaches infinity, the filters become allpass types. The low-order filters reject noise falling in the vicinity of the Nyquist frequency. However high-order filters pass all frequencies and are not frequency selective.

In a slightly different problem, a data point may be missing from a set of data and needs to be interpolated. This is equivalent to interpolating a missing data point at a sample time rather than halfway between data points as was just discussed.[6] Let us assume that equally spaced data about the missing point is used. This requires the use of only an odd number N of samples. This is equivalent to assuming that $x(p)$ is described by a $(N-1)$th-order difference equation in Table 10.6.1 and that the Nth-order difference term $\Delta^N x(0)$ is equal to zero. From Table 10.6.1, we see these

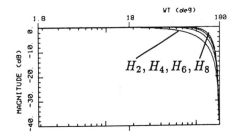

Fig. 10.6.1 Magnitude responses of Newton mid-point interpolators.

differences depend only upon order so the resulting filters do not depend upon which approximation is used unlike the previous midpoint interpolation filters. Setting these first few differences equal to zero results in

$$\Delta^2(0) = 0 = x(n+2) - 2x(n+1) + x(n)$$

$$\Delta^4(0) = 0 = x(n+4) - 4x(n+3) + 6x(n+2) - 4x(n+1) + x(n)$$

$$\Delta^6(0) = 0 = x(n+6) - 6x(n+5) + 15x(n+4) - 20x(n+3) + 15x(n+2)$$
$$- 6x(n+1) + x(n)$$

$$\Delta^8(0) = 0 = x(n+8) - 8x(n+7) + 28x(n+6) - 56x(n+5) + 70x(n+4)$$
$$- 56x(n+3) + 28x(n+2) - 8x(n+1) + x(n)$$

$$(10.6.6)$$

Shifting the samples $-N/2$ seconds produces equations having even symmetry about $n = 0$. Then solving for $x(n)$ produces

$$\Delta^2(0): \quad y(n) = \frac{1}{2}\Big[x(n+1) + x(n-1)\Big]$$

$$\Delta^4(0): \quad y(n) = \frac{1}{6}\Big[-x(n+2) + 4x(n+1) + 4x(n-1) - x(n-2)\Big]$$

$$\Delta^6(0): \quad y(n) = \frac{1}{20}\Big[x(n+3) - 6x(n+2) + 15x(n+1) + 15x(n-1) - 6x(n-2)$$
$$+ x(n-3)\Big]$$

$$\Delta^8(0): \quad y(n) = \frac{1}{70}\Big[-x(n+4) + 8x(n+3) - 28x(n+2) + 56x(n+1) + 56x(n-1)$$
$$- 28x(n-2) + 8x(n-3) - x(n-4)\Big]$$

$$(10.6.7)$$

where $y(n)$ replaces $x(n)$ which is assumed to be unknown. The transfer functions therefore equal

$$H_2(z) = \frac{1}{2}\Big[z^1 + z^{-1}\Big]$$

$$H_4(z) = \frac{1}{6}\Big[-z^2 + 4z^1 + 4z^{-1} - z^{-2}\Big]$$

$$H_6(z) = \frac{1}{20}\Big[z^3 - 6z^2 + 15z^1 + 15z^{-1} - 6z^{-2} + z^{-3}\Big]$$

$$(10.6.8)$$

$$H_8(z) = \frac{1}{70}\Big[-z^4 + 8z^3 - 28z^2 + 56z^1 + 56z^{-1} - 28z^{-2} + 8z^{-3} - z^{-4}\Big]$$

The ac steady-state gains equal

$$H_2(e^{j\omega T}) = \cos(\omega T)$$

$$H_4(e^{j\omega T}) = \frac{1}{3}\Big[4\cos(\omega T) - \cos(2\omega T)\Big]$$

$$H_6(e^{j\omega T}) = \frac{1}{10}\Big[15\cos(\omega T) - 6\cos(2\omega T) + \cos(3\omega T)\Big]$$

$$(10.6.9a–c)$$

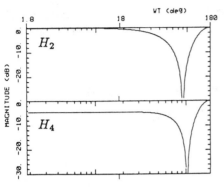 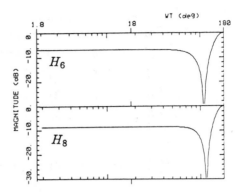

Fig. 10.6.2 Magnitude responses of missing data interpolators.

$$H_8(e^{j\omega T}) = \frac{1}{35}\Big[56\cos(\omega T) - 28\cos(2\omega T) + 8\cos(3\omega T) - \cos(4\omega T)\Big] \quad (10.6.9d)$$

setting $z = e^{j\omega T}$ in Eq. 10.6.8. These gains are plotted in Fig. 10.6.2. These are bandstop filters. They have a dc gain of unity so low-frequency signals will have missing data almost perfectly replaced. However signals with frequencies close to the Nyquist frequency have very erroneous replacement data (opposite sign, amplified level) since the Nyquist frequency gain is greater than unity. Signals at roughly one-third the sampling rate have missing values replaced by zeros.

10.6.2 DIFFERENTIATION ALGORITHMS

Another basic numerical operation involves differentiation. Differentiation algorithms can be obtained directly from the interpolation equations of Table 10.6.1 by noting that

$$x'(n+p) = x'(t)\Big|_{t=nT+pT} = \frac{dx(t)}{dt}\Big|_{t=nT+pT} = T\frac{dx(n+p)}{dp} \quad (10.6.10)$$

The Nth-order difference $\Delta^N x(0)$ is independent of p but the coefficients $C(p,k)$ depend upon p. Therefore the formulas of Table 10.6.1 can be converted into those of Table 10.6.2 by simply replacing $C(p,k)$ by $C'(p,k) = dC(p,k)/dp$ and dropping the $x(0)$ terms. For example, the first few Stirling difference equations equal

$$
\begin{aligned}
2Tx'(n+p)_2 &= C'(p,1)x(n+1) - C'(p,1)x(n-1) \\
2Tx'(n+p)_3 &= \big[C'(p,1) + C'(p,2) + C'(p+1,2)\big]x(n+1) \\
&\quad - 2\big[C'(p,2) + C'(p+1,2)\big]x(n) \\
&\quad + \big[-C'(p,1) + C'(p,2) + C'(p+1,2)\big]x(n-1)
\end{aligned}
\quad (10.6.11a\text{--}b)
$$

$$2Tx'(n+p)_4 = C'(p+1,2)x(n+2)$$
$$+ \left[C'(p,1) + C'(p,2) + C'(p+1,2) - 2C'(p+1,2)\right]x(n+1)$$
$$+ 2\left[C'(p+1,3) - C'(p,2) - C'(p+1,2)\right]x(n)$$
$$+ \left[-C'(p,1) + C'(p,2) + C'(p+1,2) + 2C'(p+1,2)\right]x(n-1)$$
$$- C'(p+1,2)x(n-2)$$

$$(10.6.11c)$$

Choosing $p = 0$ yields derivatives which are evaluated at the center of their sequences. Then Eq. 10.6.11 reduces to

$$2Tx'(n)_2 = x(n+1) - x(n-1)$$
$$2Tx'(n)_4 = \frac{1}{6}\left[-x(n+2) + 8x(n+1) - 8x(n-1) + x(n-2)\right]$$
$$2Tx'(n)_6 = \frac{1}{60}\left[x(n+3) - 9x(n+2) + 45x(n+1) - 45x(n-1)\right.$$
$$\left. + 9x(n-2) - x(n-3)\right]$$

$$(10.6.12)$$

The transfer functions of the Stirling differentiators equal

$$H_2(z) = \frac{1}{2T}\left[z^1 - z^{-1}\right]$$
$$H_4(z) = \frac{1}{12T}\left[-z^2 + 8z^1 - 8z^{-1} + z^{-2}\right]$$
$$H_6(z) = \frac{1}{120T}\left[z^3 - 9z^2 + 45z^1 - 45z^{-1} + 9z^{-2} - z^{-3}\right]$$

$$(10.6.13)$$

Then their ac steady-state gains equal

$$H_2(e^{j\omega T}) = \frac{j}{T}\left[\sin(\omega T)\right]$$
$$H_4(e^{j\omega T}) = \frac{j}{3T}\left[8\sin(\omega T) - \sin(2\omega T)\right]$$
$$H_6(e^{j\omega T}) = \frac{j}{30T}\left[45\sin(\omega T) - 9\sin(2\omega T) + \sin(3\omega T)\right]$$

$$(10.6.14)$$

letting $z = e^{j\omega T}$. The magnitude responses are plotted in Fig. 10.6.3. They have the

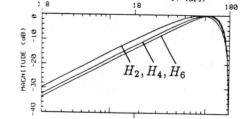

Fig. 10.6.3 Magnitude responses of Stirling differentiators.

required 20 dB/dec slope and no ripple unlike the Fourier series differentiators of Fig. 10.3.3. They have wider bandwidth than the weighted Fourier series differentiators in Fig. 10.4.3. Increasing filter order increases the differentiator bandwidth.

Other differentiators can be obtained by using the reciprocals of the integration transforms in Tables 5.2.1 and 5.2.2. To generate *nonrecursive* algorithms, we use $H_{dmn}(z)^{-1} \cong p$ transforms that have z-plane poles at only $z = 0$ (or ∞) which simply delays (or advances) the differentiator output. Therefore selecting only the $H_{00n}(z)$ and $H_{11n}(z)$ transforms, the nonrecursive differentiators which result have

$$H_{001}(z)^{-1} = \frac{1}{T}\left[1 - z^{-1}\right]$$

$$H_{002}(z)^{-1} = \frac{1}{2T}\left[3 - 4z^{-1} + z^{-2}\right]$$

(10.6.15a–b)

Table 10.6.1 A collection of numerical analysis algorithms for interpolation and extrapolation.

---

1. Newton (Newton-Gregory, forward difference)

$$x(p) = x(0) + C(p,1)\Delta x(0) + C(p,2)\Delta^2 x(0) + C(p,3)\Delta^3 x(0) + \cdots$$

(Backward difference)

$$x(p) = x(0) + C(p,1)\Delta x(-1) + C(p+1,2)\Delta^2 x(-2) + C(p+2,3)\Delta^3 x(-3) + \cdots$$

2. Gauss (Gauss-Newton, forward difference)

$$x(p) = x(0) + C(p,1,)\Delta x(0) + C(p,2)\Delta^2 x(-1) + C(p+1,3)\Delta^3 x(-1)$$
$$+ C(p+1,4)\Delta^4 x(-2) + \cdots$$

(Backward difference)

$$x(p) = x(0) + C(p,1)\Delta x(-1) + C(p+1,2)\Delta^2 x(-1) + C(p+1,3)\Delta^3 x(-2)$$
$$+ C(p+2,4)\Delta^4 x(-2) + \cdots$$

3. Stirling (Average Gauss, central difference)

$$x(p) = \frac{1}{2}\left\{2x(0) + C(p,1)\left[\Delta x(0) + \Delta x(-1)\right] + \left[C(p,2) + C(p+1,2)\right]\Delta^2 x(-1) \right.$$
$$\left. + C(p+1,3)\left[\Delta^3 x(-1) + \Delta^3 x(-2)\right] + \left[C(p+1,4) + C(p+2,4)\right]\Delta^4 x(-2) + \cdots\right\}$$

4. Bessel

$$x(p) = \frac{1}{2}\left\{\left[x(0) + x(-1)\right] + \left[C(p,1) + C(p-1,1)\right]\Delta x(0) + C(p,2)\left[\Delta^2 x(-1) + \Delta^2 x(0)\right] + \cdots\right\}$$

5. Everett

$$x(p) = -C(p-2,1)x(0) + C(p-1,1)x(1) - C(p+1,3)\Delta^2 x(-1) + C(p,3)\Delta^2 x(0) + \cdots$$

Table 10.6.2 A collection of numerical analysis
algorithms for differentiation.

1. Newton (Newton-Gregory, forward difference)

$$Tx'(p) = C'(p, 1)\Delta x(0) + C'(p, 2)\Delta^2 x(0) + C'(p, 3)\Delta^3 x(0) + \ldots$$

(Backward difference)

$$Tx'(p) = C'(p, 1)\Delta x(-1) + C'(p+1, 2)\Delta^2 x(-2) + C'(p+2, 3)\Delta^3(-3) + \ldots$$

2. Gauss (Gauss-Newton, forward difference)

$$Tx'(p) = C'(p, 1,)\Delta x(0) + C'(p, 2)\Delta^2 x(-1) + C'(p+1, 3)\Delta^3 x(-1)$$
$$+ C'(p+1, 4)\Delta^4 x(-2) + \ldots$$

(Backward difference)

$$Tx'(p) = C'(p, 1)\Delta x(-1) + C'(p+1, 2)\Delta^2 x(-1) + C'(p+1, 3)\Delta^3 x(-2)$$
$$+ C'(p+2, 4)\Delta^4 x(-2) + \ldots$$

3. Stirling (Average Gauss, central difference)

$$2Tx'(p) = C'(p, 1)[\Delta x(0) + \Delta x(-1)] + [C'(p, 2) + C'(p+1, 2)]\Delta^2 x(-1)$$
$$+ C'(p+1, 3)[\Delta^3 x(-1) + \Delta^3 x(-2)] + [C'(p+1, 4) + C'(p+2, 4)]\Delta^4 x(-2) + \ldots$$

4. Bessel

$$2Tx'(p) = C'[x(0) + x(-1)] + [C'(p, 1) + C'(p-1, 1)]\Delta x(0) + C'(p, 2)[\Delta^2 x(-1) + \Delta^2 x(0)] + \ldots$$

5. Everett

$$Tx'(p) = -C'(p-2, 1)x(0) + C'(p-1, 1)x(1) - C'(p+1, 3)\Delta^2 x(-1) + C'(p, 3)\Delta^3 x(0) + \ldots$$

Definitions: T = sample interval size, nT = nominal time value,

$p = (t - nT)/T$ = normalized time deviation (not necessarily integer),

$$C(p, k) = \frac{p(p-1)\ldots(p-k+1)}{k!} = \frac{p!}{(p-k)!k!} = \text{combinational polynomial of degree } k \text{ in } p.$$

Samples: $x(n) = x(nT)$, $x(n+p) = x((n+p)T)$
Sample deviations:

$$\Delta x(0) = x(n+1) - x(n)$$
$$\Delta^2 x(0) = \Delta[\Delta x(0)] = \Delta[x(n+1) - x(n)] = x(n+2) - 2x(n+1) + x(n)$$
$$\Delta^3 x(0) = x(n+3) - 3x(n+2) + 3x(n+1) - x(n)$$
$$\Delta^4 x(0) = x(n+4) - 4x(n+3) + 6x(n+2) - 4x(n+1) + x(n)$$
$$\Delta^q x(0) = x(n+q) - qx(n+q-1) + [q(q-1)/2!]x(n+q-2)$$
$$- [q(q-1)(q-2)/3!]x(n+q-3) + \ldots + (-1)^q x(n)$$

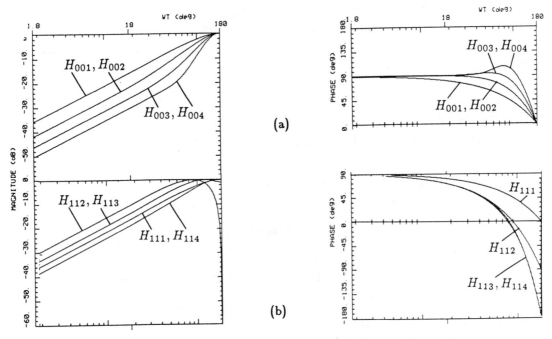

Fig. 10.6.4 Frequency responses of (a) H_{00m} and (b) H_{11m} differentiators.

$$H_{003}(z)^{-1} = \frac{1}{6T}\left[11 - 18z^{-1} + 9z^{-2} - 2z^{-3}\right]$$

$$H_{004}(z)^{-1} = \frac{1}{12T}\left[25 - 48z^{-1} + 36z^{-2} - 16z^{-3} + 3z^{-4}\right]$$

(10.6.15c–d)

from Table 5.2.1 and

$$H_{111}(z)^{-1} = \frac{1}{T}\left[z - 1\right]$$

$$H_{112}(z)^{-1} = \frac{1}{2T}\left[z - z^{-1}\right]$$

$$H_{113}(z)^{-1} = \frac{1}{6T}\left[2z + 3 - 6z^{-1} + z^{-2}\right]$$

$$H_{114}(z)^{-1} = \frac{1}{12T}\left[3z + 10 - 18z^{-1} + 6z^{-2} - z^{-3}\right]$$

(10.6.16)

from Table 5.2.2. The frequency responses for these differentiators are shown in Fig. 10.6.4. All these filters have a wider magnitude response bandwidth than the Stirling differentiators of Fig. 10.6.3. The H_{00m} filters have increased peaking at the Nyquist frequency as m increases. The H_{11m} filter respones are more linear. The H_{00m} filters maintain the 90° phase over a wider frequency range than the H_{11m} filters.

10.7 ADAPTIVE NONRECURSIVE ALGORITHMS [7]

The filters discussed thus far in this chapter have been nonadaptive. These filters are time-invariant since their $H(z)$ transfer functions have fixed coefficients which are constant with time and selected apriori. Now we consider approaches for synthesizing adaptive nonrecursive filters which are time-varying. This discussion will apply some of the adaptive fast transform filter theory developed in Chaps. 8 and 9.

10.7.1 ADAPTIVE NONRECURSIVE FILTERS

An *adaptive nonrecursive filter* or *adaptive FIR filter*, sometimes called an *adaptive linear combiner*, is shown in Fig. 10.7.1. A time domain implementation and an FFT frequency domain implementation are given. The filter input is $r(n)$, its output is $c(n)$, and its desired output is $d(n)$. The filter has internal outputs called *taps*. In the time domain form of Fig. 10.7.1a, the tap outputs are simply delayed inputs. In the frequency domain form of Fig. 10.7.1b, the tap outputs are the FFT input spectra $R(m)$ at different bin locations. In either case, the tap outputs are multiplied by the *tap weights* $\mathbf{w}_n$ and summed to form the filter output $c(n)$. The filter gain algorithm updates the $\mathbf{w}_n$ tap weights at each iteration so that the output $c(n)$ is a good approximation to the desired output $d(n)$. The adaptive FIR filter of Fig. 10.7.1a with an $\mathbf{w}_n$ impulse response is related to the time domain filter of Fig. 8.1.1c with impulse response h_T.

To express these optimum weights, we expand the vector notation of Eq. 8.1.3 as

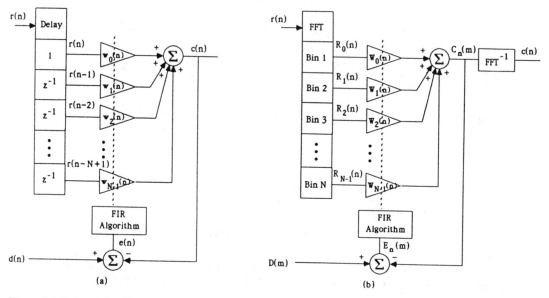

Fig. 10.7.1 Adaptive nonrecursive filter using (a) time domain and (b) frequency domain implementations.

$$\mathbf{r}_n = \left[r(n)\ r(n-1)\ \dots\ r(n-N+1)\right]^t$$

$$\mathbf{c}_n = \left[c(n)\ c(n-1)\ \dots\ c(n-N+1)\right]^t$$

$$\mathbf{d}_n = \left[d(n)\ d(n-1)\ \dots\ d(n-N+1)\right]^t$$

$$\mathbf{e}_n = \left[e(n)\ e(n-1)\ \dots\ e(n-N+1)\right]^t = \mathbf{d}_n - \mathbf{c}_n$$

$$\mathbf{w}_n = \left[w_0(n)\ w_1(n)\ \dots\ w_{N-1}(n)\right]^t$$

$$\mathbf{R}_{rr} = \begin{bmatrix} R_{rr}(0) & R_{rr}(1) & \dots & R_{rr}(N-1) \\ R_{rr}(-1) & R_{rr}(0) & \dots & R_{rr}(N-2) \\ \vdots & \vdots & \ddots & \vdots \\ R_{rr}(-N+1) & R_{rr}(-N+2) & \dots & R_{rr}(0) \end{bmatrix} = E\{\mathbf{r}_n \mathbf{r}_n^h\}$$

$$\mathbf{R}_{dr} = \left[R_{dr}(0)\ R_{dr}(1)\ \dots\ R_{dr}(N-1)\right]^t = E\{d(n)\mathbf{r}_n^*\}$$

$$(10.7.1)$$

$\mathbf{R}_{rr}$ and $\mathbf{R}_{dr}$ are the autocorrelation and cross-correlation matrices of the signals $\mathbf{r}_n$ and $\mathbf{d}_n$, respectively, where the matrix elements equal

$$\begin{aligned} R_{rr}(n,m) &= E\{r(n)r^*(m)\} = R_{rr}(n-m) \\ R_{dr}(n,m) &= E\{d(n)r^*(m)\} = R_{dr}(n-m) \end{aligned} \qquad (10.7.2)$$

The correlation matrices of Eqs. 10.7.1 and 10.7.2 assume that $\mathbf{r}_n$ and $\mathbf{d}_n$ are stationary processes. Real stationary processes have $R_{rr}(-n) = R_{rr}^*(n)$. Such matrices have equal elements on each of its N diagonals and are said to be *Toeplitz*. Nonstationary processes do not have this property.

Now the input $r(n)$, output $c(n)$, and impulse response $h_n(n)$ are related as

$$c(n) = h_n(n) * r(n) = \mathbf{r}_n^t \mathbf{w}_n = \mathbf{w}_n^t \mathbf{r}_n, \qquad C(j\omega) = H_n(j\omega)R(j\omega)$$

$$(10.7.3)$$

$$H_n(j\omega) = \sum_{m=0}^{N-1} w_m(n)z^{-m}\bigg|_{z=e^{j\omega T}}, \qquad h_n(n) = \sum_{m=0}^{N-1} w_m(n)U_0(n-m)$$

The impulse response $h_n(n)$ of this FIR filter is time-varying (see Eq. 1.11.6 where $h_m(n) \equiv h(n+1, n-m+1)$). The error $e(n)$ between the desired output $d(n)$ and the actual output $c(n)$ at iteration n and the squared error equal

$$\begin{aligned} e(n) &= d(n) - c(n) = d(n) - \mathbf{r}_n^t \mathbf{w}_n = d(n) - \mathbf{w}_n^t \mathbf{r}_n \\ |e(n)|^2 &= e^h(n)e(n) = [d(n) - \mathbf{r}_n^t \mathbf{w}_n]^h [d(n) - \mathbf{r}_n^t \mathbf{w}_n] \\ &= |d(n)|^2 + \mathbf{w}_n^h \mathbf{r}_n^* \mathbf{r}_n^t \mathbf{w}_n - d^*(n)\mathbf{r}_n^t \mathbf{w}_n - d(n)\mathbf{w}_n^h \mathbf{r}_n^* \end{aligned} \qquad (10.7.4)$$

$d(n)$, $\mathbf{r}_n$, and $\mathbf{w}_n$ can be complex-valued. The squared error is often used as a performance index. The expected value or mean of the squared error (MSE) is

$$E\{|e(n)|^2\} = E\{|d(n)|^2\} + \mathbf{w}_n^h E\{\mathbf{r}_n^* \mathbf{r}_n^t\}\mathbf{w}_n - E\{d^*(n)\mathbf{r}_n^t\}\mathbf{w}_n - \mathbf{w}_n^h E\{d(n)\mathbf{r}_n^*\} \quad (10.7.5)$$

Now Eq. 10.7.5 can be written in terms of autocorrelation and cross-correlation as

$$E\{|e(n)|^2\} = R_{dd}(0) + \mathbf{w}_n^h \mathbf{R}_{rr}^* \mathbf{w}_n - \mathbf{R}_{dr}^h \mathbf{w}_n - \mathbf{w}_n^h \mathbf{R}_{dr} \tag{10.7.6}$$

The optimum adaptive filter is the one whose tap weight vector $\mathbf{w}_n$ minimizes $E\{|e(n)|^2\}$. The expected error is a quadratic performance surface in N weight variables as shown in Fig. 10.7.2. The tap weights must be adjusted so that the filter operates at the bottom of this performance surface. When $\mathbf{r}_n$ and $\mathbf{d}_n$ are stationary processes, the shape and position of the surface remain fixed. Taking the partial derivatives of Eq. 10.7.6 with respect to the weights gives the gradient vector as

$$\nabla E_n \equiv \frac{\partial E\{|e(n)|^2\}}{\partial \mathbf{w}_n} = \left[\frac{\partial}{\partial w_0(n)} \quad \frac{\partial}{\partial w_1(n)} \quad \cdots \quad \frac{\partial}{\partial w_{N-1}(n)} \right]^t E\{|e(n)|^2\}$$

$$= \mathbf{R}_{rr} \mathbf{w}_n + \mathbf{R}_{rr}^* \mathbf{w}_n^* - \mathbf{R}_{dr}^* - \mathbf{R}_{dr} = 0 = (\mathbf{R}_{rr} \mathbf{w}_n - \mathbf{R}_{dr}) + (\mathbf{R}_{rr} \mathbf{w}_n - \mathbf{R}_{dr})^* \tag{10.7.7}$$

since $\nabla(\mathbf{w}_n^h \mathbf{R}_{rr} \mathbf{w}_n) = \mathbf{R}_{rr} \mathbf{w}_n + \mathbf{R}_{rr}^* \mathbf{w}_n^*$, $\nabla(\mathbf{R}_{dr}^h \mathbf{w}_n) = \mathbf{R}_{dr}^*$, and $\nabla(\mathbf{w}_n^h \mathbf{R}_{dr}) = \mathbf{R}_{dr}$. Setting $\nabla E_n = 0$ yields a set of N equations in N unknowns which can be solved simultaneously for the $\mathbf{w}_n$ filter weights.

Rearranging Eq. 10.7.7 shows that $\mathbf{R}_{rr} \mathbf{w}_n = \mathbf{R}_{dr}$ which is the *Wiener-Hopf* equation in discrete time. Solving for the optimum Wiener weight vector $\mathbf{w}_{opt}$ gives

$$\mathbf{w}_{opt} = [w_0(\infty) \ w_1(\infty) \ \ldots \ w_{N-1}(\infty)]^t = \mathbf{R}_{rr}^{-1} \mathbf{R}_{dr} \tag{10.7.8}$$

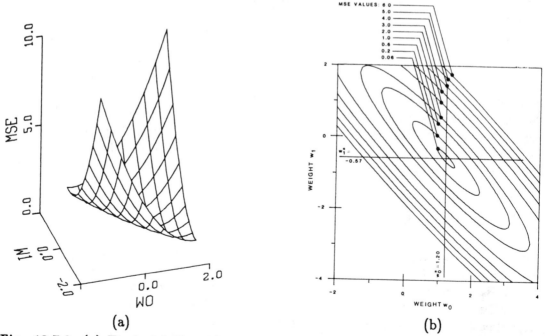

(a)   (b)

Fig. 10.7.2 (a) Typical 2-D performance surface and (b) equivalent contour plot. (From S.D. Stearns, "Fundamentals of adaptive signal processing," Sandia Report SAND84–2646, 43 p., Sandia National Laboratories, NM, June 1985.

Since the $\mathbf{R}_{rr}$ matrices are Toeplitz, they are always positive semidefinite and satisfy $\mathbf{w}^t\mathbf{R}_{rr}\mathbf{w} \geq 0$. When $\mathbf{w}^t\mathbf{R}_{rr}\mathbf{w} > 0$, the matrix is positive definite and is nonsingular so its inverse exists. Since $\mathbf{R}_{rr}$ is positive definite with probability one, it is almost always positive definite and nonsingular so $\mathbf{w}_{opt}$ is well-defined.

The optimum weight vector $\mathbf{w}_{opt}$ can also be expressed in terms of any suboptimal weight vector $\mathbf{w}_0$ as

$$\mathbf{w}_{opt} = \mathbf{w}_0 - (\mathbf{w}_0 - \mathbf{R}_{rr}^{-1}\mathbf{R}_{dr}) = \mathbf{w}_0 - \mathbf{R}_{rr}^{-1}(\mathbf{R}_{rr}\mathbf{w}_0 - \mathbf{R}_{dr})$$

$$= \mathbf{w}_0 - \frac{1}{2}\mathbf{R}_{rr}^{-1}\nabla\mathbf{E}_0 \tag{10.7.9}$$

This results by adding and subtracting $\mathbf{w}_0$ from the right side of Eq. 10.7.8, factoring out $\mathbf{R}_{rr}^{-1}$, and using the $\nabla\mathbf{E}_0$ expression of Eq. 10.7.8. This equation allows the weights to be adjusted in one step for optimum performance. The minimum error results by substituting Eq. 10.7.8 into Eq. 10.7.6 as

$$\min E\{|e(n)|^2\} = R_{dd}(0) + (\mathbf{R}_{rr}^{-1}\mathbf{R}_{dr})^t\mathbf{R}_{rr}\mathbf{w}_{opt} - 2\mathbf{R}_{dr}^t\mathbf{w}_{opt}$$

$$= R_{dd}(0) - \mathbf{R}_{dr}^t\mathbf{w}_{opt} \tag{10.7.10}$$

Therefore the filter in Fig. 10.7.1a can be viewed as a nonrecursive estimation Wiener type whose transfer function equals

$$H_{opt}(z) = \sum_{m=0}^{N-1} w_m(\infty)z^{-m} \tag{10.7.11}$$

The Wiener tap weights are given by Eq. 10.7.8 and have been obtained using the standard adaptive FIR filter derivation. However, the more general optimum filter impulse responses h_T listed in Tables 8.6.1 and 8.6.2 can be applied to FIR adaptive filters which yields a much larger variety of Class 1, 2, and 3 estimation, detection, and correlation algorithms. The optimum FIR filter weights in Fig. 10.7.1a satisfy $\mathbf{w}_{opt} = h_T u_0$ where $u_0 = [1, 0, \ldots, 0]^t$ is the unit selection vector (see Eq. 1.12.17). Therefore the weight vector $\mathbf{w}_{opt}$ equals the first column of the h_T impulse response matrix.

10.7.2 SEARCH METHODS

When the autocorrelation matrix $\mathbf{R}_{rr}$ and cross-correlation matrix $\mathbf{R}_{dr}$ are known apriori, then Eq. 10.7.8 can be evaluated to yield the optimum tap weight vector $\mathbf{w}_{opt}$. When $\mathbf{R}_{rr}$ and/or $\mathbf{R}_{dr}$ are not known, these terms must be estimated.

These estimates are generally obtained using different *search methods* or *search algorithms*. Using these methods, the $E\{|e(n)|^2\}$ performance index of Eq. 10.7.6 may be viewed as an N-dimensional parabolic surface having N independent variables (i.e., filter weights $\mathbf{w}_n$). This surface must be analyzed to determine the weight vector (Eq. 10.7.8) that globally minimizes the performance index. The process begins by choosing (or initializing) a weight vector $\mathbf{w}_0$. Then the weight vector is iteratively updated by moving $\mathbf{w}_n$ towards the global minimum $\mathbf{w}_{opt}$ of the surface.

Various search methods can be used. Random search methods select random directions to move at each iteration, evaluate the performance index for each direction (Eq. 10.7.1), and move in the direction that yields the smallest squared error value. The process repeats until there is negligible improvement in the performance index.

The gradient search methods are perhaps the most popular method. These methods compute the gradient of $E\{|e(n)|^2\}$ at the nth iteration, denoted ∇E_n, and use this information to adjust the weight vector from $\mathbf{w}_n$ to $\mathbf{w}_{n+1}$. The Newton and steepest descent methods fall in the gradient search category. The Newton method can be used when $\mathbf{R}_{rr}$ is known or estimated. Movement along the error surface is not along the gradient but along a weighted gradient. The steepest descent method can be used when only the diagonal terms of $\mathbf{R}_{rr}$ are known or estimated. Movement along the error surface is always opposite to the direction of the gradient. The weights are adjusted along the steepest descent path. These paths are shown in Fig. 10.7.3.

10.7.3 NEWTON'S METHOD AND ALGORITHMS

The *Newton* or *Newton-Raphson* method iteratively computes a root of an equation $f(w) = 0$. Expanding $f(w)$ in a Maclaurin series about $w = w_o$ gives $f(w) = f(w_o) + (w - w_o)f'(w_o) + \dots$. Truncating this series after the f' term gives a tangent approximation for $(w - w_o)$ as

$$w - w_o = \Delta w = -\frac{f(w_o)}{f'(w_o)} \tag{10.7.12}$$

when w_o is a zero or root of $f(w)$. If w_n is the nth estimate of w_o, then

$$w_{n+1} = w_n - \frac{f(w_n)}{f'(w_n)} = w_n - f'(w_n)^{-1}f(w_n) \tag{10.7.13}$$

The derivative $f'(w_n)$ is usually estimated using the Euler backward difference equation of Chap. 5.4. The iteration process continues until $|\Delta w_n| < \epsilon$ or $|f(w_n)| < \delta$. The Newton method generally converges rapidly if the initial guess w_1 is close to w_o. Otherwise convergence cannot be assured.

The Newton method can be applied to iteratively solve Eq. 10.7.7 by treating $f(\mathbf{w}_n) \equiv \nabla E_n = 0$. From Eq. 10.7.13, the recursion equation for finding the weight vector $\mathbf{w}_n$ is

$$\mathbf{w}_{n+1} = \mathbf{w}_n - \frac{1}{2}\mathbf{R}_{rr}^{-1}\nabla E_n \tag{10.7.14}$$

since $f'(\mathbf{w}_n) = \nabla(\nabla E_n) = 2\mathbf{R}_{rr}$ from Eq. 10.7.7. A more general recursion equation uses an arbitrary gain u where

$$\mathbf{w}_{n+1} = \mathbf{w}_n - u\mathbf{R}_{rr}^{-1}\nabla E_n \tag{10.7.15}$$

The convergence and stability of Eq. 10.7.15 depends upon u. In can be shown that

$$\mathbf{w}_{n+1} = (1 - 2u)^n \mathbf{w}_0 + 2u\mathbf{w}_{opt} \sum_{k=1}^{n}(1 - 2u)^k$$

$$= (1 - 2u)^n\mathbf{w}_0 + \left[1 - (1 - 2u)^n\right]\mathbf{w}_{opt} \qquad (10.7.16)$$

by recursively manipulating Eq. 10.7.15 where $\mathbf{w}_0$ is the initial weight vector. Eq. 10.7.16 shows that $\mathbf{w}_{n+1} \to \mathbf{w}_{opt}$ for $n \to \infty$ only if $|1 - 2u| < 1$. Therefore, convergence is assured when

$$0 < u < 1 \qquad (10.7.17)$$

$u < 1/2$ gives monotonic convergence of the weights but $u > 1/2$ gives oscillatory convergence. In practice, u is in the vicinity of 0.01 and convergence takes many iterations. The movement of a 2-D weight vector $\mathbf{w}_n$ from a starting point $\mathbf{w}_0 = [3, -4]^t$ to the optimum point $\mathbf{w}_{opt} = [1.20, -0.57]^t$ is shown in Fig. 10.7.3a.

The convergence time of the weights using the Newton method can be approximated. Assume that the solution $\mathbf{w}_n$ of the discrete time equation given by Eq. 10.7.16 is exponential as

$$\mathbf{w}_{n+1} = \mathbf{w}_0 e^{-n/N_w} + \mathbf{w}_{opt}(1 - e^{-n/N_w}) \qquad (10.7.18)$$

Iteration constant N_w is analogous to one time constant and equals the number of iterations needed to converge another e^{-1} factor. Expanding the exponential in a Taylor series as $e^{-x} = 1 - x + x^2/2! - \dots$, truncating after the second term, and substituting into Eq. 10.7.18 gives

$$\mathbf{w}_{n+1} \cong (1 - 1/N_w)^n\mathbf{w}_0 + \left[1 - (1 - 1/N_w)^n\right]\mathbf{w}_{opt} \qquad (10.7.19)$$

Comparing Eq. 10.7.19 with Eq. 10.7.16 and equating the coefficients of $\mathbf{w}_0$ and $\mathbf{w}_{opt}$ shows that the iteration constant N_w equals

$$N_w \cong \frac{1}{2u}, \qquad 0 < u \ll 1/2 \qquad (10.7.20)$$

If $u = 0.01$, then $N_w \cong 50$ iterations. It can also be shown that the MSE error expression can be expressed as

$$MSE(n) = MSE_{min} + (1 - 2u)^{2n} \mathbf{v}_0^t \mathbf{R}_{rr}\mathbf{v}_0 \qquad (10.7.21)$$

where $\mathbf{v}_0 \equiv \mathbf{w}_0 - \mathbf{w}_{opt}$. Therefore following the same process used in Eq. 10.7.16 to obtain Eq. 10.7.20, the MSE error has an iteration constant N_{MSE} of

$$N_{MSE} \cong \frac{1}{4u}, \qquad 0 < u \ll 1/2 \qquad (10.7.22)$$

The MSE constant N_{MSE} equals half the weight constant N_w. Thus for $u = 0.01$, the MSE error requires (115, 230, 345) iterations or (2.3, 4.6, 6.9) time constants to be within (10%, 1%, 0.1%) of the MSE_{min}, respectively.

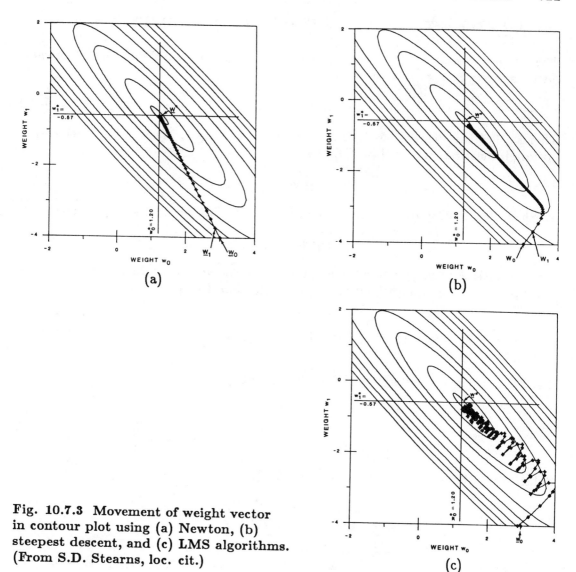

Fig. 10.7.3 Movement of weight vector in contour plot using (a) Newton, (b) steepest descent, and (c) LMS algorithms. (From S.D. Stearns, loc. cit.)

The Newton algorithm given by Eq. 10.7.15 is Class 1 since both R_{rr} and ∇E are known. When R_{rr} and/or ∇E are not known, they must be estimated by $\widehat{R}_{rr}$ and/or $\widehat{\nabla}E$ which produces Class 2 and 3 algorithms. These algorithms are obtained by simply estimating the appropriate term in Eq. 10.7.15 as:

$$\text{Class 1:} \quad w_{n+1} = w_n - uR_{rr}^{-1}\nabla E_n$$

$$\text{Class 2:} \quad w_{n+1} = w_n - u\widehat{R}_{rr}^{-1}\nabla E_n$$

$$\text{Class 2:} \quad w_{n+1} = w_n - uR_{rr}^{-1}\widehat{\nabla}E_n \qquad (10.7.23)$$

$$\text{Class 3:} \quad w_{n+1} = w_n - u\widehat{R}_{rr}^{-1}\widehat{\nabla}E_n$$

10.7.4 STEEPEST DESCENT METHOD AND ALGORITHMS

The steepest descent method moves the weights in the (opposite) direction of the gradient as shown in Fig. 10.7.3b where

$$\mathbf{w}_{n+1} = \mathbf{w}_n - \mu \nabla \mathbf{E}_n \tag{10.7.24}$$

This is called the *steepest descent* algorithm which is Class 1. When $\nabla \mathbf{E}_n$ is not known apriori, it must be estimated by $\widehat{\nabla} \mathbf{E}_n$ as

$$\mathbf{w}_{n+1} = \mathbf{w}_n - \mu \widehat{\nabla} \mathbf{E}_n \tag{10.7.25}$$

which is Class 2. Comparing Eqs. 10.7.24 and 10.7.15, we see that gain μ equals

$$\mu = u \mathbf{R}_{rr}^{-1} \tag{10.7.26}$$

μ can be estimated using the following procedure. The diagonal terms of $\mathbf{R}_{rr}$ equal the input power. If the input is white noise, then $\mathbf{R}_{rr} = \sigma^2 \mathbf{I}$ and

$$R_{rr}(n) = \frac{1}{N} E\{\mathbf{r}^h \mathbf{r}\} = \frac{1}{N} tr[\mathbf{R}_{rr}] = E\left\{ \sum_{k=0}^{N-1} r(n) r^*(k+n) \right\} \equiv \sigma^2 \tag{10.7.27}$$

Then $\mathbf{R}_{rr}^{-1}$ is equal to $(1/\sigma^2)\mathbf{I}$ and from Eq. 10.7.26,

$$\mu = \frac{u}{\sigma^2} \quad \text{(equal eigenvalues)}$$

$$\mu_{min} = \frac{u}{tr[\mathbf{R}_{rr}]} = \frac{u}{N\sigma^2} \quad \text{(unequal eigenvalues)} \tag{10.7.28}$$

Substituting the minimum μ into Eq. 10.7.25 gives

$$\mathbf{w}_{n+1} = \mathbf{w}_n - \frac{u}{N\sigma^2} \widehat{\nabla} \mathbf{E}_n \tag{10.7.29}$$

The N eigenvalues λ and eigenvectors $\mathbf{v}$ of $\mathbf{R}_{rr}$ satisfy $\mathbf{R}_{rr}\mathbf{v} = \lambda \mathbf{v}$. The eigenvalues λ of $\mathbf{R}_{rr}$ are such that the determinant or characteristic equation $|\mathbf{R}_{rr} - \lambda \mathbf{I}| = 0$. The eigenvalues of $\mathbf{R}_{rr}$ are real and positive since $\mathbf{R}_{rr}$ is almost always positive semidefinite.

10.7.5 LMS ALGORITHMS

Now an estimate of the expected $E\{|e(n)|^2\}$ error (Eq. 10.7.5) can be found using simply the actual $|e(n)|^2$ error (Eq. 10.7.4b) where

$$e(n) = d(n) - \mathbf{w}_n^t \mathbf{r}_n \tag{10.7.30}$$

Then assuming $e(n)$ is real, Eq. 10.7.7 shows that

$$\widehat{\nabla} \mathbf{E}_n = \frac{\partial \widehat{E}\{|e(n)|^2\}}{\partial \mathbf{w}_n} = \frac{\partial e^2(n)}{\partial \mathbf{w}_n} = 2e(n)\frac{\partial e(n)}{\partial \mathbf{w}_n} = -2e(n)\mathbf{r}_n \tag{10.7.31}$$

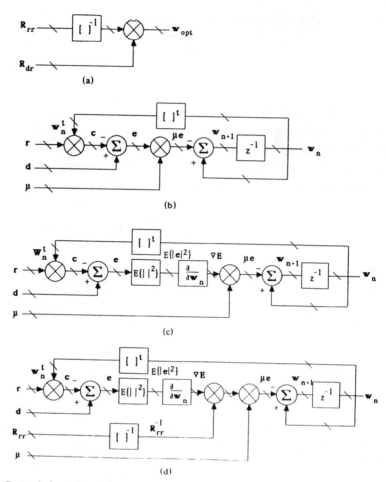

Fig. 10.7.4 Adaptive algorithms using (a) optimum (Eq. 10.7.8), (b) LMS (Eq. 10.7.38), (c) steepest descent (Eq. 10.7.24), and (d) Newton (Eq. 10.7.15) methods.

Substituting this gradient estimate into Eq. 10.7.25 gives

$$\mathbf{w}_{n+1} = \mathbf{w}_n + 2\mu e(n)\mathbf{r}_n \qquad (10.7.32)$$

which is called the *least-mean-square* or *LMS* algorithm. It is perhaps the simplest and most widely used algorithm in adaptive nonrecursive filter design.

The LMS algorithm only estimates the ∇E_n gradient vector and uses no $\mathbf{R}_{rr}^{-1}$ information. The weight vector seeks $\mathbf{w}_{opt}$ in an oscillatory manner as shown in Fig. 10.7.3c. The Newton algorithm estimates ∇E_n and uses $\mathbf{R}_{rr}$ to seek $\mathbf{w}_{opt}$ in a more monotonic manner. Convergence is faster. Combining Eqs. 10.7.15 and 10.7.31 gives

$$\mathbf{w}_{n+1} = \mathbf{w}_n + 2ue(n)\mathbf{R}_{rr}^{-1}\mathbf{r}_n \qquad (10.7.33)$$

which is called the *LMS-Newton* algorithm. The Class 2 algorithm equals

$$\mathbf{w}_{n+1} = \mathbf{w}_n + 2ue(n)(\widehat{\mathbf{R}}_{rr}^{-1})_n\,\mathbf{r}_n \qquad (10.7.34)$$

which is called the *sequential regression* or *SER* algorithm. One estimate for $(\hat{\mathbf{R}}_{rr}^{-1})_n$ is[7]

$$\hat{\mathbf{S}}_n = (\hat{\mathbf{R}}_{rr}^{-1})_{n-1}\mathbf{r}_n$$

$$(\hat{\mathbf{R}}_{rr}^{-1})_n = \frac{1}{1-\alpha}\left[(\hat{\mathbf{R}}_{rr}^{-1})_{n-1} - \alpha\frac{\hat{\mathbf{S}}_n\hat{\mathbf{S}}_n^t}{1-\alpha+\alpha\mathbf{r}_n^t\hat{\mathbf{S}}_n}\right] \qquad (10.7.35)$$

The algorithms which have been discussed in this section are summarized in Table 10.7.1 and are represented in block diagram form in Fig. 10.7.4. In order of increasing complexity, they are the optimum, LMS, steepest descent, and Newton methods. When the apriori correlations $\mathbf{R}_{rr}$ and $\mathbf{R}_{dr}$ (or gradient ∇E) are unknown, they must be estimated using estimation blocks in Fig. 10.7.4.

10.7.6 FREQUENCY DOMAIN ADAPTIVE NONRECURSIVE FILTERS

The adaptive FIR filter using a frequency domain approach is shown in Fig. 10.7.5. The input data $\mathbf{r}_n$ is converted to a generalized spectrum $\mathbf{R}_n$ by using some orthogonal discrete transform T such as Fourier, Walsh, Haar, or slant. The $\mathbf{R}_n$ spectra are adaptively filtered by the frequency domain weight vector $\mathbf{W}_n$. From Eq. 10.7.3a,

$$c_n(n) = \mathbf{w}_n^t\mathbf{r}_n = \mathbf{w}_n^t T^{-1}\mathbf{R}_n = \mathbf{W}_n^t\mathbf{R}_n \qquad (10.7.36)$$

Manipulating Eq. 10.7.36 shows that frequency domain weight vector $\mathbf{W}_n$ is related to the time domain weight vector $\mathbf{w}_n$ as

$$\mathbf{W}_n = (T^t)^{-1}\mathbf{w}_n = NT^*\mathbf{w}_n \qquad \text{or} \qquad \mathbf{w}_n = T^t\mathbf{W}_n \qquad (10.7.37)$$

where $T^t = N(T^*)^{-1}$ for unitary transforms. The time domain algorithms of Table 10.7.1 are easily converted to frequency domain algorithms using Eq. 10.7.37. For example, multiplying Eq. 10.7.32 by NT^* gives the LMS algorithm as

$$\mathbf{W}_{n+1} = \mathbf{W}_n + 2N\mu e(n)\mathbf{R}_n^* \qquad (10.7.38)$$

When the weight vector $\mathbf{w}_n$ is matched by the transform (i.e., equal to one or a few of the basis functions), the number of nonzero spectral weights in $\mathbf{W}_n$ are greatly reduced. The spectra $\mathbf{R}_n$ is much more sparse than the $\mathbf{r}_n$ matrix. Convergence of the frequency domain algorithms is faster. If there is great mismatch, then frequency domain filters provide no shorter adaptation time than time domain filters.

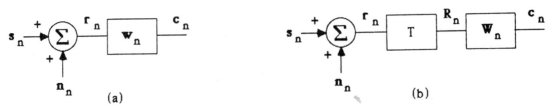

Fig. 10.7.5 Adaptive nonrecursive filter in (a) time domain and (b) frequency domain.

Table 10.7.1 Time domain algorithms for adaptive nonrecursive filters.

Filter Name	Transfer Function	Eq. Number
Newton		
Class 1:	$\mathbf{w}_{n+1} = \mathbf{w}_n - u\mathbf{R}_{rr}^{-1}\nabla\mathbf{E}_n$	(10.7.15)
Class 2:	$\mathbf{w}_{n+1} = \mathbf{w}_n - u\widehat{\mathbf{R}}_{rr}^{-1}\nabla\mathbf{E}_n$	(10.7.23b)
Class 2:	$\mathbf{w}_{n+1} = \mathbf{w}_n - u\mathbf{R}_{rr}^{-1}\widehat{\nabla}\mathbf{E}_n$	(10.7.23c)
Class 3:	$\mathbf{w}_{n+1} = \mathbf{w}_n - u\widehat{\mathbf{R}}_{rr}^{-1}\widehat{\nabla}\mathbf{E}_n$	(10.7.23d)
Steepest Descent		
Class 1:	$\mathbf{w}_{n+1} = \mathbf{w}_n - \mu\nabla\mathbf{E}_n$	(10.7.24)
Class 2:	$\mathbf{w}_{n+1} = \mathbf{w}_n - \mu\widehat{\nabla}\mathbf{E}_n$	(10.7.25)
LMS		
Class 1:	$\mathbf{w}_{n+1} = \mathbf{w}_n + 2\mu e(n)\mathbf{r}_n$	(10.7.32)
LMS-Newton		
Class 1:	$\mathbf{w}_{n+1} = \mathbf{w}_n + 2ue(n)\mathbf{R}_{rr}^{-1}\mathbf{r}_n$	(10.7.33)
Class 2:	$\mathbf{w}_{n+1} = \mathbf{w}_n + 2ue(n)(\widehat{\mathbf{R}}_{rr}^{-1})_n\,\mathbf{r}_n$	(10.7.34)

10.8 ADAPTIVE NONRECURSIVE FILTER DESIGN

To illustrate computations, we shall now design a first-order adaptive nonrecursive filter that is shown in Fig. 10.8.1. Its input is a cosine signal which is sampled at a rate of M samples per cycle ($M > 2$). The desired filter output is also a cosine. Therefore $r(n)$ and $d(n)$ equal

$$r(n) = \cos(2\pi n/M), \qquad d(n) = \cos(2\pi n/M) \tag{10.8.1}$$

The autocorrelation of $r(n)$ and the cross-correlation between $d(n)$ and $r(n)$ equal

$$E\{r(m)r^*(m-n)\} \equiv \frac{1}{M}\sum_{m=0}^{M-1}\cos(2\pi m/M)\cos(2\pi(m-n)/M) = 0.5\cos(2\pi n/M)$$

$$E\{d(m)r^*(m-n)\} \equiv \frac{1}{M}\sum_{m=0}^{M-1}\cos(2\pi m/M)\cos(2\pi(m-n)/M) = 0.5\cos(2\pi n/M)$$

$$\tag{10.8.2}$$

where $n = 0, 1, \ldots, M-1$ from Eq. 10.7.2. The expectation $E\{[\ ]\}$ of $[\ ]$ is estimated by averaging $[\ ]$ over all of the M samples. Therefore the input correlation matrix $\mathbf{R}_{rr}$ and the cross-correlation vector $\mathbf{R}_{dr}$ equal

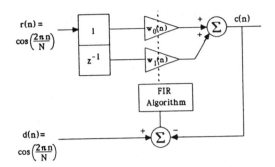

Fig. 10.8.1 First-order adaptive nonrecursive filter.

$$\mathbf{R}_{rr} = E\left\{ \begin{array}{cc} r(m)r^*(m) & r(m)r^*(m-1) \\ r(m-1)r^*(m) & r(m-1)r^*(m-1) \end{array} \right\} = 0.5 \begin{bmatrix} 1 & \cos(2\pi/M) \\ \cos(2\pi/M) & 1 \end{bmatrix}$$

$$\mathbf{R}_{dr} = E\{d(m)r^*(m) \ \ d(m)r^*(m-1)\}^t = 0.5\begin{bmatrix} 1 & \cos(2\pi/M) \end{bmatrix}^t$$

$$(10.8.3)$$

from Eq. 10.7.1. The expected squared error equals

$$E\{|e(n)|^2\} = R_{dd}(0) + \mathbf{w}_n^t \mathbf{R}_{rr}\mathbf{w}_n - 2\mathbf{R}_{rd}^t\mathbf{w}_n$$

$$= 0.5 + 0.5[w_{0n} \ w_{1n}]\begin{bmatrix} 1 & \cos(2\pi/M) \\ \cos(2\pi/M) & 1 \end{bmatrix}\begin{bmatrix} w_{0n} \\ w_{1n} \end{bmatrix} - [1 \ \cos(2\pi/M)]\begin{bmatrix} w_{0n} \\ w_{1n} \end{bmatrix}$$

$$= 0.5(w_{0n}^2 + w_{1n}^2) + w_{0n}w_{1n}\cos(2\pi/M) - [w_{0n} + w_{1n}\cos(2\pi/M)] + 0.5$$

$$(10.8.4)$$

using Eq. 10.7.6. In shorthand notation, $w_{0n} \equiv w_0(n)$ and $w_{1n} \equiv w_1(n)$. The gradient vector equals

$$\nabla \mathbf{E}_n = 2(\mathbf{R}_{rr}\mathbf{w}_n - \mathbf{R}_{dr}) = \begin{bmatrix} 1 & \cos(2\pi/M) \\ \cos(2\pi/M) & 1 \end{bmatrix}\begin{bmatrix} w_{0n} \\ w_{1n} \end{bmatrix} - \begin{bmatrix} 1 \\ \cos(2\pi/M) \end{bmatrix}$$

$$= \begin{bmatrix} w_{0n} + w_{1n}\cos(2\pi/M) - 1 \\ w_{1n} + w_{0n}\cos(2\pi/M) - \cos(2\pi/M) \end{bmatrix}$$

$$(10.8.5)$$

from Eq. 10.7.7. The optimum weight vector is

$$\mathbf{w}_{opt} = \mathbf{R}_{rr}^{-1}\mathbf{R}_{dr} = \begin{bmatrix} 1 & \cos(2\pi/M) \\ \cos(2\pi/M) & 1 \end{bmatrix}^{-1}\begin{bmatrix} 1 \\ \cos(2\pi/M) \end{bmatrix} = \begin{bmatrix} 1 \\ 0 \end{bmatrix} \qquad (10.8.6)$$

using Eq. 10.7.8.

Now suppose that the input is noisy so that $\mathbf{r}_n = \mathbf{s}_n + \mathbf{n}_n$ where the noise power $E\{|n(n)|^2\} = \sigma^2$. Then the autocorrelation and cross-correlation of Eq. 10.8.2 become

$$E\{r(m)r^*(m-n)\} \equiv \frac{1}{M}\sum_{m=0}^{M-1}[\cos(2\pi m/M) + n(m)][\cos(2\pi(m-n)/M) + n(m-n)]$$

$$= 0.5 + \sigma^2, \qquad n = 0$$

$$= 0.5\cos(2\pi/M), \qquad n = 1 \qquad (10.8.7a)$$

$$E\{d(m)r^*(m-n)\} \equiv \frac{1}{M}\sum_{m=0}^{M-1}\cos(2\pi m/M)\big[\cos(2\pi(m-n)/M)+n(m-n)\big]$$
$$= 0.5\cos(2\pi n/M) \tag{10.8.7b}$$

Therefore the correlation matrices of Eq. 10.8.3 generalize as

$$\mathbf{R}_{rr} = E\left\{\begin{array}{cc} r(m)r^*(m) & r(m)r^*(m-1) \\ r(m-1)r^*(m) & r(m-1)r^*(m-1) \end{array}\right\} = 0.5\left[\begin{array}{cc} 1+2\sigma^2 & \cos(2\pi/M) \\ \cos(2\pi/M) & 1+2\sigma^2 \end{array}\right]$$

$$\mathbf{R}_{dr} = E\big\{d(m)r^*(m) \quad d(m)r^*(m-1)\big\}^t = 0.5\big[1 \quad \cos(2\pi/M)\big]^t$$
$$\tag{10.8.8}$$

The expected squared error of Eq. 10.8.4 becomes

$$E\{|e(n)|^2\} = (0.5+\sigma^2)(w_{0n}^2+w_{1n}^2)+w_{0n}w_{1n}\cos(2\pi/M)-\big[w_{0n}+w_{1n}\cos(2\pi/M)\big]+0.5$$
$$\tag{10.8.9}$$

The noisy gradient vector of Eq. 10.8.5 becomes

$$\nabla E_n = \left[\begin{array}{c} (1+2\sigma^2)w_{0n}+w_{1n}\cos(2\pi/M)-1 \\ (1+2\sigma^2)w_{1n}+w_{0n}\cos(2\pi/M)-\cos(2\pi/M) \end{array}\right] \tag{10.8.10}$$

The optimum weight vector of Eq. 10.8.6 becomes

$$\mathbf{w}_{opt} = \left[\begin{array}{cc} 1+2\sigma^2 & \cos(2\pi/M) \\ \cos(2\pi/M) & 1+2\sigma^2 \end{array}\right]^{-1}\left[\begin{array}{c} 1 \\ \cos(2\pi/M) \end{array}\right] = \left[\begin{array}{c} \dfrac{1+2\sigma^2-\cos^2(2\pi/M)}{(1+2\sigma^2)^2-\cos^2(2\pi/M)} \\[2mm] \dfrac{2\sigma^2\cos(2\pi/M)}{(1+2\sigma^2)^2-\cos^2(2\pi/M)} \end{array}\right]$$
$$\tag{10.8.11}$$

Selecting $M = 16$ samples/cycle and a noise power $\sigma^2 = 0.01$, then Eq. 10.8.11 gives the optimum weights as

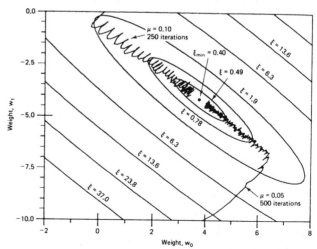

Fig. 10.8.2 Movement of weight vector in contour plot using LMS algorithm with $N = 16$ and $\sigma^2 = 0.01$. (From B. Widrow and S.D. Stearns, *Adaptive Signal Processing*, ©1985, p. 105. Reprinted by permission of Prentice-Hall, Inc., Englewood Cliffs, New Jersey.)

$$\mathbf{w}_{opt} = [0.6353, 0.3547]^t, \qquad H_{opt}(z) = 0.6363 + 0.3547z^{-1} \qquad (10.8.12)$$

When the signal is noise free, then $\sigma^2 = 0$ and the optimum weights become

$$\mathbf{w}_{opt} = [1, 0]^t, \qquad H_{opt}(z) = 1 \qquad (10.8.13)$$

The LMS algorithm performs as shown in Fig. 10.8.2 for $M = 16$ and $\sigma^2 = 0.01$. The upper path starts at $\mathbf{w}_0 = [0,0]^t$, uses gain $\mu = 0.1$, and runs for 250 iterations. The lower path starts at $\mathbf{w}_0 = [4, -10]^t$, uses gain $\mu = 0.05$, and runs for 500 iterations. Both paths converge somewhat erratically towards the optimum weight given by Eq. 10.8.11. Because the gradient estimate is noisy, the path does not always proceed in the direction of the true gradient. The smaller μ produces smaller path irregularities but convergence takes longer.

10.9 MEDIAN FILTERS

Linear FIR filters have been discussed in this chapter. Now we wish to discuss a very useful nonlinear FIR filter called the *median filter*. Median filters can be used for reducing impulsive noise while maintaining signal edge structure. Linear filters reduce impulsive noise by smoothing or averaging but, unfortunately, they also smooth signal discontinuities and therefore blur signal edges. Thus median filters can be utilized in applications where edges must be maintained like pattern recognition and CAT-scan systems. Such nonlinear filters are relatively new and have not received wide attention.[9] Nonlinear filters do not satisfy such familiar properties as superposition. Therefore new analysis methods are being developed to adequately describe these filters.

A median filter processes an odd number of samples in the following manner. It collects $2N + 1$ samples and ranks them in ascending or descending order (see Fig. 4.12.3e and discussion). The median filter output equals the median or mid-sample of this ordered sequence as

$$y(n) = \text{median of } \big[x(n-N), \ldots, x(n), \ldots, x(n+N)\big] \qquad (10.9.1)$$

This corresponds to the $(N+1)$st ordered sample. A more general nonlinear filter in

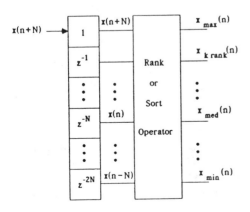

Fig. 10.9.1 Nonlinear filter having median output and kth ranked-order output.

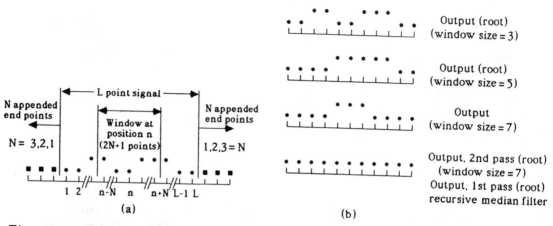

Fig. 10.9.2 (a) Median filter window $(2N+1$ points) with signal sequence (L points) and N appended end points, and (b) median filtered signal using window size of 3, 5, and 7.

this class is the *kth ranked-order filter*. It outputs the kth ordered sample as

$$y(n) = k\text{th largest value of } [x(n - N), \ldots, x(n), \ldots, x(n + N)] \qquad (10.9.2)$$

rather than the median. These $2N$-order nonlinear filters are shown in Fig. 10.9.1.

A variety of useful signals are used to characterize median filters. These include:

Constant: A sequence of $N + 1$ or more equal-valued samples.

Impulse: A set of no more than N points bounded on both sides by equal-valued sample sequences.

Edge: A sequence of monotonically rising or falling samples bounded on both sides by equal-valued samples.

Oscillation: A set of samples which do not form impulses, constants, or edges.

Root: A signal not modified by median filtering.

Impulses and oscillations are eliminated by median filtering but constants and edges are unchanged. Any signal of length $2N + 1$ that is repeatedly median filtered reduces to a root. This requires N iterations at most. In practice, the number of passes to reduce a signal to a root is much smaller. Loosely speaking, the root equals the envelope of the signal. Therefore median filters act as envelope detectors. k-th ranked-order filters act as either peak detectors $(n > N + 1)$ or valley detectors $(n < N + 1)$.

When median filtering a sequence of L samples using a $2N + 1$ window size, each end of the L sequence is appended by N constants. This is shown in Fig. 10.9.2a. The first sample of this L-sample sequence is extended to the left and the last sample of this L-sample sequence is extended to the right. Extending the sequence in both directions allows the original sequence to be median filtered to its end points without introducing edge effects. Such extended signals become roots after no more than $(L - 2)/2$ passes. Convergence to a root by repeated median filtering of a finite length signal is always

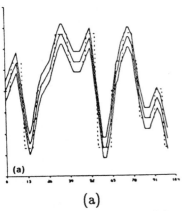

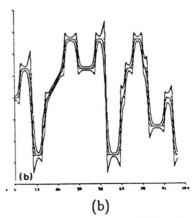

(a)   (b)

Fig. 10.9.3 Outputs of (a) averaging filter and (b) median filter for trapezoidal input signal plus impulsive noise. (From N.C. Gallagher, Jr., "Median filters: properties and applications," Rec. 16th Asilomar Conf. Circuits, Systems, and Computers, pp. 192–196, ©1982, IEEE.)

guaranteed. Any root of a certain order median filter is also a root of a lower-order median filter. For infinite length signals, oscillations can occur regardless of the number of iterations. A median filtered signal is shown in Fig. 10.9.2b using filter orders of 2, 4, and 6 or window sizes of 3, 5, and 7.

To illustrate the property that a median filter smooths a signal while retaining the edge structure, consider Fig. 10.9.3. The signal is a series of trapezoidal pulses (dashed lines) which is contaminated with impulsive noise (amplitudes $= \pm 3$ each with probability $= 0.05$). The noise has an rms value (standard derivation) of σ where $\sigma^2 = 2(3^2 \times 0.05) = 0.9$. The resulting input is lowpass filtered (using averaging of Eq. 10.2.8) in Fig. 10.9.3a and median filtered (Eq. 10.9.1) in Fig. 10.9.3b. Both filters are sixth-order and process seven samples. The mean values $E\{y\}$ of the filter outputs y are shown as the solid middle lines. The other two lines are $E\{y\} \pm \sigma y$. The edge structure of the noise-free output $E\{y\}$ is lost by the averaging filter but well maintained by the median filter.

The peak detecting property of a kth ranked-order filter is shown in Fig. 10.9.4. Here an AM envelope detector is implemented using an eighth ranked-order filter which processes nine samples. The envelope is fairly well recovered in the noise-free case shown in Fig. 10.9.4a. Slight degradation occurs for the noisy case in Fig. 10.9.4b.

Median filtering can be applied to 2-D and higher-dimensional processing. The 2-D windows can be any size and shape like a square (■), circle (●), cross (+), or an ×. Like the 1-D case, a square or circle-shaped window removes impulses, preserves edges, and tends to smooth oscillatory images. Unlike the 1-D case, multiple median filter passes need not reduce an image to a root. N-dimensional median filtering can be implemented using separable filtering (see Eq. 2.10.3). For example, 2-D median filters can be implemented using two 1-D median filters. Each appended row of the image $x(n,m)$ is 1-D median filtered to produce $x'(n,m)$. Then each column of $x'(n,m)$ is 1-D median filtered to yield $y(m,n)$. Mathematically,

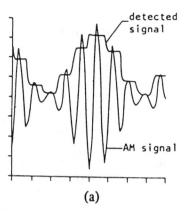
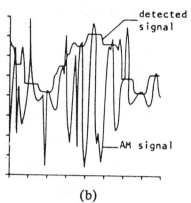

(a)   (b)

Fig. 10.9.4 Output of 8th ranked-order filter using 9 samples of (a) noise-free and (b) impulse noise contaminated AM signal. (From T.A. Nodes and N.C. Gallagher, Jr., "Median filters: some modifications and their properties," IEEE Trans. ASSP, vol. ASSP-30, no. 5, pp. 739–746, Oct., ©1982, IEEE.)

$$y(n, m) = \text{median of } \left[x'(n, m - N), \ldots, x'(n, m), \ldots, x'(n, m + N) \right]$$
$$x'(n, m) = \text{median of } \left[x(n - N, m), \ldots, x(n, m), \ldots, x(n + N, m) \right] \tag{10.9.3}$$

Nonrecursive median filters can be generalized to recursive median filters. Higher speed and improved processing are possible with IIR median filters. Their outputs equal

$$y(n) = \text{median of } \left[y(n - N), \ldots, y(n - 1), x(n), \ldots, x(n - N) \right] \tag{10.9.4}$$

Comparing Eqs. 10.9.1 and 10.9.4, the N left-most $x(n)$ input points are replaced by the N left-most $y(n)$ output points. This is illustrated in Fig. 10.9.2b. The recursive filter produces a root after only a single pass unlike the FIR filter case. Another example of IIR processing is the very noisy image and its median filtered version shown in Fig. 10.9.5. The well-defined dark and light areas and edges in Fig. 10.9.5b are unrecognizable in the unfiltered image of Fig. 10.9.5a. This result is very impressive since only a single

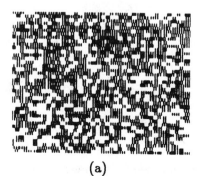
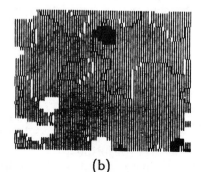

(a)   (b)

Fig. 10.9.5 (a) Input image and (b) output image of 2-D separable recursive median filter. (From N.C. Gallagher, Jr., loc. cit., ©1982, IEEE.)

pass separable median filter was used.

This short introduction to median filtering has demonstrated the great usefulness and uniqueness of nonlinear filters. Unfortunately no general theory or standard design procedures exist. Nevertheless nonlinear filters are becoming increasingly popular as their capabilities are discovered.

PROBLEMS

Design the filters in Problems 10.1–10.15 using each of the following methods ($N = 8$):
(1) Fourier series, (2) weighted Fourier series using Hamming window,
(3) frequency sampling, (4) s-to-z transforms using H_{dmn} as specified.

10.1   Several amplitude equalizer transfer functions are: (a) $[s + 2\pi(3000 \text{ Hz})]$ / $[s + 2\pi(70 \text{ Hz})]$, (b) $[s + 2\pi(10 \text{ KHz})]/[s + 2\pi(1 \text{ KHz})]^3$, (c) $14.3[s + 2\pi(200 \text{ Hz})]/[s + 2\pi(3 \text{ KHz})]$, (d) $2\pi(2.1 \text{ KHz})/[s + 2\pi(2.1 \text{ KHz})]$. Design the filters using $f_s = 20$ KHz and the H_{001} algorithm.

10.2   A wide-band lowpass voice filter is required. Design it to be 3rd order Chebyshev having 3 dB in-band ripple, a 3 dB bandwidth of 20 KHz, and a maximum low-frequency gain of unity. Use $f_s = 250$ KHz and the H_{001} algorithm.

10.3   A wide-band lowpass voice filter is required. Design it to be 4th order Chebyshev having 3 dB in-band ripple, a 3 dB bandwidth of 20 KHz, and a maximum low-frequency gain of unity. Use $f_s = 500$ KHz and the H_{111} algorithm.

10.4   A 3.2 KHz band-limiting lowpass voice filter is required. Design it to be 2nd order Butterworth having a 3 dB bandwidth of 12.8 KHz and a dc gain of unity. Use $f_s = 32$ KHz and the H_{001} algorithm.

10.5   A 1000 Bps noise suppression (i.e., data smoothing) lowpass filter is required. Design it to be 4th order Bessel having a 3 dB bandwidth of 1 KHz and unity dc gain. Use $f_s = 10$ KHz and the H_{LDI} algorithm.

10.6   A 3.5 Hz delta wave EEG lowpass filter is required. Design it to be 4th order Chebyshev having 0.5 dB in-band ripple and a 3 dB bandwidth of 3.5 Hz. Use $f_s = 20$ Hz and the H_{LDI} algorithm.

10.7   A 1 Hz phase detection filter is required. Design the filter to be 4th order Chebyshev having 1 dB in-band ripple and a 3 dB bandwidth of 1 Hz. Use $f_s = 10$ Hz and the H_{002} algorithm.

10.8   A 3.2 KHz anti-aliasing lowpass filter is required. Design it to be 3rd order Butterworth having a 3 dB bandwidth of 3.2 KHz. Use $f_s = 64$ KHz and the H_{111} algorithm.

10.9   A 941 Hz band-splitting lowpass filter is required. Design it to be 3rd order Butterworth having a 3 dB bandwidth of 941 Hz and a dc gain of 10. Use $f_s = 10$ KHz and the H_{001} algorithm.

10.10   A 2 KHz band-splitting lowpass/highpass filter is required. Design the lowpass filter to be 3rd order Butterworth having a 3 dB bandwidth of 2 KHz and unity dc gain. Use $f_s = 10$ KHz and the H_{LDI} algorithm.

10.11   A 941 Hz band-splitting lowpass filter is required. Design the lowpass to be 3rd order Butterworth having a 3 dB bandwidth of 2 KHz and unity dc gain. Use $f_s = 20$ KHz and the H_{LDI} algorithm.

10.12   Design a 5th order elliptic lowpass filter having the following parameters: $M_p = 0.28$ dB, $M_r = 50$ dB, $H_o = 20$ dB, $f_p = 1000$ Hz, and $f_r = 1556$ Hz. Use $f_s = 10$ KHz and the H_{LDI} algorithm.

10.13   Design a 4th order Bessel lowpass filter having a 3 dB bandwidth of 1 KHz and unity dc gain. Use $f_s = 20$ KHz and the H_{111} algorithm.

10.14   Design a tunable 4th order Butterworth lowpass filter having a 3 dB bandwidth that lies between 10–50 KHz. Use $f_s = 1$ MHz and the H_{001} algorithm.

10.15   Design a tunable 3rd order Bessel lowpass filter having a 3 dB bandwidth of 100 Hz and a 10:1 tuning range. Use $f_s = 25$ KHz and the H_{111} algorithm.

Design the filters in Problems 10.16–10.32 using each of the following methods (N = 8):
(1) Fourier series, (2) weighted Fourier series using Hann window,
(3) frequency sampling, (4) s-to-z transforms using H_{dmn} as specified.

10.16   Design a 4th order Chebyshev highpass filter having 3 dB passband ripple, a 3 dB corner frequency of 1 KHz, and a maximum passband gain of unity. Use $f_s = 10$ KHz and the H_{LDI} algorithm.

10.17   Design a 4th order synchronously-tuned highpass filter having a 3 dB corner frequency of 1 KHz and unity hf gain. Use $f_s = 20$ KHz and the H_{LDI} algorithm.

10.18   A Dolby B noise reduction utilizes two highpass filters. Design a 4th order Butterworth highpass filters having a 3 dB bandwidth of 1.5 KHz. Use $f_s = 25$ KHz and the H_{001} algorithm.

10.19   A F1A-weighting curve used in telephone systems using a lowpass and a highpass filter. The lowpass filter is 5th order Butterworth having unity dc gain and a 3 dB bandwidth of 2 KHz. The highpass filter is 3rd order Butterworth having unity hf gain and a 3 dB bandwidth of 620 Hz. Design the 3rd order Butterworth highpass filter. Use $f_s = 6$ KHz and the H_{001} algorithm.

10.20   Design a tunable 3rd order Chebyshev highpass filter to have a 3 dB bandwidth that lies between 2.5–12.5 KHz. Gain is arbitrary. Use $f_s = 500$ KHz and the H_{001} algorithm.

10.21   A C-weighting curve used in telephone systems can be realized using a 3rd order Butterworth highpass filter cascaded with a 4th order Chebyshev lowpass filter. The highpass filter has $f_{3dB} = 732$ Hz and $H_o = 1$. Use $f_s = 9.6$ KHz and the H_{111} algorithm.

10.22   A low-transient highpass filter has a normalized transfer function of

$$H_{CHP}(s) = \frac{s(s^2 + 0.591s + 1.177)}{(s + 0.593)(s^2 + 0.295s + 1.002)}$$

Design this filter using $f_s = 10$ Hz and the H_{001} algorithm.

10.23   A 2nd order Butterworth bandpass filter having a center frequency of 100 Hz, a 3 dB bandwidth of 10 Hz, and a mid-band gain of 10 must be designed. Its block diagram consists of two 1st order BPF stages having parameters (H_o, f_o, Q) which equal (4.44, 96.3 Hz, 14.1) and (4.44, 103.7 Hz, 14.1). Design this bandpass filter using $f_s = 1$ KHz and the H_{LDI} algorithm.

10.24   Design a 2nd order Butterworth bandpass filter having a 3 dB frequency range of 300–3200 Hz. Use $f_s = 48$ KHz and the H_{001} algorithm.

10.25   Bandpass filters used by the telegraph system have 120 Hz or 170 Hz tone spacings and are constant bandwidth. Design two 1st order bandpass filters to have a 3 dB bandwidth of 64 Hz and center frequencies of 420 Hz and 3300 Hz. Gain is arbitrary. Use $f_s = 48$ KHz and the H_{001} algorithm.

10.26   Design a 4th order Butterworth lowpass discriminator with a cut-off frequency of 500 Hz and a gain of 0.00522 volt/volt/Hz. Its block diagram realization consists of a 2nd order

LPF stage having parameters (H_o, f_o, ς) which equal $(-1, 500$ Hz, $0.924)$ cascaded with a 1st order BPF stage having parameters (H_o, f_o, ς) of $(-3.41, 500$ Hz, $0.383)$. Use $f_s = 10$ KHz and the H_{111} algorithm.

10.27   ULF direction finder systems are used in mine rescue operations. A keyed-CW transmitter connected to a large collapsible loop antenna radiates a frequency in the 900–3100 Hz range underground at each mine site. 34 frequencies are used. Above ground, a portable receiver is used to detect the location from which the desired frequency is being transmitted. Assume that the receiver uses a 1st order bandpass filter (which is insufficient). Design a bandpass filter with 2 Hz bandwidth which tunes over this frequency range. Gain is arbitrary. Use $f_s = 64$ KHz and the H_{002} algorithm.

10.28   FDM channel bands use bandpass filters having 4 KHz bandwidths with center frequencies in the range between 60-108 KHz. Design a 1st order tunable bandpass filter which tunes over this frequency range. Gain is arbitrary. Use $f_s = 500$ KHz and the H_{001} algorithm.

10.29   An acoustical data coupler uses 1170 Hz and 2125 Hz bandpass filters. The Transmit filters can be 2nd order Chebyshev having 0.5 dB ripple. Their block diagrams consist of two 1st order BPF stages having having parameters (H_o, f_o, Q) which equal $(1.41, 1.05$ or 2.00 KHz, 8.3 or $14.9)$ and $(1.41, 1.29$ or 2.24 KHz, 8.3 or $14.9)$. Design these 1170 Hz and 2125 Hz Transmit filters. Use $f_s = 25$ KHz and the H_{001} algorithm.

10.30   The Lenkurt Type 45 telephone signalling system uses a 3400/3550 Hz bandpass filter. Design it to be 2nd order Butterworth bandpass having $f_o = 3480$ Hz and $Q = 11.6$. Its block diagram consists of two 1st order BPF stages having parameters (H_o, f_o, Q) which equal $(1.41, 3.33$ KHz, $16.4)$ and $(1.41, 3.63$ KHz, $16.4)$. Use $f_s = 20$ KHz and the H_{111} algorithm.

10.31   Design a tunable 1st order bandpass filter having a variable Q and constant ω_o. Assume that its center frequency is 1 KHz and its Q must be tuned from 1 to 10. Gain is arbitrary. Use $f_s = 10$ KHz and the H_{LDI} algorithm.

10.32   Design a tunable 1st order bandpass filter having a center frequency of 1 KHz, a 2:1 tuning range, and $Q = 5$. Gain is arbitrary. Use $f_s = 10$ KHz and the H_{LDI} algorithm.

Design the filters in Problems 10.33–10.49 using each of the following methods ($N = 8$):
(1) Fourier series, (2) weighted Fourier series using 3-term Blackman window,
(3) frequency sampling, (4) s-to-z transforms using H_{dmn} as specified.

10.33   Design a 1st order Butterworth bandstop filter having a 200 Hz center frequency, $Q = 0.25$, and $H_o = 1$. Use $f_s = 4$ KHz and the H_{001} algorithm.

10.34   Design a 1st order Butterworth bandstop filter having a 200 Hz center frequency, $Q = 0.25$, and $H_o = 1$. Use $f_s = 8$ KHz and the H_{001} algorithm.

10.35   The telephone company uses 1010 Hz and 2805 Hz tones to test transmission channels. Design two 2nd order Butterworth bandstop filters having center frequencies of 1010 Hz and 2805 Hz, $Q = 10$, and $H_o = 1$. Use $f_s = 25$ KHz and the H_{111} algorithm.

10.36   Design a 2nd order Butterworth bandstop filter having a 60 Hz center frequency, $Q = 10$, and $H_o = 1$. Use $f_s = 1.2$ KHz and the H_{LDI} algorithm.

10.37   A 2600 Hz bandstop filter is required in single-frequency (SF) signalling systems used by the telephone company. A 2nd order Butterworth bandstop filter having a 3 dB band-width of 60 Hz shall be used. Its block diagram consists of two 1st order BPF stages having parameters $(H_o, f_{op}, Q_p, f_{oz})$ which equal $(1, 2577$ Hz, $61.3, 2600$ Hz) and $(1, 2623$ Hz, $61.3,$ 2600 Hz). Use $f_s = 25$ KHz and the H_{111} algorithm.

10.38 Design a 1st order Butterworth bandstop filter having a 21.6 KHz center frequency, $Q = 27$, and $H_o = 1$. Use $f_s = 64$ KHz and the H_{001} algorithm.

10.39 Hi-fi equalizers use a bank of nine filters centered from 50 Hz to 12.8 KHz (with octave spacing) to cover the audio spectrum from 20 Hz to 20 KHz. These equalizers allow the audiophile to compensate for room acoustics or speaker deficiencies, or to please his personal tastes. The equalizer provides boost and cut limits of ±12 dB. Design a biquadratic filter centered at 50 Hz with these boost and cut limits. Use $f_s = 1$ KHz and the H_{LDI} algorithm.

10.40 A low-transient 3rd order Papoulis highpass filter has a normalized transfer function of

$$H_{CHP}(s) = \frac{s(s^2 + 1.310s + 1.359)}{(s + 0.621)(s^2 + 0.691s + 0.931)}$$

Design this filter using $f_s = 10$ Hz and the H_{111} algorithm.

10.41 Design a tunable 1st order Butterworth bandstop filter having 1 Hz–30 KHz center frequency, $Q = 100$, and $H_o = 1$. Use $f_s = 1$ KHz and the H_{001} algorithm.

10.42 Design a 2nd order Butterworth bandstop filter having a center frequency of 1 KHz, $Q = 10$, and a passband gain of 10. Use $f_s = 10$ KHz and the H_{LDI} algorithm.

10.43 Design a 1st order Butterworth bandstop filter to have a $Q = 10$ and a center frequency of (a) 120 Hz and (b) 800 Hz. Use $f_s = 10$ KHz and the H_{111} algorithm.

10.44 (a) Design a 1st order allpass filter with parameters $(H_o, f_o) = (1, 318$ Hz$)$. (b) Design a 2nd order allpass filter with parameters $(H_o, f_o, \varsigma) = (1, 540$ Hz$, 0.866)$. Use $f_s = 10$ KHz and the H_{111} algorithm.

10.45 A 2nd order delay equalizer can be cascaded with a 5th order Butterworth lowpass filter to reduce its in-band delay variation. This 2nd order allpass filter has parameters $(H_o, f_o, \varsigma) = (1, 730$ Hz$, 0.866)$. Use $f_s = 15$ KHz and the H_{001} algorithm.

10.46 Design a 2600 Hz oscillator for the single-frequency in-band signalling system discussed in Problem 10.37. Use $f_s = 48$ KHz. Discuss using the H_{001} algorithm.

10.47 The Lenkurt Type 45 out-of-band signalling system utilizes a signalling oscillator. This oscillator generates tones at either 3400 Hz or 3550 Hz. Design this tunable oscillator. Use $f_s = 100$ KHz. Discuss using the H_{111} algorithm.

10.48 An electronic watch consists of a 32.786 KHz oscillator, counter, decoder/display, and battery. Design a 32.786 KHz oscillator. Use $f_s = 250$ KHz. Discuss using the H_{LDI} algorithm.

REFERENCES

1. Jong, M.T., *Methods of Discrete Signal and System Analysis*, Chap. 4, McGraw-Hill, NY, 1982.
2. Oppenheim, A.V. and R.W. Schafer, *Digital Signal Processing*, Chap. 1.6, Prentice-Hall, NJ, 1975.
3. Stearns, S.D., *Digital Signal Analysis*, Chap. 8, Hayden, NJ, 1975.
4. Tretter, S.A., *Introduction to Discrete-Time Signal Processing*, Chap. 8.8, Wiley, NY, 1976.
5. Rabiner, L.R. and B. Gold, *Theory and Application of Digital Signal Processing*, Chap. 3.17, Prentice-Hall, NJ, 1975.
6. Hamming, R.W., *Digital Filters*, 2nd ed., Chaps. 3.7 and 7.1, Prentice-Hall, NJ, 1983.
7. Stearns, S.D., "Fundamentals of adaptive signal processing," Sandia Report SAND84–2646, 43 p., Sandia Natl. Laboratory, Albuquerque, NM, June 1985.

8. Widrow, B. and S.D. Stearns, *Adaptive Signal Processing*, Chap. 6–8, Prentice-Hall, NJ, 1985.

9. Gallagher, Jr., N.C., "Median filters: properties and applications," Rec. 16th Asilomar Conf. Circuits, Systems, and Computers, pp. 192–196, 1982.

   Nodes, T.A. and N.C. Gallagher, Jr., "Median filters: some modifications and their properties," IEEE Trans. ASSP, vol. ASSP-30, no. 5, pp. 739–746, Oct. 1982.

   Arce, G.R. and N.C. Gallagher, Jr., "State description for the root-signal set of median filters," IEEE Trans. ASSP, vol. ASSP-30, no. 6, pp. 894–902, Dec. 1982.

11 RECURSIVE FILTER DESIGN

That if thou shall confess with thy mouth
the Lord Jesus and shall believe in thine heart that
God hath raised him from the dead, thou shalt be saved.
For with the heart man believeth unto righteousness;
and with the mouth confession is made unto salvation.

Romans 10:9–10

Nonrecursive or FIR filters were discussed in the last chapter. Recursive or IIR filters will be discussed in this chapter. Since low-order recursive filters have high-selectivity magnitude responses but use fewer total poles plus zeros than FIR filters, they are utilized very often. Unfortunately their transfer functions are usually more difficult to obtain than those of FIR filters. The engineer may wish to review the advantages, disadvantages, and trade-offs between recursive and nonrecursive filters outlined in Chaps. 7.3–7.4 before proceeding further.

Most recursive filters are designed using the transformations developed in Chap. 5. Some transformations like the bilinear and matched-z types are applied in the frequency domain. Other transformations like the impulse-invariant type are applied in the time domain. We shall now review these methods with emphasis on their application in design. The nonrecursive filter design methods of Chap. 10 can also be applied to recursive filters but they are less often used than those which will be discussed here (see Table 7.3.2). Adaptive IIR and Kalman filters will also be presented. The hardware and software implementation of FIR and IIR filters will be discussed in Chap. 12.

11.1 RECURSIVE FILTER PROPERTIES

A recursive digital filter is characterized by its transfer function

$$H(z) = \frac{Y(z)}{X(z)} = \frac{\sum_{k=-M}^{M} a_k z^{-k}}{\sum_{k=-N}^{N} b_k z^{-k}} = c\,\frac{\prod_{k=1}^{2M}(z - z_k)}{\prod_{k=1}^{2N}(z - p_k)} = \sum_{n=-\infty}^{\infty} h(n) z^{-n}, \qquad b_0 = 1$$

(11.1.1)

from Eq. 10.1.1. The filter has $2N$ poles and $2M$ zeros. Arranging Eq. 11.1.1 as $Y(z) = H(z)X(z)$ and taking the inverse of both sides yields

$$\sum_{k=-N}^{N} b_k y(n - k) = \sum_{k=-M}^{M} a_k x(n - k), \qquad b_0 = 1$$

(11.1.2)

737

Solving for $y(n)$ gives

$$y(n) = \sum_{k=-M}^{M} a_k x(n-k) - \sum_{\substack{k=-N \\ k \neq 0}}^{N} b_k y(n-k) \qquad (11.1.3)$$

$H(z)$ is causal if $k = 0$ rather than $-M$ and $-N$ in Eqs. 11.1.1–11.1.3. A causal $H(z)$ is absolutely stable if all of its poles p_k are inside the unit circle of the z-plane. The matrix form of a causal $y(n)$ is given by Eq. 10.1.8. A fully symmetrical form of Eq. 11.1.3 results by setting $M = N$ in which case it reduces to[1]

$$y(n) = \sum_{k=-N}^{N} \left[a_k x(n-k) - b_k y(n-k) \right], \qquad \text{set } b_0 = 0 \qquad (11.1.4)$$

The design problem in recursive filters, like nonrecursive filters, reduces to determining the impulse response coefficients $h(n)$ for all $n \geq 0$ (assuming a causal filter). However difficulty arises in obtaining the a_k and b_k values once $h(n)$ is found. To show this, consider a causal $H(z)$ described by Eq. 11.1.1. Performing long division (dividing the numerator by the denominator) yields

$$\sum_{n=0}^{\infty} h(n) z^{-n} = \frac{\sum_{k=0}^{M} a_k z^{-k}}{\sum_{k=0}^{N} b_k z^{-k}}, \qquad b_0 = 1$$

$$= z^0 a_0 + z^{-1}\{a_1 - a_0 b_1\} + z^{-2}\{(a_2 - a_0 b_2) - b_1(a_1 - a_0 b_1)\}$$
$$+ z^{-3}\{[(a_3 - a_0 b_3) - b_2(a_1 - a_0 b_1)] - b_1[(a_2 - a_0 b_2) - b_1(a_1 - a_0 b_1)]\}$$
$$+ z^{-4}\{[(a_4 - a_0 b_4) - b_3(a_1 - a_0 b_1)] - b_1\{[(a_3 - a_0 b_3) - b_2(a_1 - a_0 b_1)]$$
$$+ b_1[(a_2 - a_0 b_2) - b_1(a_1 - a_0 b_1)]\}\} + \cdots$$
$$= h(0) + h(1) z^{-1} + h(2) z^{-2} + h(3) z^{-3} + h(4) z^{-4} + \cdots \qquad (11.1.5)$$

Equating coefficients of like powers of z^{-n} leads to the set of simultaneous equations

$$h(0) = a_0$$
$$h(1) = a_1 - a_0 b_1$$
$$h(2) = (a_2 - a_0 b_2) - b_1(a_1 - a_0 b_1)$$
$$h(3) = [(a_3 - a_0 b_3) - b_2(a_1 - a_0 b_1)] - b_1[(a_2 - a_0 b_2) - b_1(a_1 - a_0 b_1)]$$
$$h(4) = [(a_4 - a_0 b_4) - b_3(a_1 - a_0 b_1)]$$
$$\qquad - b_1\{[(a_3 - a_0 b_3) - b_2(a_1 - a_0 b_1)] + b_1[(a_2 - a_0 b_2) - b_1(a_1 - a_0 b_1)]\}$$

$$\cdots \qquad (11.1.6)$$

which must be solved for a_k and b_k. Since there are only $(M + N + 1)$ independent variables (a_k and b_k excluding b_0) but an infinite number of equations (i.e., $h(n)$ values), this is an over-constrained problem with no solution. When the number of independent variables to be used is specified, then some suitable compromise in the $h(n)$ values must be made. Therefore most recursive filters have transfer functions which only approximate the ideal or desired response $H(z) = Z^{-1}[h(n)]$. Eq. 11.1.6 shows that low-order recursive filters may have long or infinite length impulse responses. IIR filters make efficient use of poles unlike FIR filters which make inefficient use of zeros.

Eq. 11.1.6 can be expressed in matrix form as

$$[h] = \left(I + [b]\right)^{-1}[a] \qquad (11.1.7)$$

This results by expressing Eq. 10.1.8 as $[y] = [a][x] - [b][y]$ and solving for $[h]$ where $[y] = [h][x]$ (see Eq. 10.1.8).

11.2 FREQUENCY TRANSFORMATION DESIGN (FREQUENCY DOMAIN ALGORITHMS)

Digital transforms directly convert analog transfer functions $H(p)$ into digital transfer functions $H(z)$ using

$$H(z) = H(p)\Big|_{p=g(z)} \qquad \text{or} \qquad H(p) = H(z)\Big|_{z=g^{-1}(p)} \qquad (11.2.1)$$

from Eq. 5.1.1. A variety of digital transforms were discussed in Chap. 5. These included the forward Euler (H_{111}), backward Euler (H_{001}), bilinear (H_{011}), modified bilinear (H_{MBDI}), lossless (H_{LDI}), optimal (H_{ODI}), and other transforms. These transforms are listed in Tables 5.2.1 and 5.2.2.

These transforms tend to produce digital filters which have time-domain responses which are similar to those for the analog filters from which they were derived. The higher the order of the transformation (i.e., the total number of poles and zeros used in $g(z)$ in Eq. 11.2.1), usually the better the match in responses. However, frequency domain responses are usually not preserved under the transformation $p = g(z)$. Only p-to-z transforms where the jv-axis of the p-plane maps onto the unit circle of the z-plane will preservation occur, and then only for magnitude response parameters like the passband and stopband ripples. The passband and stopband corner frequencies, transition bandwidth, delay, etc. are not preserved.

Even with these limitations, frequency domain transformations are the most popular method for designing recursive filters. They are simple to use and allow (unsampled) analog filter transfer functions to be directly converted into digital filter transfer functions. The classical filter nomographs of Chap. 4.2 can be used with frequency prewarping (as illustrated in Chap. 5.10) to quickly determine the required filter order.

Of the frequency transformations discussed in Chap. 5, the bilinear transform is the most popular and widely used where[2]

$$H(z) = H(p)\Big|_{p = \frac{2}{T}\frac{z-1}{z+1}}, \quad p_n = \frac{2}{v_p T}\frac{z-1}{z+1}$$

$$H(p) = H(z)\Big|_{z = \frac{1+pT/2}{1-pT/2} = \frac{1+p_n v_p T/2}{1-p_n v_p T/2}} \qquad (11.2.2)$$

(see Eqs. 5.5.7 and 5.11.9). The characteristics of this transform were discussed in Chap. 5.5. Digital filter orders using various transforms were determined in Chap. 5.10. Digital filters were designed in Chap. 5.11 following a prescribed procedure. Since all the other transformations are used in exactly the same way, we will now present several design examples using the bilinear transform, with no loss of generality. This will continue the discussion in Chaps. 5.10–5.11 which the engineer may wish to review.

EXAMPLE 11.2.1 It was determined in Example 5.10.2 that a 4th-order elliptic digital filter was needed to obtain a 0.28 dB frequency = 1 KHz and a 40 dB stopband frequency $\geq$ 2 KHz when sampling at 10 KHz. Design the filter in block diagram form and plot its frequency response.

SOLUTION From Chap. 4.2, the elliptic analog filter has a transfer function of

$$H(p_n) = \frac{k\left[p_n^2 + 2.0173^2\right]}{\left[(p_n + 0.5213)^2 + 0.4828^2\right]\left[(p_n + 0.1746)^2 + 1.0509^2\right]}$$

$$= \frac{0.1363(p_n^2 + 4.0695)}{(p_n^2 + 1.0426p_n + 0.5048)(p_n^2 + 0.3492p_n + 1.1349)} \qquad (11.2.3)$$

where $k = 0.1363$ is chosen so that $H(0) = -0.28$ dB $= 0.9683$. Its block diagram is shown in Fig. 11.2.1a. The 0.28 dB corner frequency v_p of the analog filter is computed from the frequencies ω_p and ω_s of the digital filter as

$$\frac{v_p T}{2} = \tan\frac{\omega_p T}{2} = \tan(180^\circ f_p/f_s) = \tan(180^\circ/10) = \tan(18^\circ) = 0.3249 = \frac{1}{3.078}$$

$$v_p = (2/T)\tan(180^\circ f_p/f_s) = 2(10 \ \text{KHz})\tan(18^\circ) = 2\pi(1.034 \ \text{KHz}) \qquad (11.2.4)$$

using Eq. 5.11.1b. Therefore the digital filter corner frequency of 1 KHz (or 36°) transforms to an analog filter corner frequency of 1.034 KHz (or 37.2°). They differ due to the frequency warping effects of the bilinear transform (see Fig. 5.10.3). Substituting the 0.28 dB normalized corner frequency $v_p T$ from Eq. 11.2.4 into the transformation equation of Eq. 11.2.2 shows that

$$p_n = \frac{2}{v_p T}\frac{z-1}{z+1} = 3.078\frac{z-1}{z+1} \quad \text{or} \quad z = \frac{1+pT/2}{1-pT/2} = \frac{1+0.3249p_n}{1-0.3249p_n} \qquad (11.2.5)$$

Transforming $H(p_n)$ of Eq. 11.2.3 into the digital filter gain $H(z)$ using Eq. 11.2.2 gives

$$H(z) = \frac{0.01201(z+1)^2(z^2 - 0.7981z + 1)}{(z^2 - 1.360z + 0.5133)(z^2 - 1.427z + 0.8160)} \qquad (11.2.6)$$

The block diagram of the digital filter is shown in Fig. 11.2.1b. Each analog block is transformed according to the transform relation given by Eq. 11.2.5a to obtain each digital block. Alternatively, the analog poles and zeros of each block can be converted into the digital poles and zeros using the p_n-z relation given by Eq. 11.2.5b. The dc block gains are 1 and 0.9683, respectively. The frequency responses of $H(p_n)$ and $H(z)$ are shown in Fig. 11.2.1c. The only notable difference is that the digital filter has zero gain at the Nyquist frequency due to the bilinear transform, and there is transmission zero shifting due to frequency warping. ∎

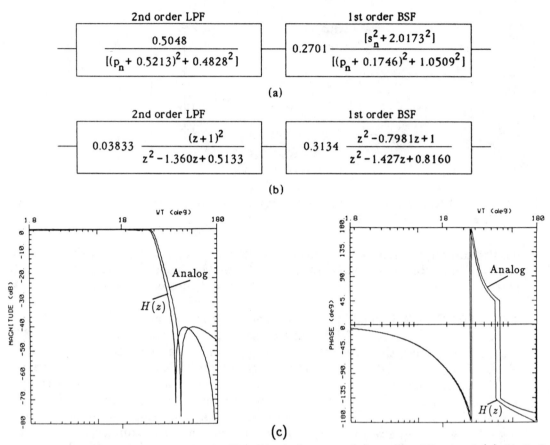

2nd order LPF

$$\dfrac{0.5048}{[(p_n + 0.5213)^2 + 0.4828^2]}$$

1st order BSF

$$0.2701 \, \dfrac{[s_n^2 + 2.0173^2]}{[(p_n + 0.1746)^2 + 1.0509^2]}$$

(a)

2nd order LPF

$$0.03833 \, \dfrac{(z+1)^2}{z^2 - 1.360z + 0.5133}$$

1st order BSF

$$0.3134 \, \dfrac{z^2 - 0.7981z + 1}{z^2 - 1.427z + 0.8160}$$

(b)

(c)

Fig. 11.2.1 Block diagram of 0.28 dB elliptic lowpass (a) analog filter and (a) digital filter of Example 11.2.1 and (c) frequency responses of both filters.

EXAMPLE 11.2.2 A digital filter must satisfy the following magnitude specifications shown in Fig. 11.2.2a: (1) dc gain = 1, (2) in-band ripple = 3 dB for $f \leq 3200$ Hz, (3) stopband rejection ≥ 40 dB for $f \geq 6400$ Hz, and (4) sampling frequency = 19.2 KHz. Determine the minimum order Butterworth, Chebyshev, elliptic, and Bessel filters needed to satisfy these requirements. Use the bilinear transform.

SOLUTION The digital filter normalized frequencies ωT must be converted into analog filter normalized frequencies vT using Eq. 5.11.1b. Here

$$v_{3dB}T/2 = \tan(\omega_{3dB}T/2) = \tan[180°(3.2 \text{ KHz})/19.2 \text{ KHz}] = \tan(30°) = 0.5774$$

$$v_{40dB}T/2 = \tan[180°(6.4 \text{ KHz})/19.2 \text{ KHz}] = \tan(60°) = 1.732$$

$$(11.2.7)$$

The normalized analog stopband frequency Ω_a equals

$$\Omega_a = \frac{v_{40dB}T}{v_{3dB}T} = \frac{2\tan(60°)}{2\tan(30°)} = 3 > \Omega_d = \frac{2(60°)}{2(30°)} = 2 \qquad (11.2.8)$$

from Table 5.10.1. It should be remembered that the analog filter frequency ratio Ω_a is always greater than the digital filter frequency ratio Ω_d (see Fig. 5.10.3)

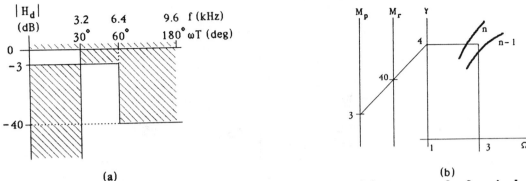

(a)   (b)

Fig. 11.2.2 (a) Magnitude response of digital filter and (b) nomograph of equivalent analog filter in Example 11.2.2.

so filter orders are reduced. Entering the data $(M_p, M_s, \gamma, \Omega_a) = (3, 40, 4, 3)$ onto the classical filter nomographs of Chap. 4.2 as shown in Fig. 11.2.2b, we find: Butterworth $n = 5$, Chebyshev $n = 4$, elliptic $n = 3$, Bessel impossible. ∎

EXAMPLE 11.2.3 In addition to the magnitude specifications of Example 11.2.2, the digital filter must satisfy the time domain specifications shown in Fig. 11.2.3. Determine whether the same filters of Example 11.2.2 can be used.

SOLUTION We shall assume that the digital filter has almost the same (but sampled) step response as the analog filter. This is true for analog step responses that are fairly smooth because the bilinear transform approximates such responses with straight-line segments connecting the samples. The step response of the selected digital filter should always be determined to verify this approximation.

The classical filter step responses are plotted in terms of normalized time $\omega_n t$ in Chap. 4.2. Since ω_{3dB} of the magnitude response is specified in Fig. 11.2.2a, normalized time $\omega_{3dB}t$ is used to label Fig. 11.2.3. The normalized step responses of the classical filters in Chap. 4.2 are transferred onto Fig. 11.2.3. For the 3rd-order elliptic filter, we use $M_p = 0.28$ dB rather than 3 dB since it can still meet the magnitude response in Fig. 11.2.2 where $\theta = 19°$. Clearly only the elliptic filter response meets the step response requirements of Fig. 11.2.3 and then only with small margin. ∎

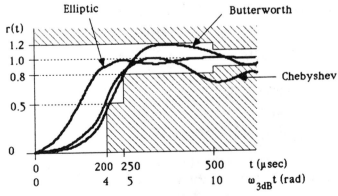

Fig. 11.2.3 Step response of digital filter in Example 11.2.3.

11.3 INVARIANT TIME DOMAIN RESPONSE DESIGN (TIME DOMAIN ALGORITHMS)

Invariant time domain response design methods were discussed in Chap. 5. These methods convert an analog filter transfer function into a digital filter transfer function such that their time domain responses $y(t)$ are identical at the sample times for identical inputs $x(t)$. This was described in block diagram of Fig. 5.12.1. Mathematically, the digital filter transfer function equals

$$H(z) = \frac{Y(z)}{X(z)} = \frac{Z[y(n)]}{Z[x(n)]} = \frac{Z\left[\mathcal{L}^{-1}\{H(s)X(s)\}_s\right]}{Z\left[\mathcal{L}^{-1}\{X(s)\}_s\right]} \qquad (11.3.1)$$

from Eq. 5.12.1. Here $x(t)$ and $y(t)$ are the analog filter input and output responses which we wish to duplicate with the sampled responses $x(n)$ and $y(n)$ of the digital filter. Impulse, step, ramp, sine, cosine, and complex exponential-invariant response designs were discussed in Chaps. 5.13–5.16. Of these, impulse-invariant response design is simple and very popular. We shall now discuss this design method in detail. The other methods follow the same procedure.

11.3.1 IMPULSE-INVARIANT TRANSFORM DESIGN

The impulse-invariant transform discussed in Chap. 5.13 preserves the impulse response $h(t)$ of the analog filter. Since $Z[U_0(t)] = 1/T$, then Eq. 11.3.1 reduces to

$$H_0(z) = TZ[h(n)] = TZ\left[\mathcal{L}^{-1}\{H(s)\}_s\right] = Z\left[\sum_{n=-\infty}^{\infty} \mathcal{L}^{-1}\{H(s - jn\omega_s)\}\right] \qquad (11.3.2)$$

as given by Eq. 5.13.1. If the analog transfer function $H(s)$ is band-limited so that $H(j\omega) = 0$ for $|\omega| \geq \omega_s/2$, there is no aliasing distortion. Then Eq. 11.3.2 shows that

$$H_0(e^{j\omega T}) = H(j\omega), \qquad |\omega| \leq \omega_s/2 \qquad (11.3.3)$$

so the digital filter has exactly the same magnitude, phase, and delay responses as the analog filter for $|\omega| \leq \omega_s/2$. When $H(j\omega)$ is not band-limited so that $H(j\omega) \neq 0$ for $|\omega| \geq \omega_s/2$ (which is almost always the case), then aliasing occurs and the frequency responses are not identical due to aliasing distortion. Nevertheless the time domain responses are identical at the sample times.

Suppose for example that $H(s)$ is causal with a transfer function given by

$$H(s) = d\,\frac{\prod_{k=1}^{M}(s + z_k)}{\prod_{k=1}^{N}(s + p_k)} = \frac{N(s)}{D(s)}, \qquad \text{Re } s > s_o \qquad (11.3.4)$$

Assuming that $N > M$ then $H(s)$ has a partial fraction expansion of

$$H(s) = \sum_{k=1}^{N} \frac{K_k}{s + p_k}, \qquad \text{Re } s > s_o \qquad (11.3.5a)$$

$$h(t) = \mathcal{L}^{-1}\big[H(s)\big] = \sum_{k=1}^{N} K_k e^{-p_k t} U_{-1}(t) \tag{11.3.5b}$$

The impulse response $h(t)$ of $H(s)$ is also given. Taking the z-transform of $h(t)$ and multiplying by T yields the impulse-invariant digital filter transfer function $H_0(z)$ as

$$H_0(z) = T Z\big[h(n)\big] = T \sum_{k=1}^{N} \frac{K_k z}{z - e^{-p_k T}} = T' \frac{\prod_{k=1}^{N}(z - c_k)}{\prod_{k=1}^{N}(z - e^{-p_k T})}, \qquad |z| > z_o \tag{11.3.6}$$

(see Table 1.9.1). Comparing Eqs. 11.3.4 and 11.3.6, we see that the transfer functions have the same number N of poles. The digital filter poles equal $z_k = e^{-p_k T}$ which are the z-transformed analog filter poles using $z = e^{sT}$. The digital filter has N zeros while the analog filter has M zeros. The zeros are not related by z-transforms unlike the poles. The analog filter must have $N > M$ or frequency aliasing distortion will be either severe ($N = M$) or intolerable ($N < M$).

Since the numerator and denominator of $H(z)$ are of equal degree, its impulse response $h(n) = 0$ for $n < 0$ so $H(z)$ is causal. Since absolutely stable and causal analog filters have poles with Re $p_k < 0$, then the digital filters have poles which satisfy $|e^{-p_k T}| < 1$ and therefore lie inside the unit circle. Thus, causal and stable analog filters transform into causal and stable digital filters.

When the analog filter magnitude response is essentially band-limited so $H(j\omega) \cong 0$ for $|\omega| \gg \omega_s/2$, there is negligible aliasing distortion. Eq. 11.3.3 shows that the digital filter frequency response is the same as the analog filter frequency response. The filter magnitude and delay characteristics as well as the nomographs of Chap. 4.2 can therefore be used directly. Thus, classical lowpass and bandpass filters such as Butterworth, Chebyshev, and elliptic filters can be easily converted using the impulse-invariant response method. This technique cannot be used for highpass, bandstop, and allpass filters whose passbands include the Nyquist frequency $\omega_N = \omega_s/2$. They must incorporate additional band-limiting above ω_N to avoid intolerable aliasing distortion.

EXAMPLE 11.3.1 Redesign the 4th-order elliptic filter of Example 11.2.1 using the impulse-invariant transform design method.

SOLUTION The transfer function $H(s)$ of the elliptic analog filter was given by Eq. 11.2.3. Expressing $H(s)$ in partial fraction expansion form yields

$$\begin{aligned}
H(s) &= \frac{0.1363\big[s_n^2 + 2.0173^2\big]}{\big[(s_n + 0.5213)^2 + 0.4828^2\big]\big[(s_n + 0.1746)^2 + 1.0509^2\big]} \\[2mm]
&= \frac{0.1129 + j0.5467}{s_n + 0.5213 + j0.4828} + \frac{0.1129 - j0.5467}{s_n + 0.5213 - j0.4828} \\[2mm]
&\quad + \frac{-0.1129 - j0.1491}{s_n + 0.1746 + j1.0509} + \frac{-0.1129 + j0.1491}{s_n + 0.1746 - j1.0509}
\end{aligned} \tag{11.3.7}$$

where $s_n = s/2\pi(1 \text{ KHz})$. The block diagram of the analog filter is shown in Fig.

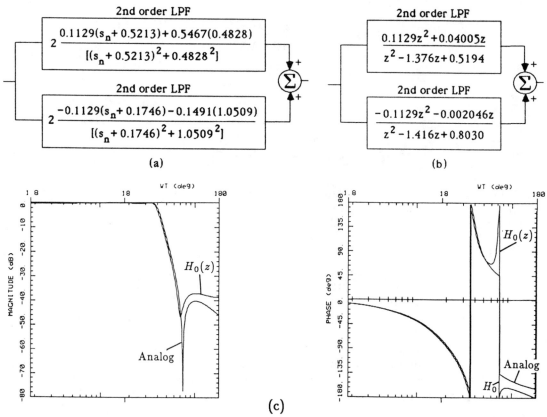

Fig. 11.3.1 Block diagram of 0.28 dB elliptic lowpass (a) analog filter and (a) digital filter of Example 11.3.1 and (c) frequency responses of both filters.

11.3.1a. Taking the inverse s-transform of $H(s)$ yields $h(t)$ as

$$\frac{1}{2\pi(1\ \text{KHz})}h(t) = 2\Big\{e^{-0.5213t_n}\Big[0.1129\cos(0.4828t_n) + 0.5467\sin(0.4828t_n)\Big]$$

$$+ e^{-0.1746t_n}\Big[-0.1129\cos(1.0509t_n) - 0.1491\sin(1.0509t_n)\Big]\Big\}U_{-1}(t)$$

$$(11.3.8)$$

where $t_n = 2\pi(1\ \text{KHz})t$. Setting $t = nT = n/(10\ \text{KHz})$ or $t_n = 0.6283n$ gives

$$Th(nT) = 4\pi\frac{1\ \text{KHz}}{10\ \text{KHz}}\Big\{0.7207^n\Big[0.1129\cos(0.3033n) + 0.5467\sin(0.3033n)\Big]$$

$$+ 0.8961^n\Big[-0.1129\cos(0.6603n) - 0.1491\sin(0.6603n)\Big]\Big\}U_{-1}(t)$$

$$(11.3.9)$$

Finding the z-transform of $Th(nT)$ using Table 1.9.1 yields $H_0(z)$ as

$$
\begin{aligned}
H_0(z) &= 1.257k\left\{ \frac{0.1129z(z - 0.6878) + 0.5467(0.2153)z}{z^2 - 1.376z + 0.5194} \right. \\
&\quad \left. + \frac{-0.1129z(z - 0.7077) - 0.1491(0.5496)z}{z^2 - 1.416z + 0.8030} \right\} \\
&= 1.257k\left\{ \frac{0.1129z^2 + 0.04005z}{z^2 - 1.376z + 0.5194} + \frac{-0.1129z^2 - 0.002046z}{z^2 - 1.416z + 0.8030} \right\} \\
&= \frac{0.04203z(z^2 - 0.6527z + 9311)}{(z^2 - 1.376z + 0.5194)(z^2 - 1.416z + 0.8030)}
\end{aligned}
$$

(11.3.10)

k is chosen so that the dc gain $H(1) = 0.9683$. The block diagram and frequency response of the filter are shown in Figs. 11.3.1b–c. The analog and digital filter responses are almost identical. The digital filter has a stopband rejection of only -37 dB rather than -40 dB due to aliasing distortion. The phase peaking around $85°$ is caused by the transmission zero not lying exactly on the unit circle. ∎

EXAMPLE 11.3.2 A digital filter must satisfy the magnitude specifications of Example 11.2.2 which were shown in Fig. 11.2.2a. Determine the minimum order Butterworth, Chebyshev, elliptic, and Bessel filters needed using the impulse-invariant response approach.

SOLUTION Eq. 11.3.3 shows that with sufficiently high sampling rate, there is negligible frequency domain aliasing so the analog and digital filter responses are identical. Since the stopband rejection ≥ 40 dB for $f \geq 6400$ Hz and the sampling frequency $f_s = 19.2$ KHz, the gain cannot exceed -40 dB above the Nyquist frequency $f_s/2 = 9.6$ KHz. Thus, there is negligible aliasing distortion. However if the sampling rate dropped by half to 9.6 KHz so the Nyquist frequency equalled 4.8 KHz, aliasing distortion would be excessive since $f_{3dB} = 3.2$ KHz.

Since the analog and digital filter responses and frequencies coincide in impulse-invariant designs, there is no frequency prewarping needed unlike the bilinear and other transforms discussed in the previous section (see Eq. 11.2.7). Therefore the normalized stopband frequency

$$
\Omega_a = \Omega_d = \frac{\omega_{40dB}T}{\omega_{3dB}T} = \frac{6.4 \text{ KHz}}{3.2 \text{ KHz}} = 2
$$

(11.3.11)

for both the analog and digital filters. Entering the data $(M_p, M_s, \gamma, \Omega_a) = (3, 40, 4, 2)$ onto the filter nomographs of Chap. 4.2 (see Fig. 11.2.2b) shows: Butterworth $n = 7$, Chebyshev $n = 5$, elliptic $n = 3$, Bessel impossible. Comparing these results with those found in Example 11.2.2, we see the filter orders needed are usually higher than those using the bilinear transform technique. This is expected because the bilinear transform always reduces or maintains filter order (see Fig. 5.10.3). ∎

EXAMPLE 11.3.3 In addition to the magnitude specifications of Example 11.3.2, the digital filter must satisfy the time domain specifications shown in Fig. 11.2.3. Determine whether the same filters of Example 11.3.2 can be used.

SOLUTION Since these are step response requirements, they would be maintained exactly if suitable analog filters were converted into digital filters using the step-invariant transform described in Chap. 5.14. However the impulse-invariant method was used which preserves the sampled impulse response of the filter. Its step response may or may not be close to that of the analog filter. This depends upon the bandwidth of the spectrum of the step (i.e., how band-limited the filter input is) relative to the filter itself. Therefore the step responses of the Butterworth, Chebyshev, and elliptic digital filters of Example 11.3.2 must actually be computed to see if they meet the requirements. ∎

11.3.2   MODIFIED IMPULSE-INVARIANT TRANSFORM DESIGN

An interesting alternative to the impulse-invariant transform is the modified impulse-invariant transform described by Antoniou.[3] Consider the analog transfer function $H(s) = N(s)/D(s)$ where $N(s)$ is its numerator and $D(s)$ is its denominator (see Eq. 11.3.4). Recall that the poles were preserved in the impulse-invariant transform but that the zeros were not. However zeros can also be preserved if we define two new analog transfer functions as

$$H_1(s) = \frac{N(0)}{N(s)}$$

$$H_2(s) = \frac{D(0)}{D(s)} \tag{11.3.12}$$

which are all-pole lowpass filters. Then the impulse-invariant response versions equal

$$H_{01}(z) = \frac{N_1(z)}{N(z)} = \frac{N_1(z)}{\prod_{k=1}^{M}(z - e^{-z_k T})}$$

$$H_{02}(z) = \frac{D_1(z)}{D(z)} = \frac{D_1(z)}{\prod_{k=1}^{N}(z - e^{-p_k T})} \tag{11.3.13}$$

using Eq. 11.3.6. Forming the ratio of these two transfer functions produces the modified impulse-invariant transform

$$H(z) = \frac{H_{02}(z)}{H_{01}(z)} = \frac{D_1(z)}{N_1(z)}\frac{N(z)}{D(z)} = \frac{D_1(z)}{N_1(z)}\frac{\prod_{k=1}^{M}(z - e^{-z_k T})}{\prod_{k=1}^{N}(z - e^{-p_k T})} \tag{11.3.14}$$

It is the product of a $D_1(z)/N_1(z)$ compensator and the matched z-transform (Eq. 11.4.1) to be discussed in the next section.

Because this method relies on closely approximating the frequency responses of both the numerator $N(s)$ and denominator $D(s)$ regardless of their orders, $H(z)$ well approximates $H(s)$ in ac steady-state. Good approximations for highpass, bandstop, and allpass filters can be obtained using the modified impulse-invariant transform. The resulting filter order increases due to the $N_1(z)$ and $D_1(z)$ terms in Eq. 11.3.14.

As long as the numerator degree is no greater than the denominator degree in $H(z)$ of Eq. 11.3.14, the filter is causal. This is always assured when $H(s)$ has $M \le N$ which is generally the case. However, the filter derived from Eq. 11.3.14 will be unstable if the zeros of $N_1(z)$ lie outside the unit circle even if the analog filter is stable. Fortunately such an unstable filter can be stabilized by cascading it with an allpass filter having

$$H_c(z) = \frac{N_2(z)}{N_2(1/z)} \tag{11.3.15}$$

where $N_2(z)$ equals the portion of $N_1(z)$ having zeros outside the unit circle. $H_c(z)$ modifies the phase and delay characteristics of the digital filter $H(z)$ but not its magnitude characteristics (see Chap. 3.12).

EXAMPLE 11.3.4 Redesign the 4th-order elliptic filter of Example 11.2.1 using the modified impulse-invariant transform method.
SOLUTION The transfer function $H(s)$ given by Eq. 11.2.3 is used to form two new transfer functions

$$H_1(s) = \frac{0.9863(2.0173^2)}{s_n^2 + 2.0173^2}$$

$$\begin{aligned} H_2(s) &= \frac{0.5729}{\left[(s_n + 0.5213)^2 + 0.4828^2\right]\left[(s_n + 0.1746)^2 + 1.0509^2\right]} \\ &= \frac{0.1814 + j0.5372}{s_n + 0.5213 + j0.4828} + \frac{0.1814 - j0.5372}{s_n + 0.5213 - j0.4828} \\ &\quad + \frac{-0.1814 - j0.1869}{s_n + 0.1746 + j1.0509} + \frac{-0.1814 + j0.1869}{s_n + 0.1746 - j1.0509} \end{aligned} \tag{11.3.16}$$

using Eq. 11.3.12 where $s_n = s/2\pi(1\ \text{KHz})$. This factoring is used in the block diagram of Fig. 11.3.2a. Converting these $H_i(s)$ into digital filter transfer functions using the impulse-invariant transform method yields

$$H_{01}(z) = \frac{1.899z}{z^2 - 0.5973z + 1}$$

$$\begin{aligned} H_{02}(z) &= 1.257\left\{ \frac{0.1814z(z - 0.6878) + 0.5372(0.2153)z}{z^2 - 1.376z + 0.5194} \right. \\ &\quad \left. + \frac{-0.1814z(z - 0.7077) - 0.1869(0.5496)z}{z^2 - 1.416z + 0.8030} \right\} \\ &= 1.257\left\{ \frac{0.1814z^2 - 0.00911z}{z^2 - 1.376z + 0.5194} + \frac{-0.1814z^2 + 0.0257z}{z^2 - 1.416z + 0.8030} \right\} \\ &= \frac{0.275z(z^2 + 0.1055z + 0.0218)}{(z^2 - 1.376z + 0.5194)(z^2 - 1.416z + 0.8030)} \end{aligned} \tag{11.3.17}$$

which have the form of Eq. 11.3.13. Forming the ratio H_{02}/H_{01} yields

$$H(z) = \frac{H_{02}(z)}{H_{01}(z)} = 0.03398\ \frac{(z^2 + 0.1055z + 0.0218)(z^2 - 0.5973z + 1)}{(z^2 - 1.376z + 0.5194)(z^2 - 1.416z + 0.8030)} \tag{11.3.18}$$

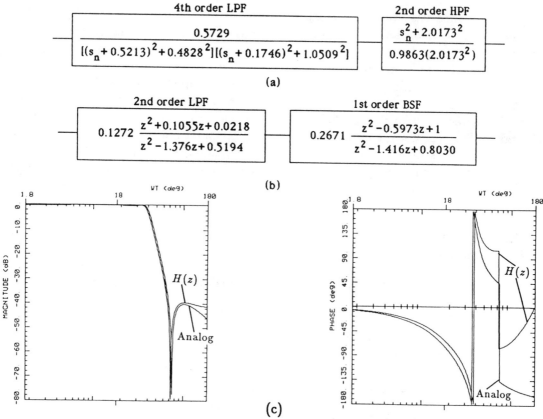

Fig. 11.3.2 Block diagram of 0.28 dB elliptic lowpass (a) analog filter and (a) digital filter of Example 11.3.2 and (c) frequency responses of both filters.

$H(z)$ has the block diagram and frequency response shown in Figs. 11.3.2b–c. The analog and digital filter magnitude responses are virtually identical. The phase responses differ only in the stopband. ∎

11.4 MATCHED Z-TRANSFORM DESIGN (POLE-ZERO MAPPING ALGORITHM)

Digital transforms directly convert (unsampled) analog filter transfer functions $H(s)$ into digital filter transfer functions $H(z)$ using Eq. 11.2.1. Assuming that $H(s)$ has the rational form given by Eq. 11.3.4, substituting the z-transform $z = e^{sT}$ or $sT = \ln z$ into Eq. 11.2.1 produces an $H(z)$ which is rational in $\ln(z)$ but irrational in z.

However a rational transfer function (in z) can be obtained by mapping the s-plane poles $s = -p_k$ and zeros $s = -z_k$ into z-plane poles $z = e^{-p_k T}$ and zeros $z = e^{-z_k T}$. Then $H(z)$ is formed as

$$H(z) = d \, \frac{\prod_{k=1}^{M}(z - e^{-z_k T})}{\prod_{k=1}^{N}(z - e^{-p_k T})}, \qquad |z| > z_o \qquad (11.4.1)$$

This is called the *matched z-transform* and was introduced in Chap. 5.1.2.[4] The modified impulse-invariant transform given by Eq. 11.3.12 involves the matched z-transform. The impulse-invariant transform given by Eq. 11.3.6 has the same denominator but a different numerator. The number of zeros is preserved under the matched z-transform unlike the impulse-invariant and modified impulse-invariant transforms.

Causal and stable analog filters always convert into causal and stable digital filters using the matched z-transform. The reasoning is identical to that following Eq. 11.3.6. The matched z-transform can be used to design highpass, bandstop, and allpass filters as well as lowpass and bandpass filters unlike the impulse-invariant transform. The matched z-transform does not preserve frequency domain characteristics such as magnitude ripple and delay nor does it preserve time domain responses. Its major advantage is ease of application.

EXAMPLE 11.4.1 Determine the transfer function of the 4th-order elliptic filter of Example 11.2.1 using the matched z-transform.

SOLUTION The transfer function $H(s)$ of the elliptic filter was given by 11.2.3. Converting the poles and zeros using the z-transform $z_k = e^{s_k T}$ yields

$$p_1,\ p_1^* = \exp\big[(-0.5213 \pm j0.4828)(2\pi(1\text{ KHz})/10\text{ KHz})\big]$$
$$= 0.7207e^{\pm j17.38°} = 0.6878 \pm j0.2153$$

$$p_2,\ p_2^* = \exp\big[(-0.1746 \pm j1.0509)(2\pi(1\text{ KHz})/10\text{ KHz})\big]$$
$$= 0.8961e^{\pm j37.83°} = 0.7077 \pm j0.5496 \tag{11.4.2}$$

$$z_1,\ z_1^* = \exp\big[(\pm j2.0173)(2\pi(1\text{ KHz})/10\text{ KHz})\big]$$
$$= e^{\pm j72.63°} = 0.2987 \pm j0.9544$$

$$z_2,\ z_2^* = \exp\big[(-\infty \pm j\pi/2)(2\pi(1\text{ KHz})/10\text{ KHz})\big] = 0$$

The digital filter transfer function therefore equals

$$H(z) = \frac{kz^2\big[(z - 0.2987)^2 + 0.9544^2\big]}{\big[(z - 0.6878)^2 + 0.2153^2\big]\big[(z - 0.7077)^2 + 0.5496^2\big]}$$
$$= \frac{0.03831z^2(z^2 - 0.5973z + 1)}{(z^2 - 1.376z + 0.5194)(z^2 - 1.416z + 0.8030)} \tag{11.4.3}$$

$k = 0.03831$ sets the dc gain $H(1) = -0.28$ dB. The block diagram and frequency response of this filter are shown in Fig. 11.4.1. The analog and digital filter magnitude responses are virtually identical. The phase responses differ only in the stopband. Comparing Figs. 11.3.2c and 11.4.1c, no difference can be seen between the modified impulse-invariant and matched z-transform filter responses. It is remarkable that the matched z-transform so well maintains the analog filter response is yet is the simplest of all these transforms to compute. ∎

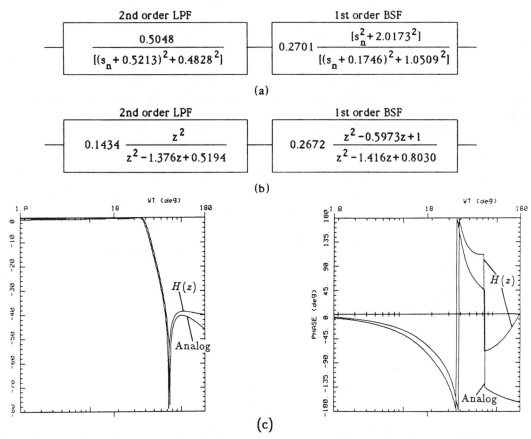

Fig. 11.4.1 Block diagram of 0.28 dB elliptic lowpass (a) analog filter and (a) digital filter of Example 11.4.1 and (c) frequency responses of both filters.

11.5 MORE NUMERICAL INTEGRATION TRANSFORMS

A number of frequency domain transforms are listed in Tables 5.2.1 and 5.2.2 and were derived using time domain integration approximations. Now more integration algorithms will be introduced which will extend Chap. 5.2.

A collection of numerical analysis algorithms for interpolation and differentiation were tabulated in Tables 10.6.1 and 10.6.2. These were used to design nonrecursive filters. For completeness, we want to continue that discussion here for integration algorithms. Since integration requires processing all inputs between the integration limits and the lower limit is usually minus infinity, infinite length memories are necessary. This is the reason for discussing integration in this recursive IIR filter chapter rather than the last chapter concerned with nonrecursive FIR filters.

11.5.1 SINGLE INTEGRATION ALGORITHMS

Integration can also be performed using the interpolation equations in Table 10.6.1. By analogy with the differentiation equation of Eq. 10.6.10,

$$y(n+p) = y(t)\Big|_{t=nT+pT} = y(n) + \int_{nT}^{nT+pT} x(t)dt = y(n) + T\int_0^p x(n+p)dp \quad (11.5.1)$$

Integrating the formulas of Table 10.6.1 as indicated by this equation leads to the integration formulas in Table 11.5.1. Using an integer value of p means that integration occurs over a region whose end-points correspond to sample points. For example, consider the first few Newton formulas in Table 11.5.1 where

$$y(n+p)_1 = y(n)_1 + px(n) + \frac{p^2}{2}\Big[x(n+1) - x(n)\Big]$$

$$y(n+p)_2 = y(n)_2 + \frac{2p^3 - 3p^2}{12}\Big[x(n+2) - 2x(n+1) + x(n)\Big]$$

$$y(n+p)_3 = y(n)_3 + \frac{p^4 - 4p^3 + 4p^2}{48}\Big[x(n+3) - 3x(n+2) + 3x(n+1) - x(n)\Big]$$
$$(11.5.2)$$

for any integer or noninteger p. Setting $p = k$ in $y(n+p)_k$ so the integral uses $(k+1)$ equally-spaced sample points including the end-points yields

$$y(n+1)_1 = y(n)_1 + \frac{T}{2}\Big[x(n+1) + x(n)\Big]$$

$$y(n+2)_2 = y(n)_2 + \frac{T}{3}\Big[x(n+2) + 4x(n+1) + x(n)\Big]$$

$$y(n+3)_3 = y(n)_3 + \frac{3T}{8}\Big[x(n+3) + 3x(n+2) + 3x(n+1) + x(n)\Big]$$

$$y(n+4)_4 = y(n)_4 + \frac{2T}{45}\Big[7x(n+4) + 32x(n+3) + 12x(n+2) + 32x(n+1) + 7x(n)\Big]$$

$$y(n+5)_5 = y(n)_5 + \frac{5T}{288}\Big[19x(n+5) + 75x(n+4) + 50x(n+3)$$

$$+ 50x(n+2) + 75x(n+1) + 19x(n)\Big]$$

$$y(n+6)_6 = y(n)_6 + \frac{T}{140}\Big[41x(n+6) + 216x(n+5) + 27x(n+4)$$

$$+ 272x(n+3) + 27x(n+2) + 216x(n+1) + 41x(n)\Big]$$
$$(11.5.3)$$

from Eq. 11.5.2. These general results are often called the *Newton-Cotes* integration formulas.[5] For $k = 1, 2,$ and 3, they are called the *trapezoid, Simpson's,* and *three-eighths* integration formulas, respectively.

Shifting time by $-k/2$ seconds produces symmetrical equations about $n = 0$. Then taking the z-transforms of Eq. 11.5.3 and forming transfer functions yields

$$H_1(z) = \frac{T}{2}\left[\frac{z^{1/2} + z^{-1/2}}{z^{1/2} - z^{-1/2}}\right] \quad\quad (11.5.4a)$$

$$H_2(z) = \frac{T}{3} \left[\frac{z^1 + 4 + z^{-1}}{z^1 - z^{-1}} \right]$$

$$H_3(z) = \frac{3T}{8} \left[\frac{z^{3/2} + 3z^{1/2} + 3z^{-1/2} + z^{-3/2}}{z^{3/2} - z^{-3/2}} \right]$$

$$H_4(z) = \frac{2T}{45} \left[\frac{7z^2 + 32z^1 + 12 + 32z^{-1} + 7z^{-2}}{z^2 - z^{-2}} \right] \tag{11.5.4b–f}$$

$$H_5(z) = \frac{5T}{288} \left[\frac{19z^{5/2} + 75z^{3/2} + 50z^{1/2} + 50z^{-1/2} + 75z^{-3/2} + 19z^{-5/2}}{z^{5/2} - z^{-5/2}} \right]$$

$$H_6(z) = \frac{T}{140} \left[\frac{41z^3 + 216z^2 + 27z^1 + 272 + 27z^{-1} + 216z^{-2} + 41z^{-3}}{z^3 - z^{-3}} \right]$$

The ac steady-state responses equal

$$H_1(e^{j\omega T}) = \frac{T}{j2} \left[\frac{\cos(\omega T/2)}{\sin(\omega T/2)} \right]$$

$$H_2(e^{j\omega T}) = \frac{T}{j3} \left[\frac{2 + \cos(\omega T)}{\sin(\omega T)} \right]$$

$$H_3(e^{j\omega T}) = \frac{3T}{j8} \left[\frac{3\cos(\omega T/2) + \cos(3\omega T/2)}{\sin(3\omega T/2)} \right]$$

$$H_4(e^{j\omega T}) = \frac{2T}{j45} \left[\frac{6 + 32\cos(\omega T) + 7\cos(2\omega T)}{\sin(2\omega T)} \right] \tag{11.5.5}$$

$$H_5(e^{j\omega T}) = \frac{5T}{j288} \left[\frac{50\cos(\omega T/2) + 75\cos(3\omega T/2) + 19\cos(5\omega T/2)}{\sin(5\omega T/2)} \right]$$

$$H_6(e^{j\omega T}) = \frac{T}{j140} \left[\frac{136 + 27\cos(\omega T) + 216\cos(2\omega T) + 41\cos(3\omega T)}{\sin(3\omega T)} \right]$$

by setting $z = e^{j\omega T}$. These gains are plotted in Fig. 11.5.1.

The ideal integrator response equals $1/j\omega$. Since the integrator gain is infinite at dc, dividing H_k given in Eq. 11.5.5 by $1/j\omega$ yields the normalized gain ratio H_{kn} as

$$H_{1n}(e^{j\omega T}) = \left[\frac{\cos(\omega T/2)}{\sin(\omega T/2)/(\omega T/2)} \right] = 1 - \frac{\omega^2}{12} + \cdots$$

$$H_{2n}(e^{j\omega T}) = \frac{1}{3} \left[\frac{2 + \cos(\omega T)}{\sin(\omega T)/(\omega T)} \right] = 1 + \frac{\omega^4}{180} + \cdots$$

$$H_{3n}(e^{j\omega T}) = \frac{1}{4} \left[\frac{3\cos(\omega T/2) + \cos(3\omega T/2)}{\sin(3\omega T/2)/(3\omega T/2)} \right] \tag{11.5.6}$$

$$H_{4n}(e^{j\omega T}) = \frac{4}{45} \left[\frac{6 + 32\cos(\omega T) + 7\cos(2\omega T)}{\sin(2\omega T)/(2\omega T)} \right]$$

Expanding numerator and denominator in power series and dividing them to form the

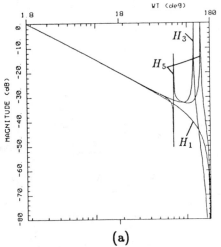
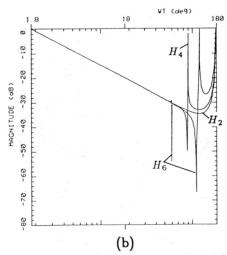

| | (a) | | (b) |

Fig. 11.5.1 Magnitude responses of Newton-Cotes integrators having gain H_n with (a) n even and (b) n odd.

power series shows that H_{kn} is maximally flat up to its $2k$-th derivative. Therefore the higher-order integrators have gains which match the ideal integrator gain $1/j\omega$ to higher frequencies.

All the H_k integrators exhibit the $1/j\omega$ response at low frequencies with a low-frequency asymptote of -20 dB/dec. They all exhibit gain peaking at integer multiples of $\omega T = 360°/k$. This is due to resonance of the k poles equally distributed at angles of $360°/k$ around the unit circle. The odd-order filters of Fig. 11.5.1a have zero gain at $\omega T = 180°$ due to the zeros at $z = -1$. The even-order filters of Fig. 11.5.1b have infinite gain peaking at $\omega T = 180°$ due to the poles at $z = -1$. Therefore odd-order filters notch at the Nyquist frequency while the even-order filters peak. None of these integrators are particularly useful because of passband gain peaking except for the trapezoidal H_1 and Simpson H_2 integrators.

EXAMPLE 11.5.1 Rather than using higher-order polynomials for approximating integrals, a low-order polynomial can be applied to adjacent intervals and the results added to obtain the entire integral value.[6] Two integrators that are well-known use the generalized trapezoid rule

$$y(n+p) - y(n) = \int_{nT}^{(n+p)T} x(t)dt$$
$$= T\big[0.5x(n) + x(n+1) + \ldots + x(n+p-1) + 0.5x(n+p)\big]$$
(11.5.7)

and the generalized Simpson's rule (use an odd number of points or even p)

$$y(n+p) - y(n)$$
$$= \frac{T}{3}\big[x(n) + 4x(n+1) + 2x(n+2) + 4x(n+3) + \ldots + 4x(n+p-1) + x(n+p)\big]$$
(11.5.8)

Determine the transfer functions of the integrators and compare with Fig. 11.5.1.
SOLUTION The z-transforms of these equations equal

$$(z^p - 1)Y_T(z) = T\left[0.5 + z^1 + z^2 + \ldots + z^{p-1} + 0.5z^p\right]$$

$$(z^p - 1)Y_S(z) = \frac{T}{3}\left[1 + 4z^1 + 2z^2 + 4z^3 + 2z^4 + \ldots + 4z^{p-1} + z^p\right]$$

(11.5.9)

Introducing a delay of $-p/2$ clocks, the transfer functions equal

$$\frac{1}{T}H_T(z) = \frac{0.5z^{p/2} + z^{p/2-1} + \ldots + z^{-(p/2-1)} + 0.5z^{-p/2}}{z^{p/2} - z^{-p/2}}$$

$$\frac{3}{T}H_S(z) = \frac{z^{p/2} + 4z^{p/2-1} + 2z^{p/2-2} + 4z^{p/2-3} + \ldots + 4z^{-(p/2-1)} + z^{-p/2}}{z^{p/2} - z^{-p/2}}$$

(11.5.10)

Setting $z = e^{j\omega T}$, the ac steady-state gains equal

$$\frac{1}{T}H_T(e^{j\omega T})$$

$$= \frac{1 + \cos(p\omega T/2) + 2\cos\left[(p-2)\omega T/2\right] + 2\cos\left[(p-4)\omega T/2)\right] + \ldots + 2\cos(2\omega T/2)}{j2\sin(p\omega T/2)}$$

$$\frac{3}{T}H_S(e^{j\omega T})$$

$$= \frac{1 + 2\cos(p\omega T/2) + 8\cos\left[(p-2)\omega T/2\right] + 4\cos\left[(p-4)\omega T/2\right] + \ldots + \cos(2\omega T/2)}{j2\sin(p\omega T/2)}$$

(11.5.11)

These gains are plotted in Fig. 11.5.2. Both sets of generalized trapezoidal and Simpson integrators have identical magnitude and phase characteristics independent of order. They are the same as the H_1 and H_2 Newton-Cotes integrators of Fig. 11.5.1. Carefully evaluating Eq. 11.5.10 shows that after pole-zero cancellations, $H_T(z) = H_1(z)$ and $H_S(z) = H_2(z)$ in Eq. 11.5.4. ∎

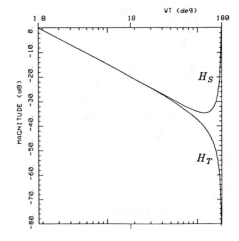

Fig. 11.5.2 Magnitude responses of generalized trapezoidal integrators and Simpson integrators.

EXAMPLE 11.5.2 The Adams-Bashforth integrators use[7]

$$y(n+1) = y(n) + \frac{T}{12}\left[23x(n) - 16x(n-1) + 5x(n-2)\right]$$

$$y(n+1) = y(n) + \frac{T}{24}\left[9x(n+1) + 19x(n) - 5x(n-1) + x(n-2)\right]$$

(11.5.12)

Determine the transfer functions.
SOLUTION Taking the z-transform yields

$$(z-1)Y(z) = \frac{T}{12}\left[23 - 16z^{-1} + 5z^{-2}\right]X(z)$$

$$(z-1)Y(z) = \frac{T}{24}\left[9z^1 + 19 - 5z^{-1} + z^{-2}\right]X(z)$$

(11.5.13)

so that

$$H_{131}(z) = \frac{T}{12}\left[\frac{23 - 16z^{-1} + 5z^{-2}}{z-1}\right]$$

$$H_{031}(z) = \frac{T}{24}\left[\frac{9z^1 + 19 - 5z^{-1} + z^{-2}}{z-1}\right]$$

(11.5.14)

These correspond to the optimum H_{131} and H_{031} integrators listed in Tables 5.2.1 and 5.2.2. They have the excellent gain responses shown in Fig. 11.5.4c. H_{031} is the best integrator. ∎

EXAMPLE 11.5.3 Some integrators find the new output $y(n+1)$ by using old values of the integral $y(n), y(n-1), y(n-2), \ldots$, new integrand values $x(n+1), x(n+2), \ldots$, and old integral values $x(n-1), x(n-2), \ldots$. For example, one type that uses three old integral values with one new and three old integrand values sets

$$y(n+1) = a_0 y(n) + a_1 y(n-1) + a_2 y(n-2)$$
$$+ T\left[b_{-1}x(n+1) + b_0 x(n) + b_1 x(n-1) + b_2 x(n-2)\right]$$

(11.5.15)

Tick developed a special case of this integration formula involving only three consecutive terms as[8]

$$y(n+1) = y(n-1) + T\left[0.3584x(n+1) + 1.2832x(n) + 0.3584x(n-1)\right] \quad (11.5.16)$$

It uses coefficients of (0.3584, 1.2832, 0.3584) rather than (0.3333, 1.3333, 0.3333) as Simpson's rule which makes the error equiripple (Chebyshev) for $0 \le \omega T/2 \le \pi/2$. The error is zero at dc. Investigate the frequency characteristics.
SOLUTION Taking the z-transform of both sides of Eq. 11.5.16 and forming the transfer function yields

$$H(z) = T\left[\frac{0.3584z + 1.2832 + 0.3584z^{-1}}{z - z^{-1}}\right]$$

(11.5.17)

Its gain in ac steady-state equals

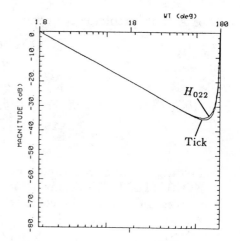

Fig. 11.5.3 Magnitude responses of Tick integrator and H_{022}.

$$H(e^{j\omega T}) = \frac{1}{j\omega}\left[\frac{0.6416 + 0.3584\cos(\omega T)}{\sin(\omega T)/(\omega T)}\right] \qquad (11.5.18)$$

which is plotted in Fig. 11.5.3.

The best integrator (maximally flat) using two poles and two zeros is $H_{022}(z)$ according to Table 5.2.1 where

$$H_{022}(z) = \frac{T}{3}\left[\frac{1 + 4z^{-1} + z^{-2}}{1 - z^{-2}}\right] \qquad (11.5.19)$$

$H_{022}(z)$ is equivalent to a 3-term Simpson integrator (cf. Eq. 11.5.4). Its response is also shown in Fig. 11.5.3. Because of its equiripple error, the Tick integrator has a slightly better response than H_{022}.   ∎

EXAMPLE 11.5.4 Predictor-corrector equations are often used when integrating $y' = f(y) = x$ over a large number of points to reduce error. The approximate value of $y(n+1)$ is obtained using the predictor equation. This value is then checked and corrected by iterative use of the corrector equation where $x(n+1) = f[y(n+1)]$. The two equations are then independent of one another. Several pairs of predictor-correctors are often used.[9] They include:

$$\text{Euler Predictor } H_{111}: \quad y(n+1) = y(n) + Tx(n)$$

$$\text{or Midvalue Predictor } H_{112}: \quad y(n+1) = y(n-1) + 2Tx(n)$$

$$\text{Trapezoidal Corrector } H_{011}: \quad y(n+1) = y(n) + \frac{T}{2}\Big[x(n+1) + x(n)\Big]$$

or Milne's predictor-corrector $(11.5.20)$

$$H_p \text{ Predictor}: \quad y(n+1) = y(n-3) + \frac{4T}{3}\Big[2x(n) - x(n-1) + 2x(n-2)\Big]$$

$$H_{022} \text{ Corrector}: \quad y(n+1) = y(n-1) + \frac{T}{3}\Big[x(n+1) + 4x(n) + x(n-1)\Big]$$

$$(11.5.21)$$

or Adam's predictor-corrector

H_{141} Predictor : $y(n+1) = y(n)$

$$+ \frac{T}{24}\Big[55x(n) - 59x(n-1) + 37x(n-2) - 9x(n-3)\Big]$$

H_{031} Corrector : $y(n+1) = y(n)$

$$+ \frac{T}{24}\Big[9x(n+1) + 19x(n) - 5x(n-1) + x(n-2)\Big]$$

$$\text{(11.5.22)}$$

Investigate their characteristics.

SOLUTION The equivalent transfer functions for the Euler and midvalue predictors and trapezoidal corrector are

$$H_{111}(z) = \frac{T}{z-1}, \qquad H_{112}(z) = \frac{2Tz}{z^2-1}, \qquad H_{011}(z) = -\frac{T}{2}\frac{z+1}{z-1} \qquad \text{(11.5.23)}$$

or for Milne's predictor-corrector

$$H_p(z) = \frac{4T}{3}\left[\frac{2 - z^{-1} + 2z^{-2}}{z - z^{-3}}\right] = \frac{4T}{3}\left[\frac{z(2z^2 - z + 2)}{z^4 - 1}\right] \qquad \text{(11.5.24)}$$

$H_{022}(z) = $ Eq. 11.5.19

or for Adam's predictor-corrector

$$H_{141}(z) = \frac{T}{24}\left[\frac{55z^3 - 59z^2 + 37z - 9}{z^3(z-1)}\right] \qquad \text{(11.5.25)}$$

$H_{031}(z) = $ Eq. 11.5.14b

Their pole-zero patterns and responses are shown in Fig. 11.5.4. H_{111}, H_{112}, and H_{011} integrate at low frequencies. The H_{112} predictor peaks at ω_N while the H_{111} predictor continues to integrate. The H_{011} corrector notches at ω_N. The Milne predictor-corrector integrates at low frequencies and peaks at ω_N. The H_p predictor also notches at $0.42\omega_N$ and peaks at $0.5\omega_N$. The Adam H_{141} predictor integrates at lower frequencies and peaks slightly around ω_N. The H_{031} corrector integrates over the entire frequency range. All the filters are causal and stable. ∎

11.5.2 MULTIPLE INTEGRATION AND DIFFERENTIATION ALGORITHMS

The algorithms of Chaps. 5.2 and 11.5.1 describe single integration. The transfer function of the digital integrator equalled

$$H_1(z) = H(p)\Big|_{p=g(z)} = \frac{1}{p}\Big|_{p=g(z)} = \frac{1}{g(z)} \qquad \text{(11.5.26)}$$

Now we want to briefly consider multiple integration. The transfer function of a kth-order integrator equals

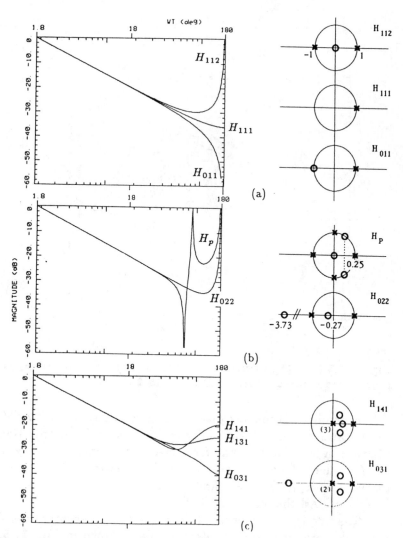

Fig. 11.5.4 Pole-zero patterns and frequency responses for (a) H_{111} Euler and H_{112} midvalue predictors and H_{011} trapezoidal corrector, (b) Milne's H_p predictor and H_{022} corrector, and (c) Adam's H_{141} predictor and H_{031} corrector.

$$H_k(z) = \frac{1}{p^k}\bigg|_{p^k = f(z)} = \frac{1}{f(z)} \tag{11.5.27}$$

If the kth-order integrator is viewed as a cascade of k first-order integrators, then its gain equals $H_1^k(z) = g(z)^{-k}$ where $g(z)$ is the first-order integrator algorithm of Eq. 11.5.26. However $H_1^k(z)$ does not give as accurate results as those given by $H_k(z) = f(z)^{-1}$ of Eq. 11.5.27 which performs k integrations in a single block. In this case, $f(z)$ is a kth-order integration algorithm.

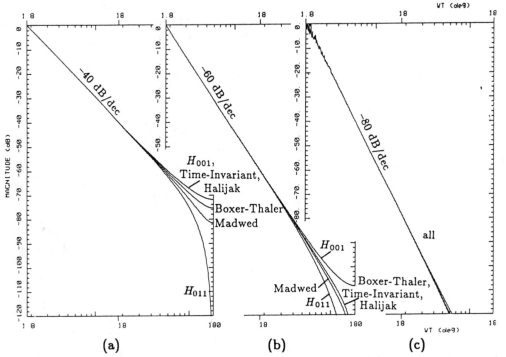

Fig. 11.5.5 Magnitude responses of (a) second-order, (b) third-order, and (c) fourth-order integrators of Table 11.5.2.

A number of these algorithms are given in Table 11.5.2.[10] These algorithms include the backward Euler H_{001}, bilinear H_{011}, time-invariant, Halijak, Boxer-Thaler, and Madwek types. Notice that for first-order integration, they all reduce to either H_{001} or H_{011} algorithms. However as the integrator order increases, they take on different forms. The H_{001} and H_{011} first-order algorithms are simply raised to the power of the integrator. All the others are not.

The frequency responses of the first-order integrators are plotted in Fig. 5.3.1 while those of the multiple-order integrators are shown in Fig. 11.5.5. Notice in Fig. 11.5.5 that the second, third, and fourth-order integrators have slopes of -40, -60, and -80 dB/dec, respectively. The best integrator is Madwed followed by Boxer-Thaler, Halijak, time-invariant, bilinear, and backward Euler. This shows that superior approximations of p^{-k} are obtained by using higher-order discrete integrators and their transforms $f(z)$ rather than cascading k first-order discrete integrators or raising $g(z)$ to the kth power.

EXAMPLE 11.5.5 Convert the kth-order integration H_{ik} algorithms of Table 11.5.2 into kth-order differentiation H_{dk} algorithms. Discuss their characteristics.
SOLUTION Simply reciprocating the algorithms of Table 11.5.2 produces an equivalent set of multiple differentiation algorithms as

$$H_{dk}(z) = \frac{1}{H_{ik}(z)} \qquad (11.5.28)$$

Letting $z = e^{j\omega T}$ in Eq. 11.5.28 shows that the magnitude and phase characteristics of $H_{dk}(e^{j\omega T})$ can be obtained from those shown in Fig. 11.5.5. Simply change the signs of the magnitudes in dB (reciprocate magnitude) and the phases. This gives a set of recursive differentiation algorithms. ∎

Table 11.5.1 A collection of numerical analysis algorithms for integration.

1. Newton (Newton-Gregory, forward difference)

$$\frac{y(p) - y(0)}{T} = px(0) + C_i(p, 1)\Delta x(0) + C_i(p, 2)\Delta^2 x(0) + C_i(p, 3)\Delta^3 x(0) + \ldots$$

(Backward difference)

$$\frac{y(p) - y(0)}{T} = px(0) + C_i(p, 1)\Delta x(-1) + C_i(p + 1, 2)\Delta^2 x(-2) + C_i(p + 2, 3)\Delta^3 x(-3) + \ldots$$

2. Gauss (Gauss-Newton, forward difference)

$$\frac{y(p) - y(0)}{T} = px(0) + C_i(p, 1,)\Delta x(0) + C_i(p, 2)\Delta^2 x(-1) + C_i(p + 1, 3)\Delta^3 x(-1)$$
$$+ C_i(p + 1, 4)\Delta^4 x(-2) + \ldots$$

(Backward difference)

$$\frac{y(p) - y(0)}{T} = px(0) + C_i(p, 1)\Delta x(-1) + C_i(p + 1, 2)\Delta^2 x(-1) + C_i(p + 1, 3)\Delta^3 x(-2)$$
$$+ C_i(p + 2, 4)\Delta^4 x(-2) + \ldots$$

3. Stirling (Average Gauss, central difference)

$$\frac{y(p) - y(0)}{T/2} = 2px(0) + C_i(p, 1)\big[\Delta x(0) + \Delta x(-1)\big] + \big[C_i(p, 2) + C_i(p + 1, 2)\big]\Delta^2 x(-1)$$
$$+ C_i(p + 1, 3)\big[\Delta^3 x(-1) + \Delta^3 x(-2)\big]$$
$$+ \big[C_i(p + 1, 4) + C_i(p + 2, 4)\big]\Delta^4 x(-2) + \ldots$$

4. Bessel

$$\frac{y(p) - y(0)}{T/2} = p\big[x(0) + x(-1)\big] + \big[C_i(p, 1) + C_i(p - 1, 1)\big]\Delta x(0) + C_i(p, 2)\big[\Delta^2 x(-1) + \Delta^2 x(0)\big] + \ldots$$

5. Everett

$$\frac{y(p) - y(0)}{T} = -C_i(p - 2, 1)x(0) + C_i(p - 1, 1)x(1) - C_i(p + 1, 3)\Delta^2 x(-1) + C_i(p, 3)\Delta^2 x(0) + \ldots$$

Definitions: T = sample interval size, nT = nominal time value,
$p = (t - nT)/T$ = normalized time deviation (not necessarily integer),

$$C(p, k) = \frac{p(p - 1)\ldots(p - k + 1)}{k!} = \frac{p!}{(p - k)!\, k!} = \text{combinational polynomial of degree } k \text{ in } p.$$

$$C_i(p, k) = \int_0^p C(p, k)\, dp = \text{integral of } C(p, k)$$

Table 11.5.1 (Continued)

$$C_i(p,1) = p^2/2, \qquad C_i(p,2) = (p^3 - 3p^2/2)/2!, \qquad C_i(p,3) = (p^4 - 4p^3 + 4p^2)/4!$$
$$C_i(p+1,1) = (p^2 + 2p)/2, \qquad C_i(p+1,2) = (2p^3 + 3p^2)/2(3!), \qquad \cdots$$
$$C_i(p+2,1) = (p^2 + 4p)/4, \qquad \cdots$$

Samples: $x(n) = x(nT)$, $x(n+p) = x((n+p)T)$
Sample deviations:

$$\Delta x(0) = x(n+1) - x(n)$$
$$\Delta^2 x(0) = \Delta[\Delta x(0)] = \Delta[x(n+1) - x(n)] = x(n+2) - 2x(n+1) + x(n)$$
$$\Delta^3 x(0) = x(n+3) - 3x(n+2) + 3x(n+1) - x(n)$$
$$\Delta^4 x(0) = x(n+4) - 4x(n+3) + 6x(n+2) - 4x(n+1) + x(n)$$
$$\Delta^q x(0) = x(n+q) - qx(n+q-1) + \left[q(q-1)/2!\right]x(n+q-2)$$
$$\qquad - \left[q(q-1)(q-2)/3!\right]x(n+q-3) + \ldots + (-1)^q x(n)$$

Table 11.5.2 A collection of first- through fourth-order integration algorithms.[10]

Type	$1/p$	$1/p^2$	$1/p^3$	$1/p^4$
Bilinear H_{011}	$\dfrac{T}{2}\dfrac{z+1}{z-1}$	$\dfrac{T^2}{4}\dfrac{(z+1)^2}{(z-1)^2}$	$\dfrac{T^3}{8}\dfrac{(z+1)^3}{(z-1)^3}$	$\dfrac{T^4}{16}\dfrac{(z+1)^4}{(z-1)^4}$
Boxer–Thaler	$\dfrac{T}{2}\dfrac{z+1}{z-1}$	$\dfrac{T^2}{12}\dfrac{z^2+10z+1}{(z-1)^2}$	$\dfrac{T^3}{2}\dfrac{z^2+z}{(z-1)^3}$	$\dfrac{T^4}{8}\dfrac{z^3+4z^2+z}{(z-1)^4} - \dfrac{T^4}{720}$
Madwed	$\dfrac{T}{2}\dfrac{z+1}{z-1}$	$\dfrac{T^2}{6}\dfrac{z^2+4z+1}{(z-1)^2}$	$\dfrac{T^3}{24}\dfrac{z^3+11z^2+11z+1}{(z-1)^3}$	$\dfrac{T^4}{120}\dfrac{z^4+26z^3+66z^2+26z+1}{(z-1)^4}$
Back. Euler H_{001}	$\dfrac{Tz}{z-1}$	$\dfrac{T^2z^2}{(z-1)^2}$	$\dfrac{T^3z^3}{(z-1)^3}$	$\dfrac{T^4z^4}{(z-1)^4}$
Time-Invariant	$\dfrac{Tz}{z-1}$	$\dfrac{T^2z}{(z-1)^2}$	$\dfrac{T^3}{2}\dfrac{z^2+z}{(z-1)^3}$	$\dfrac{T^4}{6}\dfrac{z^3+4z^2+z}{(z-1)^4}$
Halijak	$\dfrac{Tz}{z-1}$	$\dfrac{T^2z}{(z-1)^2}$	$\dfrac{T^3}{2}\dfrac{z^2+z}{(z-1)^3}$	$\dfrac{T^4}{4}\dfrac{z^3+2z^2+z}{(z-1)^4}$

11.6 ADAPTIVE RECURSIVE ALGORITHMS [11]

The filters discussed thus far in this chapter have been nonadaptive. Now we consider approaches for designing adaptive recursive filters. This will extend the discussion of adaptive nonrecursive filters in Chap. 10.7. Recall that the advantages and disadvantages of recursive and nonrecursive filters were summarized in Table 7.3.1. The table also applies to adaptive filters. Perhaps the greatest advantage of the adaptive recursive filter is its low order; its greatest disadvantage is its potential instability.

Filter stability can be guaranteed by limiting the filter weights so that the filter poles lie inside the unit circle. The performance surface of adaptive recursive filters is not generally unimodal but multimodal with local minima. Therefore it is difficult to find the global minimum and the optimum filter weights. Neither Newton or steepest descent algorithms will find global minimum on multimodal surfaces. Therefore adaptive recursive filters find limited application.

11.6.1 ADAPTIVE RECURSIVE FILTERS

An *adaptive recursive filter* or *adaptive IIR filter* is shown in Fig. 11.6.1. A time domain implementation and an FFT frequency domain implementation are shown. They consist essentially of two adaptive FIR filters (see Fig. 10.7.1) connected in a feedback configuration. One is connected in the forward path with filter weights $\mathbf{a}_n$ and the other is connected in the feedback path with filter weights $\mathbf{b}_n$. The filter input is $r(n)$, its output is $c(n)$, and its desired output is $d(n)$. The filter has internal tap outputs. In the time domain form of Fig. 11.6.1a, the tap outputs are simply delayed inputs and outputs. In the frequency domain form of Fig. 11.6.1b, the tap outputs are the FFT input spectra $R(m)$ and output spectra $C(m)$ at different bin locations. In either case, the tap outputs are multiplied by the tap weights $\mathbf{a}_n$ and $\mathbf{b}_n$ and summed to form the filter output $c(n)$. The filter gain algorithm updates the tap weights at each iteration so that the output $c(n)$ is a good approximation to the desired output $d(n)$.

To find these optimum weights, we expand the vector notation of Eq. 10.7.1 as

$$\mathbf{r}_n = \big[r(n) \ r(n-1) \ \ldots \ r(n-N+1)\big]^t$$
$$\mathbf{c}_n = \big[c(n) \ c(n-1) \ \ldots \ c(n-N+1)\big]^t$$
$$\mathbf{d}_n = \big[d(n) \ d(n-1) \ \ldots \ d(n-N+1)\big]^t$$
$$\mathbf{e}_n = \big[e(n) \ e(n-1) \ \ldots \ e(n-N+1)\big]^t = \mathbf{d}_n - \mathbf{c}_n$$
$$\mathbf{a}_n = \big[a_0(n) \ a_1(n) \ \ldots \ a_{N-1}(n)\big]^t$$
$$\mathbf{b}_n = \big[0 \ b_1(n) \ \ldots \ b_{N-1}(n)\big]^t$$
$$\mathbf{R}_{rr} = E\{\mathbf{r}_n\mathbf{r}_n^h\}, \quad \mathbf{R}_{cc} = E\{\mathbf{c}_n\mathbf{c}_n^h\}, \quad \mathbf{R}_{cr} = E\{\mathbf{c}_n\mathbf{r}_n^h\}, \quad \mathbf{R}_{dr} = E\{d(n)\mathbf{r}_n^*\}$$

$$(11.6.1)$$

Now the input $r(n)$, output $c(n)$, and impulse response $h_n(n)$ of the adaptive recursive filter are related as

$$c(n) = h_n(n) * r(n) = \mathbf{a}_n^t\mathbf{r}_n + \mathbf{b}_n^t\mathbf{c}_n, \qquad C(j\omega) = H_n(j\omega)R(j\omega)$$

$$H_n(j\omega) = \frac{\sum_{m=0}^{N-1} a_m(n)z^{-m}}{\sum_{m=0}^{N-1} b_m(n)z^{-m}}\bigg|_{z=e^{j\omega T}}, \qquad h_n(n) = \text{Eq. 11.1.6} \qquad (11.6.2)$$

The infinite number of impulse response coefficients $h_n(n)$ are related to the $2N$ filter weights $\mathbf{a}_n$ and $\mathbf{b}_n$ by Eq. 11.1.6. From Eqs. 11.1.3 and 11.6.2c, a recursive equation for implementing $H_n(z)$ is

$$c(n) = \sum_{m=0}^{N-1} a_m(n)r(n-m) + \sum_{m=1}^{N-1} b_m(n)c(n-m) = \mathbf{r}_n^t \mathbf{a}_n + \mathbf{c}_n^t \mathbf{b}_n = \mathbf{a}_n^t \mathbf{r}_n + \mathbf{b}_n^t \mathbf{c}_n$$

(11.6.3)

The error $e(n)$ between the desired output $d(n)$ and the actual output $c(n)$ at iteration n and the squared error equal

$$e(n) = d(n) - c(n) = d(n) - \mathbf{r}_n^t \mathbf{a}_n - \mathbf{c}_n^t \mathbf{b}_n = d(n) - \mathbf{a}_n^t \mathbf{r}_n - \mathbf{b}_n^t \mathbf{c}_n$$

$$|e(n)|^2 = |d(n) - \mathbf{r}_n^t \mathbf{a}_n - \mathbf{c}_n^t \mathbf{b}_n|^2$$

(11.6.4)

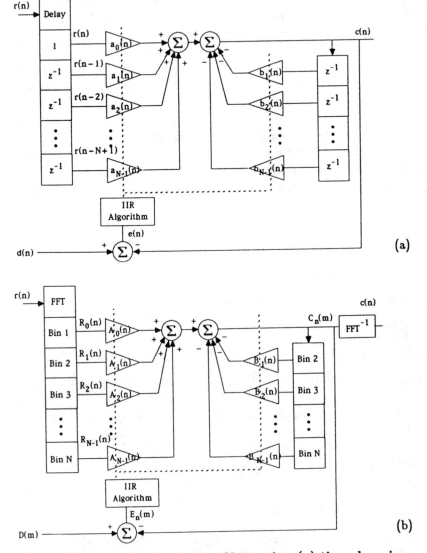

Fig. 11.6.1 Adaptive recursive filter using (a) time domain and (b) frequency domain implementations.

$d(n)$, $\mathbf{r}_n$, $\mathbf{c}_n$, and $\mathbf{w}_n$ can be complex-valued. The squared error is often used as a performance index. The expected value or mean of the squared error (MSE) is

$$E\{|e(n)|^2\} = E\{[d(n) - \mathbf{r}_n^t \mathbf{a}_n - \mathbf{c}_n^t \mathbf{b}_n][d(n) - \mathbf{r}_n^t \mathbf{a}_n - \mathbf{c}_n^t \mathbf{b}_n]^*\} \qquad (11.6.5)$$

This error involves the filter output $\mathbf{c}_n$ as well as the filter input $\mathbf{r}_n$ unlike the nonrecursive filter case in Eq. 10.7.6.

The optimum adaptive filter is the one whose filter weight vectors $\mathbf{a}_n$ and $\mathbf{b}_n$ minimizes $E\{|e(n)|^2\}$. The expected error is a nonquadratic performance surface in $2N$ variables with many local minima. The filter weights must be adjusted so that the filter operates at the global minimum (i.e. bottom) of this performance surface. Taking the partial derivatives of Eq. 11.6.5 with respect to the weights gives the gradient vector as

$$\nabla \mathbf{E}_n \equiv \left[\frac{\partial E\{|e(n)|^2\}}{\partial \mathbf{a}_n} \quad \frac{\partial E\{|e(n)|^2\}}{\partial \mathbf{b}_n} \right]^t$$

$$= \left[\frac{\partial}{\partial a_0(n)} \cdots \frac{\partial}{\partial a_{N-1}(n)} \frac{\partial}{\partial b_0(n)} \cdots \frac{\partial}{\partial b_{N-1}(n)} \right]^t E\{|e(n)|^2\} \qquad (11.6.6)$$

Evaluating Eq. 11.6.6 using Eqs. 11.6.4 and 11.6.5 shows that

$$\frac{\partial E\{|e(n)|^2\}}{\partial \mathbf{a}_n} = -E\{[d(n) - \mathbf{r}_n^t \mathbf{a}_n - \mathbf{c}_n^t \mathbf{b}_n]\mathbf{r}_n^* + \mathbf{r}_n[d(n) - \mathbf{r}_n^t \mathbf{a}_n - \mathbf{c}_n^t \mathbf{b}_n]^*\}$$

$$= (-\mathbf{R}_{dr} + \mathbf{R}_{rr}^* \mathbf{a}_n + \mathbf{R}_{cr}^* \mathbf{b}_n) + (-\mathbf{R}_{dr}^* + \mathbf{R}_{rr} \mathbf{a}_n^* + \mathbf{R}_{cr} \mathbf{b}_n^*)$$

$$\frac{\partial E\{|e(n)|^2\}}{\partial \mathbf{b}_n} = -E\{[d(n) - \mathbf{r}_n^t \mathbf{a}_n - \mathbf{c}_n^t \mathbf{b}_n]\mathbf{c}_n^* + \mathbf{c}_n[d(n) - \mathbf{r}_n^t \mathbf{a}_n - \mathbf{c}_n^t \mathbf{b}_n]^*\} \qquad (11.6.7)$$

$$= (-\mathbf{R}_{dc} + \mathbf{R}_{cr}^* \mathbf{a}_n + \mathbf{R}_{cc}^* \mathbf{b}_n) + (-\mathbf{R}_{dc}^* + \mathbf{R}_{cr} \mathbf{a}_n^* + \mathbf{R}_{cc} \mathbf{b}_n^*)$$

where the signals and filter weights can be complex-valued. For the real-valued case, combining Eqs. 11.6.6 and 11.6.7 gives the gradient vector as

$$\nabla \mathbf{E}_n = -2 \begin{bmatrix} \mathbf{R}_{dr} \\ \mathbf{R}_{dc} \end{bmatrix} + 2 \begin{bmatrix} \mathbf{R}_{rr} & \mathbf{R}_{cr} \\ \mathbf{R}_{cr} & \mathbf{R}_{cc} \end{bmatrix} \begin{bmatrix} \mathbf{a}_n \\ \mathbf{b}_n \end{bmatrix} \qquad (11.6.8)$$

Generally the autocorrelation matrix $\mathbf{R}_{rr}$ and cross-correlation matrix $\mathbf{R}_{dr}$ are known apriori. However, the autocorrelation matrix $\mathbf{R}_{cc}$ and cross-correlation matrix $\mathbf{R}_{cr}$ are not known apriori since they depend upon the filter outputs and must be estimated.

Setting $\nabla \mathbf{E}_n = \mathbf{0}$, the optimum filter weights satisfy the block matrix equation

$$\begin{bmatrix} \mathbf{R}_{dr} \\ \mathbf{R}_{dc} \end{bmatrix} = \begin{bmatrix} \mathbf{R}_{rr} & \mathbf{R}_{cr} \\ \mathbf{R}_{cr} & \mathbf{R}_{cc} \end{bmatrix} \begin{bmatrix} \mathbf{a}_{opt} \\ \mathbf{b}_{opt} \end{bmatrix} \qquad (11.6.9)$$

This yields a set of $2N$ equations in $2N$ unknowns which can be solved simultaneously for the $\mathbf{a}_{opt}$ and $\mathbf{b}_{opt}$ filter weights. Solving Eq. 11.6.9 for the optimum Wiener weight vectors $\mathbf{a}_{opt}$ and $\mathbf{b}_{opt}$ gives

$$\begin{bmatrix} \mathbf{a}_{opt} \\ \mathbf{b}_{opt} \end{bmatrix} = \begin{bmatrix} \mathbf{R}_{rr} & \mathbf{R}_{cr} \\ \mathbf{R}_{cr} & \mathbf{R}_{cc} \end{bmatrix}^{-1} \begin{bmatrix} \mathbf{R}_{dr} \\ \mathbf{R}_{dc} \end{bmatrix} \qquad (11.6.10)$$

Eq. 11.6.10 reduces to the Wiener-Hopf equation $a_{opt} = R_{rr}^{-1} R_{rd}$ as given by Eq. 10.7.8 when the filter is constrained to be nonrecursive where $b_{opt} = 0$.

The optimum weight vectors a_{opt} and b_{opt} can also be expressed in terms of any suboptimal weight vectors a_0 and b_0 as

$$\begin{bmatrix} a_{opt} \\ b_{opt} \end{bmatrix} = \begin{bmatrix} a_0 \\ b_0 \end{bmatrix} - \frac{1}{2} \begin{bmatrix} R_{rr} & R_{cr} \\ R_{cr} & R_{cc} \end{bmatrix}^{-1} \nabla E_0 \qquad (11.6.11)$$

This results by solving Eq. 11.6.8 for $[a_n, b_n]^t$ and rewriting in terms of Eq. 11.6.10. This equation allows the weights to be adjusted in one step for optimum performance. Therefore viewing Fig. 11.6.1a as a recursive estimation Wiener filter, its transfer function is given by $H_{opt}(z)$ in Eq. 11.6.2 using a_{opt} and b_{opt} from Eq. 11.6.10.

11.6.2 SEARCH METHODS AND ALGORITHMS

The *Newton* or *Newton-Raphson* method iteratively computes a root of an equation $f(w_n) \equiv \nabla E_n = 0$ as

$$w_{n+1} = w_n - f'(w_n)^{-1} f(w_n) \qquad (11.6.12)$$

when w_o is a zero or root of $f(w)$ (see Eq. 10.7.13). w_n is the nth estimate of w_o. The $f'(w_n)$ derivative equals

$$f'(w_n) = \nabla(\nabla E_n) = 2 \begin{bmatrix} R_{rr} & R_{cr} \\ R_{cr} & R_{cc} \end{bmatrix} \qquad (11.6.13)$$

from Eq. 11.6.8. Therefore from Eq. 11.6.12, the recursion equation for finding the weight vectors is (cf. Eq. 10.7.14)

$$\begin{bmatrix} a_{n+1} \\ b_{n+1} \end{bmatrix} = \begin{bmatrix} a_n \\ b_n \end{bmatrix} - \frac{1}{2} \begin{bmatrix} R_{rr} & R_{cr} \\ R_{cr} & R_{cc} \end{bmatrix}^{-1} \nabla E_n \qquad (11.6.14)$$

A more general recursion equation uses arbitrary gains u and v where (cf. Eq. 10.7.15)

$$\begin{bmatrix} a_{n+1} \\ b_{n+1} \end{bmatrix} = \begin{bmatrix} a_n \\ b_n \end{bmatrix} - \begin{bmatrix} u & 0 \\ 0 & v \end{bmatrix} \begin{bmatrix} R_{rr} & R_{cr} \\ R_{cr} & R_{cc} \end{bmatrix}^{-1} \nabla E_n \qquad (11.6.15)$$

u and v are diagonal matrices containing u and v, respectively. The a_n's are weighted by u and the b_n's are weighted by v in the constant weight case.

The *steepest descent* method moves the weights in the (opposite) direction of the gradient as (cf. Eq. 10.7.24)

$$\begin{bmatrix} a_{n+1} \\ b_{n+1} \end{bmatrix} = \begin{bmatrix} a_n \\ b_n \end{bmatrix} - \mu \nabla E_n \qquad (11.6.16)$$

μ may be a constant or a matrix. Comparing Eqs. 11.6.15 and 11.6.16, the steepest descent algorithm becomes the Newton algorithm when the gain μ equals

$$\mu = \begin{bmatrix} u & 0 \\ 0 & v \end{bmatrix} \begin{bmatrix} R_{rr} & R_{cr} \\ R_{cr} & R_{cc} \end{bmatrix}^{-1} \qquad (11.6.17)$$

Table 11.6.1 Class 1 time domain algorithms for adaptive recursive filters.

Filter Name	Transfer Function	Eq. Number
Newton		

$$\begin{bmatrix} a_{n+1} \\ b_{n+1} \end{bmatrix} = \begin{bmatrix} a_n \\ b_n \end{bmatrix} - \begin{bmatrix} u & 0 \\ 0 & v \end{bmatrix} \begin{bmatrix} R_{rr} & R_{cr} \\ R_{cr} & R_{rr} \end{bmatrix}^{-1} \nabla E_n \qquad (11.6.15)$$

Steepest Descent

$$\begin{bmatrix} a_{n+1} \\ b_{n+1} \end{bmatrix} = \begin{bmatrix} a_n \\ b_n \end{bmatrix} - \mu \nabla E_n \qquad (11.6.16)$$

LMS

$$\begin{bmatrix} a_{n+1} \\ b_{n+1} \end{bmatrix} = \begin{bmatrix} a_n \\ b_n \end{bmatrix} + 2\mu e(n) \begin{bmatrix} r_n \\ c_n \end{bmatrix} \qquad (11.6.20)$$

LMS-Newton

$$\begin{bmatrix} a_{n+1} \\ b_{n+1} \end{bmatrix} = \begin{bmatrix} a_n \\ b_n \end{bmatrix} + 2e(n) \begin{bmatrix} u & 0 \\ 0 & v \end{bmatrix} \begin{bmatrix} R_{rr} & R_{cr} \\ R_{cr} & R_{rr} \end{bmatrix}^{-1} \begin{bmatrix} r_n \\ c_n \end{bmatrix} \qquad (11.6.21)$$

The expected $E\{|e(n)|^2\}$ error of Eq. 11.6.5 can be estimated by simply using the actual $|e(n)|^2$ error of Eq. 11.6.4 where

$$e(n) = d(n) - c(n) = d(n) - a_n^t r_n - b_n^t c_n \qquad (11.6.18)$$

Then from Eq. 11.6.7 for the real-valued case,

$$\widehat{\nabla} E_n = \begin{bmatrix} \dfrac{\partial \widehat{E}\{|e(n)|^2\}}{\partial a_n} \\ \dfrac{\partial \widehat{E}\{|e(n)|^2\}}{\partial b_n} \end{bmatrix} = \begin{bmatrix} \dfrac{\partial |e(n)|^2}{\partial a_n} \\ \dfrac{\partial |e(n)|^2}{\partial b_n} \end{bmatrix} = 2e(n) \begin{bmatrix} \dfrac{\partial e(n)}{\partial a_n} \\ \dfrac{\partial e(n)}{\partial b_n} \end{bmatrix} = -2e(n) \begin{bmatrix} r_n \\ c_n \end{bmatrix} \quad (11.6.19)$$

Substituting this gradient estimate into Eq. 11.6.16 gives (cf. Eq. 10.7.32)

$$\begin{bmatrix} a_{n+1} \\ b_{n+1} \end{bmatrix} = \begin{bmatrix} a_n \\ b_n \end{bmatrix} + 2\mu e(n) \begin{bmatrix} r_n \\ c_n \end{bmatrix} \qquad (11.6.20)$$

which is called the *least-mean-square* or *LMS* algorithm.

The LMS algorithm only estimates the ∇E_n gradient vector and uses no correlation information. The weight vector seeks w_{opt} in an oscillatory manner. The Newton algorithm estimates ∇E_n and uses correlation matrices to seek $[a_{opt}, b_{opt}]^t$ in a more

monotonic manner. Convergence is faster. The *LMS-Newton* algorithm combines Eqs.
11.6.15 and 11.6.19 gives (cf. Eq. 10.7.33)

$$\begin{bmatrix} a_{n+1} \\ b_{n+1} \end{bmatrix} = \begin{bmatrix} a_n \\ b_n \end{bmatrix} + 2e(n) \begin{bmatrix} u & 0 \\ 0 & v \end{bmatrix} \begin{bmatrix} R_{rr} & R_{cr} \\ R_{cr} & R_{cc} \end{bmatrix}^{-1} \begin{bmatrix} r_n \\ c_n \end{bmatrix} \tag{11.6.21}$$

A more complicated variation of this algorithm is called the *sequential regression* or
SER algorithm. All of the Class 1 algorithms that have been discussed are summarized
in Table 11.6.1. Algorithms for other classes can easily be formed as was done with the
nonrecursive algorithms in Table 10.7.1.

11.6.3 ADAPTIVE RECURSIVE FILTER DESIGN

A design example of a second-order adaptive IIR filter $(N = 3)$ is shown in Fig. 11.6.2a.
The filter input r_n is uncorrelated white noise. The filter gain $H_n(z)$ equals

$$H_n(z) = \frac{a_0(n)}{1 - b_1(n)z^{-1} - b_2(n)z^{-2}} \tag{11.6.22}$$

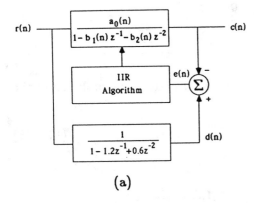

(a)

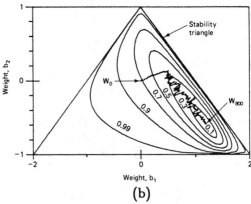

(b)

Fig. 11.6.2 (a) Second-order IIR filter
with white noise input, weight contours
for (b) LMS and (c) SER algorithms.
(From B. Widrow and S.D. Stearns,
Adaptive Signal Processing, ©1985,
p. 161. Reprinted by permission
of Prentice-Hall, Inc., Englewood
Cliffs, New Jersey.)

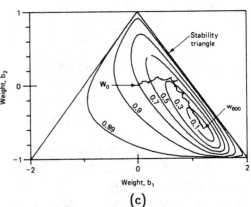

(c)

The desired output d_n is generated by the same white noise driving another filter with

$$H_{opt}(z) = \frac{1}{1 - 1.2z^{-1} + 0.6z^{-2}} \qquad (11.6.23)$$

The adaptive IIR filter will use the LMS algorithm where

$$\begin{bmatrix} \mathbf{a}_{n+1} \\ \mathbf{b}_{n+1} \end{bmatrix} = \begin{bmatrix} \mathbf{a}_n \\ \mathbf{b}_n \end{bmatrix} + 2e(n) \begin{bmatrix} \mathbf{u} & 0 \\ 0 & \mathbf{v} \end{bmatrix} \begin{bmatrix} \mathbf{r}_n \\ \mathbf{c}_n \end{bmatrix} \qquad (11.6.24)$$

$$c(n) = \mathbf{a}_n^t \mathbf{r}_n + \mathbf{b}_n^t \mathbf{c}_n$$

with starting values of $\mathbf{a}_0 = [0,0,0]^t$, $\mathbf{b}_0 = [0,0,0]^t$, $\mathbf{u} = 0.5 \, \text{diag}[1,1,1]$, and $\mathbf{v} = \text{diag}[0, 0.005, 0.0025]$. The initial output vector $\mathbf{c}_{-1} = [0,0,0]^t$ and input $\mathbf{r}_n$ is a white noise sequence. The optimum filter weights must converge to $\mathbf{a}_{opt} = [1,0,0]^t$ and $\mathbf{b}_{opt} = [0, 1.2, -0.6]^t$ comparing Eqs. 11.6.22 and 11.6.23.

The typical response of the LMS algorithm is shown in Fig. 11.6.2b for 800 iterations. The LMS contour follows a noisy steepest descent path. A SER algorithm is shown in Fig. 11.6.2c for 600 iterations. To insure IIR filter stability, $|b_2(n)| \le 1$. Also $b_2(n) < 1 + b_1(n)$ for $-2 < b_1(n) \le 0$ and $b_2(n) < 1 - b_1(n)$ for $0 \le b_1(n) < 2$. These triangular boundaries are shown in Figs. 11.6.2b–c.

11.7 KALMAN FILTERS [12]

Adaptive FIR filters were discussed in Chap. 10.7 and are very popular. The adaptive IIR filters discussed in Chap. 11.6 are not as popular because they are potentially unstable and their performance surface is not parabolic and has many local minima. The search algorithms cannot guarantee convergence to the global minimum. However, there is a type of adaptive IIR filter which does converge to a global minimum called the *Kalman filter*.

The Kalman filter can process multiple inputs and multiple outputs (MIMO) having nonstationary statistics. It can be formulated in both the time and frequency domains. It is robust and functions well even when the apriori assumptions used in its formulation are violated. Since it is a type of recursive filter, it requires little data storage. It is optimal in the MSE sense like the adaptive FIR and IIR filters. A variety of algorithms are available which improve the numerical stability of the filter. The theory can be extended to many nonlinear state and observation models.[13]

The Kalman filter is based on two additional equations (i.e., the state equation and the Kalman gain) to the one used in the adaptive FIR filter (i.e., the measurement equation). The Kalman filter is formulated assuming

$$x_{n+1} = A_n x_n + w_n \qquad (11.7.1)$$

$$y_n = C_n x_n + v_n \qquad (11.7.2)$$

$$w_n = K_n v_n \qquad (11.7.3)$$

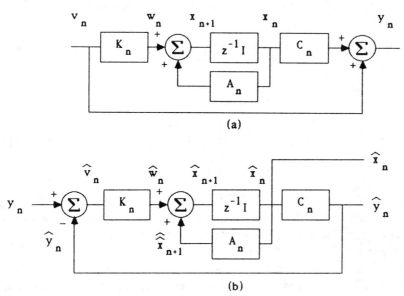

(a)

(b)

Fig. 11.7.1 (a) Time domain model of state generation and
signal observation processes, and (b) Kalman filter in time
domain.

to optimally estimate the state x_n. Eq. 11.7.1 is the *state equation*. It generates the
next state x_{n+1} from the present state x_n and process noise w_n. Eq. 11.7.2 is the
observation or *measurement equation*. It generates a noisy signal y_n from the state x_n
and measurement noise w_n. The noises w_n and v_n are in general independent. For
good Kalman filter operation however, they must not be independent but correlated as
required by Eq. 11.7.3. K_n is the *Kalman gain* which is time-varying. The model which
incorporates these three equations is shown in Fig. 11.7.1a.

The Kalman filter for optimally estimating the state x_n and the noise-free signal
y_n is shown in Fig. 11.7.1b. Comparing it with Figs. 10.7.1 and 11.6.1, we see that
the Kalman filter is IIR due to the feedback unlike the FIR filter which has no explicit
feedback. The expected state equals $E\{x_{n+1}\} = A_n E\{x_n\}$ from Eq. 11.7.1 since the
noise w_n is assumed to have zero mean. The expected measurement equals $E\{y_n\} =
C_n E\{x_n\}$ from Eq. 11.7.2 since noise v_n is also assumed to have zero mean. Therefore
the noises can be estimated as

$$\widehat{w}_n = \widehat{x}_{n+1} - \widehat{\widehat{x}}_{n+1}$$
$$\widehat{v}_n = y_n - \widehat{y}_n \qquad\qquad (11.7.4)$$
$$\widehat{w}_n = K_n \widehat{v}_n$$

where $\widehat{\widehat{x}}_{n+1} \equiv A_n \widehat{x}_n$, $\widehat{x}_n = z^{-1} I\, \widehat{x}_{n+1}$, and $\widehat{y}_n \equiv C_n \widehat{x}_n$. These six equations are implemented in Fig. 11.7.1b. Solving them simultaneously gives several recursion relations
including

$$\widehat{x}_{n+1} = A_n \widehat{x}_n + \widehat{w}_n = A_n \widehat{x}_n + K_n \widehat{v}_n$$
$$= A_n \widehat{x}_n + K_n(y_n - C_n \widehat{x}_n) \tag{11.7.5}$$
$$= (A_n - K_n C_n)\widehat{x}_n + K_n y_n$$

It is assumed that matrices A_n and C_n are known apriori but that the Kalman gain K_n is unknown and must be estimated.

We include Eq. 11.7.3 as part of the encoding model in Fig. 11.7.1a although other authors do not. Eq. 11.7.3 is necessary for the decoding model, i.e. the Kalman filter itself, in Fig. 11.7.1b. The decoding model has only one degree of freedom (i.e., measurement data y_n) because it is described by six equations in seven unknowns. Without Eq. 11.7.3, it would have two degrees of freedom and measurement noise w_n would also need to be input. This dual-input requirement is used by the adaptive noise cancellors of Chap. 9.3.

Using the MSE criterion, the error equals $\epsilon = E\{(x_{n+1} - \widehat{x}_{n+1})(x_{n+1} - \widehat{x}_{n+1})^h\}$. Then we minimize ϵ by choosing K_n such that

$$\frac{\partial \epsilon}{\partial K_n} = E\{2(x_{n+1} - \widehat{x}_{n+1})\frac{\partial \widehat{x}_{n+1}^h}{\partial K_n}\}$$
$$= E\{2[x_{n+1} - A_n \widehat{x}_n - K_n \widehat{v}_n]\widehat{v}_n^h\} = 0 \tag{11.7.6}$$

This results using the second form of Eq. 11.7.5. Rewriting Eq. 11.7.6 using correlation functions gives

$$R_{x\widehat{v}}(n+1, n) - A_n R_{\widehat{x}\widehat{v}}(n, n) = K_n R_{\widehat{v}\widehat{v}}(n, n) \tag{11.7.7}$$

The $R_{\widehat{x}\widehat{v}}$ term is zero since the estimated state and estimated noise are orthogonal. Solving for the Kalman gain K_n gives

$$K_n = R_{x\widehat{v}}(n+1, n) R_{\widehat{v}\widehat{v}}(n, n)^{-1} \tag{11.7.8}$$

The state autocorrelation $R_{xx}(n, n)$, measurement noise autocorrelation $R_{vv}(n, n)$, and state-noise cross-correlation $R_{xv}(n, n)$ are often known apriori. It can be shown that the correlations involving estimates are related to the known correlations as

$$R_{x\widehat{v}}(n+1, n) = A_n R_{xx}(n, n)C_n^t$$
$$R_{\widehat{v}\widehat{v}}(n, n) = C_n R_{xx}(n, n)C_n^t + R_{vv}(n, n) \tag{11.7.9}$$

Combining Eqs. 11.7.8 and 11.7.9, the Kalman gain equals

$$K_n = [A_n R_{xx}(n, n)C_n^t] [C_n R_{xx}(n, n)C_n^t + R_{vv}(n, n)]^{-1} \tag{11.7.10}$$

The optimum overall gain of the Kalman filter after convergence equals

$$\frac{\widehat{Y}}{Y} = [K_{opt}z^{-1}I(z^{-1}I - A)^{-1}C] [I - K_{opt}z^{-1}I(z^{-1}I - A)^{-1}C]^{-1} \tag{11.7.11}$$

When the measurement noise v_n is high (i.e., $R_{vv} \gg CR_{xx}C^t$), the data y_n is unreliable and the Kalman gain is small. In this case, the x_n estimate is based solely upon the internal state estimator of the Kalman filter since Eq. 11.7.11 reduces to

$$\frac{\hat{Y}}{Y} = K_{opt} z^{-1} I (z^{-1}I - A)^{-1} C \qquad (11.7.12)$$

When the measurement noise v_n is low (i.e., $R_{vv} \ll C_n R_{xx} C_n^t$), the data y_n is reliable and the Kalman gain is large. In this case, the x_n estimate is based solely upon the data and ignores the internal state estimator since Eq. 11.7.11 then reduces to

$$\frac{\hat{Y}}{Y} = 1 \qquad (11.7.13)$$

The Kalman filter reduces to the adaptive FIR filter when $A_n = I$ and $C_n = \Phi_{n-1}$. Setting $A_n = I$ causes the state generator to output a unit step. Φ_{n-1} is the state transition matrix needed to produce a desired output signal y_n from $x_n = U_{-1}(n)$.

For recursion purposes, several equations are used such as:

$$\hat{x}_{n+1} = A_n \hat{x}_n + K_n(y_n - C_n \hat{x}_n)$$

$$K_n = \left[A_n R_{xx}(n,n) C_n^t \right] \left[C_n R_{xx}(n,n) C_n^t + R_{vv}(n,n) \right]^{-1}$$

$$\hat{\hat{R}}_{xx}(n,n) = \hat{R}_{xx}(n,n) - \hat{R}_{xx}(n,n)C_n^t[C_n\hat{R}_{xx}(n,n)C_n^t + R_{vv}]^{-1}C_n\hat{R}_{xx}(n,n)$$

$$\hat{R}_{xx}(n+1,n+1) = A_n \hat{\hat{R}}_{xx}(n,n)A_n^t + Q_n$$

$$(11.7.14)$$

The trajectory followed by the Kalman gain to seek the global minimum MSE error is not explicitly specified in Eq. 11.7.14. The Kalman filter seeks its own trajectory unlike the FIR and IIR adaptive filters where the trajectory is explicitly defined. Since the Kalman filter and adaptive FIR filter are both derived using the same MSE input-output criterion, they must both converge to the same optimum overall gain although their formulations are different. Since the Kalman filter includes feedback, convergence may be faster but stability must be controlled by gain K_n.

Now let us modify the Kalman filter using a new viewpoint. Suppose that the time domain state model of Fig. 11.7.1a is changed into the frequency domain state model by taking the T transform of Eqs. 11.7.1–11.7.3 as

$$\begin{array}{ccc}
Tx_{n+1} = TA_n x_n + Tw_n & \longrightarrow & X_{n+1} = A_n' X_n + W_n \\
Ty_n = TC_n x_n + Tv_n & \longrightarrow & Y_n = C_n' X_n + V_n \\
Tw_n = K_n Tv_n & \longrightarrow & W_n = K_n V_n
\end{array} \qquad (11.7.15)$$

where $A_n' = TA_nT^{-1}$ and $C_n' = TC_nT^{-1}$. The primed matrices are the frequency domain equivalents of the unprimed time domain matrices. They satisfy the general relation that $H_T = Th_T T^{-1}$ from Eq. 8.1.4. The frequency domain state model is shown in Fig. 11.7.2a where the upper case variables are signal spectra rather than signals. For example, the Y_n output is the spectrum of the measurement rather that the measurement y_n itself.

The Kalman filter which estimates this spectrum $\hat{Y}_n$ is shown in Fig. 11.7.2b. Since measurement data y_n is being processed, it is converted to measurement spectrum Y_n

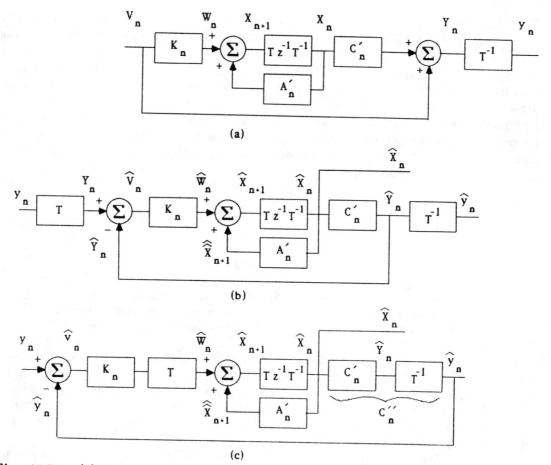

Fig. 11.7.2 (a) Frequency domain model of state generation and signal spectrum observation processes, (b) Kalman filter in frequency domain, and (c) alternative Kalman filter in frequency domain.

by the T transform block at the input. The Kalman filter outputs are estimated state spectrum $\widehat{X}_n$ and estimated signal spectrum $\widehat{Y}_n$. Passing these spectra through inverse transform blocks T^{-1} yields the time domain outputs shown in Fig. 11.7.1a.

Fig. 11.7.2c is a rearranged form of Fig. 11.7.2b and has an interesting interpretation. The T input block is moved to the summer output. The $\widehat{Y}_n$ feedback path is moved to the output of the T^{-1} block. Then the C'_n and T^{-1} blocks are combined into a single C''_n block where $C''_n = T^{-1}C'_n = C_n T^{-1}$. The output coupling matrix C''_n is the matrix of the generalized basis functions used to describe the measurement. Each column m of C''_n corresponds to one basis function f_m. The $\widehat{X}_n$ state spectrum estimates are then the generalized transform coefficients. Mathematically, the output equals

$$\text{row}_1(C''_n) = \text{row}_1(C_n T^{-1}) = [f_1(n) \ f_2(n) \ \dots \ f_N(n)]$$

$$\widehat{y}(n) = \text{row}_1(T^{-1}\widehat{Y}_n) = \text{row}_1(C''_n)\widehat{X}_n \tag{11.7.16}$$

$$= \widehat{X}_1(n)f_1(n) + \widehat{X}_2(n)f_2(n) + \dots + \widehat{X}_N(n)f_N(n)$$

Since Eqs. 11.7.15 and 11.7.16 are updated with every new measurement sample, this is a sequential or running transform system. It may also be viewed as block processing with single-sample overlap. Any T transform such as Fourier, Walsh, Haar, or slant can be used for the frequency domain Kalman filter. Proper choice of the transforms (i.e., match state and process noise to transform) may improve the convergence speed of the algorithm. This frequency domain approach allows generalized spectrum analyzers to be implemented using Kalman filters. Other unique applications are possible. This brief introduction to Kalman filters is intended only to show their relation to the matrix, FIR, and IIR filters discussed previously. For design details on Kalman filters, a variety of reference books are available.

PROBLEMS

Design these filters using each of the following methods (N = 8):
(1) impulse-invariant transform, (2) modified impulse-invariant transform,
(3) matched z-transform, (4) s-to-z transforms using H_{dmn} as specified.

11.1   Several amplitude equalizer transfer functions are: (a) $[s + 2\pi(3000 \text{ Hz})] / [s + 2\pi(70 \text{ Hz})]$, (b) $[s + 2\pi(10 \text{ KHz})]/[s + 2\pi(1 \text{ KHz})]^3$, (c) $14.3[s + 2\pi(200 \text{ Hz})]/[s + 2\pi(3 \text{ KHz})]$, (d) $2\pi(2.1 \text{ KHz})/[s + 2\pi(2.1 \text{ KHz})]$. Design the filters using $f_s = 10$ KHz and the H_{011} algorithm.

11.2   A wide-band lowpass voice filter is required. Design it to be 3rd order Chebyshev having 3 dB in-band ripple, a 3 dB bandwidth of 20 KHz, and a maximum low-frequency gain of unity. Use $f_s = 250$ KHz and the H_{011} algorithm.

11.3   A wide-band lowpass voice filter is required. Design it to be 4th order Chebyshev having 3 dB in-band ripple, a 3 dB bandwidth of 20 KHz, and a maximum low-frequency gain of unity. Use $f_s = 500$ KHz and the H_{LDI} algorithm.

11.4   A 3.2 KHz band-limiting lowpass voice filter is required. Design it to be 2nd order Butterworth having a 3 dB bandwidth of 3.2 KHz and a dc gain of unity. Use $f_s = 32$ KHz and the H_{011} algorithm.

11.5   A 1000 Bps noise suppression (i.e., data smoothing) lowpass filter is required. Design it to be 4th order Bessel having a 3 dB bandwidth of 1 KHz and unity dc gain. Use $f_s = 10$ KHz and the H_{011} algorithm.

11.6   A 3.5 Hz delta wave EEG lowpass filter is required. Design it to be 4th order Chebyshev having 0.5 dB in-band ripple and a 3 dB bandwidth of 3.5 Hz. Use $f_s = 20$ Hz and the H_{011} algorithm.

11.7   A 1 Hz phase detection filter is required. Design the filter to be 4th order Chebyshev having 1 dB in-band ripple and a 3 dB bandwidth of 1 Hz. Use $f_s = 10$ Hz and the H_{011} algorithm.

11.8   A 3.2 KHz anti-aliasing lowpass filter is required. Design it to be 3rd order Butterworth having a 3 dB bandwidth of 3.2 KHz. Use $f_s = 64$ KHz and the H_{011} algorithm.

11.9   A 941 Hz band-splitting lowpass filter is required. Design it to be 3rd order Butterworth having a 3 dB bandwidth of 941 Hz and a dc gain of 10. Use $f_s = 10$ KHz and the H_{LDI} algorithm.

11.10   A 2 KHz band-splitting lowpass/highpass filter is required. Design the lowpass filter to be 3rd order Butterworth having a 3 dB bandwidth of 2 KHz and unity dc gain. Use f_s = 10 KHz and the H_{011} algorithm.

11.11   A 941 Hz band-splitting lowpass filter is required. Design the lowpass to be 3rd order Butterworth having a 3 dB bandwidth of 2 KHz and unity dc gain. Use f_s = 20 KHz and the H_{011} algorithm.

11.12   Design a 5th order elliptic lowpass filter having the following parameters: M_p = 0.28 dB, M_r = 50 dB, H_o = 20 dB, f_p = 1000 Hz, and f_r = 1556 Hz. Use f_s = 10 KHz and the H_{011} algorithm.

11.13   Design a 4th order Bessel lowpass filter having a 3 dB bandwidth of 1 KHz and unity dc gain. Use f_s = 20 KHz and the H_{011} algorithm.

11.14   Design a tunable 4th order Butterworth lowpass filter having a 3 dB bandwidth that lies between 10–50 KHz. Use f_s = 1 MHz and the H_{011} algorithm.

11.15   Design a tunable 3rd order Bessel lowpass filter having a 3 dB bandwidth of 100 Hz and a 10:1 tuning range. Use f_s = 25 KHz and the H_{LDI} algorithm.

11.16   Design a 4th order Chebyshev highpass filter having 3 dB passband ripple, a 3 dB corner frequency of 1 KHz, and a maximum passband gain of unity. Use f_s = 10 KHz and the H_{011} algorithm.

11.17   Design a 4th order synchronously-tuned highpass filter having a 3 dB corner frequency of 1 KHz and unity hf gain. Use f_s = 20 KHz and the H_{011} algorithm.

11.18   A Dolby B noise reduction utilizes two highpass filters. Design a 4th order Butterworth highpass filters having a 3 dB bandwidth of 1.5 KHz. Use f_s = 25 KHz and the H_{LDI} algorithm.

11.19   A F1A-weighting curve used in telephone systems using a lowpass and a highpass filter. The lowpass filter is 5th order Butterworth having unity dc gain and a 3 dB bandwidth of 2 KHz. The highpass filter is 3rd order Butterworth having unity hf gain and a 3 dB bandwidth of 620 Hz. Design the 3rd order Butterworth highpass filter. Use f_s = 6 KHz and the H_{011} algorithm.

11.20   Design a tunable 3rd order Chebyshev highpass filter to have a 3 dB bandwidth that lies between 2.5–12.5 KHz. Gain is arbitrary. Use f_s = 500 KHz and the H_{011} algorithm.

11.21   A C-weighting curve used in telephone systems can be realized using a 3rd order Butterworth highpass filter cascaded with a 4th order Chebyshev lowpass filter. The highpass filter has f_{3dB} = 732 Hz and H_o = 1. Use f_s = 9.6 KHz and the H_{011} algorithm.

11.22   A low-transient highpass filter has a normalized transfer function of

$$H_{CHP}(s) = \frac{s(s^2 + 0.591s + 1.177)}{(s + 0.593)(s^2 + 0.295s + 1.002)}$$

Design this filter using f_s = 10 Hz and the H_{011} algorithm.

11.23   A 2nd order Butterworth bandpass filter having a center frequency of 100 Hz, a 3 dB bandwidth of 10 Hz, and a mid-band gain of 10 must be designed. Its block diagram consists of two 1st order BPF stages having parameters (H_o, f_o, Q) which equal (4.44, 96.3 Hz, 14.1) and (4.44, 103.7 Hz, 14.1). Design this bandpass filter using f_s = 1 KHz and the H_{011} algorithm.

11.24   Design a 2nd order Butterworth bandpass filter having a 3 dB frequency range of 300–3200 Hz. Use f_s = 48 KHz and the H_{011} algorithm.

11.25   Bandpass filters used by the telegraph system have 120 Hz or 170 Hz tone spacings and are constant bandwidth. Design two tunable 1st order bandpass filters to have a 3 dB bandwidth of 64 Hz and center frequencies of 420 Hz and 3300 Hz. Gain is arbitrary. Use $f_s = 48$ KHz and the H_{LDI} algorithm.

11.26   Design a 4th order Butterworth lowpass discriminator with a cut-off frequency of 500 Hz and a gain of 0.00522 volt/volt/Hz. Its block diagram realization consists of a 2nd order LPF stage having parameters (H_o, f_o, ς) which equal $(-1, 500$ Hz, $0.924)$ cascaded with a 1st order BPF stage having parameters (H_o, f_o, ς) of $(-3.41, 500$ Hz, $0.383)$. Use $f_s = 10$ KHz and the H_{011} algorithm.

11.27   ULF direction finder systems are used in mine rescue operations. A keyed-CW transmitter connected to a large collapsible loop antenna radiates a frequency in the 900–3100 Hz range underground at each mine site. 34 frequencies are used. Above ground, a portable receiver is used to detect the location from which the desired frequency is being transmitted. Assume that the receiver uses a 1st order bandpass filter (which is insufficient). Design a bandpass filter with 2 Hz bandwidth which tunes over this frequency range. Gain is arbitrary. Use $f_s = 64$ KHz and the H_{011} algorithm.

11.28   FDM channel bands use bandpass filters having 4 KHz bandwidths with center frequencies in the range between 60-108 KHz. Design a 1st order tunable bandpass filter which tunes over this frequency range. Gain is arbitrary. Use $f_s = 500$ KHz and the H_{LDI} algorithm.

11.29   An acoustical data coupler uses 1170 Hz and 2125 Hz bandpass filters. The Transmit filters can be 2nd order Chebyshev having 0.5 dB ripple. Their block diagrams consist of two 1st order BPF stages having having parameters (H_o, f_o, Q) which equal $(1.41, 1.05$ or 2.00 KHz, 8.3 or 14.9) and $(1.41, 1.29$ or 2.24 KHz, 8.3 or 14.9). Design these 1170 Hz and 2125 Hz Transmit filters. Use $f_s = 25$ KHz and the H_{011} algorithm.

11.30   The Lenkurt Type 45 telephone signalling system uses a 3400/3550 Hz bandpass filter. Design it to be 2nd order Butterworth bandpass having $f_o = 3480$ Hz and $Q = 11.6$. Its block diagram consists of two 1st order BPF stages having parameters (H_o, f_o, Q) which equal $(1.41, 3.33$ KHz, 16.4) and $(1.41, 3.63$ KHz, 16.4). Use $f_s = 20$ KHz and the H_{011} algorithm.

11.31   Design a tunable 1st order bandpass filter having a variable Q and constant ω_o. Assume that its center frequency is 1 KHz and its Q must be tuned from 1 to 11. Gain is arbitrary. Use $f_s = 10$ KHz and the H_{011} algorithm.

11.32   Design a tunable 1st order bandpass filter having a center frequency of 1 KHz, a 2:1 tuning range, and $Q = 5$. Gain is arbitrary. Use $f_s = 10$ KHz and the H_{011} algorithm.

11.33   Design a 1st order Butterworth bandstop filter having a 200 Hz center frequency, $Q = 0.25$, and $H_o = 1$. Use $f_s = 4$ KHz and the H_{011} algorithm.

11.34   Design a 1st order Butterworth bandstop filter having a 200 Hz center frequency, $Q = 0.25$, and $H_o = 1$. Use $f_s = 8$ KHz and the H_{LDI} algorithm.

11.35   The telephone company uses 1010 Hz and 2805 Hz tones to test transmission channels. Design two 2nd order Butterworth bandstop filters having center frequencies of 1010 Hz and 2805 Hz, $Q = 10$, and $H_o = 1$. Use $f_s = 25$ KHz and the H_{011} algorithm.

11.36   Design a 2nd order Butterworth bandstop filter having a 60 Hz center frequency, $Q = 10$, and $H_o = 1$. Use $f_s = 1.2$ KHz and the H_{011} algorithm.

11.37   A 2600 Hz bandstop filter is required in single-frequency (SF) signalling systems used by the telephone company. A 2nd order Butterworth bandstop filter having a 3 dB bandwidth of 60 Hz shall be used. Its block diagram consists of two 1st order BPF stages having parameters $(H_o, f_{op}, Q_p, f_{oz})$ which equal $(1, 2577$ Hz, 61.3, 2600 Hz) and $(1, 2623$ Hz, 61.3, 2600 Hz). Use $f_s = 25$ KHz and the H_{011} algorithm.

11.38   Design a 1st order Butterworth bandstop filter having a 21.6 KHz center frequency, $Q = 27$, and $H_o = 1$. Use $f_s = 64$ KHz and the H_{LDI} algorithm.

11.39   Hi-fi equalizers use a bank of nine filters centered from 50 Hz to 12.8 KHz (with octave spacing) to cover the audio spectrum from 20 Hz to 20 KHz. These equalizers allow the audiophile to compensate for room acoustics or speaker deficiencies, or to please his personal tastes. The equalizer provides boost and cut limits of ± 12 dB. Design a biquadratic filter centered at 50 Hz with these boost and cut limits. Use $f_s = 1$ KHz and the H_{011} algorithm.

11.40   A low-transient 3rd order Papoulis highpass filter has a normalized transfer function of

$$H_{CHP}(s) = \frac{s(s^2 + 1.310s + 1.359)}{(s + 0.621)(s^2 + 0.691s + 0.931)}$$

Design this filter using $f_s = 10$ Hz and the H_{011} algorithm.

11.41   Design a tunable 1st order Butterworth bandstop filter having 1 Hz–30 KHz center frequency, $Q = 100$, and $H_o = 1$. Use $f_s = 1$ KHz and the H_{011} algorithm.

11.42   Design a 2nd order Butterworth bandstop filter having a center frequency of 1 KHz, $Q = 10$, and a passband gain of 10. Use $f_s = 10$ KHz and the H_{011} algorithm.

11.43   Design a 1st order Butterworth bandstop filter to have a $Q = 10$ and a center frequency of (a) 120 Hz and (b) 800 Hz. Use $f_s = 10$ KHz and the H_{011} algorithm.

11.44   (a) Design a 1st order allpass filter with parameters $(H_o, f_o) = (1, 318$ Hz$)$. (b) Design a 2nd order allpass filter with parameters $(H_o, f_o, \varsigma) = (1, 540$ Hz$, 0.866)$. Use $f_s = 10$ KHz and the H_{011} algorithm.

11.45   A 2nd order delay equalizer can be cascaded with a 5th order Butterworth lowpass filter to reduce its in-band delay variation. This 2nd order allpass filter has parameters $(H_o, f_o, \varsigma) = (1, 730$ Hz$, 0.866)$. Use $f_s = 15$ KHz and the H_{011} algorithm.

11.46   Design a 2600 Hz oscillator for the single-frequency in-band signalling system discussed in Problem 11.37. Use $f_s = 48$ KHz and the H_{LDI} algorithm.

11.47   The Lenkurt Type 45 out-of-band signalling system utilizes a signalling oscillator. This oscillator generates tones at either 3400 Hz or 3550 Hz. Design this tunable oscillator. Use $f_s = 100$ KHz and the H_{011} algorithm.

11.48   An electronic watch consists of a 32.786 KHz oscillator, counter, decoder/display, and battery. Design a 32.786 KHz oscillator. Use $f_s = 250$ KHz and the H_{011} algorithm.

REFERENCES

1. Jong, M.T., *Methods of Discrete Signal and Systems Analysis*, Chap. 4, McGraw-Hill, NY, 1982.

2. Oppenheim, A.V. and R.W. Schafer, *Digital Signal Processing*, Chap. 5.2, Prentice-Hall, NJ, 1975.

3. Antoniou, A., *Digital Filters: Analysis and Design*, Chap. 7.4, McGraw-Hill, NY, 1979.

4. Ref. 3, Chap. 7.5.

5. Hamming, R.W., *Numerical Methods for Scientists and Engineers*, 2nd ed., Chap. 20.3, McGraw Hill, NY, 1973.

6. Wylie, C.R., *Advanced Engineering Mathematics*, 4th ed., Chap. 4.3, McGraw-Hill, NY, 1975.

   James, M.L., G.M. Smith, and J.C. Wolford, *Applied Numerical Methods for Digital Computations*, 2nd ed., Chap. 5.4, Harper & Row, NY, 1977.

7. Smith, J.M., *Mathematical Modeling and Digital Simulations for Engineers and Scientists*, Chap. 7.2, Wiley, NY, 1977.

8. Hamming, R.W., *Digital Filters*, 2nd ed., Chaps. 3—10, Prentice-Hall, NJ, 1983.

9. Ref. 7, Chap. 8.

10. Chang, Y.K. and P.V. Rao, "A fast and efficient digital simulation technique for control systems," Comput. & Elect. Engr., Pergamon Press, pp. 11—18, Jan. 1983.

11. Widrow, B. and S.D. Stearns, *Adaptive Signal Processing*, Chap. 8, Prentice-Hall, NJ, 1985.

12. Lindquist, C.S., "Use of signal processing algorithms in microprocessor-based control," Proc. Conf. Decision and Control, 1986.

13. Brogan, W.L. "Kalman filters – What? Why? and How?," Proc. Midwest Conf. on Circuits and Systems, 1986.

12 DIGITAL FILTER REALIZATION

> *And the Spirit and the bride (Church) say, Come.*
> *And let him that heareth say, Come.*
> *And let him that is athirst come.*
> *And whosoever will, let him take the water of life freely ...*
> *The grace of our Lord Jesus Christ be with you all. Amen.*
>
> Revelation 22:17, 21

The $H(z)$ transfer functions of recursive and nonrecursive filters were determined in the previous two chapters. Now the transfer functions will be implemented in hardware and software. This chapter will present a variety of realization forms. Their advantages and disadvantages are discussed. An elliptic filter will be implemented using the different forms to illustrate the computational complexity involved.

12.1 CANONICAL STRUCTURES

The general filter transfer function $H(z)$ equals

$$H(z) = \frac{Y(z)}{X(z)} = \frac{\sum_{k=0}^{M} a_k z^{-k}}{\sum_{k=0}^{N} b_k z^{-k}}, \qquad b_0 = 1 \qquad (12.1.1)$$

from Eq. 11.1.1. $H(z)$ is realized using delay elements (represented by z^{-1}), multipliers (the a_k's and b_k's), and adders (the summations). A block diagram realization of $H(z)$ is called a *filter structure*. $H(z)$ can be implemented by an infinite variety of structures. In general, each filter structure has different complexities, coefficient sensitivities, dynamic ranges, and other important characteristics. Based upon $H(z)$, the filter form or structure must be properly selected for good filter performance as will be discussed in Chap. 12.10.

One convenient filter classification method is based upon canonical structures. A *canonical* filter structure is one where the delay elements exactly equals the order of the transfer function.[1] The order of $H(z)$ is equal to the degree of the denominator (number of finite poles N) or the numerator (number of finite zeros M) depending upon which polynomial is larger (i.e, $\max(N, M)$). Canonical structures minimize the number of delay elements or memory locations needed for filter realization. A noncanonical structure uses more delay elements and more memory.

12.2 MULTIPLE FEEDBACK REALIZATIONS

One set of filter realizations have a *multiple feedback form*. The structure of each filter is determined by the proper factoring of the gain $H(z)$ in Eq. 12.1.1.[2] For mathematical convenience and no loss of generality, we set $N = M$ in the gain equation.

Factoring $H(z)$ into the product

$$H(z) = \sum_{k=0}^{N} a_k z^{-k} \; \frac{1}{1 + \sum_{k=1}^{N} b_k z^{-k}} = \frac{Y(z)}{W(z)} \frac{W(z)}{X(z)} \tag{12.2.1}$$

yields two equations describing the input $X(z)$, output $Y(z)$, and $W(z)$ as

$$W(z) = X(z) - \sum_{k=1}^{N} b_k \left(z^{-k} W(z) \right)$$
$$Y(z) = \sum_{k=0}^{N} a_k \left(z^{-k} W(z) \right) \tag{12.2.2}$$

Taking the inverse z-transform of Eq. 12.2.2 results in the pair of time domain equations

$$w(n) = x(n) - \sum_{k=1}^{N} b_k w(n-k)$$
$$y(n) = \sum_{k=0}^{N} a_k w(n-k) \tag{12.2.3}$$

The block diagram of this filter is shown in Fig. 12.2.1. Eq. 12.2.3a defines the nature

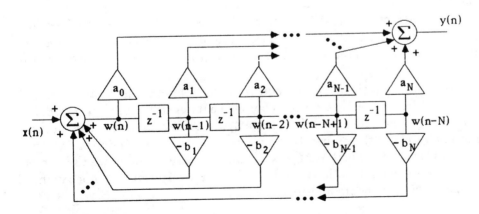

Fig. 12.2.1 1F multiple feedback filter realization.

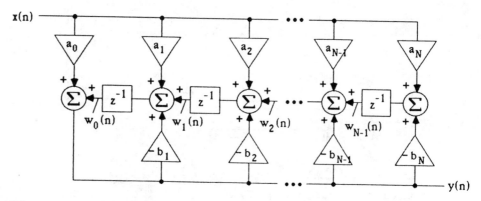

Fig. 12.2.2 2F multiple feedback filter realization.

of the feedback paths while Eq. 12.2.3b defines the nature of the feedforward paths. For brevity, we call this the first multiple feedback (1F) structure. This is a canonical filter since it requires only N delay elements.

A second multiple feedback (2F) structure results by taking the transpose of the 1F structure. This requires interchanging input and output nodes and reversing all branches so signal flow is reversed. Summing nodes become signal junction nodes and vice versa. The result is shown in Fig. 12.2.2. This structure is described by using Eq. 12.1.1 directly as

$$
\left.
\begin{aligned}
Y(z) &= \sum_{k=0}^{N} z^{-k}\Big(a_k X(z) - b_k Y(z)\Big) \\
y(n) &= \sum_{k=0}^{N} \Big(a_k x(n-k) - b_k y(n-k)\Big)
\end{aligned}
\right\}
\quad \text{reassign } b_0 = 0
\qquad (12.2.4)
$$

This is a canonical filter since only N delay elements are used.

A third multiple feedback (3F) structure results by moving the z^{-k} terms inside the parentheses of Eq. 12.2.4 as

$$
\left.
\begin{aligned}
Y(z) &= \sum_{k=0}^{N} \Big(a_k z^{-k} X(z) - b_k z^{-k} Y(z)\Big) \\
y(n) &= \sum_{k=0}^{N} \Big(a_k x(n-k) - b_k y(n-k)\Big)
\end{aligned}
\right\}
\quad \text{reassign } b_0 = 0
\qquad (12.2.5)
$$

This filter is shown in block diagram form in Fig. 12.2.3. Since it requires $2N$ delay elements, it is noncanonical.

A fourth multiple feedback (4F) structure is shown in Fig. 12.2.4. It is the transpose of the 3F structure. This is equivalent to rewriting Eq. 12.2.2 in a slightly different form as

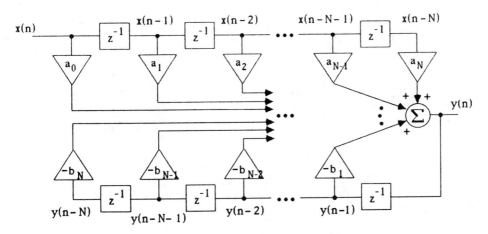

Fig. 12.2.3 3F multiple feedback filter realization.

$$W(z) = X(z) - \left(\sum_{k=1}^{N} b_k z^{-k} \right) W(z)$$

$$Y(z) = \left(\sum_{k=0}^{N} a_k z^{-k} \right) W(z)$$

(12.2.6)

The corresponding time domain equations are

$$w(n) = x(n) - \sum_{k=1}^{N} b_k w(n-k)$$

$$y(n) = \sum_{k=0}^{N} a_k w(n-k)$$

(12.2.7)

which match Eq. 12.2.3. Since this structure requires $2N$ delay elements, it is not a canonical form.

The complexities of these four structures are summarized in Table 12.10.1. The 1F and 2F forms require only N delay elements. They are canonical unlike the 3F and 4F structures which require $2N$ delay elements. All the forms require $2N + 1$ coefficient multipliers a_k's and b_k's, and $2N + 1$ registers to store the coefficients. Minimizing the number of coefficients minimizes the number of multipliers and storage registers which usually requires using minimum order filters. The 3F filter form only requires one summer while the 1F form needs two summers. The 4F form requires the most summers $(2N)$ while the 2F form requires $N + 1$ summers.

Since every delay element and coefficient is implemented using a memory location, minimizing the total number of delay elements and coefficients minimizes the memory size (software) or registers (hardware) required. Every summing node is implemented by an equation so minimizing the number of equations reduces the programming complexity, read operations, and control and sequencing needed.

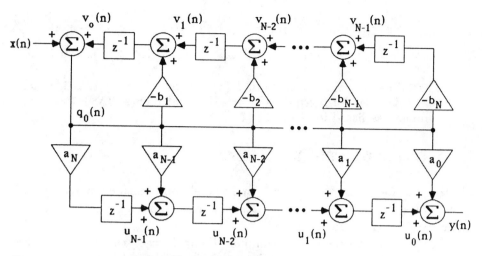

Fig. 12.2.4 4F multiple feedback filter realization.

EXAMPLE 12.2.1 The transfer function of a 4th order elliptic lowpass filter with 0.28 dB ripple and a 1 KHz bandwidth equals

$$H(z) = \frac{0.01201(z+1)^2(z^2 - 0.7981z + 1)}{(z^2 - 1.360z + 0.5133)(z^2 - 1.427z + 0.8160)} \qquad (12.2.8)$$

This $H(z)$ was determined in Example 11.2.1 using a bilinear transform. Implement $H(z)$ using a 1F multiple feedback structure.

SOLUTION This filter has the structure shown in Fig. 12.2.1. The a_k and b_k coefficients result when $H(z)$ of Eq. 12.2.8 is written in a summation form as

$$H(z) = \frac{0.01201 + 0.01443z^{-1} + 0.004850z^{-2} + 0.01443z^{-3} + 0.01201z^{-4}}{1 - 2.787z^{-1} + 3.269z^{-2} - 1.842z^{-3} + 0.4189z^{-4}}$$

$$(12.2.9)$$

Therefore matching the coefficients of Eq. 12.2.9 to those in Eq. 12.1.1 gives

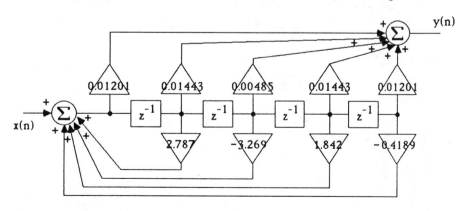

Fig. 12.2.5 1F multiple feedback filter realization of 4th order elliptic lowpass filter in Example 12.2.1.

$$a_0 = 0.01201, \quad a_1 = 0.01443, \quad a_2 = 0.004850, \quad a_3 = 0.01443, \quad a_4 = 0.01201$$

$$b_0 = 1, \quad b_1 = -2.787, \quad b_2 = 3.269, \quad b_3 = -1.842, \quad b_4 = 0.4189$$

$$(12.2.10)$$

The 1F canonical realization of $H(z)$ is shown in Fig. 12.2.5. Since $N = 4$, we see that this filter form requires $N = 4$ delay elements, $2N + 1 = 9$ multipliers, and 2 summers as listed in Table 12.10.1. ∎

12.3 BIQUADRATIC REALIZATIONS

Filters can be realized as cascade or parallel combinations of second-order subsections called *biquadratic filters* or simply *biquads*. Each biquad realizes a pair of real or complex poles and zeros simultaneously.[3]

Biquad structures are not unique. Using $N = 2$, the filters of Figs. 12.2.1–12.2.4 are easily reduced to the biquads shown in Fig. 12.3.1a–12.3.1d, respectively. The difference equations for the 1F biquad are

$$w(n) = x(n) - b_1 w(n - 1) - b_2 w(n - 2)$$
$$y(n) = a_0 w(n) + a_1 w(n - 1) + a_2 w(n - 2)$$

$$(12.3.1)$$

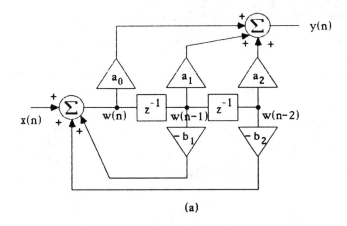

(a)

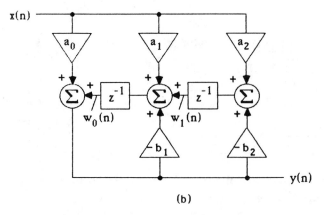

(b)

Fig. 12.3.1 (a) 1F, (b) 2F, (c) 3F, (d) 4F, and (e) 5F biquad modules for cascade filter realization.

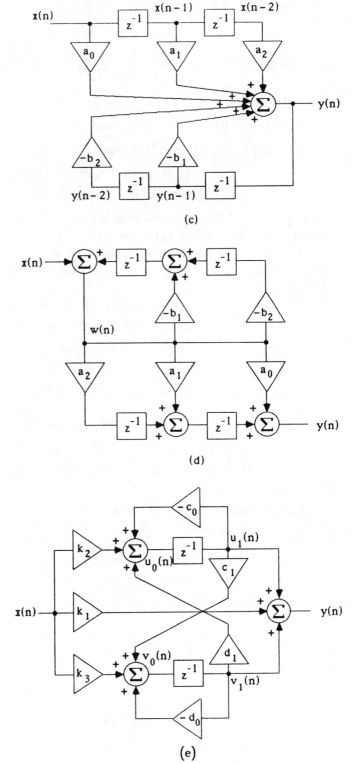

Fig. 12.3.1 (Continued)
(a) 1F, (b) 2F, (c) 3F, (d) 4F,
and (e) 5F biquad modules
for cascade filter realization.

based upon Eq. 12.2.3. The 2F biquad difference equations are

$$y(n) = a_0 x(n) + w_0(n)$$
$$w_0(n + 1) = w_1(n) + a_1 x(n) - b_1 y(n) \qquad (12.3.2)$$
$$w_1(n + 1) = a_2 x(n) - b_2 y(n)$$

based upon Eq. 12.2.4. The 3F biquad difference equations are

$$y(n) = a_0 x(n) + a_1 x(n - 1) + a_2 x(n - 2) - b_1 y(n - 1) - b_2 y(n - 2) \qquad (12.3.3)$$

from Eq. 12.2.5. Finally the 4F biquad difference equations are

$$w(n) = x(n) - b_1 w(n - 1) - b_2 w(n - 2)$$
$$y(n) = a_0 w(n) + a_1 w(n - 1) + a_2 w(n - 2) \qquad (12.3.4)$$

based upon Eq. 12.2.7. Although these various difference equation sets describe different structures, combining any equation set results in Eq. 12.3.3.

Other biquad structures are possible. The second-order canonical flow graph biquad is shown in Fig. 12.3.2. This is called the 5F biquad structure. The 5F biquad difference equations are

$$y(n) = k_1 x(n) + u_1(n) + v_1(n)$$
$$u_1(n + 1) = k_2 x(n) - c_0 u_1(n) + d_1 v_1(n) \qquad (12.3.5)$$
$$v_1(n + 1) = k_3 x(n) + c_1 u_1(n) - d_0 v_1(n)$$

Computing the transfer function of this biquad, equating it to $H(z)$ given by Eq. 12.1.1, and coefficient matching gives the relation between the a_i and b_i in terms of the k_i, c_i, and d_i as

$$a_0 = k_1, \qquad a_1 = k_1(c_0 + d_0) + k_2 + k_3,$$
$$a_2 = k_1(c_0 d_0 - c_1 d_1) + k_2(c_1 + d_0) + k_3(c_0 + d_1), \qquad (12.3.6)$$
$$b_1 = c_0 + d_0, \qquad b_2 = c_0 d_0 - c_1 d_1$$

This biquad is its own transpose.

The properties of these various biquads are listed in Table 12.10.1. 1F, 2F, and 5F biquads are canonical with two delays. The 3F biquad requires only one summer while the 1F biquad requires two summers. Although the 5F biquad is canonical, it requires the more multipliers (seven) than the other biquad filter forms (five).

12.4 CASCADE REALIZATIONS

Cascade filters have the series form shown in Fig. 12.4.1. They result when the transfer function $H(z)$ given by Eq. 12.1.1 is factored into a *product expansion* of M second-order terms as[4]

$$H(z) = \frac{\sum_{k=0}^{N} a_k z^{-k}}{\sum_{k=0}^{N} b_k z^{-k}} = \frac{\prod_{k=1}^{M}(c_{0k} + c_{1k} z^{-1} + c_{2k} z^{-2})}{\prod_{k=1}^{M}(d_{0k} + d_{1k} z^{-1} + d_{2k} z^{-2})} = \prod_{k=1}^{M} H_k(z) \qquad (12.4.1)$$

where $M = N/2$ for N even or $M = (N + 1)/2$ for N odd. The individual $H_k(z)$ terms

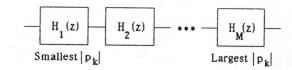

Fig. 12.4.1 Cascade filter realization.

are realized using any of the biquadratic filters discussed in the last section. A first-order section can be realized using a reduced second-order section having $a_2 = b_2 = 0$, or alternatively, as a second-order section having a single pole-zero cancellation.

The properties of the M cascaded filters constructed of 1F–5F biquad sections are listed in Table 12.10.1. The 1F, 2F, and 5F filters require the canonical number of delay elements (N) but the 3F and 4F filters require two more ($N + 2$). The 1F and 3F structures require the fewest summers ($0.5N + 1, 0.5N$). The 2F, 4F, and 5F forms require about three times as many summers. All structures require $2.5N$ multipliers except the 5F case which requires $3.5N$.

It is useful to compare these cascaded filters with the feedback filters of Table 12.10.1. Based upon total operations, the cascaded forms are slightly more complicated than the feedback forms (due primarily to more summers). For minimum complexity, the 1F feedback form requires $3N + 3$ total operations while the 1F cascade form needs $4N + 1$ operations. This is an complexity increase of about $(4N + 1)/(3N + 3) \cong 1.33 - 1/N \cong 33\%$.

One important guideline should be followed for cascade design. The filter performance depends both upon: (1) the ordering or sequence of the M sections and (2) the way in which poles and zeros are paired together to form each section. Best all around performance is usually obtained by first pairing poles and zeros that are closest together, and then ordering the sections from smallest pole magnitude to largest pole magnitude (i.e., from $|p_k| = 0$ to 1 where $|p_k|^2 = d_{2k}/d_{0k}$). This orders the stages in terms of increasing peaking in their magnitude responses to give more dynamic range.

EXAMPLE 12.4.1 Realize the $H(z)$ transfer function of the 4th order elliptic lowpass filter in Example 12.2.1 as a cascade structure. Use 1F sections.
SOLUTION The cascade block diagram of $H(z)$ is shown in Fig. 11.2.1b. This results by expressing Eq. 12.2.8 in a product form as

$$H(z) = \frac{0.01201(1 - 2z^{-1} + z^{-2})(1 - 0.7981z^{-1} + z^{-2})}{(1 - 1.360z^{-1} + 0.5133z^{-2})(1 - 1.427z^{-1} + 0.8160z^{-2})}$$

$$= \frac{0.03830(1 - 2z^{-1} + z^{-2})}{(1 - 1.360z^{-1} + 0.5133z^{-2})} \times \frac{0.3134(1 - 0.7981z^{-1} + z^{-2})}{(1 - 1.427z^{-1} + 0.8160z^{-2})} \qquad (12.4.2)$$

Using coefficient matching in Eq. 12.4.1 gives

$$c_{01} = 0.03830, \quad c_{11} = -0.07660, \quad c_{21} = 0.03830$$
$$d_{01} = 1, \quad d_{11} = -1.360, \quad d_{21} = 0.5133 \qquad (12.4.3)$$

in the first section and

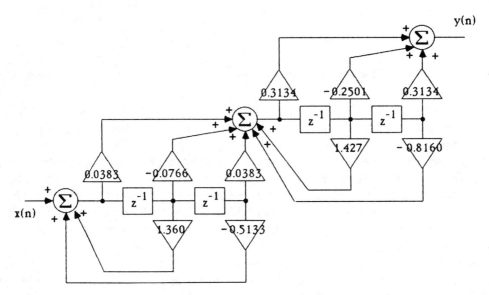

Fig. 12.4.2 1F cascade filter realization of 4th order elliptic lowpass filter in Example 12.4.1.

$$c_{02} = 0.3134, \quad c_{12} = -0.2501, \quad c_{22} = 0.3134$$
$$d_{02} = 1, \quad d_{12} = -1.427, \quad d_{22} = 0.8160 \tag{12.4.4}$$

in the second section. The resulting filter is shown in Fig. 12.4.2. The two sections are ordered so that the pole-zero pair closest each other have been grouped together in the second stage. Notice that the two input/output summers between the two biquad stages have been combined into a single summer. From Table 12.10.1 and using $N = 4$, we see that this 1F cascade realization requires $N = 4$ delay elements, $2.5N = 10$ multipliers, and $0.5N + 1 = 3$ summers. ∎

12.5 PARALLEL REALIZATIONS

Parallel filters have the parallel form shown in Fig. 12.5.1. They result when the transfer function $H(z)$ given by Eq. 12.1.1 is expressed as a *partial fraction expansion* (PFE) of M second-order terms where[5]

$$H(z) = H_0 + \sum_{k=1}^{M} H_k(z) \tag{12.5.1}$$

$M = N/2$ for N even or $M = (N + 1)/2$ for N odd. Partial fraction expansion was discussed in Chap. 1.10. The individual gain terms $H_k(z)$ equal

$$H_k(z) = \frac{c_{0k} + c_{1k}z^{-1}}{1 + d_{1k}z^{-1} + d_{2k}z^{-2}} \tag{12.5.2}$$

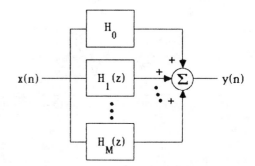

Fig. 12.5.1 Parallel filter realization.

and $H_0 = H(0)$. Such parallel structures might also be called *Foster form structures* by analogy with passive filter design.

The $H_k(z)$ terms are realized using any of the reduced biquadratic sections shown in Fig. 12.5.2. Since $H_k(z)$ does not involve a $c_{2k}z^{-2}$ term, then $c_{2k} = 0$ so these 1F–4F reduced biquads are slightly less complex than those shown in Fig. 12.3.1. The 5F biquad of Fig. 12.3.2 is unchanged. Therefore cascade and parallel filter realizations differ only in the interconnection between blocks (cf. Figs. 12.4.1 and 12.5.1) and not in the internal structure of the blocks (except $c_{2k} = 0$).

The properties of parallel filters constructed using 1F–5F reduced biquad sections are listed in Table 12.10.1. The 1F, 2F, and 5F filters require the canonical number of delay elements (N) but the 3F and 4F filters require one additional element ($N+1$). All filters require the same number ($2N+1$) of multipliers except the 5F form which needs $3N+1$. The 5F structure requires only one summer. The 1F and 3F forms require $0.5N+1$ summers while the 2F and 4F forms need almost twice as many ($N+1, N+2$).

Table 12.10.1 shows that based upon total complexity, the parallel filters have about the same complexity as their multiple feedback filter counterparts. The parallel filters are always slightly less complex that their cascade filter equivalents (due primarily to fewer multipliers and/or summers).

EXAMPLE 12.5.1 Realize the $H(z)$ transfer function of the 4th order elliptic lowpass filter in Example 12.2.1 as a parallel structure. Use 1F sections.
SOLUTION $H(z)$ must be expressed as a partial fraction expansion. From Eq. 1.10.2, this is accomplished by first expanding $H(z)/z$ as

$$\frac{H(z)}{z} = \frac{0.01201(z+1)^2(z^2 - 0.7981z + 1)}{z(z^2 - 1.360z + 0.5133)(z^2 - 1.427z + 0.8160)}$$

$$= \frac{0.02864}{z} + \frac{0.05445 - j0.3593}{z - 0.6800 - j0.2256} + \frac{0.05445 + j0.3593}{z - 0.6800 + j0.2256} \quad (12.5.3)$$

$$+ \frac{-0.06276 + j0.08907}{z - 0.7135 - j0.5540} + \frac{-0.06276 - j0.08907}{z - 0.7135 + j0.5540}$$

Combining terms gives $H(z)$ as

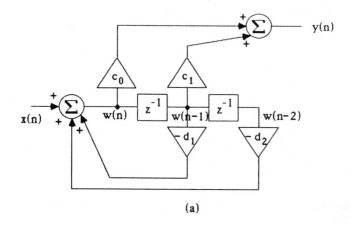

(a)

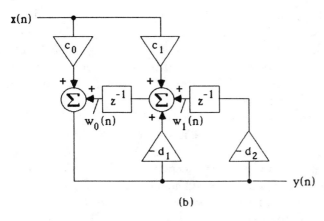

(b)

Fig. 12.5.2 (a) 1F, (b) 2F, (c) 3F, and (d) 4F biquad modules for parallel filter realization.

$$H(z) = 0.02864 + 2z \frac{0.05445(z - 0.6800) + 0.3593(0.2256)}{(z - 0.6800)^2 + 0.2256^2}$$

$$+ 2z \frac{-0.06276(z - 0.7135) - 0.08907(0.5540)}{(z - 0.7135)^2 + 0.5540^2}$$

$$= 0.02864 + \frac{0.1089 + 0.08806z^{-1}}{1 - 1.360z^{-1} + 0.5133z^{-2}} + \frac{-0.1255 - 0.009131z^{-1}}{1 - 1.427z^{-1} + 0.8160^{-2}}$$

$$(12.5.4)$$

Therefore coefficient matching in Eq. 12.5.2 gives

$$c_{01} = 0.1089, \quad c_{11} = 0.08806$$
$$d_{01} = 1, \quad d_{11} = -1.360, \quad d_{21} = 0.5133$$

$$(12.5.5)$$

in the first reduced biquad and

$$c_{02} = -0.1255, \quad c_{12} = -0.009131$$
$$d_{02} = 1, \quad d_{12} = -1.427, \quad d_{22} = 0.8160$$

$$(12.5.6)$$

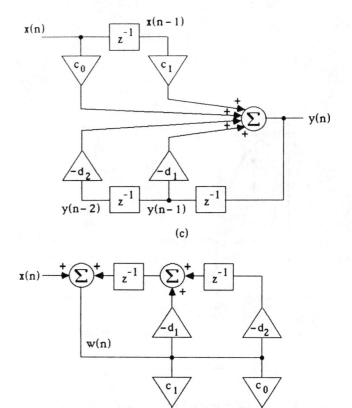

Fig. 12.5.2 (Continued)
(a) 1F, (b) 2F, (c) 3F, and
(d) 4F biquad modules for
parallel filter realization.

in the second reduced biquad. The parallel constant gain block $H_0 = 0.02864$ completes the filter realization which is shown in Fig. 12.5.3. Notice that the two summers of each 1F reduced biquad have been combined into the single summer required for the parallel realization (cf. Fig. 12.5.1). From Table 12.10.1 since $N = 4$, this 1F filter requires $N = 4$ delay elements, $2N + 1 = 9$ multipliers, and $0.5N + 1 = 3$ summers. ∎

12.6 LATTICE REALIZATIONS

Lattice filters have the lattice form shown in Fig. 12.6.1. An Nth-order lattice filter is the cascade of N first-order lattice filters. Each section is characterized by a gain k_m which is also called the reflection coefficient. There is a nonrecursive FIR form and a recursive IIR form.[6] The IIR lattice is the dual of the FIR lattice. They are canonical structures with N delays and require $2N$ multipliers and $2N$ summers.

The lattice structure converts the input sequence $x(n)$ with z-transform $X(z)$ into two other time sequences $y_m(n)$ and $u_m(n)$ with z-transforms $Y_m(z)$ and $U_m(z)$, re-

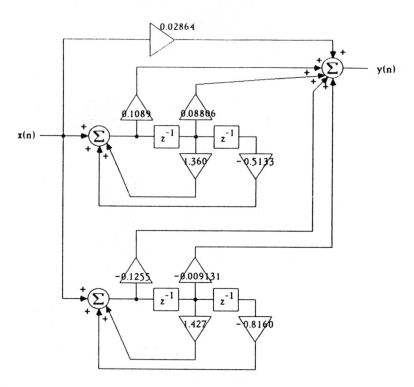

Fig. 12.5.3 1F parallel filter realization of 4th order elliptic lowpass filter in Example 12.5.1.

spectively. The upper output taps $Y_m(z)$ are sometimes called the forward prediction errors and are generated by the forward gain filter $G_m(z)$. The lower output taps $U_m(z)$ are called the backward prediction errors and are generated by the backward gain filter $H_m(z)$. The lattice filter output is taken from the upper port where $Y(z) \equiv Y_N(z)$. The lower output port $U_N(z)$ is not usually used.

The equations which describe the FIR lattice filter shown in Fig. 12.6.1a are derived as follows. The z-transformed inputs and outputs of the mth lattice stage are related in matrix form as

$$\begin{bmatrix} Y_m(z) \\ U_m(z) \end{bmatrix} = \begin{bmatrix} 1 & k_m z^{-1} \\ k_m & z^{-1} \end{bmatrix} \begin{bmatrix} Y_{m-1}(z) \\ U_{m-1}(z) \end{bmatrix} \tag{12.6.1}$$

This is analogous to a chain matrix description from two-port theory. Using Eq. 12.6.1, cascading m such stages corresponds to forming the matrix product

$$\begin{bmatrix} Y_m(z) \\ U_m(z) \end{bmatrix} = \begin{bmatrix} 1 & k_m z^{-1} \\ k_m & z^{-1} \end{bmatrix} \begin{bmatrix} 1 & k_{m-1} z^{-1} \\ k_{m-1} & z^{-1} \end{bmatrix} \cdots \begin{bmatrix} 1 & k_1 z^{-1} \\ k_1 & z^{-1} \end{bmatrix} \begin{bmatrix} Y_0(z) \\ U_0(z) \end{bmatrix}$$

$$= \begin{bmatrix} A_m(z) & B_m(z) \\ C_m(z) & D_m(z) \end{bmatrix} \begin{bmatrix} Y_0(z) \\ U_0(z) \end{bmatrix} \tag{12.6.2}$$

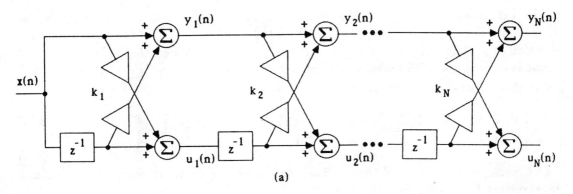

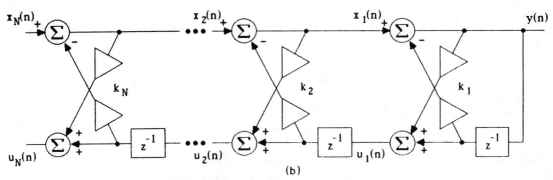

Fig. 12.6.1 (a) Nonrecursive lattice filter realization and (b) recursive lattice (all-pole) filter realization.

where the input $X(z) \equiv Y_0(z) = U_0(z)$ and the output $Y(z) \equiv Y_N(z)$. Therefore the two FIR lattice filter transfer functions between the input and the outputs of the mth section equal

$$G_m(z) \equiv \frac{Y_m(z)}{X(z)} = A_m(z) + B_m(z) = \sum_{n=0}^{m} c_{nm} z^{-n}$$

$$H_m(z) \equiv \frac{U_m(z)}{X(z)} = C_m(z) + D_m(z) = \sum_{n=0}^{m} c_{N-n,m} z^{-n} \qquad (12.6.3)$$

These two transfer functions are related as

$$H_m(z) = z^{-m} G_m(z^{-1}) \qquad (12.6.4)$$

They have pole-zero patterns which are images of one another (see Chap. 3.12). When $z = \infty$ or $z^{-1} = 0$, the upper path gain $G_N(\infty) = c_{0N} = 1$ from Fig. 12.6.1a.

An iteration process is used to determine the k_m coefficients of the lattice filter from the required FIR filter coefficients a_k in Eq. 12.1.1. The process begins at the lattice filter output and proceeds inward one stage per iteration. N iterations are necessary to compute all N gain coefficients k_m. To begin, the transfer function $H(z) = \sum_{k=0}^{N} a_k z^{-k}$

of the FIR filter to be implemented is magnitude normalized so that $H(\infty) = a_0 = 1$. This results in $H(z)/H(\infty)$ which is denoted as $G_N(z)$. Then the image transfer function $H_N(z)$ of $G_N(z)$ is computed from Eq. 12.6.4. When the FIR lattice filter input is an impulse with transform $X(z) = 1$, then the two transformed outputs equal these gains $G_N(z)$ and $H_N(z)$. Described mathematically,

$$\begin{bmatrix} Y_0(z) \\ U_0(z) \end{bmatrix} = \begin{bmatrix} X(z) \\ X(z) \end{bmatrix} = \begin{bmatrix} 1 \\ 1 \end{bmatrix} \longrightarrow \begin{bmatrix} Y_N(z) \\ U_N(z) \end{bmatrix} = \begin{bmatrix} G_N(z) \\ H_N(z) \end{bmatrix} = \begin{bmatrix} G_N(z) \\ z^{-N}G_N(z^{-1}) \end{bmatrix}$$

(12.6.5)

To compute the gains of the previous stages, we use an iteration equation obtained by inverting Eq. 12.6.1 as

$$\begin{bmatrix} G_{m-1}(z) \\ H_{m-1}(z) \end{bmatrix} = \frac{1}{1 - k_m^2} \begin{bmatrix} 1 & -k_m \\ -k_m z & z \end{bmatrix} \begin{bmatrix} G_m(z) \\ H_m(z) \end{bmatrix}$$

(12.6.6)

which always exists when $k_m \neq 1$. $G_m(z)$ and $H_m(z)$ are ascending power series in z^{-1} where

$$\begin{aligned} G_m(z) &= c_{0m} + c_{1m}z^{-1} + \ldots + c_{mm}z^{-m} \\ H_m(z) &= c_{mm} + c_{m-1,m}z^{-1} + \ldots + c_{0m}z^{-m} \end{aligned}$$

(12.6.7)

Notice that at $z = \infty$ or $z^{-1} = 0$, the gains of the lower z^{-1} branches of the FIR lattice filter in Fig. 12.6.1a equal zero. Under this condition, Eq. 12.6.7b shows that the backward gains equal

$$[H_1(\infty), H_2(\infty), \ldots, H_N(\infty)]^t = [c_{11}, c_{22}, \ldots, c_{NN}]^t = [k_1, k_2, \ldots, k_N]^t \quad (12.6.8)$$

Therefore $k_m = H_m(\infty) = c_{mm}$ is the first term in the $H_m(z)$ series or the last coefficient in the $G_m(z)$ series. Thus, the k_m's can be systematically computed beginning at the lattice output where

$$k_N = c_{NN} = H_N(z)\Big|_{z=\infty} = z^{-N}G_N(z^{-1})\Big|_{z^{-1}=0} = a_N \quad (12.6.9)$$

k_N is the coefficient of the highest order z^{-N} term in the FIR gain function in Eq. 12.1.1. When one or more of the $k_m = 1$, the recursion equation of Eq. 12.6.6 cannot be used. Then the k_m are found directly by computing $H(z)$ using flow graphs and coefficient matching (e.g., see Example 12.6.2).

EXAMPLE 12.6.1 A third-order differentiator based upon the H_{003} transform has a transfer function of

$$H(z) = \frac{11 - 18z^{-1} + 9z^{-2} - 2z^{-3}}{6} = 1.833(1 - 1.636z^{-1} + 0.8182z^{-2} - 0.1818z^{-3})$$

(12.6.10)

from Eq. 10.6.15. Implement this differentiator using an FIR lattice.

SOLUTION We begin by computing the lattice gains G_m and H_m using Eqs.

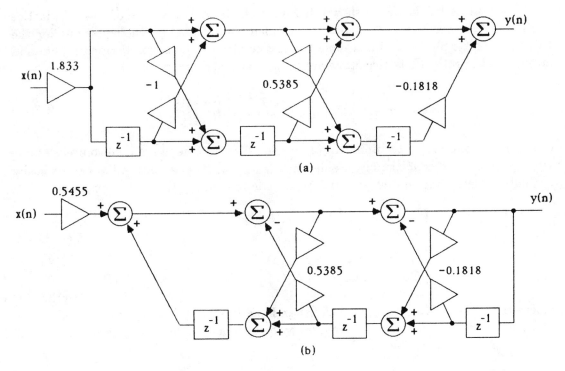

Fig. 12.6.2 (a) Nonrecursive lattice differentiator of Example 12.6.1 and (b) recursive lattice integrator of Example 12.6.2.

12.6.5 and 12.6.6 as

$$\begin{bmatrix} G_3(z) \\ H_3(z) \end{bmatrix} = \begin{bmatrix} 1 - 1.636z^{-1} + 0.8182z^{-2} - 0.1818z^{-3} \\ -0.1818 + 0.8182z^{-1} - 1.636z^{-2} + z^{-3} \end{bmatrix}$$

$$\begin{bmatrix} G_2(z) \\ H_2(z) \end{bmatrix} = \frac{1}{1 - (-0.1818)^2} \begin{bmatrix} 1 & 0.1818 \\ 0.1818z & z \end{bmatrix} \begin{bmatrix} G_3(z) \\ H_3(z) \end{bmatrix}$$

$$= \begin{bmatrix} 1 - 1.5383z^{-1} + 0.5385z^{-2} \\ 0.5385 - 1.5383z^{-1} + z^{-2} \end{bmatrix}$$

$$\begin{bmatrix} G_1(z) \\ H_1(z) \end{bmatrix} = \frac{1}{1 - 0.5384^2} \begin{bmatrix} 1 & -0.5384 \\ -0.5384z & z \end{bmatrix} \begin{bmatrix} G_2(z) \\ H_2(z) \end{bmatrix} = \begin{bmatrix} 1 - z^{-1} \\ -1 + z^{-1} \end{bmatrix}$$

(12.6.11)

The k_m's contained in Eq. 12.6.11 are determined from Eq. 12.6.8 as

$$[k_1, k_2, k_3]^t = [H_1(\infty), H_2(\infty), H_3(\infty)]^t = [-1, 0.5385, -0.1818]^t \qquad (12.6.12)$$

These are the first coefficients in the $H_m(z)$ series or the last coefficients in the $G_m(z)$ series in Eq. 12.6.11. The resulting lattice differentiator is shown in Fig. 12.6.2a. A gain constant of 1.833 is placed ahead of the differentiator for proper gain denormalization of $H(z)$ given by Eq. 12.6.10. ∎

The *all-pole* IIR lattice is shown in Fig. 12.6.1b. It is the *dual* of the FIR lattice in Fig. 12.6.1a because their equations are identical. Therefore the equations for the FIR lattice filter also describe all-pole IIR lattice filters. The z-transformed inputs and outputs of the mth IIR lattice stage are related in matrix form as

$$\begin{bmatrix} X_m(z) \\ U_m(z) \end{bmatrix} = \begin{bmatrix} 1 & k_m z^{-1} \\ k_m & z^{-1} \end{bmatrix} \begin{bmatrix} X_{m-1}(z) \\ U_{m-1}(z) \end{bmatrix} \tag{12.6.13}$$

changing output $Y_m(z)$ to input $X_m(z)$ in Eq. 12.6.1. The stage is absolutely stable when $|k_m| < 1$ and conditionally stable when $|k_m| = 1$. Therefore stability can be easily evaluated. Cascading m such stages corresponds to forming the matrix product

$$\begin{bmatrix} X_m(z) \\ U_m(z) \end{bmatrix} = \begin{bmatrix} 1 & k_m z^{-1} \\ k_m & z^{-1} \end{bmatrix} \begin{bmatrix} 1 & k_{m-1} z^{-1} \\ k_{m-1} & z^{-1} \end{bmatrix} \cdots \begin{bmatrix} 1 & k_1 z^{-1} \\ k_1 & z^{-1} \end{bmatrix} \begin{bmatrix} X_0(z) \\ U_0(z) \end{bmatrix}$$
$$= \begin{bmatrix} A_m(z) & B_m(z) \\ C_m(z) & D_m(z) \end{bmatrix} \begin{bmatrix} X_0(z) \\ U_0(z) \end{bmatrix}$$

$$\tag{12.6.14}$$

where the output $Y(z) \equiv X_0(z) = U_0(z)$ and the input $X(z) \equiv X_N(z)$. Therefore the two IIR lattice filter *reciprocal* or *inverse* transfer functions between the input and the outputs of the mth section equal

$$G_m(z) \equiv \frac{X_m(z)}{Y(z)} = A_m(z) + B_m(z) = \sum_{n=0}^{m} c_{nm} z^{-n}$$

$$H_m(z) \equiv \frac{U_m(z)}{Y(z)} = C_m(z) + D_m(z) = \sum_{n=0}^{m} c_{N-n,m} z^{-n} \tag{12.6.15}$$

These are input-output duals of the gains given by Eq. 12.6.3. These two inverse transfer functions are still related as $H_m(z) = z^{-m} G_m(z^{-1})$. They have pole-zero patterns which are images of one another. Notice that $U_m(z)/X_m(z) = z^{-m} G_m(z^{-1})/G_m(z)$ which is allpass. Therefore when $X_N(z)$ is the input, then $Y(z)$ is the all-pole IIR filter output and $U_N(z)$ is the corresponding allpass filter output. From Fig. 12.6.1b, the upper path gain $G_N(\infty) = c_{0N} = 1$. When $z = \infty$ or $z^{-1} = 0$, the gain of the lower z^{-1} branches equal zero.

The same iteration process described previously for FIR lattice filters can be used to determine the k_m coefficients from the required all-pole IIR filter coefficients b_k in Eq. 12.1.1. The process begins at the lattice filter output and proceeds inward one stage per iteration. N iterations are necessary. To begin, the reciprocal all-pole transfer function $1/H(z) = \sum_{k=0}^{N} b_k z^{-k}$ of the IIR filter to be implemented is magnitude normalized so that $H(\infty) = 1/b_0 = 1$. This results in $H(\infty)/H(z)$ which is denoted as $G_N(z)$. Then the reciprocal transfer function $H_N(z)$ of $G_N(z)$ is computed using Eq. 12.6.4. When the IIR lattice filter output is an impulse with transform $Y(z) = 1$, then the two transformed inputs equal these inverse gains $G_N(z)$ and $H_N(z)$. Described mathematically,

$$\begin{bmatrix} X_0(z) \\ U_0(z) \end{bmatrix} = \begin{bmatrix} Y(z) \\ Y(z) \end{bmatrix} = \begin{bmatrix} 1 \\ 1 \end{bmatrix} \quad \longrightarrow \quad \begin{bmatrix} X_N(z) \\ U_N(z) \end{bmatrix} = \begin{bmatrix} G_N(z) \\ H_N(z) \end{bmatrix} = \begin{bmatrix} G_N(z) \\ z^{-N}G_N(z^{-1}) \end{bmatrix}$$

$$(12.6.16)$$

To compute the gains of the previous stages, we use Eq. 12.6.6 which always has a solution when $k_m \neq 1$. Manipulating Eq. 12.6.6, a convenient recursion relation for generating only the forward gains is

$$G_{m-1}(z) = \frac{1}{1 - k_m^2}\left[G_m(z) - k_m z^{-m}G_m(z^{-1})\right] \qquad (12.6.17)$$

The k_m's can be systematically computed using $G_m(z)$ and

$$k_{m-1} = \frac{c_{m-1,m} - c_{mm}c_{1m}}{1 - c_{mm}^2} = H_{m-1}(z)\Big|_{z=\infty} = z^{-(m-1)}G_{m-1}(z^{-1})\Big|_{z^{-1}=0} \qquad (12.6.18)$$

The process begins using $k_N = b_N$ which is the coefficient of the highest order z^{-N} term in the IIR gain function.

EXAMPLE 12.6.2 The third-order integrator based upon the H_{003} transform has a transfer function of

$$H(z) = \frac{6}{11 - 18z^{-1} + 9z^{-2} - 2z^{-3}} = \frac{0.5455}{1 - 1.636z^{-1} + 0.8182z^{-2} - 0.1818z^{-3}}$$

$$(12.6.19)$$

from Table 5.2.1. Implement this all-pole integrator using an IIR lattice.
SOLUTION Since the reciprocal integrator gain $H^{-1}(z)$ of Eq. 12.6.19 matches the differentiator gain of Eq. 12.6.10, the lattice filters are exact duals and the same solution holds. Then $G_m(z)$ and $H_m(z)$ are given by Eq. 12.6.11. The k_m's are identical to those in Eq. 12.6.12 as

$$\left[k_1, k_2, k_3\right]^t = \left[-1, 0.5385, -0.1818\right]^t \qquad (12.6.20)$$

The resulting lattice integrator is shown in Fig. 12.6.2b. A gain constant of 0.5455 is placed ahead of the integrator for proper gain denormalization of $H(z)$. Using flow graph theory, the overall transfer function of this lattice can be written as

$$H(z) = \frac{0.5455}{1 + (k_1 + k_1 k_2 + k_2 k_3)z^{-1} + (k_2 + k_1 k_3 + k_1 k_2 k_3)z^{-2} + k_3 z^{-3}}$$

$$(12.6.21)$$

The denominator is the gain of the FIR lattice. Matching these z^{-m} coefficients to those of Eq. 12.6.19 gives the nonlinear set of equations involving the k_m's. Solving them simultaneously results in Eq. 12.6.20. ∎

The all-pole IIR lattice in Fig. 12.6.1b can be augmented to produce general lattices whose transfer functions have zeros as well as poles. This lattice is shown in Fig. 12.6.3. The upper $x_m(n)$ tap outputs are weighted by a set of c_m coefficients to form a new output $y'(n)$ which equals

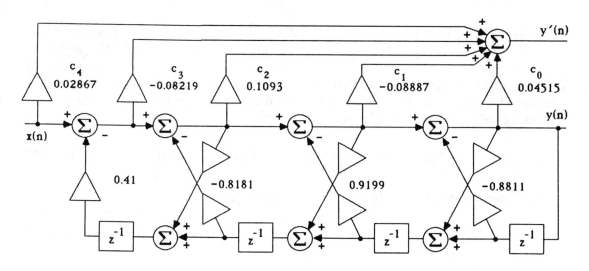

Fig. 12.6.3 Lattice filter realization of 4th order elliptic lowpass filter in Example 12.6.3.

$$y'(n) = \sum_{m=0}^{N} c_m x_m(n), \qquad x_0(n) = y(n) \tag{12.6.22}$$

Taking the z-transform of Eq. 12.6.22 gives this new transfer function as

$$\frac{Y'(z)}{X_N(z)} = \frac{Y(z)}{X_N(z)} \sum_{m=0}^{N} c_m \frac{X_m(z)}{Y(z)} = \frac{\sum_{m=0}^{N} c_m G_m(z)}{G_N(z)} = \frac{\sum_{k=0}^{N} a_k z^{-k}}{\sum_{k=0}^{N} b_k z^{-k}} \tag{12.6.23}$$

The k_m gains and $G_m(z)$ are determined in the process of matching the $\sum_{k=0}^{N} b_k z^{-k}$ denominator of the desired transfer function to $G_N(z)$. Then the required c_m coefficients are found by matching the $\sum_{k=0}^{N} a_k z^{-k}$ numerator to $\sum_{m=0}^{N} c_m G_m(z)$, and solving for the necessary c_m. This procedure is demonstrated in Example 12.6.3. From Table 12.10.1, this general lattice form requires N delays, $3N + 1$ multipliers, and $2N + 1$ summers. The total complexity is $6N + 2$ which is higher than that of the multiple feedback, cascade, and parallel filter forms.

EXAMPLE 12.6.3 Realize the 4th order elliptic lowpass filter of Example 12.2.1 as a generalized IIR lattice.
SOLUTION Since the filter $H(z)$ equals

$$
\begin{aligned}
H(z) &= \frac{0.01201 + 0.01443z^{-1} + 0.00485z^{-2} + 0.01443z^{-3} + 0.01201z^{-4}}{1 - 2.787z^{-1} + 3.269z^{-2} - 1.842z^{-3} + 0.4189z^{-4}} \\
&= \frac{c_0 + c_1 G_1(z) + c_2 G_2(z) + c_3 G_3(z) + c_4 G_4(z)}{G_4(z)}
\end{aligned}
\tag{12.6.24}
$$

then using Eq. 12.6.17, the G_m gains equal

$$G_4(z) = 1 - 2.787z^{-1} + 3.269z^{-2} - 1.842z^{-3} + 0.4189z^{-4}$$
$$G_3(z) = 1 - 2.444z^{-1} + 2.304z^{-2} - 0.8181z^{-3}$$
$$G_2(z) = 1 - 1.692z^{-1} + 0.9199z^{-2}$$
$$G_1(z) = 1 - 0.8811z^{-1}$$

(12.6.25)

k_m is determined from Eq. 12.6.18 using the last coefficient of $G_m(z)$ as

$$[k_1, k_2, k_3, k_4]^t = [-0.8811, 0.9199, -0.8181, 0.4189]^t$$

(12.6.26)

Matching numerator coefficients using Eq. 12.6.23 gives the matrix equation $a = Tc$ where

$$\begin{bmatrix} 0.01201 \\ 0.01443 \\ 0.00485 \\ 0.01443 \\ 0.01201 \end{bmatrix} = \begin{bmatrix} 1 & 1 & 1 & 1 & 1 \\ 0 & -0.8811 & -1.692 & -2.444 & -2.787 \\ 0 & 0 & 0.9199 & 2.304 & 3.269 \\ 0 & 0 & 0 & -0.8181 & -1.842 \\ 0 & 0 & 0 & 0 & 0.4189 \end{bmatrix} \begin{bmatrix} c_0 \\ c_1 \\ c_2 \\ c_3 \\ c_4 \end{bmatrix}$$

(12.6.27)

Vector a equals the numerator coefficients of the transfer function. Notice that the $(m+1)$st column of T equals the coefficients of $G_m(z)$. Solving for the c vector by inverting the matrix equation gives

$$c = T^{-1}a = \begin{bmatrix} 1 & 1.135 & 1.001 & 0.6494 & 0.2119 \\ 0 & -1.135 & -2.088 & -2.489 & -2.203 \\ 0 & 0 & 1.007 & 3.062 & 4.979 \\ 0 & 0 & 0 & -1.222 & -5.375 \\ 0 & 0 & 0 & 0 & 2.387 \end{bmatrix} \begin{bmatrix} 0.01201 \\ 0.01443 \\ 0.00485 \\ 0.01443 \\ 0.01201 \end{bmatrix} = \begin{bmatrix} 0.04515 \\ -0.08887 \\ 0.1093 \\ -0.08219 \\ 0.02867 \end{bmatrix}$$

(12.6.28)

The resulting generalized IIR lattice filter is shown in Fig. 12.6.3. From Table 12.10.1, this filter requires $N = 4$ delays, $3N + 1 = 13$ multipliers, and $2N + 1 = 9$ summers. Since the total complexity equals $6N + 2 = 26$, the general lattice filter has higher complexity than the other filter types except for wave filters. ∎

12.7 LADDER REALIZATIONS

Ladder filter realizations result when $H(z)$ is expressed as a *continued fraction expansion* (CFE).[7] The filter gain $H(z)$ of Eq. 12.1.1 can be expressed in two CFE forms as

$$H(z) = c_0 + \cfrac{1}{d_1 z + \cfrac{1}{c_1 + \cfrac{1}{\ddots + \cfrac{1}{d_N z + \cfrac{1}{c_N}}}}} \quad \text{or} \quad \cfrac{1}{c_0 + \cfrac{1}{d_1 z + \cfrac{1}{c_1 + \cfrac{1}{\ddots + \cfrac{1}{d_N z + \cfrac{1}{c_N}}}}}}$$

$$(12.7.1)$$

The first continued fraction expansion results by removing constants and poles from $H(z)$ at infinity (i.e., $z = \infty$). The ladder filter realization which results is shown in Fig. 12.7.1a. The alternative second form results by removing constants and poles of the reciprocal gain $1/H(z)$ at infinity. This ladder filter is shown in Fig. 12.7.1b. Both structures are called *Cauer 1 forms* by analogy with passive filter design. The structures differ because their branch gains have opposite directions and signs. The c_0 branch is unchanged except for sign.

The filter gain $H(z)$ can also be expanded in two other CFE forms as

$$H(z) =$$

$$c_0 + \cfrac{1}{d_1 z^{-1} + \cfrac{1}{c_1 + \cfrac{1}{\ddots + \cfrac{1}{d_N z^{-1} + \cfrac{1}{c_N}}}}} \quad \text{or} \quad \cfrac{1}{c_0 + \cfrac{1}{d_1 z^{-1} + \cfrac{1}{c_1 + \cfrac{1}{\ddots + \cfrac{1}{d_N z^{-1} + \cfrac{1}{c_N}}}}}}$$

$$(12.7.2)$$

The first continued fraction expansion results by removing constants and poles from $H(z)$ at the origin (i.e., $z = 0$ or $z^{-1} = \infty$). This ladder filter realization is shown in Fig. 12.7.2a. The alternative second form results by expanding the reciprocal gain $1/H(z)$ instead. This ladder filter is shown in Fig. 12.7.2b. These structures are called *Cauer 2 forms*. They differ because their branch gains have opposite directions and signs. The c_0 branch is unchanged except for sign. All the ladder filters of Figs. 12.7.1 and 12.7.2 are canonical with N delays. They require $2N + 1$ multipliers and $2N$ summers. From Table 12.10.1, then ladder filters have a total complexity of $5N + 1$ which is slightly higher than most other forms.

EXAMPLE 12.7.1   Realize the $H(z)$ transfer function of the 4th order elliptic lowpass filter of Example 12.2.1 having

$$H(z) = 0.01201 \, \frac{1 + 1.2015 z^{-1} + 0.4038 z^{-2} + 1.2015 z^{-3} + z^{-4}}{1 - 2.787 z^{-1} + 3.269 z^{-2} - 1.842 z^{-3} + 0.4189 z^{-4}} \quad (12.7.3)$$

as a Cauer 1 ladder filter and a Cauer 2 ladder filter.

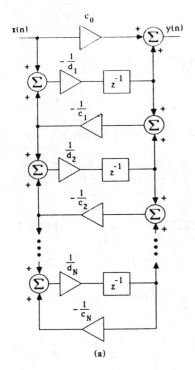

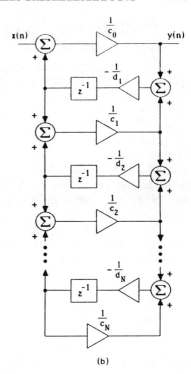

Fig. 12.7.1 Ladder filter realization using Cauer 1 form.

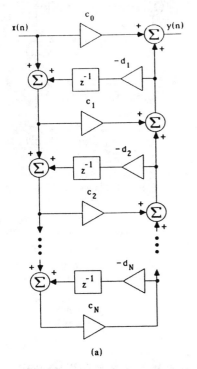

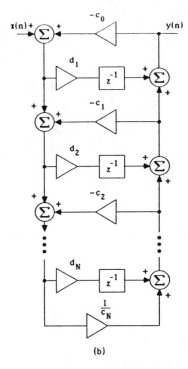

Fig. 12.7.2 Ladder filter realization using Cauer 2 form.

SOLUTION The Cauer 1 form results by expanding $H(z)$ as a continued fraction about infinity as

$$\frac{H(z)}{0.01201} = \frac{1 + 1.2015z^{-1} + 0.4038z^{-2} + 1.2015z^{-3} + z^{-4}}{1 - 2.787z^{-1} + 3.269z^{-2} - 1.842z^{-3} + 0.4189z^{-4}}$$

$$= 1 + \left\lfloor 0.2507z^{-1} + \left\lfloor -1.928 + \left\lfloor -1.052z^{-1} + \right\rfloor 1.173 \right. \right. \quad (12.7.4)$$

$$+ \left\lfloor -9.807z^{-1} + \left\lfloor -0.02066 + \left\lfloor 9.132z^{-1} + \right\rfloor 2.163 \right. \right.$$

The $\left\lfloor\right.$ symbol indicates CFE division and is used to compact the CFE equation. This realization is shown in Fig. 12.7.3a. The Cauer 2 form results by expanding $H(z)$ in a continued fraction about zero as

$$\frac{H(z)}{0.01201} = \frac{z^{-4} + 1.2015z^{-3} + 0.4038z^{-2} + 1.2015z^{-1} + 1}{0.4189z^{-4} - 1.842z^{-3}1 + 3.269z^{-2} - 2.787z^{-1} + 1} \quad (12.7.5)$$

$$= 2.387 + \left\lfloor 0.07482z^{-1} + \left\lfloor -4.346 + \left\lfloor -0.3030z^{-1} + \right\rfloor 2.782 \right. \right.$$

$$+ \left\lfloor 1.311z^{-1} + \left\lfloor -0.5779 + \left\lfloor -2.676z^{-1} + \right\rfloor 0.7540 \right. \right.$$

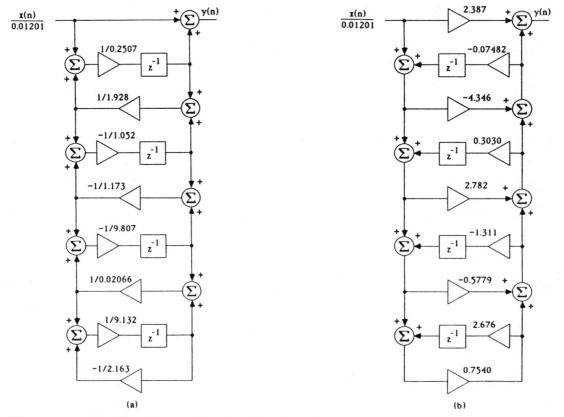

(a)   (b)

Fig. 12.7.3 Ladder filter realization of 4th order elliptic lowpass filter in Example 12.7.1 using (a) Cauer 1 form and (b) Cauer 2 form.

which is shown in Fig. 12.7.3b. The branch gains of these ladder filters are negative reciprocals of each other. Both filters are canonical since they use $N = 4$ delays. They require $2N + 1 = 9$ multipliers and $2N = 8$ summers.   ■

The ladders of Figs. 12.7.1 and 12.7.2 are usually realizable when the numerator and denominator polynomials of $H(z)$ have no missing terms, the highest degree terms differ by no more than one ($|N - M| \leq 1$), and the lowest degree terms differ by no more than one. Another ladder realization is shown in Fig. 12.7.4. This ladder filter is usually realizable when the denominator polynomial of $H(z)$ given by Eq. 12.1.1 has no missing terms and $M \leq N$. The numerator can have missing terms unlike the ladders of Figs. 12.7.1 and 12.7.2. In this new ladder, the denominator of the filter gain $H(z)$ is factored into even and odd parts as

$$H(z) = \frac{N(z)}{D(z)} = \frac{N(z)}{\text{Ev } D(z) + \text{Od } D(z)} = \frac{N_N(z)}{1 + D_N(z)} \qquad (12.7.6)$$

The numerator N_N and denominator D_N terms are related to N and D as

$$D_N(z) = \frac{\text{Od } D(z)}{\text{Ev } D(z)} \quad \text{or} \quad \frac{\text{Ev } D(z)}{\text{Od } D(z)} , \qquad N_N(z) = \frac{N(z)}{\text{Ev } D(z)} \quad \text{or} \quad \frac{N(z)}{\text{Od } D(z)} \qquad (12.7.7)$$

Using the $D_N(z)$ with the highest-order denominator, then $D_N(z)$ can be expressed in a continued fraction expansion about infinity as

$$D_N(z) = \cfrac{1}{d_1 z + \cfrac{1}{d_2 z + \cfrac{1}{d_3 z + \cfrac{1}{\ldots + \cfrac{1}{d_N z}}}}} \qquad (12.7.8)$$

Referring to Fig. 12.7.4, the d_i's are the reciprocal gain coefficients in a ladder. The c_i coefficients are determined using coefficient matching. Applying Mason's gain formula to Fig. 12.7.4, the numerator of $H(z)$ equals

$$N_N(z) = \frac{(-1)^{N+1}}{d_N z}\left[c_N \Delta_{N-1} - \frac{1}{d_{N-1} z}\left[c_{N-1}\Delta_{N-2} - \ldots - \frac{1}{d_2 z}\left[c_2 \Delta_1 - \left[\frac{c_1}{d_1 z}\right]\right]\ldots\right]\right.$$
$$(12.7.9)$$

$\Delta_k = 1 + D_k$ is the reduced filter determinant. D_k is obtained by truncating the continued fraction expansion of D_N at the kth term. The c_i coefficients are found by solving the system of equations which result where

$$N(z) = N_N(z) \text{ Ev } D(z) \quad \text{or} \quad N_N(z) \text{ Od } D(z) \qquad (12.7.10)$$

and collecting like powers of z.

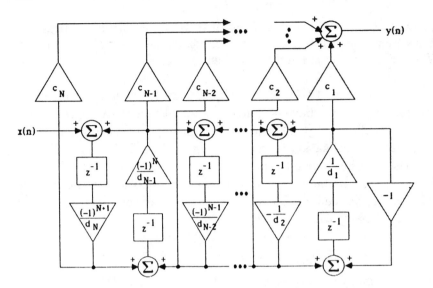

Fig. 12.7.4 Alternative ladder filter realization using Cauer 1 form.

12.8 ANALOG FILTER SIMULATIONS

The *analog filter simulation* design method utilizes the analog-to-digital transforms discussed in Chap. 5. In this approach, RLC analog filters are designed to meet the analog filter specification.[8] The Laplace transformed voltages and currents in the filter are expressed in signal flow graph form. Then the L's and C's are replaced by their digital counterparts. That is, the analog complex frequency p in every pL and pC term is replaced using any desired transform $g(z)$ from Chap. 5 where

$$p = g(z) \quad \text{or} \quad z = g^{-1}(p) \tag{12.8.1}$$

(see Eq. 5.1.1). A variety of $g(z)$ transforms are listed in Tables 5.2.1 and 5.2.2. The R's are unaffected.

To illustrate this approach, consider the passive lowpass filter shown in Fig. 12.8.1a having a voltage transfer function $H(p) = V_M(p)/V_0(p)$. It is a standard RLC ladder structure with tabulated component values for a variety of filters like Butterworth, Chebyshev, and elliptic. One possible flow graph description having voltage V_i and current I_i nodes which alternate is shown in Fig. 12.8.1b where

$$V_i = \frac{I_{i-1} - I_i}{pC_i}, \qquad I_i = \frac{V_{i+1} - V_i}{pL_i}, \qquad i = 1, \ldots, M-1 \tag{12.8.2}$$

internally and

$$I_0 = \frac{V_1 - V_0}{R_S}, \qquad I_M = \frac{V_M}{R_L} \tag{12.8.3}$$

externally at the I_0 and I_M current nodes. The input voltage node is V_0 and the output voltage node is V_M.

The branch gains $Z_i = 1/pC_i$ (or R_i) and $Y_i = 1/pL_i$ (or $1/R_i$) have units of ohms and mhos, respectively. To make the branch gains and node variables unitless while not changing the voltage and current transfer functions, the network elements are magnitude scaled by R (remember magnitude scaling requires $Z(p) = aZ_n(p/b)$ where $a = R$). This allows all the branch gains to be scaled into convenient ranges as $G = R/R_i$, $pL = pL_i/R$, and $pC = pRC_i$. Additional nodes can be inserted into the flow graph of Fig. 12.8.1b to produce the rearranged graph of Fig. 12.8.1c. Due to the leap-frog appearance of the graph, this is also called a *leap-frog filter*. This graph form easily allows us to draw its hardware implementation as shown in Fig. 12.8.1d.

The next step is to simulate and replace every $1/p$ analog integrator block in the analog filter of Fig. 12.8.1d with a $1/g(z)$ digital integrator block. This produces the digital filter in Fig. 12.8.1e having N integrators. A large variety of integrator blocks are shown in Fig. 5.9.3. In the simplest possible but seldom encountered case where $1/g(z)$ requires only a single delay and multiplier, the resulting digital filter is canonical with N delays. It requires $N + 2$ multipliers and $N + 1$ summers. The total complexity is $3N + 3$ which is the lowest in Table 12.10.1. However complexity varies with the RLC analog filter structure and the digital transform that is used.

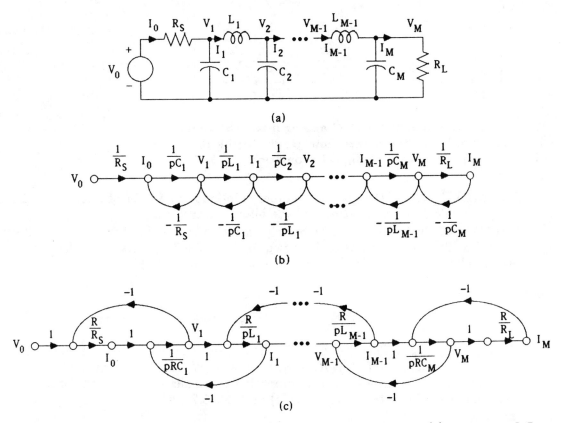

Fig. 12.8.1 (a) RLC analog filter, (b) analog filter flow graph, (c) rearranged flow graph; (d) block diagram equivalent, and (e) digital simulation on next page.

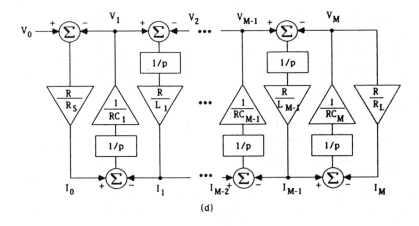

(d)

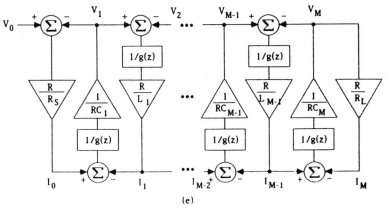

(e)

Fig. 12.8.1 (Continued) (a) *RLC* analog filter, (b) analog
filter flow graph, (c) rearranged flow graph, (d) block
diagram equivalent, and (e) digital simulation.

EXAMPLE 12.8.1 Design the 4th order elliptic lowpass filter of Example 12.2.1
using an analog filter simulation and the bilinear transform.

SOLUTION This design begins by implementing the analog lowpass filter. One
possible analog filter realization is shown in Fig. 12.8.2a.[9] This 4th order elliptic
lowpass filter has an analog transfer function (see Eq. 11.2.3)

$$H(p_n) = \frac{k(p_n^2 + 4.0695)}{p_n^4 + 1.3918p_n^3 + 2.0038p_n^2 + 1.3590p_n + 0.5729} \tag{12.8.4}$$

where $p_n = pT/v_pT$. The filter has 0.28 dB in-band ripple, a corner frequency of 1
rad/sec, and a dc gain of $R_L/(R_S + R_L) = 0.375$. We saw in Example 11.2.1 that
this unity corner frequency must be properly denormalized to the transform being
used. For the bilinear transform $H_{011}(z)$ in Table 5.2.1 where

$$p = \frac{2}{T}\frac{1 - z^{-1}}{1 + z^{-1}}, \qquad p_n = \frac{2}{v_pT}\frac{1 - z^{-1}}{1 + z^{-1}} \tag{12.8.5}$$

this denormalized corner frequency equals

$$\frac{v_p T}{2} = \tan\left(\frac{\omega_p T}{2}\right) = \tan(18°) = \frac{1}{3.078} \qquad (12.8.6)$$

from Eq. 11.2.4. This corresponds to a 0.28 dB ripple frequency of 1 KHz and a sampling frequency of 10 KHz. A block diagram of this analog filter is shown in Fig. 12.8.2b. Since the source resistor $R_S = 1\ \Omega$, we choose a magnitude scaling constant $R = 1\ \Omega$ which eliminates one multiplier in the analog filter flow graph. The bilinear transform therefore equals

$$p_n = 3.078\,\frac{1 - z^{-1}}{1 + z^{-1}}, \qquad \frac{1}{p_n} = 0.3249\,\frac{1 + z^{-1}}{1 - z^{-1}} \qquad (12.8.7)$$

combining Eqs. 12.8.5 and 12.8.6. The resulting digital filter is shown in Fig. 12.8.2c. It is obtained by replacing each $1/p_n$ analog integrator by the $0.3249(1 + z^{-1})/(1 - z^{-1})$ digital integrator. Combining Eqs. 12.8.4 and 12.8.7 gives

$$H(z) = 0.01201\,\frac{1 + 1.2015z^{-1} + 0.4038z^{-2} + 1.2015z^{-3} + z^{-4}}{1 - 2.787z^{-1} + 3.269z^{-2} - 1.842z^{-3} + 0.4189z^{-4}} \qquad (12.8.8)$$

This implementation requires 5 delays, 6 multipliers, and 11 summers.　■

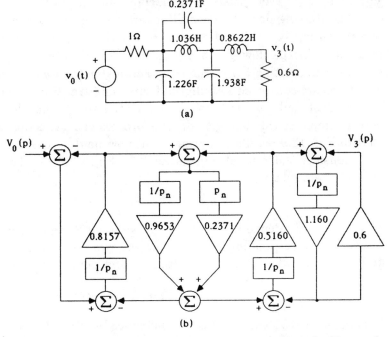

(a)

(b)

Fig. 12.8.2 4th order elliptic lowpass filter in Example 12.8.1 as (a) RLC analog filter, (b) block diagram equivalent; (c) digital simulation on next page.

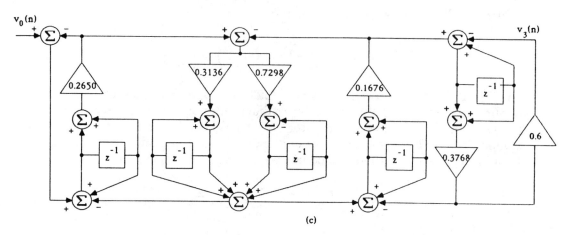

(c)

Fig. 12.8.2 (Continued) 4th order elliptic lowpass filter in Example 12.8.1 as (a) *RLC* analog filter, (b) block diagram equivalent, and (c) digital simulation.

12.9 WAVE (SCATTERING PARAMETER) REALIZATIONS

Ladder filters were designed using continued fraction expansions in Chap. 12.7. An alternative ladder filter design approach using *RLC* analog filter structures was discussed in Chap. 12.8. We now want to discuss another ladder design approach leading to the so-called *digital wave filters* or *scattering wave filters*.[10]

The starting point for such filters are still the *RLC* analog filter structures like the one shown in Fig. 12.8.1. However, rather than replacing analog integrators by digital integrators in the flow graph description of the filter, the *R*, *L*, and *C* elements are described by writing their *scattering parameter* (or *wave*) equations. Then these wave equations are converted into digital form using transforms such as the bilinear. Finally the incident and reflected waves are related to one another by reflection coefficients which completes the description. This process generates digital filters having forms like that in Fig. 12.9.6c. This approach will now be described in detail.

Consider the general impedance *Z* in Fig. 12.9.1a. The port voltage *V* and port current *I* are related as

$$V = ZI, \qquad I = \frac{V}{Z} \qquad (12.9.1)$$

Now arbitrarily define an incident voltage wave V_i and reflected voltage wave V_r. Also define the corresponding incident current wave I_i and reflected current wave I_r. In terms of these waves, the port voltage *V* and port current *I* equal

$$V = V_i + V_r, \qquad I = I_i - I_r \qquad (12.9.2)$$

V is the sum of V_i and V_r. *I* is the difference between I_i and I_r since they have opposite directions. The voltage and current waves are related by the characteristic or reference impedance Z_o as

$$V_i = Z_o I_i, \qquad V_r = Z_o I_r \qquad (12.9.3)$$

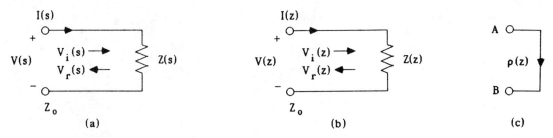

Fig. 12.9.1 Incident and reflected voltage and current waves in (a) analog port, (b) digital port, and (c) incident and reflected waves in digital port.

Combining Eqs. 12.9.2–12.9.3 and solving for V_i and V_r yields

$$V_i = \frac{V + IZ_o}{2}, \qquad V_r = \frac{V - IZ_o}{2} \tag{12.9.4}$$

V is the port voltage and I is the port current. For notational simplicity, we define a new incident wave A and a new reflected wave B as

$$A \equiv 2V_i = V + IZ_o, \qquad B \equiv 2V_r = V - IZ_o \tag{12.9.5}$$

Points of impedance imbalance or discontinuity create reflected wave B from an incident wave A. Terminating a network having reference impedance Z_o in a load Z creates reflections. Since $V = ZI$ at such a termination, then evaluating the ratio B/A from Eq. 12.9.5 yields

$$\rho(p) = \frac{B(p)}{A(p)} = \frac{V_r}{V_i} = -\frac{I_r}{I_i} = \frac{\dfrac{V}{I} - Z_o}{\dfrac{V}{I} + Z_o} = \frac{\dfrac{Z(p)}{Z_o(p)} - 1}{\dfrac{Z(p)}{Z_o(p)} + 1} \tag{12.9.6}$$

ρ is defined as the reflection coefficient at the load impedance Z. At matched terminations, $Z = Z_o$ so there are no reflections in either direction. Wave A produces **no wave B and wave B produces no wave A.**

Now convert the analog $\rho(p)$ of Eq. 12.9.6 into a digital $\rho(z)$ as

$$\rho(z) = \frac{B(z)}{A(z)} = \left. \frac{\dfrac{Z(p)}{Z_o(p)} - 1}{\dfrac{Z(p)}{Z_o(p)} + 1} \right|_{p=g(z)} \tag{12.9.7}$$

using the desired transform $p = g(z)$ as discussed in Eq. 12.8.1. The reflection coefficient $\rho(z)$ can be evaluated for each type of element used to construct RLC analog filters. Assuming that the reference impedance Z_o is resistive as R_o and that $Z(p) = R, pL,$ or $1/pC$, then Eq. 12.9.7 becomes

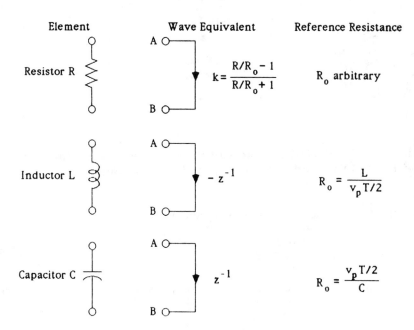

Element	Wave Equivalent	Reference Resistance

Fig. 12.9.2 Relation between incident and reflected waves in R, L, and C.

$$\rho_R(z) = \frac{\dfrac{R}{R_o} - 1}{\dfrac{R}{R_o} + 1}, \qquad \rho_L(z) = \frac{\dfrac{pL}{R_o} - 1}{\dfrac{pL}{R_o} + 1}, \qquad \rho_C(z) = \frac{\dfrac{1/pC}{R_o} - 1}{\dfrac{1/pC}{R_o} + 1} = \frac{1 - pR_oC}{1 + pR_oC}$$

$$(12.9.8)$$

Observe that the bilinear transform is matched to Eq. 12.9.8 since it has the same mathematical bilinear form where

$$p_n = \frac{2}{v_pT}\frac{1 - z^{-1}}{1 + z^{-1}}, \qquad z^{-1} = \frac{1 - p_nv_pT/2}{1 + p_nv_pT/2} = \frac{1 - pT/2}{1 + pT/2} \qquad (12.9.9)$$

from Eq. 11.2.2. Using the bilinear transform, then the reflection coefficients of Eq. 12.9.8 equal

$$\rho_R(z) = k \quad \text{for any} \quad R_o \text{ and } R$$

$$\rho_L(z) = -z^{-1} \quad \text{when} \quad R_o = \frac{L}{v_pT/2}$$

$$(12.9.10)$$

$$\rho_C(z) = z^{-1} \quad \text{when} \quad R_o = \frac{v_pT/2}{C}$$

This shows that the A and B waves are scaled versions of one another in resistors. They are delayed versions of one another in inductors (opposite sign) and capacitors (same sign). Each inductor L and each capacitor C has its own particular reference resistance R_o to produce the $\rho = \pm z^{-1}$ equivalence in Eq. 12.9.10. However, the resistor R does not require any particular R_o. Selecting $R_o = R$ is convenient because this forces

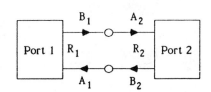

Fig. 12.9.3 Cascade connection between adaptors.

$\rho_R(z) = 0$ from Eq. 12.9.8. These results are summarized in Fig. 12.9.2. *Adaptor circuits* must be used to interconnect elements having different R_o reference resistances.

These adaptor circuits will now be described. Consider the cascade interconnection between port 1 and port 2 in Fig. 12.9.3. Since the port voltage $V_1 = V_2$ and the port current $I_1 = -I_2$, then

$$A_1 = B_2, \qquad B_1 = A_2 \tag{12.9.11}$$

The incident wave at port 1 equals the reflected wave at port 2. The reflected wave at port 1 equals the incident wave at port 2.

The series connection case is shown in Fig. 12.9.4a. The voltage and current waves must satisfy

$$V_1 + V_2 + V_3 = 0, \qquad I_1 = I_2 = I_3 \tag{12.9.12}$$

at a series connection. In terms of the waves, this is equivalent to

$$A_1 + B_1 + A_2 + B_2 + A_3 + B_3 = 0, \qquad \frac{A_1 - B_1}{R_1} = \frac{A_2 - B_2}{R_2} = \frac{A_3 - B_3}{R_3} \tag{12.9.13}$$

Solving for the reflected waves in terms of the incident waves in Eq. 12.9.13 gives

$$\begin{aligned}
B_1 &= A_1 - \beta_1(A_1 + A_2 + A_3) \\
B_2 &= A_2 - \beta_2(A_1 + A_2 + A_3) \\
B_3 &= A_2 - \beta_3(A_1 + A_2 + A_3) \\
    &= -(A_1 + A_2 + A_3) - (B_1 + B_2)
\end{aligned} \tag{12.9.14}$$

where the β_i coefficients equal

$$\beta_1 = \frac{2R_1}{R_1 + R_2 + R_3}, \qquad \beta_2 = \frac{2R_2}{R_1 + R_2 + R_3}, \qquad \beta_3 = \frac{2R_3}{R_1 + R_2 + R_3} \tag{12.9.15}$$

The parallel connection case is shown in Fig. 12.9.5a. The voltage and current waves must satisfy

$$V_1 = V_2 = V_3, \qquad I_1 + I_2 + I_3 = 0 \tag{12.9.16}$$

at a parallel junction. In terms of waves, this is equivalent to

$$A_1 + B_1 = A_2 + B_2 = A_3 + B_3, \qquad \frac{A_1 - B_1}{R_1} + \frac{A_2 - B_2}{R_2} + \frac{A_3 - B_3}{R_3} = 0 \tag{12.9.17}$$

Solving for the reflected waves in terms of the incident waves in Eq. 12.9.17 gives

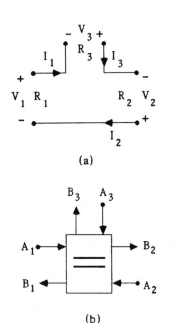

(a)

(b)

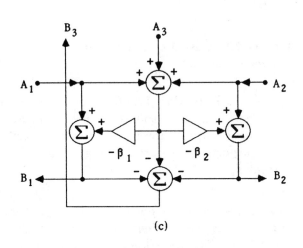

(c)

Fig. 12.9.4 (a) Series connection, (b) block diagram wave equivalent, and (c) series adaptor realization.

$$B_1 = \alpha_1 A_1 + \alpha_2 A_2 + \alpha_3 A_3 - A_1$$
$$B_2 = \alpha_1 A_1 + \alpha_2 A_2 + \alpha_3 A_3 - A_2$$
$$B_3 = \alpha_1 A_1 + \alpha_2 A_2 + \alpha_3 A_3 - A_3 \qquad (12.9.18)$$
$$= R_3(G_1 A_1 + G_2 A_2 + G_3 A_3) - R_3(G_1 B_1 + G_2 B_2)$$

where the α_i coefficients equal $(G_i = 1/R_i)$

$$\alpha_1 = \frac{2G_1}{G_1 + G_2 + G_3}, \qquad \alpha_2 = \frac{2G_2}{G_1 + G_2 + G_3}, \qquad \alpha_3 = \frac{2G_3}{G_1 + G_2 + G_3} \qquad (12.9.19)$$

A reflected wave at port k can be made *independent* of an incident wave at the same port by setting $\beta_k = 1$ for the series case and $\alpha_k = 1$ for the parallel case. Such a port is said to be *open*. For example in the port 2 open case, series connections use

$$R_2 = R_1 + R_3, \qquad \beta_1 = \frac{R_1}{R_1 + R_3}, \qquad \beta_2 = 1, \qquad \beta_3 = \frac{R_3}{R_1 + R_3} \qquad (12.9.20)$$

and parallel connections use

$$G_2 = G_1 + G_3, \qquad \alpha_1 = \frac{G_1}{G_1 + G_3}, \qquad \alpha_2 = 1, \qquad \alpha_3 = \frac{G_3}{G_1 + G_3} \qquad (12.9.21)$$

In the port 2 open case, the reflected wave B_2 is independent of the incident wave A_2 but only dependent upon incident waves A_1 and A_3.

The *RLC* ladder structure in Fig. 12.8.1a can be realized as a wave filter using N delays, $2N$ multipliers, and $5N$ summers for N even (or $5N - 1$ for N odd). The total

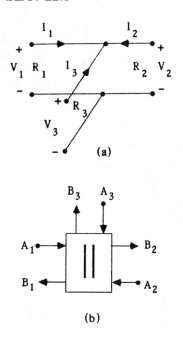

(a)

(b)

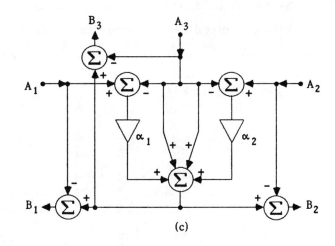

(c)

Fig. 12.9.5 (a) Parallel connection, (b) block diagram wave equivalent, and (c) parallel adaptor realization.

complexity is $8N$ (for N even) which is the highest of all the filter forms (due to the large number of summers). However the complexity varies with the RLC structure.

EXAMPLE 12.9.1 Design the 4th order elliptic lowpass filter of Example 12.2.1 using a digital wave filter.

SOLUTION This approach utilizes the tabulated RLC ladder filter used in Example 12.8.1 which is shown in Fig. 12.8.2a. The transfer function of this filter is given by Eq. 12.8.5. It has 0.28 dB in-band ripple, a ripple corner frequency of 1 rad/sec, and a dc gain of $R_L/(R_S + R_L) = 0.375$. The denormalized corner frequency for the analog RLC ladder equals $v_p T/2 = 1/3.078 = 0.3249$ from Eq. 12.8.6.

We first redraw the RLC ladder in wave filter form using series and parallel adaptor circuits. The result is shown in Figs. 12.9.6a–b. Each $R, L,$ and C element has a reference resistance R_i associated with it. Since the elements equal

$$R_S = 1, \qquad R_L = 0.6$$
$$L_1 = 1.036, \qquad L_2 = 0.8622 \tag{12.9.22}$$
$$C_1 = 1.226, \qquad C_2 = 1.938, \qquad C_3 = 0.2371$$

then using Eq. 12.9.10, their reference resistances equal

$$R_{R_S} = 1, \qquad R_{R_L} = 0.6 \qquad (\text{assumes } \rho_R = 0)$$
$$R_{L_1} = \frac{1.036}{0.3249} = 3.189, \qquad R_{L_2} = \frac{0.8622}{0.3249} = 2.654 \tag{12.9.23}$$
$$R_{C_1} = \frac{0.3249}{1.226} = 0.2650, \qquad R_{C_2} = \frac{0.3249}{1.938} = 0.1677, \qquad R_{C_3} = \frac{0.3249}{0.2371} = 1.370$$

Now we compute the port immittances and gains of the adaptors as follows.

Voltage source V_S – Source resistance R_S, Wave B_1 Input 1 (see Eq. 12.9.5):

$$B_1 = V_S - R_S I, \qquad A_1 \text{ arbitrary} \tag{12.9.24}$$

Parallel adaptor 2 (Port 2 open):

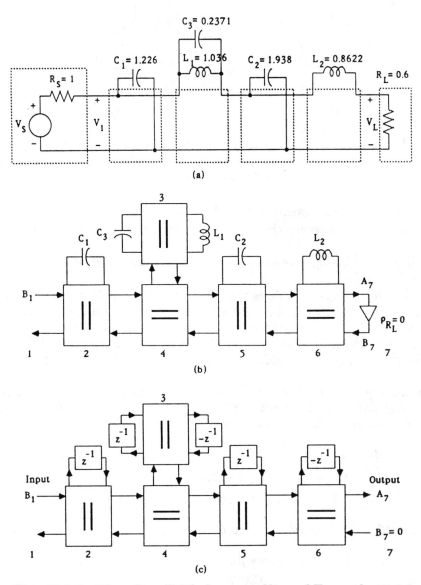

(a)

(b)

(c)

Fig. 12.9.6 4th order elliptic lowpass filter of Example 12.9.1 as (a) interconnection of series and parallel RLC elements, (b) interconnection of series and parallel adaptors, and (c) wave filter realization.

$$G_{21} = \frac{1}{R_{R_S}} = 1, \qquad G_{23} = \frac{1}{R_{C_1}} = 3.773, \qquad G_{22} = G_{21} + G_{32} = 4.773$$

$$\alpha_1 = \frac{G_{21}}{G_{21} + G_{23}} = 0.2095, \qquad \alpha_3 = \frac{G_{23}}{G_{21} + G_{23}} = 0.7905, \qquad \alpha_2 = 1$$

$$\text{(12.9.25)}$$

Parallel adaptor 3 (Port 2 open):

$$G_{31} = \frac{1}{R_{C_3}} = 0.7297, \qquad G_{33} = \frac{1}{R_{L_1}} = 0.3136, \qquad G_{32} = G_{31} + G_{32} = 1.043$$

$$\alpha_1 = \frac{G_{31}}{G_{31} + G_{33}} = 0.6994, \qquad \alpha_3 = \frac{G_{33}}{G_{31} + G_{33}} = 0.3006, \qquad \alpha_2 = 1$$

$$\text{(12.9.26)}$$

Series adaptor 4 (Port 2 open):

$$R_{41} = \frac{1}{G_{22}} = 0.2095, \qquad R_{43} = \frac{1}{G_{32}} = 0.9585, \qquad R_{42} = R_{41} + R_{43} = 1.168$$

$$\beta_1 = \frac{R_{41}}{R_{41} + R_{43}} = 0.1794, \qquad \beta_3 = \frac{R_{43}}{R_{41} + R_{43}} = 0.8206, \qquad \beta_2 = 1$$

$$\text{(12.9.27)}$$

Parallel adaptor 5 (Port 2 open):

$$G_{51} = \frac{1}{R_{42}} = 0.8562, \qquad G_{53} = \frac{1}{R_{C_2}} = 5.964, \qquad G_{52} = G_{51} + G_{53} = 6.821$$

$$\alpha_1 = \frac{G_{51}}{G_{51} + G_{53}} = 0.1255, \qquad \alpha_3 = \frac{G_{53}}{G_{51} + G_{53}} = 0.8745, \qquad \alpha_2 = 1$$

$$\text{(12.9.28)}$$

Series adaptor 6:

$$R_{61} = \frac{1}{G_{52}} = 0.1466, \qquad R_{63} = R_{L_2} = 2.654, \qquad R_{62} = R_{R_L} = 0.6$$

$$\beta_1 = \frac{2R_{61}}{R_{61} + R_{62} + R_{63}} = 0.08622, \qquad \beta_3 = \frac{2R_{63}}{R_{61} + R_{62} + R_{63}} = 1.561$$

$$\beta_2 = \frac{2R_{62}}{R_{61} + R_{62} + R_{63}} = 0.3529,$$

$$\text{(12.9.29)}$$

Load resistance R_L, Wave A_7 Output 7:

$$B_7 = \rho_{R_L} A_7 = 0 \quad \text{since} \quad \rho_{R_L} = 0 \qquad \text{(12.9.30)}$$

The resulting wave filter is shown in Fig. 12.9.6c. The input is B_1 and the output is A_7. The series and parallel adaptor blocks are each implemented as shown in Figs. 12.9.4c and 12.9.5c, respectively, using the α_i and β_i coefficients in Eqs. 12.9.25–12.9.29. This filter requires 5 delays, 10 multipliers, and 26 summers which are excessive. ∎

12.10 COMPARISON OF REALIZATIONS

To implement a digital filter $H(z)$ in either hardware or software, one of the structures of this chapter must be chosen. Seven different filter forms (multiple feedback, cascade, parallel, lattice, ladder, analog simulation, and wave) have been discussed. Since each form has a large number (if not infinite number) of implementations, a wide variety of design possibilities exist. Unfortunately, the properties of the various realization forms have not been fully interrelated so comparisons are difficult.[11] Therefore, few general statements are possible.

Some of the most important digital filter properties are:

1. *Complexity:* Related to the total number of delays, multipliers, and summers.
2. *Cost:* Proportional to complexity.
3. *Speed/Sampling Rate:* Related either to number of multipliers (no hardware multipliers) or to complexity (with hardware multipliers).
4. *Memory:* Determined by total number of delay elements and filter coefficients.
5. *Sensitivity of pole/zero locations:* Controlled by word length and arithmetic used in computations (fixed- or floating-point).
6. *Data Quantization, Coefficient Truncation, and Product Roundoff Noise:* Determined by word length.
7. *Limit cycles:* Low level oscillation which continues indefinitely due to quantization effects.
8. *Dynamic range:* Determined by word length, arithmetic used, and filter structure.

The complexity depends directly upon the number of digital elements required. Complexity depends indirectly upon filter type (lowpass, highpass, bandpass, bandstop, allpass), the filter gain characteristic (Butterworth, Chebyshev, elliptic, etc.), and the arithmetic used for computations. Table 12.10.1 shows that based upon total complexity, 1F multiple feedback structures are the simplest ($3N + 3$ total) and wave structures ($8N$ total) are the most complex. Filter cost is roughly proportional to complexity.

Speed is depended primarily by the number of multipliers. From Table 12.10.1, multiple feedback, parallel (except 5F), and ladder filters require the minimum number of multipliers ($2N + 1$). The maximum number is needed by 5F cascade ($3.5N$) and 5F parallel ($3N + 1$) filters. When the filter incorporates high-speed hardware multipliers, speed also depends upon the number of summers and the amount of memory. Then speed is proportional to total complexity. In this case, 1F multiple feedback structures are the fastest ($3N + 3$ total) and wave structures ($8N$ total) are the slowest. Parallel processing techniques can be used to produce the highest speed filters.

The memory requirements are dictated by the number of delay elements required (i.e., data storage) and the number of filter coefficients that are used (i.e., the $2N + 1$ coefficients a_i and b_i). Memory is minimized by using canonical forms (N delay elements). Almost all the filter structures of Table 12.10.1 are canonical. However, the 3F and 4F multiple feedback structures are inefficient since they require twice the number of delay elements ($2N$ delay elements and $2N + 1$ coefficients).

Table 12.10.1 Implementation complexity of digital filter structures. (*Depends upon analog filter structure.)

Chapter	Structure	Delays	Multipliers	Summers	Total
12.2	Multiple feedback 1F	N	$2N+1$	2	$3N+3$
	2F	N	$2N+1$	$N+1$	$4N+2$
	3F	$2N$	$2N+1$	1	$4N+2$
	4F	$2N$	$2N+1$	$2N$	$6N+1$
12.3	Biquad section 1F	2	5	2	9
	2F	2	5	3	10
	3F	4	5	1	10
	4F	4	5	4	13
	5F	2	7	3	12
12.4	Cascade 1F	N	$2.5N$	$0.5N+1$	$4N+1$
	2F	N	$2.5N$	$1.5N$	$5N$
	3F	$N+2$	$2.5N$	$0.5N$	$4N+2$
	4F	$N+2$	$2.5N$	$1.5N+1$	$5N+3$
	5F	N	$3.5N$	$1.5N$	$6N$
12.5	Parallel 1F	N	$2N+1$	$0.5N+1$	$3.5N+1$
	2F	N	$2N+1$	$N+1$	$4N+1$
	3F	$N+1$	$2N+1$	$0.5N+1$	$3.5N+3$
	4F	$N+1$	$2N+1$	$N+2$	$4N+4$
	5F	N	$3N+1$	1	$4N+2$
12.6	Lattice	N	$3N+1$	$2N+1$	$6N+2$
12.7	Ladder	N	$2N+1$	$2N$	$5N+1$
12.8	Analog Simulation*	N	$N+2$	$N+1$	$3N+3$
12.9	Wave*	N	$2N$	$5N$	$8N$

The sensitivity of the pole-zero locations of the filter and its response depends upon the word length (i.e., finite word size and coefficient truncation) and the type of arithmetic used in computations. Cascade, parallel, analog simulation, and wave structures have low sensitivity to coefficient truncation (or coefficient quantization) but multiple feedback and ladder structures have high sensitivity. Fixed-point arithmetic like two's complement simplifies arithmetic operations but produces high sensitivity filters. Floating-point arithmetic reduces coefficient sensitivity but requires increased hardware complexity. Cascade, parallel, and wave structures have high sensitivity to fixed-point arithmetic. Cascade and wave structures have lower sensitivity to floating-point arithmetic.

Finite word sizes are used to store and represent data, filter coefficients, and to perform products and additions. Not only does word size control the sensitivity of the filter, but it produces what may be modelled as noise in the filter. Product quantization noise is low in parallel structures, medium in cascade structures, and high in multiple feedback and ladder structures. Another problem that can exist in digital filters are limit cycles. These are low-level oscillations which continue indefinitely and are due to finite word sizes (i.e., quantization effects).

Dynamic range is also an important consideration. When there is insufficient dynamic range, signals either overflow or underflow registers and signal distortion results. Dynamic range depends upon word length, type of arithmetic used, and structure. It can be easily analyzed in cascade and parallel structures but is difficult to evaluate in other forms. Floating-point arithmetic usually increases dynamic range.

Since new filter comparisons are always being reported in the current professional journals, they must be studied to remain informed of new analytic discoveries. However, a pragmatic rather than analytic approach can be used to select the best filter realization. First, a filter is designed to meet the system requirements using several different realizations. Then these various realizations are implemented in computer software using different word lengths. The appropriate responses are then scrutinized to determine which realization has the best properties for the required application.

PROBLEMS

Design the filters in Problems 12.1–12.48 using each of the following methods: (a) Multiple feedback (1F form), (b) cascaded (2F sections), (c) parallel (3F sections), (d) lattice, (e) ladder.

Problems 12.1–12.6 describe a 1st order Butterworth lowpass filter having a unity dc gain and a normalized 3 dB corner frequency of $3.6°$.

12.1 $H_{001}(z) = \dfrac{0.0591}{1 - 0.9409z^{-1}}$

12.2 $H_{011}(z) = \dfrac{0.0304(1 + z^{-1})}{1 - 0.9391z^{-1}}$

12.3 $H_{111}(z) = \dfrac{0.0628z^{-1}}{1 - 0.9372z^{-1}}$

12.4 $H_{112}(z) = \dfrac{0.1255z^{-1}}{1 + 0.1255z^{-1} - z^{-2}}$

12.5 $H_{MBDI}(z) = \dfrac{0.03135z^{-1}(1 + z^{-1})}{1 - 0.96865z^{-1} + 0.03135z^{-2}}$

12.6 $H_{ODI}(z) = \dfrac{0.03141z^{-0.25}}{1 + 0.1255z^{-0.25} - z^{-0.5}}$

Problems 12.7–12.9 describe a 3rd order Butterworth lowpass filter having a unity dc gain and a normalized 3 dB corner frequency of 36°.

12.7 $H_{011}(z) = \dfrac{0.01809893(1+z^{-1})^3}{1 - 1.76004188z^{-1} + 1.18289326z^{-2} - 0.27805992z^{-3}}$

12.8 $H_{112}(z) = \dfrac{1.62459847z^{-3}}{1 + 2.35114101z^{-1} - 0.23606798z^{-2} - 3.07768331z^{-3} \\ \qquad + 0.23606798z^{-4} + 2.35114101z^{-5} - z^{-6}}$

12.9 $H_{ODI}(z) = \dfrac{0.03062579z^{-0.75}}{1 + 0.62573786z^{-0.25} - 2.80422607z^{-0.5} - 1.22084993z^{-0.75} \\ \qquad + 2.80422607z^{-1} + 0.62573786z^{-1.25} - z^{-1.5}}$

Problems 12.10–12.12 describe a 3rd order elliptic lowpass filter having a unity dc gain, a normalized 1.25 dB corner frequency of 3.6°, and 40 dB minimum stopband rejection.

12.10 $H_{001}(z) = \dfrac{1 - 1.94412711z^{-1} + 0.97206355z^{-2}}{1 - 2.9378776z^{-1} + 2.88042393z^{-2} - 0.94243772z^{-3}}$

12.11 $H_{111}(z) = \dfrac{0.00400999(z^{-1} - 2z^{-2} + 1.02873979z^{-3})}{1 - 2.94364489z^{-1} + 2.89189801z^{-2} - 0.94813788^{-3}}$

12.12 $H_{MBDI}(z) = \dfrac{0.00200293(z^{-1} - z^{-2} - 0.99282924z^{-3} + 1.02151228z^{-4} \\ \qquad\qquad + 0.02151228z^{-5} + 0.00717076z^{-6})}{1 - 2.97185026z^{-1} + 2.97300005z^{-2} - 1.02698560z^{-3} \\ \qquad + 0.02704305z^{-4} - 0.0011067z^{-5} + 0.00001436z^{-6}}$

Problems 12.13–12.18 describe a 1st order Butterworth highpass filter having a unity dc gain and a normalized 3 dB corner frequency of 3.6°.

12.13 $H_{001}(z) = \dfrac{0.9409(1 - z^{-1})}{1 - 0.9409z^{-1}}$

12.14 $H_{011}(z) = \dfrac{0.9695(1 - z^{-1})}{1 - 0.9390z^{-1}}$

12.15 $H_{111}(z) = \dfrac{(1 - z^{-1})}{1 - 0.9372z^{-1}}$

12.16 $H_{112}(z) = \dfrac{1 - z^{-2}}{1 + 0.1256z^{-1} - z^{-2}}$

12.17 $H_{MBDI}(z) = \dfrac{1 - z^{-1}}{1 - 0.9686z^{-1} + 0.0314z^{-2}}$

12.18 $H_{ODI}(z) = \dfrac{1 - z^{-0.5}}{1 + 0.0314z^{-0.25} - z^{-0.5}}$

Problems 12.19–12.21 describe a 3rd order Butterworth highpass filter having a unity dc gain and a normalized 3 dB corner frequency of 36°.

12.19 $H_{011}(z) = \dfrac{0.5276241(1 - 3z^{-1} + 3z^{-2} - z^{-3})}{1 - 1.7600408z^{-1} + 1.1828921z^{-2} - 0.2780595z^{-3}}$

12.20 $H_{112}(z) = \dfrac{1 - 3z^{-2} + 3z^{-4} - z^{-6}}{1 + 2.35114z^{-1} + 0.23607z^{-2} - 3.07768z^{-3} + 0.23607z^{-4}}$
$\qquad\qquad\qquad\qquad + 2.35114z^{-5} - z^{-6}$

12.21 $H_{ODI}(z) = \dfrac{1 - 3z^{-0.5} + 3z^{-1} - z^{-1.5}}{1 + 0.62574z^{-0.25} - 2.80423z^{-0.5} - 1.22085z^{-0.75} + 2.80423z^{-1}}$
$\qquad\qquad\qquad\qquad + 0.62574z^{-1.25} - z^{-1.5}$

Problems 12.22–12.24 describe a 3rd order elliptic highpass filter having a unity dc gain, a normalized 1.25 dB corner frequency of 3.6°, and 40 dB minimum stopband rejection.

12.22 $H_{001}(z) = \dfrac{1 - 3z^{-1} + 3z^{-2} - z^{-3}}{1 - 2.85036z^{-1} + 2.70861z^{-2} - 0.85779z^{-3}}$

12.23 $H_{111}(z) = \dfrac{1 - 3z^{-1} + 3.000541z^{-2} - 1.000541z^{-3}}{1 - 2.84235z^{-1} + 2.719325z^{-2} - 0.87291z^{-3}}$

12.24 $H_{MBDI}(z) = \dfrac{1 - 3z^{-1} + 3.00014z^{-2} - 0.999865z^{-3} - 0.00014z^{-4} - 0.00014z^{-5}}{1 - 2.92125z^{-1} + 2.92315z^{-2} - 1.076785z^{-3} + 0.07705z^{-4}}$
$\qquad\qquad\qquad\qquad - 0.001698z^{-5} + 0.00006628z^{-6}$

Problems 12.25–12.30 describe a 1st order Butterworth bandpass filter having a unity dc gain and a normalized 3 dB corner frequency of 3.6°.

12.25 $H_{001}(z) = \dfrac{0.044587(1 - z^{-1})}{1 - 1.464757z^{-1} + 0.710085z^{-2}}$

12.26 $H_{011}(z) = \dfrac{0.030469(1 - z^{-2})}{1 - 1.568735z^{-1} + 0.939062z^{-2}}$

12.27 $H_{111}(z) = \dfrac{0.062791(1 - z^{-1})}{1 - 1.937209z^{-1} + 1.282701z^{-2}}$

12.28 $H_{112}(z) = \dfrac{0.125581(1 - z^{-2})}{1 - 0.125581z^{-1} - 0.618034z^{-2} - 0.125581z^{-3} + z^{-4}}$

12.29 $H_{MBDI}(z) = \dfrac{0.031364(1 - z^{-2})}{1 - 1.968636z^{-1} + 1.069098z^{-2} + 0.138197z^{-3} + 0.069098z^{-4}}$

12.30 $H_{ODI}(z) = \dfrac{0.0314146(1 - z^{-2})}{1 - 0.0314146z^{-1} - 1.902011z^{-2} - 0.0314146z^{-3} + z^{-4}}$

Problems 12.31–12.33 describe a 3rd order Butterworth bandpass filter having a unity dc gain and a normalized 3 dB corner frequency of 36°.

12.31 $H_{011}(z) = \dfrac{0.000029146(1 - 3z^{-2} + 3z^{-4} - z^{-6})}{1 - 4.752454z^{-1} + 10.404607z^{-2} - 13.088947296z^{-3}}$
$\qquad\qquad\qquad\qquad + 9.977678z^{-4} - 4.370441z^{-5} + 0.881893z^{-6}$

12.32 $H_{112}(z) = \dfrac{0.0019805z^{-3}(1 - 3z^{-2} + 3z^{-4} - z^{-6})}{1 + 0.251162z^{-1} - 1.822561z^{-2} - 0.559635z^{-3}}$
$\qquad\qquad\qquad\qquad + 4.063322z^{-4} + 0.902772z^{-5} - 3.842203z^{-6}$
$\qquad\qquad\qquad\qquad - 0.902772z^{-7} + 4.063322z^{-8} + 0.559635z^{-9}$
$\qquad\qquad\qquad\qquad - 1.822561z^{-10} - 0.251162z^{-11} + z^{-12}$

12.33 $H_{ODI}(z) = \dfrac{0.000031002z^{-0.75}\left(1 - 3z^{-0.5} + 3z^{-1} - z^{-1.5}\right)}{\begin{array}{l} 1 + 0.062829z^{-0.25} - 5.704366z^{-0.5} - 0.301815z^{-0.75} \\ + 13.846402z^{-1} + 0.591901z^{-1.25} - 18.283133z^{-1.5} \\ - 0.591901z^{-1.75} + 13.846402z^{-2} + 0.301815z^{-2.25} \\ - 5.704366z^{-2.5} - 0.062829z^{-2.75} + z^{-3} \end{array}}$

Problems 12.34–12.36 describe a 3rd order elliptic bandpass filter having a unity dc gain, a normalized 1.25 dB corner frequency of 3.6°, and 40 dB minimum stopband rejection.

12.34 $H_{001}(z) = \dfrac{0.002899\big(1 - 3.957688z^{-1} + 6.611524z^{-2} - 5.828828z^{-3} + 2.718740z^{-4} - 0.543748021z^{-5}\big)}{1 - 4.4437393z^{-1} + 8.731771z^{-2} - 9.678956z^{-3} + 6.374393z^{-4} - 2.370220z^{-5} + 0.393059z^{-6}}$

12.35 $H_{111}(z) = \dfrac{0.004010z^{-1}\big(1 - 5z^{-1} + 10.719723z^{-2} - 12.159169z^{-4} + 7.249794z^{-5} - 1.810347z^{-6}\big)}{1 - 5.943645z^{-1} + 15.759308z^{-2} - 23.561726z^{-3} + 20.921646z^{-4} - 10.470511z^{-5} + 2.334847z^{-6}}$

12.36 $H_{MBDI}(z) = \dfrac{0.002003z^{-1}\left(\begin{array}{l} 1 - 4z^{-1} + 5.145368z^{-2} \\ - 5.431328z^{-3} + 4.019098z^{-5} - 0.540025z^{-6} \\ -0.174010z^{-7} - 0.019098z^{-9} - 0.000004697z^{-10}\end{array}\right)}{\begin{array}{l} 1 - 5.971850z^{-1} + 15.095846z^{-2} - 20.272236z^{-3} \\ + 14.805959z^{-4} - 5.289744z^{-5} + 0.890619z^{-6} \\ - 0.488287z^{-7} + 0.194220z^{-8} + 0.040296z^{-9} \\ + 0.013714z^{-10} + 0.001258z^{-11} + .000000325z^{-12} \end{array}}$

Problems 12.37–12.42 describe a 1st order Butterworth bandstop filter having a unity dc gain and a normalized 3 dB corner frequency of 3.6°.

12.37 $H_{001}(z) = \dfrac{1.003943 - 2z^{-1} + z^{-2}}{1.066733 - 2.062791z^{-1} + z^{-2}}$

12.38 $H_{011}(z) = \dfrac{1 - 1.996064z^{-1} + z^{-2}}{1.031426 - 1.996053z^{-1} + 0.968574z^{-2}}$

12.39 $H_{111}(z) = \dfrac{1 - 2z^{-1} + 1.003943z^{-2}}{1 - 1.937209z^{-1} + 0.941152z^{-2}}$

12.40 $H_{112}(z) = \dfrac{1 - 1.984228z^{-2} + z^{-4}}{1 + 0.125581z^{-1} - 1.984229z^{-2} - 0.125581z^{-3} + z^{-4}}$

12.41 $H_{MBDI}(z) = \dfrac{1 - 2z^{-1} + 1.000984z^{-2} + 0.001967z^{-3} + 0.000984z^{-4}}{1 - 1.968636z^{-1} + 1.000984z^{-2} - 0.029397z^{-3} + 0.000983z^{-4}}$

12.42 $H_{ODI}(z) = \dfrac{1 - 1.999013z^{-0.5} + z^{-1}}{1 + 0.0314146z^{-1} - 1.999013z^{-2} - 0.0314146z^{-3} + z^{-4}}$

Problems 12.43–12.45 describe a 3rd order Butterworth bandstop filter having a unity dc gain and a normalized 3 dB corner frequency of 36°.

12.43 $H_{011}(z) = 0.939092(1 - 4.854102z^{-1} + 10.854103z^{-2} - 13.944272z^{-3}$

$$\frac{+10.854103z^{-4} - 4.854102z^{-5} + z^{-6})}{\begin{array}{l}1 - 4.752454z^{-1} + 10.404607z^{-2} - 13.088947296z^{-3} \\ + 9.977678z^{-4} - 4.370441z^{-5} + 0.881893z^{-6}\end{array}}$$

12.44 $H_{112}(z) = 1 - 1.854102z^{-2} + 4.145898z^{-4} - 3.944272z^{-6}$

$$\frac{+4.145898z^{-8} - 1.854102z^{-10} + z^{-12}}{\begin{array}{l}1 + 0.251162z^{-1} - 1.822561z^{-2} - 0.559635z^{-3} \\ + 4.063322z^{-4} + 0.902772z^{-5} - 3.842203z^{-6} \\ - 0.902772z^{-7} + 4.063322z^{-8} + 0.559635z^{-9} \\ - 1.822561z^{-10} - 0.251162z^{-11} + z^{-12}\end{array}}$$

12.45 $H_{ODI}(z) = 1 - 5.706339z^{-0.5} + 13.854103z^{-1} - 18.294583z^{-1.5}$

$$\frac{+13.854103z^{-2} - 5.706339z^{-2.5} + z^{-3}}{\begin{array}{l}1 + 0.062829z^{-0.25} - 5.704366z^{-0.5} - 0.301815z^{-0.75} \\ + 13.846402z^{-1} + 0.591901z^{-1.25} - 18.283133z^{-1.5} \\ - 0.591901z^{-1.75} + 13.846402z^{-2} + 0.301815z^{-2.25} \\ - 5.704366z^{-2.5} - 0.062829z^{-2.75} + z^{-3}\end{array}}$$

Problems 12.46–12.48 describe a 3rd order elliptic bandstop filter having a unity dc gain, a normalized 1.25 dB corner frequency of 3.6°, and 40 dB minimum stopband rejection.

12.46 $H_{001}(z) = 13.965210(1 - 4.459047z^{-1} + 8.856994z^{-2} - 9.910822z^{-3}$

$$\frac{+6.581895z^{-4} - 2.462514z^{-5} + 0.410419z^{-6})}{\begin{array}{l}1 - 4.404593z^{-1} + 8.612865z^{-2} - 9.466534z^{-3} \\ + 6.161245z^{-4} - 2.253924z^{-5} + 0.366036z^{-6}\end{array}}$$

12.47 $H_{111}(z) = 15.0585141 - 6z^{-1} + 16.037016z^{-2} - 24.148063z^{-3}$

$$\frac{+21.580374z^{-4} - 9.004824z^{-5} + 1.810880z^{-6}}{\begin{array}{l}1 - 5.842349z^{-1} + 15.255824z^{-2} - 22.490331z^{-3} \\ + 19.720271z^{-4} - 9.762274z^{-5} + 2.160100z^{-6}\end{array}}$$

12.48 $H_{MBDI}(z) = 15.658514(1 - 6z^{-1} + 15.207431z^{-2} - 20.414859z^{-3}$

$$\frac{\begin{array}{l}+ 14.806906z^{-4} - 5.141615z^{-5} + 0.778567z^{-6} \\ - 0.470212z^{-7} + 0.198046z^{-8} + 0.035264z^{-9} \\ +0.019282z^{-10} + 0.0019795z^{-11} + 0.0003299z^{-9})\end{array}}{\begin{array}{l}1 - 5.921253z^{-1} + 14.894202z^{-2} - 20.013697z^{-3} \\ + 14.805264z^{-4} - 5.560533z^{-5} + 1.093178z^{-6} \\ - 0.518243z^{-7} + 0.199685z^{-8} + 0.022679z^{-9} \\ + 0.017899z^{-10} + 0.001604z^{-11} + 0.0000003299z^{-12}\end{array}}$$

Design the filters in Problems 12.49–12.69 using each of the following methods and the bilinear transform: (a) Analog filter simulation, (b) wave.

Problems 12.49–12.55 describe an nth order Butterworth RLC analog lowpass filter (see Fig. 12.8.1a). It has $R_S = R_L = 1$ and a 3 dB corner frequency of 1 rad/sec. Design the digital filter to have a normalized 3 dB corner frequency of 3.6^o.

	n	C_1	L_1	C_2	L_2	C_3	L_3	C_4	L_4
12.49	2	1.4142	1.4142						
12.50	3	1.0000	2.0000	1.0000					
12.51	4	0.7654	1.8478	1.8478	0.7654				
12.52	5	0.6180	1.6180	2.0000	1.6180	0.6180			
12.53	6	0.5176	1.4142	1.9319	1.9319	1.4142	0.5176		
12.54	7	0.4450	1.2470	1.8019	2.0000	1.8019	1.2470	0.4450	
12.55	8	0.3902	1.1111	1.6629	1.9616	1.9616	1.6629	1.1111	0.3902

Problems 12.56–12.61 describe an nth order 0.5 dB Chebyshev RLC analog lowpass filter (see Fig. 12.8.1a). It has $R_L = 1$ and a 0.5 dB corner frequency of 1 rad/sec. Design the digital filter to have a normalized 3 dB corner frequency of 3.6^o.

	n	R_S	C_1	L_1	C_2	L_2	C_3	L_3	C_4
12.56	2	2	0.9086	2.1030					
12.57	3	1	1.8636	1.2804	1.8636				
12.58	4	2	0.8452	2.7198	1.2383	1.9849			
12.59	5	1	1.8068	1.3025	2.6914	1.3025	1.8068		
12.60	6	2	0.8303	2.7042	1.2912	2.8721	1.2372	1.9557	
12.61	7	1	1.7896	1.2961	2.7177	1.3848	2.7177	1.2961	1.7896

Problems 12.62–12.67 describe an nth order 0.28 dB elliptic RLC analog lowpass filter (see Fig. 12.8.1a). It has $R_S = R_L = 1$ and C'_n in parallel with L_n. Its 0.28 dB corner frequency equals 1 rad/sec. Stopband rejection M_r is arbitrary. Design the digital filter to have a normalized 0.28 dB corner frequency of 3.6^o.

	n	M_r	C_1	L_1	C'_1	C_2	L_2	C'_2	C_3
12.62	3	40.84	1.2880	1.0693	0.0754	1.2280			
12.63	3	60.50	1.3324	1.1252	0.0164	1.3324			
12.64	3	81.67	1.3427	1.1382	0.0032	1.3427			
12.65	5	40.19	1.2752	1.1058	0.2288	1.7522	0.7859	0.6786	0.9789
12.66	5	60.51	1.3830	1.2252	0.0879	2.0501	1.0788	0.2390	1.2527
12.67	5	81.66	1.4280	1.2754	0.0330	2.1911	1.2166	0.0875	1.3764

Problems 12.68–12.74 describe an nth order Bessel RLC analog lowpass filter (see Fig. 12.8.1a). It has $R_S = R_L = 1$ and a 3 dB corner frequency of 1 rad/sec. Design the digital filter to have a normalized 3 dB corner frequency of 3.6^o.

	n	C_1	L_1	C_2	L_2	C_3	L_3	C_4	L_4
12.68	2	0.5755	2.1478						
12.69	3	0.3374	0.9705	2.2034					
12.70	4	0.2334	0.6725	1.0815	2.2404				
12.71	5	0.1743	0.5072	0.8040	1.1110	2.2582			
12.72	6	0.1365	0.4002	0.6392	0.8538	1.1126	2.2645		
12.73	7	0.1106	0.3259	0.5249	0.7020	0.8690	1.1052	2.2659	
12.74	8	0.0919	0.2719	0.4409	0.5936	0.7303	0.8695	1.0956	2.2656

REFERENCES

1. Antoniou, A., *Digital Filters: Analysis and Design*, Chap. 4.3, McGraw-Hill, NY, 1979.
2. Tretter, S.A., *Introduction to Discrete-Time Signal Processing*, Chap. 6.5, Wiley, NY, 1976.
3. Stanley, W.D., G.R. Dougherty, and R. Dougherty, *Digital Signal Processing*, 2nd. ed., Chap. 5.4, Reston, VA, 1984.
4. Oppenheim, A.V. and R.W. Schafer, *Digital Signal Processing*, Chap. 4.3, Prentice-Hall, NJ, 1975.
5. Rabiner, L.R. and B. Gold, *Theory and Application of Digital Signal Processing*, Chap. 2.19, Prentice-Hall, NJ, 1975.
6. Gray, A.H. and J.D. Markel, "Digital lattice and ladder filter synthesis," IEEE Trans. Audio Electroacoust., vol. AU-21, pp. 491–500, Dec. 1973.
7. Bose, N.K., *Digital Filters*, Chap. 3.7, Elsevier Science Publ., NY, 1985.
   Mitra, S.K. and R.J. Sherwood, "Canonical realizations of digital filters using the continued fraction expansion," IEEE Trans. Audio Electroacoust., vol. AU-20, pp. 185–194, Aug. 1972.
   — , "Digital ladder networks," IEEE Trans. Audio Electroacoust., vol. AU-21, pp. 30–36, Feb. 1973.
8. El-Masry, E.I. and A.A. Sakla, "Low-sensitivity digital ladder networks," Rec. 13th Asilomar Conf., pp. 273–278, Nov. 1979.
9. Zverev, A.I., *Handbook of Filter Synthesis*, 576 p., Wiley, NY, 1967.
   Hanell, G.E., *Filter Design and Evaluation*, 203 p., Van Nostrand Reinhold, NY, 1969.
10. Feltweis, A., "Digital filter structures related to classical filter networks," Arch. Eleck. Ubertragung., vol. 25, pp. 78–89, Feb. 1971.
    —, "Some principles of designing digital filters imitating classical filter structures," IEEE Trans. Circuit Theory, vol. CT-18, pp. 314–316, March 1971.
    —, "Pseudopassivity, sensitivity, and stability of wave digital filters," IEEE Trans. Circuit Theory, vol. CT-19, pp. 668-673, Nov. 1972.
11. Ref. 1, Chap. 12.11.

ANSWERS TO SELECTED PROBLEMS

Chapter 1

1.1 (a) Digital, active, LC; (e) LC, ceramic crystal.

1.4 (a) $F(j\omega) = 1/(2 - \omega^2 + j2\omega)$; (b) $\omega_n = \sqrt{2}$, slope $= -40$ dB/dec; (c) $\omega_{40dB} = 10\sqrt{2} = \omega_s$.

1.9 (a) $f(t) = U_{-1}(t+1) - U_{-1}(t-1)$; $F(s) = (e^s - e^{-s})/s = 2\sinh(s)/s$, all s.

1.14 (a) $F(j\omega) = F(s)|_{s=j\omega} = 2\sinh(s)/s|_{s=j\omega} = 2\sin(\omega)/\omega$.

1.19 (a) $f(t) = -U_{-1}(t+T/2) + 2U_{-1}(t+T/4) - 2U_{-1}(t-T/4) + U_{-1}(t-T/2)$;

$F(j\omega) = [\sin(\omega T/4) - \sin(\omega T/2)]/(\omega/2)$;

$F_p(j\omega) = 4\sum_{m=-\infty}^{\infty}(1/m)\sin(m\pi/2)U_0(\omega - m\omega_o)$, $\omega_o = 2\pi/T$;

$f_p(t) = (4/\pi)\sum_{\substack{m=1 \\ m \text{ odd}}}^{\infty}(-1)^{(m-1)/2}(1/m)\cos(m\omega_o t)$.

1.22 (c) $[f] = [e^0, e^{-0.25}, e^{-0.5}, e^{-0.75}]^t$ has $[F]_F = [2.86, 0.394 - j0.306, 0.355, 0.394 + j0.306]^t$.

1.25 (e) $F(z) = 0.5[z/(z - e^{-(2+j)})] + 0.5[z/(z + e^{-(2+j)})]$

$= z[z - e^{-2}\cos(1)]/[z^2 - 2ze^{-2}\cos(1) + e^{-4}], |z| > e^{-2}$.

1.27 (a) $f(n) = (-1/2)^n U_{-1}(n)$.

1.31 (a) $H(z) = z/[z(1+K) - 2]$; (b) $h(n) = \frac{1}{2}(1)^n U_{-1}(n)$.

1.37 (a) $H(z_1, z_2) = (1 - z_1^{-M})(1 - z_2^{-N})/(1 - z_1)(1 - z_2)$, all z_1 and z_2;

(b) $|H(e^{j\omega_1 T_1}, e^{j\omega_2 T_2})|$ in Fig. 2.10.1.

1.40 (b) $h(n) = U_{-1}(-n)$ is noncausal and conditionally stable.

1.42 (a) Root locus in Fig. 1.16.1b.

1.47 (a) $H(z) = \omega_c T(z-1)/[z^2 - 2\varsigma\omega_n z + \omega_n^2]$, $\omega_n^2 = 1 - (\omega_c T/Q) + (\omega_c T)^2$;

(c) $2(\omega_c T)/[1 + 2(\omega_c T)^2] \le Q \le 1/\omega_c T$.

1.48 (a) $E\{s(t)\} = E\{s(t) + n(t)\} = U_{-1}(t)$; $\sigma_{s+n}^2 = \sigma_s^2 + 2\rho\sigma_s\sigma_n + \sigma_n^2 = \sigma^2$.

Chapter 2

2.1 (b) $[f] = [e^0, e^{-1}, e^{-2}, e^{-3}]^t$ has $[F]_F = [1.55, 0.865 - j0.318, 0.718, 0.865 + j0.318]^t$.

2.4 (a) $[f] = [1, 1, 1, 1, -1, -1, -1, -1]^t$ has

$[F]_F = [0, 5.23e^{-j67.5°}, 0, 2.16e^{-j22.5°}, 0, 2.16e^{j22.5°}, 0, 5.23e^{j67.5°}]^t$.

2.7   (a) $[f] = [1, 0.5, 0.25, 0.13, -0.0625, -0.0313, -0.0156, -0.0078]^t$ has $[F]_C = [1.76, 1.34 - j0.735, 0.703 - j0.352, 0.781 - j0.204, 0.586, 0.781 + j0.204, 0.703 + j352, 1.34 + j0.735]^t$.

2.14 (a) $[f] = [0, 0.707, 1, 0.707, 0, 0, 0, 0]^t$ has $[F]_F = [2.41, -j2, -1, 0, -0.414, 0, -1, j2]^t$.

2.16 (b) $[F]_{Hann} = [0, -j2, j, 0, 0, 0, -j, j2]^t$.

2.20 (a) $[w]_{Hamm} = [0.086, 0.543, 1, 0.543]^t$; (b) $[f_w] = [0, 0.543, 1, 1.629]^t$.

2.22 (b) $[f] = [0, 0.25, 0.5, 0.75, 1, 0.75, 0.5, 0.25]^t$ has $[F]_S = [0.5, 0.109, 0.280, 0.061, 0, 0, 0, 0]^t$.

2.23 (a) $[f] = [0, 0.707, 1, 0.707, 0, 0, 0, 0]^t$ has $[F]_{Ha} = [0.302, 0.302, 0.177, 0, 0.177, 0.073, 0, 0]^t$.

Chapter 3

3.1   (b) $h(n) = -(-1/2)^n U_{-1}(-n - 1)$.

3.3   (a) $r(n) = (1/3)[2 + (-0.5)^n]U_{-1}(n)$.

3.7   (a) $x_1(n) = U_0(n) + (2^{n-1} - 1)U_{-1}(n)$ for $|z| > 2$;

        $x_2(n) = U_0(n) - U_{-1}(n) - 2^{n-1}U_{-1}(-n - 1)$ for $1 < |z| < 2$;

        $x_3(n) = U_0(n) + (1 - 2^{n-1})U_{-1}(-n - 1)$ for $|z| < 1$.

3.10 (b) $[H]_F = [1, 0, 0, 0]^t$ gives $[h] = (1/4)[1, 1, 1, 1]^t$.

3.13 (b) $[H]_F = [1, 0.5, 0, 0, 0, 0, 0, 0.5]^t$ gives $[h] = (1/8)[2, 1.707, 1, 0.293, 0, 0.293, 1, 1.707]^t$.

3.24 (b) $[H] = [1, 0.5, 0, 0, 0, 0, 0, 0.5]^t$ gives $[h]_P = [2, 1, 1, 2, 0, 1, 1, 0]^t$,

        $[h]_{Ha} = 0.5[3, 3, 3, 3, 1, 1, 3, -1]^t$, $[h]_S = [2, 0.88, 2, 0.89, 1.1, 0, 1.1, 0.01]^t$.

3.27 (a) $f_r(t) = U_{-1}(t - T) - U_{-1}(t - 3T)$;

        (b) $f_r(t) = U_{-2}(t) - U_{-2}(t - T) - U_{-2}(t - 2T) + U_{-2}(t - 3T)$.

3.31 (a) $f_r(t) = [\sin(\pi t)/\pi][-1/(t - T) + 1/(t - 2T)]$.

3.32 (b) $[F]_P = [0.79, 0.79, 0.56, 0.56, 0.24, 0.24, 0, 0]^t$,

        $[F]_S = [0.79, 0.40, 0.60, 0.91, 0, 0, 0.04, 0.04]^t$.

3.38 (a) $[F]_F = [2, 1 - j2, -1 - j, -1, 0, -1, -1 + j, 1 + j2]^t$;

        (b) $f_r(t) = 0.25[1 + \sqrt{5}\cos((\pi t/4) + 63.4°) - \sqrt{2}\cos((\pi t/2) - 45°) - \cos(3\pi t/4)]$

3.40 (a) $[F]_F = [1.55, 0.921e^{-j20°}, 0.718, 0.921e^{j20°}]^t$ has

        $f_r(t) = 0.388 + 0.230\cos(2\pi t + 20°) + 0.179\cos(4\pi t)$.

3.43 (a) $f_c(t) = [e^{-t} - e^{-(4-t)}]U_{-1}(t)$, $[f_c] = [0.95, 0.23, -0.23, -0.95]^t$,

        $[F_c]_F = [0, 1.18e^{-j45°}, 1.44, 1.18e^{j45°}]^t$.

3.44 (a) $[F]_{Ha} = 0.25[1, 1, 0, 0, -1, 1, 0, 0]^t$;

        (b) $f_r(t) = 0.25[Ha(0, t) + Ha(1, t) - Ha(4, t) + Ha(5, t)]$.

3.47 Droop $= 82(D/T)^2$ %, duty cycle $D/T \leq 11\%$ for droop $\leq 1\%$.

3.49 (b) For $2e^{-NT/2}/(1+\omega^2) \leq 1\%$ magnitude error, $NT \geq 10.6$ sec.

3.53 (a) Error $\cong 4/3\pi\omega_N^3 = 0.04\%$ when $\omega_N \cong 10$ and $|F| = 2/(1-\omega^2)$.

3.55 (a) $H_1(z) = (1-az)/z(z-a)$; (b) $h_1(t) = -aU_0(n-1) + (1-a^2)a^{n-2}U_{-1}(n-2)$.

Chapter 4

4.2 $(\gamma_1, \Omega_{r1}) = (2.2, 1.25)$, $(\gamma_2, \Omega_{r2}) = (4.2, 1.56)$;

Butterworth $n \geq 11$, Chebyshev $n \geq 6$, elliptic $n \geq 3$, Bessel impossible.

4.7 (a) $H(s) = K[s_n^2 + 3.522^2]/(s_n + 0.7765)[(s_n + 0.3475)^2 + 1.0859^2]$ where $s_n = s/2\pi(3.2$ KHz$)$, $\theta = 19°$.

4.11 (a) $r(t) = 0.5 + (1/\pi)\text{Si}(t-1)$, $r_{max} = r(1+\pi)$, $\gamma = 9\%$.

4.14 $H(s) = K/[(s_n + 0.5475)^2 + 0.5051^2][(s_n + 0.4438)^2 + 1.2671^2]$ where $s_n = s/2\pi(1$ KHz$)$.

4.17 (a) $H_{1,1}(s) = [1 - (s_n/2)]/[1 + (s_n/2)]$,

$H_{2,2}(s) = [1 - (s_n/2) + (s_n^2/12)]/[1 + (s_n/2) + (s_n^2/12)]$.

4.19 (a) $H_E(j\omega) = 1/(1+\omega^2)$, $H_M(j\omega) = -1/j\omega$, $H_{HR}(j\omega) = j\omega/(1+\omega^2)$;

(c),(d) $H_E(s) = 1/(1-s^2)$ is noncausal and stable; $H_M(s) = -1/s$ is causal and conditionally stable; $H_{HR}(s) = s/(1-s^2)$ is noncausal and stable.

4.24 (a) $\langle|R(j\omega)|^2\rangle = |R(j\omega)|^2 * B(j\omega) = \int_{\omega-0.05\omega_s}^{\omega+0.05\omega_s} |R(jv)|^2 * B(j(\omega-v))dv$, $\langle|S(j\omega)|^2\rangle_R \cong$ $0.1\omega_s/\omega^3$, $\langle|N(j\omega)|^2\rangle_R \cong 0.1\omega_s$; (b) $H_E(j\omega) = 1/\omega^2[\langle|S(j\omega)|^2\rangle_R + \langle|N(j\omega)|^2\rangle_R]$.

4.29 (a) $G_E(j\omega) = 1/\omega^2$, $G_M(j\omega) = -1/(1+j\omega)$, $G_{HR}(j\omega) = j\omega/(1+\omega^2 - j\omega)$.

4.34 (a) $\ln U_{-1}(t) = -100U_{-1}(-t)$ limiting to finite value, $\mathcal{F}[-100U_{-1}(-t)] = 100/j\omega$,

$\ln n(t)$ need explicit waveform; (b) $H_M(j\omega) = -100/j\omega|N_L(j\omega)|^2$.

4.39 (a) $S_C(j\omega) = \ln(1/j\omega) = -\ln\omega - j\pi/2$, $N_C(j\omega) = \ln(1e^{j \arg N}) = 0 + j \arg N(j\omega)$;

(b) $H_M(j\omega) = (-\ln\omega + j\pi/2)/|\arg N(j\omega)|^2$.

Chapter 5

5.1 (a) B: $(\gamma_1, \Omega_{a1}) = (2, 1.6)$, $(\gamma_2, \Omega_{a2}) = (4, 2.3)$, so Butterworth $n \geq 6$, Chebyshev $n \geq 4$, elliptic $n \geq 3$, Bessel impossible. (b) Bilinear has $\omega_i T = 2\tan^{-1}(0.314v_i)$,

$\Omega_{di} = \tan^{-1}(0.314\Omega_{ai})/\tan^{-1}(0.314) \leq \Omega_{ai}$, $(\gamma_1, \Omega_{d1}) = (2, 1.53)$, $(\gamma_2, \Omega_{d2}) = (4, 2.06)$.

5.5 $\omega_p T = 22.5°$, $\omega_r T = 28.9°$, $\Omega_d = 1.28$; bilinear has $\Omega_a = \tan(14.5°)/\tan(11.3°) = 1.30$;

Euler has $\Omega_a = \sin(28.9°)/\sin(22.5°) = 1.26$. Use $(M_p, M_r, \gamma, \Omega_a) = (0.28, 40, 5.18, \Omega_a)$.

5.12 (a) $H(z) = 0.01z^{-2}/(1 - 1.859z^{-1} + 0.869z^{-2})$;

(e) $H(z) = 0.01z^{-1}/(1 + 0.141z^{-0.5} - 1.9z^{-1} - 0.141z^{-1.5} + z^{-2})$.

5.15 (a) $\Delta(z) = 1 + T[1/(z-1)]$; (c) $\Delta(z) = 1 + T[(z+1)/(z-1)]$.

5.21 (a) $T = 0.0628$, $H_0(z) = 1/(1 - 0.9391z^{-1})$; $H_{0A}(z) = 0.06090/(1 - 0.9391z^{-1})$;
$H_1(z) = 0.06090z^{-1}/(1 - 0.9391z^{-1})$.

5.25 (a) $\Delta(z) = 1 - e^{-T}z^{-1}$.

5.31 $y_3(n) = 0.1585y_3(n-1) + 0.8415x_3(n-1)$.

5.33 (a) $H_{23}(z) = Tz^{-1}/(1 - z^{-1})$, $H_c(z) = [1 + (T-1)z^{-1}]/[(1+T) - z^{-1}]$;
(b) $x_1(n) = [1/(1+T)][x_1(n-1)+x(n)+(T-1)x(n-1)]$, $x_4(n) = x_1(n)-x_3(n)$; $x_2(n) = 1$ for
$x_4(n) > 0$, $x_2(n) = -1$ for $x_4(n) < 0$, $x_2(n) = 0$ for $x_4(n) = 0$; $x_3(n) = x_3(n-1)+Tx_2(n-1)$.

Chapter 6

6.1 (a) B: $(\gamma_1, \Omega_{a1}) = (2, 1/0.625)$, $(\gamma_2, \Omega_{a2}) = (4, 1/0.435)$, so Butterworth $n \geq 6$, Chebyshev
$n \geq 4$, elliptic $n \geq 3$, Bessel impossible. (b) Bilinear has $\omega_i T = 2\tan^{-1}(0.314v_i)$, $\Omega_{di} = \tan^{-1}(0.314)/\tan^{-1}(0.314\Omega_{ai})$, $(\gamma_1, \Omega_{d1}) = (2, 1/0.654)$, $(\gamma_2, \Omega_{d2}) = (4, 1/0.485)$.

6.4 $\omega_p T = 28.9°$, $\omega_r T = 22.5°$, $\Omega_d = 1.28$; bilinear has $\Omega_a = \tan(14.5°)/\tan(11.3°) = 1.30$;
Euler has $\Omega_a = \sin(28.9°)/\sin(22.5°) = 1.26$. Use $(M_p, M_r, \gamma, \Omega_a) = (0.28, 40, 5.18, \Omega_a)$.

6.11 (a) $H(z) = (1 - 2z^{-1} + z^{-2})/(1 - 1.859z^{-1} + 0.869z^{-2})$;
(e) $H(z) = (1 - 2z^{-1} + z^{-2})/(1 + 0.141z^{-0.5} - 1.9z^{-1} - 0.141z^{-1.5} + z^{-2})$.

6.12 (a) $H(z) = 0.6621(1 - z^{-1})^4/[(1 - 1.480z^{-1} + 0.5559z^{-2})(1 - 1.701z^{-1} + 0.7885z^{-2})]$.

6.14 (a) $\Delta(z) = 1 + T[1/(z-1)]$; (c) $\Delta(z) = 1 + T[(z+1)/(z-1)]$.

6.18 $y_3(n) = -0.1585y_3(n-1) + 0.8415[x_3(n) - x_3(n-1)]$.

6.20 (b) B: $(\gamma_1, \Omega_{d1}) = (2, 1.6)$, $(\gamma_2, \Omega_{d2}) = (4, 2.5)$. Solving $\omega_i - 1/\omega_i = BW_i$ gives
$(\omega_{40L}, \omega_{20L}, \omega_{3L}, \omega_o, \omega_{3U}, \omega_{20U}, \omega_{40U}) = (0.351, 0.481, 0.618, 1, 1.618, 2.081, 2.851)$. Using
Eq. 6.2.37b gives $\Omega_{a1} = \min(1.67, 1.93)$, $\Omega_{a2} = \min(2.59, 3.58)$. $(\gamma_1, \Omega_{a1}) = (2, 1.67)$, $(\gamma_2, \Omega_{a2}) = (4, 2.59)$, so Butterworth $n \geq 5$, Chebyshev $n \geq 4$, elliptic $n \geq 3$, Bessel impossible.

6.25 (a) $H(z) = 0.06746(1 - z^{-2})^2/(1 - 2.543z^{-1} + 2.816z^{-2} - 1.593z^{-3} + 0.4128z^{-4})$.

6.26 $y_3(n) = -0.1585y_3(n-1) + 0.8415[x_3(n+1) - x_3(n-1)]$.

6.28 (b) B: $(\gamma_1, \Omega_{d1}) = (2, 1/0.391)$, $(\gamma_2, \Omega_{d2}) = (4, 1/0.222)$. Solving $\omega_i - 1/\omega_i = BW_i$ gives
$(\omega_{3L}, \omega_{20L}, \omega_{40L}, \omega_o, \omega_{40U}, \omega_{20U}, \omega_{3U}) = (0.618, 0.824, 0.895, 1, 1.117, 1.214, 1.618)$. Using
Eq. 6.3.29b gives $\Omega_{a1} = \min(2.42, 2.36)$, $\Omega_{a2} = \min(4.24, 4.18)$. $(\gamma_1, \Omega_{a1}) = (2, 2.36)$, $(\gamma_2, \Omega_{a2}) = (4, 4.18)$, so Butterworth $n \geq 4$, Chebyshev $n \geq 3$, elliptic $n \geq 3$, Bessel impossible.

6.32 (a) $H(z) = (1 - 3.236z^{-1} + 4.618z^{-2} - 3.236z^{-3} + z^{-4})/(1.045 - 3.308z^{-1} + 4.616z^{-2} - 3.164z^{-3} + 0.9565z^{-4})$.

6.33 $y_3(n) = -0.1585y_3(n-1) + 0.8415[x_3(n+1) + x_3(n-1)]$.

Chapter 8

8.1 (a) $[H_E]_F = \text{diag}[0.8,0,0,0]^t$; (b) $[H_M]_F = \text{diag}[1,0,0,0]^t$, $[H_{HR}]_F = \text{diag}[0.2,0,0,0]^t$;

(c) $[H_{SS}]_F = \text{diag}[3.2,0,0,0]^t$, $[H_{SN}]_F = 0$; (e) $[c_E] = [0.0833,0.0833,0.417,$

$0.417]^t$, $[c_M] = [1,0.5,0,0.5]^t$, $[c_{HR}] = [0.5,0.33,0,0.33]^t$, $[c_{SS}] = [2,0.67,0,0.67]^t$.

8.7 (a) $[B_R] = $ block diagonal $[\; [0,\ldots,0] \; [1,\ldots,1] \; [0,\ldots,0] \;]$ with $N(1-p)/2$ zeros, Np ones

(centered about diagonal), and $N(1-p)/2$ zeros. $[S] = \text{FFT}[s]$, $[|S|^2] = [|S_0|^2,\ldots,|S_N|^2]^t$,

$\langle|S|^2\rangle = [B_R][|S|^2]$. $[R] = \text{FFT}[r]$, $[|R|^2] = [|R_0|^2,\ldots,|R_N|^2]^t$, $\langle|R|^2\rangle = [B_R][|R|^2]$.

(b) $[H_E]_F = \text{diag}[\ldots,|S_i|^2/\langle|R_i|^2\rangle,\ldots]$, $[H_M]_F = \text{diag}[\ldots,|S_i|^2/(\langle|R_i|^2\rangle - |\langle R_i\rangle|^2),\ldots]$,

$[H_{HR}]_F = \text{diag}[\ldots,S_i^*/\langle|R_i|^2\rangle,\ldots]$.

8.11 (a) $[H_E]_{Ha} = \begin{bmatrix} 0.5 & 0 & 0 & 0 \\ 0 & 0.33 & 0 & 0 \\ 0 & 0 & 0.17 & -0.17 \\ 0 & 0 & -0.17 & 0.17 \end{bmatrix}$, (b) $[H_M]_{Ha} = \begin{bmatrix} 0.5 & 0 & 0 & 0 \\ 0 & -0.25 & 0.18 & -0.18 \\ 0 & -0.18 & -0.13 & 0.13 \\ 0 & 0.18 & 0.13 & -0.13 \end{bmatrix}$.

8.19 (a) $[H_E]_P = \begin{bmatrix} 0.8 & 0 & 0 & 0 \\ 0 & 0 & 0 & 0 \\ 0 & 0 & 0 & 0 \\ 0 & 0 & 0 & 0 \end{bmatrix}$, (b) $[H_M]_P = \begin{bmatrix} 1 & 0 & 0 & 0 \\ 1 & 0 & 0 & 0 \\ 1 & 0 & 0 & 0 \end{bmatrix}$.

8.29 $\hat{E}\{SS^h\}_S = \begin{bmatrix} 1 & 2.24 & 0 & 0 \\ 2.24 & 5 & 0 & 0 \\ 0 & 0 & 0 & 0 \\ 0 & 0 & 0 & 0 \end{bmatrix}$, $\hat{E}\{NN^h\}_S = \begin{bmatrix} 4 & 0 & 3 & 0 \\ 0 & 4 & 0 & 3 \\ 3 & 0 & 4 & 0 \\ 0 & 3 & 0 & 4 \end{bmatrix}$. (a) $[H_E]_S = $

$\begin{bmatrix} 0.129 & 0.289 & -0.0968 & -0.216 \\ 0.289 & 0.645 & -0.216 & -0.484 \\ 0 & 0 & 0 & 0 \\ 0 & 0 & 0 & 0 \end{bmatrix}$, (b) $[H_M]_S = \begin{bmatrix} 0.571 & 1.28 & -0.429 & -0.958 \\ 0.571 & 1.28 & -0.429 & -0.958 \\ 0.571 & 1.28 & -0.429 & -0.958 \\ 0.571 & 1.28 & -0.429 & -0.958 \end{bmatrix}$.

8.39 (a) $[H_E]_P = \text{diag}[0.5,0.5,0,0]^t$; (b) $[H_M]_P = \text{diag}[2,-2,0,0]^t$,

$[H_I]_P = \text{diag}[0.5,-0.5,\infty,\infty]^t$, $[H_{HR}]_P = \text{diag}[1,-1,0,0]^t$.

8.41 (a) $[H_E]_S = \text{diag}[0.5,0.44,0,0.17]^t$; (b) $[H_M]_S = \text{diag}[2,-1.8,0,0.89]^t$,

$[H_{HR}]_S = \text{diag}[1,-1,0,0.75]^t$; (c) $[H_{SS}]_S = \text{diag}[0.25,-0.2,0,0]^t$, $[H_{SN}]_S = 0$.

8.55 (a) $[h_E]_{KL} = \begin{bmatrix} 0.392 & 0.238 & 0.144 & 0.087 \\ 0.238 & 0.144 & 0.087 & 0.053 \\ 0.144 & 0.087 & 0.053 & 0.032 \\ 0.087 & 0.053 & 0.032 & 0.020 \end{bmatrix}$, $[H_E]_{KL} = \text{diag}[0.608,0,0,0]^t$,

$[T]_{KL}^{-1} = \begin{bmatrix} 0.803 & 0.161 + j0.331 & -0.404 - j0.392 & -0.728 - j0.246 \\ 0.487 & -0.265 - j0.546 & 0.281 + j0.587 & 0.888 - j0.023 \\ 0.295 & 0.160 + j0.329 & 0.533 + j0.220 & 0.599 + j0.511 \\ 0.179 & -0.263 - j0.541 & 0.165 - j0.200 & -0.141 + j0.325 \end{bmatrix}$.

8.56 (a) $[h_E]_{KL} = [H_E]_{KL} = \text{diag}[0.5, 0, 0, 0]^t$, $[T]_{KL}^{-1} = [T]_{KL} = \text{diag}[1, 1, 1, 1]^t$.

Chapter 9

9.2 (a) $H_{UE}(j\omega) = 1$, $\omega \neq 2\pi(60 \text{ Hz})$ and zero otherwise;

(b) $H_{UE}(j\omega) = 1$, $\omega = 2\pi(60 \text{ Hz})$ and zero otherwise.

9.6 (a) $H_{UE}(j\omega) = \omega^2/(1 + \omega^2)$; (b) $H_{CE}(j\omega) = j\omega/(j\omega + e^{-j \arg N(j\omega)})$.

9.15 (a) $H_M(j\omega) = \sin(\omega T/2)/(\omega T/2) = 1/H_I(j\omega)$, $H_{HR}(j\omega) = 1/[H_M(j\omega) + H_I(j\omega)]$.

9.18 (a) $H_M(j\omega) = -1/j\omega(\langle |R(j\omega)|^2 \rangle - |\langle R(j\omega) \rangle|^2)$, $H_I(j\omega) = j\omega$,

$H_{HR}(j\omega) = -1/j\omega\langle |R(j\omega)|^2 \rangle$ where $\langle |R(j\omega)|^2 \rangle = |R(j\omega)|^2 * B(j\omega)$

(see solutions to Problems 4.24a and 8.7a).

9.24 (a) $H_{ss}(j\omega) = 1/j\omega(1 + \omega^2)$, $H_{nn}(j\omega) = \omega^2 e^{-j \arg N(j\omega)}/(1 + \omega^2)$, $H_{ns}(j\omega) = j\omega/(1 + \omega^2)$.

9.30 (a) $H_M(j\omega) = e^{-j\omega}/2\omega^2 = 1/H_I(j\omega)$, $H_{HR}(j\omega) = \omega^2 e^{-j\omega}/2(\omega^4 + 0.25)$.

9.32 (a) $H_M(j\omega) = 1/(1 - j\omega)(1 + j\omega)^2 = 1/H_I^*(j\omega)$, $H_{HR}(j\omega) = (1 + j\omega)(1 - j\omega)^2/[1 + (1 + \omega^2)^3]$.

(c) $H_M(s) = 1/(1 - s)(1 + s)^2 = 1/H_I(-s)$, $-1 < \text{Re } s < 1$ is noncausal and stable.

9.40 (a) $H_M(j\omega_1, j\omega_2) = e^{-j(\omega_1 + \omega_2)}/2\omega_1^2\omega_2^2 = 1/H_I(j\omega_1, j\omega_2)$,

$H_{HR}(j\omega_1, j\omega_2) = 0.5\omega_1^2\omega_2^2 e^{-j(\omega_1 + \omega_2)}/(\omega_1^4\omega_2^4 + 0.25)$.

9.42 (c) $H_M(s_1, s_2) = 1/(1 - s_1)(1 + s_1)^2(1 - s_2)(1 + s_2)^2 = 1/H_I(-s_1, -s_2)$,

$-1 < \text{Re } s_1, \text{Re } s_2 < 1$ is noncausal and stable.

9.55 (a) ASK: $H_M(j\omega) = 1/(1 - j\omega)$, $H_I(j\omega) = 1 + j\omega$, $H_{HR}(j\omega) = (1 + j\omega)/(1 + \omega^2)$.

9.58 (a) $H_M(j\omega) = 1/(1 - j\omega)(\langle |R(j\omega)|^2 \rangle - |\langle R(j\omega) \rangle|^2)$, $H_I(j\omega) = (1 + j\omega)$,

$H_{HR}(j\omega) = 1/(1 - j\omega)\langle |R(j\omega)|^2 \rangle$ (see solutions to Problems 4.24a and 8.7a).

Chapter 10

10.1   (a)-(4) $H(s) = (sT + 0.9425)/(sT + 0.02199)$ so $H(z) = (1.9425 - z^{-1})/(1.02199 - z^{-1})$.

10.8   (4) $BT = 18°$; $H(z) = 0.02951z^{-3}/[(1 - 0.6910z^{-1})(1 - 1.8455z^{-1} + 0.9410z^{-2})]$.

10.11 (4) $BT = 16.9°$; $H(z) = 0.02555z^{-1.5}/[(1 + 0.2945z^{-0.5} - z^{-1})(1 + 0.1473z^{-0.5} - 1.9132z^{-1}$

$- 0.1473z^{-1.5} + z^{-2})]$.

10.19 (4) $BT = 37.2°$; $H(z) = 0.08258(1 - z^{-1})^3/[(1 - 0.3768z^{-1})(1 - 0.6196z^{-1} + 0.2192z^{-2})]$.

10.24 (4) $f_o = 980$ Hz, $B = 2900$ Hz, $\omega_o T = 7.35°$;

$H(z) = 0.01327(1 - z^{-1})^2/(1 - 3.766z^{-1} + 5.052z^{-2} - 3.244z^{-3} + 0.8109z^{-4})$.

10.30 (4) $\omega_o T = 62.6°, Q = 11.6$;

$H(z) = 0.7888z^{-2}(1 - z^{-1})^2/(1 - 2.744z^{-1} + 4.599z^{-2} - 3.974z^{-3} + 0.9531z^{-4})$.

10.35 (4) 2805 Hz: $\omega_o T = 40.4^\circ$, $Q = 10$; $H(z) = (1 - 4z^{-1} + 6.840z^{-2} - 5.680z^{-3} + 2.016z^{-4})/$
$(1 - 3.900z^{-1} + 6.546z^{-2} - 5.349z^{-3} + 1.880z^{-4})$.

10.42 (4) $\omega_o T = 36^\circ$, $Q = 10$; $H(z) = (1 - 3.236z^{-1} + 4.618z^{-2} - 3.236z^{-3} + z^{-4})/(1 + 0.08884z^{-0.5}$
$-3.232z^{-1} - 0.2326z^{-1.5} - 4.610z^{-2} + 0.2326z^{-2.5} - 3.232z^{-3} - 0.08884z^{-3.5} + z^{-4})$.

10.47 (4) 3550 Hz: $\omega_o T = 12.78^\circ$, $Q = \infty$; $H(z) = N(z)/(1 - 1.950z^{-1} + z^{-2})$, $N(z)$ arbitrary.

Chapter 11

11.1 (a)-(4) $H(s) = (sT + 0.9425)/(sT + 0.02199)$ so $H(z) = 1.455(1 - 0.3594z^{-1})/(1 - 0.9892z^{-1})$.

11.8 (1) $h(nT) = \{0.7304^n + 0.8546^n[-\cos(15.6^\circ n) + 0.5984\sin(15.6^\circ n)]\}U_{-1}(n)$,
$H(z) = [1/(1 - 0.7304z^{-1})] - [(1 - 0.9559z^{-1})/(1 - 1.646z^{-1} + 0.7307z^{-2})]$.

(2) $0.3136H(z)$ using $H(z)$ from (1).

(3) $H(z) = z^{-3}/[(1 - 0.7304z^{-1})(1 - 1.646z^{-1} + 0.7307^{-2})]$.

(4) $BT = 18^\circ$; $H(z) = 0.003106(1 + z^{-1})^3/[(1 - 0.7265z^{-1})(1 - 1.7657z^{-1} + 0.8566^{-2})]$.

11.11 (4) $BT = 16.9^\circ$; $H(z) = 0.002620(1 + z^{-1})^3/[(1 - 0.7408z^{-1})/(1 - 1.783z^{-1} + 0.8642z^{-2})]$.

11.19 (4) $BT = 37.2^\circ$; $H(z) = 0.02225(1 - z^{-1})^3/[(1 + 0.4964z^{-1})(1 + 1.3839z^{-1} + 0.7374z^{-2})]$.

11.24 (4) $f_o = 980$ Hz, $B = 2900$ Hz, $\omega_o T = 7.35^\circ$;
$H(z) = 0.003766(1 - z^{-2})^2/(1 - 3.788z^{-1} + 5.412z^{-2} - 3.459z^{-3} + 0.8341z^{-4})$.

11.30 (4) $\omega_o T = 62.6^\circ$, $Q = 11.6$;
$H(z) = 0.1660(1 - z^{-2})^2/(1 - 1.179z^{-1} + 0.9433z^{-2} - 0.4695z^{-3} + 0.2285z^{-4})$.

11.35 (4) 2805 Hz: $\omega_o T = 40.4^\circ$, $Q = 10$; $H(z) = (1 - 3.047z^{-1} + 4.320z^{-2} - 3.047z^{-3} + z^{-4})/(1.051 - 3.122z^{-1} + 4.318z^{-2} - 2.971z^{-3} + 0.9514z^{-4})$.

11.42 (4) $\omega_o T = 36^\circ$, $Q = 10$; $H(z) = (1 - 3.236z^{-1} + 4.618z^{-2} - 3.236z^{-3} + z^{-4})/(1.045 - 3.308z^{-1} + 4.616z^{-2} - 3.164z^{-3} + 0.9565z^{-4})$.

11.47 (4) 3550 Hz: $\omega_o T = 12.78^\circ$, $Q = \infty$; $H(z) = N(z)/(1 - 1.950z^{-1} + z^{-2})$, $N(z)$ arbitrary.

Chapter 12

12.2 (a) Use $(a_0, a_1, b_1) = (0.0304, 0.0304, -0.9391)$ in Fig. 12.2.1. (e) Parallel $H_0 = 0.0304$ and
$H_1(z)$. For $H_1(z)$, use $(a_0, a_1, b_1) = (0, 0.05895, -0.9391)$ in Fig. 12.2.1.

12.4 (b) Use $(a_0, a_1, a_2, b_1, b_2) = (0, 0.1255, 0, 0.1255, -1)$ in Fig. 12.3.1b.

(d) Use $(c_1, c_2, 1/d_1, 1/d_2) = (-0.01575, 0.01575, -0.1255, 7.968)$ in Fig. 12.7.4.

12.6 Use $z^{-1/4}$ delay blocks instead of z^{-1} delay blocks (i.e. four-phase clock).

(a) $(a_0, a_1, a_2, b_1, b_2) = (0, 0.03141, 0, 0.1255, -1)$ in Fig. 12.3.1a.

(d) Use $(c_1, c_2, 1/d_1, 1/d_2) = (-0.00394, 0.00394, 0.1255, -7.968)$ in Fig. 12.7.4.

12.7 (b) Cascade $H_1(z)$ and $H_2(z)$.

For $H_1(z)$, use $(a_0, a_1, a_2, b_1, b_2) = (0.2452, 0.2452, 0, -0.5095, 0)$ in Fig. 12.3.1b.

For $H_2(z)$, use $(a_0, a_1, a_2, b_1, b_2) = (0.0738, 0.1476, 0.0738, -1.251, 0.5457)$ in Fig. 12.3.1b.

12.16 (a) Use $(a_0, a_1, a_2, b_1, b_2) = (1, 0, -1, 0.1256, -1)$ in Fig. 12.3.1a.

12.21 Use $z^{-1/4}$ delay blocks instead of z^{-1} delay blocks (i.e. four-phase clock).

(a) $(a_0, a_1, a_2, a_3, a_4, a_5, a_6) = (1, 0, -3, 0, 3, 0, -1)$ and

$(b_1, b_2, b_3, b_4, b_5, b_6) = (0.6257, -2.804, -1.221, 2.804, 0.6257, -1)$ in Fig. 12.2.1.

12.26 (b) Use $(a_0, a_1, a_2, b_1, b_2) = (0.03047, 0, 0.03047, -1.569, 0.9391)$ in Fig. 12.3.1b.

12.37 (c) Use $(a_0, a_1, a_2, b_1, b_2) = (0.9411, -1.875, 0.9374, -1.934, 0.9374)$ in Fig. 12.3.1c.

12.49 Bilinear $p_n = 31.82(1 - z^{-1}/(1 + z^{-1})$: Use gain blocks $(0.02222, 0.02222, 0, 1, 0, 0)$ in left-to-right order in Fig. 12.8.2c. Eliminate last two z^{-1} delay blocks and 0.6 gain block.

12.56 Use gain blocks $(0.5, 1.101, 0.4755, 1)$ in left-to-right order in Fig. 12.8.1d. Replace fourth branch by R/R_L.

12.62 Bilinear $p_n = 31.82(1 - z^{-1}/(1 + z^{-1})$: Use gain blocks $(0.02440, 0.02940, 0.4168, 0.02559, 1, 0)$ in left-to-right order in Fig. 12.8.2c. Eliminate last z^{-1} delay block and 0.6 gain block.

INDEX